Lecture Notes in Mechanical Engineering

Lecture Notes in Mechanical Engineering (LNME) publishes the latest developments in Mechanical Engineering—quickly, informally and with high quality. Original research reported in proceedings and post-proceedings represents the core of LNME. Volumes published in LNME embrace all aspects, subfields and new challenges of mechanical engineering. Topics in the series include:

- Engineering Design
- Machinery and Machine Elements
- Mechanical Structures and Stress Analysis
- Automotive Engineering
- Engine Technology
- Aerospace Technology and Astronautics
- Nanotechnology and Microengineering
- Control, Robotics, Mechatronics
- MEMS
- Theoretical and Applied Mechanics
- Dynamical Systems, Control
- Fluid Mechanics
- Engineering Thermodynamics, Heat and Mass Transfer
- Manufacturing
- Precision Engineering, Instrumentation, Measurement
- Materials Engineering
- Tribology and Surface Technology

To submit a proposal or request further information, please contact the Springer Editor of your location:

China: Dr. Mengchu Huang at mengchu.huang@springer.com
India: Priya Vyas at priya.vyas@springer.com
Rest of Asia, Australia, New Zealand: Swati Meherishi at swati.meherishi@springer.com
All other countries: Dr. Leontina Di Cecco at Leontina.dicecco@springer.com

To submit a proposal for a monograph, please check our Springer Tracts in Mechanical Engineering at http://www.springer.com/series/11693 or contact Leontina.dicecco@springer.com

Indexed by SCOPUS. The books of the series are submitted for indexing to Web of Science.

More information about this series at http://www.springer.com/series/11236

Dibakar Sen · Santhakumar Mohan ·
Gondi Kondaiah Ananthasuresh
Editors

Mechanism and Machine Science

Select Proceedings of Asian MMS 2018

Springer

Editors
Dibakar Sen
Department of Mechanical Engineering
Indian Institute of Science Bangalore
Bangalore, Karnataka, India

Santhakumar Mohan
Department of Mechanical Engineering
Indian Institute of Technology Palakkad
Palakkad, India

Gondi Kondaiah Ananthasuresh
Department of Mechanical Engineering
Indian Institute of Science Bangalore
Bangalore, Karnataka, India

ISSN 2195-4356 ISSN 2195-4364 (electronic)
Lecture Notes in Mechanical Engineering
ISBN 978-981-15-4476-7 ISBN 978-981-15-4477-4 (eBook)
https://doi.org/10.1007/978-981-15-4477-4

This Springer imprint is published by the registered company Springer Nature Singapore Pte Ltd.
The registered company address is: 152 Beach Road, #21-01/04 Gateway East, Singapore 189721, Singapore

Preface

International Federation for the Promotion of Mechanism and Machine Science (IFToMM) is a global organization that, as its name suggests, has been striving for over half a decade to promote the field of kinematics, mechanisms, and machines. One of the IFToMM-affiliated journals, *Mechanism and Machine Theory*, is a flagship journal of the field. IFToMM's World Congress, the biggest global gathering of the researchers in this field, is held every four years. Thus, it made sense to have a regional conference at the continent level every other year. Asian Mechanism and Machine Science (Asian MMS) grew out of that desire and began in Taipei in 2010 and subsequently held in Tokyo (2012), Tianjin (2014), Guangzhou (2016), and Bengaluru (2018). The IFToMM member organization in India, Association of Mechanisms and Machines (AMM), holds its own conference in odd years with the name National Conference on Mechanisms and Machines Conference (iNaCoMM) (the prefix "i" is added in its abbreviation as this conference attracts a few international researchers). India has a longstanding tradition in mechanisms with excellent research done in many parts of the country. Prof. J. S. Rao is a founding member of IFToMM. He had chaired IFToMM World Congress held in 1983 in New Delhi. Many Indian researchers have participated in IFToMM conferences for a long time. It is thus fitting that the 5th Asian MMS was held in Bengaluru, India, during December 17–20, 2019.

The Asian MMS 2018, although primarily intended for Asian countries, served as a global platform for the participants to exchange ideas and present their research in the fields of mechanisms and transmissions with special emphasis on industrial applications in order to stimulate the exchange of new and innovative ideas. Papers of this conference were funneled through a rigorous two-stage peer-review process. The accepted papers presented in this conference were divided into different sessions that covered the topics: Mechanism and Machine Design, Gearing and Transmissions, Actuators and Sensors, Assistive Devices, Robotics, Walking Robots, Modeling and Simulation, Dynamics, and Compliant Mechanisms.

We received 132 abstracts, which after the first stage of the review process, yielded 85 full papers, which were carefully reviewed by three reviewers per paper in a single blind-review process. There were 82 papers accepted for presentations during the conference. Finally, 76 papers were registered for the conference. In this list, there were 68 oral presentations and eight interactive presentations. The country-wise distribution of these papers was as follows: four papers from China; three from Indonesia; two papers each from Poland, Turkey, and USA; one each from Germany, Italy, Japan, Russia, and Spain. The major parts of the papers were from India: 58 papers from various regions of the country. We want to express our thanks to the reviewers who contributed to this process with their experience and expertise. Only through their effort was it possible to organize a thorough yet speedy review process.

The authors of the papers that were presented in the conference were asked to revise their papers one last time based on the feedback they would have received during the conference. Some authors preferred not to include their papers in these proceedings. We thus have 52 chapters based on the papers presented at the conference and revised afterward. We are also delighted that we are able to include extended abstracts of four plenary talks. They are deliberately in a conversational tone because they were prepared using the recorded videos.

The field of kinematics, mechanisms, and machines is considered an old hat by some. It is far from true. The field is changing with time and so are the researchers in the field. Most researchers familiar to us in this field have gradually and deftly diversified into other areas while retaining their love for geometry, kinematics, and computational methods. Design, manufacturing, controls, automation, robotics, sensors, actuators, micro and nano systems, bioX, etc., are all the allied fields that have been influencing the IFToMM community and are equally being influenced by the same community. It is thus an inclusive field and carries with it the charm of contributing its share in transforming the world since the first industrial revolution and will continue to do so in Industry 4.0. We celebrated that spirit in Asian MMS 2018 too. This book, we hope, conveys that spirit to the readers.

We thank Prof. C. Amarnath (President of AMM) and Prof. Ashitava Ghosal (a member of the Executive Council of IFToMM) who provided prescient advice throughout the preparation for the conference. Prof. G. R. Jayanth coordinated a team of students and post-docs to ensure that all was in order. We wholeheartedly thank each and every one of them for their hard work. In particular, we want to thank Mr. Safvan Palathingal, a Ph.D. student in IISc, who helped immensely with the conference and these proceedings. We also thank the Springer-Nature team for their support and patience in bringing this book to mark the Asian MMS-2018 conference.

Bangalore, India — Dibakar Sen
Palakkad, India — Santhakumar Mohan
Bangalore, India — Gondi Kondaiah Ananthasuresh
August 2019

Mechanisms for Indian Journey in Space

I am an electronics engineer working in the Indian Space Research Organization (ISRO). So, I do not pretend to know about mechanisms. I want to talk about the role of mechanisms in space research. In particular, I will touch upon the successes of mechanisms researchers in the Indian Space program and the challenges encountered by them in various space missions. Most importantly, I want to describe how mechanisms research needs to catch up with the growing demands of space missions.

One can classify space mechanisms into two groups: mechanisms needed during launching and those that are needed after that in space or extraterrestrial environments. Launch vehicle mechanisms are needed during stage separation, heat-shield separation, etc. After the launch, a spacecraft needs mechanisms for solar panel deployment, reflector deployment, gimbal pointing, antenna deployment, and so on. If we have a lander or a rover to explore Moon or Mars, many more mechanisms are needed. As you can imagine, many things need to be folded into small volume during launching and then deployed thereafter, much ingenuity is needed in designing mechanisms that enable various operations.

Another way to classify space mechanisms is based on their utility, whether it is one-time or lifetime. One-time-use mechanisms are mostly needed for deployment of various components, e.g., the solar arrays and reflectors. Momentum wheels that maintain the stability of spacecraft, scanning mechanisms, etc., need to work for long periods, say 5 to 7 years or more. Such mechanisms can be called lifetime-use mechanisms. Beyond these, we also need pointing mechanisms for articulation about one axis or two axes, say dual gimbal antenna. Solar array drive mechanisms that point to the Sun are also of the kind that must work during the entire life of the spacecraft. Special mechanisms, also of the lifetime type, are needed for landers and rovers. For example, for special doors that enable exit and entry as well as docking of a satellite or spacecraft with another. ISRO is working on such mechanisms for future explorations.

Using specially designed mechanisms, all components that will be deployed are tucked in a stowed configuration. It is like a baby in the mother's womb with legs and hands folded. After reaching space, all the necessary parts need to fold out, i.e., they are deployed one by one.

Let us not forget that space mechanisms have to work in harsh environmental conditions that include high vacuum and extreme temperatures—above 100°C on the Sun-facing side and sub-zero temperature on the rear side. You can imagine the adverse effects of the mismatch of coefficients of thermal expansion of different materials. The materials used and mechanisms built using them should be tolerant of high levels of radiation. Outgassing and re-condensation are other concerns to keep in mind. If materials give out gases or particulates, they could sit on optical components affecting their performance. Charging is another problem, and mechanisms should be designed so that they are not affected by excessive charging. And then there is a microgravity environment too. Lubricants must comply with all these environmental conditions. Cold molecular welding should also be prevented when using contact metals.

Space mechanisms ought to be designed by considering all the aforementioned conditions and yet they must be simple. They should use a minimum number of parts and have the least weight and cost. Ground handling must be easy and they should be amenable for testing on the ground. Reliability and redundancy must be ensured.

On-board power is needed to actuate space mechanisms. It could be generated from solar energy, which could be used to charge batteries. Alternatively, some mechanisms could use energy stored in springs in the stowed condition. When the mechanisms are deployed, they experience changing forces. Hence, they need control for which dampers are also needed.

Let us discuss a few examples now. In solid reflectors, the reflector surface is already formed on the ground. They just need to be deployed. In some others, the surface will be formed in space. Some need further actuation for pointing purposes. All of these operations need to be adjusted with a few arc-sec accuracy. Consider the coverage of an antenna. It should be ensured that the coverage is restricted to India and not touching upon neighboring countries. It used to be continuous coverage. Now, we have "Digital India" type coverage wherein antennas cover different zones within India. Pointing needs should be much more stringent in such cases.

Consider another example, GSAT II, where two antennas are far apart in the launching configuration but they have to sit next to each other in the deployed configuration. This needs much care in the designing mechanism. Sometimes solar arrays will have different shapes to take "satellite shadow" effects in coverage. Payload hold-down and release mechanisms call for special care. Large tubular structures need to be held fixed during launch and then deployed after reaching the orbit. Metallic elements cannot be used in certain situations to avoid interference with RF waves. The deployable antenna could weigh as much as 300 kg. Multiple antennas may need to be sequentially deployed with flatness accuracy of 100 microns. Thus, the requirements are many. So, as the satellite or spacecraft design evolves, the mechanism design has to match that.

Sail and boom mechanisms pose other challenges. They are required to balance the torque due to solar radiation pressure. ISRO has successfully launched and deployed a solar boom that is as long as 18 m. Such a long boom needs to be tucked in a small-volume canister during the launch.

Unfurlable antenna whose diameter is 6 m was successfully deployed by ISRO. As noted earlier, the antenna surface of this not formed on the ground in the stowed configuration. It is all folded in a small space. Its deployment took 20 min in space in a sequential process.

Next, I want to talk about lunar lander and rover mechanisms. They need to have semi-autonomous navigation and hazard-avoidance capabilities. Mechanisms for these have to overcome many more challenges.

One of the significant challenges of space mechanisms is their testing on the ground. Many of them cannot support their own weight because they are economically designed to operate in a microgravity environment. But they must be tested on the ground. For that, other types of mechanisms are needed for the zero-g setup. Air-bearing, balance mass mechanisms, balloon test-rigs, and water floats are some other techniques used for this purpose.

In summary, mechanisms are crucial elements in a spacecraft. They are often mission-critical elements. Mechanisms need to make the launch and aftermath safe and reliable in addition to providing all the required functionality. Yet, they must be simple. For this, they need to have good design and they must undergo extensive analysis and elaborate testing on the ground. Although it might not have been evident in the talk, the design of space mechanisms is a multidisciplinary activity. Mechanical engineers, electrical and electronics engineers, material engineers, aerospace engineers, physicists, and others must work together to ensure success. A number of biosensors are needed as we embark on human space programs. That imposes greater demands on space mechanisms. Therefore, as I said at the beginning, mechanisms need to catch up with the changing needs of space exploration.

M. Annadurai
Indian Space Research Organization (retired)
mylswamy.annadurai@gmail.com

Redundancy in Robots and Human Arm

I wish to thank Prof. T. S. Mruthyunjaya for his kind introduction and the organizers of Asian MMS, 2018, for inviting me to give this plenary lecture. This lecture deals with redundancy in robots, in the human arm, and how it is used. It focuses on the work done at the Robotics and Design Lab of IISc, although extensive research on redundancy in robots and multi-body systems has been and continues to be done in many laboratories all over the world. I will first briefly introduce the concept of redundancy and its resolution. Next, three instances of redundancy and its use, namely in a human arm and serial robots, in parallel robots that can be used for sun tracking, and in an independently actuated end-effector proposed for endoscopy, will be presented.

The Degree-of-Freedom (DoF) of a mechanism or a robot can be obtained from the well-known Grübler–Kutzbach criterion [1]. For example, in a three-link planar robot with three Rotary (R) joints (a 3R robot), the Grübler–Kutzbach criterion gives the DoF as 3. This implies that the end-effector of a 3R-robot can be positioned and oriented arbitrarily in a plane by rotating the three joints. To obtain the desired position and orientation, one needs to solve a set of three nonlinear transcendental equations relating the position and orientation of the end-effector with the three R joint rotations. It can be shown that there exist two possible sets of joint angles that give the desired end-effector position and orientation [2]. If, however, the orientation of the end-effector is not of interest and one attempts to obtain the joint variables for a desired position of the end-effector, one cannot obtain unique or finitely many sets of joint variables. This is termed as *redundant* and, in such mechanisms and robots, there are more joints than required to achieve a task. Several such redundant robots have been designed and fabricated. In biological systems, such as in a human arm or a snake, redundancy is ubiquitous. To operate a redundant robot, one needs to impose additional constraints and choose one of the solutions from the infinitely many possible solutions—this is termed as *resolution of redundancy*. Extensive research has been done over the years and worldwide on the resolution of redundancy (see, for example, [3–11]). In this talk, I present how redundancy can be usefully utilized through three case studies.

Resolution of Redundancy in the Human Arm and Hyper-Redundant Robots

One of the approaches to obtain a unique set of joint variables is to optimize the function of the joint variables. In this context, researchers have proposed minimization of a) rotations at the joints [2], b) joint velocities [1, 2] or joint torques, c) obstacle avoidance, d) maximization of a manipulability index, etc. (see [4]). More recently redundancy was resolved using a backbone curve [7] and by minimizing the linear velocity of the far end of a link [8]. The author and his students have used the idea of tractrix curve [5], to obtain realistic and natural-looking motions of a highly redundant robot [8, 9]. It is shown that the motions of the links die down as one moves from the head to the tail and one can extend the approach to obstacle avoidance [10].

In an older work [5], it was proposed that the redundancy can be used to make the velocity distribution at all points of the workspace of the robot isotropic. The human arm is known to be redundant with seven DoFs—a three-DoF joint at the shoulder, a two-DoF joint at the elbow, and a two-DoF joint at the wrist. Since only six DoFs are required for positioning and orienting the hand, the human arm is redundant. In a recent work, extensive experiments with point-to-point reaching tasks were done by adult human subjects. These experiments [9, 10] consist of reaching a random target (spaced 45° apart) from an initial location and had three phases—an initial phase without any perturbation, a second phase where a lateral force was applied by a robot, and finally, a third phase where the lateral force was switched off. In the initial phase, the recorded hand trajectories were found to be mostly straight and when the lateral force was applied, the hand trajectory deviated significantly from the straight path, and over trials, the trajectories became straighter. When the lateral force was switched off, the curvature of the hand trajectory reverses demonstrating the subjects learn how to adapt to the lateral force. Using statistical analyses, the researchers demonstrated that those subjects who explore the redundancy present in the arm learn to adapt to the external force faster. It was also observed that the largest errors in the trajectories are not the same in all directions—the error distribution is close to an ellipse with the largest error along $\sim 100°$ and smallest along $\sim 225°$. As the subjects learn, the eccentricity of the ellipse reduces. However, they never become circular (or isotropic) as possible in redundant serial robots. Taken together, these new results suggest one possible use of redundancy in the human arm is for learning to adapt to external perturbations and to homogenize the workspace.

Sun Tracking Using Parallel Robots

One approach to harvest incident solar energy is by reflecting the incident beam to a central tower receiver where it can be converted to electricity. In this approach, a large number of mirrors (upwards of 5000), also called heliostats, track the sun accurately (within 2 mrad pointing accuracy) while focusing the incident energy onto a distant receiver (up to 1 km away) [14]. The main advantages of this approach are that energy conversion efficiency is larger than current photo-voltaic systems and the energy can be stored in thermal form without the use of batteries. The sun moves very slowly ($\sim 180°$ in 12 hours) in the East-West direction and with seasons, it also moves in a North-South direction. A two-DoF mechanism is thus required for sun tracking and there exist two main sun tracking mechanisms—the Azimuth-Elevation (Az-El) and the Target-Aligned (T-A) drives [14]. Due to the need to track the sun in the presence of winds and other disturbances, the heliostats use significant structural material. In a recently completed Indo-US funded project, two novel redundant three DoF parallel robots, namely the well-known 3-RPS and the 3-UPU robots (R, P, S, and U denote rotary, prismatic, spherical, and a universal joint, respectively). In both these configurations, the mirror is supported at three connection points (unlike at a single point in the AZ-El and T-A configurations) and one can save between 15%–60% in structural material to achieve the same pointing accuracy [15,16]. Additionally, since linear drives (P oint) are used, one can do away with expensive and accurate speed reductions systems. Finally, in the case of the 3-UPU configuration, the mechanism can be electronically switched to the Az-El and T-A configurations when required to improve efficiency [17]. This research is expected to reduce the cost of heliostats which forms a significant component of the total cost of a concentrated solar power system.

Redundancy and Soft Robots

A typical endoscope is about 12 mm in diameter, about 1.5 m long and at the end carries an imaging system and, in modern endoscopes, a channel through which a catheter with an end-effector can be deployed. In modern endoscopes, in addition to imaging the Gastro-Intestinal (GI) tract, the end-effector can be used for taking biopsy samples or performing simple surgeries. The main difficulty is that image changes when the distal end is moved toward the point of surgery as there is no independent control of the end-effector. In a recently completed project, funded by the Robert Bosch Foundation, we have developed an independently actuated end-effector. The main challenge was to select very small actuators (< 4.5 mm to fit in the available channel) and toward this end, we have developed Miniaturized Pneumatically "Artificial" Muscles (MPAMs). The MPAMs consist of a silicone tube braided with nylon fibers. When the silicone tube is pressurized with air, the MPAMs contract by up to 20% and act like muscles. We have developed an independently actuated end-effector with three MPAMs and it has been demonstrated that the actuator is able to achieve the required motion [18]. Additionally, a new improved model which better predicts the motion of the MPAM [19, 20] and the redundant end-effector has been developed.

Summary

This talk included the work done by the research students in the Robotics and Design Lab in the last few years. New and novel findings on how exploitation of redundancy in the human arm can help in learning external perturbations are shown. It is shown that using redundancy in a special way can generate realistic and natural-looking motion in hyper-redundant and "snake" robots and avoid obstacles. Redundancy in parallel robots helps in reducing structural weight in sun-tracking mechanisms based on parallel robots thereby creating the possibility of making heliostats less expensive. Finally, the design and modeling of a miniature end-effector actuated using flexible actuators were presented.

Ashitava Ghosal
Mechanical Engineering
Indian Institute of Science, Bengaluru
asitava@iisc.ac.in

Acknowledgement

The contents of this talk are based on the work of several research students in the Robotics and Design Laboratory at the Mechanical Engineering Department, IISc, Bangalore. In particular, I wish to thank Ashith Shyam R. B., Ashwin K. P., Midhun S. Menon, Puneet Singh, and V. C. Ravi.

References

1. G. Gogu, Mobility of mechanisms: a critical review, *Mechanism and Machine Theory* 40 (9) pp. 1068–1097, 2005.
2. A. Ghosal, *Robotics: Fundamental Concepts and Analysis* (see Chapter 3), Oxford University Press, New Delhi, 2006.
3. C. A. Klein and C. H. Huang, Review of the pseudo-inverse for the control of kinematically redundant manipulators, *IEEE Trans. on Systems, Man, and Cybernetics*, SMC-**13**(3), pp. 245–250, 1983.
4. T. Yoshikawa, Analysis and control of robotic manipulators with redundancy, *In Robotics Research: The First International Symposium*, Eds. M. Brady and R. Paul, Cambridge, Massachusetts, MIT Press, pp. 735–748, 1984.
5. Y. Nakamura, *Advanced Robotics: Redundancy and Resolution, Addison-Wesley*, 1991.
6. G. S. Chirikjian and J. W. Burdick, A modal approach to hyper-redundant manipulator kinematics, *IEEE Trans. on Robotics and Automation*, **10**(3), pp. 343–354, 1994.
7. D. Reznik and V. Lumelsky, Sensor based motion planning in three dimensions for a highly redundant snake robot, *Advanced Robotics*, **9**(3), pp. 255–280, 1995.

8. S. Sreenivasan, P. Goel and A. Ghosal, A real-time algorithm for simulation of flexible objects and hyper-redundant manipulators, *Mechanism and Machine Theory*, **45**, pp. 454–466, 2010.
9. V. C. Ravi, S. Rakshit and A. Ghosal, A, Redundancy resolution using tractrix: Simulations and experiments, Trans. *ASME, Journal of Mechanisms and Robotics*, **2**, pp. 031013–1, 2010.
10. M. S. Menon, V. C. Ravi and A. Ghosal, A. Trajectory planning and obstacle avoidance for hyper-redundant serial robots, *Trans. ASME, Journal of Mechanisms and Robotics*, **9**, pp. 0445055–7, 2017.
11. A. Ghosal and B. Roth, A new approach for kinematic resolution of redundancy. *International Journal of Robotics Research*, **7**(2), pp. 22–35, 1988.
12. P. Singh, S. Jana, A. Ghosal and A. Murthy, Exploration of joint redundancy but not task space variability facilitates supervised motor learning, *Proceedings of the National Academy of Sciences of USA*, **113**(50), pp. 14414–14419, 2016.
13. P. Singh, *The Role of Basal Ganglia and Redundancy in Supervised Motor Learning*, Ph.D. Thesis, Indian Institute of Science, 2017.
14. L. Vant-Hull, *Concentrating solar power technology: Principles, developments and applications*, Woodhead Publishing, Cambridge, UK, Ch. 8, pp. 240–283, 2012.
15. Ashith Shyam R. B. and A. Ghosal, Three-degree-of-freedom parallel manipulator to track the sun for concentrated solar power systems, *Chinese Journal of Mechanical Engineering*, **28**(4), pp. 793–800, 2015.
16. Ashith Shyam R. B., M. Acharya and A. Ghosal, A heliostat based on a three degree-of-freedom parallel manipulator, *Solar Energy*, **157**, pp. 672–686, 2017.
17. Ashith Shyam R. B. and A. Ghosal, Path planning of a 3-UPU wrist parallel manipulator for sun tracking in central receiver tower systems, *Mechanism and Machine Theory*, **119**, pp. 130–141, 2018.
18. Ashwin, K. P., Jose Don P. and A. Ghosal, Modeling and analysis of a flexible end-effector for actuating endoscopic catheters, *Proc. of 14th World Congress in Mechanism and Machine Science 2015*, Taipei, Taiwan, 2015.
19. Ashwin, K. P. and A. Ghosal, Mathematical model for pressure-deformation relationship on miniaturized McKibben actuators, *Proc. of iNaCoMM 2017*, Mumbai, India, 2017.
20. Ashwin, K. P. and A. Ghosal, A survey of static modeling of miniaturized pneumatic artificial muscles with new model and experimental results, *Trans. ASME, Applied Mechanics Reviews*, **70**(4), pp. 040802, 2018.

In Search of Fundamentally Robust and Intrinsically Complex Designs in Pursuit of High-Performance Mechanisms and Machine Design

With an eye toward improving performance and achieving simplicity in machine and mechanism design, one must delve into intrinsically complex processes. I will describe six topics from my past research, which may be cohesively connected with the common theme of finding simplicity amid complexity. I would like to begin with the following quote:

> Conventional wisdom can be like concrete; it is good to walk on, and it gets you places fast. But nothing beautiful, rare or worthwhile will grow in it. However, if it is formed to cup a seedling in a nurturing environment, growth is but certain and its fruits are sure to follow.

This was something I had written when I was in my late forties; I continue to believe in it till this day. Over the years, I am pleased to see many younger colleagues subscribe to it and grow in their careers. At this point, I wish to recall a friend, a mentor, and an accomplished engineer, Mr. Dimitry Grabbe (1928–2011). He was the Director of Applied Technology at AMP Inc., having to his credit more than 500 patents in areas such as machine design, semiconductor packaging, optoelectronic connector design, and printed-circuit-board technology. Among his awards were ASME Leonardo da Vinci and IEEE Components, Packaging, and Manufacturing Technology Awards. He had been an integral part of the US space exploration and had founded the Maine Research Corporation. My association with him greatly influenced my thinking, and thus shaping my early career.

Let me begin with my first topic, the discovery and development of the primal equations in designing four-bar crank-rocker linkages. Much work exists on this topic. Extensively reported in the master's thesis of my student, Mark J. Erickson, these equations offer profound simplicity and elegance, providing a lot of insights, while revealing families of solutions for a whole host of mobility synthesis problems:

$$sin\left(\frac{\Delta\theta_4}{2}\right)sin(\mu_m) = \frac{R_2}{R_4} \quad \& \quad sin(\theta_{4m})sin\left(\frac{\Delta\mu}{2}\right) = \frac{R_2}{R_4}$$
$$sin\left(\frac{\Delta\theta_3}{2}\right)sin(\mu_m) = \frac{R_2}{R_3} \quad \& \quad sin\,(\theta_{3m})sin\left(\frac{\Delta\mu}{2}\right) = \frac{R_2}{R_3}$$
$$sin\left(\frac{\Delta\theta_4}{2}\right)sin(\mu_{3m}) = \frac{R_2}{R_1} \quad \& \quad sin(\theta_{4m})sin\left(\frac{\Delta\theta_3}{2}\right) = \frac{R_2}{R_1}$$

These equations involve 10 variables: four link lengths, $R_1$to R_4, the mean and range values of three angles, i.e., the coupler angle, θ_3, the rocker angle, θ_4, and the transmission angle, μ. One of the equations was found using symbolic mathematics, and then the rest were subsequently derived. With four free variables to choose, there exist 210 combinations, with a few eliminated as being redundant or not satisfying Grashof's criterion. The equations made it easy to rule out impossible user specifications and also to obtain possible solutions. They also led to a graphical method to locate the ground pivots. There are other interpretations and uses of these equations, e.g., obtaining a simple harmonic motion of the output member when the input crank is considered to be relatively much smaller.

Let me move on to my second topic: the mechanical advantage of Single-Input Port and Multiple-Output Port (SIPMOP) mechanisms. For example, a hand-held device, called the *elastrator* [1], is used to stretch a rather stiff elastic band into a nearly square-shaped opening. This means that the force applied by the hand stretches the band in two orthogonal directions simultaneously. Therefore, the device requires one input force and exerts two output forces. This system is analogous to two springs in series acted upon by a single force. We found that its net Mechanical Advantage (MA) is governed by an equation [1], where MA1 and MA2 are component mechanical advantages evaluated in orthogonal directions 1 and 2, respectively, assuming only a single output force in action with a single input force:

$$\frac{1}{MA} = \frac{1}{MA_1} + \frac{1}{MA_2} \Rightarrow MA = \frac{MA_1 MA_2}{MA_1 + MA_2}$$

Two fundamental observations emerge: (i) MA will be lower than even the lower of MA_1 and MA_2, and (ii) it will converge to the lower value only if the higher value tends to infinity. Having recognized this, we were able to effectively design the elastrator (and other similar mechanisms) for a high mechanical advantage. Furthermore, the use of a SIPMOP allows us to synthesize a mechanism with a nonlinear output function. For example, a single-degree-of-freedom SIPMOP mechanism, symmetric about two planes, with two linear springs in orthogonal directions, is studied for a mechanical advantage as well as energy absorption. The energy vs. displacement profiles can be nonlinear concave or convex, or even linear. The linear energy case is very interesting in that it produces a nonlinear, constant-force mechanism wherein the force remains a constant over a finite displacement. Such a constant-force mechanism was used to design an accelerated pavement testing machine, to support the research of a civil engineering faculty member, Dr. Thomas White, at Purdue University.

Compliant mechanisms, which derive some or all of their functionality from the deformation of their flexible segments, are often multiple-output mechanisms. So, the SIPMOP concepts we discussed apply to them as well. The third topic we will consider is a compliant mechanism, again from the viewpoint of mechanical advantage. A fully compliant crimping mechanism was developed at AMP Inc. It was desired that its mechanical advantage increases monotonically with displacement. But even with a dozen variant prototypes tested, the mechanical advantage showed a declining trend with displacement. The subject has been thoroughly analyzed and the following conclusions are drawn: two significant factors affecting mechanical advantage are (i) the structural configuration of the Pseudo-Rigid-Body Model (PRBM) [2] and (ii) the energy stored in compliant elements of the mechanism. Several case studies examined suggest that minimizing the latter contribution relative to that of an optimized structural configuration will improve the mechanical advantage of a compliant mechanism. Therefore, its effect cannot be neglected. This crimping mechanism became an obsession and, in part, paved the way for research on compliant mechanisms design.

Let me dwell a little longer on constant-force mechanisms, as the fourth topic I want to discuss. A notable development with John Jenuwine and Jeff Albers, master's students at Purdue University, was the design of a constant-force accelerated pavement testing machine. Morgan Murphy and Larry Howell, Ph.D. students at Purdue, and I considered a variety of compliant mechanism configurations for generating constant force [3]. We utilized various segment types, from rigid to compliant, and joint types from revolute and prismatic to flexures to obtain a wide range of them. Some used cam-like surfaces in a vein to reduce the number of segments involved. The concept of generating constant force has led to a number of interesting applications. Today, we can accomplish it much more easily using simple beam segments. What looked complex at the beginning became simple as we continued to explore various aspects of the problem.

The fifth topic I wish to touch upon is the design of compliant mechanisms for generating arbitrary force–displacement profiles. Since the early days of the AMP Inc. crimping mechanism, we have come a long way in our understanding about compliant mechanisms. An important development has been the ability to represent compliant mechanisms with Pseudo-Rigid-Body Models (PRBM). The classical work of Bishopp and Drucker [4] considers the analysis of large elastic deformation of a cantilever beam with an end load, using elliptic integral solutions. Using this as a benchmark, a striking observation here has been that the locus of the tip can be approximated using a PRBM, with a pseudo-rigid-body angle in the order of 75° corresponding to a beam end location relative error less than 0.5%. This is true even for very large variations in the ratio of the transverse to the axial beam end loads. It enabled us to define terms such as characteristic pivot, radius, and radius factor, and allowing very interesting, hitherto unknown revelations regarding large-deflection beam behavior and properties. For instance, a nearly linear relation was found between the beam end force and pseudo-rigid-body angle. The PRBM has been found to be effective and accurate for analysis and design of compliant mechanisms, from simple hand tools and exercisers to more complex applications in MEMS, biomedical devices, and more. Among those who have been actively engaged in compliant mechanisms research, the works of Larry Howell and G. K. Ananthasuresh have been momentous.

I will finish with a brief discussion of the sixth topic: minimization of indeterminacy in array-like supporting structures using compliant mechanisms. Imagine long beams on multiple supports. A mattress with multiple springs is a good practical example. Many similar applications are statically indeterminate structures, and pose problems with their analysis and design. There is a nice way to alleviate this indeterminacy, if we were to replace a large number of these supports as constant-force supports made possible by compliant mechanisms. Imagine several supports on a single girder, all but two are constant-force compliant supports, designed for known force. Then, the unknown reaction forces at the two "rigid" supports may be quickly evaluated from the loading. In such a setting, I believe the accuracy of the positioning of the intermediate constant-force supports, and their engagement with the girder may not quite be as critical, as long as a preload is maintained, allowing much freedom in arranging the supports with resulting cost benefits. I leave you with these thoughts to imagine many other interesting possibilities.

As you saw, the six intrinsically complex topics we dealt with aimed at achieving simplicity. A major theme that emerged from these is that of compliant mechanisms. Let me define a compliant mechanism, while contrasting it with elastic or flexible mechanisms.

> ...elastic or flexible mechanisms are necessarily those that undergo elastic deformation. Their response may or may not be vibratory. In this context, a high-speed mechanism may be regarded as elastic or flexible. The Webster's definition of the word 'compliant', on the other hand is 'readily disposed to comply. SUBMISSIVE'...

V. K. Stokes of General Electric made a Visiting Fellow presentation, entitled "Thermoplastics in Load-bearing Applications: New Opportunities for Engineering Science" in the School of Chemical Engineering at Purdue University in 1987. He said, "...time has arrived for more challenging uses for thermoplastic materials in load-bearing applications ..." He cited weight reduction, ease of fabrication of complex shapes, and cost reduction resulting from functional integration as advantages. He added, "Lack of development of the design methodology was currently a barrier in their application." He foresaw such developments as new opportunities for engineering science. Today, "compliant mechanism design" has met that challenge, with momentous, unending applications in the aerospace industry, the automotive industry, the healthcare industry, and many more. Clearly, there is no limit to the application areas, when they offer numerous and lucrative benefits, and the complex problems are rendered simpler to solve, and easier to understand, apply, and design.

Departing from the "conventional wisdom" and leaning heavily toward inclusion of compliance in design, provided the "seedling in a nurturing environment," and the "fruits" are now beginning to be evident.

Ashok Midha
Mechanical and Aerospace Engineering, Missouri
University of Science and Technology, Rolla, Missouri, USA
midha@mst.edu

References

1. Midha, A., Hall, A. S., Her, I., and Bubel, G. M. (1984). Mechanical advantage of single-input and multiple-output ports mechanical device. *Journal of Mechanisms, Transmissions, and Automation in Design*, **106**(4), 462–469.
2. Howell, L. L., and Midha, A. (1995). Parametric deflection approximations for end-loaded, large-deflection beams in compliant mechanisms. *Journal of mechanical Design*, **117**(1), 156–165.
3. Murphy, M. D., Midha, and Howell, L. L. (1997). The Topological Synthesis of Compliant Mechanisms. *Mechanism and Machine Theory*, **31**(2), 1996, 185–199.
4. Bisshopp, K. E., & Drucker, D. C. (1945). Large deflection of cantilever beams. *Quarterly of Applied Mathematics*, **3**(3), 272–275.

Cell-sized Molecular Robots and Beyond

In this presentation, I will talk about the development of molecular robots. What is a robot? In general, while answering this question, people tend to think of humanoid robots, such as Atlas or Asimo. However, a textbook on robot-engineering gives you the following answer: In a robot, there are sensors for receiving external information, processors for processing the information, actuators for acting upon the world with processor-generated signals, and body for constructing these three devices into one individual structure. In other words, robots need not just be humanoids; rather, autonomous vehicles, planetary probes, and artificial-intelligence speakers can all be regarded as robots. In fact, these machines are expanding the range of application of robotics from our daily life to the most advanced scientific explorations.

Let us imagine that you dropped a coin into the gap between the closets. How do you pick that coin back? You might want to have a picking tool. What if the coin goes farther behind the closet? Also, what if the coins are much smaller, and the gaps are small and complex like mazes? In such a case, what is needed to look for coins behind hidden mazes that are invisible to humans might be a small *robot*. Even though Prof. Richard Feynman said, "There is plenty of room at the bottom", in his famous 1959 lecture, how much can we miniaturize a robot to combine the four aforementioned elements?

The size of a molecule is measured in angstroms and nanometres, and a molecule is a minimal unit that can be freely manufactured by considering the energy levels. Chemists, whose mission is to design and synthesize new molecules, have created molecular sensors, molecular processors, and molecular actuators. With the progress of polymer and supramolecular chemistry, it is also possible to construct a larger-body device sufficient to install these devices. Then, can we make a molecular robot by combining these devices? No, it is not that straightforward. Robots cannot be constructed by simply putting the devices together. It is essential to set up a common protocol for information communication so that the devices can be connected to each other by wiring with molecules.

What exactly is a common-molecular-information-communication protocol? We do this by adopting a DNA molecule as the molecule used for communication. It is a long-chain polymer called "deoxyribonucleic acid" that is used to store genetic information in living organisms. In recent times, with the development of molecular biology, DNA sequences consisting of ATGC combinations can be designed freely and can be synthesized quickly and inexpensively ($0.2/base). DNA structural nanotechnology, proposed by Ned Seeman, makes it possible to create the desired shape using the designed DNA sequence's hybridization. The complementary strand-formation reaction of DNA continues until reaching the most stable structure among all combinations of the sequences in solution through a brute-force approach. Based on this technology, many studies that produced DNA switches, DNA circuits, and DNA molecular computers have been reported. We thought that if a molecular sensor, molecular processor, and molecular actuator could be wired in the molecular body using reactions of DNA hybridization, then a molecular robot could be constructed. In the end, our aim is to create artificial white blood cells.

White blood cells recognize molecules released by bacteria, determine their orientation, and move by themselves to chase and capture the bacteria. If an artifact is created having such a series of mechanisms, we would be able to realize a molecular robot that mimics its counterparts in a living cell. In the future, such robots will be able to recognize, manipulate, or eliminate targets in a microenvironment that is difficult to reach and even difficult to observe with a microscope. At present, we are in the process of building up an autonomously moving cell-sized *amoeba-type molecular robot* [1] and hereby I describe our activities.

First, as a body of a molecular robot, a lipid-bilayer vesicle with similar size and structure as living cells, a micrometer-sized *liposome*, is used. This capsule-like structure, however, causes random membrane fluctuation due to Brownian motion but does not move. Therefore, we introduce an actuator inside its body. The motor protein *kinesin* is adopted as a molecular actuator. This molecule walks on a rail-like rigid cytoskeleton-protein assembly called *microtubule* (diameter: 25 nm, length: several micrometers or more) using energy dissipation in the process of hydrolysis of ATP to ADP. On a glass substrate, a gliding motion of the microtubules, a reconstituted actuator system, can be observed when kinesin is placed and microtubules are added on it. This molecular actuator was introduced inside the membrane of the liposome and aimed to realize the amoeba-like movement by generating the deformation behavior from the inside. Liposomes are usually spherical. By fixing the kinesin on the inside of this liposome membrane and sliding the rigid microtubule on it, it is expected that the soft membrane will be pushed out locally to realize the deformation mechanism.

There were many technical difficulties in integrating the aforementioned molecular devices, but, by focusing on a physical property called the *phase separation of the lipid membrane*, it was finally observed that the membrane of the liposome was continuously deformed repeatedly by the motion of the actuator. The lifespan is approximately 80 min, and the movement is stopped depending on whether the microtubules are broken or degraded, or the availability of the ATPs. In the end, the body returns to a spherical shape. However, what is important for robots is controllability. How can we control this kind of artificial cell-like structure that moves freely? *E. coli*, a type of bacteria, switches between two phases moving in a random direction and stopping in place when searching for food. Taking this switching as a hint, we designed a *sensor/processor* for stopping/moving the prototype artificial cells by molecular signals

We focused on the part that connects the kinesin and the membrane and designed a *molecular clutch* by using the DNA molecule. Similar to a clutch in an automobile, when a molecular clutch is engaged, the power of the (kinesin) motor is transmitted to the body (membrane), and if the clutch is disengaged, the motor idles (just on the microtubule), and the body does not move. The clutch was designed to realize the switching by the input of a regulatory DNA molecule with a specific sequence. By using light-responsive DNA, the following system was realized: when the light was externally irradiated, a control "start" DNA molecule was generated and the clutch was turned on, and then the movement was initiated. When the control DNA coded the stop signal, the clutch turned off. We call this system as the prototype amoeba-type molecular robot. Currently, start and stop are one way, but it is possible to repeat the mode change (shape-shifting and stop), as with *E. coli*, by using photoresponsive molecules with different wavelength responses.

So, what is missing in the amoeba-type molecular robot to behave "smarter", like real white blood cells, in the future? First, it is its lack of movement with a sense of direction. Currently, the direction in which the individual actuators move inside the body is random, and the direction of the movement of the whole body cannot be determined. Therefore, we are considering bundling actuators to the system. Kakugo and colleagues [2] created a gliding assay system in which the microtubule actuator is modified with single-stranded DNA. By adding DNA molecules containing their complementary sequence, hundreds of microtubules can be moved into a bundle. This swarm-type collective motion was broken up again by the input of another type of DNA molecule.

We are now conducting research to introduce this system of bundle microtubule actuators to align the direction of movement according to the input signal of DNA molecules inside an amoeba-type molecular robot. Also, the research to program DNA molecule circuits that can control the order and timing of the ON/OFF signal in advance has also started.

The next important development is the input of the molecular signal, from the outside. White blood cells are thought to change their direction of movement by receiving bacterial molecules locally. What about amoeba-like molecular robots?

DNA clutch molecules are now present inside the robot at a concentration of 800 nM. Now, let us consider introducing a control DNA molecule from outside. The DNA hydrophilic molecule is blocked by the oily membrane of a Giant Unilamellar Vesicle (GUV), and even if the DNA is modified with hydrophobic molecules, the introduced amount of DNA can be estimated to be less than 1 nM. We are now also struggling with developing special channels that enable certain DNA molecules to penetrate—but before that, what do we usually do when a required signal is weak? Yes, we try to amplify the signal. We have developed a system that can amplify a DNA molecule of a specific sequence isothermally inside the GUV.

Finally, how do we power the molecular robot? In our prototype, although the deformation of the membrane can be realized, it does not have enough power to pull the body attached to the surface of the substrate for the crawling movement. The trial calculation requires at least a 100-fold increase in the traction force on the membrane. Therefore, we are also considering strengthening motor proteins to increase the power.

If these improvements come true, it will result in a structure that can be called as a smart molecular robot, which can tackle more-advanced challenges than fetching the coin from behind the closet. Although it might still sound like Sci-Fi, molecular robot development is being carried out in a way compatible with that of molecular machines, DNA, and proteins used by living cells, which are actively used in the field of molecular/synthetic biology. Molecular robots, such as the amoeba-type, will be researched at a similar scale to that of living cells. If their circuits do not directly interfere with living gene expression systems, they can be used to monitor and care for the cellular health of humans who cannot directly observe the state of all cells in cell culture. Environmental monitoring or agriculture support are also possible applications. By using DNA molecular circuits that can be designed, even by undergraduate students, through free software, their complicated motion will be programmed in a manner that is autonomous, but different from that of living organisms.

I hope that molecular robots, artifacts that are good neighbors to living cells, will be expanded into a new field of micromachines in the future.

Shin-ichiro M. Nomura
Department of Molecular Robotics
Tohoku University, Japan
nomura@molbot.mech.tohoku.ac.jp

References

1. Y. Sato, Y. Hiratsuka, I. Kawamata, S. Murata, and S. M. Nomura, (2017). Micrometer-sized Molecular Robot Changes its Shape in Response to Signal Molecules. *Science Robotics*, **2**, eaal3735.
2. J. J. Keya, R. Suzuki, A. M. R. Kabir, D. Inoue, H. Asanuma, K. Sada, H. hess, A. Kuzuya, and A. Kakugo (2018). DNA-assisted Swarm Control in a Biomolecular Motor System. *Nature Communications*, **9**: 453.

Contents

About the Editors

Dibakar Sen is a Professor in the Centre for Product Design and Manufacturing and Department of Mechanical Engineering, Indian Institute of Science (IISc), Bangalore. He has done his MSc and Ph.D. in engineering from IISc in 1993 and 1997 respectively, and has worked in ISRO as an Engineer before joining IISc . His research interests include computer aided analysis and synthesis of mechanisms. geometric and topological modeling, digital human modeling, computer aided ergonomy, and advanced prosthetic systems design. He developed multiple products in the agriculture, disability and medical domains; he holds 3 patents and has 6 patents pending, published about 35 papers in international journals and about 65 in Conferences.

Santhakumar Mohan is currently an Associate Professor in the Department of Mechanical Engineering, Indian Institute of Technology (IIT) Palakkad. After his ME in Mechanical Engineering from Government College of Technology, Coimbatore, he went on to do his Ph.D. in Engineering Design from IIT Madras. He has worked in IIT Indore as an Associate Professor before joining IIT Palakkad. Dr. Mohan has received many awards over the years, including research fellowships from the Alexander von Humboldt Foundation (Germany), Erasmus+ (France), Korea Robotics Society (South Korea), KAIST University/National Research Foundation (South Korea), and AICTE (India). His research interests include parallel robotic platforms, assistive and rehabilitation robots, field robotics, mobile manipulators and underwater robotic systems. He holds 4 patents and has published 15 book chapters, 53 journal publications, and 65 papers in various international conferences.

Gondi Kondaiah Ananthasuresh is Professor and currently Chair of Mechanical Engineering, Indian Institute of Science, Bengaluru. After his Ph.D. in Mechanical Engineering from University of Michigan, Ann Arbor (USA), he was a post-doctoral research associate in the Microsystems Technology Laboratory, Massachusetts Institute of Technology (MIT), Cambridge, USA. His research areas include compliant mechanisms, topology optimization, microelectromechanical systems, kinematics, and cell mechanics. Prof. Ananthasuresh has been the recipient of the Shanti Swarup Bhatnagar Prize, Swarnajayanthi Fellowship, and Abdul

Kalam Technology Innovation Fellowship by the Government of India; the Ralph R. Teetor Educational Award by the Society of Automotive Engineers, and the National Science Foundation (NSF) Faculty Early Career Award in recognition of his academic excellence. He serves on the editorial board of many prestigious journals, holds 8 patents, and has published 2 textbooks, 4 edited books, 14 book chapters, 100 journal articles and 166 conference papers.

Multi Operation Modes of 4-CRU Parallel Mechanism For 3D-Printing Building

Pradiktio Putrayudanto, Latifah Nurahmi, and Guowu Wei

Abstract This paper deals with the operation modes analysis of a 4-CRU parallel mechanism for 3D printing building as a new technology for constructing the sustainable houses. The analysis is based upon an algebraic approach, namely the Euler parameters quaternion and primary decomposition. The 4-CRU parallel mechanism is composed of rectangular base and platform that are connected by four identical CRU legs. Each leg consists of cylindrical (C), revolute (R) and universal (U) joints. Dimensions of rectangular base and platform are the design parameters and can be varied. Initially, the parallel mechanism is described by a set of four constraint equations that define the motion of CRU legs. Then the primary decomposition is computed over a set of constraint equations and it turns out that the 4-CRU parallel mechanism has three distinct operation modes, namely the Schönflies mode, the reversed Schönflies mode and additional modes. The Degree-Of-Freedom (DOF) of the additional mode can be either 4-DOF or even degenerate into 3-DOF, depending on the ratio of the platform to the base. This additional mode is also a transition mode for the mechanism to switch from the Schönflies mode to the reversed Schönflies mode, and vice versa. This condition makes the 4-CRU parallel mechanism become reconfigurable manipulator as long as the additional modes exist. A mock-up of 4-CRU parallel mechanism has been fabricated to depict the Schönflies mode, the reversed Schönflies mode and the additional modes performed by the moving platform.

Keywords Parallel mechanism · Operation modes · Primary decomposition · Schönflies motion

P. Putrayudanto · L. Nurahmi (✉)
Department of Mechanical Engineering, Institut Teknologi Sepuluh Nopember, Surabaya, Indonesia
e-mail: latifah.nurahmi@me.its.ac.id

P. Putrayudanto
e-mail: pradiktio.putrayudanto16@mhs.me.its.ac.id

G. Wei
School of Computing, Science & Engineering, University of Salford, Manchester, UK
e-mail: g.wei@salford.ac.uk

D. Sen et al. (eds.), *Mechanism and Machine Science*, Lecture Notes in Mechanical Engineering, https://doi.org/10.1007/978-981-15-4477-4_1

1 Introduction

3D-printing technology is the process of making an object based on the formed digital model by depositing the material in layers [1]. It was first developed in Japan in 1984 and became the starting point of the additive manufacturing process [2]. Due to its potential benefits, 3D printing technology has also been developed and implemented in the construction sector. It is shown by the increasing number of buildings that were created by the 3D-Printing Building Technology, for example is a 400 ft^2 house that was constructed in 24 h by the Russian Robot Company, Apis Cor [1]. The 3D-printing building technology offers many benefits over conventional development process, namely, lower costs than traditional construction methods, environmentally friendly construction process, time savings and reduced number of injuries [3].

Spatial parallel robot with less than 6-DOF is being developed in many sectors such as industrial, medical or academics [4, 5]. Spatial parallel robot with less than 6-DOF has some additional movement, simple structure design and use less actuators [6–8]. These factors lead to cost reduction in overall usage of spatial parallel robot. In general, spatial parallel robot with less than 6-DOF has one base, one moving platform and three or more identical limbs. With a particular geometrical condition, those limbs create an overconstrained mechanism [6, 9]. One kind of those mechanisms is 4-DOF robot manifold, named the Schönflies Mode. In [10], the dynamics behaviour of the 4-DOF parallel mechanisms was thoroughly investigated. By using the Study kinematics, the operation mode analysis and self motion of the 2-RUU and 4-RUU parallel mechanisms were provided in [11, 12]. It turns out that the operation modes of the 4-RUU parallel mechanism can be derived from the decomposition of the two 2-RUU parallel mechanisms. By using the Cayley parametrization, Coste et al. [13] determined the number and type of operation modes of the 4-UPU parallel mechanism.

In this paper, the operation modes of the 4-CRU parallel mechanism related to design parameters are discussed. Operation modes that can be performed by 4-CRU parallel mechanism are the Schönflies mode (S_1), the reversed Schönflies mode (S_2) and additional mode (AM). Since the 4-CRU parallel mechanism has more than one operation mode, there exist transitional configurations that allow the moving platform to switch from one operation mode to another. This paper is organized as follows: Sect. 2 addresses the architecture of the 4-CRU parallel mechanism. The targeted application of the 4-CRU parallel mechanism under study is provided in Sect. 3. Section 4 presents kinematic analysis of the 4-CRU parallel mechanism and Sect. 5 reveals relationship among design parameters and additional modes.

2 Architecture of a 4-CRU Parallel Mechanism

The 4-CRU parallel mechanism consists of one rectangular prism as a base, one rectangular moving platform and four identical CRU legs each connects the frame rod to the moving platform. The mock-up is shown in Fig. 1.

The origin O of the fixed frame (Σ_0), as shown in Fig. 2, is located at the centre of lower base and composed of (x, y, z) axes. The origin P of the moving frame (Σ_1) is located at the centre of moving platform and composed of (u, v, w) axes. The side lengths of the 4-CRU parallel mechanism base are $2a$ and $2b$, and the side lengths of the 4-CRU parallel mechanism moving platform are $2c$ and $2d$. The vertical distance between point A_i and the ground is l_i. Hence, there are four design parameters, namely, *a, b, c, d* and four design variables, namely, l_1, l_2, l_3, l_4.

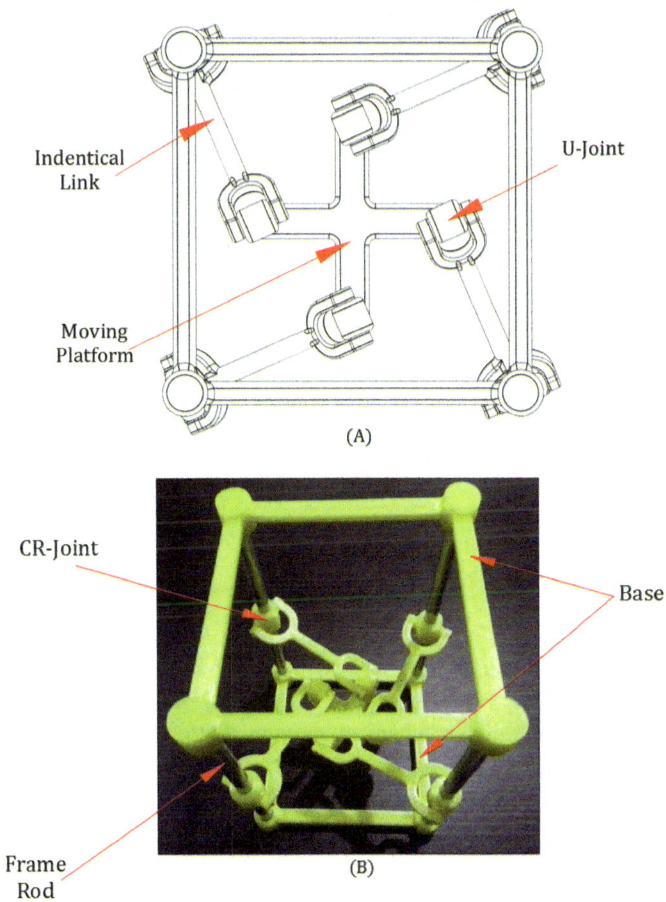

Fig. 1 Mock-up of a 4-CRU parallel mechanism **a** top view, **b** mock-up

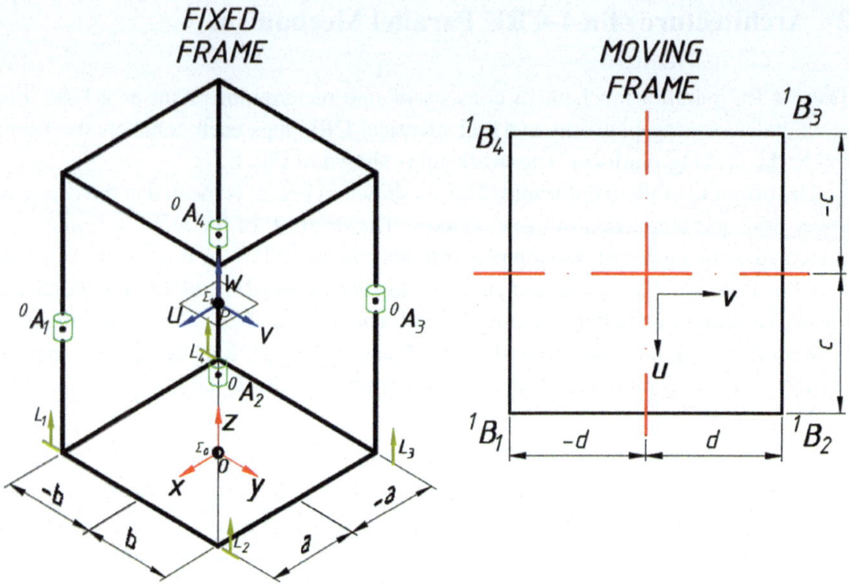

Fig. 2 The position of points A_i and B_i

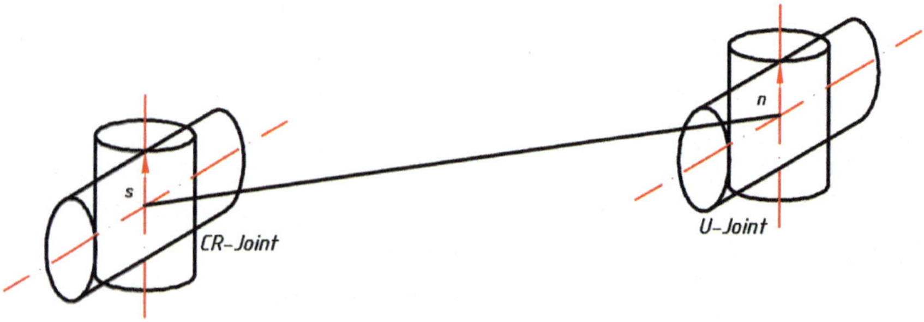

Fig. 3 Directions of universal joint (**s** and **n**)

The vertices of 4-CRU parallel mechanism base are denoted by points A_i ($i = 1, \ldots, 4$) and their coordinates with respect to the fixed frame are $^0\mathbf{A}_1 = [1, a, -b, l_1]^T$, $^0\mathbf{A}_2 = [1, a, b, l_2]^T$, $^0\mathbf{A}_3 = [1, -a, b, l_3]^T$, $^0\mathbf{A}_4 = [1, -a, -b, l_4]^T$. While the vertices of the 4-CRU parallel mechanism moving platform are denoted by points B_i ($i = 1, \ldots, 4$) and their coordinates with respect to the moving frame are $^1\mathbf{B}_1 = [1, c, -d, 0]^T$, $^1\mathbf{B}_2 = [1, c, d, l0]^T$, $^1\mathbf{B}_3 = [1, -c, d, 0]^T$, $^1\mathbf{B}_4 = [1, -a, -d, 0]^T$. Figure 3 shows the directions of **s**-axis of cylindrical link, $^0\mathbf{s}_i = [0, 0, 0, 1]^T$, and an **n**-axis of revolute joint on the moving platform, $^1\mathbf{n}_i = [0, 0, 0, 1]^T$ [14, 15].

3 Application of the 4-CRU Parallel Mechanism for 3D Printing Building

3D printing is a process in which physical objects are created by depositing materials in layers. As the technology continues to grow, 3D printing technology can be used to make simple parts to complex building shapes. This is a promising technique that may lead to revolutionize construction industry in near future.

When printing with non-symmetrical nozzle, the thickness of printed parts should be maintained equal which means that the 3D printing should not only perform three translational motions but also possess the orienting function. As a consequence, the 3D printing mechanism can be realized by the Schönflies parallel mechanism. The 4-CRU parallel mechanism is selected to carry out the task of 3D printing buildings as shown in Fig. 4. The 3D printing mechanism should be able to adapt to different building sizes thus the dimensions of base and moving platform can be reconfigured. This reconfiguration will affect the number and type of operation modes which becomes the utmost concern of this paper.

Fig. 4 Application for 3D printing building

4 Kinematic Analysis of the 4-CRU Parallel Mechanism

4.1 Constraint Equations

Constraint equations are derived whose solutions define the movement of the moving platform with respect to the fixed frame (Σ_0). The Euler parameter quaternion transformation is used to transform the vector point of ${}^1\mathbf{B}_i$ and $\mathbf{n}$ expressed in the moving frame (Σ_1) into ${}^0\mathbf{B}_i$ and $\mathbf{N}$ expressed in the fixed frame (Σ_0), as follows:

$$\mathbf{T} = \begin{bmatrix} x_0^2 + x_1^2 + x_2^2 + x_3^2 & 0_{3\times1}^T \\ \mathbf{d} & \mathbf{R} \end{bmatrix} \tag{1}$$

$$\mathbf{R} = \begin{bmatrix} x_0^2 + x_1^2 - x_2^2 - x_3^2 & 2(x_1x_2 - x_0x_3) & 2(x_1x_3 + x_0x_2) \\ 2(x_1x_2 - x_0x_3) & x_0^2 - x_1^2 + x_2^2 - x_3^2 & 2(x_2x_3 - x_0x_1) \\ 2(x_1x_3 - x_0x_2) & 2(x_2x_3 + x_0x_1) & x_0^2 - x_1^2 - x_2^2 + x_3^2 \end{bmatrix} \tag{2}$$

$$\mathbf{d} = \begin{bmatrix} X \\ Y \\ Z \end{bmatrix} \tag{3}$$

Parameters x_0, x_1, x_2, x_3 in the homogenous transformation matrix above represent the rotation of a rigid body [16, 17]. The Euler parameter should always satisfy $x_0^2 + x_1^2 + x_2^2 + x_3^2 = 1$ to remove singular parametrization. Vector $\mathbf{d}$ represents the translation movement which consists of X,Y and Z. It describes translation movement along x-axis, y-axis and z-axis, respectively.

Using the homogenous transformation matrix, location of point B_i and vector $\mathbf{N}$ with respect the fixed frame system coordinate (Σ_0) can be written as ${}^0\mathbf{B}_i = \mathbf{T}\,{}^1\mathbf{B}_i$ ($i = 1, \ldots, 4$) and $\mathbf{N} = \mathbf{T}\,\mathbf{n}$. Vector between point A_i and B_i can be formulated as follows: $\overrightarrow{ab_i} = {}^0\mathbf{B}_\mathrm{i} - {}^0\mathbf{A}_i$ ($i = 1, \ldots, 4$). Constraint equations (f_i) are obtained by eliminating the denominator from triple scalar product multiplication. The constraint equation represents three vectors ($\overrightarrow{ab_i}, \mathbf{s}_i, \mathbf{N}$) that are always lying on the same plane [18, 19] as shown in Fig. 5. Hence,

$$f{\cdot}i = \mathbf{s}_i{\cdot}(\overrightarrow{ab_i} \times \mathbf{N}) \tag{4}$$

After applying (4), we obtain four constraint equations as follows:

$$\begin{aligned} f_1 :& Xx_0x_1 - Xx_2x_3 + Yx_0x_2 + Yx_1x_3 - ax_0x_1 + ax_2x_3 + bx_0x_2 + bx_1x_3 \\ &+ cx_0x_1 + cx_2x_3 - dx_0x_2 + dx_1x_3 = 0 \end{aligned} \tag{5}$$

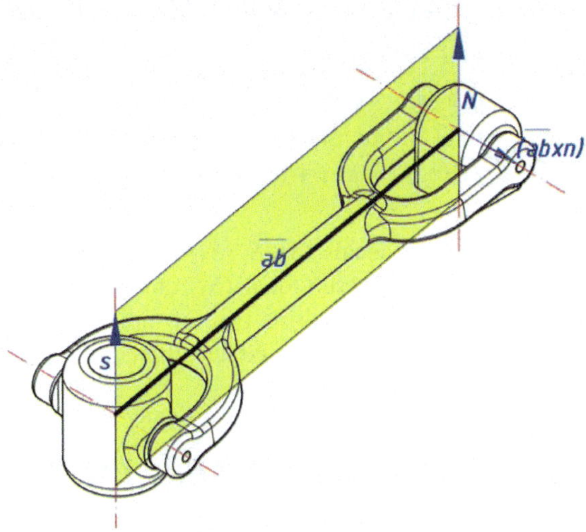

Fig. 5 Representation of constraint equations

$$f_2 : Xx_0x_1 - Xx_2x_3 + Yx_0x_2 + Yx_1x_3 - ax_0x_1 + ax_2x_3 - bx_0x_2 - bx_1x_3 + cx_0x_1 + cx_2x_3 + dx_0x_2 - dx_1x_3 = 0 \quad (6)$$

$$f_3 : Xx_0x_1 - Xx_2x_3 + Yx_0x_2 + Yx_1x_3 + ax_0x_1 - ax_2x_3 - bx_0x_2 - bx_1x_3 - cx_0x_1 - cx_2x_3 + dx_0x_2 - dx_1x_3 = 0 \quad (7)$$

$$f_4 : Xx_0x_1 - Xx_2x_3 + Yx_0x_2 + Yx_1x_3 + ax_0x_1 - ax_2x_3 + bx_0x_2 + bx_1x_3 - cx_0x_1 - cx_2x_3 - dx_0x_2 + dx_1x_3 = 0 \quad (8)$$

4.2 Derivation of Operation Modes

The constraint equations written in Eqs. (5)–(8) are collected as a polynomial ideal in the ring $(x_0, x_1, x_2, x_3, X, Y, Z)$ with the design parameters (a, b, c, d). The type of operation mode thus is identified by decomposing that ideal through primary decomposition [13]. It produces three sub-ideals, as follows:

$$S_1 : \langle x_1, x_2 \rangle \quad (9)$$

$$S_2 : \langle x_0, x_3 \rangle \quad (10)$$

$$S_3 :\langle(b^2c^2 - c^2d^2)X^2 + (a^2d^2 - c^2d^2)Y^2, (ab - ad - bc + cd)x_0^2 + (ab + ad + bc + cd)x_3^2, (ab + ad - bc - cd)x_1^2 + (ab - ad + bc - cd)x_2^2, (a - c) x_0x_1 + (-a - c)x_2x_3, (b - d)x_0x_2 - (a + c)x_1x_3, (-ad + cd)Yx_1 + (bc - cd)Xx_2, (ad - cd)Yx_0 + (bc + cd)Xx_3, (bc - cd)Xx_0 - (ad + cd) Yx_3, (bc + cd)Xx_1 + (ad + cd)Yx_2\rangle \tag{11}$$

The three sub-ideals (S_1, S_2, S_3) represent operation modes of 4-CRU parallel mechanism. The third sub-ideal, (S_3), contains nine equations. To physically interpret the operation modes with its type of movement, we substitute Eqs. (9)–(11) into Eq. (1) which will be discussed hereafter.

4.3 Operation Modes of 4-CRU PM

The transformation matrix of the first operation mode characterized in Eq. (9) is

$$\mathbf{T}_{S_1} = \begin{bmatrix} 1 & 0 & 0 & 0 \\ X & \dfrac{x_0^2 - x_3^2}{x_0^2 + x_3^2} & \dfrac{2x_0x_3}{x_0^2 + x_3^2} & 0 \\ Y & \dfrac{2x_0x_3}{x_0^2 + x_3^2} & \dfrac{-x_0^2 + x_3^2}{x_0^2 + x_3^2} & 0 \\ Z & 0 & 0 & 1 \end{bmatrix} \tag{12}$$

This transformation matrix indicates that the platform has 4-DOF motion. It consists of 1-DOF rotation about the z-axis and 3-DOF translation. The orientation of moving platform in this operation mode is parallel to the fixed frame (Σ_0). This operation mode is illustrated in Fig. 6 and called the Schönflies mode (S_1).

The transformation matrix of the second condition from Eq. (10) is

$$\mathbf{T}_{S_2} = \begin{bmatrix} 1 & 0 & 0 & 0 \\ X & \dfrac{x_1^2 - x_2^2}{x_1^2 + x_2^2} & \dfrac{2x_1x_2}{x_1^2 + x_2^2} & 0 \\ Y & \dfrac{2x_1x_2}{x_1^2 + x_2^2} & \dfrac{-x_1^2 + x_2^2}{x_1^2 + x_2^2} & 0 \\ Z & 0 & 0 & -1 \end{bmatrix} \tag{13}$$

The analysis shows that the platform is reversed about an axis parallel to the xy-plane by 180° from the identity condition. The identity condition occurs when the moving frame and the fixed frame are coincident ($\Sigma_0 \equiv \Sigma_1$) and the transformation matrix becomes the identity matrix. The moving platform has 4-DOF with 3-DOF translation and 1-DOF rotation about z-axis. This operation mode is illustrated in Fig. 7 and called the reversed Schönflies mode (S_2).

Fig. 6 Schönflies mode (S_1)

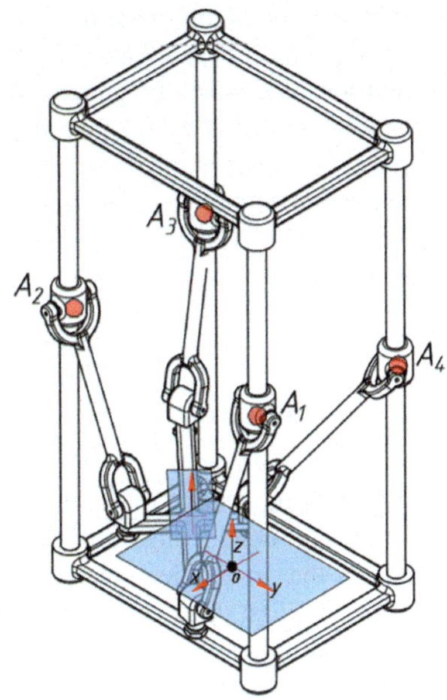

Fig. 7 Reversed Schönflies mode (S_2)

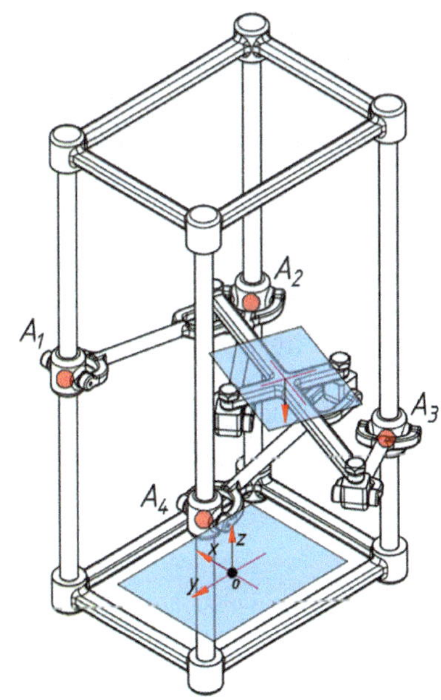

The last operation mode defined in Eq. (11) is described by nine equations. An operation mode besides 4-DOF operation modes (reversed Schönflies mode and Schönflies mode) will be generated if all equations in Eq. (11) are satisfied. Those equations consist of design parameters a, b, c, d and variables $(X, Y, x_0, x_1, x_2, x_3)$. Variable Z does not appear which indicates that translation movement along z-axis is an independent movement. Accordingly, this operation mode is called "Additional Mode 1 (AM_1)". This operation mode is degenerate mode in which its DOF is lower than 4-DOF.

4.4 Additional Mode 1 (AM_1)

The type of movement of Additional Mode 1 (AM_1) is investigated by identifying Eq. (11) and conducting kinematic interpretation. Each equation has two or more variables that make the equation represent dependent movement. We only need to consider the first three equations from Eq. (11) because the others are only another form of multiplication. Hence, the simplified equations of Additional Mode 1 (AM_1) are obtained as follows:

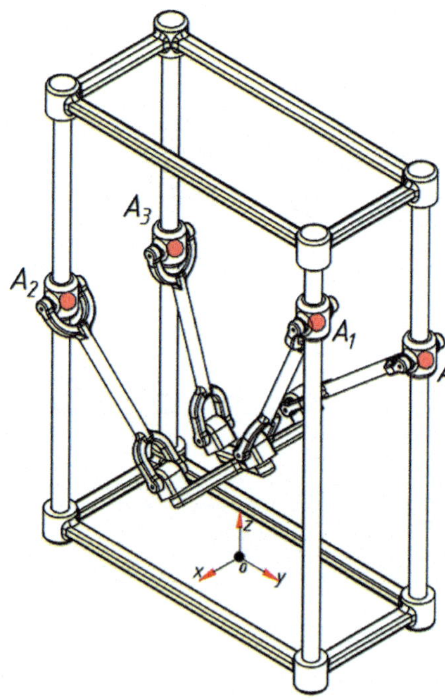

Fig. 8 Additional mode (AM_1)

$$X = Y\sqrt{\frac{-d^2(a-c)(a+c)}{c^2(b-d)(d+b)}} \tag{14}$$

$$x_3 = x_0\sqrt{\frac{(c-a)(b-d)}{(a+c)(d+b)}} \tag{15}$$

$$x_2 = x_1\sqrt{\frac{(c-a)(b+d)}{(a+c)(d+b)}} \tag{16}$$

Based on Eqs. (14)–(16), the Additional Mode 1 (AM_1) has 3-DOF movement. It consists of 1-DOF rotation about x-axis, 1-DOF translation along z-axis and 1-DOF dependent translation on xy-plane. In this operation mode, the moving platform is not parallel to the base and tilted within 0°–180°, as shown in Fig. 8.

5 Relation Among Design Parameters and Additional Modes

According to Eqs. (14)–(16), additional mode will occur in real space if and only if $(c-a)(b-d) \geq 0$ is satisfied. There are three conditions which satisfy this inequality, $(c-a)(b-d) \geq 0$.

The first condition is $(c = a)$ and $(b \neq d)$. It generates three operation modes, consisting of the Schönflies mode (S_1), the reversed Schönflies mode (S_2) and additional mode 2 (AM_2). This additional mode has 3-DOF movement consisting of 2-DOF translation on yz-plane and 1-DOF translation movement along x-axis. The following is the transformation matrix of first condition $(c = a)$ and $(b \neq d)$ at additional mode 2 (AM_2):

$$\mathbf{T}_{AM_2} = \begin{bmatrix} 1 & 0 & 0 & 0 \\ 0 & 1 & 0 & 0 \\ Y & 0 & \frac{x_0^2 - x_1^2}{x_0^2 + x_1^2} & \frac{-2x_0x_1}{x_0^2 + x_1^2} \\ Z & 0 & \frac{2x_0x_1}{x_0^2 + x_1^2} & \frac{x_0^2 - x_1^2}{x_0^2 + x_1^2} \end{bmatrix} \tag{17}$$

The second condition of design parameter is $(c \neq a)$ and $(b = d)$. It generates three operation modes, consisting of the Schönflies mode (S_1), the reversed Schönflies mode (S_2) and additional mode 3 (AM_3). This additional mode has 3-DOF movement consisting of 2-DOF translation on xz-plane and 1-DOF translation along y-axis. The following is the transformation matrix of the second relation $(c \neq a)$ and $(b = d)$ at additional mode 3 (AM_3):

$$\mathbf{T}_{AM_3} = \begin{bmatrix} 1 & 0 & 0 & 0 \\ X & \frac{x_0^2 - x_2^2}{x_0^2 + x_2^2} & 0 & \frac{2x_0x_2}{x_0^2 + x_2^2} \\ 0 & 0 & 1 & 0 \\ Z & \frac{-2x_0x_2}{x_0^2 + x_2^2} & 0 & \frac{x_0^2 - x_2^2}{x_0^2 + x_2^2} \end{bmatrix} \tag{18}$$

The third condition of design parameter is $(c = a)$ and $(b = d)$. It generates three operation modes, consisting of the Schönflies mode (S_1), the reversed Schönflies mode (S_2) and additional mode 4 (AM_4). This additional mode has 4-DOF movement consisting of 2-DOF translation on yz-plane with translation movement along y-axis depends on x-axis and 2-DOF rotation. The following is the transformation matrix of the third condition $(c = a)$ and $(b = d)$ at additional mode 4 (AM_4).

$$\mathbf{T}_{AM_4} = \begin{bmatrix} 1 & 0 & 0 & 0 \\ \frac{-Yx_2}{x_1} & \frac{x_0^2 + x_1^2 - x_2^2}{x_0^2 + x_1^2 + x_2^2} & \frac{2x_1x_2}{x_0^2 + x_1^2 + x_2^2} & \frac{2x_0x_2}{x_0^2 + x_1^2 + x_2^2} \\ Y & \frac{2x_1x_2}{x_0^2 + x_1^2 + x_2^2} & \frac{x_0^2 - x_1^2 + x_2^2}{x_0^2 + x_1^2 + x_2^2} & \frac{-2x_0x_1}{x_0^2 + x_1^2 + x_2^2} \\ Z & \frac{-2x_0x_2}{x_0^2 + x_1^2 + x_2^2} & \frac{2x_0x_1}{x_0^2 + x_1^2 + x_2^2} & \frac{x_0^2 - x_1^2 - x_2^2}{x_0^2 + x_1^2 + x_2^2} \end{bmatrix} \tag{19}$$

6 Conclusion

In this paper, the method of algebraic approach was applied to define operation modes of the 4-CRU parallel mechanism. The constraint equations corresponding to design parameters are derived and the primary decomposition is computed. It shows that the 4-CRU parallel mechanism has three operation modes, namely, the Schönflies Mode (S_1), the Reversed Schönflies Mode (S_2) and Additional Mode (AM). The motion type of each operation mode performed by the 4-CRU parallel mechanism was identified. It turns out that there is a specific condition between the design parameter and the additional mode that should be fulfilled, thus different additional modes can be generated. The first design parameter condition is $(c = a)$ and $(b \neq d)$. It generates Additional Mode 2 (AM_2) with 1-DOF rotation about the x-axis and 2-DOF translation on the yz-plane. The second design parameter condition is $(c \neq a)$ and $(b = d)$. It produces Additional Mode 3 (AM_3) with 1-DOF rotation about the y-axis and 2-DOF translation on the xz-plane. The third design parameter condition is $(c = a)$ and $(b = d)$. It generates Additional Mode 4 (AM_4) which consists of 2-DOF rotation and 2-DOF translation on the yz-plane with translation movement along the y-axis depending on x-axis.

Acknowledgements This work was supported by the Royal Academy of Engineering under the Industry Academia Partnership Programme—17/18 scheme, Grant No. IAPP1\100109.

References

1. Sakin M, Kiroglu Y (2017) 3d printing of buildings: construction of the sustainable houses of the future by bim. Energy Procedia 10:702–711
2. Pirjan A, Mihaela Petrosanu D (2013) The impact of 3d-printing technology on the society and economy. J Inf Syst Oper Manage 7:163–173
3. Hager I, Golonka A, Putanowicz R (2016) 3d printing of buildings and building components as the future of sustainable construction? Procedia Eng 151:292–299
4. Kiguchi K, Liyanage M (2008) A study of a 4dof upper-limb power-assist intelligent exoskeleton with visual information for perception-assist. In: 2008 IEEE international conference on robotics and automation, pp 3666–3671
5. Li Y, Ma Y, Liu S, Luo Z, Mei J, Huang T, Chetwynd D (2014) Integrated design of a 4-dof high-speed pick-and-place parallel robot. CIRP Ann Manuf Technol 63:185–188
6. Zhao TS, Dai JS, Huang Z (2002) geometric analysis of overconstrained parallel manipulators with three and four degrees of freedom. JSME Int J Ser C Mech Syst Mach Elem Manuf 45:730–740
7. Kong X, Gosselin C, Richard P-L (2007) Type synthesis of parallel mechanisms with multiple operation modes. J Mech Des 129:595–601
8. Kong X (2013) Type synthesis of 3-dof parallel manipulators with both a planar operation mode and a spatial translational mode. J Mech Robot 5(4)
9. Kong X (2014) Reconfiguration analysis of a 3-dof parallel mechanism using euler parameter quaternios and algebraic geometry method. Mech Mach Theory 74:188–201
10. Gosselin C (2009) Compact dynamic models for the Tripteron and Quadrupteron parallel manipulators. Proc Inst Mech Eng Part I: J Syst Control Eng 1–12
11. Nurahmi L, Caro S, Wenger P (2015) Operation modes and self-motions of a 2-ruu parallel manipulator. In: Recent advances in mechanism design for robotics, pp 417–426
12. Nurahmi L, Caro S, Wenger P, Schadlbauer J, Husty M (2016) Reconfiguration analysis of 4-ruu parallel manipulator. J Mech Mach Theory 96:269–289
13. Coste M, Demdah KM (2015) Extra modes of operation and self-motions in manipulators designed for schoenflies motion. J Mech Robot 7:97–113
14. Tsai LW (1999) Robot analysis: the mechanics of serial and parallel manipulators. John Wiley & Sons Inc
15. Wei G, Dai JS (2014) Reconfigurable and deployable platonic mechanism with a variable revolute joint. Adv Robot Kinemat 485–495
16. Kong X (2016) Reconfiguration analysis of a 4-dof 3-rer parallel manipulator with equilateral triangular base and moving platform. Mech Mach Theory 98:180–189
17. Spring KW (1986) Euler parameters and the use of quaternion algebra in the manipulation of finite rotations. Mech Mach Theory 21:365–373
18. Riley KF, Hobson MP, Bence SJ (2006) Mathematical methods for physics and engineering. Cambridge University Press
19. Nurahmi L, Solichin M (2017) Motion type of 3-rps parallel manipulator for ankle rehabilitation device. In: 2017 international conference on advanced mechatronics, intelligent manufacture, and industrial automation (ICAMIMIA)

References

Manipulator Design for a Haptic System with Improved Performance

Veysel Sekendiz, İbrahimcan Görgülü, Mehmet Görkem Karabulut, Gökhan Kiper, and Mehmet İsmet Can Dede

Abstract This paper presents the work carried out to improve the design of an existing kinesthetic haptic device. The proposed improvement is designated for enhancing this device's impedance width which is a common metric in performance evaluation of haptic devices. In this study, kinematic design optimization, static balancing, constructional design enhancement, and actuation system design studies are presented.

Keywords Haptic device · Parallel manipulator · Impedance width · Transparency · Manipulator stiffness

1 Introduction

A teleoperation system is composed of a master system on the human operator site, a slave system on the task site, and a communication infrastructure between these two systems. Generally, the master system is distant from the task site and human operator accomplishes the task using the master system by sending commands to use the slave system. Usually, the mentioned commands are motion commands and the master system is responsible for acquiring these motion commands from the human operator. Depending on the necessities of the task, the mentioned motion commands

V. Sekendiz · İ. Görgülü · M. G. Karabulut · G. Kiper · M. İ. C. Dede (✉)
Department of Mechanical Engineering, İzmir Institute of Technology, Urla, İzmir, Turkey
e-mail: candede@iyte.edu.tr

V. Sekendiz
e-mail: mehmetsekendiz@iyte.edu.tr

İ. Görgülü
e-mail: ibrahimcangorgulu@iyte.edu.tr

M. G. Karabulut
e-mail: mehmetkarabulut@iyte.edu.tr

G. Kiper
e-mail: gokhankiper@iyte.edu.tr

D. Sen et al. (eds.), *Mechanism and Machine Science*, Lecture Notes in Mechanical Engineering, https://doi.org/10.1007/978-981-15-4477-4_2

are acquired by sensing generally the hand motion and sometimes sensing the foot, leg, eye, or total body movement of the human operator.

In the teleoperation system, if the slave system's physical interaction information with the task site is to be sent to the master system as force feedback, then the master system is required to transmit this fed-back force information to the human operator. The devices that both acquire the motion of the desired limbs of the human operator and provide the force transmission to the human operator are called kinesthetic haptic devices.

This study is part of work done for the project titled "design of a Haptic system with Improved performance via developing its new Sub-Systems (HISS)". The HISS project aims to propose a new mechanism, new sub-components, and model improvements in desktop haptic device technology. The manipulator is to serve as the master device in the precision required teleoperation tasks. The manipulator to be used is selected as the R-CUBE mechanism [1] which is a 3-degree-of-freedom 3-legged translational parallel manipulator with 5R limbs where the revolute joint at the base has parallel axis to the next revolute joint axis and the remaining 3 axes are parallel to each other and orthogonal to the first two. The 3 revolute joint axes or the 3 limbs on the base are mutually perpendicular to each other. While the R-CUBE mechanism has a translational mobile platform, a wrist mechanism can be installed on the platform in order to control the orientation as well. This kinematic architecture has been previously used for a haptic device (Fig. 1) [2]. In this initial design, one of the ground-fixed joints of the parallelogram of each leg is coupled with a direct-drive actuation system and the other one is coupled with an angular position sensor. The links are manufactured from an Aluminum alloy and the joint structures are built by using double bearings. On the mobile platform, depending on the need, a handle or

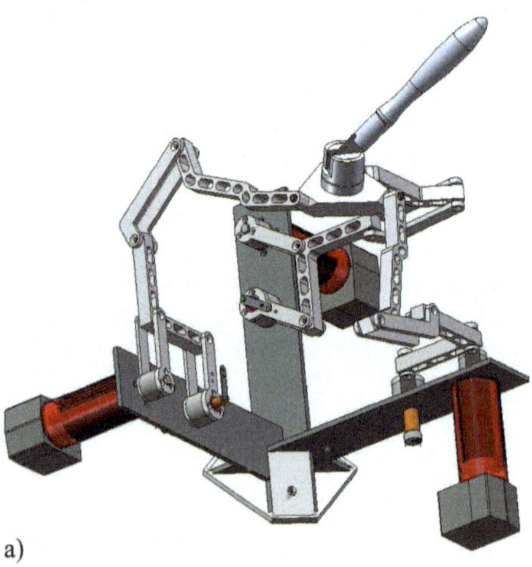
a)

b)

Fig. 1 HIPHAD v1.0. **a** CAD model, **b** real model [3]

a passive wrist mechanism can be placed to change the degree-of-freedom of measuring the pose of the handle from 3 to 6. Also, the manipulator can be re-oriented from the orientation that appears in the CAD drawing in Fig. 1a to the orientation in which one of the axes of operation is aligned with the gravity vector as shown in Fig. 1b.

The performance of a kinesthetic haptic device can be assessed by a number of different metrics. In [4], these metrics are categorized into three parts as unpowered, powered, and control system properties. Nevertheless, the overall design of a kinesthetic device is evaluated based on its impedance width and transparency. Impedance width which was introduced in [5] is the difference between the minimum and maximum mechanical impedance that can be displayed by the haptic device. The minimum impedance is a measure of resistance to free motion that is felt by the human operator when he/she tries to back-drive the haptic device. This performance is affected adversely by the friction at the joints and the moved mass/inertia. The maximum impedance is the measure of the maximum resistance that can be displayed by the device to the back-driving motion of the human operator. Based on the impedance width performance criterion, the major problems of the initial design can be listed as follows:

(1) links are constructed to have relatively larger mass and inertia, which results in increasing the minimum impedance.
(2) direct-drive actuation limits the maximum torque that can be generated in a limited area, which results in decreasing the maximum impedance.
(3) joint structures are constructed by using low-friction bearings, which results in decreasing the maximum impedance since the stability of the device when displaying larger impedance is jeopardized by minimizing the physical damping [5].
(4) total number of links are more than necessary because of the use of the parallelograms, which increases the minimum impedance.
(5) link lengths are not optimized for maximizing the stiffness, which affects both minimum and maximum impedance adversely.
(6) the manipulator is not passively balanced for gravitational effects, which limits the performance of the actuators that are working against gravitational effects thus, decreases the maximum impedance.

A measurement of the impedance characteristics of the initial design was presented in [6]. This study presents the work done in mechanical and actuation aspects to improve impedance width performance of the R-CUBE kinematic architecture to be used as a high-performance haptic device. The design studies presented in this paper are listed as the kinematic design, static balancing, constructional design, and semi-active actuator design.

2 Manipulator Architecture and Kinematic Design

Figure 2 depicts the components and kinematic parameters of the modified version of the R-CUBE mechanism. The parallelogram loops in version 1 (Fig. 1) do not exist in version 2. Also, the link lengths are modified as explained in further sections.

The manipulator comprises identical legs attached to 3 mutually orthogonal axes. Joints A_i, C_i, D_i are revolute joints, whereas joints B_i are universal joints for $i = 1, 2, 3$. Link length parameters are depicted in Fig. 2. A fixed frame O(X, Y, Z) is attached to the fixed link and a moving frame P(x, y, z) is attached to the mobile platform. Since the manipulator is a translational manipulator, the P(x, y, z) frame axes remain parallel to the O(X, Y, Z) frame axes at all times. Due to the kinematic architecture of the manipulator, $B_1C_1D_1$, $B_2C_2D_2$, and $B_3C_3D_3$ define planes attached to the mobile platform perpendicular to x-, y-, and z-axes, respectively. The locations of these planes with respect to the O(X, Y, Z) frame are determined by the A_iB_i links. Therefore, Leg 1 ($A_1B_1C_1D_1$) is responsible for the motion of the platform along the x-axis and similarly, Leg 2 and Leg 3 are responsible for the motion of the platform along the y- and z-axes, respectively. As a result of this architecture, the workspace of the end-effector point P is cubic.

The forward/inverse kinematics formulation of the R-cube manipulator is very simple and very few link lengths need to be determined for the rigid-body kinematics, while several other link lengths are arbitrary for kinematic relations assuming rigid links and joints. It is desired that the end-effector point P is at the center of the cubic workspace when A_1B_1, A_2B_2, and A_3B_3 are perpendicular to the X-, Y-, and

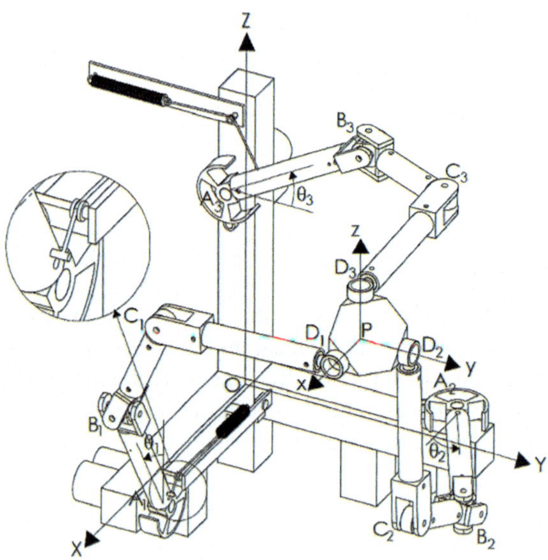

$$|OA_1| = |OA_2| = |OA_3| = a$$

$$|A_1B_1| = |A_2B_2| = |A_3B_3| = b$$

$$|B_1C_1| = |B_2C_2| = |B_3C_3| = c$$

$$|C_1D_1| = |C_2D_2| = |C_3D_3| = d$$

$$|D_1P| = |D_2P| = |D_3P| = e$$

Fig. 2 HISS kinematic parameters

Z-axes, respectively. This is the home position of the manipulator. Actuated joint angles θ_i are measured from the A_iB_i locations at the home position as depicted in Fig. 2. In order for point P to be at the center of the cubic workspace at the home position, $|A_iB_i| = b = a - e$. The forward kinematics equations are easily obtained as follows:

$$P_x = b(1 + \sin\theta_1) \tag{1}$$

$$P_y = b(1 + \sin\theta_2) \tag{2}$$

$$P_z = b(1 + \sin\theta_3) \tag{3}$$

Consequently, the inverse kinematics equations are obtained as

$$\theta_1 = \sin^{-1}\left(\frac{P_x}{b} - 1\right) \tag{4}$$

$$\theta_2 = \sin^{-1}\left(\frac{P_y}{b} - 1\right) \tag{5}$$

$$\theta_3 = \sin^{-1}\left(\frac{P_z}{b} - 1\right) \tag{6}$$

Accordingly, the forward kinematics Jacobian matrix $\widehat{J}$ is obtained as

$$\widehat{J} = \begin{bmatrix} b\cos\theta_1 & 0 & 0 \\ 0 & b\cos\theta_2 & 0 \\ 0 & 0 & b\cos\theta_3 \end{bmatrix} \tag{7}$$

Since HISS is designed to be used as a haptic device, the dimensional synthesis of the modified R-CUBE mechanism is conducted in accordance with the haptic performance metrics. The performance of a haptic manipulator is determined by its mechanical impedance characteristics since it designates the quality of the reflected force/torque to the user. The range between the maximum and minimum value determines the range of forces/torques to be rendered, while the frequency range indicates the diversity of the forces/torques which is felt by the user. Hence, higher range in both the impedance and frequency is needed to enhance the perception of reality and quality of the performance. However, impedance and the frequency range depend on the performance of kinematics, stiffness, and dynamics characteristics which are highly contradicted with each other. Higher stiffness is required to render higher impedances. However, increasing the stiffness performance is achieved by driving the manipulator to singular poses and/or increasing the thickness of the links. In both cases, either kinematic performance is decreased in terms of manipulability or dynamic performance is decreased which reduces the performance of minimum impedance. Therefore, an optimization is required to determine the kinematic parameters by virtue of kinematics, stiffness, and dynamic performance metrics. Since the relationship between these metrics is non-linear, evolutionary or artificial intelligence based optimization algorithms are needed. In this study, the multi-objective genetic algorithm is adopted for optimization. In order to construct the objective functions, kinematic, stiffness, and dynamic models are procured. In kinematic performance evaluation, manipulability [7] and condition number [8] are the optimization parameters for achieving singularity avoidance and isotropic motion resolution distribution within the workspace. Next, the stiffness model is obtained by using the virtual joint method [9] and represented by means of a stiffness matrix in the Cartesian space, $\widehat{K}_C$. The computation is achieved by the multiplication of $\widehat{K}_C = (\widehat{J}_\theta \widehat{K}_\theta^{-1} \widehat{J}_\theta^T)^{-1}$, where $\widehat{J}_\theta$ is the Jacobian matrix including the virtual joint variables and $\widehat{K}_\theta$ denotes the stiffness matrix in joint space which only depends on the geometry and mechanical properties of the links. The determinant of the stiffness matrix, which is pose dependent, is computed in order to obtain the measure of rigidity as a scalar metric [10]. Finally, the determinant of the inertia matrix of the modified R-CUBE is calculated in order to obtain the mass distribution with respect to pose changes. The aim is to increase the manipulability and stiffness while decreasing the inertia and condition number. The objective functions to be minimized are constructed as follows:

$$O_1 = \left|\widehat{J}\right|^{-1} \tag{8}$$

$$O_2 = \left\|\widehat{J}\right\| \left\|\widehat{J}^{-1}\right\| \tag{9}$$

$$O_3 = \left|\widehat{M}_C\right| \tag{10}$$

$$O_4 = \left|\widehat{K}_C\right|^{-1} \tag{11}$$

where | | is the determinant and || || is the 2-norm operator and $\widehat{M}_C$ and $\widehat{K}_C$ are the pose dependent inertia and stiffness matrices in Cartesian space. O_1, O_2, O_3, and O_4 are the inverse of manipulability index, condition number, the determinant of the inertia matrix, and the inverse of the determinant of the stiffness matrix, respectively. The links are assumed to be made of carbon fiber composite tubes with 10 mm outer and 6 mm inner diameter, hence associated mass and stiffness values are used. Constraint equations are introduced to the optimization algorithm in order to avoid singularity. An extra constraint is added for ergonomics so that the hand of the user does not collide with any of the links. The optimization is conducted by Pareto-front approach. Hence, a set of optimum solutions is obtained and the one which has the highest impedance and natural frequency is selected. Optimum solution parameters are as $b = 111.7$ mm, $c = 74.1$ mm, $d = 121.8$ mm. In this case, the maximum range of the input joints is $\pm 32.5°$ and the values of objective functions are computed as $O_1 = 10.638$, $O_2 = 10.63$, $O_3 = 51 \times 10^{-4}$, and $O_4 = 8.821 \times 10^{-7}$.

3 Constructional Design and Balancing

It is important to have a lightweight and high stiffness manipulator to have high transparency and high frequency range in impedance type of haptic device. Therefore, carbon fiber composite tubes and aluminum connections are selected as the materials for the links (Fig. 3).

When using a manipulator as a master device, the operator should not feel the weight and inertial effects of the links. For gravity compensation, springs are designed. The weights of the moving platform and all links that move in the same z-level with the moving platform are lumped to Leg 3, which is responsible for the z-axis motion and balanced with a single spring attached to the driving link A_3B_3 of Leg 3. The potential energy of the platform and links B_3C_3 and C_3D_3 change together, so their masses are added up and lumped to joint B_3. Also, to simplify the balancing solution, half of the masses of links B_1C_1, C_1D_1, B_2C_2, and C_2D_2 are

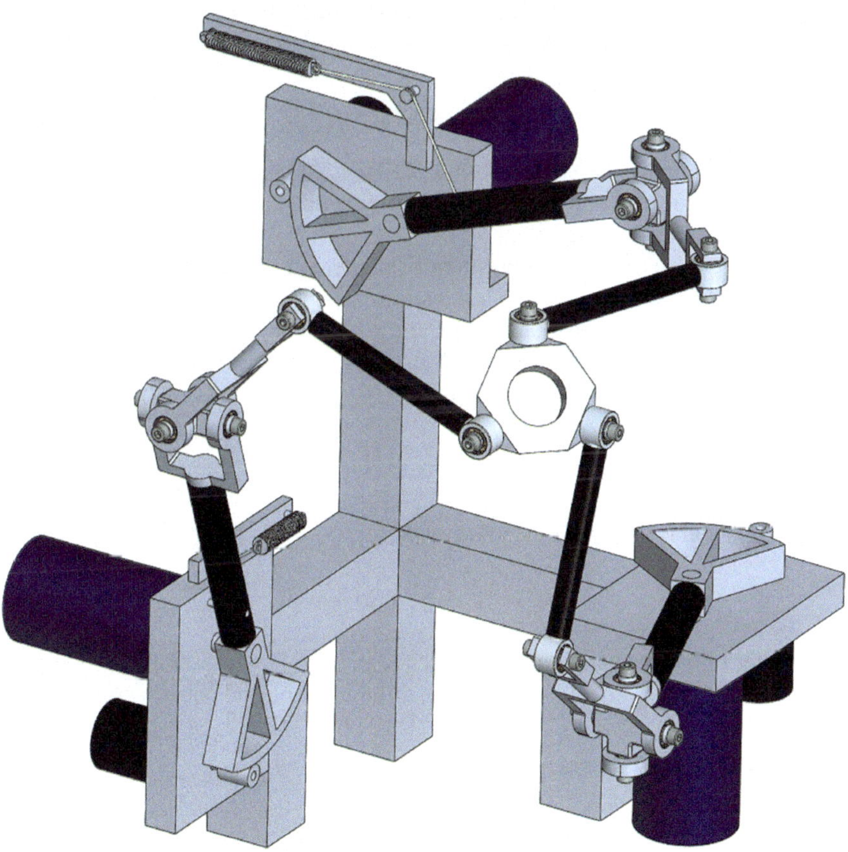

Fig. 3 HISS CAD design

Fig. 4 Spring balancing

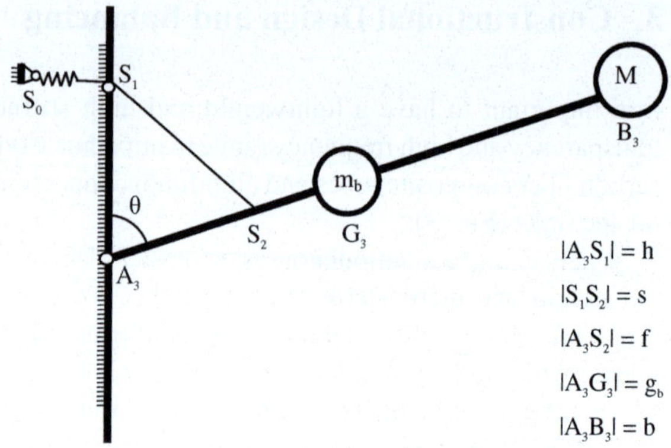

added to the aforementioned lumped mass as well. The reason for taking the half of mass for these links is that the mass centers of these links are approximately at the middle and hence the z-motion of these centers of masses is approximately half of that of the platform. Let M denote the whole lumped mass. Then

$$M = m_P + m_c + m_d + 2(m_c/2) + 2(m_d/2) = m_P + 2m_c + 2m_d \qquad (12)$$

where m_c, m_d, and m_P are masses of the platform, link B_iC_i, and link C_iD_i, respectively. Also considering the mass m_b of the driving link A_3B_3, the inverted pendulum shown in Fig. 4 needs to be balanced.

The spring attachment point location or the length of the string attached to the spring should be adjusted such that the spring is unloaded when $\theta = 0$, i.e., when link A_3B_3 is at upright position. $|A_3S_2| = f$, $|A_3S_1| = h$, and spring constant k are design parameters. The total potential energy of the masses and the spring can be written as

$$U = m_b g g_b \cos\theta + Mgb\cos\theta + \frac{1}{2}ks^2 \qquad (13)$$

Cosine theorem in triangle $S_1S_2A_3$ reads

$$s^2 = f^2 + h^2 - 2fh\cos\theta \qquad (14)$$

In order to have constant potential energy, its variation with respect to θ should be zero. Substituting Eq. (14) in Eq. (13), differentiating and equating to zero:

$$\frac{\partial U}{\partial \theta} = (-m_b g g_b - Mgb + kfh)\sin\theta = 0 \Rightarrow kfh = (m_b g_b + Mb)g \qquad (15)$$

All terms on the right-hand side of Eq. (15) are known. First for chosen f and h, the desired k value is determined. Then a tension spring is manufactured according to the desired k value. The spring constant of the manufactured spring is then determined

Fig. 5 Spring balanced HIPHAD v1.0

with several tests. The measured spring constant is usually different from the desired value. f or h value is adjusted according to the measured k value.

The only remaining unbalanced link is link A_1B_1. This link can also be balanced individually with a similar approach. The balance springs for the new design are illustrated in Figs. 2 and 3. To test the balancing performance, High Precision Haptic Device (HIPHAD) version 1 shown in Fig. 1 is balanced with two springs (Fig. 5). The tests show that the moving platform remains in static balance wherever the operator positions it when the actuators are not active.

4 Semi-active Actuator Design

A haptic device does not only work to display constraints to stimulate the human operator but it also produces motion of the handle to stimulate effects such as surface roughness or slip conditions. In the previous version of the device, the actuation system was composed of a direct-drive system that included a BLDC motor. An actuation system composed in this way or by combining an electrical motor with a reduction system such as Capstan drive enables to display both constraints and motion at the handle of the device. However, although these actuation systems are designed for producing motion, applying brakes with these actuation systems to display constraints has a limiting factor. This limiting factor is the size of the actuator. In order to display a larger amount of static forces, larger electric motors are required which degrade the dynamic performance of the system.

In haptic device development, MR-fluid based brake systems can produce controllable viscosity-based damping force. A design of an MR-brake based haptic device is introduced in [11] and a multi-degree-of-freedom haptic device was built by using an MR-fluid brake [12]. Some researchers even used the MR-brakes with the conventional motor-capstan drive systems [13] to increase the z-width of the device by increasing the maximum stiffness of the system.

A Magneto-Rheological (MR) fluid based semi-active actuation system or in other words, an MR-fluid based brake system is designed to resolve this problem of displaying a larger amount of static forces without jeopardizing the dynamic performance of the system. There are two major disadvantages of such systems;

(1) it cannot be the sole actuation system in a haptic device since the device is required to display motion like in the solution given in [14].
(2) the conventional design of a rotary MR-fluid based brake applies brake in both directions when it is activated. This results in an undesired stiction feeling displayed to the human user.

The design proposed has two rotary MR-brakes assembled on top of each other to constrain the motion of a common shaft in between them in opposite directions. In this way, decoupled way of braking is possible and thus, the stiction problem is resolved [15]. The design and its performance evaluation is provided in [16] and presented in Fig. 6.

The key element in the design is the one-way clutch or in other words, one-way bearing, which constrains the motion in one direction and lets it free in the other. This new MR-brake design has two directional brake conditions as presented in Fig. 7:

(1) The rotation in the clockwise direction:
  - Top rotating part of the MR-brake and the shaft are coupled
  - The human operator feels the braking torque with the activation of the top part of the MR-brake
  - Free rotation is in the reverse direction, even when the brake is activated for the top part.
(2) The rotation in the counterclockwise direction:
  - Bottom rotating part of the MR-brake and the shaft are coupled
  - The human operator feels the braking torque with the activation of the bottom part of the MR-brake
  - Free rotation is in the reverse direction, even when the brake is activated for the bottom part.

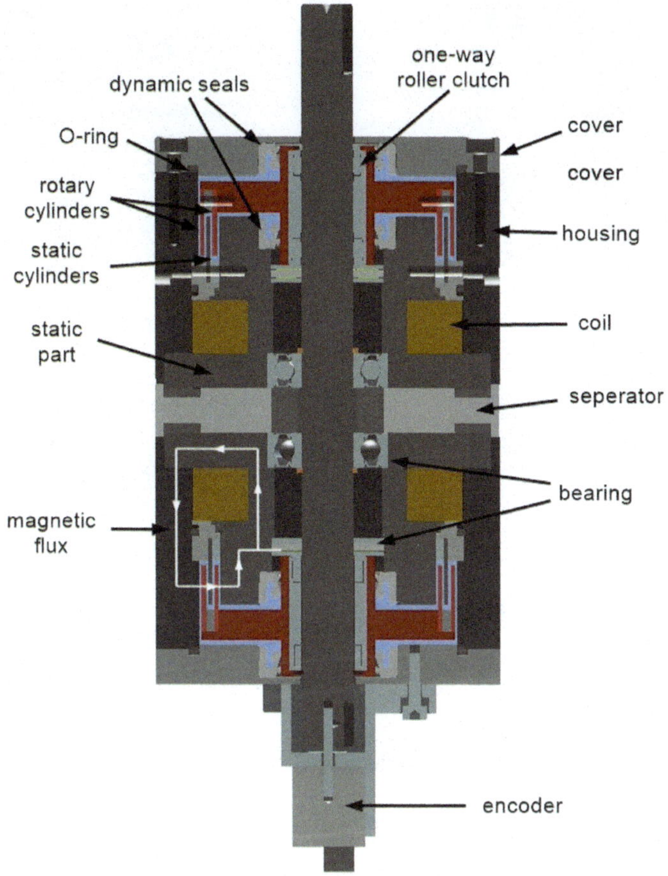

Fig. 6 Design details of the bi-directional MR-brake system [16]

As a result of this design, the brake can be applied in only one direction and even motion in both directions can be constrained simultaneously at different braking torques. The results from an experimental study presented in Fig. 8 clearly indicates that when the MR-brake constrains the motion in one direction, the user is free to move in the other direction.

The experimental test setup in Fig. 8 is composed of a handle attached to the MR-brake via a torque sensor. The user moves the handle in both directions during the test. The plot on the bottom right corner is for the measured torques by the torque sensor as the user moves the handle. After a while, the MR-brake is activated in one direction then, it is activated in the other direction and finally, they are both in off condition to terminate the experiment. It is clearly observed from this plot that the user's motion is constrained by a maximum torque when he tries to move in the direction of the constraint and when he moves in the other direction, the measured torque is minimized. This result proves the working principle of the bi-directional

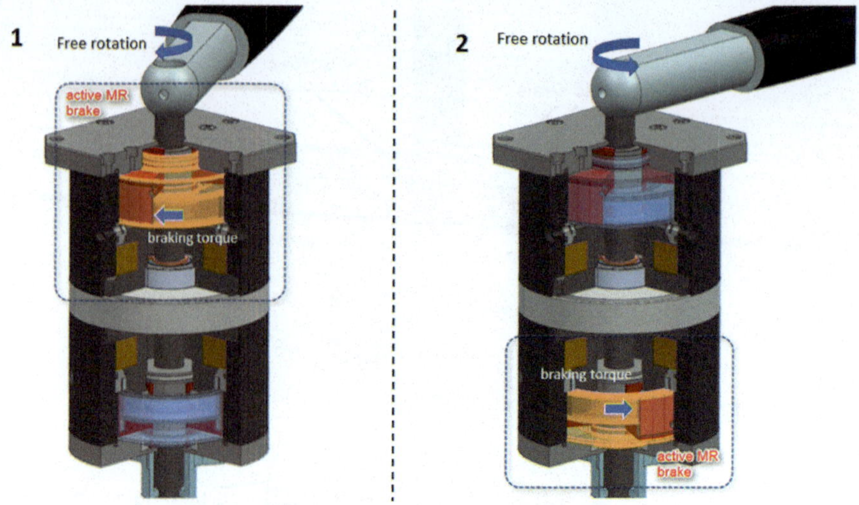

Fig. 7 Bi-directional operation of the new MR-brake system

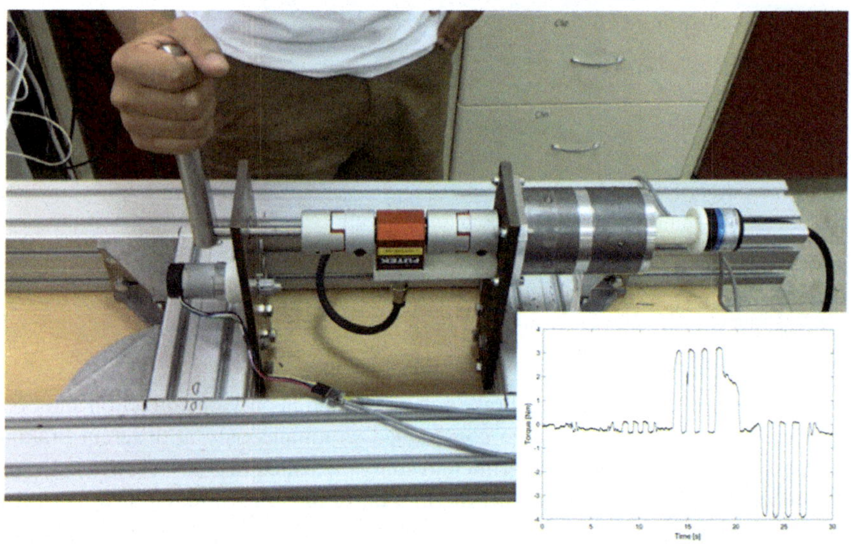

Fig. 8 User evaluation test setup and the result

MR-brake system. A study is carried out to characterize the developed system. The results of this characterization are presented in Table 1.

Table 1 Bi-directional MR-brake specifications [16]

Category	Specification
Product mass	3.57 kg
Braking torque @ 2 A	3.8 N m
Off-state torque	0.15–0.4 N m
Bandwidth	63 rad/s
External diameter	80 mm
Length	124 mm
Coil wire dia.	0.5 mm
Max. operational current	2 A
Number of turns (coil)	450
Magneto-rheological material	MRF-122EG
Magnetic material	AISI 1008 steel
Non-magnetic material	Stainless steel and aluminum

5 Conclusions

This study presents the mechanical design of a manipulator to be used for a haptic system that can acquire accurate motion demand even when it is providing force feedback and that has enhanced impedance width and transparency performance. First, the kinematic parameters are optimally designed according to kinematic, stiffness, and dynamic performance indices. Then a lightweight and statically balanced constructional design is presented. Finally, the semi-active actuator design is given. Manufacturing and tests of the prototype of the designed manipulator are the future works for this study.

Acknowledgements This work is supported by The Scientific and Technological Research Council of Turkey via grant number 117M405.

References

1. Li W, Gao F, Zhang J (2005) R-CUBE, a decoupled parallel manipulator only with revolute joints. Mech Mach Theory 40:467–473
2. Bilgincan T, Gezgin E, Dede MİC (2010) Integration of the hybrid-structure haptic interface HIPHAD v1.0. In: Proceedings of international symposium mechanism and machine theory, Izmir, Turkey, pp 267–284
3. Bilgincan T (2010) Design of a six degree-of-freedom haptic hybrid platform manipulator. MSc thesis, İzmir Institute of Technology, İzmir, Turkey, May 2010
4. Samur E (2012) Performance metrics for haptic interfaces. Springer
5. Colgate JE, Brown JB (1994) Factors affecting the z-width of a haptic display. In: Proceedings of IEEE international conference on robotics and automation, pp 3205–3210

6. Mobedi E, Görgülü İ, Dede MİC (2018) Experimental evaluation of actuation and sensing capabilities of a haptic device, accepted to be presented in European Conference on Mechanism Science (EUCOMES), Aachen, Germany
7. Yoshikawa T (1985) Manipulability of robotic mechanisms. Int J Robot Res 4(2):3–9
8. Salisbury JK, Craig JJ (1982) Articulated hands: force control and kinematic issues. Int J Robot Res 1(1):4–17
9. Gosselin C (1990) Stiffness mapping for parallel manipulators. IEEE Trans Robot Autom 6(3):377–382
10. Carbone G, Ceccarelli M (2010) Comparison of indices for stiffness performance evaluation. Front Mech Eng 5(3):270–278
11. Liu B, Li W, Kosasih PB, Zhang X (2006) Development of an MR-brake-based haptic device. Smart Mater Struct 15(6):1960
12. Senkal D, Gurocak H (2009) Spherical brake with MR fluid as multi degree of freedom actuator for haptics. J Intell Mater Syst Struct 20(18):2149–2160
13. Samur E, Flaction L, Spaelter U, Bleuler H, Hellier D, Ourselin S (2008) A haptic interface with motor/brake system for colonoscopy simulation. In: Symposium on haptic interfaces for virtual environment and teleoperator systems, haptics 2008, pp 477–478. IEEE
14. Rossa C, Lozada J, Micaelli A (2014) Design and control of a dual unidirectional brake hybrid actuation system for haptic devices. IEEE Trans Haptics 7(4):442–453
15. Karabulut M, Taner B, Dede M (2015) MR Sıvısı ile Çalışan Haptik Kol Tasarımı-design of MR-fluid based Haptic Paddle, TOK 2015 Bildiri Kitabı, pp 867–872
16. Karabulut MG, Dede MIC (2018) Design and experimental validation of an MR-fluid based brake for use in haptics. In: Proceedings of actuators 2018, Bremen, Germany

A Continuously Variable Transmission System Designed for Human–Robot Interfaces

Emir Mobedi and Mehmet İsmet Can Dede

Abstract Within a predefined limit, continuously variable transmission (CVT) systems can continuously vary the power transmission ratio. The transmission in CVTs is achieved via friction, belt or gear systems. If CVT designs can incorporate back-drivability, independent output position and impedance variation, shock absorbtion, and low mass and inertia, they can be employed in human–robot interfaces. Among various types of CVT designs, the two-cone drive CVT designs have a major drawback since the output torque and position cannot be changed independent of each other. The friction wheel used in this design does not have a holonomic motion capability and causes this inconvenience. In order to overcome this problem, a sphere is used in this work for the CVT design as the transmission element. In addition, it is stated in the literature that common CVT drive systems do not have the capability to be used in cyclic bidirectional motion. In the presented CVT design, a second sphere is added to the system with two springs from the lower part of the cones for pretension in order to solve the bidirectional transmission problem. In this paper, the working principle and conceptual design details of the novel two-cone CVT drive are presented. Experimental results showed that the novel CVT has the capacity to transmit bidirectional power with some accuracy.

Keywords Continuously variable transmission · Human–robot interaction · Haptics

1 Introduction

For many decades, robots' primary employment has been in industrial tasks such as polishing, welding, pick, and place. They are developed to follow only the predefined trajectory with high position accuracy and they can be named as conventional industrial robots. Even if they had the capability to achieve some tasks, they were

E. Mobedi · M. İ. C. Dede (✉)
Mechanical Engineering Department, Izmir Institute of Technology, Urla, İzmir, Turkey
e-mail: candede@iyte.edu.tr

E. Mobedi
e-mail: emirmobedi@iyte.edu.tr

D. Sen et al. (eds.), *Mechanism and Machine Science*, Lecture Notes in Mechanical Engineering, https://doi.org/10.1007/978-981-15-4477-4_3

not allowed to work with the human operators side by side. This limitation was due to the safety reasons as well as the absence of advanced sensory systems. As a consequence, the conventional industrial robots are designed to operate in working cells or workshops inside the protective fences.

In recent years, a new generation of robots named collaborative robots have been produced to work alongside with humans in order to increase the efficiency of the work. These robots are developed under ISO 10218-1 standard to avoid possible damages to the human coworker who is present in the work cell of the robot. According to the fifth clause (5.10.5) of this standard, robots with inherent force exertion limitations can work alongside the human operators while in continuous contact with them since these robots have the capability of avoiding the application of excessive forces to the operator. This capability can be implemented in the robots by control and/or mechanical design. The CVT design presented in this paper is a mechanical solution for constructing the active joint structure of a robot by following the fifth part of the standard, which enables the user to work with the robot in continuous contact situation.

Considering the joint structures of collaborative robots, designers should consider some additional design criteria including backdrivability, mechanical impedance, workspace, low mass and inertia, variation of impedance without disturbing the motion, and shock absorbtion [1]. Good backdrivability is achieved when the resistance of the robot to the motion input by the human from the end-effector is minimized [2]. When there is a physical interaction, the robot is expected to be backdrivable so that the human–robot interaction happens in a two-way stream. The aforementioned resistance is defined as mechanical impedance which relates the induced force to the subsequent velocity for a robot. It should be the maximum for enhancing accuracy and minimum for enhancing the backdrivability. Thus, the problem is narrowed down to the adjustment of the robot's impedance either mechanically or by control. A solution to this has been proposed by designing variable impedance actuators and joint structures.

Joints configured with antagonistic springs and motors are listed in the literature as variable impedance joints [3]. The primary advantage is that without the need of using a current control for the motor, joint stiffness and position are changed independent of each other by the help of this mechanical stiffness device [4]. However, in this design, the workspace is limited because of the limits of the used servomotors.

Another design is antagonistic springs with independent motors [4]. In this setup, there are two motors which are used for stiffness variation and joint position variation. Accordingly, the mechanism is capable of altering the joint motion and stiffness independently. The device absorbs shocks with the help of a spring which is located in between the joint and the base. However, the motor that controls the stiffness is positioned at the joint and this causes an increase in the joint structure's mass and inertia, which results in the degradation of the backdrivability capability.

Among the possible CVT designs, planetary gear-based CVT drives are capable of altering the stiffness mechanically [5]. In addition, it is claimed that it is able to provide protection inherently against overloading. Also, the design is claimed to be simpler than spring-based variable-stiffness joints. Hence, it is easy to carry out the force and kinematic analysis of the system. On the other hand, stiffness of the joint and its position cannot be varied independent of each other since output resistance is an input to the system. In addition, since the mechanism is produced with the planetary gear system, the inertia of the gears increases the inertia felt by the user in addition to the increased friction effects while backdriving the system from the end-effector.

Another well-known CVT drive is developed by the use of dry friction. The first robot that uses this working principle, Cobot [6], was designed for haptic applications. In this design, there are CVT wheels, which are coupled with the proximal link and the power cylinder. When stiffness variation is required, the power cylinder is rotated and CVT wheels are steered depending on the desired stiffness. Therefore, stiffness variation is achieved by altering the angle between the power cylinder and CVT wheels. However, the power cylinder makes the design bulky because of its footprint. Spherical continuously variable transmission (S-CVT) is yet another CVT design example [7]. In S-CVT system, two input discs, two output discs, and one variator having friction surfaces are connected with a sphere in order to adjust the stiffness of the output discs by the help of friction force. Input and output discs have point-type of contact with the sphere. By adjusting the location of the rotation axis of the sphere, output velocity and output torque are altered.

Finally, CVT cone drives are used to change the output torque in a continuous way. In this drive system, two cones are located parallel to each other. The transmission between the cones is realized with the help of dry friction, belt or gear [8]. Two problems are detected in the literature related to dry-friction two-cone drive systems. Firstly, the transmission occurs only in one way and output torque cannot be changed without altering the output position. These issues limit the usage of two-cone CVT systems for the human–robot interface. Hence, in this study, the working principle, conceptual design details, and experimental validation results of a new two-cone CVT drive system are presented to propose a solution to the aforementioned problems.

2 Operation of the New CVT Design

In two-cone CVTs with friction drive, there are two cones, one being the input and the other being the output cone. There is also a wheel that transmits the moment from the input cone to the output cone. Moment and speed transmission levels are altered by changing the ratio of the radii between the input and output cones by varying the point of contact of the wheel. The wheel that is used for transmission of the moment does not have holonomic motion capability and thus, output position and torque cannot be controlled independently. Hence, this defect limits the usage of the two-cone CVT drive as a joint structure of the human–robot interfaces. The first modification to overcome this problem is made by using a sphere, which is capable of executing holonomic motion (Fig. 1). Having the capability of holonomic motion output position and torque can now be controlled independent of each other. Friction surfaces are created on the input and output cones to avoid the application of a large

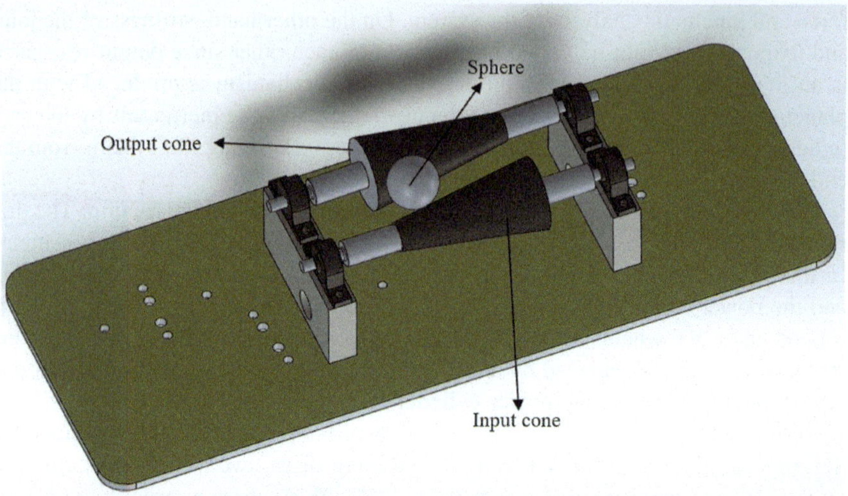

Fig. 1 A 3D view of the new CVT design with minimal components

amount of normal force for high moment transmission. At this point, a limitation of the continuously variable transmission system should be mentioned. This limitation occurs because the sphere and the cone form a point-type of contact for ideally rigid structures. As a result of this, the transmitted torque is limited. However, in the application, neoprene material is used for the friction surface which was also used in HAPKIT for constructing the friction drive [9]. The use of this soft material changes the contact type from an ideal point-type of contact to an ellipse-shaped contact. This ellipse-shaped contact area enlarges with respect to the applied normal force creating a larger contact surface. In this way, the aforementioned limitation on the limitation of the torque transmission is avoided (Fig. 2).

The location of the spheres with respect to the input and output cones axes is critical since the sphere should be moved with ease along the cones to change the transmission ratio. Locating the sphere's center on the plane formed by the axes of the input and output cones prevents the possibility of the holonomic motion. The mathematical proof of the reason is presented as follows:

$$\vec{M}_1 = r\vec{u}_2^{(0)} \times F_{t1}\vec{u}_3^{(0)} = F_{t1}r\vec{u}_1^{(0)} \tag{1}$$

$$\vec{M}_2 = -r\vec{u}_2^{(0)} \times F_{t2}\vec{u}_3^{(0)} = -F_{t2}r\vec{u}_1^{(0)} \tag{2}$$

Friction forces (F_{t1} and F_{t2}) are in the same direction and have the same magnitude in this ideal representation of the system. Therefore, the sphere cannot roll around the $\vec{u}_1^{(0)}$ axis and it needs to slip between the two cones when the transmission ratio is required to be changed. On the contrary, this slippage is an undesired effect since it affects the transmission adversely. As a solution to this problem, the sphere is located

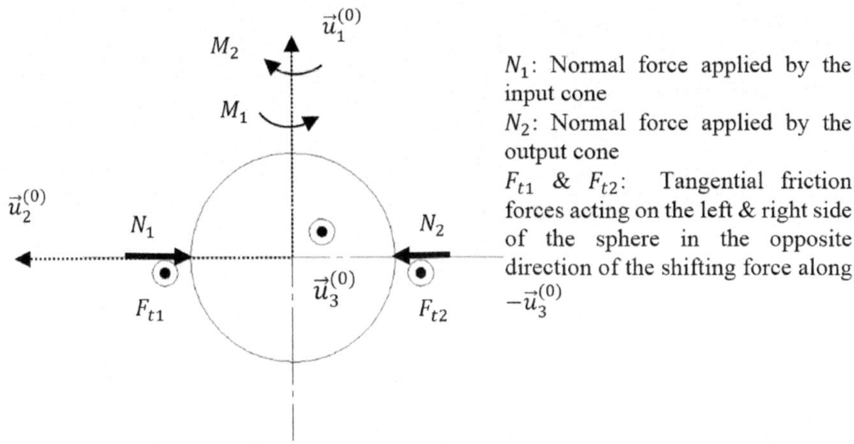

Fig. 2 Forces and moments acting on the sphere when its center is located on the plane formed by the axes of the input and output cones

above the input and output cones and as a result of this, it rolls while it moves to change the transmission ratio. The next set of equations which are based on Fig. 3 proves this phenomenon.

The angle between the frames $\mathcal{F}_0$ and $\mathcal{F}_1$ and $\mathcal{F}_0$ and $\mathcal{F}_2$ are defined as $\sphericalangle \vec{u}_1^{(0)} \rightarrow \vec{u}_1^{(1)} = \frac{\pi}{2} - \beta_1$; $\sphericalangle \vec{u}_1^{(0)} \rightarrow \vec{u}_1^{(2)} = \frac{\pi}{2} + \beta_2$. Based on this definition, the moments that the friction forces in the opposite direction of the shifting force are calculated as follows:

$$\vec{M}_1 = r\vec{u}_2^{(1)} \times F_{t1}\vec{u}_3^{(1)} = F_{t1}r\vec{u}_1^{(1)} \tag{3}$$

$$\vec{M}_2 = r\vec{u}_2^{(2)} \times F_{t2}\vec{u}_3^{(2)} = F_{t2}r\vec{u}_1^{(2)} \tag{4}$$

The above vector equations can be resolved in the inertial frame $\mathcal{F}_0$ as presented in Eqs. 5 and 6. It should be noted that $\widehat{C}^{(0,i)}$ is the transformation matrix between the inertial frame and the ith frame and $\widetilde{u}_3$ resembles the skew symmetric matrix form of the $\widetilde{u}_3$ column matrix.

$$\overline{M}_1^{(0)} = \widehat{C}^{(0,1)}\overline{M}_1^{(1)} = F_{t1}re^{\widetilde{u}_3(90-\beta_1)}\overline{u}_1 = F_{t1}r[\overline{u}_1 \sin\beta_1 + \overline{u}_2 \cos\beta_1] \tag{5}$$

$$\overline{M}_2^{(0)} = \widehat{C}^{(0,2)}\overline{M}_2^{(2)} = F_{t2}re^{\widetilde{u}_3(90+\beta_2)}\overline{u}_1 = F_{t2}r[-\overline{u}_1 \sin\beta_2 + \overline{u}_2 \cos\beta_2] \tag{6}$$

The $\vec{u}_2$ component of the moments in Eqs. 5 and 6 makes the sphere roll on the cones. The $\vec{u}_1$ components of the moments are in the opposite directions which results in no rotation about the $\vec{u}_1$ axis. As a consequence of this, the spherical transmission

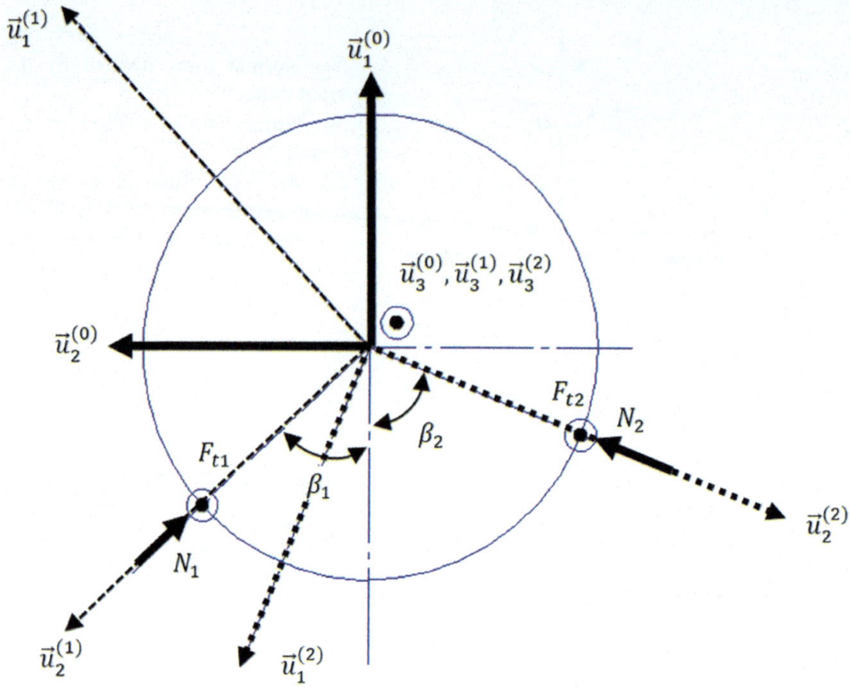

Fig. 3 Moments and forces acting on the sphere when it is located above the input and output cones

element can be moved without the need of any slippage and this is the reason why the sphere must be located above the cones.

Nevertheless, relocation of the sphere above the cones does not solve the problem fully. Figures 4 and 5 are drawn to present the new problems. In Fig. 4, the human operator manipulates the output cone via a handle attached on the output cone.

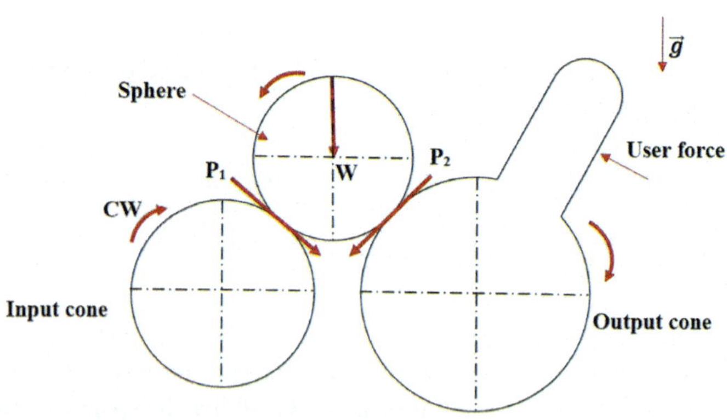

Fig. 4 Input moment applied in the CW direction

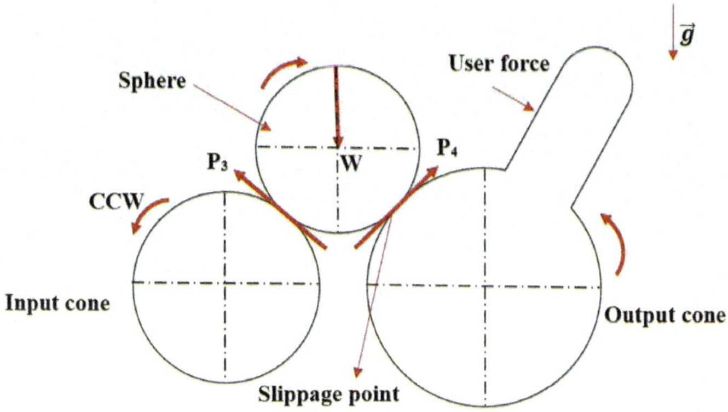

Fig. 5 Input moment applied in the CCW direction

The transmission achieved as the input is in CW direction is shown in Fig. 4 and CCW direction I shown in Fig. 5. The friction forces (P_1, P_2) on the sphere, which are required for the moment transmission, are in the downward direction for CW input (Fig. 4). However, for the CCW input, the friction forces (P_3, P_4) are in the upward direction (Fig. 5). When the friction forces are in the downward direction, the sphere is pulled down and the normal force increases, which guarantees the moment transmission at higher values. However, when the friction forces are upward, the sphere is pushed away from the cones and the normal forces decrease, which decreases the maximum amount of transmissible moment. In Figs. 4 and 5, W is the weight of the sphere. To overcome this problem, a second sphere below the cones is integrated to the design. Additionally, two springs with pretensions are used to apply forces (F_1, F_2) to each sphere in order to generate equal amount of normal force on each sphere (Fig. 6).

3 Proof of the Concept Design

The design details of this new CVT are explained in this section in order to be used for manufacturing a prototype for proof of the design tests (Fig. 7).

First, the carriage system is described in Fig. 8. This carriage system carries the spheres.

The carriage system design consists of 8 parts which are compression springs (**6**), support elements (**5**), 4 pieces of steel pins (**4**) and bearings (**3**), 2 pieces of steel spheres (**1, 2**), and the carriage (**8**). The springs' pretension should be adjusted carefully during the assembly process of the carriage. The carriage system is mounted on the screw-nut (**11**) and linear guides (**9**) as a next step (Fig. 9). In Fig. 9, the other components of the design are listed as cones (**10**), bedding pieces for the cones (**12**), and the carriage's linear guides (**13**).

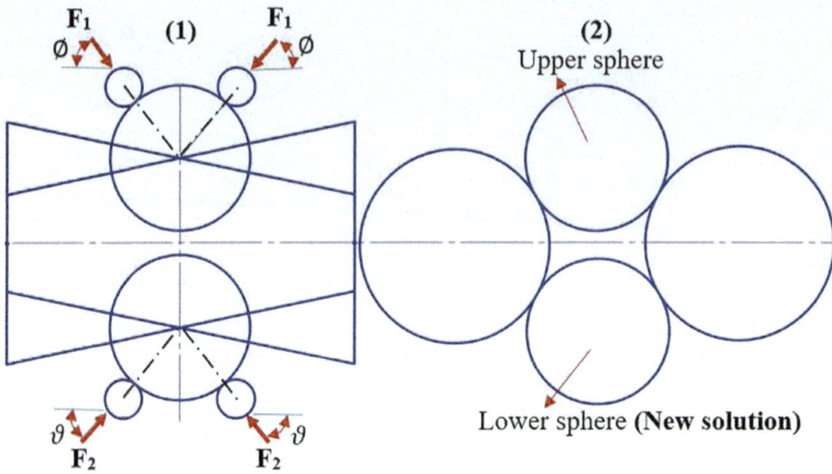

Fig. 6 (1) Spring pretension forces acting on the spheres (2) Location of the upper and lower spheres with respect to the cones

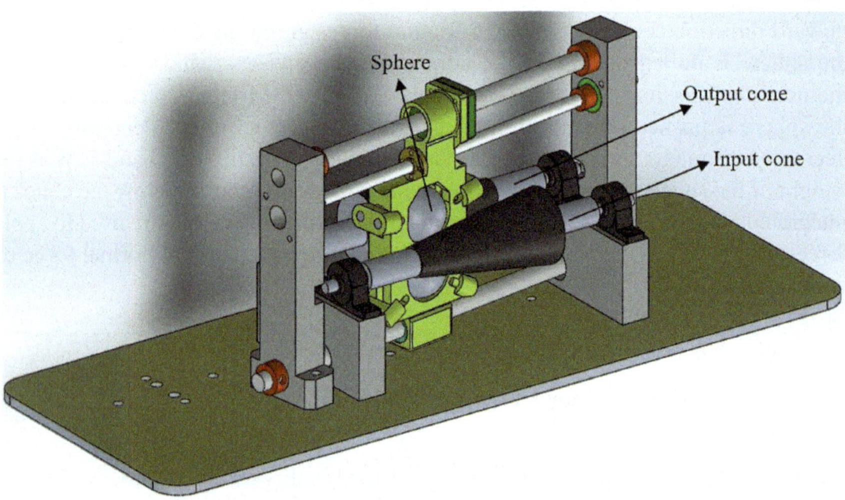

Fig. 7 3D view of the new CVT with its complete set of components

The maximum allowable torque transmission is regulated by adjusting the pretension of the springs (**6**). The normal force is decreased by decreasing the pretension of the spring in the assembly process to adjust lower maximum transmission torques. Hence, if this maximum torque limit is violated, slippage occurs between the spheres and the cones and the torques beyond the set limit are not transmitted. Hence, a safety feature is included in the design by the use of this property.

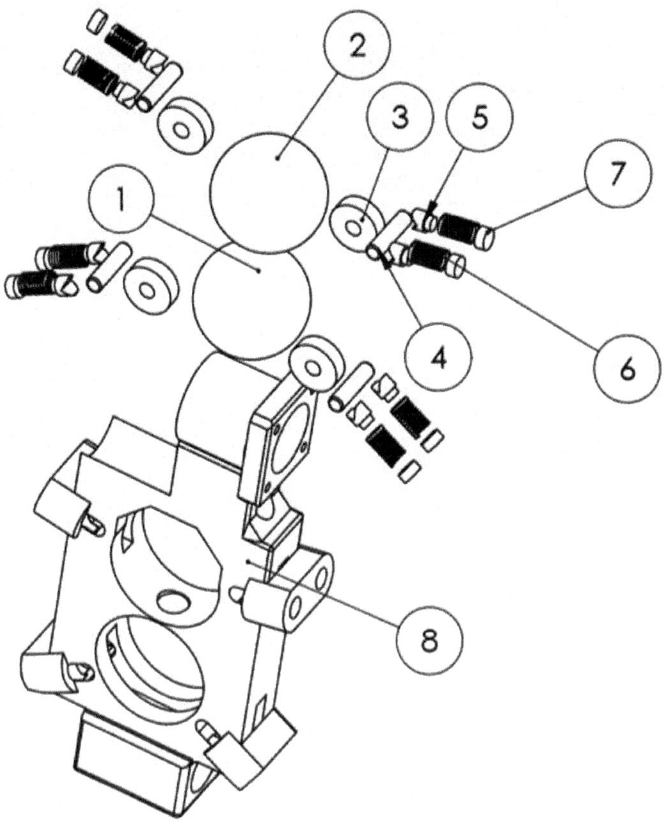

Fig. 8 3D view of the components of the carriage system

If the system is arranged so that the spheres are aligned along the vertical axes, then the compression force (F_2) on the below sphere has to be higher than the compression force (F_1) on the above sphere in order to overcome the gravitational effects. As a result of this adjustment, normal forces between the cones and the spheres are set to be equal to each other, and the same type of transmission in two directions is made possible.

Since bearings are used to transmit spring force to the spheres, this minimizes the induced friction in the assembly of the sphere. Consequently, faster response times can be reached during the variation of the transmission ratio.

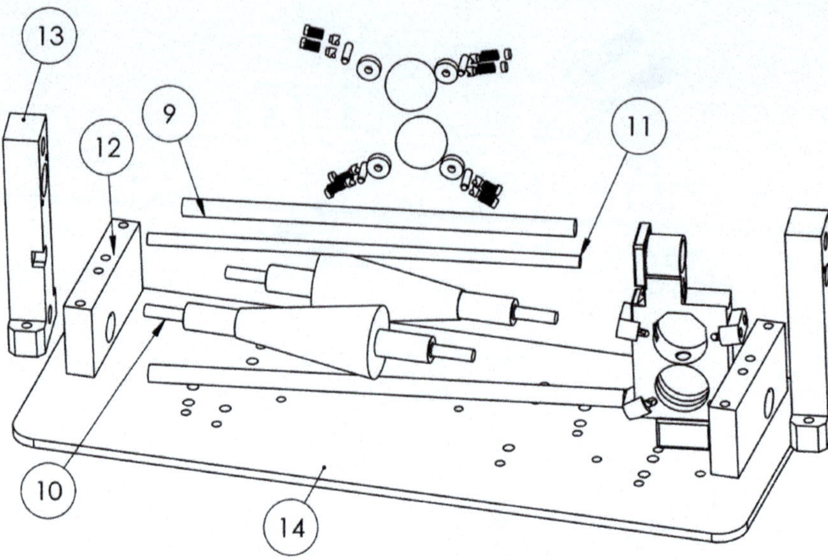

Fig. 9 3D view of the new CVT design with all of its components

4 Experimental Results

The proof of concept design is investigated via an experiment. In this experiment, torque transmission during CW and CCW inputs are observed to see if they are equal to each other for the same amount of input moments. Six different locations of the cone that corresponds to different test diameters are initially selected. The selected locations' diameters are measured on the cones with the help of a digital caliper (Mitutoyo). Input torque of 72 mNm is applied by hanging a calibrated mass on the input cone from a measured moment arm length. An encoder (Fig. 10) with 0.3515625° resolution is assembled on the input cone to observe if any slippage occurs between the input cone and the sphere. Output cone is coupled with a force/torque sensor (ATI, Nano 17), which is presented in Fig. 11, in which only the torque along the axis of rotation of the cone is measured. The output cone is mechanically fixed to the inertial frame in order to constrain its rotation. The results are presented as follows:

In Table 1, the output torques for the same input torques in both CW and CCW directions are measured for each test point. According to these results, the torque transmission in both directions was accomplished.

Nevertheless, there are relatively small inconsistencies between the output torque measured in CW and CCW directions at the same test point. Also, the inconsistency between the direction of operation is not always in the favor of one direction. For example, the CW output torque is higher than the CCW only in the first, third, fourth, and sixth test locations. Therefore, it can be concluded that the inconsistency is not due to the concept that is tested here but due to the measurement inconsistencies and

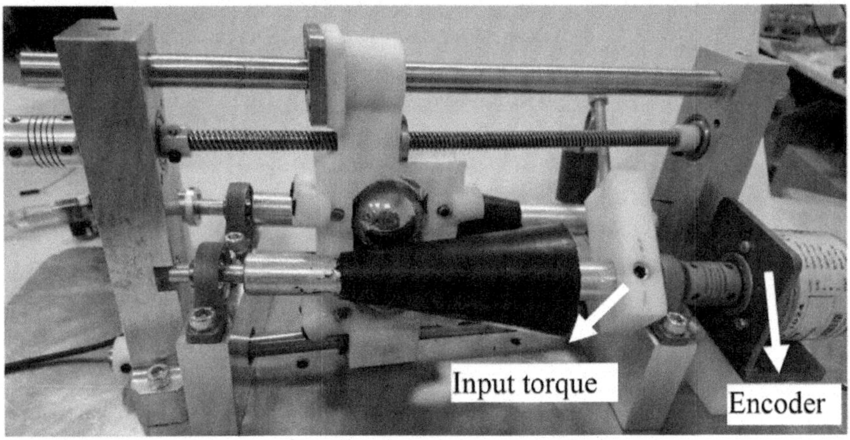

Fig. 10 Experimental setup right side view

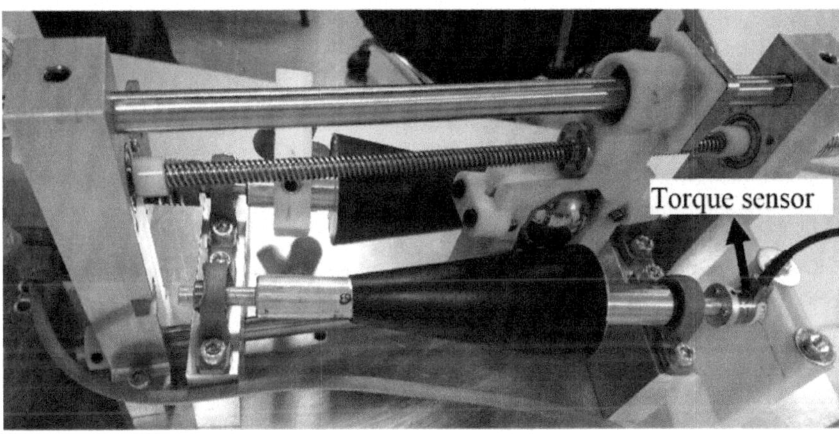

Fig. 11 Experimental setup left side view

manufacturing errors. Especially, the manufacturing errors affecting the smoothness of the surfaces and the errors when placing the friction surfaces on the cones may have affected these results.

5 Conclusion

A new CVT system is introduced in this study to be used for human–robot interaction systems. The primary novelty of the design presented in this work is the modification of the transmission element from a wheel to two spheres. The experimental

Table 1 Experimentally obtained output torque results

r_{out}/r_{in}	Input torque (mNm)	Direction	Output torque (mNm)
0.56	72	CW	33.39
		CCW	32.22
0.75	72	CW	40.17
		CCW	43.36
0.95	72	CW	52.52
		CCW	51.98
1.07	72	CW	66.57
		CCW	65.17
1.31	72	CW	68.21
		CCW	73.80
1.36	72	CW	85.66
		CCW	80.40

results with the proof of concept prototype have showed that bidirectional torque transmission was made possible with this design. Also, with this design, the user is able to alter the backdrivability level by ensuring safe physical interaction which is one of the most critical issues of the human–robot interfaces. As a future work, the proposed design will be optimized to be used for haptic device designs for varying the impedance range of the device.

Acknowledgements This work is supported in part by The Scientific and Technological Research Council of Turkey via grant number 117M405.

References

1. Wolf S, Grioli G, Eiberger O, Friedel W, Grebenstein M, Höppner H, Burdet E, Caldwell DG et al (2016) Variable stiffness actuators: review on design and components. IEEE Trans Mechatron 21(5):2418–2430
2. Ishida T, Takanishi A (2006) A robot actuator development with high backdrivability. In: IEEE conference on automation and mechatronics, pp 1–6
3. Migliore SA, Brown EA, DeWeerthgn SP (2005) Biologically inspired joint stiffness control. In: ICRA proceedings of the IEEE international conference on robotics and automation, pp 4508–4513
4. Vanderborght B et al (2013) Variable impedance actuators: a review. J Robot Auton Syst. In: ICRA proceedings of the IEEE international conference on robotics and automation, pp 1601–1614
5. Ivanov KS, Ualiev G, Tultaev B (2014) Kinematic and force analysis of robot with adaptive electric drives. Appl Mech Mater. J Robot Auton Syst 555:273–280
6. Faulring EL, Colgate JE, Peshkin MA (2006) The cobotic hand controller: design, control and performance of a novel haptic display. Int J Robot Res 25(11):1099–1119

7. Kim J (2002) Design and analysis of a spherical continuously variable transmission. J Mech Des 124(1):21–29
8. Sclater N, Chironis NP (2001) Mechanism and mechanical devices sourcebook. The McGraw-Hill Companies
9. Martinez MO, Morimoto TK, Taylor AT et al (2016) 3D printed haptic devices for educational applications. In: IEEE haptics symposium, pp 126–133

Cable Driven Parallel Robot with Big Interference-Free Workspace

Vincentius A. Handojo, Adlina T. Syamlan, Latifah Nurahmi, Bambang Pramujati, Mohamad N. Tamara, and Unggul Wasiwitono

Abstract Current evacuation processes of natural disaster pose difficulties, since there are limitations from rescue team as well as workspace provided by existing tools. Cable driven parallel robot (CDPR) is a robot designed to overcome these limitations. It provides a large workspace with high mobility. Accordingly, this research aims to determine the geometric, static, and kinematic models of suspended CDPR consisting of 8 cables for search-and-rescue operation. The dimensions of mobile platform (length, width, and height) and cable arrangements are computed to achieve the largest interference-free workspace. The arrangements are varied based on the combination of anchor points and exit points. The optimum designs are obtained and presented based on three case studies by varying the external forces. The case where the CDPR is subjected to minimal external wrenches performs maximum workspace about 64.8% of the desired workspace.

Keywords Cable driven parallel robot · Wrench feasible workspace · Twist feasible workspace · Cable interferences · Optimum designs

---

V. A. Handojo · A. T. Syamlan · L. Nurahmi (✉) · B. Pramujati · U. Wasiwitono
Department of Mechanical Engineering, Institut Sepuluh Nopember, Surabaya, Indonesia
e-mail: latifah.nurahmi@me.its.ac.id

V. A. Handojo
e-mail: vincentius.adiyanto.handoyo@gmail.com

A. T. Syamlan
e-mail: syamlan.adlina@gmail.com

B. Pramujati
e-mail: pramujati@me.its.ac.id

U. Wasiwitono
e-mail: unggul@me.its.ac.id

M. N. Tamara
Politeknik Elektronik Negeri Surabaya, Kampus PENS Surabaya, Surabaya, Indonesia
e-mail: nasir_meka@eepis-its.edu

D. Sen et al. (eds.), *Mechanism and Machine Science*, Lecture Notes in Mechanical Engineering, https://doi.org/10.1007/978-981-15-4477-4_4

1 Introduction

Earthquakes are the result of sudden energy releases in a form of seismic waves from tectonic plates. Regardless of the epicenter, an earthquake with ahuge amount of energy might be powerful enough to destroy a city on catastrophical proportions. Most city is densely populated and due to land availability, most buildings are growing on height which drives an increase in fatalities. The evacuation process may take weeks, even months since mostly rescue teams rely on heavy machineries such as excavators. It creates the need for an efficient and modern solution, such as cable driven parallel robot (CDPR).

Research on parallel mechanism for both academic and industrial purposes dates back to 1950s when the hexapod was built for testing purpose by Dunlop tyres [1]. Since then, researches on parallel mechanisms have been conducted for their potential application. Recently, the rigid legs connecting the base to the mobile platform are replaced by cables, which is widely known as Cable Driven Parallel Robot. CDPR is a blend of parallel mechanism with mechanical properties of cables, leading to a flexible mechanism. Cables can withstand high tensions, which leads to a high payload ratio. In addition to its outstanding weight bearing abilities, CDPR also has advantages of operating on dynamic trajectories possed by parallel mechanism. Compared to rigid legs, cables may generate a larger workspace since rigid legs can only retract, therefore, consuming space [2]. Based on this fact, CDPR is a perfect fit as a solution for search-and-rescue operations [3].

Currently, CDPR has been studied to serve wide ranges of applications. For instance, the use of CDPR as cargo and pallet handler for logistic purposes. Three or more telescoping masts attached to deck are used to deploy cables controlled by series of sensor. The system is robust enough to both load and unload cargo regardless of pitch–yaw–roll movement of the ship due to sea condition [4]. In addition to pallet handler, CDPR has been developed for large industrial structure, such as performing painting and sandblasting tasks in offshore wind turbine [5]. Researchers also started to look beyond the conventional fixed frame CDPR to reconfigurable ones, such as for aircraft maintenance [6] or large structures in general [7]. Industrial applications sometimes require the platform to be able to twist and change orientation of the payload. The ability to twist is strictly related to the dimensions of gearbox, actuators, and winches. To better address this issue, Gagliardini *et al.* developed a solution of Twist Feasible Workspace [8, 9] by extending the idea of Wrench Feasible from static analysis to kinematic analysis.

The vast applications of CDPR also drive analysis in mainly the biggest drawbacks of CDPR: cable collision. Cable collision may occur both between cables and its surroundings. Blanchet and Merlet developed an algorithm by using swept base and shortest distance algorithm to overcome the issue of cable collision [10]. Furthermore, real environments may generate disturbance to the end effector, which cannot be neglected. Therefore, disturbance robustness and workspace generation are critical to CDPR. Bosscher developed a new method of the Specified Robustness Workspace [11].

Accordingly, this research aims to determine the geometric, static, and kinematic models of suspended CDPR consisting of 8 cables for search-and-rescue operation. Analysis of Wrench Feasible Workspace, Twist Feasible Workspace, Capacity Margin, and Cable Interferences are addressed to generate optimum design(s) of the CDPR.

2 Static and Kinematic Models

Statics. Static Model of CDPR is an analysis carried out on platform at idle state. Let us consider a CDPR platform connected by 8 cables to the body frame, as shown in Fig. 1. Both the base and the platform have a Cartesian coordinate system denoted by $\mathfrak{F}_b$ axes for base frame and $\mathfrak{F}_p$ for platform. $\mathfrak{F}_b$ of (x, y, z) axes is positioned on the bottom-center of the body frame origin O and $\mathfrak{F}_p$ of (u, v, w) axes is positioned at the middle-center of the platform with origin P. Orientation of platform $\mathfrak{F}_p$ is expressed with respect to fixed frame $\mathfrak{F}_b$. It states that the fixed frame of CDPR is set as reference of -platform movement. Therefore, rotation matrix $\mathbf{R}$ is applied to the platform, expressed as follows:

$$\begin{aligned}\mathbf{R}(\phi, \theta, \psi) &= \mathbf{R}_z(\phi)\mathbf{R}_y(\theta)\mathbf{R}_x(\psi)\\ &= \begin{bmatrix} c(\phi)c(\theta) & c(\phi)s(\theta)s(\psi) - s(\phi)c(\psi) & c(\phi)s(\theta)c(\psi) + s(\phi)s(\psi) \\ s(\phi)c(\theta) & s(\phi)s(\theta)s(\psi) - c(\phi)c(\psi) & s(\phi)s(\theta)c(\psi) + c(\phi)s(\psi) \\ -s(\theta) & c(\theta)s(\psi) & c(\theta)c(\psi) \end{bmatrix}\end{aligned} \tag{1}$$

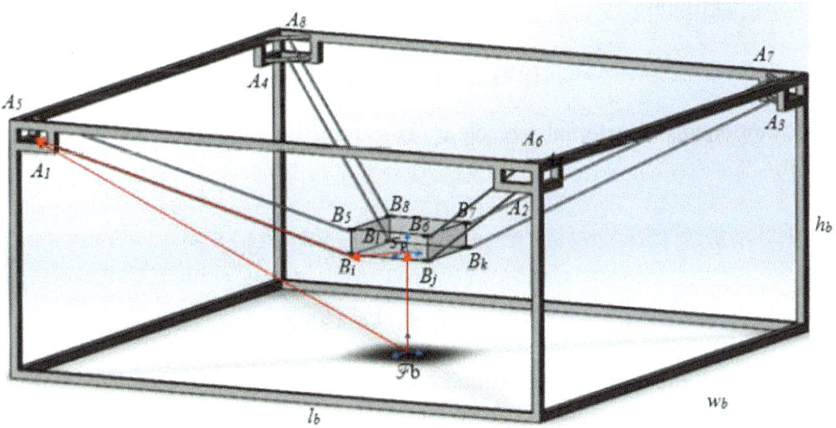

Fig. 1 Schematic of suspended CDPR

Vector position of point P expressed in the fixed frame $\mathfrak{F}_b$ is denoted by vector $\mathbf{p}$. Vector position of exit point A_i for $i = 1, \ldots, 4$, expressed in the fixed frame $\mathfrak{F}_b$ is denoted as position vector $\mathbf{a}_i^b$. Vector position of anchor point B_j for $j = 1, \ldots, 8$ expressed in the platform frame $\mathfrak{F}_p$ from the platform is denoted as position vector $\mathbf{a}_j^p$. jth cable connects point A_i to B_i with a certain tension is considered as a straight line $(A_i B_j)$. $\mathbf{l}_j$ is a vector of jth cable from B_j to A_i, or mathematically expressed as

$$\mathbf{l}_j = \mathbf{a}_i^b - \mathbf{p} - \mathbf{R}\mathbf{b}_j^p, \ i = 1, \ldots, 4, \ j = 1, \ldots, 8 \tag{2}$$

The unit vector can be expressed as

$$\hat{\mathbf{l}}_j = \frac{\mathbf{l}_j}{\|\mathbf{l}_j\|_2} \tag{3}$$

where $\|\mathbf{l}_j\|_2$ is the Euclidean norm of $\mathbf{l}_j$. Static equilibrium equation used for Fig. 1 is expressed as follows:

$$\mathbf{W}\tau + \mathbf{w}_e = 0 \tag{4}$$

$\mathbf{W}$ is a wrench matrix, which consists of forces and moments unit vectors exerted by m cables to the platform, that can be mathematically expressed as follows:

$$\mathbf{W} = \begin{bmatrix} \hat{\mathbf{l}}_1 & \ldots & \hat{\mathbf{l}}_j \\ \mathbf{R}\mathbf{b}_1^p \times \hat{\mathbf{l}}_1 & \ldots & \mathbf{R}\mathbf{b}_m^p \times \hat{\mathbf{l}}_j \end{bmatrix} \tag{5}$$

τ is a vector consisting of m cable tensions, which is defined as $\tau = \begin{bmatrix} \tau_1 \ \tau_2 \ \ldots \ \tau_m \end{bmatrix}$. $\mathbf{w}_e$ is an external wrench applied to the platform. It consists of external forces and moments as follows:

$$\mathbf{w}_e = \begin{bmatrix} \mathbf{f} \ \mathbf{m} \end{bmatrix}^T = \begin{bmatrix} f_x \ f_y \ f_z \ m_x \ m_y \ m_z \end{bmatrix}^T \tag{6}$$

The components of external wrench are constrained within given boundary as follows:

$$f_{\min} \leq f_x, f_y, f_z \leq f_{\max} \tag{7}$$

$$m_{\min} \leq m_x, m_y, m_z \leq m_{\max} \tag{8}$$

When $m > 6$, the system is classified as underdetermined problem. The reason is that m tensions from unknown cables can only be solved by 6 equations, resulted in infinite solutions. To solve this problem, Moore–Penrose pseudoinverse $\mathbf{W}^\dagger$ and null space $\mathbf{n}$ are introduced and stated as follows [12]:

$$\tau = \tau_n + \tau_0 = \mathbf{W}^{\dagger}\mathbf{W}_e + \lambda \boldsymbol{n} \quad \tau_n \geq 0 \tag{9}$$

Kinematics. Kinematic model of CDPR is an analysis to determine velocity of winding and unwinding of the cables, which is stated in the below equation:

$$\mathbf{Jt} = \dot{\mathbf{l}} \tag{10}$$

where $\mathbf{J}$ = Jacobian Matrix, $\dot{\mathbf{l}}$ = Cable Velocity, $\mathbf{t}$ = twist (velocity of the mobile platform). Jacobian matrix is obtained by transposing wrench matrix, $\mathbf{W}$, as follows:

$$\mathbf{J} = \mathbf{W}^T \tag{11}$$

Twists of mobile platform comprise two components, namely linear velocity ($\mathbf{v}$) and angular velocity ($\boldsymbol{\omega}$), as follows [13]:

$$\mathbf{t} = \begin{bmatrix} \mathbf{v} \ \omega \end{bmatrix}^T \tag{12}$$

2.1 Wrench Feasible Workspace

The equilibrium of mobile platform defined by Eq. (4) can only be achieved if and only if the position, or a pose, of the mobile platform is within the Wrench Feasible Workspace. A pose is said to achieve equilibrium if all assigned wrench $\mathbf{T}$ can be stabilized by available wrench $\mathbf{A}$. The available wrench $\mathbf{A}$ depends on the architecture of cables attached to the mobile platform and its tension limits. Wrench $\mathbf{T}$ also depends on the application of external wrench of the CDPR. All wrenches can be achieved when satisfying the following conditions:

$$\mathbf{T} \subseteq \mathbf{A} \tag{13}$$

where available wrench set $\mathbf{A}$ is defined as a group of wrench from cable $\boldsymbol{\tau}_j\mathbf{w}_j$ and wrench set $\mathbf{T}$ is a group of external wrench $\mathbf{w}_e$.

Figure 2 shows example of three-cables-planar-mechanism with its maximum and minimum tension τ_j of each cables represented in free body diagram (FBD). External Wrench $\mathbf{T}$ cannot be compared directly with the maximum and minimum tension of the cables because each matrix has different dimensions. Taking that into account, the tension limits of each cable are to be multiplied by wrench matrix $\mathbf{W}$, which resulted in $\mathbf{A}$ along with its dimension as shown in Fig. 2. $\Delta\tau = \tau_{\max} - \tau_{\min}$ represents the tension limits of each cables. A set of $\mathbf{T}$ inside a set $\mathbf{A}$ resulted in compliance of Eq. (4) [12].

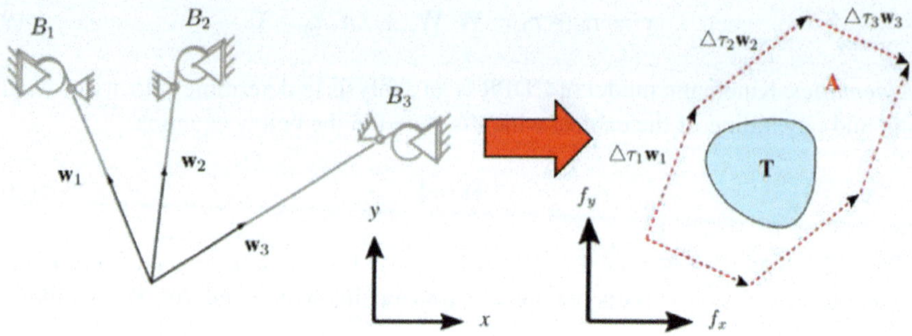

Fig. 2 Wrench feasible workspace of planar CDPR

2.2 *Twist Feasible Workspace*

The velocity of the mobile platform should be within a set of range as stated below:

$$v_{\min} \leq v_x, v_y, v_z \leq v_{\max} \tag{14}$$

$$\omega_{\min} \leq \omega_x, \omega_y, \omega_z \leq \omega_{\max} \tag{15}$$

where v_{min} and v_{max} are lower and upper limits of mobile platform linear velocity, whereas ω_{min} and ω_{max} are lower and upper limits of mobile platform angular velocity. Maximum and minimum velocity of the cables are defined as $\dot{l}_{\max}$ and $\dot{l}_{\min}$, respectively, where $\dot{l}_{\min} = -\dot{l}_{\max}$. A set of cable velocities intended for winding and unwinding is defined as Available Cable Velocity Set $\dot{l}_a$, as follows:

$$\dot{l}_a = \{\dot{l} \mid \dot{l}_{\min} \leq \dot{l}_j \leq \dot{l}_{\max}, j = 1, \ldots, m\} \tag{16}$$

A pose is said to be twist feasible if the velocity of each cable is inside available cable velocity set [12].

2.3 *Capacity Margin*

Capacity margin is an index to prove Eq. (13), which is proposed by Guay et al. [14]. It measures minimum degree of constraint satisfaction at which a set of task wrench **T** is inside the available wrench **A**, as follows:

$$s = \min(\min\ s_{u,y}), u = 1, \ldots, n\ y = 1, \ldots, m \tag{17}$$

where degree of constraint satisfaction $s_{u,y}$ states the distance at the far end of degree $\mathbf{w}_e$ from **T** to the cable surface from **A**. Positive sign in s indicates that the Eq. (13)

is fulfilled, or when **T** is inside **A**. Conversely, the negative sign of s indicates that one of the degree **T** is outside **A**. s has value of zero when one of degree **T** is at the boundary of **A** [14].

3 Interferences

Cable-to-Cable Interference. Cable-to-cable interference occurs when cables connecting the base to the mobile platform collide with one another, which leads to platform instability. Interferences might cause a pose of mobile platform being rejected as available workspace. One approach to determine interference is by calculating the shortest distance between each cable. Let us consider cables $A_i B_i$ and $A_j B_j$ as a line segment with $M_j M_i$ as the shortest distance between two cables, as shown in Fig. 3.

The line segment can be formulated as $\vec{OM} = \vec{OA} + l\vec{AB}$ where $\vec{OA}_i$ is position vector of anchor point A_i with respect to the fixed frame. $\vec{QB}_i$ is position vector of exit point B_i with respect to the moving frame. l is fraction of cable length within 0 to 1, and $\vec{OQ}$ is position vector of mobile platform origin. Using loop vector method, it yields

$$\vec{OM}_i = \vec{OA}_i + \vec{A_i B_i} l_i \tag{18}$$

$$\vec{OM}_j = \vec{OA}_j + \vec{A_j B_j} l_j \tag{19}$$

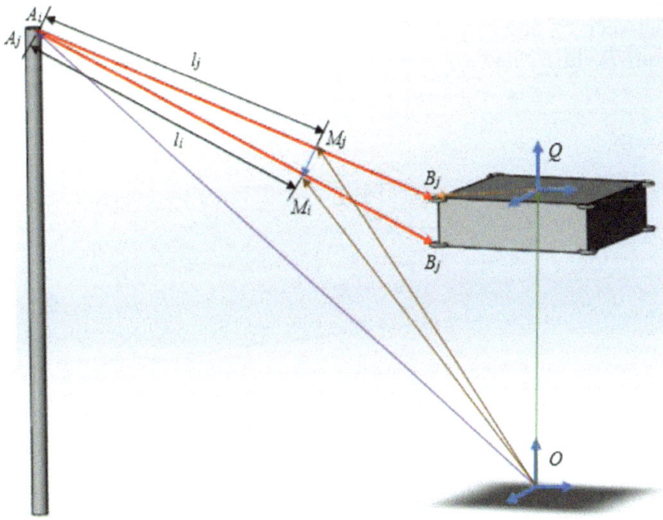

Fig. 3 Cable-to-cable interference

$$\vec{AB_i} = \vec{OQ_i} + (\mathbf{R} \cdot \vec{QB_i} - \vec{OA_i}) \tag{20}$$
$$\vec{AB_i} = \vec{OQ_j} + (\mathbf{R} \cdot \vec{QB_j} - \vec{OA_j}) \tag{21}$$

Substituting Eqs. (20) and (21) to Eqs. (18) and (19) for each cable leads to

$$\vec{OM_i} = \vec{OA_i} + \vec{OQ_i}l_i + (\mathbf{R} \cdot \vec{QB_i} - \vec{OA_i})l_i \tag{22}$$
$$\vec{OM_j} = \vec{OA_j} + \vec{OQ_j}l_i + (\mathbf{R} \cdot \vec{QB_j} - \vec{OA_j})l_j \tag{23}$$

By merging i and j cables using Eqs. (22) and (24), vector M_jM_i is obtained as follows:

$$\vec{M_iM_j} = \vec{A_jA_i} + \vec{OQ}(l_i - l_j) + (\mathbf{R} \cdot \vec{QB_i} - \vec{OA_i}l_i + (\mathbf{R} \cdot \vec{QB_i} - \vec{OA_i})l_j \tag{24}$$

Hence, the closest distance between i and j cables is

$$d_{i,j} = \left\| \vec{M_jM_i} \right\|_2 \tag{25}$$

In this research, a pose is rejected when $d_{i,j} = 0$, which means cables i and j collide [15].

Cable-to-Platform Interference. The idea of cable-to-platform interference analysis is to convert each side of the platform into a plane and then determine the shortest distance intercepting vertex at the mobile platform to the plane. In this research, platform vertex and anchor points are located at the same position.

Figure 4 shows the schematic diagram of interference between cable $\vec{P_0P_1}$ and the upper side of the mobile platform. P_0 is the exit point of the CDPR, while P_1 is the anchor point. P_0P_1 intercepts the sides of the platform at point *P(t)*, where *t* defines segment between P_0 and *P(t)*. **d** is a vector from point *P(t)* to B_j. $\boldsymbol{e}_i$ and $\boldsymbol{e}_j$ are unit vectors from B_j to B_k and B_k to B_l, respectively. The normal vector **c** of plane is

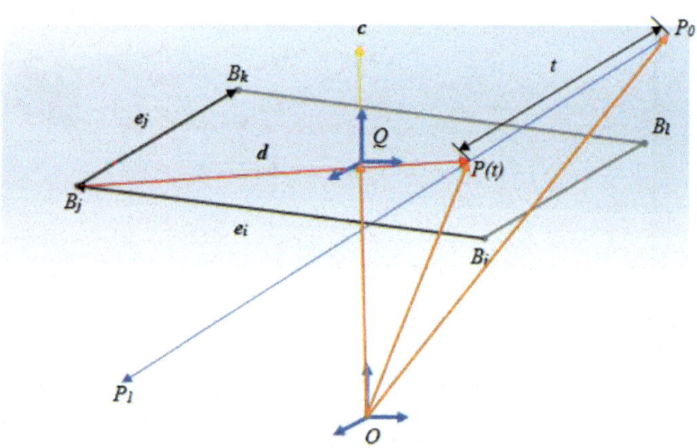

Fig. 4 Cable–platform interference

stated as follows:

$$\mathbf{c} = \mathbf{e}_i \times \mathbf{e}_j \tag{26}$$

Using the geometry of CDPR, with $\mathbf{R}$ as rotation matrix, B_j, B_k, and B_l are converted to

$$B_j = \vec{OQ} + \mathbf{R} \cdot \vec{QB_j} \tag{27}$$
$$B_k = \vec{OQ} + \mathbf{R} \cdot \vec{QB_k} \tag{28}$$
$$B_l = \vec{OQ} + \mathbf{R} \cdot \vec{QB_l} \tag{29}$$

$\mathbf{e}_i$ and $\mathbf{e}_j$ are substituted to Eqs. (27)–(29), which become

$$\mathbf{e}_i = \vec{B_j B_k} = \mathbf{R} \cdot (\vec{QB_k} - \vec{QB_j}) \tag{30}$$
$$\mathbf{e}_j = \vec{B_k B_l} = \mathbf{R} \cdot (\vec{QB_l} - \vec{QB_k}) \tag{31}$$

P(t) is obtained using loop vector as follows:

$$\vec{OP}(t) = \vec{OP_0} + \vec{P_0P_1} \cdot t \tag{32}$$
$$\vec{OP}(t) = \vec{OQ} + \mathbf{R} \cdot \vec{QB_k} + B_k P(t) \tag{33}$$

Equation (32) is substituted to Eq. (33) yields

$$\mathbf{d} = B_k P(t) = OP_0 + P_0P_1.t - (OQ + \boldsymbol{R}.QB_k) \tag{34}$$

Position of *P(t)* can be calculated using Eqs. (35) and (36) leads to

$$\mathbf{d}^T \cdot \mathbf{c} = 0 \tag{35}$$
$$(\vec{OP_0} + \vec{P_0P_1} \cdot t - (\vec{OQ} + \mathbf{R} \cdot \vec{QB_k}))^T \cdot (\boldsymbol{e}_i \times \boldsymbol{e}_j) = 0 \tag{36}$$

Substituting Eqs. (30) and (31) to Eqs. (35) and (36) to obtain *P(t)* :

$$(\vec{OP_0} + \vec{P_0P_1} \cdot t - (\vec{OQ} + \mathbf{R} \cdot \vec{QB_k}))^T \cdot (\mathbf{R} \cdot (\vec{QB_k} - \vec{QB_j}) \times \mathbf{R} \cdot (\vec{QB_l} - \vec{QB_k})) = 0 \tag{37}$$

Cables will collide with the mobile platform when *P(t)* is inside the mobile platform, or can be mathematically expressed as

$$B_{lx} \leq P(t)_x \leq B_{jx} \tag{38}$$
$$B_{ly} \leq P(t)_y \leq B_{jy} \tag{39}$$
$$B_{lz} \leq P(t)_z \leq B_{jx} \tag{40}$$

where *x,y,z* are the orientation axes [16].

4 Design Problem Formulation

The parameters considered in this paper consists of desired workspace, allowable tension, external forces, coordinates of exit points A_i and anchor points B_i, variation of cable arrangements, and mobile platform dimension. Eight cables with suspended configuration will connect the mobile platform to the fixed frame, as shown in Fig. 1.

In this research, cable tension is limited within 1–1000 N. Allowable cable velocity $\dot{l}_j$ is $-1.7\ \text{m/s} \leq \dot{l}_a \leq 1.7\ \text{m/s}$ where positive and negative signs signify the direction of velocity. Total platform mass and payload are 100 kg.

Let us consider the width (w_b), length (l_b), height (h_b) of the base as 100 m, 70 m, and 30 m respectively, as shown in Fig. 1. Hence, the coordinates of anchor points A_i with respect to the fixed frame are

$$\begin{aligned} \mathbf{a}_1^b &= (35, -50, 29.7)^T \ \mathbf{a}_2^b = (35, 50, 29.7)^T \\ \mathbf{a}_3^b &= (-35, 50, 29.7)^T \ \mathbf{a}_4^b = (-35, -50, 29.7)^T \end{aligned} \tag{41}$$

The CDPR is designed to have a rectangular platform with l_p long, w_p wide, and h_p tall, as shown in Fig. 5. Length of l_p is equal to width of w_p with its range 0.3–0.7 m by 10 times iteration. Height of h_p is also varied from 0.1 to 0.5 m by 10 times iteration. Hence, the coordinates of exit points B_i with respect to the moving frame are

$$\begin{aligned} \mathbf{b}_1^p &= \left(\frac{w_p}{2}, -\frac{l_p}{2}, 0\right)^T \ \mathbf{b}_2^p = \left(\frac{w_p}{2}, \frac{l_p}{2}, 0\right)^T \\ \mathbf{b}_3^p &= \left(-\frac{w_p}{2}, \frac{l_p}{2}, 0\right)^T \ \mathbf{b}_4^p = \left(-\frac{w_p}{2}, -\frac{l_p}{2}, 0\right)^T \\ \mathbf{b}_5^p &= \left(\frac{w_p}{2}, -\frac{l_p}{2}, \frac{h_p}{2}\right)^T \ \mathbf{b}_6^p = \left(\frac{w_p}{2}, \frac{l_p}{2}, \frac{h_p}{2}\right)^T \\ \mathbf{b}_7^p &= \left(-\frac{w_p}{2}, \frac{l_p}{2}, \frac{h_p}{2}\right)^T \ \mathbf{b}_8^p = \left(-\frac{w_p}{2}, -\frac{l_p}{2}, \frac{h_p}{2}\right)^T \end{aligned} \tag{42}$$

Exit point A_i is connected to anchor point B_j but is restricted to one cable per point. Exit points of A_1, A_2, A_3, A_4 are connected to anchor points B_5, B_6, B_7, B_8. Exit points of A_1, A_2, A_3, A_4 are also to be varied into anchor points B_1, B_2, B_3, B_4. The number of cable arrangements generated is $4! = 24$. The new design consists of cable arrangement and dimensions (width, length, and height) of the mobile platform. The number of all possible design(s) are obtained by multiplying the number of cable arrangements with number of variation from dimensions of the platform. The desired workspace volume of CDPR will be discretized into 1000 poses. Designs with a certain platform dimension and cable arrangements are expected to meet all poses.

The total feasible designs with the combinations of workspace volume and cable arrangements are 2400 possibilities, all of which will be compared to deter-

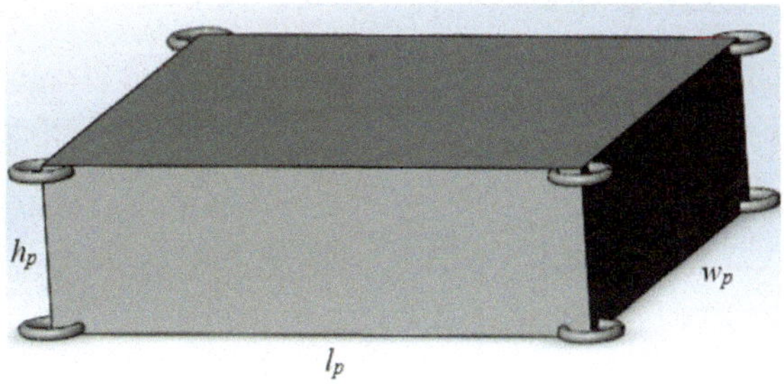

Fig. 5 Mobile platform dimension

mine the optimum workspace. The optimum workspace is obtained by intersecting wrench feasible workspace, twist feasible workspace, cable-to-cable interference-free workspace, and cable-to-platform interference workspace. Wrench feasible workspace (WFW) is obtained by calculating both geometric from Eq. (2) and static from Eq. (9). All cables tension values of every test pose will be compared with the maximum and minimum tension limits of the cable. A pose is said to be feasible if it complies with Eq. (13). Twist feasible workspace (TFW) is obtained by calculating kinematic model from Eq. (10). The twist is firstly calculated, then compared with the pre-determined allowable available cable velocity set. A pose is said to be twist feasible when it is inside the limits of available cable velocity set.

Cable-to-cable interference-free workspace (CCIFW) is a method to determine the shortest distance between cables to prevent collision. It is calculated using Eq. (25). In this research, a pose is rejected when the shortest distance is equal to 0m, due to the assumption that the cables are considered as a line with no finite radius, ignoring the cable properties. Cable-to-platform interference-free workspace (CPIFW) is a method to validate collision between cables and platform by using Eqs. (37)–(40). The optimum workspace for each case study is obtained from the intersection of all four generated workspaces, expressed as follows:

$$\text{optimum workspace} = WFW \cap TFW \cap CCIFW \cap CPIFW \quad (43)$$

5 Case Studies

Case 1: On Platform Center of Mass. The external wrench used for Case 1 is 100 kg, including weight of the mobile platform and payload. Two designs as shown in Figs. 6a and 7a generate the biggest workspace of 64.8%. The biggest attainable workspace of both designs is shown in Figs. 6a and 7a, respectively.

Holes are seen at the bottom of the optimum workspace since the poses in the area do not fulfill twist feasible workspace and allowable cable velocity $\dot{l}_a$.

Case 2: On Platform Center of Mass with External Forces. The external wrench used for Case 2 is 100 kg including the weight of the mobile platform and payload on the center of mass of the mobile platform, with an additional wind force of

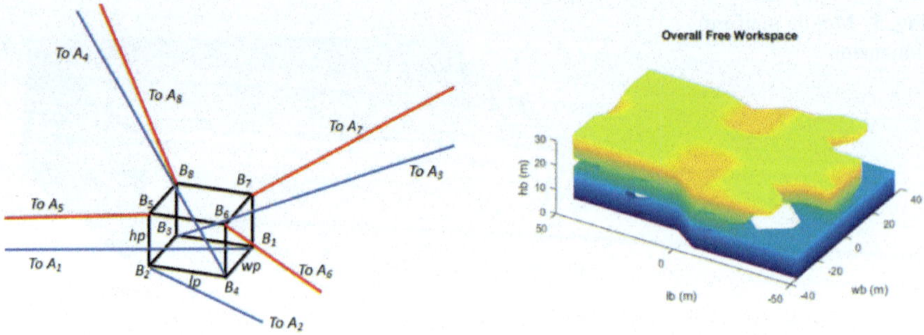

(a) Cable Arrangement (b) Optimum Workspace

Fig. 6 Optimum design 1 of Case 1

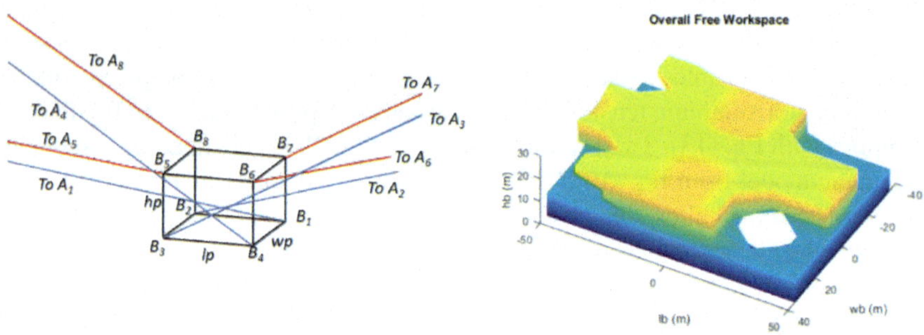

(a) Cable Arrangement (b) Optimum Workspace

Fig. 7 Optimum design 2 of Case 1

150 N about x and y axes. Seven designs are obtained which can generate the biggest workspace of 19.6% of the desired workspace. The cable arrangement among seven design is identical. One of seven optimum designs is shown in Fig. 8a, where $l_p = w_p = 0.48$ m and $h_p = 0.10$ m as shown in Fig. 8b. There is a significant reduction in workspace volume between Case 1 and Case 2, with a gap of 45.2%. It indicates that the robot work on the environment with weaker winds to generate a bigger workspace.

Case 3: Offset from Platform Center of Mass with External Forces. The external wrench used in Case 3 is 100 kg including the weight of the mobile platform and payload, l_x and l_y offset from the center of mass of the mobile platform. These external wrenches are applied with offset l_x and l_y from the center of mass of the mobile platform. Let us consider $l_x = l_y = 5$ mm. It produces moments in x and y axes which are obtained by multiplying payload with the offset distances, becomes 5 Nmm.

Four designs are obtained which can generate the biggest workspace of 10.6%. The cable arrangement among four designs is identical. One of four optimum designs is shown in Fig. 9a, where $l_p = w_p = 0.30$ m and $h_p = 0.46$ m as shown in Fig. 9b.

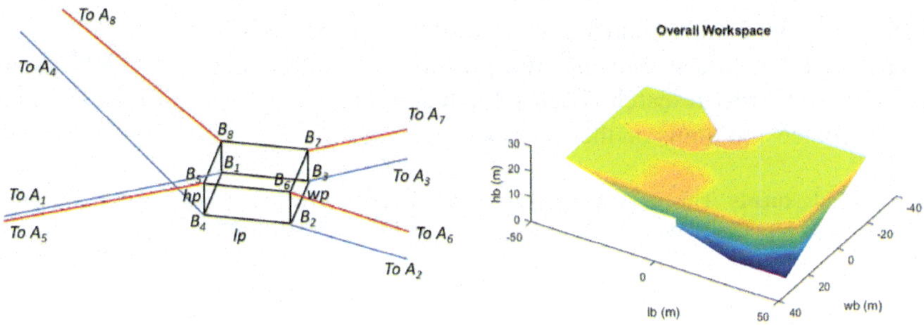

(a) Cable Arrangement (b) Optimum Workspace

Fig. 8 Optimum designs 1 of Case 2

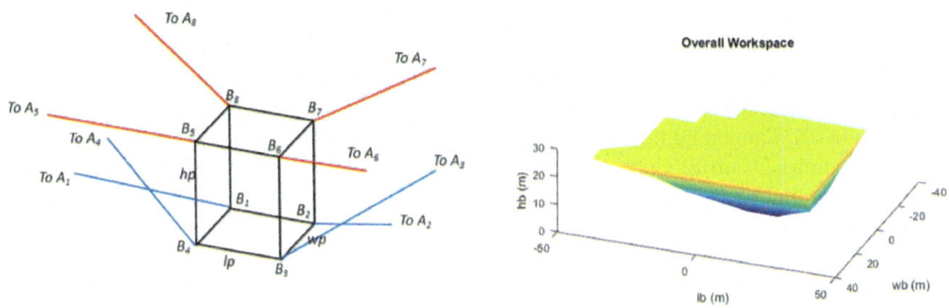

(a) Cable Arrangement (b) Optimum Workspace

Fig. 9 Optimum designs 1 of Case 3

There is a slight decrease in workspace volume between Case 2 and Case 3, with a gap of 9%. It indicates that the load position should be considered to increase the workspace.

6 Conclusions

This paper presents the optimum designs of CDPR by maximizing the overall workspace generated by CDPR when subjected to payload and external wrenches. The analysis used to determine the overall workspace is wrench feasible workspace, twist feasible workspace based on the kinematic equilibrium, cable-to-cable interference, and cable-to-platform interference. A suspended CDPR with 8 cables is used, with exit points denoted by A_i and anchor points denoted by B_i. Variations are conducted on the dimensions of the rectangular platform and cable arrangements. The location of payload and number of external forces applied to the platform affect the overall workspace of the CDPR. The biggest overall workspace of 64.8% in Case 1 is decreased to 19.6% in Case 2 and the lowest overall workspace of 10.6% is achieved in Case 3. Since this research neglects material properties of cables, a throughout analysis of the influence of material properties on interferences in CDPR should be studied. Further research is needed to determine the influence of maximum and minimum tension limits on the workspace volume.

Acknowledgements This work was supported by Ministry of Research and Higher Education of Indonesia, under the scheme of Internal Collaboration Research 2018–2020.

References

1. Gough V, Whitewall S (1962) Universal tyre test machine. In: Proceedings of the FISITA 9th international technical congress. pp. 117–137
2. Gosselin C (2013) Cable-driven parallel mechanism: state of the art. Bullet JSME, 1–17
3. Caro S, Martin A, Design and modelling of cable driven parallel robots: state of the art and open issues. Lecture Notes
4. Holland CS, Cannon DJ (2004) Cable array robot for material handling
5. Gagliardini L, Caro S, Gouttefarde M, Wenger P, Girin A (2014) Optimal design of cable driven parallel robots for large industrial structures. In: International conference on robotics and automation (ICRA)
6. Nguyen DQ, Goutterfarde M (2014) Study of reconfigurable suspended cable-driven parallel robots for aircraft maintenance. In: International conference on intelligent robots and systems (IROS)
7. Nguyen DQ, Goutterfarde M, Company O, Pierrot F (2014) On the analysis of large-dimension reconfigurable suspended cable driven parallel robots. In: International conference on robotics and automation (ICRA)
8. Gagliardini L, Caro S, Gouttefarde M (2015) Dimensioning of cable-driven parallel robots actuators, gearboxes and winches according to the twist feasible workspace. In: International conference on automation science and engineering (CASE)
9. Gagliardini L, Caro S, Gouttefarde M, Girin A (2016) Discrete reconfiguration planning for cable-driven parallel robots. J Mech Mach Theory 100:313–317
10. Blanchet L, Merlet J-P (2014) Interference detection for cable driven parallel robot (CDPRS). In: Advanced intelligent mechatronics, pp. 1413–1418
11. Bosscher PM (2004) Disturbance robustness measures and wrench-feasible workspace generation techniques for cable-driven robots. PhD thesis, Georgia Institute of Technology
12. Nurahmi L, Pramujati B, Caro S, Hanafie J (2017) Dimension synthesis of suspended eight cables-driven parallel robot for search-and-rescue operation. In: ICAMIMIA
13. Hanafie J, Nurahmi L, Caro S, Pramujati B (2017) Design optimization of spatial four cables suspend cable driven parallel robot for rapid life-scan. In: ICOME
14. Guay F, Cardou P, Ruiz AL, Caro S (2011) Measuring how well a structure supports varying external wrenches. In: New advances in mechanisms, transmission and applications, pp. 385–392
15. Blanchet L, Merlet JP (2014) Interference detection for cable-driven parallel robots (CDPRS). In: International conference on advanced intelligent mechatronics (AIM)
16. Cardou P, Perreault S, Gosselin CM, Otis MJD (2010) Geometrical determination of the interference-free constant-orientation workspace of parallel cable driven mechanisms. J Mech Robot 2(3):1–9

Stiffness Distribution of 3-RPS Parallel Manipulator Based on the Base-and-platform Configuration

Latifah Nurahmi, Rath Kautsar, and Ranjan Jha

Abstract Rehabilitation of ankle injuries must be under supervision of a therapist to prevent severe injuries. However, relying on a therapist is not consistent and accurate. In the present days, parallel manipulators play a key role as an ankle rehabilitation device to overcome the problem imposed by a therapist. The ankle rehabilitation device should be able to sustain the external forces and moments exerted from the patient's feet. The external forces and moments applied to the platform can cause deflections, either translation or rotational deflections. These deflections depend mostly upon the stiffness properties of the device. The stiffness property is one of crucial factors to be considered in designing the ankle rehabilitation device. Therefore, this paper aims to analyse and establish the static and stiffness model of the 3-RPS parallel manipulator which is employed as an ankle rehabilitation device. Each RPS leg applies one constraint force and one actuation force intersecting the centre of spherical joint. By collecting all constraint forces and actuation forces, the Jacobian matrix can be formulated according to the Screw theory and Quaternion parameters. Influences of actuators and base-and-platform configuration are identified by using stiffness index, namely eigenvalues of the stiffness matrix. The base-and-platform configuration is described by the distance of each joint to the origin of coordinate frame and the angle among three legs. For given external wrenches, numerical simulations have been performed to determine the optimal distributions of actuators and base-and-platform configuration by taking into account the stiffness.

L. Nurahmi (✉) · R. Kautsar
Department of Mechanical Engineering, Institut Teknologi Sepuluh Nopember, Surabaya 60111, Indonesia
e-mail: latifah.nurahmi@me.its.ac.id

R. Kautsar
e-mail: aida@me.its.ac.id

R. Jha
BioMedical Instrumentation Division, CSIR-Central Scientific Instruments Organisation Sector 30-C, Chandigarh 160030, India
e-mail: ranjan.jha@csio.res.in

D. Sen et al. (eds.), *Mechanism and Machine Science*, Lecture Notes in Mechanical Engineering, https://doi.org/10.1007/978-981-15-4477-4_5

Keywords Stiffness · Statics · Design parameters · Actuation · Constraint · 3-RPS

1 Introduction

The ankle sprains are the common injuries that occur in the population, especially sport players. Rehabilitation of ankle injuries must be under the supervision of a therapist to prevent severe injuries. However, relying on a therapist is not consistent and accurate. The ankle rehabilitation devices are developed to help faster recovery of ankle injuries and become one of the solutions to overcome the lack of therapist number. Parallel robots are widely used for the ankle rehabilitation device due to its simple structure [1]. A base and a moving platform, joined by rigid links or flexible links, are the main components of a parallel robot [2].

The important thing in the ankle rehabilitation devices is the range of motion (ROM) measurement [3]. This measurement aims to find an appropriate size of device component. Improper size may cause harmful effect to the patient. The 6-dof parallel manipulator is also employed in [4] because it is sufficient to provide ankle joint motion. Unfortunately, the centre of rotations of previous parallel manipulators do not perfectly match with the patients' ankles.

Compliant ankle rehabilitation robot CARR with three rotational dof was studied. This study was conducted by varying the workspace and torque actuation capacity considering patients' physical constraints [5]. The study of soft parallel robot (SPR) using modified genetic algorithm for ankle rehabilitation device was conducted in [6]. Aside of parallel manipulator, series of actuators with both low and high stiffness springs for force transmission was developed as an ankle rehabilitation device in [7]. The studies of stiffness model were performed by several researchers by using Screw theory [8], strain energy [9], and force-torque sensor [10].

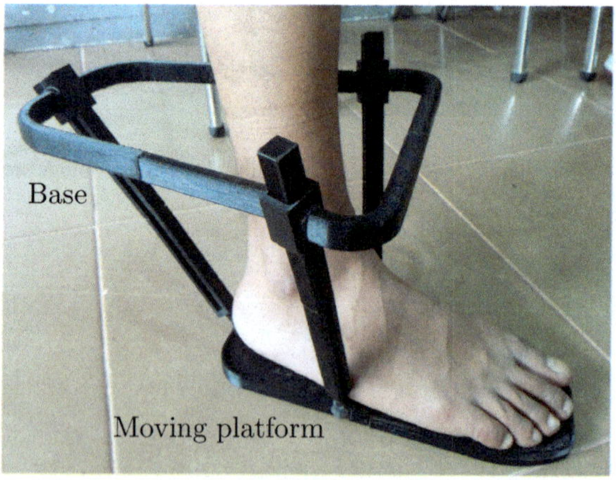

Fig. 1 Application for ankle rehabilitation device

Fig. 2 Necessary motion provided by ankle rehabilitation device

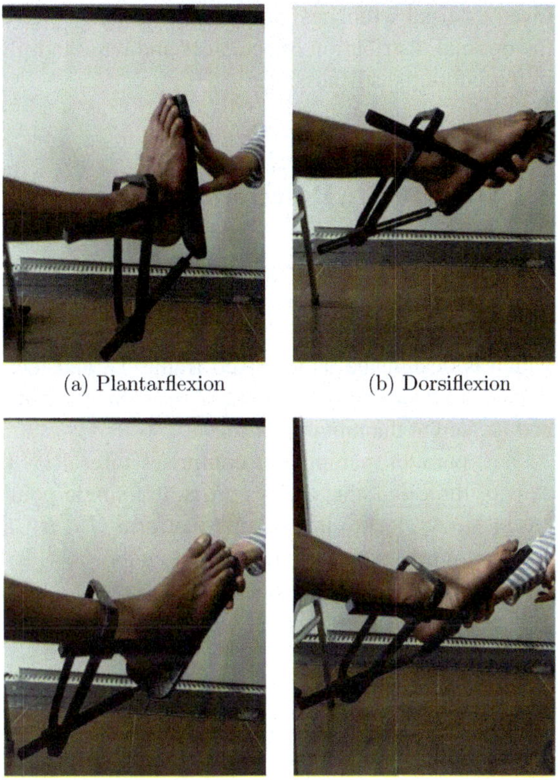

(a) Plantarflexion (b) Dorsiflexion

(c) Supination (d) Pronation

The 3-RPS parallel manipulator in which its centre of rotations is constructed close to the centre of rotations of patients' ankles has been developed in [11], as shown in Fig. 1. The base and moving platform have the shape of an isosceles triangle base. The base is hooked to the patient's leg and the platform is placed on the patient's foot palm to follow its motion as a therapeutic device. Figure 2 shows the manipulator under study performing dorsiflexion, plantarflexion, supination, and pronation.

Accordingly, this paper aims to analyse and establish the Jacobian and stiffness model of the ankle rehabilitation device. The device is realized based on a 3-RPS parallel manipulator comprising three RPS legs where three axes of revolute joint cross at a single point. Each RPS leg applies one constraint force. Since the prismatic joint is actuated, one actuation force is applied on the moving platform passing through the centre of spherical joint. By collecting all constraint forces and actuations forces, the wrench matrix of the manipulator under study is formulated based on the Screw theory and Quaternion parameters. Influences of actuators and base-and-platform configuration are identified by using stiffness index, namely eigenvalues of the stiffness matrix. The base-and-platform configuration is described by the distance of each joint to the origin of the coordinate frame and the angle among three legs. For given external wrenches, numerical simulations have been performed to determine the optimal distribution of actuators and base-and-platform configuration by taking into account the stiffness.

2 Architecture Description of Manipulator

Figure 3 depicts the schematic diagram of a 3-RPS parallel manipulator developed in this paper. Let us assume that three legs of the manipulator can be moved freely with respect to the vertical axis. The origin O belongs to the coordinate system (x, y, z) which is designated as the fixed frame Σ_0 and located at the base. The origin P belongs to coordinate system (u, v, w) which is illustrated as the moving frame Σ_1 and located at the moving platform.

This parallel manipulator comprises three RPS kinematic chains in which the axes of three R-joints always cross at a single point. Three RPS legs connect the base (A_1, A_2, A_3) to the moving platform (B_1, B_2, B_3) and each is composed of a revolute, a prismatic, and a spherical joints.

Let the first points A_1 and B_1 be on the x-axis and u-axis, respectively, hence $\varepsilon_1 = 0°$. Therefore, points A_i, B_i, and $\mathbf{s}_i$ can be determined as

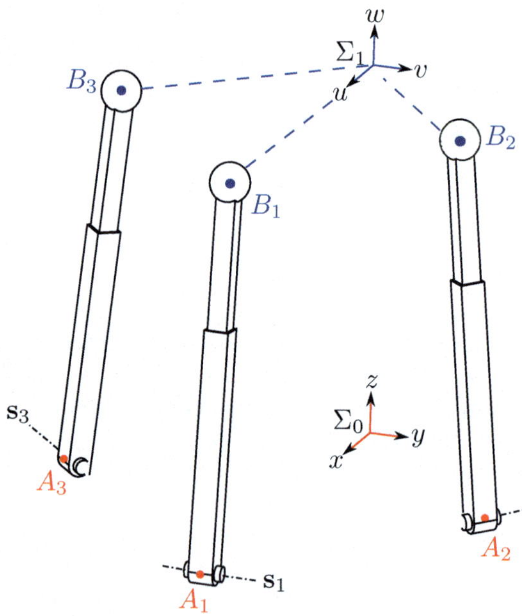

Fig. 3 Schematic of the 3-RPS manipulator

$$\begin{aligned}\mathbf{r}^0_{A_1} &= \begin{bmatrix}1 \; a_1 \; 0 \; 0\end{bmatrix}^T,\\ \mathbf{r}^0_{A_2} &= \begin{bmatrix}1 \; \cos(\varepsilon_2)a_2 \; \sin(\varepsilon_2)a_2 \; 0\end{bmatrix}^T,\\ \mathbf{r}^0_{A_3} &= \begin{bmatrix}1 \; \cos(\varepsilon_3)a_3 \; \sin(\varepsilon_3)a_3 \; 0\end{bmatrix}^T.\end{aligned} \tag{1}$$

$$\begin{aligned}\mathbf{r}^1_{B_1} &= \begin{bmatrix}1 \; b_1 \; 0 \; 0\end{bmatrix}^T,\\ \mathbf{r}^1_{B_2} &= \begin{bmatrix}1 \; \cos(\varepsilon_2)b_2 \; \sin(\varepsilon_2)b_2 \; 0\end{bmatrix}^T,\\ \mathbf{r}^1_{B_3} &= \begin{bmatrix}1 \; \cos(\varepsilon_3)b_3 \; \sin(\varepsilon_3)b_3 \; 0\end{bmatrix}^T.\end{aligned} \tag{2}$$

$$\begin{aligned}\mathbf{s}_1 &= \begin{bmatrix}0 \; 0 \; 1 \; 0\end{bmatrix}^T,\\ \mathbf{s}_2 &= \begin{bmatrix}0 \; -\sin(\varepsilon_2) \; \cos(\varepsilon_2) \; 0\end{bmatrix}^T,\\ \mathbf{s}_3 &= \begin{bmatrix}0 \; -\sin(\varepsilon_3) \; \cos(\varepsilon_3) \; 0\end{bmatrix}^T.\end{aligned} \tag{3}$$

The relation among design parameters $(a_1, a_2, a_3, b_1, b_2, b_3, \epsilon_1, \epsilon_2, \epsilon_3)$ can be obtained by using design methodology established by Nurahmi et al. [12], as follows:

$$\begin{aligned}a_1 &= -a_3\frac{\cos(\varepsilon_2-\varepsilon_3)}{\cos(\varepsilon_2)}, \quad a_2 = a_3\frac{\cos(\varepsilon_3)}{\cos(\varepsilon_2)},\\ b_1 &= -b_3\frac{\cos(\varepsilon_2-\varepsilon_3)}{\cos(\varepsilon_2)}, \quad b_2 = b_3\frac{\cos(\varepsilon_3)}{\cos(\varepsilon_2)},\\ \varepsilon_2 &\neq \pm\frac{\pi}{2} \text{ or } \varepsilon_2 \neq \pm\frac{3\pi}{2}.\end{aligned} \tag{4}$$

Thus, the number of design parameters drop into 4 design parameters, namely $(a_3, b_3, \varepsilon_2, \varepsilon_3)$.

3 Screw-Based Jacobian Analysis

The Jacobian matrix maps the input and output velocities of the manipulators. In this paper, the Screw theory is employed to derive the Jacobian matrix of the 3-RPS parallel manipulator.

3.1 Screw Theory

The screw coordinates are the coordinate system which is built from a screw. The coordinates of a unit screw ($\hat{\$}$) are described by a pair of vectors (Roth 1984) as follows:

$$\hat{\$} = \begin{bmatrix} \mathbf{u} \\ \mathbf{r} \times \mathbf{u} + \lambda \mathbf{u} \end{bmatrix} \tag{5}$$

where $\mathbf{u}$ is a unit vector which shows the direction of $\hat{\$}$, $\mathbf{r}$ is the position vector of any point on the screw axis, λ is a pitch which is the ratio of translational to rotation, $\mathbf{r} \times \mathbf{u}$ is the moment of the screw axis. Revolute joint, a joint which causes only rotational motion, has zero pitch ($\lambda = 0$), hence the unit screw becomes

$$\hat{\$}_0 = \begin{bmatrix} \mathbf{u} \\ \mathbf{r} \times \mathbf{u} \end{bmatrix}. \tag{6}$$

While the prismatic joint has infinite pitch ($\lambda = \infty$) which only has translation motion, the unit screw is define as

$$\hat{\$}_\infty = \begin{bmatrix} \mathbf{0} \\ \mathbf{u} \end{bmatrix}. \tag{7}$$

3.2 Coordinate Transformation

The first step to derive the Jacobian matrix is to carry out the coordinate transformation. The coordinates of point B_i should be transformed into the fixed frame coordinate by using the transformation matrix $\mathbf{M}$, as follows:

$$\mathbf{M} = \begin{pmatrix} x_0^2 + x_1^2 + x_2^2 + x_3^2 & \mathbf{0}_{3\times 1}^T \\ \mathbf{d} & \mathbf{R} \end{pmatrix}. \tag{8}$$

$\mathbf{R}$ is a rotational matrix containing the Quaternions parametrization and $\mathbf{d}$ is the point-displacement vector as follows:

$$\mathbf{R} = \begin{pmatrix} \alpha & 2(x_1x_2 - x_0x_3) & 2(x_1x_3 + x_0x_2) \\ 2(x_1x_2 + x_0x_3) & \beta & 2(x_2x_3 - x_0x_1) \\ 2(x_1x_3 - x_0x_2) & 2(x_2x_3 + x_0x_1) & \gamma \end{pmatrix}$$

$$\begin{aligned} \alpha &= x_0^2 + x_1^2 - x_2^2 - x_3^2 \\ \beta &= x_0^2 - x_1^2 + x_2^2 - x_3^2 \\ \gamma &= x_0^2 - x_1^2 - x_2^2 + x_3^2 \end{aligned} \tag{9}$$

$$\mathbf{d} = \begin{pmatrix} X \\ Y \\ Z \end{pmatrix}.$$

The parameters x_0, x_1, x_2, x_3, which appear in matrix $\mathbf{R}$, are defined as Quaternion parameters of the rotation. They are used to represent a spatial Euclidean displacement and should fulfil the following inequality:

$$x_0^2 + x_1^2 + x_2^2 + x_3^2 - 1 = 0. \tag{10}$$

This condition should always be fulfilled. By using transformation matrix, point B_i can be written with respect to fixed frame as

$$\mathbf{r}^0_{B_i} = \mathbf{M}\,\mathbf{r}^1_{B_i}. \tag{11}$$

Hence, the vector parallel to the direction of prismatic joint can be defined as $\overrightarrow{A_iB_i} = \mathbf{r}^0_{A_i} - \mathbf{r}^0_{B_i}$. Its unit vector yields:

$$\hat{ab_i} = \frac{\overrightarrow{A_iB_i}}{||\overrightarrow{A_iB_i}||}. \tag{12}$$

The coordinates of point B_i and the unit vector ab_i will be used to derive the expression of Jacobian matrix in the following.

3.3 Jacobian of 3-RPS Parallel Manipulator

Since the manipulator is composed of RPS leg, one constraint force $\mathbf{F}_{ci}$ is exerted by each leg as illustrated in Fig. 4 as a purple arrow. It crosses point B_i of the spherical joint, which is orthogonal to the actuated prismatic joint and parallel to the axis $\mathbf{s}_i$ of the revolute joint. Prismatic joint in each leg is motorized which results in one actuation force $\mathbf{F}_{ai}$ whose axis is parallel to the direction of the corresponding actuated joint $\hat{ab_i}$ as derived in Eq. (12). Both forces correspond to the zero-pitch wrench, as follows:

$$\mathbf{F}_{ci} = \begin{bmatrix} \mathbf{r}^0_{B_i} \times \hat{ab}_i \\ \hat{ab}_i \end{bmatrix}, \quad \mathbf{F}_{ai} = \begin{bmatrix} \mathbf{r}^0_{B_i} \times \mathbf{s}_i \\ \mathbf{s}_i \end{bmatrix}, \quad i = 1, 2, 3. \tag{13}$$

By collecting all the actuation forces and constraint forces, the Jacobian matrix of the corresponding manipulator is expressed as

$$\mathbf{J} = \left[\mathbf{F}_{a1}\ \mathbf{F}_{a2}\ \mathbf{F}_{a3}\ \mathbf{F}_{c1}\ \mathbf{F}_{c2}\ \mathbf{F}_{c3}\right]^T. \tag{14}$$

This matrix is also called in [13, 14] as an extended Jacobian matrix. This matrix is composed of actuation Jacobian and constraint Jacobian, which are expressed as $\mathbf{J} = \left[\mathbf{F}_{ai}\right]^T$ and $\mathbf{J} = \left[\mathbf{F}_{ci}\right]^T$, respectively.

4 Stiffness Model Identification

Every rigid body may perform infinitesimal deflections due to specific loads. The spatial relation of force-deflection can be linearly described as a 6×6 symmetric positive semi-definite matrix, called stiffness matrix. This matrix relates the deflections of moving platform to an external wrenches applied on the moving platform. For the lower-mobility parallel manipulator where the moving platform exhibiting less than 6-dof, the overall stiffness is a result of a combination between the constraint and actuation wrenches.

Accordingly, the formulation of stiffness matrix for the 3-RPS parallel manipulator is

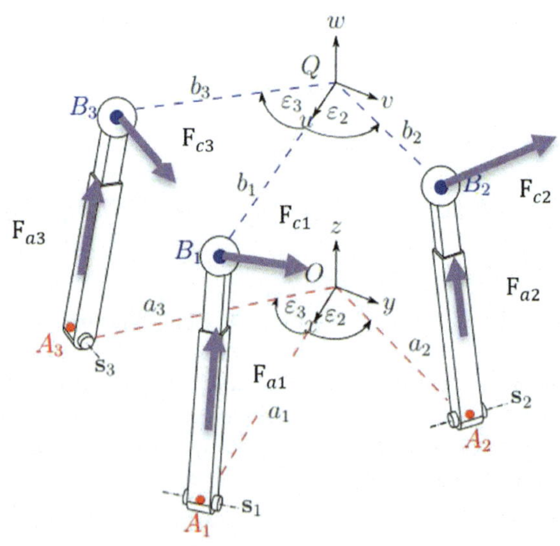

Fig. 4 Actuation and constraint forces

$$\mathbf{K} = \mathbf{J}^T \chi \mathbf{J} \tag{15}$$

where χ is a diagonal matrix that consists of $\chi = \text{diag}\left[k_{a_i}, k_{ci}\right]$, $i = 1, 2, 3$. The stiffness matrix **K** can be decomposed into two parts. The first part describes the stiffness of manipulator due to the forces exerted by the actuators. The second part illustrates the manipulator stiffness due to the forces that are inherently induced by the manipulator architecture. These two stiffness can be expressed as follows:

$$\mathbf{K}_a = \mathbf{J}_a^T \chi_a \mathbf{J}_a, \tag{16}$$

$$\mathbf{K}_c = \mathbf{J}_c^T \chi_c \mathbf{J}_c. \tag{17}$$

4.1 *Stiffness Subject to Actuations*

The prismatic joint in each leg is actuated. The value of each motor stiffness is collected as a diagonal matrix χ_a. The diagonal matrix χ_a is 3×3 matrix that shows the value of actuator stiffness k_{ai}. The value of k_{ai} is obtained from the motor stiffness mounted in each leg. The selected actuator for the 3-RPS parallel manipulator is a lead screw motor of 20 mm in diameter and its minimum stiffness is 0.3 kN/μm.

4.2 *Stiffness Subject to Leg Constraints*

The constraint stiffness depends on the three constraint forces ($\mathbf{F}_{ci}$) induced by the manipulator architecture whose axes are parallel to $\mathbf{s}_i$ (revolute joint axes). These constraint forces may produce deflections to the manipulator (Δx_{ci}) and it depends on the leg stiffness (k_{ci}) (Fig. 5). The determination of leg stiffness is shown in Fig. 6 and they are collected into a diagonal matrix χ_c.

Each leg of RPS is subjected to tension-compression and bending induced by the constraint force. Thus, the leg stiffness is given by the following formulation:

$$k_{ci} = \frac{EA}{L} + \frac{EI}{L^3}, \quad i = 1, 2, 3. \tag{18}$$

The material used to fabricate the 3-RPS parallel manipulator is PLA with Young's modulus $E = 3500$ MPa, the diameter and length of revolute joint are 15 mm and 28.4 mm, respectively.

5 Stiffness Distribution

According to the analysis in Sect. 2, there are four design parameters that are not specified and synthesized, namely $a_3, b_3, \varepsilon_2, \varepsilon_3$. The value of one design parameter can be changed to another, therefore one of the parameters is assigned to be $a_3 =$ 100 mm.

Let us consider the value of b_3 is less than a_3 and positive-definite. The angle parameters are bounded within: $\varepsilon_2 = [0, 90°]$ and $\varepsilon_3 = [-90°, 0]$.

The moving platform is assumed to be working at the altitude 100 mm. The rotational motion defined by the azimuth angle ϕ has a range of values $[-180°, 180°]$. While the rotational motion defined by the tilt angle θ has a range of values $[0, 180°]$.

Variations of these parameters result in the magnitude changes of stiffness matrix. The stiffness distribution of the 3-RPS parallel manipulator can be evaluated by extracting the eigenvalue (λ) of the stiffness matrix.

Figure 6 shows the distribution of the manipulator stiffness at $b_3 = 25$, 50, 75, 100 mm with the ε_2 and ε_3 values being varied. At $b_3 = 100$ mm, the largest λ presents the maximum stiffness value when $\varepsilon_2 = 5°$ and $\varepsilon_3 = -5°$ with eigenvalues 5.5×10^5 Nmm.

The largest λ when $b_3 = 75$ mm is 1.8×10^4 Nmm, where the two legs are assembled with $\varepsilon_2 = [60°, 80°]$ and $\varepsilon_3 = [0°, -15°]$. At $b_3 = 50$ mm, largest λ is when the two legs are assembled with $\varepsilon_2 = [70°, 80°]$ and $\varepsilon_3 = [0°, -20°]$, in which the eigenvalues reaches $\lambda = 10{,}000$ Nmm. The stiffness decreases significantly when the angle parameter $\varepsilon_2, \varepsilon_3$ are respectively over 30°.

By taking into consideration of the value of the design parameters represented in Fig. 1 and the design relation as in Eq. (4), one case study that maximum stiffness can be selected as shown in Table 1. This dimension shows that the platform size is

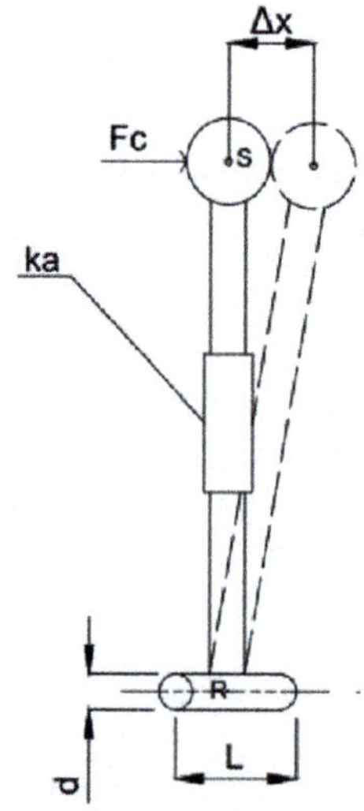

Fig. 5 Deflection of each leg

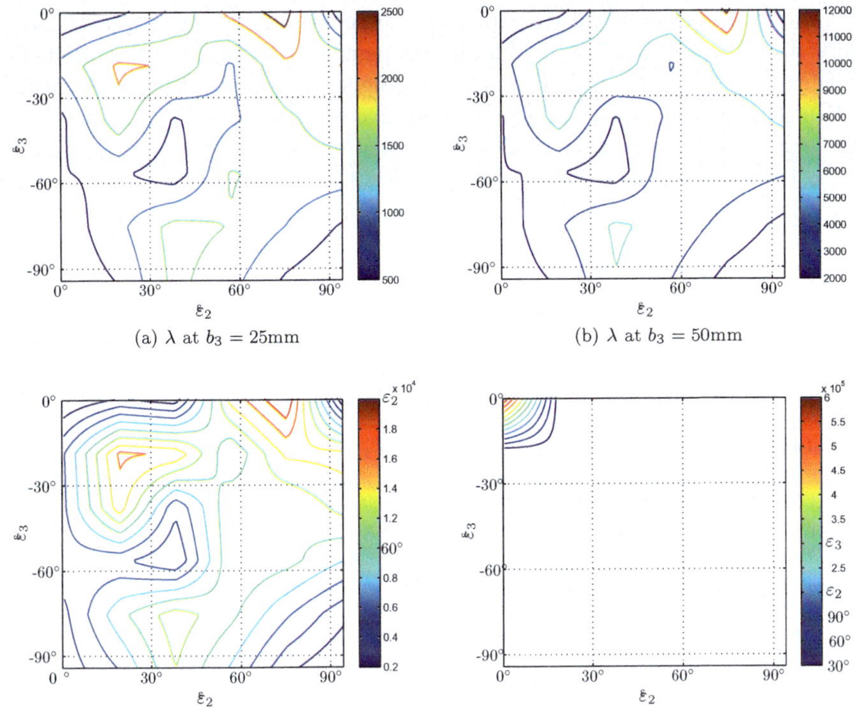

(a) λ at $b_3 = 25$mm

(b) λ at $b_3 = 50$mm

(c) λ at $b_3 = 75$mm

(d) λ at $b_3 = 100$mm

Fig. 6 Stiffness distribution in terms of the design parameters b_3, ε_2, ε_3

Table 1 Selected design parameters

ε_2	ε_3	a_1 (mm)	a_2 (mm)	a_3 (mm)	b_1 (mm)	b_2 (mm)	b_3 (mm)
50°	−10°	78	153	100	60	115	75

0.75 smaller than the base size. The second leg is assembled 50° from the x-axis, while the third leg is assembled −10° from the x-axis.

6 Conclusions

Jacobian analysis and stiffness distribution of the developed 3-RPS parallel manipulator for ankle rehabilitation device have been presented in this paper. This manipulator consists of three RPS legs in which the three axes of revolute joints always intersect at one point. Due to the structure of RPS leg, each leg applies one constraint force parallel to the axis of the revolute joint. The prismatic joint is actuated, thus an actuation force is applied onto the moving platform. The Jacobian matrix is formulated by collecting all constraint and actuation forces, based on the Screw theory and Quaternion parameters. The influence of the design parameters of base and platform to the manipulator stiffness can be identified by using stiffness index, namely eigenvalue of stiffness matrix. Numerical simulations have been performed and it shows that the maximum stiffness is obtained when the platform is 0.75 times smaller than the base. Eventually, the new 3-RPS parallel manipulator with new dimension of design parameters based on its stiffness value can be generated.

References

1. Dasgupta B, Mruthyunjaya TS (2000) Stewart platform manipulator: a review. Mech Mach Theory 35:15–40
2. Jamwal PK, Xie SQ, Tsoi YH, Aw KC (2010) Forward kinematics modelling of a parallel ankle rehabilitation robot using modified fuzzy interference. Mech Mach Theory 45:1537–1554
3. Chen G, Zhou Z, Wang N, Wang Q (2017) Range-of-motion measurement with therapist-joined method for robot-assisted ankle stretching. Robot Auton Syst 94:34–42
4. Yonezawa T, Nomura K, Onodera T, Ding M, Mizoguchi H, Takemura H (2014) Development and performance evaluation of parallel link type human ankle rehabilitation assistive device. In: IEEE international conference on robotics and biomimetics, pp 802–807
5. Zhang M, Cao J, Zhu G, Miao Q, Zeng X, Xie SQ (2017) Reconfigurable workspace and torque capacity of a compliant ankle rehabilitation robot (CARR). Robot Auton Syst 98:213–221
6. Jamwal PK, Xie S, Aw KC (2009) Kinematic design optimization of a parallel ankle rehabilitation robot using modified genetic algorithm. Robot Auton Syst 57:1018–1027
7. Chen G, Qi P, Guo Z, Yu H (2016) Mechanical design and evaluation of a compact portable knee-ankle-foot robot for gait rehabilitation. Mech Mach Theory 103:51–64
8. Xu Q, Li Y (2008) An investigation on mobility and stiffness of a 3-DOF translational parallel manipulator via screw theory. Robot Comput Integr Manuf 24:402–414
9. Cao WA, Ding H (2018) A method for stiffness modelling of 3R2T overconstrained parallel robot mechanisms based on scrw theory and strain energy, vol 52, pp 10–29
10. Lu Y, Chen L, Wang P, Zhang B, Zhao Y, Hu B (2013) Static and stiffness analysis of a novel six-component force/torque sensor with 3-RPPS compliant parallel structure, vol 62, pp 99–111
11. Nurahmi L, Solichin M, Harnany D, Kurniawan A (2017) Dimension synthesis of 3-RPS parallel manipulator with intersecting R-axes for ankle rehabilitation device. In: 18th international conference on advanced robotics (ICAR). China, Hong Kong
12. Nurahmi L, Caro S, Wenger P (2015) Design of 3-RPS parallel manipulator based on operation mode. In: 14th IFToMM world congress in mechanism and machine science, Taipei, Taiwan
13. Amine S, Masouleh MT, Caro S, Wenger P, Gosselin C (2012) Singularity conditions of 3T1R parallel manipulators with identical limb structures. ASME J Mech Robot 4:011011-1–011011-11
14. Amine S, Masouleh MT, Caro S, Wenger P, Gosselin C (2012) Singularity analysis of 3T2R parallel mechanisms using Grassmann-Cayley algebra and Grassmann geometry. Mech Mach Theory 52:326–340

A Geometric Method for Non-singular Path-planning in the Constant Orientation Workspace of a Stewart Platform Manipulator

Prem Kumar Prasad and Sandipan Bandyopadhyay

Abstract This paper presents a novel geometric solution to the problem of finding singularity-free paths joining two arbitrary points in the constant orientation workspace of a semi-regular Stewart platform manipulator. The formulation builds upon the known closed-form expression for the gain-type singularity surface of the manipulator. Using a rational parametrisation of the surface, it computes the geodesic curve on this surface, connecting the projections of the two given points on this surface. A sequence of spheres is then constructed in such a manner that each sphere is tangential to a previous one as well as the singularity surface, at a point on the said geodesic curve. Thus the geodesic curve acts as a guide, over which the singularity-free sphere is rolled, till it reaches its destination. Multiple methods for computing such sequences of spheres are presented and compared with the help of a numerical example. Finally, a sequence of line segments connecting the centres of the spheres is constructed, which connects the two given points via a provably singularity-free path.

Keywords Stewart platform manipulator · Singularity-free path-planning · Constant orientation singularity surface · Geodesic curve

1 Introduction

Path-planning of parallel manipulators is known to be a hard problem in robotics. One of the major impediments is the existence of gain-type singularities inside the workspace. These singularities occur on non-linear surfaces of complicated shapes, and any feasible path between two points in the workspace has to ensure that there is no intersection with these surfaces at *any* point on the path. In the context of the

P. K. Prasad · S. Bandyopadhyay (✉)
Indian Institute of Technology Madras, Chennai, India
e-mail: sandipan@iitm.ac.in

P. K. Prasad
e-mail: kumar.prasad.prem@gmail.com

D. Sen et al. (eds.), *Mechanism and Machine Science*, Lecture Notes in Mechanical Engineering, https://doi.org/10.1007/978-981-15-4477-4_6

Stewart platform manipulator, Dasgupta and Mruthyunjaya [1] presented a method by characterising the singularities via the ill-conditioning of certain matrices. Dash et al. [2] reported a numerical technique of clustering points around the singularity surface, representing them as obstacles and then avoiding these obstacles to find a path. Sen et al. [3] used the calculus of variations to reduce the problem of path-planning of a Stewart platform manipulator to a boundary value problem (BVP).

While these methods achieved reasonable success in their objectives, they suffered from various problems, such as computational complexity in the case of [2], and the difficulty in proving the existence as well as the uniqueness of the solution of a BVP in the case of [3]. More importantly, none of these methods could be proven to be fool-proof. In contrast, the method presented in this paper is based on geometry, and the paths it produces have the advantage of being certifiably free of singularities. It builds on three important theoretical developments, namely, the determination of the singularity manifold of the semi-regular Stewart platform manipulator (SRSPM) in the closed form [4]; identification of a rational parameterisation of the said manifold by Coste and Moussa [5]; and finally, computation of the singularity-free spheres (SFS) in the constant orientation workspace of the SRSPM [6]. Using these known results, the problem of path-planning is reduced to two problems in geometry: finding the geodesic curve on the gain-type singularity surface between two given points, and finding an SFS that is tangential to a given sphere as well as the singularity surface at a specified point. These two geometric building-blocks are used to construct four different variants of a method to demonstrate the concept. These are applied to the numerical example concerning an SRSPM, and the results obtained are compared critically, in terms of lengths of the path obtained, CPU-time consumed etc.

The approach presented is applicable to Stewart platform manipulators of any architecture, and other manipulators. However, the singularity surface of the manipulator must be known in closed-form, and it should be amenable to algebraic parametrisation for the method to work. Finally, in this approach, too, a BVP needs to be solved to compute the geodesic curve. The computational success of this step is not always guaranteed.

The rest of the paper is organised as follows: the mathematical approach to the problem is explained in Sect. 2. Numerical results are presented in Sect. 3, followed by a discussion of the same in Sect. 4. Finally, the conclusions are presented in Sect. 5.

2 Mathematical Formulation

The mathematical formulation and algorithms developed for finding the singularity-free path in the translational (i.e., constant orientation) workspace is described in this section. The steps are first described in brief, followed by the mathematical details.

2.1 *Overview of the Steps Followed in Finding A Singularity-Free Path, Given a Source and a Destination Point*

The task is to find a path, which is C^0-continuous and also connects the two given points, while not touching a given surface (i.e., the singularity surface of the SRSPM— see Fig. 1) at any point. The steps follow:

1. Check if the source and the destination points can be connected by a line segment which does not intersect the singularity surface at any point. The problem is solved trivially if such a path exists.
2. If the above fails, then construct the singularity-free spheres (SFS) centred around the source and the destination spheres, following the procedure given in [6]. These spheres are termed as s_1 and s_n, respectively, in Fig. 2. The points of tangency of these spheres with the singularity surface also define the respective projections of the terminal points onto the said surface.

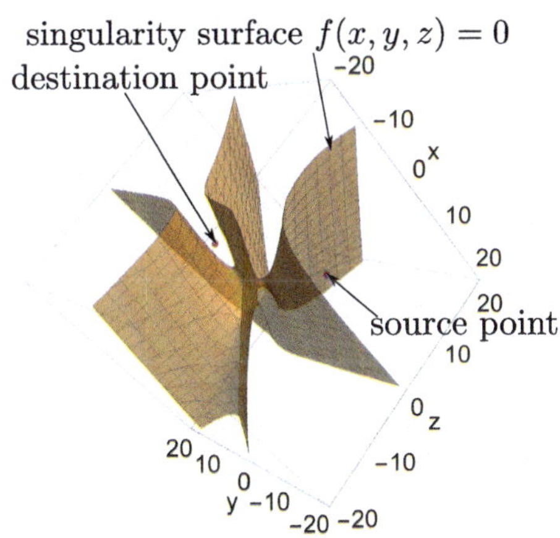

Fig. 1 Source and destination points

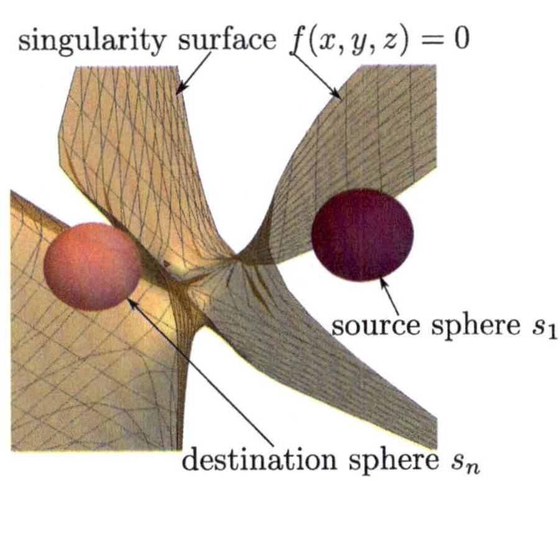

Fig. 2 Source and destination spheres

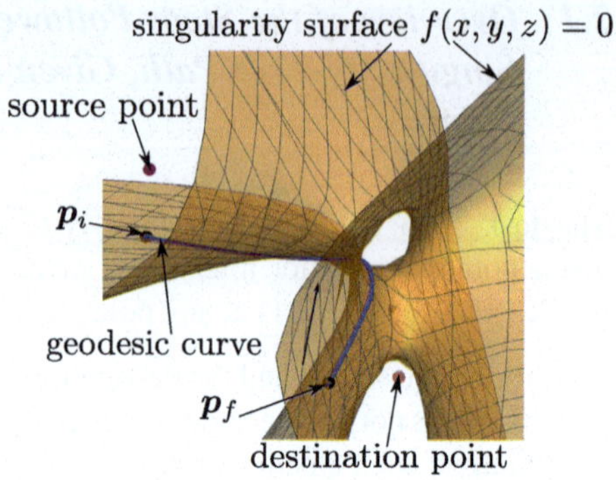

Fig. 3 Geodesic curve found on the singularity surface shown along with the source and destination points

3. Find the geodesic curve shown in Fig. 3 on the singularity surface connecting the points of tangency of s_1 and s_n with the singularity manifold.
4. Find a sequence of spheres, starting from either only s_1 or s_n, or both, such that each consecutive sphere is tangent to its predecessor as well as to the singularity surface at a point on the geodesic curve. The sequence terminates when starting from s_1 (s_n), one finds a sphere that interferes with s_n (s_1) (see Fig. 9), or starting from both sides, two of them interfere somewhere in the middle of the path (see Fig. 11). The algorithm fails if no such sequence can be found.
5. Join the source point of the path to its destination point, by a set of connected line segments passing through the centres of the sequence of spheres.

Multiple possibilities exist in implementing Step 4, of which four variants are studied in this paper. These are described below.

2.2 *Methods to Compute the Sequence of Singularity-Free Spheres*

Two strategies of choosing an appropriate 'step size' while computing the successive spheres along the geodesic curve are implemented, in the unidirectional as well as the bidirectional mode, leading to a total of four different methods.

2.2.1 Method A: Unidirectional Progression in Uniform Steps

Let, the geodesic curve be denoted by $\boldsymbol{\gamma}(t)$, where $t \in [0, 1]$ is the *path parameter*. The parameter t is divided into N equal segments of size δt for the sake of computation. The sequence of SFS is advanced from s_1 by an SFS which is tangential to s_1 as well as the singularity surface at $t = \delta t$. The mathematical formulation for the same is given in Sect. 2.4. However, this sphere is not inserted in the sequence immediately, as the search proceeds to the next SFS tangential to the singularity

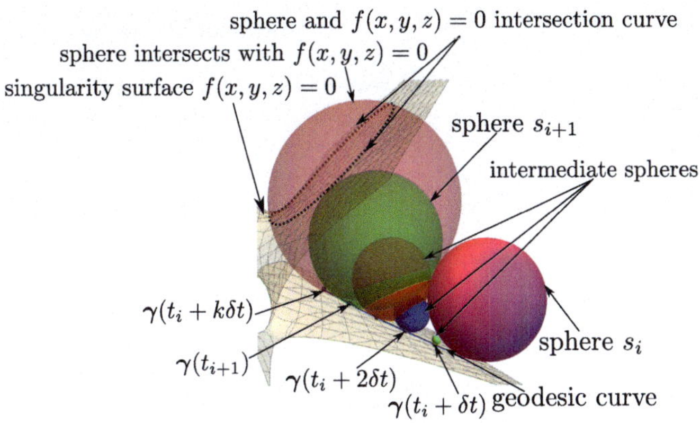

Fig. 4 SFS s_{i+1} found by method A for a known SFS s_i

surface at $t = 2\delta t, \ldots (k-1)\delta t$, till the sphere which maintains tangency with s_1 as well as the singularity surface at $t = k\delta t$ is no longer an SFS, i.e., it intersects the singularity surface at some other point(s). The sphere at $t = (k-1)\delta t$ becomes the next sphere in the sequence, i.e., s_2, and the process continues to find the next SFS, with s_2 as the updated 'starting' sphere. This is done to increase the size of the SFSs in the sequence, which helps the path in maintaining a higher distance from the singularity surface. The process is continued, till an SFS of this sequence interferes with the destination sphere, s_n. A flowchart[1] explaining method A has been presented in Fig. 6, and the case of intermediate spheres has been shown in Fig. 4.

2.2.2 Method B: Bidirectional Progression in Uniform Steps

Method B is the bidirectional version of method A. Two independent sequences of SFSs are propagated out of s_1 and s_n, and the algorithm terminates successfully, if an SFS from one sequence interferes with another from the other sequences.

2.2.3 Method C: Unidirectional Progression with Adaptive Step Size

In method C, the initial step size is set at 1, i.e., starting at s_1, the SFS tangential to s_1 as well as the singularity surface at $\boldsymbol{\gamma}(1)$ is determined. If no such SFS exists, then t is halved and the SFS tangential to s_1 and the singularity surface at $\boldsymbol{\gamma}(1/2)$ is computed. Again, if no (real) SFS is found, the parameter t is halved from its current value and the process is repeated, till it returns a new SFS, i.e., the second sphere in the sequence, denoted by s_2, which is tangential to s_1, as well as the singularity surface, at $t = t_2$ (say). The algorithm now resumes with s_2 as the current sphere,

[1] Flowcharts for the other methods could not be included for the want of space.

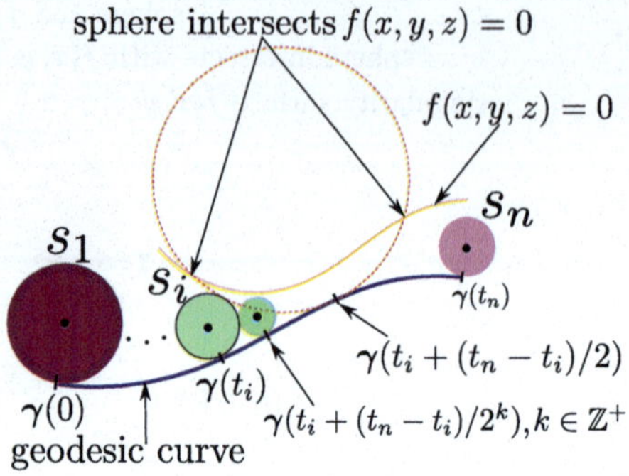

Fig. 5 SFS s_{i+1} found by method C for a known SFS s_i

and with $t \in [t_2, 1]$. This process is illustrated in the general case in Fig. 5, where one attempts to find an SFS tangential to s_i as well as the singularity surface at the point $\gamma\left(t_i + \frac{t_n - t_i}{2}\right)$. It can be seen that such a sphere exists, but it intersects the singularity surface elsewhere, hence does not qualify as an SFS. The process is, therefore, continued, till one finds an SFS at $\gamma\left(t_i + \frac{t_n - t_i}{2^k}\right)$, for $k \in \mathbb{Z}^+$, satisfying all the Requirements; hence, it is inserted in the sequence as s_{i+1}.

2.2.4 Method D: Bidirectional Progression with Adaptive Step Size

Method D is the bidirectional counterpart of method C; hence, further details pertaining to it are skipped.

These methods are now applied to the specific case of an SRSPM.

2.3 Geometry of the SRSPM

The geometrical model of the SRSPM is shown in Fig. 7. The six actuated prismatic joints are represented by the rectangular blocks. The global reference frame $\{1\}$ is attached at $\boldsymbol{o}_1$, i.e., the centroid of the base platform. The axis $\boldsymbol{X}_1$ passes through the vertex $\boldsymbol{b}_1$. Similarly, the reference frame $\{2\}$ is attached at the centroid of the moving platform, $\boldsymbol{o}_2$, and the axis $\boldsymbol{X}_2$ passes through $\boldsymbol{a}_1$. The axes $\boldsymbol{Z}_1$ and $\boldsymbol{Z}_2$ are perpendicular to the base and the moving platforms, respectively. The radii of the circumcircles passing through the vertices of the base and the moving platforms are r_b and r_a, respectively. Angular spacings between the adjacent pairs of legs are $2\gamma_a$ and $2\gamma_b$ for the moving and base platforms, respectively. Figure 8 shows the angular spacing for the moving platform of the SRSPM.

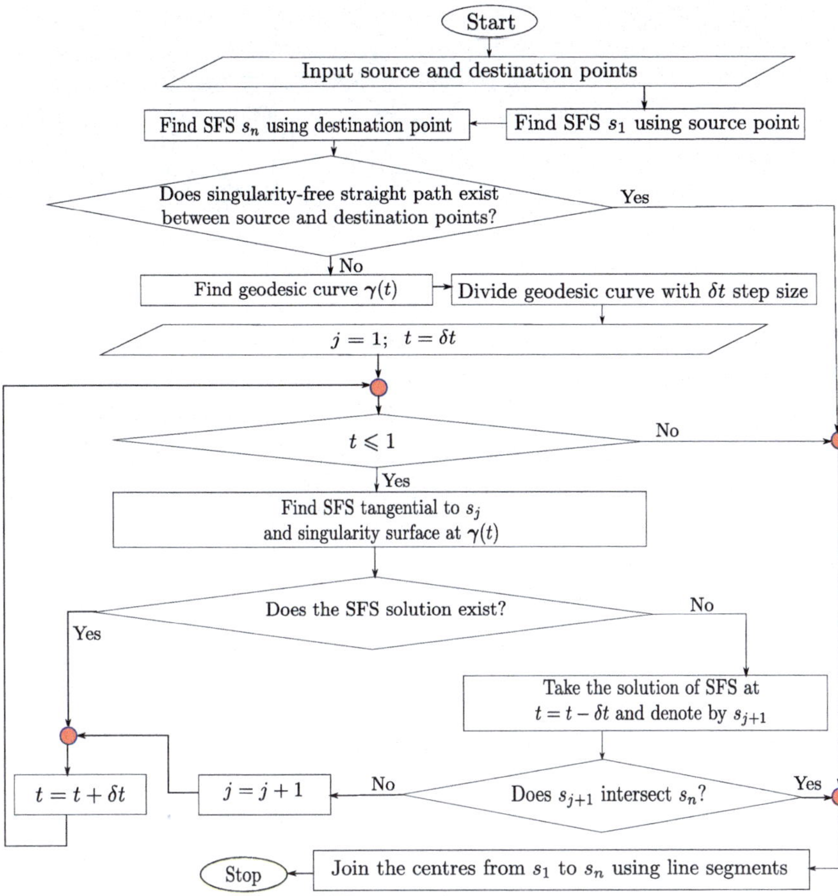

Fig. 6 Flowchart for method A

2.4 *Derivation of the Singularity-Free Sphere Which Is Tangential to both the Singularity Surface and a Given Sphere*

The position kinematics of the SRSPM is governed by the constraint equations $\boldsymbol{f}(\boldsymbol{\theta},\ \boldsymbol{x}) = \boldsymbol{0}$, where $\boldsymbol{x}$ is the vector of the *task space* (i.e., output) variables, and $\boldsymbol{\theta}$ is the vector of the active joint variables. Gain-type singularities occur when the forward kinematic equations have repeated solutions [7–9]. Following the *implicit function theorem*, it can be said that Jacobian of $\boldsymbol{f}$ with respect to $\boldsymbol{x}$, i.e., $\boldsymbol{J}_{fx}$, will always be singular at gain-type singularities. Therefore, the condition for gain-type singularity is given by:

$$S(\boldsymbol{\theta}, \boldsymbol{x}) \triangleq \det(\boldsymbol{J}_{fx}) = 0, \tag{1}$$

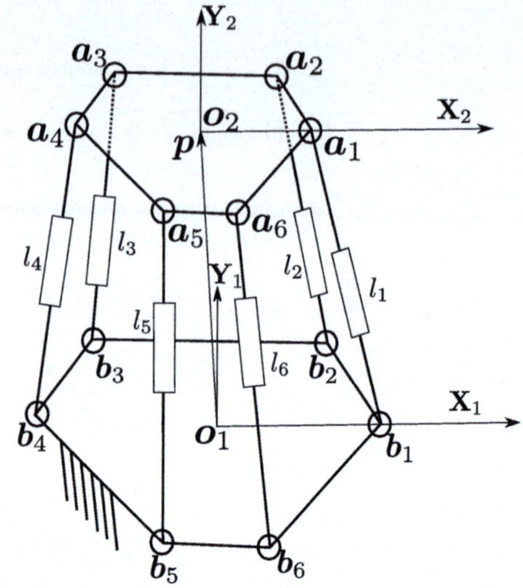

Fig. 7 Geometric model of SRSPM

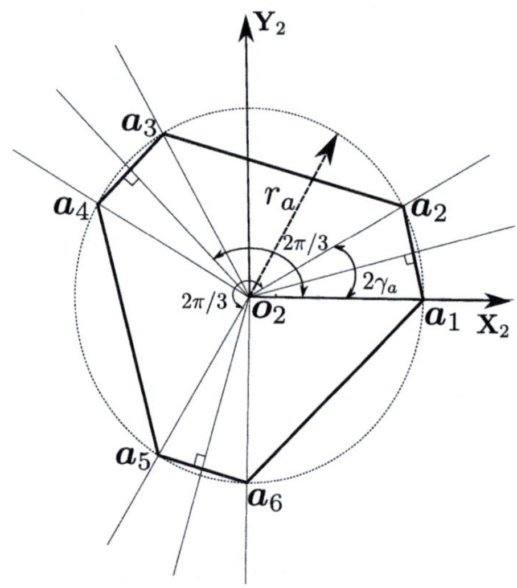

Fig. 8 Geometric description of the moving plate

For a *fixed orientation* of the moving platform, the singularity manifold of the SRSPM is of degree 3 in the coordinates $x,\ y,\ z$ of the centroid of the moving platform, $\boldsymbol{p}$:

$$\begin{aligned} S(\boldsymbol{\theta}, \boldsymbol{x}) = a_0 &+ a_1 x + a_2 x^2 + a_3 y + a_4 xy + a_5 y^2 + a_6 z + a_7 xz + a_8 x^2 z \\ &+ a_9 yz + a_{10} xyz + a_{11} y^2 z + a_{12} z^2 + a_{13} xz^2 + a_{14} yz^2 + a_{15} z^3. \end{aligned} \tag{2}$$

The coefficients $a_i \in \mathbb{R}$ depend on the orientation of the moving platform, i.e., Rodrigues' parameters c_1, c_2, c_3, and the architecture parameters of the manipulator, namely, γ_a, γ_b, r_a and r_b. The equation of the known sphere (denoted by s_g) of radius r_0 and centre $[x_0,\ y_0,\ z_0]^\top$ is given by:

$$g(x, y, z) = (x - x_0)^2 + (y - y_0)^2 + (z - z_0)^2 - r_0^2 = 0. \tag{3}$$

The desired sphere s_t has a radius of r, and is centred at $\boldsymbol{c} = [x_c, y_c, z_c]^\top$. Its equation is given by:

$$g_1(x, y, z) = (x - x_c)^2 + (y - y_c)^2 + (z - z_c)^2 - r^2 = 0. \tag{4}$$

The points, $\boldsymbol{p}_1 = [x_p,\ y_p,\ z_p]^\top$ and $\boldsymbol{p}_2 = [x_s,\ y_s,\ z_s]^\top$ are the tangency points of s_t with s_g and singularity surface, respectively. At the points $\boldsymbol{p}_1$ and $\boldsymbol{p}_2$ the normals to the sphere s_t should align with the normal of the sphere s_g and that of the singularity surface, respectively. The unknown variables at this stage are $x_c, y_c, z_c, x_s, y_s, z_s, x_p, y_p, z_p$ and r. More equations in these variables are derived as follows.

At the point $\boldsymbol{p}_1$ the gradients of g and g_1 align and the following tangency condition is obtained, of which only two scalar equations are linearly independent:

$$\nabla g \times \nabla g_1 = \mathbf{0} \Rightarrow h_{1i} = 0,\ i = 1,\ 2,\ 3,\ \text{where,} \tag{5}$$

$$h_{11} = y_0 z_c - y_p z_c - y_c z_0 + y_p z_0 + y_c z_p - y_0 z_p, \tag{6a}$$

$$h_{12} = x_0 z_c - x_p z_c - x_c z_0 + x_p z_0 + x_c z_p - x_0 z_p, \tag{6b}$$

$$h_{13} = z_0 z_c - z_p z_c - z_c z_0 + z_p z_0 + z_{c_2} z_p - z_0 z_p. \tag{6c}$$

The total degree of the equations $h_{1i} = 0, (i = 1,\ 2,\ 3)$ is two. At the point $\boldsymbol{p}_2$, the gradients of S and g_1 align and give the following tangency conditions from which two independent equations are obtained:

$$\nabla S \times \nabla g_1 = \mathbf{0} \Rightarrow h_{2i} = 0,\ \text{for } i = 1,\ 2,\ 3. \tag{7}$$

The total degree of $h_{2i} = 0,\ (i = 1,\ 2,\ 3)$ is three in $x_s,\ y_s,\ z_s,\ x_c,\ y_c$ and z_c. The respective highest degrees of these variables are provided in Table 1 for $h_{2i} = 0$, $(i = 1,\ 2,\ 3)$.

Table 1 Structure of the equations $h_{2i}, i = 1,\ 2,\ 3$

Variables	x_c	y_c	z_c	x_s	y_s	z_s
h_{21}	1	1	0	2	2	2
h_{22}	1	0	1	3	2	3
h_{23}	0	1	1	2	3	3

Both Eqs. (3) and (4) are satisfied by $\boldsymbol{p}_1$, leading to the following equations:

$$g(x_p, y_p, z_p) = (x_p - x_0)^2 + (y_p - y_0)^2 + (z_p - z_0)^2 - r_0^2 = 0, \quad (8)$$

$$g_1(x_p, y_p, z_p) = (x_p - x_c)^2 + (y_p - y_c)^2 + (z_p - z_c)^2 - r^2 = 0. \quad (9)$$

Equation (4) is satisfied by $\boldsymbol{p}_2$, which results in:

$$g_1(x_s, y_s, z_s) = (x_s - x_c)^2 + (y_s - y_c)^2 + (z_s - z_c)^2 - r^2 = 0. \quad (10)$$

The singularity surface described in Eq. (2) contains $\boldsymbol{p}_2$, leading to:

$$\begin{aligned} &a_0 + a_1 x_s + a_2 x_s^2 + a_3 y_s + a_4 x_s y_s + a_5 y_s^2 + a_6 z_s \\ &+ a_7 x_s z_s + a_8 x_s^2 z_s + a_9 y_s z_s + a_{10} x_s y_s z_s \\ &+ a_{11} y_s^2 z_s + a_{12} z_s^2 + a_{13} x_s z_s^2 + a_{14} y_s z_s^2 + a_{15} z_s^3 = 0. \end{aligned} \quad (11)$$

Finally, in ten unknowns, namely, $x_c, y_c, z_c, x_s, y_s, z_s, x_p, y_p, z_p$ and r, eight independent equations are obtained. Knowing a point on the singularity surface (or more specifically, on the geodesic curve $\boldsymbol{\gamma}(t)$ in this case), where the sphere s_t is tangent to the singularity surface, i.e., knowing x_s, y_s, z_s, Eq. (11) gets eliminated. The number of independent equations gets reduced to seven which can be solved to get the seven remaining variables: $x_p, y_p, z_p, x_c, y_c, z_c$ and r. In this case, the *Bézout number* signifying the maximum number of (complex) solutions to the system of seven polynomial equations is computed to be 288. The solution which leads to an s_t that does not intersect with the singularity surface and also has the largest radius among all the (real) solutions obtained is chosen as the final solution.

2.5 Finding the Geodesic Curve on the Singularity Surface

The details of finding the geodesic curve on the singularity manifold are described in this section. Let, the initial point be $\boldsymbol{p}_i$ and final point be $\boldsymbol{p}_f$ on the singularity surface. The surface admits a rational parameterisation in $(u(t), v(t))$ as shown by Coste and Moussa [5]. The geodesic on the singularity manifold is obtained as a solution to the following differential equations expressed in terms of $\boldsymbol{r}_u = \frac{\partial r}{\partial u}, \boldsymbol{r}_v = \frac{\partial r}{\partial v}, E = \boldsymbol{r}_u \cdot \boldsymbol{r}_u$, $F = \boldsymbol{r}_u \cdot \boldsymbol{r}_v$ and $G = \boldsymbol{r}_u \cdot \boldsymbol{r}_v$: (see, e.g., [10], pp. 182–185)

$$\frac{\mathrm{d}}{\mathrm{d}t}\left(E\frac{\mathrm{d}u}{\mathrm{d}t} + F\frac{\mathrm{d}v}{\mathrm{d}t}\right) - \frac{1}{2}\left(\left(\frac{\partial E}{\partial u}\frac{\mathrm{d}u}{\mathrm{d}t}\right)^2 + 2\frac{\partial F}{\partial u}\frac{\mathrm{d}u}{\mathrm{d}t}\frac{\mathrm{d}v}{\mathrm{d}t} + \left(\frac{\partial G}{\partial u}\frac{\mathrm{d}v}{\mathrm{d}t}\right)^2\right) = 0, \quad (12)$$

$$\frac{\mathrm{d}}{\mathrm{d}t}\left(F\frac{\mathrm{d}u}{\mathrm{d}t} + G\frac{\mathrm{d}v}{\mathrm{d}t}\right) - \frac{1}{2}\left(\left(\frac{\partial E}{\partial v}\frac{\mathrm{d}u}{\mathrm{d}t}\right)^2 + 2\frac{\partial F}{\partial v}\frac{\mathrm{d}u}{\mathrm{d}t}\frac{\mathrm{d}v}{\mathrm{d}t} + \left(\frac{\partial G}{\partial v}\frac{\mathrm{d}v}{\mathrm{d}t}\right)^2\right) = 0. \quad (13)$$

Solution of Eqs. (12, 13) with the boundary conditions $\boldsymbol{\gamma}(0) = \boldsymbol{p}_i$ and $\boldsymbol{\gamma}(1) = \boldsymbol{p}_f$ results in the geodesic curve $\boldsymbol{\gamma}(t)$.

3 Numerical Example

In this section, different methods proposed in Sect. 2.2 are demonstrated with the help of a numerical example. For a fair comparison of the different methods, a constant set of orientation parameters, as well as initial and final positions are used. The problem is solved in `Mathematica` [11], using the default precision of the system. `Mathematica` internal commands like `NSolve` have been used to solve the polynomial Eqs. (5, 7, 8, 9, 10) corresponding to the tangency conditions described in Sect. 2.4. The internal command `NDSolve` has been used to solve the BVP as described in Sect. 2.5. All the solutions obtained using `NSolve` of `Mathematica` have been validated by substituting back in the original equations and confirming that the norms of the residues were less than a predefined tolerance, $\epsilon = 10^{-10}$. All the computations are performed on an Intel® Core™ i7–4770 CPU running at 3.4 GHz.

The numerical values of the architecture parameters chosen are $\gamma_b = \frac{\pi}{6}$ rad, $\gamma_a = \frac{\pi}{4}$ rad, $r_a = \frac{1}{3}$ and $r_b = 1$. All the distances are scaled with respect to the base circum-radius r_b; hence, the length units are not specified. The orientation parameters are chosen to be $c_1 = 1$, $c_2 = 1$ and $c_3 = -\frac{1}{10}$. The source and destination points chosen are $[10,\ \frac{3}{2},\ 10]^\top$ and $[-14,\ -10,\ -3]^\top$, respectively. The time taken to compute the geodesic curve is nearly 14 min, which is a common preparatory step for the different methods of finding the sequence of spheres, which follow. The additional time taken in each method is listed in Table 2.

The path obtained by joining the centres of the SFSs by line segments are shown in Figs. 9, 11, 13 and 15 along with the SFSs for methods A, B, C and D, respectively. The paths obtained are shown in Figs. 10, 12, 14 and 16 along with the singularity surface for the methods A, B, C and D, respectively.

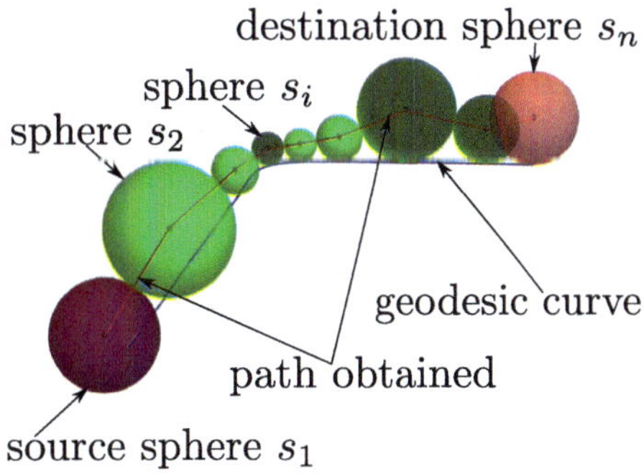

Fig. 9 The sequence of SFSs obtained from Method A

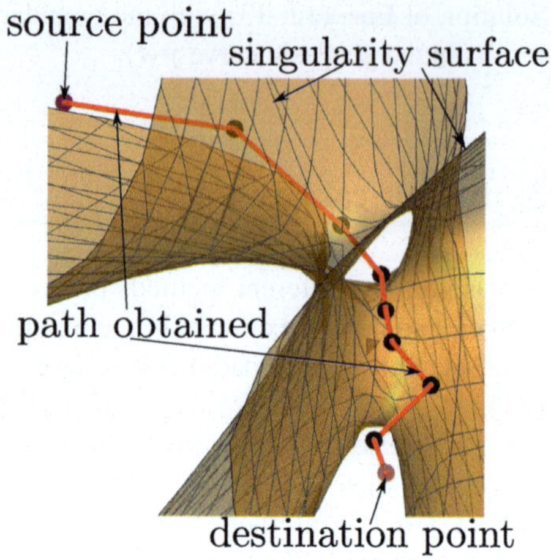

Fig. 10 The singularity-free path obtained from Method A

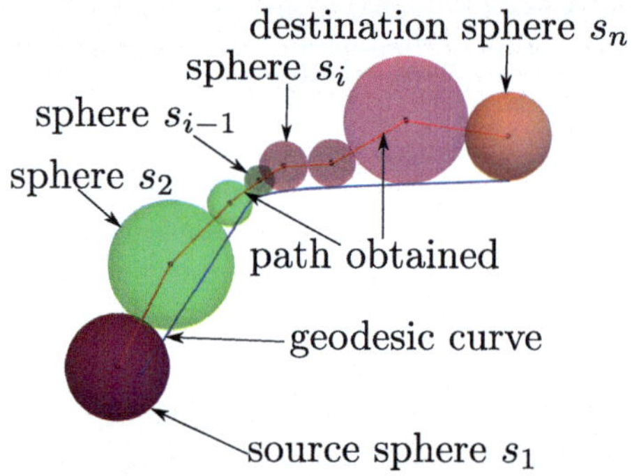

Fig. 11 The sequence of SFSs obtained from Method B

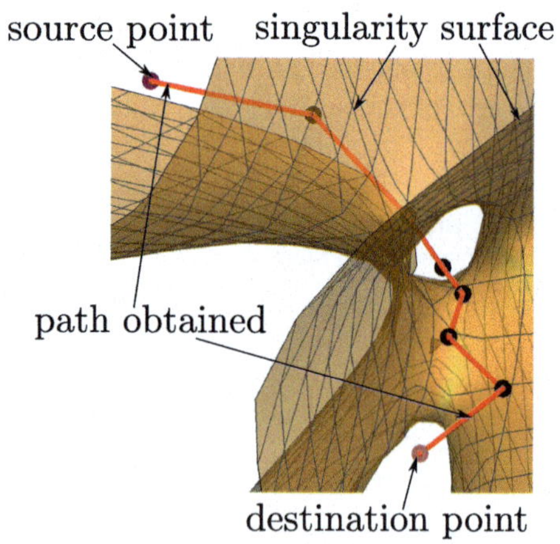

Fig. 12 The singularity-free path obtained from Method B

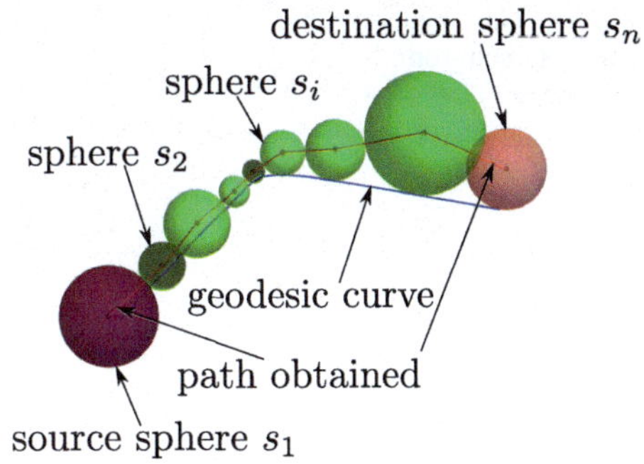

Fig. 13 The sequence of SFSs obtained from Method C

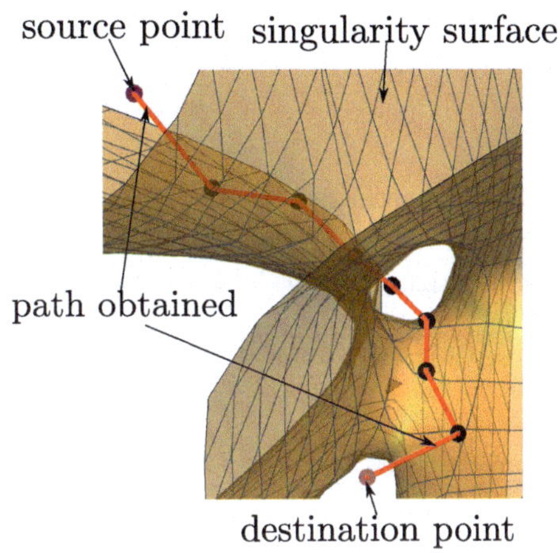

Fig. 14 The singularity-free path obtained from Method C

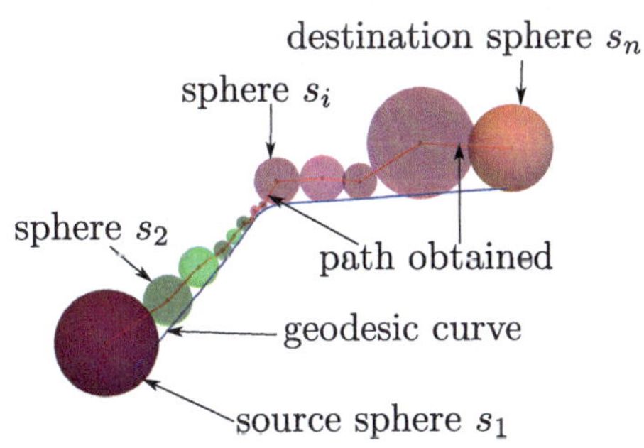

Fig. 15 The sequence of SFSs obtained from Method D

4 Discussions

The SFSs obtained from the different methods show significant variations in their radii. As expected, all the SFS radii are observed to decrease as the geodesic curve passes through a narrow 'hole' in the singularity surface. In case of method A and method B, it is the crucial part to determine the step size δt along the geodesic curve

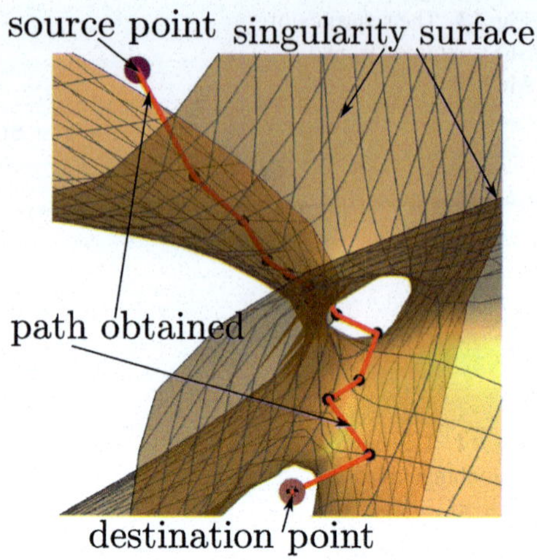

Fig. 16 The singularity-free path obtained from Method D

Table 2 Comparison of the different methods

Measure	A	B	C	D
Smallest sphere radius	0.996	0.996	0.791	0.072
Closest distance of path from $f = 0$	0.921	0.921	0.765	0.072
Number of SFSs	9	8	9	16
Computational time (s)	57.4	48.0	43.2	79.8
Length of path obtained	32.075	32.391	32.242	32.676

which should not be *too large*, as that would be likely to result in no real solution for the next SFS in the sequence. It should also not be *very small* as then would require more computational time, and also could result in smaller spheres, which could bring the resulting path very close to the singularity surface. In contrast, the main advantage of method C and method D is that no initial specification of the step size along the geodesic path is required. Consequently, the path obtained in each method is *unique*.

For the step size $\delta t = 0.029$ for methods A and B and for the example case demonstrated in Sect. 3, a number of performance indices have been documented in Table 2 for all the methods. The computational times listed in Table 2 each are the averages of ten different runs. It can be seen in Table 2 that the smallest SFS radii are the same in the methods A and B. It is of the same order in method C, but one order of magnitude smaller in the method D. This signifies that methods A and B have an advantage over methods C and D as bigger the sphere better may be the

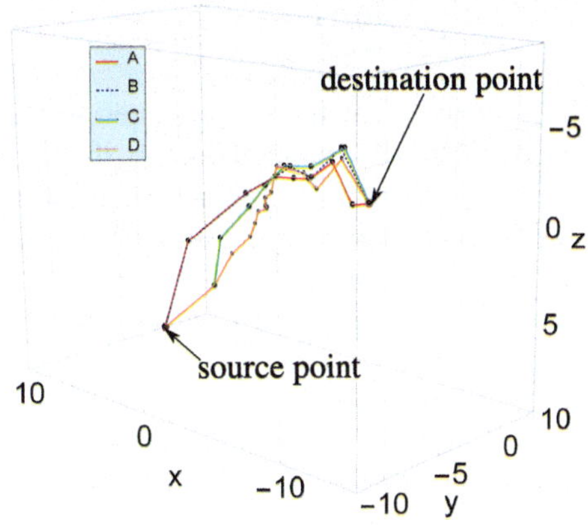

Fig. 17 Path obtained from methods A, B, C and D

possibility of the path staying away from the singularity manifold. Farther the path remains from the singularity surface the better it may be considered, though it is understood that no monotonic relation may exist between the physical distance from the singularity surface and the near-singular behaviour of the manipulator. The radii of the SFSs do indicate the distance from the singularity surface, but there exists the possibility of points on the path being closer to the singularity surface at some other points on the line segments connecting the centres of the sphere, than at the centres themselves. To check this, the paths were divided into 100 segments in each case and the minimum distance of the points from the singularity surface was computed. Among all the distances computed for a particular method, the minimum distance obtained is shown in Table 2. This measure indicates that in the method D the path lies closest to the singularity surface, as can be seen in Fig. 15. The number of SFSs along the geodesic curve found are far less for methods A, B and C as compared to method D, which in a way indicates smaller spheres and lesser distance from the singularity manifold.

In terms of computational time, method C has a significant advantage over all the other methods. In all the methods, path length covered is approximately the same but the method A seems to have a minor advantage in this regard. Figure 17 shows all the paths obtained together. It is easy to see that the initial path segments in the methods A and B are identical, as expected. In the methods C and D, the match is restricted only to the first segment.

It may be noted that the performance measures presented here are not necessarily generic, as they have been computed based on a single example. In a more rigorous study involving a large number of cases the outcome may vary significantly.

5 Conclusions

In this paper, a method is presented for finding a path free of gain-type singularity for the SRSPM. The method uses the knowledge of the closed-form expression of its singularity surface in $\mathbb{R}^3$ for a constant orientation of the moving platform. Using the parametric form of the singularity manifold, a geodesic curve is obtained on it. Different methods have been proposed to obtain piece-wise linear paths using the knowledge of the geodesic curve. These are illustrated using a numerical example and the results are compared.

To the best of knowledge of the authors, the methods proposed in this paper for finding a singularity-free path for the SRSPM are novel. While they have been proposed in the context of the SRSPM, in principle they hold for any form of Stewart platform manipulator, or indeed, any other manipulator, whose singularity manifold is known in closed form, and amenable to a rational parametrisation.

In the future, the obtained path would be smoothened via blending at the centres of the SFSs to allow a physical robot to trace it. Also, the formulation would be extended to include general spatial motions in $\mathbb{SE}(3)$.

Acknowledgements The authors would like to thank Mr. Anirban Nag, Doctoral scholar and Ms. Shivani Guptasarma, Dual Degree student, both in the Department of Engineering Design, Indian Institute of Technology Madras, for reading this paper thoroughly and suggesting improvements. The first author would also like to thank Council of Scientific & Industrial Research (CSIR) for the scholarship provided under the grant number: 09/084(0697)/2017-EMR-I.

References

1. Dasgupta B, Mruthyunjaya T (1998) Singularity-free path planning for the Stewart platform manipulator. Mech Mach Theory 33(6):711–725
2. Dash AK, Chen IM, Yeo SH, Yang G (2003) Singularity-free path planning of parallel manipulators using clustering algorithm and line geometry. In: 2003 IEEE international conference on robotics and automation (Cat. No.03CH37422), vol 1, pp 761–766
3. Sen S, Dasgupta B, Mallik AK (2003) Variational approach for singularity-free path-planning of parallel manipulators. Mech Mach Theory 38:1165–1183
4. Bandyopadhyay S, Ghosal A (2004) Analysis of configuration space singularities of closed-loop mechanisms and parallel manipulators. Mech Mach Theory 39:519–544
5. Coste M, Moussa S (2015) On the rationality of the singularity locus of a Gough-Stewart platform biplanar case. Mech Mach Theory 87:82–92
6. Nag A, Reddy V, Agarwal S, Bandyopadhyay S (2018) Identifying singularity-free spheres in the position workspace of Semi-regular Stewart Platform Manipulators. In: Lenarčič J, Merlet JB (eds.) Advances in Robot Kinematics 2016, Cham, pp 421–430. Springer International Publishing
7. Bandyopadhyay S, Ghosal A (2006) Geometric characterization and parametric representation of the singularity manifold of a 6–6 Stewart platform manipulator. Mech Mach Theory 41(11):1377–1400
8. Srivatsan RA, Bandyopadhyay S (2014) Determination of the safe working zone of a parallel manipulator. In: Thomas F, Perez Gracia A (eds) Computational kinematics: proceedings of the

6th international workshop on computational kinematics (CK2013), Dordrecht, pp 201–208. Springer Netherlands
9. Ghosal A (2006) Robotics: fundamental concept and analysis. Oxford University Press, India, New-Delhi
10. Somasundaram D (2010) Differential geometry: a first course. Alpha Science International Limited, New-Delhi
11. Wolfram Research Inc (2016) Mathematica, Version 11.0. Champaign, IL

A Review on Dynamic Balancing and Link Shape Synthesis of Planar Mechanisms

Sajjan Singh Bajiya, Kailash Chaudhary, and Himanshu Chaudhary

Abstract This paper reviews the various methods developed for balancing of the planar mechanisms and synthesizing the link shapes. The methods discussed in this paper are used for complete force balance, complete force and moment balance, partial force and moment balance as well as for the link shape synthesis of different planar mechanisms. The concepts, applications, and limitations of various methods are discussed and reviewed from the available literature in the area of mechanism balancing. The better understanding of available methods will definitely help the researchers working in this area in analyzing the current practices and in developing the new methods.

Keywords Shaking force and moment · Driving torque · Mass redistribution · Counterweight · Topology · Optimization

1 Introduction

Several review papers throw light on the quantum of work carried out on the balancing of the mechanisms especially dynamic balancing [1–7]. Shaking moment and driving torque get increases due to balancing of shaking force alone. Review of the literature regarding the complete balancing, i.e., shaking force and the shaking moment states that there will be no clear-cut method for static and dynamic balancing.

S. S. Bajiya (✉)
Department of Mechanical Engineering, University College of Engineering and Technology, Bikaner, India
e-mail: sajjansinghbajiya@gmail.com

K. Chaudhary
Department of Mechanical Engineering, MBM Engineering College, JNVU Jodhpur, Jodhpur, India
e-mail: k.chaudhary.mech@jnvu.edu.in

H. Chaudhary
Department of Mechanical Engineering, MNIT Jaipur, Jaipur, India
e-mail: hchaudhary.mech@mnit.ac.in

D. Sen et al. (eds.), *Mechanism and Machine Science*, Lecture Notes in Mechanical Engineering, https://doi.org/10.1007/978-981-15-4477-4_7

Several works of the literature are very specific and applicable to the specific mechanism. A generalized methodology is not available which can be applied to all types of mechanisms, particularly multi-loop mechanisms. Therefore, optimal balancing methods are explored. The literature based on the methods developed for optimal balancing of force and moment is reviewed including the various optimization methods for simultaneous minimization of the shaking force and the shaking moment. However, the methods used for balancing the flexible mechanisms are excluded in this paper. The methods used for the link shape synthesis are also reviewed and it is concluded that an initial design domain is required to start the procedure and may not always relate the final link shapes with the dynamic balance of the mechanisms. The limitations of all these methods are also discussed which justifies the future research directions.

The resultant forces, i.e., shaking force and shaking moment are transmitted to the frame due to moving links of the mechanism. For a mechanism having n moving links, shaking force and shaking moment are given by

$$\vec{\mathrm{F}}_{sh} = -\sum_{i=1}^{n} m_i \vec{a}_i Z$$

$$\vec{\mathrm{n}}_{sh} = -\sum_{i=1}^{n} \vec{r}_i \times m_i \vec{a}_i + \vec{n}_i Z$$

where m_i and a_i are mass and acceleration of the mass center of ith link, respectively. r_i is the position vector of mass center while n_i is inertia moment due to angular acceleration $\dot{\omega}_i$ and is given by

$$n_i = I_i \dot{\omega}_i + \omega_i \times I_i \omega_i Z$$

in which I_i is the inertia tensor.

2 Complete Shaking Force Balancing

The complete balancing of shaking force known as static balancing requires the total center of mass to be fixed for a mechanism. The two common approaches used to achieve this are the redistribution of the link masses and the use of the counterweights for the mechanism links. The analytical methods have been developed to trace and keep the total mass center of the mechanism fixed. Shchepetilnikov [8] presented the method of "*Principal Vectors*" in which the position of the mass center is described using the vectors that are directed along the links of the mechanism. Similarly, the "*Method of Linearly Independent Vectors*" was presented for the complete force balancing of four-bar as well as six-bar planar mechanisms with arbitrary link mass distribution (Fig. 1) [9]. The balancing conditions are presented for the internal mass

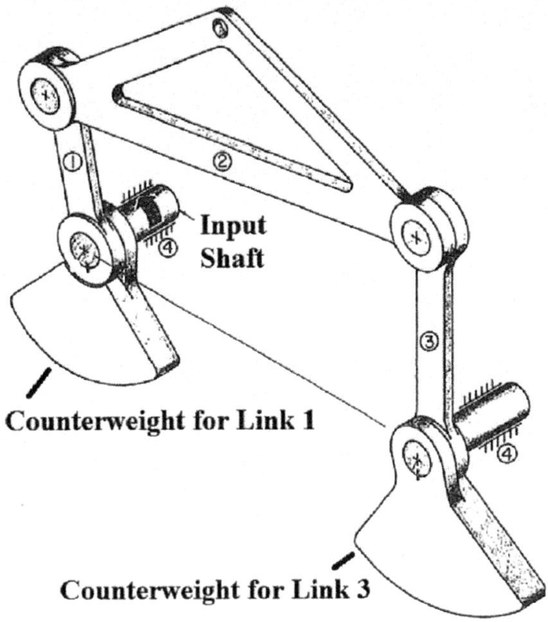

Fig. 1 Complete force balancing of planar four-bar mechanism using counterweights [9]

redistribution and for the counterweight addition. In this method, the links masses are redistributed in such a way that it eliminates the time-dependent terms coefficients in an equation representing the trajectory of the total center of mass of the mechanism.

According to the conservation of linear momentum law, when the resultant force F is zero, linear momentum p is conserved.

$$\frac{\mathrm{d}p}{\mathrm{d}t} = F = 0Z$$

The shaking force of the mechanism will get vanished when linear momentum p remains constant for any kind of motion at all times. The linear momentum for a completely balanced mechanism is given as follows:

$$p = \sum_{i=1}^{n} m_i \dot{r}_i = \mathrm{Const}Z$$

with i standing for the total number of links in a mechanism, m_i is mass of the link, r_i *is* the position vector of its center of mass, and $\dot{r}_i$ is the time-dependent term.

The elimination of time-dependent term coefficients results in a fixed center of mass of the mechanism and thus the complete shaking force balancing is achieved. This method was extended in [10] and it was proved that the counterweights required for the complete force balance of an n-link planar mechanism are half of the total number of the links. The "*Contour Theorem*" was developed to differentiate between the mechanisms which can be fully force balanced and those which cannot. Contour theorem examines the nature of the paths from the individual links to the ground. It was found that the pinned planar mechanisms can always be force balanced as they do not have a time-dependent coefficient in the center of mass equation. Based on the same approach, the complete force balancing conditions were presented for various types of planar mechanisms with multi-degrees of freedom in [11, 12]. The counterweights are used to balance the mechanism containing both the revolute and the prismatic joints. This method is used to decide the number of the counterweights required for complete force balancing and for the selection of the links to which the counterweight is to be attached. As an extension of the method proposed in [9], Kochev [13] used the ordinary vector algebra to derive the conditions for full force balancing in planar mechanisms. Thus, the linear balancing conditions in the Cartesian form are presented in this method. A computer program is also developed to fully balance the shaking force in the mechanisms based on the *Method of Linearly Independent Vectors* which also controls the increment in the shaking moment through properly designing the counterweights [14]. In another approach, the balancing conditions are presented to reduce the root-mean-square (RMS) and maximum value of the shaking force in the mechanisms which is known as the best uniform balancing [15].

The shaking force in a press machine can be minimized by designing a cam mechanism [16]. Adjusting Kinematics Parameter (AKP) approach is used to achieve force balancing along with the trajectory tracking in a five-bar real-time controllable (RTC) mechanism [17, 18]. Similar to the robot manipulators, the RTC mechanism is driven by the servomotors and can be scheduled and planned in real time. The AKP approach was found better than the counterweight method as far as the reduction in the servomotors torques and joint forces are concerned.

The effect of the complete force balance on the other dynamic properties was studied in [1, 19]. It was found that the complete force balance increases the shaking moment and driving torque in a mechanism. Therefore, only force balancing is not useful and the moment balance is also needed to balance the mechanism completely.

3 Complete Shaking Force and Shaking Moment Balancing

To achieve the dynamic balance in the mechanisms, the shaking moment is completely balanced by eliminating the angular momentum of the moving links along with the full force balance. The complete elimination of the angular momentum by adding the counterweights and/or using the mass distribution of link is not possible [4]. Therefore, the shaking moment is generally reduced by adding disk counterweights [10, 20–22], cam-actuated oscillating counterweight [23], physical pendulum [24], counter-rotating disks [3], inertia counterweight [25, 26], geared counterweights [27, 28], duplicate mechanism [29], and moment balancing idler loops [30]. Moore [31] presented the different sets of the design parameters that dynamically balance a four-bar mechanism without using the counter-rotations.

Berkof [24] developed a method in which the coupler mass is dynamically substituted by the concentrated masses located at the rotating links to completely balance the force and moment in the mechanisms. In this method, the coupler is treated as a massless link having two concentrated masses and thus only the rotating links need to be balanced for complete balancing of the mechanism. Also, to obtain complete balancing of the planar mechanisms, dynamic replacement of point masses for the moving links described as the mass flow concept can be used [28].

As an extension of the method of linearly independent vectors for full force balance, Elliot and Tesar [32, 33] presented a theory to balance shaking force and moment in the complex planar mechanisms. Esat and Bahai [27] extended the method developed by Tepper and Lowen [10] and found that for a fully force balanced mechanism, the moment can be completely eliminated using the geared counter-inertias (Fig. 2). Kochev [34] suggested the balancing of the shaking moment in a force balanced mechanism by prescribing the input speed fluctuations using noncircular gears.

The cam mechanism was used to fully balance the shaking force and moment in the planar mechanisms in [1, 23] (Fig. 3). The reduction of the inertia forces is achieved by using a cam-counterweight arrangement where the cam-driven masses keep the mechanism's center of mass fixed.

Feng [20, 21] developed a method for complete force and moment balance of the planar mechanisms combining mass redistribution and addition of the geared inertia counterweights (Fig. 4). The analytical conditions for complete force and moment balance for 17 types of eight-bar mechanism and 26 types of four-, five-, and six-bar mechanism with prismatic pairs are presented in this method. Arakelian and Smith [5, 29] proposed a method to completely balance the planar mechanisms using the counterweights connected through toothed-belt transmission system and gears. This arrangement generates an equal and opposite movement to the mechanism's center of mass movement and hence completely balances the mechanism (Fig. 5). Similarly, a method to improve the balancing of a double slider-crank mechanical system was presented in [35] (Fig. 6). The shaking force is balanced in this method by using the

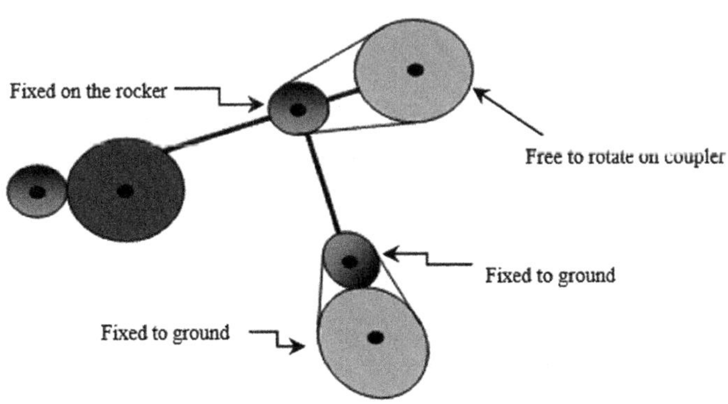

Fig. 2 Complete balancing of planar four-bar mechanism using geared counter-inertias [27]

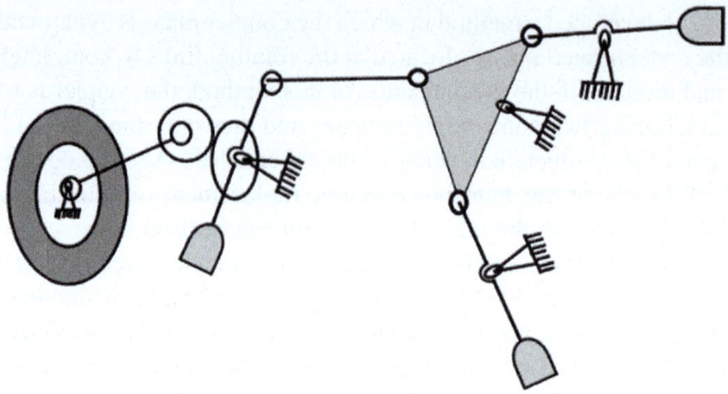

Fig. 3 Dynamic balancing of planar six-bar mechanism using cam-operated counterweight [1, 23]

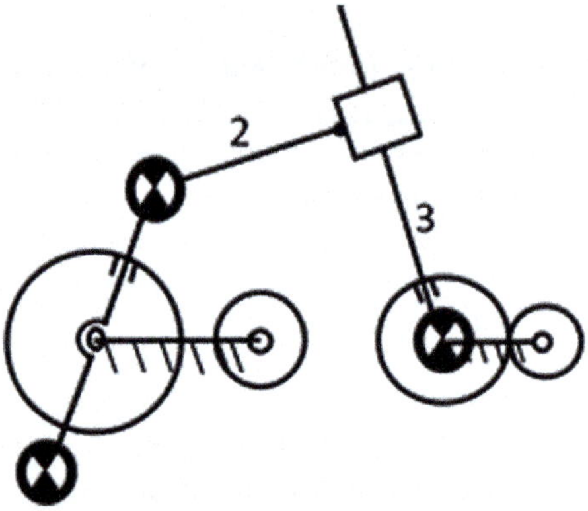

Fig. 4 Complete shaking force and shaking moment balancing through mass redistribution and addition of inertia counterweights [20]

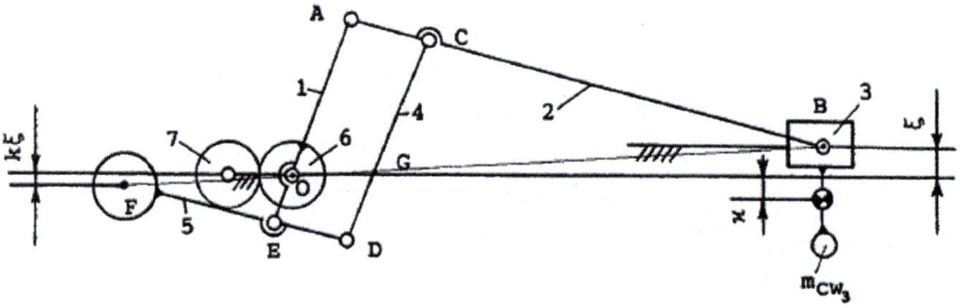

Fig. 5 Complete balancing of slider-crank mechanism based on the pantograph copying properties [29]

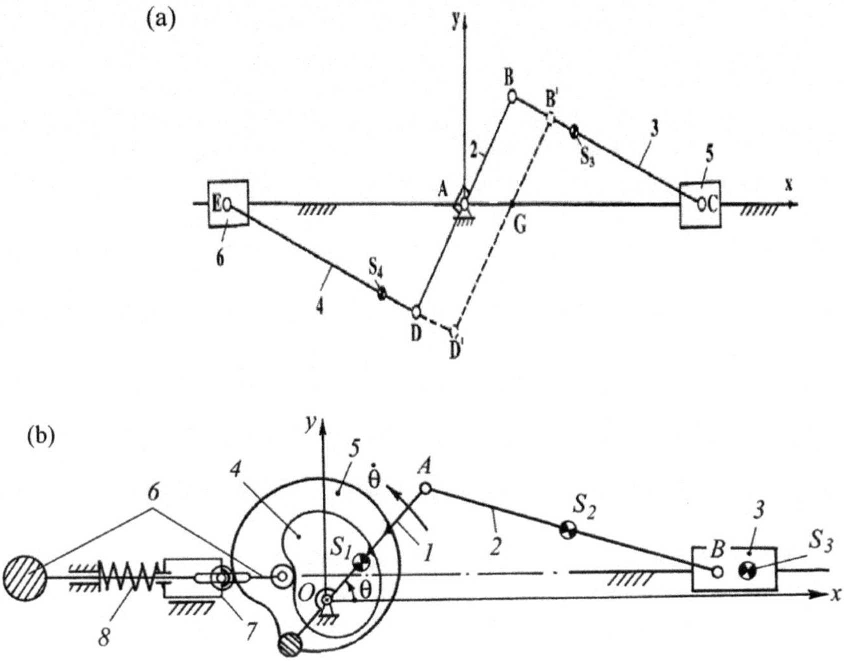

Fig. 6 **a** Self-balanced slider-crank mechanism [35], **b** balanced and torque compensated off-set slider-crank mechanism

two slider-crank mechanisms having equal and opposite movements. The shaking moment is balanced in this mechanism by the design modification of the second connecting rod of the double slider-crank mechanical system. Based on the same approach, Arakelian and Briot [36] used counterweight with the cam mechanism to balance a slider-crank mechanism. In this mechanism, the torque reduction is achieved through the spring that is used to maintain contact with the balancing cam mechanism.

Bagci [30, 37] solved the balancing problem of planar mechanisms containing multiple prismatic pairs that cannot be fully force balanced using the balancing criteria presented in [10]. He termed it as "force transmission irregularities" and used the "idler loop" concept with counter-rotating disks to completely balance the force and moment in the mechanism. Dresig and Dien [38] proposed a method to completely balance the shaking force and moment using a single rigid body known as the *balancing body*. In this method, the inertia forces and moments both for the mechanism and the balancing body are eliminated by controlling the motion of the balanced body.

An analytical method was proposed for static and dynamic balancing in the mechanisms in [39]. In this method, the balancing conditions are presented by the algebraic equations using joint angular velocities and complex variables. Thus, by using computer algebra, the necessary and sufficient conditions are derived to completely balance the shaking force and moment in the mechanism.

Considering the elasticity of mechanism links, kineto-elastodynamic inertia forces are the additional inertia forces developed by the acceleration field resulting from the vibration of the links. The contribution of the kineto-elastodynamic inertia forces should be taken into account toward the shaking force and moment while balancing planar mechanisms. The effects of the dynamic balancing on the elastodynamic properties are investigated in [40–43]. It is found that these methods increase the overall mass, space, cost, and complexity in the mechanisms, and most of them are applicable to the simple mechanisms only. Hence, complete force and moment balancing are achieved by producing shaking moment that counteracts the moment in the original mechanism.

4 Partial Shaking Force and Shaking Moment Balancing

Instead of complete balancing of shaking force and shaking moment independently, minimization of them simultaneously is more useful from the design point of view. The optimization methods used to simultaneously minimize the shaking force and moment in planar mechanism can be categorized as

4.1 Method of Harmonic Balancing

A method based on the harmonic analysis is used to balance the harmonics of the shaking forces and moments in which the forces and moments are formulated using the Gaussian least-square formulation and Fourier series [44]. Han [15] presented a least-square approach to balance the complex mechanisms in which a counterweight connected to the input shaft results in the first harmonic balancing of the inertia induced forces and moments. Stevensen [45] presented a method for the complete balancing of a harmonic unbalance of the machine, including inertia forces, moments of the inertia forces, and inertia torques, utilizing six weights on three shafts in the machine parallel to the coordinate axes and rotating at the speed of the harmonic. Similarly, two Oldham couplings were used to balance the second harmonics of the shaking force and shaking moment in [46]. This type of balancer runs at the primary speed of the machine whereas the Lanchester balancer runs at double the primary speed to get the same balancing effect. A method was developed based on the harmonic balancing to find the region boundaries to locate the additional shafts in [47]. Arakelian and Dahan [48] presented a method to achieve the first harmonic balancing of the shaking moment by using the counterweight connected to the input shaft in such a way that the counterweight rotation axis remains at an offset to the input shaft axis (Fig. 7).

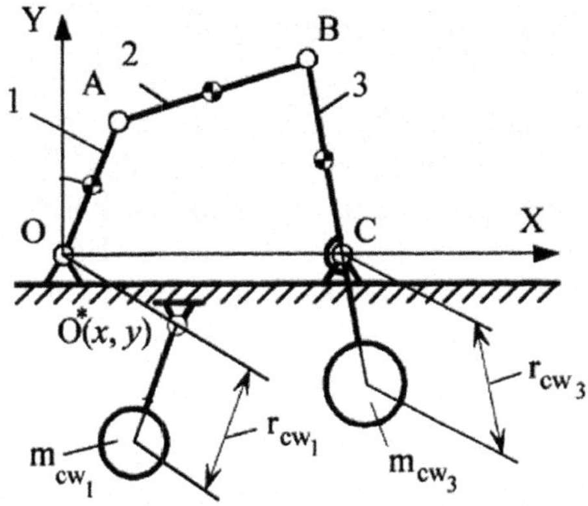

Fig. 7 First harmonic balancing of the shaking moment in planar four-bar mechanism [48]

4.2 Extension of Method of Linearly Independent Vectors

Based on the fact that in a force balanced mechanism, the shaking moment reduces to a pure torque, Berkof and Lowen [49, 50] developed the theory of shaking moment optimization. In this method, RMS value of the shaking moment is minimized for a force balanced in-line planar four-bar mechanism. The link mass distribution ratios as the functions of the link length ratios are defined as the design variables. This method was the extension of the previously developed force balancing method [9].

The angular momentum principle is used to define the shaking moment of a force balanced four-bar mechanism. It was shown that the shaking moment cannot be completely balanced without using additional links, but a partial moment balance may be obtained by the desired internal mass rearrangement. Carson and Stephens [51] extended this method by considering feasible limits for the link parameters. The optimum mass distribution is found for the mechanism links to balance it, and the kinematic design criterion is satisfied by fixing the link length ratios. Similarly, Haines [52] optimized the RMS values of the shaking moment and the driving torque for a force balanced mechanism by constraining the link parameters within the physical limits.

Tepper and Lowen [53] presented a method for partial force balancing of four-bar mechanisms, which allows control over the increase in ground bearing forces. They presented a trade-off method, wherein the RMS shaking force of constant input speed, a general four-bar mechanism is optimized while the increase of the RMS ground bearing forces is limited using Lagrange multiplier formulation. They presented an optimization method that represents feasible trade-off techniques for optimizing the RMS shaking force of a four-bar mechanism while keeping the RMS ground bearing forces lower than those of the fully force balanced mechanism.

4.3 Optimum Design of Counterweights

A partial force balancing method for a planar four-bar mechanism was developed in [25, 26] which allows the prescribed maximum shaking force. The simultaneous minimization of shaking moment, input torque, and bearing forces is achieved by using the three-counterweight technique (Fig. 8). This method determines the parameters of the three counterweights that must be attached to the input, coupler, and output links, respectively.

Demeulenaere [54–57] developed a convex optimization technique for planar mechanisms to determine the counterweight parameters for its balancing. This technique is used to design point-mass counterweights and minimum-inertia counterweights for the planar mechanisms. The counterweight parameters are chosen as the design variables while the upper and lower limits on mass and center of mass coordinates are presented as the convex constraint functions. The ratio of RMS values of the optimized dynamic force to the original dynamic force known as *balancing effect index* is defined as the objective function. Verschuure [58] extended this method and introduced a set of auxiliary parameters, μ-parameters, that are a function of the classical mass parameters to describe a counterweight. Due to the introduction of these parameters, an analytical derivation of the mechanism kinematics and dynamics is not necessary for this method (Fig. 9).

Mechanism balancing problem was formulated as a multi-objective optimization problem by Farmani [59] based on the concepts of the physical pendulum and inertia counterweights developed in [24] which completely balance all linear and rotary mass effects. The thickness of counterweights and disks for both input and output links of a planar four-bar mechanism are considered as the objective functions and

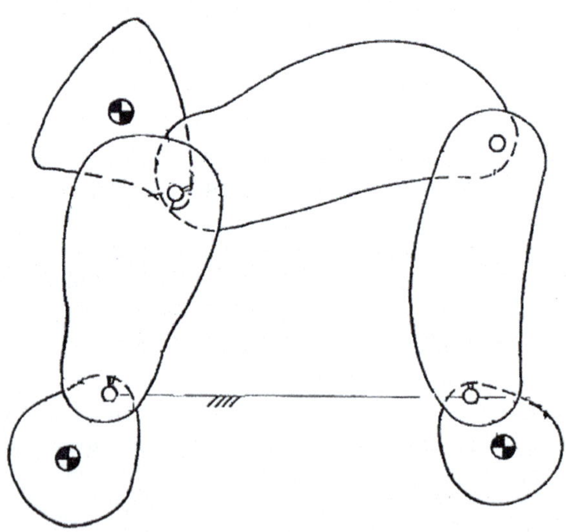

Fig. 8 Balancing of planar four-bar mechanism using three counterweights [25, 26]

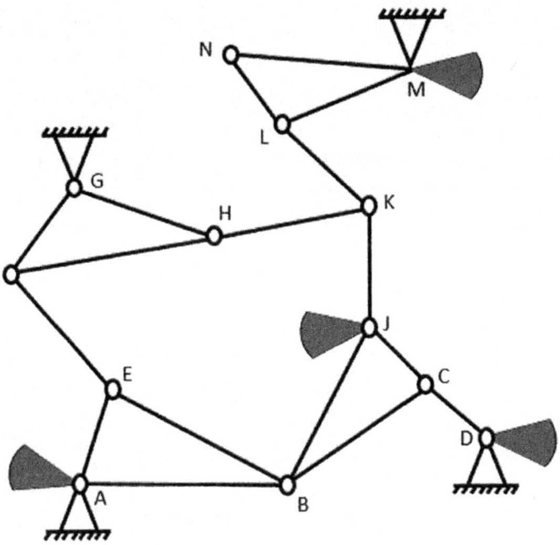

Fig. 9 Partial balancing of planar ten-bar mechanism using sector-type counterweights [58]

are minimized using Particle Swarm Optimization (PSO) and Genetic Algorithm (GA). The multiple optimum solutions were found for the optimization problem and then a fuzzy decision maker was used to select the best solution.

4.4 Optimum Mass Distribution Using Equimomental System of Point Masses

A design procedure for the optimum mass distribution was developed by Gill and Freudenstein [60, 61] for high-speed spherical four-bar mechanism links. This analysis is based on a quadratic-programming technique that allows an optimum trade-off between shaking forces, shaking moments, input-torque fluctuation, and bearing reactions. They replaced the mass distribution of a rigid link with a fixed point by four equivalent point masses to simplify inertia-force calculations. Rahman [62] extended this method to the general spatial mechanisms. The previous method was limited to one particular type, i.e., spherical mechanism. Lee and Cheng [63] used a two-point-mass model for combined balancing of shaking force, shaking moment, and torque by developing the optimality criterion using Lagrangian and Newtonian formulations. They formulated an optimization problem for the mechanism balancing and solved the same by Heuristic Optimization Technique of Lee and Freudenstein [64, 65]. This method presents the trade-offs among shaking force, shaking moment, input-torque fluctuations, and bearing reactions by redistributing the link masses and adding counterweights. Thus, the optimum balancing of the mechanism can be achieved in the design stage. The two-point-mass model developed in [66] for momentum conservation was adopted to model the inertial properties of the mechanism.

Chaudhary and Saha [67–69] solved the mechanism balancing problem as an optimization problem by finding the optimum mass distribution of the mechanism links along with constraining the mechanism parameters within the feasible limits. Thus, by optimizing links inertial properties, i.e., mass, location of center of mass, and moment of inertia, the shaking force and moment are minimized simultaneously. In this method, the inertia properties of the mechanism are represented by the equimomental point-mass systems that are dynamically equivalent to the rigid moving links of the mechanisms. The equimomental point-mass systems are used to identify the design variables and to formulate the constraints for the optimization problem formulation. They proposed three equimomental point masses for each link of the planar mechanisms [70]. To determine the shaking force and the shaking moment, the equimomental point-mass parameters are used to formulate the dynamic equations of motion for the mechanisms. An optimization problem is formulated for finding the optimum mass distribution of the mechanism links that minimizes the shaking force and shaking moment. The point-mass parameters are defined as the design variables and limits on links masses, and inertias are presented as the constraints for the optimization problem. The objective function is defined as the weighted sum of the dimensionless RMS values of the shaking force and shaking moment calculated using Newton–Euler equations of motions.

5 Methods for Link Shape Synthesis

To carry various loads acting on links, link shapes are to be decided once optimum inertial properties of a balanced mechanism are obtained. In the literature, several methods are suggested to find the mechanism link shapes for specified inertial properties of the mechanism that requires an initial design domain to start with. The methods for link shape synthesis available in the literature may be categorized as

5.1 Small Element Superposing Method

Feng [71] developed a Small Element Superposing Method (SESM) to find link shapes. This method discretizes the initial assumed shape of the link into small mass elements and locates these mass elements systematically along the link length (Fig. 10). The design equations corresponding to the optimized mass, center of mass location, and inertia are satisfied to form the link by superposing these small mass elements.

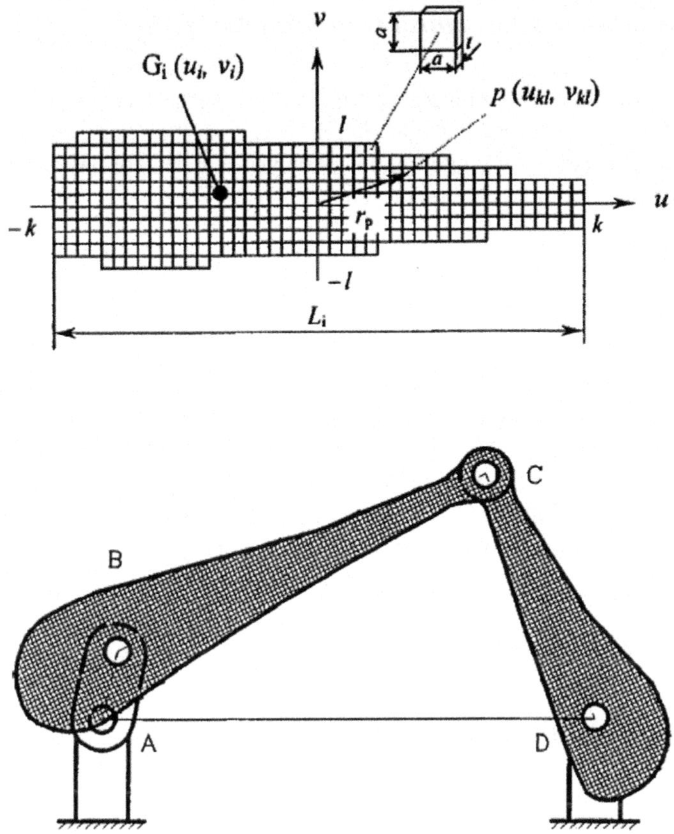

Fig. 10 Small element superposing method for finding link shape [71]

5.2 *Voxel-Based Discretization*

In the convex optimization method developed by Demeulenaere [54–57], designing point-mass counterweights and minimum-inertia counterweights was reformulated as a convex nonlinear optimization problem. This method was used to optimally design the counterweights used for the balancing purpose.

For planar mechanisms, a particular counterweight shape is enforced through additional constraints. The counterweight shapes considered were point mass, minimum-inertia counterweights, and sector-type counterweights. Based on this approach Verschuure [58] used a voxel-based discretization to describe the final counterweight shape, inspired by the typical discretizations used in the area of topology optimization.

5.3 Evolutionary Structural Optimization Method

This method is used to optimize the shape of the rotating machinery shaft. In this method ineffective material is gradually removed from the design domain [72–74]. This method executes the Finite Element Analysis (FEA) with finite-size elements in each iteration to find the optimum shape.

5.4 Gradient-Based Method

This method is used to synthesize the link shapes. In this method, considering the total volume of all the links as a constraint function, external work done by a given external force is maximized [75]. The external work done for the mechanism defined as the objective function satisfies the requirement of the kinetic energy of the mechanism for an assigned time interval. The initial design domain for the link is considered as the design variable and is optimized to get the optimum link shape (Fig. 11).

5.5 Topology Optimization Method

A fully compliant mechanism is a single entity of an elastic material continuum which requires determination of its geometrical form. It is challenging to design a compliant mechanism as a single body satisfying all kinematic and dynamic requirements. Shape synthesis of these mechanisms are required to consider constraints such as self-intersections, avoiding intersections with other segments, and restrictions on the available space and material. The link shapes are also found through the topology optimization based on parametric curves [76] and non-intersecting closed polygons [77]. In this approach, the vertices of the control polygon that define parametric curves for links are used to identify the design space. Based on FEA, this formulation considers the constraints to avoid self-intersections and intersections with other segments. It also ensures the optimum utilization of the available space and material. The required inertial properties of the design domain are achieved through optimizing the vertices of the control polygon as shown in Fig. 12.

5.6 Shapes for Specific Path and Motion

Identification of the feasible material domain associated with the link geometries helps to find the link shapes for the interference-free motion [78]. Three main infeasibilities, i.e., structural, kinematic, and geometric infeasibilities are identified so that the actual mechanism can be fabricated by their removal. This method focuses on

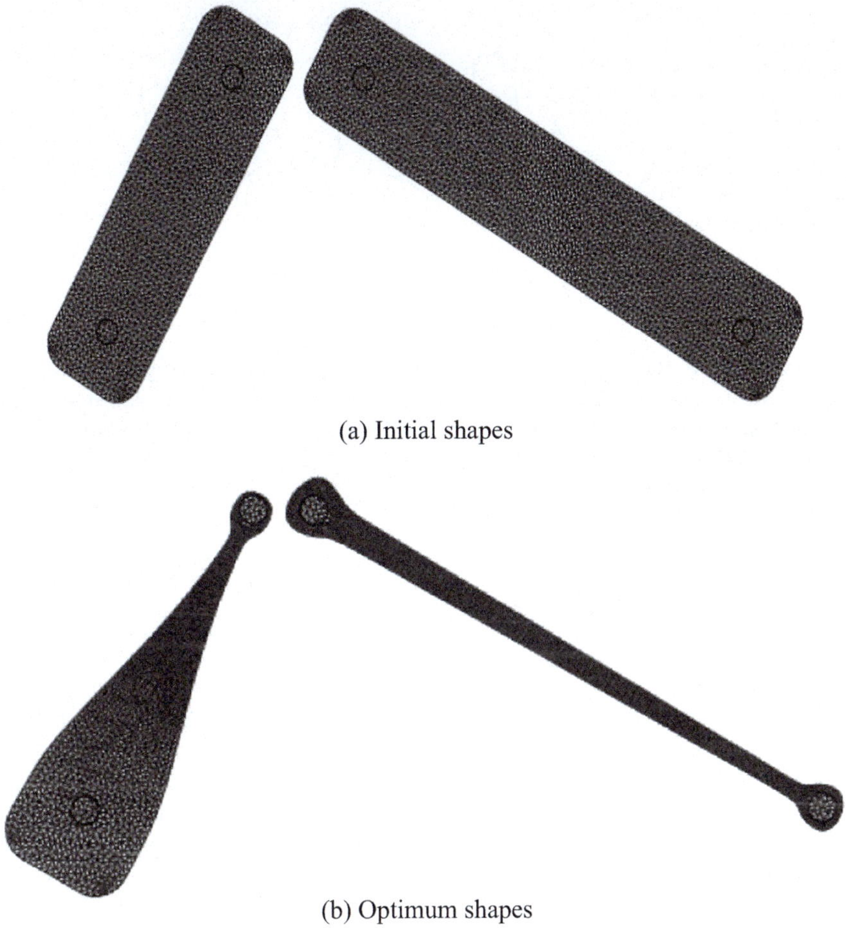

Fig. 11 Optimum shape for piston-crank mechanism using gradient-based method [75]

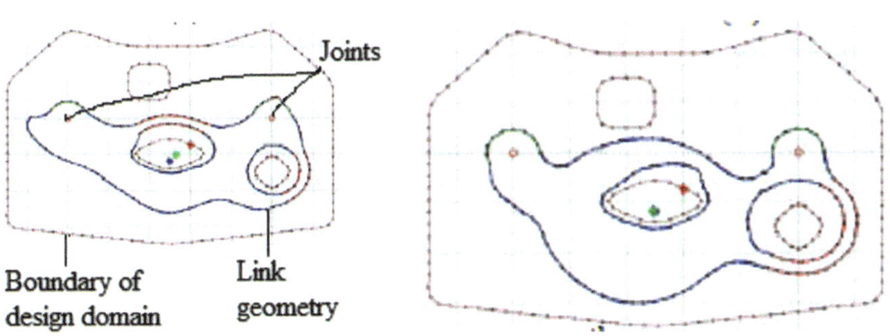

Fig. 12 Geometry synthesis through topology optimization [77]

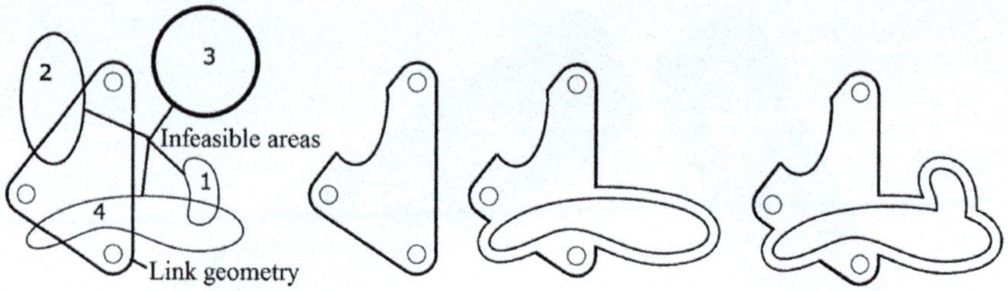

Fig. 13 Determination of feasible geometry [78]

interference-free motion with prespecified link shapes while topology methods [76, 77] synthesize the link shape by controlling the vertices positions of closed parametric curves. Similarly, shape optimization can also be completed using graphical and analytical techniques by generating a specified path or motion using the mechanism's dimensional synthesis [79, 80] (Fig. 13).

5.7 Shapes for Given Inertial Parameters

The shapes can be constructed using CAD for the given inertial parameters of the mechanism links as well as the given kinematic structure [81–84]. This two-stage optimization method includes the dynamic balancing and the shape synthesis of mechanism links. The benefit associated with this method is that the links of the balanced mechanism are of less and uniform thickness while the traditionally balanced mechanisms using force and inertia counterweights are of large thickness and radius compared to the original link parameters. Also, no predefined shapes or design domain is required in the proposed method. The resulting stresses for links of the balanced mechanism can be calculated at the weakest sections under external loads.

6 Conclusions

Techniques of shaking force balancing and its effect on other dynamic quantities are reviewed in Sect. 2. It was found that only shaking force balancing is not useful as it increases shaking moment and driving torque in the mechanism. Review of the literature regarding the complete balancing of the shaking force and the shaking moment in Sect. 3 states that there will be no clear-cut method for static and dynamic balancing. Several works of literature are very specific and applicable to the specific mechanism. A generalized methodology is not available which can be applied to all types of mechanisms, particularly multi-loop mechanisms. Therefore, optimal balancing methods are explored. The literature based on the methods developed for optimal balancing of force and moment is reviewed in Sect. 4 including the various optimization methods for simultaneous minimization of the shaking force and the shaking moment. In Sect. 5, the methods used for the link shape synthesis are reviewed and it is concluded that these methods require an initial design domain to start the procedure and do not relate the final link shapes with the dynamic balance of the mechanisms. The limitations of all these methods are also discussed which justifies the research gap.

References

1. Kamenskii VA (1968) On the questions of the balancing of plane linkages. J Mech 3(4):303–322
2. Lowen GG, Berkof RS (1968) Survey of investigations into the balancing of linkages. Mech Mach Theory 3(4):221–231
3. Lowen GG, Tepper FR, Berkof RS (1983) Balancing of linkages—an update. Mech Mach Theory 18(3):213–220
4. Kochev IS (2000) General theory of complete shaking moment balancing of planar linkages: a critical review. Mech Mach Theory 35:1501–1514
5. Arakelian VH, Smith MR (2005) Shaking force and shaking moment balancing of mechanisms: a historical review with new examples. ASME J Mech Des 127:334–339
6. Wijk VV, Herder JL, Demeulenaere B (2009) Comparison of various dynamic balancing principles regarding additional mass and additional inertia. ASME J Mech Robot 1:041006-1–9
7. Arakelian V, Briot S (2015) Balancing of linkages and robot manipulators—advanced methods with illustrative examples. Springer International Publishing
8. Shchepetilnikov VA (1968) The determination of the mass centres of mechanisms in connection with the problem of mechanism balancing. Mech Mach Theory 3(4):367–389
9. Berkof RS, Lowen GG (1969) A New method for completely force balancing simple linkages. ASME J Eng Ind 91(1):21–26
10. Tepper FR, Lowen GG (1972) General theorems concerning full force balancing of planar linkages by internal mass redistribution. ASME J Eng for Ind 94(3):789–796
11. Walker MJ, Oldham K (1978) A General theory of force balancing using counterweights. Mech Mach Theory 13:175–185
12. Walker MJ, Oldham K (1979) Extensions to the theory of balancing frame forces in planar linkages. Mech Mach Theory 14:201–207
13. Kochev IS (1988) A new general method for full force balancing of planar linkages. Mech Mach Theory 23(6):475–480
14. Smith M (1975) Dynamic analysis and balancing of linkages with interactive computer graphics. Comput Aided Des 7(1):15–19
15. Han C (1967) Balancing of high speed machinery. ASME J Eng Ind 89(1):111–117
16. Chiou ST, Davies TH (1997) Partial cancellation of shaking force harmonics by cam modification. Proc Inst Mech Eng. Part C: J Mech Eng Sci 211:253–263
17. Ouyang PR, Li Q, Zhang WJ (2003) Integrated design of robotic mechanisms for force balancing and trajectory tracking. Mechatronics 13:887–905
18. Ouyang PR, Zhang WJ (2005) Force balancing of robotic mechanisms based on adjustment of kinematic parameters. ASME J Mech Des 127(3):433–440
19. Lowen GG, Tepper FR, Berkof RS (1974) The quantitative influence of complete force balancing on the forces and moments of certain families of four-bar linkages. Mech Mach Theory 9:299–323
20. Feng G (1990) Complete shaking force and shaking moment balancing of 26 types of four-, five- and six-bar linkages with prismatic pairs. Mech Mach Theory 25(2):183–192
21. Feng G (1991) Complete shaking force and shaking moment balancing of 17 types of eight-bar linkages only with revolute pairs. Mech Mach Theory 26(2):197–206

22. Chiou ST, Bai GJ, Chang WK (1998) Optimum balancing designs of the drag-link drive of mechanical presses for precision cutting. Int J Mach Tools Manuf 38(3):131–141
23. Kamenskii VA (1968) On the problem of the number of counterweights in the balancing of plane linkages. J Mech 3(4):323–333
24. Berkof RS (1973) Complete force and moment balancing of inline four-bar linkage. Mech Mach Theory 8:397–410
25. Tricamo SJ, Lowen GG (1983) A novel method for prescribing the maximum shaking force of a four-bar linkage with flexibility in counterweight design. J Mech, Trans, Autom Des 105:511–519
26. Tricamo SJ, Lowen GG (1983) Simultaneous optimization of dynamic reactions of a four-bar linkage with prescribed maximum shaking force. J Mech, Trans, Autom Des 105:520–525
27. Esat I, Bahai H (1999) A theory of complete force and moment balancing of planar linkage mechanisms. Mech Mach Theory 34:903–922
28. Ye Z, Smith MR (1994) Complete balancing of planar linkages by an equivalent method. Mech Mach Theory 29(5):701–712
29. Arakelian VH, Smith MR (1999) Complete shaking force and shaking moment balancing of linkages. Mech Mach Theory 34:1141–1153
30. Bagci C (1982) Complete shaking force and shaking moment balancing of link mechanisms using balancing idler loops. ASME J Mech Des 104:482–493
31. Moore B (2009) Dynamic balancing of linkages by algebraic methods. PhD thesis, Research Institute for Symbolic Computation, Johannes-Kepler University Linz, Austria
32. Elliott JL, Tesar D (1977) The theory of torque, shaking force, and shaking moment balancing of four link mechanisms. ASME J Eng Ind 99(3):715–722
33. Elliott JL, Tesar D (1982) A general mass balancing method for complex planar mechanisms. Mech Mach Theory 17(2):153–172
34. Kochev IS (1992) Active balancing of the frame shaking moment in high speed planar machines. Mech Mach Theory 27(1):53–58
35. Arakelian VH (2006) Shaking moment cancellation of self-balanced slider-crank mechanical systems by means of optimum mass redistribution. Mech Res Commun 33:846–850
36. Arakelian VH, Briot S (2010) Simultaneous inertia force/moment balancing and torque compensation of slider-crank mechanisms. Mech Res Commun 37:265–269
37. Bagci C (1979) Shaking force balancing of planar linkages with force transmission irregularities using balancing idler loops. Mech Mach Theory 14(4):267–284
38. Dresig H, Dien NP (2011) Complete shaking force and shaking moment balancing of mechanisms using a moving rigid body. Tech Mech 31(2):121–131
39. Gosselin CM, Moore B, Schicho J (2009) Dynamic balancing of planar mechanisms using toric geometry. J Symb Comput 44:1346–1358
40. Raghu E, Balasubramonian A (1990) Experimental study on the elastodynamic behavior of the unbalanced and the counterweighted four-bar mechanisms. J Mech Des 112(3):271–277
41. Xi F, Sinatra R (1997) Effect of dynamic balancing on four-bar linkage vibrations. Mech Mach Theory 32(6):715–728
42. Martini A, Troncossi M, Rivola A (2013) Elastodynamic effects of mass-balancing: experimental investigation of a four-bar linkage. Adv Mech Eng 1–10
43. Martini A, Troncossi M, Carricato M, Rivola A (2014) Elastodynamic behavior of balanced closed-loop mechanisms: numerical analysis of a four-bar linkage. Meccanica 49(3):601–614
44. Norton RL (2011) Kinematics and dynamics of machinery. Tata McGraw Hill Education Private Limited, New Delhi, India
45. Stevensen EN (1973) Balancing of machines. ASME J Eng Ind, 650–656
46. Tsai L (1984) Oldham-coupling second-harmonic balancer. ASME J Mech, Trans, Autom Des 106(3):285–290
47. Davies TH, Niu GH (1994) On the retrospective balancing of installed planar mechanisms. Proc Inst Mech Eng. Part C: J Mech Eng Sci 208:39–45
48. Arakelian V, Dahan M (2001) Partial shaking moment balancing of fully force balanced linkages. Mech Mach Theory 36:1241–1252

49. Berkof RS, Lowen GG (1971) Theory of shaking moment optimization of forced-balanced four-bar linkages. ASME J Eng Ind 93B(1):53–60
50. Lowen GG, Berkof RS (1971) Determination of force-balanced four-bar linkages with optimum shaking moment characteristics. J Eng Ind 39–46
51. Carson WL, Stephenes JM (1978) Feasible parameter design spaces for force and root-mean-square moment balancing an in-line 4r 4-bar synthesized for kinematic criteria. Mech Mach Theory 13:649–658
52. Haines RS (1981) Minimum RMS shaking moment or driving torque of a force-balanced 4-bar linkage using feasible counterweights. Mech Mach Theory 16(3):185–195
53. Tepper FR, Lowen GG (1975) Shaking force optimization of four-bar linkage with adjustable constraints on ground bearing forces. ASME J Eng Ind, 643–651
54. Demeulenaere B (2004) Dynamic balancing of reciprocating machinery with application to weaving machines. PhD thesis, Department of Mechanical Engineering, University of Leuven, Belgium
55. Demeulenaere B, Aertbelien E, Verschuure M, Swevers J, Schutter JD (2006) Ultimate limits for counterweight balancing of crank-rocker four-bar linkages. ASME J Mech Des 128:1272–1284
56. Demeulenaere B, Swevers J, Schutter JD (2004) A convex optimization framework for dynamic balancing of planar linkages. Proc. ISMA 1997–2010
57. Demeulenaere B, Verschuure M, Swevers J, Schutter JD (2010) A general and numerically efficient framework to design sector-type and cylindrical counterweights for balancing of planar linkages. ASME J Mech Des 132:011002(1–10)
58. Verschuure M (2009) Counterweight balancing of mechanisms using convex optimization techniques. PhD thesis, Department of Mechanical Engineering, University of Leuven, Belgium
59. Farmani MR, Jaamialahmadi A, Babaie M (2011) Multiobjective optimization for force and moment balance of a four-bar linkage using evolutionary algorithms. J Mech Sci Technol 25(12):2971–2977
60. Gill GS, Freudenstein F (1983) Minimization of inertia-induced forces in spherical four-bar mechanisms. Part 1: the general spherical four-bar linkage. ASME J Mech, Trans, Autom Des 105:471–477
61. Gill GS, Freudenstein F (1983) Minimization of inertia-induced forces in spherical four-bar mechanisms. Part 2: wobble-plate engines. ASME J Mech, Trans, Autom Des 105:478–483
62. Rahman S (1996) Reduction of inertia-induced forces in a generalized spatial mechanism. PhD thesis, Department of Mechanical Engineering, The New Jersey Institute of Technology
63. Lee TW, Cheng C (1984) Optimum balancing of combined shaking force, shaking moment, and torque fluctuations in high speed linkages. ASME J Mech, Trans, Autom Des 106(2):242–251
64. Lee TW, Freudenstein F (1976) Heuristic combinatorial optimization in the kinematic design of mechanisms; Part 1, theory. ASME J Eng Ind 98(4):1277–1280
65. Lee TW, Freudenstein F (1976) Heuristic combinatorial optimization in the kinematic design of mechanisms; part 2, application. ASME J Eng Ind 98(4):1281–1284
66. Wiederrich JL, Roth B (1976) Momentum balancing of four-bar linkages. ASME J Eng Ind 98(4):1289–1295
67. Chaudhary H, Saha SK (2007) Balancing of four-bar linkages using maximum recursive dynamic algorithm. Mech Mach Theory 42(2).216–232
68. Chaudhary H, Saha SK (2008) Balancing of shaking forces and shaking moments for planar mechanisms using the equimomental systems. Mech Mach Theory 43:310–334
69. Chaudhary H, Saha SK (2009) Dynamics and balancing of multibody systems. Springer-Verlag, Berlin
70. Chaudhary H, Saha SK (2006) Equimomental system and its applications. In: Proceedings of 8th biennial ASME Conference on Engineering Systems Design and Analysis, Torino, Italy, 2006
71. Feng B, Morita N, Torii T (2002) A new optimization method for dynamic design of planar linkage with clearances at joints. ASME J Mech Des 124:68–73
72. Xie YM, Steven GP (1993) A simple evolutionary procedure for structural optimization. Comput Struct 49(5):885–896

73. Xie YM, Steven GP (1996) Evolutionary structural optimization for dynamic problems. Comput Struct 58(6):1067–1073
74. Kim Y, Tan A, Yang B, Kim W, Choi B, An Y (2007) Optimum shape design of rotating shaft by ESO method. J Mech Sci Technol 21:1039–1047
75. Azegami H, Zhou L, Umemura K, Kondo N (2013) Shape optimization for a link mechanism. Struct Multidiscip Optim 48(1):115–125
76. Xu D, Ananthasuresh GK (2003) Freeform skeletal shape optimization of compliant mechanisms. ASME J Mech Des 125:253–261
77. Yoganand G, Sen D (2011) Link geometry synthesis for prescribed inertia. In: Proceedings of 15th national conference on machines and mechanisms, IIT Madras, India
78. Sen D, Chowdhury S, Pandey SR (2004) Geometric design of interference-free planar linkages. Mech Mach Theory 39:737–759
79. Sandor GN, Erdman AG (1984) Advanced Mechanism Design – Analysis and Synthesis, vol 2. Prentice Hall, New Jersey
80. Freudenstein F (2010) Approximate synthesis of four-bar linkages. Resonance 15(8):740–767
81. Chaudhary K, Chaudhary H (2015) Optimal dynamic balancing and shape synthesis of links in planar mechanisms. Mech Mach Theory 93:127–146
82. Chaudhary K, Chaudhary H (2014) Dynamic balancing of planar mechanisms using genetic algorithm. J Mech Sci Technol 28(10):4213–4220
83. Chaudhary K, Chaudhary H (2015) Optimal design of planar slider-crank mechanism using teaching-learning-based optimization algorithm. J Mech Sci Technol 29(12):5189–5198
84. Chaudhary K, Chaudhary H (2016) Optimal dynamic design of planar mechanisms using teaching-learning-based optimization algorithm. Proc Inst Mech Eng, Part C: J Mech Eng Sci 230(19):3442–3456

Canonical Decomposition of the Instantaneous Kinematics of Darboux Frame

K. Rama Krishna

Abstract This paper gives a screw theoretic insight into the instantaneous kinematics of the Darboux frame (D-frame) moving on a smooth surface. It is established that the velocity state of D-frame is described by a three screw system. The principal screws of this system depend only upon the local shape of the surface. The twist of the D-frame is decomposed into twisting about surface normal line and a screw which is a ruling of a cylindroid. The properties of this cylindroid are studied. It is also shown that the twist of the D-frame can be obtained by a linear combination of pure rotations about three axes. This result is used to arrive at an equivalent 3R (revolute) serial chain whose end-effector is the D-frame. An illustrative numerical example is presented.

Keywords Surface curvature · Screw · Equivalent linkage

1 Introduction

In the study of the differential geometry of smooth spatial curves, the use of the Frenet-Serret frame (F-S-frame) [1] and its instantaneous kinematics is an indispensable tool. Analogously, the local differential geometry of a smooth surface at a point is captured by the kinematics of a geometric construct called the Darboux frame (D-frame) [1]. The instantaneous kinematics of a rigid body is elegantly studied via screw theory [2] which casts lines, screws, and twists in a single framework. The motion of the F-S-frame has been analyzed using screw theoretic concepts to some extent in the context of the striction curve of an axode surface.

Dooner [3] introduced an osculating cylinder to a spatial curve whose radius is referred to as "radius of torsure". It is claimed there that the instantaneous screw axis (ISA) of the twist of the F-S-frame is the axis of this cylinder. The locus of the axes of such osculating cylinders for the family of ∞^1 geodesics at a point on a smooth

K. Rama Krishna (✉)
Department of Mechanical Engineering, Indian Institute of Technology Delhi, New Delhi, India
e-mail: ram.itbhu@gmail.com

D. Sen et al. (eds.), *Mechanism and Machine Science*, Lecture Notes in Mechanical Engineering, https://doi.org/10.1007/978-981-15-4477-4_8

surface is termed “cylindroid of torsure”. The analysis does not use D-frame since it is limited to geodesics on a surface.

The present work for the first time in literature comprehensively studies the instantaneous kinematics of the D-frame from a screw theoretic point of view. The formulation of the twist of the D-frame and its primary partitioning into two twists is presented in Sect. 2. The cylindroid component of the primary decomposition is studied in Sect. 3. Complete decomposition and equivalent serial chain substitute-connection producing the twist of the D-frame is presented in Sect. 4.

2 Kinematics of D-Frame

The motion of the D-frame and the local Differential geometry of the surface are interconnected. Let C be a smooth curve on a smooth surface S_f (Fig. 1). At a point P on curve C, imagining the surface as a boundary of a solid object, let n be the surface normal line directed into the material side of the surface. Let T be the line passing through point P and tangential to the curve C at point P. Line p passes through the point P and is such that Tnp is a right-handed orthogonal triad of these three lines. This rigid Tnp construct with the origin P being a point on the surface S_f is called the D-frame. Without loss of generality, let us assume that the origin of the global fixed coordinate system is such that point P is at this origin instantaneously. The lines T, n and p are written in the dual-vector form as $T = \underline{T} + \epsilon\underline{0}$, $n = \underline{n} + \epsilon\underline{0}$, and $p = \underline{p} + \epsilon\underline{0}$, respectively. $\underline{T}$, $\underline{n}$, and $\underline{p}$ are unit vectors.

2.1 Twist of D-Frame

As the point P moves along the curve C, the motion of the D-frame evolves and the directions of T, n, and p are parametrized. We assume here that the point P moves with unit speed on the surface S_f, i.e., $ds/dt = 1$, where s is the arc length

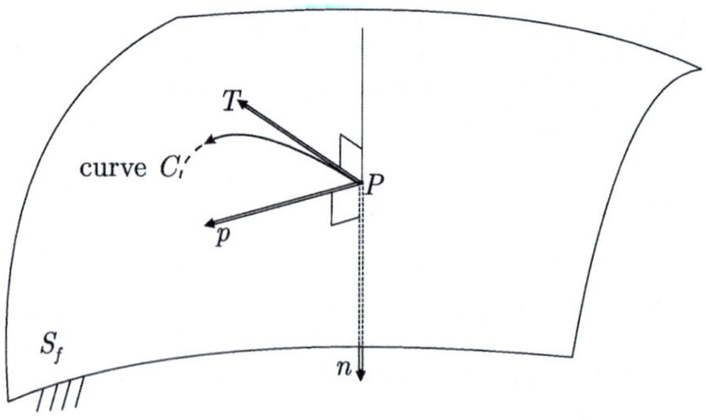

Fig. 1 D-frame moving along curve C at point P on surface

parameter of the curve C. Let the instantaneous twist of the D-frame be labeled $\$_{\mathrm{D}}$. The angular velocity ($\underline{\omega}$) of the D-frame is written [1] as

$$\underline{\omega} = t\underline{T} + \gamma\underline{n} + k\underline{p} \tag{1}$$

where t, γ, and k are the geodesic torsion, geodesic curvature, and normal curvature, respectively, of the curve C at point P. Since the point P moves with unit speed in the direction of $\underline{T}$, its velocity $\underline{v}$ is written as

$$\underline{v} = \underline{T} \tag{2}$$

Then, using Eqs. (1) and (2), the twist of the D-frame can be written in the dual-vector form as

$$\$_{\mathrm{D}} = t\underline{T} + \gamma\underline{n} + k\underline{p} + \epsilon\underline{T} \tag{3}$$

It is worth noticing that while the *first-order differential geometric properties of the curve C at a point P instantiate the Tnp frame, the second-order differential geometric properties of the curve C relate to the first-order kinematics of the D-frame.*

2.2 *Primary Decomposition of $\$_D$*

Now let us partition the twist $\$_{\mathrm{D}}$ of the D-frame formulated above in Eq. (3) into two twists as follows:

$$\$_{\mathrm{D}} = \$_1 + \$_2 \tag{4}$$

where

$$\$_1 = \gamma\underline{n} + \epsilon\underline{0} \tag{5}$$
$$\$_2 = t\underline{T} + k\underline{p} + \epsilon\underline{T} \tag{6}$$

The angular velocities $\underline{\omega}_1$ and $\underline{\omega}_2$ corresponding to the twists $\$_1$ and $\$_2$, respectively, are

$$\underline{\omega}_1 = \gamma\underline{n} \tag{7}$$
$$\underline{\omega}_2 = t\underline{T} + k\underline{p} \tag{8}$$

The pitches h_1 and h_2 of the twits $\$_1$ and $\$_2$ are obtained as

$$h_1 = 0 \tag{9}$$

$$h_2 = \frac{\underline{\omega}_2 \cdot \underline{v}_2}{\underline{\omega}_2 \cdot \underline{\omega}_2} = \frac{(t\underline{T} + k\underline{p}) \cdot (\underline{T})}{(t\underline{T} + k\underline{p}) \cdot (t\underline{T} + k\underline{p})} = \frac{t}{t^2 + k^2} \tag{10}$$

The ISAs corresponding to the twists $\$_1$ and $\$_2$ are labeled L_1 and L_2, respectively, and are obtained in dual-vector form as

$$L_1 = \underline{n} + \epsilon\underline{0} \tag{11}$$

$$L_2 = \frac{1}{\sqrt{t^2 + k^2}}\left\{t\underline{T} + k\underline{p} + \epsilon\left(\frac{k^2}{t^2 + k^2}\underline{T} - \frac{kt}{t^2 + k^2}\underline{p}\right)\right\} \tag{12}$$

From Eqs. (5), (7), (9), and (11), it is evident that the twist $\$_1$ is a twist about the zero pitch screw $\hat{\$}_1$ written as

$$\hat{\$}_1 = \underline{n} + \epsilon\underline{0} \tag{13}$$

which also happens to be the surface normal line n. The amplitude of this twist is equal to the geodesic curvature (γ) of the curve C at point P. The other partitioned twist $\$_2$, is analyzed in the next section.

3 Cylindroid of $\$_2$

Firstly, it can be observed from Eqs. (11) and (12) that the screw reciprocal product of the lines L_1 and L_2 is zero. Also, L_1 and L_2 are orthogonal to each other. Hence, it can be concluded that the shortest distance between the two lines is zero. This means that the *ISA of* $\$_2$ *intersects the ISA of* $\$_1$ *(which also happens to be n) orthogonally*. Using Eq. (12), the z-coordinate of the location of ISA from point P on the line n is obtained as

$$z = \frac{k}{t^2 + k^2} \tag{14}$$

The twist $\$_2$ and ISA L_2 are functions of two parameters, namely, curvature k and geodesic torsion t. These quantities vary according to the tangential direction $\underline{T}$ at the point P on the surface. Let the X-axis and Y-axis of the global fixed coordinated system be along the first and second principal normal curvature directions, respectively, in the tangent plane as shown in Fig. 2. Z-axis points along $\underline{n}$. $\underline{T}$ makes an angle θ with the positive X-axis. The Euler and Bonnet's (also attributed to Bertand) formulas [1] for normal curvature (k) and geodesic torsion (t), respectively, in the direction of $\underline{T}$ are given by

$$k = k_1 \cos^2\theta + k_2 \sin^2\theta \tag{15}$$

$$t = (k_2 - k_1)\sin\theta\cos\theta \tag{16}$$

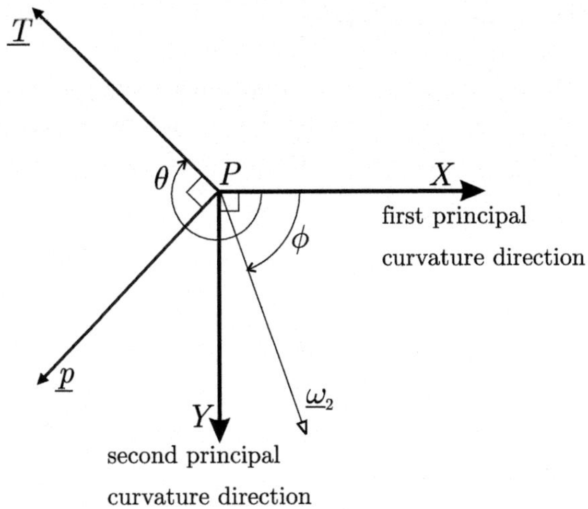

Fig. 2 Relative orientation of various axes as seen in the tangent plane. $\underline{n}$ points into the page and is not shown

where k_1 and k_2 are the principal normal curvatures ($k_1 > k_2$ algebraically for all types of surfaces). The positive sense of the angle θ is taken to be the direction of n. Since k and t are functions of a single parameter θ (for given principal normal curvatures), the ISA L_2 too varies according to only the angle θ.

From the above arguments, we can conclude that the locus of L_2 is a ruled surface with its rulings, parallel to the tangent plane to the surface at point P, and orthogonal to the n-line at P. Such a surface is called a *right conoid* or *cylindroid* [2] whose *nodal axis is the surface normal line* n. The properties of this cylindroid are studied below.

3.1 Geometry of the Cylindroid

Using $\underline{T} = \cos\theta\hat{i} + \sin\theta\hat{j}$ and $\underline{p} = \cos(\theta - \pi/2)\hat{i} + \sin(\theta - \pi/2)\hat{j}$ in Eq. (8) gives:

$$\omega_2 = k_2 \sin\theta\hat{i} - k_1 \cos\theta\hat{j} \tag{17}$$

Substituting Eqs. (15) and (16) in Eqs. (10) and (14) gives

$$h_2 = \frac{(k_2 - k_1)\sin\theta\cos\theta}{k_1^2\cos^2\theta + k_2^2\sin^2\theta} \tag{18}$$

$$z = \frac{k_1\cos^2\theta + k_2\sin^2\theta}{k_1^2\cos^2\theta + k_2^2\sin^2\theta} \tag{19}$$

3.1.1 Principal Screws

We find the maximum and minimum values of the pitch given by Eq. (18). The screws $\hat{\$}_\alpha$ and $\hat{\$}_\beta$ corresponding to maximum and minimum pitches, respectively, are known as *principal screws*. Using Eq. (18), vanishing of the first derivative of h_2 with respect to θ gives the condition

$$k_1^2 \cos^2\theta = k_2^2 \sin^2\theta \tag{20}$$

which implies $k_1 \cos\theta = \pm k_2 \sin\theta$. The pitches of the principal screws, labeled $h_{2\alpha}$ (minimum) and $h_{2\beta}$ (maximum), are obtained as

$$h_{2\alpha} = -\frac{(k_1 - k_2)}{2k_1k_2} \tag{21}$$

$$h_{2\beta} = \frac{(k_1 - k_2)}{2k_1k_2} \tag{22}$$

The principal screws are located at a z-coordiante $z_{\alpha,\beta}$ given by:

$$z_{\alpha,\beta} = \frac{1}{2}\left(\frac{1}{k_1} + \frac{1}{k_2}\right) = \frac{(k_2 + k_1)}{2k_1k_2} = \frac{\text{Mean Curvature}}{\text{Gaussian Curvature}} \tag{23}$$

The *central plane of the cylindroid* containing the principal screws is parallel to the tangent plane (here XY-plane) of the surface and is located at a distance equal to the mean radius of curvature at the point on the surface.

Using Eq. (17), it immediately follows that the ISAs of the principal screws $\hat{\$}_\alpha$ and $\hat{\$}_\beta$ are mutually orthogonal and each one makes an angle $\pi/4$ with the principal normal curvature directions; the directions of the *principal screws bisect the principal curvature directions and coincide along the directions of maximum and minimum geodesic torsion.*

3.1.2 Extent of Cylindroid

The extent of the cylindroid is determined from the extreme values of z given by Eq. (19). The condition for extreme values of z are obtained as

$$\sin\theta \cos\theta = 0 \tag{24}$$

Let ϕ be the angle made by $\underline{\omega}_2$ with positive X-axis (Fig. 2). Using the above equation, the minimum and maximum values of z and the corresponding direction of the ISA given by angle ϕ, are obtained as below. For a convex surface we obtain

$$\text{when} \quad \theta = 0\text{:} \quad z_{\min} = \frac{1}{k_1} \quad \phi = \frac{3\pi}{2} \tag{25}$$

$$\text{when} \quad \theta = \frac{\pi}{2}\text{:} \quad z_{\max} = \frac{1}{k_2} \quad \phi = 0 \tag{26}$$

$$\text{when} \quad \theta = \pi\text{:} \quad z_{\min} = \frac{1}{k_1} \quad \phi = \frac{\pi}{2} \tag{27}$$

$$\text{when} \quad \theta = \frac{3\pi}{2}\text{:} \quad z_{\max} = \frac{1}{k_2} \quad \phi = \pi \tag{28}$$

From Eqs. (18) and (24), it is evident that the condition for extreme values of z, renders the pitch of the corresponding screws as zero. *This means the screws of the cylindroid at the extreme values of z along the nodal axis are pure rotations*. Let $\hat{\$}_a$ and $\hat{\$}_b$ be two screws of zero pitch corresponding to $\phi = \pi/2$ and $\phi = 0$, respectively. The ISAs of $\hat{\$}_a$ and $\hat{\$}_b$ are then orthogonal to the normal planes containing the maximum and minimum normal curvature directions, respectively. The use of $\hat{\$}_a$ and $\hat{\$}_b$ is made in the following sections.

4 Total Decomposition of $\$_{\mathrm{D}}$

The above study of the twist $\$_2$ established the *two-system* (linear two-system of screws) inherent in it. So the twist $\$_2$ can be written as a linear combination of the two independent screws of the cylindroid. If the principal screws are chosen for this purpose, then

$$\$_2 = \lambda_\alpha \hat{\$}_\alpha + \mu_\beta \hat{\$}_\beta \tag{29}$$

From Eqs. (4), (13) and (29) we have

$$\$_{\mathrm{D}} = \gamma \hat{\$}_1 + \lambda_\alpha \hat{\$}_\alpha + \mu_\beta \hat{\$}_\beta \tag{30}$$

The screw $\hat{\$}_1$ is along the nodal axis of the cylindroid of the twist $\$_2$. The three screws $\hat{\$}_1$, $\hat{\$}_\alpha$, and $\hat{\$}_\beta$ are three mutually orthogonal and concurrent screws of different pitches. Such a triad of screws are eligible to be named as the *principal screws of the three-system*. Hence, we can say that the *instantaneous kinematics of the D-frame belongs to a three-system whose principal screws are all identifiable once the surface normal and principal normal curvature directions and values are known at a point on the surface.*

Special case: $k_1 = k_2$
In this case, the pitches of the principal screws of the cylindroid are equal and zero, i.e., $h_{2\alpha} = h_{2\beta} = 0$. Also, we have $h_1 = 0$. This is a case of the *second special*

three-system enunciated by Hunt [2] wherein the spherical motion of the D-frame is described. This makes sense because the D-frame moves on a locally spherical surface.

4.1 Canonical Decomposition and Substitute-Connection

In another way of decomposing $\$_2$ other than principal screws, we employ the zero pitch screws $\hat{\$}_a$ and $\hat{\$}_b$ described in Sect. 3.1. Using these, the twist $\$_2$ can be written as

$$\$_2 = \lambda_a \hat{\$}_a + \mu_b \hat{\$}_b \tag{31}$$

From Eqs. (4), (13) and (31) we have

$$\$_{\rm D} = \gamma \hat{\$}_1 + \lambda_a \hat{\$}_a + \mu_b \hat{\$}_b \tag{32}$$

The three screws $\hat{\$}_1$, $\hat{\$}_a$, and $\hat{\$}_b$ have zero pitch and are mutually orthogonal. The ISA of $\hat{\$}_1$ is common perpendicular to the axes of $\hat{\$}_a$, and $\hat{\$}_b$. The two ISAs of $\hat{\$}_a$, and $\hat{\$}_b$ are

(a) orthogonal to the normal planes $\pi_{\rm max}$ and $\pi_{\rm min}$ (Fig. 3) containing the maximum and minimum principal normal curvature directions, respectively, at point P on the surface
(b) intersecting the surface normal line n orthogonally at distances of first and second principal normal radius of curvature ($1/k_1$ and $1/k_2$), respectively.

From the above discussion, we have that the instantaneous twist $\$_{\rm D}$ of the D-frame is a combination of three rotations whose axes are mutually orthogonal and these axes are readily obtained from the information about the surface normal line and the locations of two centers of curvature (C_a and C_b in Fig. 3) in the two normal planes containing the maximum and minimum principal normal curvature directions.

A mechanism interpretation of the above kinematic result is an equivalent spatial 3R (revolute) substitute-connection for the instantaneous motion of the D-frame shown in Fig. 3. The end-effector of the serial 3R manipulator is the D-frame whose arbitrary twist can be generated for different angular velocities of the three revolute joints. The connection is substitute in the sense that the instantaneous kinematic behavior of the D-frame constrained by the surface can also be achieved by considering the D-frame as an end-effector of a 3R chain without the presence of the surface. The *decomposition of the twist of the D-frame into three rotations about axes which can be attributed uniquely to the local geometric shape of the surface is indeed canonical*. In this 3R manipulator, there is a clear decoupling between the joints in affecting the geodesic curvature and normal curvature of the curve C traced on the surface instantaneously. *While rotation about* $\hat{\$}_1$ *solely determines the geodesic curvature, rotation about* $\hat{\$}_a$, and $\hat{\$}_b$ *determines the normal curvature*.

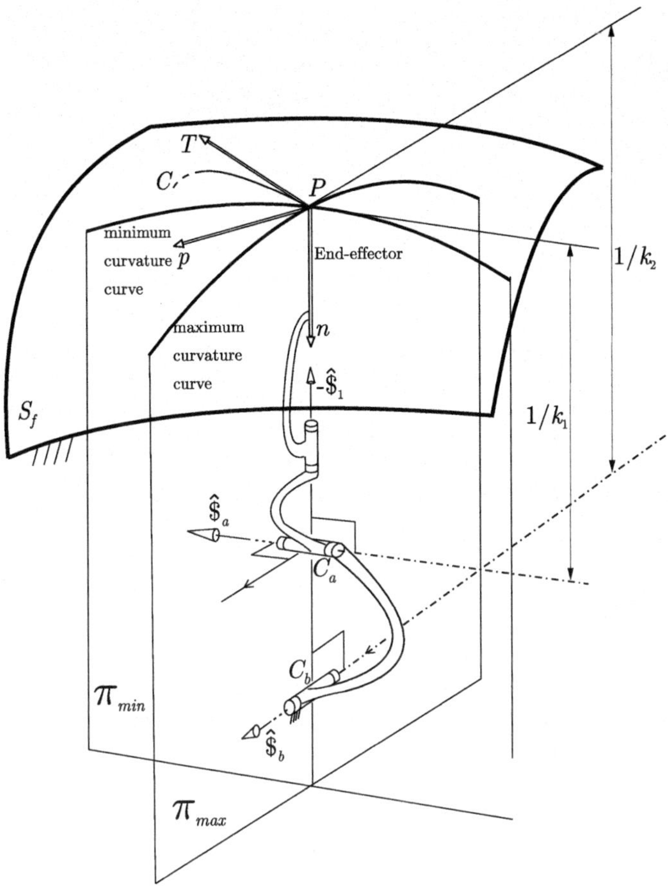

Fig. 3 3R substitute-connection for the instantaneous kinematics of D-frame

4.2 Numerical Example

In Fig. 4, the surface patch is represented using NURBS. The surface normal, principal normal curvatures, and directions are computed. This information is used to find the principal screws (Fig. 4a) of the three-system which are shown in red ($-\hat{\$}_1$), green ($\hat{\$}_\alpha$), and blue ($\hat{\$}_\beta$). The blue and green screws span a cylindroid which is also shown in the figure. The three zero pitch screws for the 3R substitute-connection are shown in Fig. 4b.

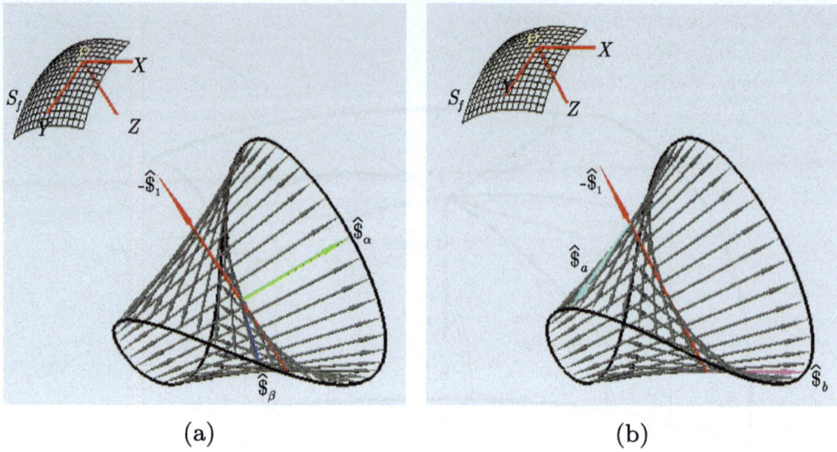

Fig. 4 Screws basis for twist of D-frame at point P on surface S_f. **a** Principal screws, **b** zero pitch screw basis

5 Discussion

Consider the case when $\gamma = 0$ in the analysis described in Sect. 4. In this case the D-frame moves along those curves for which the geodesic curvature is zero. These curves are none other than the geodesics because *geodesic curvature of a geodesic is zero* [4]. Hence, only the cylindroid of the twist $\$_2$ captures the instantaneous kinematics of the D-frame. We have seen in Sect. 3 that the twist $\$_2$ contains only normal curvature and geodesic torsion terms. It is known in differential geometry that the geodesic torsion of all the curves in a given tangential direction is the same and equal to the torsion of the geodesic curve [4]. This means that the *cylindroid of the twist* $\$_2$ *remains the same immaterial of whether the D-frame moves along a geodesic or any general curve on the surface*. Hence, the "cylindroid of torsure" proposed in [3] need not be associated with only the family of geodesics at a point on the surface and, in fact, associated with the family of all the curves passing through a point on the surface.

In the 3R substitution connection discussed in Sect. 4.1, let us assume there is no unit speed restriction on the end-effector point P. It is known in the studies on the instantaneous kinematics of serial manipulators [5] that for a general three degrees-of-freedom manipulator, for the sum of the squares of the three joint angular velocities being unity, the tip of the linear velocity of the end-effector point lies on an ellipsoid. However, in the present case, it can be shown that the tip of the velocity vector of point P lies on an ellipse which lies on the tangent plane to the surface and the major axis of the ellipse is along the second principal normal curvature direction. The degeneracy of the ellipsoid to ellipse is due to the following reason. The end-effector point P lies on a line n (screw of zero pitch) which is reciprocal to the screws (also of zero pitch here) of the twists $\hat{\$}_a$, $\hat{\$}_b$ and $\hat{\$}_1$; this happens when the 3R manipulator is in a special configuration with the point P at the *workspace boundary*. Hence, the end-effector point cannot move along the surface normal direction. The insight is that, for the 3R manipulator, the surface under consideration is a workspace boundary and the end-effector point traces different curves locally on the workspace boundary as the three joints are moved at different angular velocities.

6 Conclusions

Twist of the D-frame is obtained as a linear combination of three screws. The spatial arrangement of these three axes and their pitches are readily obtained using the normal line, principal normal curvature directions, and principal normal curvature information at a point of the surface. The geodesic curvature is a result of a twist about one of the screws and the normal curvature of the curve on a surface is determined by twists about two other screws. The instantaneous point-path of the origin of the D-frame lies on the workspace boundary of a spatial 3R serial manipulator.

References

1. Brand L (1947) Vector and tensor analysis. Wiley, New York, NY
2. Hunt KH (1978) Kinematic geometry of mechanisms. Wiley, New York, NY
3. Dooner DB (2003) Introducing radius of torsure and cylindroid of torsure. J Robot Syst 20(8):429–436
4. Kreyszig E (1991) Differential geometry. Dover Publishers, New York, NY
5. Ghosal A (2006) Robotics: fundamental concepts and analysis. Oxford University Press, New Delhi

Multibody Modeling of Direct and Cross-Axis Seat to Head Transmissibility of the Seated Human Body Supported with Backrest and Exposed to Vertical Vibrations

Raj Desai, Anirban Guha, and P. Seshu

Abstract Drivers of the road and off-road vehicles are continuously exposed to low-frequency whole-body vibration of significant magnitudes arising from tire/track-terrain interactions. The human body models reported in the literature for analyzing these are usually of lumped parameter category having only a few degrees of freedom. Most of them are also one-dimensional models, i.e., they allow an analysis of vibration only in the vertical direction. The highly complex motion that the human body is subjected to in real life requires at least a two-dimensional model. This paper presents a multibody human body model of 12 degrees of freedom that represents a seated human with backrest support. The motion has been assumed to occur in only the sagittal plane allowing vibration analysis in two dimensions. The modeling of the interconnection of masses using rotational and translational springs and dampers and the contact with backrest gives a better simulation of forces transmitted to the upper and lower torso. A sensitivity study to find the effect of critical model parameters on the frequencies of peak moduli helps in identifying the section of the human body which should be given appropriate support for improving comfort. This can lead to the re-design of the seat by considering the effect on different sections of the human body. This is expected to be a better approach than the current practice of considering the overall effect on the human body.

Keywords Whole-body vibration · Genetic algorithm · Multibody modeling

R. Desai (✉) · A. Guha
Department of Mechanical Engineering, Indian Institute of Technology, Mumbai, India
e-mail: rajdesai@iitb.ac.in

A. Guha
e-mail: anirbanguha@iitb.ac.in

P. Seshu
Department of Mechanical Engineering, Indian Institute of Technology, Dharwad, India
e-mail: seshu.iitb@gmail.com

D. Sen et al. (eds.), *Mechanism and Machine Science*, Lecture Notes in Mechanical Engineering, https://doi.org/10.1007/978-981-15-4477-4_9

1 Introduction

Drivers of vehicles are continuously exposed to low-frequency whole-body vibration of significant magnitudes arising from tire/track-terrain interactions. Studies have shown a high presence of back disorders such as disc degeneration and muscular pain among the occupational vehicle drivers exposed to whole-body vibration. The threat of vibration exposure depends on three different factors: time span, amplitude, and frequency of the absorbed vibration. In the design of the vehicle field, designers and manufacturers need to understand how the vibrational signals are damped or amplified by the human-seat cushion interfaces, seat suspension, vehicle suspension, etc. in order to improve human physiological comfort. The human body response is studied using a lumped parameter model [1, 2], multibody models [3], finite element models [4, 5], and most recently, matrix models [6, 7]. Most of the human body models reported in the literature are of the lumped parameter type which considers the human body as many concentrated rigid masses representing different body parts having few degrees of freedom and is extensively studied for the study of the response of the seated body. The analysis is very simple, easy method in obtaining the equation of motion, and matches the experimental results. But with the limitation that it has been used for one-dimensional vibrational analysis only. Patil [8] used a lumped parameter model to optimize the tractor suspension system. Chiang [9] studied responses of pregnant passengers exposed to vertical vibrations due to road roughness and suggested a suitable selection of a sitting position in a vehicle that can reduce her level of discomfort. In finite element-based models, the human body is modeled using the beam, spring, and mass elements to model spine vertebra, muscles, viscera, head, pelvis, and buttocks tissue. Such a model can be used to study the biodynamic response of the human body and is used in predicting injuries to passengers. Multibody human models are made of arbitrary shaped rigid bodies interconnected by kinematic joints such as pin, ball-socket joints, and translational-rotational spring-damper joint. Multibody models usually treat the spine as an interconnected structure with rigid masses representing the vertebral bodies, and deformable flexible elements represent intervertebral discs. Cho [10] developed a nine degree of freedom (DoF) multibody model and was able to depict the vertical motion of the head. Kim [11] developed biomechanical models with different structures of upper and lower body and arrived at the best model to represent the biodynamic response. Lumbar support plays an important role in decreasing low back pain and maintaining proper natural shape spine. Diana [12] investigated the effectiveness of lumbar support in changing the lumbar lordosis angle of the spine and pelvic posture with different thickness supports. Lumbar support towards motorcyclists also helps them reducing the discomfort level and preventing them from muscle fatigue during the prolonged riding process [13]. Lumbar flattening during sitting was also reduced with a lumbar support pillow [14]. Jonathan [15] studied the effect of different lumbar support conditions and found that convex cushion made up of soft foam was very effective in reducing seat to head transmissibility (STHT) and improving comfort level. Thus, a proper understanding of forces transmitted through the backrest is needed in order to make an optimal design of lumbar support.

Considering that automotive seats are designed with an inclined backrest to provide physiologically comfortable and controlled sitting posture, a detailed 12 DoF seated human body model with inclined backrest support is developed to study vertical and fore-and-aft STHT responses of the human body head exposed to vertical excitation. The highly complex motion that the human body is subjected to in real life requires at least a two-dimensional model. The majority of the reported models did not include cross-axis fore-aft STHT transmissibility responses. The human head has translation and rotation in the sagittal plane. Thus, the head is modeled as a rigid mass with 3° of freedom. A multi-objective genetic algorithm and subsequently a global criterion-based minimization scheme have been used to fine-tune a total of twenty-six model parameters consisting of spring and damper coefficients. Thereafter, parameter sensitivity analysis has been done to find the critical parameters which have a high effect on STHT. The variation of forces transmitted to the human body at the backrest and neck region as a result of frequency changes is also presented.

2 Multibody Biodynamic Modeling

A two-dimensional 12 DoF multibody biomechanical model of the seated human body with inclined back support is shown in Fig. 1. The model consists of four rigid masses 1–4 representing thigh-pelvis, lower torso, upper torso, and head, respectively. Each joint J_i consists of rotational and translational spring-damper as shown in Fig. 2. Also, Joint 3 is modeled differently, having parameters in x and z direction. When there are vibrations coming either in vertical or fore-aft, the head will have vertical, fore-aft, and rotational motion. The human body contact with seat is modeled as linear spring-dampers at thighs, pelvis, and lumbar region of the spine. The spring-dampers at backrest contact at c_3 are inclined at an angle alpha (α)—called as backrest inclination angle. Backrest forces are modeled in such a way that due to vertical vibrations of the seat, forces generated in both x-z directions at the lumbar region are coupled. This causes the vertical and fore-aft motion of the head as shown in Fig. 5 and Eq. 8. Horizontal forces generated by backrest in x direction causes fore-aft movement of the head. Experimental data reported by Wang [16] is used for parameter optimization in which the accelerometer was mounted at the head at the location (x_m, z_m). This has been considered for the estimation of the transfer function given by Eqs. 11 and 12 (Figs. 3 and 4).

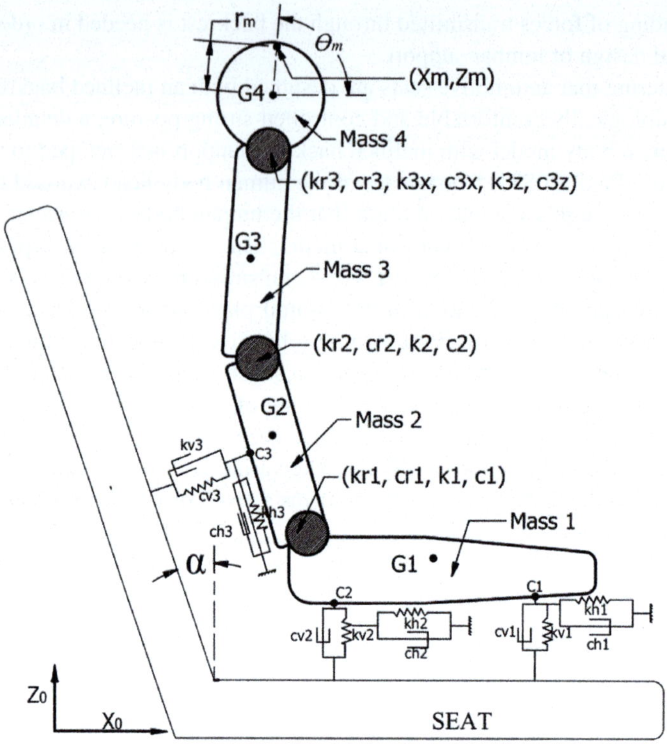

Fig. 1 12 DoF human body model

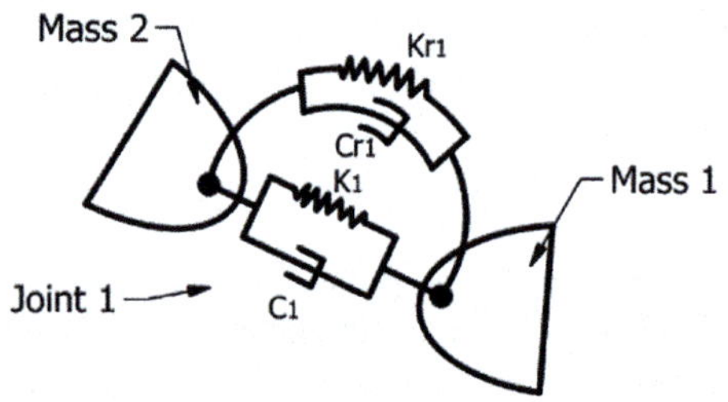

Fig. 2 Joint 1 representation

2.1 Equations of Motion

The detailed equation of motion for mass 1 is shown below. The newtonian approach is used to derive the equations of motion. For mass 1 from the free body diagram shown in Fig. 3 the equation of motion for mass 1 can be written as

$$m_1\ddot{x}_1 = -f_{1x} - f_{c1x} - f_{c2x} \tag{1}$$

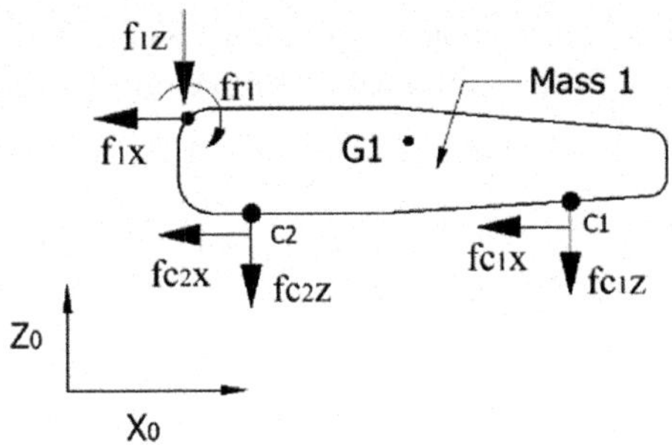

Fig. 3 Free body diagram mass 1

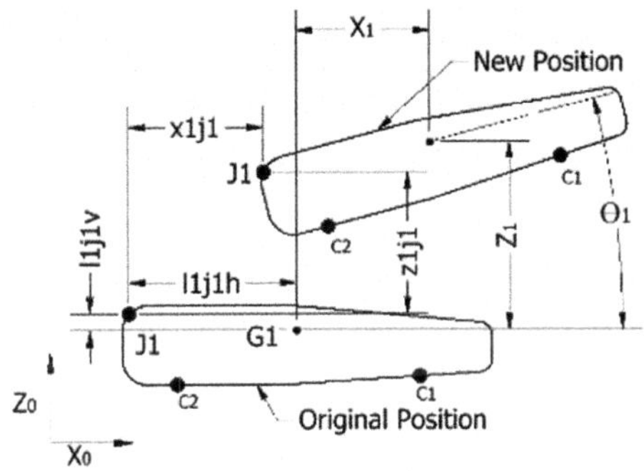

Fig. 4 Mass1 displaced and rotated

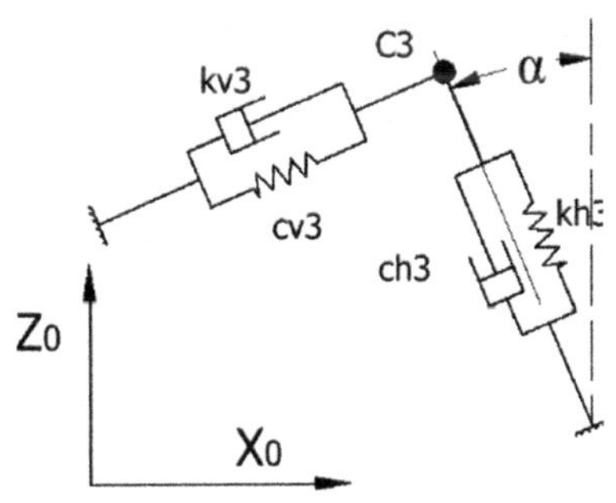

Fig. 5 Backrest support representation

$$m_1\ddot{z}_1 = -f_{1z} - f_{c1z} - f_{c2z} \tag{2}$$

$$I_1\ddot{\theta}_1 = -f_{1x}l_{1j1v} - f_{1z}l_{1j1h} - f_{c1z}l_{1c1h} - f_{c1x}l_{1c1v} - f_{c2z}l_{1c2h} - f_{c2x}l_{1c2v} - f_{r1} \tag{3}$$

Each mass can have translation and rotation. Figure 4 shows details of joint coordinates for mass 1 and the corresponding forces are shown in Eq. 4.

$$\begin{aligned} x_{j11} &= x_1 - l_{1j1v}\theta_1 \\ z_{j11} &= z_1 + l_{1j1h}\theta_1 \\ f_{r1} &= k_{r1}(\theta_1 - \theta_2) + c_{r1}(\dot{\theta}_1 - \dot{\theta}_2) \\ f_{1x} &= k_1(x_{1j1} - x_{2j1}) + c_1(\dot{x}_{1j1} - \dot{x}_{2j1}) \\ f_{1z} &= k_1(z_{1j1} - z_{2j1}) + c_1(\dot{z}_{1j1} - \dot{z}_{2j1}) \\ f_{c1x} &= k_{h1}(x_{c1} - x_0) + c_{h1}(\dot{x}_{c1} - \dot{x}_0) \\ f_{c1z} &= k_{v1}(z_{c1} - z_0) + c_{v1}(\ddot{z}_{c1} - \ddot{z}_0) \\ f_{c2x} &= k_{h2}(x_{c2} - x_0) + c_{h2}(\dot{x}_{c2} - \dot{x}_0) \\ f_{c2z} &= k_{v2}(z_{c2} - z_0) + c_{v2}(\ddot{z}_{c2} - \ddot{z}_0) \end{aligned} \tag{4}$$

For mass 2 the equation of motion is shown in Eqs. 5–7.

$$m_2\ddot{x}_2 = f_{1x} - f_{2x} - f_{c3x} \tag{5}$$

$$m_2\ddot{z}_2 = f_{1z} - f_{2z} - f_{c3z} \tag{6}$$

$$I_2\ddot{\theta}_2 = f_{1x}l_{2j1v} + f_{1z}l_{2j1h} - f_{2x}l_{2j2v} - f_{2z}l_{2j2h} - f_{c3x}l_{3c3v} - f_{c3z}l_{2c3h} + f_{r1} - f_{r2} \tag{7}$$

Similarly, for mass 3 and 4, equations of motion are derived. Backrest supporting the upper body and forces applied are modeled in x and z directions given by Eq. 8

$$\begin{aligned} \begin{bmatrix} f_{c3x} \\ f_{c3z} \end{bmatrix} = \begin{bmatrix} \cos\alpha & \sin\alpha \\ \sin\alpha & -\cos\alpha \end{bmatrix} \begin{bmatrix} k_{v3}\cos\alpha & k_{v3}\sin\alpha \\ k_{h3}\sin\alpha & -k_{h3}\cos\alpha \end{bmatrix} \begin{bmatrix} x_{c3} \\ z_{c3} - z_0 \end{bmatrix} + \\ \begin{bmatrix} \cos\alpha & \sin\alpha \\ \sin\alpha & -\cos\alpha \end{bmatrix} \begin{bmatrix} c_{v3}\cos\alpha & c_{v3}\sin\alpha \\ c_{h3}\sin\alpha & -c_{h3}\cos\alpha \end{bmatrix} \begin{bmatrix} \dot{x}_{c3} \\ \dot{z}_{c3} - \dot{z}_0 \end{bmatrix} \end{aligned} \tag{8}$$

Forces at the neck, modeled in x-z direction, are given by Eq. 9

$$\begin{aligned} f_{3x} &= k_{3x}\left(x_{3j3} - x_{4j3}\right) + c_{3x}\left(\dot{x}_{3j3} - \dot{x}_{4j3}\right) \\ f_{3z} &= k_{3z}\left(z_{3j3} - z_{4j3}\right) + c_{3z}\left(\dot{z}_{3j3} - \dot{z}_{4j3}\right) \end{aligned} \tag{9}$$

STHT is considered at the location of the accelerometer on the head and is given by Eq. 10.

$$\begin{aligned} x_m &= x_4 - r_m \sin\theta_m \theta_4 \\ z_m &= z_4 + r_m \cos\theta_m \theta_4 \end{aligned} \tag{10}$$

The above equations for masses 1–4 can be rewritten in the form as
$[M]\ddot{x} + [C]\dot{x} + [K]x = FZ$

where

$[M]$ Mass matrix (size 12 × 12)
$[K]$ Spring Stiffness matrix (size 12 × 12)
$[C]$ Damping coefficient matrix (size 12 × 12)
$[F]$ Input excitation force matrix (size 12 × 1).

$\ddot{x}$, $\dot{x}$, x represents the output acceleration, velocity, and displacement vector of each body (size 12 × 1). Frequency-domain (FD) method is mainly suitable for linear systems with harmonic excitations as input. Using Laplace transform, equations of motion are transformed from the time domain to the frequency domain to match frequency-domain experimental data. The solution takes into account only the steady-state response of the system as. Seat to head transmissibility defined as a non-dimensional ratio of the response amplitude displacement, velocity or acceleration of a system in steady-state vibration. The excitation amplitude is expressed as a function of the vibration frequency. From the measurement of transmissibility, the researcher is interested in knowing which frequencies resonate (peak in the transmissibility) and how much damped is each resonance over the entire frequency range. The amplitude of the peak in response is inversely proportional to the damping parameter. The effect of the model parameters on the transmissibility ratio helps one to identify the critical parameters which can be given more importance in order to improve human comfort. The peak in the transmissibility shows vibration at that frequency is amplified by a gradual accumulation of stored energy due to repeated extension and squeezing of buttock tissue. The STHT transfer function for head motion during direct vertical (DV) vibrations is defined in Eq. 11.

$$\text{DV STHT} = \frac{z_m}{z_0} = \frac{\dot{z}_m}{\dot{z}_0} = \frac{\ddot{z}_m}{\ddot{z}_0} \tag{11}$$

Similarly, there is fore-aft motion to head due to vertical vibrations at the seat. This is called cross-axis fore-aft (CFA) STHT and is given by Eq. 12.

$$\text{CFA STHT} = \frac{x_m}{z_0} = \frac{\dot{x}_m}{\dot{z}_0} = \frac{\ddot{x}_m}{\ddot{z}_0} \tag{12}$$

2.2 *Parameters Identification*

The parameters for the proposed 12 DoF model are identified through the minimization of an error difference function of the model response and target experimental values. Multi-objective GA is employed to determine the human model parameters imposing limit bound constraints on the model parameters. The objective function is selected to minimize the squared sum of errors associated with DV and CFA functions between the model computed and the target experimental values. Two objective functions associated with the DV STHT and CFA STHT are defined by Eq. 13 where f in the equation represents frequency. The subscripts M and E represent the model and experimental data, respectively.

$$J_1 = \sum_{i=1}^{N} (\text{DV STHT}_{Mz}(f_i) - \text{DV STHT}_{Ez}(f_i))^2$$

$$J_2 = \sum_{i=1}^{N} (\text{CFA STHT}_{Mz}(f_i) - \text{CFA STHT}_{Ez}(f_i))^2 \quad (13)$$

A flowchart explaining the parameter optimization process using multi-objective GA is shown in Fig. 6.

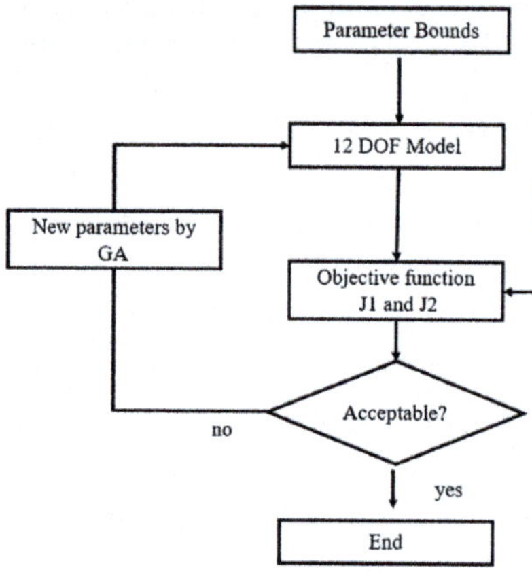

Fig. 6 Flowchart for model parameter search

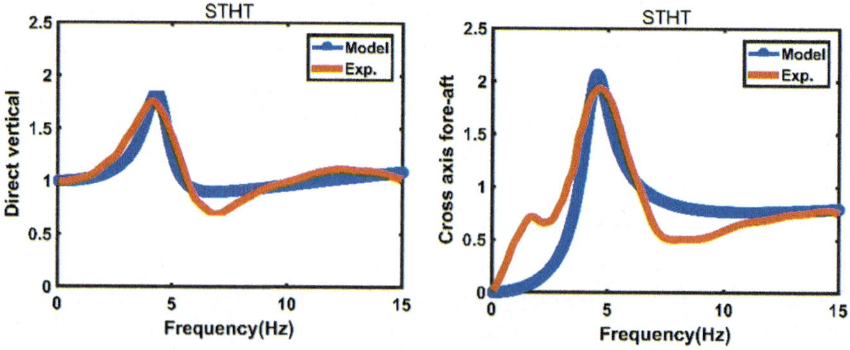

Fig. 7 Experimental versus human body model seated with inclined backrest ($\alpha = 24°$)

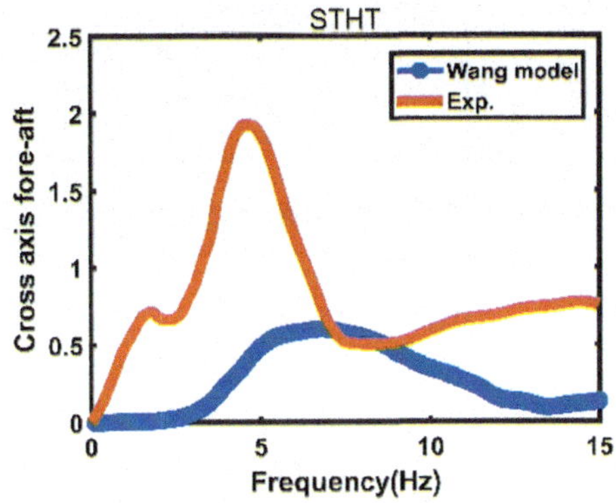

Fig. 8 Cross axis fore-aft STHT by 5 DOF model [16]

To solve the above multi-objective problem, the global criterion method is considered here as stated by Rao [17]. The overall error is defined as a combination of both errors as observed in Eq. 14.

$$J_{\text{Model}} = \sum_{i=1}^{2} \left[\frac{J_i^* - J_i}{J_i^*} \right]^2 \tag{14}$$

where, $J_i^*, i = 1$ and 2 are the minimum errors when the model is optimized with multi-objective GA. At this stage, having estimated the J_i^*, again using single-objective GA, the function error for two cases will be minimized simultaneously by using Eq. 14. The mass and inertia values for different human body parts are obtained from the literature [18, 19]. Considering 78 kg total mass of the human body, using mass ratio and corresponding radius of gyration values, the mass and inertia of different body parts are shown in Table 1. Mass of legs not considered in the model. The values obtained after the global criterion method, are the final coefficients for the 12-DoF model.

Geometric parameters of C.G and joint point coordinates are presented in Table 1. The units are as follows: Geometric parameter of mass center of gravity (C.G) and joint point coordinates in mm, mass in kg, inertia in kgm^2, spring stiffness in N/m, damping in Ns/m, rotational stiffness Nmrad^{-1}, and rotational damping Nmsrad^{-1}.

The optimized model parameters are shown in Table 2.

2.3 *Evaluation of Model*

The goodness of fit (GoF) criteria is used to estimate the model accuracy. It is evaluated by taking the ratio of root mean square (RMS) error to the mean value and is calculated by Eq. 15 [20]:

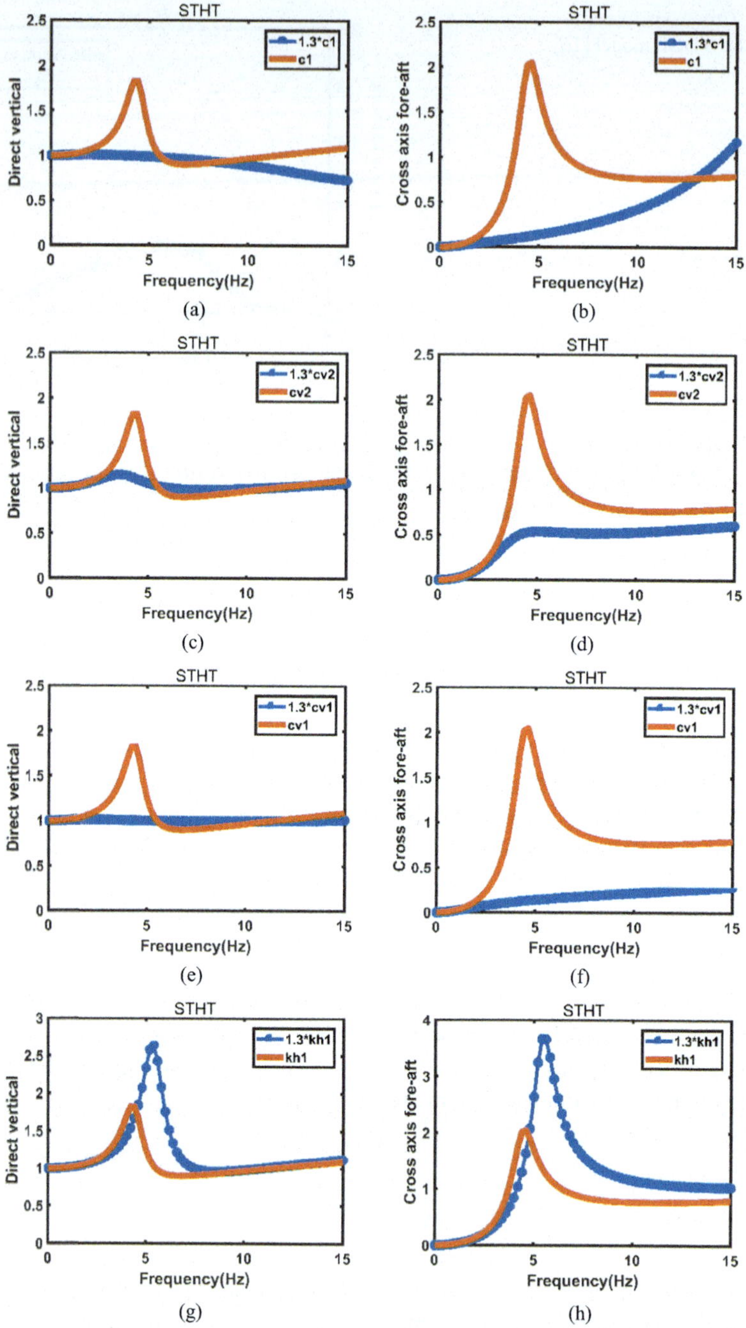

Fig. 9 Changes in DV STHT and CFA STHT resulting from changes in the optimized model parameters

Table 1 The geometric parameter of C.G, joint point, mass, and inertia data

	Mass	Inertia	Position (x)	Position (z)
Thighs-pelvis	26.67	0.58	95	83.5
Lower torso	10.84	0.45	−48	247
Upper torso	24.64	0.36	21	540
Head	6.31	0.022	100	790
Joint 1			−44	139
Joint 2			−17	386
Joint 3			53	706
C_1			173	0
C_2			0	0
C_3			−80	230

Table 2 Optimized model parameters

Parameter	Value	Parameter	Value	Parameter	Value
k_{h1}	30583414.14	k_{3z}	205138.32	c_{v3}	19710053.54
k_{v1}	386847.83	k_{r1}	486967.39	c_1	73968.34
k_{h2}	992590.84	k_{r2}	263647.82	c_2	106243.69
k_{v2}	148010.25	k_{r3}	63271.70	c_{3x}	145250.40
k_{h3}	27851689.87	c_{h1}	183002.95	c_{3z}	154585.23
k_{v3}	22423242.65	c_{v1}	153857.85	c_{r1}	138491.41
k_1	31966.19	c_{h2}	216455.30	c_{r2}	117606.41
k_2	355903.55	c_{v2}	120239.24	c_{r3}	167194.99
k_{3x}	265878.46	c_{h3}	22226895.30		

$$\varepsilon = 1 - \frac{\sqrt{\sum (\tau_e - \tau_m)^2/(N-2)}}{\sum \tau_e/N} \tag{15}$$

Here τ_e is test datum (experimental values), τ_m is the results obtained from the model and N is the number of test data points used for comparison. When ε is equal to 1, the fit of the model predicted results to the test data is very accurate. In this research study, the prediction of CFA STHT for the 12 DoF multibody model has been compared with a 5 DoF model reported by Wang [16].

2.4 *Parameter Sensitivity*

The effect of changes in the optimized model parameters (stiffness and damping coefficients) on the resonance frequency and the response magnitudes at resonance in both the DV and CFA STHT were studied. For a biodynamic response, each model parameter was allowed to change by an increased amount of 30% from the optimized value while the other parameters were fixed at the optimized values. This helps in identifying the parameters of the model producing resonance behavior and more critical to vibrations. A matrix for the parameter sensitivity analysis of the body model is taken from research work by Zheng [3] and is shown in Eq. 16.

$$P_{ij} = \frac{\Delta R_i / R_i}{\Delta^j \psi / ^j \psi} \tag{16}$$

In Eq. 16 ψ ={k_{h1}, k_{v1}, k_{h2}, k_{v2}, k_{h3}, k_{v3}, k_1, k_2, k_{3x}, k_{3z}, k_{r1}, k_{r2}, k_{r3}, c_{h1}, c_{v1}, c_{h2}, c_{v2}, c_{h3}, c_{v3}, c_1, c_2, c_{3x}, c_{3z}, c_{r1}, c_{r2}, c_{r3}} are the model parameters; $^j\psi$ and Δ $^j\psi$ represent the jth (j = 1, 2, …, 26) parameter of vector ψ at its optimized value and small perturbation of the parameter, respectively; R_i represents the response to be studied, with $i = 1$ being peak modulus of the DV STHT, $i = 2$ the corresponding resonance frequency, $i = 3$ peak modulus of CFA STHT and $i = 4$ the corresponding resonance frequency computed with the set of model parameters at their optimized values; ΔR_i are the differences of the above model parameters between using the optimized and new set of parameters with perturbation. With this new set of model parameters, the peak DV STHT, CFA STHT, and the corresponding frequencies can be computed as $R_{\mathrm{i}}^{\mathrm{new}}$ and hence ΔR_i can be calculated as $\Delta R_i = R_i^{\mathrm{new}} - R_i$. The greater the value of P_{ij} the more sensitive the resonance behavior of R_i to the parameter $^j\psi$.

Direct Vertical seat to head transmissibility—Changes in the magnitude and resonance frequency of the STHT at the resonance of the model response as a result of 30% changes in the optimized model parameters are given in Table 3 The table shows that the highest changes in both peak modulus and resonance frequency occurred with changes in c_1, c_{v1}, k_{h1}, and c_{v2}.

Cross-axis fore-aft seat to head transmissibility—Similarly changes in the magnitude and resonance frequency of the CFA STHT at the resonance of the model response as a result of 30% changes in the optimized model parameters are presented in Table 3. The table presents that the highest changes in the peak modulus occurred with changes in c_{v1}, k_{h1}, c_{v2}, and c_1.

2.5 Backrest and Neck Forces

Low back pain and neck injuries are common among vehicle drivers. These forces and moments are transferred to the intervertebral disc, the vertebral body, and the facet joints of the human spine. The intervertebral disc is a viscoelastic-fluid material and its mechanical properties are dependent on the rate of loading. Thus, a proper understanding of how these forces are transmitted and how they vary with frequency is important in order to have a better prediction of the dynamics of the human body. This can lead to the design of proper sitting conditions in order to improve driving

Table 3 Parameter sensitivity

	$i=1$	$i=2$	$i=3$	$i=4$		$i=1$	$i=2$	$i=3$	$i=4$
k_{h1}	0.312	0.173	0.569	0.155	c_{h1}	−0.069	−0.003	−0.145	0.009
k_{v1}	−0.124	0.090	−0.249	0.120	c_{v1}	−0.322	−0.385	−0.615	1.656
k_{h2}	0.008	0.007	0.015	0.003	c_{h2}	−0.078	−0.003	−0.164	0.013
k_{v2}	−0.015	0.013	−0.031	0.013	c_{v2}	−0.271	−0.130	−0.504	1.656
k_{h3}	0.000	0.000	0.000	0.000	c_{h3}	−0.003	0.000	−0.006	0.000
k_{v3}	0.000	0.000	0.001	0.000	c_{v3}	0.026	0.003	0.047	0.000
k_1	0.027	−0.023	0.056	−0.025	c_1	−0.325	−0.452	−0.308	1.656
k_2	0.000	0.000	0.000	0.000	c_2	−0.002	0.000	−0.003	0.000
k_{3x}	0.000	0.000	0.000	0.000	c_{3x}	0.000	0.000	0.000	0.000
k_{3z}	0.000	0.000	0.000	0.000	c_{3z}	0.000	0.000	0.000	0.000
k_{r1}	−0.002	0.000	−0.004	0.000	c_{r1}	−0.018	−0.003	−0.033	−0.003
k_{r2}	0.000	0.000	0.000	0.000	c_{r2}	0.000	0.000	0.000	0.000
k_{r3}	0.000	0.000	0.000	0.000	c_{r3}	0.000	0.000	0.000	0.000

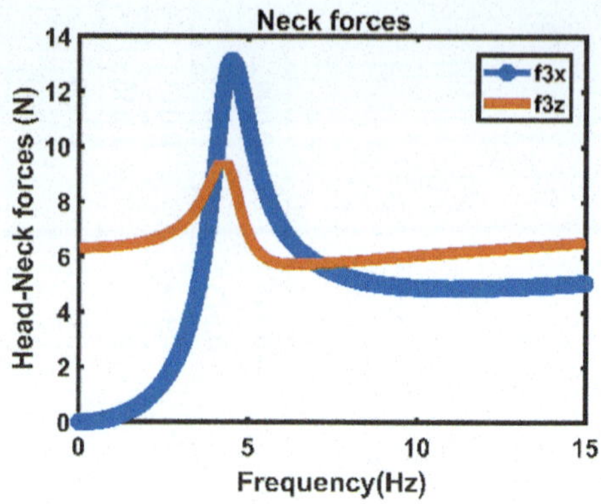

Fig. 10 Neck forces variation

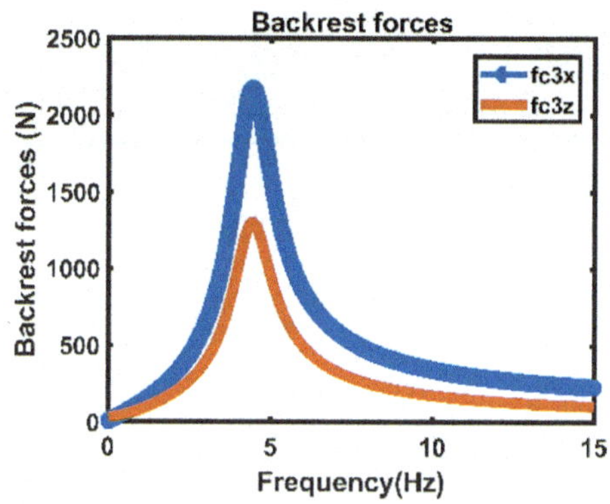

Fig. 11 Backrest forces variation

comfort. Using Eqs. 8 and 9 the forces in the lower lumbar region and cervical spine are computed and shown in Figs. 10 and 11.

3 Results

A total of 26 model parameters—consisting of translational and rotational spring-damper values—are optimized. Bounds on each model parameter are obtained from the literature. Experimental data have been compared with simulation results Fig. 7. It is seen that the two biodynamic functions of the model have a trend similar to that of the experimental values.

The performance of the 5 DoF model reported in the literature to evaluate CFA STHT is shown in Fig. 8. This shows the superiority of the 12 DoF model over the 5 DoF model.

GoF is used to compare the performance of both the models by comparing them with experimental results. The results are shown in Table 4.

Table 4 Model comparison for CFA STHT

Human body model	12 DOF	5 DOF [16]
GoF (ε)	0.9076	0.1523

The plot showing the changes in the top four most critical parameters is shown in Fig. 9. The greatest changes in resonance frequency resulted from changes in c_{v1}, c_{v2}, c_1, and k_{h1}. Seated human body contact with seat at pelvis (c_{v2} parameter) and pelvis-upper torso translational damping parameter (c_1) had an overall greatest effect on DV and CFA STHT. This was as expected because the force from the vehicle is transmitted through the pelvis and the upper portion of the human body tries to rotate about the pelvis. Thus, this is a critical region for analyzing the human body subjected to vehicle generated vibrations. Also, the horizontal shearing effect of k_{h1}. and vertical compression effect of c_{v1} at thigh's contact with seat has a considerable effect on DV and CFA STHT. Figure 9e, f show that a 30% increase in c_{v1} results in the elimination of peaks for DV and CFA STHT and a significant reduction in the fore-aft motion of the head. Figure 9g, h show that a 30% increase in k_{h1} parameter results in a substantial increase in peak amplitude of head motion. Thus, for improving driver comfort, the focus should be given to seat to pelvis and thigh contact properties. By providing appropriate cushion support at this region, head movement in both x and z directions can be reduced.

Considering acceleration amplitude of 1 m/s^2 variation of neck joint and backrest-lower lumbar contact forces resulting from changes in vibration frequency is shown in Figs. 10 and 11. The forces shown in Fig. 10 are internally generated forces between the upper torso and head, whereas, the forces shown in Fig. 11 are transmitted to the lower lumbar region due to contact with the backrest. Neck and backrest forces are more predominating in x direction (fore-aft) in comparison with z direction (vertical). The forces transmitted are seen to be highest in the range 4–6 Hz.

4 Conclusion

A detailed 12 DoF multibody biomechanical model of the human body in a sitting posture with backrest support is presented in this paper. Parameter identification has been done by using experimental data reported in the literature. A goodness of fit analysis shows the superiority of the model over a 5 DoF model reported by Wang [16]. Parameter sensitivity analysis shows that horizontal shearing k_{h1}, vertical compression c_{v1}, rotational c_1 are high at the pelvis and vertical compression effect of c_{v1} at thigh contact with seat has more influence on DV and CFA STHT. The amount of forces transmitted to the human body at the backrest and neck region as a result of frequency variation is presented. It is seen that forces transmitted are highest at 4–6 Hz. The head motion in both vertical and fore-aft direction can be predicted accurately thus giving a better understanding of how the human body responds to external excitations—particularly forces which are transmitted to the lower lumbar region of the spine or neck. Future work can focus on coupling this model with a vehicle model to find the best seat cushion and seat suspension parameters and analyze human body response on different road conditions. The model is also flexible enough to be used to analyze various postures of drivers by changing joint locations and the back support inclination angle.

References

1. Muksian R, Nash CD (1974) A model for the response of seated humans to sinusoidal displacements of the seat. J Biomech 7(3):209–215
2. Wei L, Griffin MJ (1998) Mathematical models for the apparent mass of the seated human body exposed to vertical vibration. J Sound Vib 212(5):855–874
3. Zheng G, Qiu Y, Griffin MJ (2011) An analytic model of the in-line and cross-axis apparent mass of the seated human body exposed to vertical vibration with and without a backrest. J Sound Vib 330(26):6509–6525
4. Zhang X, Qiu Y, Griffin MJ (2015) Developing a simplified finite element model of a car seat with occupant for predicting vibration transmissibility in the vertical direction. Ergonomics 58(7):1220–1231
5. Kitazaki S, Griffin MJ (1997) A modal analysis of whole-body vertical vibration, using a finite element model of the human body. J Sound Vib 200(1):83–103
6. Marzbanrad J, Afkar A (2013) A biomechanical model as a seated human body for calculation of vertical vibration transmissibility using a genetic algorithm. J Mech Med Biol 13(04):1350053
7. Marzbanrad J, Jamali S, Afkar A (2016) Calculation of APMS and STHT responses in three translational axis vibration for human body using biomechanical modeling and matrix method. Comput Res Prog Appl Sci Eng 2(4):150–157
8. Patil MK, Palanichamy MS, Ghista DN (1978) Man-tractor system dynamics: towards a better suspension system for human ride comfort. J Biomech 11(8–9):397–406
9. Liang CC, Chiang CF, Nguyen TG (2007) Biodynamic responses of seated pregnant subjects exposed to vertical vibrations in driving conditions. Veh Syst Dyn 45(11):1017–1049
10. Cho Y, Yoon YS (2001) Biomechanical model of human on seat with backrest for evaluating ride quality. Int J Ind Ergon 27(5):331–345
11. Kim TH, Kim YT, Yoon YS (2005) Development of a biomechanical model of the human body in a sitting posture with vibration transmissibility in the vertical direction. Int J Ind Ergon 35(9):817–829
12. Carvalho DE, Callaghan JP (2012) Influence of automobile seat lumbar support prominence on spine and pelvic postures: a radiological investigation. Appl Ergon 43(5):876–882
13. Shafieia UKM, Karuppiaha K, Tamrina SBM (2015) The effectiveness of new model of motorcycle seat with built-in lumbar support J Teknol 77(27):97–103
14. Grondin DE, Triano JJ, Tran S, Soave D (2013) The effect of a lumbar support pillow on lumbar posture and comfort during a prolonged seated task. Chiropr Man Ther 21(1):21
15. DeShaw J, Rahmatalla S (2016) Effect of lumbar support on human-head movement and discomfort in whole-body vibration. Occup Ergon 13(1):3–14
16. Wang W (2006) A study of force-motion and vibration transmission properties of seated body under vertical vibration and effects of sitting posture. PhD diss., Concordia University
17. Rao SS (1996) Engineering Optimization. John Wiley & Sons, New York, pp 779–783
18. Gągorowski A (2010) Simulation study on stiffness of suspension seat in the aspect of the vibration assessment affecting a vehicle driver. Logistics and Transport 11(2):55–62
19. Winter DA (2009) Biomechanics and motor control of human movement. John Wiley & Sons, pp 82–84
20. Wong JY (1989) Terramechanics and off-road vehicles. Elsevier Science Publishing Company, New York

MechAnalyzer: Software to Teach Kinematics Concepts Related to Cams, Gears, and Instantaneous Center

R. Dikshithaa, Siddhant Jain, Janani Swaminathan, Rajeevlochana G. Chittawadigi, and Subir Kumar Saha

Abstract Theory of Machines and Mechanisms is the foundation for the world's race toward automation. Simple mechanisms, cams, and gears find applications in all such automated systems. Therefore, a thorough understanding of these basics is quite essential. The authors are a part of the development team of MechAnalyzer, a software aimed at easing the teaching and learning of concepts related to mechanisms. In this paper, several new modules are reported. Cam module lets user select properties of a cam-follower mechanism and the steps required to draw cam profile is animated and shown to the user. Similarly, concepts related to gear profiles and meshing of gears are illustrated in gears module. Another module is presented which lets user understand the velocity analysis of planar mechanisms using the Instantaneous Center (IC) method. All these modules and the earlier set of modules to teach kinematics of mechanisms are readily available for free at http://www.roboanalyzer.com/mechanalyzer.html.

Keywords Mechanisms · Velocity analysis · Instantaneous center · Cam follower · Cam profile · Gear profile · Involute gears

R. Dikshithaa · R. G. Chittawadigi (✉)
Department of Mechanical Engineering, Amrita School of Engineering, Amrita Vishwa Vidyapeetham, Bengaluru, India
e-mail: rg_chittawadigi@blr.amrita.edu

R. Dikshithaa
e-mail: rdikshithaa97@gmail.com

S. Jain · J. Swaminathan
Department of Mechanical Engineering, National Institute of Technology Karnataka, Surathkal, India
e-mail: siddhant.jain150397@gmail.com

J. Swaminathan
e-mail: janjanamayee@gmail.com

S. K. Saha
Department of Mechanical Engineering, Indian Institute of Technology, New Delhi, India
e-mail: saha@mech.iitd.ac.in

D. Sen et al. (eds.), *Mechanism and Machine Science*, Lecture Notes in Mechanical Engineering, https://doi.org/10.1007/978-981-15-4477-4_10

1 Introduction

Study of motion of machine components or members (i.e., links or joints) neglecting the cause of motion is known as kinematics of mechanism. Kinematic analysis and static force analysis of mechanisms play an important role in the field of machine design. Understanding the graphical methods and the behavior of the mechanisms is very important. Thus, learning and understanding of these mechanisms becomes very essential. In the initial stages, it becomes very difficult for the students to imagine the working of these models, hence visualizations of these results in faster learning. As there is an increase in the usage of mechanisms in industries, it has become very essential for the effective learning of mechanisms in engineering.

The concepts in the theory of machines involves a lot of visualization and many graphical methods are employed to solve problems, as an alternative to analytical methods. It becomes a challenge for both, the teacher and the student, to teach as well as to understand them. Physical models can be built to illustrate the motion class, but that will be a tedious and time-consuming task; it also limits the study of other possibilities of these mechanisms. However, in a software, the students can see different outputs based on their input parameters and make the entire learning process fun. Also, in case of graphical methods, it takes too long to solve each problem, which becomes a constraint on solving a variety of problems.

MechAnalyzer is a 3D model-based learning software. It was developed with the intention of serving as a teaching aid for the basic concepts which are a part of undergraduate courses related to Theory of Machines. It is a Windows desktop application developed using Visual C# and OpenTK (a wrapper of OpenGL). The graphs are plotted in an open source library called ZedGraph. The active development of this software began in the year 2013 and is evolving since then. Versions 1 and 2 of MechAnalyzer were reported in [1] and had forward kinematics analysis of four-bar, slider-crank, and five-bar mechanisms which included the animation of motion and the resulting graph results were plotted. In version 3 of MechAnalyzer [2] the software was completely revamped and was set on the lines of RoboAnalyzer [3], another educational software developed by the last two authors. The user interface was made similar to that of RoboAnalyzer. Many mechanisms with their variants were added in this version such as Whitworth quick return, double slider, four-bar quick return, cam, gear, steering, pantograph, and wiper mechanism. This version lacked graph plots which were taken up in the next version.

Velocity and Acceleration Diagram (VAD) module was added in version 4.0 of MechAnalyzer [4]. It carried out the position, velocity, and acceleration analysis of planar mechanisms using graphical methods. The stepwise animation for both velocity and acceleration can be shown. The analysis was done for the inversions of mechanisms, if present. Static Force Analysis (SFA) module was added in version 4.1 of MechAnalyzer [5]. It could perform static force analysis of four-bar, slider-crank, Whitworth quick return, crank-slotted, and four-bar quick return mechanisms. The required input parameters were taken from the user and further, the force polygons and free body diagrams were drawn in a stepwise animation, the way it is done by a teacher on a blackboard. In this paper, concepts related to gears, cams, and Instantaneous Center (IC) methods are reported.

At present, there are a few commercially available software that show us the animation of gears and the transmission forces between them. For example, MechDesigner [6], Linkage [7], and Artas [8]. But these are not available for free. A few websites have gear generator [9] applications that can be used to achieve these tasks. However, it works only for users with an active Internet connection. However, to the best of the knowledge of the authors the available software do not show the animation of gear profile form based on the users' input parameters. Hence, a gear profile module to show the animation of the different gear profiles and a gear module that clearly shows the meshing of involute gears and has the ability to solve simple numerical problems has been added to MechAnalyzer software. Similarly, to the best knowledge of the authors, there are no simulation or educational software that show the stepwise animation of displacement, velocity, and acceleration graph of follower motion and also the stepwise animation of drawing cam profile. These have been developed as a new cam module in MechAnalyzer. This enables the faster learning of problems and also allows the user to solve a variety of problems. One way of performing velocity analysis of planar mechanisms is by finding the instantaneous centers between the links. These instantaneous centers may be found by analytical or graphical methods. Graphical approach gives a better insight into the mechanism and thus helps in better understanding. But it suffers from the disadvantage of being tedious and time consuming. This module aims to tackle such problems, and provides a computer-based approach.

In this paper, an overview of the new modules reported in this paper is given. Cam module is explained in Sect. 2, followed by gear profile module in Sect. 3. Meshing of involute gears is covered in Sect. 4, followed by Instantaneous Center method in Sect. 5 and then the conclusions.

2 Overview of Cam Module

Cams are generally used to move a link or a combination of links (known as follower), which are further used to control certain tasks such as the flow of fuel in an internal combustion engine using cam-value mechanisms. The cam may be rotating or reciprocating whereas the follower may be rotating, reciprocating, or oscillating. Complicated output motions which are otherwise difficult to achieve can easily be produced with the help of cams.

It is a difficult task for the teachers to teach the topic on cams. The drawing of cam displacement diagrams for different types of motion and the cam profile for different types of followers is hard to understand without any visual aid. To make it easier for both students and teachers, this module has been developed.

The advantages this software will provide to students and teachers are as follows:

- View the stepwise procedure for drawing the displacement diagram,
- View the stepwise procedure for drawing the cam profile,
- View the rotation of cam along with the motion of the follower.

Cam module has been developed as a part of the MechAnalyzer software. It has a separate tab on the same windows as that of the VAD and SFA module. It has been developed as a Windows Form Application using Visual C# as the programming language in Microsoft Visual Studio. An open source library called ZedGraph has been used for drawing displacement, velocity, and acceleration graphs and also to show the animation of the cam profile.

This module consists of four tabs to display graphs, draw displacement curve, cam profile, and to rotate the cam. This module includes the four different types of motions of follower such as the simple harmonic motion, cycloidal motion, constant acceleration motion, and constant velocity motion. The displacement, velocity, and acceleration graphs with respect to these motions are displayed. Ascent angle, dwell angle, descent angle, height, base circle radius, offset, roller radius are taken as input parameters. The user is allowed to select the type of follower, either knife edge or roller follower and also the type of ascent and descent motion of follower. Graphs based on these parameters are displayed. Figure 1 shows the tab that displays position, velocity, and acceleration graphs based on user input.

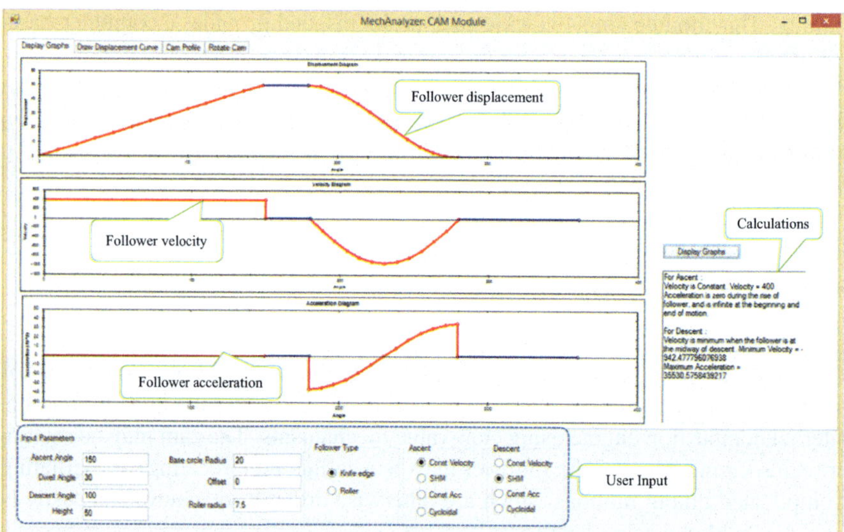

Fig. 1 User interface of cam module showing follower motion plots. Follower has constant velocity during ascent and simple harmonic motion (SHM) during descent

2.1 *Equations of Displacement, Velocity, Acceleration of Motions*

Constant Velocity: If the follower is moving with constant velocity, it means that the displacement of the follower is proportional to the cam angular displacement and the slope of displacement curve is constant. If the motion of the follower were a straight line, it would have equal displacements in equal units of time, i.e., has uniform velocity throughout. The acceleration, except at the end and beginning of rise and fall would be zero. Figure 1 shows constant velocity motion graph during the ascent of the follower.

Simple Harmonic Motion: The follower displacement is a simple harmonic function with the initial and final velocities as zero. However, there is an abrupt change in acceleration during the initial and final positions. An example is shown for the descent of the follower in Fig. 1.

Constant Acceleration Motion: The velocity increases at a uniform rate during the first half of the motion and decreases at a uniform rate during the second half of the motion. The acceleration is constant and positive throughout the first half of the motion, as shown in f, and is constant and negative throughout the second half. Figure 2 shows constant acceleration follower motion during ascent.

Cycloidal Motion: A cycloid is the locus of a point on a circle rolling on a straight line. By far, this is the best motion profile for a follower as it does not have any abrupt change of acceleration, i.e., jerk. Follower motion plots with cycloidal motion are shown for the descent in Fig. 2.

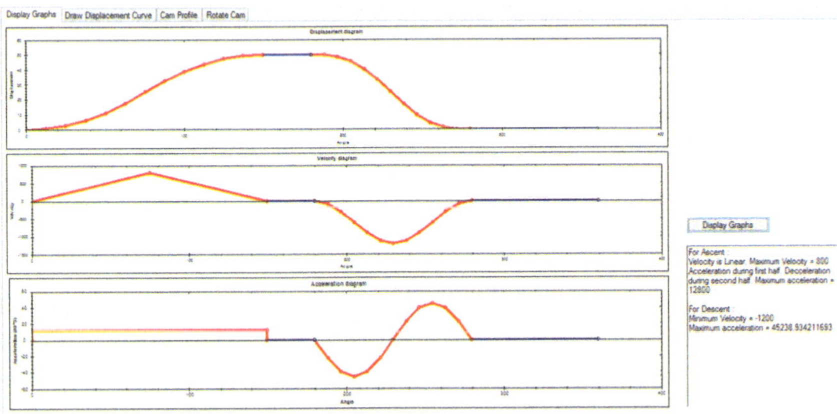

Fig. 2 Follower with constant acceleration in ascent and cycloidal motion in descent

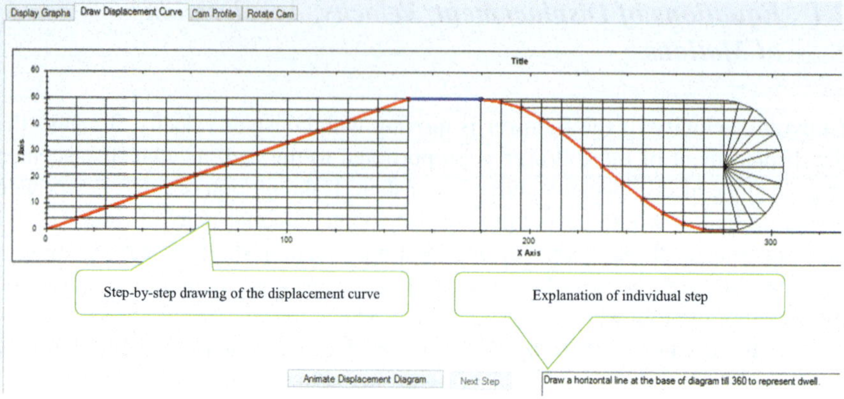

Fig. 3 Displacement curve for constant velocity in ascent and SHM in descent

2.2 Displacement Curve

A stepwise animation of the displacement curve drawn is shown in the displacement curve tab of the cam module. On clicking the given button, the drawing of the displacement curve starts and on clicking the "Next Step" button the next step to be followed to continue drawing the displacement curve is shown along with a message which displays what has been done in that step. A screenshot of the user interface displacement curve is shown in Fig. 3. Using this feature, the teacher can explain the steps required to draw the displacement curve for a follower motion, which is otherwise very tedious to do on a blackboard. Moreover, with just a few mouse clicks, the different combinations of ascent and descent motion can be tried in the classroom. Later, the students can also play around the options and understand the steps required to draw, at their own pace.

2.3 Cam Profile

There are two graph areas on this tab. The first graph area is just to display the follower displacement curve while the other is to display the step-by-step procedure for drawing the cam profile. The displacement curve is displayed because the construction of the cam profile requires measurements from the displacement curve. The steps involved in the drawing of the cam profile involves drawing a circle and dividing the angles of it for ascent, descent, and dwell. Thereafter radial lines are drawn for a radial cam and tangential lines are drawn for a cam with offset. Corresponding distances are measured from the displacement curve and marked on the radial or tangential lines. If the follower is of roller type, some more steps are involved to compensate the roller radius. All these steps are time consuming. To show all combinations of the

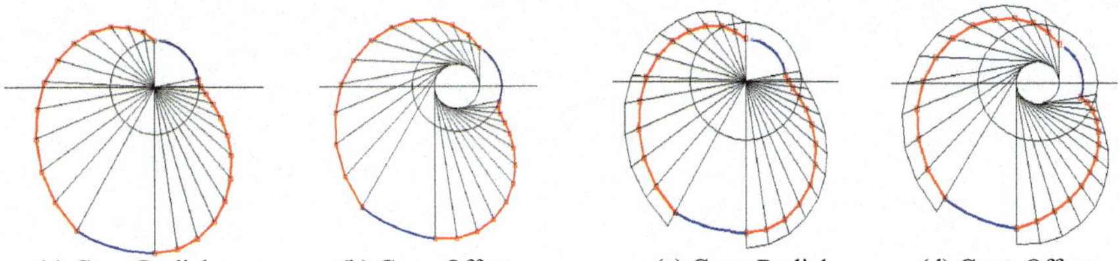

(a) Cam: Radial Follower: Knife-edge
(b) Cam: Offset Follower: Knife-edge
(c) Cam: Radial Follower: Roller
(d) Cam: Offset Follower: Roller

Fig. 4 Cam profile drawing for various combinations of cam and follower

cam profile, it is very difficult in limited time. Hence, the steps required are animated in the cam profile tab. The final drawing step for various combinations are shown in Fig. 4, for follower displacement used in Fig. 3.

2.4 Rotate Cam

This tab shows the animation of rotation of the cam along with the follower's rise and fall motion. The user can pause the rotation at any time and continue it from the paused position. One important application of this feature is that students can understand why the lines are drawn radial in case of a radial cam and tangential in case of an offset cam. When the animation is paused, the direction of the follower motion is along the radial and tangential lines, respectively, for those cases. These aspects are otherwise very difficult to explain and learn by just looking at blackboard or textbook figures. Screenshots for these examples are illustrated in Fig. 5.

3 Overview of Gear Profile Module

Two curves of any shape that fulfill the law of gearing can be used as the profiles of teeth. In other words, an arbitrary shape of one of the mating teeth can be taken and applying the law of gearing the shape of the other can be determined. Such gears are said to have conjugate teeth. However, it will be very difficult to manufacture such gears and the cost will be high. Thus, there arises the need to standardize gear teeth. Common forms of teeth are cycloidal profile teeth and involute profile teeth. Before, students understand the geometry of the teeth it becomes necessary for them to understand the basic profile of the teeth. Drawing these profiles on board can be attempted for one set of parameters. However, if the students have to be taught various possibilities with these input parameters, it becomes difficult and time consuming. Hence, this gear profile module has been developed to give students a better understanding of the gear profile with animation. It also shows the different behavior of the profile based on the input. It has been developed as a Windows

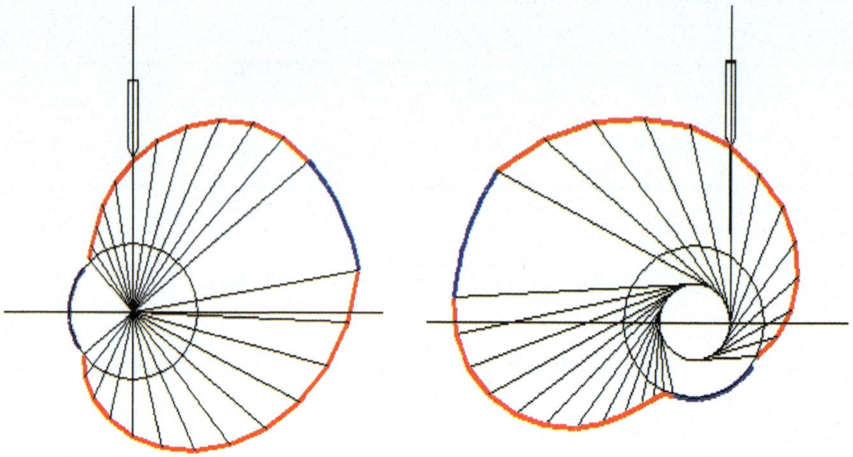

(a) Cam: Radial
Follower: Knife-edge

(b) Cam: Offset
Follower: Knife-edge

Fig. 5 Screenshots during animation of follower displacement with rotation of cam

Form Application using Visual C# as the programming language in Microsoft Visual Studio. An open source library called ZedGraph has been used for drawing the bodies/objects and to show the animation. Primary objectives of this module are

- Better understanding of basic gear profiles;
- Visual aid helps the students learn and understand the profile better;
- Allows the user to visualize all possible outputs for a given input.

The common profiles that are taught with regard to gear profile are cycloidal, hypocycloid, epicycloid, and involute profiles, all of which have certain input parameters that can be varied and accordingly the profile varies. The user interface of a gear profile is shown in Fig. 6, where each profile has a separate tab.

3.1 Cycloid Profile

A cycloid is the locus of a point on the circumference of a circle that rolls without slipping on a fixed straight line. The parametric equation for a cycloid is

$$x = r(t - \sin t) \tag{1}$$

$$y = r(1 - \cos t) \tag{2}$$

where t is the parameter, representing time. Radius (r) is given as an input parameter and the speed of the animation can also be varied. Figure 7 shows screenshots during

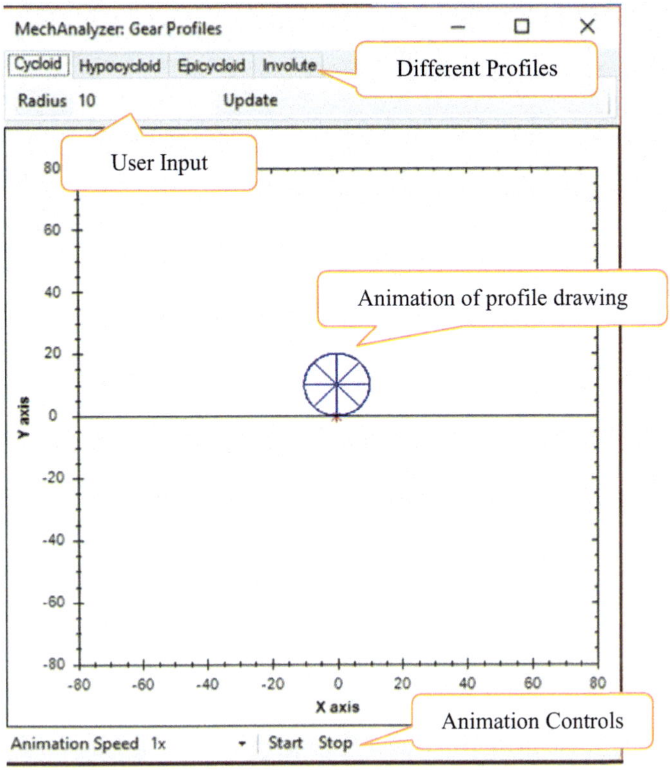

Fig. 6 User interface of gear profile module

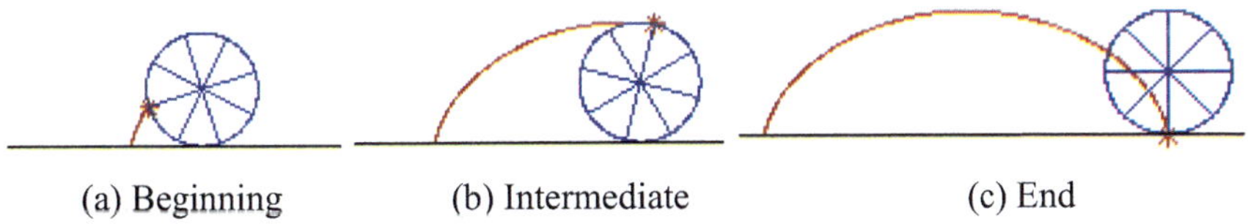

Fig. 7 Screenshots during animation of cycloid profile

the animation of drawing of cycloid profile. It is to be noted that the trace of the point is shown in "red" color. This profile is also taught during drawing courses such as Engineering Drawing and Machine Drawing and the software can also be used for effective explanation.

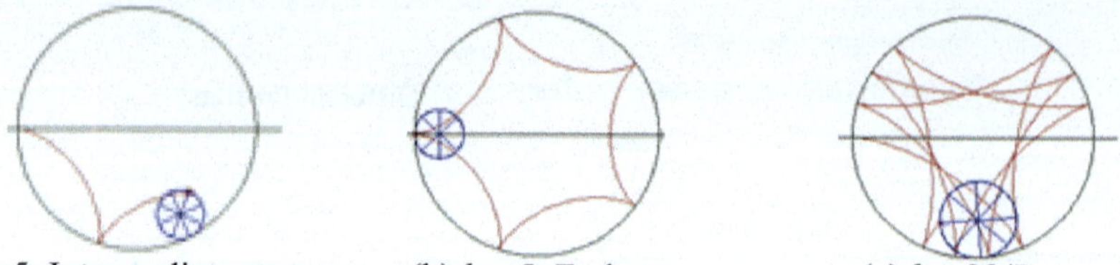

(a) $k = 5$: Intermediate step (b) $k = 5$: End step (c) $k = 22/7$ (irrational number)

Fig. 8 Animation of drawing of hypocycloid profile

3.2 *Hypocycloid Profile*

A hypocycloid is the locus of a point on the circumference of a circle that rolls without slipping inside the circumference of another circle.

The parametric equation for a hypocycloid is

$$x = (R - r)\cos\theta + r\cos\left(\frac{R - r}{r}\theta\right) \tag{3}$$

$$y = (R - r)\sin\theta - r\sin\left(\frac{R - r}{r}\theta\right) \tag{4}$$

where R is the fixed circle radius and r is the rolling circle radius. If k is defined as a variable such that $k = \frac{R}{r}$, then value of k gives the number of cusps. Several interesting shapes of hypocycloid profiles are also possible as illustrated in Fig. 8.

3.3 *Epicycloid Profile*

An epicycloid is the locus of a point on the circumference of a circle that rolls without slipping on the circumference of another circle. The parametric equation for an epicycloid is

$$x = (R + r)\cos\theta - r\cos\left(\frac{R + r}{r}\theta\right) \tag{5}$$

$$y = (R + r)\sin\theta - r\sin\left(\frac{R + r}{r}\theta\right) \tag{6}$$

where R is the fixed circle radius, r is the rolling circle radius, and $k = \frac{R}{r}$ is the number of cusps. Screenshots during the animation of the epicycloid profile are given in Fig. 9.

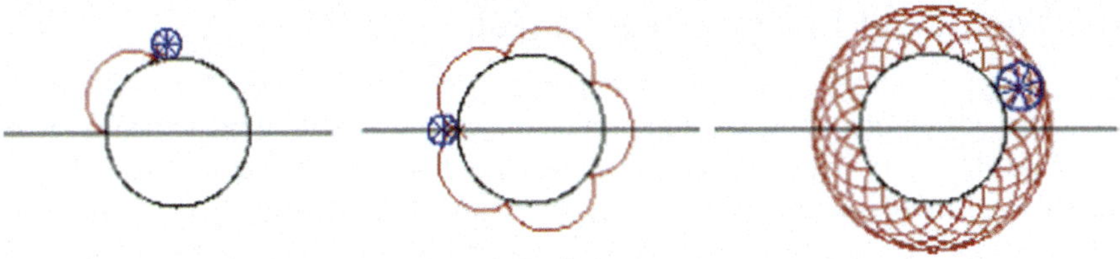

(a) $k = 5$: Intermediate step (b) $k = 5$: End step (c) $k = 22/7$ (irrational number)

Fig. 9 Animation of drawing of epicycloid profile

3.4 *Involute Profile*

An involute is defined as the locus of a point on a straight line which rolls without slipping on the circumference of a circle. Also, it is the path traced out by the end of a piece of taut cord being unwound from the circumference of a circle. The x and y coordinates of the involute profile are defined as

$$x = (\text{radius} * \cos\theta) + j * \cos(\text{angle} + 90) * \text{pi}/180.0 \tag{7}$$

$$y = (\text{radius} * \sin\theta) + j * \sin(\text{angle} + 90) * \text{pi}/180.0 \tag{8}$$

where θ is the initial angle from which tangents are being started, they keep decrementing by a standard angle (i.e., given by 360°/total number of divisions), and j is the counter for angle increment. Screenshots of the instances during the animation are shown in Fig. 10. Similar to the cycloidal profile, the involute profile is also taught and practiced in Engineering Drawing and Machine Drawing courses and hence the implementation presented here can be readily used in such courses.

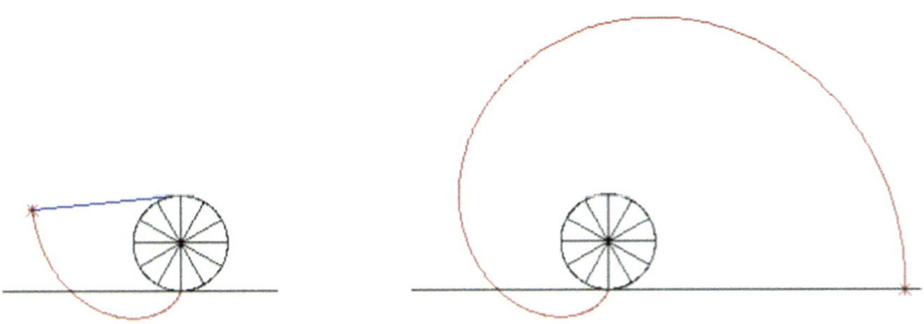

(a) Intermediate step (b) End step

Fig. 10 Animation of drawing of involute profile

4 Overview of Involute Gear Module

Gears have been one of the most important and also a difficult topic to be explained on board. It is very important for the students to have a complete understanding of the basic terminologies of gears. It takes more time for the teachers to draw the exact replica of the gear tooth on the board and explain them. Concepts like meshing of gears, movement of point of contact becomes tough to be explained on board. This is where the necessity of developing this module was envisaged, where the basic terminologies of gear could be explained in an effective way. The objectives of this module are

- To develop the interface for better and faster understanding of basic gear terminologies;
- To animate the movement of two gears;
- To solve basic problems based on gears.

The gear module has been developed on the same software architecture as that of the gear profile module. The main user interface of the involute gear module is shown in Fig. 11.

This module has two different ways of accepting the input parameters based on which the geometry of the gears is generated. There are buttons, each for the basic terminologies of gears which, when clicked, display the curve or path on the gears. There is also a textbox that displays the definition of the terminology also the different possible output values for the given input, based on the inbuilt calculations. These calculations can be used by students to verify if a problem solved by them by hand is correct or not.

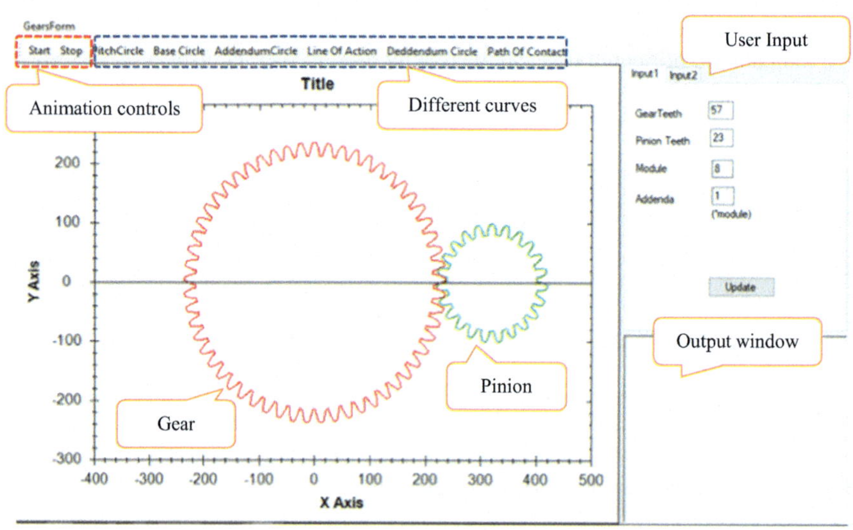

Fig. 11 User interface for involute gears module

4.1 Formulation of Gear Tooth

An involute is the path traced out by the end of a piece of taut cord being unwound from the circumference of a circle. The circle from which it is unwound is called the base circle. For a given set of input values, the curves corresponding to a tooth of each gear are drawn by adapting the formulations reported in [10]. To obtain the remaining teeth, each point in these curves of the first tooth is rotated as many times as the number of teeth in that gear. The expressions of the coordinates after rotation are obtained by multiplication with a rotation matrix, i.e.,

$$x = x_1 \cos\theta - y_1 \sin\theta \tag{9}$$

$$y = x_1 \sin\theta + y_1 \cos\theta \tag{10}$$

where x_1 and y_1 are the coordinates of each point on the first tooth and θ is the rotation angle, whose initial angle is (360°/total number of teeth on the gear) and is later incremented till all the teeth are generated.

To achieve animation, all these points are stored in a list and sent to another method or function for the rotation of the gear. By now, each point on the gear is defined uniquely, therefore to get the complete rotation of the gear these points are further rotated by a rotation matrix, whose equations are again given by Eqs. (9) and (10), but the rotation angle changes each time. To maintain constant velocity, the rotation angle of gear and the pinion differs by the velocity ratio. In a similar way, the other gear is modeled and animated. To achieve proper meshing at the beginning of the animation, the pitch point is assumed to be on the center-to-center line between the two gears. This is achieved by rotation of both wheels by a certain angle. Due to space constraints, a detailed explanation of the formulations and implementation are not included in this paper.

5 Overview of Instantaneous Center (IC) Module

Velocity analysis of planar mechanisms can be performed using the analytical method (whenever a closed-form solution exists) or using graphical methods. Among graphical methods, the relative velocity vector-based approach is one of the most common methods, which is already present in MechAnalyzer software. Another graphical method is the usage of Instantaneous Centers of a set of two links and determining relative velocities between them. This method consists of the identification of all possible instantaneous centers between the links in a given mechanism and measuring distances between some of these centers and using them in calculations. Ideally, any four-link or six-link planar mechanism with single degree-of-freedom can be solved using this method. However, in a classroom environment, solving a variety of problems requires time and effort. Hence, Instantaneous Center (IC) method module has been implemented in MechAnalyzer software.

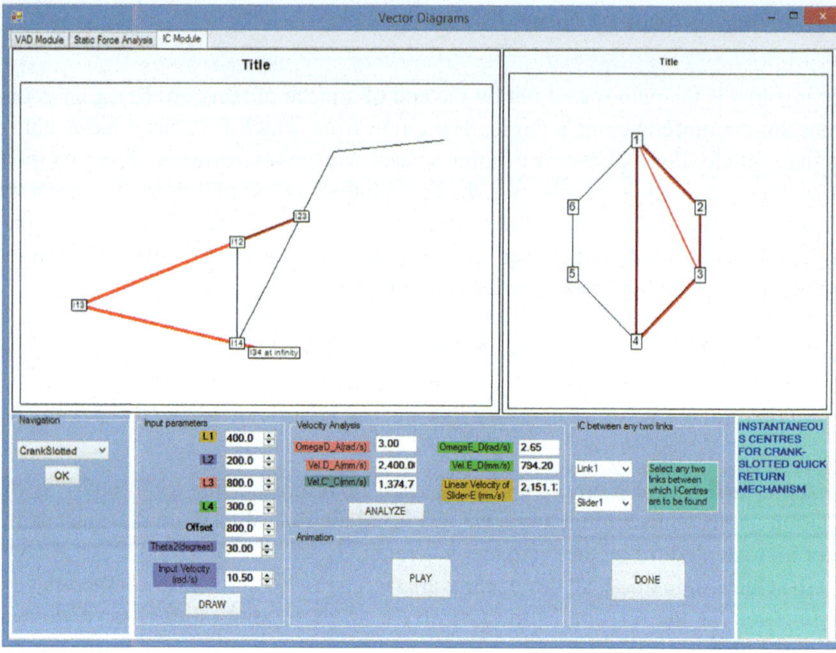

Fig. 12 User interface Instantaneous Center (IC) module

A stepwise animation of finding the center for all combinations of links is shown. It is used to draw vector diagrams for finding instantaneous centers between links for common mechanisms such as four-bar, slider-crank, and slotted-lever quick return mechanism. It provides the option of observing the behavior of a mechanism over a complete cycle of operations as per user's input, rather than at only a few instantaneous positions. It also makes possible parametric studies of these mechanisms to improve performance and analysis. The calculations of velocity analysis are also displayed to the user and hence it can be used as a validation tool after manual calculations by the user. In addition, it also allows the animation of steps to determine the instantaneous center between a set of two links. For example, in a crank-slotted quick return mechanism, the IC between the crank and the first slider is obtained as illustrated in Fig. 12.

6 Conclusion

Various concepts in kinematics require visualization of motion of the mechanisms for effective education. To aid the process, MechAnalyzer software has been developed. It has been developed in a modular way so that newer modules can be added with lesser efforts. In this paper, modules related to cam analysis, gear profiles, involute gear meshing, and velocity analysis of planar mechanisms using Instantaneous Center (IC) method have been reported. These modules are developed with functionality that has a step-by-step procedure to replicate the blackboard teaching methodology. One major advantage is that various inputs can be used in the software and their effects can be seen and learnt in a fast, easy, and productive manner.

References

1. Hampali S, Chittawadigi RG, Saha SK (2015) MechAnalyzer: 3D model based mechanism learning software. In: Proceedings of the 14th world congress in mechanism and machine science
2. Lokesh R, Chittawadigi RG, Saha SK (2015) MechAnalyzer: 3D simulation software to teach kinematics of machines. In: Proceedings of the 2nd International and 17th national conference on machines and mechanisms
3. Othayoth RS, Chittawadigi RG, Joshi RP, Saha SK (2017) Robot kinematics made easy using RoboAnalyzer software. Comput Appl Eng Educ 25(5):669–680
4. Verma SK, Kumar R, Chittawadigi RG, Saha SK (2016) Kinematic analysis of mechanisms using velocity and acceleration diagrams (VAD) module in MechAnalyzer software. In: Proceedings of the 8th Asian conference on multibody dynamics
5. Verma SK, Swaminathan J, Chittawadigi RG, Saha SK (2016) Static force analysis of planar mechanisms in MechAnalyzer software. In: Proceedings of the XII IPRoMM national workshop on industrial problems on machines & mechanisms
6. MechDesigner (2017) https://www.psmotion.com/. Accessed June 2017
7. SAM Mechanism design by Artas Engineering Software (2017) https://www.artas.nl/en/. Accessed June 2017
8. Linkage Mechanism Designer and Simulator (2017) http://blog.rectorsquid.com/linkage-mechanism-designer-and-simulator/. Accessed July 2017
9. Gear generator (2017) https://geargenerator.com. Accessed July 2017
10. Reyes O, Rebolledo A, Sanchez G (2008) Algorithm to describe the ideal spur Gear profile. In: Proceedings of the world congress on engineering vol II

Modeling the Effects of Linear and Torsional Spring Based Passive Assistance on Human Gait

Rajbeer Singh Anand, Neeta Kanekar, and Anirban Guha

Abstract Older adults and people with various neurological diseases need some external assistance even during normal walking. Amongst various factors, due to a decrease in the muscle strength, they are not able to produce a large force that significantly affects the pre-swing phase of their gait cycle, an important determinant of walking speed. While there is significant work done on active assistance for walking, almost no work has been reported on the effect of torsional spring and combinations of torsional and linear springs as a mode of passive assistance. OpenSim was used to model the effects of linear and torsional spring-based passive assistance on the total metabolic energy and ankle moment during human gait. A total of six conditions were simulated out of which five gave satisfactory findings. Particularly, a combination of ankle torsional and gastrocnemius path springs showed the largest reduction in the peak total metabolic energy and peak plantar flexor moment during the pre-swing phase of the gait cycle, suggesting that such passive assistance could be used as an effective mode of support for walking.

Keywords Gait · Modeling · Passive device · Torsional spring · Linear spring

1 Introduction

Neuromuscular system changes go hand in hand with advancing age and include altered muscle strength, joint mobility, balance, and cardiovascular fitness [1–10]. Due to these changes, both the performance of activities and mobility can get deteriorated. Older adults are severely affected by reduced mobility and it increases their risk of disability, morbidity, and mortality [11]. Gait mechanics is impacted by

R. S. Anand (✉) · A. Guha
Department of Mechanical Engineering, Indian Institute of Technology, Bombay, India
e-mail: rajbeer2007@iitb.ac.in

A. Guha
e-mail: anirbanguha@iitb.ac.in

N. Kanekar
Department of Biosciences and Bioengineering, Indian Institute of Technology, Bombay, India
e-mail: nkanekar@iitb.ac.in

D. Sen et al. (eds.), *Mechanism and Machine Science*, Lecture Notes in Mechanical Engineering, https://doi.org/10.1007/978-981-15-4477-4_11

age-related changes in the neuromuscular system, which is thought to take part in reducing mobility and functional independence [12]. Older adults and people with various neurological diseases need some external assistance even during normal walking. Amongst various factors, due to a decrease in the muscle strength, they are not able to produce a large force which to a great extent affects the pre-swing phase of their gait cycle. The pre-swing phase is generally observed during 50–60% of the gait cycle and helps in propelling the body forward and in the transfer of body weight from one leg to the other [13].

Generally, the external assistance given to people with neurological diseases is active assistance. These are the devices that have an actuation system or are automated in any sense. A few of the active assistive devices developed are: BLEEX [14], XOS [15], HULC [16], and HAL [17]. In order to improve load-carrying capacity for soldiers, DARPA since 2000 has developed BLEEX and XOS. Human Universal Load Carrier (HULC) is a second generation prototype of BLEEX designed by Lockheed Martin. HAL was developed by Japan, which was the first commercial exoskeleton. HAL helps people with disabilities to walk at a speed of 4 km/h independently and to go up and down the stairs easily. However, these devices typically use a huge battery source which makes the device bulky and very expensive.

A passive assistive device does not need any external source of power but rather uses springs, dampers or various materials that have the ability to store energy cultivated by human motion and to use this energy as and when required to support a posture or a movement. A few passive assistive devices are mentioned here. A passive orthosis linkage [18] was developed by Berkelman et al. for foot drop condition in patients. Exoskeleton with clutch and spring system was attempted by Collins et al. [19] for human gait. A novel three degrees of freedom ankle exoskeleton [20] was developed to strengthen the subject's muscle power in military missions. These make a very efficient choice for an assistive device.

While there is significant work reported on active assistance for walking, there is very limited literature available for passive assistance. Almost no work has been reported on the effect of torsional spring as a mode of passive assistance, and a combination of torsional and linear springs. The purpose of this study was, therefore, to model the effects of passive assistance on human gait. Particularly, torsional and linear springs of different stiffness constants were used either alone or in different combinations to represent different simulation conditions.

2 Methods

Open source software OpenSim was used to carry out simulations for gait using a two dimensional model of the human musculoskeletal system (Fig. 1). Computer simulations of bipedal walking were generated with body segments' motion occurring in the sagittal plane. The subject in this study had a weight of 72.6 kg and a standing height of 1.8 m and was made to walk at a comfortable speed of 1.2 m/s on the treadmill. The initial condition was defined using the marker data for kinematics

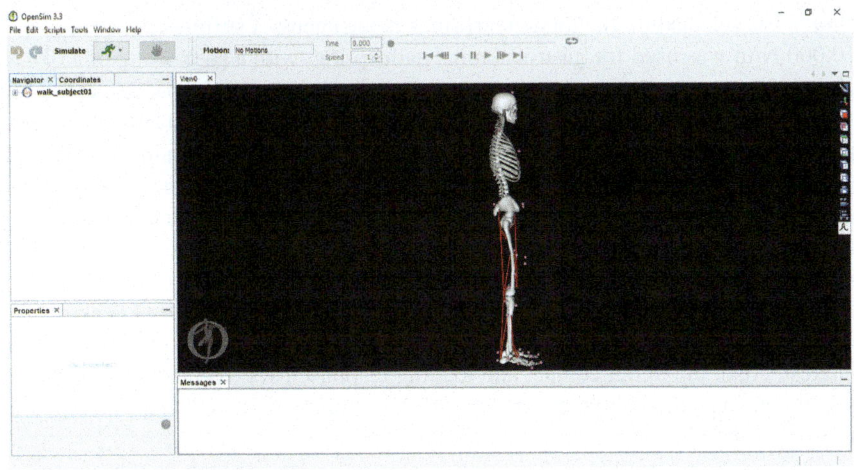

Fig. 1 OpenSim interface

in healthy young subjects. The effect of linear and torsional spring-based passive assistance, provided at both the lower limbs, on the total metabolic energy and joint moments during human gait were studied. The torsional spring employed at the ankle had a stiffness constant of 1 Nm/° which was also the case for the torsional spring at the knee. The linear spring replicating the gastrocnemius muscle (gastrocnemius path spring) had a stiffness of 10,000 N/m, and the linear spring replicating the soleus muscle (soleus path spring), 11,000 N/m. Normal walking condition (no springs) was used as a control condition.

For determining the stiffness constant of the springs, the total metabolic energy (TME) and the right ankle moment were taken into account. The total metabolic energy [21] was identified as the muscle energy expenditure in the lower limb of the human body. This includes the gastrocnemius, soleus, tibialis anterior, and iliopsoas muscles, which help in gait. The right ankle moment in the sagittal plane was also used for determining spring constants. It concerns with rotation about the ankle in the sagittal plane and represents the tendency for the ankle to go either into dorsiflexion or plantarflexion. The reduction in peak TME and peak ankle moment during the pre-swing phase represented the effectiveness of passive assistance. The six conditions simulated in the study were: torsional springs being used at the ankles, torsional springs at the knees, gastrocnemius path springs at both the legs, soleus path springs at both the legs, combination of ankle torsional and gastrocnemius path springs, and combination of ankle torsional and soleus path springs. For torsional springs at the ankles, a stiffness constant of $k = 1$ Nm/° was chosen in reference to Hong et al. [20]. For torsional springs at the knees, different spring constants were implemented but the TME exceeded for all the cases as compared to the TME for the control condition, which was not desirable. This means that muscles were expending more energy when the torsional spring was implemented at the knee. Therefore, for the sake of comparison, the spring constant identical to that used at the ankles ($k = 1$

Nm/°) was selected for the torsional springs at the knees. A stiffness constant of $k =$ 10,000 N/m was used for gastrocnemius path spring, with a preloading of 1107 N, after implementing different spring constants on the basis of the parameters: TME and right ankle moment, which were found optimal for $k = 10{,}000$ N/m. For soleus path spring, a stiffness constant of $k = 11{,}000$ N/m (with a preloading of 1052 N) was found to be optimal on the basis of TME and right ankle moment. And for the combinations of torsional springs at the ankles and path springs at the legs, the same spring constants were used as mentioned above.

The pre-swing phase, which is generally between 50 and 60% of the gait cycle, was considered to be the main focus of the study. The peak TME and peak plantar flexor moment of the right ankle during the pre-swing phase were the parameters of interest. Also, the initial contact and loading response (first 10% of the gait cycle), and the swing phase (between 60 and 100% of the gait cycle) were included as additional criteria for describing the results.

3 Results

During the pre-swing phase of the gait cycle, the peak TME was observed to be 1152.076 W/kg and the right ankle had 130,263 Nm of peak plantar flexor moment in the normal walking (control) condition. The first condition with torsional springs being used at both the ankles gave a peak TME value of 831.818 W/kg during the pre-swing phase (Fig. 2). This brought about a 27.8% reduction in the peak TME value as compared to the control condition.

Also, the peak plantar flexor moment for the right ankle (Fig. 3), obtained for the pre-swing phase, was found to be 123,124 Nm for the condition of torsional springs at both ankles. This resulted in a 5.48% reduction in the peak plantar flexor moment as compared to the control condition (130,263 Nm).

Torsional springs placed at the knees of both the legs, the second condition, produced a peak TME of 1402.852 W/kg (Fig. 4), which was 21.77% higher than that of the control condition (1152.076 W/kg), in the pre-swing phase.

The third condition was that of the lower limbs loaded with linear springs replicating gastrocnemius muscles, which gave a peak TME of 1184.802 W/kg during the pre-swing phase (Fig. 5). And thus a small increment of 2.84% was seen in the peak TME as compared to the control condition. The peak plantar flexor moment of the right ankle for this condition was observed to be 100,431 Nm (Fig. 6), which reduced by 22.9% from that obtained for the control condition (130,263 Nm).

The fourth condition saw the linear springs replicating soleus muscles being used on both the limbs. The TME for this condition was 1026.828 W/kg (Fig. 7), which resulted in a 10.87% reduction as compared to the control condition (1152.076 W/kg). The peak plantar flexor moment of the right ankle for this condition was 118,695 Nm (Fig. 8), which was an 8.88% reduction as compared to the control condition (130,263 Nm).

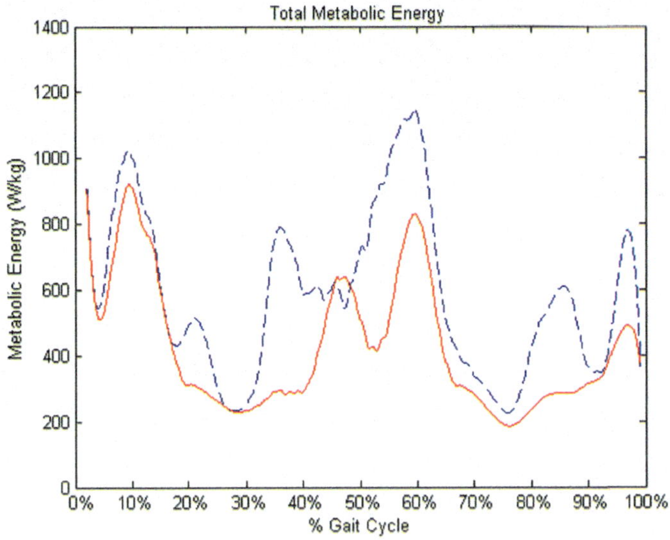

Fig. 2 Total metabolic energy during the gait cycle for the control condition (blue) and for the ankle torsional springs condition (red)

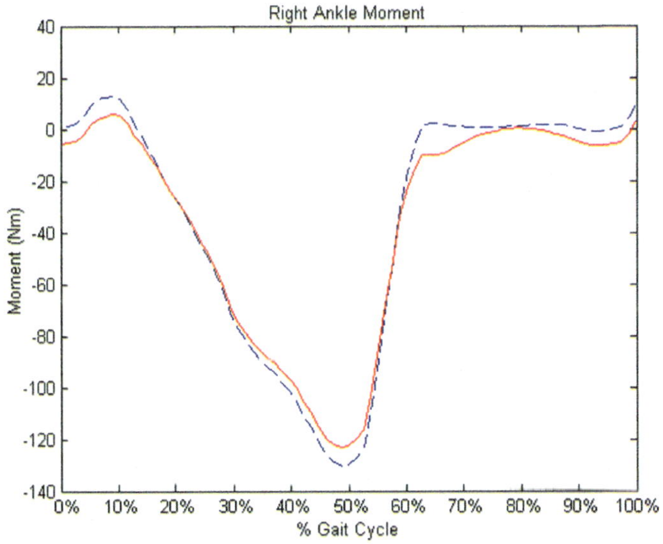

Fig. 3 Right ankle moment during the gait cycle for the control condition (blue) and for the ankle torsional springs condition (red)

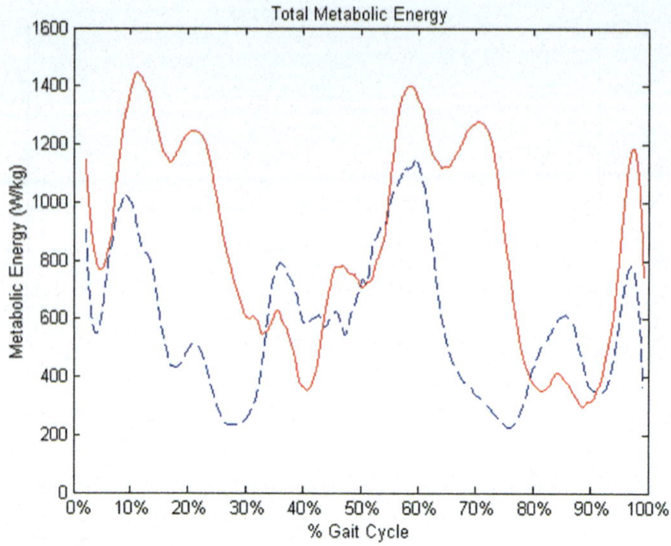

Fig. 4 Total metabolic energy during the gait cycle for the control condition (blue) and for the knee torsional springs condition (red)

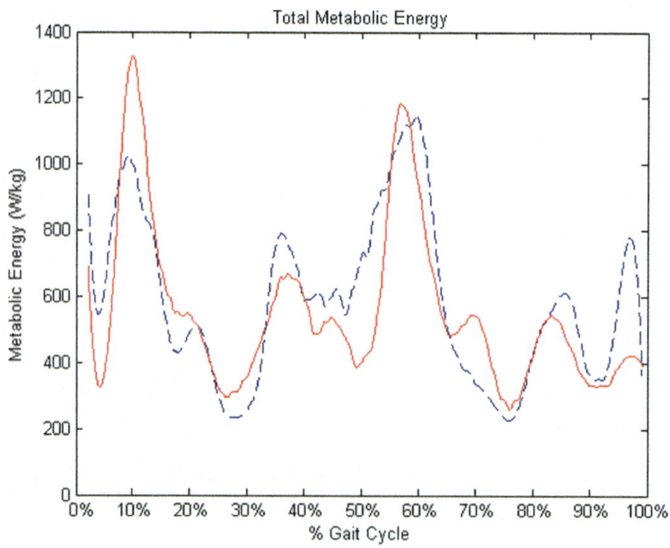

Fig. 5 Total metabolic energy during the gait cycle for the control condition (blue) and for the gastrocnemius path springs condition (red)

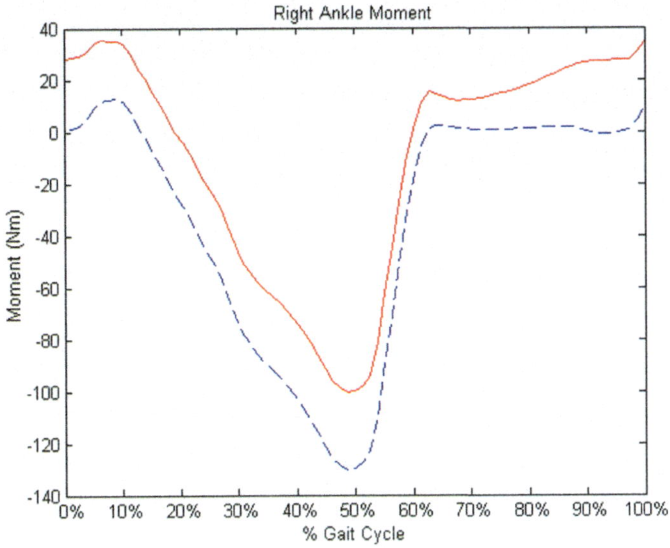

Fig. 6 Right ankle moment during the gait cycle for the control condition (blue) and for the gastrocnemius path springs condition (red)

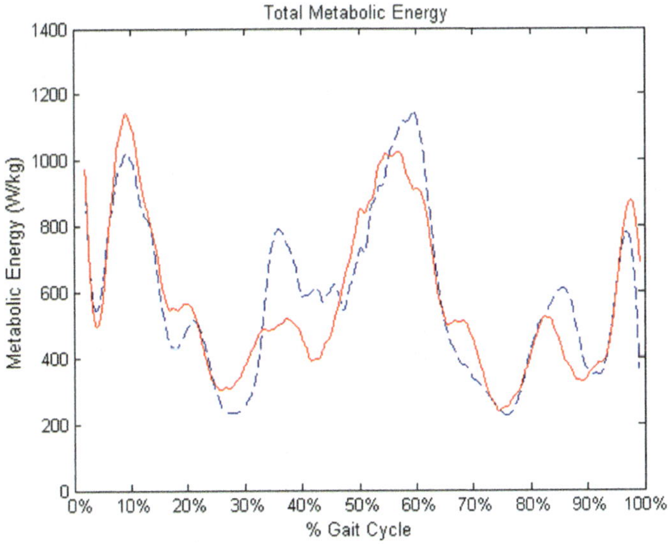

Fig. 7 Total metabolic energy during the gait cycle for the control condition (blue) and for the soleus path springs condition (red)

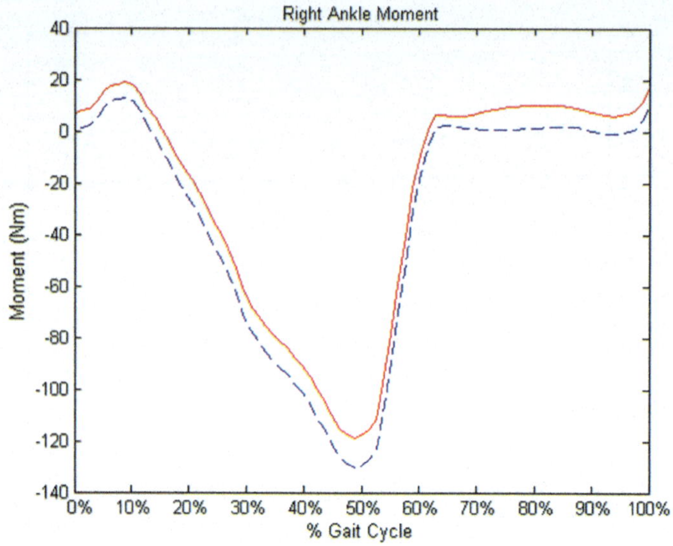

Fig. 8 Right ankle moment during the gait cycle for the control condition (blue) and for the soleus path springs condition (red)

Based on the above results, a combination of ankle torsional and gastrocnemius path springs was implemented. This was the fifth condition in which, the peak TME was found to be 719.252 W/kg (Fig. 9). This resulted in a reduction of 37.57% as compared to the control condition (1152.076 W/kg) during the pre-swing phase. The peak plantar flexor moment of the right ankle for this condition was 93,292 Nm (Fig. 10). This was a 28.38% reduction in the peak moment from that of the control condition (130,263 Nm).

The sixth condition was a combination of ankle torsional and soleus path springs which resulted in the peak TME of 835.547 W/kg (Fig. 11), a reduction of 27.47% as compared to the control condition (1152.076 W/kg).

For this condition, the peak plantar flexor moment of the right ankle during the pre-swing phase was found to be 111,556 Nm, a 14.36% reduction in the moment as compared to the control condition (130,263 Nm) (Fig. 12).

4 Discussion

The objective of the present study was to model the effects of passive assistance on human gait. Torsional and linear springs were used either alone or in different combinations to simulate different conditions of passive assistance. A reduction in peak TME and peak ankle moment during the pre-swing phase represented the effectiveness of passive assistance.

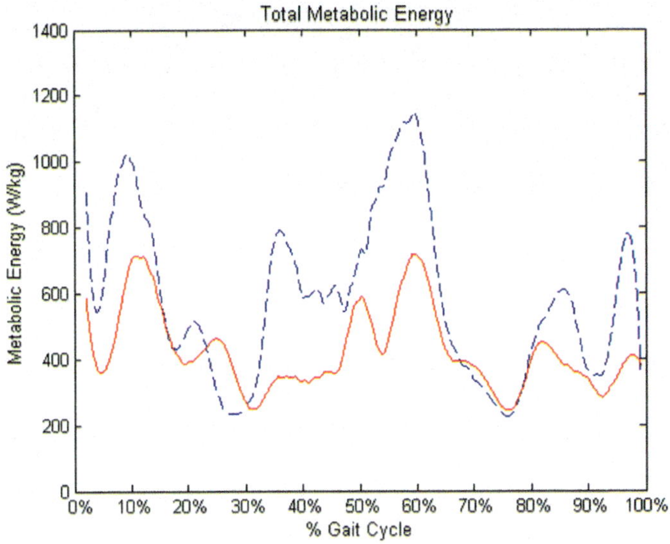

Fig. 9 Total metabolic energy during the gait cycle for the control condition (blue) and for the combination of ankle torsional springs and gastrocnemius path springs condition (red)

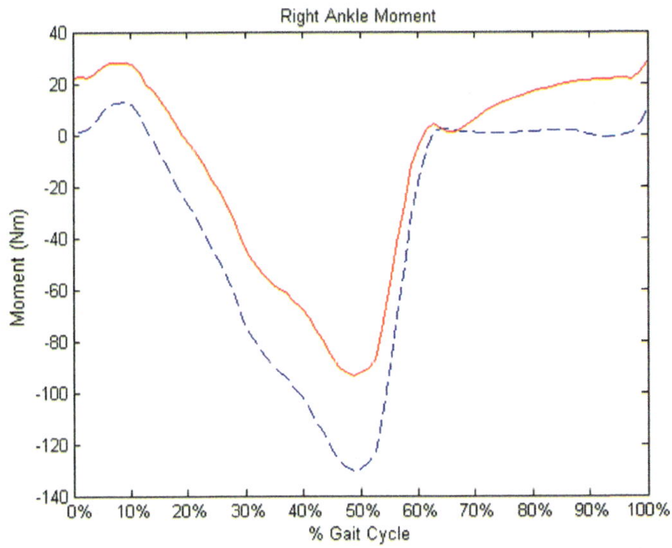

Fig. 10 Right ankle moment during the gait cycle for the control condition (blue) and for the combination of ankle torsional springs and gastrocnemius path springs condition (red)

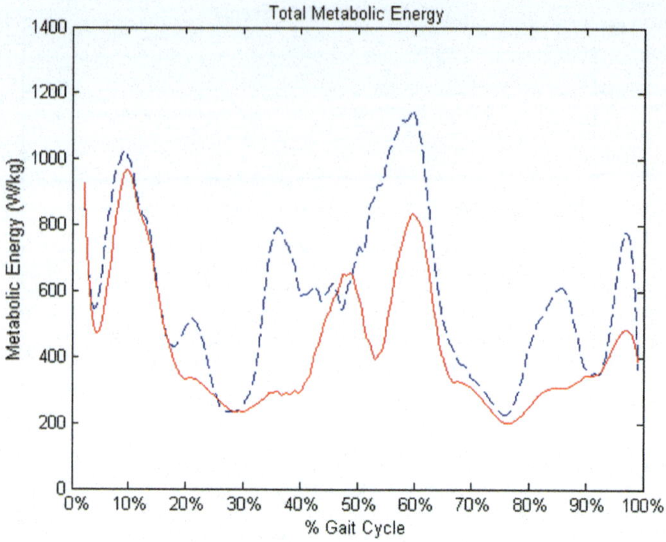

Fig. 11 Total metabolic energy during the gait cycle for the control condition (blue) and for the combination of ankle torsional springs and soleus path springs condition (red)

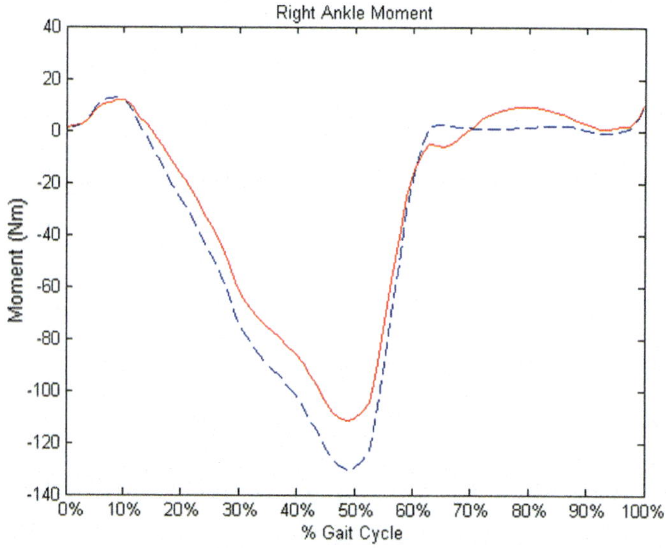

Fig. 12 Right ankle moment during the gait cycle for the control condition (blue) and for the combination of ankle torsional springs and soleus path springs condition (red)

According to the results, the first condition with torsional ankle springs resulted in a peak TME reduction of 27.8% and the TME reduction was generally observed for the whole gait cycle as compared to the normal gait (control condition). This was a favorable condition for the human gait as the energy expenditure, in this case, was lower than that for the control condition. As far as the right ankle peak plantar flexor moment for the pre-swing phase is concerned, a small reduction of 5.48% was observed as compared to normal gait. Also, the moments during the loading response phase and swing phase were within the limit of the moments seen for the normal gait. In the second condition, torsional springs were used at the knees; the TME for this case exceeded the TME in normal gait condition for nearly the whole of the gait cycle, which means large energy expenditure was seen for the muscles of the lower limb in this case. Therefore, the second condition was not explored further. The third condition was the case where linear springs replicating gastrocnemius muscles were used at both the limbs. The peak TME during the pre-swing phase in this case almost remained the same as the normal gait but an increase in TME was seen during the end of loading response and the starting of midstance phase (the midstance phase is between 10 and 30% of the gait cycle). Also, the peak plantar flexor moment during pre-swing was greatly reduced, but the dorsiflexor moments during the loading response and the swing phase exceeded the limits as compared to the normal gait. In the fourth condition, results for linear springs replicating soleus muscles were investigated, and a decrease in the peak TME during the pre-swing phase was observed. Also, an overall trend of reduced TME was observed across the gait cycle. For the peak plantar flexor moment during pre-swing, a reduction was seen compared to the normal gait and generally a good trend for the whole gait cycle was observed. Therefore, an encouraging result was seen in the fourth condition for both the parameters. Therefore, based on the above-mentioned outcomes, combinations of springs were implemented for the last two conditions. The fifth condition consisted of torsional springs at the ankles and linear springs replicating gastrocnemius muscles on the legs. A large reduction in peak TME was observed during the pre-swing phase and generally, a highly reduced TME was seen across the gait cycle. Also, a large reduction in the peak plantar flexor moment was observed during the pre-swing phase, and generally, the moments were in the normal range throughout the gait cycle. However, a further drop in the dorsiflexor moment during the loading response phase and the swing phase would be desirable. The sixth condition included a combination of torsional ankle springs and soleus path springs. The peak TME for this condition also saw a large reduction during the pre-swing phase, and an overall decrease in TME was observed for the whole gait cycle. Also, the peak plantar flexor moment for the right ankle decreased by 14.36% as compared to normal gait and generally followed a good trend for the whole gait cycle.

Other than the second condition, every other condition looked promising according to the parameters of total metabolic energy and right ankle moment. Pre-swing is a critical phase of the gait cycle and one of the most important phases studied to evaluate the effectiveness of any gait assistive device. Other phases can also be of importance depending on the requirements, for example, for the loading response and the swing phases of gait, a plantar flexor moment can cause improper heel strike and foot drag, respectively, which can result in stumbling and a fall for the subject. However, for the above conditions, no such issues were found. For the third and fifth conditions, there was dorsiflexor moment during the loading response phase and the swing phase but the reduction in the peak plantar flexor moment during pre-swing was also high as compared to control condition. The fifth condition seems the most promising as this particular combination of springs gave the largest reduction in peak TME and peak plantar flexor moment during the pre-swing phase, however, it also resulted in an increase in dorsiflexor moment beyond the normal range during the swing phase. As such, an optimization of this condition can be explored in the future. The reduction in peak TME and peak plantar flexor moment during the pre-swing phase was lower for the sixth condition as compared to the fifth. However, both, TME and ankle moments remained in the normal range for the rest of the gait cycle.

5 Conclusion

In this paper, a study was carried out to model the effect of torsional and linear springs, used at the lower limbs, on the human gait. OpenSim was used to carry out the simulations. Six different conditions were simulated on the basis of the springs being used, their combinations, and their placement on the lower limb. Out of the six, five conditions showed a positive effect on gait based on the following parameters: total metabolic energy and right ankle moment. The peak total metabolic energy and the peak plantar flexor moment during the pre-swing phase of the gait reduced for nearly all the five conditions as compared to the normal gait (control) condition. Also, for most of the conditions both total metabolic energy and right ankle moment gave favorable results for other phases of the gait cycle as well. Different conditions can be preferred for different situations or different requirements during gait.

Passive assistance can be preferred over active assistance due to the lightweight structure of passive devices. They help in maintaining or improving muscle strength and mobility in the subjects, because a fair bit of human activity is involved while using passive assistance. They are cheaper and occupy lesser space as compared to active assistive devices. No battery or external source is required for them. They are also energy and gait efficient.

Interaction effects were not considered for the current study. The above results were for a young subject's data; a new study has to be carried out for older adults which can be a part of future investigation. Also, other parameters, in addition to those used in this study, can be explored.

References

1. Chale-Rush A, Guralnik JM, Walkup MP et al (2010) Relationship between physical functioning and physical activity in the lifestyle interventions and independence for elders pilot. J Am Geriatr Soc 58:1918–1924
2. Judge JO, Davis III RB, Ounpuu S (1996) Step length reductions in advanced age: the role of ankle and hip kinetics. J Gerontol Med Sci 51A(6):M3O3–M312
3. Graf A, Judge JO, Ounpuu S, Thelen DG (2005) The effect of walking speed on lower extremity joint powers among elderly adults who exhibit low physical performance. Arch Phys Med Rehabil 11:2177–2183
4. Kang HG, Dingwell JB (2000) Dynamics and stability of muscle activations during walking in healthy young and older adults. J Biomech 42:2231–2237
5. Karamanidis K, Arampatzis A (2005) Mechanical and morphological properties of different muscle-tendon units in the lower extremity and running mechanics: effect of aging and physical activity. J Exp Biol 208:3907–3923
6. Kerrigan DC, Lee LW, Nieto TJ, Markman JD, Collins JJ, Riley PO (2000) Kinetic alterations independent of walking speed in elderly fallers. Arch Phys Med Rehabil 81:730–735
7. Kerrigan DC, Lee LW, Collins JJ, Riley PO, Lipsitz LA (2001) Reduced hip extension during walking: healthy elderly and fallers versus young adults. Arch Phys Med Rehabil 82:26–30
8. Koster A, Penninx BW, Newman AB (2007) Lifestyle factors and incident mobility limitation in obese and non-obese older adults. Obesity (Silver Spring) 15:3122–3132
9. Murray MP, Duthie Jr EH, Gambert SR, Sepic SB, Mollinger LA (1985) Age-related differences in knee muscle strength in normal women. J Gerontol 40:275–280
10. Sepic SB, Murray MP, Mollinger LA, Spurr GB, Gardner GM (1986) Strength and range of motion in the ankle in two age groups of men and women. Am J Phys Med 65:75–84
11. Perera S, Patel KV, Rosano C et al (2016) Gait speed predicts incident disability: a pooled analysis. J Gerontol A Biol Sci Med Sci 71:63–71
12. McGibbon CA (2003) Toward a better understanding of gait changes with age and disablement: neuromuscular adaptation. Exerc Sport Sci Rev 31:102–108
13. Perry J, Burnfield JM (2010) Gait analysis: normal and pathological function. SLACK Incorporated
14. Zoss A, Kazerooni H (2006) Design of an electrically actuated lower extremity exoskeleton. Adv Robot 20(9):967–988
15. Wang ZX (2010) Raytheon launched XOS2 second generation exoskeleton device. Light Weapons 23:24–44
16. Spencer A (2010) The report of production display for "HULC" made in Lockheed Martin. Available: http://www.lockheedmartin.com. Accessed May 2013
17. Veneman JF, Ekkelenkamp R, Kruidhof R, et al (2005) Design of a series elastic- and Bowden cable based actuation system for use as torque-actuator in exoskeleton-type training. In: Proceedings of the 9th international conference on rehabilitation robotics, 28 June–1 July, Chicago, IL, USA
18. Berkelman P, Rossi P, Lu T, Ma J (2007) Passive orthosis linkage for locomotor rehabilitation. In: Proceedings of the 2007 IEEE 10th international conference on rehabilitation robotics, June 12–15, Noordwijk, The Netherlands
19. Collins SH, Wiggin MB, Sawicki GS (2015) Reducing the energy cost of human walking using an unpowered exoskeleton. Nature 522
20. Hong MB, Shin YJ, Wang JH (2014) Novel three-DOF ankle mechanism for lower-limb exoskeleton: kinematic analysis and design of passive-type ankle module. In: IEEE/RSJ international conference on intelligent robots and systems (IROS 2014) Sept 14–18, 2014, Chicago, IL, USA
21. Umberger BR, Gerritsen KGM, Martin PE (2003) A model of human muscle energy expenditure. Comput Methods Biomech Biomed Eng 6(2):99–111

Harvesting Energy from Chaotic Vibration

Jin Xie, Jian-qi Zhang, and Zhao-hui Liu

Abstract This paper focuses on harvesting energy from chaotic vibration by attaching piezoelectric patches at the end of a cantilever beam. For more efficiently harvesting energy, the cantilever beam should be bistable and the majority of the trajectories should be of large amplitude orbit. The optimization problem of parameters is discussed.

Keywords Harvesting energy · Chaotic vibration · Piezoelectric cantilever beam

1 Introduction

Vibrations arising from mechanical engineering are classified into three catalogues: regular vibration, random vibration and chaotic vibration. A well-known example of regular vibration is harmonic oscillation which is governed by determined dynamic equations; The random vibrations, in which no periodic trajectory can be found on a short time scale, are caused by some random components, whereas, chaotic vibration exhibits irregular and aperiodic trajectory under determined dynamic equations without any random components. The differences between the three kinds of vibration are detailed in Ref. [1].

J. Xie (✉) · J. Zhang · Z. Liu
School of Mechanical Engineering, Southwest Jiaotong University, Chengdu, China
e-mail: xj_6302@263.net

J. Zhang
e-mail: 1064245151@qq.com

Z. Liu
e-mail: zhhliu@home.swjtu.edu.cn

D. Sen et al. (eds.), *Mechanism and Machine Science*, Lecture Notes in Mechanical Engineering, https://doi.org/10.1007/978-981-15-4477-4_12

In the field of mechanism analysis and designing, chaotic motion has attracted much attention since later of the last century. Chaotic motions are observed in linkages [2–5], cams [6, 7], gears and gear trains [8–12]. Additionally, chaos can also be observed in large-sized mechanical equipment, such as vehicle [13], rotor/bearing [14], and wind turbine [15], and same kinds of the mechanical system in which there are some nonlinear elements, for examples, nonlinear elastic or spring elements, nonlinear dapping, such as stick-slip friction, backlash, play, or bilinear springs.

Vibration energy harvesting, which generates electrical energy from ambient vibration, is considered a key technique in the development of autonomous sensing nodes and microsystem with an extended lifetime in unattended operation [16]. Due to the availability of piezoelectric materials of good performances and ease of application, piezoelectric energy harvesting has received the most attention [17, 18].

Typically, a piezoelectric energy harvester is constructed by attaching piezoelectric patches at the end of a cantilever beam clamped onto the vibrating host part. In order to remedy the limitations on the broadband performance characteristics of energy harvester based on resonant interaction, intentional introduction of nonlinear into the design of energy harvester is done by attaching one magnet to the tip of cantilever beam while another magnet is fixed in the reference frame [19]. Such piezomagnetoelastic structure usually has two potential minima [20], which are named as double-well potential [21], and more, the harvester is bistable. Obviously, the oscillation of the cantilever beam in a bistable energy harvester that is determined by the external environmental excitation as well. Lots of researchers have reported their work on the performance of bistable energy harvester in response to harmonic excitation and random excitation. Under the high lever harmonic excitation, the cantilever beam yield large-orbit, and larger power output over a wide bandwidth. It is noted that the large-orbit branch is accompanied by a chaotic attractor and small branches of undesired intrawell oscillations [22]. While the responses of bistable energy harvester to random excitation are quite different from that to harmonic excitation. Under random excitation, the power generation of a bistable energy harvester can be enhanced by making it vibrate around both stable equilibrium points [23] or combining the harmonic excitation to the random force [24].

As yet, to the best of our knowledge, less research has been conducted on the performance of bistable energy harvester subjected to chaotic excitation. In this regard, this paper briefly presents exploratory work and addresses the case that a bistable energy harvester is attached onto the Duffing oscillator, which is the commonest example in nonlinear oscillation texts to demonstrate chaos. The rest of this paper is organized as follows: In Sect. 2, the model of Duffing oscillator with and without the energy harvester is presented, and the simulations are carried out. Based on the results, the performance of the system is discussed. In Sect. 3, the interwell jumping frequency ratio (IJFR) is introduced; based on which the optimization method for gaining the best harvesting efficiency is proposed in Sect. 4. Conclusions are drawn in Sect. 5.

2 The Bistable Energy Harvester for Chaotic Vibration

2.1 Setup of the Energy Harvester

As mentioned in the last section, there are many mechanisms and mechanical equipment exhibiting chaotic motion. For the sake of simplification, here we attach a bistable energy harvester onto the Duffing oscillator. The system is depicted in Fig. 1.

As a whole, it can be regarded as a vibrating system with three degrees of freedom. The base is driven by a harmonic force $F_e = A_0 \cos(\Omega\tau)$; Under some parameters of the harmonic excitation, the nonlinearity of the spring with cubic stiffness and negative linear stiffness leads the oscillator to chaotically vibrate; then the cantilever beam vibrates subjected to not only the chaotic excitation of the oscillator but the repulsive force acting between the magnets as well.

2.2 Modelling and Simulation

Duffing oscillator. The derivation of the dynamics equation for the Duffing oscillator is available in Ref. [25], it is not provided in this paper. For completeness of this paper, it is reproduced as the following

$$\begin{cases} \dot{x}_3 = x_4 \\ \dot{x}_4 = -2\xi x_4 + \delta x_3 - \eta x_3^3 - A_0 \cos(\Omega\tau) \end{cases} \tag{1}$$

Equation 1 is solved numerically using *ode45* command of MATLAB®. The bifurcation diagram of displacement of the oscillator with respect to the frequency of harmonic excitation is depicted in Fig. 2, in which it is shown that the oscillator is in chaotic vibration as $A_0 = 0.825$, and the frequency Ω being in the interval [0.84 0.98], [1.04 1.11] and [1.49 1.7].

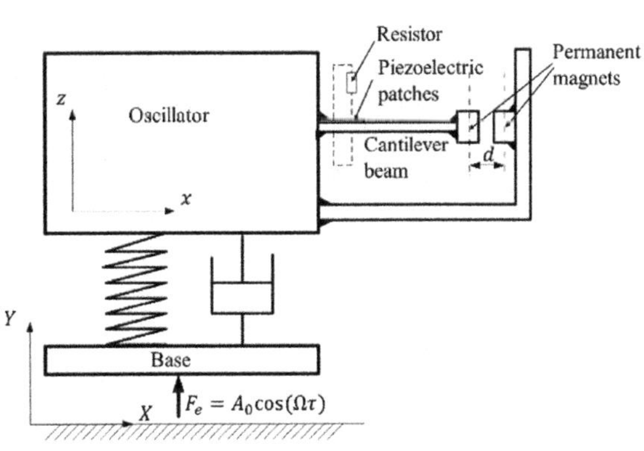

Fig. 1 Schematic of the bistable energy harvester attached onto the Duffing oscillator

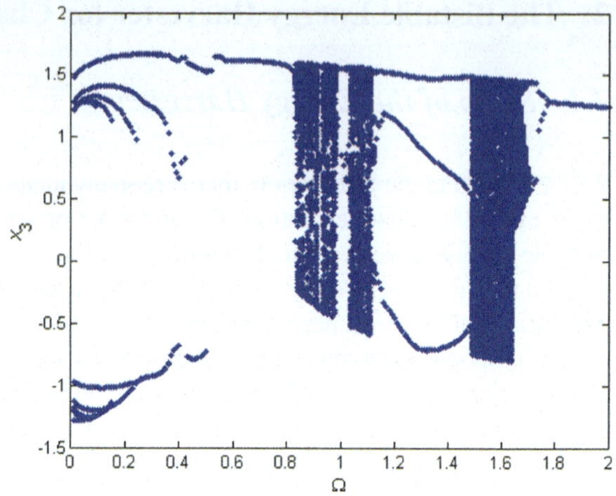

Fig. 2 Bifurcation diagram of displacement of oscillator w.r.t. frequency ($A_0 = 0.825$)

Duffing oscillator with energy harvester. Employing classical Euler-Bernoulli theory, Lagrange dynamic formulation, piezoelectric constitutive equation and Kirchoff's voltage law, the dimensionless dynamics equation of the system depicted in Fig. 1 can be written as [26]

$$\begin{cases} \dot{x}_1 = x_2 \\ \dot{x}_2 = -2\xi_1 x_2 - (1-\kappa_1)x_1 - \kappa_3 x_1^3 + \vartheta^2 x_5 + \mu x_4 - \beta_1 x_3 + \chi_1 x_3^3 \\ \dot{x}_3 = x_4 \\ \dot{x}_4 = -2\xi_2 x_4 + \beta_2 x_3 - \chi_2 x_3^3 - f_1 \varepsilon x_2 - f\cos(\omega\tau) \\ \dot{x}_5 = -x_2 - \alpha x_5 \end{cases} \tag{2}$$

Equation 2 describes a mechanical-electric-magnetic coupling inhere in the system. The first two equations result from the dynamics analysis of the cantilever beam, where κ_1 and κ_3 consider the contribution of the distance d shown in Fig. 1; the third and fourth equations result from that of the oscillator; and the fifth results from that of an electric circuit, where α considers the contributions of the resistor R_L.

The results of simulation using Eq. 2 support that even though there is a little change in the configuration of the oscillator, attaching the energy harvester to the oscillator has less impact on the motion of oscillator since the mass of the cantilever beam is much smaller than the mass of oscillator. Therefore, both the cantilever beam and the oscillator are in chaotic vibration simultaneously.

Due to that, the piezoelectric patch is attached to the clamped end of the cantilever beam, the vibration of a cantilever beam in the frame oxz affixed to the oscillator is most concerned. As depicted in Fig. 1, the configuration of the energy harvester is the same as that for harmonic or random excitation in most literature. But it is subjected to chaotic excitation, which is of broadband, and energy is distributed over a wide spectrum of frequencies. For more efficient transduction under the chaotic excitement, the energy harvester should be bistable at least [27]. To this end, the

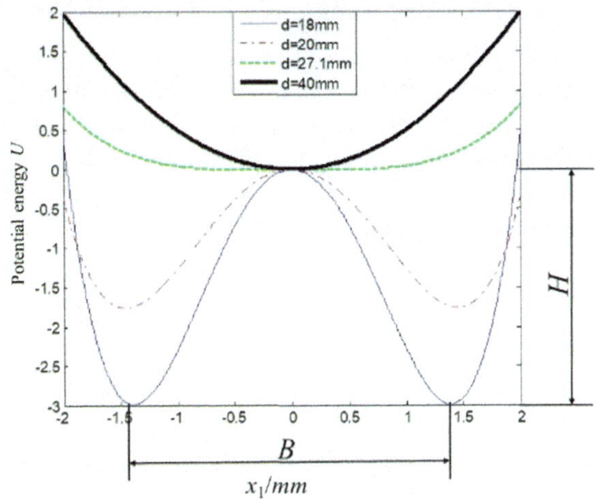

Fig. 3 Potential function for different distance d

parameters of the energy harvester should be carefully designed. For example, distance d shown in Fig. 1 should meet some kinds of requirements that are derived in the following procedure.

It starts with the analysis of the potential function of cantilever beam and magnetic field. From the first two equations in Eq. 2, the potential function can be written as

$$U = -(\kappa_1 - 1)\frac{x_1^2}{2} + \kappa_3\frac{x_1^4}{4} \tag{3}$$

where both κ_1 and κ_3 are the function of the distance d as previously mentioned. Taking different values of d, the potential energy can be calculated using Eq. 3, and the results are plotted in Fig. 3.

From Fig. 3, it can be seen that when $d \geq 27.1$ mm, the potential energy has unique minimal value at (0, 0), and when $d < 27.1$ mm, the potential energy has two minima, which form two potential wells separated by a potential barrier. Theoretically, the barrier corresponds to an unstable equilibrium point of system, and each well corresponds to a stable equilibrium point. With these evidences, we arrive at the conclusion that $d < 27.1$ mm is the necessary condition for the bistable energy harvester. Of cause, this value of d will be varied if the other parameters of the energy harvester is varied causing κ_1, κ_3 to be varied. In Fig. 3, the symbols, B and H, are used to indicate the separation distance between the wells and the height of potential barrier, respectively, when $d = 18$ mm.

3 Harvesting Efficiency of Chaotic Vibration

The two stable equilibrium points of the system derived from the potential function can be intuitively depicted in Fig. 4. The two stable equilibrium points correspond, respectively, to the two deflected positions of the cantilever beam where the repulsive magnetic force between the two permanent magnets is balanced by the longitudinal stiffness of the cantilever beam. And the unstable point is the symmetrical position of the cantilever beam.

The behaviour of the bistable energy harvester is such that the tip of the cantilever beam can either oscillate around one stable equilibrium point, visiting only one well, or jump between the two wells, depending upon not only the lever of the excitation, but values of the energy harvester parameters and the initial condition as well.

The primary interest in energy harvesting is the voltage output of the convertor connected to the piezoelectric patches across the resistor R_L. The average power harvested is given as

$$P = \frac{V_{\text{rms}}^2}{R_L} \tag{4}$$

where V_{rms} is the root mean square value of alternating-current voltage.

Assuming that the cantilever beam jumps between the two well under the exciting of chaotic vibration, the waveform of V_{rms} is certainly irregular, and the average power P maybe approaches to zero if the piezoelectric patch is connected to convertor with rectifying circuits [28]. Such a situation can be averted if the level of jumping between two wells is high enough [29]. This requires that not only cantilever beam works on large amplitude orbit, but has a higher percentage of such motion. The interwell jumping frequency ratio (IJFR) Φ is here introduced to represent the percentage.

Once jumping is counted when the tip displacement of the cantilever beam, x_1, goes from top to bottom, or vice versa, in the diagram of x_1 as a function of time, as depicted in Fig. 5, and once interwell jumping is counted when Δx_1 is larger than the separation distance between the wells, B. IJFR, Φ, is defined as

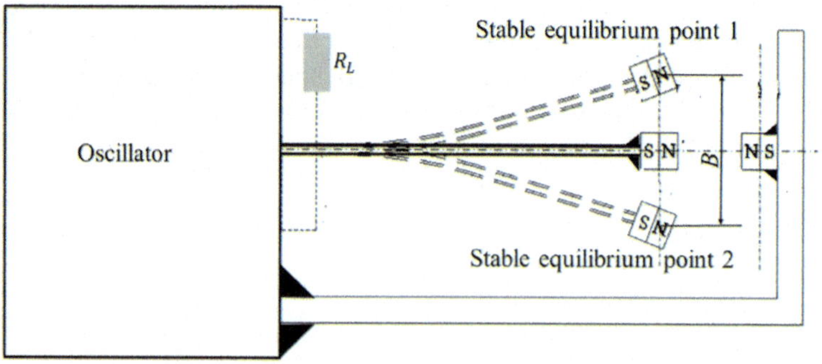

Fig. 4 The two stable equilibrium points of the energy harvester

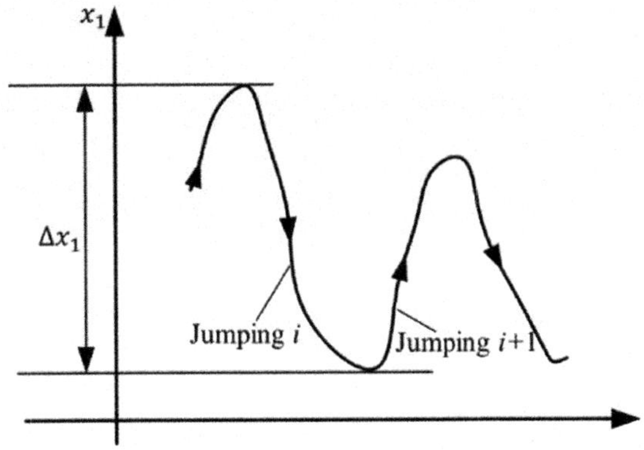

Fig. 5 Counting of Jumping

$$\Phi = \frac{\text{The Number of Interwell Jumping}}{\text{The Total Number of Jumping}} \times 100\% \tag{5}$$

Figure 6 shows the variation of IJFR Φ with respect to different distances d.

From Fig. 6, it can be seen that when $d \leq 17$ mm, Φ is equal to zero. Then $d > 17$ mm is the crucial condition for the bistable energy harvester to harvest chaotic vibration energy efficiently; When $d > 17$ mm, IJFR fluctuates, and arrives at its peak $\Phi = 95\%$ as $d = 23$ mm. This provides a clue for optimizing the parameters of the system for higher harvesting efficiency. To help understand what happens as $d = 23$ mm, the time series displacement of the tip of the cantilever beam and its phase portraits are presented in Fig. 7.

Figure 7 shows that the absolute majority of the trajectories are of large amplitude orbit even though the initial point is close to the stable equilibrium point 1, labelled as well 1 in Fig. 7b. This supports that under chaotic excitement, the bistable energy harvester can reach the highest harvesting efficiency when $\Phi = 95\%$, which will be involved in the objective function for optimization of the energy harvester in the next

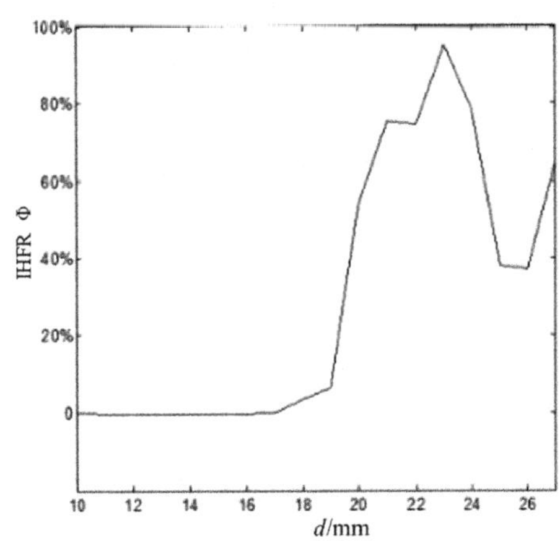

Fig. 6 IJFR for different values of d

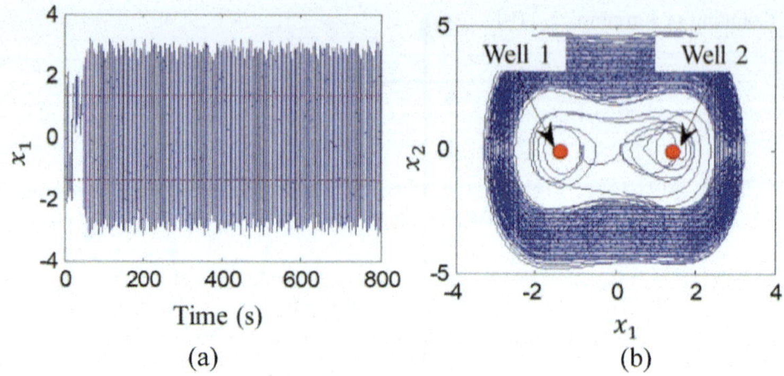

Fig. 7 Time series displacement and phase portraits of tip of the cantilever beam for $d = 23$ mm

section. The advantage of this measure to enhance harvest energy efficiency over the technique to control the harvester working on a particular large amplitude orbit [30] is that it needs not to input any additional energy into the system.

4 Optimization Design of Bistable Energy Harvester for Chaotic Vibration

This section discusses the problem of the optimal design of the harvesting system.

The set of optimal objectives includes maximizing the average power harvested in Eq. (4) and satisfying IJFR $\Phi \geq 95\%$. Since the performance of the bistable energy harvester under the excitation of chaotic vibration depends upon many factors as mentioned in previous sections, there are many design parameters to be optimized. For simplicity, we choose, here, the distance d and resistor R_L as the variables. The constraints should be established based on the dynamics analysis. For example, based on the analysis in Sects. 2 and 3, the constraint for d can be written as $17 \leq d \leq 27.1$, and the constraint for R_L is $1 < R_L < 1000$ taking the real circumstance into consideration.

This optimization problem is complex and nonlinear. The solution can be obtained by employing Monte Carlo method. Firstly, generate 100 grid points under the limitation for each variables using unbiased random number generator; Then find out those points satisfying IJFR $\Phi \geq 95\%$ among the 100 grid points; Finally, find out the point not only satisfying IJFR $\Phi \geq 95\%$ but making the average power harvested maximized. The following will demonstrate the details of this procedure.

From Fig. 2, it has been seen that the oscillator is in chaotic vibration as $A_0 = 0.825$, and the frequency Ω being in interval [1.49 1.7]. With the values of d and R_L on each of grid points generated by unbiased random number generator and $A_0 = 0.825$, $\Omega \in [1.49\ 1.7]$, the simulations of dynamics Eq. 2 are conducted, and a cloud of points satisfying IJFR $\Phi \geq 95\%$ are obtained. The cloud of points as $A_0 = 0.825$, $\Omega = 1.7$ is presented in Fig. 8, where the points are assigned with blue.

Next using Eq. (4), the average power harvested, P, for each point satisfying IJFR $\Phi \geq 95\%$ can be obtained, furthermore, the largest value, $P_{\max}$, and the values of relevant points can be found, as depicted in Fig. 9.

With Fig. 9, one is able to find the maximum average power harvested and the values of relevant parameters of harvesting system. For example, as $A_0 = 0.825$, $\Omega =$

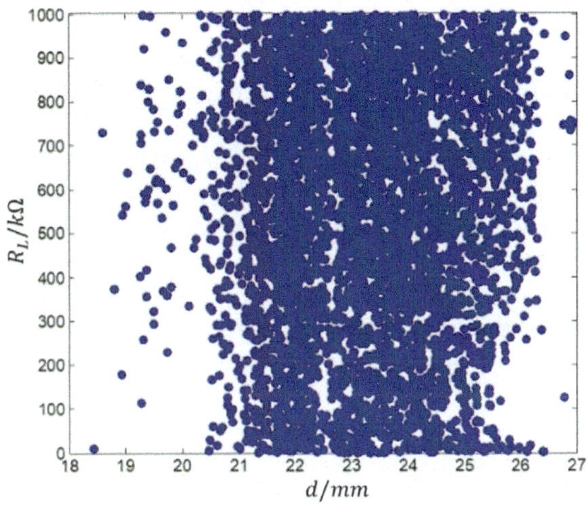

Fig. 8 Cloud of points satisfying IJFR $\Phi \geq 95\%$ as $A_0 = 0.825$, $\Omega = 1.7$

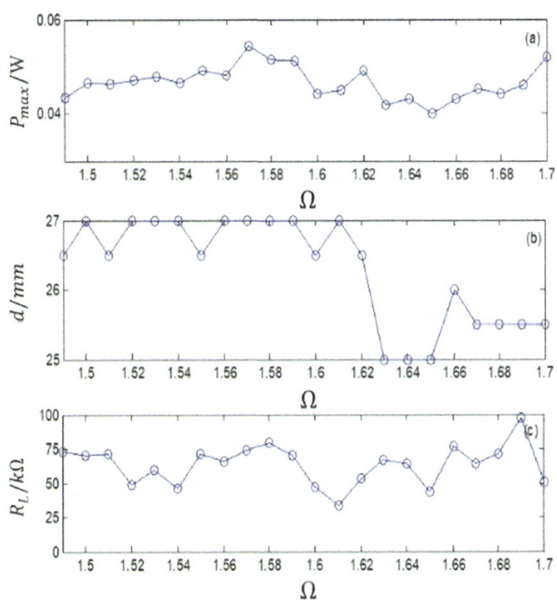

Fig. 9 Maximized average power harvested and relevant parameters w.r.t. frequency ($A_0 = 0.825$)

1.52, the maximized average power harvested can reach to 0.046(W) provided that $R_L = 50\ \text{k}\Omega, d = 27$ mm. In addition to this, it can be found, from Fig. 9, that when $A_0 = 0.825$, $\Omega \in [1.49\ 1.7]$, the value of d should be in the interval [25 27] (mm) and the value of R_L should be in the interval [45 80] (kΩ) for the purpose of gaining the best performance of the vibration energy harvester.

5 Conclusions and Future Work

This paper focuses on designing an energy harvester for chaotic vibration. The piezoelectric patches are attached at the end of a cantilever beam, which is clamped onto the Duffing oscillator, which exhibits chaotic motion.

Since chaotic vibration is irregular and the energy is distributed over a wide spectrum of frequencies, the energy harvester should be designed to meet the requirement that not only cantilever beam works on large amplitude orbit, but has a higher percentage of such motion. To this end, the harvester should be bistable and interwell jumping should be at a higher level. Interwall jumping frequency ratio (IJFR) is introduced, which can be regarded as a new index for harvesting efficiency. The optimization procedure based on the Monte Carlo method is illustrated.

Much work remains before the proposed system is implemented in a realistic harvesting environment. The system involves many parameters contributing to the performance of the system to a lesser or greater extent, a thorough study on these parameters is needed. The algorithm of optimization is time-consuming and needs to be further modified.

Acknowledgements The authors would like to acknowledge the financial support of NSFC (National Natural Science Foundation of China) under the Grant No. 51575457.

References

1. Kautz R (2011) Chaos the science of predictable random motion. Oxford University Press
2. Seneviratne LD, Earles SWE (1992) Chaotic behavior exhibited during contact loss in a clearance joint of a four-bar mechanism. Mech Mach Theory 27(3):307–321
3. Belato D, Weber HI et al (2001) Chaotic vibrations of a nonideal electro-mechanical system. Int J Solids Struct 38(10–13):1699–1706
4. Chang Z, Zhang C et al (2001) Bifurcation and chaos in linkage with a clearance. J Mech Strength 23(1):77–79
5. Wei D, Wang Y, Jiang T et al (2017) Chaos vibration of pinion and rack steering trapezoidal mechanism containing two clearances. Mech Syst Signal Process 92:146–155
6. Taborda JA, Santini S et al (2009) Active chaos control of a cam-follower impacting system using FPIC technique. Proc IFAC 42(7):327–332
7. Zhou J, Wang X et al (2015) Nonlinear dynamic characteristics of a quasi-zero stiffness vibration isolator with cam–roller–spring mechanisms. J Sound Vib 346:53–69

8. Chang-Jian C-W, Chang S-M (2011) Bifurcation and chaos analysis of spur gear pair with and without nonlinear suspension. Nonlinear Anal Real World Appl 12(2):979–989
9. Farshidianfar A, Saghafi A (2014) Identification and control of chaos in nonlinear gear dynamic systems using Melnikov analysis. Phys Lett A 378(46):3457–3463
10. Saghafi A, Farshidianfar A (2016) An analytical study of controlling chaotic dynamics in a spur gear system. Mech Mach Theory 96:179–191
11. Wang J, Zheng J et al (2012) An analytical study of bifurcation and chaos in a spur gear pair with sliding friction. Procedia Eng 31:563–570
12. Xiang L, Gao N et al (2018) Dynamic analysis of a planetary gear system with multiple nonlinear parameters. J Comput Appl Math 327:325–340
13. Zhu Q, Ishitobi M (2006) Chaotic vibration of a nonlinear full-vehicle model. Int J Solids Struct 43(3):747–759
14. Ehich FF (1991) Observation of chaotic vibration phenomena in high speed rotor dynamics. ASME J Vibrat Acoust 113:50–57
15. Barszcz AT et al (2015) Modelling of a chaotic load of wind turbines drivetrain. Mech Syst Signal Process 54–55:491–505
16. Hadas Z, Singule V et al (2010) Development of energy harvesting sources for remote applications as mechatronic systems. In: Proceeding of 14th international power electronics and motion control conference
17. Priya S (2007) Advances in energy harvesting using low profile piezoelectric transducers. J Electroceram 19:167–184
18. Cook-Chennault KA, Thambi N et al (2008) Powering MEMS portable devices: a review of non-regenerative and regenerative power supply systems with emphasis on piezoelectric energy harvesting system. Smart Mater Struct 17(4):043001
19. Wickenheiser AM, Garcia E (2010) Broadband vibration-based energy harvesting improvement through frequency up-conversion by magnetic rectification. Smart Mater Struct 19(6):065020
20. Eeturk A, Hoffmann J, Inman DJ (2009) A piezomagnetoelastic structure for broadband vibration energy harvesting. Appl Phys Lett 94(25):254102
21. Robinson RC (2004) An introduction to dynamical systems: continuous and discrete. Pearson Education, Inc
22. Eriture A, Inman DJ (2011) Piezoelectric energy harvesting. Wiley
23. De Paula AS, Inman DJ et al (2015) Energy harvesting in a nonlinear piezomagnetoelastic beam subjected to random excitation. Mech Syst Signal Process 54–55:405–416
24. Foupouapouognigni O, Nono Dueyou Buckjohn C et al (2018) Hybrid electromagnetic and piezoelectric vibration energy harvester with Gaussian white noise excitation. Phys A Stat Mech Appl 509:346–360
25. Guckenheimer J, holmes P (1983) Nonlinear oscillations, dynamical systems and bifurcations in vector field. Springer, New York
26. Zhang J-Q (2017) Design and analysis of energy harvesting with bistable piezoelectric cantilever beam for Duffing oscillator. M. E. Thesis, Southwest Jiaotong University, Chengdu, China (in Chinese)
27. Tran N, Ghayesh MH, Arjomandi M (2018) Ambient vibration energy harvesters: a review on nonlinear techniques for performance enhancement. Int J Eng Sci 127:162–185
28. Lan C, Tang L, Harne RL (2018) Comparative methods to assess harmonic response of nonlinear piezoelectric energy harvesters interfaced with AC and DC circuits. J Sound Vib 421:61–78
29. Daqaq MF, Masana R et al (2014) On the role of nonlinearities in vibratory energy harvesting: a critical review and discussion. ASME Appl Mech Rev 66:040801-1–040801-23
30. Geiyer D, Kauffman JL (2015) Chaotification as a means of broadband energy harvesting with piezoelectric materials. ASME J Vibrat Acoust 137:051005-1–051005-8

Parameter Estimation and Calibration of Parallel Mechanism Based Robots

Tanvi Verma, Gaurav Bhutani, T. A. Dwarakanath, and Dinesh Biswas

Abstract This paper deals with the calibration of a Hybrid Robot which is a combination of serial and parallel mechanisms having five degrees of freedom. In this paper, a new methodology has been introduced to determine the kinematics of such robots to ease the complexities. A reference frame has been defined on the moving platform which helps to bifurcate the hybrid mechanism for analysis. Calibration of the PKM is complex compared to serial robots because of the passive joints present in the PKM. The calibration of the parallel mechanism part of the hybrid robot has been presented here. Sensitivity analysis of the kinematic parameters of the PKM has been performed to identify the most significant parameters to influence change in the joint variable. Generalized calibration algorithm has been developed using the least square approximation method which requires less number of poses (only 8 poses). The calibration procedure presents an approach which does not involve any external measurement or calibrating devices. The experimental results are validated through a simulation model and hence can be effectively used to calibrate the hybrid robot.

Keywords Parallel kinematic mechanism · Parameter estimation · Sensitivity analysis · Calibration

T. Verma (✉)
Indira Gandhi Delhi Technical University for Women, New Delhi, India
e-mail: vtanvi13@gmail.com

G. Bhutani · T. A. Dwarakanath · D. Biswas
Division of Remote Handling & Robotics, BARC, Mumbai 400085, India
e-mail: bhutani@barc.gov.in

T. A. Dwarakanath
e-mail: tad@barc.gov.in

D. Biswas
e-mail: dcb@barc.gov.in

D. Sen et al. (eds.), *Mechanism and Machine Science*, Lecture Notes in Mechanical Engineering, https://doi.org/10.1007/978-981-15-4477-4_13

1 Introduction

Parallel Kinematic Mechanisms (PKMs) have significant advantages over serial robots because of their unique capabilities like high flexibility, high stiffness, elevated accuracy, dynamic loading, precision and low cost of manufacturing. Several PKM structures have been proposed for various applications like machine tools, pick and place high-velocity manipulators, flight simulators, medical robots, etc. and the application space is expanding every year. These applications possess some strict requirements in terms of accuracy and high speed [1, 2]. So, absolute positioning accuracy is an important issue in utilizing the PKMs in many of the applications. After manufacturing the parts and undertaking assembly of the robotic mechanism, the actual dimensions of the mechanism differ from theoretical dimensions. All the dimensions of kinematic parameters cannot be measured after assembly for accounting the difference in the kinematic equations. Also, the theoretical kinematic models are built with the notion of exactness in constraints. In practice the resulting assembly deviates from the theoretical constraints. To accommodate these changes, the practical (real) model may have to introduce additional parameters and may have to introduce finite small values to account the change from the ideal assumed parameters. The additional parameters and dimensional deviations alter the positional accuracy of the robot. Hence the calibration is needed. It has been well understood that kinematic calibration of PKMs is a practical and economical method to enhance positional accuracy. The implementation of the calibration process can be carried out through four sequential steps of modeling, measurement, parameter identification, and implementation in such a way that the kinematic model matches completely with the real model. The measurement of dimensions of PKMs through ball bar or CMM after iteration of calibration is time consuming and vulnerable to human errors. Thus, parameter identification becomes an important step in context of time and cost without compromising the accuracy of the real system.

In the recent past, several kinematic calibration approaches have been developed to improve the accuracy of PKMs. Han Sung Kim [3] used direct kinematics to calibrate the 3 DOF parallel robots and improved accuracy from 534 to 128 μm which is still not an acceptable accuracy for such high-speed robots because it hampers repeatability at a later stage. Shaoping Bai et al. [4] described a method to calibrate the medical parallel robot by developing an error model and measurement by optical sensors, where accuracy to reach a pose was 0.1 mm after calibration. Tian Huang et al. [5] proposed a method of creating an identification matrix of the geometric parameters for checking the flatness of plane and straightness of the two orthogonal axes parallel to the plane of the platform. This method was applicable to PKMs having less than 6 DOF by checking 25 poses in the first level of calibration. Many researchers have worked on two basic approaches like least square approximation method and Ridge estimation method for calibration of geometric parameters. However, these methods are applied in less than 6 DOF parallel robots [6]. Researchers used least square approximation method by checking the results for 50 poses, which was laborious and highly time consuming [7]. A control algorithm of calibration was developed by them to calibrate the parallel robots which inculcated both inverse and direct kinematics of robots using high-frequency optical sensors and CMM [8, 9]. However, for some of the parallel robots, it is complicated to determine the direct kinematics. So, this is not the efficient way to calibrate parallel robots [10]. Pujun Bai et al. [11] placed LASER sensors on the platform of the parallel robot to avoid random location errors using non-linear optimization technique. Maurine et al. [12] used distance-based approach to calibrate the delta robot. It was only valid for 3 DOF mechanisms. Many researchers worked on the error compensation approach for calibration of parallel mechanisms. This calibration approach was mainly applied to 6 DOF Hexaglide parallel manipulators wherein 42 geometric parameters were improved by applying least square method on 25 configurations. The authors used LASER tracker which increased the overall cost of manipulator; but positional accuracy of 0.03 mm was achieved after 7 levels of calibration process [13]. Parameter identification is very crucial but it takes lot of effort and is a very time-consuming step of calibration [14]. Several researchers used fuzzy logic and genetic algorithms to identify the actual values of the parameters; but these algorithms depend only on the specific design of the robot and are not generalized [15, 16].

This paper deals with the improved methodology for calibration of a 5 DOF Hybrid Robot as shown in Fig. 1. A hybrid robot is a combination of a serial and a parallel robot. A parallel module is a spatial parallel kinematic mechanism and has three degrees of freedom. The serial link has two degrees of freedom. In a parallel mechanism module, the legs are connected through universal joints with a

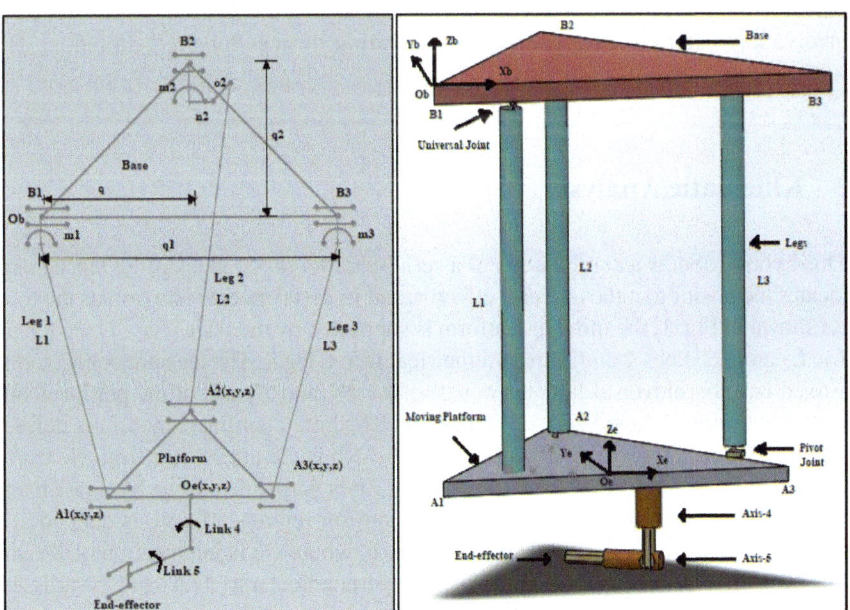

Fig. 1 3D model of hybrid mechanism

fixed base and through pivot joints at the platform side. The construction is such that the mobility of the mechanism is hard to determine as Kutzbach or Grubler's criteria cannot be applied due to the redundancy issue. Neither DKP nor IKP of the hybrid system is simple. Resorting to either Direct Kinematic Problem (DKP) or Inverse Kinematic problem (IKP) may not be possible. To ease the complexities of the system, a methodology is formulated wherein a reference frame defined at a point "Oe" is introduced on the moving platform. This reference frame helps to divide the hybrid mechanism for analysis into two steps. This paper presents the methodology designed to determine the kinematics of such a robot. Inaccuracy in the serial robot is caused due to accumulation of errors in each of its links whereas inaccuracy in the parallel robot is caused due to the averaging of errors. It is always hard to determine the reasons behind the errors in a parallel robot. Therefore, the second section of this paper deals with the sensitivity analysis of the geometric parameters present in the parallel module on the joint space accuracy. The impact of every parameter on the kinematic transformation needs to be determined before parameter identification. The third section presents the methodology to perform the parameter estimation for calibration using the least square approximation method. The results of the calibration procedure are validated using a simulated model developed in a 3D modeling software. Since all the kinematic parameters cannot be estimated in one step, the stage-wise parameter estimation and calibration procedure are established. The fourth section presents the trajectories having 8 poses, which calibrates the robot in minimum time. In the last section, the calibration results and its validation have been presented. Joint space accuracy is shown to be improved after using calibrated kinematic parameters. Therefore, the paper presents an approach, which does not involve any external measurement or calibrating devices for the calibration. The method is cost and time efficient.

2 Kinematic Analysis

This hybrid robot is a combination of a serial-parallel robot, the parallel mechanism locates the position of the tool end-effector, and its serial mechanism orients the tool. As shown in Fig. 1, the moving platform is supported by three legs having leg length L_1, L_2 and L_3. Legs 1 and 3 are symmetrical w. r. t. leg 2. The legs are connected to a fixed base by universal joints denoted as B_1, B_2, and B_3 and at the platform side with pivot joints denoted as A_1, A_2, and A_3. The base coordinate system is defined as Ob (Xb, Yb, Zb), where Ob is the point B_1. Xb is the unit vector from B_1 to B_3. Yb is perpendicular to B_1B_3 and towards B_2. Zb is perpendicular to Xb and Yb and pointing upwards. As discussed earlier, a platform reference frame is defined at a point "Oe (Xe, Ye, Ze)" as shown in Fig. 2a, b, where Oe is intersection of Xe and Ye, Xe is the unit vector from A_1 to A_3, Ye is perpendicular to A_1A_3 and towards A_2, Ze is perpendicular to Xe and Ye and pointing upwards. The serial module has two links, namely, Axis-4 and Axis-5 which gives two rotations. The tool end-effector is attached to serial links. q, q_1, and q_2 are used to determine the leg connection

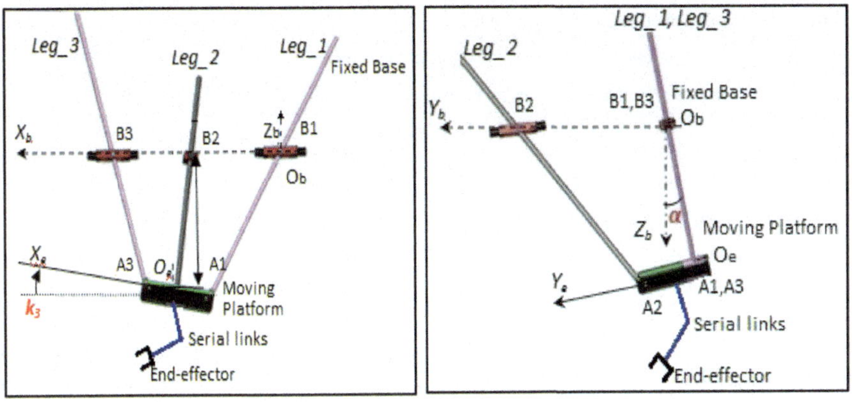

Fig. 2 **a** Front view of the hybrid robot and **b** right view of the hybrid robot

point at the base. m_1, m_2, and m_3 are the offsets between two axis rotation of the universal joint on legs 1, 2, and 3. To incorporate for the errors in the leg connection points, leg 2 offsets n_2 and o_2 are introduced which are in Xb and Yb axes direction, respectively. Legs L_1, L_2, and L_3 are having active prismatic joints due to which the mechanism can be positioned at various positions. Also, L_2 has an additional passive twist joint.

The definition of the point "Oe" is made on the assumption that the passive joint axis passing through A_1 and A_3 are in line with each other and are perpendicular to the passive joint axis at A_2. The reference frame defined at "Oe" is made common between the parallel and serial portion of the hybrid robot. This reference frame helps to divide the hybrid mechanism for the analysis. This leads us to solve the inverse kinematics of the parallel portion and direct kinematics of the serial portion on the hybrid mechanism. Thus, we can easily determine the joint space and the tool tip point at the specified location (Fig. 3; Table 1).

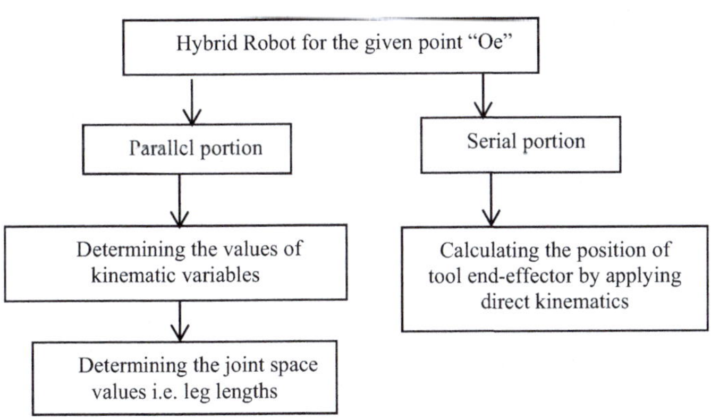

Fig. 3 Block diagram of the procedure of finding kinematics of the Hybrid robot

Table 1 Variables used in inverse kinematics

Variables	Unit	Physical meaning
α	Radian	The angle between A_1A_3 and the plane XbYb
$k3$	Radian	The angle between B_1A_1 and the plane XbYb

For the parallel portion, the base connection points of all the three legs w. r. t. base frame are given by B_i where $i = 1, 2, 3$. The leg connection points of all the three legs at the platform side are given by A_i where $i = 1, 2, 3$. The posture of the point "Oe" w. r. t. base frame is known and the objective is to determine the corresponding joint space values (L_i), i.e., leg lengths of the three legs and the corresponding passive joints values (Δ_{ij}). The variables (α, $k3$) are used to define the relation between the base frame and the platform frame "Oe". The methodology to solve the inverse kinematics is (a) to calculate the variables from the given posture of point "Oe", (b) to obtain the relation of the frame "Oe" w. r. t. the base frame from the given posture of point "Oe", and (c) to calculate the leg connection points at the platform side w. r. t. the base frame and thus calculate the leg lengths for the given point "Oe". The detailed procedure is described below:

The relationship of Frame "Oe" w. r. t. the base frame is fully described with the help of variables and the known point "Oe" and the variables are also defined using the given point "Oe". It can be shown as

$$ {}^{\text{base}}_{\text{Oe}}T = \begin{bmatrix} Ck3 & 0 & Sk3 & \text{Oe}_{x\_\text{base}} \\ S\alpha.Sk3 & C\alpha & -S\alpha.Ck3 & \text{Oe}_{y\_\text{base}} \\ -C\alpha.Sk3 & S\alpha & C\alpha.Ck3 & \text{Oe}_{z\_\text{base}} \\ 0 & 0 & 0 & 1 \end{bmatrix} \tag{1} $$

where

$\text{Oe}_{x\_\text{base}}$, $\text{Oe}_{y\_\text{base}}$ and $\text{Oe}_{z\_\text{base}}$ are the x, y, and z coordinates of the given point "Oe",

α and $k3$ are the variables,

S represents the sine function,

C represents the cosine function,

Ob represents the base frame,

Oe represents the Oe frame present on the platform,

$$ \alpha = \tan^{-1}\left(\frac{-{}^{\text{base}}_{y}\text{Oe}}{{}^{\text{base}}_{z}\text{Oe}}\right) \tag{2} $$

To define the value of variable $k3$, one assumption has to be made that B_2A_2 Oe is in the same plane which is perpendicular to Xe, which means B_2 Oe $\perp$ Xe.

$$ k3 = \tan^{-1}\left(\frac{m_1}{m}\right) \tag{3} $$

where
$m_1 = (\mathrm{Oe}_{x\_\mathrm{base}} - B_2 x)$,
$m = (\mathrm{Oe}_{z\_\mathrm{base}} - B_2 z).\ \mathrm{C}\alpha - (\mathrm{Oe}_{y\_\mathrm{base}} - B_2 y)\ \mathrm{S}\alpha$.

After calculating the variables and transformation matrix, we can determine the leg connection points at the platform side w. r. t. the base frame (${}^{\mathrm{base}}_{A1,A2,A3}P$) and thus calculate the leg lengths:

$$ {}^{\mathrm{base}}_{A1,A2,A3}P = {}^{\mathrm{base}}_{\mathrm{Oe}}T \cdot {}^{\mathrm{Oe}}_{A1,A2,A3}P \tag{4}$$

The leg lengths (L_i) of all the 3 legs are determined by:

$$ L_i = |B_i - A_i| \quad \text{where } i = 1, 2, 3 \tag{5}$$

Now, for the serial portion, point "Oe" is the input task space. It is relatively easy to determine the position of the end-effector w. r. t. the base using forward kinematics. In this hybrid robot, there are two rotation movements of the two axes (axis-4 and axis-5) and one translation movement. Axis-4 is having rotation about Z-axis (Θ_4). Axis-5 is having rotation about X-axis (Θ_5). Then a fixed translation movement takes place from axis-5 towards the tool end-effector. After applying forward kinematics, the position of the end-effector w. r. t. the base for the given point "Oe" can be expressed as

$$ {}^{\mathrm{base}}_{\mathrm{tool}}P = {}^{\mathrm{base}}_{\mathrm{Oe}}T \cdot {}^{\mathrm{Oe}}_{\mathrm{Axis-4}}R \cdot {}^{\mathrm{Axis-4}}_{\mathrm{Axis-5}}R \cdot {}^{\mathrm{Axis-5}}_{\mathrm{tool}}T \tag{6}$$

Hence, the kinematics of the hybrid robot is easily understood by the geometry of the robot. The combination of the forward and inverse kinematics can easily define the joint space and position of the end-effector.

3 Calibration Algorithm

Parallel mechanisms are generally regarded as being highly accurate due to non-cumulative joint errors but to remove the cumulative errors due to passive joints is itself a big task compared to serial mechanisms [17]. So, parameter estimation of PKMs needs more time and better techniques as compared to serial mechanisms. Here, parameter estimation is a very crucial step while calibrating any kind of robot. Calibration is a combination of sensitivity analysis and parameter estimation.

In this hybrid robot, there are 17 geometric parameters which can influence the accuracy of the tool end-effector. To understand the nature of the parameters on the joint space, we first do the sensitivity analysis. After understanding the sensitivity of every parameter, the parameter estimation algorithm using the least square approximation method is applied. This is the classical approach for estimating the values of the parameters. The reason behind choosing this technique is that (a) it does not require any velocity and time measurement for any poses, (b) it only needs partial differentiation values for any pose which is easy to determine with the help of software module in a few seconds, (c) it does not need any measurement devices and sensors fitted with the mechanism, (d) it only needs engineering drawing values of parameters, and (e) it is simple to understand and results obtained are nearly 98% accurate.

3.1 Sensitivity Analysis

Sensitivity analysis is the study of how the uncertainty in the output (joint space) of a mathematical model or system (numerical or otherwise) can be apportioned to different sources of uncertainty in its inputs (kinematic parameters). In case of calibrating models with a large number of parameters, a primary sensitivity test can ease the calibration stage by focusing on the sensitive parameters. This analysis is performed by putting all the kinematic parameter values same as the theoretical values except for one kinematic parameter. The 17 parameters which influence the joint space accuracy are, namely, q, q_1, q_2, m_1, m_2, m_3, n_2, o_2, $A_1(x, y, z)$, $A_2(x, y, z)$, and $A_3(x, y, z)$ (as shown in Fig. 2).

Generally, it is concluded that it is not necessary that all the geometric parameters influence all the joint space. Leg lengths (L_1, L_2, and L_3) are a function of several geometric parameters (17 in this case) and the given point "Oe".

Influence of all geometric parameters is summarized in Table 2 and graphs are generated to demonstrate the impact of parameters on the leg lengths to achieve a desired trajectory.

Graphs in Figs. 4, 5, and 6 show the impact of deviation in the kinematic parameters on the particular leg lengths while the platform is moving in a trajectory along the X-axis. The blue-colored curves show the nominal values of the leg lengths (based on theoretical values) and the red colored curves show the actual values of the leg lengths (based on the deviated parameters). All dimensions are in mm.

Case—1 Impact of deviation in the kinematic parameters on leg length L_1 while platform is moving along the X-axis

The parameters and its contribution in affecting the leg lengths L_1 are mentioned in Table 3.

Case—2 Impact of deviation in the kinematic parameters on leg length L_2 while platform is moving along the X-axis

The parameters and its contribution in affecting the leg lengths L_2 are mentioned in Table 4.

Case—3 Impact of deviation in kinematic parameters on leg length L_3 while the platform is moving along the X-axis

Table 2 Influence of geometric/kinematic parameters on the leg lengths

L_i	m_1	m_2	m_3	q_1	q	q_2	n_2	o_2	A_{1x}	A_{1y}	A_{1z}	A_{2x}	A_{2y}	A_{2z}	A_{3x}	A_{3y}	A_{3z}
L_1	√				√	√	√	√	√	√	√		√				
L_2		√			√	√	√	√				√	√	√			
L_3			√	√	√	√	√	√					√		√	√	√

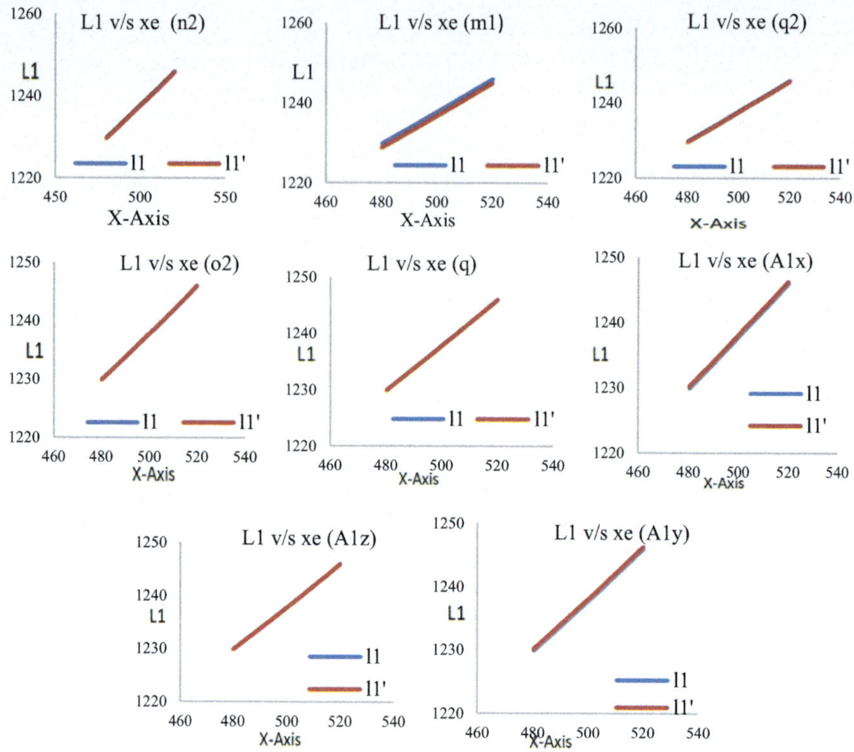

Fig. 4 Graphs showing the impact of various kinematic parameters on leg length L_1 while platform is moving along the X-axis

The parameters and its contribution in affecting the leg lengths L_3 are mentioned in Table 5.

The observations from the graphs are summarized as below:

The graphs in the Figs. 4, 5 and 6 illustrate that m_1, m_2, and m_3 reduce L_1, L_2, and L_3, respectively. A positive change in $A_1(x, y, z)$ increases L_1 and vice versa. A positive changes in $A_3(x, y, z)$ decreases L_3. A_2x has negligible effect on any of the leg lengths. A positive change in A_2z decreases L_2. A positive change in q_1 increases L_3 and vice versa. All the leg lengths are influenced by q, q_2, A_2y, n_2, and o_2. A positive change in n_2 decreases L_1 and L_2 but increases L_3. There is negligible effect on leg lengths due to any change in o_2. For a positive change in q, L_1 is increased but L_2 and .. are decreased. A positive change in q_2 decreases L_1 but increases L_2 and L_3. For a positive change in A_2y, L_1 is increased but L_2 and L_3 are decreased.

After the above mathematical analysis and observations, the joint space functions (leg lengths) can be written as a function of the following kinematic parameters:

$L_1 = f\ (m_1, A_1x, A_1y, A_1z, q, q_2, n_2, o_2, A_2y,$ Point Oe).

$L_2 = f\ (m_2, A_2x, A_2y, A_2z, q, q_2, n_2, o_2,$ Point Oe).

$L_3 = f\ (m_3, q_1, A_3x, A_3y, A_3z, q_2, q, n_2, o_2, A_2y,$ Point Oe).

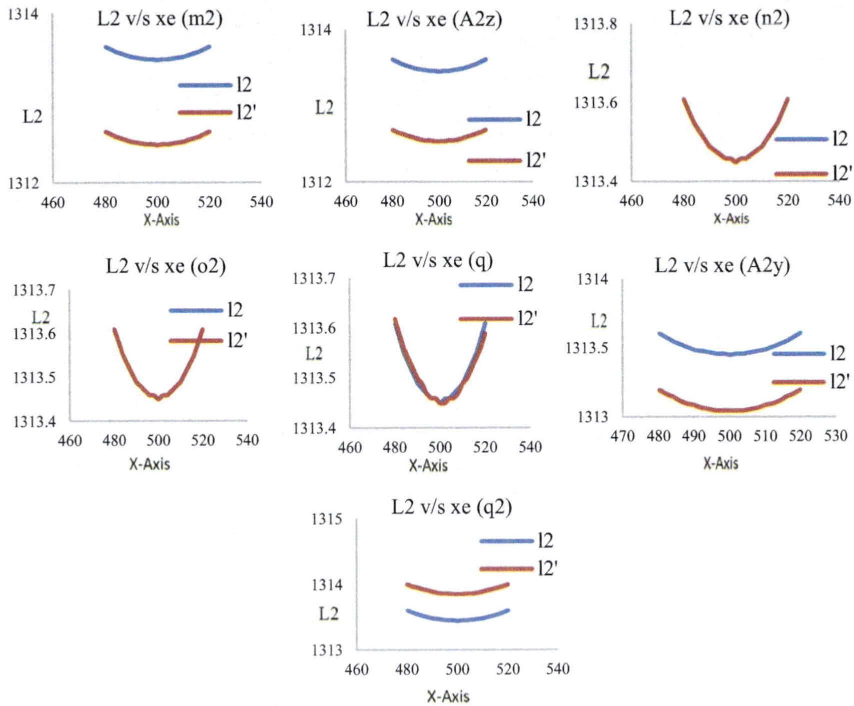

Fig. 5 Graphs showing the impact of various kinematic parameters on leg length L_2 while the platform is moving along the X-axis

3.2 *Parameter Estimation*

The least-squares algorithm is implemented for approximating a given function by a weighted sum of other functions. The best approximation can be defined as that which minimizes the difference between the original function and the approximation. It is a classical approach because it requires leg lengths for the given set of poses. For a least-squares algorithm, the quality of the approximation is measured in terms of the squared differences between the two values (actual and nominal). In this hybrid robot, the least square approximation algorithm is applied as follows:

The theoretical leg length is a function of the nominal values of kinematic parameters given in the engineering drawing and it can be expressed as

$$L_i = f(P_i) \tag{7}$$

The actual leg length is a function of the actual values of kinematic parameters and it can also be expressed as

$$L_i' = f(P_i') \tag{8}$$

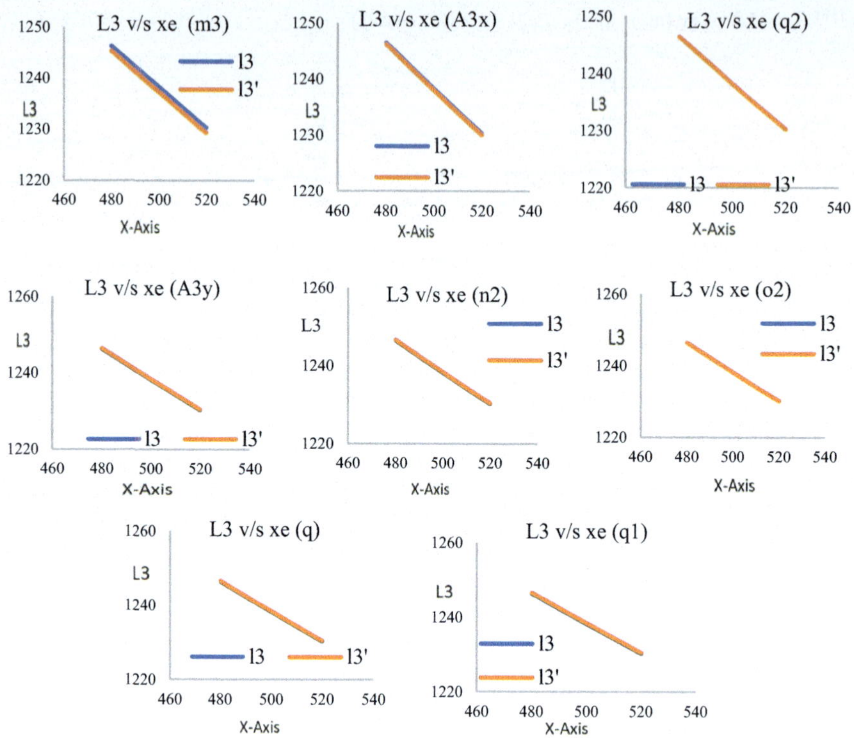

Fig. 6 Graphs showing the impact of various kinematic parameters on leg length L_3 while the platform is moving along the X-axis

Table 3 Nature of parameters affecting the leg length L_1

Nature	Parameters
Major parameters	m_1, A_1x, A_1z
Minor parameters	n_2, o_2, q, q_2, A_1y

Table 4 Nature of parameters affecting the leg length L_2

Nature	Parameters
Major parameters	m_2, A_2z, q_2, A_2y
Minor parameters	n_2, o_2, q

Table 5 Nature of parameters affecting the leg length L_3

Nature	Parameters
Major parameters	m^3
Minor parameters	n_2, o_2, q, A_3z, q_2, A_3y, A_3x

From the difference of Eqs. (7) and (8), we get;

$$(L_i' - L_i) = f(P_i + \Delta P_i) \tag{9}$$

where

L_i = leg length of leg i and $i = 1, 2, 3$ at nominal values of parameters.

P_i = nominal values of parameters of leg i given in the engineering drawing data.

L_i' = actual leg length of leg i and $i = 1, 2, 3$ at actual values of parameters during manufacturing.

P_i' = actual values of the parameters after manufacturing/assembly.

The generalized equation to determine the change in parameter in the hybrid robot is given as

$$\begin{bmatrix} \Delta L_1 \\ \Delta L_2 \\ \Delta L_3 \end{bmatrix} = \begin{bmatrix} \frac{\partial L_1}{\partial P_1} & \frac{\partial L_1}{\partial P_2} & \cdots\cdots\cdots\cdots & \frac{\partial L_1}{\partial P_n} \\ \frac{\partial L_2}{\partial P_1} & \frac{\partial L_2}{\partial P_2} & \cdots\cdots\cdots\cdots & \frac{\partial L_2}{\partial P_n} \\ \frac{\partial L_3}{\partial P_1} & \frac{\partial L_3}{\partial P_2} & \cdots\cdots\cdots\cdots & \frac{\partial L_3}{\partial P_n} \end{bmatrix} \begin{bmatrix} \Delta P_1 \\ \Delta P_2 \\ \cdots\cdot \\ \Delta P_n \end{bmatrix} \tag{10}$$

where

ΔL_i = difference between the measured and nominal (ideal) leg lengths.

$\frac{\partial Li}{\partial P_i}$ = partial differentiation values of the leg length w. r. t. any parameter at any point "Oe" considering theoretical values of parameters.

ΔP_i = change in the value of the kinematic parameter. This value is determined using *Mathematica* Software.

To analyze and validate the results of calibration, a 3D model of the mechanism having actual kinematic parameters is made in modeling software; and the leg lengths are measured at different coordinates of point "Oe". The calibration results converge to the actual kinematic parameters as in the 3D model. The stage-wise parameter estimation is carried out to calibrate all 17 kinematic parameters. Sometimes, due to different order of partial differentiation values, all parameters cannot be calibrated in one step. So, stages have been categorized. Hence, the three stages of calibration are as follows:

1. Ist stage—8 parameters like $m_1, m_2, m_3, q_1, A_1x, A_1z, A_3x$, and A_3z are calibrated using minimum 8 poses. As we know from sensitivity analysis, the nature of parameters while moving along the X and Z direction is nearly same and these parameters are influenced by only one leg length. Hence, calibration of these parameters is considered in the Ist stage of calibration. Two constraints are to be taken care of: (a) A_1z and A_3z are same after calibration and (b) m_1, m_2, and m_3 should be greater than zero or positive value.
2. IInd stage—sensitivity analysis of the parameters of leg 2 and the parameters along Y-axis are completely different. Their contribution is about 10^{-5} order (negligible) and it becomes highly sensitive to deal with them. So, the parameters like A_1y, A_2x, A_3y, and A_2z are considered for parameter estimation in the IInd stage analysis. The equations are decoupled and the order of the partial

differentiation values are quite negligible. So, 8 poses are required to calibrate these parameters. The constraint which has to be taken care of is that value of A_1y is equal to A_3y after calibration.

3. IIIrd stage—after calibrating the 12 parameters in the above two stages, if the error still exists in the leg lengths (between measured and calculated value) then 5 parameters, i.e., n_2, o_2, q, q_2, and A_2y are calibrated. Hence, only 8 poses are required to calibrate these 17 parameters.

4 Best Selection of Trajectory for Calibration

Calibration not only depends on the parameter estimation but also on the poses that are selected for calibrating the robot. The selection of poses is completely dependent on the nature of geometric parameters on the joint space values (leg lengths) while moving along the X, Y, and Z axes individually. Sometimes, the impact on the leg lengths while moving along the X and Z axes is much influential as compared to that of Y-axis. The selection of poses is a very important task. As a result, many trajectories have been analyzed for calibrating the kinematic parameters accurately. The following are some trajectories that are used while calibrating the robot with the help of a selection of poses:

1. Rectangular-based trajectory having eccentricity along X-axis:
   In this trajectory, 8 poses are selected in such a way that the corners of the rectangle and its mirror image about the XY plane with eccentricity along X-axis is as shown in Fig. 7a. Keeping the value of the Y and Z coordinates almost same, only the X-coordinate of the poses is largely influenced. As a result, only those parameters which are linearly dependent are correctly estimated. This selection criterion is only applicable for linearly dependent kinematic parameters. By using this trajectory in the calibration procedure, it is seen that the joint space accuracy improves from 1100 to 100 μm.
2. Cylindrical-based trajectory having eccentricity along X and Y axes:
   In this trajectory, poses are selected at the top and bottom circumference of the cylinder having eccentricity along X and Y axes as shown in Fig. 7b. It is concluded that a combination of slight change in X and Y coordinates of the poses only influences the linearly dependent parameters with some minimal deviation in the non-linear parameters. By using this trajectory in the calibration procedure, it is seen that the joint space accuracy improves from 1100 to 90 μm.
3. Spherical-based trajectory:
   In this trajectory, poses are selected at the surface of sphere of radius "r" as shown in Fig. 7c. In this, Y and Z coordinates of the poses are equally changed with gradual change in the X-coordinate. This trajectory estimates the linearly dependent parameters with 98% accuracy and non-linear parameters with 5–10% accuracy. By using this trajectory in the calibration procedure, it is seen that the joint space accuracy improves from 1100 to 50 μm.

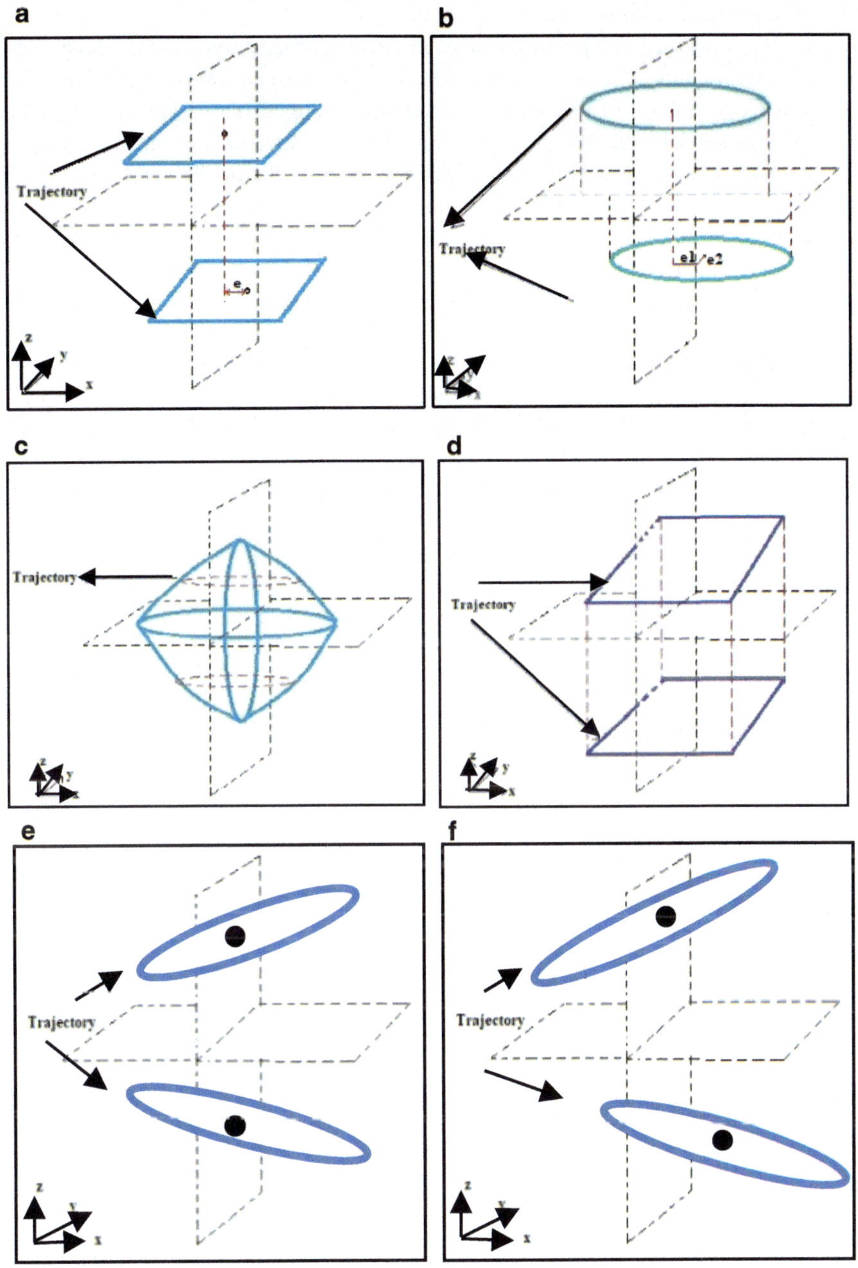

Fig. 7 **a** Rectangular-based trajectory. **b** Cylindrical-based trajectory. **c** Spherical-based trajectory. **d** Oblique parallelogram-based trajectory. **e** Oblique circular-based trajectory. **f** Oblique circular-based trajectory having eccentricity along X-axis

4. Oblique parallelogram-based trajectory:
   In this trajectory, the poses are selected at the corner points of the oblique parallelogram and its mirror image about the XY plane as shown in Fig. 7d. X, Y, and Z coordinates of the poses are gradually changed. This trajectory estimates the linearly dependent parameters with 98% accuracy and non-linear parameters with 15–20% accuracy. By using this trajectory in the calibration procedure, it is seen that the joint space accuracy improves from 1100 to 50–60 μm.
5. Oblique circular-based trajectory:
   In this trajectory, the poses are selected at the circumference of the oblique circle and its mirror image about the XY plane as shown in Fig. 7e. It is a combination of a gradual change in the X and Z coordinates of the poses with constant change in the Y-coordinate. This trajectory estimates the linearly dependent parameters with 98% accuracy and non-linear parameters with 95% accuracy. By using this trajectory in the calibration procedure, it is seen that the joint space accuracy improves from 1100 to 30–40 μm.
6. Oblique circular-based trajectory having eccentricity along X-axis:
   In this trajectory, the poses are selected at the circumference of the oblique circle and its mirror image about the XY plane with eccentricity along the X-axis as shown in Fig. 7f. Here, excessive change in X-coordinate of the poses estimates non-linearly dependent parameters with 99.9% accuracy and hampers the linear parameter estimation. By using this trajectory in the calibration procedure, it is seen that the joint space accuracy improves from 1100 to 90 μm.

Therefore, it is concluded that "Oblique circular-based trajectory" is the most suitable and effective trajectory for calibrating the parallel portion of the hybrid robot, achieving joint space accuracy of 30–40 μm.

5 Results

The calibration algorithm of the parallel portion of the hybrid robot is presented here. Using simulated model, tests are performed to validate this calibration algorithm with the help of a 3D solid model designed in solid modeling software. The sensitivity of the kinematic parameters is well analyzed to estimate the actual value of parameters. Minimum poses based on different trajectories (8 poses) are categorized to estimate the parameters in three stages analysis. Result obtained from the simulated model of the Ist stage analysis for estimating the parameters is shown in Table 6. Joint space accuracy (Leg lengths accuracy) has been improved from 6000 to 30–40 μm to reach the desired position in two calibration iterations only.

The results presented in Table 7 conclude that there is a drastic improvement in the deviation of leg lengths after calibration. This has also been verified by the simulated model.

Table 6 Result of the estimation of actual values of parameters in Ist stage calibration (using only 8 poses) obtained from the simulated model

Kinematic parameters	Engineering drawing values of kinematic parameters	Actual values of kinematic parameters (through least square analysis)
m_1	0	0.453733
m_2	0	1.00144
m_3	0	0.765878
q_1	1000.53299	1000.65639
A_3x	196.806	195.06067
A_1x	−194.806	−194.624766
A_1z	584	583.836
A_3z	584	583.386

All dimensions are in mm

Table 7 Improvement in joint space accuracy (using 8 poses only)

Initial error in leg lengths (mm) before calibration			After calibration, final error in leg lengths (mm)		
dL_1	dL_2	dL_3	dL_1'	dL_2'	dL_3'
−6.44	−2	3.53	0	−0.06	−0.03
−5.98	−1.9	3.09	0.01	−0.07	−0.03
−6.32	0.87	3.43	0	−0.04	−0.02
−5.92	0.74	3.01	0.01	−0.06	−0.03
−5.88	−1.86	2.95	0.01	−0.05	−0.04
−5.67	−1.82	2.75	0.02	−0.04	−0.04
−6.04	0.77	3.13	0.01	−0.07	−0.02
−5.82	0.7	2.9	0.01	−0.07	−0.03

6 Conclusion

This paper presented a calibration methodology that requires data like poses and leg lengths only. In this, kinematics with bifurcation of the parallel-serial robot has helped the calibration process to do parameter estimation without any ambiguity. The paper presented the sensitivity analysis, which gives an estimation of the kinematic parameters largely influencing the joint space parameters. The paper presented the best and suitable trajectory for calibrating this parallel portion of the hybrid robot. The joint space accuracy (Leg lengths accuracy) has remarkably improved from 6000 to 30–40 μm to reach a desired position. The results have been validated through a simulation model.

References

1. Bi ZM, Lang SYT, Orban P, Verner M (2007) An integrated design tool box for tripod-based parallel kinematic machines. ASME J Mech Des 129(8):799–807
2. Bi ZM, Jin Y, Gibson R, McTotal P (2009) Kinematics of parallel kinematic machine exechon. In: Proceedings of the 2009 IEEE international conference on information and automation, China
3. Kim HS (2005) Kinematic calibration of a cartesian parallel manipulator. Int J Control Autom Syst 3(3):453–460
4. Bai S, Teo MY (2003) Kinematic calibration and pose measurement of a medical parallel manipulator by optical position sensors. J Robot Syst 20(4):201–209 (2003)
5. Huang T, Chetwynd DG, Whitehouse DJ, Wang J (2005) A general and novel approach for parameter identification of 6-DOF parallel kinematic machines. Mech Mach Theory 40:219–239
6. Liu Y, Wu J, Wang L, Wang J (2016) Parameter identification algorithm of kinematic calibration in parallel manipulators. Adv Mech Eng 8(9):1–16
7. Zhang JF, Chen QH, Wu CY (2014) Kinematic calibration of a 2-DOF translational parallel manipulator. Adv Robot 28:707–714
8. Nahvi A, Hollerbach JM, Hayward V (1994) Calibration of a parallel robot using multiple kinematic closed loops. IEEE
9. Song Y, Zhang J, Lian B, Sun T (2016) Kinematic calibration of 5-D of parallel kinematic machine. Precis Eng 45:242–261
10. Majarena AC, Santolaria J, Samper D, Aguilar JJ (2011) Modelling and calibration of parallel mechanisms using linear optical sensors and a coordinate measuring machine. Meas Sci Technol 22
11. Bai P, Mei J, Huang T, Chetwynd DG (2016) Kinematic calibration of Delta robot using distance measurements. Proc IMechE Part C: J Mech Eng Sci 230(3):414–424
12. Maurine P, Dombre E (1996) A calibration procedure for the parallel robot delta 4. In: Proceedings of the 1996 IEEE international conference on robotics and automation Minneapolis, Minnesota
13. Guo J, Wang D, Fan R, Chen W, Zhao G (2017) Kinematic calibration and error compensation of a hexaglide parallel manipulator. Proc IMechE Part B: J Eng Manuf 1–11
14. Liu Y, Jun W, Wang L, Wang J (2016) Parameter identification algorithm of kinematic calibration in parallel manipulators. Adv Mech Eng 8(9):1–16
15. Bai Y, Zhuang H and Wang D (2008) Calibration of parallel machine tools using fuzzy interpolation method. In: Proceedings of the 2008 IEEE international conference on technologies for practical robot applications, Woburn, MA, 10–12 Nov 2008. IEEE, New York
16. Fan C, Zhao G, Zhao J (2015) Calibration of a parallel mechanism in a serial-parallel polishing machine tool based on genetic algorithm. Int J Adv Manuf Tech 81:27–37
17. Vischer P, Clavel R (1998) Kinematic calibration of parallel Delta Robot. Robotica 16:207–218

Modeling of Piezoelectric Energy Harvester for Medical Applications Using Intelligent Optimization Techniques

P. Mangaiyarkarasi and P. Lakshmi

Abstract Piezoelectric energy harvester is developed and optimized based on Intelligent Optimization Techniques (Bat algorithm (BA), Grey Wolf Optimization (GWO), and Hybrid Grey Wolf Optimizer tuned Bat Algorithm (HGWOBA)) for medical applications. The structure is modeled by utilizing analytical equations to identify the voltage and power generated by the piezoelectric device and to use them for medical applications. These analytical equations of voltage and power are utilized as the objective function of the intelligent techniques to enhance the harvested voltage and power by optimizing the piezoelectric parameters. In this work, pacemaker and hearing aids are chosen as the medical applications, for which, the piezoelectric device is modeled and designed with respect to intelligent optimization techniques. Therefore, the design variables of the piezoelectric harvester are optimized and designed for medical applications. Moreover, the harvesting efficiency, stress, frequency, and load impedance of the piezoelectric device is calculated. Finally, with the results obtained, a comparative study is analyzed between the optimization techniques. Among them, HGWOBA is more efficient in comparison to BA and GWO, which has improved the efficiency of the unimorph piezoelectric harvester from 48.78 to 68.96%, which is designed for medical applications.

Keywords Analytical equations · Fitness function · Hybrid grey wolf optimizer tuned bat algorithm · Pacemaker · Hearing aids

1 Introduction

Implantable medical devices have obtained an appreciable implication in enhancing the lifespan of patients. The devices that are presently utilized in various portions of the human body are cardiac pacemakers [1], hearing aids [2], cardioverter-defibrillator [3], intrauterine devices [4], etc. as illustrated in Figs. 1, 2, 3, and 4. The

P. Mangaiyarkarasi (✉) · P. Lakshmi
Anna University, CEG Campus, Chennai, India
e-mail: mangaipadma@gmail.com

P. Lakshmi
e-mail: p_lakshmi@annauniv.edu

D. Sen et al. (eds.), *Mechanism and Machine Science*, Lecture Notes in Mechanical Engineering, https://doi.org/10.1007/978-981-15-4477-4_14

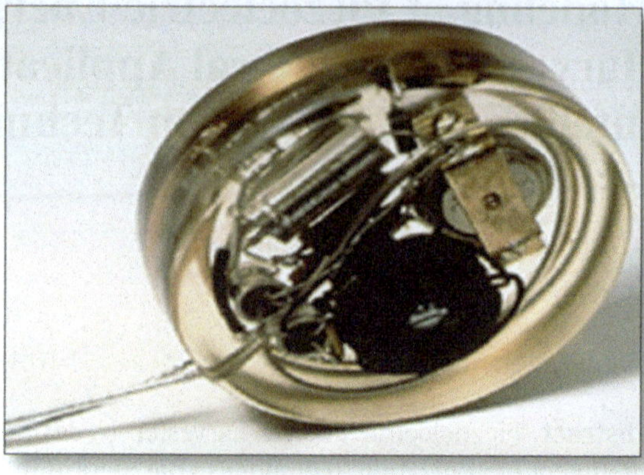

Fig. 1 First artificial cardiac pacemaker

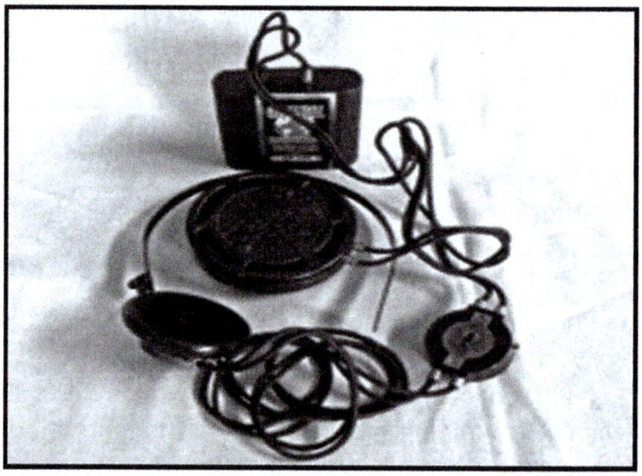

Fig. 2 Carbon hearing aid

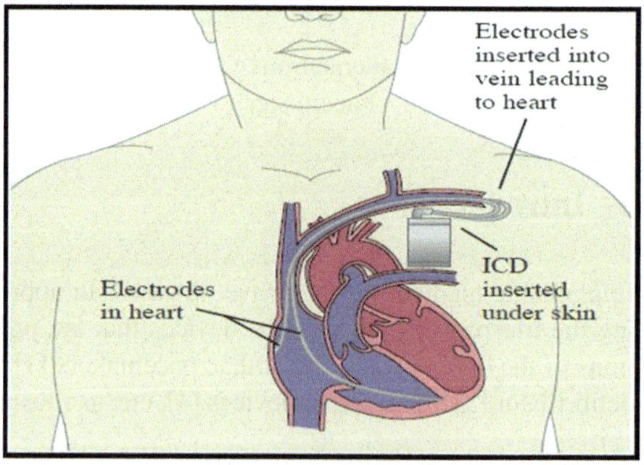

Fig. 3 Implantable cardioverter-defibrillator

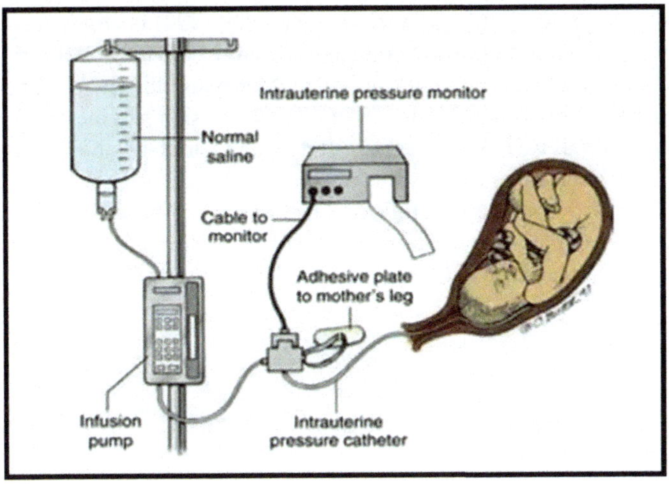

Fig. 4 Intrauterine pressure catheter

bio-medical device can contribute real-time treatment and examination for numerous diseases linked with our body. Because of the decreased lifetime of batteries and requirement of additional surgeries for battery replacement, large efforts are taken to eliminate the need for batteries. Piezoelectric energy-harvesting devices are incorporated inside the human body to transform mechanical energy into electric energy [5, 6].

In recent years, analysts have proposed more techniques to boost up the life period of the human body by implanting medical devices, such as enhancing the biocompatibility of the medical devices through power supply [7], embedding piezoelectric energy harvester with an elastic magnifier for pacemakers [8], supplying energy for implantable remote-miniaturized devices [9], and optimizing piezoelectric speaker for hearing aids [10]. From the survey, the proposed approaches require more time to select the appropriate design variables of the energy harvester. To avoid the above-mentioned issues, some of the intelligent optimization techniques such as Genetic Algorithm [11], Bat Algorithm [12], Grey Wolf Optimization [13], Hybrid Grey Wolf Optimizer tuned Genetic Algorithm [14], and Hybrid Bat Algorithm [15] are utilized to optimize the design variables of the piezoelectric harvesters, to generate maximum efficiency, and to power medical devices. Various approaches have been analyzed for the optimization of piezoelectric harvesters and have been presented. Sunithamani et al. discussed the PZT length optimization [16] and performance of MEMS piezoelectric harvesters [17]. Sunithamani et al. conducted experimental analysis of piezoelectric harvester with various substrate thickness and proof mass shapes [18]. Ye et al. [19] utilized GA technique to improve the power generated from traffic vibrations. Nabavi and Zhang [20] utilized GA to enhance the voltage generated from piezoelectric harvesters. Lakshmi and Khan [21] used GA to enhance the stability of the system.

In this work, Euler–Bernoulli analytical modeling [22] is chosen and the equations are derived to determine the voltage and power produced by the piezoelectric energy harvester. Intelligent optimization techniques use the fitness function, from the derived analytical equations, by optimizing the design parameters of the piezoelectric harvester. Based on these optimized design variables, the energy harvester is designed for pacemaker and hearing aids.

This paper is formulated as follows. The analytical modeling and intelligent optimization techniques are explained in Sect. 2. Section 3 discusses the medical applications. Simulation results are explained in Sect. 4. Conclusion is made in Sect. 5.

2 Analytical Modeling-Intelligent Optimization Techniques

2.1 Analytical Modeling of Piezoelectric Energy Harvesters

As stated in IEEE standard on piezoelectricity, the voltage generated is proportional to stress, which is given by [23]

$$V = g_{31} t_{\text{pi}} \sigma_{\text{pi}} \tag{1}$$

where V—harvested voltage, g_{31}—piezoelectric coefficient, t_{pi}—piezoelectric thickness, σ_{pi}—stress applied. The strain and stress generated is given by

$$\varepsilon_{\text{pi}} = \sigma_{\text{pi}} / Y_{\text{pi}} \tag{2}$$

where ε_{pi}—strain and Y_{pi}—Young's modulus of the piezoelectric material.

Euler–Bernoulli model is used in this work [22] in which the piezoelectric layer and substrate bend about a common neutral axis. This neutral axis is determined by a modulus weight algorithm [24]. The neutral axis equation is given as

$$z_s = \frac{\frac{t_{\text{pi}}}{2} t_{\text{pi}} \frac{Y_{\text{pi}}}{Y_{\text{be}}} + \left(t_{\text{pi}} + \frac{t_{\text{be}}}{2}\right)}{t_{pi} \frac{Y_{\text{pi}}}{Y_{\text{be}}} + t_{\text{be}}} \tag{3}$$

where z_s—neutral axis of the beam, t_{be}—thickness of the beam, and Y_{be} is the young's modulus of the beam.

The structure of piezoelectric harvester [11] is shown in Fig. 5. The voltage is calculated as

$$\varepsilon_{\text{pi}} = -\frac{M}{(Y_{\text{pi}} I_{\text{pi}} + Y_{\text{be}} I_{\text{be}})} \left(z_s - \frac{t_{\text{pi}}}{2}\right) \tag{4}$$

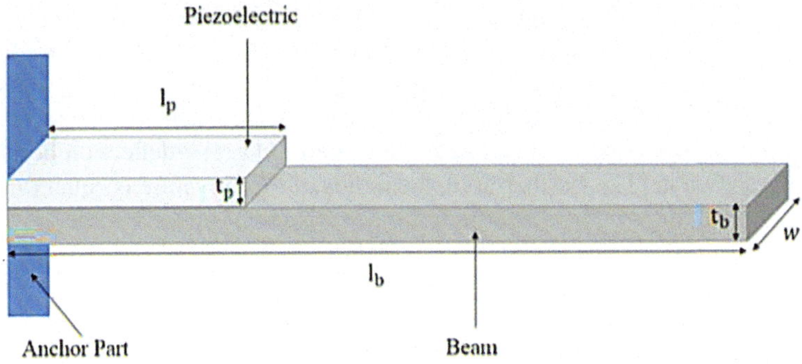

Fig. 5 Structure of piezoelectric energy harvester

where M—actuating momentum. I_{pi} and I_{be}—moments of inertia of the piezoelectric film and beam.

$$I_{\text{pi}} = \frac{1}{3} w_{\text{pb}} \left[z_s^3 - \left(z_s - t_{\text{pi}} \right)^3 \right] \tag{5}$$

$$I_{\text{be}} = \frac{1}{3} w_{\text{pb}} \left[\left(t_{\text{pi}} + t_{\text{be}} - z_s \right)^3 + \left(z_s - t_{\text{pi}} \right)^3 \right] \tag{6}$$

where w_{pb}—width of the piezoelectric or beam layer.

Substituting Eqs. (5) and (6) into (4),

$$\varepsilon_{\text{pi}} = \frac{6MY_{\text{be}}t_{\text{be}}\left(t_{\text{pi}} + t_{\text{be}}\right)}{w_{\text{pb}}\left[Y_{\text{pi}}^2 t_{\text{pi}}^4 + Y_{\text{be}}^2 t_{\text{be}}^4 + 2Y_{\text{pi}}t_{\text{pi}}Y_{\text{be}}t_{\text{be}}\left(2t_{\text{pi}}^2 + 3t_{\text{pi}}t_{\text{be}} + 2t_{\text{be}}^2\right)\right]} \tag{7}$$

Finally, as given in [11], the harvested voltage from the piezoelectric device is given as

$$V = -\frac{6g_{31}M\varphi(1+T)}{w_{\text{pb}}t_{\text{pi}}\left[1 + \varphi^2 T^2 + 2\varphi\left(2 + 3T + 2T^2\right)\right]} \tag{8}$$

where $\varphi = \frac{Y_{\text{be}}t_{\text{be}}}{Y_{\text{pi}}t_{\text{pi}}}$

$T = \frac{t_{\text{be}}}{t_{\text{pi}}}$

The actuating momentum is calculated as

$$M = (I_{\text{be}}Y_{\text{be}})\hat{n} \tag{9}$$

where $\hat{n}$ is the average radius of curvature. The governing equation of the beam [22] is determined as

$$\rho A \frac{\partial^4 b(x,t)}{\partial t^4} + Y_{\text{be}} I_{\text{be}} \frac{\partial^4 b(x,t)}{\partial x^4} = F(t) \tag{10}$$

where b—displacement of the beam, ρ—density of the beam, A—cross-sectional area, and $F(t)$—external force applied to the beam, $b(x, t)$—deflection besides the Z-axis as shown in [11]. Furthermore, the radius of the curvature is equal to (11):

$$n(x,t) = \frac{\partial^2 b(x,t)}{\partial x^2} \tag{11}$$

The average radius of curvature is calculated in (12):

$$\hat{n} = \frac{1}{l_{\text{pi}}} \int_0^{l_{\text{pi}}} n(x,t)\text{d}x \tag{12}$$

where l_{pi}—length of the piezoelectric film.

The power generated by the piezoelectric device [22] is

$$P = \sum_{i=0}^{m} \frac{v_s^2 R_L}{(R_L + R_S)^2 m} \tag{13}$$

where v_s—source voltage, R_L and R_s—load impedance and source impedance, m—number of time steps, respectively.

Equations (8) and (13) represent the voltage and power equations of the piezoelectric harvester.

Stress, Natural Frequency, Load Impedance, and Harvester Efficiency

The stress, natural frequency, load impedance, and harvester efficiency are the electrical and mechanical characteristics of the piezoelectric device.

Stress is computed in [11]. Natural frequency is calculated as

$$\omega_n = 2\pi f \tag{14}$$

where ω_n—natural frequency

Load impedance enhances the power produced by the piezoelectric device [25].

$$R = \frac{t_{\text{pi}}}{w_{\text{pb}} l_{\text{be}} \varepsilon_{33}} \tag{15}$$

where R—resistive load, l_{be}—beam length, and ε_{33}—permittivity.

Harvester efficiency is estimated to identify the effectiveness of the piezoelectric device [26].

$$\eta_{\text{harv}} = \frac{\text{electrical energy}}{\text{work done by applied force}} * 100 \tag{16}$$

where η_{harv}—harvesting efficiency of the piezoelectric harvester.

2.2 Piezoelectric Energy Harvester-Based Intelligent Optimization Techniques

BA, GWO, and HGWOBA are the intelligent optimization techniques carried out in multi-objective mode to increase the voltage and power, harvested from the piezoelectric harvester, and to power medical devices (pacemaker and hearing aids).

Bat Algorithm (BA)—It is a high-level method, which determines its solution, by sensing the distance of the prey (solution), through its echolocation behavior. The flowchart of BA is given in [12]. Here, Eqs. (8) and (13) are utilized as the BA's objective function. Firstly, the bat population, frequency, pulse rate, and loudness parameters are initialized [12]. The solutions are found for the initial set of values, by applying fitness function. New set of results are generated by adjusting the variables. Finally, BA correlates the present results with the earlier results, and it exhibits the optimal output [12]. By this process, the BA technique is executed, to optimize the design variables of the harvester, for the increment of performance of the piezoelectric device.

Grey Wolf Optimization (GWO)—It is an exclusive, high-level method, where the grey wolves are divided into four types: they are alpha, beta, delta, and omega [13]. The essential methods of GWO are (a) trailing, hounding, and impending the target; (b) following, surrounding, and tormenting the target until it stops moving; and (c) assaulting the target. The flowchart of GWO is given in [13]. In this context, the objective function (fitness function) of GWO is derived from Eqs. (8) and (13). The fitness function of grey wolves is calculated in an iterative system, where it correlates the current output with the earlier output. The positions of the grey wolves are updated based upon its output, and eventually the optimal results are presented. Through this process, the dimensions of the piezoelectric and beam structure are optimized by the GWO technique, for the increment of effectiveness of the unimorph piezoelectric harvester.

Hybrid Grey Wolf Optimizer tuned Bat Algorithm (HGWOBA)—It is the proposed optimization technique, in this research work, to optimize the design variables of the unimorph piezoelectric energy harvester. The flowchart of HGWOBA is given in Fig. 6. In this context, the BA parameters are optimized by the GWO parameters to produce an optimized solution [14]. The objective function is derived from Eqs. (8) and (13). The parameters of HGWOBA are population size, vector coefficients, frequency, pulse rate, and loudness rate, and they are initialized. The primary set of results is produced from these variables. At first, the GWO and BA parameters are applied to the entire population. Then, the population is segregated into sub-populations [14]. The frequency rate is applied to sub-population and the pulse rate, and loudness parameter is applied to the entire population. Eventually, the solutions are updated, and new set of solutions are produced. The current (present) solution is

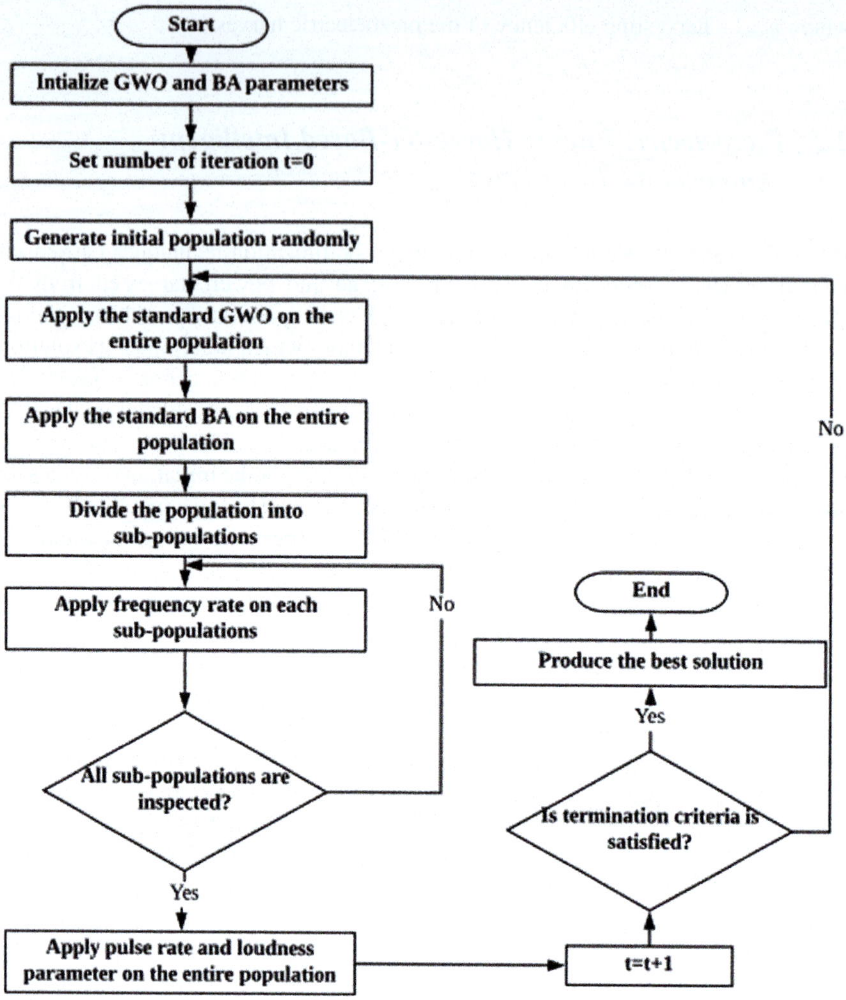

Fig. 6 Flowchart of HGWOBA

compared with the earlier solution. If the results are not satisfied, then it repeats the process, otherwise, the best solution is produced to enhance the performance of the piezoelectric device.

3 Medical Applications—Pacemaker and Hearing Aids

Medical applications, such as pacemaker and hearing aids, are powered based on the dimensions of the piezoelectric energy harvester and optimized utilizing intelligent optimization techniques. Normally, the power requirement of a pacemaker is about 10 μW [27] and voltage consumption is about 1.5–120 V [28], respectively, whereas hearing aid consumes power and voltage of about 10 mW [29] and 1.3 V, respectively [30]. Material used in this work is PZT-5A. Hearing aids and pacemakers are powered from the output voltage and power generated by the piezoelectric device. The artificial cardiac pacemaker is shown in Fig. 7 [1] and hearing aids using the electric output from piezoelectric energy harvester are shown in Fig. 8 [31].

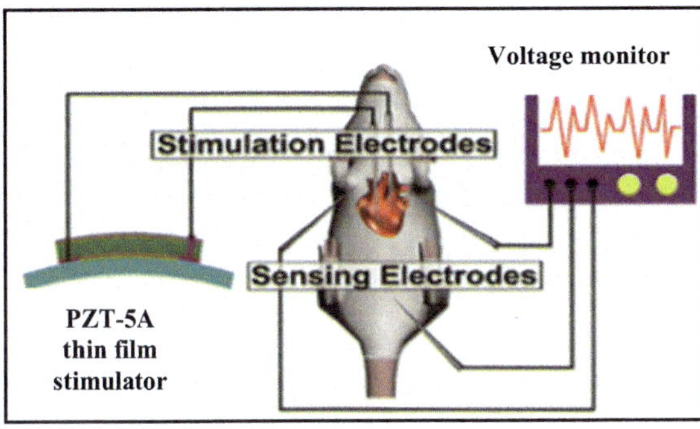

Fig. 7 Artificial cardiac pacemaker using the electric output from the PZT-5A thin film

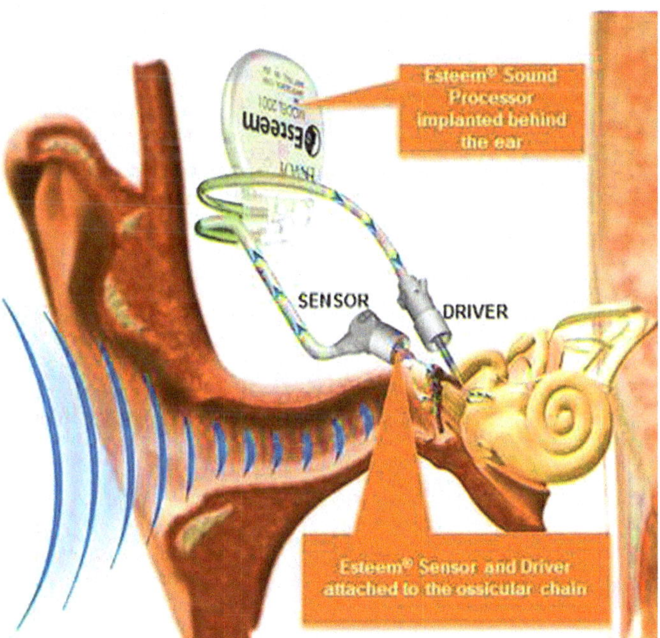

Fig. 8 Hearing aid using the electric output from the piezoelectric sensor

4 Results and Discussions

The design of piezoelectric harvester is established utilizing MATLAB. The substrate is made of aluminum and piezoelectric material used is Lead Zirconate Titanate (PZT-5A). Figure 5 shows the structure of piezoelectric energy harvester. The parameters in Table 1 [11] are used to determine the voltage and power generated by the piezoelectric energy harvester.

Figures 9, 10, and 11 show the un-optimized results of the unimorph harvester. The voltage, power, natural frequency, stress, and efficiency generated by the piezoelectric

Table 1 Specifications of piezoelectric energy harvester [11]

Design variables	Explanation	Value
l_{be}	Length of the beam	3 mm
t_{be}	Thickness of the beam	200 μm
Y_{be}	Young's modulus of beam	70 GPa
ρ_{be}	Density of beam	2700 kg/m^3
l_{pi}	Length of the PZT layer	3 mm
t_{pi}	Thickness of the PZT layer	50 μm
Y_{pi}	Young's modulus of PZT	50 GPa
ρ_{pi}	Density of the PZT layer	7600 kg/m^3
g_{31}	PZT coefficient	−9.5e−3 V*m/N
w_{pb}	Width of the Piezoelectric or beam layer	1 mm

Fig. 9 Voltage response

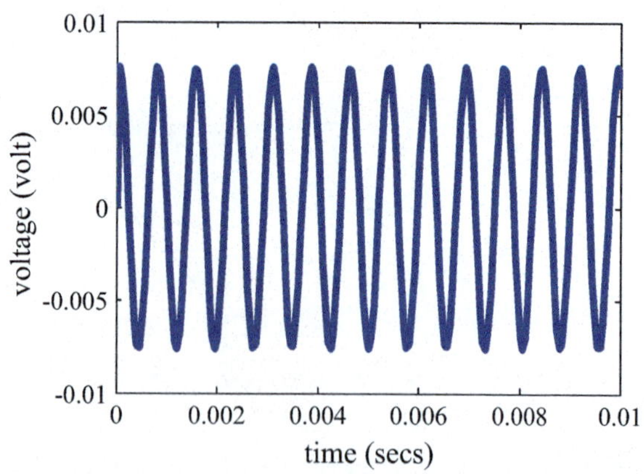

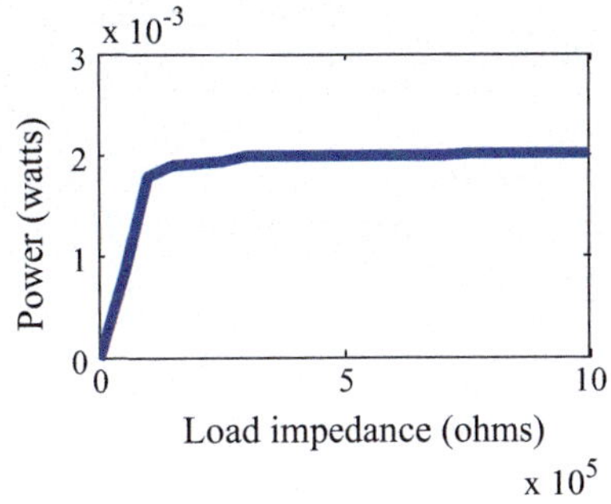

Fig. 10 Power response—load impedance

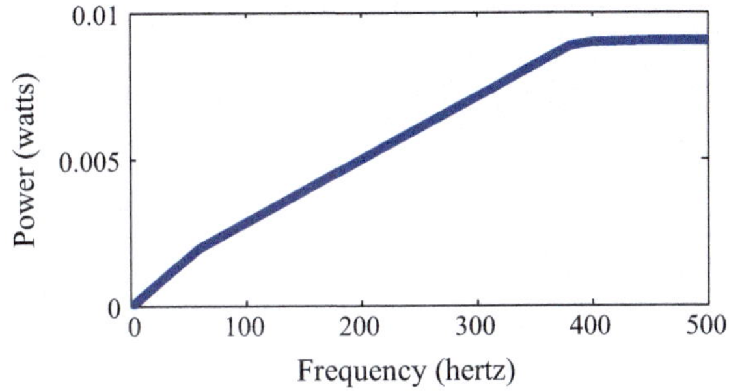

Fig. 11 Power response—natural frequency

structure using analytical modeling from Figs. 9, 10, and 11 are **7.6 mV** [11], **2.2 mW, at 120 kΩ, 65 Hz, 1.65 MPa and 24.9%** [26], respectively.

The voltage and power generated are improved by using the intelligent optimization techniques, which are performed by optimizing five physical dimension variables with lower and upper bound limits [11] as given in Table 2. It is performed four times with various population sizes (5, 15, 30, and 45) [11]. Here, OPT1, OPT2, OPT3, and OPT4 specify the optimized energy harvester for four optimizations [11].

Since the un-optimized results from Figs. 9, 10, and 11 are lower, optimization techniques are used to improve the results by optimizing the design variables of the energy harvester.

The intelligent optimization techniques are established in MATLAB software, in multi-objective mode, to improve the amount of voltage and power generated by the energy harvester by optimizing the parameters of the piezoelectric and beam structure. The objective function and constraints of the optimization techniques [11] are given below:

Table 2 Lower and upper boundary limits of the unimorph piezoelectric structure [11]

Parameters	l_{be} (μm)	t_{be} (μm)	l_{pi} (μm)	t_{pi} (μm)	w_{pb} (μm)
Range	[100, 5000]	[100, 200]	[500, 5000]	[10, 50]	[500, 2500]

$$\begin{aligned}&\text{Maximize: Voltage, Power}\\&\text{Based on: design rules of the optimizable parameters}\end{aligned} \quad (17)$$

The five variables of the unimorph piezoelectric energy harvester, which is optimized, are beam length (l_{be}), piezoelectric length (l_{pi}), beam thickness (t_{be}), piezoelectric thickness (t_{pi}), and width of the beam or piezoelectric layer (w_{pb}).

The optimization techniques are executed four times (OPT1, OPT2, OPT3, and OPT4) with four population sizes (5, 15, 30, and 45) [11]. The lower and upper boundary limits [11] and the optimized dimensions of the harvester for GA [11], BA, GWO, and HGWOBA are given in Tables 2, 3, 4, and 5, respectively. With respect to the optimized parameters, the voltage, power, stress, natural frequency, and harvester efficiency are determined.

Table 3 BA optimized parameters

PS [11]	OPT	l_{be} (μm)	t_{be} (μm)	l_{pi} (μm)	t_{pi} (μm)	w_{pb} (μm)
5	OPT1	2450	190	2350	30	935
15	OPT2	4250	130	250	50	980
30	OPT3	4950	115	260	30	870
45	OPT4	4850	120	360	30	1600

Table 4 GWO optimized parameters

PS [11]	OPT	l_{be} (μm)	t_{be} (μm)	l_{pi} (μm)	t_{pi} (μm)	w_{pb} (μm)
5	OPT1	3000	170	2150	25	915
15	OPT2	2200	120	257	50	976
30	OPT3	2926	125	550	20	866
45	OPT4	2895	140	430	20	1700

Table 5 HGWOBA optimized parameters

PS [11]	OPT	l_{be} (μm)	t_{be} (μm)	l_{pi} (μm)	t_{pi} (μm)	w_{pb} (μm)
5	OPT1	4350	200	2356	38	920
15	OPT2	2350	180	254	50	910
30	OPT3	2330	120	670	40	880
45	OPT4	2300	118	450	40	1800

PS—population size [11]

4.1 BA, GWO, and HGWOBA—Voltage, Power, Frequency, and Stress

Generated Voltage Versus Time

From Fig. 12 and with the optimized dimensions from Tables 3, 4, and 5, the voltage generated using **HGWOBA (OPT4-60.1 mV) is more in comparison to GA (OPT4-27.81 mV)** [11], **GWO (OPT4-36.2 mV), and BA (OPT4-32.1 mV)** and efficient to power cardiac pacemakers and hearing aids.

Generated Power Versus Load impedance

See Fig. 13.

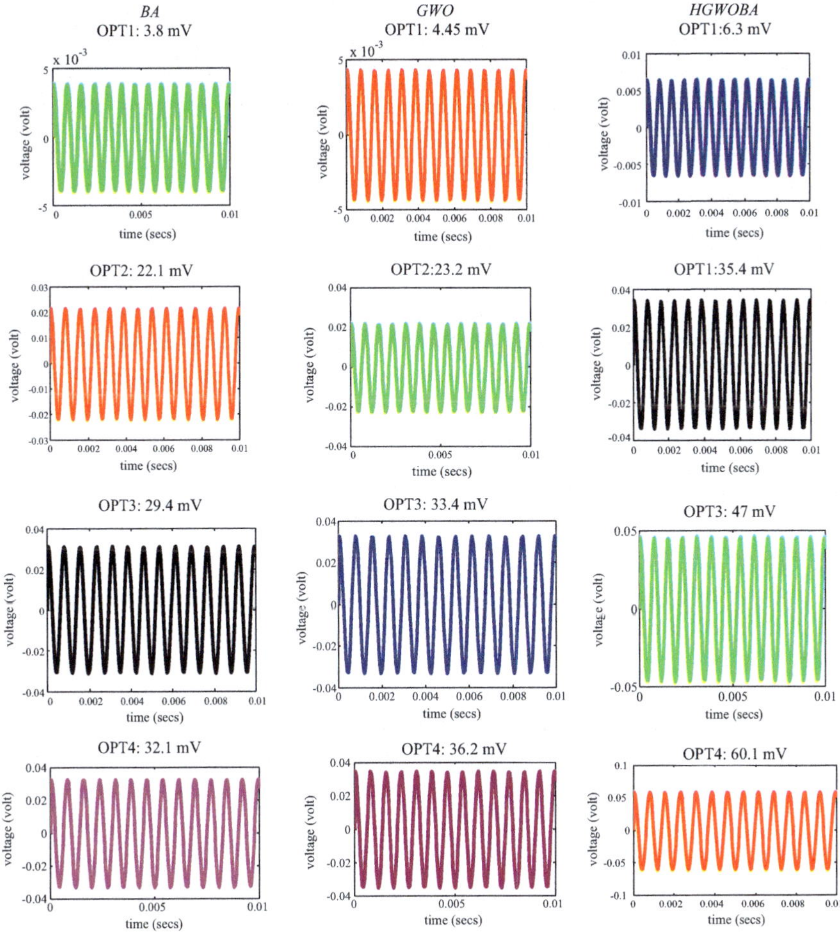

Fig. 12 Voltage response of BA, GWO, and HGWOBA

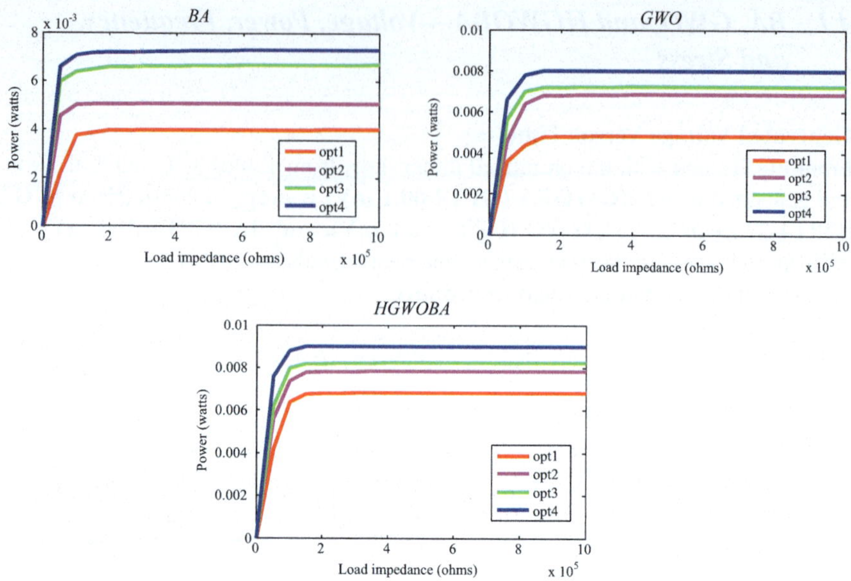

Fig. 13 Power versus load impedance—BA, GWO, and HGWOBA

Figure 13 shows that the power harvested from piezoelectric harvester corresponding to the load impedance, using **HGWOBA (OPT4-10 mW at 180 kΩ) is more than GWO (OPT4-8.04 mW at 160 kΩ) and BA (OPT4-7.25 mW at 145 kΩ)**, which is more efficient to power cardiac pacemakers and hearing aids.

Generated Power Versus Frequency

Figure 14 and the optimized physical dimensions from Tables 3, 4, and 5 confirm that the natural frequency generated by the piezoelectric harvester corresponding to the generated power using **HGWOBA (OPT4-190 Hz at 10 mW) is more than GWO (OPT4-172 Hz at 8.04 mW) and BA (OPT4-168 Hz at 7.25 mW)**, when executed for four optimizations, at four different population sizes [11], which is more efficient to power cardiac pacemakers and hearing aids.

Stress Analysis

Table 6 shows the stress analysis for BA, GWO, and HGWOBA.

HGWOBA generates more stress, when compared to GWO and BA. The maximum stress is produced at OPT4: BA-9.86 MPa, GWO-10.88 MPa, and HGWOBA-14.1 MPa, respectively, which is used to power cardiac pacemakers and hearing aids.

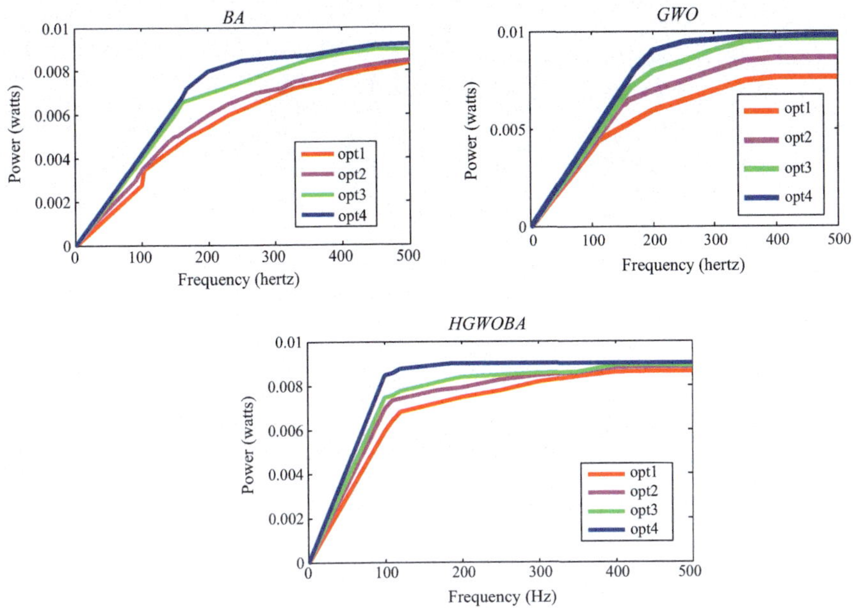

Fig. 14 Power versus frequency—BA, GWO, and HGWOBA

Table 6 Comparison of stress for BA, GWO, and HGWOBA

PS [11]	OPT	Stress (MPa)		
		BA	GWO	HGWOBA
5	OPT1	4.5	5.3	9.2
15	OPT2	5.6	8.7	11.56
30	OPT3	7.2	9.6	12.8
45	**OPT4**	**9.86**	**10.88**	**14.1**

4.2 Run Time for BA, GWO, and HGWOBA

Figure 15 proves that the run time taken by **HGWOBA (1.3 min) is very less when compared to GWO (1.5 min) and BA (1.67 min)**.

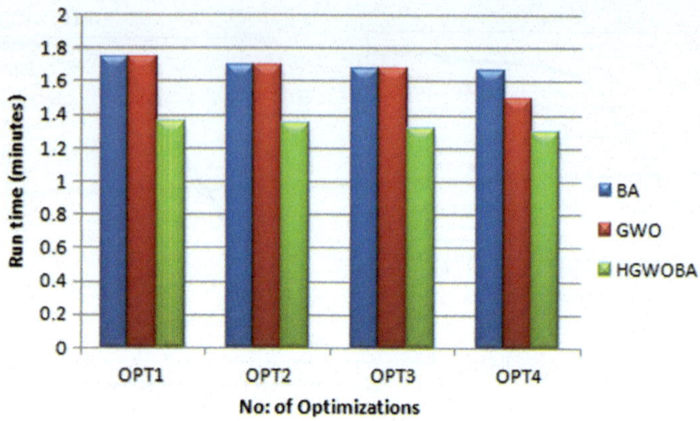

Fig. 15 Comparison of run time between BA, GWO, and HGWOBA

4.3 Comparison of Harvester Efficiency for BA, GWO, HGWOBA

Figure 16 proves that **HGWOBA produces more efficiency (68.96%), when compared to GWO (58.9%) and BA (48.78%)**. Therefore, the above results prove that HGWOBA is more efficient than BA and GWO, in all aspects, to power cardiac pacemakers and hearing aids.

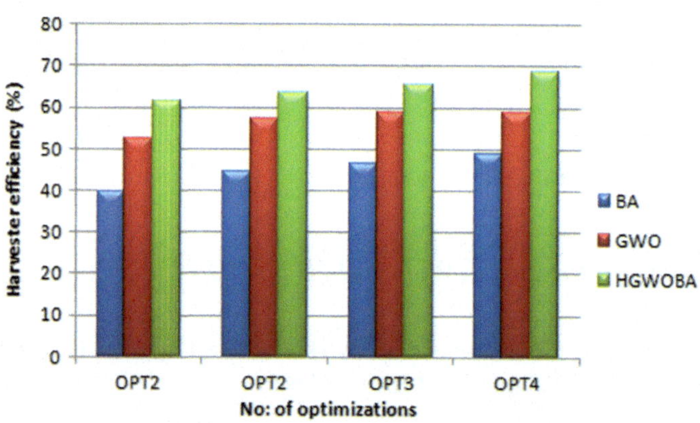

Fig. 16 Comparison of harvester efficiency between BA, GWO, and HGWOBA

5 Conclusion

Analytical Modeling of unimorph piezoelectric structure using intelligent optimization techniques is examined to determine the power and voltage from piezoelectric cantilever for medical applications. Intelligent optimization techniques (BA, GWO, and HGWOBA) are utilized for optimizing the dimensions of the piezoelectric energy harvester to yield maximum power and voltage for medical applications, i.e., pacemaker and hearing aids. Moreover, the un-optimized and optimized results are compared, and the results procured justify that **HGWOBA is efficient (68.96%) to power the medical devices**, when compared to the other optimization techniques and the un-optimized results in all conditions.

References

1. Hwang GT, Byun M, Jeong CK, Lee KJ (2015) Flexible piezoelectric thin-film energy harvesters and nanosensors for biomedical applications. J Adv Healthcare Mater 4(5):646–658
2. Curran JR, Galster JA (2013) The master hearing aid. Trends Amplification 17(2):108–134
3. https://www.chss.org.uk/heart-information-and-support/about-your-heart-condition/common-heart-conditions/heart-arrhythmias-2/icds-implantable-cardioverter-defibrillators/
4. https://medicalfoxx.com/amnioinfusion.html
5. Mitcheson PD, Green TC, Yeatman EM (2007) Power processing circuits for electromagnetic, electrostatic and piezoelectric inertial energy scavengers. Microsyst Technol 13(11):1629–1635
6. Song H, Kumar P, Maurya D (2017) Ultra-low resonant piezoelectric MEMS energy harvester with high power density. J Microelectromech Syst 26(6):1226–1234
7. Amar AB, Kouki AB, Cao H (2015) Power approaches for implantable medical devices. Sensors 15(11):28889–28914
8. Galbier AC, Karami MA (2016) A bistable piezoelectric energy harvester with an elastic magnifier for applications in medical pacemakers. In: ASME conference on smart materials, adaptive structures and intelligent systems, vol 2, pp 1–9
9. Kim A, Ochoa M, Rahimi R, Ziaie B (2015) New and emerging energy sources for implantable wireless microdevices. IEEE Access 3:89–98
10. Martins GC, Nunes PR, Cordioli JA (2014) On the optimization of a piezoelectric speaker for hearing aid application through multi-physical FE models. In: International conference on engineering optimization, pp 317–322
11. Nabavi S, Zhang L (2017) Design and optimization of piezoelectric MEMS vibration energy harvesters based on genetic algorithm. IEEE Sens J 17(22):7372–7382
12. Tupia M, Cueva R, Guanira M (2018) A bat algorithm for job scheduling in ceramics production lines. In: International conference on infocom technologies and unmanned systems, pp 266–270
13. Mirjalili S, Mirjalili SM, Lewis A (2014) Grey wolf optimizer. Adv Eng Softw 69:46–61
14. Tawhid MA, Ali AF (2017) A hybrid grey wolf optimizer and genetic algorithm for minimizing potential energy function. Memetic Comp 9(4):347–359
15. Liang H, Liu Y, Shen Y (2018) A hybrid bat algorithm for economic dispatch with random wind power. IEEE Trans Power Syst 33(5):5052–5061
16. Sunithamani S, Lakshmi P, Eba Flora T (2014) PZT length optimization of MEMS piezoelectric energy harvester with a non-traditional cross section: simulation study, Microsyst Technol 20(12):2165–2171
17. Sunithamani S, Lakshmi P (2015) Simulation study on performance of MEMS piezoelectric energy harvester with optimized substrate to piezoelectric thickness ratio. Microsyst Technol 21(4):733–738

18. Sunithamani S, Lakshmi P (2017) Experimental study and analysis of unimorph piezoelectric energy harvester with different substrate thickness and different proof mass shapes. Microsyst Technol 23(7):2421–2430
19. Ye G, Yan J, Wong ZJ, Soga K, Seshia A (2009) Optimization of a piezoelectric system for energy harvesting from traffic vibrations. In: IEEE international ultrasonics symposium, pp 759–762
20. Nabavi S, Zhang L (2016) MEMS piezoelectric energy harvester design and optimization based on genetic algorithm. In: IEEE international ultrasonics symposium proceedings, pp 1–4
21. Lakshmi P, Khan MA (2000) Stability enhancement of a multimachine power system using fuzzy logic based power system stabilizer tuned through genetic algorithm. Int J Electr Power Energy Syst 22:137–145
22. Eggborn T (2003) Analytical models to predict power harvesting with piezoelectric materials. Master thesis, Virginia Polytechnic Institute and State University, May 2003
23. Meitzler AH, Tiersten HF, Warner AW, Berlincourt D, Couqin GA, Welsh FS III (1988) IEEE standard on piezoelectricity. Society
24. Hibbeler RC (1997) Mechanics of materials. Prentice Hall
25. Moura S (2006) Design of an energy harvester. ME128 Project 3c, Department of Mechanical Engineering, University of California, Berkeley
26. Athavale AA (2015) An analytical model for piezoelectric unimorph cantilever subjected to an impulse load Master of Science. Graduate School-New Brunswick Rutgers, The State University of New Jersey, May 2015
27. Davis S (2014) Energy harvesting poised to eliminate pacemaker battery. http://www.powerelectronics.com/energy-harvesting/energy-harvesting-poisedeliminate-pacemaker-battery. Accessed Sep 2014
28. https://en.wikipedia.org/wiki/Artificial_cardiac_pacemaker
29. Gaspar PD, Felizardo V (2013) Energy-harvesting methods for medical devices. Rechargeable Sens Netw 327–356
30. Yasui H, Kurosawa M, Morita T, Kanda T, Higuchi T (1998) Piezoelectric transducer for a hearing aid using PZT thin film. J Acoust Soc Am 103(5)
31. Anderson L (2014) The Esteem® hearing implant. Audiology Online, Article 12780. http://www.audiologyonline.com. Accessed July 2014

A Modular End Effector for Sample Collection from Extraterrestrial Terrain

Rippudaman Singh, Abhinandan Kapoor, Gaurav Sharma, Abhishek Kumar, A. Shankara, and K. A. Keshava Murthy

Abstract In the present work, a modular end effector has been conceptually designed to facilitate excavation using the gripper of the robotic arm of a proposed future rover. The advantage of the modular end effector is that it uses a common robotic arm for carrying out surface and subsurface collection using the respective active and passive modules thereby reducing the mass associated with independent multiple arms used for carrying out specific excavation. The design of the different modules of the collection mechanism is based on the mass of the sample and the interaction of excavation modules with the terrain. To provide contamination-free samples, each sample is required to be encapsulated in the very unit where it is collected and then placed inside the storage container for earth return. A rapid prototype of the mechanism has been developed and functionally demonstrated over lunar soil simulant. This paper contains the details of design, analysis, fabrication and tests carried out on a modular end effector to verify its functional performance as a sample collection mechanism. The present work provides a theoretical and experimental groundwork to design sample return mechanisms for future extraterrestrial missions. The calculations are done using MATLAB, the modelling using Siemens NX and the designed components are rapidly prototyped.

Keywords Sample return · Excavation · End effector · Design · Analysis · Test

---

R. Singh (✉)
SMBS (School of Mechanical and Building Science), Vellore Institute of Technology, Chennai, India
e-mail: rippudamans@gmail.com

A. Kapoor · G. Sharma · A. Kumar · A. Shankara · K. A. Keshava Murthy
UR Rao Satellite Centre, Old Airport Road, Bengaluru, India
e-mail: akapoor@isac.gov.in

G. Sharma
e-mail: gsrs@isac.gov.in

D. Sen et al. (eds.), *Mechanism and Machine Science*, Lecture Notes in Mechanical Engineering, https://doi.org/10.1007/978-981-15-4477-4_15

1 Introduction

It is important to know the properties of extraterrestrial bodies in order to realize its potential for future colonization [1]. There are multiple ways in which the properties of the celestial body can be studied, viz. Fly-by, touch and go, and surface collection. Amongst these, the surface collection provides an opportunity to maximize the scientific findings by carrying out in situ tests on multiple locations or rigorous tests on returned samples. In situ tests are specific and avoid the complexity and cost associated with a return mission whereas sample return mission reduces mass associated with in situ instrumentation and provides the flexibility to do multiple tests by different teams on the return samples even as the test procedures improve in the future [2]. As the technology is not looking into before, the modular concept is a different perspective and an addition to the sample return technology. Some of the scientific studies of soil include (Previously done):

- High voltage breakdown and electrical insulation characteristics
- The Soil Mechanics Experiment
- The solar wind Experiment
- The passive seismic Experiment
- Lunar Dust Detector.

Currently, the only successful sample return missions that have been conducted are the Unmanned Lunar missions by the Late Soviet Union and the Manned Apollo space missions by the United States of America.

The importance of sample return over in situ, advantages previous missions, how it would help humanity—Overall the motivation for the project is learned from the extensive study conducted by Scott Sandford in his "*The power of sample return missions—Stardust and Hyabusa*" where he investigates the use of samples brought from the extraterrestrial surface to Earth and its use in the present and in the future, He also discusses the advantages sample return and the disadvantages of in situ which were a highly valuable resource in understanding the problem [2].

The importance of sample isolation and the need for sample return of uncontaminated soil samples, the procedure of sterilization is understood from "*Sample Acquisition and Caching using Detachable Scoops for Mars Sample Return*" where P. Younse and his team discusses the importance of bringing back large volumes of unhindered sample from the surface of Mars, they also share a prototype model of a rover which focuses on collection method and a suggestion for a modular model. They also mentioned the sterilization processes they would use to keep the samples uncontaminated which are what we would follow as well [3].

All excavation calculations required for lunar surface regolith are referred from "*Calculation of Excavation Force for ISRU on Lunar Surface*" [5]. In situ, and findings are obtained from the Luna 20 report of the configuration of the soil sample.

And finally, the types of technology used, the need for newer technology, the drawbacks and limitations of existing technology are realized from "*Developing Sample Return Technology using the Earth's Moon as a Testing Ground*" [1].

The sterilization and the importance of unhindered samples are understood from the study conducted by *Mr. J. Eigenbrode* and his team on "*A Field-Based Cleaning Protocol for Sampling Devices Used in Life-Detection Studies*" [4].

The focus of the present work is on unmanned sample collection using the rover's robotic arm and its containment in storage containers housed inside the ascent unit of the Lander for earth return. A modular end effector has been conceptually designed to facilitate excavation using the gripper of the robotic arm of a proposed future rover. The two active modules mounted over the chassis of the rover, one for surface and other for subsurface excavation house the prime movers for operating the passive excavation cum collection modules that, in turn, are housed inside the storage container of the Lander. The arm's movement is used to latch the arm's gripper with the active module, open the storage container, latch with the passive module, excavate the sample, encapsulate it, place the passive module into the storage container and finally close the storage container.

2 Mechanism Details

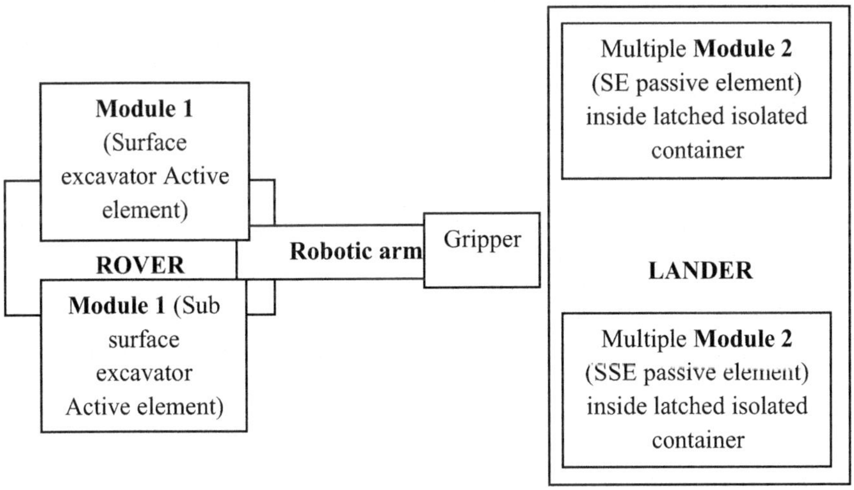

Following is the working principle (Refer above block diagram while going through the mechanism):

1. Robotic Arm (RA) is commanded to pick up the Module 1 (either Surface excavator (SE) or Subsurface Excavator (SSE) Active element) available on the rover with the help of gripping mechanism on RA and then RA is steered towards the Lander and it is commanded to open the door latch of housing for picking up the module 2 (either SE or SSE).

2. Then RA picks up the module 2 and the desired operation of surface soil collection using scoops or in-depth soil collection using drills is performed.
3. After this, the module 2 is retracted back from the soil and is kept inside the module 2 housing using RA. Then the door is closed using RA. This completes the sampling at one site.
4. After this, the rover with the other set of drills and scoops is hovered on to other sampling site and perform the similar operation.
5. The individuality in the approach of this mechanism design is that sampling can be done from multiple sites without any contamination, thereby, maintaining the virginity of the soil which has been a major issue in the earlier sample return missions. Also, the maximum volume of soil can be retained in the drill housing by the ingenious design of a cover closing mechanism after the soil sampling is done. The overall operation sequence shown below in Fig. 1 shows surface and in-depth excavation using scoop and drill, respectively.

2.1 *The Operation Sequence*

See Fig. 1.

2.2 *Passive Element Details*

The volume of the collectors is determined by the amount of sample which has to be collected per container. It was assumed that the amount of sample to be collected by the surface excavator to be 200 g but the volume of the container is designed for 300 g at 100% capacity; this is to make sure that the sample collection of 200 g can easily be achieved with about 67% volume occupied by the sample collection unit.

In the case of Subsurface excavator, a similar principle is followed but the volume here would be cylindrical.

2.2.1 For Surface Excavator (Fig. 2)

Mass of sample to be collected assumed as 300 g as mentioned above.

Density of extraterrestrial soil [5]: 1680 kg/m^3.

Hence the volume of the sample to be collected: 178,571 mm^3.

Total volume of sample which can be collected using SE

The surface excavator is a combination of a Cuboid and a Prism.

Dimensions of the Cuboid considered with both the "Scoops" enclosed: $66 \times 44 \times 35$ mm^3 = 101,640 mm^3.

Dimension of the prism: ½ * 66 * 50 * 44 mm^3 = 72,600 mm^3.

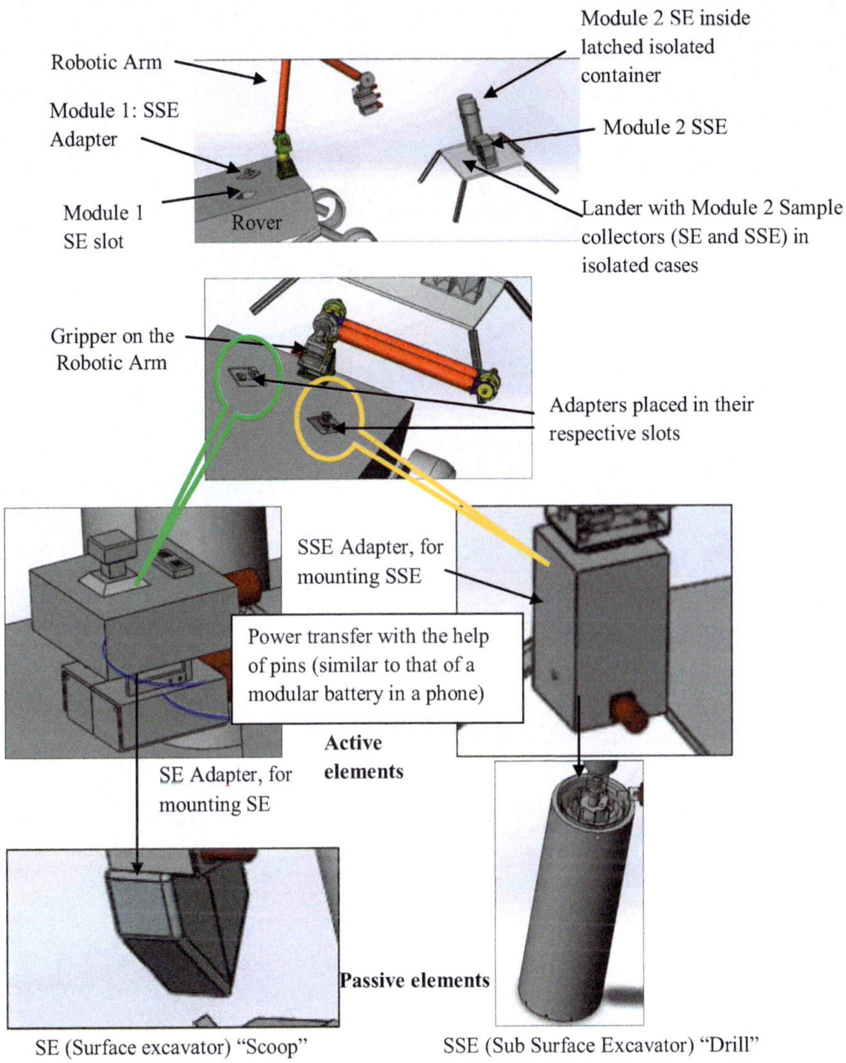

Fig. 1 The operation sequence

Hence total actual volume = 101,640 + 72,600 = 171,240 mm^3.

2.2.2 For Subsurface Excavator (Fig. 3)

This passive module is to collect the volume of the sample over a depth, the mass to be collected is assumed to be about 300 g. Hence as stated by the above calculations, the required volume of sample for 300 g is 178,571 mm^3.

Fig. 2 Surface excavator with the above-mentioned dimensions

Fig. 3 Subsurface excavator with below-mentioned dimensions

Actual volume which can be collected: 382569.75 mm^3 with cylindrical dimensions of radius 28.5 mm and a height of 150 mm for full capacity.

Volume for 50% capacity is 191,284 mm^3.

Therefore, in about 50% of the capacity of the Subsurface excavator, it would be able to collect 300 g.

Titanium has been chosen as the material for the passive elements as they come in contact with the soil of the extraterrestrial surface. Titanium is considered to be one of the most ideal metals which react to the external environment very rarely. The surface property of the material is also very hard which implies that the bits of the metal would not contaminate the soil sample.

2.3 *Active Element Details (Fig. 4a, b)*

2.3.1 Surface Excavator

The mechanism is actuated by 3 motors, 2 for its locking mechanism and one to provide the opening and closing mechanism. The opening and closing mechanism is performed with the help of gears in the ratio of 1:3.6. There are two reasons as to why such a reduction is required

(1) To make sure that the Scoops rotate slowly when compared to the motor shaft rotation
(2) To make sure enough torque is transferred with a smaller motor for the Surface Excavation process.

The gear attached to the primary shaft (Smaller one as depicted in the Fig. 4) rotates and at the same time rotates two adjacent gears, one which is directly connected to the Scoop and the other smaller gear which is responsible to rotate another gear as shown in the opposite direction in order to obtain a claw-like kinematic movement.

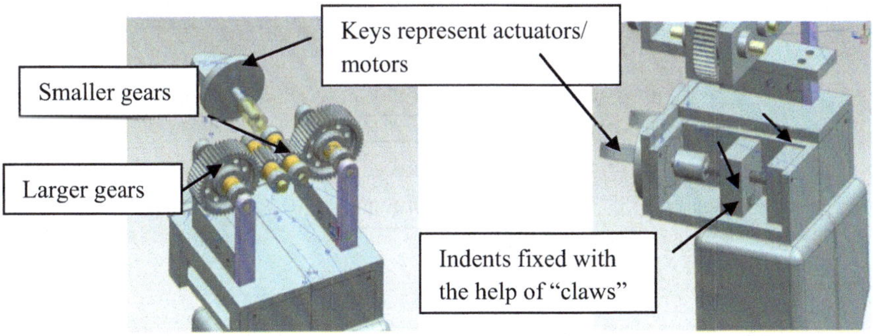

Fig. 4 **a** Rotation motion provided by the motor, in turn, rotates each of the passive elements, in turn, rotates each of the passive elements. **b** Indents locked by the "claw" in the active element

The passive element (Surface excavator/scoop) is attached to the active element via an indent projecting out of the passive element. The claws responsible to hold the indents are a part of the active element powered by an actuator on both the sides, each of these locks is individually powered by a separate motor.

2.3.2 Subsurface Excavator (Figs. 5a, b, and 6a, b)

The problem to overcome, regarding the subsurface excavator, was the effective retainment of the sample in the desired volume after collection. Present methods don't deal with effective retention of samples after collection as there is a sample loss during retention. In this paper, an innovative method of sample collection and retention is presented. Concepts were proposed with a different type of locking mechanism, but they require an additional actuator to make it possible. A separate actuator in the passive element, with this the practicality of power transfer was in question; hence the locking had to be done from the active element. Finally, the proposed concept was concluded as it does the job with one less actuator.

The active element of the subsurface excavator consists of two actuators one to provide the necessary torque to rotate the cylinder (Subsurface excavator) and the other to lock the cylinder in its place and to make sure that the cylinder does not fall off (All the DOF apart from rotation along its axis are fixed). The torque is transmitted through a double D slot and the subsurface excavator is intact in its place due to the clamping force provided by the slider which is actuated by a motor.

The passive element has a Ratchet and Pawl Mechanism (Fig. 7) which allows it to be used to excavate if it is rotated in the anticlockwise direction, while clockwise direction makes the centre shaft move freely which allows the shutter to close the gap, isolating the sample collected inside and the external environment.

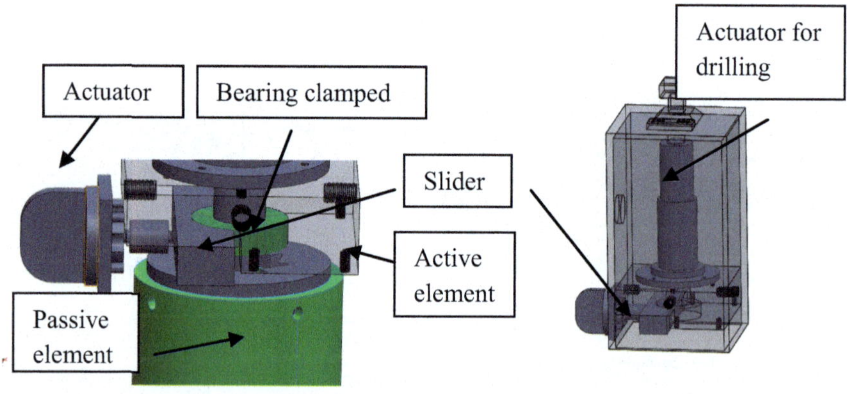

Fig. 5 **a** Clamping between the active and the passive elements via slider. **b** Active element of subsurface excavator

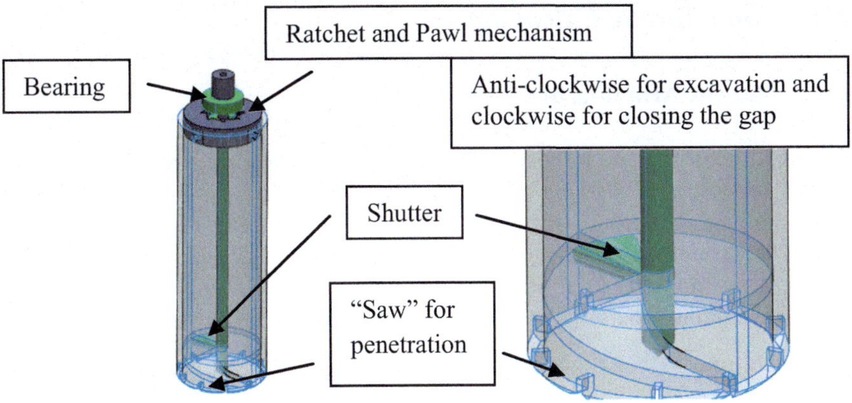

Fig. 6 **a** Passive element for collecting samples once over a depth. **b** The shutter which closes the gap the required sample is collected

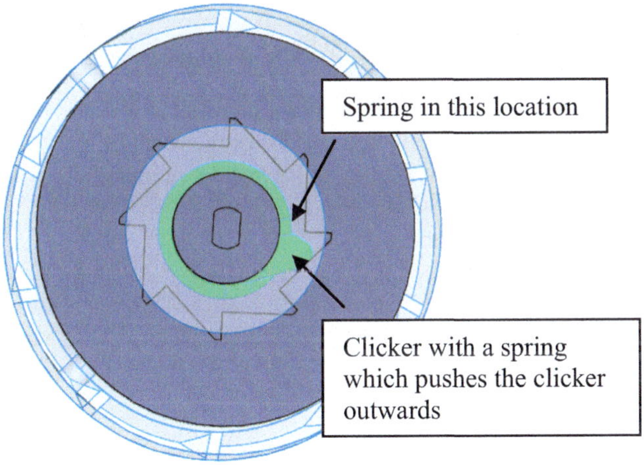

Fig. 7 Ratchet and Pawl Mechanism. Anticlockwise the whole container moves as the clicker engages while the clockwise clicker is free and hence only the centre shaft moves

The passive element also has saw-like edges on the bottom to help with the penetration in the soil.

2.4 *Motor Torque Calculations*

2.4.1 Robotic Arm (Fig. 8)

For the torques required to operate the motors of the Robotic arm on the extraterrestrial surface, the following equations were used. The equations are derived by following basic moment balance multiplied by a certain factor.

For the complete mechanism to work 4 degrees of freedom is required in the robotic arm. Motor 1, 2, 3: have a rotation about the Z-axis while Motor 4 has rotation about the Y-axis.

Nomenclature

Mo1	Mass of motor 1
Mo2	Mass of motor 2
Mo3	Mass of motor 3
M1	Mass of module 1-Active element
M2	Mass of module 2-Passive element
L1	Length of tube 1 (As shown in the figure)
L2	Length of tube 2
L3	Length of the module 1
x3	The distance of CG from the top surface of the module 1
x4	The distance of CG from the top surface of the module 2
ρ	Density of tube
R	Outer diameter of the tube
r	Inner diameter of the tube

Moment equation for:

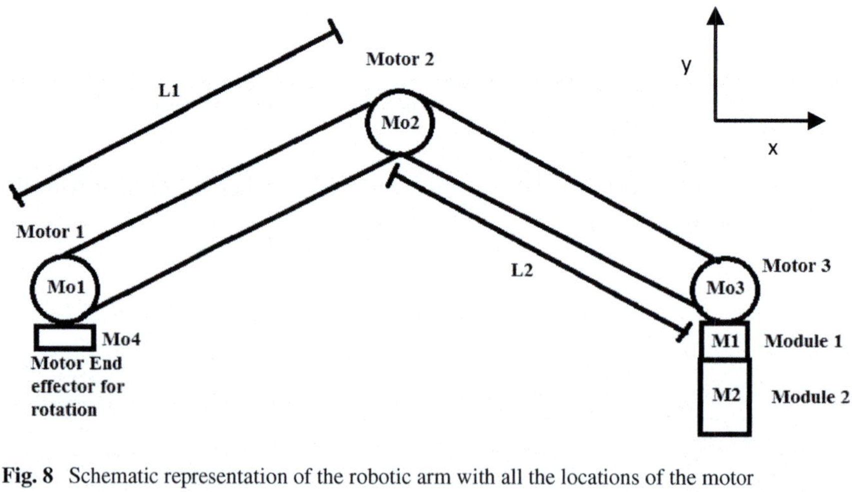

Fig. 8 Schematic representation of the robotic arm with all the locations of the motor

Motor 1 torque:

$$\tau 1 = (\frac{\pi\rho 1L1\left(R1^2 - r1^2\right)L1}{2} + Mo2L1 + \pi\rho 2L2\left(R2^2 - r2^2\right)\left(L1 + \frac{L2}{2}\right) + Mo3(L1 + L2) + M1(x3 + L2 + L1) + (M2 + S)(x4 + L3 + L2 + L1))g \quad (1)$$

Motor 2 torque:

$$\tau 2 = (\pi\rho 2L2\left(R2^2 - r2^2\right)\left(\frac{L2}{2}\right) + M1(x3 + L2) + (M2 + S)(x4 + L3 + L2))g \quad (2)$$

Motor 3 torque:

$$\tau 3 = (M1(x3) + (M2 + S)(x4 + L3))g \quad (3)$$

2.4.2 Excavation Torque Calculation on an Extraterrestrial Surface [5]

$$T_x = -F_{\text{blade}} \sin\alpha + P_p \cos(\alpha - \delta) + F_{\text{side}} \cos\beta + (W_b/g)a_h \quad (4)$$

$$T_y = F_{\text{blade}} \cos\alpha + W_b + P_p \sin(\alpha - \delta) + F_{\text{side}} \sin\beta + (W_b/g)a_v \quad (5)$$

Hence total excavation force T:

$$T = \sqrt{T_x^2 + T_y^2} \quad (6)$$

For Subsurface excavator T_y is considered multiplied by the distance of the centre of the ramp to obtain the torque and T_x is considered for the amount of push force required to go forth with the excavation process. In the case of Surface excavator T is considered the force required for the excavation process. The complete calculations with an explanation are given in the reference mentioned, the appendix below shows the calculations in the form of a MATLAB code. For robotic arm 6 Nm Brushed DC motor for Motor—1, 2, 3, 1 Nm Brushed DC Motor for drilling operation and 0.8 Nm Brushed DC Motor for scoop opening and clamping mechanism. These were chosen based on the availability of motors in the organization for the rapid prototype. The calculation results are given in Tables 1, 2 and 3.

Table 1 Motor torque required in robotic arm

Motors	Torque from calculations for g = 1.6 (Nm)	Multiplying by a factor 2 (Nm)
Motor 1	3.14	6.3
Motor 2	1.136	2.27
Motor 3	0.16	0.33

Table 2 Torque required in an active element for surface excavator

Motor	Torque required by the actuators at g = 1.6 (Nm)	Multiplying by a factor 2 (Nm)
Motor for open and close	0.136	0.27
Motor to lock indents (assuming no jamming)	2.39e−5	4.78e−5

Table 3 Active element for subsurface excavator

Motor	Actuators at g = 1.6	Multiplying by a factor 2
Drilling	1.34 Nm	3.68 Nm
Clamping force required	4.5 N	9 N

2.5 *Rapid Prototyping*

Scale 1:1, Pen for size reference.

The components were made to demonstrate the Kinematics of the Mechanism, and to have a better understanding of how the mechanism works (Figs. 9a, b, 10a, b, 11a, b).

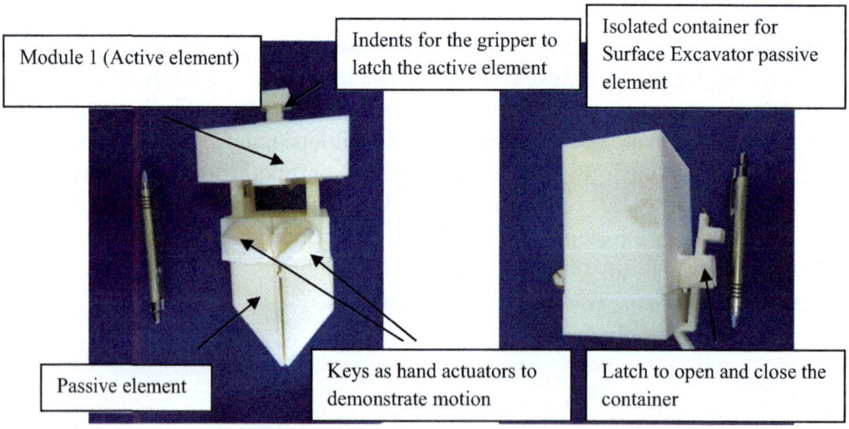

Fig. 9 **a** Module 1 attached. **b** Isolated container which would be on the lander (SE)

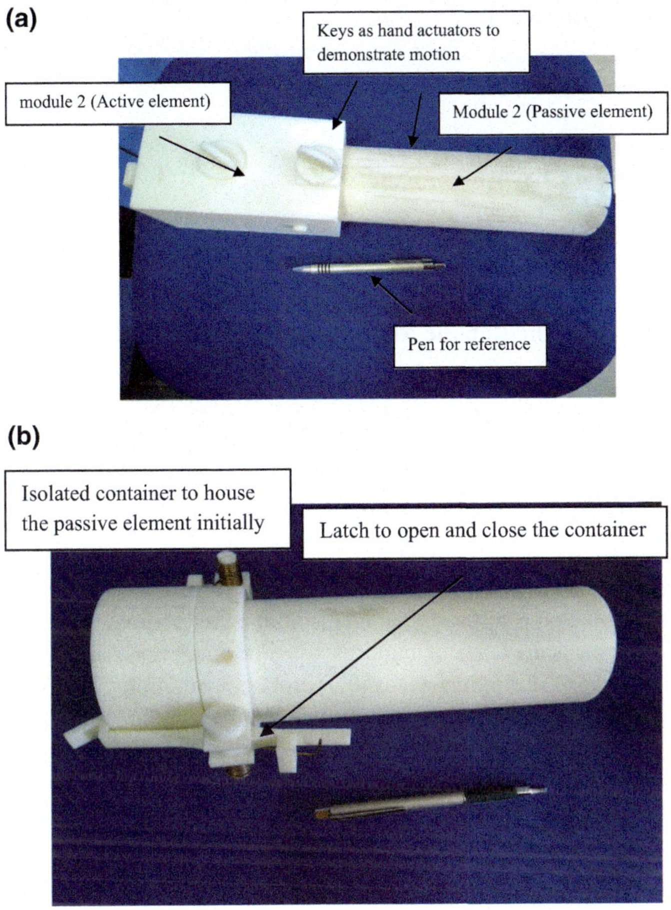

Fig. 10 **a** Module 2 attached. **b** Isolated container which would be on the lander (for SSE)

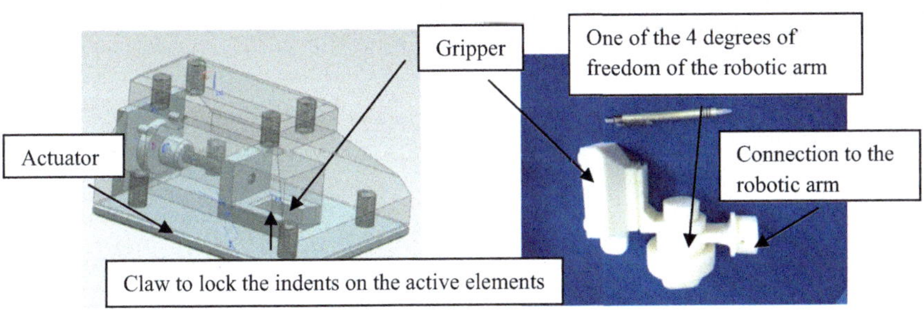

Fig. 11 **a** Gripper attached to the robotic arm. **b** Gripper and the robotic arm interface

2.5.1 Surface Excavator

See Fig. 9a, b.

2.5.2 Subsurface Excavator

See Fig. 10a, b.

2.5.3 Gripper on the Robotic Arm

See Fig. 11a, b.

3 Conclusion and Future Scope

- Currently, the rapid prototype model has been developed, certain experiments have been performed on a soil simulant and from these experiments it was concluded that the subsurface excavator requires some design tweaks in order to achieve greater efficiency during the excavation process while in the case of the surface excavator the experiments showed positive results. An ingenious end effector to bring contamination-free samples from extraterrestrial surface is being presented.
- Detailed design, Space compatible hardware realization, testing and automation to be carried out in the future.

Acknowledgements Firstly, I would like to thank Mr. MP Singh, URSC, for providing me the opportunity to work with excellent scientists at the U R Rao Satellite Centre, Bangalore. I would also like to thank my internal guide Dr. Harish R, SMBS VIT Chennai, for providing support throughout my project period.

References

1. Developing sample return technology using the earth's moon as a testing ground (2012)
2. Neal CR, Shearer CK, Wadwha M, Borg L, Jolliff B, Treiman A, Sandford SA (2005) The power of sample return missions—Stardust and Hayabusa
3. Younse P, Stroupe A, Huntsberger T, Garrett M, Eigenbrode JL, Benning LG, Fogel M, Steele A, Team Sample acquisition and caching using detachable scoops for mars sample return
4. Eigenbrode J, Benning LG, Maule J, Wainwright N, Steele A, Amundsen HEF, and the AMASE (2006) team (submitted) A field-based cleaning protocol for sampling devices used in life-detection studies. Astrobiology, 2008
5. (David) Zeng X, Avenue E, Burnoski L, Agui H, Wilkinson A (2007) Calculation of excavation force for ISRU on lunar surface

Kinematic Modeling of Walking Mechanism

Dishant Kavathia and Jatin Dave

Abstract This paper presents an approach to design and optimize the planar walking mechanism that enables fast locomotion along with minimal actuation. The design of walking mechanism is based on the symmetrical foot-point path to get the required stability for walking on an even terrain and simplify its motion control. A systematic methodology is used for the concept generation of walking mechanisms to achieve these objectives. A brief study of a wide range of walking mechanisms developed in the past is used to develop Computer-aided Design (CAD) model of a single Degree of freedom (DOF) planar walking mechanism. An analytical method for dimensional synthesis based on path generation is used to get dimensions of linkages. Additionally, geometric constraints of motion are used to anticipate free variables to solve for unknown terms in the equation. However, the inclusion of the control system required for simulation in the CAD software is not feasible. Construction stages are shown to aid the development of the simulation model. This is accomplished by interfacing Solidworks and MATLAB/Simulink environments. This paper, in short, is an effort to develop simulation-based methods for optimizing planar mechanisms.

Keywords Path generation · Dimensional synthesis · Geometric constraint programming · Computer-aided design · Simulink model · MATLAB analysis

1 Introduction

As far as mobility and energy efficiency is concerned, walking mechanisms are conceivably better than the conventional wheeled or tracked vehicles for a wide range of terrains. The main advantage of legged locomotion is the use of isolated footholds developed by imitating legged animals in nature. In the course of recent years, researchers have endeavored to build walking machines that are fit for carrying out specific tasks on even terrains. However, many of these walking mechanisms

D. Kavathia · J. Dave (✉)
Institute of Technology, Nirma University, Ahmedabad, India
e-mail: Jatin.dave@nirmauni.ac.in

D. Kavathia
e-mail: 14bme037@nirmauni.ac.in

D. Sen et al. (eds.), *Mechanism and Machine Science*, Lecture Notes in Mechanical Engineering, https://doi.org/10.1007/978-981-15-4477-4_16

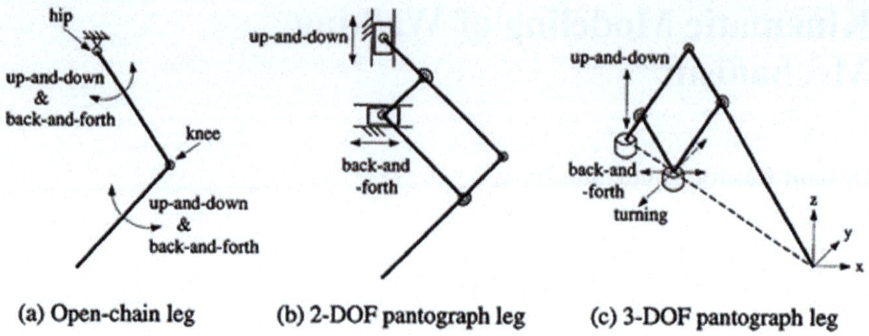

Fig. 1 Different type of open-chain walking mechanisms [4]

were either laboratory scaled machines or not useful in terms of utility and practical applications. Due to the large number of complexity involved in legged locomotion, further research is required in multiple domains of legged locomotion system to adhere to the practical needs of such applications. A wide range of walking mechanisms has been proposed and reported in the literature over the past few decades. Based on the desired objectives and walking terrain, machines up to eight legs and a total of eighteen active degrees of freedom have been considered [1]. The GE Quadruped [2] and Carnegie Mellon University's Hexapod [3] have two DOFs on each leg with a total of eight and twelve active DOFs, respectively. These mechanisms used open-chain linkages for walking on uneven terrain as they are able to provide a mechanical advantage but are less of practical importance taking into account lesser strength and efficiency. On the other hand, a closed-loop linkage can plausibly be treated consistently in Multibody Dynamics (MBD) method enabling facile analysis of kinematic parameters such as velocity, acceleration, and determinant [4] (Fig. 1).

In order to design an energy-efficient walking mechanism, Gravitationally Decoupled Actuation (GDA) system needs to be employed to have individual control over three mutually perpendicular directions, i.e., back-and-forth, up-and-down, and turning motion by different actuators [5]. Walking mechanism equipped with a perfect gravitationally decoupled actuation system can generate a straight horizontal foot-point path with ideally no energy loss in the complete cycle.

In Fig. 2a [5], the power requirement at each joint is non-zero (a1 and a2) since joints are not under gravitationally decoupled design. However, in walking mechanism shown in Fig. 2b [5], the net power generated is theoretically zero as it is designed as per the concept of gravitationally decoupled design. Hence, it can be said that walking mechanism with GDA has higher energy efficiency than the mechanism without GDA [2], but it is still fundamentally limited by control technology to equip the GDA mechanism for a well-structured environment. This limits its practical application and can only be used as a theoretical base to maximize energy efficiency. Moreover in the GDA mechanism, each actuator has to be independently controlled, leading to a multi DOF mechanism and it would also require high-gain feedback to cancel the natural dynamics of the machine itself. The control of such under-actuated

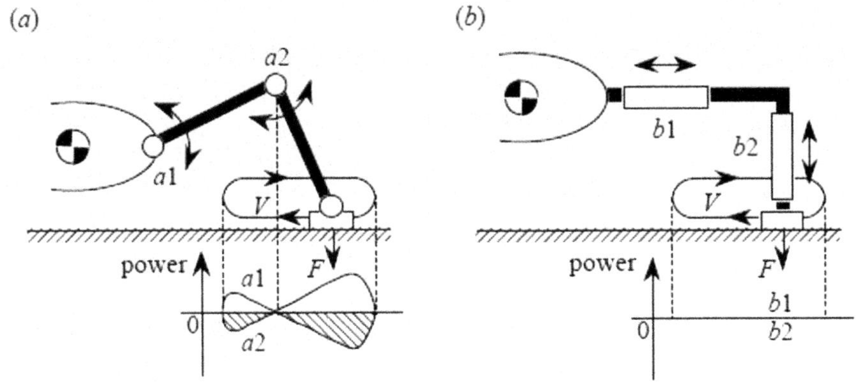

Fig. 2 **a** Non-GDA mechanism. **b** GDA mechanism [5]

mechanisms is an open and interesting problem in the control system with relatively few general principles. Hence, a non-GDA mechanism is developed in further work as the scope of current research is limited to single DOF mechanisms and to simplify the control system by directly obtaining a block diagram with the help of CAD model.

In order to obtain dimensions of various linkages, dimensional synthesis for path generation has been performed. Analytical synthesis for two or three precision points is relatively simple and equations can be solved using computer programming. In this paper, synthesis is carried out for two precision points and Chebyshev's theorem has been employed for minimum structural error. A non-linear system of equations has also been considered by researchers [6]. Further, the solution of dynamics forces and trajectories for foot and joints of the multi-legged robot is available in [7–9] for closed-chain and open-chain type leg mechanism if walking robot.

Using simulation tools [10], mathematical equations are coded and analysis has been performed. In this work, the CAD geometry is prepared after obtaining dimensions from synthesis and integrated with the suitable interfacing options of the simulation environment.

2 Dimensional Synthesis for Path Generation

The subsequent Fig. 3 shows a four-bar mechanism to discuss dimensional synthesis using the path generation method for two precision point case where the coupler point is moving from P1 to P2.

A loop closure equation is shown in Eq. (1) considering the rotation of crank O_2A_1 by an angle β_2 from the initial position in a clockwise direction

$$\overrightarrow{O_2O_1} + \overrightarrow{A_1P_1} + \overrightarrow{P_1P_2} = \overrightarrow{O_2A_2} + \overrightarrow{A_2P_2} \tag{1}$$

Fig. 3 Schematic diagram for two-position synthesis

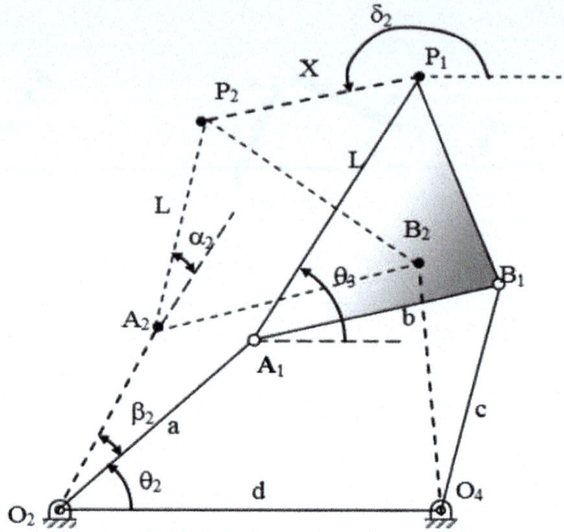

which is the vector addition of the left dyad of mechanism for two precision point synthesis for path generation.

Writing in terms of link lengths and angles, the equation given as

$$Ae^{j(\beta_2+\theta_2)} + Le^{j(\theta_3+\alpha_2)} = Xe^{j\delta_2} + Ae^{j\theta_2} + Le^{j\theta_3}. \tag{2}$$

These results in two scalar equations in term of eight factors θ_2, β_2, A, L, θ_3, α_2, δ_2, and X. Of these, α_2, X and δ_2 are characterized based on the solution from Chebyshev's equation, while θ_2, θ_3, β_2 are considered as free choices (anticipated quantities based on geometric constraints) and A and L remain unknown terms to be solved. Similarly, the vector loop equation for the right dyad is solved. The solution of dimensions also depends on the initial guess values taken for the variables.

Expanding Eq. (2) using Euler's identity,

$$\begin{aligned} &A(\cos\theta_2 + j\sin\theta_2)(1 - \cos\beta_2 - j\sin\beta_2) + L(\cos\theta_3 + j\sin\theta_3) \\ &(1 - \cos\alpha_2 - j\sin\alpha_2) + X(\cos\delta_2 + \sin\delta_2) = 0 \end{aligned} \tag{3}$$

Now, separating the real part and imaginary part of the Eq. (3) and equating it to zero, we get

Real part:

$$\begin{aligned} &A\cos\theta_2(1 - \cos\beta_2) + A\sin\theta_2\sin\beta_2 + L\cos\theta_3(1 - \cos\alpha_2) \\ &+ L\sin\theta_3\sin\alpha_2 + X\cos\delta_2 = 0 \end{aligned} \tag{4}$$

Imaginary part:

$$A \sin\theta_2(1 - \cos\beta_2) + A\cos\theta_2 \sin\beta_2 + L\sin\theta_3(1 - \cos\alpha_2)$$
$$+ L\sin\theta_3 \sin\alpha_2 + X\sin\delta_2 = 0 \quad (5)$$

Both these equations when solved will give a set of values for the unknown terms A and L, i.e., dimensions of linkages. The value for physical parameters such as angle traversed by crank length (α_2), angle made by P_1P_2 w.r.t X-axis (δ_2), and distance traversed by coupler point (X) in Eqs. (4) and (5) have been decided based on Chebyshev's theorem, which is been used to minimize structural error which distributes coupler points evenly to minimize the deviation of actual coupler curve w.r.t desired one. Moreover, with the help of geometric constraints value of θ_2, β_2, θ_3 is obtained and used as free choices to solve the equation.

In the scope of current work dimensional synthesis based on path generation for 3 precision points has been incorporated. For 3 precision point case 12 variables are to be solved from 4 sets of equation, 2 sets of equation from simplified vector loop equation of P_1P_2 and other from P_2P_3, i.e., displacement of the coupler point while traversing path from P_2 to P_3. The solution for 4 sets of equation using the above analytical technique will be time-consuming and cumbersome. Hence, Geometric Constraint Programming (GCP) in which parametric features of CAD tools are used for the graphical imposition of governing mathematical equations using geometric constraints available in the software has been used. This results in a specific set of values to be taken as free choices, then the equations are programmed in MATLAB to obtain dimensions of walking mechanism and to execute its gait analysis.

Hence, these procedures ensure that the mechanism follows the desired path and free from branching defects. The image on the left shown in Fig. 4 depicts that with the use of GCP, the inverse of the mechanism about all the three precision point converges to a single location. Moreover, it is useful to obtain the dimension and orientation of the fixed link. The image on right in Fig. 4 shows the path or coupler

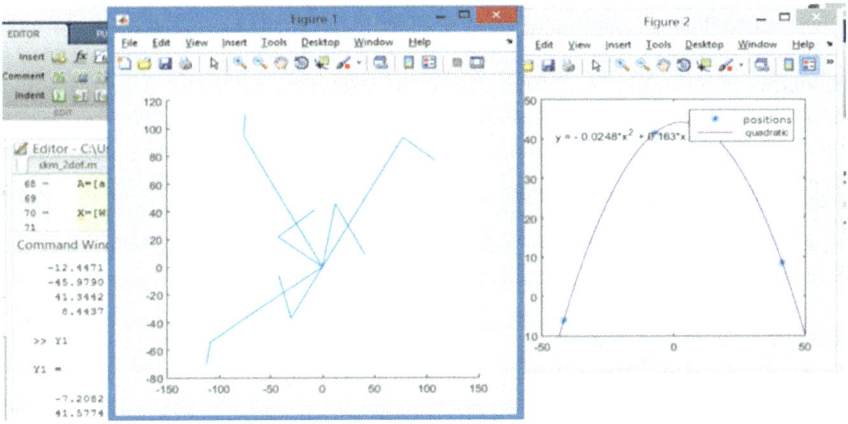

Fig. 4 MATLAB solution for path synthesis

curve of walking mechanism along its 3 precision points and the quadratic equation which resembles a walking gait.

The observation made under the solution of foot trajectory and mechanism motion generation also satisfies the criteria of not consideration of output link with the two rockers and keeping the length of a crack small compared to coupler and follower given by [4].

The evaluation criterion is also inspired by the study of passive dynamic walking which employs a non-GDA mechanism to propel the mechanism by its own weight, does not require a motor and can completely work on mechanical components which is also the motivation behind the current work. Passive dynamic walking has a limitation that it does not have defined walking gait, so control or simulation of the model becomes difficult. Hence, the alternative direction is to advance a single DOF mechanism propelled by a single motor and synthesize it using dimensional synthesis for path generation and incorporating GCP to follow the defined gait.

3 Conceptual Design of Walking Mechanism

An innovative approach to obtain a mathematical model of the walking mechanism involves interfacing MATLAB environment directly to CAD tool such as Solidworks. This method requires dimensional modeling of mechanism in CAD tool which then can be directly converted to a Simulink model using "export to SimMechanics" link obtained after the interface. Both first and second generation models can be obtained from this method and by integrating various Simulink blocks, results for peak crank torque, angular velocity, and acceleration are obtained. The CAD model is shown in Fig. 5 which imported as an XML and implemented in MATLAB/Simulink environment. The block diagram of Simulink is illustrated in Fig. 6.

In order to get a six-bar mechanism with symmetrical coupler curve against the four-bar mechanism, additional dyad is added to a four-bar mechanism to form an embedded regular or skew pantograph. As a result, the coupler curve generated at the output point is amplified from that of the four-bar mechanism. Moreover, for a similar coupler curve in the four-bar mechanism, the linkages would be too large to handle; hence, the six-bar mechanism provides more compactness and control of the mechanism.

For this compound mechanism, one end of the pantograph is connected to and driven by the coupler point C of four-bar linkage, while the other end of pantograph point D is fixed for walking on flat ground. The motion at point C is provided by rotation of the crank and is amplified through the pantograph by a factor of −N at the foot-point E, where N = −DF/CB, here negative sign refers to the inverted shape of the curve at point E as compared to C. The advantage of this compounded mechanism over four-bar mechanism is to get an amplified symmetrical coupler curve at the foot-point.

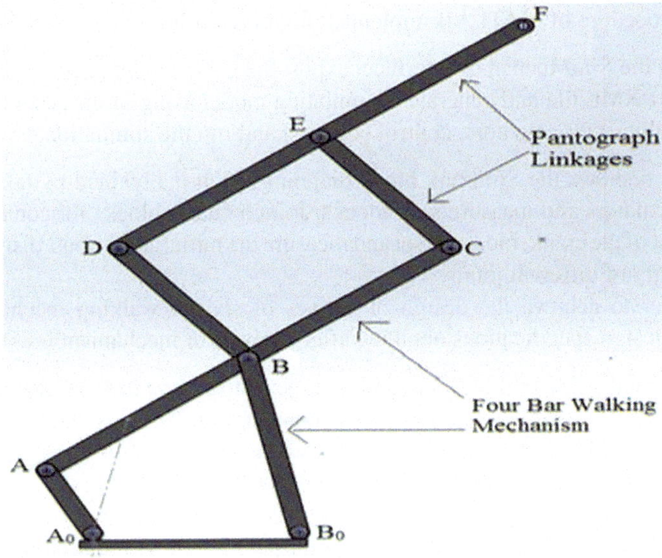

Fig. 5 CAD model of six-bar mechanism

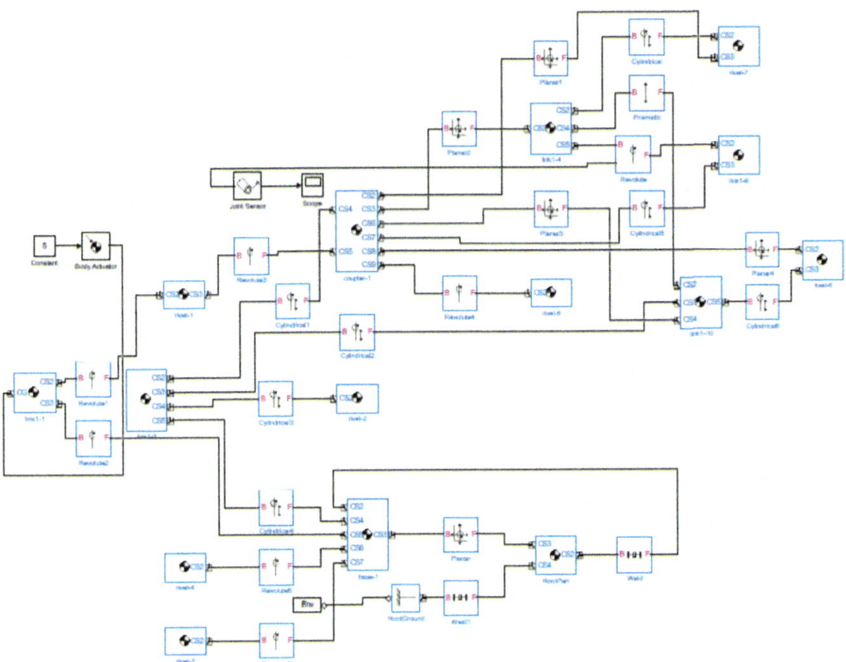

Fig. 6 Simulink block diagram of six-bar mechanism

The procedure of MATLAB implementation is as follows:

1. Using the SimMechanics interface, export the CAD model in XML format.
2. Import XML file and generate a simulation model using SimMechanics.
3. Add blocks for actuators, control equation and run the simulation.

Figure 6 shows the Simulink block diagram which is obtained by taking signal source, actuators and measurement blocks, Joint actuator blocks for controlling the movement of the crank, blocks to set and measure the initial conditions, displacement, and velocity of different joints.

In order to achieve the design objectives of six-bar walking mechanism, the required design specifications obtained after analysis of mechanism is as follows:

1. Horizontal stride (sh) of 0.3 m and a vertical stride (sv) of 0.2 m is required for the back-and-forth, as well as up-and-down motions.
2. The required maximum walking speed for the mechanism is 0.3 m/sec which is equivalent to the angular speed of 30 rpm for the crank.
3. There should be no interaction between propelling and non-propelling path of walking mechanism. Also, the foot-point path height is required to be at least 0.03 m so that any small obstacle can be easily avoided without requiring the actuation of any DOF motion.

4 Conclusions

This paper is a simulation-based approach for understanding planar walking mechanisms with data-driven method for its kinematic synthesis. The paper shows the approach of connecting CAD software with MATLAB/Simulink to obtain dimensions and trajectory of the walking mechanism. In this study, friction, inertia, and dynamic coupling effect are all neglected. However, these effects are required to be considered for the practical application of a walking mechanism in any environment. Hence, the primary focus of further research would be to enhance capabilities for dynamic analysis of the mechanism using the discussed methodology and achieve design objectives.

Acknowledgements I would like to thank Prof. Jatin Dave for guiding me in this research work. Also, I would like to thank the Mechanical department of Nirma University for allowing me to use CAD facility and for providing financial, as well as moral support to complete the research work.

References

1. Morecki, Waldron KJ (1997) Human and machine locomotion. Springer
2. Mosher RS (1969) Exploring the potential of a quadruped. SAE Trans 78(2):836–843 Papers 690150–690349
3. Sutherland IE, Ullner MK (1984) Footprints in the asphalt. Int J Robot Res 3(2):29–36
4. Shieh W-B (1996) Design and optimization of planar leg mechanisms featuring symmetrical foot-point paths. PhD thesis, Institute for Systems Research
5. Arikawa K, Hirose S (2007) Mechanical design of walking machines. Phil Trans R Soc A. https://doi.org/10.1098/rsta.2006.1888
6. Kinzel EC, Schmiedeler JP, Pennock GR (2007) Function generation with finitely separated precision points using Geometric constraint programming. ASME J Mech Des 129:1185–1190
7. Chen X, Wang L, Ye X, Wang G, Wang H (2013) Prototype development and gait planning of biologically inspired multi-legged crablike robot. Mechatronics 23:429–444
8. Orin DE, McGhee RB, Vukobratović M, Hartoch G, Kinematic and kinetic analysis of open-chain linkages utilizing Newton-Euler methods. https://doi.org/10.1016/0025-5564(79)90104-4
9. Ghassaei A (2011) The design and optimization of a crank-based leg mechanism. PhD thesis, Pomona College Department of Physics and Astronomy, Claremont, CA
10. Cekus D, Posiadała B, Waryś P (2014) Integration of modelling in solidworks and matlab/simulink environments. Arch Mech Eng 61(1). https://doi.org/10.2478/meceng-2014-0003

Contact Analysis of a Gear Pair Using Linear Complementarity

Mangesh Pathak, Manish Gautam, and Sourav Rakshit

Abstract The current computation models for gear contact analysis and wear prediction are mostly based on finite element analysis which consumes a lot of time and effort. In this paper, an alternate model to spur gear contact analysis is developed using linear complementarity. A linear complementarity solver computes the contact forces between meshing teeth along the path of contact. From the contact forces, sliding wear in gear tooth is predicted. Archard's wear model is used for the wear prediction. The results of linear complementarity and finite element model are compared for both contact forces and sliding wear. For identical meshing gear pair, the linear complementarity model consumes much less computation time than the finite element model.

Keywords Contact force · Spur gear · Sliding wear · Lemke's algorithm · Archard's wear equation

1 Introduction

Gears are used in number of applications like transmission of power, reduction in speed, etc. Contact analysis is an important part for evaluation of safety, life, and integrity of gears. Normally, two types of analysis models are being followed currently: (1) Lumped Parameter Model (2) Finite Element Model [1], [2]. In this paper, we take an alternative approach for gear contact analysis using linear complementarity [3]. In this work, the contact between a pair of mating spur gear teeth is formulated as a linear complementarity problem (LCP). LCP is used as a time-stepping method to solve multibody contact-impact problems [4]. Gear train is a multibody system with multiple contact configurations. Even in simple case of two meshing gears, multiple points of contact are involved. It becomes difficult to analyze the system with equations of motion alone. Linear complementarity method uses complemen-

M. Pathak (✉) · M. Gautam · S. Rakshit
Indian Institute of Technology, Chennai, India
e-mail: 1993mangesh@gmail.com

D. Sen et al. (eds.), *Mechanism and Machine Science*, Lecture Notes in Mechanical Engineering, https://doi.org/10.1007/978-981-15-4477-4_17

tary variables. In contact problems, normal contact velocity and normal contact force are complementary to each other, i.e., when $F > 0$, $\upsilon = 0$, and when $\upsilon > 0$, $F = 0$ where F is normal contact force and υ is normal contact velocity. Using such complementary pairs, the formulation of multiple contact problems becomes simpler.

In this paper, the linear complementarity problem is formulated for the meshing spur gear pair and is solved for contact forces. Using contact forces, the contact pressures are calculated to predict the sliding wear in gear tooth using Archard's wear model [5]. Finally, the results from linear complementarity model and finite element model are compared. In this work, rigid body model is considered. In future, model will be extended for deformable bodies.

2 Related Work

There are two main areas of research related to our work: linear complementarity and gear wear analysis.

Lotstedt [6] was the first to formulate multiple contact problems with friction using quadratic programming. Baraff [7] developed an algorithm for frictional contact problems based on Dantzig's algorithm which is used for solving linear complementarity problems (LCP). Stewart and Trinkle [8] formulated position-based time-stepping formulation for solving multibody contact-impact problems using Lemke's algorithm [3]. This formulation adopted the polyhedral approximation of friction cone which approximates the true friction cone to a desired accuracy. Anitescu and Potra [9] also developed a formulation based on explicit Euler integration and polyhedral approximation of friction cone. This is a modified version of Stewart and Trinkle's formulation in which the contact velocities are complementary variables rather than the displacements.

Ku [10], [11] studied the tribology of gears and presented the factors affecting gear life. Dudley [12] has given the wear failure modes which occur frequently in gear operations. Wu and Cheng [13] analyzed the sliding wear in spur gears by using their sliding wear model [14] based on Archard's wear approach and derived the formulae for wear rate and gear tooth wear. Flodin and Andersson [5] performed wear simulations based on a generalized Archard's wear equations for both spur and helical gears. This wear model of Flodin and Andersson does not incorporate any tooth profile errors and deflections. Brauer and Andersson [15] developed a wear model with mixed finite element and analytical approach for spur gears. In this model, finite element method is used to determine contact forces and analytical expressions based on Hertz theory are used to calculate contact pressures and widths. Bajpai et al. [16] proposed a similar surface wear prediction methodology by using finite elements-based contact model in conjunction with Archard's wear model for spur and helical gears. This wear model incorporated tooth profile errors and deflections. Similar model was developed by Yuksel and Kahraman [17] for tooth profile wear in planetary gear sets. Following that, a surface wear model was developed for hypoid gear pairs by Park and Kahraman [18].

3 Method

The method for contact analysis of two gears consists of the kinematic model for determining the contact points along the path of contact, and the linear complementarity analysis to get the contact forces at contact points along the line of contact.

3.1 The Kinematic Model

For a pair of mating spur gears, contact points always lie on the path of contact. This path of contact is fixed for a given gear pair owing to the property of involute profile of gear tooth. Sometimes there is only one tooth in contact and other times there are two teeth in contact (Fig. 1). Thus the contact ratio, i.e., number of teeth in contact over the whole path varies between 1 and 2. At the start of meshing, two pair contacts are established, one at lowest point of tooth contact (LPTC: point A') and the other at highest point of single tooth contact (HPSTC: point C') (Fig. 2). As the meshing of gears progress, the leading pair leaves the contact at highest point of tooth contact (HPTC: point D') of pinion, and hence only one pair is in contact from lowest point of single tooth contact (LPSTC: point B') to HPSTC (C') till another incoming pair establishes the contact at A' to initiate two pairs contact again. The expressions for these limiting radii for single gear tooth contact and double gear teeth contact are given as [2], [19],

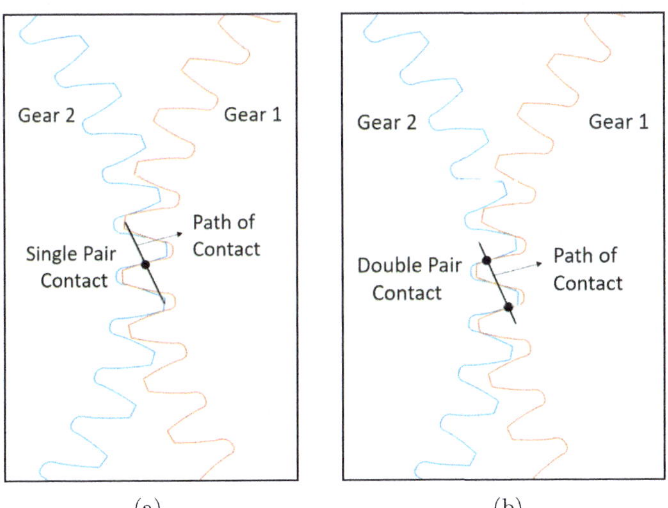

Fig. 1 Meshing spur gears **a** single pair contact, **b** double pair contact

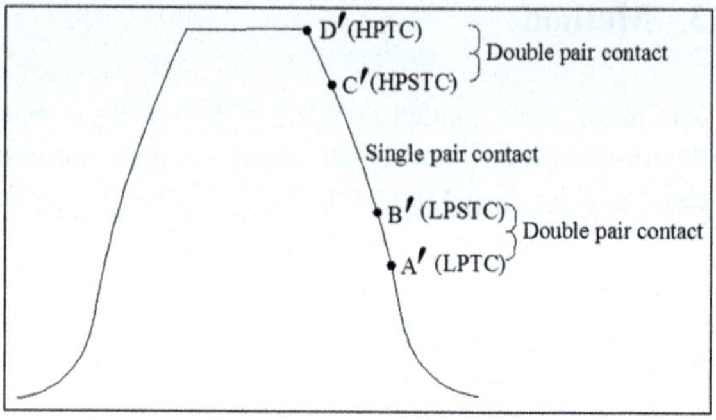

Fig. 2 LPTC, LPSTC, HPSTC, and HPTC points on gear profile

$$r_{HPTC} = r_{a1} \tag{1}$$

$$r_{HPSTC} = \sqrt{r_{b1}^2 + (P_b - AD + \sqrt{r_{a1}^2 - r_{b1}^2})^2} \tag{2}$$

$$r_{LPSTC} = \sqrt{r_{b1}^2 + (\sqrt{r_{a1}^2 - r_{b1}^2} - P_b)^2} \tag{3}$$

$$r_{LPTC} = \sqrt{r_{b1}^2 + (\sqrt{r_{a1}^2 - r_{b1}^2} - AD)^2} \tag{4}$$

$$AD = \sqrt{r_{a1}^2 - r_{b1}^2} + \sqrt{r_{a2}^2 - r_{b2}^2} - a \sin \alpha \tag{5}$$

where r_{a1} and r_{a2} are the tip circle radii of pinion and gear, r_{b1} and r_{b2} are the base circle radii of pinion and gear, P_b is base pitch, a is centre distance, α is pressure angle, and $r_{HPTC}, r_{HPSTC}, r_{LPSTC}, r_{LPTC}$ are the respective radii of points HPTC, HPSTC, LPSTC and LPTC on gear tooth.

The kinematic model for meshing spur gears is developed in MATLAB. The equations for the spur gear involute profile and trochoid fillet profile are used for the generation of exact gear tooth geometry [20].

3.2 *Linear Complementarity Model*

Linear complementarity problem consists of a set of non-negative variables that are pairwise complementary (i.e., either one variable is positive and the other is zero or vice versa) and are related by a set of linear equations. According to Anitescu-Potra's model [9], for every contact point j, the normal contact velocity v_j and the normal contact force F_j form a complementary pair, i.e., $F_j \geq 0, \quad v_j \geq 0, \quad \text{and} \quad F_j v_j = 0$. This statement, which implies that when $F_j > 0, v_j = 0$, and when $v_j > 0, F_j = 0$, is also represented as $0 \leq F_j \perp v_j \geq 0$. n be the number of contact points, and since v_j and F_j are linearly related,

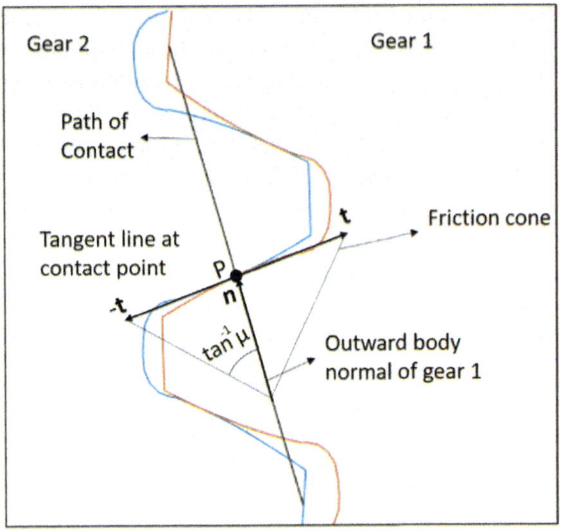

Fig. 3 Friction cone for a pair of meshing gear teeth at contact point

$$\sum_{j=1}^{n} F_j v_j = \mathbf{F}^T \mathbf{v} = \mathbf{F}^T(\mathbf{AF} + \mathbf{b}) = 0 \tag{6}$$

where $\mathbf{A} \in \mathbb{R}^{n \times n}$ is a positive semidefinite matrix having information on body mass and inertia. All external and body forces are grouped in the vector $\mathbf{b} \in \mathbb{R}^n$.

If friction is included, a tangential component of force and velocity at every contact point also need to be considered. The magnitude of the friction force at contact point j is linearly related to the normal force by Coulomb's law. But this lead to non-linear complementarity problems which are difficult to solve. Hence the polyhedral approximation of the friction cone is considered. For two-dimensional case, the tangent space is a line separating two bodies at contact point (Fig. 3) [8], [9], [21].

The Anitescu-Potra's model [9] gives a complementarity formulation for rigid body contact problems with dynamic friction that can be solved using Lemke's algorithm [3]. In Anitescu-Potra's model, at each contact point j, the complementary constraints are

$$0 \le v_{n_j} \perp p_{n_j} \ge 0 \tag{7}$$

$$\mathbf{0} \le \boldsymbol{\rho}_j = \lambda_j \mathbf{e}_j + \mathbf{W}_{f_j}^T \mathbf{v} \perp \mathbf{p}_{f_j} \ge \mathbf{0} \tag{8}$$

$$0 \le \zeta_j = \mu p_{n_j} - \mathbf{e}_j^T \mathbf{p}_{f_j} \perp \lambda_j \ge 0 \tag{9}$$

where $\perp$ indicates the complementary variables, v_{n_j} $(= \mathbf{c}_{n_j}^T \mathbf{v})$ is the normal velocity between the two bodies at contact point j, p_{n_j} is the normal component of the contact impulse, $\mathbf{p}_{f_j}$ is the friction component of the contact impulse, $\mathbf{e}_j \in \mathbb{R}^D$ is a vector of ones, D is the number of edges of the friction pyramid, λ_j is a variable that approximates the magnitude of the sliding velocity, μ is the Coulomb friction coefficient, n_d is the dimension of the configuration space, and $\mathbf{W}_{f_j} \in \mathbb{R}^{n_d \times D}$ is the friction contact wrench matrix that transforms the generalized velocity $\mathbf{v} \in \mathbb{R}^{n_d}$ of the body along the tangential component of the friction cone. For a two-dimensional problem ($n_d = 3$, two translations and one rotation), at each contact point j, there are only two directions $\{\mathbf{t}_j, -\mathbf{t}_j\}$ of the friction component of the contact impulse, each perpendicular to the outward contact normal $\mathbf{c}_{n_j}$ (Fig. 3). The friction contact wrench matrix $\mathbf{W}_{f_j}$ is $\begin{bmatrix} \mathbf{t}_j & -\mathbf{t}_j \\ \mathbf{r}_j \otimes \mathbf{t}_j & -\mathbf{r}_j \otimes \mathbf{t}_j \end{bmatrix}$, where $\mathbf{r}_j$ is the vector from the body centroid to contact point j, and $\otimes$ is the two-dimensional equivalent of cross-product. Thus Anitescu-Potra's linear complementarity model [9], [21] for multiple contact-impact problems can be written as

$$0 \leq \begin{bmatrix} \tilde{\mathbf{A}}_{nn} & \tilde{\mathbf{A}}_{nf} & \mathbf{0} \\ \tilde{\mathbf{A}}_{fn} & \tilde{\mathbf{A}}_{ff} & \tilde{\mathbf{E}} \\ \tilde{\boldsymbol{\mu}} & -\tilde{\mathbf{E}}^T & \mathbf{0} \end{bmatrix} \begin{bmatrix} \tilde{\mathbf{p}}_n^t \\ \tilde{\mathbf{p}}_f^t \\ \tilde{\boldsymbol{\lambda}}^t \end{bmatrix} + \begin{bmatrix} \tilde{\mathbf{b}}_n^t \\ \tilde{\mathbf{b}}_f^t \\ \mathbf{0} \end{bmatrix} = \begin{bmatrix} \tilde{\mathbf{v}}_n^t \\ \tilde{\boldsymbol{\rho}}^t \\ \tilde{\boldsymbol{\xi}}^t \end{bmatrix} \perp \begin{bmatrix} \tilde{\mathbf{p}}_n^t \\ \tilde{\mathbf{p}}_f^t \\ \tilde{\boldsymbol{\lambda}}^t \end{bmatrix} \geq 0 \tag{10}$$

where superscript t indicates value at time t. $\tilde{\mathbf{p}}_n = \left[p_{n_1}, \ldots, p_{n_K}\right]^T$, $\tilde{\mathbf{p}}_f = \left[\mathbf{p}_{f_1}, \ldots, \mathbf{p}_{f_K}\right]^T$, and $\tilde{\boldsymbol{\lambda}} = [\lambda_1, \ldots, \lambda_K]^T$ are the concatenated normal contact impulses, friction impulses, and λs, respectively, for all the K contact points in the assembly, and $\tilde{\mathbf{A}}_{nn} = \tilde{\mathbf{W}}_n^T \tilde{\mathbf{M}}^{-1} \tilde{\mathbf{W}}_n$,
$\tilde{\mathbf{A}}_{nf} = \tilde{\mathbf{W}}_n^T \tilde{\mathbf{M}}^{-1} \tilde{\mathbf{W}}_f$,
$\tilde{\mathbf{A}}_{ff} = \tilde{\mathbf{W}}_f^T \tilde{\mathbf{M}}^{-1} \tilde{\mathbf{W}}_f$,
$\tilde{\mathbf{A}}_{fn} = \tilde{\mathbf{W}}_f^T \tilde{\mathbf{M}}^{-1} \tilde{\mathbf{W}}_n$,
$\tilde{\mathbf{b}}_f = \tilde{\mathbf{W}}_f^T \left(\tilde{\mathbf{M}}^{-1} \left(\tilde{\mathbf{F}}_b + \tilde{\mathbf{F}}_{ext}\right) \Delta t + \tilde{\mathbf{v}}^{t-1}\right)$ and
$\tilde{\mathbf{b}}_n = \tilde{\mathbf{W}}_n^T \left(\tilde{\mathbf{M}}^{-1} \left(\tilde{\mathbf{F}}_b + \tilde{\mathbf{F}}_{ext}\right) \Delta t + \tilde{\mathbf{v}}^{t-1}\right)$.
$\tilde{\mathbf{M}}$ is the concatenated mass matrix, $\tilde{\mathbf{W}}_n = \left[\mathbf{W}_{n_1}, \ldots, \mathbf{W}_{n_K}\right]$ is the concatenated normal contact wrench matrix where $\mathbf{W}_{n_j}^T = \left[\mathbf{c}_{n_j}^T \quad \mathbf{r}_j \otimes \mathbf{c}_{n_j}\right]$, $\tilde{\mathbf{W}}_f = \left[\mathbf{W}_{f_1}, \ldots, \mathbf{W}_{f_K}\right]$ is the concatenated friction wrench matrix, $\tilde{\boldsymbol{\mu}} = diag\,[\mu_1, \ldots, \mu_K]$, $\tilde{\mathbf{E}} = diag\,[\mathbf{e}_1, \ldots, \mathbf{e}_K]$, $\tilde{\mathbf{v}}_n = \left[v_{n_1}, \ldots, v_{n_K}\right]^T$, $\tilde{\boldsymbol{\rho}} = \left[\rho_1, \ldots, \rho_K\right]^T$, $\tilde{\boldsymbol{\xi}} = [\zeta_1, \ldots, \zeta_K]^T$, $\tilde{\mathbf{v}}^t$ is the concatenated generalized velocity at time t, Δt is the time step of simulation, and $\tilde{\mathbf{F}}_b$ and $\tilde{\mathbf{F}}_{ext}$ are the concatenated body and external forces, respectively, for all the parts.

In this problem, it is considered that gears are running in steady state, i.e., running at constant velocity and transmitting constant torque. This ensures static equilibrium condition at all contact configurations. Hence, the contact configurations are independent of each other as meshing progresses along the line of contact, and doing simultaneous static contact analysis at contact points is equivalent to continuous contact analysis. During meshing of gears, four possible contact conditions can be found. The free body diagrams for each of the contact condition is shown in Fig. 4. On driving gear, input torque is applied by some driving agency like motor. The

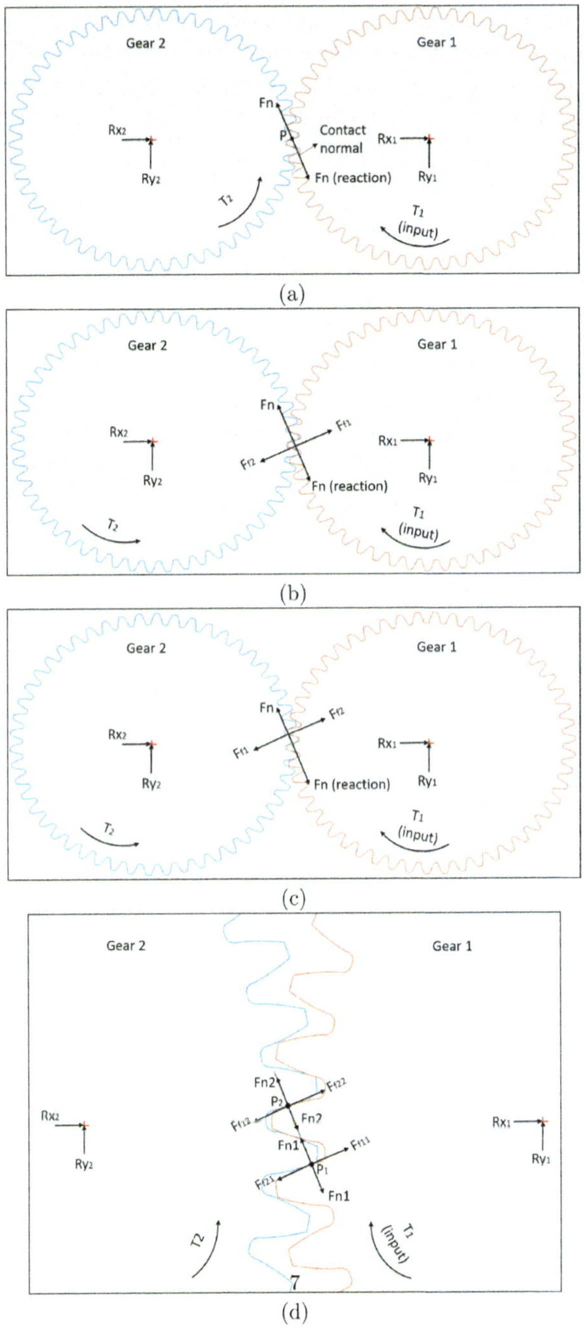

Fig. 4 Free body diagram for meshing gears **a** at pitch point **b** at single pair contact on the path of approach side **c** at single pair contact on the path of recess side **d** at double pair contact

torque is transmitted to the driven gear through contact force. An equal magnitude force will act in opposite direction on driving gear as a reaction. Friction force acts along the line perpendicular to the line of action. The direction of friction force gets reversed at pitch point [22]. For case (a) in Fig. 4, the equilibrium equations for gear 1 are

$$F_{nx} + R_{x1} = 0 \tag{11}$$

$$F_{ny} + R_{y1} = 0 \tag{12}$$

$$F_n(r_{b1}) - T_1 = 0 \tag{13}$$

where F_{nx} and F_{ny} are the x and y-components of the normal contact force F_n, R_x, and R_y are the x and y-components of the reaction forces acting on bearings of gears, T_1 and T_2 are the respective torque values acting on gear 1 and 2, r_{b1} and r_{b2} are the respective base circle radii values of gear 1 and 2. Similarly, the equilibrium equations for gear 2 can be obtained. It can be seen that the reaction forces on bearings and the torque acting on the gear can be found. For case (b) in Fig. 4, the equilibrium equations for gear 1 are

$$F_{f1x} + F_{nx} + R_{x1} = 0 \tag{14}$$

$$F_{f1y} + F_{ny} + R_{y1} = 0 \tag{15}$$

$$-F_{f1}(r_{f1}) + F_n(r_{b1}) - T_1 = 0 \tag{16}$$

where F_{f1} and F_{f2} are the friction forces acting on gear 1 and 2, respectively, r_{f1} and r_{f2} are the respective moment arms of the friction force acting on gear 1 and 2. Similarly, the equilibrium equations for gear 2 can be obtained. Since there is sliding, by Coulomb's friction law, $F_f = \mu F_n$. Thus, the reaction forces on bearings and the torque acting on the gear can be found. Case (c) can be solved in the similar way. For case (d) in Fig. 4, the equilibrium equations for gear 1 are

$$F_{f11x} + F_{f12x} + F_{n1x} + F_{n2x} + R_{x1} = 0 \tag{17}$$

$$F_{f11y} + F_{f12y} + F_{n1y} + F_{n2y} + R_{y1} = 0 \tag{18}$$

$$-F_{f11}(r_{f11}) + F_{f12}(r_{f12}) + (F_{n1} + F_{n2})(r_{b1}) - T_1 = 0 \tag{19}$$

where F_{n1} and F_{n2} are the normal contact forces at contact points P1 and P2, respectively, F_{f11} and F_{f21} are the friction forces acting on gear 1 and 2 at contact point P1, F_{f12} and F_{f22} are the friction forces acting on gear 1 and 2 at contact point P2, r_{f11} and r_{f21} are the respective moment arms of the friction force acting on gear 1 and 2 at contact point P1, r_{f12} and r_{f22} are the respective moment arms of the friction force acting on gear 1 and 2 at contact point P2. Similarly, the equilibrium equations for gear 2 can be obtained. Here there are 4 unknowns R_{x1}, R_{y1}, F_{n1}, F_{n2} and equations available are 3. Thus, the reaction forces on bearings and the torque acting

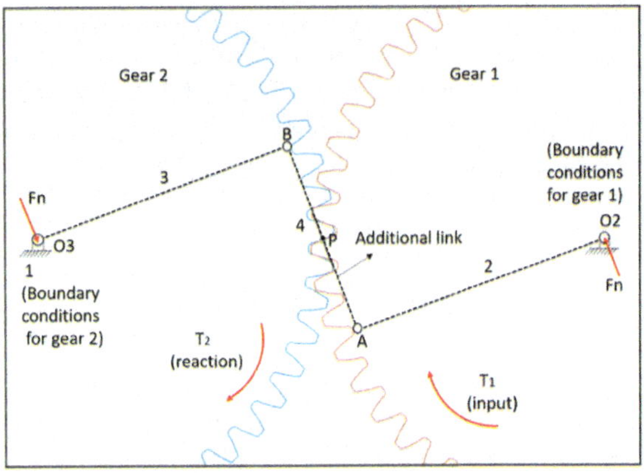

Fig. 5 Instantaneous load and boundary conditions for LCP model of meshing gears

on the gear cannot be found in this case. Thus, this problem needs an alternative method other than elementary mechanics. In Fig. 5, an equivalent mechanism with lower pairs is presented for meshing gears. It can be observed that the contact force always act along the link 4. This generates the reaction forces (of same magnitude as contact force) at bearings (at O2 and O3) and reaction torque on gear 2, which are treated as boundary conditions. Thus, Fig. 5 gives the instantaneous load and boundary conditions for meshing gears. The magnitude of these forces is constant as the input torque is constant. The external force matrix $\tilde{\mathbf{F}}_{ext}$ for gear 1 in our LCP model is given as $\left[-F_{n_x} \quad F_{n_y} \quad -T_1\right]^T$ and for gear 2 as $\left[F_{n_x} \quad -F_{n_y} \quad -T_2\right]^T$. F_{n_x} and F_{n_y} are the x and y-components of the normal contact force F_n, T_1, and T_2 are the respective torque values acting on gear 1 and 2. The standard sign conventions are followed here.

After each time step Δt, we update the positions of gears and use the LCP (Eqs. (6)–(10)) to find the contact forces at the current time step. PATH solver [23] is used to solve the LCP.

3.3 Calculation of Wear

After obtaining contact forces, contact pressure values are computed using expression [24],

$$p_H = 2F_t/\pi a_H \tag{20}$$

where p_H is the local contact pressure, F_t is the load transmitted at a particular point, a_H is the semi-hertzian contact width. Archard's wear model [5] is applied to obtain the wear depth for each point on the gear tooth flank. Archard's wear equations give relationship between wear, sliding distance, and contact pressure.

$$h/S = k_w p_H \quad (21)$$

where h is the wear depth, S is the sliding distance the point is sliding against the interacting surface, p_H is the local contact pressure and k_w is the dimensional wear constant.

4 Results and Discussion

The gear parameters considered for generating gear geometry are given in Table 1. The values of radii for points (LPTC, LPSTC, HPSTC, and HPTC) on tooth profile are calculated from (Eqs. (1) to (5)) and are tabulated in Table 2. The normal and tangent unit vectors at the contact point are given in Table 3.

We have compared our linear complementarity analysis with finite element simulation of the same pair of spur gears in contact using ANSYS software [25]. In finite element analysis, the gear contact problem is considered as a plane stress problem. The meshed model of gear pair is shown in Fig. 6. Fine meshing is done near tooth contact and fillet area to capture stress concentration effects in the contact zone. The

Table 1 Details of gear pair

Parameter	Pinion	Gear
Profile	Involute	Involute
Number of teeth	50	50
Module (mm)	2.5	2.5
Pressure angle (degree)	20	20
Pitch circle diameter (mm)	125	125
Tip circle diameter (mm)	130	130
Base circle diameter (mm)	117.46	117.46
Mass (mm^2)	12271.85	12271.85
Polar moment of inertia (mm^4)	23968449.81	23968449.81
Face width (mm)	15	15
Modulus of elasticity (N/mm^2)	206000	206000
Poisson's ratio	0.3	0.3
Co-ordinates of centroid	[125; 0]	[0; 0]
Contact ratio	1.75	
Wear coefficient ($mm^2 N^{-1}$)	$5e - 10$	
Torque applied on pinion (Nmm)	37500 (clockwise)	
Pinion speed (rpm)	1500	
Coulomb friction coefficient	0.2	

Table 2 Typical points on gear tooth profile

Point	Radii (mm)
Highest point of tooth contact (HPTC)	65
Highest point of single tooth contact (HPSTC)	62.815
Lowest point of single tooth contact (LPSTC)	62.196
Lowest point of tooth contact (LPTC)	60.592

Table 3 Calculated vectors

Parameter	Pinion	Gear
Outward normal vector(c_n)	[−0.3420; 0.9397]	[0.3420; −0.9397]
Tangent vector(t)	[0.9397; 0.3420]	[−0.9397; −0.3420]

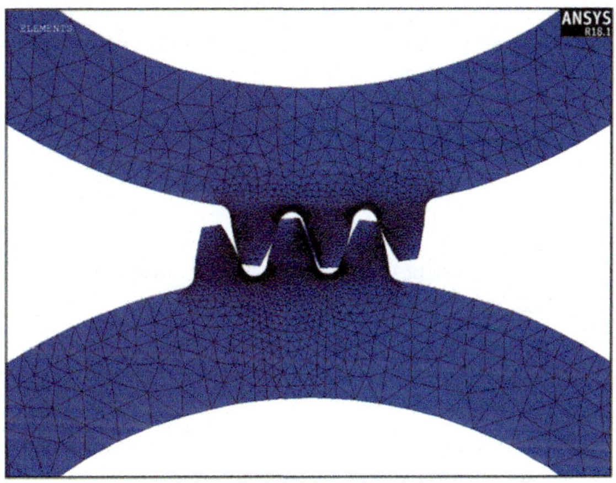

Fig. 6 Meshed model of gear pair from finite element analysis

Table 4 Mesh details of gear pair

1	Type of element	Plane183, Targe169, Conta172
2	Number of elements	87440
3	Number of nodes	166372

mesh details of gear pair are given in Table 4. In both linear complementarity and finite element models, multiple static contact analysis are performed for various contact points between HPTC and LPTC. For points from LPTC to LPSTC and HPSTC to HPTC, double contact pair is considered. For points from LPSTC to HPSTC, single contact pair is considered. In linear complementarity model, the contact force is shared equally between gear teeth during double pair contact (Fig. 7a). During single pair contact, the single gear tooth takes up all the contact load. The expression for sliding velocity in gear pair is given by [5],

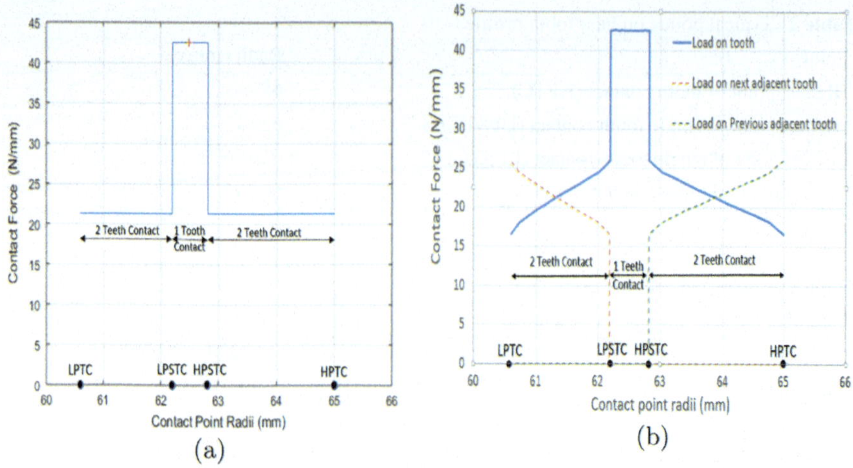

Fig. 7 Load on pinion teeth per unit face width vs contact point radii, **a** linear complementarity model **b** finite element model

$$v_s = (w_p + w_g)y \tag{22}$$

where v_s is sliding velocity, w_p and w_g are magnitudes of angular velocities of pinion and gear, respectively, and y is the distance along line of action to the arbitrary point chosen from the pitch point. Using the contact forces and sliding velocities, sliding wear is predicted for different points on the gear tooth profile (Fig. 9a). In finite element analysis model, the contact load shared between gear teeth during double pair contact is not equal (Fig. 7b). Thus the load sharing ratio is not constant during double pair contact (Fig. 8b). The wear depth versus contact point radii plot for finite element model is shown in (Fig. 9b).

It is observed that in linear complementarity model, contact load during double pair contact is half of the full load whereas in finite element model this value is in between 40 and 60% of the full load. This is may be due to the rigid bodies considered in linear complementarity model and deformable bodies in finite element model. The sliding wear calculated in both the models using Archard's wear equations [5], give sudden increase in wear near LPSTC and HPSTC points due to sudden change in the load. The wear at pitch point is zero due to pure rolling. The wear predicted goes on increasing as we go away from the pitch point due to increasing value of sliding velocity (Eq. (22)). In linear complementarity model, wear plots are steeper in the extreme regions of the gear tooth profile (i.e., near LPTC and HPTC points) because the contact forces are higher as compared to the contact forces from finite element model. For single pair contact region (i.e., LPSTC to HPSTC) both models give same results. The computation time required for the analysis of each contact point in linear complementarity model (0.04 s) is very little compared to that of finite element model (6 min). The configuration details of systems used are mentioned in Table 5.

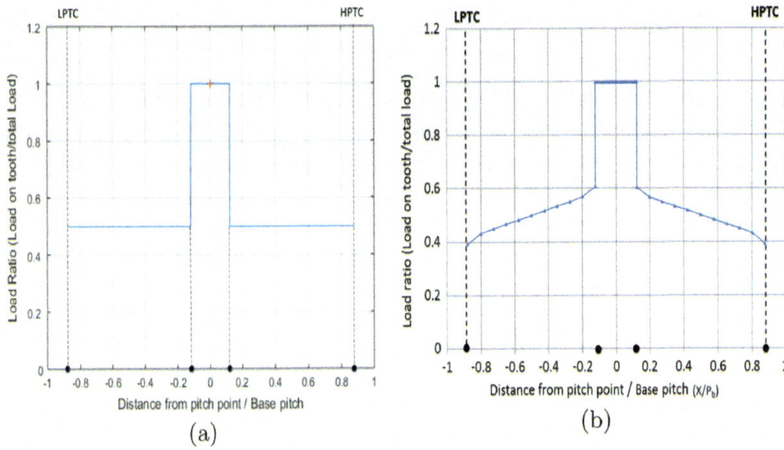

Fig. 8 Load ratio versus distance from pitch point, **a** linear complementarity model **b** finite element model

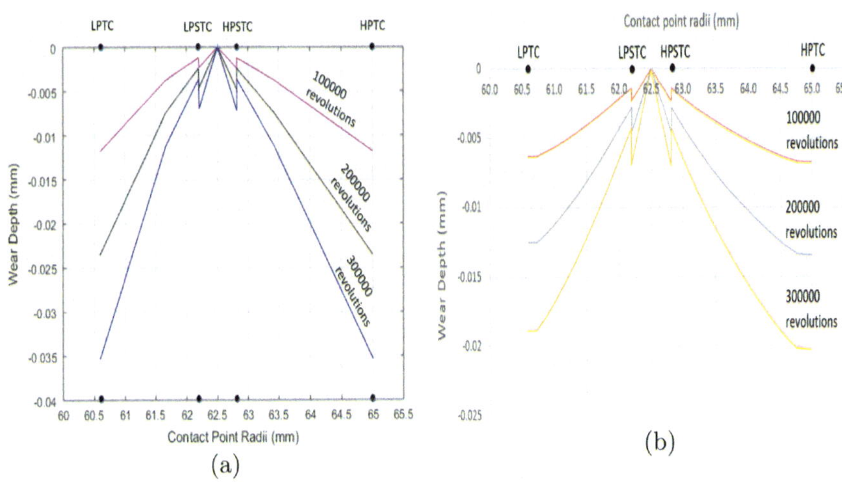

Fig. 9 Wear depth versus contact point radii, **a** linear complementarity model **b** finite element model

Table 5 Details of system configurations used

Configuration	Linear complementarity model	Finite element model
RAM (GB)	16	128
Processor	Intel core i5	Intel core i5
OS name	Windows 10	Windows XP

5 Conclusion

In present work, attempt is made to use the linear complementarity as an alternate method to finite element analysis for the analysis of gear pair contact. The computation time required with linear complementarity model is much less than that with finite element model. The model developed here using linear complementarity principle is meant for rigid bodies only whereas finite element model incorporates the elasticity of bodies. Future work will concentrate on the extension of linear complementarity model to deformable bodies [26].

References

1. Ambarisha V, Parker R (2017) Nonlinear dynamics of planetary gears using analytical and finite element models. J Sound Vib 302:577–595
2. Thirumurugan R, Muthuveerappan G (2010) Maximum fillet stress analysis based on load sharing in normal contact ratio spur gear drives. Mech Based Des Struct Machines 38:204–226
3. Cottle R, Pang J, Stone R (1992) The linear complementarity problem. Soc Ind Appl Math
4. Stewart D (2000) Rigid-body dynamics with friction and impact. Soc Ind Appl Math Rev 42:3–39
5. Flodin A, Andersson S (1997) Simulation of mild wear in spur gears. Wear 207:16–23
6. Lotstedt P (1984) Numerical simulation of time-dependent contact and friction problems in rigid body mechanics. SIAM J Sci Stat Comput 5:370–393
7. Baraff D (1994) Fast contact force computation for non-penetrating rigid bodies. In: computer graphics proceedings, annual conference series, pp 23–34
8. Stewart D, Trinkle J (1996) An implicit time-stepping scheme for rigid body dynamics with inelastic collisions and coulomb friction. Int J Numer Meth Eng 39:2673–2691
9. Anitescu M, Potra F (1997) Formulating dynamic multi-rigid-body contact problems with friction as solvable linear complementarity problems. Nonlinear Dyn 14:231–247
10. Ku P (1974) Tribology of gears and splines. Proceedings tribology workshop, pp 402–427
11. Ku P (1975) Gear failure modes—importance of lubrication and mechanics. ASLE Trans 19:239–249
12. Dudley D (1980) Gear wear. In: Peterson MB, Winer WO (eds), ASME wear control handbook, pp 755–830
13. Wu S, Cheng H (1993) Sliding wear calculation in spur gears. J Tribol 115:493–500
14. Wu S, Cheng H (1991) A sliding wear model for partial-ehl contacts. J Tribol 113:134–141
15. Brauer J, Andersson S (2003) Simulation of wear in gears with flank interferencea mixed fe and analytical approach. Wear 254:1216–1232
16. Bajpai P, Kahraman A, Anderson N (2004) A surface wear prediction methodology for parallel-axis gear pairs. J Tribol 126:597–605
17. Yuksel C, Kahraman A (2004) Dynamic tooth loads of planetary gear sets having tooth profile wear. Mech Mach Theory 39:695–715
18. Park D, Kahraman A (2009) A surface wear model for hypoid gear pairs. Wear 267:1595–1604
19. Hwang S, Lee J, Lee D, Han S, Lee K (2013) Contact stress analysis for a pair of mating gears. Math Comput Model 57:40–49
20. Buckingham E (1963) Analytical mechanics of gears. Dover Publication, New York
21. Lloyd J (2005) Fast implementation of lemkes algorithm for rigid body contact simulation. In: Proceedings of the 2005 IEEE international conference on robotics and automation, pp 4538–4543

22. Howard I, Jia S, Wang J (2001) The dynamic modelling of a spur gear in mesh including friction and a crack. Mech Syst Sig Process 15:831–853
23. Dirkse S, Ferris M (1995) The path solver: a non-monotone stabilization scheme for mixed complementarity problems. Optim Meth Softw 5:123–156
24. Maitra G (1994) Handbook of gear design. Tata McGraw-Hill Publication, New Delhi
25. Stolarski T, Nakasone Y, Yoshimoto S (2006) Engineering analysis with ANSYS software. Butterworth-Heinemann
26. Song P, Pang J, Kumar V (2004) A semi-implicit time-stepping model for frictional compliant contact problems. Int J Numer Meth Eng 60:2231–2261

A Model-View-Controller-Based Software Approach for the Interactive Design of Planar Mechanisms

Burkhard Corves, Mathias Huesing, and Mario Müller

Abstract Nowadays, mechanism design is no longer a process based on pen and paper. Due to the increased complexity of mechanisms, a more effective proceeding is required. Since the appearance of increasingly powerful computers, the number of software tools for mechanism design had likewise increased. This huge number of software solutions can be divided into two categories—geometric software products and dedicated mechanism design software. Since most of these software solutions have disadvantages concerning the function range or usability, a new software product is under development by the Institute of Mechanism Theory, Machine Dynamics and Robotics (IGMR) of RWTH Aachen University. The consideration of new software engineering methods and software design processes allow a continuous implementation of functionalities that are not yet state of the art. One of these functionalities is the combination of analytical and numerical approaches within the kinematical analysis of planar mechanisms. Besides presenting the combination of analytical and numerical approaches, this paper will further outline the motivation to develop a new software solution. In addition, the software architecture as well as the software engineering process will be described.

Keywords Kinematic analysis and design of mechanisms · Combination of analytical and numerical calculation methods · Model-view-controller-based software approach · Scrum for software development

B. Corves (✉) · M. Huesing · M. Müller
IGMR, RWTH Aachen University, Aachen, Germany
e-mail: corves@igmr.rwth-aachen.de

M. Huesing
e-mail: huesing@igmr.rwth-aachen.de

M. Müller
e-mail: mueller@igmr.rwth-aachen.de

D. Sen et al. (eds.), *Mechanism and Machine Science*, Lecture Notes in Mechanical Engineering, https://doi.org/10.1007/978-981-15-4477-4_18

1 Introduction

The Institute of Mechanism Theory, Machine Dynamics and Robotics (IGMR) of RWTH Aachen University has a long tradition regarding the analysis, synthesis, and design of planar, spherical and spatial mechanisms [1]. Within the last decades, the use of mechanisms design software has played an ever-increasing role both in student education and as a tool for the design of mechanisms [2–4]. Such software can be divided into two different groups: general geometric software products (e.g., Cinderella [5] and GeoGebra [6]) and dedicated mechanism design software [4, 7]. General geometry software allows using graphical methods of mechanism theory in a dynamic way, but it also requires explicit knowledge in mechanism design in order to perform analysis or synthesis tasks [8]. In contrast, mechanism design software requires less explicit knowledge in mechanism design and supports the design process with special functionalities in the fields of mechanism synthesis and analysis. On the other hand, dedicated mechanism software often comes along with deficits regarding ease of use and is therefore restricted when it comes to understanding its functionalities intuitively especially if the design engineer has limited mechanism design background.

Based on these deficits, a new software for the development of planar mechanisms—Mechanism Developer (MechDev)—is under development at IGMR. It aims to be easy to handle and at the same time allows complex analysis and synthesis based on a Model-View-Controller approach [9]. To outline the need for action, the motivation for the development of a new software tool is described within this paper precisely. To allow a dynamical extension of the software, a special software architecture based on plug-ins is defined. One of these plug-ins is the kinematical analysis plug-in. It combines analytical and numerical computation methods. The basic approach of this functionality will be presented within this paper. This approach distinguishes between numerically and analytically computable loops of a kinematical structure. It is based on a topological analysis in order to allow an optimized numerical calculation for a given mechanism. As mentioned before, MechDev is designed to be a software extendable with different functionalities, which may not have been envisioned yet. To meet such a demand, a special software architecture is required and will be described in detail within this paper.

2 Motivation

As mentioned before, a huge number of software solutions for the synthesis and analysis of mechanisms were developed during the last decades. Some of them are solutions for a specific synthesis or analysis task; others claim to be a universal tool for the mechanisms design process. By analyzing the state of the art of software solutions, most of these solutions neither include all functionalities that are required for the mechanism design process nor enable adequate software handling.

Software such as SAM [10] or ASOM [11] aim to be an industrial solution for the kinematic and kinetostatic analysis of planar mechanisms. Both software solutions provide features in design, simulation, and optimization of mechanisms. Since both software solutions do not support cam mechanisms, they are no universal tools for the design of planar mechanisms. MechAnalyzer is a software developed for effective teaching and learning mechanisms-related courses [12]. It provides the forward kinematics for more than ten mechanism structures. However, the number of mechanism structures is limited and the preinstalled structures cannot be extended by the user. GENESYS is a software developed by the former IFG of the Leibniz University in Hannover, Germany [13, 14]. The functionalities of this software increased during several years of development. It provides methods for the synthesis and kinematical as well as kinetostatical analysis of planar mechanisms. Since this also includes cam mechanisms, a powerful tool for industrial and academic customers has been developed. The lack of this software is not the functionality but the fact that its handling is neither intuitive nor user friendly. Further powerful tools are OPTIMUS MOTUS [15] and MechDesigner [16]. Especially within the field of cam design, both solutions are outstanding. The disadvantage of both software solutions is that they are very expensive. In summary, there is a need for a mechanism design software that is easy, quick, and intuitive to handle and that provides the necessary knowledge of mechanism theory.

Therefore, the IGMR decided to develop a new software that aims to the requirements listed in Fig. 1.

To meet these requirements, the following questions have to be answered:

- How must a software tool be designed that it can be optimally integrated into the workflow of a design engineer?
- How can repetitive steps in the workflow be shortened and simplified?
- What level of support allows each user to benefit the most from the software?

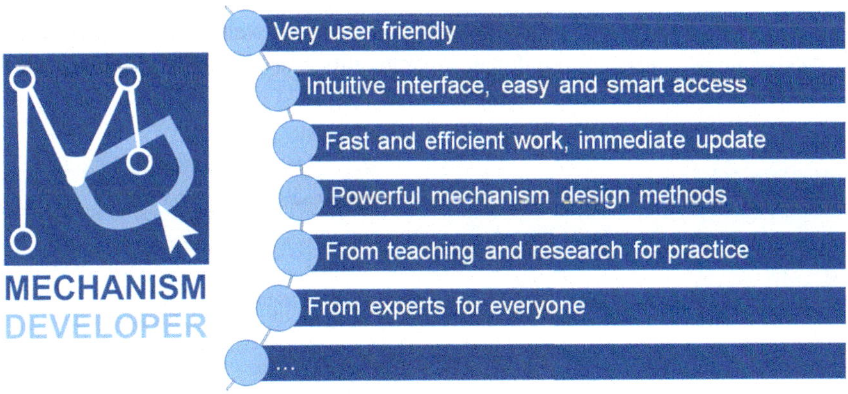

Fig. 1 Requirements of the new mechanism development software MechDev

- How must a computational algorithm look like to support an optimal dynamically interactive working method?

Some of these questions will be answered within this paper and some were already answered by Mannheim [17] or Müller et al. [9].

3 Software Architecture

As already mentioned, the essential requirement is that the software tool should contain powerful mechanism design methods as well as an intuitive and user-friendly interface. Changing requirements with regards to user interfaces and new research results are leading to new mechanism design methods. So the software has to evolve continually. To provide such a continuous evolution, a special software architecture has to be considered. Software architecture itself deals with the high-level structure of a software system. It is the first artifact that can be analyzed to determine how well quality attributes are being achieved [18].

A software architecture that allows a continuous evolution of the user interface as well as the continuous extension in functionality such as new mechanism design methods is the Model-View-Controller (MVC) architecture. This architecture allows a capsulation of each part such as model, view, and control. By this capsulation, each part can be modified independently. That is why MVC is recommended especially for user-interactive systems [19–21]. The model includes the design intelligence and provides the data of the system. The view is the interface to the user and visualizes the data, especially input and output data. The controller takes part in the interaction with the user and forwards the changes to the model. Figure 2 demonstrates the user interaction with an MVC-based software.

The main advantage of the software architecture is that the design intelligence as well as the data model is independent of the user interface and both are loosely coupled by the controller. Each part can be modified or even replaced individually. In case of a mechanisms development software, this leads to the fact that for instance

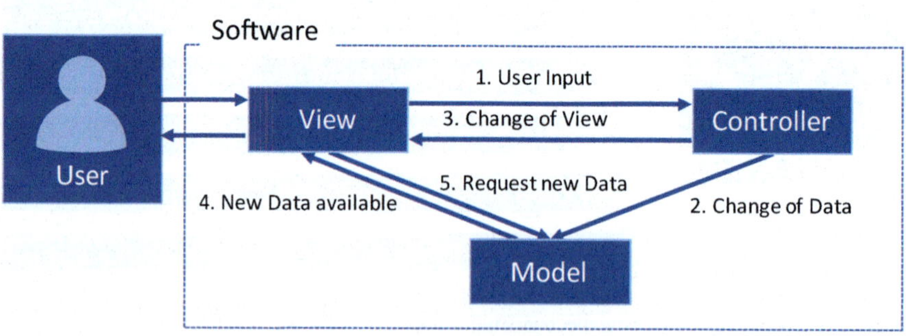

Fig. 2 Interaction of the components in an MVC architecture based on [19]

Table 1 Possible plug-ins for the different MVC components within the software MechDev

Model	View	Controller
Data Model	2-Dimensional View	Provided by RCP
Kinematical Analysis	Analysis Diagrams	
Kinetostatical Analysis	Interactive Diagrams	
Structural Synthesis(*)	Properties Window	
Dimensional Synthesis(*)	Console	
...	...	

Yet to be unimplemented plug-ins are marked by (*).

the methods for the analysis and synthesis can be updated without changing the user interface and vice versa.

To add functions dynamically, the modules of the MVC architecture consist of different plug-ins. The software MechDev is based on the Rich Client Platform (RCP) of Java. This platform provides the possibility to add additional functionalities and update methods via plug-ins. Table 1 shows examples of possible plug-ins. Plug-ins that are not yet implemented are marked by (*). The data model builds the base of the model. It includes the data structure and represents how a mechanism is composited within the software. The data model as well as the kinematical analysis is based on the dissertation of Mannheim [17]. The plug-in concerning the kinetostatic analysis is based on the Newton–Euler equations. The other plug-ins are not yet implemented but can be developed independently based on the data model plug-in.

The user interface of MechDev is shown in Fig. 3. It shows that all View plug-ins mentioned in Table 1 are implemented. The 2-Dimensional View shows an animated sketch of the mechanisms and allows to visualize the motion of the mechanism. Besides, the results of the kinetostatic analysis are shown as arrows indicating joint forces and torques. The analysis diagram sub-window allows the user to visualize data calculated by the model. In addition, the interactive diagram, for instance, allows editing the actuation function of a servo drive. The properties window in the lower left of Fig. 3 shows the properties of the selected element such as the name and its position. It also allows a user interaction, for instance, to rename or relocate the selected element. The console displays feedback for instance if the mechanism cannot be calculated.

The view components can be rearranged individually and modified easily due to the decoupling of view and model. The controller component of the MVC architecture itself is provided by the RCP framework.

Since the capsulation into model view and control allows a modification of every plug-in, the software can be updated and thus improved continuously. Such an agile updating process can be carried out in sprints via the Scrum framework (see Fig. 4).

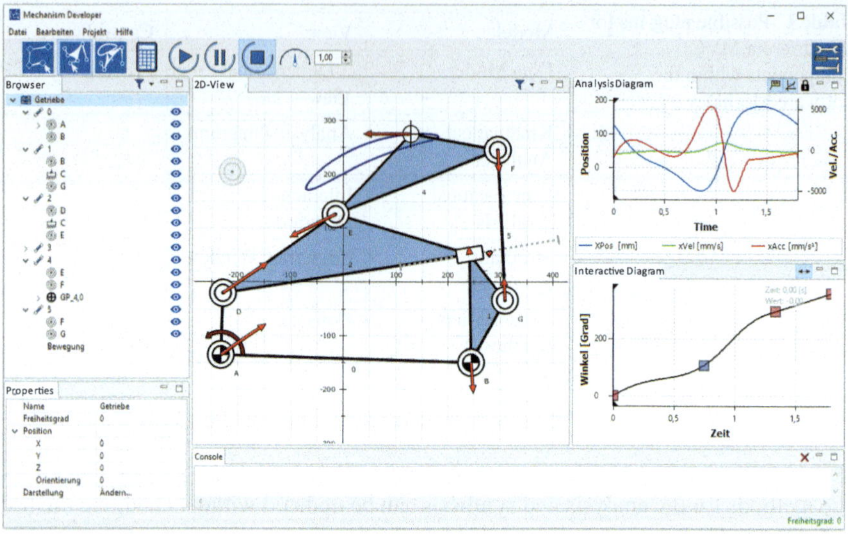

Fig. 3 Interface of the mechanism design software MechDev

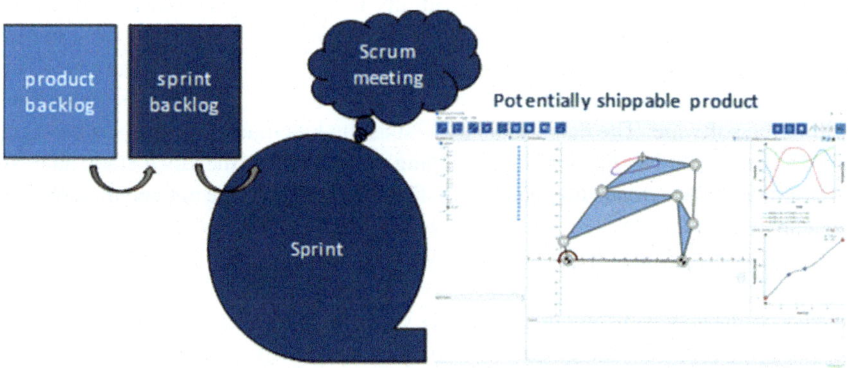

Fig. 4 Sketch of the Scrum process for agile software development

Scrum is an agile method for software development [22]. The software development process is divided into different sprints. A sprint itself is a period in which a useable and potentially shippable product is created (i.e., a new software version). Therefore, product requirements are taken from a product backlog into the backlog of the actual sprint. The product backlog is a list of all initial requirements for the software. Within the duration of the sprint, these requirements have to be implemented on the software. The team responsible for the implementation meets in so-called Scrum meetings. Within these meetings, each team member has to inform the others what he completed and what he plans to complete. Through these regular sprint meetings, each team member always is up-to-date and can react to new circumstances [23].

4 Kinematic Analysis

The kinematic analysis is one of the major requirements for the mechanism design process. The analysis aims to calculate the motion of the mechanism. In MechDev the calculation is realized in a separated plug-in and based on [17]. This special approach differentiates between numerical and analytical computations. Whenever a mechanism or only just a part of it can be calculated analytically, an analytical approach is used. In all other cases, numerical approaches have to be applied. By doing so, the kinematic analysis is as performant as possible. For the decision whether an analytical approach is possible or a numerical approach has to be utilized, the topology of a mechanism has to be analyzed. Therefore, the mechanism has to be fractionated into different loops during the topological analysis. This approach as well as the analytical and numerical analysis is shown within the following sections. The kinematic analysis is described with the help of the academic example shown in Fig. 5.

4.1 Topological Analysis

As mentioned before, the topological analysis leads to numerically and analytically computable fragments—the so-called loops. Figure 5 shows an 8-bar mechanism that can be separated into three independent loops (I–III). One of these loops can

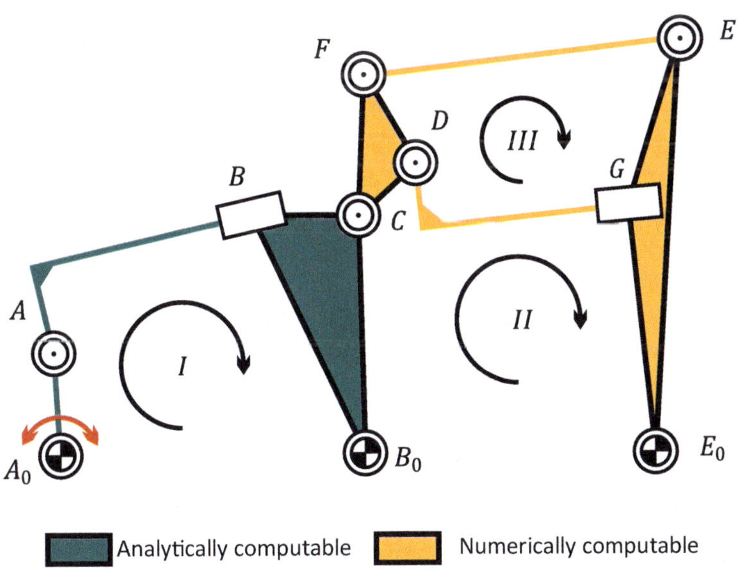

Fig. 5 Mechanism with analytical and numerical computable loops

be calculated analytically and the other two have to be calculated numerically. To determine these loops, the mechanism has to be transferred into a topological graph. The nodes of the graph represent the links of the mechanism. The vertices respectively represent the joints. Based on graph theory, the different independent loops can be found by a depth-first search starting with the frame. The depth-first search is one of the most fundamental and important ways to analyze graphs [24].

The first loop consists of the four joints A_0, A, B, and B_0 as well as their connecting links. Loop II consists of five joints B_0, C, D, G, E_0, and their connecting links. The last loop III contains the remaining joints E and F as well as the joints D and G of loop II.

To decide whether a loop can be calculated analytically, the elements of each independent loop with an unknown position have to be determined. In loop I, the joints A_0 and B_0 have a known position since they are located in the frame and do not change their position over time. In addition, the location of joint A is known because the link $\overline{A_0A}$ is actuated with a defined motion. The only unknown joint in loop I is B. Thus, the kinematical properties of this joint can be determined analytically.

Every other loop contains at least two joints with unknown kinematic properties. Loop II for instance contains the frame joints B_0 and E_0. Besides, the motion of joint C can be determined by calculating loop I. The remaining joints D and G still have an unknown motion, so do the joints of loop III. Loops with more than one unknown joint cannot be calculated analytically and have to be calculated numerically instead.

4.2 Analytical Approach

The previous section dealt with the detection of analytical and numerical solvable loops. This section will show how loop I can be calculated analytically using the approach of the German guideline VDI 2142 [25]. This guideline describes a modular, analytical computation method for dyads consisting of revolute and prismatic joints via complex numbers. The calculation will be described by the example given in Fig. 6.

The kinematic properties (position, velocity, and acceleration) of the two joints D_1 and D_2 are known and represented by the complex numbers z_{D_1} and z_{D_2}. Besides, the constant relative position of the revolute joint D_1 with respect to the prismatic joint S_1 is given by the complex number z_1. The complex number z_2 is known as well. It represents the position of the thrust axis of the prismatic joint D_2 relative to joint S_1. This information can be used to calculate the unknown angle of the thrust axis ψ and the stroke s of joint S_1.

The basic loop equation is represented by (1). The right-hand side of the equation describes the path from the origin to D_1 via D_2 and S_1. Since the constant complex vectors z_1 and z_2 as well as the stroke s are relative to the prismatic joint S_1, they have to be rotated by the angle ψ.

$$z_{D_1} = z_{D_2} + e^{i\psi}(z_1 + z_2 + s) \quad (1)$$

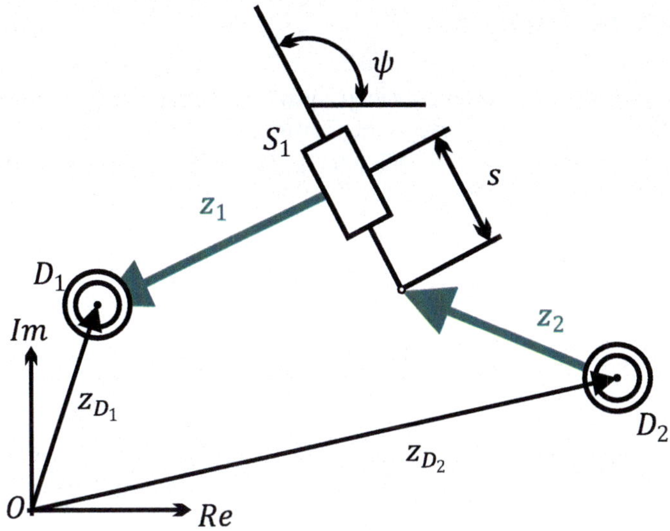

Fig. 6 Dyad consisting of two revolute joints and one prismatic joint

This problem describes geometrically the determination of a tangent to a circle starting from a certain point. To solve this problem, Eq. (1) can be rearranged as follows:

$$w = z_1 + z_2 \tag{2}$$

$$z_{D_1} - z_{D_2} = z_d = e^{i\psi}(w + s) \tag{3}$$

The absolute value of z_d can be squared and rearranged so that the unknown stroke s and the angle ψ can be determined as

$$|z_d|^2 = (\mathrm{Re}(w) + s)^2 + \mathrm{Im}^2(w) \tag{4}$$

$$s = -Re(w) + K \cdot \sqrt{|z_d|^2 - \mathrm{Im}^2(w)} \tag{5}$$

$$\psi_1 = \arg\left(\frac{z_d}{s + w}\right) \tag{6}$$

Since there are always two possible tangents to a circle, the parameter K in Eq. (5) determines which tangent to the circle is chosen for the calculation. K can have the values $+1$ or -1. The calculation of other dyads as well as the calculation of the velocity and acceleration will not be part of this paper. It can be taken from [17] or [25].

4.3 Numerical Approach

The numerical approach is based on loop equations. Each joint has a transformation matrix ${}^{i-1}\boldsymbol{T}_i$. This matrix represents the transformation between link $i-1$ and link i and depends on the position of each joint q_i. For revolute joints, q_i is the angular displacement between the two linkages. That leads to the transformation matrix of a revolute joint $\mathbf{T}_{\mathrm{R}}$ as follows:

$$\boldsymbol{T}_R(q_i) = \begin{pmatrix} \cos(q_i) & -\sin(q_i) & 0 & 0 \\ \sin(q_i) & \cos(q_i) & 0 & 0 \\ 0 & 0 & 1 & 0 \\ 0 & 0 & 0 & 1 \end{pmatrix} \tag{7}$$

For a prismatic joint, q_i represents the linear displacement. Its transformation matrix is given in Eq. (8). The transformation matrix $\boldsymbol{T}_P$ depends on the angle of the thrust axis ψ relative to the previous link.

$$T_P(q_i) = \begin{pmatrix} 1 & 0 & 0 & \cos(\psi)\cdot q_i \\ 0 & 1 & 0 & \sin(\psi)\cdot q_i \\ 0 & 0 & 1 & 0 \\ 0 & 0 & 0 & 1 \end{pmatrix} \tag{8}$$

Since the beginning and ending of a closed loop is the same, the product of every transformation matrix of one loop with n transformations will end in the identity matrix $\boldsymbol{I}$. These transformations contain the joint transformations as well as the link transformations. The transformation matrix of a link contains its geometrical information. So for one loop, the zeros of a function $\boldsymbol{F}(q)$ can be formulated as follows:

$$\boldsymbol{T}_{ges} = \prod_{i=1}^{n} {}^{i-1}\boldsymbol{T}_i(q_i) = \boldsymbol{I} \tag{9}$$

$$\boldsymbol{F}(\vec{q}) = \boldsymbol{T}_{ges}(\vec{q}) - \boldsymbol{I} \overset{!}{=} \underline{0} \tag{10}$$

Since only one solution is required and an initial solution to the problem is always given, Newton's method is a suitable method to find the unknown joint positions q_i. Therefore, a Taylor series can be used to approximate the function up to the first-order derivative:

$$\boldsymbol{F}(\vec{q}) = \boldsymbol{T}^*_{ges} - \boldsymbol{I} + \sum_{i=0}^{n} \left(\frac{\partial \boldsymbol{T}_{ges}}{\partial q_i}\right)^* \Delta q_i^* + \boldsymbol{Res}(\vec{q}, \vec{q}^*) \tag{11}$$

The variables marked with an asterisk (*) are variables of the previous iteration. The joint position of the actual iteration $\vec{q}$ can be calculated by the determination of $\Delta\vec{q}^*$ since the following equation is valid:

$$\vec{q} = \vec{q}^* + \Delta\vec{q}^* \tag{12}$$

Also for the numerical approach, the determination of velocities and acceleration will not be part of this paper and can be taken from [17].

5 Conclusion

The presented paper outlined the motivation to develop a new mechanism design software. Even if there is an enormous number of existing software, the ideal design software does not yet exist. The IGMR of the RWTH Aachen University started the software project MechDev, that aims to develop such a functional as well as ergonomic software tool. The appearance as well as the interaction with the user interfaces change during the software lifecycle. Good software has to follow these changes. To allow such changes MVC is used to capsulate the user interface from the model. The capsulation into plug-ins allows applying new methods in mechanism theory and visual elements to the software. One of these methods is the kinematical analysis explained in this paper. This approach combines analytical and numerical calculation methods in order to provide an efficient analysis of the mechanism.

References

1. Corves B (2016) The first hundred years of mechanism science at RWTH Aachen University. In: López-Cajún C, Ceccarelli M (eds) Explorations in the history of machines and mechanisms. Springer International Publishing, Cham, pp 51–64. ISBN 978-3-319-31182-1
2. Corves B (2004) Computer-aided lectures and exercises: graphical analysis and synthesis in mechanism theory. In: Huang T (ed) 11th world congress in mechanism and machine science, Tianjin, China
3. Kurtenbach S, Mannheim T, Hüsing M, Corves B (2013) Content and realization of mechanism theory at RWTH Aachen University. In: Proceedings of ISEMMS 2013, pp 39–46
4. Hüsing M, Braune R, Corves B (2014) Dimensional synthesis with mechanism processing strategy. In: Flores P, Viadero F (eds) Proceedings of EUCOMES 2014. Springer, pp 139–146
5. Kortenkamp U (2017) Cinderella. https://www.cinderella.de. Accessed 06 Nov 2017
6. International GeoGebra Institute. GeoGebra. https://www.geogebra.org. Accessed 19 Feb 2019
7. Petuya V, Macho E, Altuzarra O, Pinto C, Hernandez A (2014) Educational software tools for the kinematic analysis of mechanisms. Comput Appl Eng Educ 22(1):72–86. https://doi.org/10.1002/cae.20532
8. Corves B, Riedel M, Hüsing, M (2011) Descriptive and intuitive mechanism design and synthesis using geometry-based computer-aided methods. In: 13th world congress in mechanism and machine science, Guanajuato, México

9. Müller M, Mannheim T-P, Hüsing M, Corves B (2018) MechDev—a new software for developing planar mechanisms. In: Kecskeméthy A et al (eds) Third conference on interdisciplinary applications in kinematics, Lima, Peru
10. Artas Engineering Software. SAM. https://www.artas.nl/en/. Accessed 19 Feb 2019
11. info-key. ASOM. https://info-key.de/asom/. Accessed 19 Feb 2019
12. Saha SK. MechAnalyzer. http://www.roboanalyzer.com/mechanalyzer.html. Accessed 19 Feb 2019
13. Braune R (1996) Das Projekt GENESYS. In: Verein Deutscher Ingenieure (ed) Kurvengetriebe, Gelenkgetriebe, gesteuerte Antriebe. VDI Verlag, Düsseldorf, pp 17–46. ISBN 9783180912813
14. Braune R. GENESYS, https://www.getriebe-software-genesys.de. Accessed 19 Feb 2019
15. Nolte R. OPTIMUS MOTUS. https://nolte-nc-kurventechnik.homepage.t-online.de/en/index.html. Accessed 19 Feb 2019
16. PSMotion Ltd. MechDesigner. https://www.psmotion.com/. Accessed 19 Feb 2019
17. Mannheim T-P. Kombination analytischer und numerischer Berechnungsmethoden als Grundlage eines Softwaresystems zum Getriebeentwurf, Dissertation. RWTH Aachen University, Institut für Getriebetechnik und Maschinendynamik, Aachen
18. Bass L, Clements P, Kazman R (2010) Software architecture in practice, 2nd edn, vol 14. Addison-Wesley, Boston, Mass. [u.a.]. ISBN 0-321-15495-9
19. Goll J (2011) Methoden und Architekturen der Softwaretechnik. Vieweg+Teubner Verlag, Wiesbaden. ISBN 3834881643
20. Lucia A de, Ferrucci F (eds) (2013) Software engineering. International summer schools ISSE 2009–2011. Springer, Berlin. ISBN 978-3-642-36053-4
21. Noble J (ed) (2013) Transactions on pattern languages of programming III. Springer, Berlin. ISBN 978-3-642-38675-6
22. Adi P (2015) Scrum method implementation in a software development project management. Int J Adv Comput Sci Appl 6(9):198–204. https://doi.org/10.14569/IJACSA.2015.060927
23. Gonçalves L (2018) Scrum. Control Manag Rev 62(4):40–42. https://doi.org/10.1007/s12176-018-0020-3
24. Ahn H-K, Shin C-S (2014) Algorithms and computation. Springer International Publishing, Cham. ISBN 978-3-319-13074-3
25. VDI 2142. Blatt 2. Auslegung ebener Kurvengetriebe: Berechnungsmodule für Kurven- und Koppelgetriebe. Beuth Verlag GmbH, Verein Deutscher Ingenieure, Berlin

Design Issues for a Walking-Flying Robot

Marco Ceccarelli, Daniele Cafolla, Matteo Russo, and Giuseppe Carbone

Abstract In this paper, a design problem is attached to develop a robotic platform with hybrid mobility capability by combining locomotion and flight operations. Such a complex mobile robot can be required in applications in which one motion capability is not enough to ensure the success of an assigned task, such as in damaged cultural heritage sites, risky industrial environments, or accidented power plants where several kinds of obstacles cannot be overcome by locomotion and flying is not always possible. Design requirements are analyzed in order to identify the peculiarities of combining locomotion and flight in a suitable robotic platform with modular mechanism design that is based on parallel architecture for leg structures and propeller plate for drone flight. Design issues are discussed in terms of size, features, component assembly, user-oriented operation, sensors, and equipment for assigned tasks. A specific example is reported as referring to the design experience of HB platform that has been developed at LARM for applications in Cultural heritage sites both for inspection and intervention. A prototype is illustrated to show the feasibility of the proposed integrated mechanism design with suitable operation capabilities.

Keywords Robotics · Mobile robots · Robot design · Mechanism design · Mechanism analysis

---

M. Ceccarelli (✉)
Laboratory of Robot Mechatronics, Department of Industrial Engineering, University of Rome Tor Vergata, Roma, Italy
e-mail: marco.ceccarelli@uniroma2.it

D. Cafolla
Bioengineering-Biomechatronics Lab, IRCCS Neuromed, Pozzilli, Italy
e-mail: d.cafolla@infordata.net

M. Russo
Faculty of Engineering, Organization University of Nottingham, Nottingham, UK
e-mail: matteo.russo@nottingham.ac.uk

G. Carbone
Department of Mechanical, Energy and Management Engineering, University of Calabria, Rende, Italy
e-mail: giuseppe.carbone@unical.it

D. Sen et al. (eds.), *Mechanism and Machine Science*, Lecture Notes in Mechanical Engineering, https://doi.org/10.1007/978-981-15-4477-4_19

1 Introduction

Usually, robotic mobility is addressed with systems separately for locomotion and flight [1–3]. The first type of systems includes wheeled, crawling, and legged solutions, with a large variety of mechanical designs, as pointed out for example in [3]. For these systems, locomotion is addressed for motion planning with obstacle avoidance even just for movement aims. In addition, mobile robots are used for transportation purposes with payload equipment to pursue system tasks. However, locomotion is focused on ground environments for the motion on surfaces that can be, or not be, flat and known a priori.

Flying mobile robots are developed recently with drone structures, with a variety of solutions for aerial applications, as for example reported in [4, 5]. Usually they have space mobility in 3D space with a small payload equipment mainly for visual inspection and monitoring. Solutions are designed at different dimensional scales and with multiple propellers.

Very few designs have been conceived to combine the two systems in a fully integrated solution and they are still at a laboratory development stage, according to the authors' knowledge. Some solutions are available, even in the market, as a fairly simple collaboration of the two systems, with the ground mobile robot as a platform for the drone system. Interesting examples of robot designs that integrate flying and walking modes can be found in [6, 7]. These robots are characterized by a bioinspired foldable structure that can be used both to fly and to walk, but they are not flexible enough to operate independently in real tasks and their mechanical structure and sensors are very limited.

In this paper, following experiences with an HB design that was developed at LARM in Cassino [8–11], the problem of conceiving a walking-flying robot is attached to outline requirements and concepts for a fully integrated design, by merging the mobility capabilities of locomotion and flight for a mobile robot that is useful in a wide range of applications, as well as in particular environments and tasks where each of the two mobile solutions cannot succeed.

2 Requirements for Full Mobility

Full mobility can be understood as the capability to move in any environment with no limits by obstacles or any other constraint in the environment space. Such full mobility requires capability not only to move on the ground or in the space but also to combine locomotion and flight in an integrated way. In addition, even task goals can require such full mobility integration due to both motion planning and task peculiarities. The above general considerations are summarized in the scheme of Fig. 1, where it is indicated that locomotion mobility can be performed in ground environments independently of the flight mobility for aerial operation but by having

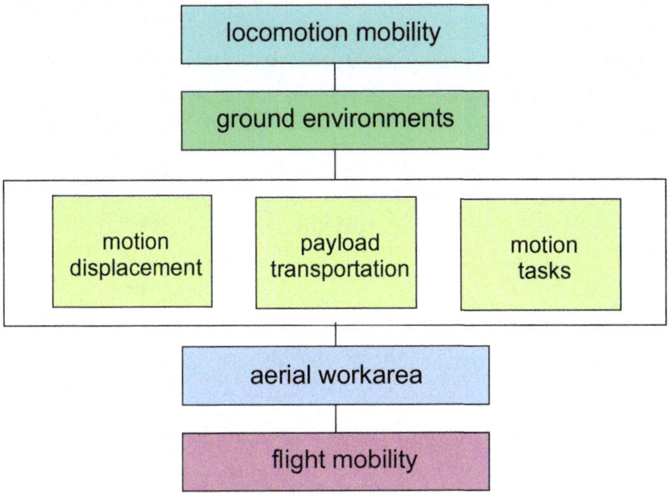

Fig. 1 A scheme for general mobility features

in common or sharing the task goals in terms of motion and transportation issues that may require their concurrent action.

In Fig. 2, requirements are summarized as design issues as coming from locomotion and flight modes separately, but with aspects whose synergy can be considered essential to achieve full mobility with the abovementioned characteristics. In particular, a structure design can be determined as a function of the locomotion mechanisms (crawlers, wheels, or legs) and the flight propellers (in number, material, and shape). Specific design solutions for those components can shape the frame of the overall system but require specific mechanism designs that can be properly defined with simulation for operation performance, action transmission, and strength resistance. The actuation is required with proper power levels and compact sizes to still ensure proper ranges of motion, even according to task needs. The locomotion structure should be designed with proper sizes and characteristics as a function of similar design issues for the flight structure, and vice versa in order to avoid an overestimation or oversizing of one system over the other.

As indicated in Fig. 2, design issues include also a selection of proper sensors for internal control and environment detection that can influence design choices for mechanism solution or can be constrained by the structural solutions. The full-mobility design will require sensors that can detect 3D motion both with vision and other measures with numerical data. The 3D motion performance in speed and acceleration should not limit the sensing capabilities both for the motion control and task operation. Additional equipment can be installed onboard for prescribed tasks as tools, sensors, and other subsystems with an on-demand design. The control system will satisfy the operation regulation of locomotion and flight separately as well as in integrated simultaneous operation.

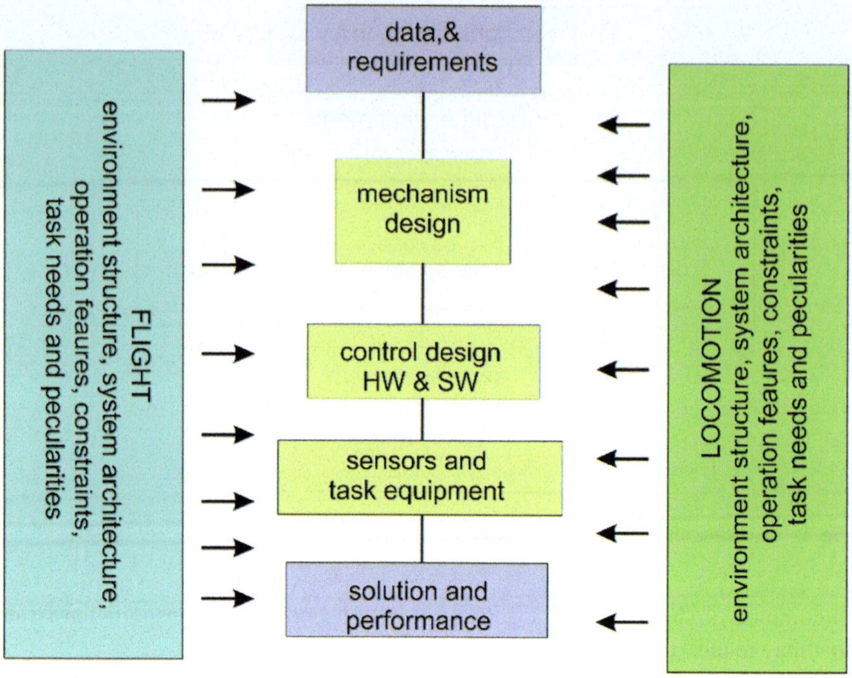

Fig. 2 Design issues for integrated locomotion-flight mobility

The main mobility features of the locomotion and flight modes are summarized in Table 1. The flight mode is characterized by aerial movement speed and by a 3D motion capability that is not limited by ground surfaces. Furthermore, it can be used for aerial monitoring of wide areas through cameras and sensors, thanks to the increased vision horizons. The ground locomotion can be used in clustered spaces with the nearest proximity to the areas under examination. Furthermore, it is characterized by smaller power consumption and stable static balance while walking. Moreover, the combination of locomotors and flying systems can be used to lift an increased payload, to recover dynamically from unexpected obstacles, and to overcome obstacles that are unsurmountable for the ground locomotion module only with optimized power consumption.

Table 1 Mobility features

Flight	Ground locomotion	Locomotion + flight
Aerial speed	Object proximity	Maximum payload
3D motion	Long battery life	Improved balancing
Aerial monitoring	Stable positioning	Obstacle avoidance

Summarizing, it is expected that the design of a locomotion system will give requirements and constraints to the flight system, and vice versa. The specific numerical values of requirements and design issues can be determined by the expected operation of the full mobile system and its main task goals.

3 Conceptual Solutions

A conceptual design for a walking-flying robot can be conceived as either a modular or an integrated system. As shown in Fig. 3, a modular solution is characterized by several modules which are mechatronics-based independent from each other. Each module is able to achieve a single task that is controlled by an independent control architecture (walking for the locomotion module, flying for the flight module, and sensing/manipulation with additional onboard equipment). A central control unit communicates and coordinates module control units in order to achieve system integration. When using a modular approach, each module sees the other subsystems as a black box with a limited number of inputs and outputs only. Thus, each system is easily replaceable with different systems with similar functionality or different performance capabilities. However, full integration of the system for an optimized performance requires proper adjustments that can be difficult in preserving the simple architecture of the overall system.

On the other hand, an integrated solution is characterized by a single mechatronic system where flying, walking, and sensing are embedded, as in the conceptual scheme in Fig. 4. This solution is controlled by a single unit that can operate all the functions of the structure, although the control architecture is considerably more complex than the one of the modular solution.

In Fig. 5, a conceptual solution of a specific solution is presented with modular design for a platform as a 4-leg, 4-propeller walking-flying robot. The proposed design consists of the main platform body, which contains the control units and sensors and is connected to the flight module with flight propellers and the locomotion module with 4-leg walking system. The legged locomotion can be considered more

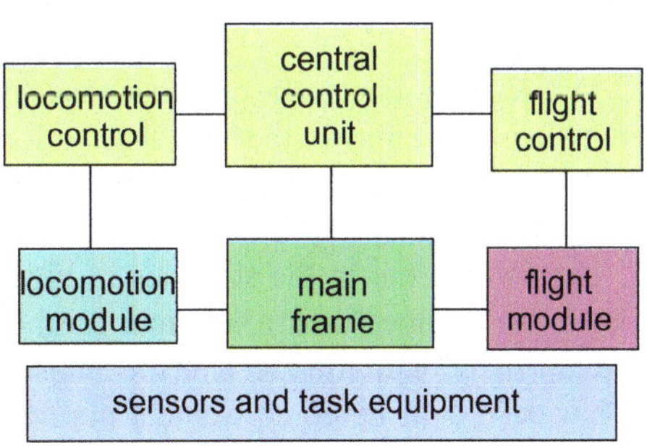

Fig. 3 A modular design for walking-flying robots

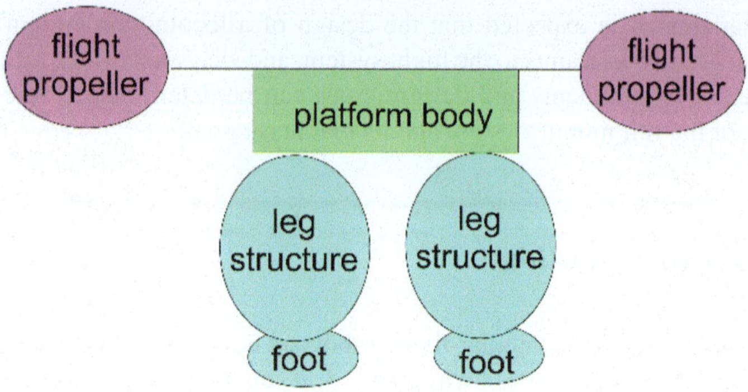

Fig. 4 An integrated design for walking-flying robots

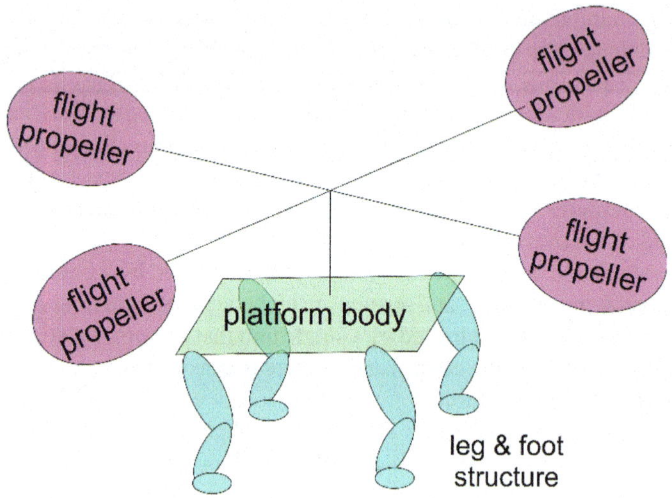

Fig. 5 A conceptual design for a 4-leg, 4-propeller walking-flying robot

convenient in unstructured environments to have the possibility to exploit the walking versatility.

A 4-leg solution can be considered as with the minimum number of legs, although more legs will improve the load distribution on the ground and the system robustness both in structure and gait performance whereas a more complicated design will be required for the system and its control.

As pointed out in the schemes in Figs. 4 and 5, the design solution will much depend on the mechanism solutions for the units and the assembly will have a proper overall shape with the mechanisms chosen for leg structure. In particular, in Fig. 4 it has been considered convenient not to refer to any specific topology in order to indicate the wide well-established solutions for mobile robots, whereas in Fig. 5 an anthropomorphic structure of the legs is drawn to indicate the possibility of the humanoid solution too. However, mechanism solutions can be considered also with the parallel architecture of hybrid solution for combining advantages and reducing drawbacks of specific designs, as for example combining leg mechanisms with wheeled feet.

Summarizing, a combination of locomotion systems with a flight module expands the mobility ranges giving the possibility both of near-motion and panoramic views as well as to overpass constraints of one or the other single mobility solution.

4 Applications

The main applications for mobile robots with flight and ground locomotion capability are in service tasks, such as surveillance and inspection of environments that are difficult or dangerous to reach for human beings. Furthermore, it is possible to add on-board instruments for small intervention and maintenance tasks, especially for examples in nuclear plant environments and Cultural heritage frames. Another important field of application is in case of a natural disaster, when a human being might be unable to reach and inspect areas of natural or urban environments.

The dual nature of the proposed walking-flying robot makes it able to work both in clustered spaces, where a drone could easily collide with obstacles, and in open spaces that are unreachable through ground locomotion. The locomotion-flying robot design allows high flexibility with on-board equipment, which can include cameras, a lidar, infrared sensors, and other kinds of vision systems, and even small instruments for interventions, such as robotic arms with interchangeable tools.

A walking-flying robot can perform many other services when equipped with additional systems for specific tasks or even for manipulation abilities, such as for monitoring and intervention in bridge structures or in aerial cable lines.

The flying capability can be also exploited not really for flying but for floating extra load capability by the locomotor unit in application areas where flying is not permitted either by the environment conditions or even by law restrictions. Such a floating operation of the flight module can also be used to help the stability of the walking in areas that in general are not suitable for walking locomotion.

5 An Example

An illustrative example is reported by the direct experience of the authors at LARM in developing a full mobile platform for operations within Cultural heritage frames.

The HB platform has been designed with a modular design of locomotor and drone for a synergic operation in tasks for inspection and small interventions in restoration activities, [8–11]. A prototype as in version HB III has been designed and built with a structure merging the locomotor frame with the drone frame in a unique plate giving a fully integrated system.

Figure 6 shows two conceptual solutions that have been worked by referring to the above schemes in Figs. 4 and 5. In particular Fig. 6a refers to a modular design in which the units are indicated by their plate frames: one for the drone module, one for sensor and equipment module for a customized task, and one for the legged locomotor unit. In Fig. 6b the abovementioned plate frames are merged in one platform plate hosting both the flight propellers and the legged system with the equipment and sensors on board of it. The first solution with modules can be considered more flexible in being adaptable with more activity in design and operation regulation to a variety of tasks with specific features whereas the second one with only one robot body can be understood as a universal-like system. In addition, in both solutions in Fig. 6 the locomotion unit is designed with legs having parallel architectures that will ensure high payload when a suitable mechanism solution will also give a proper large workspace of the foot extremity.

The HB III prototype in Fig. 7 refers to a solution with an integrated design with one platform body as an evolution from previous solutions that looked at the modular design. It is designed with a size of 757 × 757 × 470 mm and weighs 2.85 kg for applications in Cultural heritage frames with relatively small structures both for limiting the encumbrance and the effect of its action in the working area. It has been assembled by using 3D printed parts, commercial actuators, control boards, and sensors also to achieve a user-oriented design and operation. Its main technical specifications are summarized in Table 2.

The HB III platform design in Fig. 7 is characterized by

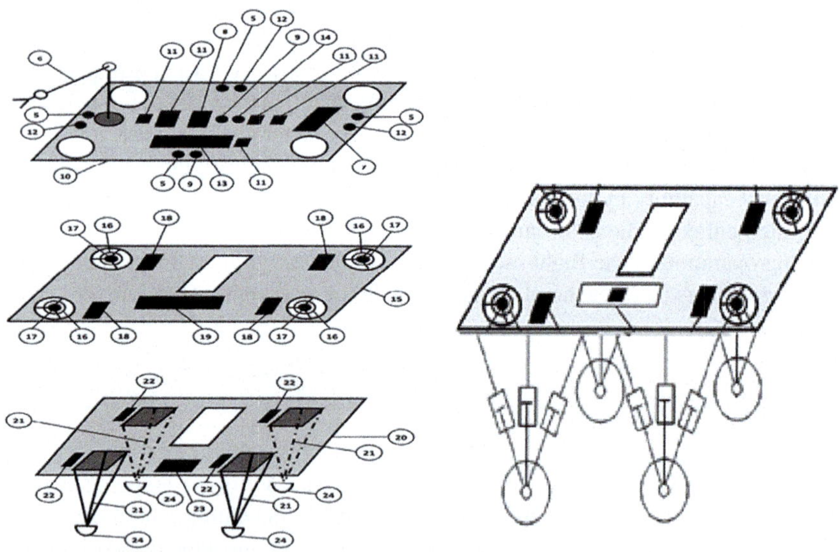

a) b)

Fig. 6 A conceptual design of a walking-flying service robot for cultural heritage frames: **a** a patent scheme; **b** a kinematic design

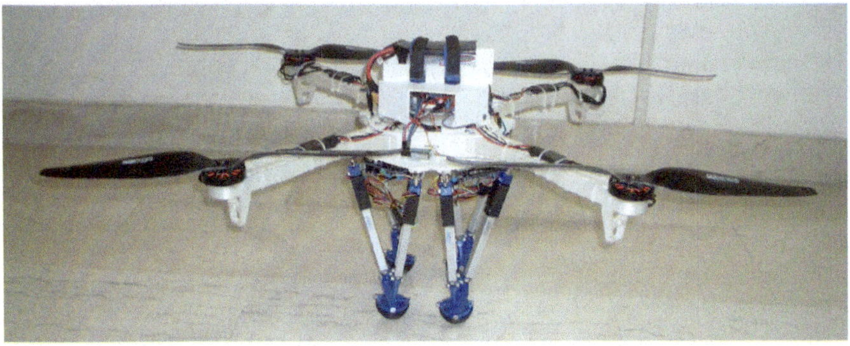

Fig. 7 A lab prototype of the HB robotic platforms

Table 2 Main design characteristics of the HB III prototype in Fig. 7

Feature	Ground locomotion	Flight mode
Maximum power consumption	60.0 W	400.0 W
Payload capability	120.0 N	40.0 N
Working distance	100.0 m	50.0 m
Battery life	360 min	30 min
Speed	0.1 m/s	8 m/s

- a platform structure with merged modules for locomotor, drone, and main frame;
- plate design in size and shape with the capability to host a variety of sensors both in the upper and bottom free surfaces as a function of the specific tasks for the application;
- design solutions and component selections with low-cost features and user-oriented operation according to service application and expected users;
- leg mechanism with a tripod parallel architecture;
- the drone frame with a quadcopter structure with light arms.

The leg architecture is a 4 degrees-of-freedom mechanism that is based on a 3-SPR parallel manipulator that has been developed at LARM initially for the Humanoid design project. It consists of a base platform fixed to a waist/trunk, three limbs connected to the platform with a spherical joint with a varying length by a linear servomotor, and a moving foot-plate with all the limbs converging to its central point, thanks to a specific connecting linkage.

The drone frame has been designed with a light solid structure although reconfigurable mechanisms could be included in the arm design to redirect the action of the propellers for a more flexible flight.

The operation feasibility of the HB III prototype has been validated with experimental tests both in indoor and outdoor environments for ground, aerial, and combined mobility, as reported in the examples in Fig. 8.

a)

b)

c)

Fig. 8 Snapshots of operation tests of the lab HB prototype in Fig. 7: **a** walking; **b** flying; **c** flying-assisted walking

In particular, snapshots of a walking operation test in Fig. 8a show HB III in its home position and in two different steps of a walking operation in a grass field. In Fig. 8b, the prototype is shown while it is ready to fly, taking flight, and flying in open space. A flying-assisted walking for improving payload and motion balancing is reported in Fig. 8c, in home position, during a step, and in the final position within a narrow indoor space.

The HB platform design has been useful to implement the design concepts of a fully mobility robotic platform for service applications with a direct experience and feasibility of a design development with user-oriented solutions.

6 Conclusions

This paper presents considerations for designing mobile service robots with flying and walking capabilities in an integrated design. Requirements and design issues are discussed to propose either a modular design or a fully integrated design of a system with a legged walking robot and a drone. The feasibility of the conceptual design is illustrated by referring to the HB platform design that the authors have recently developed for tasks in Cultural heritage frames.

References

1. International Organization for Standardization (2017) Robots and robotic devices—vocabulary. ISO 8373:2012
2. International Federation of Robotics (2017) World robotics 2017 service robots
3. Ceccarelli M, Kececi EF (eds) Designs and prototypes of mobile robots. Robotics engineering book series. ASME Press. ASME, New York
4. Bachrach A, He R, Roy N (2009) Autonomous flight in unknown indoor environments. Int J Micro Air Veh 1(4):217–228
5. Bircher A, Alexis K, Burri M, Oettershagen P, Omari S, Mantel T, Siegwart R (2015) Structural inspection path planning via iterative viewpoint resampling with application to aerial robotics. In: Proceedings of IEEE international conference on robotics & automation (ICRA 2015), Seattle, pp 6423–6430
6. Daler L, Mintchev S, Stefanini C, Floreano D (2015) A bioinspired multi-modal flying and walking robot. Bioinspiration Biomim 10(1):016005
7. Daler L, Lecoeur J, Hählen PB, Floreano D (2013) A flying robot with adaptive morphology for multi-modal locomotion. In: Proceedings of IEEE/RSJ international conference on intelligent robots and systems (IROS 2013), Tokyo, pp 1361–1366
8. Ceccarelli M, Cafolla D, Carbone G, Russo M (2016) Device with legs and helices. Italian Patent no 102016000103321
9. Ceccarelli M, Cigola M (2012) Service robots for restoration of goods of cultural heritage. Chapter 12. Service robots and robotics-design and application, engineering science reference. IGI Global, Hershey, pp 213–228
10. Ceccarelli M, Cafolla D, Russo M, Carbone G (2017) Design and construction of a demonstrative heritagebot platform. In advances in service and industrial robotics, mechanisms and machine science. Springer, Cham, vol 49, pp 355–362
11. Ceccarelli M, Cafolla D, Carbone G, Russo M, Cigola M, Senatore JL, Gallozzi A, Di Maccio R, Ferrante F, Bolici F, Supino S, Colella N, Bianchi M, Intrisano C, Recinto G, Micheli AP, Vi stocco D, Nuccio MR, Porcelli M (2017) In: Proceedings of HeritageBot service robot assisting in cultural heritage, robotic computing (IRC). IEEE first international workshop on robotic computing for cultural heritage (IRCCH 2017), Taichung City, pp 440–445

Indoor Robot Navigation and Mapping Using Sensory Fusion

Giorgio Ballestin and Teresa Zielinska

Abstract A mobile robot needs to perform localization and navigation tasks. In indoor environments, global localization sensors such as GPS cannot be used. Thus the robot has to build an environmental map using other available sensors (e.g., laser scanner, stereo camera, and bumpers) and localize itself on this map. The goals of this work are (i) to design and implement sensory fusion (available sensors are laser scanner, stereo camera, bumpers, and odometry) during creation of an environmental map; (ii) to perform obstacle detection during navigation; and (iii) to perform goal-driven navigation with obstacle avoidance. The work has been carried out on a Seekur Jr. robot by Adept MobileRobots, which in the state of the art uses most often the laser scanner alone. The proposed approach is a novelty technique to obtain a 3D enviromental map by merging the precision of the laser scanner with the capability of obtaining dense spatial information of the cameras.

Keywords Mobile robots · 3D reconstruction · Stereo vision

1 Introduction

In the past years a lot of research has been produced in the field of autonomous robots. These robots are often useful in many fields such as space industry, for planetary rovers, and even as domestic robots. The evolution of robotics is gradually allowing the utilization of robots not only in industrial environments, but also in people's daily lives, performing menial or difficult tasks.

G. Ballestin (✉)
Department of Informatics, Bioengineering, Robotics, and System Engineering (DIBRIS), University of Genoa, Genova, Italy
e-mail: giorgio.ballestin@gmail.com

T. Zielinska
Faculty of Power and Aerospace Engineering, Warsaw University of Technology, Warszawa, Poland

D. Sen et al. (eds.), *Mechanism and Machine Science*, Lecture Notes in Mechanical Engineering, https://doi.org/10.1007/978-981-15-4477-4_20

One of the most important characteristics of autonomous working robots is their ability to navigate in the surroundings. To accomplish this task, the robot must be well aware of the environment in order to make a proper action planning using the sensor data available as feedback. Sometimes it is also required to achieve such knowledge in unknown and unstructured environments, with possible changes of the scene over time, like doors being opened or chairs being moved. In this scenario, the robot must be able to create a map of the surrounding environment and localize itself inside it. In order to perform complex tasks, a high-level representation of the environment is needed, carrying as much information as possible.

Sensory fusion is widely applicated in contemporary robotics, both with sensors of the same kind [1–3], for example by using multi-resolutional sensors [4], and with different complementary types [5–8]. Applications of sensor fusion on mobile robots often involve obtaining the localization from several sensors by feature matching from an existing database, or using SLAM techniques and probabilistic approaches. Some approaches derive from information theory, using neural networks, Dempster–Shafer reasoning [9], or fuzzy logic [10, 11]. Our objective is to obtain a 3D representation of the environment using cheap cameras, reducing the noise, thanks to the precision of the laser scanner. Point clouds are usually obtained by using RGB-D devices (i.e., kinect); the cheaper counterpart is using a stereo rig of high-resolution images. The aim of this project is to use low-resolution cameras instead.

The state of the art for 3D reconstruction using stereo vision is very well defined [12–14]: the general method to obtain a 3D point cloud from a pair of images involves finding the correspondence of the same feature in two (or more) different views, and then deriving metric distance information from the fact that the rays ideally intersect into a unique point [15]. The process requires a precise model of the cameras' geometry: multi-view and epipolar geometry are in fact considered fundamental aspects of the computer vision field [15–17].

Some issues commonly related to 3D reconstruction can be related to occlusions, texturelessness of the images, and difference in the images due to perspective variation. Occlusions will happen when a feature is missing in one of the views because it is covered by something else, and thus prevents a proper matching and triangulation. A solution has been proposed in a past research [18] by proposing a new way to generate the disparity space image (DSI), using correlation rather than subtraction. Texturelessness of the images is the lack of image variation (e.g., blank walls), which severely increases the uncertainty of a correct match. The perspective variation poses a similar threat, increasing the difficulty of finding matches.

Different approaches to reconstructing the scene involve choosing different algorithms to match the features from the stereo pair, by using different criteria and cost functions. Choosing a different algorithm has repercussions on the quality of the obtained disparity image and therefore of the final point cloud [19].

The aim of this study is to implement sensory fusion techniques on the robot Seekur Jr. This research proposes an approach of fusing the vision information with the environmental 2D maps created using a laser scanner, to also obtain a 3D map.

Most of the works with the Seekur Jr, in the past literature, use the laser scanner only [20], thus it would be of interest to integrate new types of sensory feedback to the robot.

2 Apparatus

The Seekur Jr. robot can be programmed with MobileRobots APIs, which are the C++ ARIA and ARNL libraries. The 2D laser scanner mounted on the Seekur Jr is a Sick LMS100, with a field of view of 270° and an angular resolution of 0.50°. The reading frequency can be set to 25 or 50 Hz. The statistical error of the laser measurement is 12 mm. The camera used is the MobileRanger C3D Stereo Camera, from Focus Robotics. It is a black-and-white stereo camera with a $752X480$ (WVGA) resolution. The environment used for experiments is an unorganized indoor laboratory classroom, where several tables and chairs were moved to obtain different scenarios. In order to simplify the data gathering process, the robot will stop the motion when collecting the data. Once the computations are done, the motion will start again.

The Seekur Jr. robot is able to build detailed maps of the environment using the laser scanner and the software provided by the manufacturer—Mapper3. The robot has to be manually driven around the room, by either using the remote controller or through the MobileEyes program connecting through ArNetworking connection protocol. The connection to the robot is operated by using a Secure Shell protocol (SSH).

Once the mapping has been completed, a scan log file will be generated, containing the readings taken by the laser scanner during the motion at different time instants, in polar coordinates: for each reading, there will be 540 values which are the distances measured every 0.5° arcs of the full 270° field of view. The robot will take a new reading after moving a threshold distance or rotating by a threshold angle.

The scan log file can then be converted into a *.map* file, by merging all the readings and discarding outliers. One limitation of the 2D laser scanner is that objects such

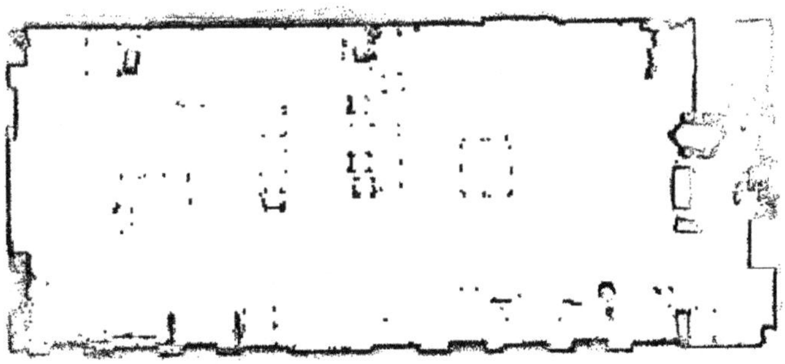

Fig. 1 The map obtained with the laser scanner

as tables or chairs are depicted as sparse dots, as the scanner takes the readings on a plane at a specific height, and therefore only detects their legs (Fig. 1). In short, any object with a small horizontal section at that height is problematic for the laser mapping process.

3 Our Approach

The first step of this research was to automatically build the map by implementing an exploration algorithm capable of successfully navigating through all the room, and then processing said map to better depict the navigable space.

After the 2D map has been obtained, the robot will be able to precisely localize itself using the laser scanner. The localization is performed by means of the $ArLocalizationTask$ class of the ARNL library, which runs an asynchronous thread that uses the Monte Carlo Localization Algorithm to precisely ($\pm$ 5cm) obtain the robot's pose using the laser data.

After the 2D map is computed, observation points are processed to collect the vision data. To obtain a good 3D representation of the scene, all the obstacles must appear at least once in the visual dataset. Some restrictions were also added to make this project compatible with another research.

The restrictions were that (i) Each observation point had to lie on the intersection of a grid with 50 cm cell size, to limit the perspective distortion between images taken from adjacent nodes; (ii) Observation points had to be part of a connected graph, thus there cannot be split groups of points, and (iii) Observation points had to not lie in unreachable areas, nor too close to obstacles, as the robot must be able to turn while taking pictures.

Once the observation points are computed, they are added in the 2D map as goal items, and the robot can collect the pictures. For each stereo pair, the corresponding point cloud is computed from a 3D reconstruction process. The point clouds from each each stereo pair are converted into global coordinates by exploiting the laser localization system, and merged together to obtain a 3D representation of the environment.

3.1 Exploring the Environment

The robot needs to be able to autonomously navigate in the room. Ideally, each area of the room should be visited only once, and there should not be unmapped areas when the algorithm stops. The approach chosen is simple: the robot moves around the boundary of the room by staying close to the right wall (or any object attached to it) until it returns to the starting position. The choice derives from the fact that the field of view (270°) and range (20 m) of the Seekur Jr laser scanner are both very

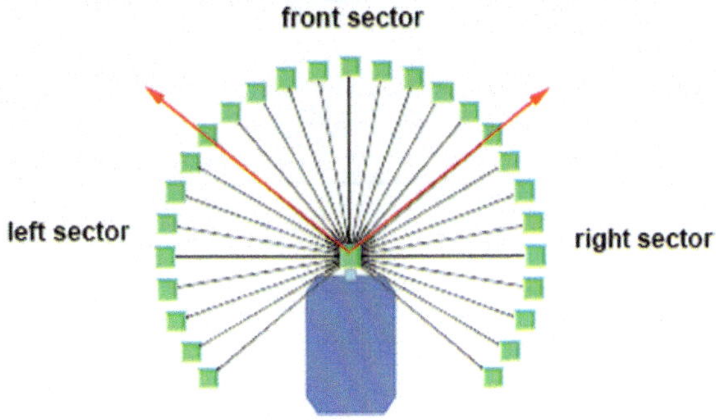

Fig. 2 Subdivision of the laser scan field of view

large for indoor environments, and thus performing one loop is enough to map all the elements of the room in most situations.

The control algorithm implemented therefore is a reactive control which only uses the current readings from the laser scanner to decide the direction of the upcoming movement. The laser readings have been grouped into three sectors (Fig. 2): the first 90° on the left belong to the left sector, the next 90° belong to the front sector, and the last 90° belong to the right sector. Since the angle 0° is the one pointing forward, the three sectors are from −135° to −45° for the left sector, from −45° to 45° for the front sector, and from 45° to 135° for the right sector.

During the motion, the closest measurement will be selected from each sector; if the value is lower than a certain threshold, that sector will be marked as occupied. The threshold has to be adjusted with respect to the size of the robot, it has to be wide enough to allow safe maneuvers. Once the data from the three sectors has been obtained, the robot will have three boolean values that represent the presence or absence of an obstacle on the front, right, or left side; thus 2^3 situations are possible (Fig. 3).

To complete the loop of the room, the robot can either do a clockwise motion by sticking to the obstacle to the left, or a counter-clockwise motion by sticking to the obstacle on the right. As a personal choice, the control algorithm was set to follow a counter-clockwise motion.

- **State *a***: all three sectors are occupied. This means the robot reached a dead end, and must retreat to continue the navigation. Since the robot is skid steering and can rotate around it's axis, by rotating right it will eventually reach state *f* or *g*, where it is able to continue the motion.
- **State *b***: the robot must rotate to the right to keep following the counter-clockwise motion.
- **State *c***: also in this case the robot must rotate to the right to keep following the counter-clockwise motion.
- **State *d***: in this situation the robot has obstacles on both sides, but nothing in front of it; therefore, it should keep moving forward.
- **State *e***: this situation is similar to the state c; the following action needed is still a rotation to the right.

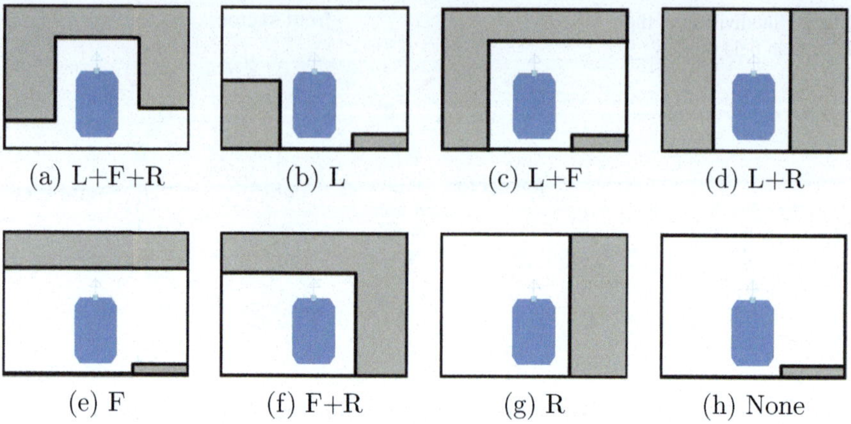

Fig. 3 Possible positions of obstacles. F = Front, L = Left, R = Right

Table 1 Truth table of the reactive control: R is a right turn, F is forward movement

Left sector	1	1	1	1	0	0	0	0
Front sector	1	0	1	0	1	1	0	0
Right sector	1	0	0	1	0	1	1	0
Action taken	R	R	R	F	R	R	F	R

- **State *f***: in this case the robot reached a corner, and must continue to the left. It is possible to either rotate to the left or to the right: in both situations, the robot state will eventually reach the state *g*.
- **State *g***: similar to the case of *d*, the robot can just keep moving forward.
- **State *h***: this situation is similar to the states *b* or *c*. By simply rotating right the robot will manage to keep its counter-clockwise motion.

It can be noted from the summarizing decision table (Table 1) how any decision is either a rotation to the right or a forward translation.

This is due to the fact that the left rotation can also be accomplished by rotating right by the complementary angle: the next reached state will be the same, although rotating on the right will require more time as the angle will be greater. Left turns only happen in state *f*, which most of the times happens when the robot reaches a corner of the room (for square-shaped rooms): turning right instead of left in those situations assures that the robot takes a full view of the room, giving some degree of redundancy in the spotted landmarks which helps the matching process later on. Sometimes, however, a left turn can still be required when there is not enough space to do a right turn.

3.2 Processing the Map and Collecting the Vision Data

After the exploration has been performed, a first raw representation of the environment will be available. In the current state, the map has a very good definition of the external walls, but has very sparse information about the middle (Fig. 4). In particular, objects difficult to detect for the laser scanner are represented by a group of sparse dots.

We thus developed a program that extracts a bounded navigation region. The approach used is that whenever a pair of points have a relative distance smaller than the size of the robot (plus some margin to tackle the uncertainty), it is connected by a forbidden line. This is due to the fact that the robot would be unable to navigate between them anyway; also, if two points are very close together, it is likely that they are part of the same object.

The number of points of a map can quickly escalate to several thousands even for small rooms, usually in the 10^4 order. By considering the connections between points A,B and B,A for a single connection, when having just two points, only one line has to be added; with three points, three lines are needed. By adding points the number of lines needed will always increase by the amount of points that were already present. Therefore, if there are $n > 1$ points close together, the triangular number k of lines between them can be obtained with Eq. 1:

$$k = \sum_{i=1}^{n-1} i = 1 + \cdots + n - 1 = \frac{n(n-1)}{2} \tag{1}$$

For computational efficiency, the choice was to downsample the map by grouping points that were very close together into clusters, and then only connect the clusters. It is then possible to extract the inner surface that represents the navigable space with a simple reachability check from the origin, since it is known that it always lies within the navigable space.

To obtain a good 3D representation of the scene, all the obstacles must appear at least once in the visual dataset. By using the map obtained in the previous step, a set of goal positions have been procedurally defined. As already stated the observation points have to be placed on the intersections of a grid with 50 cm cell size, and they must not lie in unreachable areas nor too close to obstacles.

When all the observation points have been defined, it is then possible to drive the robot to collect the images. Each goal can be singularly selected for navigation, or a global tour can also be used.

When the robot has reached the needed goal, it will collect both a laser reading and several images, using procedural filenames. Since the laser localization provides the pose of the robot in absolute coordinates, this information is embedded inside the filename of each image as well.

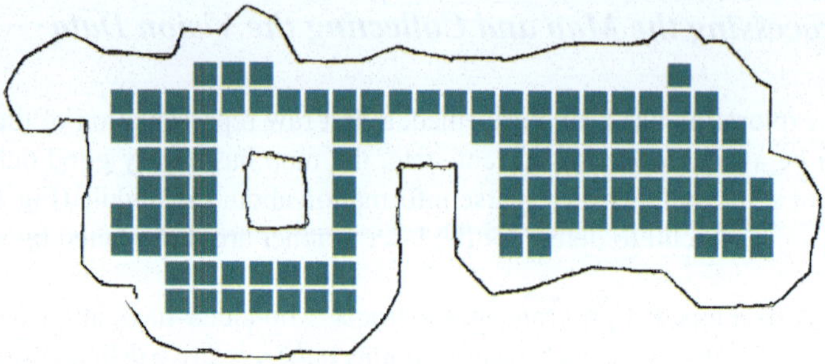

Fig. 4 Definition of observation points

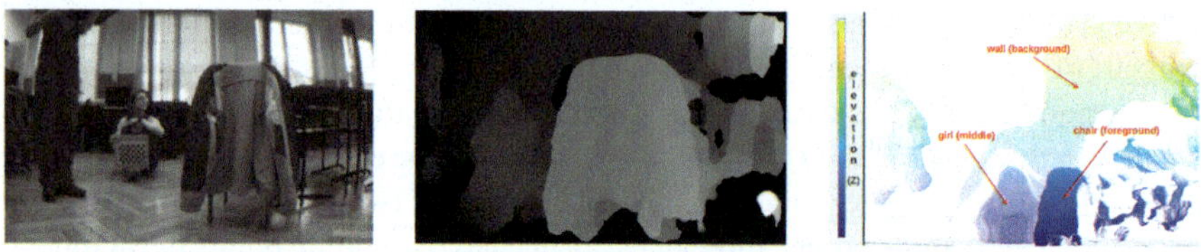

(a) Raw image (b) Disparity map (c) 3D reconstruction

Fig. 5 3D reconstruction process

3.3 Calibrating the Camera and Reconstructing the Environment

The MobileRanger cameras have a built-in software which is able to provide the disparity map, processed directly with the embedded nDepth processor. Several different settings of parameters have been tested with the built-in block matching and calibration of the cameras; however the result obtained was always very noisy. Each computed depth presented considerable error with respect to the real distances (well over 70 cm of error). For this reason, the stereo camera has been calibrated again. The calibration has been performed by taking 36 pictures of a checkerboard with 25 mm wide cells from different angles, and using the OpenCV function $stereoCalibrate()$.

After the calibration, the precision has been evaluated again, using the semi-global block matching algorithms. The depth computed for the obstacles has been compared to the real distances, measured both with the laser scanner and manually, and provided reasonable accuracy (10 ± 2cm of error). The measures have been performed on 10 different images where flat obstacles were placed at different intervals (Fig. 5).

The obtained point cloud has been filtered from the points with a computed distance smaller than 40 cm or greater than 4.5 m, to remove outliers.

Fig. 6 Orientation of the two reference frames

3.4 Fusing Sensor Data

Combining the data from the laser scanner with the data obtained from the stereo camera requires both data types to be in the same format, as by default the reference frames of the two sensors are different (Fig. 6).

The laser scanner and the stereo camera are fixed to the chassis of the robot, thus the relative distances between sensors are known. Assuming the aligment error in the robot manufacturing is trivial, the two reference frames are aligned ($X_L = Z_C$, $Y_L = -X_C, Z_L = -Y_C$), thus a simple change of variables can be used for simplicity. After the coordinate transformation, the point cloud can be directly compared with the laser data. Each stereo pair is collected together with a laser reading: since the laser readings are much more precise compared to the cameras, they have been used to remove part of the outliers.

The 3D points which had an X and Y coordinate too far away from the laser readings have therefore been removed. The minimum proximity to a laser reading to mark a 3D point as valid can be defined with a threshold value. Once a point cloud has successfully been computed and filtered from the outliers, it can then be converted into global coordinates. Each image is tied with the global position, which is embedded inside the filename. Each 3D point can thus be converted into global coordinates simply by adding the X and Y values of the robot's pose in that specific observation point.

Since the coordinate system of this 3D representation of the environment is the same as the one used to create the 2D map, the two maps are practically overlapped: the localization obtained from the laser scanner will also apply for the 3D environment.

4 Results

The program has been tested on five different layouts of a room approximately $7x15$ m wide with several obstacles. The objects have been arranged in a way to make it as hard as possible for the robot to navigate, with many narrow passages and elements hard to detect such as chairs and tables.

The exploration algorithm has always managed to successfully navigate in the room and return to starting position (e.g., Fig. 7). The reaction time of the robot when approaching obstacles was always fast enough to let it avoid them accordingly, thus the robot never ended up hitting an obstacle.

The map processing algorithms took less than 5 ms to read the map generated from the exploration. The 2D map of the room obtained in the experiments was composed by an average of $10k \pm 1k$ points. Execution time is tied to the size of the room, but without loss of generality, most indoor environments should be processed by the whole pipeline in under one second. The processing speed is thus suitable for real-time applications.

The produced final 2D map models (e.g., Fig. 8) are reasonably precise with respect to the real measurements, as the laser scanner has a statistical error of 12 mm

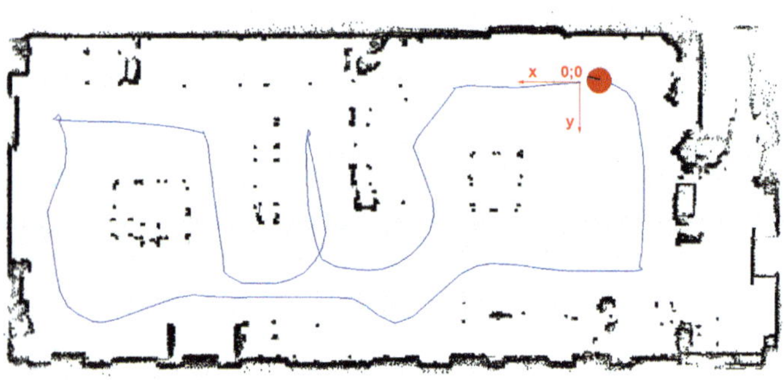

Fig. 7 Path followed by the Seekur Jr during the exploration

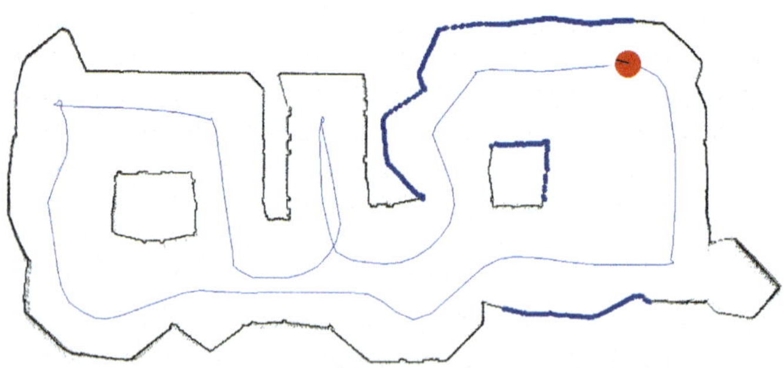

Fig. 8 The extracted navigable space. Blue dots represent the laser scanner readings on the displayed position of the robot

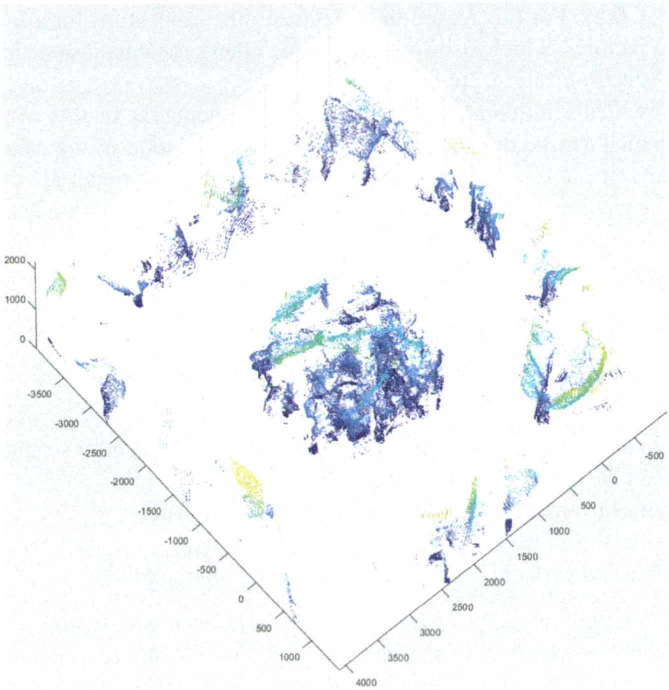

Fig. 9 The point cloud obtained after merging 20 different views. The shape distinguishable in the middle is a table, while the elements in the boundaries are walls

(specified by its manufacturer), which can be considered sufficient for a robot of the size of the Seekur Jr. The localization, being based on the laser scanner data, had an average error of 5 ± 1 cm, also including the mapping error. The error has been measured by placing markers on the ground at a measured distance from the robot, and commanding the robot to move by that distance, and then measuring the drift. The processed map is considered more suitable for the development of alternative navigation algorithms (e.g., using cell decomposition approaches, Voronoi graphs, and rapidly exploring random trees), if the one provided by ARNL does not fit the needs: this is due to the area occupied by obstacles being removed by the connected region representing the navigable space. In the current state of the art, there have been little to no researches aiming at automatically processing the maps obtained with Seekur Jr. robots rather than manually editing them, thus making this project a good starting point for further work with this robot.

After merging the point clouds obtained from different stereo pairs, the overall point cloud has been downsampled to reduce the number of points by a factor of 10. The navigable area is polished from outliers, and the obstacles are correctly distanced, even though some noise is still present (e.g., Fig. 9). The computational time of the vision system is the bottleneck of the whole pipeline: when using the onboard processor, the time required from taking the pictures to the final reconstruction is on average 1 ± 0.5 s. For this reason in this project, the robot stops for a few seconds to take the pictures, which also simplifies some other problematics (such as motion blur, data synchronization, etc.).

The presented solution can be improved by enhancing the quality of the disparity map, which can be obtained by increasing the resolution of the cameras used. Nevertheless, it is a viable solution to automatically obtain a rough 3D shape of the surrounding environment using cheap cameras, while being able to navigate inside it.

5 Conclusions

The scope of the work thus is

- To obtain a 2D map by exploring the environment with the laser scanner;
- To process the obtained map to collect the vision data;
- To identify the objects in the environment by using vision;
- To combine the above information to create a 3D map;
- To localize and perform goal-driven navigation with the robot.

Navigation and mapping for mobile robots have been widely researched in the past: the contribution of this paper is to show a possible implementation approach taking into account the limitations of the Seekur Jr. robot. Our approach is to obtain a 3D enviromental map by first obtaining a 2D map using a laser scanner, and later enriching such map with the point clouds obtained from the stereo vision system. We have extracted the navigable space from the 2D laser maps, which could be useful for implementing new navigation techniques on the used robot.

The project was implemented on a Seekur Jr. robot by MobileRobots: the program of automatic exploration and data collection for mapping has been described. A reactive control algorithm has been designed using the ARIA library to explore the indoor scene by following the walls, and gathering the required data. Two representations of an environment have been obtained, both 2D and 3D; methods to process and polish both maps have been implemented. In the former map, a better representation of the navigable space has been extracted, which was needed to automatically define waypoints for the vision data retrieval; in the latter, stereo vision methodologies have been applied, together with sensor fusion, to obtain a point cloud representation of the room. The laser data has been used to limit the noise during the stereo vision 3D reconstruction.

Navigation in the obtained maps is possible through a user interface implemented through the ArNetworking library. The robot is therefore able to either be teleoperated or to perform goal-driven navigation with obstacle avoidance, by means of the ARNL localization and path planning methods.

Overall, the sensory data fusion between the laser scanner and the vision system allowed a better representation of the environment than the one obtainable with only one type of sensor. Without the vision system, the map obtained with the laser scanner would be restricted to a 2D map. On the other hand, the laser scanner data is needed to merge the point clouds from different points of view and to remove some of the outliers. Without the laser data, merging different views would require the implementation of a robust landmark matching vision system, which could be difficult to perform with lower quality pictures. Also, the perspective difference between different views would represent a big issue for landmarks matching. The presence of the laser scanner therefore allows the utilization of very cheap cameras while still maintaining an acceptable degree of precision in the 3D reconstruction. The presented solution is thus suitable for situations where the goal is to cheaply obtain a general reconstruction of an area (e.g., an office).

The presented solutions can be applied in any scenario where automatic mapping is preferred, either because the presence of human personnel is problematic (e.g., hazardous or unreachable environments, such as radioactive or extraterrestrial) or simply for convenience reasons.

Future development ideas could consist in matching the vision data (e.g., with SIFT descriptors) to obtain a more polished version of the final 3D model. Hopefully, the information gathered in this project will help future development of software for the Seekur Jr. robot.

References

1. Elfes A (1987) Sonar-based real-world mapping and navigation. IEEE J Robo Autom 3(3):249–265
2. Jesus F, Ventura R (2013) Simultaneous localization and mapping for tracked wheel robots combining monocular and stereo vision. J Autom Mob Robo Intell Syst 7
3. Sola J, Monin A, Devy M, Vidal-Calleja T (2008) Fusing monocular information in multicamera slam. IEEE Trans Robo 24(5):958–968
4. Çiftçioğlu Ö, Sariyildiz S (2012) Data sensor fusion for autonomous robotics. In: Proceedings of the serial and parallel robot manipulators-Kinematics. Dynamics, Control and Optimization, InTech, London
5. Kam M, Zhu X, Kalata P (1997) Sensor fusion for mobile robot navigation. In: Proceedings of the IEEE, vol. 85(1), pp 108–119
6. Ashokaraj I, Tsourdos A, Silson P, White B (2004) Mobile robot localisation and navigation using multi-sensor fusion via interval analysis and ukf. In: Proceedings of the 2004 towards autonomous robotic systems (TAROS). University of Essex, Colchester, UK
7. Lee TE (2014) Multiple sensor indoor mapping using a mobile robot. CS 229 Machine Learning final project
8. Sutas A (2014) Delivery bot: indoor mobile robot navigation with sensory fusion localization in the appearance space. thesis, University of Edinburgh
9. Pagac D, Nebot EM, Durrant-Whyte H (1998) An evidential approach to map-building for autonomous vehicles. IEEE Trans Robot Autom 14(4):623–629
10. Saeedi J, Faez K (2012) Infrared and visible image fusion using fuzzy logic and population-based optimization. Appl Soft Comput 12(3):1041–1054
11. Dubois DJ (1980) Fuzzy sets and systems: theory and applications, vol 144. Academic press
12. Oleari F, Rizzini DL, Caselli S (2013) A low-cost stereo system for 3d object recognition. In: IEEE international conference on intelligent computer communication and processing (ICCP)
13. Achmad MH, Findari WS, Ann NQ, Pebrianti D, Daud MR (2016) Stereo camera-based 3d object reconstruction utilizing semi-global matching algorithm. In: International conference on science and technology-computer (ICST), pp 194–199, IEEE

14. Klimentjew D, Stroh A, Zhang H, Zhang J (2008) 3d real-time stereo reconstruction for small humanoid robots. In Proceedings of the 3rd CAS symposium on robotics and manufacturing technology
15. Hartley RI, Sturm P (1997) Triangulation. Comput Vis Image Underst 68(2):146–157
16. Faugeras O, Luong Q-T, Papadopoulo T (2004) The geometry of multiple images: the laws that govern the formation of multiple images of a scene and some of their applications. MIT press
17. Hartley R, Zisserman A (2003) Multiple view geometry in computer vision. Cambridge University Press
18. Intille S, Bobick A (1994) Disparity-space images and large occlusion stereo. Comput Vis-ECCV 94:179–186
19. Szeliski R, Scharstein D (2004) Sampling the disparity space image. IEEE Trans Pattern Anal Mach Intell 26(3):419–425
20. Genevois T, Zielińska T (2014) A simple and efficient implementation of ekf-based slam relying on laser scanner in complex indoor environment. J Autom Mob Robot Intell Syst 8

The Kinematics and Dynamics of Engine Start Systems

Madhusudan Raghavan, Norman K. Bucknor, and Vatche Donikian

Abstract Interest in low-voltage hybrid systems has grown because of the potential for significant fuel economy gains at a relatively low cost. Start/stop technology in hybrid powertrains is an economical technical approach in this general area of powertrain refinement. A new engine start mechanism concept is presented herein. It can change its operating state between two modes, respectively, involving a geared or belted connection to the crankshaft. This novel starter device enables an ultrafast start as well as a 5–10% fuel economy improvement, depending on voltage and power levels implemented. We also present a detailed engine cranking model developed in AMESim that distinguishes between motoring and firing events to get an accurate representation of engine dynamic torque response during engine start/stop events.

Keywords Engine model · Cranking · Start/stop

1 Introduction

A recent IEO report [1] forecasts a 20% increase in the world population over the next two decades, accompanied by a 40% increase in energy use. Fossil fuels are still expected to supply over 75% of world energy in 2040. Petroleum remains the world's largest energy source, and worldwide energy-related CO_2 emissions would rise from 34 billion metric tons per year in 2015 to 43 billion metric tons per year in 2040, a 29% increase. To address this future scenario having a high population, urbanization, and increased energy usage and emissions, we look for novel automotive propulsion architectures with improved efficiency. The problem of designing

M. Raghavan (✉) · N. K. Bucknor
Propulsion Systems Research Laboratory, Warren, USA
e-mail: madhu.raghavan@gm.com

N. K. Bucknor
e-mail: norman.k.bucknor@gm.com

V. Donikian
Department of Mechanical Engineering, University of California, Irvine, USA
e-mail: vdonikia@uci.edu

D. Sen et al. (eds.), *Mechanism and Machine Science*, Lecture Notes in Mechanical Engineering, https://doi.org/10.1007/978-981-15-4477-4_21

ultracompact transmissions is also a valuable aspect of this research as it frees up room for the insertion of electrification components within the packaging envelopes of existing propulsion systems.

1.1 Previous Work

Start/stop systems save fuel by switching off the engine when vehicle velocity is zero. This is particularly effective in traffic lights and in traffic jams in urban environments, when most vehicle engines remain in the idle mode and continue to burn fuel while pumping out emissions. Start/stop systems are typically calibrated to quickly restart the engine when the driver lifts his foot off the brake in preparation to begin driving away.

Wellmann et al. [2] compare such idle stop systems and lay out the specifications for the associated driveline components to ensure acceptable drive quality. Chen et al. [3] study the fuel consumption of a small car on various certification drive cycles and show a nearly 4% improvement with the installation of a start/stop system. Beer and Teulings [4] investigate the applicability of an early injection and spark approach to refining engine starts for an idle stop system. Heinzmann et al. [5] explore the use of start/stop systems along with idle sailing on a motorcycle operated over various driving cycles and suggest a nearly 30% fuel economy improvement potential. Whitall [6] demonstrates that the classical city/highway cycle combination used for fuel economy label certification perhaps does not do adequate justice to technologies such as start/stop systems which are designed for maximum impact in urban use. The city/highway combination cycle underestimates the actual benefits obtained on city-centric cycles. Bishop et al. [7] present results on a powerful belted alternator starter system with a nearly 6% fuel economy improvement in city operation and a 4% improvement in highway operation. Hawkins et al. [8] describe a system that uses a 115 V architecture with a Li-ion battery to execute a belt-based start/stop and recuperation/boost system for a 2.4 L 4-cylinder gasoline engine. Duan et al. [9] describe the use of a novel transmission clutch which remains ON in its passive state but turns OFF when energized. Such a clutch is able to keep the transmission in gear in preparation for a quick launch when coupled to an engine with a start/stop system. This arrangement eliminates the need for an electric auxiliary pump to pressurize the clutches when the engine is switched off.

2 Electromechanical Starter System

In the present work, we introduce a new electromechanical starter device that has dual-mode operation. It can selectively be connected to the crankshaft via a gearset or a belt, for different preferred operating states, as will be described in the following. This unique combination of elements allows both efficient start/stop operations of the

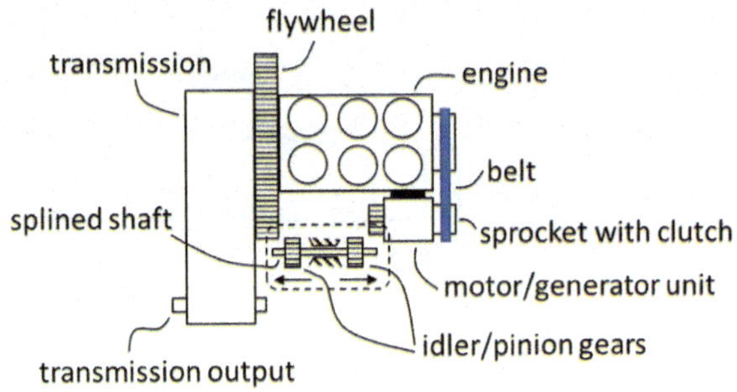

Fig. 1 Engine with motor/generator for geared/belted start

engine as well as mild hybridization of the powertrain for improved fuel economy. The system of interest in the present study is shown in Fig. 1. It shows a top view of a propulsion system, comprised of an engine and a transmission.

A four-quadrant electric machine labeled as a motor/generator unit is mounted on the engine. One end of this machine is connected to a sprocket via an electric clutch. There is a similar sprocket on the crankshaft and a belt running over these two sprockets that provides a mechanical connection between the motor/generator and the crankshaft when the electric clutch is engaged. The motor/generator is designed so that its rotor shaft is accessible on both sides. The power electronics controlling the motor/generator are arranged like a donut on one side, leaving room for the rotor shaft to come through. The rotor shaft on the side opposite to the belt and electric clutch has a gear mounted on it. The gear–link combination shown in the portion of Fig. 1 indicated by the rectangle with a dashed line boundary, serves the purpose of connecting the gear on the motor/generator shaft to the crankshaft, when suitably actuated. Additional details of this gear–link combination are shown in Fig. 2.

The device of Fig. 2 includes two solenoids which serve as actuators, moving two idler/pinion gears into position when required. Like-colored actuators act upon like-colored pinions, blue on blue and red on red. This entire arrangement is perhaps better understood by referring to the stick diagram of Fig. 3, wherein the kinematic

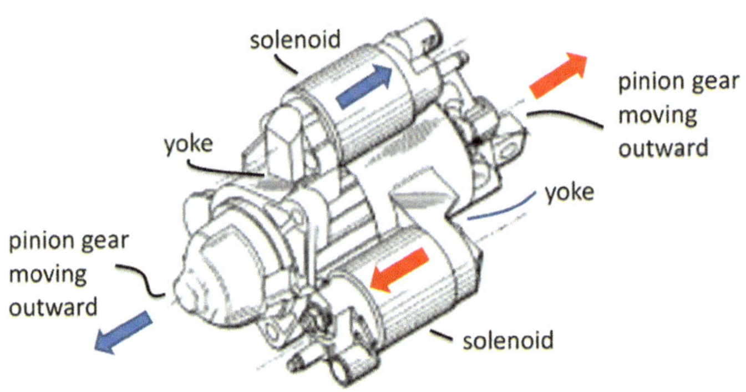

Fig. 2 Actuation device for novel electromechanical starter

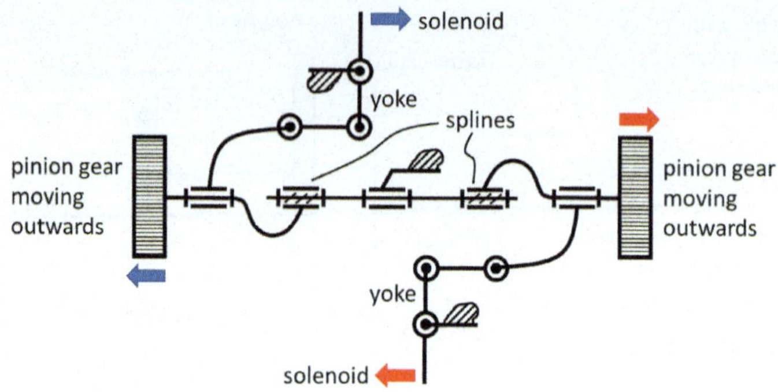

Fig. 3 Stick diagram of electromechanical starter

details are clearly represented. When the solenoids are not energized, the pinion gears are in their retracted positions, held in place by springs (which are not shown in the figure). Upon actuation of the solenoids, each pinion gear slides outwards as shown by the arrows. The blue pinion gear engages with a corresponding gear on the crankshaft flywheel. The red pinion gear engages with the prior-mentioned gear on the motor/generator rotor shaft. The pinion gears are respectively mounted on coaxial shafts which in turn are splined to a central shaft that is free to rotate. In this manner this mechanism arrangement is able to establish a geared connection between the motor/generator and the crankshaft, when required, with the motor/generator serving the purpose of the conventional starter motor. Once the engine has started and is running at an rpm in excess of the motor/generator rpm, passive overrunning clutches similar to those used on conventional starters allow the pinion gears to disengage from their respective gear connections and retract to their normal position.

It is easy to see that the above described arrangement enables the selective connection of the motor/generator to the crankshaft either via the belt connection at one end of the motor/generator or via the geared connection at the opposite end. Each connection provides a different geometric ratio between motor/generator rpm and the engine rpm, thus enabling optimal use of the engine-motor/generator combination depending on vehicle operating conditions. Additional details of this starter device may be found in Namuduri et al. [10].

2.1 *Experimental Results*

The speed ratio in effect when the motor/generator connects to the crankshaft via the belt is in the neighborhood of around 3:1. The speed ratio in effect when the motor/generator connects to the crankshaft via the geared arrangement is much higher and more typical of the conventional starter speed ratio. Thus, it makes sense to start the engine using the geared connection and then to switch to the belted mode of operation for normal hybrid functions during driving. These include torque addition

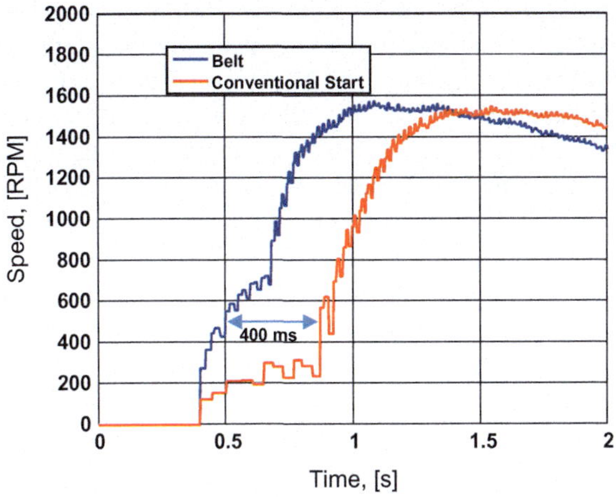

Fig. 4 Engine starting times comparison

on the driveline when a burst of acceleration is needed and regenerative braking on the driveline when the vehicle is decelerating. This regenerated energy may be stored in a battery for reuse during torque addition or engine start events thus contributing to an overall improved fuel economy for the vehicle. Depending on the voltage and energy specifications of the battery and the motor/generator unit and the vehicle mass (and hence available kinematic energy for regenerative capture), fuel economy improvements in the 5–10% range are achievable.

The belted connection may also be used for engine starts when operating conditions and ambient conditions are suitable. Very cold weather and humidity may cause belt slip and are to be avoided. Nevertheless, we have found experimentally that these belted starts are in fact superior to geared starts via conventional starter motors because the motor/generator may be designed to have considerably higher power and torque as compared to conventional starter motors. Figure 4 shows a comparison between a belted start using such a motor/generator versus a geared start using a conventional starter motor. In this case, the belted start was able to reach a threshold of 500 rpm approximately 400 ms faster than the conventional starter, thereby enabling a smoother, quicker, and nearly imperceptible start. This is a further benefit of the proposed system, in addition to the abovementioned fuel economy improvement potential.

3 Modeling Engine Starts

We have been studying quick engine starts and restarts on a 2.0 L turbocharged engine coupled to a GF6 six-speed automatic transmission. We use a Simulink model of a conventional powertrain incorporating an engine model developed in AMESim. In order to accurately model a vehicle powertrain during start-stop events, an engine model that can simulate crankshaft torque under cranking and firing conditions is necessary. Our starting point was a Simulink model which included an engine-starter system which incorporated dynamics of the starter pinion gear engagement, as well as the other elements of the drivetrain such as torque converter and alternator. For the engine start case, the vehicle is assumed stationary and we only need to model the torque load created by these elements as they have elements that rotate with the engine. Instead of using a Simulink engine model, we inserted an AMESim-based engine model, packaged as an s-function for use in the Simulink model. The engine model receives the throttle command, the engine firing command, the load torque, and the starter's cranking torque from Simulink, and in return sends the engine indicated torque, crank angle position, accumulated crank angle, and engine speed. The Simulink model is shown in Fig. 5.

The white block is the s-function interface module between Simulink and the AMESim model. The engine model layout in AMESim is shown below in Fig. 6. The primary components of the model are (a) a cylinder pressure sub-model utilizing motoring and firing cylinder pressure curve data as a function of crank angle and (b) a slider-crank sub-model representing the rotational dynamics of the engine

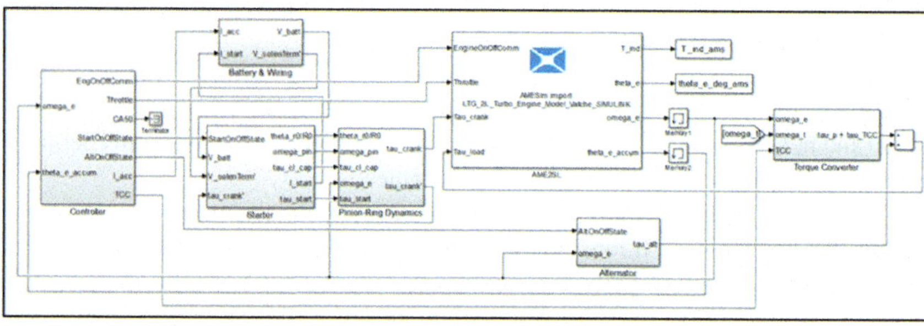

Fig. 5 Starter-engine cranking model in Simulink

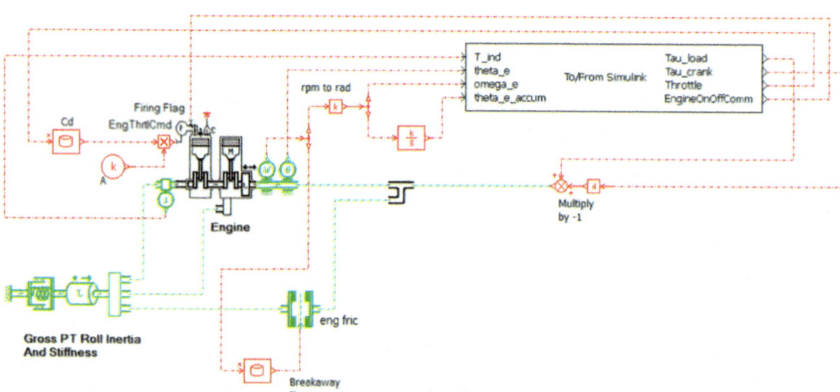

Fig. 6 Engine model in AMESim

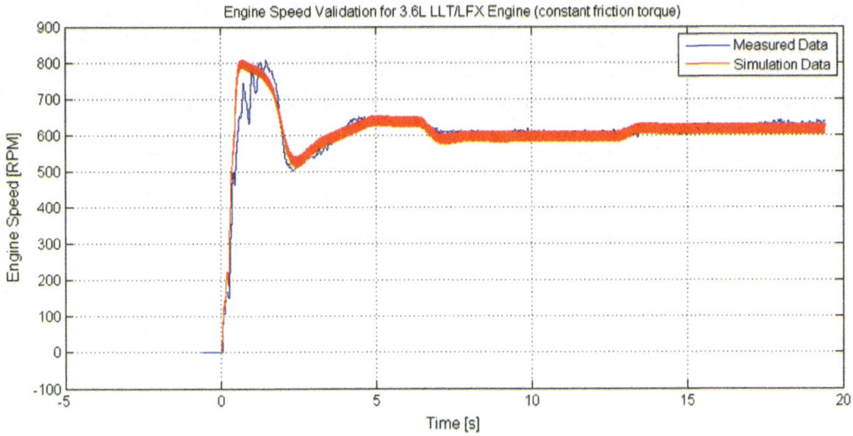

Fig. 7 Engine speed comparison for 3.6 L V6 engine validation

mechanism. The higher friction torque seen when trying to start a stationary engine is represented by a friction-torque versus engine speed curve. During the cranking phase, cylinder pressure is represented by the motoring pressure data, with some additional gas dynamics physics modeling at the speed below which the motoring data is not available. The cylinder pressure in each cylinder is modeled as a function of the crank angle. Once the ignition ('firing') flag is set high, the engine model transitions to the firing pressure data.

The AMESim engine model is convenient since replacing one engine with another requires changing a smaller set of parameters in the model. The cranking model was validated with measured data from two different engines. The comparison of engine speed (simulation versus experiment) for a 3.6 L V6 engine is shown in Fig. 7. The simulation is able to follow the trends of the measured data well. There are certain instances when the simulation is not able to exactly follow the measured data, because there are certain changes and events that occur in a real engine start up that are not captured in the model. A close-up plot of the cranking event is shown in Fig. 8.

It can be seen in the plot above that the simulation model achieves a somewhat higher firing speed than the actual engine. A similar start-up test was conducted for a 1.4 L turbocharged 4-cylinder engine. The comparison of engine speeds for this test is shown in Fig. 9. Once again, the simulation data is able to follow the measured data reasonably well over the course of the event. A close-up of the cranking, in Fig. 10, again shows some differences between the two sets of data, as in Fig. 8.

Figure 10 shows that, similar to the V6 model, the simulation cranks the engine up to speed faster than the measured data. These differences can be attributed to a number of factors, including simplified dissipative effects in the engine model, real-world engine control differences (spark timing, cam-phasing, and throttle setting), and the real versus modeled engine load. The initial spark timing and cam-phasing can be set but the model does not modify these parameters during the crank-to-run event.

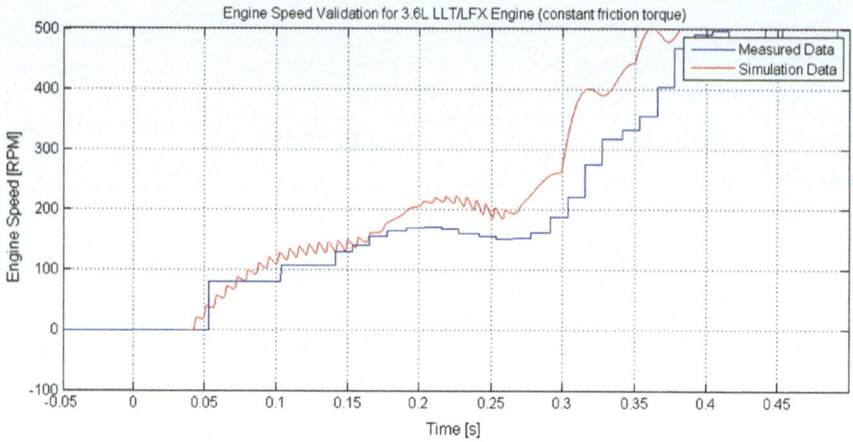

Fig. 8 Engine speed close-up for 3.6 L V6 validation

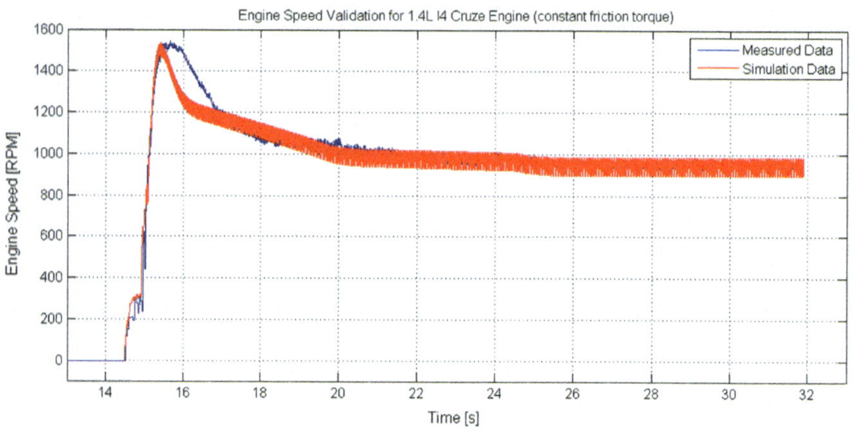

Fig. 9 Engine speed comparison for 1.4 L 4-cylinder validation

Overall, the results show that the cranking model is well correlated with measured data and can be used for detailed vehicle simulations.

The experimental data in Figs. 4, 7, 8, 9, and 10 were gathered by logging signals from an instrumented engine control module (ECM) using INCA software from ETAS.

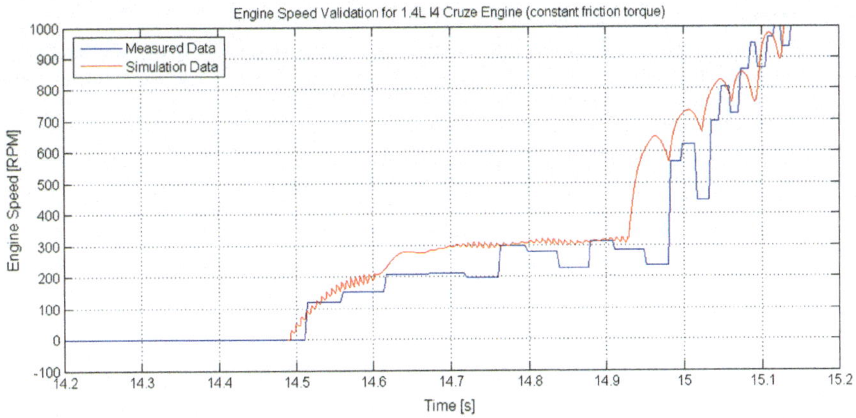

Fig. 10 Engine speed close-up for 1.4 L 4-cylinder validation

4 Summary and Conclusions

Interest in low-voltage hybrid systems has grown because of the potential for significant fuel economy gains at a relatively low cost. Start/stop systems offer an economical pathway to addressing the problems associated with increased emissions due to urbanization, traffic jams, and increasing travel. The novel electromechanical device presented in this paper, when operating in geared mode, is capable of achieving quick engine starts in such start/stop equipped engines. Additionally, this dual-mode device allows the powertrain to switch to belted operation to accomplish hybrid functions such as regenerative braking and driveline torque boosting. In fact, a 5–10% fuel economy improvement may be achieved, depending on voltage and power levels implemented. We have also presented a detailed engine cranking model developed in AMESim that distinguishes between motoring and firing events to get an accurate representation of engine dynamic torque response during engine start/stop events.

References

1. International Energy Outlook (2016) DOE/EIA-0484. https://www.eia.gov/outlooks/ieo/pdf/0484(2016).pdf
2. Wellmann T, Govindswamy K, Tomazic D (2013) Integration of engine start/stop systems with emphasis on NVH and launch behavior. SAE Int J Engines 6(2):1368–1378. https://doi.org/10.4271/2013-01-1899
3. Chen H, Zuo C, Yuan Y. Control strategy research of an engine smart start/stop system for a micro car. SAE 2013-01-0585
4. Beer J, Teulings W. Optimized start strategy for stop/start operation of a micro-hybrid vehicle. SAE 2007-01-0298
5. Heinzmann B, Scholz SRP, and Anantha P (2013) Start/Stop strategies for two-wheelers in the emerging markets. SAE Int J Passeng Cars Mech Syst 6(3):1595–1604. https://doi.org/10.

4271/2013-32-9125
6. Whitall I. Off-cycle fuel consumption evaluation of stop-start systems. SAE 2012-01-1601
7. Bishop J, Nedungadi A, Ostrowski G, Surampudi B, Armiroli P, Taspinar E. An engine start/stop system for improved fuel economy. SAE 2007-01-1777
8. Hawkins S, Billotto F, Cottrell D, Houtman A et al (2012) Development of general motors' eAssist powertrain. SAE Int J Alt Power 1(1). https://doi.org/10.4271/2012-01-1039
9. Duan C, Samie F, Hebbale K, Li D et al (2015) Normally-engaged dual-piston clutch for engine stop-start application. SAE technical paper 2015-01-1141. https://doi.org/10.4271/2015-01-1141
10. Namuduri C, Rajala A, Holmes A, Nehl T, Atluri V, Raghavan M (2016) Hybrid powertrain with mechatronics actuator assembly and method of controlling the same. US Patent 9,481,357, Nov 1, 2016

Synthesis and Analysis of Jansen's Leg-Based Mechanism for Gait Rehabilitation

D. S. Mohan Varma

Abstract Theo Jansen built wind-powered walking mechanisms or strandbeasts using a basic 12-linked leg mechanism which has a single degree of freedom (DoF). In this work, the Jansen leg mechanism is used as a start point in an optimization routine to arrive at a single DoF mechanism to trace the ankle trajectory. Optimization and forward dynamic analysis methodology of the mechanism are presented. This preliminary work can be used in the design of rehabilitation devices and assistive exoskeletons.

Keywords Strandbeast · Optimization · Human walking · Forward dynamic analysis · Kane's method.

1 Introduction

Robotic exoskeletons such as Lokomat, LOPES, ALEX, etc., have been designed to enable the gait training of persons with walking disabilities. However, the high cost of these devices makes them unaffordable for most clinics in India. Six bar, 10-bar, and cam-based designs have been used by researchers to synthesize ankle trajectory tracing mechanisms and to make preliminary designs of gait training devices [1–4]. In this work, optimization is used to vary the standard link lengths of Theo Jansen mechanism to arrive at a modified Theo Jansen mechanism that can trace the ankle trajectory. Additionally, forward dynamic analysis methodology of the Jansen's leg mechanism is also described.

Theo Jansen's strandbeasts are build for translational motion. Almost all of these "walking" mechanisms have more than one leg and can walk independently. The goal of this work is not to design an independent walking mechanism but to design a mechanism that can assist or support in the gait training process where, typically, a

D. S. M. Varma (✉)
School of Mechanical Engineering, Centre for Biomaterials,
Cellular and Molecular Theranostics, Vellore Institute of Technology, Vellore, India
e-mail: mohanvarma@vit.ac.in

D. Sen et al. (eds.), *Mechanism and Machine Science*, Lecture Notes
in Mechanical Engineering, https://doi.org/10.1007/978-981-15-4477-4_22

group of physiotherapists are involved—one enabling the movement of ankle while two others hold the patient to prevent imbalance and fall. While the body weight can be supported by a harness, repetitive and consistent movement of the ankle along a predefined trajectory can be achieved via a mechanism. In this work, Jansen's leg is used as a start point in the optimization to arrive at an ankle trajectory tracing mechanism. Although tracing ankle trajectory is the primary goal of most preliminary designs, additional mechanisms might be required for support at the knee and ankle [2, 3] in some cases. Jansen leg mechanism has motion resembling the human leg and has 12 links connecting the crank (at the hip) to the end-effector (at the ankle). Therefore, in this work Jansen's leg is used as a start point in the optimization. This design could possibly enable additional mechanisms/supports to be included/mounted at strategic locations as per the requirement in later stages of the design.

Powered gait training orthosis [5–7] such as Lokomat, active leg exoskeleton (ALEX), lower extremity powered exoskeleton (LOPES), etc., are available in the Western market. However, these devices are expensive and out of reach for most clinics in India. Some researchers have developed simpler mechanically driven designs of gait rehabilitation devices. Shao et al. [1] used a two degree of freedom (DoF), seven-bar crank-slider mechanism as the start point to design simplistic rehabilitation robot. Ji and Manna [8] synthesized an ankle-tracing mechanism using a four-bar crank-rocker mechanism as the start point in the optimization. Manna et al. [3] used a combination of planetary gears and cams to achieve timing and coordination in their gait rehabilitation system. Tsuge and McCarthy [2] designed a single DoF linkage system to guide natural ankle trajectory. They included a cam-driven mechanism to control the foot orientation. In this work, the single DoF Jansen leg is used as a start point in the optimization to arrive at a mechanism that can trace the ankle trajectory.

Patnaik and Umanand [9] presented a bond graph approach for the forward kinematics and inverse dynamics analysis of the Theo Jansen mechanism. Nansai et al. used projection method to study the dynamics [10] and stability [11] of a four-legged Theo Jansen leg-based platform. They did not study the application of Jansen leg for gait training applications. In this work the step-by-step procedure for the forward dynamic analysis of the Jansen's leg mechanism using Kane's method is presented. Kane's method [12] is preferred by many researchers for forward dynamics analysis as the vectorial approach is less cumbersome (as opposed to Newton–Euler or Lagrange's equations). Although the step-by-step procedure to obtain equation of motion (EoM) using Kane's method is described well in the literature, in most cases, a symbolic manipulator would be required to derive the EoM which would then be solved using a numerical solver. The methodology presented in this work can be used to directly code and solve the EoM without the need for a symbolic manipulator.

2 Methodology

Jansen's leg has 12 links connected by revolute joints and is a one DoF mechanism (Fig. 1). A single actuator can be used to propel the leg for walking. In order to use this mechanism for rehabilitation, the motion of Jansen's leg during one cycle should

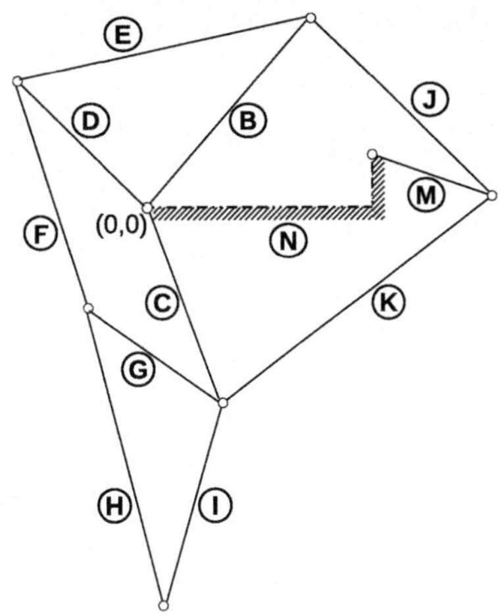

Fig. 1 Theo Jansen mechanism

resemble that of the human leg. In fact, if Jansen's leg can be modified such that a point on the end-effector traces ankle trajectory with respect to the hip during gait, the mechanism can potentially be used to design a gait training device. The optimization methodology used to achieve this is described below.

2.1 *Optimization*

Experimental data for normal walking is obtained from the literature [13]. Using the hip and ankle marker data, an envelop which defines the boundary of closed loop ankle trajectory is constructed (Fig. 2). Normal walking data for 10 trials is used. The objective function (to be minimized) during optimization is the error between trajectory of a point on the mechanism and the target trajectory envelop defined from the experimental data. The objective function is defined clearly below. Link lengths, a scaling factor, and coordinates of a point on the end-effector are used as the design variables in the optimization. Constraints on the link lengths ensure that each of the three four-bar mechanisms in the Jansen's leg mechanism are feasible and remain Grashof. The constraints also ensure that the link which is the crank (i.e., the smallest) at the beginning of optimization remains the crank after optimization. Equations describing the objective function and constraints are given below.

The design variables are the x and y lengths of the fixed link—x_N, y_N; link lengths (see Fig. 1)—$l_M, l_J, l_B, l_K, l_C, l_D, l_E, l_F, l_G, l_H, l_I$; coordinates of a point in link G (expressed in the links local coordinate system)—x_G, y_G; and S the scaling factor. The objective function is defined as $f = \sum w_i |E_i|$, where $i = 1, \ldots, 8$. Each of the E_is is the error between the extremes (minimum and maximum x and y coordinates) of the trajectory traced by the mechanism, and the corresponding extreme values of the inner and outer envelops. For instance, $|E_1|$ is the difference between the smallest

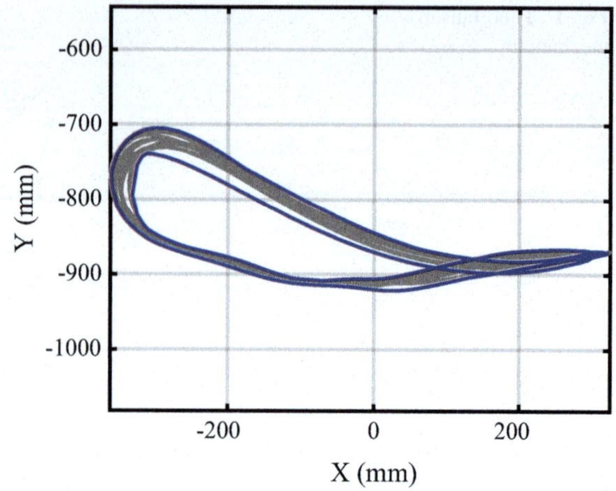

Fig. 2 Experimental data from [13]

x coordinate of the ankle trajectory $x_{\min}$ and the corresponding value of the inner envelop of experimental data $x^{in}_{\min}$. If the smallest $x_{\min}$ is within the envelop, i.e., $x_{\min} < x^{in}_{\min}$, weightage factor w_1 is zero. The objective function is the summation of eight such terms which ensure that the four extreme points—$x_{\min}$, $x_{\max}$, $y_{\min}$ and $y_{\max}$—of the ankle trajectory lie within the envelop of experimental data.

The constraints ensure that the four-bar mechanisms are feasible and remain Grashof with the same crank at the end of optimization. For the four-bar formed by links *N*-*M*-*J*-*B*, the constraints are

$$l_M < l_N \tag{1}$$

$$l_M < l_J \tag{2}$$

$$l_M < l_B \tag{3}$$

$$|l_B - l_J| < l_N + l_M < l_B + l_J \tag{4}$$

and

$$|l_B - l_J| < |l_N - l_M| < l_B + l_J. \tag{5}$$

The constraints given by Eqs. 1–3 ensure that the link *M* remains the crank, constraints given by Eqs. 4 and 5 ensure that the mechanism is feasible. For the four-bar formed by links *N*-*M*-*K*-*C*, the constraint given by Eq. 1 and

$$l_M < l_K \tag{6}$$

$$l_M < l_C \tag{7}$$

ensure that the link M remains the crank and constraints

$$|l_K - l_C| < l_N + l_M < l_K + l_C \tag{8}$$

and

$$|l_K - l_C| < |l_N - l_M| < l_K + l_C \tag{9}$$

ensure that the four-bar mechanism is feasible. Similarly, for four-bar formed by links C-D-F-G, constraints

$$l_G \le l_C \tag{10}$$

$$l_G \le l_D \tag{11}$$

$$l_G \le l_F \tag{12}$$

ensure that the link G remains the crank and constraints

$$|l_F - l_G| \le l_D + l_C \le l_F + l_G \tag{13}$$

and

$$|l_F - l_G| \le |l_D - l_C| \le l_F + l_G \tag{14}$$

ensure that the four-bar is feasible, i.e., can be formed. Note that in the constraint equations for four-bar mechanism formed by links C-D-F-G, a $\le$ is used as opposed to inequalities in constraint equations for four-bar mechanisms formed by links N-M-K-C and links N-M-J-B where "<" is used. This was done intentionally, as it was seen that the optimization routine was unable to converge to a solution otherwise.

For the closed loop triangles formed by links B-D-E and G-H-I, the constraints,

$$l_B < l_D + l_E \tag{15}$$

and

$$l_G < l_H + l_I \tag{16}$$

ensure that the triangles are feasible. Additional constraints that the four-bar mechanisms remain Grashof are also included. Lower and upper bounds of 1 and 200 cm are used for each of the link lengths. For the coordinates x_N, y_N, x_G, and y_G, bounds of (−200, 200) cm are used and (0.25, 2.5) are used as the bounds for the scaling factor S. Initial link lengths of the Theo Jansen mechanism are taken as $x_N = 38$ cm, $y_N = 7.8$ cm, $l_M = 15$ cm, $l_J = 50$ cm, $l_B = 41.5$ cm, $l_K = 61.9$ cm, $l_C = 39.3$ cm, $l_D = 40.1$ cm, $l_E = 55.8$ cm, $l_F = 39.4$ cm, $l_G = 36.7$ cm, $l_H = 65.7$ cm, and $l_I = 49$ cm. Coordinates of a point in link G (expressed in the local coordinate system) x_G, y_G and S the scaling factor are obtained by trial and error as 15.8 cm, −1.44 cm, and 1.5, respectively.

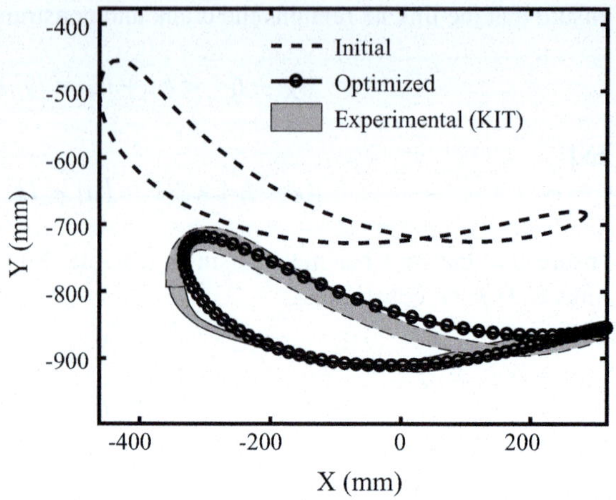

Fig. 3 Ankle trajectory obtained using optimization of the Jansen leg mechanism

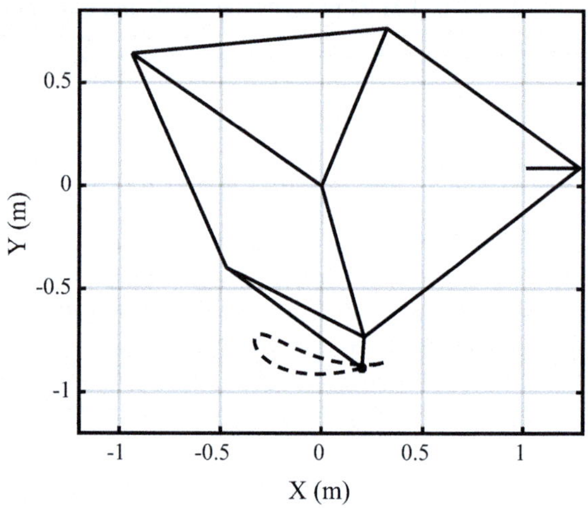

Fig. 4 Modified Theo Jansen mechanism

MATLAB's "fmincon" function is used for the optimization. The goal was to obtain a solution that can trace the ankle trajectory. If a "global" solution is to be sought, the problem definition should include all constraints which would take into account kinematic and dynamic factors, specifications on the stresses, and many other factors that would be clear only after obtaining inputs from clinicians and/or users. In the current stage of the design, using the default optimization algorithms provided by MATLAB, a preliminary solution is obtained. The results from optimization are $x_N = 53.6$ cm, $y_N = 4.6$ cm, $l_M = 14.13$ cm, $l_J = 62.1$ cm, $l_B = 43.8$ cm, $l_K = 71.2$ cm, $l_C = 40.2$ cm, $l_D = 60.1$ cm, $l_E = 66.8$ cm, $l_F = 60.12$ cm, $l_G = 40.2$ cm, $l_H = 74.1$ cm, $l_I = 43.14$ cm, $x_G = 14.7$ cm, $y_G = -3.31$ cm, and $S = 1.89$. The initial and final trajectory traced by the point on the mechanism are shown in Fig. 3. The modified Theo Jansen mechanism is shown in Fig. 4.

These preliminary results show that the Theo Jansen mechanism can be modified to trace the trajectory of human ankle during walking. The mechanism in its current form (Fig. 4) would need a work space of 2 m high and 2.5 m wide. Although the design needs to be simplified further, these preliminary results show that the mechanism can be potentially used to design devices for rehabilitation and gait training purposes and in the development of assistive exoskeletons.

Although the trajectory traced by the mechanism has a better match with the experimental data as opposed to the initial trajectory, several discrepancies still exist between the final trajectory and the experimental data. These differences can be attributed to the objective function. The objective function ensures that the extreme (maximum and minimum x and y coordinates) of the traced trajectory are within the envelop. The underlying assumption was that achieving this would ensure that the entire trajectory would lie in the bounds. However, the results indicate that the definition of the objective function needs to be revised.

2.2 *Forward Dynamic Analysis*

Kane's method [12] is chosen to perform the forward dynamic analysis as the vectorial approach makes the method a more suitable choice for computer implementation. Kane's method is based on the principle of virtual powers or Jourdain's principle which states that: *A constrained mechanical system performs motions such that the total virtual power δP of the constraint force and torque is zero.* In the case of Theo Jansen mechanism, Kane's equation is given by

$$\vec{T}.\frac{\partial \vec{\omega}_M}{\partial \dot{\theta}_M} + \sum_P \vec{F}_{gP}.\frac{\partial \vec{V}_{PO}}{\partial \dot{\theta}_M} + \sum_P \vec{F}^*_{PO}.\frac{\partial \vec{V}_{PO}}{\partial \dot{\theta}_M} + \sum_P \vec{T}^*_P.\frac{\partial \vec{\omega}_P}{\partial \dot{\theta}_M} = 0 \quad (17)$$

where $\vec{T}$ is the torque applied on the link M. $\vec{\omega}_M$ (= $[0; 0; \dot{\theta}_M]$) is the angular velocity of link M. $\dot{\theta}_M$ is the chosen generalized coordinate. $\vec{F}_{gP}$ is the gravity force acting at the CoM of link P, where $P = B, C, D, E, F, G, H, I, J, K$, and M. $\vec{V}_{PO}$ is the velocity of CoM of the link P and $\vec{\omega}_P$ is the angular velocity of the link P. $\vec{F}^*_{PO}$ and $\vec{T}^*_P$ are the inertia force and inertia torque of link P given by

$$\vec{F}^*_{PO} = -m_P \vec{A}_{PO} \quad (18)$$

and

$$\vec{T}^*_P = -\mathbf{I}_P \vec{\alpha}_P \quad (19)$$

where m_P is the mass of link P, $\vec{A}_{PO}$ is the acceleration of CoM of link P, $\mathbf{I}_P$ is the moment of inertia of link P, and $\vec{\alpha}_P$ is the angular acceleration of link P. Note that all the quantities (in Eq. 17 and in the following equations) are expressed with respect to the ground fixed coordinate system N and are expressed in N.

Since the mechanism has a single DoF, Kane's equation gives a single second-order differential equation (Eq. 17) in vector form. For the forward dynamics analysis of the mechanism using MATLAB, the function "ode45" is used. This function uses initial (i.e., at time t = 0) angular orientation (θ_M) and angular velocity ($\dot{\theta}_M$) values as the inputs and estimates θ_M and $\dot{\theta}_M$ for the given time span. Using state-space representation, the second-order differential equation is written as

$$\begin{aligned} y_1 &= \theta_M \\ y_2 &= \dot{\theta}_M \end{aligned} \tag{20}$$

and

$$\begin{aligned} \dot{y}_1 &= y_2 \\ \dot{y}_2 &= \ddot{\theta}_M = f(y_1, y_2). \end{aligned} \tag{21}$$

The $\ddot{\theta}_M = f(y_1, y_2)$ is obtained from Kane's equation (Eq. 17) as described below:

Components of CoM accelerations: The acceleration of CoM $\vec{A}_{PO}$ of any link P in a kinematic chain connected by revolute joints can be written as

$$\vec{A}_{PO} = \frac{d^2\vec{r}_{PO}}{dt^2} = \sum_{Q=N}^{P} \vec{\alpha}_Q \text{ x } \vec{r}_{QP} + \sum_{Q=N}^{P} \vec{\omega}_Q \text{ x } (\vec{\omega}_Q \text{ x } \vec{r}_{QP}) \tag{22}$$

where

$$\vec{r}_{PO} = \sum_{Q=N}^{P} \vec{r}_{QP} \tag{23}$$

and Q varies from N (the ground link), along the kinematic chain, to the link P which contains the point PO. The vectors $\vec{r}_{QP}$ are vectors on links along the kinematic chain starting from origin on link N to the point PO on link P. Equation 22 can be rewritten as

$$\vec{A}_{PO} = \vec{A}^{\alpha}_{PO} + \vec{A}^{\omega}_{PO} \tag{24}$$

where

$$\vec{A}^{\alpha}_{PO} = \sum_{Q=N}^{P} \vec{\alpha}_Q \text{ x } \vec{r}_{QP} = \sum_{Q=N}^{P} \begin{bmatrix} -r_{QPy} \\ r_{QPx} \\ 0 \end{bmatrix} \ddot{\theta}_Q \tag{25}$$

and

$$\vec{A}^{\omega}_{PO} = \sum_{Q=N}^{P} \vec{\omega}_Q \text{ x } (\vec{\omega}_Q \text{ x } \vec{r}_{QP}) = \sum_{Q=N}^{P} \begin{bmatrix} -r_{QPx} \\ -r_{QPy} \\ 0 \end{bmatrix} \dot{\theta}^2_Q. \tag{26}$$

Using Eq. (24) in Eq. (18)

$$\vec{F}^{*\alpha}_{PO} = -m_P \vec{A}^{\alpha}_{PO} \text{ and } \vec{F}^{*\omega}_{PO} = -m_P \vec{A}^{\omega}_{PO}. \tag{27}$$

Using Eq. (27) in Eq. 17 and rearranging the terms, we get

$$-\sum_P m_P \left[\frac{\partial \vec{V}_{PO}}{\partial \dot{\theta}_M}\right]^T \sum_{Q=N}^{P} \begin{bmatrix} -r_{QPy} \\ r_{QPx} \\ 0 \end{bmatrix} \ddot{\theta}_Q - \sum_P \left[\frac{\partial \vec{\omega}_P}{\partial \dot{\theta}_M}\right]^T \begin{bmatrix} 0 \\ 0 \\ I_{Pz} \end{bmatrix} \ddot{\theta}_P = -\sum_P \left[\frac{\partial \vec{V}_{PO}}{\partial \dot{\theta}_M}\right]^T \vec{F}^{*\omega}_{PO} - \left[\frac{\partial \vec{\omega}_M}{\partial \dot{\theta}_M}\right]^T \vec{T} - \sum_P \left[\frac{\partial \vec{V}_{PO}}{\partial \dot{\theta}_M}\right]^T \vec{F}_{gP}. \tag{28}$$

Kinematic analysis: The angular velocity (acceleration) of any link in a kinematic chain connected by revolute joints can be written as a linear combination of angular velocities (accelerations) of the other links. In other words, the angular velocity of any link $\dot{\theta}_P$ in the mechanism is linearly dependent on $\dot{\theta}_M$ and angular acceleration of any link $\ddot{\theta}_P$ is linearly dependent on $\ddot{\theta}_M$. These equations can be written as

$$\begin{aligned} \dot{\theta}_P &= k_P \dot{\theta}_M \\ \ddot{\theta}_P &= k_P \ddot{\theta}_M + s_P. \end{aligned} \tag{29}$$

Equation 28 can now be rewritten as

$$\begin{aligned} &-\sum_P m_P \left[\frac{\partial \vec{V}_P}{\partial \dot{\theta}_M}\right]^T \sum_{Q=\mathbf{N}}^{\mathbf{P}} \begin{bmatrix} -r_{QPy} \\ r_{QPx} \\ 0 \end{bmatrix} (k_Q \ddot{\theta}_M + s_Q) \\ &-\sum_P \left[\frac{\partial \vec{\omega}_P}{\partial \dot{\theta}_M}\right]^T \begin{bmatrix} 0 \\ 0 \\ I_{Pz} \end{bmatrix} (k_P \ddot{\theta}_M + s_P) \\ &= -\sum_P \left[\frac{\partial \vec{V}_P}{\partial \dot{\theta}_M}\right]^T \vec{F}^{*\omega}_P - \left[\frac{\partial \vec{\omega}_M}{\partial \dot{\theta}_M}\right]^T \vec{T} - \sum_P \left[\frac{\partial \vec{V}_P}{\partial \dot{\theta}_M}\right]^T \vec{F}_{gP}. \end{aligned} \tag{30}$$

Algorithm for forward dynamic analysis: MATLAB's Runge–Kutta algorithm ("ode45") requires the initial values, i.e., $y_1 = \theta_M$ and $y_2 = \dot{\theta}_M$ at time $t = 0$ and the handle to a function that can be used to determine $\dot{y}_1 = y_2$ and $\dot{y}_2 = \ddot{\theta}_M$. The general algorithm to determine $\ddot{\theta}_M$ using Kane's method is as follows:

Step 1: Obtain the initial values $\theta_M(0)$ and $\dot{\theta}_M(0)$ at time $t = 0$.
Step 2: Determine all angular positions θ_P using θ_M and the link lengths. Here, $P = B, C, D, E, F, G, H, I, J, K$, and M with respect to N.
Step 3: Determine transformation matrices $\mathbf{T}_P$ of all links with respect to N.
Step 4: Determine CoM positions of all links using Eq. 23

$$\vec{r}_{PO} = \sum_{Q=N}^{P} \vec{r}_{QP} = \sum_{Q=N}^{P} \mathbf{T}_Q \vec{r}^{\,Q}_{QP}. \tag{31}$$

Q varies from N, the ground link, to the link P which contains the point PO. The vectors $\vec{r}_{QP}$ are vectors on links along the kinematic chain starting from origin on link N to the point PO on link P. $\vec{r}^{\,Q}_{QP}$ is the same vector expressed in the body-fixed coordinate system.

Step 5: Determine k_Ps for all the links such that, $\dot{\theta}_P = k_P \dot{\theta}_M$.

Step 6: Form the angular velocity vectors of all the links- $\vec{\omega}_P = k_P \left[0; 0; \dot{\theta}_M\right]$.

Step 7: Determine the CoM velocities of all the links:

$$\vec{V}_{PO} = \frac{\mathrm{d}\vec{r}_{PO}}{\mathrm{d}t} = \sum_{Q=N}^{P} \vec{\omega}_Q \times \vec{r}_{QP}. \tag{32}$$

Step 8: Redo steps 5 till 7, using $\dot{\theta}_M = 1$. The CoM velocities thus obtained are the partial derivatives with respect to $\dot{\theta}_M$, $\frac{\partial \vec{\omega}_P}{\partial \dot{\theta}_M}$, and $\frac{\partial \vec{V}_P}{\partial \dot{\theta}_M}$.

Step 9: Determine $\vec{A}^{\omega}_{PO}$ using Eq. 26 and $\vec{F}^{*\omega}_{PO}$ using Eq. 27.

Step 10: Determine the right-hand side of Eq. 30.

$$RHS = \sum_{P} \left[\frac{\partial \vec{V}_P}{\partial \dot{\theta}_M}\right]^T \vec{F}^{*\omega}_P - \left[\frac{\partial \vec{\omega}_M}{\partial \dot{\theta}_M}\right]^T \vec{T} - \sum_{P} \left[\frac{\partial \vec{V}_P}{\partial \dot{\theta}_M}\right]^T \vec{F}_{gP}. \tag{33}$$

Step 11: Determine s_P (Eq. 29) of all the links.

Step 12: Determine left-hand side of Eq. 30 using $\ddot{\theta}_M = 0$:

$$\begin{aligned} LHS_{res} = &-\sum_{P} m_P \left[\frac{\partial \vec{V}_P}{\partial \dot{\theta}_M}\right]^T \sum_{Q=\mathbf{N}}^{\mathbf{P}} \begin{bmatrix} -r_{QPy} \\ r_{QPx} \\ 0 \end{bmatrix} s_Q \\ &-\sum_{P} \left[\frac{\partial \vec{\omega}_P}{\partial \dot{\theta}_M}\right]^T \begin{bmatrix} 0 \\ 0 \\ I_{Pz} \end{bmatrix} s_P. \end{aligned} \tag{34}$$

Step 13: Determine left-hand side of Eq. 30 using $\ddot{\theta}_M = 1$ and all $s_P = 0$:

$$\begin{aligned} LHS_c = &-\sum_{P} m_P \left[\frac{\partial \vec{V}_P}{\partial \dot{\theta}_M}\right]^T \sum_{Q=\mathbf{N}}^{\mathbf{P}} \begin{bmatrix} -r_{QPy} \\ r_{QPx} \\ 0 \end{bmatrix} k_Q \\ &-\sum_{P} \left[\frac{\partial \vec{\omega}_P}{\partial \dot{\theta}_M}\right]^T \begin{bmatrix} 0 \\ 0 \\ I_{Pz} \end{bmatrix} k_P. \end{aligned} \tag{35}$$

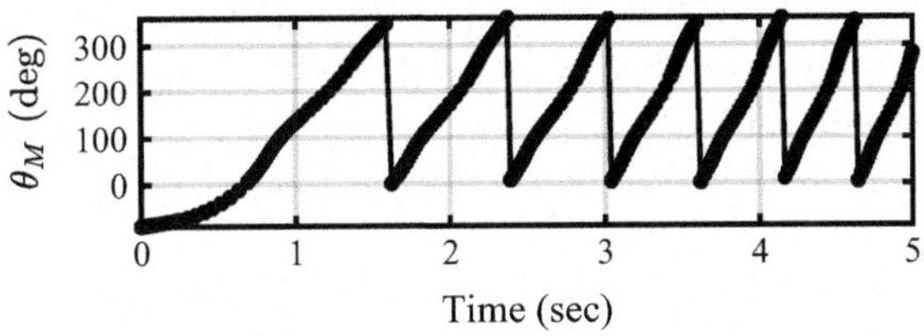

Fig. 5 Crank angle versus time for the forward dynamics simulation of the modified Theo Jansen mechanism

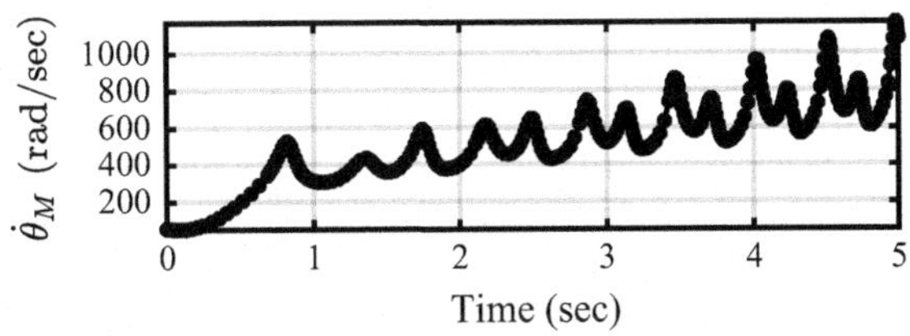

Fig. 6 Angular velocity of the crank versus time for the forward dynamics simulation of the modified Theo Jansen mechanism

Step 14: Determine $\ddot{\theta}_M = (RHS - LHS_{res})/LHS_c$.

The above methodology/algorithm can be used to codify Kane's method for forward dynamics analysis of Jansen's leg mechanism. In comparison, Newton's method would require writing the equilibrium equations of all 12 links (3 per link), 5 loop closure equations (2 per loop), and the constraint equations for all the revolute joints. This could become extremely cumbersome and difficult to solve, even on a computer. Use of Lagrange's method would require a symbolic manipulator to find the partial derivatives with respect to the generalized coordinate q, $\dot{q}$, and time. Kane's method presents the single scalar equation of motion using the principal of virtual powers in a vectorial (codable) form. Using Kane's equation and the above methodology, computer implementation of forward dynamic analysis of Theo Jansen mechanism can be made relatively simpler and easier to handle.

The forward dynamic simulation was carried out for time = 0–5 s with a constant torque of 10 N m and initial values of θ_M and $\dot{\theta}_M$ as $-90°$ and 1 rad/sec, respectively. The results (Figs. 5 and 6) show that for a constant torque the angular velocity of the crank increases exponentially. For the stable operation of the mechanism, either a controller should be included (as in [10]) or the optimization problem statement should be expanded to include additional parameters and constraints that ensures constant velocity operation. Future work will include more optimization trails which would incorporate dynamic analysis as well.

3 Conclusion

The optimization and dynamic analysis of the mechanism described in this work can be used as a basis for the design and analysis of the gait rehabilitation device and/or assistive exoskeletons. Furthermore, the forward dynamic analysis methodology (Kane's method) can be used as a basis for computational modeling of Jansen's leg-based rehabilitation devices and for legged walking platforms or other robotic manipulators. In this work, a step-by-step algorithm to codify and solve the EoM of Theo Jansen mechanism using Kane's method is presented. Typically, a symbolic manipulator would be required to derive the EoM which would then be solved using a numerical method. The methodology presented here can be used to directly code and solve the EoM without the need for a symbolic manipulator.

While Theo Jansen mechanism has been used by others in the design of mobile platforms, etc., there are no papers that attempted to used Theo Jansen mechanism to trace ankle trajectory to arrive at a mechanism for gait training. Although this has been attempted here, there are limitations that are still to be addressed. One of the major limitations of the current design is that the size of the mechanism is not practical. The preliminary design obtained in this work needs further changes to ensure compact design and adaptability to persons of different heights which could be two of the primary challenges to be addressed. Definition of the objective function in the optimization problem statement needs to be further refined to ensure that the entire trajectory fits into the envelop of experimental data and to ensure constant velocity operation. Additionally, timing control should be incorporated to ensure a match with the gait cycle of the user. Analyses described in this work will be used as a basis to make further design modifications to the mechanism in order to arrive at a simplistic and affordable gait training device.

References

1. Shao Y, Xiang Z, Liu H, Li L (2016) Conceptual design and dimensional synthesis of cam-linkage mechanisms for gait rehabilitation. Mech Mach Theory
2. Tsuge BY, McCarthy JM (2016) An adjustable single degree-of-freedom system to guide natural walking movement for rehabilitation. J Med Dev
3. Manna YA, Du F, Ji Z (2011) Motion control and coordination in the development of a conceptual gait rehabilitation system. J Mech Eng Sci 19(9):2482–2487
4. Tsuge BY, Plecnik MM, McCarthy (2016) Homotopy directed optimization to design a six-bar linkage for a lower limb with a natural ankle trajectory. J Mech Robot
5. Banala S, Kim S, Agrawal S, Scholz J (2009) Robot assisted gait training with active leg exoskeleton (alex). IEEE Trans Neural Syst Rehabil Eng 17:2–8
6. Jezernik S, Colombo G, Keller T, Frueh H, Morari M (2003) Robotic orthosis lokomat: a rehabilitation and research tool. Int Neuromodul Soc 6(2):108–115
7. Van Asseldonk EHF, Ekkelenkamp R, Veneman JF, Van der Helm FCT, Van der Kooij H (2007) Selective control of a subtask of walking in a robotic gait trainer (lopes). In: Proceedings of the IEEE 10th international conference on rehabilitation robotics, Noordwijk, The Netherlands, June 2007
8. Ji Z, Manna Y (2008) Synthesis of a pattern generation mechanism for gait rehabilitation. J Med Dev
9. Patnaik L, Umanand L (2016) Kinematics and dynamics of Jansen leg mechanism: a bond graph approach. Simul Model Pract Theory 60:160–169
10. Nansai S, Elara MR, Iwase M (2013) Dynamic analysis and modeling of Jansen mechanism. Procedia Eng 64:1562–1571

11. Nansai S, Rojas N, Elara MR, Sosa R, Iwase M (2015) A novel approach to gait synchronization and transition for reconfigurable walking platforms. Digital Commun Netw 1:141–151
12. Kane TR, Levinson DA (2005) Dyn Theory Appl. The Internet-First University Press
13. Mandery C, Terlemez O, Do M, Vahrenkamp N, Asfour T (2015) The kit whole-body human motion database. In: International conference on advanced robotics, pp 329–336

Textile DOBBY Mechanism Synthesized as a Mechanical D Flip-Flop

C. Amarnath and Ankit Mehta

Abstract There are several devices and machinery which are products of ingenious thought and subsequent refinement. Textile machinery specifically those used in the small scale sector in developing nations are one example of such an evolution. Here we examine the possibilities of enhancement of the performance of the DOBBY—a device used in weaving machines which are attached to a loom for weaving small patterns. The DOBBY (and the Jacquard) is the forerunner of modern digital systems and uses reciprocating "knives" for actuating the "healds" through a set of hooks. The hooks are brought into engagement and disengagement with the reciprocating knives depending upon the position of "feelers" that sense holes on punched cards. This ingenious device has served the handloom and power loom sector well. There is a limit, however, on operating speeds and thus throughput because of the dynamics of the "make and break" contact between the knives and hooks. The knife and hook engage with a jerk that causes unwanted dynamic forces and vibration which in turn is reflected in the weave. An attempt has been made to eliminate the "knife and hook" design through use of a circuit composed of "mechanical logic gates" to eliminate the "make and break" and maintain a continuous set of derivatives of motion in the critical portions of the activity of the device. The number of moving links in the mechanism has increased as a consequence of this artifice. The DOBBY can be recognized to be a "D Flip-Flop" in computer parlance by casting and examining its Truth Table. The elements in the circuit that was arrived at are AND and OR mechanical gates which had been synthesized over the past few years. The use of mechanisms with lower pairs ensured that the devices exhibit smooth motion. A working model was made and demonstrated.

Keywords Textile machinery · Dobby · Mechanical logic gates · Logic circuits

---

C. Amarnath (✉)
IIT Bombay, Mumbai, India
e-mail: c.amarnath@gmail.com

A. Mehta
Idea Forge, Mumbai, India
e-mail: ankit@ideaforge.com

D. Sen et al. (eds.), *Mechanism and Machine Science*, Lecture Notes in Mechanical Engineering, https://doi.org/10.1007/978-981-15-4477-4_23

1 Introduction

There exist a large number of systems in which the input, output, and other links have only two states of interest—ON or OFF, 0 or 1, OPEN or CLOSE, etc. The other intermediate positions of the system are not of relevance. Such Binary systems have been done extensively used in the field of Electronics and Computer Science.

In electronics, the states of interest are the LOW and HIGH voltages in a circuit, in Logic FALSE and TRUE are the two states of interest. Similarly it can be observed in mechanical systems that certain links of interest (like the input and output links) are required to occupy only two finitely separated positions of interest. Such systems will be referred to as Mechanical Binary Systems.

Boolean Variables 0 and 1 are used in electronics to represent LOW and HIGH voltages, respectively; in logic they represent FALSE and TRUE states, respectively. In the Mechanical domain 0 and 1 can be used to represent two finitely separated positions of the links that are of interest to us.

2 Form Closed Logic Gates

Amarnath [1–8] has developed two-input logic gates by using a two-degree of freedom seven-bar linkage in a novel way. A 2-d.o.f system is required since we are looking for two-input and single-output gates. By a suitable choice of the link lengths, it is possible to use such mechanisms for providing the "AND/NAND" and "OR/NOR" logic.

2.1 AND Gate

Figure 1a shows a seven-link mechanism proposed earlier [1, 2], with two sliders A and B as inputs. These inputs occupy two finitely separated positions, designated as "0" and "1". Points C1, C2, C3, and C4 are the positions of the joint C for the four possible input combinations (A, B) = (0, 0), (0, 1), (1, 0), and (1, 1), respectively. The pivot R on output slider Z is the center of a circle passing through C1, C2, and C3. With pivot R so chosen, output slider Z occupies position "1" only when both inputs A and B are "1", otherwise slider Z is at position "0". The companion Fig. 1b shows the same gate, but with rotary inputs and outputs, the synthesis procedure being similar to the earlier. A NAND gate can be obtained from the AND gate of Fig. 1b, simply by locating the fixed pivot Q of the output link at Q′ as shown in Fig. 1b. Hence an input which causes a clockwise rotation of the output link of the AND gate will result in an anti-clockwise rotation of the output link of the NAND gate.

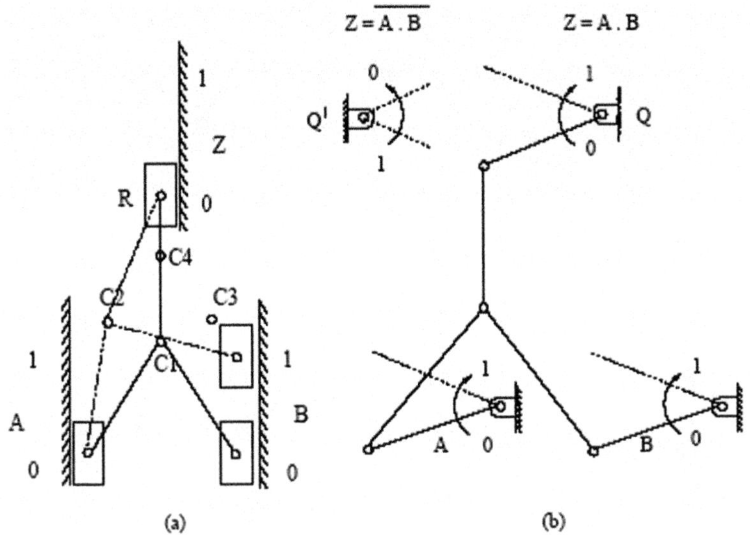

Fig. 1 Mechanical AND gate [1, 2]

The devices in Fig. 1 are approximate devices, in the sense that a transition from (A, B) = (0, 0) to (A, B) = (0, 1), or (1, 0) results in a slight motion of output link Z.

2.2 *OR Gate*

Figure 2 shows a seven-link mechanism with sliders A and B as the two inputs. Points C1, C2, C3, and C4 are the positions of joint C for the four possible input combinations. Here pivot R is at the center of the arc passing through C2, C3, and C4. Thus the output slider occupies position "0" only when both inputs are "0" and is at "1" otherwise there by providing the OR logic.

3 Dobby Mechanism

The dobby mechanism enables weaving looms to create sheds in the longitudinal threads (wrap) for the insertion of the cross threads (weft) by the means of the shuttle. The shed is formed by pulling up selected threads in a specific pattern with a frame called the heald. The heald movement is enabled by the central drive.

In a simple weave, the shed opens only for a single pass of the shuttle (from now on called a pick), as in Fig. 3a. The shed may sometimes be required to be open for more than one pick in more complex weaving patterns (Fig. 3b) Thus the

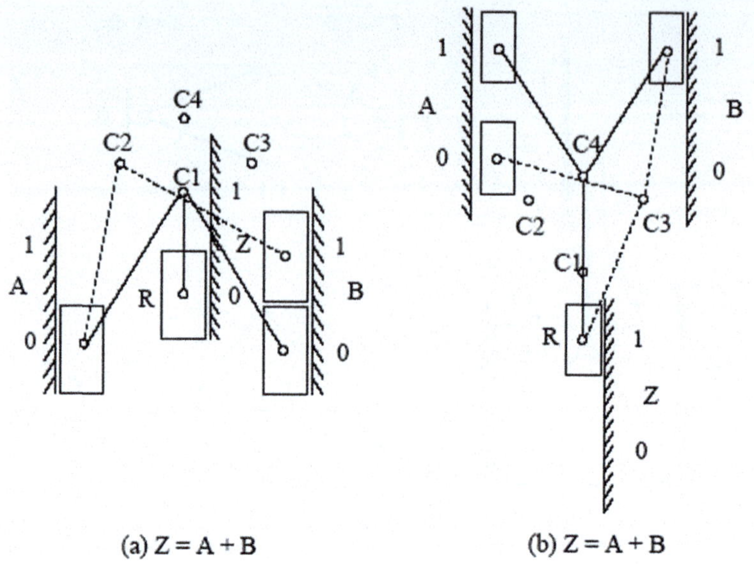

Fig. 2 Mechanical OR gate [1, 2]

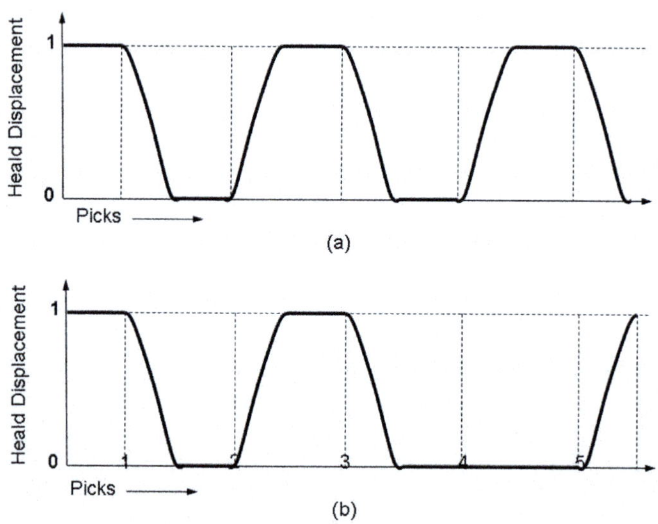

Fig. 3 Heald displacement versus picks graph

heald should stay in the raised or lowered position for the requisite number of picks, irrespective of the central drive's operation.

The Keighley dobby mechanism [3] is one of the mechanisms that are capable of achieving the kind of motion that is desired by the weaving operation as described

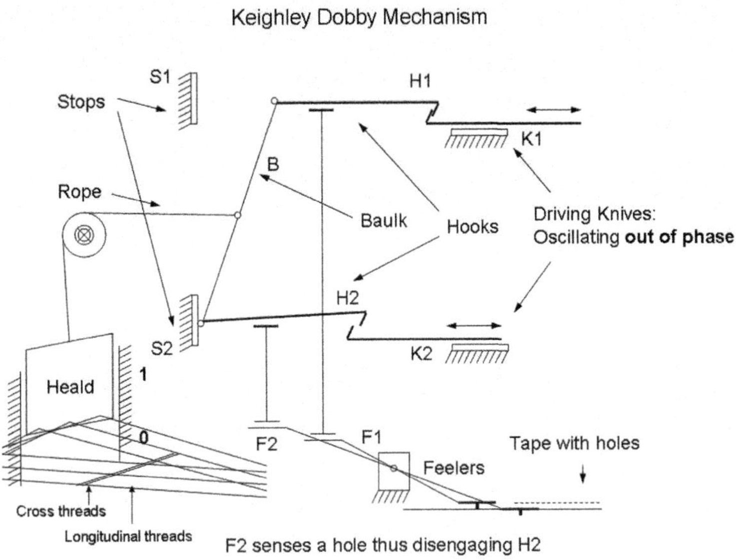

Fig. 4 Keighley dobby mechanism with the heald at the upper position

above. Figure 4 shows the Keighley dobby mechanism with the heald in the upper position and Fig. 5 shows the heald in the lower position. The Keighley dobby takes its inputs from a tape with the holes punched in the desired pattern. The tape continuously rolls on an indexing mechanism, and at a fixed time interval, which is synchronized with the indexing; the feelers of the sensor feel the tape. The presence of a hole in the tape would lead to the raising of the hooks which in turn leads to the disengagement of the driving knives from the hooks. When both hooks are engaged the heald stays in the upper position and when both are disengaged the heald stays in the lower position. If one of the hooks is engaged the heald will move from one position to the other.

By making holes on the tape according to the desired weaving pattern, it is ensured that the heald stays in the upper position or the lower position for the requisite number of picks.

To analyze the Keighley dobby mechanism, (Fig. 6) the finitely separated positions of the links of interest, i.e., the knives, the sensor, and the heald, in the dobby mechanism have been marked as 0 and 1. The feelers and the tape assembly is termed as the sensing module with output 0 or 1.

The Keighley dobby mechanism uses the physical engaging and disengaging of the hooks as a means of following the sensor input. This physical engaging and disengaging leads to a limitation on the loom speed due to unfavorable dynamic effects. Thus a need is felt to improvise upon the Keighley dobby and come up with a faster loom.

The various inputs and the output of a dobby mechanism are stated in the figure below (Fig. 7).

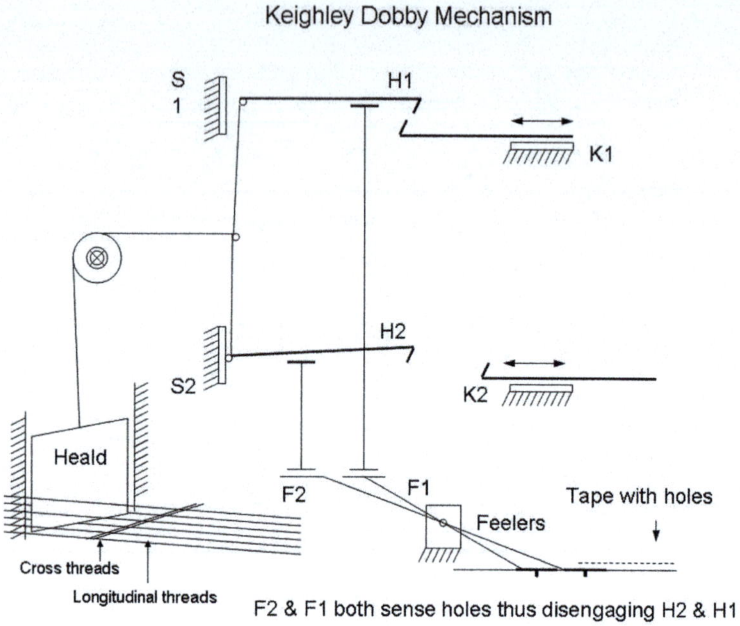

Fig. 5 Keighley dobby mechanism with the heald at the lower position

Fig. 6 Keighly dobby mechanism with logic state representations

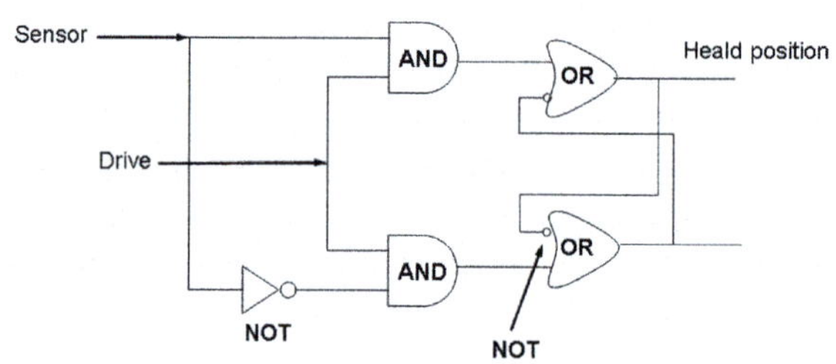

Fig. 7 Dobby mechanism represented as a black box

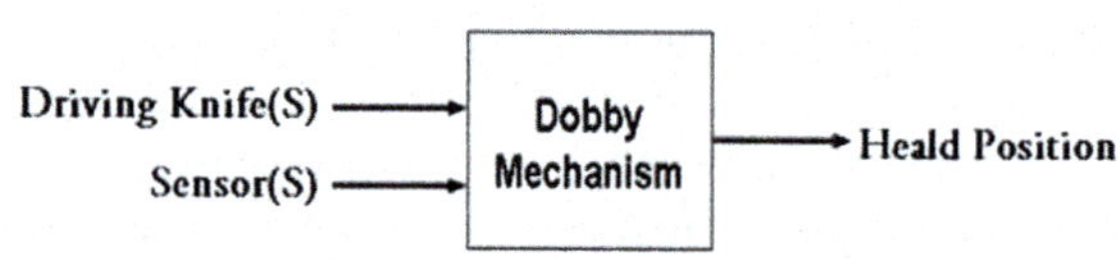

The logic state representations and the truth table are given in Tables 1 and 2.

From the truth table, it can be noted that the logic of operation of the dobby is similar to the operation [3, 4] of a clocked D flip-flop. The sensor becomes the input D and the drive is the clock for the D flip-flop.

The logic circuit of the dobby mechanism is implemented using the seven-bar AND and OR gates. The following sections describe the design of the new dobby mechanism, and the performance of a working model built.

Table 1 Dobby mechanism's inputs and output tasks and their logic state representations

Task	Task purpose	0 logic state	1 logic state
Drive(s)	Powers the movement of the heald	Ready to raise the heald from lower position	Ready to lower the heald from raised position
Sensing(s)	Determines which state the heald is to stay in	Heald lowering command	Heald raising command
Heald positioning	Determines the weave style	Lower heald position	Raised heald position

Table 2 Truth table for a dobby mechanism

Sensing	Drive	Heald positioning
0	0	0
0	1	0
1	0	1
1	1	1

3.1 *An Implementation of the Logic Circuit Using Form Closed Gates*

The mechanism of Fig. 8 consists of two seven-bar AND gates and two seven-bar OR gates. The above figure only shows the schematic arrangement of the gates. It is important to note that in this implementation of the dobby mechanism there is no physical engaging and disengaging occurring as in the conventional dobby and thus is expected to perform smoothly at higher speeds of operation. Due to absence of the knives we call the new mechanism as the "Knifeless dobby mechanism". The following section describes the design of a working model.

3.2 *Working Model of the Alternative Dobby Mechanism*

A working model of the Knifeless dobby mechanism suggested in Sect. 3.1 was fabricated. The actual implemented model has a few changes from the proposed schematic design used in the Sect. 3.1. Instead of the seven-bar OR gate as shown to be useful in Fig. 8 we used the seven-bar AND gate in the inverted position to implement the OR logic. Due to symmetry reasons the AND gate, placed geometrically in the opposite direction as shown in Fig. 9, was used as an OR gate in the Knifeless dobby.

The actual model was implemented in two planes with one pair of the input AND gates operating in one layer and the other pair of OR gates operation in another plane. This was done in order to reduce its planar dimensions. The device was demonstrated under the name of "Knifeless Dobby" (Figs. 10, and 11) at the India ITME fair in Mumbai.

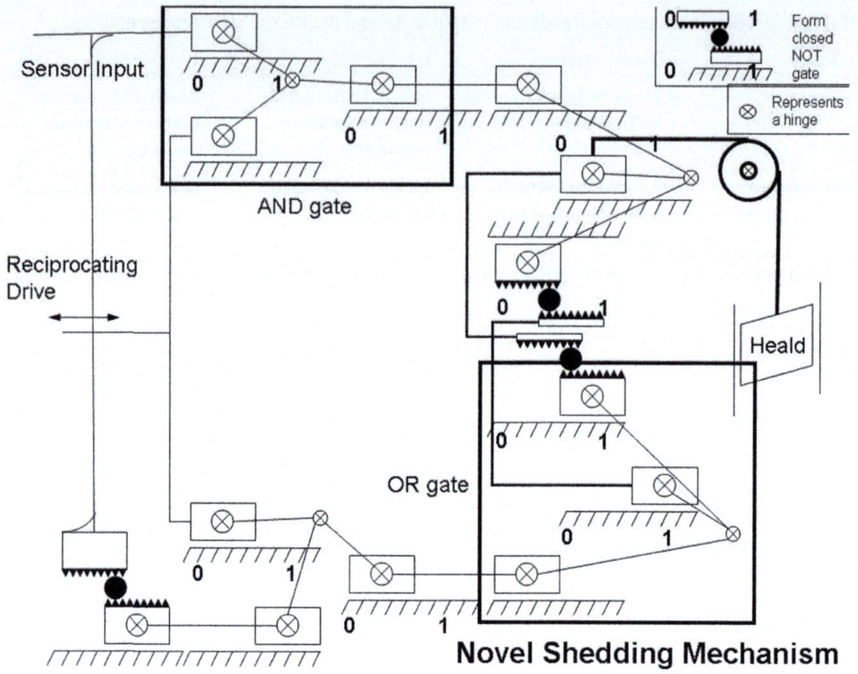

Fig. 8 The novel dobby mechanism

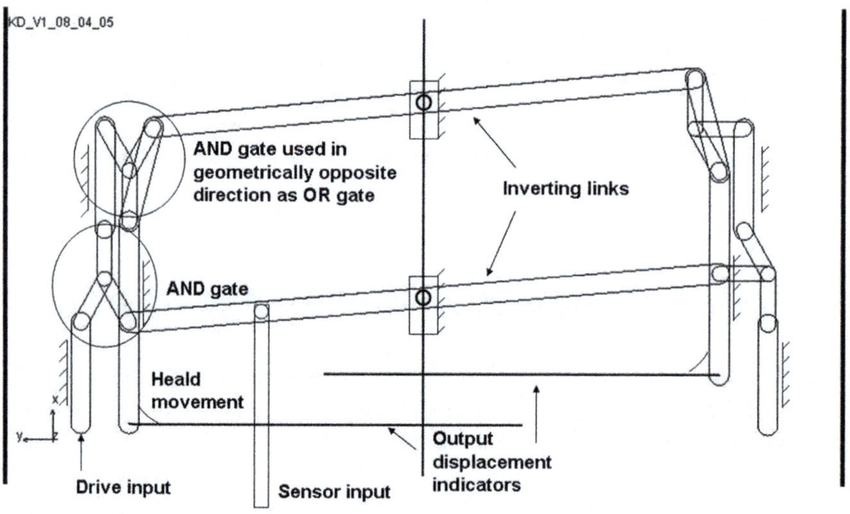

Fig. 9 Theoretical assembly of the new dobby

Fig. 10 The inside view of the dobby, also showing the drive motor and the sensor scale used for demonstration purposes

Fig. 11 Side view of the Knifeless dobby

The working model of the Knifeless dobby performed according to the expected logic. Theoretically structural error—if large—would result in the model misbehaving when the input motions are out of sync but the Knifeless dobby mechanism built did not exhibit a tendency to lock up in actual practice due to minimization of structural error.

4 Conclusion

It has been shown and demonstrated that a dobby based on mechanical logic gates is feasible. A prototype (if built) could demonstrate with certainty the ability of these devices to perform at high speed.

The need to design an exact form closed AND device was felt while exploring the Knifeless dobby mechanism. It was felt that it may be necessary to device new AND gates which would not suffer from high structural error or could be exact ones—though like exact mechanisms they could gain or lose mobility.

Acknowledgements The authors would like to acknowledge the help by Tina Primrose in preparing the paper.

References

1. Amarnath C (1989) Personal inventions notebook. Robotics Laboratory, IIT Bombay. http://www.me.iitb.ac.in/~amarnath/mechanical_gates.html
2. Amarnath C, Deo HV (2003) Logic gates based on bar link mechanisms. In: Proceedings the 11th world congress in mechanism and machine science, Tianjin, China
3. Talukdar MK, Sriramulu PK, Ajgaonkar DB (1998) Weaving-machine-mechanisms-management. Mahajan Publishers, p 259
4. Mottershead A (1973) Electronic devices and circuits—an introduction. Prentice-Hall, India, pp 477–478
5. Amarnath C, Deo HV (2003) Synthesis of circuit breaker mechanisms based on mechanical logic gates. In: Proceedings of the 11th world congress in mechanism and machine science, Tianjin, China
6. Sapre SS (1991) Mechanical digital logic systems. Master's disseration, IIT Bombay
7. Deo HV (2001) Application of mechanical logic gates to switchgear design. Bachelor's thesis, IIT Bombay
8. Shigley JE, Uicker JJ (1995) Theory of machines and mechanisms, 2nd edn. McGraw-Hill, pp 181–188

Design of Oblique Leaf-Spring Suspension Mechanism for Heavy Vehicles

Jing-Shan Zhao, Hong-Wei Song, Yun Zhang, and Xiang Liu

Abstract The ride comfort, driving flexibility and stability, and operation safety all directly or indirectly depend on the suspension mechanism. Therefore a suspension system defines, in a certain degree, the vehicle's performance. Traditional longitudinal spring suspension system occupies a large space which is not beneficial either to enlarging the steering angle or to the layout of vehicle chassis. This paper therefore proposes a suspension mechanism within which there are four identical single leaf-springs assembled in parallel. Thus the suspension forms a statically indeterminate oblique structure which improves the stiffness, strength, and reliability while reducing its total mass. The oblique arrangement of the leaf-springs allows the suspension to occupy smaller space which therefore provides larger steering angle for a vehicle. Experiments with real vehicles proved that this kind of suspension should also improve the dynamic performance as it reduces the suspension mass due to the redundant leaf-springs. Therefore, the oblique leaf-spring suspension mechanism can enhance the ride comfort, handling flexibility and stability, and driving safety of a vehicle.

Keywords Parallel leaf-spring · Suspension mechanism · Vehicle performance · Handling stability

1 Introduction

It is the role of an automotive suspension to ensure that the elastic connections between wheels and axles, and vehicle frame could bear the load, moderate the impact, attenuate the vibration, and adjust the vehicle's position in driving process [1]. Suspension is an important part of chassis [2] to directly determine the stability and smoothness of a vehicle in operation. Generally suspensions can be divided into

J.-S. Zhao (✉) · H.-W. Song · Y. Zhang · X. Liu
Department of Mechanical Engineering, Tsinghua University, Beijing 100084, People's Republic of China
e-mail: jingshanzhao@mail.tsinghua.edu.cn

Y. Zhang · X. Liu
Advanced Engineering of Dongfeng Commercial Vehicle Technical Center, Wuhan, China

D. Sen et al. (eds.), *Mechanism and Machine Science*, Lecture Notes in Mechanical Engineering, https://doi.org/10.1007/978-981-15-4477-4_24

non-independent suspensions, independent suspensions [3], and semi-independent suspensions. Nowadays, the widely used independent suspensions are MacPherson suspensions, double wishbone suspensions, and multilink suspensions [4]. Stiffness is an important parameter to measure the mechanical properties of a suspension. Verros and Natsiavas [5] presented a method to analyze the damping and stiffness [6–8] of nonlinear models. For non-independent suspensions, both sides of the two wheels are connected by one integral axle. Wheels and axles are connected to the whole frame by elastic components. Gerrard [9] considered that the suspension system is a single elastic system connecting the wheel carrier to the ground and that the elastic system can be represented in its entirety by single stiffness matrix. The non-independent suspension has the advantages of a large carrying capacity, high reliability, simple structure, and low production cost. The structure of an independent suspension is characterized by distinct connections for each wheel to the frame through elastic components [10]. Recently and Sharma [11] discussed a primary suspension system that could isolate cabin from the vibrations of vehicle's chassis resulting from road irregularities [12]. Most of the commercial vehicles [13] usually use multiple leaf-springs for their non-independent suspensions to attenuate the shocking forces from the bumpy road. The multiple leaf-springs are arranged along the longitudinal direction of the vehicle's frame. One end of the multiple leaf-spring is hinged with the frame and the other one is hinged with the frame through a swing rod [14]. The main disadvantages of these suspensions are that large deformations of the longitudinal multiple leaf-springs resulting from an external load often lead to the change of wheel alignment, which will induce a tilt of the vehicle, an abnormal wear of wheels and even an instability of the vehicle. As the longitudinal multiple leaf-springs form a series structure, the failure of any springs will cause catastrophe of the vehicle. So, the stability of the traditional multiple leaf-springs is relatively low. During steering, there must be enough space to avoid structural interference. An increase in the lateral distance between the vehicle frame and longitudinal multiple leaf-spring will deteriorate the vehicle's anti-roll ability [15]. Zhao et al. [16, 17] proposed a compliant independent suspension by using flexible plate-form links with an anisotropic elasticity. They first investigated R-type kinematic chains to generate planar motions and then discussed the equivalent substitution of a flexible link with an infinite-R kinematic chain. The innovative design of an independent suspension [18], besides eliminating the shambling shocks induced by the jumping of the wheels, can also decrease wear of the tires. Therefore, these kinds of independent suspension can be used as either front or rear suspension. The steering system [3] accompanying the rectilinear moving suspensions [19] greatly improves the ride and handling properties of advanced vehicles [20, 21].

In this paper, a generic novel oblique leaf-spring suspension is proposed for heavy-duty commercial vehicles. The oblique leaf-spring allows a larger turning space while maintaining the identical steering alignment properties of the usual vehicles. As a result, it fundamentally reduces the abnormal wear and even tearing of tires. The oblique leaf-spring suspension accompanied by the steering system [22] should improve the ride comfort, handling stability, and driving safety of heavy-duty vehicles.

2 Statics of Oblique Leaf-Spring

The oblique leaf-spring suspension consists of four pieces of single leaf-springs, trunnions, knuckle scaffold, and single leaf-springs which are symmetrical about plane π_1 and plane π_2 simultaneously as shown in Fig. 1.

Assuming that the four springs in parallel uniformly bear the external loads, including vertical force, lateral force, longitudinal force, vertical torque M_Z, lateral torque M_Y, and longitudinal torque M_X. The basic parameters of the oblique leaf-spring suspension are illustrated in Fig. 2.

The initial tilting angle of the curved leaf-spring, δ, shown in Fig. 2 is

$$\delta = \arcsin \frac{b - a}{2l} \tag{1}$$

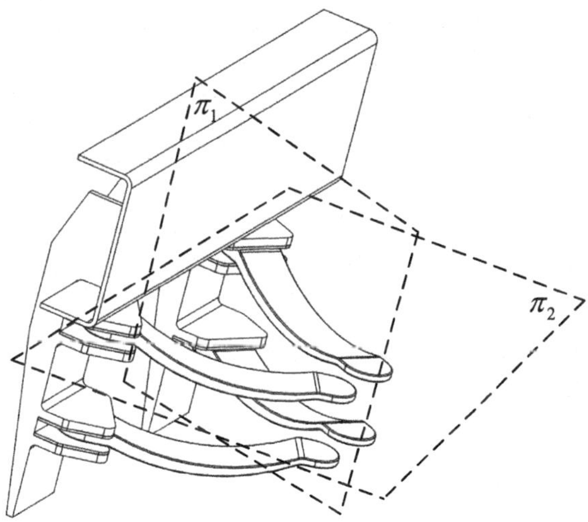

Fig. 1 Configuration of the oblique leaf-spring suspension mechanism

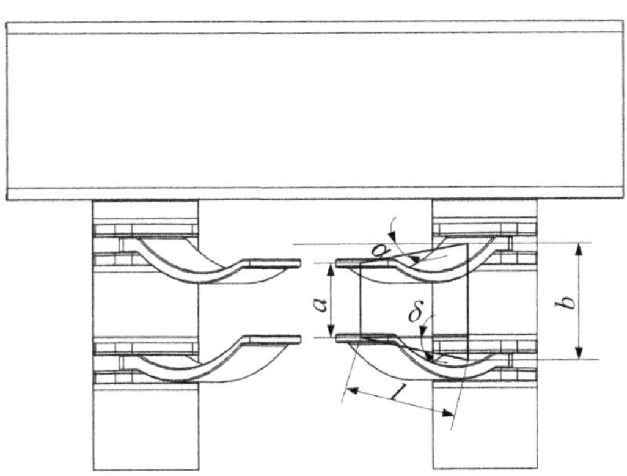

Fig. 2 Size parameters of the suspension

where a is the vertical distance between two layers of springs at the connection with steering knuckle, b is the vertical distance between the connections at the frame for the two layers of springs, l is the length of the spring, α is the angle between the spring arrangement direction and the direction of plane π_1, δ is the angle between the spring direction and plane π_2.

As the functions of the suspension are already known, the values of the parameters need to be identified based on the common use.

2.1 Load Analysis

This section mainly describes the load of a leaf-spring, and the stress diagram illustrated in Fig. 3 with two coordinate systems of $oxyz$ and $ox_1y_1z_1$ at the end of each spring connecting the knuckle.

The $x(x_1)$-axis is perpendicular to the spring arrangement's direction, whereas the y-axis's direction is collinear to the spring arrangement. Furthermore the y-axis is along the connecting direction of the spring with the frame or knuckle, and z-axis is then determined in accordance to the right-hand rule. Moreover y_1 locates in the horizontal plane, the subtended angle between y_1 and y is δ. As a result, z_1-axis is along the vertical direction, thus the angle between z_1- and z-axes is δ.

Assume that a vector r_1 is expressed by $r_1 = \left(x_1\ y_1\ z_1\right)^T$ in the $ox_1y_1z_1$ coordinate system, the same vector r is, in the coordinate system $oxyz$, given by

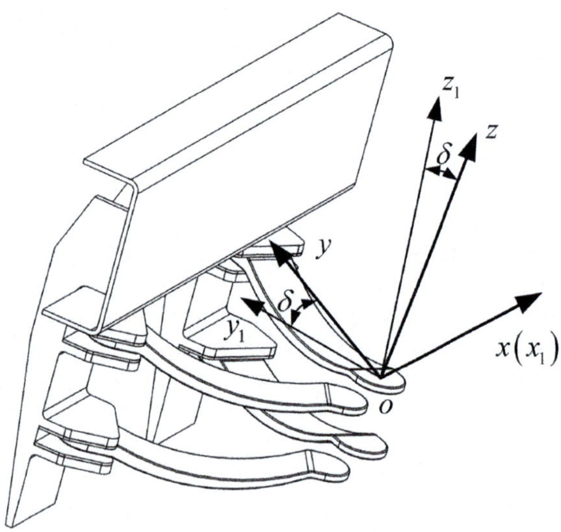

Fig. 3 Coordinate system of spring end

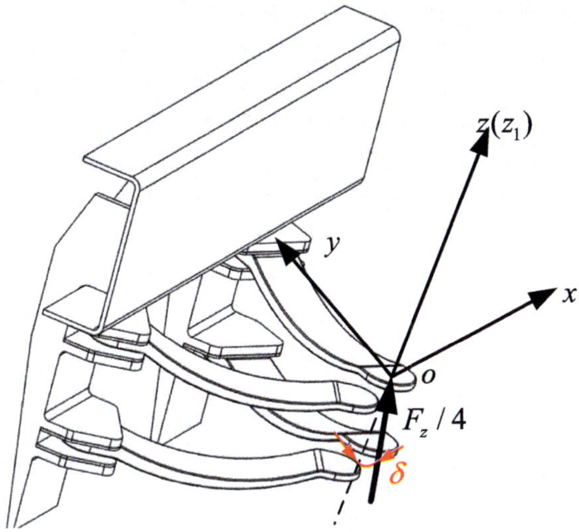

Fig. 4 Vertical load on each spring

$r = \left(x\ y\ z\right)^T$, then $r = T_C r_1$ where $T_C = \begin{bmatrix} 1 & 0 & 0 \\ 0 & \cos\delta & \sin\delta \\ 0 & \sin\delta & \cos\delta \end{bmatrix}$ is the transformation matrix from frame $ox_1y_1z_1$ to frame $oxyz$.

2.1.1 Analysis of Vertical Force

Assuming that the external load from vertical direction is F_z, therefore the vertical force on each leaf-spring is $F_z/4$, in the direction of z_1, shown in Fig. 4.

The vertical force in the coordinate system of $oxyz$ is

$$F_{\text{vertical}} = \left[0\ \tfrac{F_z}{4}\sin\delta\ \tfrac{F_z}{4}\cos\delta\right] \quad (2)$$

Putting δ into Eq. (2), we get the force along the spring, $F_{\text{vertical}y} = \frac{F_z}{4}\frac{b-a}{2l}$, and the force perpendicular to the spring, $F_{\text{vertical}z} = \frac{F_z}{4}\sqrt{1-\left(\frac{b-a}{2l}\right)^2}$.

2.1.2 Analysis of Lateral Force

The lateral load is F_y, therefore the lateral force of each spring is $F_y/4$, as shown in Fig. 5.

The lateral force in the coordinate frame $ox_1y_1z_1$ is $F_{\text{lateral}l} = \left[-\frac{F_y}{4}\cos\alpha\ \frac{F_y}{4}\sin\alpha\ 0\right]$. According to the transformation matrix T_C, the lateral force in the coordinate frame $oxyz$ is $F_{\text{lateral}} = \left[-\frac{F_y}{4}\cos\alpha\ \frac{F_y}{4}\cos\delta\sin\alpha\ \frac{F_y}{4}\sin\delta\sin\alpha\right]$,

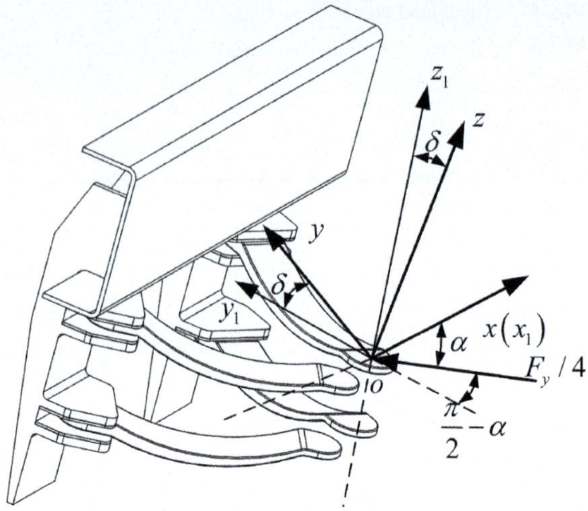

Fig. 5 Lateral load on each leaf-spring

substituting δ into which presents the force on the spring resulting from the lateral load.

The longitudinal force of the spring is

$$F_{\text{lateral}x} = -\frac{F_y}{4}\cos\alpha \tag{3}$$

The force along the direction of spring arrangement is

$$F_{\text{lateral}y} = \frac{F_y \sin\alpha}{4}\sqrt{1-\left(\frac{b-a}{2l}\right)^2} \tag{4}$$

The normal force of the spring is

$$F_{\text{lateral}z} = \frac{F_y \sin\alpha}{4}\frac{b-a}{2l} \tag{5}$$

2.1.3 Analysis of Longitudinal Force

Suppose the longitudinal load is F_x, then the component of each spring is $F_x/4$. The diagram of load on one leaf-spring is illustrated in Fig. 6.

The longitudinal force in the $ox_1y_1z_1$ coordinate frame can be similarly obtained

$$F_{\text{longitudinal}l} = \left[\tfrac{F_x}{4}\sin\alpha\ \tfrac{F_x}{4}\cos\alpha\ 0\right] \tag{6}$$

Fig. 6 Longitudinal load on each leaf-spring

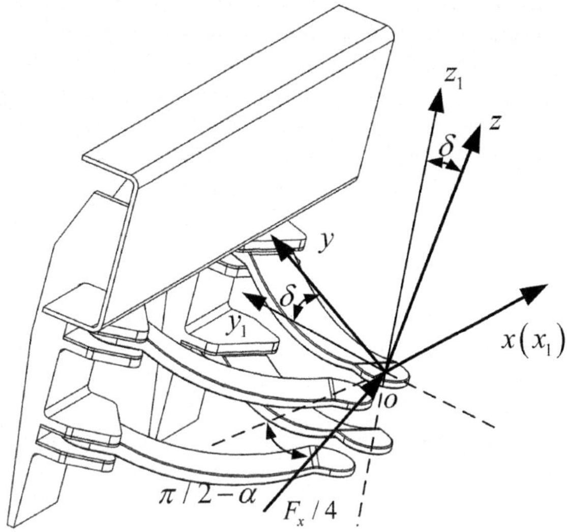

Using transmission matrix T_C, we have $F_{\text{longitudinal}} = T_C F_{\text{longitudinal}l}$ and then obtain the longitudinal force in the $oxyz$ coordinate frame:

$$F_{\text{longitudinal}} = \left[\,\frac{F_x}{4}\sin\alpha \;\; \frac{F_x}{4}cos\delta\cos\alpha \;\; \frac{F_x}{4}\sin\delta\cos\alpha\,\right] \tag{7}$$

Substituting δ into Eq. (7) leads to

$$F_{\text{longitudinal}x} = \frac{F_x}{4}\cos\alpha \tag{8}$$

$$F_{\text{longitudinal}y} = \frac{F_x\cos\alpha}{4}\sqrt{1-\left(\frac{b-a}{2l}\right)^2} \tag{9}$$

$$F_{\text{lateral}\,z} = \frac{F_x\cos\alpha}{4}\frac{b-a}{2l} \tag{10}$$

where $F_{\text{longitudinal}x}$ is the force along the longitudinal direction of the vehicle, $F_{\text{longitudinal}y}$ is the lateral force along the arrangement direction of the spring, and $F_{\text{lateral}\,z}$ represents the vertical force which is perpendicular to the leaf-spring.

2.2 *Torque Analysis*

In the coordinate system shown in Fig. 7, x-axis is perpendicular to the arrangement direction of the spring; y-axis is along the horizontal direction of the spring's arrangement. In addition, z-axis stands for vertical direction.

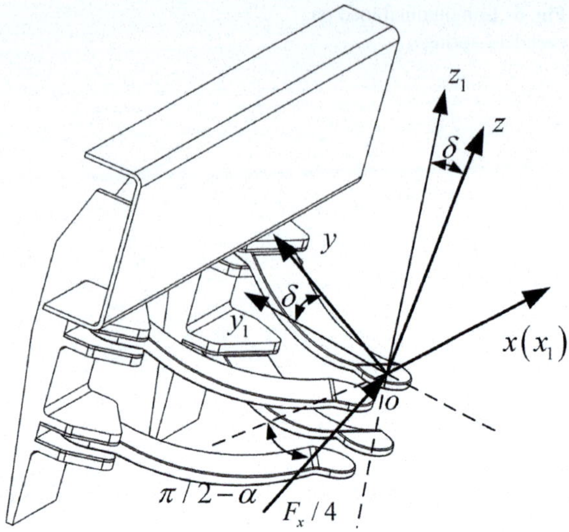

Fig. 7 Coordinate on the knuckle

2.2.1 Analysis of Longitudinal Torque

The longitudinal torque is generated by the vertical load shown in Fig. 8.

$M_X = F_Z l_0$ where l_0 is the distance between action point of the load and geometry center of the knuckle. The leaf-springs uniformity bear the longitudinal moment, so that the loading torque of each spring is $M_X/4$, the torque about the x-axis is $M_{\text{longitudinal } x} = \frac{M_X}{4}\sin\alpha$, and the torque about y-axis is $M_{\text{longitudinal } y} = \frac{M_X}{4}\cos\alpha$.

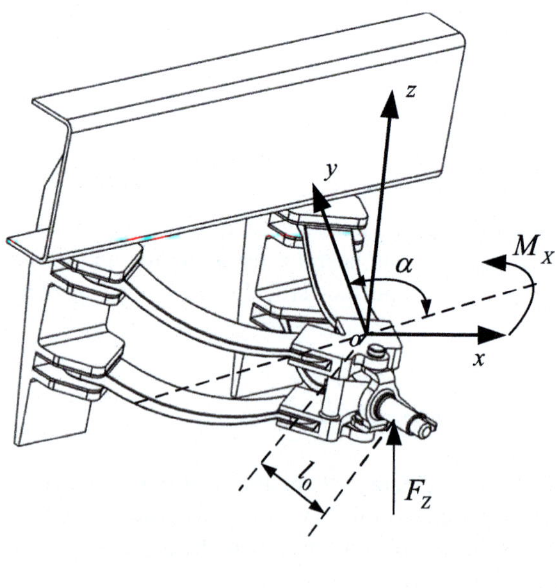

Fig. 8 Longitudinal torque

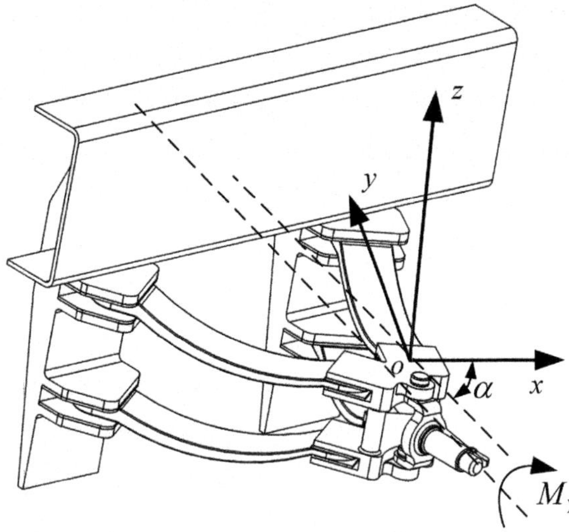

Fig. 9 Longitudinal torque

2.2.2 Analysis of Lateral Torque

Lateral torque is caused by accelerating or braking the vehicles, which is shown in Fig. 9.

In addition, F_X is the longitudinal load, R is the radius of the wheel $M_Y = F_X R$. The torque of each spring is $M_Y/4$, the torque around the x-axis is $M_{\text{side }x} = \frac{M_Y}{4}\cos\alpha$, and the torque around the y-axis is $M_{\text{side }y} = \frac{M_Y}{4}\sin\alpha$.

2.2.3 Analysis of Vertical Torque

The moment M_Z about the vertical direction is caused by turning vehicles as shown in Fig. 10.

$$M_Z = \frac{\mu}{3}\sqrt{\frac{G_t^3}{p}} \tag{11}$$

where μ stands for the sliding friction coefficient under normal conditions and $\mu = 0.6$ [23] for rubber bushing, G_t is the vertical load of the tire, and P is the pressure of the tire.

The vertical load is distributed on four parallel leaf-springs, therefore, the torque of each spring is $M_Z/4$, the diagram of vertical torque is described in Fig. 11.

Vertical torque M_1 is around the direction of y_1, the torque M_2 about the z-axis has been decomposed into M_1 and M_2 as illustrated in Fig. 11 where y_1-axis is along the arrangement direction of the spring,z_1-axis is perpendicular to y_1- and x-axis, simultaneously.

Fig. 10 Vertical torque

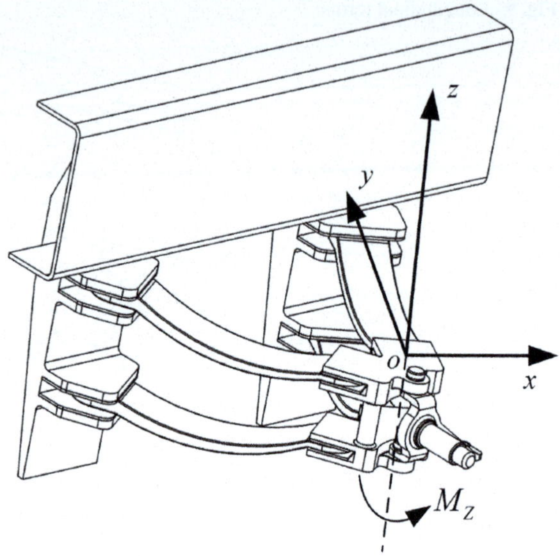

Fig. 11 Vertical torque

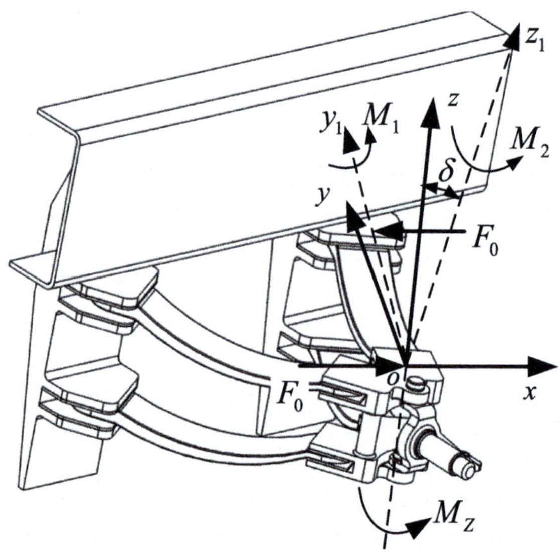

$$M_1 = \frac{M_Z}{4}\sin\delta = \frac{M_Z}{4}\frac{b-a}{2l} \tag{12}$$

$$M_2 = \frac{M_Z}{4}\cos\delta = \frac{M_Z}{4}\sqrt{1-\left(\frac{b-a}{2l}\right)^2} \tag{13}$$

The moment around the y_1-axis is M_1 which makes the spring warping.

$$M_1 = \frac{M_Z}{4}\frac{b-a}{2l} \tag{14}$$

The torque around the z_1-axis is

$$M_2 = \frac{M_Z}{4}\sqrt{1-\left(\frac{b-a}{2l}\right)^2} \tag{15}$$

The load of the spring along x-axis, F_0, is

$$F_0 = \frac{M_Z}{4l}\sqrt{1-\left(\frac{b-a}{2l}\right)^2} \tag{16}$$

After describing the forces and torques of the single leaf-spring and the knuckle, we apply CAE software to analyze the relationship between the load and deformation of this oblique leaf-spring suspension to optimize the design.

3 Computer Simulation and Discussion

3.1 Reconstruction of One Leaf-Spring

The suspension consists of four parallel leaf-springs, each of which bears the longitudinal torque, so that the torque of each spring is $M_X/4$ illustrated in Fig. 8. The torque around the x-axis is $M_{\text{longitudinal } x} = \frac{M_X}{4}\sin\alpha$ which will force the spring to bend, as shown in Fig. 12a. The torque around the y-axis is $M_{\text{longitudinal } y} = \frac{M_X}{4}\cos\alpha$ which will force the spring warping, and which is represented in Fig. 12b.

Each of the spring is permanently connected with the frame by one end, the other end is a ring joint connecting the knuckle. The leaf-spring are made of 52CrMnBA material. In order to get the optimum curvature radius for a leaf-spring, we set a CAE model to analyze it. Considering the difficulty of processing in actual production, the leaf-spring adopts a circular arc curved shape. Finite element software is used to analyze the leaf-spring with radii of 170 mm, 200 mm, 300 mm, 400 mm, 500 mm, 600 mm, 700 mm, 800 mm, 900 mm, 1000 mm, 1500 mm, 2000 mm and an infinite radius for straight leaf-spring, respectively. We analyze the maximum stress on leaf-spring for different radii. The simulation results of stress analysis are shown in Fig. 13.

The analysis of straight leaf-spring is illustrated in Fig. 14.

The maximum stresses of different radii are illustrated in Table 1.

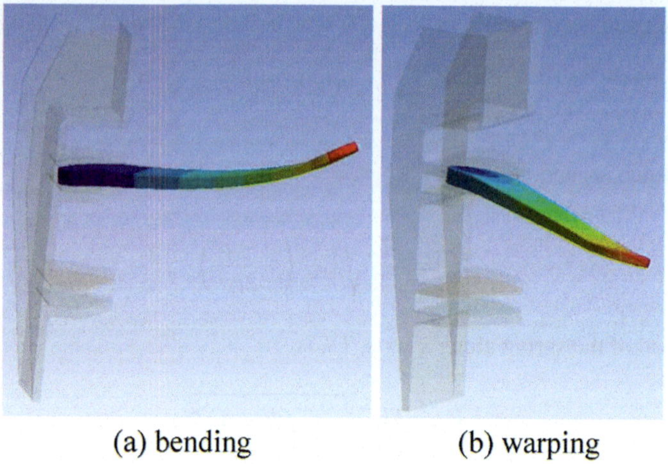

Fig. 12 Bending and warping

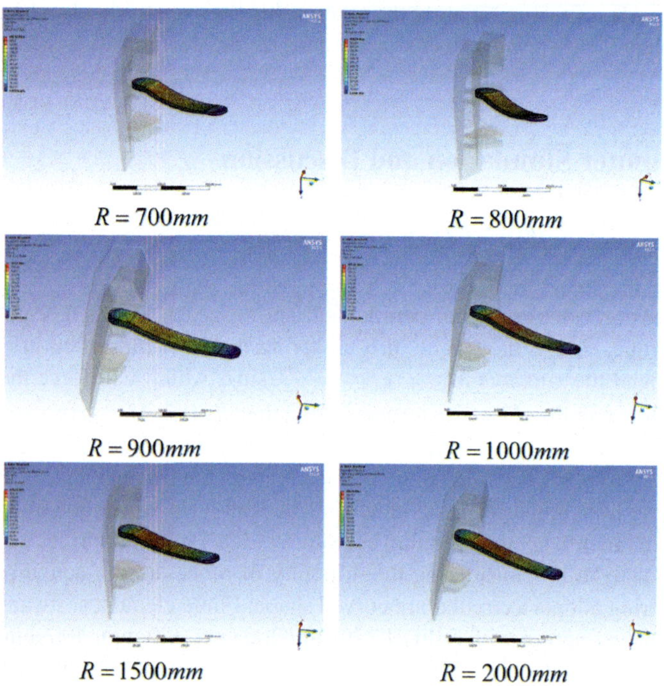

Fig. 13 Stress analysis of different radii

Fig. 14 Stress distribution on the straight leaf-spring

Table 1 Maximum stress of leaf-springs with different radii

Radius R (mm)	170	200	300	400	500	600
Maximum stress (MPa)	7069.8	1356.2	656.31	564.96	508.37	510.63
Radius R (mm)	700	800	900	1000	1500	2000
Maximum stress (MPa)	495.52	498.36	494.8	459.44	446.15	439.49

According to the data analysis, the maximum stress decreases with the increase of the radius of a single leaf-spring, and it ultimately tends to be a constant value, as the hatched areas shown in Fig. 15.

The stiffness of leaf-spring with different radii is illustrated in Table 2.

The fitting curve in Fig. 16 represents the relationship between the stiffness and the radius of a single leaf-spring. A small radius change causes a large change in stiffness of a single leaf-spring. With the increase of the radius of a single leaf-spring, the stiffness tends to be a constant value, which is shown in Fig. 16.

Fig. 15 Maximum stress and the radius of the leaf-spring

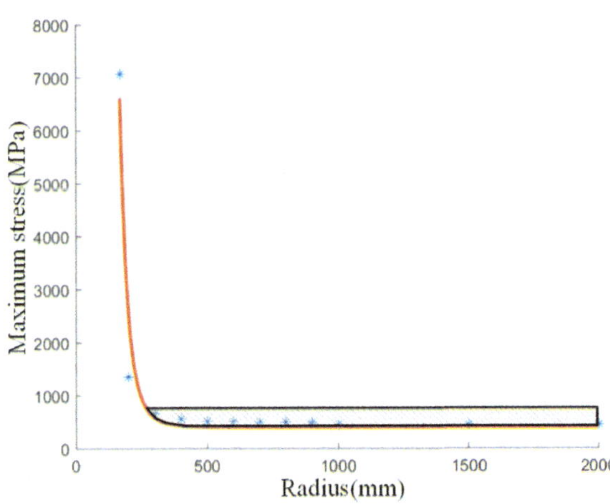

Table 2 Stiffness of leaf-spring with different radii

Radius R (mm)	170	200	300	400	500	600
Stiffness (N/mm)	37.92	135.54	197.48	208.40	213.24	216.84
Radius R (mm)	700	800	900	1000	1500	2000
Stiffness (N/mm)	216.99	218.89	219.86	220.41	220.55	223.49

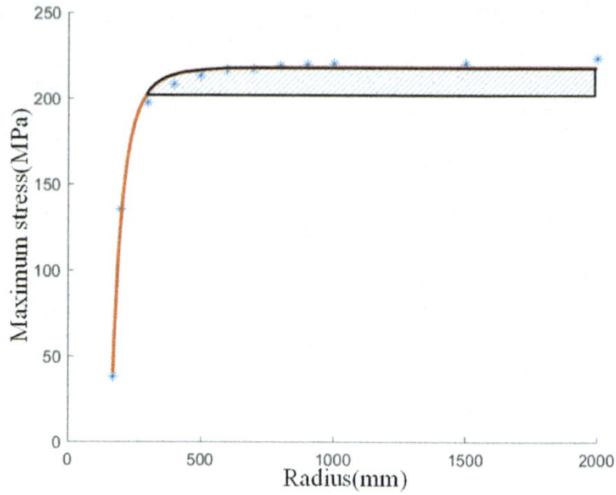

Fig. 16 Stiffness and radius

It has uneven loading problem when we use the traditional multiple leaf-springs. Multiple leaf-springs usually contain more than four pieces of leaf-springs. As the length of every piece is different, it may increase the stress concentration. Sometimes, stress concentration may cause a certain leaf-spring to damage easily. Therefore single leaf-spring tends to replace the traditional multiple leaf-springs. On the one hand, it can avoid the stress concentration, and on the other hand the self-weight of the suspension will be lighter than that of the multiple leaf-springs. However, only one single leaf-spring has lower strength compared with the multiple leaf-springs, therefore we try to use four parallel single leaf-springs to replace the existing multiple leaf-springs in order to get a longer service life and also to reduce the total mass of the suspension.

3.2 *Deformation Compliance of Leaf-Spring Suspension Assembly*

To simulate the equivalent stress, stiffness, and loading status under different loads, we suppose that the value of the vertical static payload is 3 tons. As a result, the vertical load equals $F_z = 30{,}000\,\text{N}$ when we take $g = 10\,\text{N/kg}$ for the sake of safety. The maximum lateral force is $F_y = \lambda F_z$ and the maximum longitudinal force is expressed by $F_x = \lambda F_z$ where $\lambda = 0.6$ which is the load distribution coefficient during braking [24]. The three orthogonal loads are shown in Fig. 17.

The maximum lateral force and maximum longitudinal force are $F_y = 18{,}000\,\text{N}$ and $F_x = 18{,}000\,\text{N}$, respectively. The steering resisting torque is

$$M_z = \frac{\mu}{3}\sqrt{\frac{G_t^3}{p}} \tag{16}$$

where μ is the coefficient of rolling resistance between the tire and the pavement under normal conditions, $\mu = 0.6$ [23], G_t is the vertical load on the tire, p is the tire pressure, and the steering moment is $M_z = 2013.47\,\text{N m}$.

Since the driving condition is different, the stress state of suspension is also different. The most complex state is when braking and steering happen at the same time. The stress of the leaf-spring is analyzed by only considering the vertical load along z-axis while the vehicle is driving across a bumpy road. When the vehicle is braking, both the vertical load along the z-axis and longitudinal load along the x-axis coexist. We analyze the deformation of every parallel single leaf-spring under the driving conditions when braking and steering are happened at the same time. The rotational torques of the suspension about the x-, y- and z-axes satisfy that $M_x = M_y = M_z = 2014\,\text{N m}$. The applied forces and applied torques in three directions at the steering knuckle are shown in Table 3.

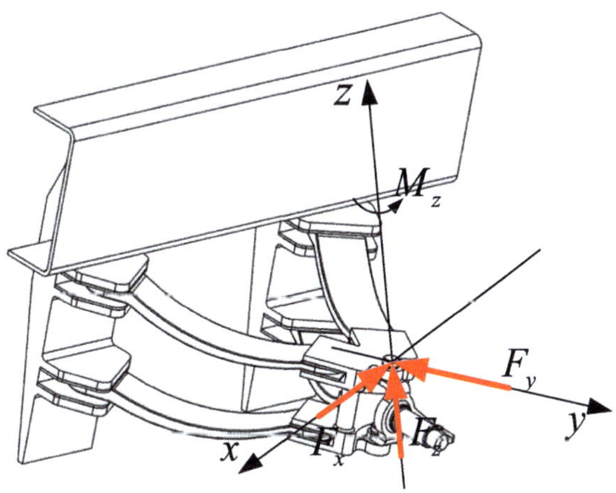

Fig. 17 Stress status of suspension

Table 3 Forces and torques

Load	Longitudinal force	Lateral force	Vertical force	Longitudinal torque	Lateral torque	Vertical torque
Value	18 kN	18 kN	30 kN	2014 N m	2014 N m	2014 N m

When braking and steering are simultaneously occurred, the total deformation of the parallel single leaf-spring suspension is 3.6488 mm according to the CAE simulation, which is shown in Fig. 18.

Therefore there are six-dimensional forces acting on the knuckle. The deformation of every spring is illustrated in Fig. 19.

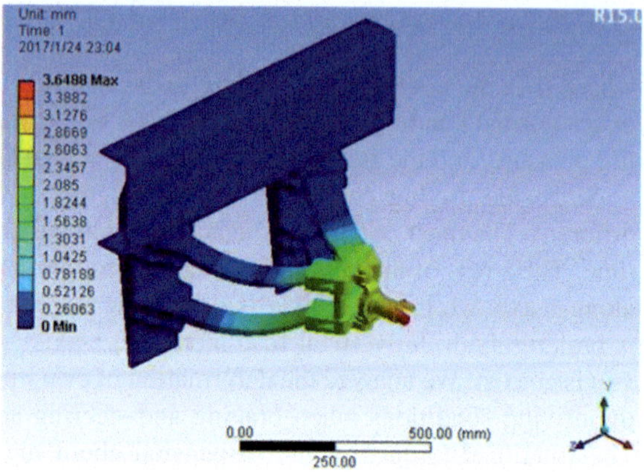

Fig. 18 Total deformation of the suspension

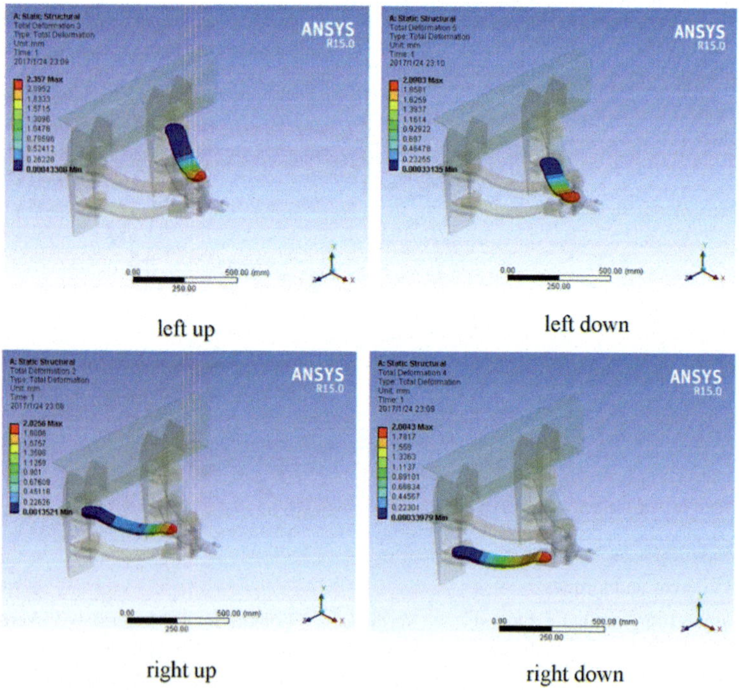

Fig. 19 Deformation of every spring

Table 4 Deformation of every spring

Spring	Left up	Left down	Right up	Right down
Deformation	2.357 mm	2.090 mm	2.026 mm	2.004 mm

Deformation of each spring is illustrated in Table 4.

When there are some forces and torques exerted on the steering knuckle, the maximum stress is less than 900 MPa and the total deformation is less than 5 mm. Therefore, the simulation results comply with design requirements.

3.3 *Stress and Strain Optimization*

Assume that the vertical load is $F_z = 30\,\text{kN}$, lateral load is $F_y = 30\,\text{kN}$, the longitudinal load is $F_x = 30\,\text{kN}$, and the steering torque about the z-axis is $M_z = 2014\,\text{N m}$. We simulate the deformation and stiffness of the suspension (Fig. 20).

The stiffness can be calculated after knowing the maximum deformation under the applied load. Table 5 lists the stiffness in six directions.

The analysis shows that the stiffness along the y-direction is large enough, which means that the suspension has a good resistance to lateral deformations. The stiffness in the x-direction is 333.3 N/mm and the stiffness in the z-direction is 355.8 N/mm. Therefore, the vehicle has good ride comfort. The rotational stiffnesses of the suspension about the x-, y- and z-axes are relatively larger which indicates that the anti-torsion and the resistance of the suspension to bending are both good.

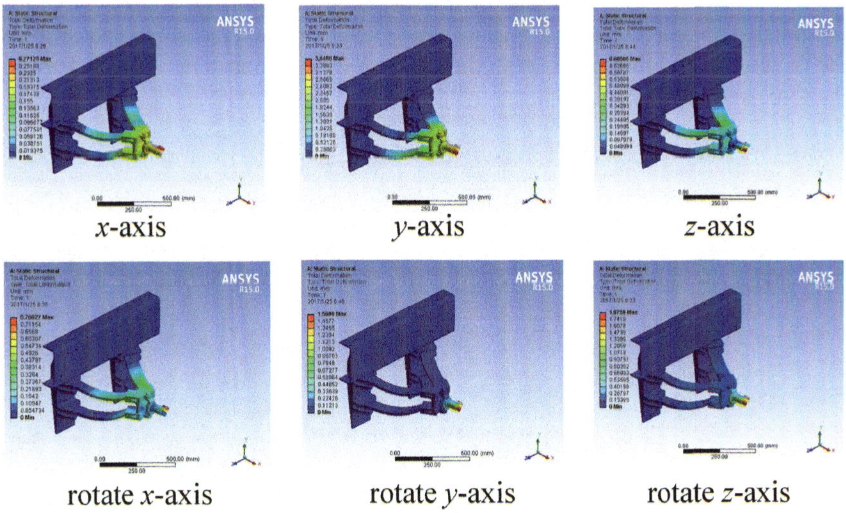

Fig. 20 Deformations in six directions

Table 5 Directional stiffness of the suspension

	x-axis	y-axis	z-axis	Rotate x-axis	Rotate y-axis	Rotate z-axis
Deformation	90.0 mm	6.6 mm	84.3 mm	0.015 rad	0.006 rad	0.024 rad
Stiffness	333.3 N/mm	4545.5 N/mm	355.8 N/mm	1.34×10^5 N m/rad	3.36×10^5 N m/rad	0.84×10^5 N m/rad

4 Prototype Manufacturing and Elementary Experiments

According to the design we proposed, we manufactured a prototype to test the mechanical properties of the oblique leaf-spring suspension and install the prototype on the fixture.

In order to get the static stiffness of the suspension system, we produced a jig to connect multiple-purpose test station (MTS) and the oblique leaf-spring system in Fig. 21. The jig is used to connect the knuckle and the actuators with connecting piece. It has three sides. Each side connects to its individual hydraulic cylinder of MTS. The force from the horizontal hydraulic cylinder represents the load along the y-direction, the vertical hydraulic one stands for the load along the z-direction.

To investigate the idea, we performed the experiment illustrated in Fig. 22. As it is shown that the oblique leaf-spring is assembled on MTS. The vertical static stiffness and lateral static stiffness of this oblique leaf-spring suspension are the main data. Use vertical and lateral hydraulic cylinders to exert force to the knuckle, respectively. We collect the data of force and displacement in order to calculate the stiffness.

The experimental results of the static load test are shown in Figs. 23 and 24. The vertical stiffness test of leaf-spring is measured by the static load exerted by vertical hydraulic cylinder. The hydraulic cylinder exerted the tensile force slowly up to 4 tons along the z-direction, and the data collection equipment measured and collected the forces at the testing point. With these data, we extrapolated the relation between the force and the deformation:

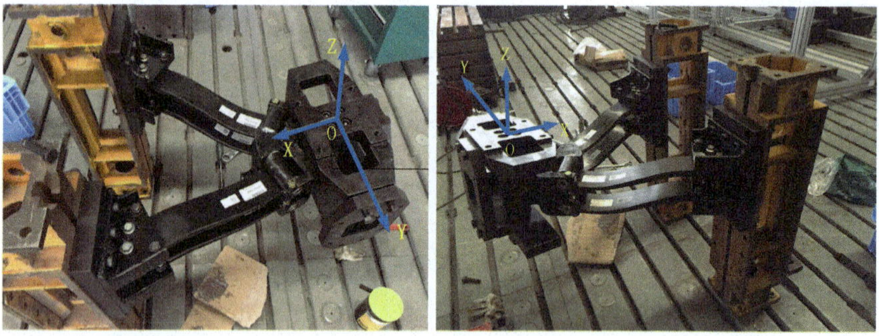

Fig. 21 Coordinate system on the knuckle

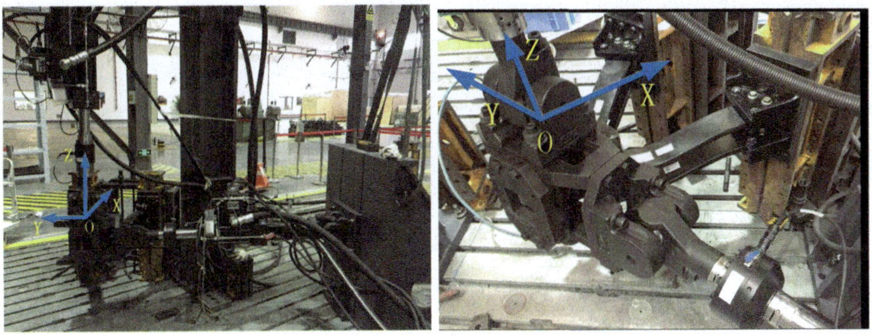

Fig. 22 Assembly drawing of hydraulic cylinder piston

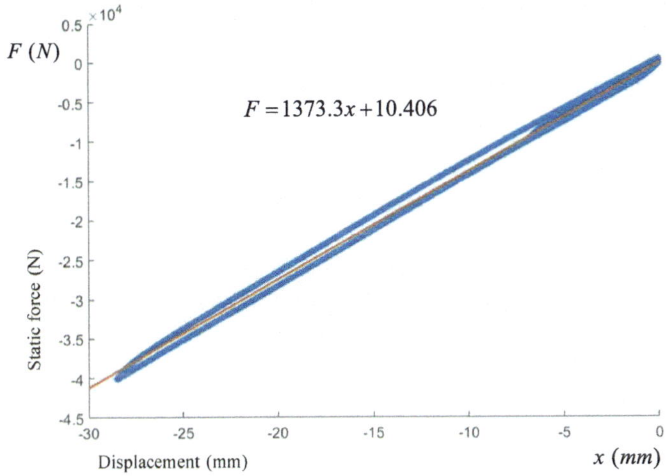

Fig. 23 Vertical force and displacement

$$F = 1373.3x + 10.406 \tag{20}$$

where x represents the deformation and F is the force.

The lateral hydraulic cylinder is used to load the tensile force from 0 to 1.5 tons along the y-direction. The lateral static stiffness could be obtained from the extrapolation:

$$F = 6323.1x \tag{21}$$

Therefore, an accurate and simple model of oblique leaf-spring is imminently developed. These experiments involve the effects of different excitation loads, excitation stresses, and displacement on the leaf-spring. The slope of the fitted line stands for the static stiffness of oblique leaf-spring suspension. The vertical stiffness

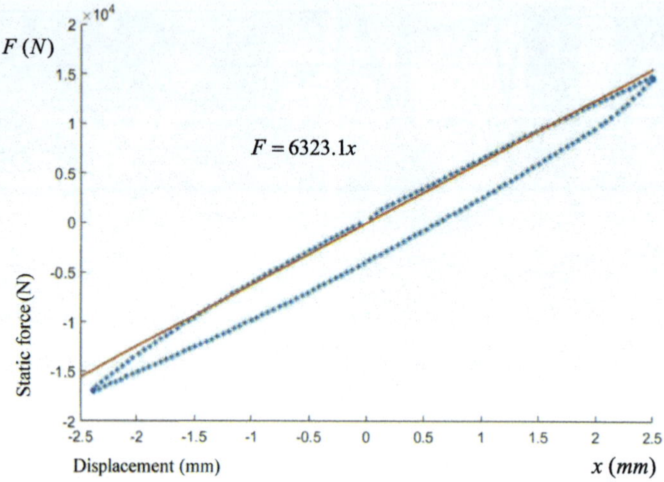

Fig. 24 Lateral force and displacement

is 1373.3 N/mm within the displacement of 30 mm in total and the lateral stiffness is 6323.1 N/mm within the largest deformation of 5 mm which is particularly well for independent suspension for heavy vehicles. There is a supporting bar in Fig. 25, when the deformation of the vertical direction reaches the maximum point, the supporting bar will take in action. It will make the deformation change slowly, which means the stiffness of the oblique leaf-spring suspension will have nonlinear variation.

Assume n_1 and n_2 are the vibrational frequencies of front suspension and rear suspension, and $n_1 = \frac{1}{2\pi}\sqrt{\frac{c_1}{m_1}}$ and $n_2 = \frac{1}{2\pi}\sqrt{\frac{c_2}{m_2}}$. The relation between the load on the leaf-spring and stiffness of the linear suspension are $f_{c1} = \frac{m_1 g}{c_1}$, $f_{c2} = \frac{m_2 g}{c_2}$, and c_1 and c_2 are the stiffnesses of the front and rear suspensions, we can get the result of the relations between the vibrational frequency and the stiffness [25].

The matching of the front and rear suspension's vibrational frequency and the static deflection influence the driving comfort a lot. Comparing the different load between the front and rear suspensions, the static deflection of the front suspension should be always larger than the rear suspension for commercial vehicles if we want to get a good ride comfort. For real applications, the stiffness of the parallel oblique leaf-spring suspension could satisfy the actual needs of the multiple-axle heavy commercial vehicles.

The oblique leaf-spring suspension prototype has already been manufactured and installed to the front suspension of Dongfeng Commercial Automobile illustrated in Fig. 25. One end of the oblique leaf-spring connects the knuckle scaffold through trunnions, the other end of the leaf-spring is permanently fixed with the frame.

Traditional spring is hinged at both ends, one is connected with the frame and the other end is connected with the swing rod. Oblique leaf-spring suspension structure is quite different from the exiting leaf-spring suspensions in that it contains four parallel single leaf-springs. One end of each leaf-spring connects the frame and the

Fig. 25 Installation to the front suspension

other end connects the knuckle scaffold. Comparing with the multiple leaf-spring suspension, this design can avoid a stress concentration, furthermore an appropriate connection at both ends of the spring ensures an appropriate stiffness. As a result, it has a very wide application prospect especially in the front suspension of heavy vehicles.

5 Conclusions

This paper proposed an oblique leaf-spring suspension system. It contains four identical single leaf-springs assembled in parallel. As a result, the suspension forms a statically indeterminate structure of which stiffness, strength, and reliability can meet the need of the vehicles, while reducing the total mass of the vehicles. Accordingly, the redundant arrangement of the oblique leaf-spring suspension improves the wheel alignment's parameters while occupying a smaller space. Therefore it provides larger steering angle for the vehicle. The oblique leaf-spring suspension can improve both the ride comfort, and the handling flexibility and stability of the vehicles. Besides, it could also ameliorate the motion smoothness, control balance and the driving safety of the commercial vehicles.

Acknowledgements This research was supported by the Major Scientific and Technological Innovation Project of Hubei Province of China (2014AAA003).

References

1. Moon Il-D, Yoon H-S, Oh C-Y (2006) A flexible multi-body dynamic model for analyzing the hysteretic characteristics and the dynamic stress of a taper leaf-spring. J Mech Sci Technol 20(10):1638–1645
2. Crews JH, Mattson MG, Buckner GD (2011) Multi-objective control optimization for semi-active vehicle suspensions. J Sound Vib 330(23):5502–5516
3. Németh B, Fényes D, Gáspár P (2016) Independent wheel steering control design based on variable-geometry suspension. IFAC Papers On Line 49(11):426–431
4. Smith JH (2002) An introduction to modern vehicle design. Butterworth Heinemann, pp 285–288
5. Verros G, Nastiavas S, Papadimitriou C (2005) Design optimization of quarter-car models with passive and semi-active suspensions under random road excitation. J Vib Control 11(11):581–606
6. Zhao JS, Liu X, Feng ZJ, Dai JS (2012) Stiffness of a rectilinear suspension with automatic length compensation branches. Mech Mach Theory 56(1):99–122
7. Knapczyk J, Maniowski M (2006) Stiffness synthesis of a five-rod suspension for given load–displacement characteristics. Proc Inst Mech Eng Part D: J Automob Eng 220(7):879–888
8. Mirza N, Hussain K, Day AJ, Klaps J (2008) Effect of component stiffness and deformation on vehicle lateral drift during braking. Proc Inst Mech Eng Part K: J Multi-Body Dyn 223(1):9–22
9. Gerrard MB (2005) A tool for the analysis and the design of compliant suspension linkages, the equivalent elastic mechanism. SAE technical paper, United States, 17–19 January 2005
10. Trikande MW, Jagirdar VV, Sujithkumar M (2014) Evaluation of semi-active suspension control strategies for 8 × 8 armoured vehicle using stochastic road profile inputs. In: IFAC proceedings, Kanpur, India, 13–15 March 2014
11. Sharma SK, Pare V, Chouksey M, Rawal BR (2016) Numerical studies using full car model for combined primary and cabin suspension. Procedia Technol (23):171–178
12. Chen MZQ, Hu Y, Li C, Chen G (2014) Semi-active suspension with semi-active inerter and semi-active damper. IFAC Proc 47(3):11225–11230
13. Güvenc BA, Kural E, Keşli B, Gülbudak K, Güngor S, Kanbolat A (2006) Semi active suspension control system development for a light commercial vehicle. IFAC Proc 39(16):391–396
14. Sanger DJ, Chen JQ, Zhang SJ (2000) A general method for the stiffness analysis of manipulator mechanisms. Proc Inst Mech Eng Part C J Mech Eng Sci 214(5):673–685
15. Dhananjay Rao K (2014) Modeling simulation and control of semi active suspension system for automobiles under MATLAB Simulink using PID Controller. In: IFAC proceedings, Kanpur, India, 13–15 March 2014
16. Zhao JS, Chu FL, Feng ZJ (2009) Synthesis of a rear wheel suspension mechanism with pure rectilinear motion. ASME J Mech Des 131(10):101007/1–9
17. Zhao JS, Chu FL, Feng ZJ (2008) Synthesis of rectilinear motion generating spatial mechanism with application to automotive suspension. ASME J Mech Des 130(6):876–884
18. Zhao JS, Jin HZ, Zhao S, Chu FL, Feng ZJ (2010) Innovative concept design of an independent front suspension and steering system. Proc Inst Mech Eng Part D: J Automob Eng 224(10):1487–1499

19. Kang J (2008) Forced vibration analysis for dynamic coupling of finite-element spring-reduced suspension system. Proc Inst Mech Eng Part C J Mech Eng Sci 222(3):349–356
20. Gillespie TD (1985) Heavy truck ride. SAE Paper, United States, paper no. 850001
21. Tsampardoukas G, Stammers CW, Guglielmino E (2007) Semi-active control of a passenger vehicle for improved ride and handling. Proc Inst Mech Eng Part D: J Automob Eng 222(3):325–352
22. Mántaras DA, Luque P, Vera C (2004) Development and validation of a three-dimensional kinematic model for the McPherson steering and suspension mechanisms. Mech Mach Theory 39(6):603–619
23. Osipenko MA, Nyashin Y, Rudakov RN (2003) A contact problem in the theory of leaf spring bending. Int J Solids Struct 40:3129–3136
24. Yang Q (2006) Analysis on braking force distribution coefficient of low-speed truck. Tractor Farm Transp 33(2):66–68
25. Zhang H (1987) Automobile design. China Machine Press 7(36):155–161

Review of Structural Synthesis Algorithms for Epicyclic Gear Trains

V. R. Shanmukhasundaram, Y. V. D. Rao, and S. P. Regalla

Abstract The kinematic structure of an epicyclic gear train (EGT) contains the information regarding which link is connected to which other link and by what type of joint. In accordance with the graph theoretic framework for EGTs, its kinematic structure can be represented using a simple graph model. During the conceptual stage, structural synthesis of EGTs is performed, and it is aimed at the enumeration of all non-isomorphic graphs (concepts), for a given number of links and DOF. All the candidate graphs are generated by performing some combinatorial operations on a set of parent graphs. Basically, the structural synthesis approaches in the mechanisms literature can be grouped into three categories, namely, recursive, non-recursive and acyclic graph-based schemes, respectively. In this work, the strategies for generation of EGTs, adopted by different researchers and spanning all the three categories, are reviewed. Moreover, the different graph representations encountered in synthesis of EGTs and the generation algorithms are lucidly explained using examples. A recent trend is the development of synthesis methods of EGTs inspired from established methods of those of linkage-type kinematic chains.

Keywords Mechanism design · Graph theory · Acyclic graphs · Trees · Gear trains

1 Introduction

Graph theory has been traditionally used in the conceptual stage of mechanism design following the works of Freudenstein and his coworkers [1, 2]. This work concerns about epicyclic gear trains (EGTs), a type of complex planar mechanism, widely used as a vital component of mechanical transmission finding applications in different

V. R. Shanmukhasundaram (✉) · Y. V. D. Rao · S. P. Regalla
Department of Mechanical Engineering, BITS Pilani, Hyderabad Campus, Hyderabad 500078, Telangana, India
e-mail: shanmukhasundaram89@gmail.com

Y. V. D. Rao
e-mail: yvdrao@gmail.com

S. P. Regalla
e-mail: spr.bits@gmail.com

D. Sen et al. (eds.), *Mechanism and Machine Science*, Lecture Notes in Mechanical Engineering, https://doi.org/10.1007/978-981-15-4477-4_25

industries such as automobiles, machine tools and wind energy systems. By applying graph theory during the conceptual stage, it is possible for an EGT designer to generate an atlas of feasible kinematic structures.

The application of graph theory to the modeling of kinematic structure of EGTs was an outcome of the work by Buchsbaum and Freudenstein [3]. In another paper by Freudenstein [4], the principles of rotation and displacement graphs of EGTs are given. The graph theoretic framework enabled in the development of structural synthesis algorithms for enumerating candidate graphs of EGTs of a certain complexity [5–9]. Generally, for a given number of links and DOF, the task of synthesis is aimed at obtaining the complete set of rotation as well as displacement graphs.

The first step of any structural synthesis procedure is the process of generation wherein all the candidate EGT chains are built from parent graphs. This is typically realized by means of a graph theory-based algorithm. The next step is isomorphism detection aimed at eliminating isomorphic graphs, all of which have the same kinematic structure. Lastly, the non-isomorphic graphs are checked for those corresponding to degenerate EGT chains and then eliminated. An EGT chain is called degenerate or rigid if any of its subchains (subgraphs) have zero DOF based on the *Grübler* criterion.

A review of generation methods adopted by different researchers during the last 50 years is presented in this work. Basically, the generation methods can be grouped into three categories, namely, recursive, non-recursive and acyclic graph-based schemes. The recursive as well as non-recursive schemes enumerate the rotation graphs firstly and then generate displacement graphs. On the other hand, acyclic graph-based methods enumerate displacement graphs directly [10–16]. The issue of isomorphism test, degeneracy detection and synthesis results of EGTs is well explained in other works [17–24] and therefore will not be dealt here.

This chapter is organized in the following manner. In Sect. 2, the originally formulated simple graph model for EGTs is explained along with the concepts of rotation and displacement graphs. Section 3 deals with the generation algorithms explained with the help of suitable examples. Section 4 concludes the chapter with salient features of this work. Also offered are recommendations for future research directions in synthesis of EGTs.

2 Geared Kinematic Chains and EGTs—Excerpts from Buchsbaum and Freudenstein (1970) [3]

The general degree of freedom (DOF) equation for planar kinematic chains is given by

$$F = 3(N - J - 1) + \sum f_i \tag{1}$$

where (F) denotes the DOF of the kinematic chain, N being the number of links, J denoting the number of pair connections or joints, and f_i is the DOF of joint (J_i). The general DOF equation is also applicable for geared kinematic chains (GKCs) which are characterized by the simultaneous presence of turning and gear pair connections. For a GKC, J is the sum of J_T turning pairs ($f_i = 1$ per pair) and J_G gear pairs (f_i = 2 per pair). Upon substituting these, the DOF equation for GKC is obtained as

$$F = 3N - 3(J_T + J_G) - 3 + ((1 \times J_T) + (2 \times J_G)) \tag{2}$$

An important consideration for GKCs is that the number of gear pair connections is always lesser than that of turning pairs and it is based on the following inequality:

$$J_T \geq J_G + 1 \tag{3}$$

EGTs represent a class of GKCs for which the set of turning pair edges is always a spanning tree:

$$J_T = N - 1 \tag{4}$$

By comparing Eqs. (2) and (4), the relationship between number of links (N) and number of gear pair edges (J_G) for an F-DOF EGT is obtained as

$$J_G = (N - 1 - F) \tag{5}$$

2.1 *Simple Graph Model for EGTs of One DOF*

The graph of an EGT serves as an isomorphic substitute for kinematic structure of same. The links and joints are modelled as vertices and edges, respectively, of a simple graph. The edges are coloured so as to distinguish turning pairs from gear pairs. As apparent from Eqs. (2) to (5), the graph of an N-link one DOF EGT consists of N vertices and $(2N - 3)$ edges. Out of the $(2N - 3)$ edges, $(N - 2)$ of them are gear pairs and the remaining $(N - 1)$ are turning pairs.

The EGT graph is a closed graph, in which at least two edges are incident to each of its vertices. It is also incomplete in the sense that not every pair of vertices is connected; mathematically this can be expressed as

$$(N - 1) + (N - 2) + \text{Unconnected vertex pairs} = N\boldsymbol{C}_2 \tag{6}$$

2.2 Additional Restrictions—Concept of Fundamental Circuit

For a simple graph of N vertices and $(2N - 3)$ edges, ideally, with $(N - 1)$ turning pair edges and $(N - 2)$ gear pair edges, the relative distribution of the two kinds of edges can result in $(2N - 3)\mathbf{C}_{(N-1)}$ or $(2N - 3)\mathbf{C}_{(N-2)}$ combinations. Not every possible combination can be a legitimate EGT graph unless it satisfies some restrictions which are based on practical design considerations, as imposed in the original framework [3], and which are as follows:

- There can be no circuit formed exclusively by turning pair edges.
- Every vertex must have at least one incident edge, which is a turning pair.
- The deletion of a gear pair edge cannot cause an increase in number of components. In other words, no gear pair edge can be a bridge or cut edge.

From the above considerations, it is clear that the subgraph of an EGT formed by deleting all its gear paired edges cannot be disjoint and therefore is a spanning tree (also evident from Eq. (4)). As a consequence, a gear pair edge incident at any pair of vertices (multiple edges between same vertices not allowed) belonging to the spanning tree gives rise to a fundamental circuit (FC). Figure 1a depicts a graph of a 5-link EGT, where turning pair edges are thin and that of gear pair are heavy. A spanning tree as shown in Fig. 1b holds all the turning pair edges together. The FCs with respect to the three gear pairs are shown in Fig. 1c–e.

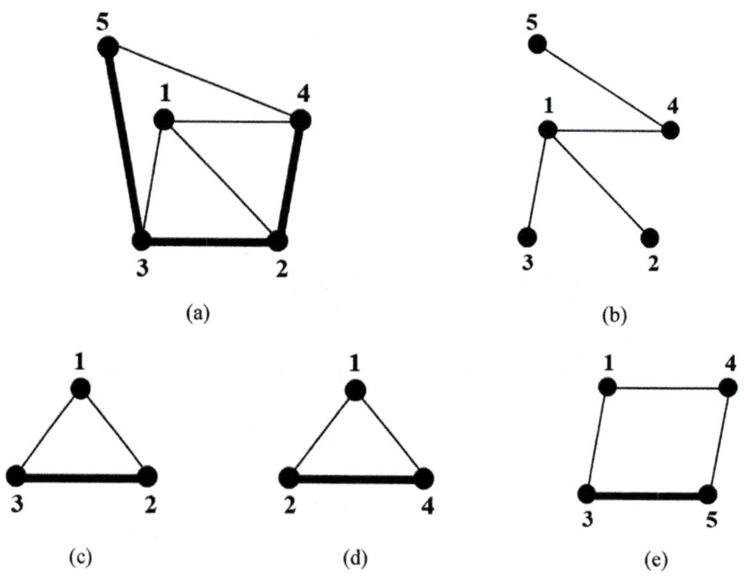

Fig. 1 **a** Simple graph model of a 5-link EGT, **b** its spanning tree obtained by the elimination of gear pair edges, **c–e** fundamental circuits with respect to the gear pair edges

2.3 *Concepts of Rotation and Displacement Graphs as in Freudenstein (1971) [4]*

It can be recalled that every FC in the EGT graph is composed of a gear pair edge along with a unique path of turning pair edges. Each turning pair edge is labelled to indicate a level corresponding to the position of its axis in space. There exists a transfer vertex (TV) in every FC such that all the turning pair edge/s on one side of the TV possesses the same level and those on the other side are at a differently labelled level.

The TV also represents the carrier link. A labelled EGT graph enables in the determination of linear displacement equations and it is also referred to as a displacement graph. A rotation graph contains the information of gear pair edges and their TVs. Using a rotation graph, rotational displacement equations can be derived.

Considering the example of 5-link EGT of Fig. 1, for FCs 1-2-3 and 1-2-4 the TV is undoubtedly vertex-1. Whereas for the FC 1-3-5-4, the TV can be either vertex-1 or vertex-4. This results in two different rotation graphs as shown in Table 1. Alongside the rotation graph, the possible labelled graph is also tabulated. The edges are labelled such that the TV details in the corresponding rotation graph are preserved. For both the rotation graphs, only one labelled graph/displacement graph is possible, as apparent from Table 1.

Table 1 Rotation graphs derived from the unlabelled graph of Fig. 1 along with their labelled graphs

Rotation graph derivable from unlabelled graph of Fig. 1	Labelled graph with respect to the rotation graph
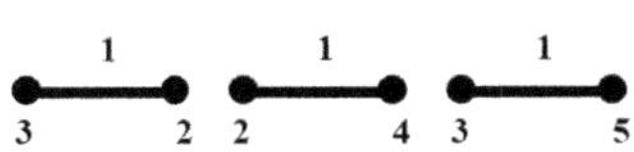	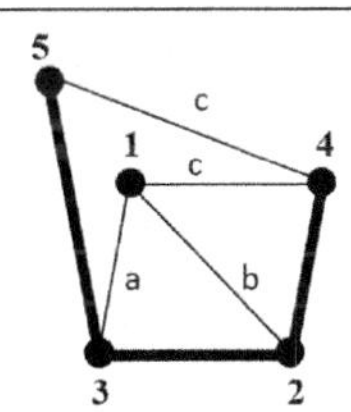
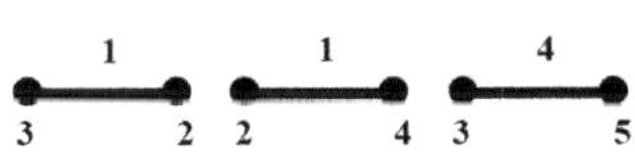	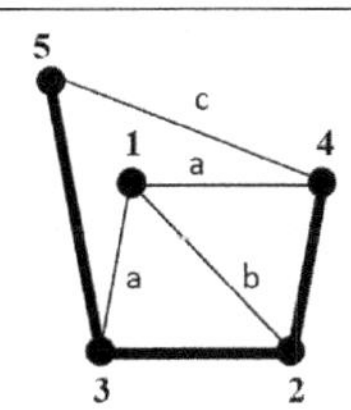

3 Structural Synthesis of EGTs

The structural synthesis of EGTs, for a given number of links and DOF, is focused on enumerating all the non-isomorphic displacement graphs [5]. There exist two different approaches presented in the mechanisms literature: the first approach involves generation of all rotation graphs followed by generation of displacement graphs with respect to every non-isomorphic rotation graph; whereas the second approach directly enumerates all the displacement graphs from a set of parent graphs. Naturally, test for isomorphism is inevitable in the overall synthesis procedure for either of the approaches.

For the former case, firstly, a set of parent graphs are subjected to a graph theory-based synthesis algorithm (either recursive or non-recursive) so as to arrive at all the combinatorially possible rotation graphs. Then, those rotation graphs that contain degenerate structure/s as subgraphs are eliminated. Finally, displacement graphs are generated with respect to every non-degenerate rotation graph. The overall process is depicted in Fig. 2. According to the algorithm of the latter case, displacement graphs are directly obtained from parents known us acyclic graphs, followed by test for degeneracy as illustrated in Fig. 3.

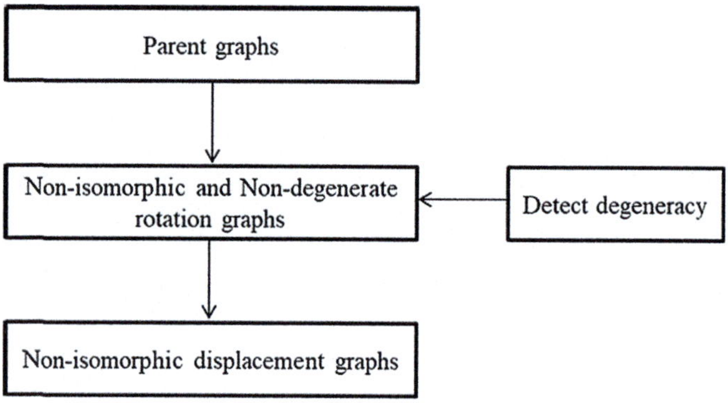

Fig. 2 The process of displacement graphs enumeration also including the generation of rotation graphs

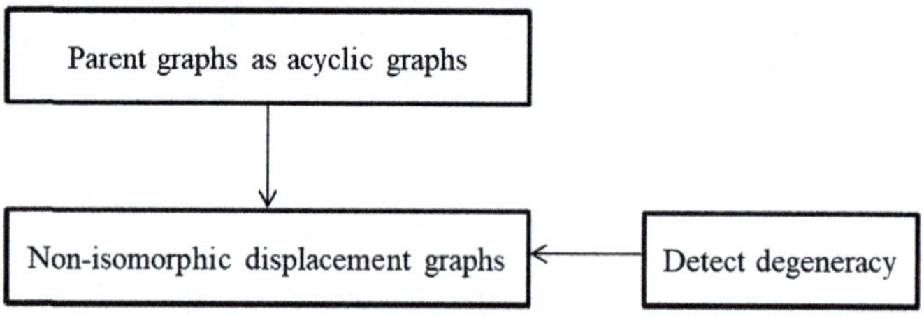

Fig. 3 The process of displacement graph generation using acyclic graphs as parents

3.1 Non-recursive Scheme

This method was first introduced by [3]. Subsequent works that used this method are those of [5, 8, 20]. The first step involves the enumeration of all the permissible non-coloured graphs whose edge set has no distinction between turning pair and gear pair edges. These are nothing but all the simple and closed graphs for N vertices and $(2N-3)$ edges (i.e. for enumerating N-link 1 DOF EGTs). All such graphs can either be synthesized or taken from a graph library database. Further, those graphs possessing cut vertices or in other words, showing a graph connectivity of 1, are taken out of consideration at this stage.

The second step involves the determination of all structurally distinct ways of distributing $(N-2)$ gear pair edges among the $(2N-3)$edges with a condition that the remaining edges (i.e. turning pair edges) constitute a spanning tree. An isomorphism test needs to be performed at this stage to eliminate duplicates thereby arriving at the set of all non-isomorphic geared graphs.

The third step is aimed at the enumeration of rotation graphs that can be derived for a certain gear pair edge distribution of a parent graph (geared graph). This is achieved by selecting TV details for all the FCs associated with the geared graph. In an FC there can be more than one possible TV, depending upon the labelling of turning pair edges. Therefore, for the same geared graph, choice of different TVs for FCs will lead to different rotation graphs possibilities. In order to arrive at the exhaustive possibilities of rotation graphs, this needs to be performed for all geared graphs enumerated earlier. This step is culminated with an isomorphism check so as to arrive at all N-link rotationally non-isomorphic graphs. Degeneracy check is also carried out to ensure that only non-degenerate graphs remain.

The fourth and last step is dedicated to the generation of displacement graphs with respect to every rotation graph. This is realized by assigning edge levels in all possible ways such that the TV details in the rotation graph are preserved. Isomorphic displacement graphs are possible only if they are generated from a common rotation graph and if detected they must be discarded. Hence, the final set of EGT graphs are each rotationally as well as displacement non-isomorphic as depicted in the overall synthesis process in Fig. 4.

The non-recursive scheme is elucidated using an example aimed at the enumeration of all 4-link EGTs, in the form of displacement graphs. In this case the parent graph of 4 vertices ($N = 4$) and 5 edges ($2N - 3 = 5$) is considered and there exists only one such graph which is shown in Fig. 5a. For the 4-vertex parent graph, gear pair edges and turning pair edges can be distinguished resulting in $\mathbf{C}_{(2N-3)}^{(N-2)}$ or $\mathbf{C}_{(2N-3)}^{(N-1)}$ combinations (10 precisely) of geared graphs which are illustrated in Fig. 5b–k. Of these 10 combinations, the two graphs of Fig. 5g, j need to be excluded as they contain turning pair edges giving rise to a loop. Of the remaining eight geared graphs, the four graphs of Fig. 5b–e are isomorphic to each other. Similarly, the two graphs Fig. 5f, k are isomorphic and so is the pair of graphs of Fig. 5h, i.

If one representative is considered out of each set of isomorphic graphs, then effectively 3 non-isomorphic graphs remain. Now, the next step is to identify the

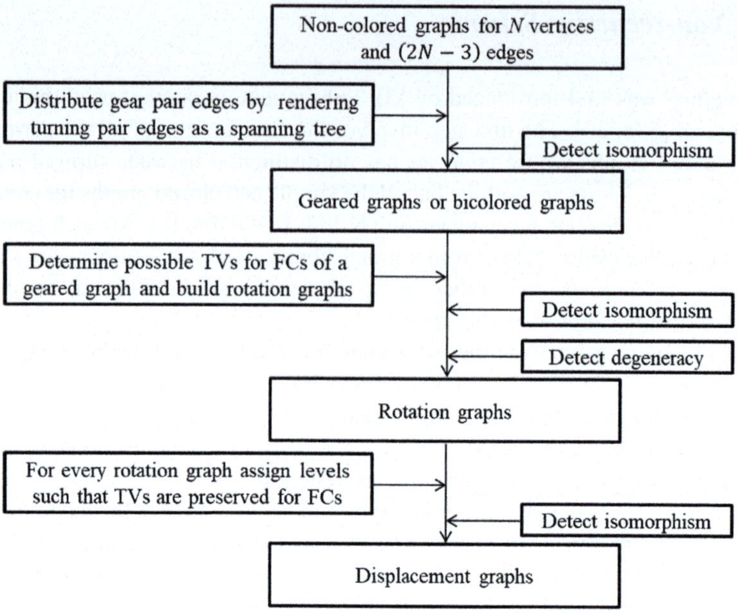

Fig. 4 The overall process of enumeration of EGTs using a non-recursive scheme

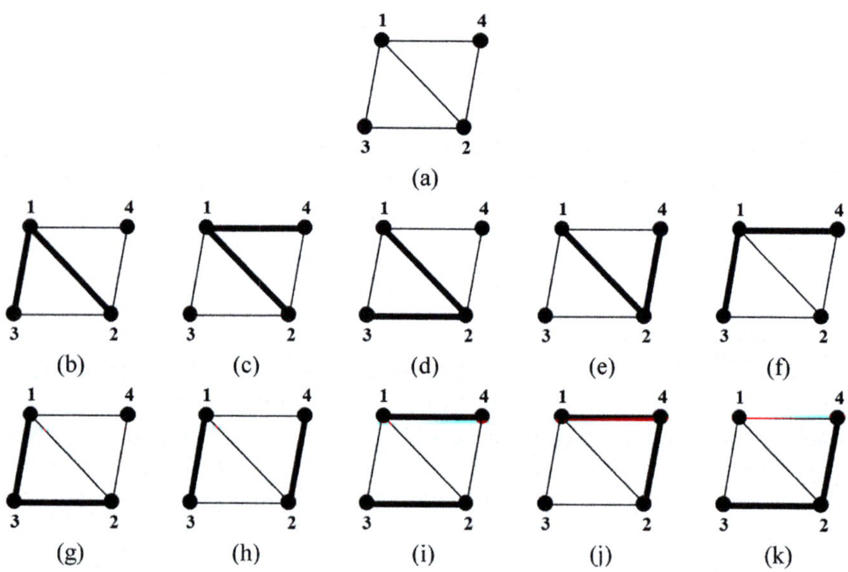

Fig. 5 **a** Parent graph for 4 vertices, **b**–**k** combinations of geared graphs with two gear pair edges and three turning pair edges with respect to the parent graph

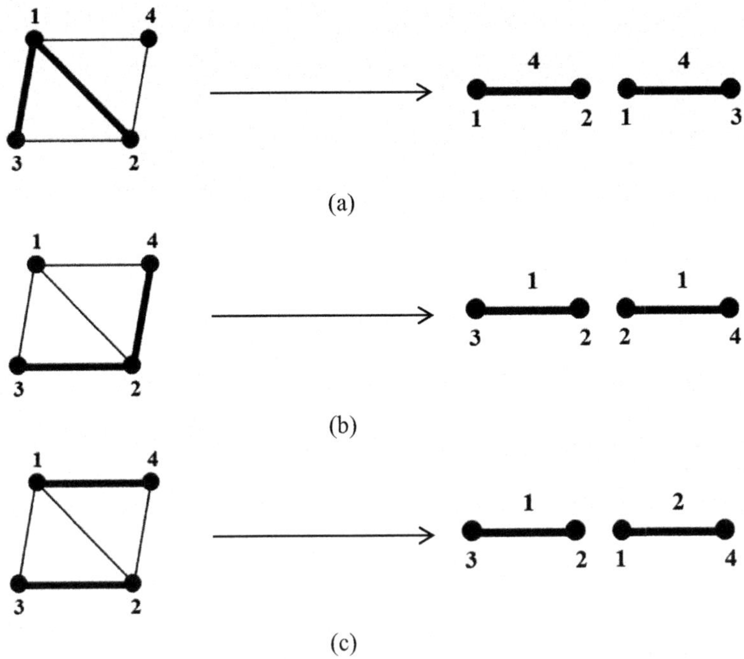

Fig. 6 **a–c** Three 4-vertex non-isomorphic geared graphs along with the derived rotation graphs

ways of assigning TVs to FCs of all the 3 geared graphs thereby leading to generation of rotation graphs; this is illustrated in Fig. 6.

For the geared graph in Fig. 6a, vertex-2 cannot be a TV for FC 1-3-2-4 since it would necessitate more than 2 axis labels for its turning pair edges. Therefore for both the FCs 1-3-2-4 and 4-1-2, vertex-4 is chosen as TV. For the geared graphs in Fig. 6b, c the choice of TVs is straightforward, leading to the derivation of respective rotation graphs. It can be observed that the rotation graph derived in Fig. 6a is isomorphic to that of (b) despite them being derived from different geared graphs. Therefore, in effect there exist two rotationally non-isomorphic graphs. Now, with respect to these rotation graphs, 3 non-isomorphic displacement graphs are derived as shown in Table 2.

3.2 *Recursive Scheme*

This procedure was introduced by Tsai [6] and is also employed by further researches due to [7, 9]. The idea behind this method is to generate N-link rotation graphs from the set of all $(N-1)$-link graphs. This is realized by adding a graph component consisting of a turning pair edge and a gear pair edge that are incident to each other, and whose free ends being fused to a vertex pair of a $(N-1)$-link EGT graph.

Table 2 Displacement graphs derived from the 2 rotationally non-isomorphic graphs

4-link rotation graph	Displacement graphs derived
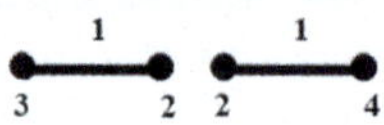	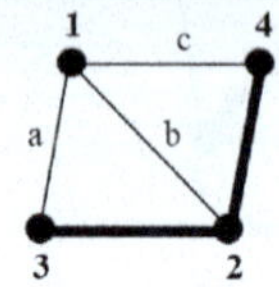 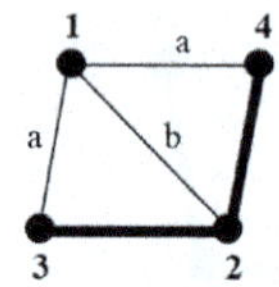
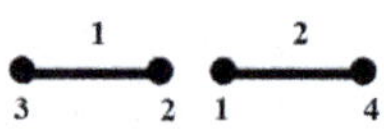	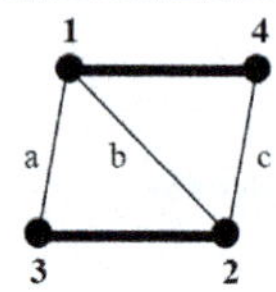

After ascertaining the TV of the FC with respect to the added gear pair, the rotation graph information of the *N*-link EGT is stored. This process is repeated for each of the $(N - 1)\boldsymbol{C}_2$vertex pairs and in this way to all of the distinct $(N - 1)$-link EGTs. Those *N*-link EGT graphs generated in this manner, that violate fundamental rules are eliminated. Finally, an isomorphism test is then conducted among all the generated graphs thereby resulting in all non-isomorphic *N*-link rotation graphs. Then, with respect to every rotation graph, displacement graphs are generated in the usual manner by assigning levels. The overall schematic for recursive scheme is shown in Fig. 7. Unlike the non-recursive scheme, here the test for degeneracy is not required since all the rotation graphs generated using the recursive scheme are anyway free from degenerate structure/s.

Table 3 shows a 4-link unlabelled graph, in which a turning pair edge and gear pair edge can be added to $4\mathbf{C}_2$ different pairs of vertices: (1, 2); (1, 3); (1, 4); (2, 3); (2, 4); (3, 4). To any pair of vertices, the link consisting of the turning and gear pair edges can be added in 2 ways resulting in a total of 12 possibilities.

3.3 *Acyclic Graph-Based Approaches for Enumeration of Displacement Graphs*

The constituents of an acyclic graph are single edges and/or polygons [12]. A Polygon of *N* vertices represents the set of like-labelled edges whose topological structure can be any tree for *N* vertices in which every one of the $(N - 1)$ edges is labelled with the same level (Table 4). The equivalence of a polygon and a like-labelled tree is

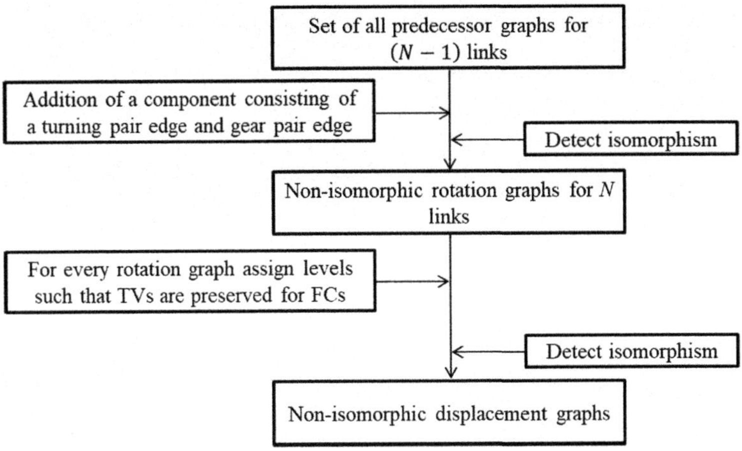

Fig. 7 The overall process of enumeration of EGTs using a recursive scheme

justified by the fact that both of them have the same NC_2 paths connecting all pairs of vertices. A polygon of 3 vertices is equivalent to the only 3-vertex tree with both its edges labelled alike. A polygon of 4 vertices is equivalent to the 2 possible 4-vertex trees with all its edges labelled alike. Similarly, a 5-vertex polygon represents the 3 possible 5-vertex trees with all its edges labelled same. In any acyclic graph, the number of edge labels equals the total number of single edges and polygons. The minimum and maximum bound for the number of edge labels of the spanning tree corresponding to an acyclic graph is given by

$$2 \leq \text{Number of edge labels in any acyclic graph} \leq (N - 1) \tag{7}$$

For the case of 3-vertex acyclic graph, there is only a single possibility and it is given in Fig. 8a where both the single edges correspond to different labels. There exist three different 4-vertex acyclic graphs as shown in Fig. 8b–d. In this way, set of connected single edges and/or polygons correspond to different edge labelled possibilities of a tree where polygons are used in the place of like-labelled subgraphs.

It should be noted here that, the edge labelled trees of different topologies can correspond to the same acyclic graph as shown in Table 5. We are interested in every possible edge labelled tree for N vertices and at this point it is assured that all these possibilities are nothing but a superset of acyclic graphs for the same N vertices.

Generation of Acyclic graphs

Similar to trees, all acyclic graphs from $(N - 1)$-vertex can be converted to N-vertex acyclic graphs in a recursive manner by adding a single edge with a label. This new labelled edge can be incident to any of the $(N - 1)$ vertices of the predecessor. Below mentioned is a general method for generating all N-vertex acyclic graphs for a given $(N - 1)$-vertex acyclic graph:

Table 3 The recursive generation of 5-link EGT graphs from a 4-link EGT graph

Addition of a graph component consisting of a turning pair and a gear pair to a 4-link EGT graph	Graph component incident to vertex pair	Generated 5-link graphs	
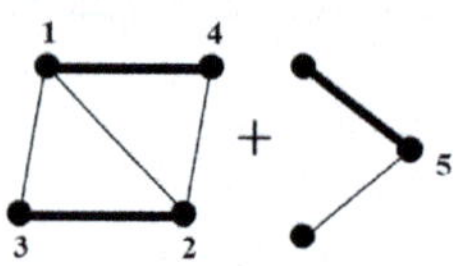	1-2	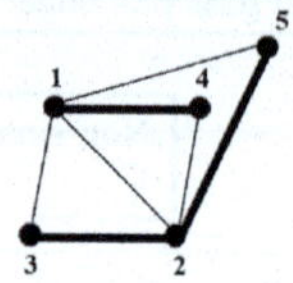	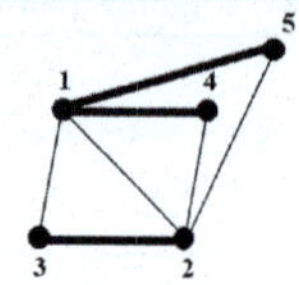
	1-3	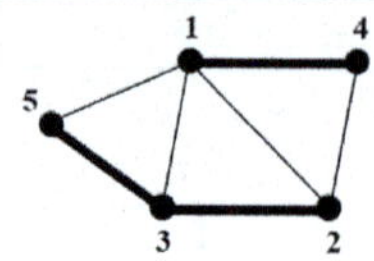	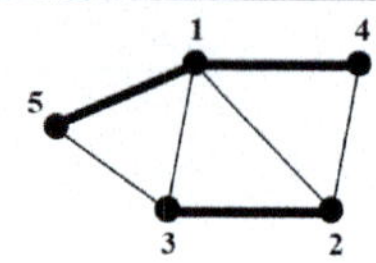
	1-4	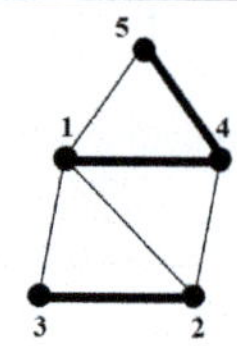	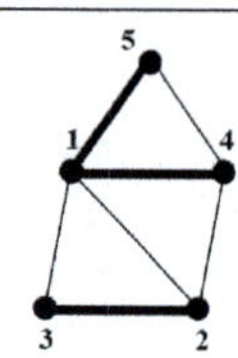
	2-3	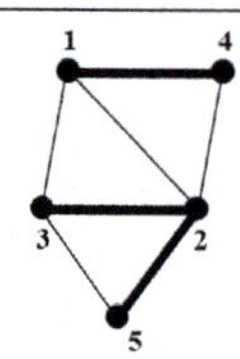	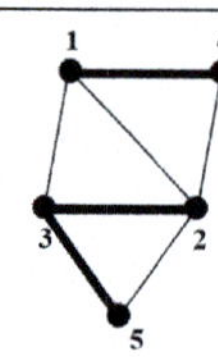
	2-4	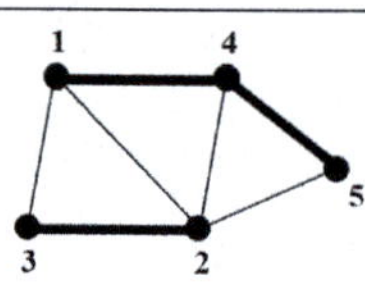	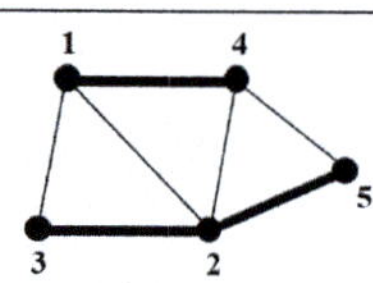
	3-4	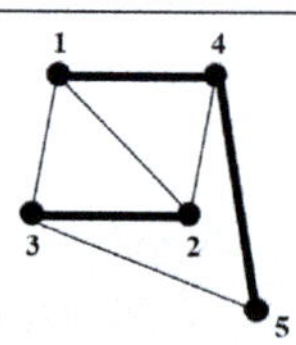	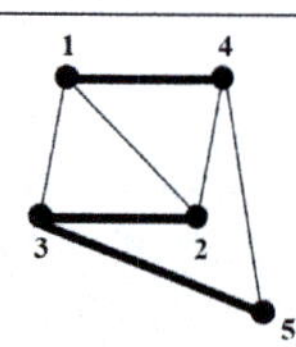

1. Determine those set of vertices that occupy similar topological position in the given acyclic graph which is to be evolved. In this way partition the vertex set of parent graph, into equivalent classes.
2. Add the new edge to any one representative vertex belonging to an equivalent class and in this way to all the representative vertices of corresponding equivalent classes one at a time. The idea is to use symmetry in the graph.
3. The label of the new edge can either be that of any of the edges already incident on that particular vertex (i.e. to which the new edge is being added) or an

Table 4 Polygons and their like-labelled trees

Polygons of Acyclic graphs	Equivalent Labelled Trees
	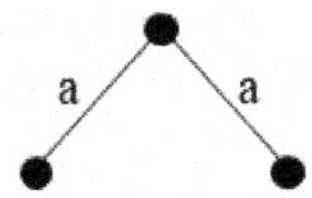
	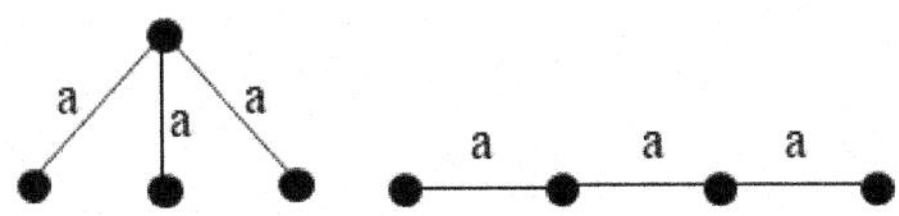
	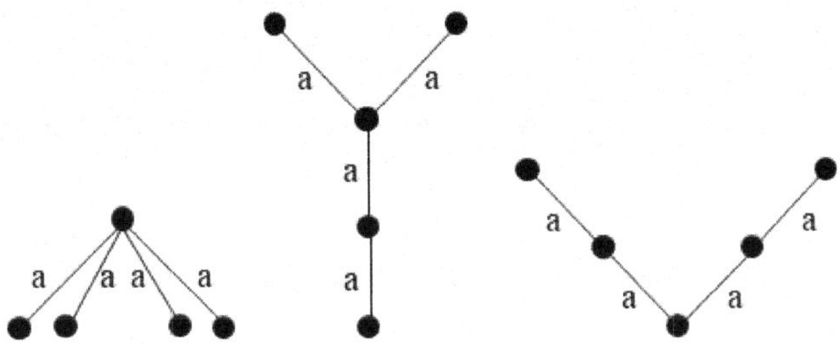

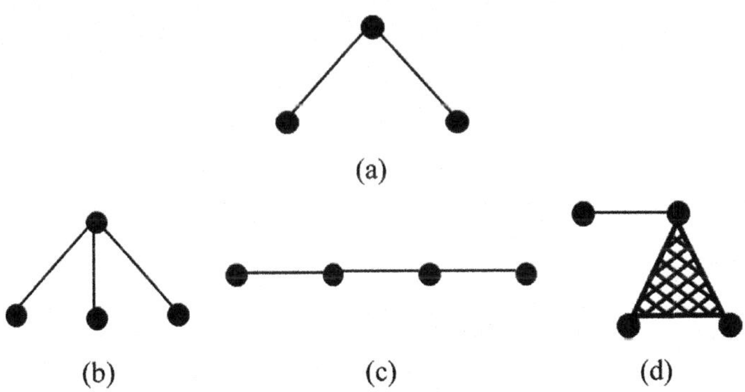

Fig. 8 **a** The only 3-vertex acyclic graph, **b–d** the three different 4-vertex acyclic graphs

entirely different label not existing in the predecessor graph. Corresponding to all possibilities, the N-vertex acyclic graphs are generated.

4. Finally, isomorphism test is to be conducted among all the generated N-vertex acyclic graphs to remove the duplicates.

Example Generation of 6 vertex acyclic graphs from a 5 vertex acyclic graph (Fig. 9).

Vertex set of this acyclic graph can be grouped into Equivalence classes—(1) (23) (45). We consider vertices 1, 2, 4 as the representatives from their respective

Table 5 Examples of different structured edge labelled trees for a common acyclic graph

Different structured Edge labelled Trees	Common Acyclic Graph
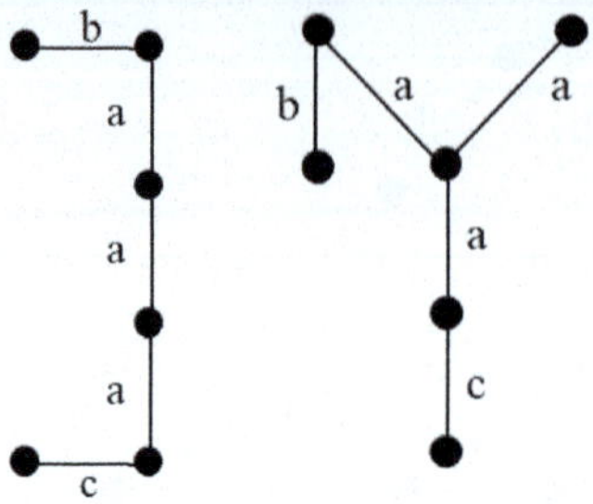	
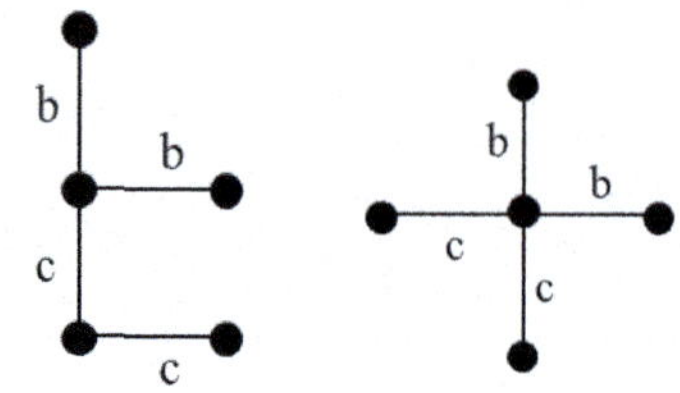	

Fig. 9 A 5-vertex acyclic graph

equivalent classes. In this given graph edges 1-4 and 1-5 possess different edge labels, respectively, and the polygon 1-2-3 possesses a third label.

At vertex 1—There exist edges belonging to all the three different labels of the graph. The new edge when incident to vertex-1 can possess any of the three labels or an entirely new label. So, in total there exist four possible 6-vertex acyclic graphs which can be generated by adding an edge at vertex-1. The four acyclic graphs are shown in Fig. 10. Out of the four graphs, the ones generated due to the new edge label same as that of edges 1-4 or 1-5 are topologically same because vertices 4 and 5 belong to the same equivalent class of the parent 5-vertex graph. This is evident from Fig. 10b, c.

At vertex 2—There exists only the polygon 1-2-3 incident at vertex 2. The new edge (to be incident at 2) can either have a same label as that of polygon 1-2-3 or possess a different label. If the new edge is of the same label as that of the polygon 1-2-3, the order of the polygon increases by one resulting in a four vertex polygon. This

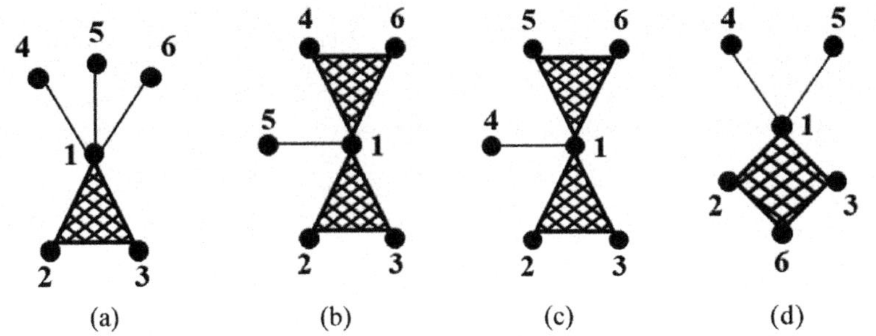

Fig. 10 Generated 6-vertex acyclic graphs due to addition of a labelled edge at vertex-1

is because polygon in itself is a tree with all like-labelled edges and when another like-labelled new edge is incident then the order of the tree and therefore the polygon also increases by one. The two possibilities are depicted in Fig. 11.

At vertex 4—The only edge incident at 4 is edge 1-4. The new edge can be of the same label as edge 1-4 or otherwise. If the new edge is of the same label as 1-4 then it gives rise to a three-sided polygon. Both the possibilities are mentioned below in Fig. 12.

Some of the graphs generated this way are isomorphs and those need be eliminated. In this work the above-mentioned procedure was followed and acyclic graphs with

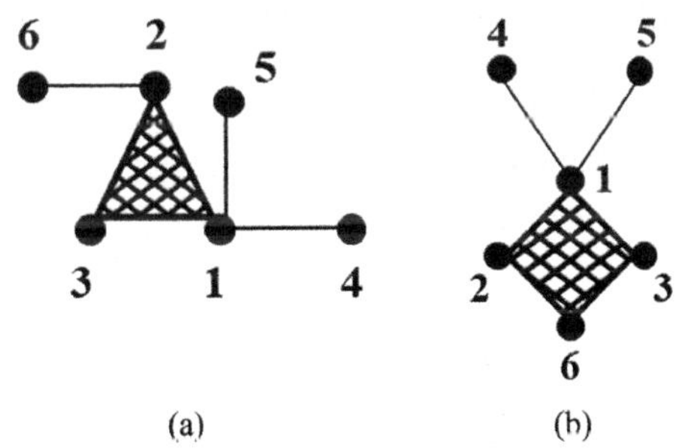

Fig. 11 Generated 6-vertex acyclic graphs due to addition of a labelled edge at vertex-2

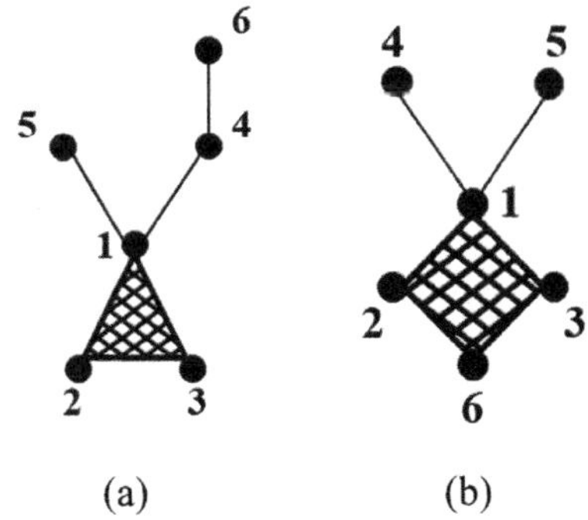

Fig. 12 Generated 6-link acyclic graphs due to addition of a labelled edge at vertex-4

Table 6 Enumeration results for acyclic graphs with and up to 8 vertices

Number of vertices	Number of non-isomorphic acyclic graphs
3	1
4	3
5	8
6	21
7	58
8	164

up to 8 vertices, is enumerated successfully and whose results are shown in Table 6. The present enumeration results for acyclic graphs is in agreement with that of given by [16].

3.3.1 Enumeration of Displacement Graphs Using Acyclic Graph-Based Approaches

Method of Addition of Labelled Turning Pair Edge and a Gear Pair Edge to Predecessor Graph [15]

This work was proposed by Hsu et al. [15]. Accordingly, an acyclic graph together with the connected gear pair edges constitutes a displacement graph of an EGT. The enumeration of displacement graphs for *N*-link, *F*-DOF is carried out from the set of all the $(N - 1)$-link, *F*-DOF displacement graphs. This is achieved by the addition of a labelled turning pair edge and a gear pair edge to a pair of vertices of $(N - 1)$-link graph. This operation is performed to all the $(N - 1)\mathbf{C}_2$ pairs of vertices. While adding the component consisting of turning and gear pair edges, different edge labels of the added turning pair edge needs to be considered and it is illustrated with an example of generation of all 4-link displacement graphs from the only 3-link graph. Here, in the parent 3-link EGT (Table 7), there exist three (i.e.$3\mathbf{C}_2$) possible pairs of vertices for the addition of a turning pair edge and a gear pair edge and they are (1, 2); (1, 3); (2, 3).

Case of graph component added to the pair of vertices {(1, 2); (1, 3)}
If the new turning pair edge is 1-4 (incident at vertex-1) and the new gear pair edge is 2-4 (incident at vertex-2), then the resulting 4-link displacement graphs can take two forms: the label of turning pair edge 1-4 being same as that of 1-3 (as depicted by polygon 1-3-4 in the acyclic graph) and the label of 1-4 is different to that of edges 1-2 and 1-3.

If the new turning pair edge is 2-4 (incident at vertex-2) and gear pair edge is 1-4 (incident at vertex-1) then it results in a single 4-link displacement graph having all the turning pair edge labels, as different (Table 7).

Table 7 Generation of 4-link displacement graphs from the only 3-link EGT based on method of Hsu et al. [15]

Parent 3-link EGT displacement graph in acyclic graph representation	Addition of turning and gear pair edges to a particular pair of vertices	Generated 4-link displacement graphs in acyclic graph representation	Whether a valid graph satisfying fundamental circuit rule
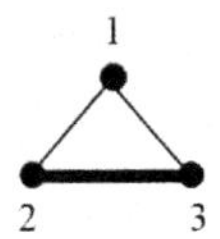	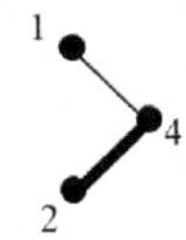	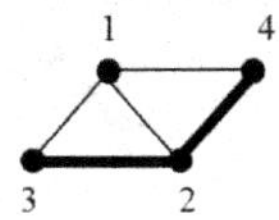	Yes
		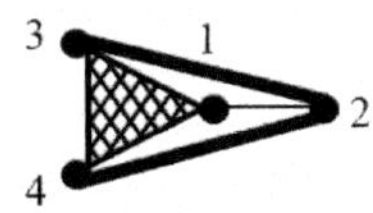	Yes
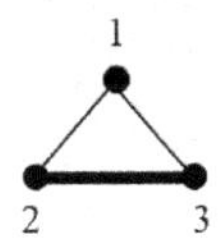	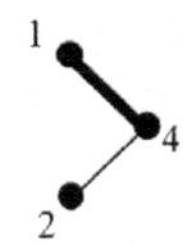	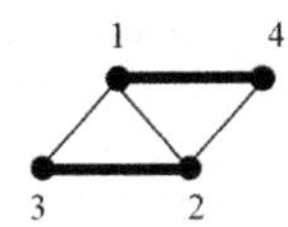	Yes
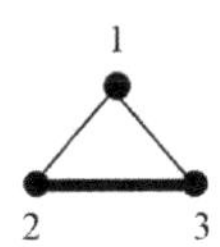	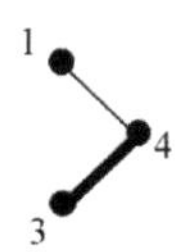	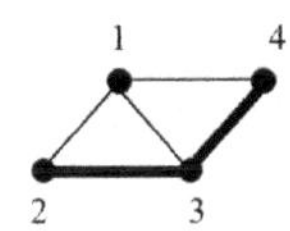	Yes
		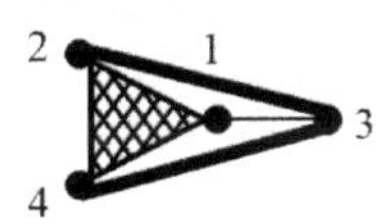	Yes
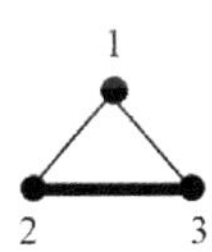	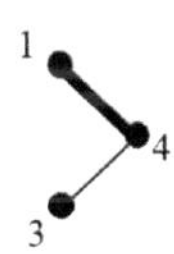	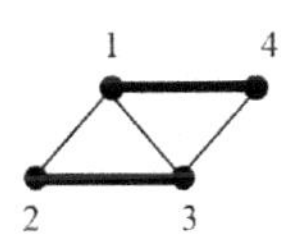	Yes
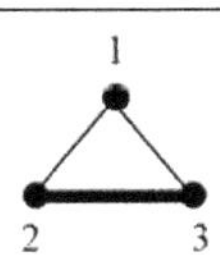	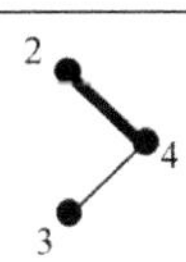	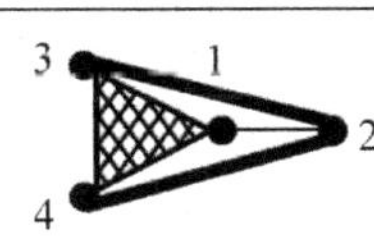	Yes
		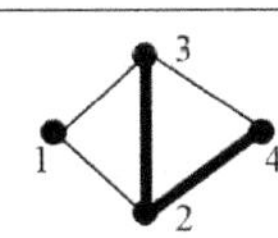	No

(continued)

Table 7 (continued)

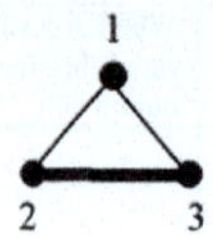	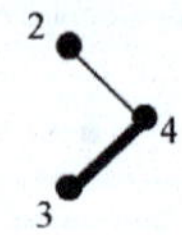	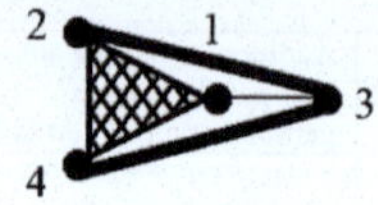	Yes
		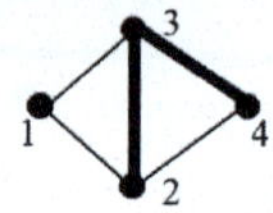	No

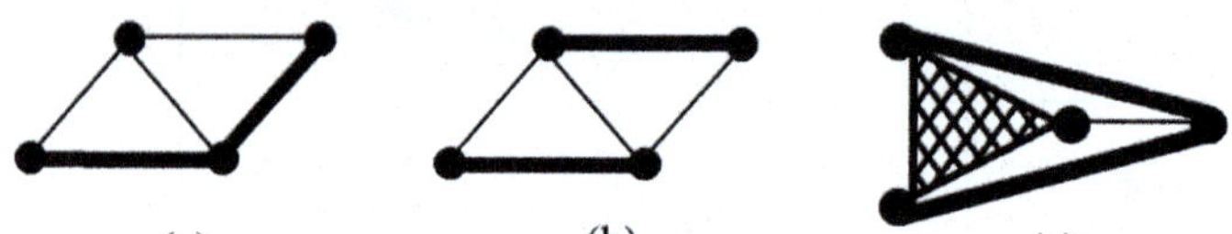

(a) (b) (c)

Fig. 13 The three 4-link non-isomorphic displacement graphs in their acyclic graph representation

The same set of graphs will also be generated if the vertex pair is (1, 3), owing to the symmetry inherent in the 3-link EGT.

Case of graph component added to the pair of vertices (2, 3)
If the new turning pair edge is 3-4 (incident at vertex-3) and the new gear pair edge is 2-4 (incident at vertex-2), then it results in the generation of two possibilities which include a displacement graph consisting of polygon 1-3-4 owing to same labelled edges 1-3 and 1-4, and another graph having edges 1-3 and 1-4 labelled differently and it is invalid because three different edge labels exist in FC 2-1-3-4 corresponding to edges 1-2, 1-3 and 3-4.

If the turning and gear pair edges are interchanged (i.e. 2-4 as turning pair edge and 3-4 as gear pair edge) it gives rise to the same set of graphs as previously (Table 7), because of the symmetric location of links 2 and 3.

Among the all generated displacement graphs, after checking for isomorphism, three 4-link displacement graphs result and they are shown in Fig. 13.

Method of Admissible Graphs Hsu 1993 [13]

Admissible graphs of EGTs are those obtained by transforming all the gear pair edges into revolute pair edges of an EGT in its acyclic graph representation. Conversely, EGTs can very well be enumerated given the set of all the admissible graphs, as parent graphs. Hsu [13] synthesized all the non-fractionated admissible graphs using the published atlases of contracted graphs. From a given N-vertex admissible graph, $(N - 2 - F)$ gear pair edges can be distributed by replacing that many revolute edges, such that no two FCs have a common gear pair edge thereby resulting in

Table 8 Set of 5-vertex admissible graphs along with the derived displacement graphs

5-vertex admissible graphs	5-link non-isomorphic displacement graphs
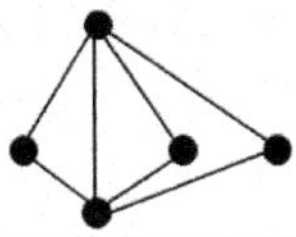	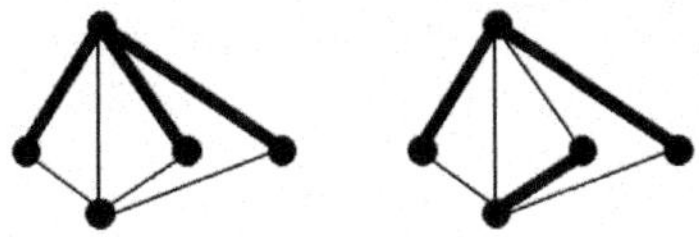
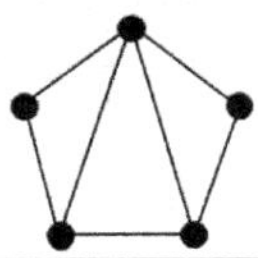	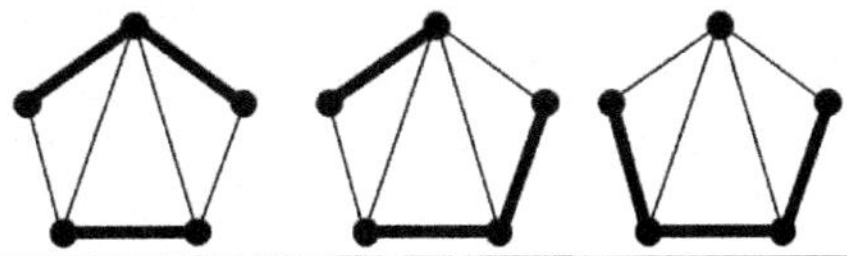

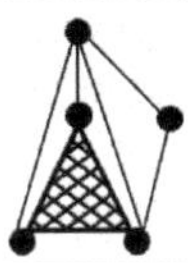	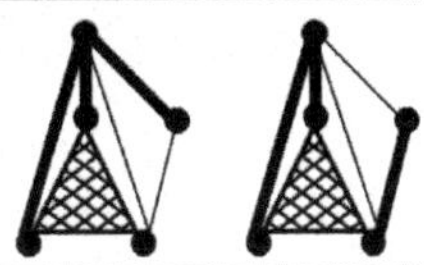
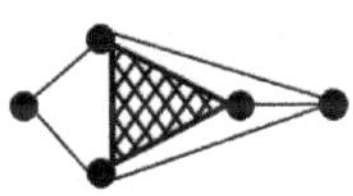	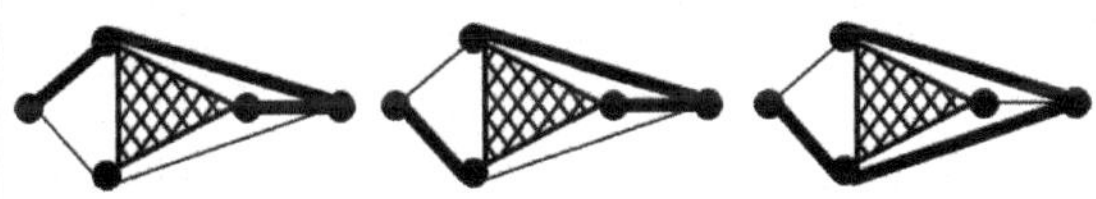

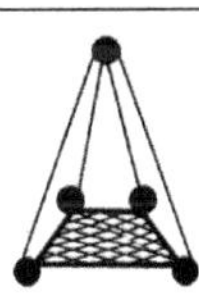	

N-link, *F*-DOF displacement graphs which will necessarily contain an acyclic graph (connected graph of all labelled turning pair edges), as the spanning tree. This combinatorial procedure will yield isomorphic graphs which need to be eliminated in order to arrive at the set of all *N*-link displacement graphs with respect to the parent admissible graph. Table 8 contains all the 13 displacement graphs of 5-link and 1 DOF which are all derived from 7 admissible graphs for five vertices.

Table 9 A 5-vertex acyclic graph along with its feasible pairs of vertices for addition of gear pairs

Acyclic graph	Feasible pair of vertices for FCs
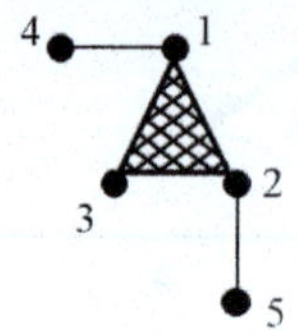	(1, 5) (2, 4) (3, 4) (3, 5) (4, 5)

Method of Acyclic Graphs as Parents by Hsu and Hsu [16]

An N-vertex acyclic graph can be converted into displacement graphs of N-link EGTs of 1 DOF by addition of $(N - 2)$ gear pair edges. Every gear pair edge must give rise to a feasible FC, complying with the fundamental rules of the EGT graph. If there exist k such feasible pairs of vertices of the given N-vertex acyclic graph, which are suitable for the addition of gear pairs, then it is possible to obtain all the $(k)\mathbf{C}_{(N-2)}$ combinations of displacement graphs. As an example, a 5-vertex acyclic graph is shown in Table 9 along with the feasible pairs of vertices which are suitable for addition of a gear pair edge to form a FC. There exist 5 feasible pairs of vertices for the acyclic graph given in Table 9 and combining those 3 at a time will result in 10 different 5-link displacement graphs of 1 DOF and are tabulated in Table 10.

3.4 Synthesis of EGTs from Basic Kinematic Chains (BKCs)

A BKC can be considered as a complex mechanical system consisting of simple joints, multiple joints and turning pair connections existing between different links. A multiple joint is formed when more than 2 links are incident at same. In the graph representation of a BKC, links are modelled as vertices and turning pair connections as edges. In order to model a multiple joint, the works of Ding et al. [25–27] proposed the use of solid as well as hollow vertices belonging to the vertex set of their graph model of BKC. A solid vertex would represent a link and the hollow vertex together with the edges incident on same would represent a multiple joint. An 8-link BKC is shown in Fig. 14a, where link-1, link-5 and link-6 are incident at a common multiple joint marked as "a". In the graph representation shown in Fig. 14b, the links sharing the multiple joint "a" are incident at a hollow vertex which is marked as "9(a)".

The representation of EGT in a graph model requires the information of labels of the turning pair edges thereby resulting in a displacement graph. In the work of Yang et al. [17], they have modified their bicolored (i.e. solid and hollow) topological graph they had developed previously to include geared pair edges also (i.e. in addition to turning pair edges). The edge labelling of turning pairs is also effected by making all like-labelled edges incident at a hollow vertex. This way, similar to acyclic graph

Table 10 The ten displacement graphs derived by combining 3 FCs at a time for the graph of Table 9

$5C_3$ combinations of FCs	Displacement graph
{(1, 5); (2, 4); (3, 4)}	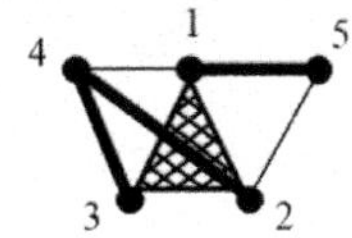
{(1, 5); (2, 4); (3, 5)}	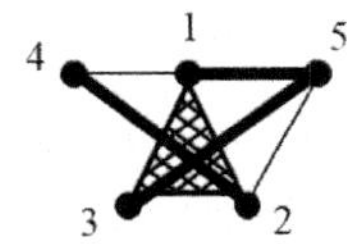
{(1, 5); (2, 4); (4, 5)}	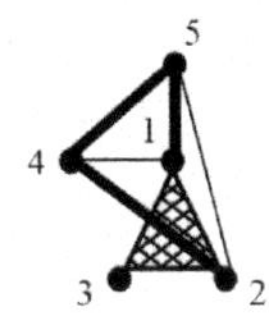
{(1, 5); (3, 4); (3, 5)}	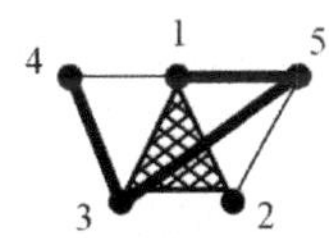
{(1, 5); (3, 4); (4, 5)}	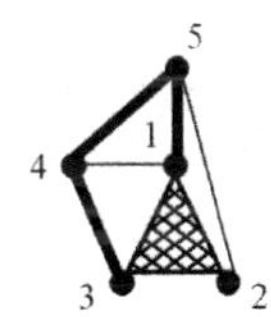
{(1, 5); (3, 5), (4, 5)}	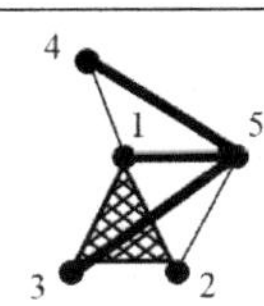

(continued)

and at the same time unlike conventional graph representation, there is no necessity to mention the label beside the turning pair edges. A Simpson gear train is shown in the EGT graph representation of Yang and Ding [18] wherein gear pair edges are 2-3, 2-4, 3-5, 5-6. The turning pair edges formed by links: 1, 4, 5, and 6 are all same labelled and are therefore connected by a hollow vertex-7.

In the works by Souza et al. [28, 29], in order to enumerate EGTs for a certain number of links, firstly, all the non-fractionated BKCs corresponding to Variety V

Table 10 (continued)

{(1, 5); (3, 4); (3, 5)}	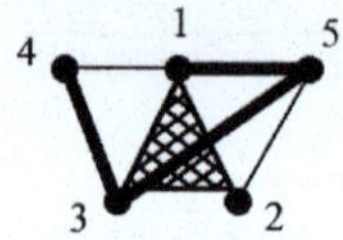
{(2, 4); (3, 4); (3, 5)}	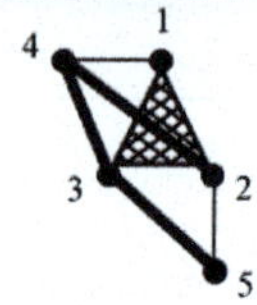
{(2, 4); (3, 4); (4, 5)}	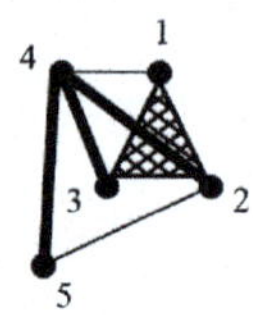
{(3, 4); (3, 5); (4, 5)}	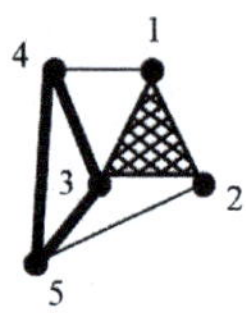

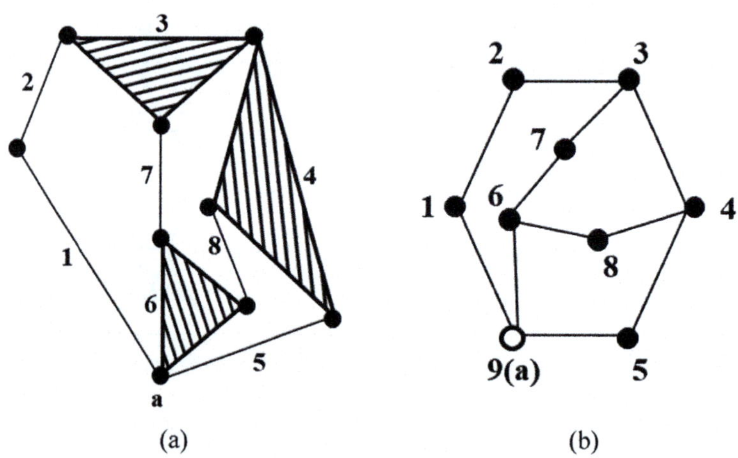

Fig. 14 **a** A 8-link basic kinematic chain with a multiple joint marked as "a" **b** its corresponding bicolored graph representation given by Ding et al. [25]

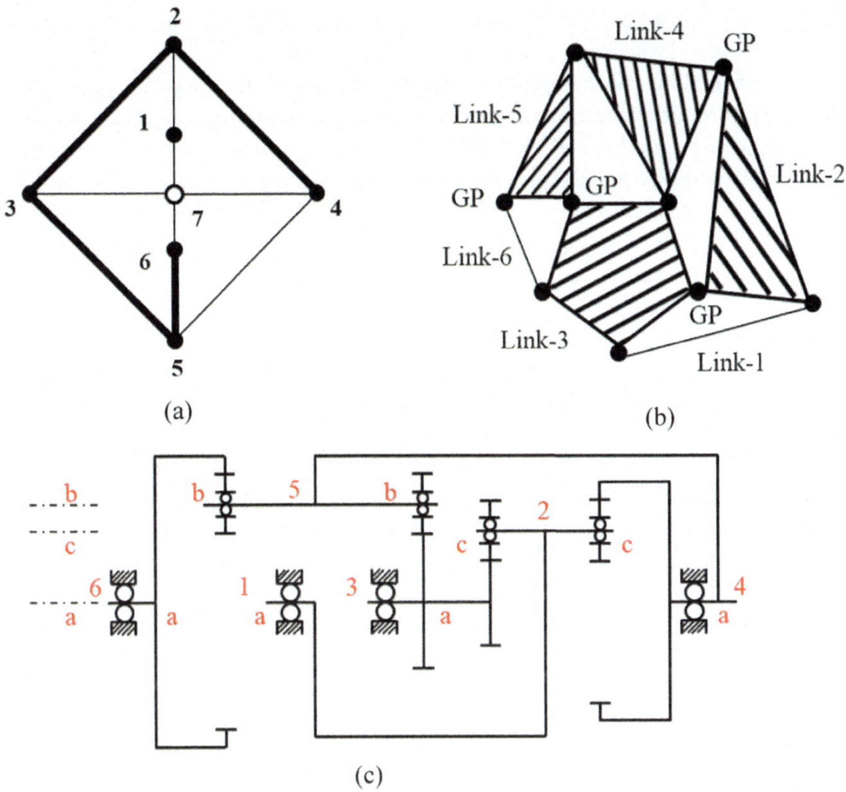

Fig. 15 **a** Representation of a 6-link EGT (Simpson gear train) given by Yang et al. [17], **b** BKC representation of the same 6-link EGT with particular vertices marked as GP denoting gear pair connections, **c** Schematic of Simpson gear train

= 0, order of screw system $\lambda = 2$ and DOF = 1 are enumerated. Then, it is shown that EGTs can be enumerated by colouring the vertices of a given BKC. In Fig. 15b, one such BKC is shown in which certain vertices are coloured as GP denoting the gear pair connection with respect to the EGT graph of Fig. 15a. Also shown is the corresponding schematic in Fig. 15c.

4 Conclusions

In this chapter, all the major graph theoretic methods for generation of EGTs were reviewed. It was seen that the ultimate aim of any structural synthesis algorithm for EGTs for a given number links and DOF is to enumerate all the non-isomorphic displacement graphs. Certain approaches enumerate the set of all rotation graphs, in an intermediate step before eventually generating the displacement graphs. However, the acyclic graph-based methods, directly generate the set of displacement graphs (i.e. in one step) [10–16].

A recent trend has been to synthesize EGTs from BKCs (i.e. as parents). Displacement graphs require labelling the turning pair edges and in this regard the adoption of multiple joint graphs of BKCs to EGTs is a great step towards realization of its automated synthesis for higher number of links [17, 20]. It is worthy to note that, in the graph representation of an EGT given by Ding and coworkers [25–27], the labelling of two or more turning pair edges with the same axis is effected with the help of a multiple joint in an equivalent BKC.

The correspondence between edge labelled trees and acyclic graphs is explained in this work and the basis of acyclic graph representation used by Hsu and coworkers is clarified. It was seen that the labelled turning pair edges in a displacement graph can be modelled by an acyclic graph. The structure of acyclic graphs bears similarities to those of "Assur groups", "Baranov trusses" etc. which are well known in the mechanisms literature for their use in synthesis of BKCs. Taking inspirations from such well-established methods, future synthesis algorithms for EGTs using acyclic graphs as parents is promising. The idea can be to generate rotation graphs from acyclic graphs as parents followed by generation of displacement graphs with respect to each rotation graph as per the method used in [23].

Generally speaking, structural synthesis is aimed at obtaining the complete set of EGTs for a given number of links and DOF. However, future works can concentrate on developing methods to synthesize EGTs with particular structural features [30]. In this regard, classification of rotation graphs based on certain similarities in their graph structure is essential.

It is hoped that this review will benefit researchers and motivate them to come up with novel synthesis methods for EGTs, using graph theory.

Acknowledgements VRS would like to thank Prof. Varadaraju Dharanipragada for his help and valuable insights on this work.

References

1. Freudenstein F, Dobrjanskyj L (1966) On a theory for the type synthesis of mechanisms. In: Görtler H (ed) Applied mechanics. Springer, Berlin, Heidelberg, pp 420–428
2. Dobrjanskyj L, Freudenstein F (1967) Some applications of graph theory to the structural analysis of mechanisms. J Manuf Sci Eng 89(1):153–158
3. Buchsbaum F, Freudenstein F (1970) Synthesis of kinematic structure of geared kinematic chains and other mechanisms. J Mech 5(3):357–392
4. Freudenstein F (1971) An application of Boolean algebra to the motion of epicyclic drives. J Eng Ind 93(1):176–182
5. Ravisankar R, Mruthyunjaya TS (1985) Computerized synthesis of the structure of geared kinematic chains. Mech Mach Theory 20(5):367–387
6. Tsai LW (1987) An application of the linkage characteristic polynomial to the topological synthesis of epicyclic gear train. ASME J Mech Transm Autom Des 109(3):329–336

7. Kim JU, Kwak BM (1990) Application of edge permutation group to structural synthesis of epicyclic gear trains. Mech Mach Theory 25(5):563–574
8. Shin JK, Krishnamurthy S (1993) Standard code technique in the enumeration of epicyclic gear trains. Mech Mach Theory 28(3):347–355
9. Rao YVD, Rao AC (2008) Generation of epicyclic gear trains of one degree of freedom. J Mech Des 130(5):052604
10. Hsu CH, Lam KT (1992) A new graph representation for the automatic kinematic analysis of planetary spur-gear trains. ASME J Mech Des 114(1):196–200
11. Hsu CH (1992) A graph notation for the kinematic analysis of differential gear trains. J Frankl Inst 329(5):859–867
12. Hsu CH, Lam KT (1993) Automatic analysis of kinematic structure of planetary gear trains. J Mech Des 115(3):631–638
13. Hsu CH (1993) Synthesis of kinematic structure of planetary gear trains by admissible graph method. J Frankl Inst 330(5):913–927
14. Hsu CH (1993) A graph representation for the structural synthesis of geared kinematic chains. J Frankl Inst 330(1):131–143
15. Hsu CH, Lam KT, Yin YL (1994) Automatic synthesis of displacement graphs for Planetary gear trains. Math Comput Model 19(11):67–81
16. Hsu CH, Hsu JJ (1997) An efficient methodology for the structural synthesis of geared kinematic chains. Mech Mach Theory 32(8):957–973
17. Yang W, Ding H, Zi B, Zhang D (2018) New graph representation for planetary gear trains. J Mech Des 140(1):012303
18. Yang W, Ding H (2018) Automatic detection of degenerate planetary gear trains with different degree of freedoms. Appl Math Model 64:320–332
19. Yang W, Ding H (2018) The perimeter loop-based method for the automatic isomorphism detection in planetary gear trains. J Mech Des 140(12):123302
20. Yang W, Ding H (2019) The complete set of one-degree-of-freedom planetary gear trains with up to nine links. J Mech Des 141(4):043301
21. Rai RK, Punjabi S (2019) A new algorithm of links labelling for the isomorphism detection of various kinematic chains using binary code. Mech Mach Theory 131:1–32
22. Kamesh VV, Rao KM, Rao ABS (2017) Detection of degenerate structure in single degree-of-freedom planetary gear trains. J Mech Des 139(8):083302
23. Shanmukhasundaram VR, Rao YVD, Regalla SP (2019) Enumeration of displacement graphs of epicyclic gear train from a given rotation graph using concept of building of kinematic units. Mech Mach Theory 134:393–424
24. Shanmukhasundaram VR, Rao YVD, Regalla SP (2019) Algorithms for detection of degenerate structure in epicyclic gear trains using graph theory. J Braz Soc Mech Sci Eng 41(11):496
25. Ding H, Zhao J, Huang Z (2009) Unified topological representation models of planar kinematic chains. J Mech Des 131(11):114503
26. Ding H, Yang W, Huang P, Kecskemethy A (2013) Automatic structural synthesis of planar multiple joint kinematic chains. J Mech Des 135(9):091007
27. Ding H, Yang W, Zi B, Kecskemethy A (2016) The family of planar kinematic chains with two multiple joints. Mech Mach Theory 99:103–116
28. Souza MBD, Vieira RDS, Martins D (2018) Synthesis of epicyclic gear trains with one and two degrees of freedom from kinematic chains belonging to minimal sets. Int J Mech Robot Syst 4(4):383–400
29. Souza MBD, Vieira RDS, Martins D (2017) Enumeration of kinematic chains with zero variety for epicyclic gear trains with one and two degrees of freedom. In: International symposium on multibody systems and mechatronics. Springer, Cham
30. Ding H, Cai C (2020) Patent analysis and structural synthesis of epicyclic gear trains used in automatic transmissions. Appl Sci 10(1):82

Vision-Based Grasp Planning Based on Grasp Quality Metrics and Its Hardware Implementation

Roshan Kumar Hota, Aditya Negi, and C. S. Kumar

Abstract This work presents grasp planning on everyday objects using vision. The hand considered is a one degree-of-freedom parallel jaw gripper of Mitsubishi Movemaster robot. Candidate grasping points are chosen on the object and a grasp matrix is computed for the grasp. The grasp matrix can be used to computationally determine a force-closure grasp feasibility. For selecting the candidate grasping points, image of the object is used. Three quality metrics based on different physical notions of quality of grasp are computed. The first quality measure tells how far a grasp is from violating the friction limits, the second gives the worst case performance of the force-closure for all external wrenches, and the third tells how well the object is enclosed from all directions. The main contribution of the paper is to compare grasps based on different quality measures and understand their physical interpretation.

Keywords Grasp planning · Grasp quality metrics · Computer vision · Linear-inequality · Force-closure

1 Introduction

Robots with the ability to grasp and manipulate in human and industrial environment are going to see great utility. Challenges for better grasping and manipulation systems span mechanical design, controls, planning and intelligence, perception, and sensors. The importance of research in the field is evidenced by the keen interest shown by many companies like Google [1] and Amazon [2] (amazon picking challenge). If we consider grasping task in a household environment, there are objects of different shapes and sizes. Some objects have special features built in to facilitate grasping.

R. K. Hota (✉) · A. Negi · C. S. Kumar
Mechanical Engineering Department, Indian Institute of Technology, Kharagpur, India
e-mail: roshanhota@iitkgp.ac.in

C. S. Kumar
e-mail: kumar@mech.iitkgp.ac.in

D. Sen et al. (eds.), *Mechanism and Machine Science*, Lecture Notes in Mechanical Engineering, https://doi.org/10.1007/978-981-15-4477-4_26

A water jug, for example, has a handle without which it would be very difficult to grasp it. A robot should be able to exploit such features of objects.

One of the first studies in grasping and manipulation was done by Salisbury and Craig [3] where they have discussed force application capabilities of fingers of the robotic hands. Salisbury and Roth in [4] have developed the condition for force-closure on an object. Force-closure refers to a condition where the contact forces created by the fingers can resist any arbitrary external wrench. Murray et al. in [5] further developed the mathematical notation for the constraints for force-closure. Various solution methods for the constraints have been proposed. We have used linear inequality as proposed by Prattichizzo and Trinkle in [6]. Another approach to grasping objects is caging [7]. Caging is a method in which the object is allowed to move to some degree yet it cannot move arbitrarily far because it is "caged" by the fingers. Caging, however, is not suitable if the object has to be manipulated or a task has to be performed. We, therefore, work with force-closure in this paper. The fingers can produce any desired wrench on the object if the object is held in force-closure.

Both caging and grasping are based on mechanics and geometrical constraints and are analytical approaches to grasp planning. Besides these, there are empirical approaches that learn grasps from human demonstration. Researchers in this method typically use data gloves to input human demonstration and imitate based on joint angles [8, 9] or task wrench space [10]. We use the analytical method in this work because it has some distinct advantages over the empirical learning-based algorithms. While the learning-based method generates new grasps from the demonstrated grasps, the analytical method we have used generates grasps from the mechanics of contact and grasping. This allows us to have an insight based on physics and geometry which is not available in learning-based methods. The physical insight also allows us to define grasp quality metrics quantifying a particular behavior of the grasp. Also, the learning-based methods requires an initial phase of training which is completely avoided in the analytical method.

We have considered grasp planning using vision. We have presented the complete process of perception to grasp planning and execution with a focus on grasp planning. We have used everyday objects like water bottle in our work. The quality of a force-closure grasp can be based on many different physical notions. We compare the grasp planning based on three different notions from the works in [6] and [11]. The first measure is how far the contact force is from violating the friction cone. Farther away the contact force from violating the friction cone, better is the grasp. Second is the measure of worst-case performance of the grasp to resist external wrenches. Worst-case performance of various grasps are compared. The grasp with better worst-case performance is considered better. Third measure is called the volume of wrench space. It tells how well the object is bounded from all sides. We show the three measures for all the grasps and discuss which grasp quality metric can be chosen in a particular situation. The main contribution of the paper lies in the physical interpretations of the various grasp quality measures and to demonstrate through examples how the measures can be used in different situations.

2 Mathematical Formulation

In this section we develop the measures for goodness of the grasp. The computations are based on the grasp matrix (**G**).

Consider the object twist ($\boldsymbol{\nu}$) to be the vector of the velocity of the origin of frame {**O**} and its angular velocity (Fig. 1).

$$\nu = \begin{bmatrix} \mathbf{v} \\ \omega \end{bmatrix} \tag{1}$$

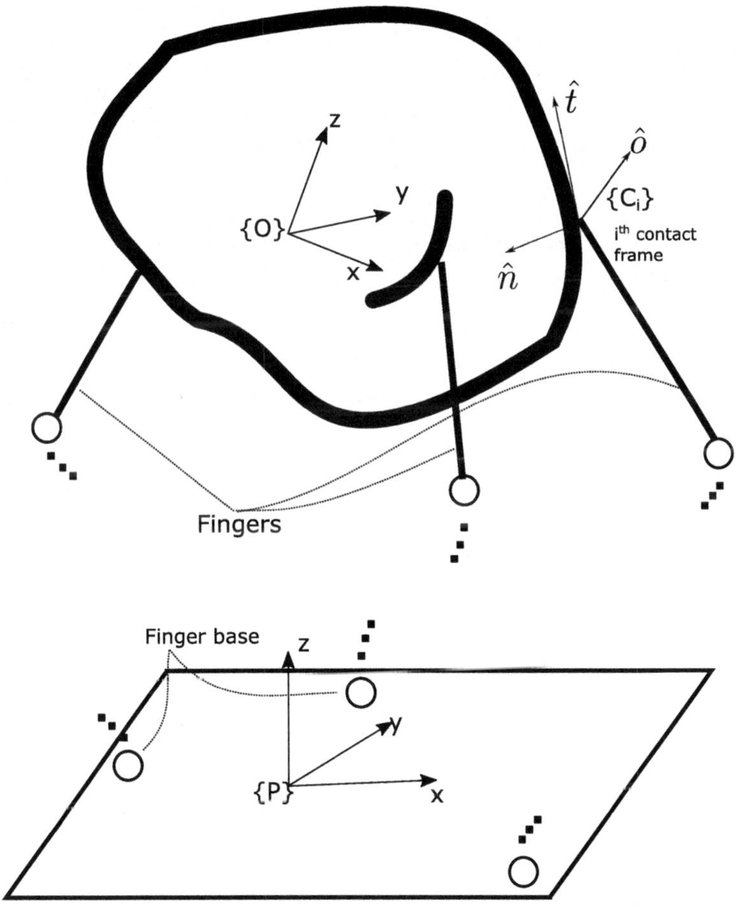

Fig. 1 A representation of the object and the contact points along with the fingers and the palm. The palm frame {P}, the object frame {O}, and the contact frame for one contact $\{C_i\}$ is shown

Here, $\mathbf{v}$ is the velocity of the origin of frame $\{\mathbf{O}\}$ and ω is the angular velocity of the frame. In a similar manner, the twist, $\boldsymbol{\nu}_{c_i}$, of the contact frame $\{\mathbf{C}\}_i$ expressed in the palm frame $\{\mathbf{P}\}$, is given by the vector of the velocity of origin of frame $\{\mathbf{C}\}_i$ and its angular velocity (ω_{c_i}).

$$\boldsymbol{\nu}_{c_i} = \begin{bmatrix} \mathbf{v}_{c_i} \\ \omega_{c_i} \end{bmatrix} \tag{2}$$

where

$$\begin{aligned} \mathbf{v}_{c_i} &= \mathbf{v} + \omega \times \mathbf{r}_i \\ \omega_{c_i} &= \omega \end{aligned} \tag{3}$$

The object twist ($\boldsymbol{\nu}$) and the contact twist at the ith contact ($\boldsymbol{\nu}_{c_i}$) are related as follows:

$$\begin{aligned} \boldsymbol{\nu}_{c_i} &= \mathbf{P}_i^T \boldsymbol{\nu} \\ where\ \mathbf{P}_i &= \begin{bmatrix} \mathbf{I}_{3\times3} & \mathbf{0}_{3\times3} \\ -\mathbf{S}(\boldsymbol{r}_i) & \mathbf{I}_{3\times3} \end{bmatrix} \in \mathbb{R}^{6\times6} \end{aligned} \tag{4}$$

Here, $\mathbf{I}_{3\times3} \in \mathbb{R}^{3\times3}$ is an identity matrix and the term $S(\mathbf{r})$ is the skew of a vector used for performing cross product. For $\mathbf{r} = [r_x \;\; r_y \;\; r_z]^T$ it is given by

$$\mathbf{S}(\mathbf{r}) = \begin{bmatrix} 0 & -r_z & r_y \\ r_z & 0 & -r_x \\ -r_y & r_x & 0 \end{bmatrix} \tag{5}$$

Equation (4) evaluates the contact twist in frame $\{\mathbf{P}\}$. The contact twist ($\boldsymbol{\nu}_{c_i}$) has to be expressed in the contact frame $\{\mathbf{C}\}_i$ so that it can be easily used for identifying feasible and infeasible grasp. The contact frame at the ith contact is chosen such that the x-axis is along the normal pointing toward the object or into the object. The other two axes are in the tangential plane to form a right-handed coordinate system. The contact frame is shown in Fig. 1. A rotation matrix $\mathbf{R}_i \in \mathbb{R}^{3\times3}$ between the palm and the contact frame can be written as $\mathbf{R}_i = [\hat{n} \;\; \hat{t} \;\; \hat{o}]^T$. Here, $\hat{n}$ is the unit vector along the normal at i^{th} which is also the x-direction of the contact frame. This normal direction can be obtained by image processing. The other two directions can be obtained to satisfy the right-handed system. For obtaining the tangential and orthogonal direction of the contact frame, we make use of an arbitrary unit vector, ($\hat{\mathbf{a}}$), different from $\hat{n}$. We perform the following computations for obtaining the other two directions:

$$\hat{\boldsymbol{t}} = \hat{\boldsymbol{n}} \times \hat{\boldsymbol{a}} / \left\| \hat{\boldsymbol{n}} \times \hat{\boldsymbol{a}} \right\| \tag{6}$$

$$\hat{\boldsymbol{o}} = \hat{\boldsymbol{n}} \times \hat{\boldsymbol{t}} / \left\| \hat{\boldsymbol{n}} \times \hat{\boldsymbol{t}} \right\| \tag{7}$$

Further

$$\mathbf{R}_{i,blk} = Blockdiag(\mathbf{R}_i, \mathbf{R}_i) = \begin{bmatrix} \mathbf{R}_i & \mathbf{0} \\ \mathbf{0} & \mathbf{R}_i \end{bmatrix} \in \mathbb{R}^{6\times 6}$$
$$\tilde{\boldsymbol{G}}_i^T = \mathbf{R}_{i,blk}^T \mathbf{P}_i^T \tag{8}$$

$$\tilde{\boldsymbol{G}}^T = \begin{pmatrix} \tilde{\boldsymbol{G}}_1^T \\ \tilde{\boldsymbol{G}}_2^T \\ \tilde{\boldsymbol{G}}_3^T \end{pmatrix} \tag{9}$$

The matrix $\tilde{\boldsymbol{G}}^T \in \mathbb{R}^{6n_c\times 6}$, that we call partial grasp map, maps the object twist expressed in palm frame to the contact twist expressed in contact frame. Here, n_c is the number of contacts. These computations can be made for a contact point on the object or on the finger. For a contact point on the object:

$$\boldsymbol{\nu}_{c,obj} = \tilde{\boldsymbol{G}}^T \boldsymbol{\nu} \tag{10}$$

We can consider two contact frames both of whose origin is at the contact point. One frame is attached to the object and the other frame is attached to the finger. To write the components of motion that are transferred between the finger and the object, we have to consider the nature of the contact. Three kinds of contact are point contact without friction, point contact with friction or hard finger (HF) contact, and soft finger contact. We consider the point contact with friction (HF) contact model. A HF contact transmits the three linear velocity components but does not transmit the rotational ones. We use a matrix **H**, that we call the selection matrix, to select the transmitted motion components. The selection matrix for HF contact is given by

$$\boldsymbol{H}_i = [\boldsymbol{I}_{3\times 3} \quad \mathbf{0}_{3\times 3}] \in \mathbb{R}^{3\times 6} \tag{11}$$

Selection matrix for all the n_c contacts will be

$$\boldsymbol{H} = Blockdiag(\boldsymbol{H}_1, \boldsymbol{H}_2, \ldots.\boldsymbol{H}_{n_c}) \in \mathbb{R}^{3n_c\times 6n_c} \tag{12}$$

The expression for complete grasp matrix (G) is

$$\boldsymbol{G}^T = \boldsymbol{H}\,\tilde{\boldsymbol{G}}^T \tag{13}$$

To transform the object twist into the contact twist transmitted from the finger to the object ($\boldsymbol{\nu}_{cc,obj}$), we use the transpose of the complete grasp matrix. Since only the translational components of motion get transmitted from the finger to the object $\boldsymbol{\nu}_{cc,obj} \in \mathbb{R}^{3n_c\times 1}$. The transmitted twist obtained in Eq. 14 is expressed in the contact frame.

$$\boldsymbol{\nu}_{cc,obj} = \mathbf{G}^T \boldsymbol{\nu} \tag{14}$$

Equation 14 relates the twists. From this equation we can find the relation between the forces by using the principle of virtual work. Equation 15 relates the external wrench $\mathbf{w} \in \mathbb{R}^{6\times1}$ applied to the object and contact forces $\lambda \in \mathbb{R}^{3n_c\times1}$.

$$-\mathbf{w} = \mathbf{G}\boldsymbol{\lambda} \tag{15}$$

Each contact force has three components, the normal force and the two tangential forces. The contact forces are expressed in the contact frame (C_i). A set of three terms in ˘ represents the components of the contact force along the axes of the contact frame at a particular contact location. The first of the three terms represents the normal contact force. Since a finger can only push the object and not pull it, this contact force should be positive. The other two components are frictional forces that should be within the friction limits. We apply these conditions computationally while checking for force-closure. There are the two conditions that we need to check for force-closure. The first condition is that any wrench can be created on the object and the second is that this wrench should be created for a valid contact force. The first condition translates into the condition on the rank of the grasp matrix (**G**). The grasp matrix should be full rank, that is its rank should be six. The second condition is that grasp matrix should have a nontrivial null space. And the null vectors should be valid contact forces. In other words, we should have $\mathbf{G}$˘ $= 0$ for a valid ˘. And, for the contact force to be valid we should have the normal components of the contact force positive and the tangential components within the friction limits. The friction limits are given by Coulomb's law with a friction coefficient that we have chosen to be $(\mu = 0.6)$.

$$\sqrt{\lambda_{3i-1}^2 + \lambda_{3i}^2} \leq \mu\lambda_{3i-2}, \text{ where } i = 1, 2, 3 \tag{16}$$

The condition for nontrivial null-space on **G** means that the contacts can squeeze the objects arbitrarily hard. Thus, it can also resist any external wrench on the object. All of the conditions above are posed as a linear programming problem. The result of the linear programming problem tells us if a feasible ˘ is possible. It also gives a measure 'd' [6] that is the measure of the 'distance' of the contact forces from violating the friction limits. Therefore, we use 'd' as a measure of the quality of the grasp.

Additionally, we use two other measures for goodness of grasp. One is the minimum eigenvalue and the other is the volume of wrench space measure.

2.1 *Smallest Singular Value*

For a grasp matrix $\mathbf{G} \in \mathbb{R}^{m\times n}$, the singular values of **G** are the n non-negative square roots, σ_i, of the eigenvalues of $\mathbf{G}^T\mathbf{G}$. The smallest singular value gives an estimate of the firmness of the grasp.

The measure of grasp quality is defined as

$$\delta(\mathbf{G}) = \sigma_{\min}(\mathbf{G}) \tag{17}$$

We say that $\mathbf{G}_1 < \mathbf{G}_2$ if and only if $\delta(\mathbf{G}_1) < \delta(\mathbf{G}_2)$. It is noted that although the worst-case analysis of the grasp is given by $\delta(\mathbf{G})$, the stability of the grasp is not ensured by this quality measure. With this measure, different grasps based on their worst performance and the grasp with 'best' worst performance is chosen.

2.2 *Volume in Wrench Space*

Let $\sigma_1, \sigma_2, \ldots, \sigma_6$ be the singular values of $\mathbf{G}$. Then $\beta > 0$ exists such that

$$\nu(\mathbf{G}) = \beta(\sigma_1\sigma_2 \ldots \sigma_6) \tag{18}$$

The volume measure provides a better estimate as it takes into account all the singular values and not just the minimum. It gives a measure of how well the object is bounded on all sides by the finger contacts.

3 Method

We acquire the image, perform image processing on it to get the boundary, and then find out the normal at the boundary point. Figure 2 shows the setup used to acquire the images. To find out the normal we interpolate a few (say 20) points on the boundary with a line. Then we find the direction vector perpendicular to the line. The normal vector at the point on the boundary which is the middle of the selected points is the obtained normal vector.

We use this method to find the normal vector at the candidate grasping points. We select the candidate grasping points as points on the opposing surfaces of the object. This kind of grasping points is appropriate for a two-fingered hand or a parallel plate gripper. In our work we consider the Movemaster robot which has a two jaw gripper. Therefore this selection of contact points is appropriate for our work. Additionally we use two nearby contact points at each surface to simulate a patch contact instead of a point contact. The knowledge of contact locations and contact normals are used to find the grasp matrix (G). The flow of steps is shown in Fig. 3.

We use everyday objects for our experiments. To simplify the perception problem, we have chosen dark colored objects kept in a white background. This enables us to easily find out the boundary of the object and further find out the normal vectors.

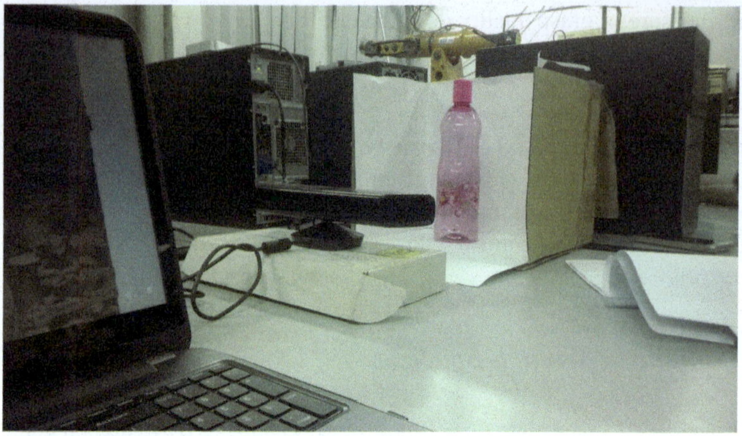

Fig. 2 Setup for acquiring the image of the objects. Kinect is used to acquire the images

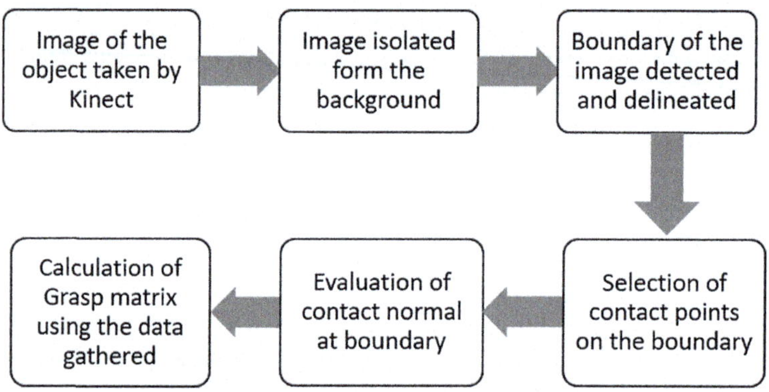

Fig. 3 Evaluating the grasp matrix of an object

3.1 The Movemaster Robot

Movemaster II RM-501 model manufactured by Mitsubishi Electric Corporation is a five degrees of freedom, vertical multi-joint type manipulator arm. It has five revolute joints actuated by DC servomotors rated DC 24 V and 15 W output. The resolution of the joints coordinates at the waist, shoulder and elbow coordinates is 0.025, and of the wrist is 0.075. Thus the movement of every degree of freedom is set by an integer representing a unit of the coordinate resolution. Feedback on every degree of freedom of the robot is obtained by means of optical incremental encoders. It has a maximum synthesis speed of 400 mm/s (wrist tool surface) and gives a position repeat accuracy of 0.5 mm. Permissible handling weight is limited to 1.2 kg, including that of the gripper.

Movemaster is commanded from a program written the National Instrument's LabView interface. In our work we keep the object at a known location with respect to the robot and compute the joint angles to reach to the object from the kinematics of the robot. We have used the robot to verify that the grasp planning is effective. The use of the robot can be integrated with object location detection and automatic reach and grasp action.

4 Results

Different household objects were chosen and grasp synthesis was done on their images. We present here two representative objects. The aim is to compare and contrast the parameters used to define the quality of the grasp. Two contact points are selected sequentially from top to bottom in the image of the object. The points selected are such that they lie on a horizontal line, i.e., it is presumed that the grasp attempt will be made in a select fashion. The parameters are then graphically represented with the image of the object.

The MATLAB code to determine the optimal grasp using the three measures was run for the two objects. At this stage in the project relatively simpler shapes are tested. Figure 4 depicts two of the objects.

Following is the grasp analysis of the objects 1 (bottle) and 2 (apple). For every object, position of all the grasps is depicted by marking the contact points and the normal to the object surface on its processed image followed by the plots of the value of grasping parameters corresponding to each grasp.

The observation (Fig. 5) is that the three parameters do not give the same result, though it can be commented that the conclusions reached by the three measures are not very different.

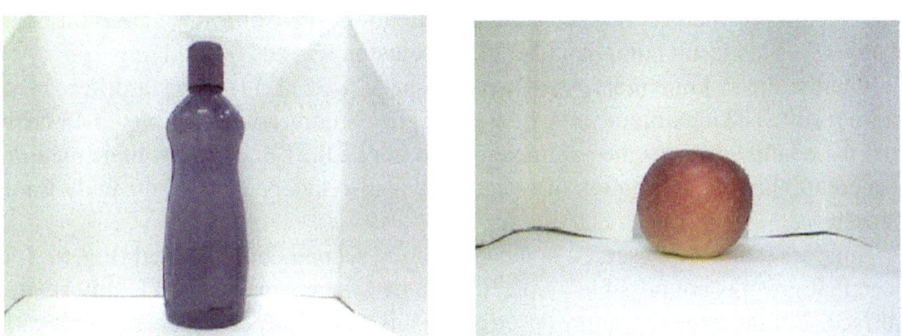

(a) Bottle (b) Apple

Fig. 4 The objects used for grasp analysis

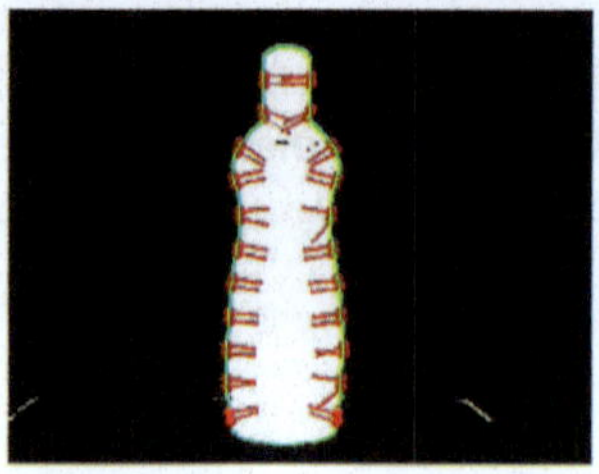

(a) Object with all grasps marked

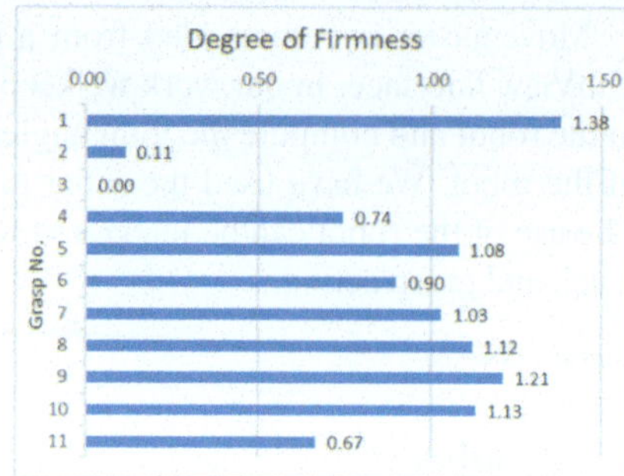

(b) Degree of Firmness, d

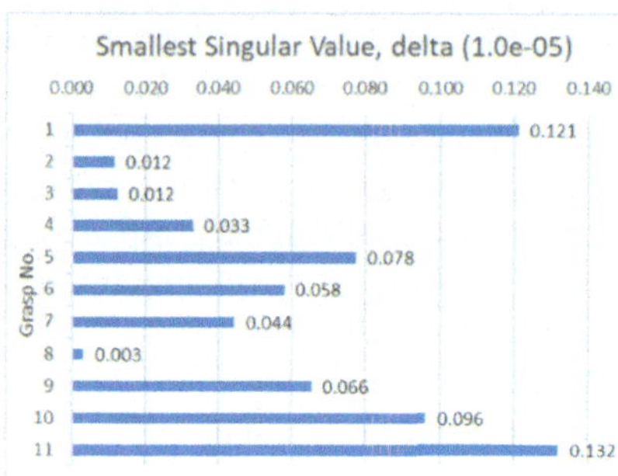

(c) Smallest Singular Value, δ

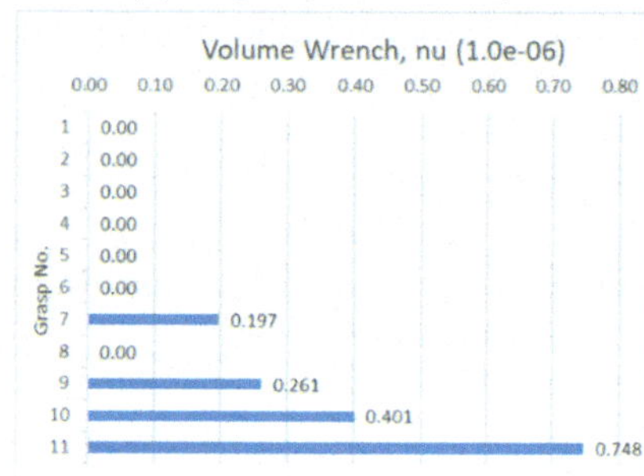

(d) Volume of Wrench, ν

Fig. 5 Analysis of grasps on the first object. The contact locations are numbered from top to bottom

The first measure, d (Fig. 5b), utilizes optimization of the wrench applied by the hand to maintain equilibrium. It is at a value of zero at the second and last grasps. The value of d is dependent on the choice of the origin of the body frame.

The smallest singular value, δ (Fig. 5c), gives a worst-case analysis of the grasp for which it is evaluated. It is seen that the maximum value of δ is achieved for grasp number 5. This result complies with the conclusion reached by studying d.

Finally, the volume measure of wrench space, ν (Fig. 5d), gives a different but close result. The maximum value of ν is for grasp number 4. It is to be remembered that the results given by the parameters need not be the same. The volume measure is a comprehensive evaluation of a grasp and is also independent of the body frame chosen.

Similar results are observed in the case of the second object as well (Fig. 6). It is seen that grasp is not possible by placing the two fingers/grippers on highly slanted surface. Although the procedure described in this report only considers the projection of an object and not the actual 3D geometry, it provides good results.

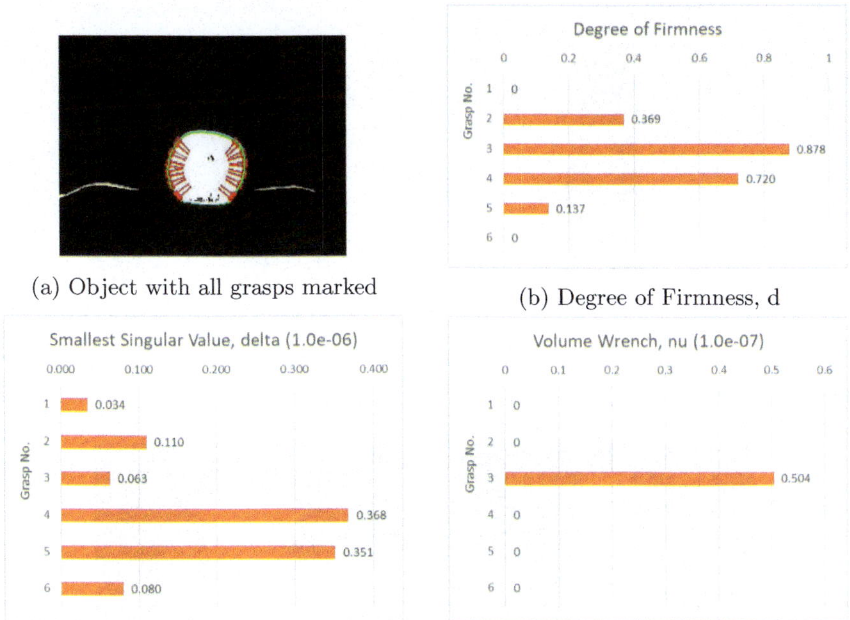

(a) Object with all grasps marked

(b) Degree of Firmness, d

(c) Smallest Singular Value, δ

10

(d) Volume of Wrench, ν

Fig. 6 Analysis of grasps on the second object. The points are numbered from top to bottom

4.1 Hardware Implementation

We perform the hardware implementation on an object to test if the grasp plans generated by us are effective or not. The test object is grasped in (a) upright and (b) upside-down orientations. The analysis and results follow.

As can be seen from Figs. 7, 8, and 9, the measure, degree of firmness, is maximum at the straight edges and reduces at the slant edges, which is expected. The grasps below are made at various of these locations. The control method for grasp execution is open loop. The built in gripper close command is used for this purpose. The object is shown grasped in two orientations—one upside-down of the other. This is to show the possibility of bringing in the contribution of gravity in the grasping process. It is intuitive to see that the case (b) of Fig. 8 will be easier to grasp under gravity (Fig. 10).

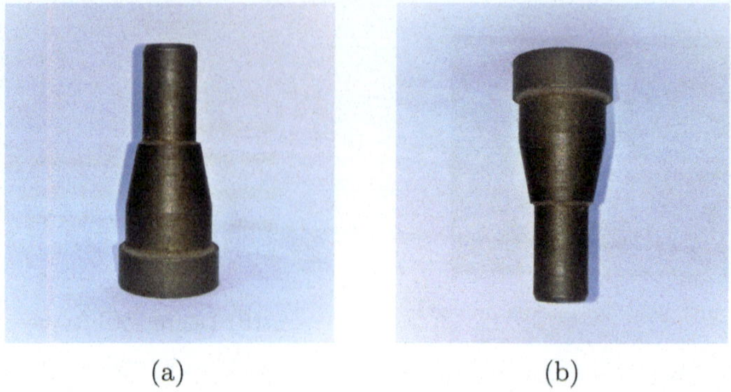

Fig. 7 Test object

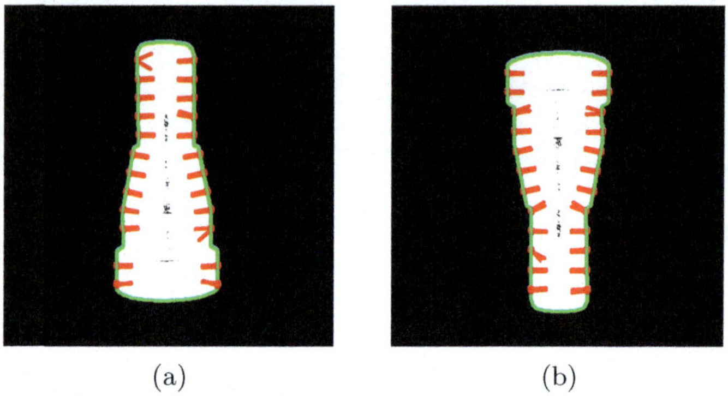

Fig. 8 Grasp analysis on test object

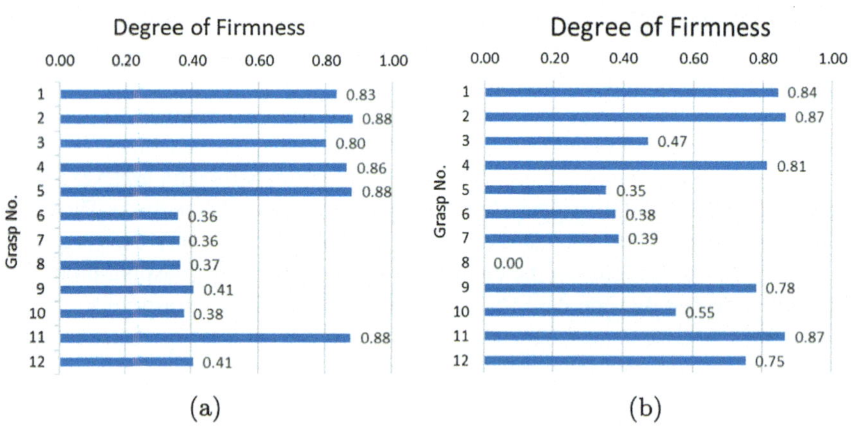

Fig. 9 Degree of firmness of grasps on the object

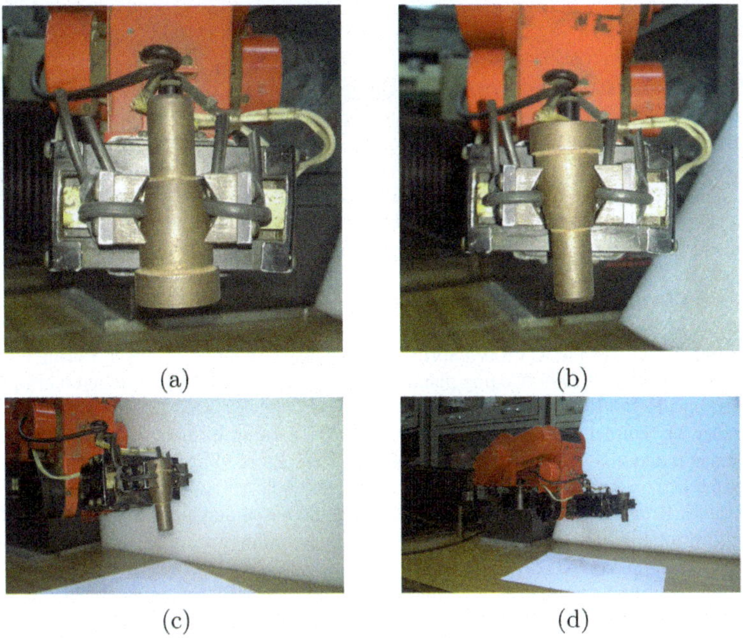

Fig. 10 Physical execution of grasp

5 Conclusions

We have done grasp synthesis for objects using vision data. We detect the normal at the boundary of the object and find out the grasp matrix corresponding to a contact configuration. Grasp matrix has been utilized to determine the feasibility of the grasp. We have evaluated three measures to determine the goodness of the grasp. The first measure, d, tells us the feasibility as well as the goodness while the smallest singular value (δ) and volume of wrench space (ν) give us the worst-case analysis and goodness of the envelope, respectively. For the two representative objects, we evaluated these measures and found that the feasibility of the grasp shown by the measure, d, can be verified by intuition. The second and third measure suggest the best grasp according to the worst-case analysis and enveloping analysis. We have tested the feasibility of the grasp on an object by lifting it with a robot having a parallel plate gripper.

References

1. Levine S, Pastor P, Krizhevsky A, Ibarz J, Quillen D (2017) Learning hand-eye coordination for robotic grasping with deep learning and large-scale data collection. Int J Robot Res 37:421–436
2. Correll N, Bekris KE, Berenson D, Brock O, Causo A, Hauser K, Okada K, Rodriguez A, Romano JM, Wurman PR (2018) Analysis and observations from the first amazon picking challenge. IEEE Trans Autom Sci Eng 15:172–188
3. Salisbury JK, Craig JJ (1982) Articulated hands. Int J Robot Res 1:4–17
4. Salisbury JK, Roth B (1983) Kinematic and force analysis of articulated mechanical hands. J Mech Transm Autom Des 105(1):35–41
5. Murray RM, Li Z, Sastry SS, Sastry SS (1994) A mathematical introduction to robotic manipulation. CRC Press
6. Prattichizzo D, Trinkle JC (2008) Springer handbook of robotics. Springer, Berlin, p 671–700
7. Rimon E, Blake A (1999) Caging planar bodies by one-parameter two-fingered gripping systems. Int J Robot Res 18(3):299–318
8. Fischer M, van der Smagt P, Hirzinger G (1998) Learning techniques in a dataglove based telemanipulation system for the dlr hand. In: 1998 Proceedings of the IEEE International Conference on Robotics and Automation, vol 2, pp 1603–1608, IEEE
9. Ekvall S, Kragic D (2004) Interactive grasp learning based on human demonstration. In: Proceedings of the ICRA'04. 2004 IEEE international conference on robotics and automation, 2004, vol 4, pp 3519–3524, IEEE
10. Aleotti J, Caselli S (2008) Programming task-oriented grasps by demonstration in virtual reality. In: Proceedings of IEEE/RSJ international conference on intelligent robots and systems, WS on grasp and task learning by imitation
11. Li Z, Sastry SS (1988) Task-oriented optimal grasping by multifingered robot hands. IEEE J Robot Autom 4(1):32–44

Forward Kinematics of Cable-Driven Continuum Robot Using Optimization Method

K. P. Ashwin and Ashitava Ghosal

Abstract An elephant trunk robot is a continuum robot consisting of a flexible backbone actuated by two pairs of cables (or tendons) offset by a distance and along the circumference of the backbone. By pulling the cables, the continuum robot assumes the shape of an arc of a circle in 3D space. In the literature, forward kinematics of the robot has been developed using differential geometry of 3D curves. In this paper, we show that forward kinematics can also be obtained by discretizing along the length of the robot, with each discrete element modeled as a four-bar mechanism, and using a minimization approach. Each of the four-bar mechanism is assumed to be rigid with constant link length. It is shown that minimizing the angle made by the coupler link with the fixed link of the parallel linkage results in a profile which is numerically same as that derived from differential geometry for a segment of the cable-driven robot in 2D. The results obtained for actuation in 2D are extended to 3D, using two four-bar mechanisms actuated by the two sets of cables, and it is shown to match the analytical formulation available in the literature. The proposed method of using a sequence of four-bar mechanisms opens up a new perspective in modeling the forward kinematics of cable-driven continuum robots.

Keywords Elephant trunk robot · Flexible robots · Four-bar mechanism · 2D and 3D motion

1 Introduction

Continuum robots are generally serial robots which use flexible links instead of rigid links. Since the degree of freedom of such robots are infinite, they possess all the advantages of a hyper-redundant robot. A detailed review on the design, history,

K. P. Ashwin · A. Ghosal (✉)
Indian Institute of Science, Bengaluru, India
e-mail: asitava@iisc.ac.in

K. P. Ashwin
e-mail: ashwinkp@iisc.ac.in

D. Sen et al. (eds.), *Mechanism and Machine Science*, Lecture Notes in Mechanical Engineering, https://doi.org/10.1007/978-981-15-4477-4_27

Fig. 1 Elephant trunk robot [3]

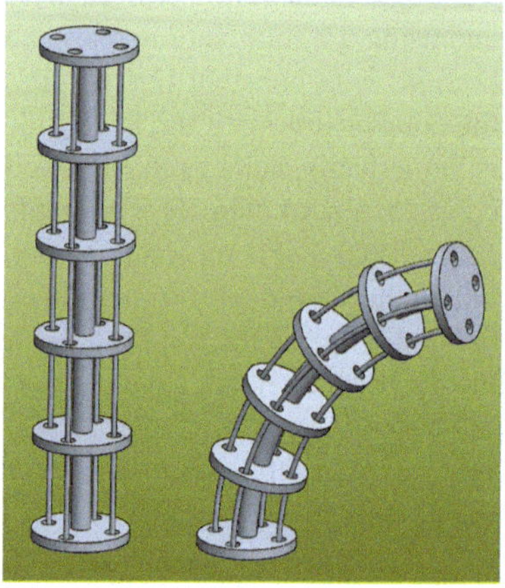

Fig. 2 Elephant trunk robot configuration

and applications of continuum robots can be found in [1, 2]. One of the earliest continuum robot developed was the Rice/Clemson 'elephant-trunk' robot shown in [3] (refer Fig. 1). The robot which resembles a spinal column structure consists of a flexible backbone whose pose can be adjusted by pulling a set of cables attached to it (refer Fig. 2). The robot is made of multiple segments and each segment can be actuated using a set of four cables which are placed 90° apart from each other. Guide disk attached at regular intervals on the backbone routes the cables (tendons) from the base of the robot till the tip of the segment so that the separation between the cables are maintained. Detailed analysis of the workspace of the robot can be found in Ref. [4]. Many advanced flexible robots are modifications of this original concept, with difference mostly in the actuation mechanism (see [5, 6]).

The forward kinematics of the cable-driven continuum robot is to find the pose of the robot when the lengths of cables are given. In Ref. [7] Gravagne and Walker derived analytical expressions for the forward kinematics of elephant trunk robot using concepts from differential geometry. The co-ordinates of a point on the robot is determined using two parameters—α which is angle of rotation of the plane that contains the final configuration of the robot and about the initial axis of the robot[1] and the angle $\beta(s)$ made by the tangent of backbone with the initial axis (which varies along the length of the deformed backbone). In this paper, we show that the forward kinematics solutions derived in the literature can also be obtained by solving an optimization problem. While extending the differential geometry-based analytical methods to solve the model could be nontrivial for uneven spacing between the cables and for robots with general cable routing, the method proposed in this paper could provide a simpler alternative for tackling such complexities. In Sect. 2, the forward kinematics model for single cable actuation and the resulting configuration in 2D plane is discussed. Extension of the method to out-of-plane bending of the actuator is shown in Sect. 3. Finally, conclusions are presented in Sect. 4.

2 Forward Kinematics Model for 2D Continuum Robot

In a cable-driven continuum robot with two pairs of cables, if the two opposing cables in one of the pairs (cables which are 180° apart) are actuated, the robot will be restricted to move within a 2D plane. Among the opposing cable duo, only one cable is pulled at a time, since otherwise, the robot can buckle. In this section, the formulation of forward kinematics of 2D planar continuum robot is discussed.

The proposed formulation assumes the profile of the continuum robot as a set of connected 4-bar parallel linkages. To this end, the robot is discretized into n number of segments along the backbone of the continuum robot as shown in Fig. 3.

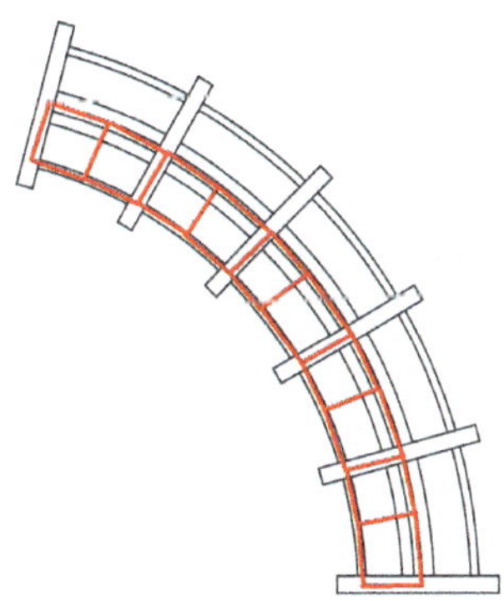
Fig. 3 Discretization of robot in 2D plane

[1]The configuration of the robot will always be in a plane and is shown to take the shape of an arc of a circle.

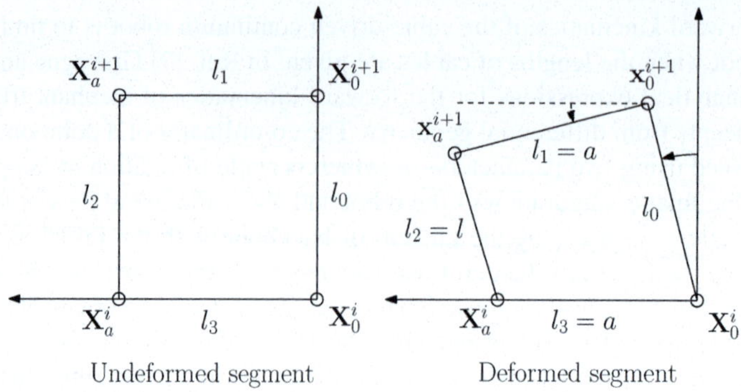

Fig. 4 Nomenclature for single segment

In this approach, only one cable from the opposing cable pair is considered, as it can be seen in the later part of the section that the other cable does not have any role in manipulation of the robot in the quadrants containing the actuated cable.[2] With reference to Fig. 4, one segment of the continuum robot can be approximated as a 4-bar linkage, with the first crank, coupler, second crank, and fixed link having lengths l_0, l_1, l_2, and l_3, respectively. Since the cables pass through holes present in the guide disks, the lengths $l_1 = l_3 = a$ where a is the constant spacing between the backbone and the cables. For the robot of length L_0 measured along the backbone, the length of the first crank $l_0 = \frac{L_0}{n}$. When the cables are actuated to a final length of $L_a = L_0 - \Delta L_a$ where ΔL_a is the prescribed change in length of the cable, the length of the second crank $l_2 = l = \frac{L_a}{n}$. From the loop-closure constraint equations of the 4-bar linkage located at the base of the continuum robot, we get

$$f(\theta, \beta) = l_0^2 + 2a^2 - l^2 + 2al_0 \sin(\theta - \beta) - 2a\,(l_0 \sin\theta + a\cos\beta) = 0 \quad (1)$$

where θ is the angle made by the first crank and the axis of backbone while β is the coupler angle measured relative to the fixed link. Plotting the values of θ and β, we can see from Fig. 5 that a minimum exists for $\beta - \theta$ curve for a given set of length parameters.

Differentiating the above equation with respect to θ and setting $\frac{d\beta}{d\theta} = 0$ and simplifying, we get

$$\cos(\theta - \beta) = \cos\theta$$

From the above equation, we can get the value of crank angle (and subsequently a configuration of the linkage) where the rate of change of coupler angle with respect to the crank angle is zero. This is given by

[2]The opposing cable is used to retract the robot to its original position if the backbone material is not perfectly elastic, and for manipulation in the opposing direction of the first cable.

Fig. 5 β versus θ

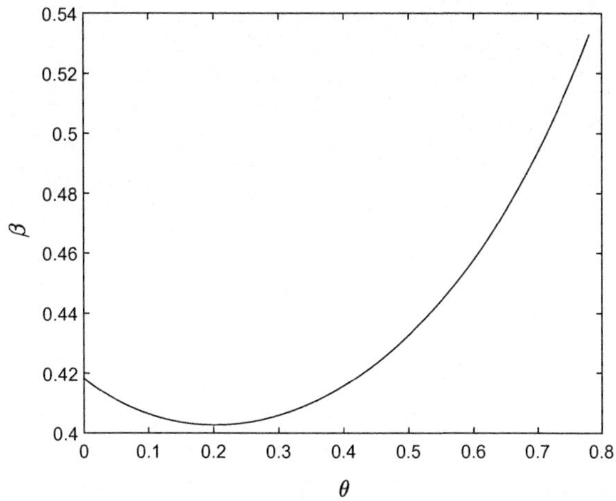

$$\beta = 2\left(\theta - k\pi\right) \tag{2}$$

where k is an integer.

In the initial un-actuated configuration, the lengths of link are $l_0 = l_2 = \frac{L_0}{n}$. In this position, the initial coupler angle is zero. After actuation, the nearest minimum for the coupler angle (from zero value) appears when $k = 0$, or $\beta = 2\theta$ for small displacements of the first crank. In what follows, it is shown that the configuration which produces minimum change of coupler angle is the same as the configuration obtained analytically by Gravagne and Walker [7].

Substituting $\theta = \frac{\beta}{2}$ in Eq. (1) and simplifying, we get

$$-l^2 + \left(l_0 - 2a\sin\left(\frac{\beta}{2}\right)\right)^2 = 0 \tag{3}$$

Since l takes only positive values, we get

$$\sin\frac{\beta}{2} = \frac{l_0 - l}{2a} \tag{4}$$

The assembled discrete segments represent the robotic continuum for large values of n. In that case, the right-hand side of the above equation will be very small. For small angles, we get

$$\beta = \frac{l_0 - l}{a} \tag{5}$$

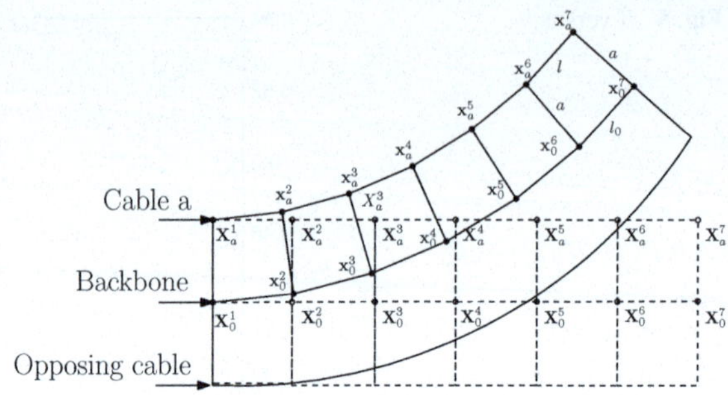

Fig. 6 Assembly of 4-bar linkages for iteration procedure

The value of coupler angle for the distal tip, which is same as the angle $\beta(L_0)$ in the derivation shown in [7], can be found by multiplying the above expression for β with the total number of segments n. The value of coupler angle so obtained,

$$\beta(L_0) = n\frac{l_0 - l}{a} = \frac{\Delta L_a}{a} \tag{6}$$

is same as the value obtained analytically in [7].

Since it has been shown that the forward kinematics of continuum robot can be obtained by minimizing the coupler angle of approximated 4-bar parallel linkages, the following formulation may be used to implement the optimization procedure.

Let the co-ordinates of the ends of the coupler attached to the first crank and the second crank in the undeformed position be denoted by $\mathbf{X}_0^{i+1}$ and $\mathbf{X}_a^{i+1}$, respectively (see Fig. 6). The corresponding deformed positions are denoted by $\mathbf{x}_0^{i+1}$ and $\mathbf{x}_a^{i+1}$, respectively. Similarly, the co-ordinates of the ends of the fixed link attached to the first crank and the second crank are $\mathbf{X}_0^{i}$ and $\mathbf{X}_a^{i}$, respectively. For the segment at the base of robot $i = 1$ and $i = n$ at the tip. With the above, the optimization problem may be formulated as

$$\underset{\mathbf{x}_b^{i+1},\mathbf{x}_a^{i+1}}{\arg\min} \quad \arccos\left(\left(\frac{\mathbf{X}_0^i - \mathbf{X}_a^i}{\|\mathbf{X}_0^i - \mathbf{X}_a^i\|}\right)\cdot\left(\frac{\mathbf{x}_0^{i+1} - \mathbf{x}_a^{i+1}}{\|\mathbf{x}_0^{i+1} - \mathbf{x}_a^{i+1}\|}\right)\right)$$

Subject to

$$\begin{aligned} \|\mathbf{x}_0^{i+1} - \mathbf{X}_0^i\| &= l_0 \\ \|\mathbf{x}_a^{i+1} - \mathbf{X}_a^i\| &= l \\ \|\mathbf{x}_0^{i+1} - \mathbf{x}_a^{i+1}\| &= a \end{aligned} \tag{7}$$

Given data: $\mathbf{X}_0^i$, $\mathbf{X}_0^{i+1}$, $\mathbf{X}_a^i$, $\mathbf{X}_a^{i+1}$, l_0, l, a.

The solution to the above optimization problem gives the co-ordinates of tips $\mathbf{x}_0^{i+1}$ and $\mathbf{x}_a^{i+1}$ and the iterative method starts from the base segment and proceeds towards the tip of the robot. With reference to Eq. (4), in order to ensure that the right-hand side of the equation is always less than 1, the value of n may be chosen such that

$$n > \frac{L_0 - L_a}{2a} \tag{8}$$

Figure 7 shows the profile plotted for a continuum robot using the solution obtained from analytical method as well as the discretized optimization method solved using *fmincon* in Matlab. The solution, as proved before, is exact except for the tolerances in the numerical procedure. Figure 8 shows that the profile remains similar with different number of segments.

In the next section, we extend the approach to 3D continuum robots.

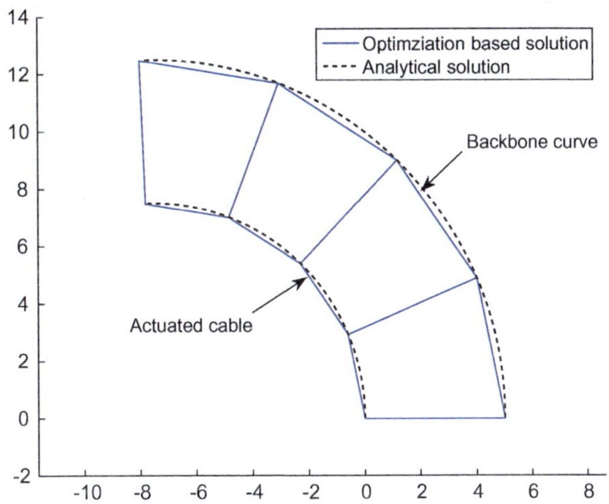

Fig. 7 Comparison between analytical and optimization-based solutions

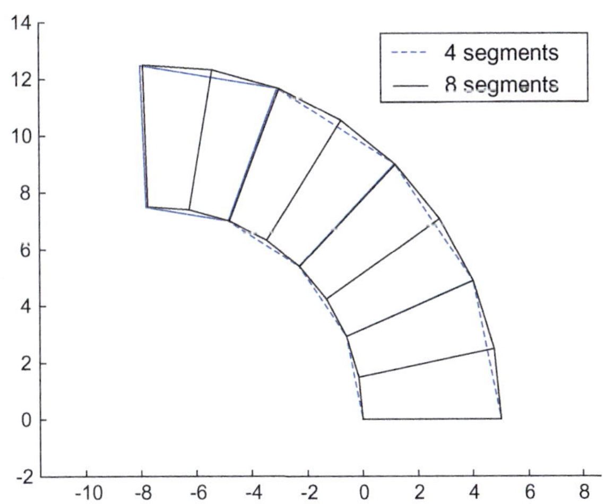

Fig. 8 Solution with increased number of segments

3 Forward Kinematics Model for Continuum Robot in 3D

If both pairs of cables are actuated simultaneously, the continuum robot will assume a deformed pose in 3D space. For simultaneous actuation, the discretization of robot is carried out as shown in Fig. 9. Here, the backbone along with the actuating cables which are placed at an angle $\phi = 90°$ apart can be considered as a set of two adjoined 4-bar linkages (a 7-bar linkage as shown in Fig. 10). The two quadrilaterals have the vertices $(\mathbf{X}_0^i, \mathbf{X}_0^{i+1}, \mathbf{X}_a^{i+1}, \mathbf{X}_a^i)$ and $(\mathbf{X}_0^i, \mathbf{X}_0^{i+1}, \mathbf{X}_b^{i+1}, \mathbf{X}_b^i)$ initially. These quantities change to $(\mathbf{X}_0^i, \mathbf{x}_0^{i+1}, \mathbf{x}_a^{i+1}, \mathbf{X}_a^i)$ and $(\mathbf{X}_0^i, \mathbf{x}_0^{i+1}, \mathbf{x}_b^{i+1}, \mathbf{X}_b^i)$ after actuation. The subscripts a and b are used to denote the respective values for the linkages formed by cable pairs a and b, respectively. The coupler angles may be defined as β_a and β_b for the two 4-bar linkages.

Since the deformation of the robot is commutative—i.e., the actuation is the same irrespective of the order of actuating the cables—the final pose will be the one where coupler angles of both the linkages are minimized simultaneously and independently. Hence, the pose can be determined from solving the optimization problem

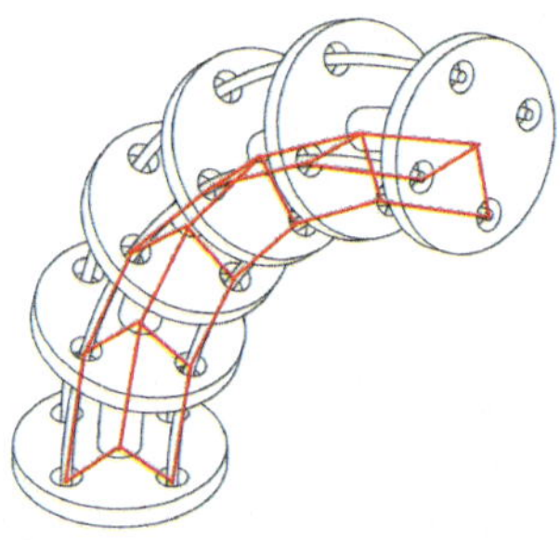

Fig. 9 Discretization of robot in 3D

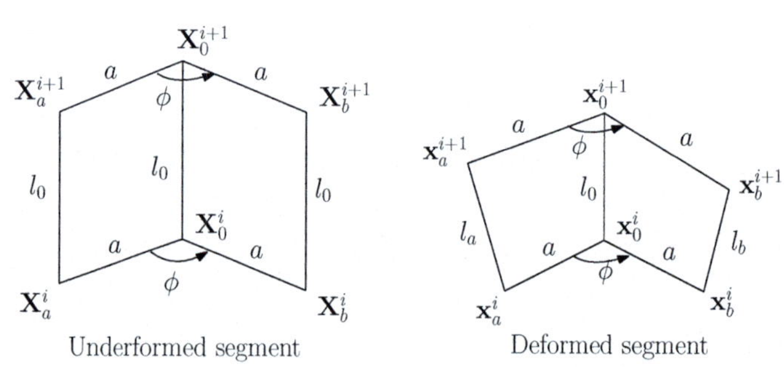

Fig. 10 Nomenclature of segment in 3D

$$\underset{\mathbf{x}_b^{i+1},\mathbf{x}_a^{i+1}}{\arg\min} \left[\arccos\left(\left(\frac{\mathbf{X}_0^i-\mathbf{X}_a^i}{\|\mathbf{X}_0^i-\mathbf{X}_a^i\|}\right)\cdot\left(\frac{\mathbf{x}_0^{i+1}-\mathbf{x}_a^{i+1}}{\|\mathbf{x}_0^{i+1}-\mathbf{x}_a^{i+1}\|}\right)\right)\right]^2 + \left[\arccos\left(\left(\frac{\mathbf{X}_0^i-\mathbf{X}_b^i}{\|\mathbf{X}_0^i-\mathbf{X}_b^i\|}\right)\cdot\left(\frac{\mathbf{x}_0^{i+1}-\mathbf{x}_b^{i+1}}{\|\mathbf{x}_0^{i+1}-\mathbf{x}_b^{i+1}\|}\right)\right)\right]^2 \tag{9}$$

Subject to

$$\begin{gathered}
\|\mathbf{x}_0^{i+1}-\mathbf{X}_0^i\| = l_0 \\
\|\mathbf{x}_a^{i+1}-\mathbf{X}_a^i\| = l_a \\
\|\mathbf{x}_b^{i+1}-\mathbf{X}_b^i\| = l_b \\
\|\mathbf{x}_0^{i+1}-\mathbf{x}_a^{i+1}\| = a \\
\|\mathbf{x}_0^{i+1}-\mathbf{x}_b^{i+1}\| = a \\
\arccos\left(\left(\frac{\mathbf{x}_0^i-\mathbf{x}_a^i}{\|\mathbf{x}_0^i-\mathbf{x}_a^i\|}\right)\cdot\left(\frac{\mathbf{x}_0^i-\mathbf{x}_b^i}{\|\mathbf{x}_0^i-\mathbf{x}_b^i\|}\right)\right) - \arccos\left(\left(\frac{\mathbf{X}_0^i-\mathbf{X}_a^i}{\|\mathbf{X}_0^i-\mathbf{X}_a^i\|}\right)\cdot\left(\frac{\mathbf{X}_0^i-\mathbf{X}_b^i}{\|\mathbf{X}_0^i-\mathbf{X}_b^i\|}\right)\right) = 0
\end{gathered} \tag{10}$$

Given data: $\mathbf{X}_0^i, \mathbf{X}_0^{i+1}, \mathbf{X}_a^i, \mathbf{X}_a^{i+1}, \mathbf{X}_b^i, \mathbf{X}_b^{i+1}, l_0, l_a, l_b, a$

where $l_a = \dfrac{L_a}{n}$, $l_b = \dfrac{L_b}{n}$ and the last equality constraint ensures that the separation between the cables remains constant even after deformation.

In Figs. 11 and 12, the pose of the robot with two cables actuated by equal amounts as well as by different values are shown. As is the case with the 2D, analytical solution and optimization-based solution varies only by the numerical error induced due to the optimization scheme.

To summarize, we have shown that the forward kinematics of a cable-driven continuum robot, with a fixed tendon spacing, can be calculated using an optimization-based approach. The actuating cable of the continuum robot as well as the backbone are discretized to a number of segments and the two are considered as the two cranks of a 4-bar linkage. The constant spacing between the cable and the backbone separates the cranks and forms the fixed link as well as the coupler. Minimizing the coupler angle from the stationary initial position based on the constraints imposed by the cable lengths, backbone length as well as the constant spacing between the cable and backbone, a unique pose is obtained for the robot. This unique pose is shown to match the solution obtained using differential geometric approach [7].

It is worth noting that the approach is applicable only if the given lengths are constant. For a robot which does not have guiding disks to keep a constant cable spacing, a will vary and this approach cannot be used in the present form. For the case where the cables are passed through guiding disks with holes drilled with different spacings, the cables will not drift in circumferential direction. In reference [8], forward kinematics of such a system with general tendon routing is posed using a set of differential equations based on Cosserat theory. However, the expressions are non-intuitive and are applicable to tendon routes that can be analytically expressed. In our optimization-based method proposed in this paper, the same would translate to adding specific constraints to the problem, such as:

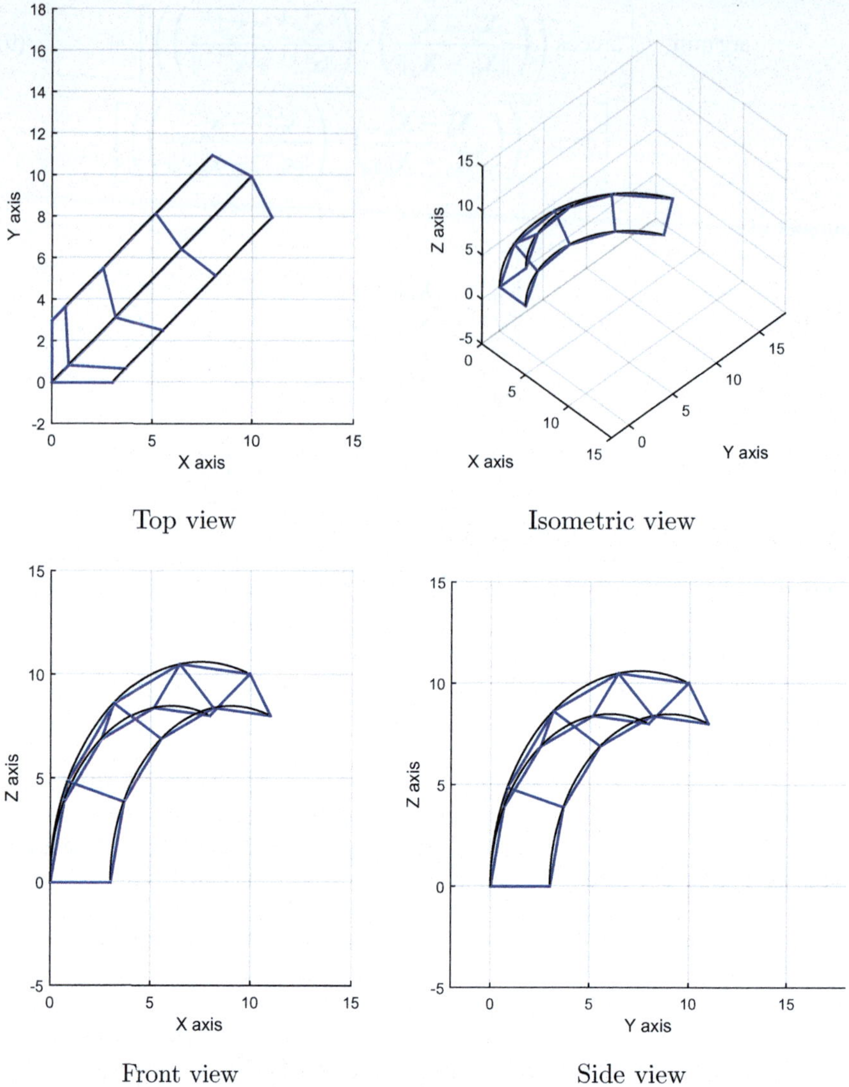

Fig. 11 Profile of robot with cables actuated by equal displacements (black lines for analytical solution and blue lines for optimization-based solution)

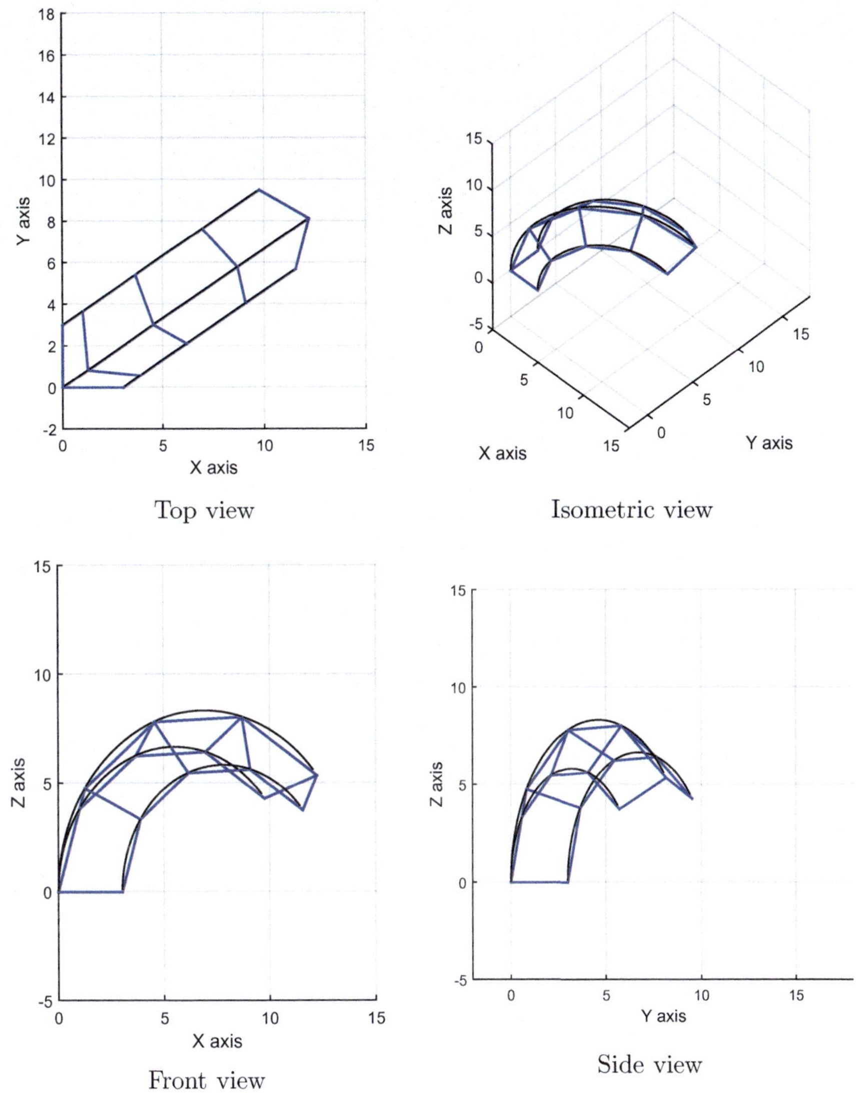

Fig. 12 Profile of robot with cables actuated by different displacements (black lines for analytical solution and blue lines for optimization-based solution)

$$\|\mathbf{x}_0^{i+1} - \mathbf{x}_a^{i+1}\|_j = a_j \quad (11)$$

$$\|\mathbf{x}_0^{i+1} - \mathbf{x}_b^{i+1}\|_j = a_j \quad (12)$$

$$\left\{\arccos\left(\left(\frac{\mathbf{x}_0^i - \mathbf{x}_a^i}{\|\mathbf{x}_0^i - \mathbf{x}_a^i\|}\right)\cdot\left(\frac{\mathbf{x}_0^i - \mathbf{x}_b^i}{\|\mathbf{x}_0^i - \mathbf{x}_b^i\|}\right)\right) - \right. \quad (13)$$

$$\left. \arccos\left(\left(\frac{\mathbf{X}_0^i - \mathbf{X}_a^i}{\|\mathbf{X}_0^i - \mathbf{X}_a^i\|}\right)\cdot\left(\frac{\mathbf{X}_0^i - \mathbf{X}_b^i}{\|\mathbf{X}_0^i - \mathbf{X}_b^i\|}\right)\right)\right\}_j = \phi_j \quad (14)$$

where $(\cdot)_j$ represent the specified values at segment index j. The above optimization problem can again be solved numerically using *fmincon* provided by Matlab.

4 Conclusions

The forward kinematics of cable-driven continuum robots such as the well-known Rice/Clemson elephant trunk has been analytically solved by Gravagne and Walker based on differential geometry. It is shown in this paper that the same can also be achieved using optimization methods. The continuum robot is discretized into segments with each segment represented as a 4-bar parallelogram linkage for planar manipulation. From the initial horizontal position of the coupler, a unique pose of the 4-bar linkage is obtained by minimizing the angle made by a coupler with the fixed link, while maintaining the constraints of link lengths. When the iterative process is continued for the entire robot, the resulting pose is shown to be the same as the one derived using the traditional differential geometric methods. A minimum number of segments required for the procedure is also suggested based on the derivation and the method shows consistency with varying number of segments. The concept is directly extended to 3D using a 7-bar linkage or two adjoined 4-bar linkages and with an extra constraint to maintain the constant angle between the cables. The numerical procedure and analytical solution are shown to match each other as in the 2D case.

The method suggested in this paper suggests a new perspective in modeling the forward kinematics of cable-driven continuum robots. The proposed method is easy to apply to robots having uneven separation between the cables and also for robots with generalized cable routing—for which analytical methods could be a daunting task. Research work exploring these areas is ongoing.

References

1. Walker ID (2013) Continuous backbone 'continuum' robot manipulators. ISRN Robot 2013:19
2. Robinson G, Davies JBC (1999) Continuum robots—a state of the art. In: Proceedings of the 1999 IEEE international conference on robotics and automation, pp 2849–2854
3. Walker ID, Hannan MW (1999) A novel 'elephant's trunk' robot. In: Proceedings of IEEE/ASME international conference on advanced intelligent mechatronics, pp 410–415

4. Cao K, Kang R, Wang J, Song Z, Dai J (2015) Kinematic model and workspace analysis of tendon-driven continuum robots. In: Proceedings of 14th IFToMM world congress, pp 640–644
5. Festo Didactic Inc., 607 Industrial Way West, Eaton-580 town N.J. 07724. www.festo.com. Accessed 23 June 2018
6. Xu K, Simaan N (2008) An investigation of the intrinsic force sensing capabilities of continuum robots. IEEE Trans Robot 24(3):576–587
7. Gravagne IA, Walker ID (2000) On the kinematics of remotely-actuated continuum robots. In: Proceedings of IEEE international conference on robotics & automation, pp 2544–2550 (2000)
8. Rucker DC, Webster RJ III (2011) Statics and dynamics of continuum robots with general tendon routing and external loading. IEEE Trans Robot 27(6):1033–1044

An Efficient Methodology to Determine the Usability of Exoskeleton to Control a Serial Manipulator

K. Arun, Abijith P. Nair, Rajeevlochana G. Chittawadigi, K. Venkatesh, and Srinivas Patnaik

Abstract In the field of robotics, several instances exist when a robot has to be controlled or jogged to move to a required configuration. It can be done using input devices such as a teach pendant, which requires the user to be well experienced in handling it. Another intuitive way to control a physical robot is by using an exoskeleton, a device that can be worn on a human arm which records the motion of the human which can be fed into the robot controller. In this paper, an efficient methodology to determine the usability of the exoskeleton to control a serial manipulator is presented. First, the CAD model of the exoskeleton is used to extract its Denavit–Hartenberg (DH) parameters using an analytical method developed by one of the authors. The obtained DH parameters and the CAD files of the exoskeleton are then imported in the Virtual Robot Module of RoboAnalyzer software, which acts as a server application. The virtual exoskeleton can be moved by giving the motion of the joints and hence acts as an intermediate step to validate whether the joint angles obtained from the sensors of the exoskeleton are correct or not, which otherwise requires trial and error strategy. Once the joint angles are found to be correct and free of noise, they can be used to control virtual and physical robot manipulators. In this paper, ABB IRB140 has been controlled both virtually and physically through an exoskeleton developed at ABB Ability Innovation Center.

K. Arun · A. P. Nair · R. G. Chittawadigi (✉)
Deptartment of Mechanical Engineering, Amrita School of Engineering, Amrita Vishwa Vidyapeetham, Bengaluru, India
e-mail: rg_chittawadigi@blr.amrita.edu

K. Arun
e-mail: arunkvsp@gmail.com

A. P. Nair
e-mail: apn126@gmail.com

K. Venkatesh · S. Patnaik
ABB Ability Innovation Center, Bengaluru, India
e-mail: venkatesh.k@in.abb.com

S. Patnaik
e-mail: srinivas.patnaik@in.abb.com

D. Sen et al. (eds.), *Mechanism and Machine Science*, Lecture Notes in Mechanical Engineering, https://doi.org/10.1007/978-981-15-4477-4_28

Keywords Master-slave manipulator · Exoskeleton · Simulation · Serial robot · DH extraction

1 Introduction

The subject of robotics is relevant in today's engineering curriculum because of the robot's ability to perform tireless and dangerous jobs. A robot is only meaningful when it is meant to relieve a human worker from doing a boring, unpleasant, or a very precise job. It is a machine capable of carrying out a complex series of actions autonomously. Robots can be controlled through an external control device called a controller or the control may be embedded within. A robot controller is a combination of hardware and software to program and controls the motion of a combination of different joints in single or multiple robots to place the robot manipulator in the user-defined position and orientation. Industrial robots usually consist of a jointed arm (multi-linked chain) and an end-effector to manipulate objects. The robots can be further classified into serial robots and parallel robots. Serial robots are the most common industrial robots which are designed as a series of links connected by motor-actuated joints that extend from a fixed base to an end-effector.

These manipulators can be used to perform tasks that are generally defined in the Cartesian space, whereas the joints or actuators of the robot are controlled in the joint space. They can be controlled electronically, using a device called a teach pendant or through the use of switches, joysticks, anthropomorphic harness, or an end-effector grip controlled through an offline-programmed computer. The teach pendant controls the manipulator using buttons which allows the user to increase or decrease the values of the robot joint angles or move the robot in the Cartesian space directly along the principal directions. However, in this method of controlling a robot with a joystick, operator response is limited and there is a high probability of the operator becoming disoriented. This is especially critical when the operator is handling sensitive equipment such as nuclear waste, hazardous chemicals, or during bomb disposal. It also lacks force feedback and does not have configuration feedback.

An alternate method to control a robot intuitively is by using an exoskeleton. An exoskeleton is a wearable device that works in tandem with the user. It is placed on the user's body and acts as amplifiers that augment, reinforce, or restore human performance and can be made out of rigid materials such as metal or carbon fiber, or they can be made entirely out of soft and elastic parts. They can be mobile or fixed/suspended (usually for tele-operation or rehabilitation) and can cover the entire body, just the upper or lower extremities, or even a specific body segment such as the ankle or the hip. This has many applications especially in remote or tele-operation of robots for use in situations deemed harmful for human life. For example, such an exoskeleton can be used to remotely diffuse a bomb. It can also be used to remotely operate a robot to handle dangerous materials in chemical factories. A generic term used for such tasks is master-slave manipulation, where an exoskeleton acts as a master and the robot acts as a slave, which mimics the motion of the master. Several researchers have developed exoskeleton worn on the human arm to control serial robots. While some use only kinematic mapping [1], others also have force-feedback sensors to let the user feel forces felt by the end-effector and thus act in a more intuitive manner [2]. A thorough overview of their applications in industries is given in [3].

A physical exoskeleton is generally connected to a computer to obtain joint angle values from the sensors mounted on the exoskeleton. The values obtained may have noise and jitter due to several reasons such as interference, error in mounting the sensor, etc. [4]. Also, the mounting of the sensor on a joint axis might have a basis shift, i.e., the zero angle corresponding to the DH frames attached to the links of the corresponding joint might be different from that obtained from the joint sensor. These problems should be identified and resolved before the exoskeleton can be used to control a physical robot. An efficient methodology is explained in this paper which acts as a solution to such problems.

A brief overview of robot kinematics is presented in Sect. 2, followed by the extraction of DH parameters from the CAD model of the exoskeleton in Sect. 3. The proposed methodology is explained in Sect. 4 where the extracted DH parameters and a CAD model of the robot are used in the instrumentation of the joint angle values obtained and validate them by mapping motion to an exoskeleton model in RoboAnalyzer, co-developed by one of the authors. Section 5 reports the application of exoskeleton in controlling a virtual and physical robot, followed by the conclusions.

2 Robot Kinematics

Kinematics is the study of the motion of bodies without consideration of the forces or moments that cause their motion or caused due to motion. Robot kinematics is concerned with the relationship between the individual joints of the robot manipulator and the position and the orientation, collectively referred to as robot pose, of the tool or end-effector. It is broadly classified as forward and inverse kinematics.

Forward kinematics is used to determine the pose of the end-effector, given the values for the joint variables of the robot. The joint variables are the angles between the links in the case of revolute or rotational joints, and the link extension in the case of prismatic or sliding joints, respectively. Inverse kinematics, on the other hand, deals with the determination of joint values for a required robot pose. The knowledge of both forward kinematics and inverse kinematics is required for the master-slave manipulation of an exoskeleton and serial robot.

2.1 Denavit and Hartenberg Parameters

The architecture of any serial robot can be represented using the Denavit–Hartenberg (DH) parameters, which is a commonly used convention. A coordinate frame (Frame 1) is attached to the base link (Link 0) of the robot such that its Z-axis (Z_1) is along the first joint axis. The second joint axis, i.e., Z_2, is represented as a line in 3D space. Four parameters are required to reach from Frame 1 to the line, as illustrated in Fig. 1a. The parameters are defined in Table 1.

Once the second joint axis is reached, a frame (Frame 2) is attached at the point physically located on the first link (Link 1). The procedure is continued for the remaining joints from the base to the end-effector link. Hence, using four parameters for each joint, one can reach from the base link to the end-effector link (Link n), by attaching n DH frames, where n refers to the degree-of-freedom (DOF) of the serial robot. Rules on attaching the coordinate frames, referred to as DH frames, are outlined in [5], and are skipped here for the sake of brevity. An example of DH frames attached to the first three links of a robot is shown in Fig. 1b.

Homogenous Transformation Matrix (HTM) is used to represent the position and orientation of a coordinate frame with respect to another coordinate frame. Referring to Fig. 1b and by using the DH parameters convention, the HTM of Frame $i + 1$ with respect to Frame i is obtained as

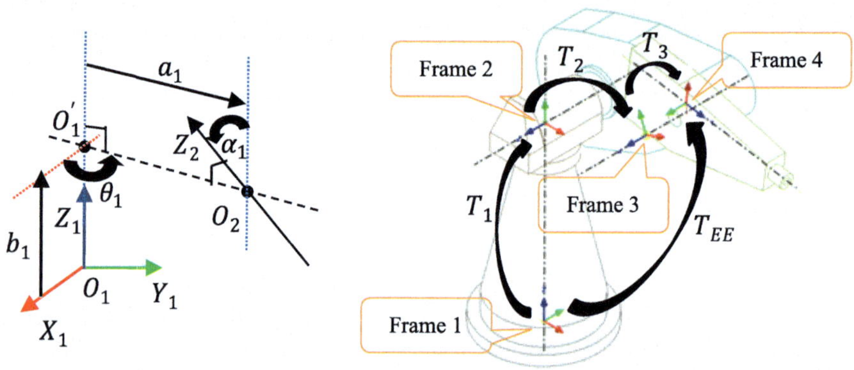

(a) Line represented in a frame (b) DH Frames attached on links

Fig. 1 Denavit–Hartenberg (DH) parameters

Table 1 Definitions of DH parameters

Parameter	Description
Joint offset (b_1)	Linear distance from O_1 to O_1' along Z_1
Joint angle (θ_1)	Angle from X_1 to common normal (dashed line) about Z_1
Link length (a_1)	Linear distance from O_1' to O_2 along common normal
Twist angle (α_1)	Angle from Z_1 (a line parallel to it) and the line (Z_2) about common normal

$$T_i = \begin{bmatrix} \cos\theta_i & -\sin\theta_i\cos\alpha_i & \sin\theta_i\sin\alpha_i & \alpha_i\cos\theta_i \\ \sin\theta_i & \cos\theta_i\cos\alpha_i & -\cos\theta_i\sin\alpha_i & \alpha_i\sin\theta_i \\ 0 & \sin\alpha_i & \cos\alpha_i & b_i \\ 0 & 0 & 0 & 1 \end{bmatrix} \quad (1)$$

The forward kinematics is obtained by multiplication of individual HTMs from the base to the end-effector frames as

$$T_{EE} = T_1 * T_2 * T_3 \ldots * T_n \quad (2)$$

2.2 *Master-Slave Manipulation*

A typical master-slave manipulation is where an exoskeleton acts as a master and a serial robot as a slave. The motion of the tip of the exoskeleton is mapped onto the motion of the end-effector of the robot. The desired motion could be one-to-one mapping of the master motion or it could be scaled up or down to suitable workspace. Various steps involved in the master-slave manipulation are illustrated in Fig. 2.

As evident from the steps in Fig. 2, it is imperative that the DH parameters of both exoskeleton and robot are to be known to perform forward kinematics of master (exoskeleton) and inverse kinematics of slave (robot). These topics are covered in the following sections.

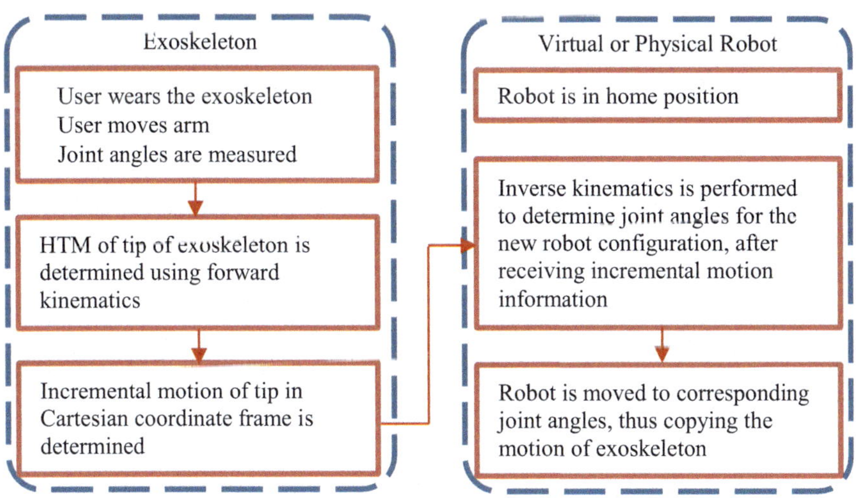

Fig. 2 Steps in the master-slave manipulation scheme

3 Extraction of DH Parameters

The Denavit–Hartenberg (DH) parameters of the exoskeleton have to be known. They can be obtained by drawing the exoskeleton on a paper, locating the joint axes and measuring the angles and distances between the joint axes in the physical prototype. However, these steps are possible only when the physical prototype is available. These steps can also be performed on the 3D CAD assembly of the exoskeleton in CAD software such as Autodesk Inventor. The third author has proposed an analytical method to extract DH parameters from a serial robot in [6]. It involved the creation of an add-in/plugin for Autodesk Inventor which could read the joint axes and determine the parameters programmatically. This add-in has been extensively used by the third author and others to extract the DH parameters of industrial robots from their CAD models and have exported the 3D model into the RoboAnalyzer software. A recent work which explains how DH parameters of a serial robot can be extracted to include in a simulation environment is reported in [7]. Though the proposed work is demonstrated using Autodesk Inventor software, it can also be carried out in other commercial software such as SOLIDWORKS, SpaceClaim, CATIA, etc.

3.1 DH Parameters of Exoskeleton

A physical prototype of an exoskeleton, developed by an earlier set of student interns, was available to make it more functional. It had just the links and joints and did not have encoders. The 3D CAD model of the exoskeleton was then developed in Autodesk Inventor software and the procedure mentioned in [7] was used to extract the DH parameters of the exoskeleton using the CAD model itself. A screenshot of the Autodesk Inventor add-in [6] and the CAD model of the exoskeleton is shown in Fig. 3.

The extracted DH parameters of the exoskeleton are reported in Table 2. Note that these parameters if had to be determined from a physical prototype would be very difficult as measuring the distances and imagining how those dimensions are related requires more experience and understanding of the spatial geometry.

The CAD files of individual links of the exoskeleton are further exported as STL (stereolithography) files, as reported in [7]. The DH parameters along with the STL files are then imported in Virtual Robot Module (VRM) of RoboAnalyzer. The CAD model of exoskeleton plays an important role in the instrumentation of the physical prototype of the exoskeleton as discussed in Sect. 4.

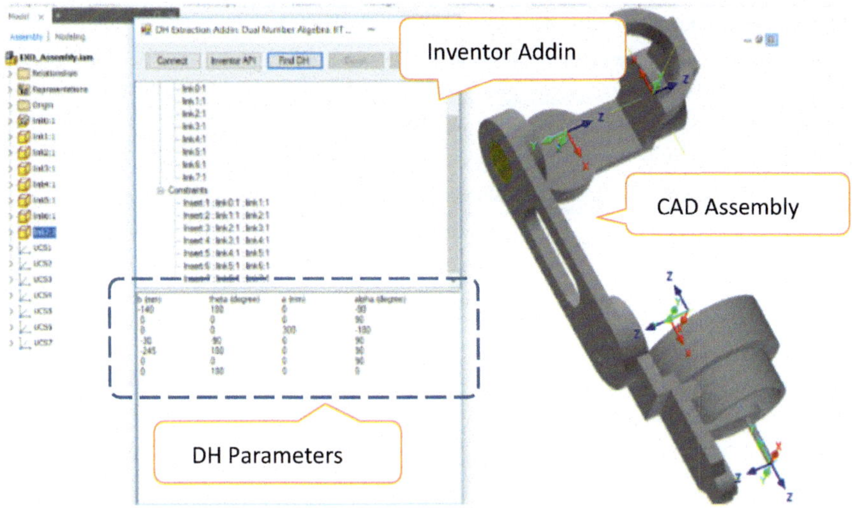

Fig. 3 DH parameters extraction of exoskeleton from the CAD model

Table 2 DH parameters of an exoskeleton

Joint number	Joint offset (b_i) (mm)	Joint angle (θ_i) (degrees)	Link length (a_i) (mm)	Twist angle (α_i) (degrees)
Joint 1	−140	Joint variable	0	−90
Joint 2	0	Joint variable	0	90
Joint 3	0	Joint variable	300	180
Joint 4	−30	Joint variable	0	90
Joint 5	−245	Joint variable	0	90
Joint 6	0	Joint variable	0	90
Joint 7	0	Joint variable	0	0

3.2 *DH Parameters of ABB IRB140*

ABB Innovation Center has ABB IRB140 robot manipulator which the exoskeleton had to control. Since the CAD model of this robot was not already present in the robot library of VRM (that has more than 20 models from various robot manufacturers), the CAD model of the robot was downloaded from CAD repository of ABB's public website [8] and the same procedure explained in the previous section was followed to extract its DH parameters and also prepare its CAD model for VRM in RoboAnalyzer. The extracted DH parameters are reported in Table 3 and exactly match with the important length dimensions on the specification sheet of the robot, given in Fig. 4.

The extracted DH parameters are required for the mapping of the motion between the exoskeleton and the robot, virtual at first and then the physical robot that is

Table 3 DH parameters of ABB IRB140

Joint number	Joint offset (b_i) (mm)	Joint angle (θ_i) (degrees)	Link length (a_i) (mm)	Twist angle (α_i) (degrees)
Joint 1	352	Joint variable	70	90
Joint 2	0	Joint variable	360	0
Joint 3	0	Joint variable	0	−90
Joint 4	380	Joint variable	0	−90
Joint 5	0	Joint variable	0	90
Joint 6	65	Joint variable	0	0

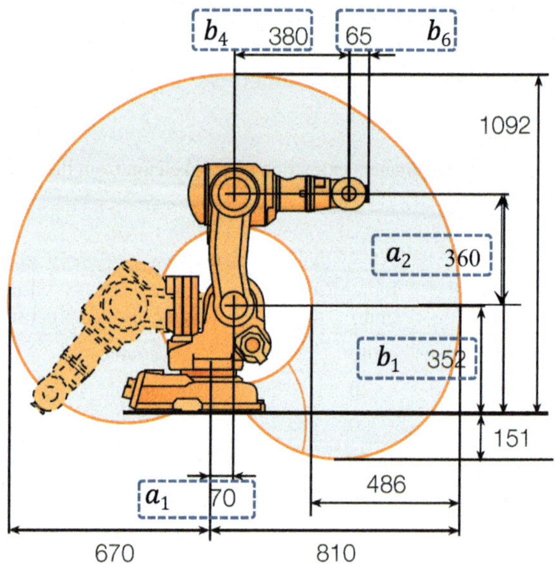

Fig. 4 Specifications of ABB IRB140 robot arm [8]

implemented in Sect. 5, where the imported CAD model in VRM is also used for visualization of the robot motion.

4 Instrumentation of Exoskeleton

The prototype of the exoskeleton had to be fitted with joint angle measurement sensors such as rotary potentiometers. To achieve this, a series of sequential steps were followed as explained in this section.

4.1 Test Setup with Rotary Encoders

A test setup consisting of a breadboard and seven inexpensive rotary potentiometers were connected as shown in Fig. 5a to an Arduino Uno board, whose schematic circuit diagram is in Fig. 5b. Since the output from the rotary potentiometers is in the form of analog signals, the wires carrying the data from the wiper pin is input to the analog pins of the Arduino board.

The potentiometers which work as a voltage divider circuit havea linear output of voltage measured across its end, depending on the angle by which the knob of the potentiometer is rotated. The angles were read by an Arduino controller and fed through serial (USB) communication into a software application in a computer, which maps each angle measured to a joint in the CAD model of the exoskeleton model in the VRM application, as illustrated in Fig. 6.

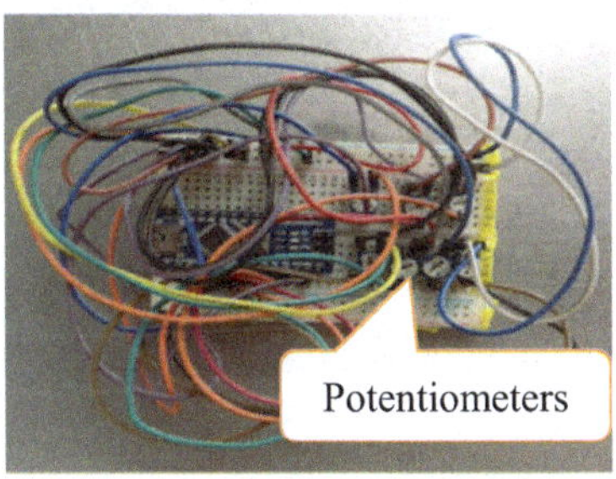

(a) Breadboard with rotary potentiometers

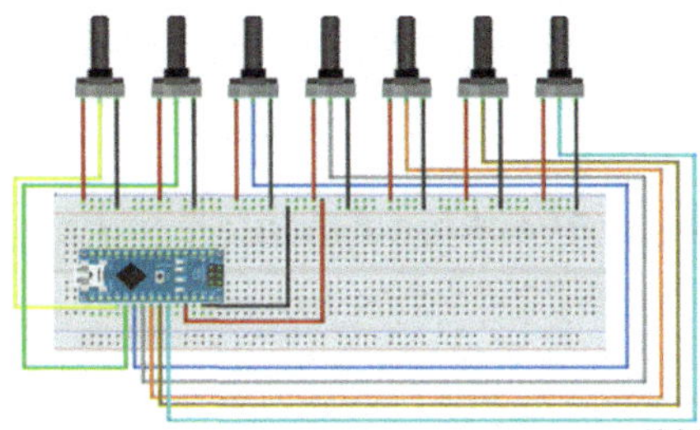

(b) Schematic circuit diagram with Arduino Uno

Fig. 5 Test setup with rotary encoders

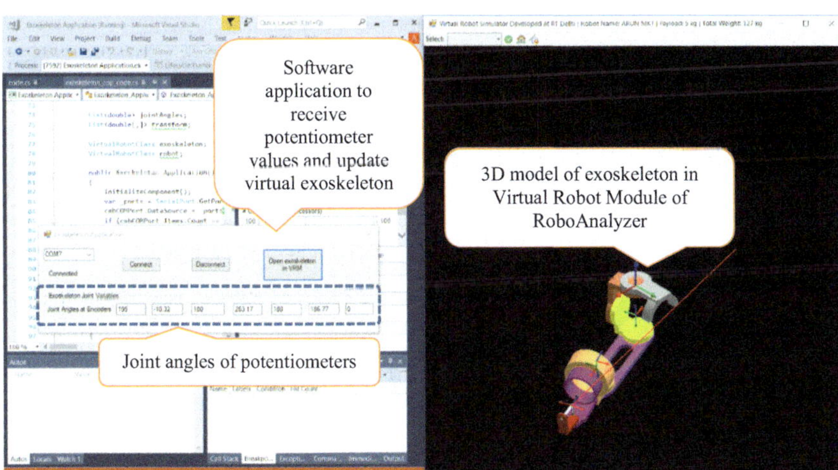

Fig. 6 Software application to validate angles measured by potentiometers

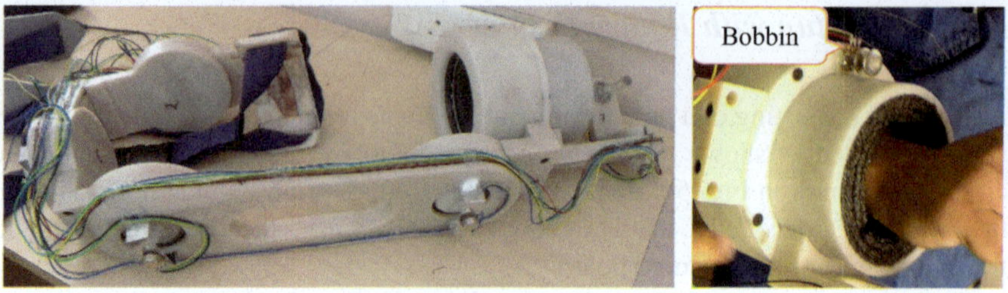

(a) Metallic clamps and potentiometers at each joint (b) Bobbin for fifth axis

Fig. 7 Prototype of exoskeleton fitted with clamps and potentiometers

Thus, the rotation of the encoders provided joint level jogging of the virtual exoskeleton. This exercise was essential to establish the connection and the workflow for mapping potentiometers motion to individual joints in virtual exoskeleton which would then be used in the context of a physical exoskeleton.

4.2 Attaching Encoders to Physical Prototype

The physical prototype was prepared by assembling individual links created in a CNC machine and had bearings at the joints. There was sufficient place for mounting of potentiometers. Metallic clamps were used to house a potentiometer on one of the links of a joint and the rotor of the potentiometer on the other link of the joint, allowing the rotor to rotate whenever the joint was rotated. The rotary potentiometer used to extract the angular information produces values from 0 to 1023 which are then converted into angles between 0° and 275°. The prototype with metallic clamps and potentiometer is shown in Fig. 7a. For the fifth joint axis which corresponds to the "roll" motion at the wrist of the exoskeleton, attaching a clamp to house a conventional potentiometer was not possible due to the shape of the links connected to the joint. To overcome this, a bobbin was attached to one of the links of this joint and a tight tread was used to rotate along with the other joint, thus resulting in the rotation of a potentiometer attached to the bobbin, as shown in Fig. 7b. The bobbin has to be mounted with good care to ensure it does not come out during the operation. A better approach for this joint would be to use a resistive strip used in [1], which will be attempted in the future.

4.3 Mapping of Measured Joint Angles and Virtual Model

A challenge in placing the metallic clamps and the potentiometer is setting the zero angle of the joint with the zero angle of the potentiometer. As it was difficult to align both of them, the clamps and potentiometers were attached such that there is good rotation range available at each joint. The angles were measured from the potentiometers and sent to the virtual exoskeleton model in VRM. Starting from the first joint, the angle measured was offset with some angle such that the physical and virtual model correspond to the same joint angle. The exercise was carried out for all the joints, and offset angles to be added for the joints from the base to the last link are $-80°$, $-20°$, $+180°$, $+150°$, $+180°$, $-10°$, and $0°$, respectively.

5 Control of Robot

A serial robot has motors or actuators at its joints which can be controlled to perform various tasks. Generally, these tasks are defined in the Cartesian space whereas the joints can be controlled in joint space. To provide the relationship between the two spaces, robot kinematics is used, which is defined briefly in Sect. 2. The exoskeleton prototype has been used to control a virtual robot in both simulation and actual scenarios, as explained in this section.

5.1 Control of Robot in Virtual Robot Module

Referring to Fig. 2, the tip of the exoskeleton is determined using forward kinematics. At every instant, the incremental motion of this point in the Cartesian space is determined, in the three translation axes and three rotational axes (Euler angles). These incremental motions are then provided to the virtual robot model using a method proposed in [9, 10] which uses the inverse Jacobian method. This method has been implemented in VRM, which returns the increments in each joint value of the robot model. Thus, the incremental motion of exoskeleton (master) in Cartesian space is mapped to incremental joint angle values of the robot (slave). One can also use the closed-form inverse kinematics formulations to obtain the possible joint angles required to reach the next robot configuration and choose the joint angles solutions that are closest to the current values of joint angles.

An image of an operator wearing exoskeleton and controlling a robot model in VRM is shown in Fig. 8. The master-slave manipulation was carried out to achieve the motion of the end-effector of the virtual robot. As the length of the links is fixed, the exoskeleton resulted in a slight erroneous motion as the joint axes of the prototype and the human operator were not exactly aligned. Also, due to the clamps at joints, some motion on the shoulder joints was restricted. There is however scope for improvements in the design of the exoskeleton which will be attempted in the future.

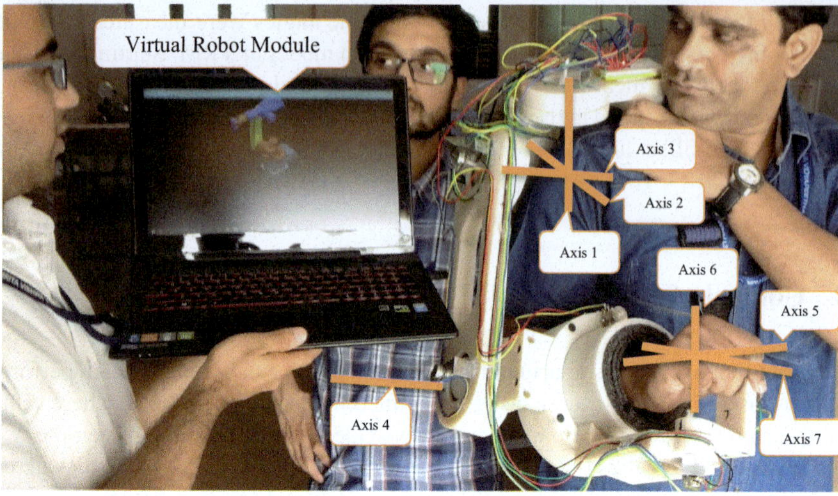

Fig. 8 Master-slave manipulation of a robot in the Virtual Robot Module (VRM)

5.2 Control of Robot in ABB RobotStudio

RobotStudio is a proprietary software developed by ABB that can be used to perform offline simulation of various robot models present in it. Once an offline program or a set of robot motions are found satisfactory, it can be sent to the robot controller, which then would move a physical robot in the same way. Robot simulation software like this help engineers learn robot programming and test their program on a virtual robot, before running on an actual one, thus ensuring safety and improved productivity.

RobotStudio can be programmed using RAPID language and has some native support to control a virtual robot in its environment from different programming interfaces. An example is its ability to listen to communication received on a User Datagram Protocol (UDP) and move the robot accordingly.

The software application explained in Sect. 5.1 was further developed to send the joint angles of the robot model (ABB IRB140) as data packets in UDP on a particular port, which were received by the RobotStudio application running on the same computer. The joint angles received by RobotStudio were then used to update the joint angles of the virtual robot in it. Thus, the Cartesian motion of the exoskeleton was mapped to the motion of the virtual robot in RobotStudio. Alternatively, one can also attempt to supply incremental motion in the Cartesian space to the RobotStudio application, which can move the virtual robot accordingly.

A screenshot of ABB RobotStudio and Virtual Robot Module (VRM) showing robot motion for operator moving the exoskeleton is shown in Fig. 9. The motion of the virtual robot in RobotStudio was also sent further to a physical IRB140 robot and the robot motion achieved was satisfactory. The master-slave manipulation of real robot is shown in Fig. 10. As there was a face plate on the end-effector, there was not much scope for the control of the orientation of the robot. Several attempts

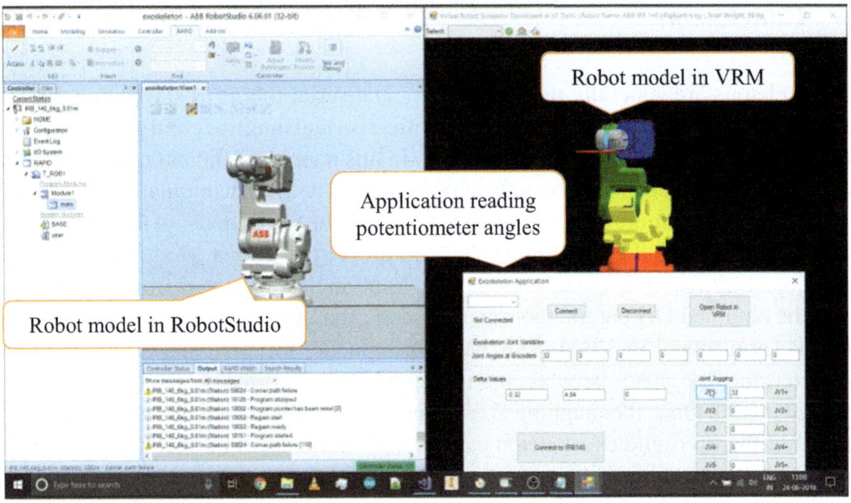

Fig. 9 Movement of ABB IRB140 robot in virtual environments

Fig. 10 Master-slave manipulation of physical ABB IRB140 robot

on controlling the position of the end-effector of the robot, without changing the orientation were made. However, there is large scope for improvement which will be attempted in the future.

6 Conclusion

Exoskeletons are generally used as masters and serial robots as slaves in a master-slave manipulation scheme, which allows effective and intuitive control of robots that are located remotely or in hazardous areas. In this paper, an efficient methodology to determine the usability of an exoskeleton to control a serial manipulator is proposed. It consists of determining the DH parameters from the CAD model of the exoskeleton and then preparing the CAD model of it in the Virtual Robot Module (VRM), co-developed by one of the authors. Further, joint angles measured from exoskeleton and the obtained DH parameters can be used to map the exact motion of the virtual exoskeleton model in VRM. This helps in reducing the time and efforts required to determine the offset between the measured angle and the one at the joint in the virtual model. Once the mapping is established, the incremental motion in Cartesian space of the tip of the exoskeleton can be supplied as input in VRM to achieve the corresponding motion of the virtual robot. Further, the same can be sent to specific applications like ABB RobotStudio to control a virtual robot in it. Thereafter, one can also control a physical robot. Thus, the workflow explained in this paper can be used for effective master-slave manipulation.

References

1. Mukherjee S, Zubair M, Suthar B, Kansal S (2013) Exoskeleton for tele-operation of industrial robot. In: Proceedings of conference on advances in robotics
2. Zubair M, Suthar B, Kansal S, Mukherjee S (2014) Haptics exoskeleton for tele-operation of industrial robot. In: Proceedings of the 3rd joint international conference on multibody dynamics and the 7th Asian conference on multibody dynamics
3. de Looze MP, Bosch T, Krause F, Stadler KS, O'Sullivan LW (2016) Exoskeletons for industrial application and their potential effects on physical work load. J Ergon 59(5):671–681
4. Rawal S, Kansal S, Zubair M, Suthar B, Mukherjee S (2016) Jitter removal in KUKA KR-5 using modified Kalman filter while tele-operating with exoskeleton. In: Proceedings of the Asian conference on multibody dynamics
5. Saha SK (2014) Introduction to robotics. Tata McGraw-Hill Publications
6. Rajeevlochana CG, Saha SK, Kumar S (2012) Automatic extraction of DH parameters of serial manipulators using line geometry. In: Proceedings of 2nd joint international conference on multibody system dynamics
7. Patwardhan A, Prakash A, Chittawadigi RG (2018) Kinematic analysis and development of simulation software for Nex Dexter robotic manipulator. In: Proceedings of international conference on robotics and smart manufacturing
8. ABB IRB140 Datasheet. https://new.abb.com/products/robotics/industrial-robots/irb-140/irb-140-data. Accessed June 2018
9. Sadanand ROM, Chittawadigi RG, Saha SK (2013) Virtual robot simulation in RoboAnalyzer. In: Proceedings of the 1st international & 16th national conference on machines and mechanisms
10. Sadanand ROM, Chittawadigi RG, Joshi R, Saha SK (2015) Virtual robots module: an effective visualization tool for robotics toolbox. In: Proceedings of the 2nd international conference on advances in robotics

Simultaneous Control of Passive Stiffness Characteristics and Position with Elastically Constrained Underactuated Mechanisms

Takeshi Tawada, Nobuyuki Iwatsuki, and Ikuma Ikeda

Abstract Passive compliance is a necessary feature to ensure safety of home robots and medical robots working in human daily life. Such kinds of robots are desired to adjust their stiffness characteristics according to the task at hand and the surrounding environment. In this paper, we present an approach to control both the position and passive stiffness of output links of elastically constrained underactuated link mechanisms. Specifically, we consider a planar parallel link mechanism with nine degrees-of-freedom (DoF) consisting of six active revolute pairs with rotary actuators, eighteen passive revolute pairs, and nineteen links. Six of the passive revolute pairs are elastically constrained by torsional coil springs, thus the mechanism has passive compliance. We propose an optimization procedure to control the output position and multi-directional non-linear stiffness characteristics simultaneously by determining input actuator angles. The validity of the proposed analysis and optimization procedure is also experimentally corroborated.

Keywords Passive compliance · Underactuated mechanism · Non-linear stiffness

1 Introduction

Nowadays, robots are required to flexibly perform tasks, such as grasping and assembling soft or fragile objects. For general-purpose robots that perform a number of such tasks, it is necessary to have appropriate flexibility according to the object to be treated and the ambient environment. Especially for robots performing everyday activities around humans, adding compliance characteristics is important to ensure safety.

One of the approaches to provide flexibility to mechanisms is to add passive compliance through the use of elastic elements. Elastically constrained underactuated

T. Tawada (✉)
Graduate Student, School of Engineering, Tokyo Institute of Technology, Tokyo, Japan
e-mail: tawada.t.aa@m.titech.ac.jp

N. Iwatsuki · I. Ikeda
School of Engineering, Tokyo Institute of Technology, Tokyo, Japan

D. Sen et al. (eds.), *Mechanism and Machine Science*, Lecture Notes in Mechanical Engineering, https://doi.org/10.1007/978-981-15-4477-4_29

link mechanisms have been studied to add desired passive stiffness characteristics to the mechanisms, e.g., a rolling contact joint with elastic band and its application in a three DoF parallel mechanism [1], a grasping mechanism with designed stiffness [2], a remote center of compliance device with 3-RRR link mechanism [3], and an application of an elastic serial link mechanism to peg-in-hole problem [4]. However, the desired stiffness characteristics of these mechanisms are predetermined and set in the design stage, and so are not suitable for general-purpose robots that must be adaptive, that must modify their stiffness characteristics while performing diverse tasks by adjusting the actuator input angles on the fly.

In previous research, a planar parallel manipulator with elastic elements with five DoF that can adjust its passive stiffness [5], and a planar link mechanism with nine DoF and six actuators that can generate desired non-linear passive stiffness characteristics in one direction [6] were proposed. However, some tasks require the adjustment of multi-directional stiffness characteristics. Moreover, we need to precisely control the position of the output point in no-load states at the same time so as to use the mechanism as a planer manipulator.

Therefore, in this research, we have tried to control simultaneously multi-directional passive stiffness characteristics and the output position of an elastically constrained underactuated link mechanism. The proposed method is applied to the modified nine DoF and six actuators mechanism shown in Fig. 1. An experimental validation was carried out with a prototype.

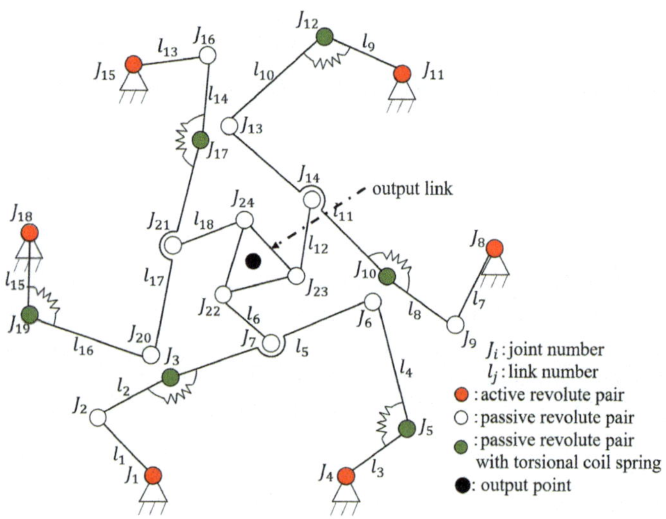

Fig. 1 A planar parallel link mechanism with nine DoF and six actuators

2 Kinetostatic Analysis

2.1 Target Mechanism

The target mechanism shown in Fig. 1 consists of the three sub-chains shown in Fig. 2, arranged rotationally symmetrically. Let the position and the angle of each joint be $J_i(X_i, Y_i)$ and θ_i, respectively. $\boldsymbol{x}(x, y)$ and ψ are the position and orientation of the output link. J_1 and J_4 are fixed joints. J_3 and J_5 are elastically constrained by torsional coil springs. The length of each link is represented by l_j. Since this mechanism has nine DoF, three DoF remain after the six input actuator angles are determined. Therefore, three parameters $\boldsymbol{\theta}_C$ should be assumed to specify the configuration. We chose $[\theta_3\ \theta_{10}\ \theta_{17}]^T$ as $\boldsymbol{\theta}_C$ and refer to them as configuration determinants.

2.2 Forward Statics

Let $\boldsymbol{F} = [F_x\ F_y\ M_\psi]^T$ be the external force and torque applied to the output point. When the actuator input angles $\boldsymbol{\theta}_A = [\theta_1\ \theta_4\ \theta_8\ \theta_{11}\ \theta_{15}\ \theta_{18}]^T$ and $\boldsymbol{\theta}_C$ are specified, $\boldsymbol{F}$ can be calculated according to the following steps:

Step 1: Kinematics Given the configuration determinants, $\boldsymbol{\theta}_C$, the positions and angles of all joints are obtained kinematically.

Step 2: Derive the Closed-Loop Equation By considering routes from the fixed joints to the output point on the sub-chain shown in Fig. 2, we obtain two vector equations as

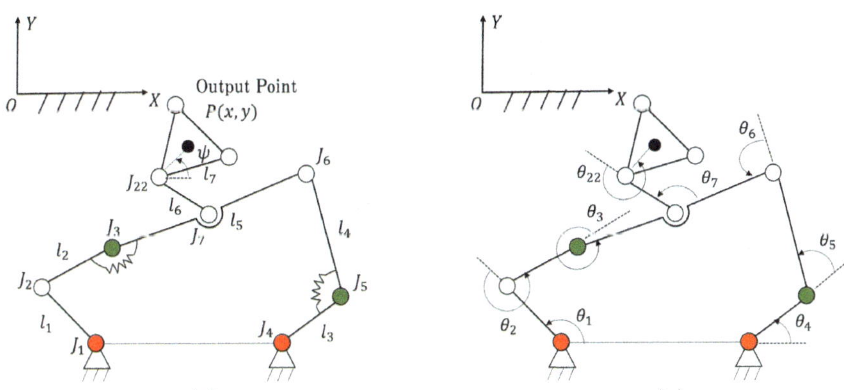

Fig. 2 Mechanism parameters of the subchains

$$\overrightarrow{OJ_1} + \overrightarrow{J_1J_2} + \overrightarrow{J_2J_3} + \overrightarrow{J_3J_7} + \overrightarrow{J_7J_{22}} = \overrightarrow{OP} + \overrightarrow{PJ_{22}} \quad (1)$$

$$\overrightarrow{OJ_4} + \overrightarrow{J_4J_5} + \overrightarrow{J_5J_6} + \overrightarrow{J_6J_7} + \overrightarrow{J_7J_{22}} = \overrightarrow{OP} + \overrightarrow{PJ_{22}}. \quad (2)$$

From the geometric relation among the joint angles, we obtain

$$\theta_1 + \theta_2 + \theta_3 - \theta_4 - \theta_5 - \theta_6 + \pi = 0 \quad (3)$$

$$\theta_1 + \theta_2 + \theta_3 + \theta_7 + \theta_{22} = \psi. \quad (4)$$

Similarly for the remaining two sub-chains, the following equations hold.

$$\overrightarrow{OJ_8} + \overrightarrow{J_8J_9} + \overrightarrow{J_9J_{10}} + \overrightarrow{J_{10}J_{14}} + \overrightarrow{J_{14}J_{23}} = \overrightarrow{OP} + \overrightarrow{PJ_{23}} \quad (5)$$

$$\overrightarrow{OJ_{11}} + \overrightarrow{J_{11}J_{12}} + \overrightarrow{J_{12}J_{13}} + \overrightarrow{J_{13}J_{14}} + \overrightarrow{J_{14}J_{23}} = \overrightarrow{OP} + \overrightarrow{PJ_{23}} \quad (6)$$

$$\theta_8 + \theta_9 + \theta_{10} - \theta_{11} - \theta_{12} - \theta_{13} + \pi = 0 \quad (7)$$

$$\theta_8 + \theta_9 + \theta_{10} + \theta_{14} + \theta_{23} = \psi + \frac{2}{3}\pi \quad (8)$$

$$\overrightarrow{OJ_{15}} + \overrightarrow{J_{15}J_{16}} + \overrightarrow{J_{16}J_{17}} + \overrightarrow{J_{17}J_{21}} + \overrightarrow{J_{21}J_{24}} = \overrightarrow{OP} + \overrightarrow{PJ_{24}} \quad (9)$$

$$\overrightarrow{OJ_{18}} + \overrightarrow{J_{18}J_{19}} + \overrightarrow{J_{19}J_{20}} + \overrightarrow{J_{20}J_{21}} + \overrightarrow{J_{21}J_{24}} = \overrightarrow{OP} + \overrightarrow{PJ_{24}} \quad (10)$$

$$\theta_{15} + \theta_{16} + \theta_{17} - \theta_{18} - \theta_{19} - \theta_{20} + \pi = 0 \quad (11)$$

$$\theta_{15} + \theta_{16} + \theta_{17} + \theta_{21} + \theta_{24} = \psi + \frac{4}{3}\pi \quad (12)$$

Since the number of variables $\boldsymbol{\theta}$, $\boldsymbol{x}$, ψ is 27 and the total DoF of the mechanism is 9, the closed-loop equation can be derived as

$$\boldsymbol{\Phi}(\boldsymbol{\theta}, \boldsymbol{x}, \psi) = 0, \quad (13)$$

by summarizing 18 independent equations above.

Step 3: Calculate the External Force By differentiating the closed-loop equation, we obtain

$$\frac{\partial \boldsymbol{\Phi}}{\partial \boldsymbol{\theta}}\delta\boldsymbol{\theta} + \frac{\partial \boldsymbol{\Phi}}{\partial \boldsymbol{x}}\delta\boldsymbol{x} + \frac{\partial \boldsymbol{\Phi}}{\partial \psi}\delta\psi = 0. \quad (14)$$

with the angles of passive joints $\boldsymbol{\theta}_P = [\theta_2\ \theta_3\ \theta_5\ \theta_6\ \theta_7\ \theta_9\ \theta_{10}\ \theta_{12}\ \theta_{13}\ \theta_{14}\ \theta_{16}\ \theta_{17}\ \theta_{19}\ \theta_{20}\ \theta_{21}\ \theta_{22}\ \theta_{23}\ \theta_{24}]^T$, Eq. (14) can be written as

$$\frac{\partial \boldsymbol{\Phi}}{\partial \boldsymbol{\theta}_A}\delta\boldsymbol{\theta}_A + \frac{\partial \boldsymbol{\Phi}}{\partial \boldsymbol{\theta}_P}\delta\boldsymbol{\theta}_P + \frac{\partial \boldsymbol{\Phi}}{\partial \boldsymbol{x}}\delta\boldsymbol{x} + \frac{\partial \boldsymbol{\Phi}}{\partial \psi}\delta\psi = 0. \quad (15)$$

By means of the principle of virtual work, the relation between the external force $\boldsymbol{F}$ and internal torque $\boldsymbol{\tau}$ of each passive joint generated by torsional springs, is derived as

$$\delta\boldsymbol{\theta}_P{}^T\boldsymbol{\tau} = \begin{bmatrix} \delta\boldsymbol{x} \\ \delta\psi \end{bmatrix}^T \boldsymbol{F}. \tag{16}$$

Here, note that the torque of the 12 passive joints with no elastically elements should be zero since friction is not considered. Equations (15), (16) along with the fixed actuator inputs lead to

$$\boldsymbol{F} = -\begin{bmatrix} \frac{\partial\boldsymbol{\Phi}}{\partial\boldsymbol{x}} & \frac{\partial\boldsymbol{\Phi}}{\partial\psi} \end{bmatrix}^T \frac{\partial\boldsymbol{\Phi}}{\partial\boldsymbol{\theta}_P}^{-T}\boldsymbol{\tau}. \tag{17}$$

2.3 Inverse Statics

When $\boldsymbol{\theta}_A$ and the external force are given, configuration determinants $\boldsymbol{\theta}_C$ in state of static equilibrium can be calculated according to the following steps:

Step 1: Jacobian Matrix Let $\boldsymbol{\theta}_V = [\theta_2\ \theta_5\ \theta_6\ \theta_7\ \theta_9\ \theta_{12}\ \theta_{13}\ \theta_{14}\ \theta_{16}\ \theta_{19}\ \theta_{20}\ \theta_{21}\ \ \theta_{22}\ \theta_{23}\ \theta_{24}]^T$ be the angles of passive joints excluding $\boldsymbol{\theta}_C$, then Eq. (15) can be written as

$$\frac{\partial\boldsymbol{\Phi}}{\partial\boldsymbol{\theta}_C}\delta\boldsymbol{\theta}_C + \frac{\partial\boldsymbol{\Phi}}{\partial\boldsymbol{\theta}_V}\delta\boldsymbol{\theta}_V + \frac{\partial\boldsymbol{\Phi}}{\partial\boldsymbol{x}}\delta\boldsymbol{x} + \frac{\partial\boldsymbol{\Phi}}{\partial\psi}\delta\psi = 0. \tag{18}$$

Total differentiation of the external force becomes

$$\delta\boldsymbol{F} = \frac{\partial\boldsymbol{F}}{\partial\boldsymbol{\theta}_C}\delta\boldsymbol{\theta}_C + \frac{\partial\boldsymbol{F}}{\partial\boldsymbol{\theta}_V}\delta\boldsymbol{\theta}_V + \frac{\partial\boldsymbol{F}}{\partial\boldsymbol{x}}\delta\boldsymbol{x} + \frac{\partial\boldsymbol{F}}{\partial\psi}\delta\psi, \tag{19}$$

where partial differentiation of the external force by an arbitrary variable i is derived from Eq. (17) as

$$\begin{aligned}\frac{\partial\boldsymbol{F}}{\partial i} = &-\left(\frac{\partial}{\partial i}\begin{bmatrix} \frac{\partial\boldsymbol{\Phi}}{\partial\boldsymbol{x}} & \frac{\partial\boldsymbol{\Phi}}{\partial\psi} \end{bmatrix}^T\right)\frac{\partial\boldsymbol{\Phi}}{\partial\boldsymbol{\theta}_P}^{-T}\boldsymbol{\tau} + \begin{bmatrix} \frac{\partial\boldsymbol{\Phi}}{\partial\boldsymbol{x}} & \frac{\partial\boldsymbol{\Phi}}{\partial\psi} \end{bmatrix}^T \frac{\partial\boldsymbol{\Phi}}{\partial\boldsymbol{\theta}_P}^{-T}\left(\frac{\partial}{\partial i}\frac{\partial\boldsymbol{\Phi}}{\partial\boldsymbol{\theta}_P}\right)\frac{\partial\boldsymbol{\Phi}}{\partial\boldsymbol{\theta}_P}^{-T}\boldsymbol{\tau} \\ &-\begin{bmatrix} \frac{\partial\boldsymbol{\Phi}}{\partial\boldsymbol{x}} & \frac{\partial\boldsymbol{\Phi}}{\partial\psi} \end{bmatrix}^T \frac{\partial\boldsymbol{\Phi}}{\partial\boldsymbol{\theta}_P}^{-T}\frac{\partial\boldsymbol{\tau}}{\partial i}.\end{aligned} \tag{20}$$

From Eqs. (18) and (19), the Jacobian matrix, Jac, is obtained as

$$\begin{aligned}\delta\boldsymbol{F} &= \left(\frac{\partial\boldsymbol{F}}{\partial\boldsymbol{\theta}_C} - \begin{bmatrix} \frac{\partial\boldsymbol{F}}{\partial\boldsymbol{\theta}_V} & \frac{\partial\boldsymbol{F}}{\partial\boldsymbol{x}} & \frac{\partial\boldsymbol{F}}{\partial\psi} \end{bmatrix}\begin{bmatrix} \frac{\partial\boldsymbol{\Phi}}{\partial\boldsymbol{\theta}_V} & \frac{\partial\boldsymbol{\Phi}}{\partial\boldsymbol{x}} & \frac{\partial\boldsymbol{\Phi}}{\partial\psi} \end{bmatrix}^{-1}\frac{\partial\boldsymbol{\Phi}}{\partial\boldsymbol{\theta}_C}\right)\delta\boldsymbol{\theta}_C && (21) \\ &= \text{Jac}\ \delta\boldsymbol{\theta}_C. && (22)\end{aligned}$$

Step 2: Calculate the Configuration Determinants We obtain $\boldsymbol{\theta}_C$ in state of static equilibrium for a specified external force $\boldsymbol{F}_S$ by solving the non-linear equation,

$$\boldsymbol{F}_S - \boldsymbol{F} = \boldsymbol{0} \tag{23}$$

for $\boldsymbol{\theta}_C$ by means of the Newton–Raphson method. In each iteration step n, $\boldsymbol{F}$ should be calculated by forward statics and $\boldsymbol{\theta}_C$ updated as

$$\boldsymbol{\theta}_{C n+1} = \boldsymbol{\theta}_{C n} + \mathrm{Jac}^{-1} (\boldsymbol{F}_S - \boldsymbol{F}) \,. \tag{24}$$

3 Optimization of the Stiffness Characteristics

In this section, we propose an optimization procedure to control the output position and multi-directional non-linear stiffness characteristics of the target mechanism.

3.1 Optimization Procedure

When the proposed mechanism is used as a two DoF planar manipulator, the output position, (x, y), in the no-load state must be controlled precisely. On the other hand, passive stiffness characteristics need only be approximately optimized. Therefore, two actuator inputs $\boldsymbol{\theta}_\alpha$ are chosen for precise positioning, and remaining four actuator inputs $\boldsymbol{\theta}_\beta$ are used for optimization of the stiffness characteristics, which is implemented as a non-linear least squares problem. Moreover, these two objectives are carried out simultaneously.

Definition of Stiffness Characteristics When an external force is applied in a specified direction $\boldsymbol{t}_i$, let ξ be the $\boldsymbol{t}_i$-directional displacement from the no-load state; i.e.,

$$\xi = \boldsymbol{t}^T (\boldsymbol{x} - \boldsymbol{x}_0) \tag{25}$$

where $\boldsymbol{x}_0$ denotes the output position in the no-load state. Then define the stiffness characteristics as the $|\boldsymbol{F}| - \xi$ relationship. Figure 3 displays the example of an ideal

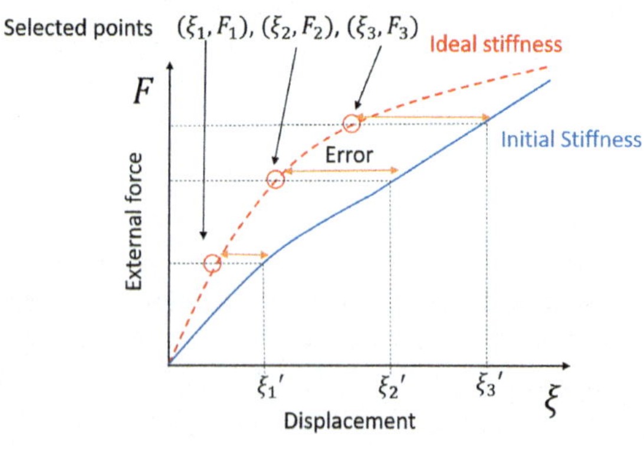

Fig. 3 Initial and ideal stiffness characteristics

stiffness characteristics curve. We choose m number of points (ξ_i, F_i) on the ideal curves and set the objective function S as the sum of square errors in the force-directional displacements, i.e.,

$$S(\boldsymbol{\theta}_A) = \sum_{i=1}^{m} (\xi_i - \xi_i')^2, \tag{26}$$

where ξ_i' is the $\boldsymbol{t}_i$-directional displacement from the no-load state with the current actuator input. In the case of multi-directional optimization, several directions $\boldsymbol{t}_{i,\cdots,m}$ should be chosen.

3.2 Optimization Algorithm

Figure 4 contains a flowchart of the optimization procedure. In the first iteration, precise positioning using $\boldsymbol{\theta}_\alpha$ is done by the Newton–Raphson method. $\boldsymbol{x}_S$ denotes the specified output position in the no-load state. The Jacobian matrix, $\frac{\partial \boldsymbol{x}}{\partial \boldsymbol{\theta}_\alpha}$, can be derived as

$$\begin{bmatrix} \frac{\partial \boldsymbol{\theta}_P}{\partial \boldsymbol{\theta}_A} \\ \frac{\partial \boldsymbol{x}}{\partial \boldsymbol{\theta}_A} \\ \frac{\partial \boldsymbol{\psi}}{\partial \boldsymbol{\theta}_A} \end{bmatrix} = - \begin{bmatrix} \frac{\partial \boldsymbol{\Phi}}{\partial \boldsymbol{\theta}_P} & \frac{\partial \boldsymbol{\Phi}}{\partial \boldsymbol{x}} & \frac{\partial \boldsymbol{\Phi}}{\partial \boldsymbol{\psi}} \\ \frac{\partial \boldsymbol{F}}{\partial \boldsymbol{\theta}_P} & \frac{\partial \boldsymbol{F}}{\partial \boldsymbol{x}} & \frac{\partial \boldsymbol{F}}{\partial \boldsymbol{\psi}} \end{bmatrix}^{-1} \begin{bmatrix} \frac{\partial \boldsymbol{\Phi}}{\partial \boldsymbol{\theta}_A} \\ \frac{\partial \boldsymbol{F}}{\partial \boldsymbol{\theta}_A} \end{bmatrix}, \tag{27}$$

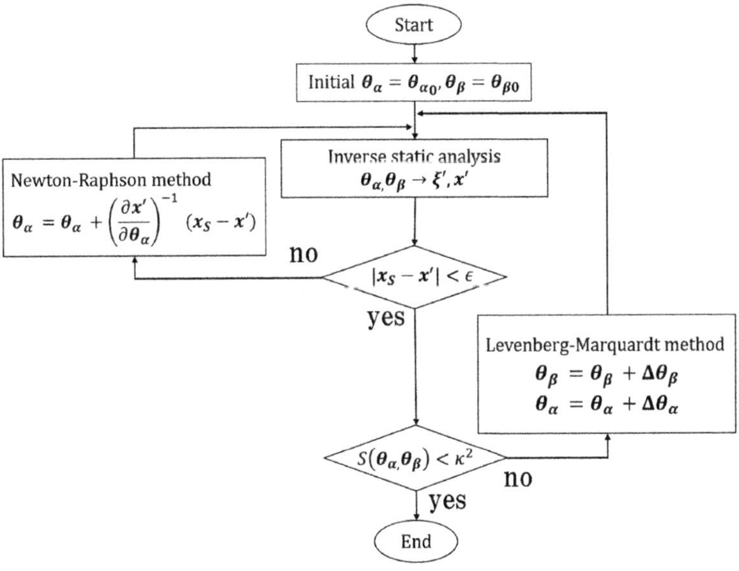

Fig. 4 Flowchart of optimization

from Eqs. (18) and (19). When the positioning accuracy ϵ is satisfied, $\boldsymbol{\theta}_\alpha$ and $\boldsymbol{\theta}_\beta$ will be updated by the Levenberg–Marquardt method (LM) to decrease the objective function. In this phase, the conditional expression for maintaining positioning,

$$\frac{\partial \boldsymbol{x}'}{\partial \boldsymbol{\theta}_\alpha}\Delta\boldsymbol{\theta}_\alpha + \frac{\partial \boldsymbol{x}'}{\partial \boldsymbol{\theta}_\beta}\Delta\boldsymbol{\theta}_\beta = \boldsymbol{0}, \tag{28}$$

is considered. The Jacobian matrix is derived considering not only $\Delta\boldsymbol{\theta}_\beta$, but also $\Delta\boldsymbol{\theta}_\alpha$, thus

$$\frac{\partial \boldsymbol{\xi}'}{\partial \boldsymbol{\theta}_\alpha}\Delta\boldsymbol{\theta}_\alpha + \frac{\partial \boldsymbol{\xi}'}{\partial \boldsymbol{\theta}_\beta}\Delta\boldsymbol{\theta}_\beta = \mathrm{Jac}\Delta\boldsymbol{\theta}_\beta \tag{29}$$

is obtained, where

$$\frac{\partial \boldsymbol{\xi}'}{\partial \boldsymbol{\theta}_A} = \begin{bmatrix} \boldsymbol{t}_1 \frac{\partial x}{\partial \boldsymbol{\theta}_A}{}_{F=F_1} \\ \vdots \\ \boldsymbol{t}_m \frac{\partial x}{\partial \boldsymbol{\theta}_A}{}_{F=F_m} \end{bmatrix}. \tag{30}$$

$\boldsymbol{x}'$ and $\boldsymbol{\xi}'$ denote the position in the no-load state and the vector with i-th component ξ_i' in the current iteration step, respectively. From Eqs. (28), (29) and the update rule of LM, the incremental, $\Delta\boldsymbol{\theta}_\alpha$ and $\Delta\boldsymbol{\theta}_\beta$, and the Jacobian matrix are determined as

$$\mathrm{Jac} = \frac{\partial \boldsymbol{\xi}'}{\partial \boldsymbol{\theta}_\beta} - \frac{\partial \boldsymbol{\xi}'}{\partial \boldsymbol{\theta}_\alpha}\left(\frac{\partial \boldsymbol{x}'}{\partial \boldsymbol{\theta}_\alpha}\right)^{-1}\frac{\partial \boldsymbol{x}'}{\partial \boldsymbol{\theta}_\beta}, \tag{31}$$

$$\Delta\boldsymbol{\theta}_\beta = \left[\mathrm{Jac}^T\mathrm{Jac} + \lambda\,\mathrm{diag}\left(\mathrm{Jac}^T\mathrm{Jac}\right)\right]^{-1}\mathrm{Jac}^T\left(\boldsymbol{\xi} - \boldsymbol{\xi}'\right), \tag{32}$$

$$\Delta\boldsymbol{\theta}_\alpha = -\left(\frac{\partial \boldsymbol{x}'}{\partial \boldsymbol{\theta}_\alpha}\right)^{-1}\frac{\partial \boldsymbol{x}'}{\partial \boldsymbol{\theta}_\beta}\Delta\boldsymbol{\theta}_\beta, \tag{33}$$

where λ is the damping factor. Once the positioning accuracy is violated due to the accumulated error, the positioning by the Newton–Raphson method is performed again. When the positioning accuracy and the tolerance κ^2 of the objective function are both satisfied, iteration ends.

3.3 Application Examples

We show two examples that optimize the stiffness characteristics in multiple directions while positioning at the origin. The proposed optimization process is carried out with respect to two different desired characteristics patterns—A and B. As shown in Fig. 5, pattern A is set for use in people-rich environment so that softening spring characteristics along three directions are necessary for safety. Pattern B is for grasp-

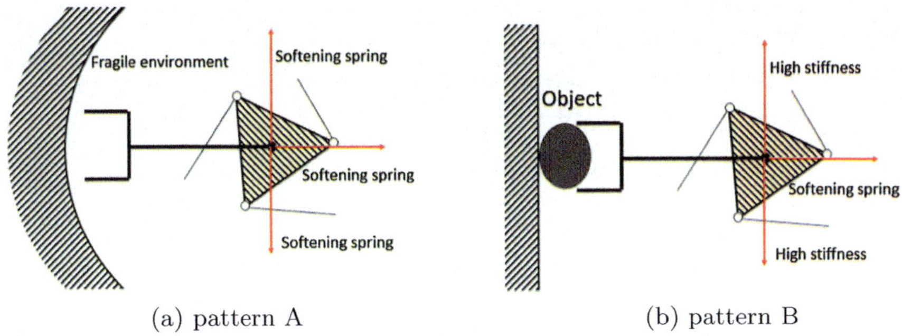

Fig. 5 Conceptual diagram of stiffness pattern A and B

Table 1 Selected points of stiffness pattern A

Force direction	Displacement $\boldsymbol{\xi}$[m]			External force $\boldsymbol{F}$[N]		
$-x$	ξ_1	ξ_2	ξ_3	F_1	F_2	F_3
	0.005	0.01	0.025	1.0	1.6	2.4
x	ξ_4	ξ_5	ξ_6	F_4	F_5	F_6
	0.005	0.01	0.025	1.0	1.6	2.4
$-y$	ξ_7	ξ_8	ξ_9	F_7	F_8	F_9
	0.005	0.01	0.025	1.0	1.6	2.4

Table 2 Selected points of stiffness pattern B

Force direction	Displacement $\boldsymbol{\xi}$[m]			External force $\boldsymbol{F}$[N]		
$-x$	ξ_1	ξ_2	ξ_3	F_1	F_2	F_3
	0.003	0.006	0.009	0.8	1.6	2.4
x	ξ_4	ξ_5	ξ_6	F_4	F_5	F_6
	0.003	0.006	0.009	0.8	1.6	2.4
$-y$	ξ_7	ξ_8	ξ_9	F_7	F_8	F_9
	0.005	0.012	0.025	1.0	1.6	2.4

ing fragile objects, thus softening spring characteristics along the direction of motion is desired for safety, and high stiffness along the directions of perpendicular thereto is also needed for positioning accuracy. Tables 1 and 2 show the selected points in the ideal curve for each pattern.

The mechanical dimensions used in the application examples and the experiment in Sect. 4 are listed in Table 3. Table 4 shows the initial and final values of the actuator input angles. The positioning accuracy is set at 0.0001 m, and the tolerance κ, at 0.001 m. Figure 6 shows the transition of the objective function. Here, it can be confirmed that the objective function is remarkably reduced. The observed vibration

Table 3 Mechanism dimensions

Link length [m]							
l_1	l_2	l_3	l_4	l_5	l_6	l_7	
0.10	0.10	0.10	0.20	0.25	0.10	0.10	
Spring Constant [Nm/rad]		Natural angle [deg]		Joint coordinates [m]			
k_1	k_2	ϕ_1	ϕ_2	X_1	Y_1	X_4	Y_4
0.188	0.188	180	135	−0.125	−0.325	0.125	−0.325

Table 4 The initial and final values of actuator angles

	θ_1	θ_4	θ_8	θ_{11}	θ_{15}	θ_{18}
Initial angle [°]	120	40	230	160	330	300
Final angle (pattern A)	73.4	64.1	196.0	176.3	312.7	307.1
Final angle (pattern B)	94.5	56.3	215.2	175.1	352.2	276.3

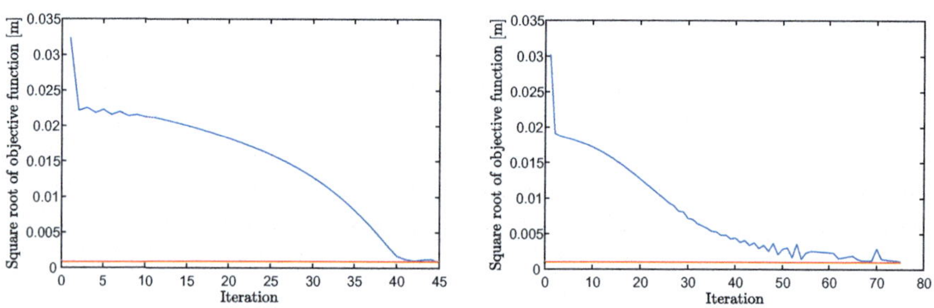

(a) pattern A (b) pattern B

Fig. 6 Decrease of the objective function

of the objective function is considered to be an influence of positioning. Figures 7 and 8 show the optimized stiffness characteristics. The red circles denote the given selected points. Here, we can visually confirm that the desired stiffness characteristics is approximately achieved. These facts show that the proposed procedure achieves the simultaneous control of stiffness characteristics in multiple directions and of position.

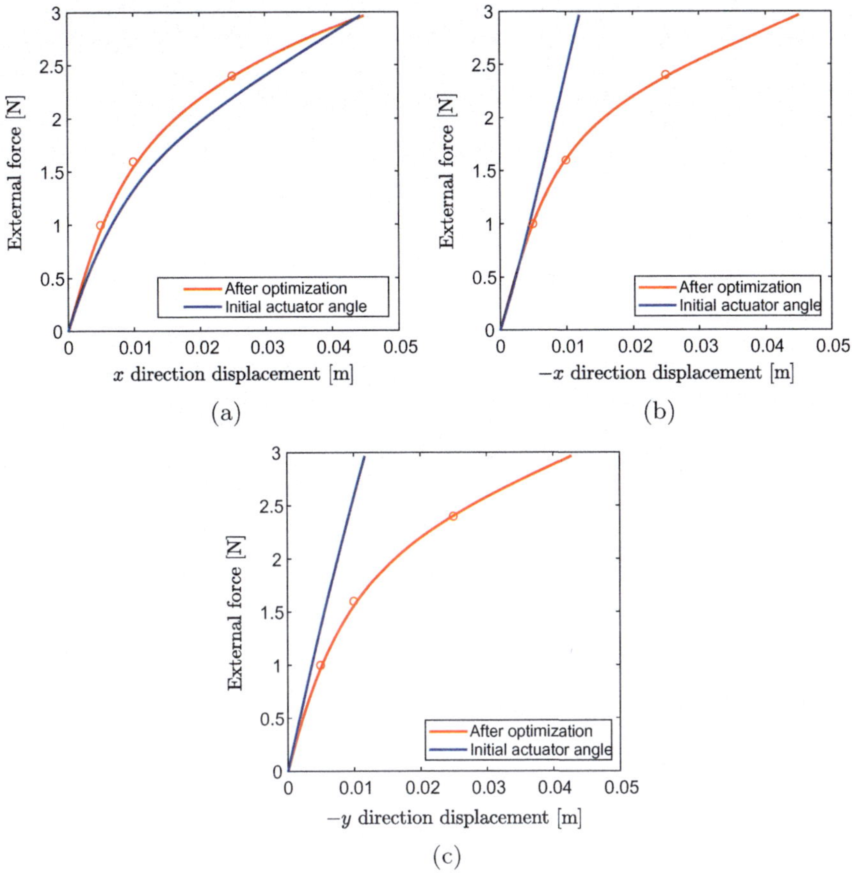

Fig. 7 Stiffness characteristics (pattern A)

4 Experiment

4.1 Experimental Setup

Figure 9 shows a prototype of nine DoF and six actuators mechanism. The mechanism dimensions are the same as in Table 3. For the active revolute pairs, DC motors with encoders with resolution of 512 pulses per revolution and harmonic drives that have reduction ratio of 100 or 50 are mounted. Figure 10 shows a passive joint with a torsional coil spring. The spring is fixed with a plate that has a V-groove. The prototype has a structure that prevents interference by height differences of each subchain.

As shown in Fig. 11, an external force is applied along a certain direction by pressing a force gauge against the output point using a feed mechanism. A ball

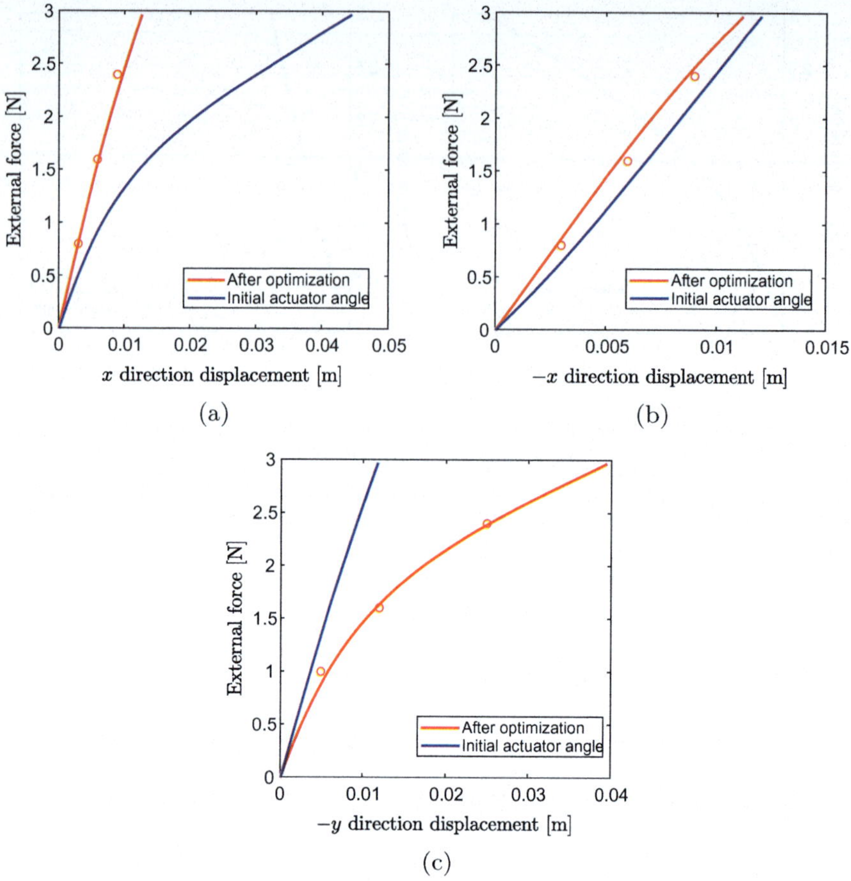

Fig. 8 Stiffness characteristics (pattern B)

bearing is attached at the output point so that the direction of external force does not change by friction. The stiffness characteristics is measured as a force-displacement relationship determined from the feed amount and the measured force.

4.2 *Experimental Results*

Figure 12 presents the measured stiffness characteristics neglecting the positioning error due to the mechanical tolerances. The analyzed stiffness characteristics agree well with the measured. Note that non-linear stiffness characteristics of the softening spring and hardening spring are generated well.

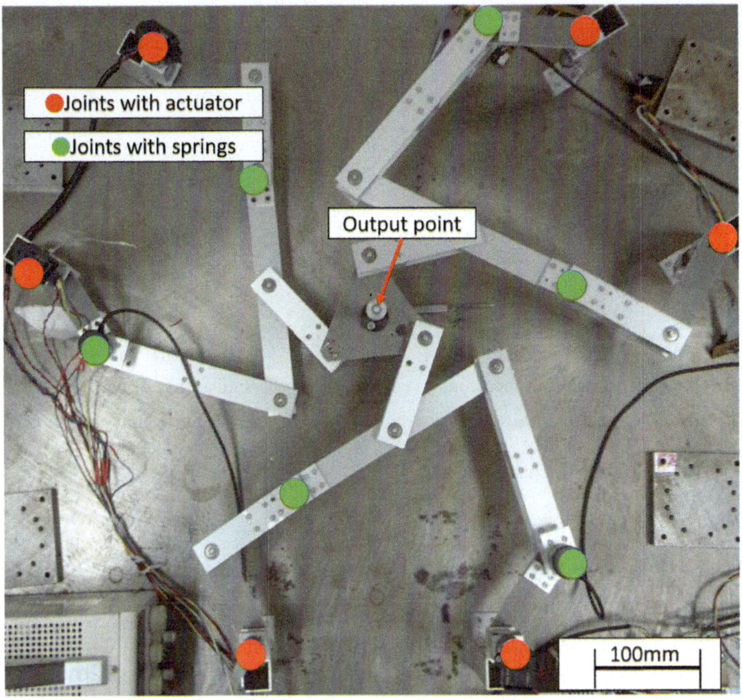

Fig. 9 A prototype of nine DoF and six actuators mechanism

Fig. 10 Joint with torsional spring

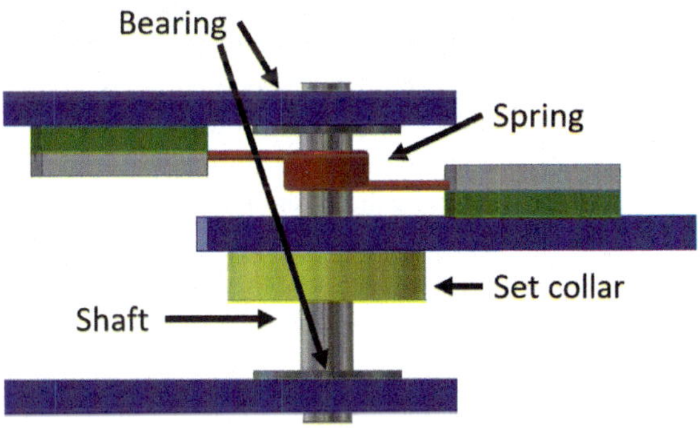

Fig. 11 Force gauge feed mechanism

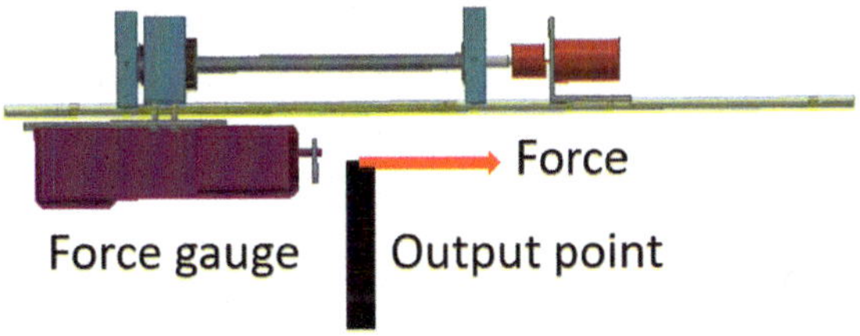

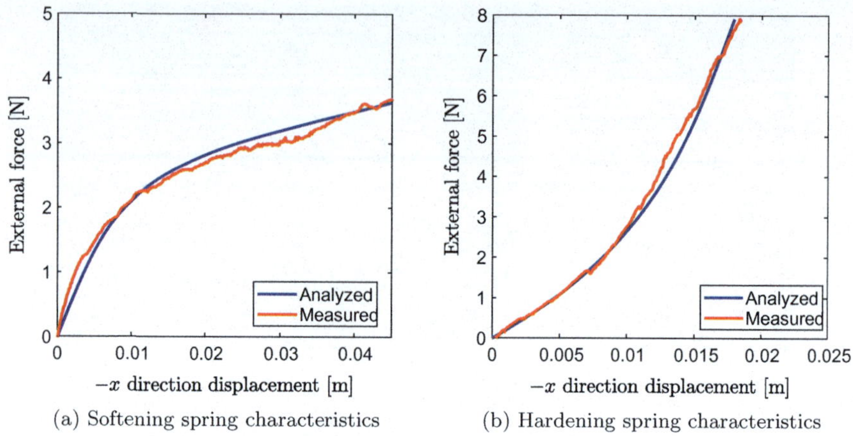

Fig. 12 Measured stiffness characteristics

5 Conclusions

A method of simultaneously controlling both the passive stiffness characteristics and the position in the no-load state is proposed and applied to an elastically constrained underactuated mechanism with nine DoF and six actuators. In summary,

(1) A new optimization procedure to calculate input joint motions that generate both precise output motion of the output link in the no-load state and multi-directional non-linear stiffness is proposed and formulated.
(2) The proposed optimization procedure is applied in two different examples, revealing that the output position and multi-directional stiffness characteristics are successively optimized with the procedure.
(3) The experimental validation of the proposed method is carried out with a prototype. The analyzed stiffness characteristics agree well with the measured. It is thus confirmed that the proposed methods are effective and useful.

References

1. Jeanneau A, Herder J, Laliberte T, Gosselin C (2004) A compliant rolling contact joint and its application in a 3-DOF planar parallel mechanism with kinematic analysis. In: Proceedings of the ASME design engineering technical conferences, pp 689–698
2. Stavenuiter R, Birglen L, Herder J (2017) A planar underactuated grasper with adjustable compliance. Mech Mach Theory 112(5):295–306
3. Kim W, Byung-Ju Y, Whang C (2000) Rcc characteristics of planar/spherical three degree-of-freedom parallel mechanisms with joint compliances. J Mech Des 122(1):10–16

4. Yun S (2008) Compliant manipulation for peg-in-hole: is passive compliance a key to learn contact motion?. In: Proceedings of the IEEE international conference on robotics and automation, pp 1647–1652
5. Iwatsuki N, Kotte T, Morikawa K (2010) Simultaneous control of the motion and stiffness of redundant closed-loop link mechanisms with elastic elements. J Mech Sci Technol 24(1):285–288
6. Iwatsuki N, Terashima N (2010) Flexibility control of redundant closed-loop mechanisms with elastic elements. In: CD-ROM Proceedings of the 1st IFToMM Asian conference on mechanism and machine science, pp 1–6

A Numerical Algorithm to Reconstruct the Missing Points from the Membrane Wings of a Bat via the Computerized Motion Capture System

Sudeep Kumar Singh and Jing-Shan Zhao

Abstract The bat is an efficient flyer with flapping membrane wings but the measurement of aerodynamic forces is the major challenge. The motion of the bat's wing is very complex which comprises translation as well as rotation. During the bat's flight, the wing membrane will deform in the downstroke and upstroke process of wings. The investigation of the detailed kinematics of short-nosed fruit bat (*Cynopterus sphinx*) was conducted. We performed high-speed recordings of the kinematics to obtain 3D reconstructions of membrane wing movements. Living bats were used for performing the experiments. We used the motion capture system, *OptiTrack*, to measure the data set on the flapping membrane wings of the bat and finally rebuild the bat's flight in the form of soft wings through this experiment. However, due to the complex kinematics of the bat's wing, there are some periods when many points on the wing cannot be tracked by the motion capture system as the wing flaps. To rebuild the wing profile, a detailed analysis of every point is required. Therefore, this paper proposes numerical algorithms and *MATLAB* programs to recover the missing data from the wing during the flight of a bat. It is the purpose to find efficient calculation and analysis procedures and to provide ideas and techniques for the analysis of all marked points on the membrane wings of the bat.

Keywords Bat flight · Bionic robot · Reconstruction of the wing structure

1 Introduction

Bats are the only mammals that can naturally fly. They evolved the ability of powered flight more than 50 million years ago. Bats are more maneuverable than other birds and are a proficient flyer for heavy loading transportation at a very slow speed. Taking this opportunity many countries and regions, such as the United States, Europe, and

S. K. Singh · J.-S. Zhao (✉)
Department of Mechanical Engineering, Lee Shau Kee Science & Technology Building, Tsinghua University, Room A1033/A413, Beijing 100084, China
e-mail: jingshanzhao@mail.tsinghua.edu.cn

D. Sen et al. (eds.), *Mechanism and Machine Science*, Lecture Notes in Mechanical Engineering, https://doi.org/10.1007/978-981-15-4477-4_30

other developed regions, have set up research institutions one after another for engaging in flapping-wing flight investigation. Birds, insects, and bats, they all have different airfoils, different flight characteristics, and skills, but they all use flapping-wing flying methods, with varying superiority and efficiency of flapping-wing flights [1–3]. Being large in population, the birds, insects, and bats have become the research object for flapping-wing flight and due to this, corresponding birds and insect flapping-wing aircraft were born [4–8].

Different from birds and insects, wings of bats are a kind of elastic membrane. During the flight, the airfoil deforms greatly. It has not only the characteristics and advantages of bird flight and insect flight but also high flight efficiency [9]. Bats have the ability to fly at low speed and low altitudes, and some small bats have the ability to hover [10]. Compared to insects, hummingbirds, and other hovering creatures [11, 12], the flapping frequency of bats is very low. The bats' flight possesses excellent efficiency and remarkable maneuverability too. The hovering flights of bats are 40% and 60% less expensive metabolically than that of hawk moths and hummingbirds, respectively, for the similar body size range of large insects, small birds, and bats [13–15].

Although research on the bat was started during the eighteenth century, research on bat flight gradually gained attention in recent decades only [16–19]. In recent years, domestic and foreign scholars have gradually begun to pay more attention to bat flight research [20–24]. The analysis of the motion of bats is also one of the foci of this research. As the availability of high quality (especially speed) cameras for motion capture are increasing with the development of new technologies, the interest in quantifying animal locomotion from multiple view videos are also on the rise. Despite this interest, 3D motion reconstruction methods mainly focus on the capture and reconstruction of human motion [25–28]. So, there are very few established methods for the study of animal locomotion and flights of creatures.

The flight of bats is a very complex phenomenon. The wings of a bat are elastic membranes and have many joints like the human palm and fingers. The mechanical features of wing skin play a significant role in determining aerodynamic characteristics of the wings. The integrated motions at the many hand joints produce complex and functionally versatile dynamic wing conformations [29]. During the flying process, the wings fold, unfold, and flap. So many parts of the wing get hidden which cannot be captured by any camera even in the surrounding of multiple cameras. For the reconstruction of wings and body kinematics of a bat, we need the coordinates of all the points of wings and body. So there is the requirement to discover the coordinates of missing points from the wing.

Our contribution in this paper is a computer-based algorithm for discovering the missing points from the point cloud obtained from the motion capture (mocap) system. This will help to reconstruct the smooth flying process of the bat.

2 The Experimental Setup

The motion capture system was used to collect the position information of the marking points on the wings of bats. The experimental data were processed by numerical analysis. The reconstruction of bat wing membranes was implemented using *MATLAB,* which can provide the basis for the design of the bat's wing bionic mechanism. The specific points on the bat's wings were selected as the research object. These points were tracked during the flight of the bat in this experiment.

*2.1 Motion Capture System (***OptiTrack***)*

The motion capture system consists of two parts: hardware and software. The hardware parts contain 12 high-speed infrared cameras, PRIME 17 W, a concentrator, lines, and camera calibration tools. Cameras can capture a maximum of 360 films per second. All the cameras were powered by a concentrator through lines. In the software parts, *Motive* was used. The various tracking applications of motion capture system can be controlled by software, *Motive*. It calibrates and configures the mocap systems, as well as offers interfaces for capturing and processing of 3D data. The captured data either can be recorded or live-streamed into other channels. The main components of the hardware and the setup of the motion capture system are shown in Fig. 1.

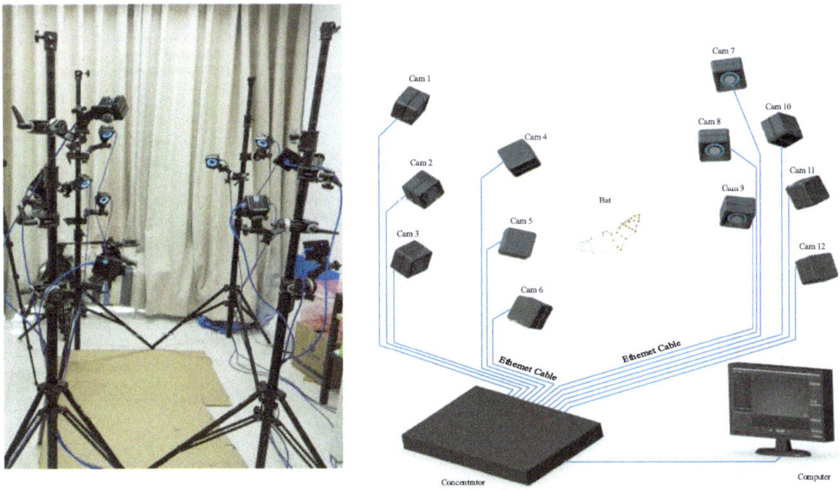

Fig. 1 Schematic diagram of the experimental setup, *OptiTrack* motion capture system

2.2 *The Principle of the Experiment*

The *OptiTrack* motion capture system uses passive infrared positioning technology. When any point of the space can be seen by two or more cameras simultaneously, the picture of that point can be captured by the system. Schematics are shown in Fig. 2. Optical motion capture systems utilize multiple 2D images from each camera to compute, or reconstruct, corresponding 3D coordinates. The camera has a frequency of 360 FPS and the bat has around 10 Hz (found by observation). So for one wingbeat, 36 frames (images) can be captured by the motion capture system.

3 Method of the Experiment

The experiments were performed at the Guangdong Institute of Applied Biological Resources, Guangzhou, PR China. A closed house was constructed to perform the experiments on living bats, such that the bat could not fly away. Five short-nosed fruit bats (*Cynopterus sphinx*) were used for the experiments. Markers were pasted on the wings and body of the bat. Once the camera was ready at their positions synchronized with the computer, the bats were allowed to fly in the house. The steps of the experiments are divided into the following categories.

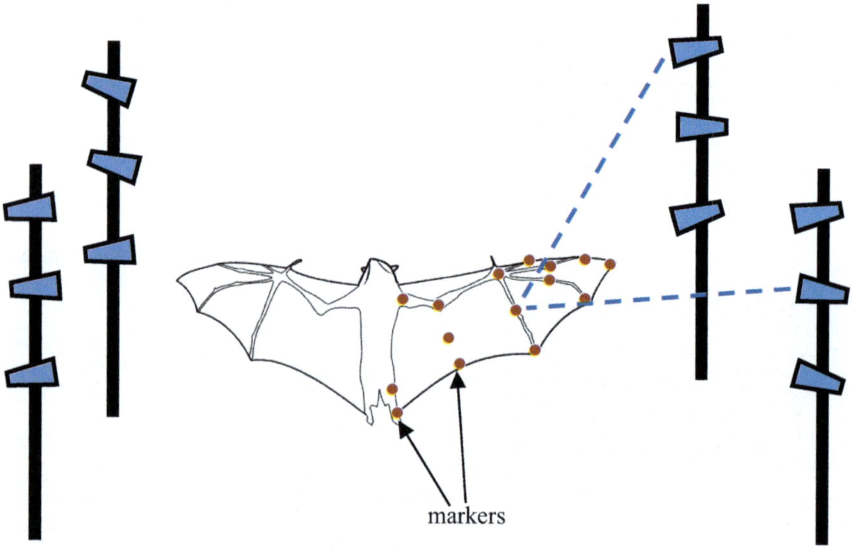

Fig. 2 Schematic of the working principle

3.1 Preparing the Capture Area

For the best tracking results, it is required to prepare and clean up the capture environment. We removed all the unnecessary obstacles that could block the camera views. We tried to keep the floor and surrounding non-reflective since LED illumination from cameras will reflect off of it. We used rubber mats to cover the reflective area. We either removed all the items with reflective surfaces or covered them with non-reflective materials in order to avoid extraneous reflections.

3.2 Placing and Aiming Cameras

The optical motion capture system utilizes multiple 2D images from each camera to compute, or reconstruct, corresponding 3D coordinates. Therefore, for the best tracking results, the cameras must be placed in such a way that each of them captures the unique vantage of the target capture area. The cameras must be mounted securely onto the stable structures (e.g. truss system) so that they do not move throughout the capture. The cameras are placed in such a way that their views overlap the most in the region where most of the capture will take place. Simultaneously it was also ensured that each point in the capture volume must be visible by a minimum of two cameras at all times. Even the small movement of any camera after calibration must be re-calibrated.

3.3 Camera Calibration

All the cameras must be first calibrated to track the 3D points. From the calibration process, *Motive* computes location and alignment of each camera as well as amounts of lens distortions in captured images. *Motive* establishes a 3D capture volume by using the calibration data, and within this volume, motion tracking is accomplished. Then setting of the ground plane was done which is the last step of the calibration process. Basically, Motive sets up a virtual 3D coordinate system through the calibration process which is used during tracking of points.

3.4 Pasting the Markers

The retro-reflective markers were placed onto the membrane wings and body of the bat. Biological glue was used for the safety of the membrane wings of the living bat. The markers must be attached securely. The selected points were mostly the joints on the membrane wings of the bat. Motive derives the position of each body segment from related markers that we place on the membrane wing and body. It is important to prevent the shifting of markers as much as possible.

Now the bat with marked points on the wings and body is ready to fly within the experimental volume of the created house. So when a bat flies, the marked points from the wings and body can be tracked and stored. The stored data will be used later for the post-processing.

4 Initial Processing of the Data

The captured data from the mocap system has no information about the marked points, i.e. which data is for which marked points. So we used an open software *MeshLab* for the same. The different processes used for getting the exact coordinates of all the available marked points are explained below.

4.1 Extracting the Data

The bagged data from the experiment were extracted and saved in the *Excel* format. This data contains all the available coordinates without any information on marked points. From this data first we chose the best 5 continuous wingbeats, 180 frames, using the software and hardware part of the *OptiTrack*. As this format of data cannot be used in *MeshLab*, a *MATLAB* program was used to convert this format of data to text format. The output of the program gave us as many numbers of files as the frames in 5 wingbeats, i.e. 180 files, in text format. Every file contains the coordinates of the available marked points from respective frames but again without any information of marked points.

4.2 Finding the Coordinates of Available Marked Points Using MeshLab

The text data obtained from the output of the *MATLAB* program were presented to the *MeshLab* one by one, which presents the point cloud as shown in Fig. 4. Now by using a couple of steps, the 3D coordinates are got with the clear information of marked points (i.e. we know the marked points corresponding to the coordinates). These coordinates are now saved as *.xyz* file format.

5 Restoring the Lost Data

The 3-dimensionally complex wing motion of a bat results in substantial self-occlusion, which makes direct tracking of marker positions tedious and often impossible. Among such markers, some can be discovered by using a numerical algorithm and some others can be discovered by using parameter fitting through *MATLAB*.

5.1 Numerical Algorithm

The data obtained from the experiment for n number of points are shown in Fig. 3. The black dots 11, 21, 31, …., n1 and p1 are one group of data, and the gray dots 12, 22, 32, …., n2 are the second group of data, respectively, indicating that the two groups of marked points are different. Actually, the first group of data belongs to the first frame and the second group of data to the next frame. Here the coordinate of point p2, which belongs to the second frame, is missing. At the moment, our job is to find out the coordinates of the unmeasured (missing) data point p2, which is indicated by a yellow dot, through the two sets of data. First, we used the data of n known points to find the coordinates of p2 and hence the position vector of the same.

$$\begin{cases}\Delta x_1 = x_{12} - x_{11}\\ \Delta y_1 = y_{12} - y_{11}\\ \Delta z_1 = z_{12} - z_{11}\end{cases} \quad \begin{cases}\Delta x_2 = x_{22} - x_{21}\\ \Delta y_2 = y_{22} - y_{21}\\ \Delta z_2 = z_{22} - z_{21}\end{cases} \quad \begin{matrix}\cdots\cdots\\ \cdots\cdots\\ \cdots\cdots\end{matrix} \quad \begin{cases}\Delta x_n = x_{n2} - x_{n1}\\ \Delta y_n = y_{n2} - y_{n1}\\ \Delta z_n = z_{n2} - z_{n1}\end{cases}$$

These groups of data obtained from n points were used as the basis for calculating the unknown data points.

Using the distance formula, we found the distances between the point p1 and the points 11, 21, 31, ……., n1, respectively:

$$a_{pi} = \sqrt{(x_{p1} - x_{i1})^2 + (y_{p1} - y_{i1})^2 + (z_{p1} - z_{i1})^2}, i = 1, 2, 3, \ldots, n \qquad (1)$$

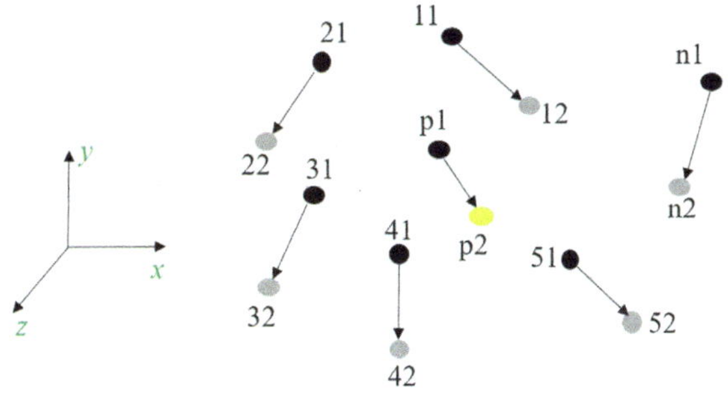

Fig. 3 The schematic of a set of points for restoring the coordinates of the unknown point

Summation:

$$a = \sum_{i=1}^{n} a_{pi} \tag{2}$$

All the points are related to each other so we are using the concept of the influence factor, which is as follows.

$$\begin{cases} k_1 = \dfrac{a - a_{p1}}{(n-1)a} \\ k_2 = \dfrac{a - a_{p2}}{(n-1)a} \\ \ldots\ldots\ldots\ldots\ldots \\ \ldots\ldots\ldots\ldots\ldots \\ k_n = \dfrac{a - a_{pn}}{(n-1)a} \end{cases} \tag{3}$$

Combining the above expressions, we can get the coordinates of point p2, which are as follows.

$$\begin{cases} x_{p2} = x_{p1} + k_1\Delta x_1 + k_2\Delta x_2 + \ldots\ldots\ldots\ldots + k_n\Delta x_n \\ y_{p2} = y_{p1} + k_1\Delta y_1 + k_2\Delta y_2 + \ldots\ldots\ldots\ldots + k_n\Delta y_n \\ z_{p2} = z_{p1} + k_1\Delta z_1 + k_2\Delta z_2 + \ldots\ldots\ldots\ldots + k_n\Delta z_n \end{cases} \tag{4}$$

In this example, we used n known points to find one unknown point. So we can use this algorithm as a general technique. In the process of actual data processing, this paper uses six known points to solve unknown points. A *MATLAB* program was used to implement this algorithm. For proof of our algorithm, we chose one known point as an unknown point and applied this algorithm. The result obtained was within the acceptable limit. The maximum error was $\pm$ 1.37% in the *x*-coordinate. Remember, there are several principles to follow when choosing known points. First, the known points must be selected on the wing in the same side of the unknown point, that is, it must not be selected on the wing of the other side of the body; second, try to select the points that are similar to the unknown point; and third, select the point as near as possible around the unknown point. Avoid being completely on the other side.

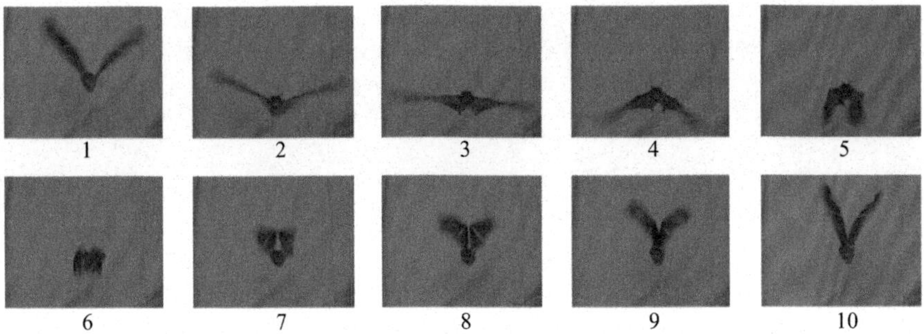

Fig. 4 Schematic of the screenshots of the flight of a bat for one wingbeat

5.2 *Parameter Fitting*

Arbitrarily ten frames were selected from one wingbeat of a flying bat, which is shown in Fig. 4. From the figure, we can see that when the bat flies to the time interval represented by 7, 8, and 9 frames, a large number of marked points on the bat wings are blocked due to the folding of the wings. This results in the inaccessibility of data. As it was not possible to find enough known points near the missing data point, the missing data points could not be recovered using the numerical algorithm discussed in the previous technique. So there is the requirement of another technique. For a marked point on a bat's wing, its position has continuity in the time dimension. Therefore, a single marked point on a bat body in one cycle can be taken as an object, and the position of this point at every instant can be obtained by parameter fitting using *MATLAB*. So if there are any missing data points, it can be recovered. This process can be continued for all the other points from one frame to restore the missing data.

In the measured point cloud, we designated several points as the research object and named them distinctly as shown in Fig. 5. The following ten points are used as samples in this research analysis. The points marked in Fig. 1–2, 3–4, 5–6, 7–8, and 9–10 are symmetrical about the midline of the body of the bat. Here points 1 and 2 are at the end of the third finger of the wings, points 9 and 10 at the end of the fourth finger, points 5 and 6 at the end of the fifth finger, points 3 and 4 at the joint of the intersection of five fingers of the wing, whereas points 7 and 8 are on the feet of the bat.

The data obtained by the preliminary processing is about the 3D coordinates of the marking points at different times. Since the infrared camera used in the motion capture system captures the frame at the rate of 360 FPS, and the bat flaps with the frequency of 10 Hz approximately (found with observation), we have selected 36 frames of data, that is, about one wingbeat. The time interval between the two frames is 1/360 s. In Fig. 6, the coordinates and trajectory of point 1 are shown which is for one complete cycle. It shows there are many intermediate points that are missing. Obviously, the missing data will be obtained in the subsequent calculations.

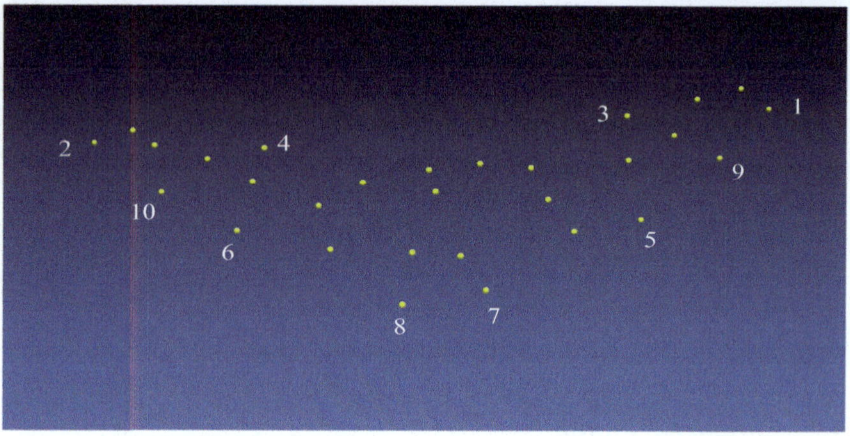

Fig. 5 Point cloud with named markers

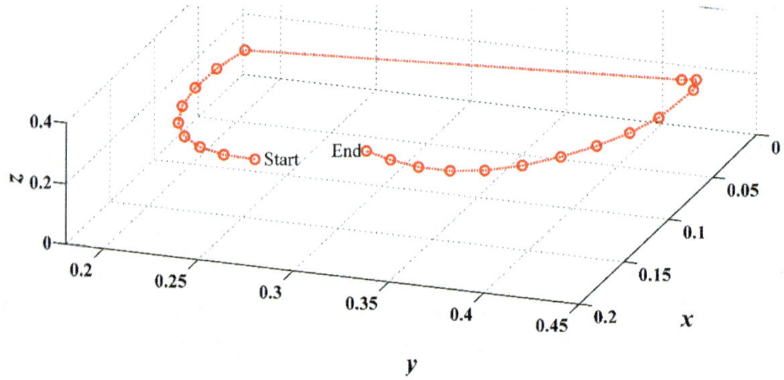

Fig. 6 Coordinates and trajectory of point 1 for one wingbeat

The coordinate data of a point from five wingbeats were presented to *MATLAB*. We used point 1, as shown in Fig. 5, for our analysis. The curve fitting tool was used to fit the presented data. In the process of fitting the data, the Fourier series fitting was used for this paper. After analyzing a different number of terms for the Fourier series, the final selected number of terms for the *x*-, *y*-, and *z*-coordinates were 7, 6, and 7, respectively. Figure 7, a 2D curve whose coordinates were simulated individually with time t for one wingbeat of the bat's flight, shows the *x*-, *y*-, and *z*-coordinates of point 1. The fitting parameter equations used are as follows:

$$\begin{aligned} x_1 = {} & -0.1965 + 0.068\cos(10.4t) + 0.193\sin(10.4t) + 0.064\cos(20.8t) + 0.077\sin(20.8t) \\ & + 0.062\cos(31.2t) + 0.037\sin(31.2t) + 0.062\cos(41.6t) + 0.004\sin(41.6t) + 0.064 cos(52t) \\ & - 0.023 sin(52t) + 0.017 cos(62.4t) - 0.120\sin(62.4t) + 0.030\cos(72.8t) + 0.023\sin(72.8t). \end{aligned}$$

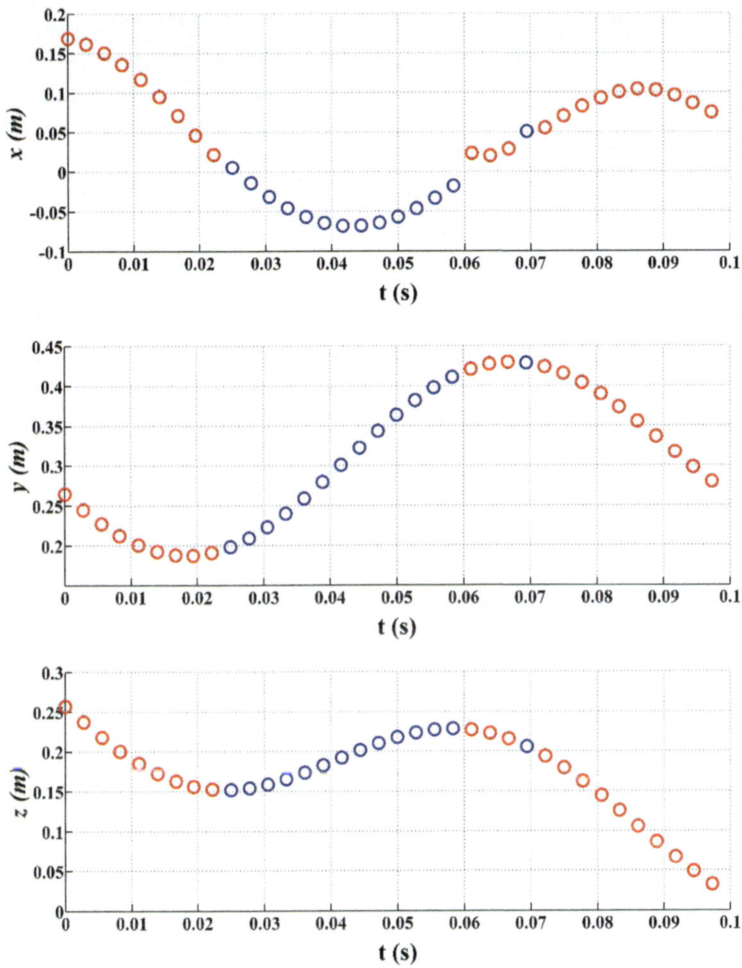

Fig. 7 Fitting curves of 3D coordinates of point 1 with time t for one wingbeat

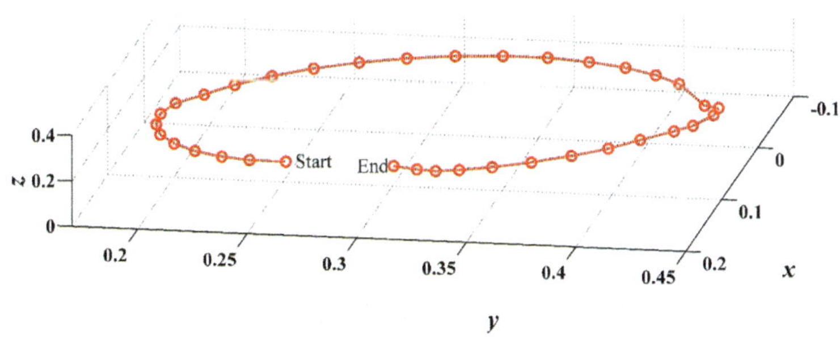

Fig. 8 The simulated coordinates and trajectory of point 1 for one wingbeat

$$y_1 = 0.4028 - 0.00071\cos(11.11t) - 0.106\sin(11.11t) - 0.03818\cos(22.22t) - 0.0356sin(22.22t)$$
$$-0.0271\cos(33.33t) - 0.0111\sin(33.33t) - 0.02355\cos(44.44t) + 0.0017\sin(44.44t)$$
$$-0.01805\cos(55.55t) + 0.0011\sin(55.55t) - 0.02711\cos(66.66t) - 0.0889\sin(66.66t).$$

$$z_1 = -0.254 + 0.1624\cos(9.423t) + 0.364\sin(9.423t) + 0.1961\cos(18.846t) + 0.1364sin(18.846t)$$
$$+0.1777\cos(28.269t) - 0.0155\sin(28.269t) + 0.0775\cos(37.692t) - 0.1011\sin(37.692t)$$
$$-0.01497\cos(47.115t) - 0.08017\sin(47.115t) - 0.0202\cos(56.538t) - 0.01805\sin(56.538t)$$
$$-0.06686\cos(65.961t) - 0.06671\sin(65.961t). \quad (5)$$

We can see from Fig. 6, the coordinates of the data points at a number of instants are missing in the middle. We can enter the required time values into Eq. (5) to obtain the corresponding data point coordinates. A *MATLAB* program was used to recover the missing data by using the parameter equation obtained above. The restored coordinates *x, y,* and *z* are separately shown in Fig. 7 by the blue dots.

6 Results

By using the above-demonstrated technique, we can obtain the coordinates of missing points from the wing and body of the bat or any other moving object. For the validation of the algorithm, we applied this technique to the coordinates of a known point and compared the obtained results.

The actual coordinates of a point obtained from the data of the experiment:
x: –0.1452, ***y***: 0.2110, ***z***: 0.2840
The calculated coordinates of the same point using the numerical algorithm:
x: –0.1432, ***y***: 0.2101, ***z***: 0.2828
The percentage error in the calculated coordinates:
x coordinate: $\pm$ | [{–0.1452 – (–0.1432)}/(–0.1452)]| $\times$ 100% $= \pm$ 1.377%
y coordinate: $\pm$ | [(0.2110–0.2101)/0.2110]| $\times$ 100% $= \pm$ 0.427%
z coordinate: $\pm$ | [(0.2840–0.2828)/0.2840]| $\times$ 100% $= \pm$ 0.423%

Apart from this, we also traced point 1 during one cycle of the bat flight by using parameter fitting to obtain the space curve shown in Fig. 8. In Fig. 8, the start and end indicate the 1st and 36th frames of one wingbeat. From the figure, it is also clear that the space curve is continuous during the flying process in one wingbeat, which is the requirement for the reconstruction of the membrane wings of the bat.

7 Conclusion

A novel computerized method was established to track the 3D coordinates of any type of complex motion. The method was verified by comparing the experimental results with simulated results for the motion of wings of short-nosed fruit bat (*Cynopterus sphinx*). The accuracies of the comparison are in good agreement with each other. Using this method, any number of marked points on the wing of a bat (or any other flapping vehicle) can be tracked and the final structure of the wing of the bat can be reconstructed.

We generated the parameter equation for the position of the membrane wings in 3D space through the fitting, which is given in equation number 5. The derivatives of this equation can be used to find the velocity and acceleration of the corresponding points. Further, it can be used to calculate the aerodynamic forces acting on the membrane wings of the bat. The kinematic analysis is the requirement of every newly designed mechanism or vice versa. So this technique can be applied to the kinematic analysis of any other flapping vehicles or mechanisms.

References

1. Bennett AG, Obye RC, Jeglum PM (1979) Swimming and flying in nature. Environ Biol Fishes 4:95
2. Hedrick TL, Cheng B, Deng X (2009) Wingbeat time and the scaling of passive rotational damping in flapping flight. Science 324:252–255
3. Usherwood JR (2016) Physiological, aerodynamic and geometric constraints of flapping account for bird gaits, and bounding and flap-gliding flight strategies. J Theor Biol 408:42–52
4. Fenelon MAA, Furukawa T (2010) Design of an active flapping wing mechanism and a micro aerial vehicle using a rotary actuator. Mech Mach Theory 45:137–146
5. Keennon M, Klingebiel K, Won H (2012) Development of the nano hummingbird: a tailless flapping wing micro air vehicle. In: 50th AIAA aerospace science meeting including the new horizons forum aerospace exposition. pp 1–24
6. Yang L-J, Esakki B, Chandrasekhar U, Hung K-C, Cheng C-M (2015) Practical flapping mechanisms for 20 cm-span micro air vehicles. Int J Micro Air Veh 7:181–202
7. Jung HK, Choi JS, Wang C, Park GJ (2015) Analysis and fabrication of unconventional flapping wing air vehicles. Int J Micro Air Veh. 7:71–88
8. Yang W, Wang L, Song B (2018) Dove: a biomimetic flapping-wing micro air vehicle. Int J Micro Air Veh 10:70–84
9. Swartz SM, Groves MS, Kim HD, Walsh WR (1996) Mechanical properties of bat wing membrane skin. J Zool 239:357–378
10. Norberg UML (1976) Aerodynamics of hovering flight in the long-eared bat *Plecotus Auritus*. J Exp Biol 65:459–470
11. Ellington CP (1984) The aerodynamics of hovering insect flight. IV. Aeorodynamic mechanisms. Philos Trans R Soc B Biol Sci 305:79–113
12. Altshuler DL, Dudley R, Ellington CP (2004) Aerodynamic forces of revolving hummingbird wings and wing models. J Zool 264:327–332
13. Winter Y (1998) Energetic cost of hovering flight in a nectar-feeding bat measured with fast-response respirometry. J Comp Physiol B Biochem Syst Environ Physiol 168:434–444
14. Voigt CC, Winter Y (1999) Energetic cost of hovering flight in nectar-feeding bats (Phyllostomidae: Glossophaginae) and its scaling in moths, birds and bats. J Comp Physiol–B Biochem Syst Environ Physiol 169:38–48
15. Winter Y, Von Helversen O (1998) The energy cost of flight: do small bats fly more cheaply than birds? J Comp Physiol B Biochem Syst Environ Physiol 168:105–111
16. Tian X, Iriarte-Diaz J, Middleton K, Galvao R, Israeli E, Roemer A, Sullivan A, Song A, Swartz S, Breuer K (2006) Direct measurements of the kinematics and dynamics of bat flight. Bioinspir. and Biomim. 1:S10–S18

17. Bergou AJ, Swartz S, Breuer K, Taubin G (2011) 3D reconstruction of bat flight kinematics from sparse multiple views. Proc IEEE Int Conf Comput Vis 1618–1625
18. Hedenström A, Johansson LC, Wolf M, von Busse R, Winter Y, Spedding GR (2007) Bat flight genenrates complex aerodynamic tracks. Nature 316:894–898
19. Muijres FT, Johansson LC, Barfield R, Wolf M, Spedding GR, Hedenström A (2008) Leading-edge vortex improves lift in slow-flying bats. Science 319:1250–1253
20. Hubel TY, Riskin DK, Swartz SM, Breuer KS (2010) Wake structure and wing kinematics: the flight of the lesser dog-faced fruit bat, *Cynopterus brachyotis*. J Exp Biol 213:3427–3440
21. Ramezani A, Chung S-J, Hutchinson S (2017) A biomimetic robotic platform to study flight specializations of bats. *Sci Robot.* **2**, eaal2505
22. Bahlman JW, Swartz SM, Breuer KS (2013) Design and characterization of a multi-articulated robotic bat wing. Bioinspir. Biomim. 8:016009
23. Tokita M, Abe T, Suzuki K (2012) The developmental basis of bat wing muscle. Nat. Commun. 3:1302–1309
24. MacAyeal LC, Riskin DK, Swartz SM, Breuer KS (2011) Climbing flight performance and load carrying in lesser dog-faced fruit bats (*Cynopterus brachyotis*). J Exp Biol 214:786–793
25. Hua G, Wu Y (2007) A decentralized probabilistic approach to articulated body tracking. Comput Vis Image Underst 108:272–283
26. Deutscher J, Reid I (2005) Articulated body motion capture by stochastic search. Int J Comput Vis 61:185–205
27. Wang JM, Fleet DJ, Hertzmann A (2008) Gaussian process dynamical models for human motion. IEEE Trans Pattern Anal Mach Intell 30:283–298
28. Stenger B, Mendonca PRS, Cipolla R (2001) Model-based 3D tracking of an articulated hand. In: Proceedings of the 2001 IEEE Computer Society Conference on Computation Visual Pattern Recognition CVPR 2001. **2**, II-310-II–315
29. Swartz S, Iriarte-Diaz J, Riskin D, Tian X, Song A, Breuer K (2007) Wing structure and the aerodynamic basis of flight in bats. In: 45th AIAA Aerospace Sciences Meeting Exhibiton pp 1–10

Dynamic Analysis of a Legged Lander System

Umesh Singh, Manish Trikha, and Ashitava Ghosal

Abstract Interplanetary missions have increased in the recent years, with spacecrafts landing successfully on Moon and Mars. The entry, descent, and landing on another planet is still a challenging job, since earlier few missions were unsuccessful. The lander is a mechanical system which is a combination of links and joints which provides the soft-landing platform for the spacecraft. The lander needs to be designed such that it can absorb the maximum impact energy and provide stable landing. Dynamic analysis of a lander is required in order to understand the behavior of impact forces on the lander. The main objective of this paper is to analyze an existing legged lander system configuration which will be capable of withstanding and attenuating the impact force. ADAMS simulation of a single-legged system is carried out for a range of terminal velocities. The simulation results are compared with the results of a single-legged system drop test.

Keywords Impact · Kinematics · Damper · Lander · Dynamics

1 Introduction

The curiosity of humans to explore other planets of our solar system has motivated interplanetary missions. Many spacecraft have been sent by different countries for the exploration of other solar system planets. Luna1 was the first orbiter sent by USSR and then it was followed by many missions such as MESSENGER (Mercury), Akatsuki (Venus), Luna10, Lunar orbiter1, Apollo and Chandrayaan1 (moon), Mars, Viking and Mars Orbiter Mission, etc., (mars), Juno (Jupiter), Cassini (Saturn), and Rosetta(comet). These spacecrafts have landed on the surface to understand the

U. Singh (✉) · M. Trikha
U R Rao Satellite Centre, Indian Space Research Organisation, Bangalore, India
e-mail: usingh@ursc.gov.in

M. Trikha
e-mail: mtrikha-hsfc@isro.gov.in

A. Ghosal
Department of Mechanical Engineering, Indian Institute of Science, Bangalore, India
e-mail: asitava@iisc.ac.in

D. Sen et al. (eds.), *Mechanism and Machine Science*, Lecture Notes in Mechanical Engineering, https://doi.org/10.1007/978-981-15-4477-4_31

physical properties of the soil and to explore the possibilities to sustain life. Different landing techniques like airbag, Sky Crane, and legged lander system [1] have been designed and used by space agencies according to different impact attenuating systems for the safe landing. Airbag landing technique was used for landing of SPIRIT and OPPORTUNITY on Mars. Sky Crane was used for the landing of CURIOSITY on Mars. A legged lander system was used for the landing of APOLLO (on Moon) and VIKING (on Mars). The legged landing gear is generally used to land where the atmosphere is not present and space organizations from the USA, USSR, JAPAN, and ESA have already sent their lander to the other terrestrial body. Surveyor-7 was the first unmanned legged lander sent by NASA which successfully landed on the Moon in 1968. Apollo-11 [11] lunar module was the first manned mission which landed on the Moon in 1969. Viking-1 was the first unmanned legged lander by NASA which successfully landed on Mars in 1976. Philae Lander was the first to land on a comet by ESA in 2014. Apollo lunar landings and Viking Mars landings missions are the most popular missions in the space era. The systems and the mission designed techniques for Apollo and Viking missions are the foundation for many missions now.

Even though previous landers have safely landed and the missions were successful due to the environmental challenges and landing uncertainties, there is significant interest in the scientific communities to understand more about the kinematics, dynamics, and landing stabilities of the legged landers. During landing the exact impact height, lander orientation, boulders, and craters and in some cases about wind speed and wind direction where the atmosphere present is uncertain. This makes landing very challenging and so giving opportunities to design reliable, robust, and safe lander for future missions. For legged landers, rock hazards are one of the largest challenges. Legged landers built so far have had from 20 to 30 cm of ground clearance (after leg stroke for landing load attenuation). Terminal descent thrusters cannot spend any more than a few hundreds of milliseconds within a meter or so of the surface without digging trenches, launching small rocks into the landing gear, and producing destabilizing ground effect backpressure on the bottom of the lander. For this reason, legged landers with integrated propulsion systems approach the ground at relatively high speed (2.4 m/s) that has the converse effect of increasing susceptibility to slope-induced tip-over hazards.

Many researcher are working today to study the impact interaction phenomena of the legged lander with the soil of planets (see, for example, [2–5, 12]). Several researchers are working on different energy dissipation techniques [1] during landing and some are working on the study of stable landing [6–8].

The landing gear design requires the understanding of kinematics and dynamics of the lander and its interaction with the soil during landing. The impact force transmitted to the lander structure depends critically on the contact stiffness of the lunar soil. Some of the recent literature talks about the contact force model [1, 5] but determining contact stiffness for the lunar soil still remains a challenge. To develop an accurate contact force model, experimental test data is required. In the absence of experimental data and an accurate contact model for the lander and lunar soil, we have used a Hertzian contact model between the metal footpad and the lunar soil which is assumed to be having an elastic behavior. The objective of this paper is to understand the kinematics and dynamics of the legged lander system. One-leg model is generated in CAD software and dynamic analysis is done to study the dynamics of the lander. ADAMS simulation of a single-legged system is carried out and the impact force transmitted to the system is determined. Later simulation studies are done for a range of velocities for two different drop heights. Results are compared with the actual test data from a prototype lander. It is shown that the average impact force is found to be about 11,000 N and 11,060 N from simulations and test data gives a value of about 11,295 m and 10,090 N for 300 mm and 800 mm drop height, respectively.

2 Legged Lander Systems and Mathematical Modeling

The legged lander system is generally used where low impact energy during landing is required. It is also known as a soft lander. Impact velocity is reduced during landing by using thruster firing and at a certain height the thruster is switched off and the lander makes free fall on the ground. Free-fall height is fixed and sensed by an onboard sensor and so kinetic energy during impact is approximately known. A Legged lander can be of 3, 4, 5-legged configuration system or with more legs. From a stability point of view, minimum of three legs are required but four-legged lander system is chosen so that energy absorption by each leg will be reduced and for a more stable design. Five legs and more can be chosen for landing, but more number of legs will increase the lander mass and reduce the payload mass. Due to the launcher restriction of the total payload mass, we need to design the lander with less mass, more stiff and stable. By analyzing all input requirements and taking stability as the main criteria for lander design, we have chosen four-legged lander system for analysis. The four-legged lander system was earlier used for most of the interplanetary missions by different space agencies and it was successful.

A single leg of four-legged lander system is modeled in CAD software. The single leg of a lander was modeled with four links: (1) cylinder, (2) piston, (3) knee, and (4) deck. There are four joints, namely piston-cylinder (slider joint), cylinder-deck (revolute joint), knee-deck (revolute joint), and between piston and knee (revolute joint).

The legged landing gear is an arrangement of primary and secondary struts (see Fig. 1) arrangement which is fixed at the bottom of the lander structure. Figures 2 and 3 also show the legged landing gear in two other views.

The primary strut is the piston-cylinder arrangement filled with honeycomb damper inside which absorbs the landing impact energy by crushing the honeycomb. The secondary strut is a hollow rod which allows the primary strut piston to move inside the cylinder during impact and crush the honeycomb to absorb the impact energy and to take the bending loads during impact. This lander is equipped with the lander footpad which provides the bearing area for landing during landing. The energy is transferred from the footpad to the primary and secondary strut and then to the structure during landing.

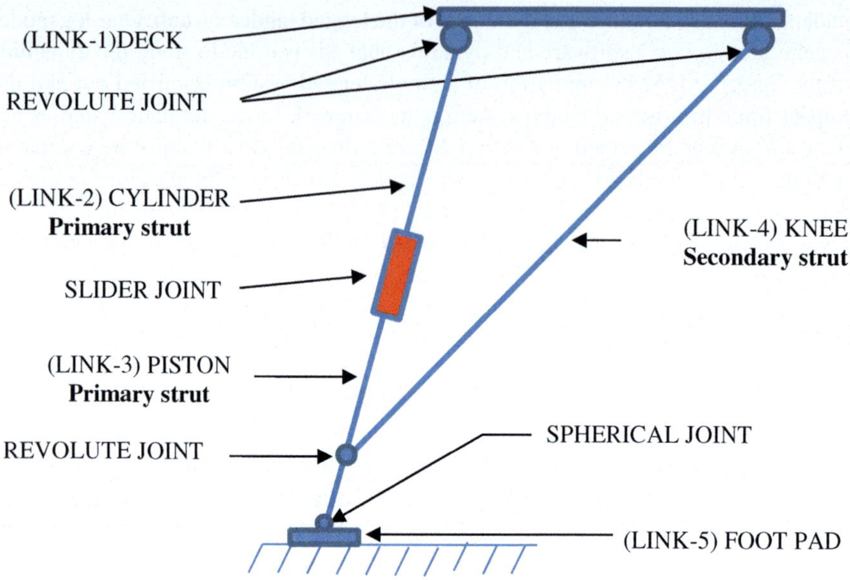

Fig. 1 Mathematical model of a single leg of the lander system

Fig. 2 Side view

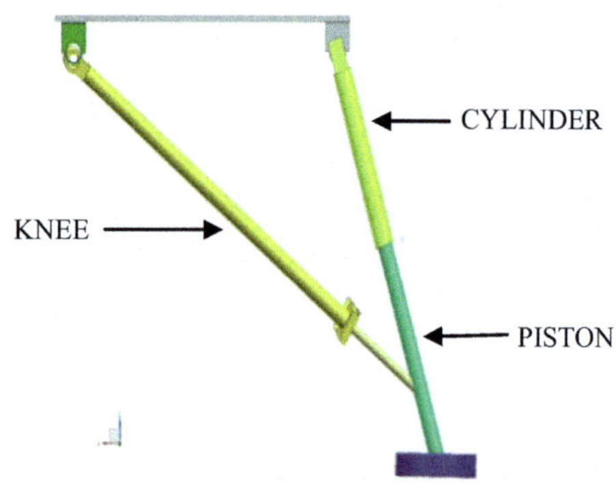

Fig. 3 Front view

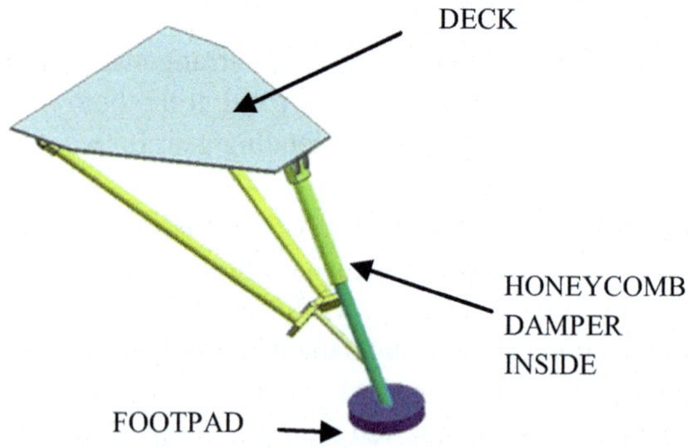

The following assumptions are made in the mathematical model:

- Links are assumed to be rigid.
- Contact between lander footpad and surface is assumed as Hertzian contact.
- The impact surface is made of aluminum alloy.

The mathematical model consists of four links and four joints as shown in Fig. 1. The degree-of-freedom (DOF) of the model is calculated using the well-known Grubler–Kutzbach criterion

$$\mathrm{DOF} = 3(N - J - 1) + \sum F_i \tag{1}$$

where N is the number of links, J is the number of joints, and F_i is the degree-of-reedom at joint i. In this case $N = 4$, $J = 4$, and F_i is 1 for all joints. This gives a DoF of 1.

The primary leg (cylinder, piston, and footpad) are made up of aluminum alloy and the secondary leg (knee) is made up of titanium alloy. An aluminum honeycomb core is used in the primary strut as a damper for energy absorption. The density of aluminum alloy and titanium alloy used is 2.8 g/cm^3 and 4.4 g/cm^{3,} respectively.

3 Simulation

The simulations of a single leg of the lander system are carried out for different parameters using ADAMS software. The details of a single leg of the lander system are given in Table 1. Two different heights are considered for the free-fall impact simulations and dynamic analysis is carried out for 300 and 800 mm drop height. The contact stiffness (K) between the footpad and the impact surface is calculated using Hertzian contact theory as follows:

$$K = 2 * a * E^{\circ} \tag{2}$$

Table 1 Single leg lander detail

Entity	Value
Total mass of lander	131 kg
Mass of the single-leg system	8.5 kg
Maximum vertical velocity	3 m/s
Contact stiffness	3.6 E +006 N/mm
Damping	<0.1% of contact stiffness N/m^2/cm
Coefficient of friction	0.15

where "*a*" is equal to the radius of the footpad and $1/E^{\circ} = (1-\nu_1^2/E^1) + (1-\nu_2^2/E^2)$. "$E^{\circ}$" is equivalent Young's modulus during contact; "ν" is equal to poisons ratio of the material. For simulating contact force in ADAMS, contact stiffness and damping ratio are required. Since the present lander impact experiment was done on an aluminum alloy metal surface, there is a metal to metal (Al–Al) contact and this contact is best modeled by using a Hertzian contact model [9, 1]. The aluminum honeycomb damper is modeled as a spring with force deflection characteristics (Fig. 4).

Figure 5 shows the contact force during the impact at the surface and Fig. 6 shows the force at cylinder and deck joint during impact. Figure 7 shows the displacement of the piston inside the cylinder during impact. The simulation results for 800 mm and 300 mm height are shown in Figs. 5, 6, 7, 8, 9, and 10, respectively. The comparison of simulation results for 300 and 800 mm is given in Table 2.

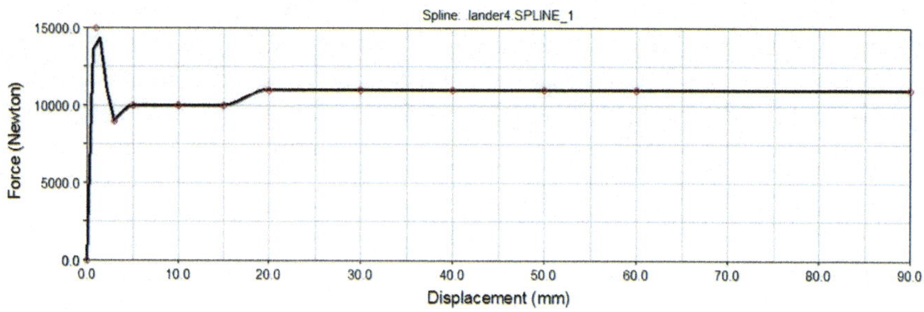

Fig. 4 Spline force used to model honeycomb damper

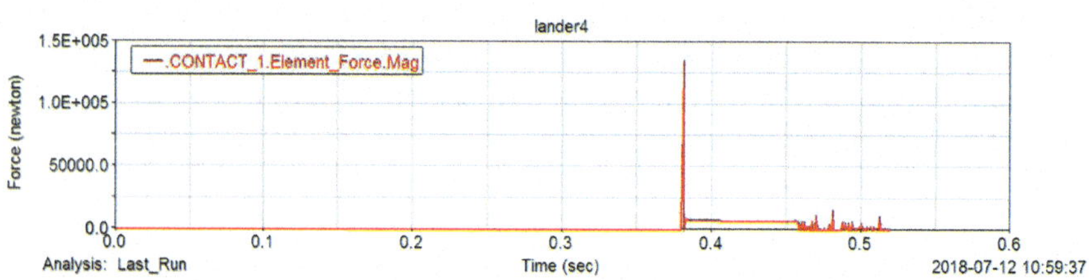

Fig. 5 Contact force for 800 mm height

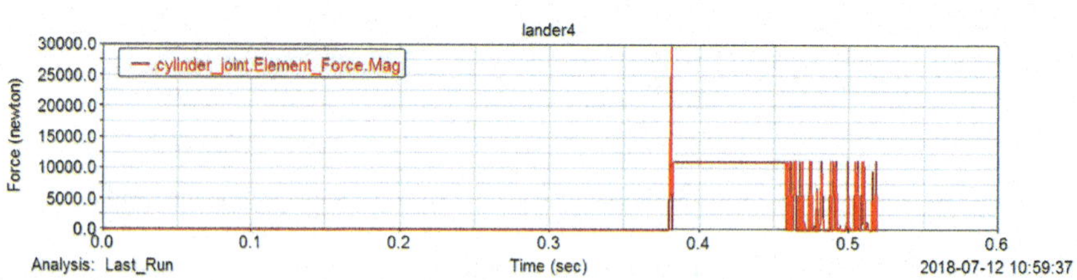

Fig. 6 Cylinder joint force for 800 mm height

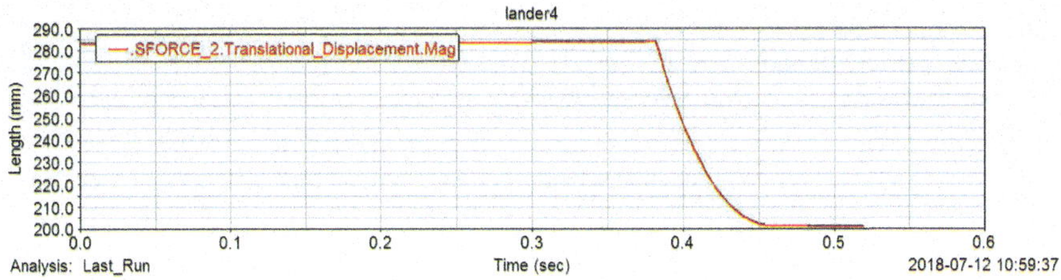

Fig. 7 Piston displacement during impact for 800 mm height

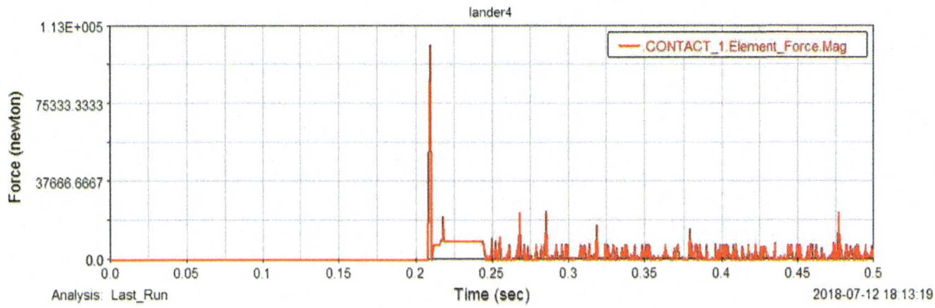

Fig. 8 Contact force for 300 mm height

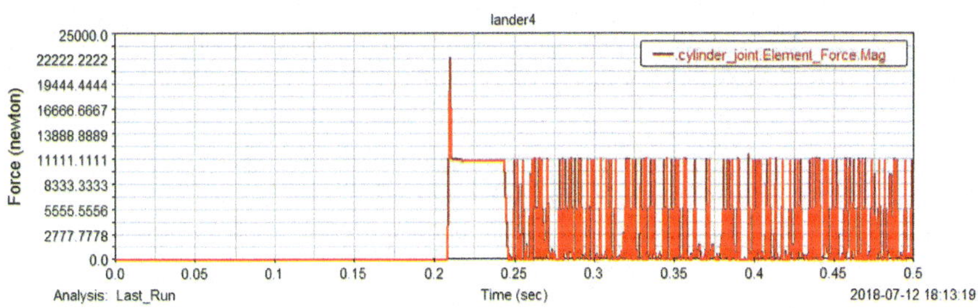

Fig. 9 Cylinder joint force for 300 mm height

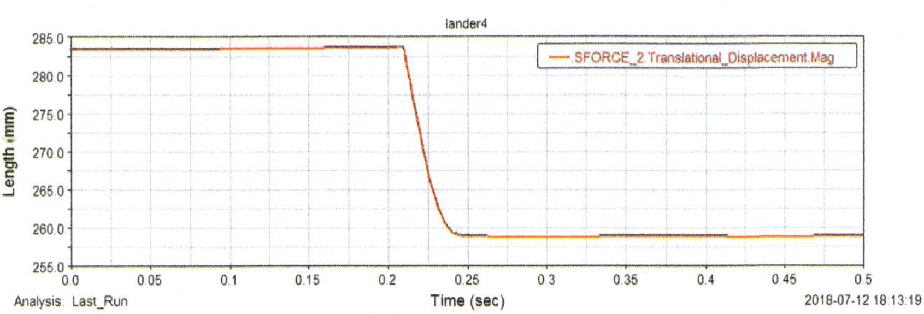

Fig. 10 Piston movement for 300 mm height

Table 2 Comparisons of simulation data for 800 and 300 mm drop height

Description		800 mm	300 mm
Contact force (N)	Max	1,35,400	1,03,390
	Avg.	7104.5	8,858
Cylinder joint force (N)	Max	29,700	22,358
	Avg.	11,059	11,000
Knee joint force (N)	Max	1,08,000	54,965
	Avg.	5,800	3,547
Knee piston joint force (N)	Max	1,12,040	59,022
	Avg.	5,600	3,300
Piston-cylinder joint force (N)	Max	27,880	18,622
	Avg.	1,525	1,103
Piston displacement(mm)		81.45	24.51
Impact duration(ms)		68.2	38.6

4 Comparisons with Test Data

The validation of simulation results is done by comparison with drop test data (Ref. [10]). The test is done on a single leg of the lander system for two different heights, i.e., 300 and 800 mm. A force measuring device (FMD) is used to measure the impact force. The force is measured at the cylinder to the deck joint. For dropping the lander from a specific height, a guide rod was used. The total mass during the drop test including the leg and support equipment is around 138 kg. Figure 11 shows the real arrangement of the test model of the lander single leg.

A comparison of the magnitude of force at the cylinder and deck interface for test data and simulation data is shown in Table 3.

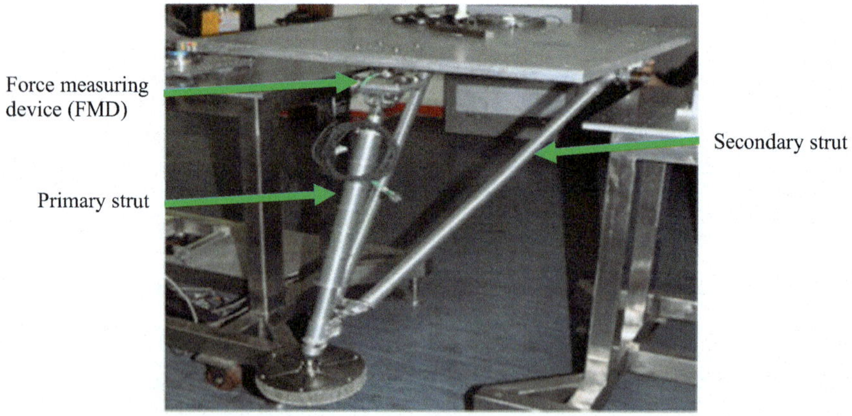

Fig. 11 Test model of a single leg of the lander (from Ref. [10])

Table 3 Comparisons of test and simulation results for cylinder joint force

Test no.	Height (mm)	Avg. force (N) from test	Avg. force (N) from simulation	% error
1	300	11293.84	11,000	2.60
2	800	10088.51	11,059	9.62

Table 4 Comparisons of test and simulation data of piston displacement

Test no.	Height (mm)	Piston movement (mm)-test	Piston movement (mm)-simulation	% error
1	300	23.6	24.51	3.8
2	800	96.5	81.45	15.6

A comparison of piston displacement after impact for simulation result and test results is given in Table 4.

5 Conclusions

In this paper dynamic analysis of a legged lander system is performed to understand the kinematics and dynamics of the system. An existing legged lander system is considered for analysis. A CAD model of a single leg of four-legged lander system is prepared and dynamic simulations are done in ADAMS® software. A case study of the impact of the lander for different heights is done. Spline force function is used to model honeycomb damper which has nicely simulated the damping force. Simulation and test results for lander impact are compared for the cylinder joint force at the deck for two different heights. The results are within a 10% error for the resultant force at the deck interface except for the initial spike seen in the simulation result. Piston movement comparison between test and simulation result is done for both the heights and the error is within 16%. The dynamic study will serve as an initial baseline for further investigation. We also plan to analyze different configurations and different impact surfaces and stresses generated from the impact.

Acknowledgements We are very thankful to Shankara A. and Gaurav Sharma of the Indian Space Research Organisation (ISRO) for their kind support to provide necessary help at various stages of the realization of this paper. We also thank Shubhakant Swain, CSM software for providing support in ADAMS simulations.

References

1. Lavender RE (1964) Touchdown dynamics analysis of spacecraft for soft lunar landing. In: NASA technical note NASA TN D-2001
2. Xu L, Nie H, Wan J, Lin Q, Chen J (2011) Analysis of landing impact performance for Lunar Lander based on flexible body. In: IEEE international conference on computer science and automation engineering
3. Sharma G, Shankara A, Nagaraj BP, Sureshakumar HN (2015) Stability simulation of scaled rigid lunar lander model under different loading condition. In: 9th national symposium and exhibition on aerospace and related mechanisms, ISRO satellite centre, Bengaluru, India, January 30–31, 2015
4. JinBao C, Hong N, JunLin W, Qing L (2014) Investigation on landing impact dynamic and low-gravity experiments for deep space lander. Mech Astron 57(10):1987–1997
5. Huang B, Jiang Z, Lin P, Ling D (2015) Research on impact process of lander footpad against simulant lunar soils. Hindawi Publishing Corporation, Shock and Vibration, Volume 2015, Article ID 658386
6. Liu Y, Song S, Li M, Wang C (2017) Landing stability analysis for lunar landers using computer simulation experiments. Int J Adv Robot Syst 2017:1–15
7. Lin Q, Kang Z, Ren J, Zhao Q, Nie H (2014) Impact analysis of Lunar lander soft landing performance caused by the body gravity centerline shift. Proc Amer Soc Civil Eng. https://doi.org/10.1061/(asce)as.1943-5525.0000441
8. Fachbereich Produktionstechnik V (2015) Touchdown dynamics and the probability of terrain related failure of planetary landing systems. Tag der mundlichen prufung 20
9. Giesbers J (2012) Contact mechanics in MSC ADAMS. Bachelor Thesis, University of Twente
10. Sharma G, Nagraj BP (2017) Chandrayaan-2 static and dynamic of lunar landing leg from launch to touchdown. ISRO internal document No. ISRO-ISAC-CHANDRAYAAN-2-AR-1258
11. Rogers WF (1972) Apollo experience report-Lunar module landing gear subsystem. In: NASA technical note, Report no NASA TN D -6850
12. Weixiong RG (2017) Preliminary design of reusable lunar lander landing system. Space engineering. Master's Thesis, Luleå University of Technology, Sweden

Development and Demonstration of a Wheat Harvester for Small Farms of India

Harsh Modi and Vineet Vashista

Abstract In this project, we focus on developing wheat harvester for small farms of India. Farming has become efficient today since farmers have started adopting modern technologies such as tractors, threshers, etc. Still, farmers having small farms in India cannot afford such machines. Also, these machines are not designed to work on small farms. So, in small farms, manual labor is used for basic farming operations such as harvesting. It costs farmers concerning both, the money and the time. Also, people working as labor in such farms develop health issues such as back pain as they need to sit in an awkward position on the ground while harvesting. In our research, we have noticed that around 20 people are required to harvest the wheat in around 0.25-acre land in 1 day. So, this wheat harvester has the potential to reduce the labor force required tremendously.

Keywords Chainsaw · Scythe · Shear strength · Wheat straw · Pedaling rate

1 Introduction

Wheat is the second largest agricultural crop in India by economic value, as per the year 2009 data [1]. Its production in the year 2017 has been 98.510 million metric tons in India [2]. One of the vital tasks in the production of wheat is harvesting. The technology of harvesting has improved (e.g. combine harvester) and has made the production of wheat efficient in the developed countries [3]. But in the developing countries like India, some farmers are still very far from adopting such technology.

The average farm size of India in the years 2010–11 was 1.15 ha (2.84 acres) [4] compared to about 427 acres [5] in the USA at that time. Also, the average farm size has been reducing each year in India [6]. The area held by marginal and small farms (less than 2 ha or 4.94 acres) is about 44.58% [7] of the total operational agricultural

H. Modi · V. Vashista (✉)
Indian Institute of Technology Gandhinagar, Ahmedabad, India
e-mail: vineet.vashista@iitgn.ac.in

H. Modi
e-mail: modi.harsh@btech2014.iitgn.ac.in

D. Sen et al. (eds.), *Mechanism and Machine Science*, Lecture Notes in Mechanical Engineering, https://doi.org/10.1007/978-981-15-4477-4_32

land holdings of India. So, the technology designed to work in big farms, which are generally found in countries like the USA, is not efficient in small farms of India.

Currently, farmers having such small farms in India use manual harvesting method, which involves sitting down on the ground and cutting the wheat with the use of a sickle. This process is time-consuming and requires huge labor force. So, the idea behind this project is to solve the problems of such farmers, who cannot use big machinery on their farms. It was an attempt to make the wheat harvester as affordable as possible to farmers, with minimal or no usage of an external power supply.

2 Existing Solutions

There have been many technological innovations which have made wheat harvesting easier. But there are some problems with such technologies. Combine harvesters [3], which do reaping, threshing, and winnowing in a single process, are costly products, and they are suitable for big farms. Image of such combine harvester can be seen in Fig. 1. For small farms, machines as shown in Figs. 2 and 3 exist. But they require an internal combustion engine to power them, which makes them costly [8]. Also for the machine shown in Fig. 2, it can cause back pain to the person carrying the machine. So the idea behind this project was to make a wheat harvester, which uses no external power or minimum possible external power, and can be very affordable to farmers. Also, it should be suitable for farmers having a farm size of around 0.25 acres.

Fig. 1 Combine harvester (*Image source* https://pixabay.com/en/harvester-wheat-field-crop-heaven-2526334)

Fig. 2 Small internal combustion engine powered wheat harvester (*Image source* https://youtu.be/sjasNdqqaBY)

Fig. 3 Small internal combustion engine powered wheat harvester (*Image source* https://youtu.be/knNBGHMjfvQ)

3 Methodology

The project started with an exploration of the different existing solutions of small-scale wheat harvesters. Initially, the wheat harvester without any external power was conceptualized. The design was prepared considering the learning from the previous project [9]. In this project, the primary cutter was chosen to be chainsaw (considering the requirement of continuous cutting) and no other similar mechanisms

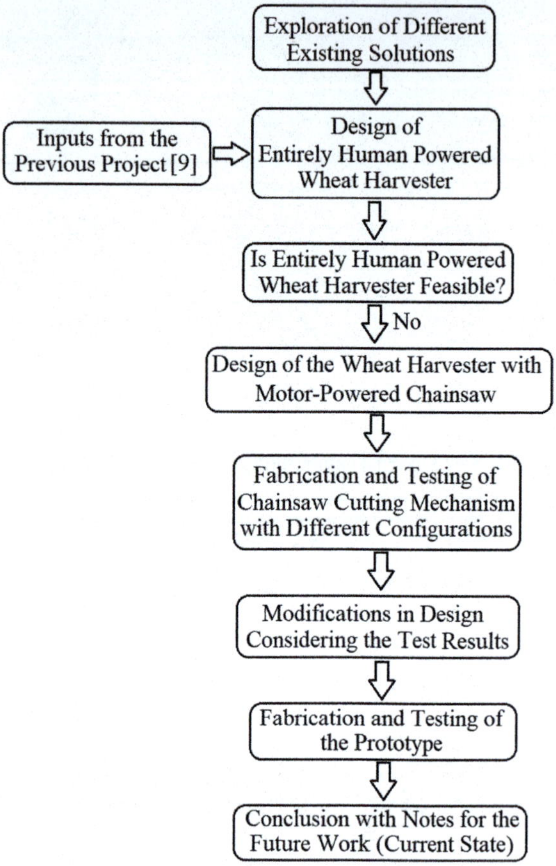

Fig. 4 Flowchart of the project

were explored. The feasibility calculations for such an entirely human-powered wheat harvester were done, and it was determined that the entirely human-powered wheat harvester is not possible. Afterwards, the wheat harvester with the use of minimal external power requirement (i.e. DC motor) was designed. Along with this, the primary cutting mechanism (i.e. chainsaw) was also tested with different configurations. After the testing and design modifications, the prototype was fabricated and tested. The flowchart of the overall methodology can be found in Fig. 4.

3.1 Chainsaw as a Primary Cutter

This project has been inherited from the earlier project [9]. It was a human-powered wheat harvester, which used scythe as a primary cutter. It was utilizing quick return mechanism for the movement of the scythe blade. The mechanism also had a return

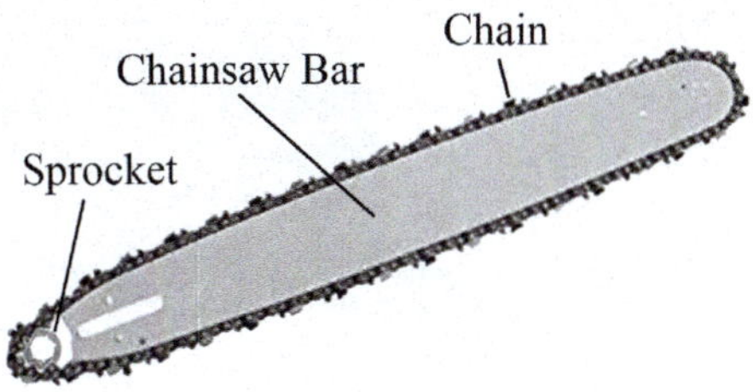

Fig. 5 Chainsaw bar and chain (*Image source* https://goo.gl/Z9qQs9)

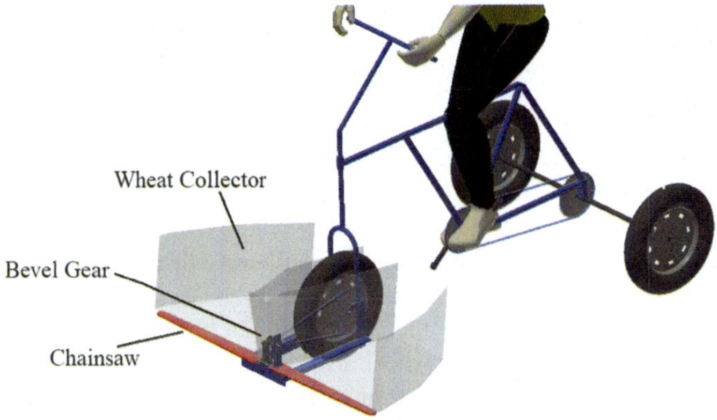

Fig. 6 CAD model of entirely human-powered wheat harvester

stroke of the scythe in which the cutting was not happening. Geneva mechanism was used for non-continuous movement of the vehicle, so that vehicle does not move forward during the time of return stroke. It was not efficient as it required acceleration and de-acceleration of the vehicle.

So in this project, it was desired to use a cutting mechanism which can cut the wheat continuously. After brainstorming, we came up with the idea of using a chainsaw as a primary cutter. The IC engine powered chainsaws have been used widely for cutting the wood. So, the idea was that the chainsaw mechanism would be capable of cutting less stronger wheat with very less requirement of power (Fig. 5).

3.2 Concept of Entirely Human-Powered Wheat Harvester

The ultimate goal of this project was to make the wheat harvester as affordable as possible to the farmers. So, the first idea was to make it entirely human-powered. The CAD model of such conceptual wheat harvester was prepared as shown in Fig. 6.

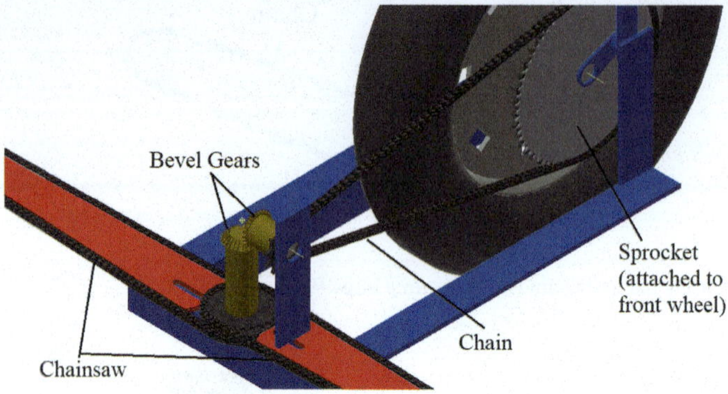

Fig. 7 Mechanism which powers the chainsaws

As per the design, the person sitting on the seat does the pedaling. The pedaling rotates the sprocket attached to the pedals. The rotation of the sprocket causes the chain attached to it to rotate and the chain would eventually rotate the sprocket attached to the rear axle of the harvester. This will move the harvester in the forward direction.

The mechanism which powers the chainsaws can be seen in Fig. 7. Due to the motion of the harvester, the front wheel also rotates with the same rotation speed at which the rear wheels rotate. The front wheel rotates the sprocket attached to it, and the rotation is transferred to chainsaw sprockets via chain and bevel gears. The desired gear ratio can also be achieved here using sprockets having a certain number of teeth.

Before moving ahead with the concept, the power available by pedaling and the power required for the wheat harvesting were compared.

1. Shear strength of wheat straw $\tau = 29.57\,\text{MPa}^1 = 29.57 \cdot 10^6\,\text{Pa}$ [10][1],
2. Pedaling power (Optimum)[2] $P_{\text{pedal}} = 60\,\text{W}$ [11],
3. Pedaling rate (Optimum)[2] $\omega = 50\,\text{RPM} = 5.23\,\text{rad}/s$ [11],
4. External diameter of wheat straw[3] $D_{\text{out}} = 3.7\,\text{mm} = 3.7 \cdot 10^{-3}\,\text{m}$,
5. Internal diameter of wheat straw[3] $D_{\text{in}} = 1.9\,\text{mm} = 1.9 \cdot 10^{-3}\,\text{m}$,
6. Number of wheat straws[3] $N = 26/\text{m}$,
7. Gear Ratio $G = \frac{44}{18}(\frac{\text{Input}}{\text{Output}})$,

[1]The shear stress was taken from the reference [10] for 2 straws and 27% moisture content. This was the largest value of the shear stress mentioned in the reference [10]. It was taken to ensure that the machine is designed to the highest load possible. It was also assumed that the shear stress values mentioned in the reference [10], which were determined from the region other than this project's region, would be applicable in determining the required power. Also, the experiment done in the reference [10] has taken the wheat straws with the average cross-sectional area of $2.32\,\text{mm}^2$, where the cross-sectional area of the wheat straws in this project's scope was 7.91mm^2. In the further calculations, the force is determined by Force = Shear Stress × Cross-sectional Area, which takes the cross-sectional area of the wheat straw into account. So the shear stress values were assumed to be independent of the cross-sectional area.

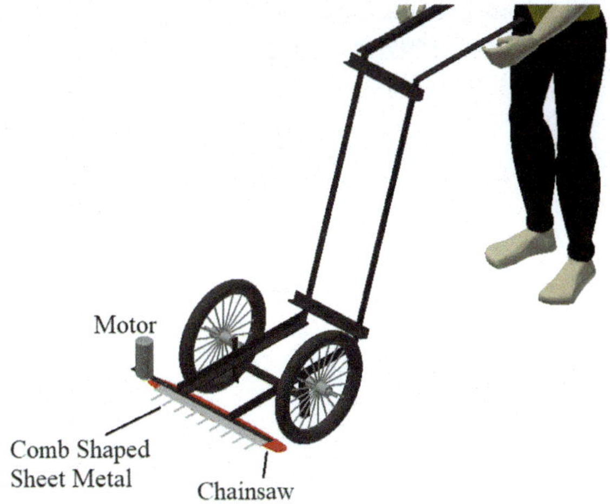

Fig. 8 CAD model of wheat harvester with motor-powered chainsaw

8. Diameter of the sprocket driving chainsaw $D_{\text{chainsaw}} = 3\,\text{cm} = 3 \cdot 10^{-2}\,\text{m}$,

$$\begin{aligned}\text{Force required } \mathrm{F}_{\text{Required}} &= \tau \times \frac{\pi}{4} \times (D_{\text{out}}{}^{2} - D_{\text{in}}{}^{2}) \times \mathrm{N} \\ &= 6.09 \cdot 10^{3}\,\mathrm{N}\end{aligned} \tag{1}$$

$$\begin{aligned}\text{Power required } \mathrm{P}_{\text{Required}} &= \mathrm{F}_{\text{Required}} \times \frac{D_{\text{chainsaw}}}{2} \times G \times \omega \\ &= 1168.61\,\mathrm{W}\end{aligned} \tag{2}$$

$$\text{So, } \mathrm{P}_{\text{required}} > \mathrm{P}_{\text{pedal}} \tag{3}$$

From the above calculations, it was evident that just the pedal power will not be sufficient to drive the chainsaw. This is also not considering frictional losses and power required to drive the vehicle.

3.3 *Wheat Harvester with Motor-Powered Chainsaw*

As the entirely human-powered wheat harvester was not feasible, it was decided to use a small motor to drive the chainsaw. The CAD model of the wheat harvester with a motor-powered chainsaw is shown in Fig. 8.

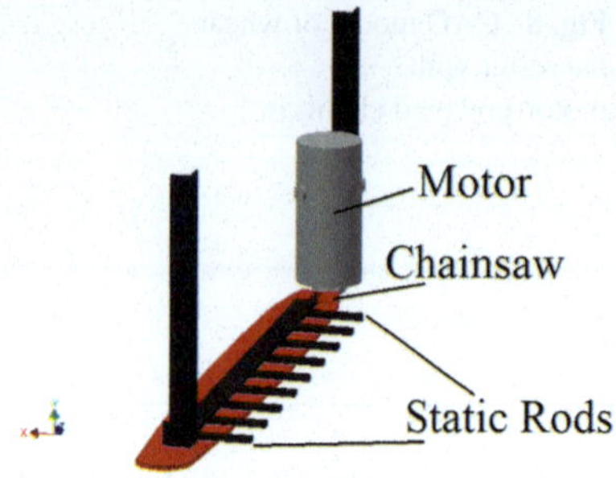

Fig. 9 Testing setup having static rods above chainsaw

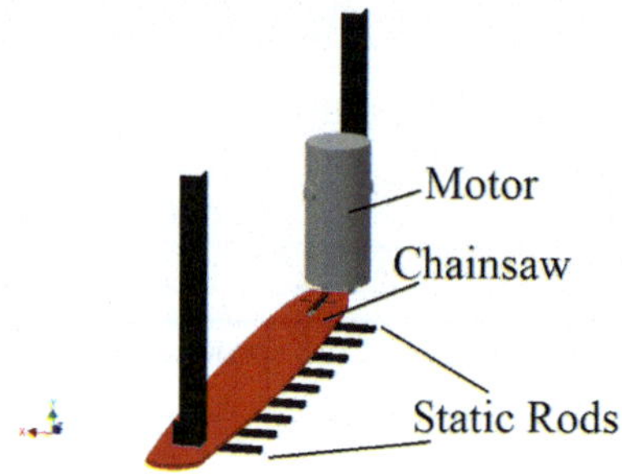

Fig. 10 Testing setup having static rods below chainsaw

As the chainsaw was being driven by a motor, it was not required to have a vehicle on which a person can sit and is driven by pedals. Instead, it was decided to make a cart which can be pushed by hands. Also, in the entirely human-powered wheat harvester, two chainsaws with approximately 0.5 m span each were used side by side considering the width of the vehicle. Now, in order to reduce the power required, only one chainsaw has been used. Though this would reduce the area covered by the wheat harvester,it would be much better than cutting the wheat manually. The hardware specifications of the prototype are explained in Sect. 6.

This design consists of a chainsaw in the front, which is driven by the motor placed on one side of the chainsaw. The motor can be powered by a 12 V battery. The battery used in our prototype was having the capacity of 7000 mAh. The wheat harvester can be operated like a lawn mower by a single person. The static rods are put in the front in order to avoid wheat straws from bending sideways due to the motion of the chainsaw's chain. There were two options for the placement of these rods as shown in Figs. 9 and 10. The decision on which one to use was taken by an experiment explained in the next section.

Fig. 11 Testing with setup having static rods above chainsaw

4 Testing and Analysis

4.1 Feasibility Test of Chainsaw as a Primary Cutter

In order to check the feasibility of using the chainsaw as a primary cutter, testing setup was manufactured as per the design shown in Figs. 9 and 10. Both the testing setups were tested on a wheat farm. The testing setup having static rods below chainsaw proved to be better than the testing setup having static rods above chainsaw. As shown in Fig. 11, the wheat was getting stuck in between static rods and chainsaw after getting cut in the testing setup having static rods above the chainsaw. This was not the problem with the testing setup having static rods below the chainsaw. So, in the final prototype, the design as shown in Fig. 10 was used. The full video of tests can be seen at [12, 13].

The average current drawn during the test was around 3.5 A with a peak of about 5.2 A at the starting. So, considering the capacity of 7000 mAh of the battery used in the prototype, wheat harvester should be able to last for 2 h after the full charge.

4.2 Final Prototype Testing

By the time the final prototype was ready, the harvesting season of wheat was over. So, the final prototype was tested on the grass instead of wheat. Also, instead of static

Fig. 12 Final prototype of wheat harvester

rods used in testing setup, sheet metal cut in the shape of the comb was used. This is considering the fact that with the sheet metal, different spacing between comb teeth and different thickness of the comb tooth can be obtained. The important observations made during the test are explained below:

1. The sheet metal used in the test, for avoiding wheat straws from bending sideways, was very thin and was not strong enough. It was getting bent sometimes. So, thicker and reinforced sheet metal should be used in future prototypes.
2. Sometimes, the grass was getting bent in the forward direction and therefore chainsaw's chain could not make contact with grass to cut it. Still, it is yet to be determined if it happens with the real wheat also.
3. The already available motor was used in order to avoid the purchase of a new motor just for the concept proofing. So the motor used here was not chosen as per the requirements, and hence it was not expected to work flawlessly. Due to this, the motor was stalling sometimes. In the future, a proper motor can be chosen for the full-scale testing of the prototype (Fig. 12).

The full video of the final prototype test with the explanation of different parts can be seen at [14].

5 Conclusion and Future Work

The objective of this project was to do the concept proofing of such a small-scale wheat harvester, which does not require the heavy-duty engine to power it. It was desired that the wheat harvester should be able to do the cutting of the wheat straws. The motor used in the prototype was not selected as per the requirement (i.e. already available motor was used in order to avoid the purchase of a new motor just for the concept proofing) and hence it was not expected that the wheat harvester will work flawlessly. After the test, it was determined that this wheat harvester concept has the potential to replace the labor-intensive and time-intensive task of harvesting the wheat manually, still being very affordable. The cost of preparing the prototype was around Rs. 5000. So, if produced in batch, it can be manufactured at a cheaper cost. Some future work required is listed below:

1. The final prototype is to be tested in the real wheat farm in order to determine if the wheat straws bend in the forward direction.
2. It can be tested how much area the wheat harvester can cover in one charge cycle of the battery.
3. Some mechanism is to be developed for collecting the wheat after it is cut.
4. The prototype should be more robust in order to tackle the harsh agricultural conditions.
5. The project was targeted toward the crop of wheat only. But, it can be determined if the same mechanism can be utilized for other crops such as rice, millet, etc. as the shear strength and structure of these crops would be different than wheat.

6 Hardware Specifications of the Parts Used in the Prototype (Wheat Harvester with Motor-Powered Chainsaw)

See Table 1.

Table 1 Hardware specifications of the parts used in the prototype

Serial no.	Description of the part	Specifications
1	Chainsaw	MAF 22 in. chainsaw bar and chainsaw
2	Motor	Robokits RMCS-2002 (DC servo motor 24 V 100 W with 2000 PPR encoder)[a]
3	Battery	Exide 12 V 7 Ah battery
4	Wheels (2)	16 in. solid tube wheels (make not known)

[a]The already available motor was used to avoid the purchase of a new motor just for the prototyping. The final model does not require a servo motor or PPR encoder. Also, the wattage rating can be chosen for further experimentation to meet the requirement of the wheat harvester

References

1. Wikipedia (2018) Agriculture in India. https://en.wikipedia.org/wiki/Agriculture_in_India#Major_crops_and_yields. Accessed June 28 2018
2. Index Mundi (2018) India wheat production by year. https://www.indexmundi.com/agriculture/?country=in&commodity=wheat&graph=production. Accessed June 28 2018
3. Wikipedia (2018) Combine harvester. https://en.wikipedia.org/wiki/Combine_harvester. Accessed Sept 11 2018
4. Agricultural Census Division—Department of Agriculture & Cooperation—National Informatics Centre (2010–11) Reports—table 11: average size of operational holding for all social groups. http://agcensus.dacnet.nic.in/nationalholdingtype.aspx. Accessed June 28 2018
5. United States Department of Agriculture (2018) Farms and land in farms 2017 summary. https://downloads.usda.library.cornell.edu/usda-esmis/files/5712m6524/3j333500g/ht24wn24x/FarmLandIn-02-16-2018.pdf. Accessed June 28 2018
6. Agricultural Census Division—Department of Agriculture & Cooperation—National Informatics Centre (2010–11) Graphs—average size of operational holdings as per different agriculture censuses. http://agcensus.dacnet.nic.in/nationalholdingtype.aspx. Accessed June 28 2018
7. Agricultural Census Division—Department of Agriculture & Cooperation—National Informatics Centre (2010–11) Reports—table 14: all India number and area of operational holdings by size classes (all social groups). http://agcensus.dacnet.nic.in/nationalT1sizeclass.aspx. Accessed June 28 2018
8. Alibaba (2018) Qualified mini rice and wheat cutting machine/rice wheat reaper machine for sale. https://www.alibaba.com/trade/search?fsb=y&IndexArea=product_en&CatId=&SearchText=mini+rice+and+wheat+cutting. Accessed Sept 14 2018
9. Naji A, Modi H, Raut P, Dabhi P, Gayatri S, Nirmal T, Vakharia V, Kumar R Final project report : synthesis and analysis of mechanisms. Unpublished report. https://drive.google.com/file/d/1wEYvymXEMzHNzA9ja-q6r8pBskVj1dro/view
10. Kushwaha RL, Vaishnav AS, Zoerb GC (1983) Shear strength of wheat straw. Can Agricul Eng 25(2):163–166
11. Tiwari P, Gite L, Pandey M, Shrivastava A (2011) Pedal power for occupational activities: effect of power output and pedalling rate on physiological responses. Int J Ind Ergon 41(3):261–267
12. Modi H (2018) Testing setup having static rods above chainsaw. https://youtu.be/XhwIapKMlPE. Accessed June 28 2018
13. Modi H (2018) Testing setup having static rods below chainsaw. https://youtu.be/ozIkVxX4y5o. Accessed June 28 2018
14. Modi H (2018) Testing final prototype of affordable wheat harvester. https://youtu.be/rGcJLf89Xag. Accessed June 28 2018

Design of Passive Compliant Constant-Force Mechanism

Nianfeng Wang, Jianliang Zhang, and Xianmin Zhang

Abstract Force regulation is a big challenge when robot end-effector interacts with the unknown environment. To satisfy such requirement, kinds of constant-force mechanisms (CFMs) are designed. Compared with the active CFMs, the passive CFMs get rid of introducing complex sensors and advanced control algorithms. Chained beam-constraint model (CBCM) can be used to model the large deflection problems efficiently. Based on CBCM, this paper explores the buckling properties of flexure beam and then presents one kind of CFM. Such CFM obtains the constant-force output by utilizing the flexible fixed-guided beam, which undergoes its first buckling mode. Finite element analysis (FEA) and experiment are also given to prove the designed CFM.

Keywords CBCM · Fixed-guided beam · Buckling mode

1 Introduction

Constant-force mechanism (CFM) can provide a nearly constant output force over a range of input displacements. Unlike conventional elastic structures such as linear spring, the output force of CFM is not proportional to the input displacement [1]; its stiffness maintains zero regardless of the input displacement. Hence, CFM is also termed as zero-stiffness or quasi-zero-stiffness mechanisms. When the CFM interacts with the unknown environment, the reaction force remains nearly constant no matter how complex the environment is. This property enables CFM to avoid vibration, keeping both the operated objects and actuators safe and reliable. Thus, CFMs are

N. Wang (✉) · J. Zhang · X. Zhang
School -1, South China University of Technology, Guangzhou, China
e-mail: menfwang@scut.edu.cn

J. Zhang
e-mail: jl-zhang@outlook.com

X. Zhang
e-mail: zhangxm@scut.edu.cn

D. Sen et al. (eds.), *Mechanism and Machine Science*, Lecture Notes in Mechanical Engineering, https://doi.org/10.1007/978-981-15-4477-4_33

used in various domains, such as energy absorption [2], output protection in both macro-scale and micro-scale applications [3, 4], statically balancing devices [5].

Kinds of methods have been put forward to design CFMs. Generally, they are divided into two types. The first type is active CFM, which generates constant output by process control [6]. Normally, it sets up a force feedback system by force sensors and control algorithms. Although many precise sensors and advanced algorithms have been developed to make the control process faster and more reliable [7–9], the design of active CFM is complicated and costs a lot. The second type is passive CFM, which can be subdivided into two types, rigid passive CFM and compliant passive CFM.

Normally, rigid passive CFM consists of linear springs or the C-shape constant-force springs. It achieves constant force output by cleverly changing the direction and magnitude of the linear spring reaction force. To improve the accuracy of output force and make smaller the mechanism' size, Lambert and Herder [3] proposed an adjustable double-slider constant-force mechanism. Nathan [5] proposed a parallelogram four-bar hinged-lever constant-force mechanism, aimed to support a mass stably. This kind of CFM is easy to model and design, but it cannot get rid of the conventional mechanisms' disadvantages such as friction, wear, and assembly [10]. The compliant CFM comprises of flexible part such as thin beam or compliant joint. It uses the flexible components to store energy and thus dynamically balance the energy storage and release. It can overcome the shortcomings of rigid passive CFM, with no friction, no backlash, low cost, good compatibility, etc. Many scholars designed curved-beam or composite beam to achieve constant output through shape optimization and topology. Lan et al. [11] utilized the modeling of slender beams and one designing concept of rigid passive CFM to develop curved beam for adaptive robot end-effector operations. Pedersen et al. [12] developed a constant-force actuator that can achieve constant output by topology optimization. The structure of the CFM designed by optimization is usually strange and thus it is hard to fabricate and assemble. Furthermore, the formation of modeling the beam and shape optimization are complex [10]. Another method of designing compliant passive CFM is stiffness combination, means designing a zero-stiffness mechanism by combining the positive-stiffness and negative-stiffness mechanism. Hao [13] developed an adjustable constant-force gripper by combining bistable beams and linear spring. A micrometer was applied to change the output force accurately by adjusting the preloading of linear spring. Using a similar method, CC Lan [1] designed an adjustable CFM for adaptive end-effector operations. It is worth noting that the negative-stiffness part, which is normally produced by bistable beam, needs preloading before the designed mechanism reaches the constant-force region.

This paper aims to develop a CFM by exploring the buckling properties of thin beam. We find that in a certain range, the greater the initial rotated angle, the more level the initial part of force-displacement curve. If the curve is level enough, it will become a quasi-straight horizontal line. Due to large deflection of flexible parts occurred in the process, we mainly use the chained beam-constraint model (CBCM) to formulate the output. In addition, the relevant simulated analysis method is mentioned. We fabricate a prototype and verify it experimentally.

2 Analytic Basis of the Design

2.1 CBCM

Many methods have been proposed to model the deflection of flexible parts in compliant mechanisms, e.g., the elliptic integral method (EIS), the pseudo-rigid-body method (PRBM), the finite element analysis (FEA). The EIS represents the exact solution to the Bernoulli–Euler beam equation but the derivation is complicated. Only when the end slope and load parameters are given, the solution is straightforward [14]. The principle of the PRBM is to simulate the deformation of the flexible member in the mechanism by using a rigid body member with an equal force–deformation relationship, so that the flexible member can be analyzed using the theory of conventional rigid members [14]. But the accuracy of PRBM is not always high enough and it is not suitable when the axial force is the main force. FEA is a commonly used analysis method. When using it, the continuous complex overall structure is often dispersed into a finite number of unit cells by meshing, and the physical state characteristics of the entire structure are obtained by analysis and state transfer of each unit [14]. However, it is slightly difficult to be used in designing. Therefore, it is still necessary to find a method that is accurate, simple, and widely available.

CBCM divides a flexible beam into several elements, and then uses beam-constraint model (BCM) method to model each element. This method can also analyze the stress distribution and deflection curve during beam deformation [15]. BCM is a method developed to accurately predict the behavior of a flexible beam within its deflection range. It has been demonstrated that when defection is within 10% of the length of the flexible beam, BCM can accurately capture the relevant nonlinear information. More importantly, BCM is a simple, parameterized, and closed mathematical model [16]. The merits of CBCM are as follows. 1. Because each BCM segment can accurately formulate the deflection, CBCM requires fewer elements than other methods with the same desired accuracy. 2. The maximum compressive load of CBCM is larger than that of BCM, so it can predict complex deflection with higher buckling modes. 3. When the axial load is significant, CBCM is more suitable than other methods that ignore axial strain (such as EIS).

Figure 1a shows the discretization of a beam. In this model, the transverse force applied at its free end is F_0, the axial force is P_0, and the moment is M_0; parameters θ_0, X_0, and Y_0 denote the tip slope, the horizontal, and vertical tip deflections of the beam, respectively, if the cantilever beam is equally divided into N elements. For the $i-1$th ($1 \leq i \leq$ N) element, its local coordinate system is established at the free end of the $i-1$th element, i.e., node $i-1$ is the coordinate origin of the ith element. Node 0 represents the fixed end of the beam, node N represents the free end of the beam. Figure 1b shows the schematic of the ith element,where f_i, p_i, and m_i denote the normalized transverse load, axial load, and moment while δx_i, δy_i, and α_i denote the normalized axial displacement, transverse displacement, and end slope, respectively. The load and deflection parameters are normalized with respect to the beam parameters as [15]

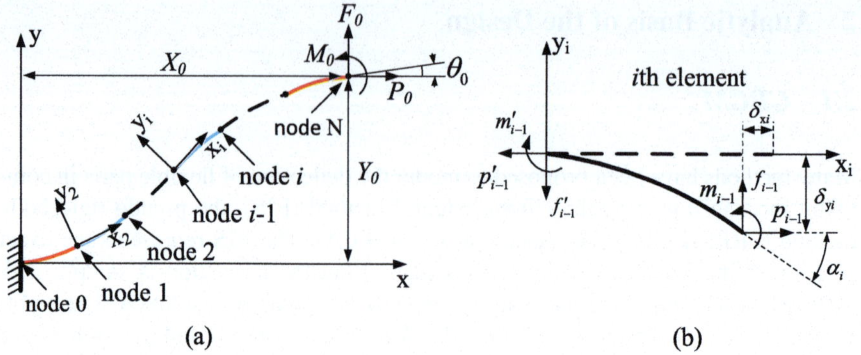

Fig. 1 Discretization of beam

$$f_i = \frac{F_i L_i^2}{EI},\ p_i = \frac{P_i L_i^2}{EI},\ m_i = \frac{M_i L_i}{EI},\ \delta_{xi} = \frac{U_{xi}}{L_i},\ \delta_{yi} = \frac{U_{yi}}{L_i}$$
$$f_0 = \frac{F_0 L^2}{EI} = N^2 f_1,\ p_0 = \frac{P_0 L^2}{EI} = N^2 p_1,\ m_0 = \frac{M_0 L}{EI} = N m_N \tag{1}$$

where L_i denotes the length of the ith element, if the beam is divided equally, $L_i = L/N$. I denotes the cross-sectional moment of inertia and it is a constant because the section of the beam is considered to be uniform. Through the formulation of BCM [15], the load equilibrium equations of the ith element can be expressed as

$$\begin{bmatrix} f'_{i-1} \\ p'_{i-1} \\ m'_{i-1} \end{bmatrix} = \begin{bmatrix} 1 & 0 & 0 \\ 0 & 1 & 0 \\ (1+\delta_{xi}) & -\delta_{yi} & 1 \end{bmatrix} \begin{bmatrix} f_i \\ p_i \\ m_i \end{bmatrix} \tag{2}$$

where f'_{i-1}, p'_{i-1} and m'_{i-1} are the loads transmitted to $i-1$th element from the ith element. Due to the deformation of the $i-1$th element, there is a slope between the local coordinate system of the ith element and the end of the $i-1$th element, so the f_{i-1}, p_{i-1}, m_{i-1} can be expressed as

$$\begin{bmatrix} f_{i-1} \\ p_{i-1} \\ m_{i-1} \end{bmatrix} = \begin{bmatrix} \cos\alpha_{i-1} & -\sin\alpha_{i-1} & 0 \\ \sin\alpha_{i-1} & \cos\alpha_{i-1} & 0 \\ 0 & 0 & 1 \end{bmatrix} \begin{bmatrix} f'_{i-1} \\ p'_{i-1} \\ m'_{i-1} \end{bmatrix} \tag{3}$$

The rotation of the coordinate frame of the ith element can be expressed as

$$\theta_i = \sum_{1}^{i-1} \alpha_i \tag{4}$$

where $i = 2,3,\ldots,N$. Hence, Eq. (2) can be rewritten as

$$\begin{bmatrix} \cos\theta_i & -\sin\theta_i & 0 \\ \sin\theta_i & \cos\theta_i & 0 \\ (1+\delta_{xi}) & \text{-}\delta_{yi} & 1 \end{bmatrix} \begin{bmatrix} f_i \\ p_i \\ m_i \end{bmatrix} = \begin{bmatrix} f_1 \\ p_1 \\ m_0 \end{bmatrix} \tag{5}$$

We can get the geometric constrains of the entire beam as follow:

$$\sum_{i=1}^{N-1} \left[\begin{bmatrix} \cos\theta_i & -\sin\theta_i \\ \sin\theta i & \cos\theta_i \end{bmatrix} \begin{bmatrix} L_i(1+\delta_{xi}) \\ L_i\delta_{yi} \end{bmatrix} \right] = \begin{bmatrix} X_0 \\ Y_0 \end{bmatrix} \tag{6}$$

$$\theta_N + \alpha_N = \theta_0 \tag{7}$$

The geometric constraint Eqs. (6)–(7), the loads equilibrium Eq. (5), and the BCM formulations [15] constitute the chained beam-constraint model. For load parameters P_0, F_0, M_0 and displacement parameters X_0, Y_0, θ_0, any three of them can be calculated by CBCM if the other three parameters are given.

2.2 Formulation and Optimization

Figure 2 shows the schematic of a fixed-guided cantilever beam, whose fixed end is point A and free end is point B. And F, δ denote the reaction force and displacement of point B. The initial slope of the entire beam is θ. Other parameters are the same as Sect. 2.1. Hence, some deformation parameters of the beam can be written as

$$\begin{aligned} X_0 &= L_{AB} - \delta\sin\theta \\ Y_0 &= -\delta\cos\theta \\ \theta_0 &= 0 \end{aligned} \tag{8}$$

And the reaction force can be expressed by F_0, P_0

$$F = F_0\cos\theta + P_0\sin\theta \tag{9}$$

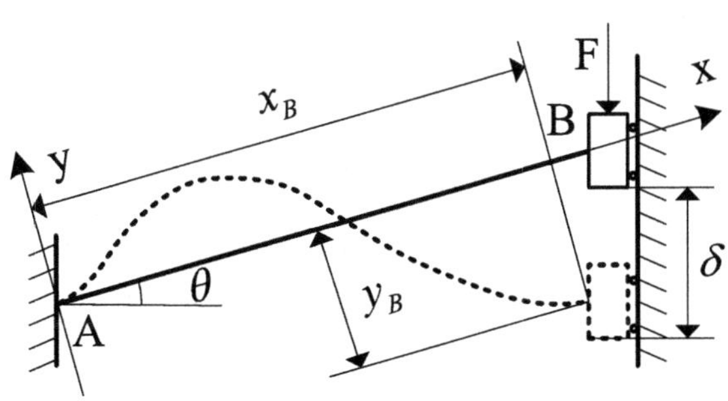

Fig. 2 Schematic of a fixed-guided beam

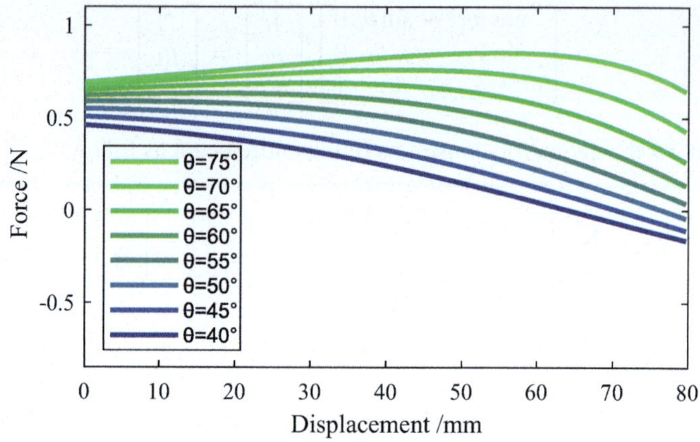

Fig. 3 The force-displacement curve corresponding to different initial slope θ

On the basis of CBCM, we explore the force–displacement relationship of the same fixed-guided beam when it goes though different initial slope θ. The material we used is stainless steel SUS304, whose Young's modulus is 193 GPa, with Poisson's ratio $\mu = 0.3$. The in-plane thickness $T = 0.1$mm, out-of-plane thickness $H = 1.5$ mm, length of entire beam $L = 100$ mm. In addition, the initial slope increases from 40 to 75° with an increment of 5°.

As Fig. 3 shows, for a certain initial slope θ, the force will not change too much in the initial part of the force-displacement curve. In fact, all the beam parameters such as L, T, and H would inflect the relationships between output force and displacement. Hence, there exists an optimal beam parameter set that enables the initial part of the force-displacement curve to be as level as possible. If the initial part of the curve is level enough, it will become a transverse quasi-straight line, meaning that the output force will remain quasi-constant regardless of the input displacement.

We allow that the output of constant force fluctuates within 5%. To simplify the optimization, some parameters of the fixed-guided beam are given as follow: beam length $L = 100$ mm, in-plane thickness $T = 0.1$ mm, fabricated material is SUS304. Particle swarm optimization (PSO) algorithm is applied to optimize initial slope θ and out-of-plane thickness H. The number of particles is 20 and the number of iterations is 300; both the acceleration factors C1, C2 = 2. To make faster the calculating speed, the restrictions are as follows:

$$\begin{cases} 50^\circ \le \theta \le 80^\circ \\ 10 \le H \le 20 \end{cases}$$

The optimal set of the parameters found by the PSO is given as $\theta = 64^\circ$, $H = 13$ mm. After getting the optimal set, we use FEA and experiments to verify the optimization result.

2.3 FEA Verification

When we use FEA to simulate the optimization result, we have to consider the bulking mode of the thin beam when it undergoes different deformation. Otherwise, the nonlinear deformation of beam would easily converge to higher order buckling mode, which will lead to erroneous simulation result.

The FEA of flexible beam needs to be performed in two steps. The first step is to analyze the buckling mode. In this step, the buckling modal information of the beam can be obtained by loading with a small deformation, so that the critical load under the corresponding constraint condition can be obtained. Figure 4a shows the initial shape of the beam, and (b)–(d) are the deformation shapes of the first to third buckling modes (evenly enlarged 8 times) and their corresponding eigenvalues. The second step is post-buckling analysis. By introducing the defect information obtained in the first step, we can use Riks method or general static method to get the FEA result by loading displacement. It should be noted that we have to open the nonlinear large deformation option during this step. Additionally, we have to verify the FEA result by judging whether the deflection shape in FEA matches the experiment one. If it does not, some setting parameters in step two need modifying. Figure 5a show the rational deflection shape in FEA while (b) shows the erroneous one. It can be easily found that the thin beam under first buckling mode is under this situation.

(a) Initial shape

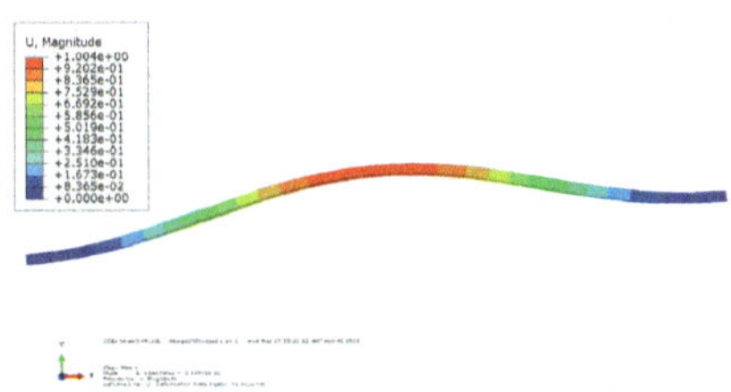

(b) First buckling mode

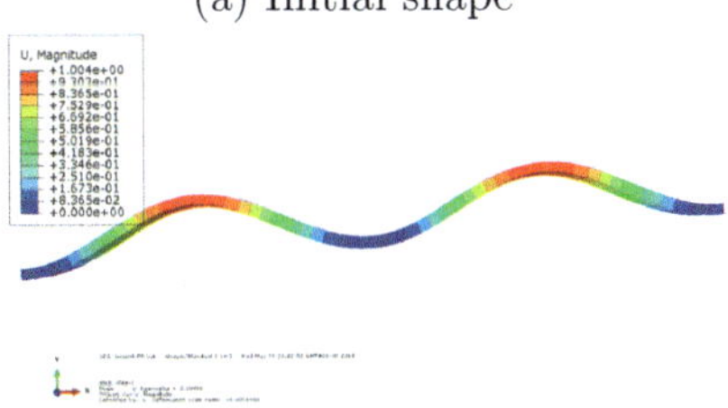

(c) Second buckling mode

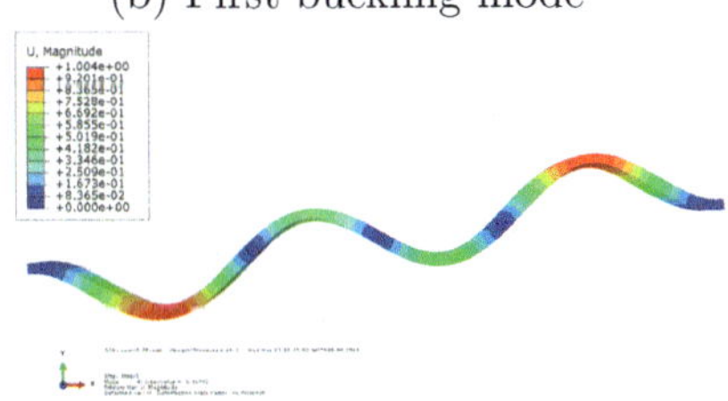

(d) Third buckling mode

Fig. 4 Initial shape and different buckling mode of the beam

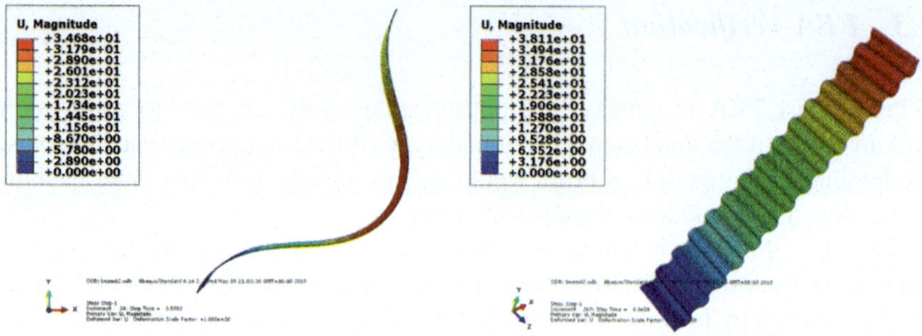

(a) Rational deflection shape (b) Erroneous deflection shape

Fig. 5 Deflection shape in FEA

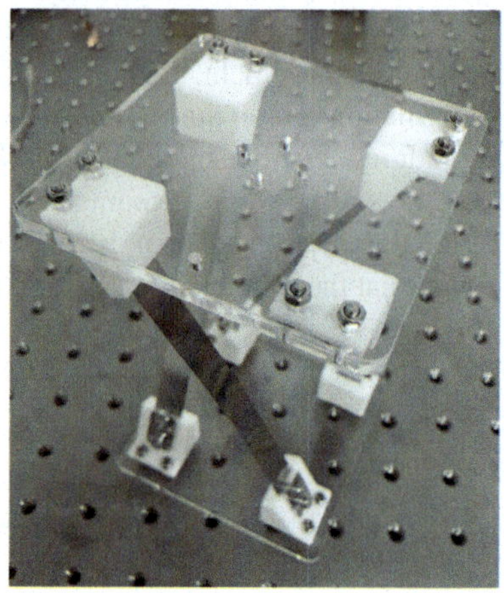

Fig. 6 CFM prototype

3 Experiment Prototype

3.1 Experiment

We design a CFM utilizing the optimal set of beam's parameters mentioned in Sect. 2.2, whose prototype is shown in Fig. 6. It mainly comprises four spirally distributed beams. To verify it experimentally, we design a testing device shown in Fig. 7. The device consists of a ball screw slide rail, a 57 stepper motor, a force sensors, and some mounts. The ball screw slide rail is driven by a 57 stepper motor to generate linear displacement. The force sensor is connected with the CFM's base, to measures the reaction force when the CFM goes through different positions. The position control of the ball screw slide rail, data acquisition, observation, and storage of the force sensor are all processed by a computer. The computer is not shown in the figure. The ball screw slide type is HBX1204-90, with a positioning accuracy of 0.03 mm and a minimum displacement speed of 0.0625 mm/s. Force sensor adopts

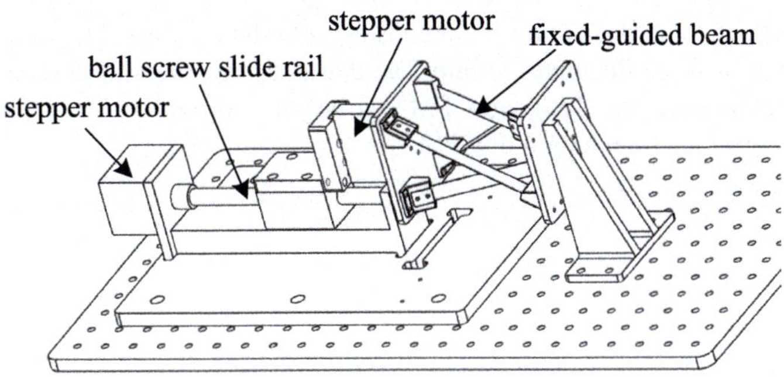

Fig. 7 Testing device

Fig. 8 Results

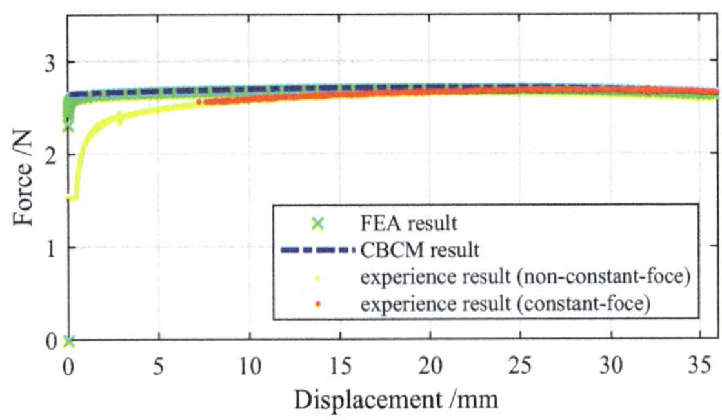

SMT series sensor of American Interface Company, whose measuring range is [−10 N, +10 N], nonlinearity is 0.05%, precision is up to 0.005 N, sampling frequency is up to 160 Hz, with ten times of range overload protection .

3.2 *Result*

The results shown in Fig. 8 indicate that the FEA result (green curve) and the CBCM modeling result (blue curve) are consistent, and the experiment result (yellow and red curve) is consistent with the FEA and CBCM results in the latter part, but not in the initial part. Such phenomenon results from the processing of thin beams. The flexible beams used in this experiment was obtained by laser cutting. As we all know, an important processing requirement in sheet metal processing is flatness. The beam we used is so thin that it would generate plastic deformation during processing or preservation. Therefore, in the initial stage of the experiment, the beam is not an expected straight beam but a curved beam actually. As a result, in the beginning of the experiment, the actual output force is smaller than the expected one.

The experiment result consists of non-constant-force region and constant-force region, showed as a yellow curve and red curve in Fig. 8. The yellow one indicates that the output force does not remain constant. On the contrary, the red one means that the output force fluctuates within 5%, that is to say, the output force is quasi-constant. Moreover, the designed CFM can provide 27 mm constant-force stroke, 21% as long as the total length of CFM.

4 Conclusion

This paper has proposed one constant-force mechanism. Unlike the other design methods, we do not have to use the complicated algorithms to model the compliant part or assemble CFM with multiple compliant parts. With the help of CBCM, we can model the thin beams easily and also get the optimal beam parameters by PSO. Also, we propose the corresponding FEA to simulate the optimization result and then verify it by experiment. The result shows that the designed CFM has good behavior of constant-force output.

Acknowledgements The authors would like to gratefully acknowledge the reviewers' comments. This work is supported by National Natural Science Foundation of China (Grant Nos. 51575187, U1713207), the Fundamental Research Funds for the Central University, Science and Technology Planning Project of Guangdong Province (Grant No. 2017A010102005), Natural Science Foundation of Guangdong Province (S2013030013355).

References

1. Chen Y-H, Lan C-C (2012) An adjustable constant-force mechanism for adaptive end-effector operations. J Mech Des 134(3):031005
2. Jenuwine J, Midha A (1994) Synthesis of single-input and multiple-output port mechanisms with springs for specified energy absorption. J Mech Des 116(3):937–943
3. Lambert P, Herder JL (2017) An adjustable constant force mechanism using pin joints and springs. In: New trends in mechanism and machine science. pp 453–461
4. Wang P, Xu Q (2017) Design of a flexure-based constant-force xy precision positioning stage. Mech Mach Theory 108:1–13
5. Nathan R (1985) A constant force generation mechanism. J Mech Trans Autom Des 107(4):508–512
6. Xu Q (2013) Adaptive discrete-time sliding mode impedance control of a piezoelectric micro-gripper. Ieee Trans Robot 29(3):663–673
7. Liang W, Song Y, Lv H, et al (2010) A novel control method for robotic belt grinding based on SVM and PSO algorithm. In: Intelligent computation technology and automation. pp 258–261
8. Kiguchi K, Fukuda T (1997) Intelligent position/force controller for industrial robot manipulators-application of fuzzy neural networks. Ieee Trans Ind Electron 44(6):753–761
9. Pliego-Jiménez J, Arteaga-Pérez MA (2015) Adaptive position/force control for robot manipulators in contact with a rigid surface with uncertain parameters. Eur J Control 22:1–12
10. Wang P, Xu Q (2018) Design and modeling of constant-force mechanisms: a survey. Mech Mach Theory 119:1–21
11. Lan C-C, Wang J-H, Chen Y-H (2010) A compliant constant-force mechanism for adaptive robot end-effector operations. In: Robotics and Automation. pp 2131–2136

12. Pedersen C, Fleck N, Ananthasuresh G (2006) Design of a compliant mechanism to modify an actuator characteristic to deliver a constant output force. J Mech Des 128(5):1101–1112
13. Hao G, Mullins J, Cronin K (2017) Simplified modelling and development of a bi-directionally adjustable constant-force compliant gripper. Proc Inst Mech Eng Part C J Mech Eng Sci 231(11):2110–2123
14. Howell LL (2001) Compliant mechanisms. John Wiley & Sons, America
15. Ma F, Chen G (2016) Modeling large planar deflections of flexible beams in compliant mechanisms using chained beam-constraint-model. J Mech Robot 8(2):021018
16. Awtar S, Slocum AH, Sevincer E (2007) Characteristics of beam-based flexure modules. J Mech Des 129(6):625–639

A Neurodynamic Approach to Stabilization of a 10 DOF Biped Mechanism Using Reinforcement Learning

Aditya Kameswara Rao Nandula, Sudhir Raj, A. K. Deb, and C. S. Kumar

Abstract In this paper, we propose a Reinforcement Learning (RL) based approach to stabilizing and control of a biped robot. Bipeds have complex requirements of stabilization and gait planning in multiple degrees of freedom. A reinforcement learning strategy is presented in this work for the stabilization of a single leg as well as a biped to create a learned behavior of a system. In this paper, each leg of the Humanoid biped robot is approximated as a double inverted pendulum, and its static stabilization is studied in the sagittal plane. The equations of motion are derived using Lagrange's formulation method. An equivalent Humanoid robot single leg and biped model developed in Gazebo. Through Robotic Operating System (ROS), a reinforcement learning based control algorithm was developed for static stabilization, and the simulation was carried out on the Gazebo model. A total of 1458 states are used for training the reinforcement learning algorithm.

Keywords Humanoid · Biped robot · Reinforcement learning

1 Introduction

Reinforcement Learning (RL) is finding considerable attention in robotics in the past few years. In computer science, the origin of reinforcement learning is deeply-rooted, although RL approaches are similar like neurodynamic programming and adaptive

A. K. R. Nandula (✉)
Advanced Technology Development Centre, Indian Institute of Technology Kharagpur, Kharagpur, India
e-mail: akraonandula@gmail.com

A. K. Deb
Electrical Engineering Department, Indian Institute of Technology Kharagpur, Kharagpur, India
e-mail: alokkanti@ee.iitkgp.ac.in

S. Raj · C. S. Kumar
Mechanical Engineering Department, Indian Institute of Technology Kharagpur, Kharagpur, India
e-mail: sudhir.mechanical@gmail.com

C. S. Kumar
e-mail: kumar@mech.iitkgp.ac.in

D. Sen et al. (eds.), *Mechanism and Machine Science*, Lecture Notes in Mechanical Engineering, https://doi.org/10.1007/978-981-15-4477-4_34

dynamic programming discussed in literature by Werbos [1–3], Bertsekas et al. [4–6], and others from optimal control field [7]. The RL can use as an on-line controller-based approach [8] and fuzzy was an off-line controller-based approach and before experimenting data is collected. To maximize the reward, RL uses a mapping between situations and actions. The RL approach did not require a pre-collected data for stabilization humanoid biped robot. The humanoid biped leg approximated as an inverted double pendulum and RL scheme modeled for stabilization by using its non-linear motion equation.

Oh et al. [9] proposed a method to acquire center of mass trajectory using a specified zero moment point (ZMP) in real-time. The entire biped locomotion pattern separated into periodic and transient walking stages. In every stage of locomotion, we can obtain a solution in analytical form for center of mass for a specified zero moment point based on inverted pendulum model. Here humanoid robot dynamic model is assumed as inverted pendulum model. ZMP limitations for the biped robot are evaluated in the motion of inverted pendulum model. The model of inverted pendulum with center of gravity motion is considered as of biped robot. This approach for humanoid walking was proposed by Tang and Er [10]. Thant et al. [11] proposed a "cubic spline interpolation based trajectory planning," which demonstrates the smooth trajectory of walking biped robot. In continuous action domains, the self-scaling RL algorithm developed by Franklin et al. [12].

Conventionally, balancing, and gait generation uses motion balance control and pattern generation. The pattern generation of walking conducted offline using the stability principle with a ZMP, and a human-like two-stage gait generation and gesture stabilization control employ an IPM control for the real-time alteration of joint positions during humanoid biped robot walking, using all sensors data available. Our system implements control of the biped robot system through an equivalent multi-degree of freedom inverted pendulum control problem. The multi-degree of inverted pendulum problem through its challenging control regime and similarity to some extent with bipedal walking has therefore been used as an initial framework to study the efficacy of various control schemes. In particular, the study explores modeling, control by linear as well as non-linear control methods and a suitable variant of neuro-dynamic control through reinforcement learning (RL) proposed for implementation. In this paper, we describe the study of the control problems associated with biped static stabilization and experimental verification of stabilizing control schemes on 10 DOF biped robot, which considered as a double inverted pendulum.

2 Kinematic Modeling of IIT Kharagpur 10 DOF Humanoid Biped Robot

We modeled and fabricated a humanoid biped platform developed at IIT Kharagpur namely KYZR. The 10 degrees of freedom (DOF) humanoid biped robot is one of its kind in India. It has 5 DOF in each leg, with a height of 66 cm (up to hip). Each leg consists of the foot, calf, thigh and hip link's which performs motions like a human leg. Each leg has five joints namely, a hip roll, hip pitch, knee, foot pitch, foot roll. The joint actuators are the smart actuators with a controller, speed reducer, and a driver. The motors can also read their current position and speed. The mechanical body is made up of aluminum and in-house 3D printed plastic parts connecting the links and motors. The overall weight of the robotic leg will be 6 kg approx. The robot is designed to perform a simple bio-mechanically compatible closed-loop and semi-generative gaits and motion sequences. This humanoid biped robot was designed to work in any terrain. Figure 1a shows the kinematic diagram and Fig. 1b shows the existing hardware of the 10 DOF humanoid biped robot platform in Robotics and intelligent systems lab, IIT Kharagpur.

In general, any robot can be described kinematically using the Denavit - Hartenberg (D - H) representation by stating the values of four quantities for each link pair. These four quantities are the joint angle (θ_i), twist angle (α_i), joint Offset (d_i) and link length (a_i). D - H Parameter of right and left leg of a bipedal robot made at IIT Kharagpur are shown in Tables 1 and 2.

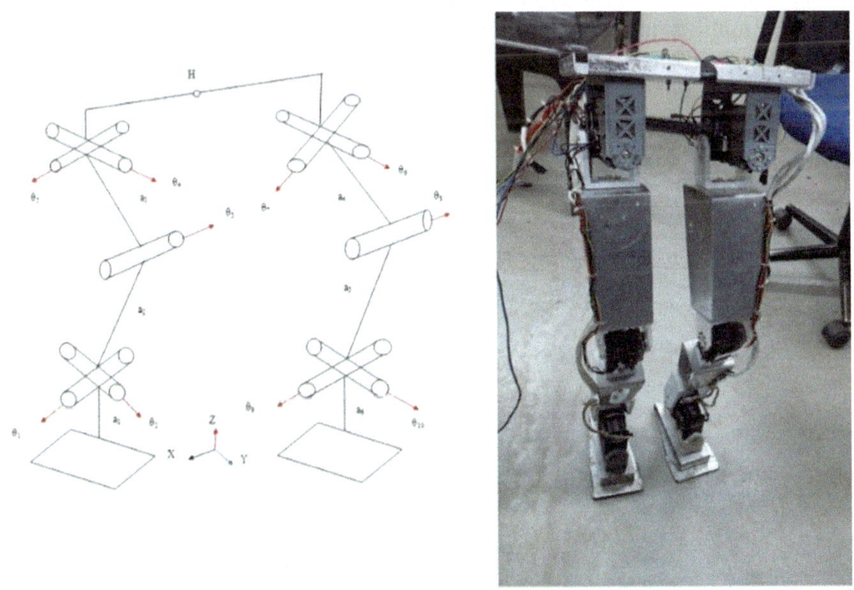

(a) Kinematic Diagram (b) Hardware

Fig. 1 **a** Kinematic diagram, **b** Hardware

Table 1 Right leg D - H parameters

Links (i)	θ_i (deg)	d_i (cm)	a_i (cm)	α_i (deg)
1	θ_{10}	$d_1 = 0$	$a_1 = 0$	0
2	θ_9	$d_2 = 0$	$a_2 = -25.5$	−90
3	θ_8	$d_3 = 0$	$a_3 = -24$	−180
4	θ_7	$d_4 = 0$	$a_4 = 0$	−90
5	θ_6	$d_5 = 0$	$a_5 = 0$	−90

Table 2 Left leg D - H parameters

Links (i)	θ_i (deg)	d_i (cm)	a_i (cm)	α_i (deg)
6	θ_5	$d_5 = 0$	$a_5 = 0$	−90
7	θ_4	$d_4 = 0$	$a_4 = 0$	−90
8	θ_3	$d_3 = 0$	$a_3 = -24$	−180
9	θ_2	$d_2 = 0$	$a_2 = -25.5$	−90
10	θ_1	$d_1 = 0$	$a_1 = 0$	0

3 Experimental Setup

The implementation carried out on a humanoid biped platform developed at IIT Kharagpur namely KYZR. For humanoid biped robot stabilization in the sagittal plane, a simplified motion dynamics of a leg approximated as a sagittal-plane body dynamics of a double inverted pendulum and extension and flexion torques at the ankle and knee considered as shown in Fig. 2. The assumptions considered are as follows

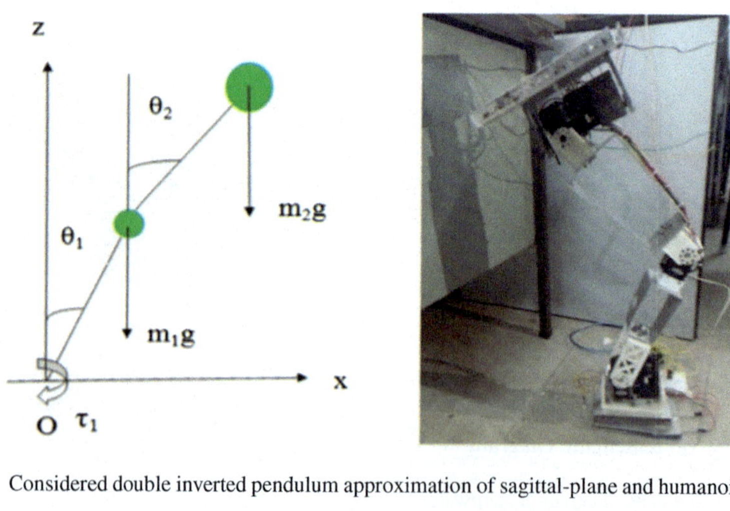

Fig. 2 Considered double inverted pendulum approximation of sagittal-plane and humanoid biped leg

(a) Body is supposed to be rigid and, m_2 is mass of body
(b) The foot was static at the origin 'O' and while for joint 1 damping and stiffness coefficients are c_1, k_1 for the motion contact part.
(c) At the knee, the corresponding mass is m_1.

where

(a) a and b are the calf and thigh links lengths with mass m_1 and m_2.
(b) θ_1 and θ_2 are the ankle and knee joint angle.

The Lagrange's modeling uses generalized coordinates which makes systems analysis simple, and this method eliminates the constraining forces from the motion equations. Modeling based on Euler-Lagrange derived by specifying a Lagrange's, as variance of potential and kinetic energy of the system. This resulted system with differential nonlinear equations consisting of three 2nd order equations as explained below.

Using Lagrange's equation, the torque τ_i is calculated as follows

$$\tau_i = \frac{d}{dt}\left(\frac{\partial L}{\partial \dot{\theta}_i}\right) - \frac{\partial L}{\partial \theta_i} \tag{1}$$

where, i is link number $i = 1, 2$.

Lagrange Function $= L =$ Ki−Po
Total Kinetic Energy of the system (Ki)
Total Potential Energy of the system (Po)

$$\begin{aligned} L = {} & \frac{1}{2}(m_1 + m_2)a^2\dot{\theta}_1^2 + \frac{1}{2}m_2 b^2\dot{\theta}_2^2 + m_2 ab\dot{\theta}_1\dot{\theta}_2\cos(\theta_1 - \theta_2) \\ & - (m_1 + m_2)ga\cos\theta_1 - m_2 g l_2 \cos\theta_2 \end{aligned} \tag{2}$$

Based on the above Eqs. (1) and (2) the torques τ_1 and τ_2 are calculated and used for the development of the RL based control algorithm.

4 Control Architecture

Reinforcement learning is a basic problem in artificial intelligence. Generally, in an unknown situation, the agent attempts to act. It interacts with environment by performing some activities which change the state of the environment. After each action in the environment, a reward is awarded to agent. The main aim of agent was to get maximum aggregate reward from the environment.

Associative Search Element (ASE) is to select the right or left action of control. From $X(t) = [x_1(t), x_2(t), \ldots x_{n\text{-}1}(t), x_n(t)]$ as the vector input and output element $y(t)$ obtained as in Eq. (3)

$$y(t) = f\left[\sum_{i-1}^{n} w_i(t) * x_i(t) + n(t)\right] \tag{3}$$

where $n(t)$—random real variable and f in equation as a threshold function and d as probability density function.

$f(x) = -1$, if $x \leq 0$ (left action control)

$+1$, if $x \geq 0$ (right action control) over discrete-time the weights w_i change as in Eq. (4), $1 \leq i \leq n$,

$$w_i(t+1) = \alpha * r(t) * e_i(t) + w_i(t) \tag{4}$$

where

$e_i(t)$ = eligibility for ith pathway input at t.
α = defining the w_i rate of change, positive constant.
$r(t)$ = at t, real reinforcement value.

The final control input is given in Eq. (5)

$$F = \pm y(t) F_{\max} \tag{5}$$

where the value of maximum force applied is $F_{\max}$.

ROS offers some features, including device control low-level and hardware interface. At runtime, ROS has dispersed structure with procedures that permit executables to independently designed and lightly coupled. This framework is flexible to implement in C++, Python, and Lisp programming languages. The Humanoid biped robot leg motors are interfaced with ROS Software using a Dynamixel USB to RS485 converter module. To send and receive the joint angle through it to the joint actuators.

Experimental verification of stabilization on a humanoid biped robot platform for the control of biped legs was carried out using the Robot Operating System (ROS). Each leg has five actuators of which only two are used in the experiment to show the working of Reinforcement learning based neurodynamic schemes in sagittal plane stabilization. A simplified two degree of freedom equivalent to each leg considered.

The typical architecture developed for ROS based control of the Humanoid robotic leg shown in Fig. 3.

Figure 4 shows typical experimental flow chart of the approach.

The humanoid robot leg controlled using a control algorithm. Initially, the algorithm sets the Humanoid leg to standing in a home position. We developed new algorithms in ROS to control and obtain their angular position at that instance. At this point, if the Humanoid leg is deviated from its position using joint slider control out of stable Zone. The developed reinforcement learning based control algorithm will stabilize it with a control signal.

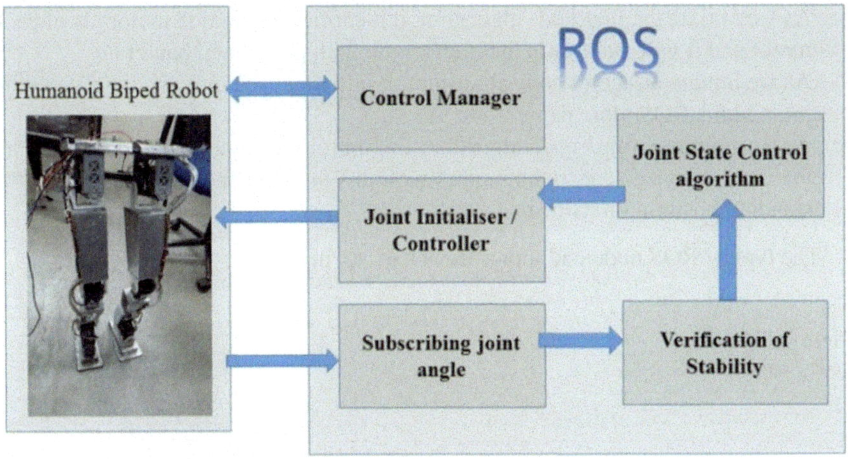

Fig. 3 Architecture developed for ROS based control of the humanoid robotic leg

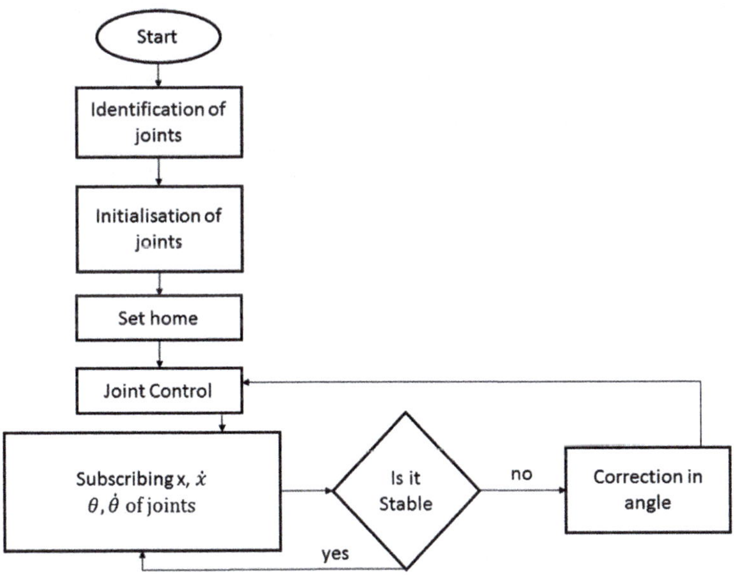

Fig. 4 Experimental method flow chart

In ROS, three different nodes are created for joint control and subscribe to joint parameter data. The flowchart in Fig. 5 displays the different nodes created and communication flow to joints to attain real-time control. When the hardware interfaced with the system, the following steps are followed to make the humanoid leg actuators active.

1. As we initiate the first node algorithm, it identifies the unique motor ids of each motor and it will make sure that motors are with predefined baud rate.
2. As we initiate a second node algorithm, it initializes the motors associated with joint with a ROS topic name.
3. As we initiate a third node algorithm to control the joint angles of each joint, the control signal sent to the motors with this the Humanoid leg will retain the standing position on its own.

The typical ROS node and topics flowchart are in Fig. 5.

Fig. 5 Flowchart of nodes/topics in ROS

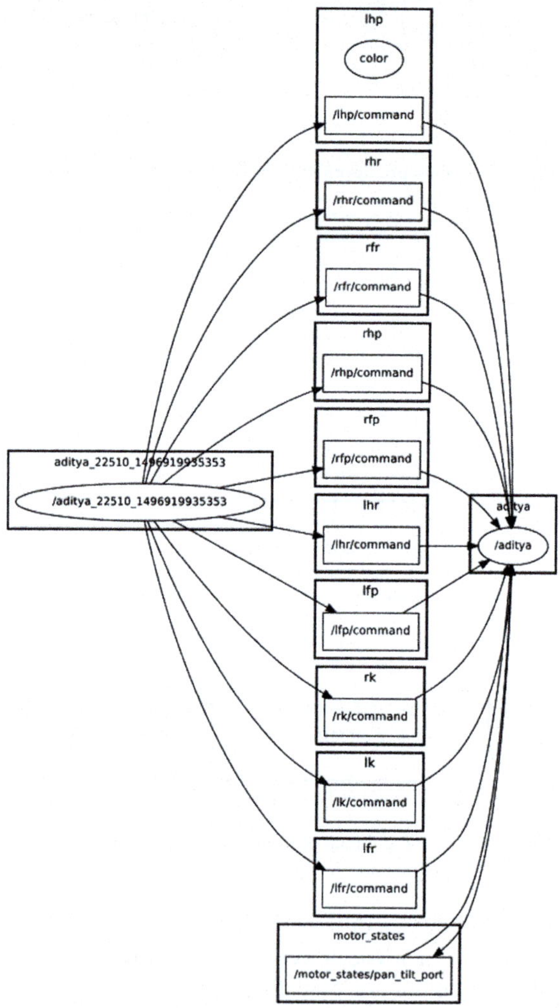

5 Simulation

The 10 DOF equivalent humanoid biped robot model for single leg and biped was developed in Gazebo. The gazebo is an open-source robot simulator. In this, it is easy to design robots (using URDF or SDF formats) and, we can rapidly test algorithms using the Gazebo-ROS control interface or other programming languages, train Artificial Intelligence (AI) system using real situations and test the performance regressively. The Gazebo offers the ability to precisely and competently simulate a population of the robot in complex indoor and outdoor environmental conditions. The Gazebo supports different sensors such as pressure sensors, joint state sensors, imu e.t.c.

The single-leg and biped Gazebo model was shown in Fig. 6a, b.

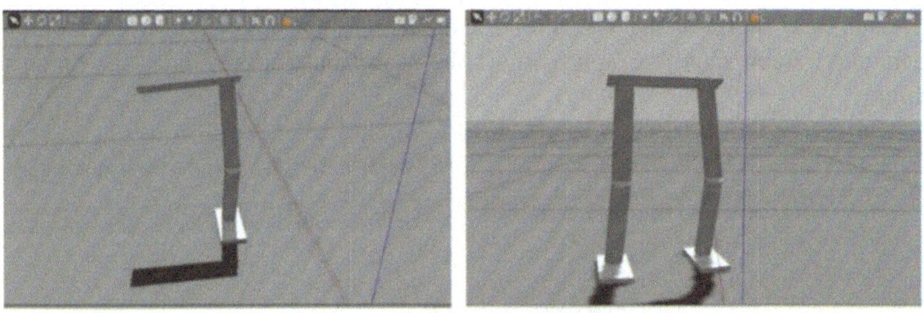

(a) Single leg equivalent model (b) Double leg equivalent model

Fig. 6 **a** Single leg equivalent model, **b** Double leg equivalent model

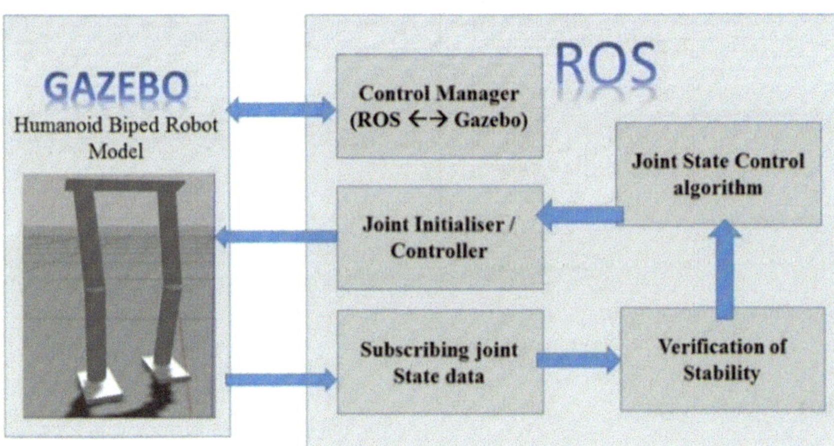

Fig. 7 ROS-Gazebo control block diagram

The ROS-Gazebo control architecture/interface provides the feasibility to test the algorithms in different scenarios. The ROS-Gazebo control block diagram shown in Fig. 7.

The Gazebo model is equivalent to the real model. We carried out the initial trials of the reinforcement learning based neuro-dynamic control algorithm developed in ROS on this Gazebo model and implemented the same on the hardware. The node flowchart shown in Fig. 8 shows the communication between the ROS and Gazebo simulator for a single leg and biped simulations.

For stabilization of biped using reinforcement learning based on neurodynamic control, the displacement and velocity were assumed as three states. The angle of the first link was partitioned as six states. There were three states each was partitioned for the angular velocity of first link, the angular velocity and angle of second link. A sum of states 1458 was considered for the reinforcement learning model. This training leads to the stabilization of the humanoid biped.

The state-space boxes for reinforcement learning used for inverted double pendulum stabilization of approximate the humanoid biped single-leg shown in Table 3.

The total no. of boxes considered are $3 \times 3 \times 6 \times 3 \times 3 \times 3 = 1458$.

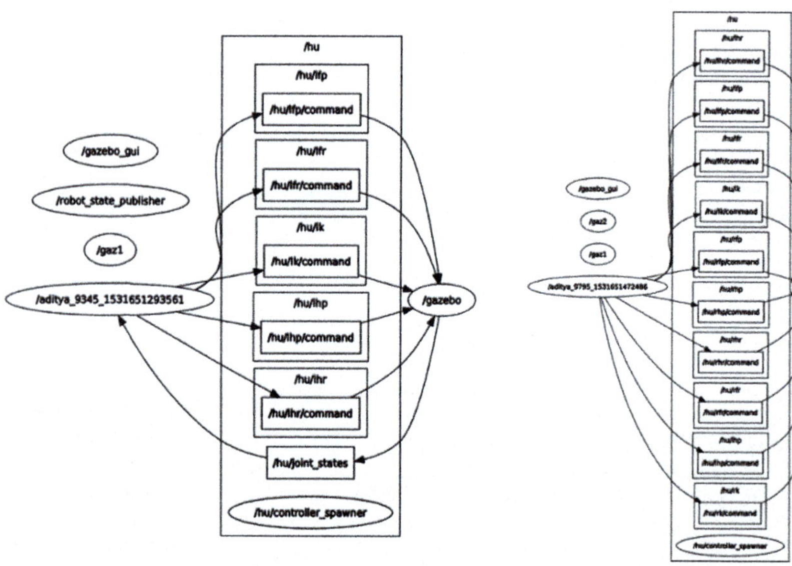

(a) Single leg(5DOF) (b) Double leg (10DOF)

Fig. 8 Node flowchart

Table 3 Reinforcement learning boxes

x (m)	<−2.40		−2.40, 2.40		>2.40	
$\dot{x}$ (m/sec)	<−0.50		−50, 50		>50	
θ_1(Radian)	<−0.21	−0.21, −0.11	−0.11, 0	0,0.11	0.11, 0.21	>0.21
$\dot{\theta}_1$(Radian/sec)	<−2.09		−2.09, 2.09		>2.09	
θ_2(Radian)	<−0.104		−0.104, 0.104		>0.104	
$\dot{\theta}_2$ (Radian/sec)	<−1.04		−1.04, 1.04		>1.04	

6 Results

Using the state-space data, an algorithm was developed in ROS with python. Initially, the single-leg or biped set to its home position, and it deviates from its position. Then the static stabilization algorithm is started to stabilize the biped or single leg.

Initially, the algorithm simulated on the Gazebo model and, we performed the stabilization control on the hardware. Some of the poses while experimenting on a single leg stabilization on the equivalent Gazebo model and real hardware is shown in Fig. 9.

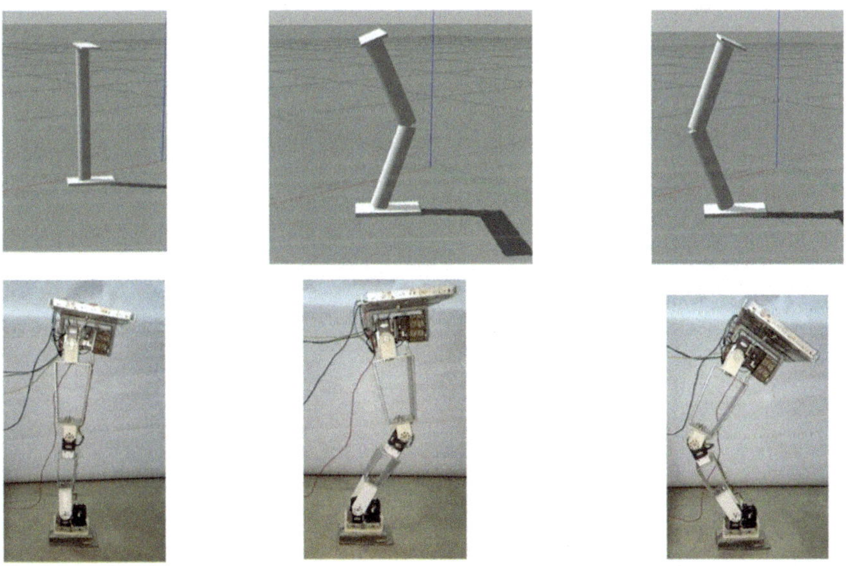

Fig. 9 Postures of humanoid single leg while experimenting in Gazebo and on real hardware

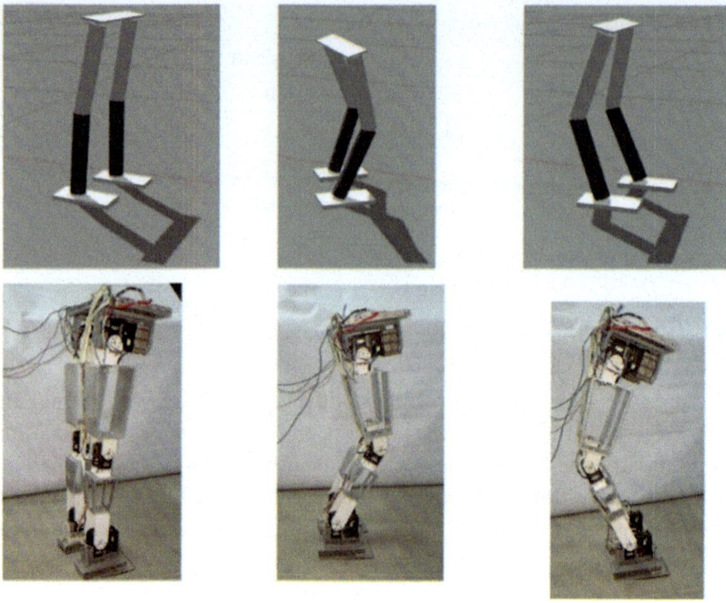

Fig. 10 Postures of humanoid double leg while experimenting in Gazebo and on real hardware

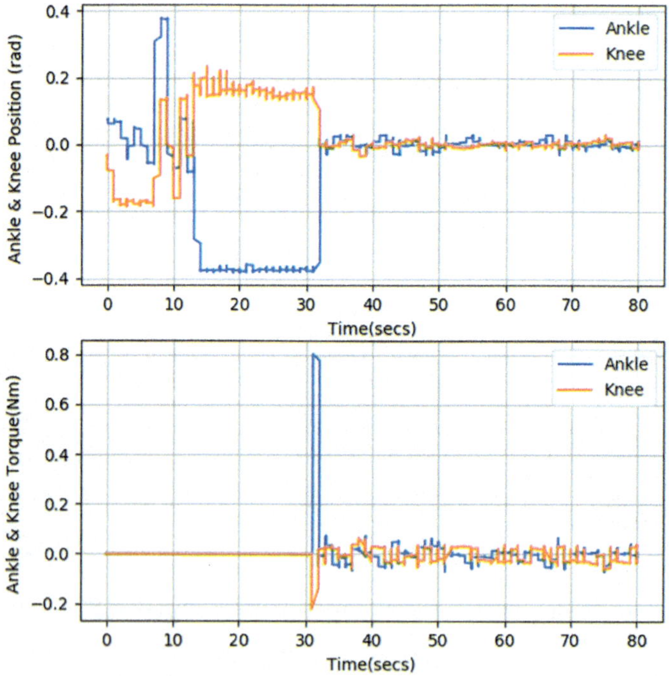

Fig. 11 Time versus position and torque plots of left leg

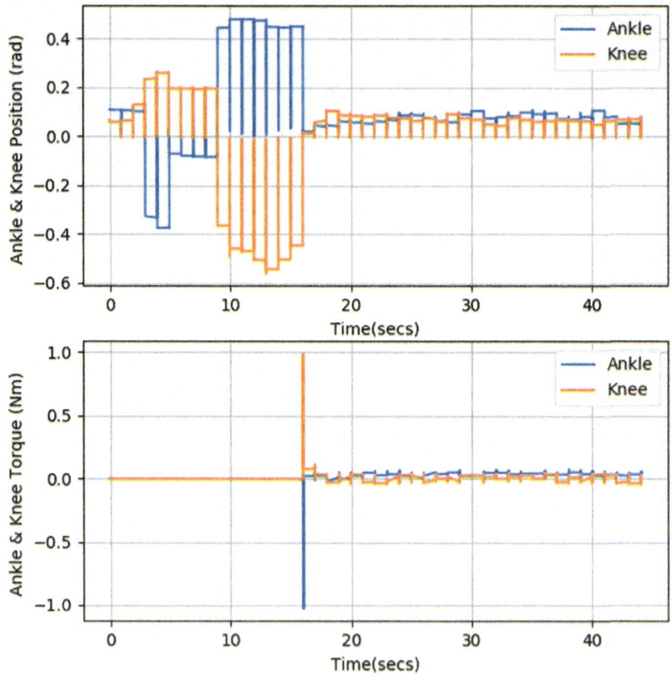

Fig. 12 Time versus position and torque plots of right leg

By considering the results above, the modified algorithm for the double leg is implemented on the developed equivalent Gazebo model of biped and on real hardware. Some of the postures in experimental results are shown in Fig. 10.

Initially, single-leg or biped set to home position, and it deviated from its position to some extreme position, and then the stabilization algorithm started. Then it starts stabilizing itself. The joint parameter data (i.e., time, position, velocity, torque) was recorded to a file in every experiment performed. Figure 11 shows the plot for time versus position and torque for the ankle and knee joints of the left leg.

Figure 12 shows the plot for time versus position and torque for the ankle and knee joints of the right leg.

7 Conclusion

Validation of static stabilization on a 10 DOF Humanoid biped robot was carried out. The results show the proposed Reinforcement Learning algorithm achieves promising stability in the static mode balancing of the robot. Therefore, the static version of the RL stabilization algorithm has been demonstrated. It reveals new compiled angles that were stable and not trained before, indicating good performance of the method used.

Acknowledgements The authors gratefully acknowledge the research and financial support from IIT Kharagpur in Centre for Robotics.

References

1. Werbos P (1992b) Neurocontrol and supervised learning: an overview and evaluation. In: White, Sofge (eds) Handbook of intelligent control. V.N. Reinhold, New York, pp. 65–89
2. Werbos P (2004) 1 ADP: goals, opportunities, and principles. In: SiBarto J, Powell W, Wunsch D (eds) Handbook of learning and approximate dynamic programming. Wiley-IEEE Press, Piscataway, NJ, pp 3–44
3. Werbos P (2009) Intelligence in the brain: a theory of how it works and how to build it. Neural Netw 22(3):200–212. Goal-Directed neural system
4. Bertsekas D (2009) Neuro-dynamic programming. In: Encyclopedia of optimization, pp 2555–2560
5. Bertsekas D (2010a) Approximate policy iteration: a survey and some new methods. http://stuff.mit.edu/people/dimitrib/API_Survey.pdf
6. Bertsekas D, Tsitsiklis J (1996) Neuro-dynamic programming. Athena Scientific
7. Khan SG, Herrmann G, Lewis FL, Pipe T, Melhuish C Reinforcement learning and optimal adaptive control: An overview and implementation examples
8. Raj S (2016) Reinforcement learning based controller for stabilization of double inverted pendulum. In: 1st IEEE international conference on power electronics, intelligent control and energy systems (ICPEICES-2016)
9. Yonghwan Oh, Kyung-ho A, Doikkimand chang wan K (2006) An analytical method to generate walking pattern of humanoid robot. In: IEEE industrial electronics, IECON 2006, pp 4159–4164
10. Tang Z, Joo Er M (2007) Humanoid 3D gait generation based on inverted pendulum model. In: 22nd IEEE international symposium on intelligent control, Singapore, 1–3 October 2007
11. Thant AA, Aye KK (2009) Application of cubic spline interpolation to walking patterns of biped robot. World Academy of Science, Engineering and Technology
12. Benbrahim H, Franklin JA (1997) Biped dynamic walking using Reinforcement learning. Robot Auton Syst 22:283–302

Human-Centric Optimal Design of Biomimetic Exosuit for Loaded Walking: A Simulation Study

Karthick Ganesan and Abhishek Gupta

Abstract Exoskeletons can augment the capacity of humans and restore mobility. Design of exoskeletons has been primarily experimental. The effect of different actuation and control strategies are learned from experiments. Human device interaction can be considered in the design phase using biomechanical simulations and it can provide optimal designs. In this study, we use musculoskeletal modeling and simulation to find optimal actuator configuration of exosuits for reducing the metabolic cost of walking with heavy loads. Computed muscle control algorithm was used to compute the muscle excitations and actuator controls required to achieve experimental kinematics. A muscle energy expenditure model was used to estimate the reduction in metabolic cost with each actuator configuration. Results show that biarticular actuator configurations can provide higher metabolic cost reduction compared to uniarticular configurations.

Keywords Exoskeleton · Musculoskeletal modeling · Metabolic cost · Human–robot interaction

1 Introduction

In India 26.8 million (2.21%) persons suffer from disabilities among which 5.36 million (20%) suffer from locomotor disabilities [1]. Wearable robotic devices can restore and enhance their mobility. Powered prostheses, a class of wearable robotic devices, have enabled amputees to walk near normal [2, 3]. Exoskeletons, another class of wearable robotic devices, have enabled patients with spinal cord injury [4] and hemiparetic post-stroke patients [5] to walk better. Exoskeletons can also assist soldiers, firefighters, and laborers in lifting heavy loads.

The main objective of exoskeletons is to reduce the metabolic cost. Fully passive [6], power tethered [7], and autonomous exoskeletons [8] have reduced the metabolic

K. Ganesan (✉) · A. Gupta
Department of Mechanical Engineering, Indian Institute of Technology Bombay, Mumbai, India
e-mail: karthick.ganesan@iitb.ac.in

A. Gupta
e-mail: abhi.gupta@iitb.ac.in

D. Sen et al. (eds.), *Mechanism and Machine Science*, Lecture Notes in Mechanical Engineering, https://doi.org/10.1007/978-981-15-4477-4_35

cost of normal walking. Exoskeletons have also reduced the metabolic cost in loaded walking [9] and in running [10]. They have reduced the energy cost of walking of hemiparetic post-stroke patients [5].

Design and development of exoskeletons are mostly experimental in nature and human device interaction is not considered during the design phase. For example, optimal stiffness for a minimum metabolic cost for a passive elastic ankle exoskeleton was found through experiments [6]. Experiments were used to understand the effect of increased assistance on the metabolic cost in a multiarticular powered soft exosuit [11]. Musculoskeletal simulations also have been used to explain the mechanisms underlying the reduction in metabolic cost with exoskeletons. A simple lumped muscle-tendon model of ankle joint revealed optimum stiffness for a passive elastic exoskeleton mentioned earlier [12]. More complex EMG-driven musculoskeletal models have explained the unexpected effects of exoskeleton assistance for ankle on metabolic rate during walking [13] and hopping [14].

However these studies were carried out after experiments with the device. Human–device interaction can be considered in the design phase using biomechanical simulations. This will speed up the design and development process, save expenses, and provide optimal exoskeleton designs. The important questions for designers are which joint to provide assistance, when and how much assistance is to be provided. Assistance has been provided individually to ankle, knee, and hip joints and also to multiple joints. Different designs have reported different reductions in metabolic cost. It is difficult to compare these designs because of variations in parameters such as device mass and control strategies.

Recent studies have used musculoskeletal simulations to answer these questions for running [15] and loaded walking [16]. For loaded walking, seven assistive devices were considered namely hip flexion, hip extension, hip abduction, knee flexion, knee extension, ankle plantar flexion, and ankle dorsiflexion. It was found that hip abduction device has the greatest potential to reduce the metabolic cost. These studies revealed that the assistance moment pattern does not necessarily have to follow the net joint moment pattern and assistive device on a particular joint can affect the muscle activities of other joints. These findings are in line with experimental findings. In our group, we have used musculoskeletal simulations to inspire the design of novel active transtibial prosthesis [17].

One of the limitations of the study of exoskeletons with loaded walking [16] is that the exoskeleton was modeled using ideal actuators directly providing torque at the joints. Recent exoskeleton designs use soft robotic technology wherein force is applied along a path similar to muscles. These soft exosuits provide various advantages such as lightweight, more comfort, and the force can be applied across multiple joints similar to biarticular muscles in the human body. Biarticular muscles could provide energy efficiency [18].

The objective of this study is to find how exosuits with different actuator configurations in parallel to different muscles affect the metabolic cost of loaded walking through musculoskeletal simulations. We hypothesized that biarticular actuators can provide higher metabolic cost reduction compared to the corresponding uniarticular actuators. Also, we hypothesized that placing actuators in parallel to muscles compared to directly actuating at the joints may provide different results in optimization.

2 Methods

2.1 *Musculoskeletal Model*

We used OpenSim (version 3.3), an open-source biomechanical modeling software [19], for modeling and simulations. We used the same musculoskeletal model used in a previous study of loaded walking with assistive devices [16]. The three-dimensional musculoskeletal model has 39 degrees of freedom. Eight of the degrees of freedom which are not very important for the study were locked: bilateral ankle eversion, toe flexion, wrist flexion, and wrist deviation. In effect, each leg in the model has five degrees of freedom: hip flexion/extension, hip abduction/adduction, hip internal/external rotation, knee flexion/extension, and ankle plantar flexion/dorsiflexion and they are shown in Fig. 1. It has 80 muscle-tendon units for actuating the lower limb and 17 ideal torque actuators for actuating the upper limb [20]. Muscles were modeled by Hill type equilibrium model [21]. Musculotendon parameters are based on measurements of 21 cadavers and magnetic resonance images of 24 young healthy individuals. For the simulations of loaded walking, a 38 kg of load was added. The load was modeled as a hollow cylindrical channel with uniform density and welded to the torso.

2.2 *Exosuit Model*

Exosuit actuators were modeled as ideal force-producing elements producing force along the prescribed path. It was implemented using PathActuator in OpenSim. It applies controllable tension along the geometry path [19]. The force produced is given by

$$F = x * F_{opt} \tag{1}$$

where x is the control signal to the actuator, varying from 0 to 1 and F_{opt} is the optimal force, the maximum force that the actuator can produce. Path of the actuator can be prescribed by specifying a set of points. Ten different actuator configurations were considered. They were to mimic the soleus, gastrocnemius, tibialis anterior, biceps femoris short head, biceps femoris long head, psoas, gluteus maximus, rectus femoris,

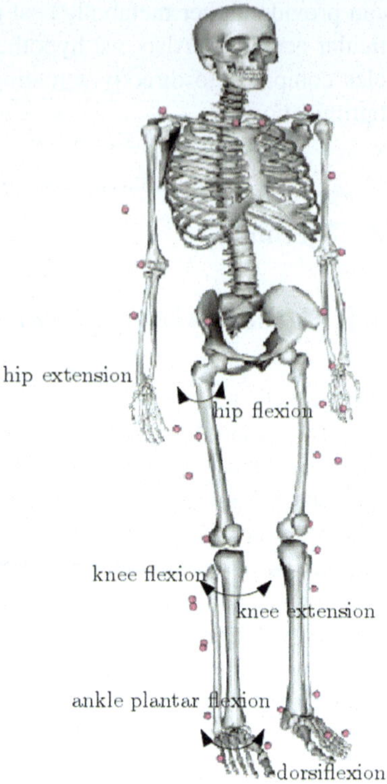

Fig. 1 Musculoskeletal model and its important degrees of freedom

and vasti muscles. The function of these muscles and their origin and insertion segments are given in Table 1.

Muscles like soleus are uniarticular, span a single joint whereas muscles like gastrocnemius are biarticular, and span two joints. The path of each actuator was defined by only two points, origin and insertion. No via points or wrapping objects were considered. The attachment points of the actuators were similar to the muscles except that these points were chosen such that they lie outside the body based on marker locations on the model. Path of all the actuators is shown in Fig. 2, although during simulations only one actuator was added at a time. The optimal force of the path actuators was set at a very high value of 30000 N so that the actuator can produce the maximum possible force.

2.3 *Simulations*

We used the scaled, mass adjusted model and experimental gait data with load, without exosuit of one subject supplemented with the literature [16] for the simulations. We used the Computed Muscle Control (CMC) algorithm [22] as depicted in Fig. 3

Table 1 Path of the actuators

S. No.	Actuator path	Function	Origin	Insertion
1	Soleus (SOL)	Ankle plantar flexion	Shank	Foot
2	Gastrocnemius (GAS)	Ankle plantar flexion, knee flexion	Thigh	Foot
3	Tibialis Anterior (TA)	Ankle dorsiflexion	Shank	Foot
4	Biceps Femoris Short Head (BFSH)	Knee flexion	Thigh	Shank
5	Biceps Femoris Long Head (BFLH)	Knee flexion, hip extension	Pelvis	Shank
6	Psoas	Hip flexion	Pelvis	Thigh
7	Gluteus Maximus (GMax)	Hip extension	Pelvis	Thigh
8	Rectus Femoris (RF)	Hip flexion, knee extension	Pelvis	Shank
9	Vasti (VAS)	Knee extension	Thigh	Shank
10	Gluteus Medius (GMed)	Hip abduction	Pelvis	Thigh

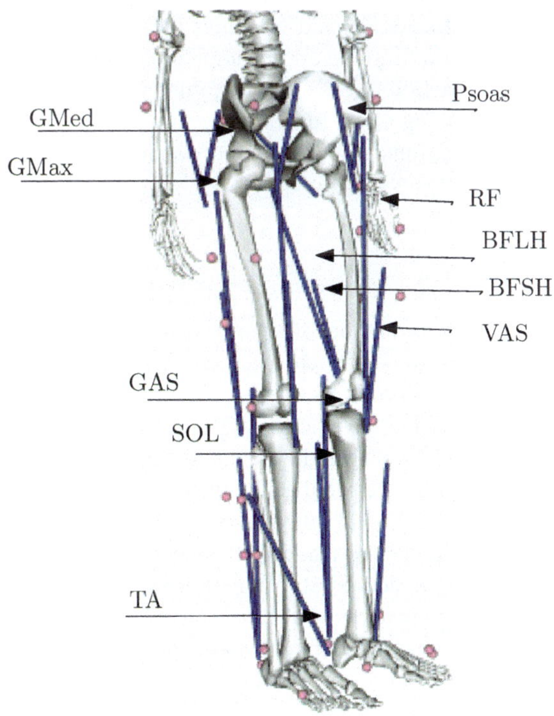

Fig. 2 Path of different actuators. In simulations only one actuator was added at a time

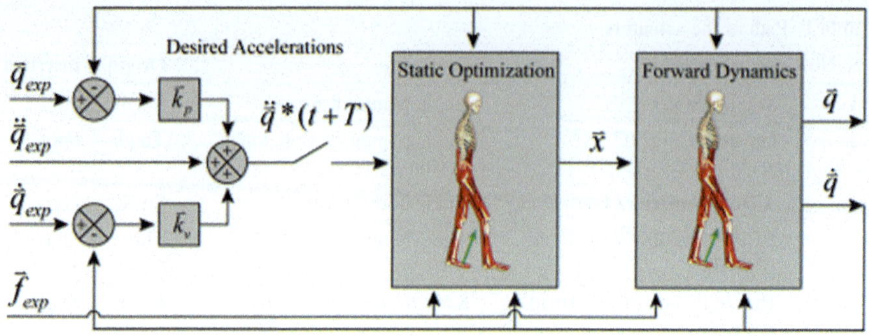

Fig. 3 Schematic of the Computed Muscle Control (CMC) algorithm [19]

in OpenSim 3.3 to estimate the muscle and actuator controls. Desired model accelerations ($\ddot{q}^*$) that will drive the model toward experimentally measured kinematics are calculated using proportional plus derivative (PD) control on the difference between experimental generalized coordinates ($\boldsymbol{q}_{exp}$), velocities $\left(\dot{\boldsymbol{q}}_{exp}\right)$, and model generalized coordinates ($\boldsymbol{q}$) and velocities ($\dot{\boldsymbol{q}}$). Muscle redundancy is resolved using static optimization. In static optimization model kinematics and experimentally measured ground reaction forces are used to find the muscle excitations and actuator controls by minimizing a cost function.

The cost function is the sum of squared actuator controls as given in Eq. (2).

$$J = \sum_{i=1}^{n_x} x_i^2 \tag{2}$$

where n_x is the number of controls in the model and x_i is the ith control. The optimization is performed at each specified time step. The actuator controls include muscle excitations and controls for reserve and residual actuators, and controls for additional actuators such as path actuator. Reserve actuators produce a small amount of torque about a joint when model's actuators are not able to produce a needed force at an instant in time. Residual actuators account for the dynamic inconsistencies between experimental data and the musculoskeletal model. Reserve actuators for hip, knee, and ankle joints were added. Residual actuators for the six degrees of freedom of pelvis were added. First, simulation was performed without any exosuit actuators. Then ten simulations were performed with one exosuit actuator (for both legs) at a time, with the same experimental data.

2.4 Metabolic Cost Estimation

Metabolic cost was estimated from the simulations using a muscle energy expenditure model [23] with an orderly recruitment model to determine the ratio of slow- to fast-twitch fibers that were excited at each instant of the simulation [24]. Umberger2010MuscleMetabolicsProbe in OpenSim 3.3 was used for this purpose. In brief, in Umberger's metabolic model, total muscle energy expenditure rate was calculated as a sum of activation heat rate, maintenance heat rate, shortening/lengthening heat rate, and mechanical work rate of muscle contractile element. Each term was modeled using empirical relations derived from various experiments. A basal rate of 1.2 W/kg was added to account for whole-body metabolism. This was integrated over the gait cycle to get the total metabolic energy.

3 Results

3.1 Metabolic Cost

Individual muscle contribution to total metabolic cost can provide insight into which muscle or joint has to be the target for assistance. From the simulation of loaded walking without any actuator, the individual muscle contribution as a percentage of total metabolic energy excluding basal energy is shown in Fig. 4. Gluteus medius

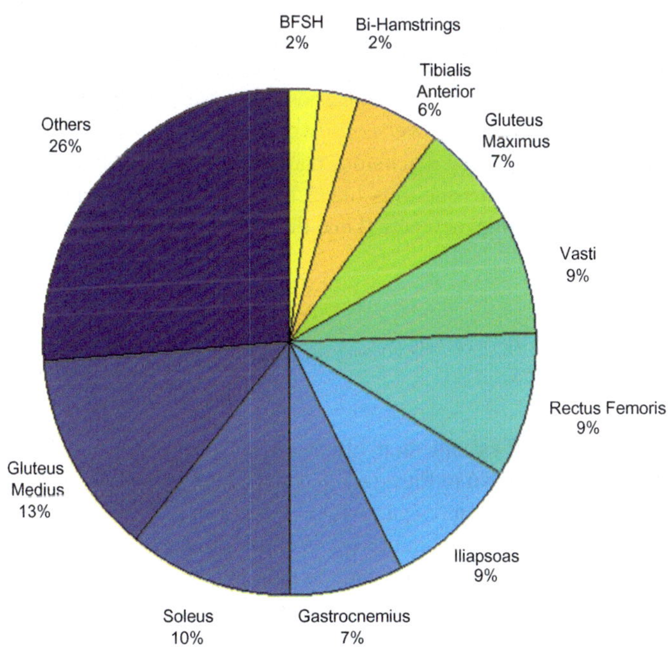

Fig. 4 Individual muscles contribution to metabolic cost excluding basal cost for walking with load, without any actuator

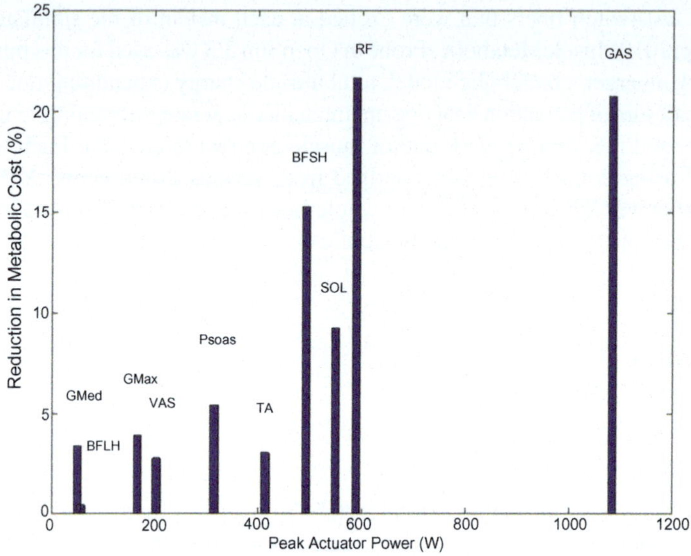

Fig. 5 Percentage reduction in metabolic cost versus peak actuator power for different muscle-like actuators. Simulations included one actuator at a time. For abbreviations refer Table 1

contributed the largest followed by soleus, iliopsoas, gastrocnemius. and rectus femoris.

The percentage reduction in metabolic cost for different actuator configurations and their corresponding peak power is shown in Fig. 5. Biarticular rectus femoris-like actuator (hip flexion-knee extension) had the highest metabolic cost reduction followed by biarticular gastrocnemius-like actuator (knee flexion-ankle plantar flexion), uniarticular BFSH-like actuator (knee flexion), and soleus-like actuator (ankle plantar flexion).

3.2 *Muscle Forces*

The actuator force profiles and their effect on muscle forces of the right leg for different actuators are shown in Figs. 6, 7, 8, 9, and 10. Only a few important muscle forces are shown. In these figures gait cycle is from opposite toe off to opposite toe off, because of the experimental data used in simulations.

Soleus-like actuator. Soleus-like actuator only reduced the force of soleus muscle as shown in Fig. 6. It had insignificant effect on other muscles. Gastrocnemius is an ankle plantar flexor as well as knee flexor. Although the actuator could replace the force of gastrocnemius at the ankle, gastrocnemius's force is required at the knee joint. So the gastrocnemius force is not influenced significantly by the soleus actuator.

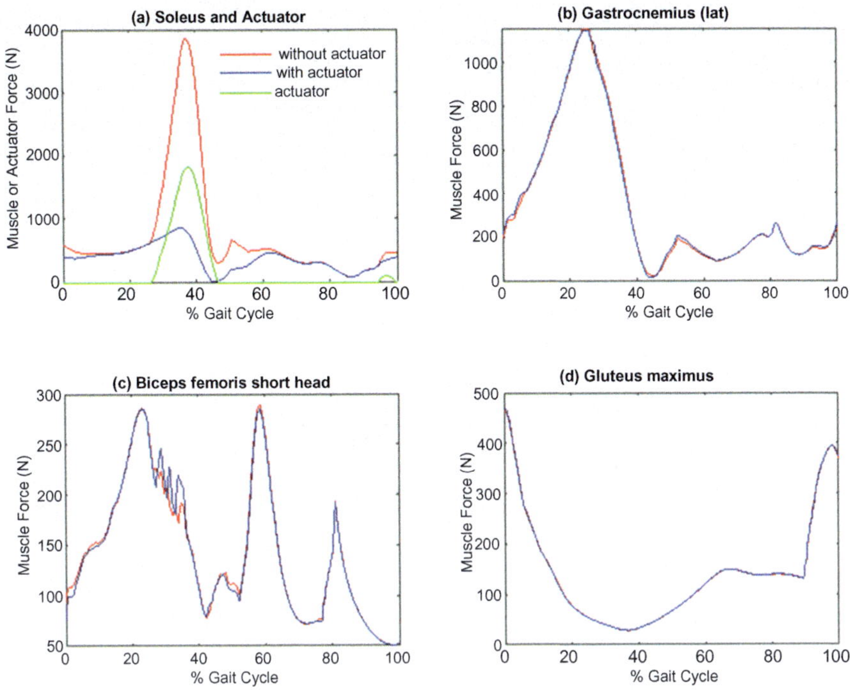

Fig. 6 Forces in important muscles with soleus-like actuator. Actuator force (green), without actuator (red), with actuator (blue)

Gastrocnemius-like actuator. However when the actuator parallel to gastrocnemius (biarticular) is used, it produces a moment at the knee as well as ankle joints. Hence it reduced the force of gastrocnemius, BFSH which is a knee flexor, and soleus as shown in Fig. 7. This resulted in greater metabolic reduction compared to soleus-like actuator. Gastrocnemius and soleus combined contribute around 18% toward the total metabolic cost excluding basal cost (Fig. 4). A recent experimental study showed that biarticular actuator with spring for an ankle-foot exoskeleton could reduce higher metabolic cost compared to monoarticular and biarticular actuators without spring [25].

BFSH-like actuator. BFSH-like actuator assists knee flexion. It replaced the function of gastrocnemius muscle at the knee and also BFSH muscle. In effect gastrocnemius muscle force reduced as shown in Fig. 8. Since the gastrocnemius muscle's contribution to ankle joint was also reduced, soleus muscle force increased to compensate for this. This resulted in less metabolic cost reduction compared to biarticular gastrocnemius-like actuator.

Psoas-like actuator. Psoas-like actuator reduced the peak force of both iliacus and psoas which are uniarticular hip flexors as shown in Fig. 9. Peak force of rectus femoris was reduced only slightly because it reduced knee extension moment as well.

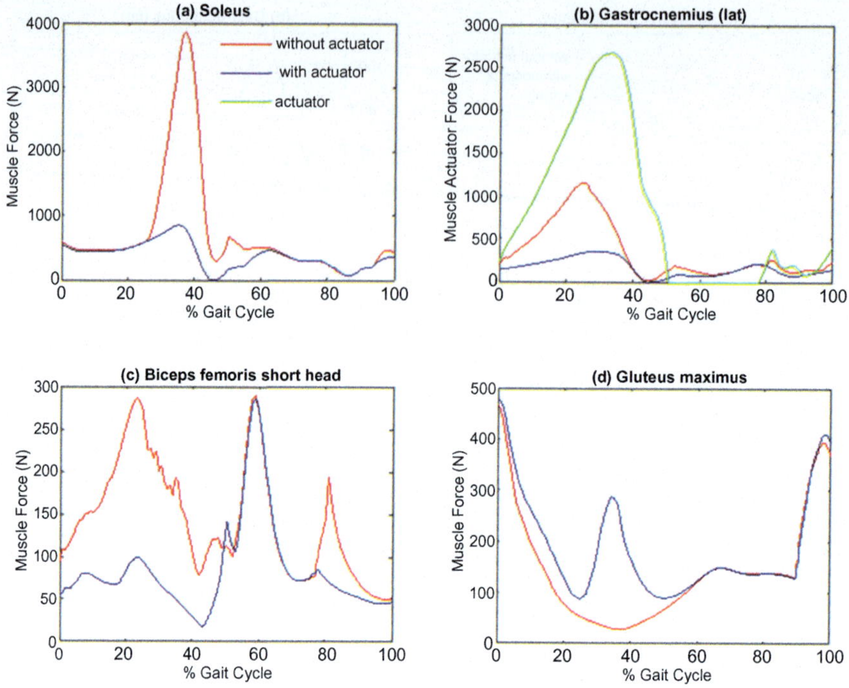

Fig. 7 Forces in important muscles with gastrocnemius-like actuator. Actuator force (green), without actuator (red), with actuator (blue)

This resulted in reduction of peak force of gastrocnemius. To compensate for this there was increase in soleus force during the same period.

Rectus femoris-like actuator. Rectus femoris-like actuator reduced the activity of rectus femoris muscle as shown in Fig. 10. It also reduced the activity of psoas during 40–60% of the gait cycle. However it increased the peak force of vasti muscles. Iliopsoas, rectus femoris, and vasti combined contribute around 26% toward total metabolic cost excluding basal cost (Fig. 4). This actuator reduced the muscle metabolic energy of all of these muscles resulting in the highest metabolic cost reduction.

4 Discussion

When driven by actuators directly connected to the joints, hip flexion actuator had the highest metabolic reduction followed by hip abduction actuator [16]. With path actuators, psoas-like actuator (hip flexion) and gluteus medius-like actuator (hip abduction) did not produce the highest metabolic savings. It is because the chosen

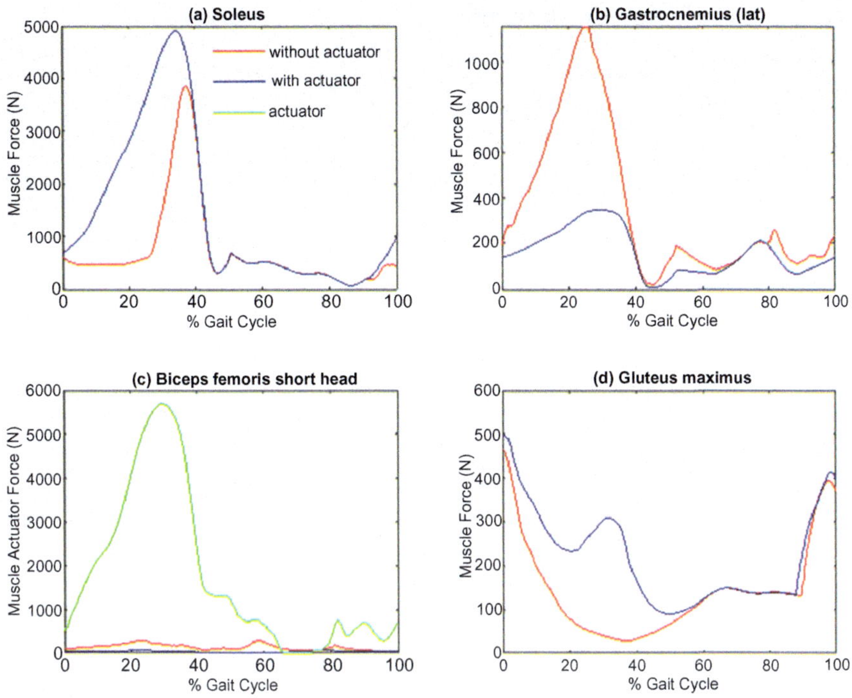

Fig. 8 Forces in important muscles with BFSH-like actuator. Actuator force (green), without actuator (red), with actuator (blue)

attachment points may result in moment arm characteristics which are not economical. The metabolic reductions caused by other path actuators were similar to driving with actuators directly connected to the joint. BFLH-like actuator also did not produce significant metabolic reduction for the same reason.

There was no significant change in the gastrocnemius muscle force with soleus actuator. However experimental evidence shows that gastrocnemius EMG activity is reduced by monoarticular ankle exoskeletons [6, 25]. Previous simulation study [16] reported on average, across subjects and trials, 12% reduction in peak ankle plantar flexion moment of medial gastrocnemius with monoarticular ankle plantar flexion device. Further analysis of their results revealed that this reduction is subject- and trial-specific and for the subject and trial data used in this study, no significant change in gastrocnemius muscle force was found. Muscle compensations observed with monoarticular actuators were similar to that reported by [16].

With BFSH-like actuator, the metabolic savings were still higher compared to soleus-like actuator despite increased soleus force. This was due to the reduction in soleus muscle metabolic energy probably because of reduced fiber velocities. Although the actuator's path is similar to BFSH muscle, it could reduce the force of gastrocnemius muscle. So in this case the joint which was given assistance is more important than the path of the actuator.

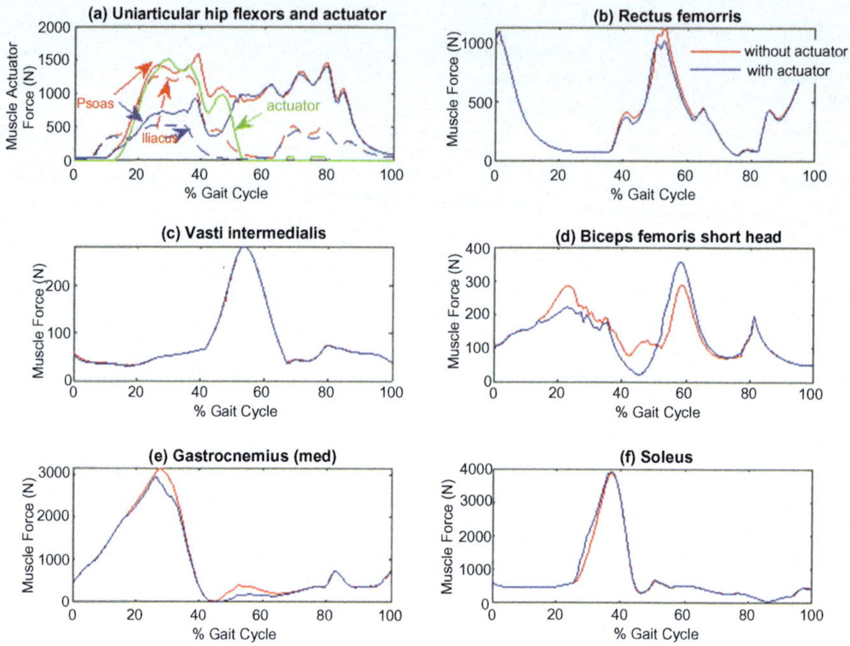

Fig. 9 Forces in important muscles with psoas-like actuator. **a** solid—psoas, dashed—iliacus, actuator force (green), **a**–**f** without actuator (red), with actuator (blue)

One of the limitations of this study is that the actuator paths were modeled using only origin and insertion points. Via points and wrapping were not included. Origin and insertion points were approximate based on marker locations. For more realistic simulations, locations of these attachment points have to be measured experimentally. To consider the wrapping of the actuator over the body, more complex models will be required.

This study provided insights into how different actuator configurations affect the metabolic cost of loaded walking. In future we will consider modified metabolic cost functions and cost functions other than metabolic cost. In the current study we used the experimental data without any actuator. In future we will perform predictive simulations without using experimental data. We will also extend this to the design and control of other wearable devices such as prostheses.

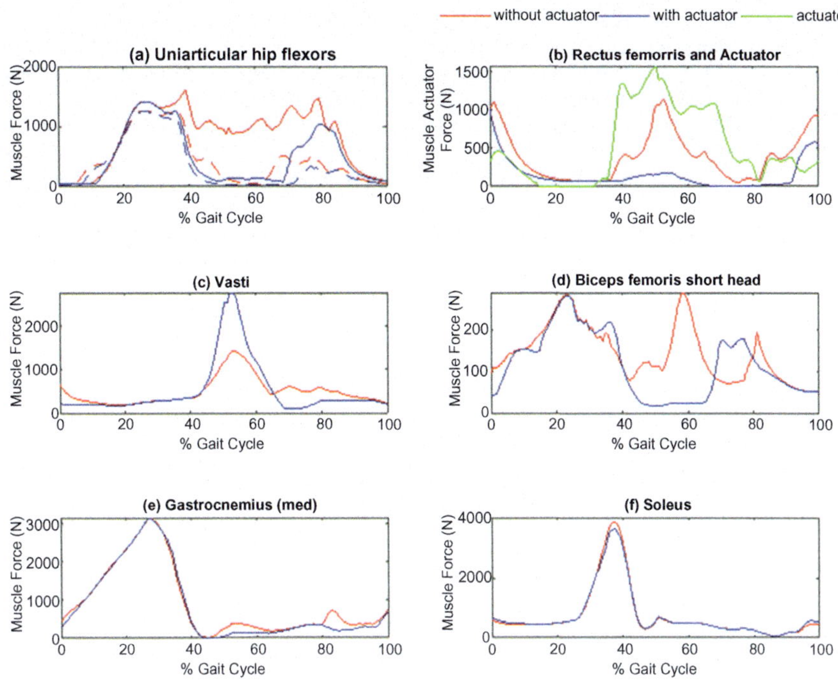

Fig. 10 Forces in important muscles with RF-like actuator. **a** solid—psoas, dashed—iliacus, **b** actuator force (green), **a–f** without actuator (red), with actuator (blue)

Acknowledgements We thank the OpenSim team for making the models and experimental data available as open source.

References

1. "Disabled persons in India: A statistical profile", Social Statistics Division. Ministry of Statistics and Programme Implementation, Government of India, 2016
2. Martinez-Villalpando EC, Herr H (2009) Agonist-antagonist active knee prosthesis: a preliminary study in level-ground walking. J Rehabil Res Dev 46(3):361–373
3. Herr HM, Grabowski AM (2012) Bionic ankle-foot prosthesis normalizes walking gait for persons with leg amputation. Proc R Soc B Biol Sci 279(1728):457–464
4. Esquenazi A, Talaty M, Packel A, Saulino M (2012) The Rewalk powered exoskeleton to restore ambulatory function to individuals with thoracic-level motor-complete spinal cord injury. Am J Phys Med Rehabil 91(11):911–921
5. Awad LN et al (2017) A soft robotic exosuit improves walking in patients after stroke. Sci Transl Med 9(400)
6. Collins SH, Wiggin MB, Sawicki GS (2015) Reducing the energy cost of human walking using an unpowered exoskeleton. Nature 522(7555):212–215
7. Malcolm P, Derave W, Galle S, De Clercq D (2013) A simple exoskeleton that assists plantarflexion can reduce the metabolic cost of human walking. PLoS One 8(2)

8. Mooney LM, Rouse EJ, Herr HM (2014) Autonomous exoskeleton reduces metabolic cost of human walking. J Neuroeng Rehabil 11(1):1–5
9. Mooney LM et al (2014) Autonomous exoskeleton reduces metabolic cost of human walking during load carriage. Journal of Neuroengineering and Rehabilitation 11(1):80
10. Lee G et al (2017) Reducing the metabolic cost of running with a tethered soft exosuit. Sci Robot 2(6):6708
11. Quinlivan BT et al (2017) Assistance magnitude versus metabolic cost reductions for a tethered multiarticular soft exosuit. Sci Robot 2(2), 4416
12. Sawicki GS, Khan NS (2016) A simple model to estimate plantarflexor muscle-tendon mechanics and energetics during walking with elastic ankle exoskeletons. IEEE Trans Biomed Eng 63(5):914–923
13. Jackson RW, Dembia CL, Delp SL, Collins SH (2017) Muscle–tendon mechanics explain unexpected effects of exoskeleton assistance on metabolic rate during walking. J Exp Biol 220(11):2082–2095
14. Farris DJ, Hicks JL, Delp SL, Sawicki GS (2014) Musculoskeletal modelling deconstructs the paradoxical effects of elastic ankle exoskeletons on plantar-flexor mechanics and energetics during hopping. J Exp Biol 217(22):4018–4028
15. Uchida TK, Seth A, Pouya S, Dembia CL, Hicks JL, Delp SL (2016) Simulating ideal assistive devices to reduce the metabolic cost of running. PLoS One
16. Dembia CL, Silder A, Uchida TK, Hicks JL, Delp SL (2017) Simulating ideal assistive devices to reduce the metabolic cost of walking with heavy loads. PLoS One 12(7)
17. Joshi S, Gupta A (2016) Conceptual design of an active transtibial prosthesis based on expected joint and muscle forces in a unilateral transtibial amputee: a modelling study. In: Proceedings of the ASME international mechanical engineering congress and exposition, vol 3, p 11
18. Wells RP (1988) Mechanical energy costs of human movement: An approach to evaluating the transfer possibilities of two-joint muscles. J Biomech 21(11):955–964
19. Delp SL et al (2007) OpenSim: Open-source software to create and analyze dynamic simulations of movement. IEEE Trans Biomed Eng 54(11):1940–1950
20. Rajagopal A, Dembia C, DeMers M, Delp D, Hicks J, Delp S (2016) Full body musculoskeletal model for muscle- driven simulation of human gait. IEEE Trans Biomed Eng 63(10):2068–2079
21. Millard M, Uchida T, Seth A, Delp SL (2013) Flexing computational muscle: modeling and simulation of musculotendon dynamics. J Biomech Eng 135(2):21005
22. Thelen DG, Anderson FC (2006) Using computed muscle control to generate forward dynamic simulations of human walking from experimental data. J Biomech 39(6):1107–1115
23. Umberger BR, Gerritsen KG, Martin PE (2003) A model of human muscle energy expenditure. Comput Methods Biomech Biomed Eng 6(2):99–111
24. Uchida TK, Hicks JL, Dembia CL, Delp SL (2016) Stretching your energetic budget: How tendon compliance affects the metabolic cost of running. PLoS One 11(3)
25. Malcolm P, Galle S, Derave W, de Clercq D (201) Bi-articular knee-ankle-foot exoskeleton produces higher metabolic cost reduction than weight-matched mono-articular exoskeleton. Front Neurosci 12

Damping of Wind-Induced Galloping Oscillations of Solar Trackers

Ninad H. Watwe and V. Kartik

Abstract A solar tracker is a machine that uses a mechanical prime mover to tilt solar photo voltaic panels from east to west during the day to track the sun for maximizing the production of PV power by keeping the incident angle of sun-rays perpendicular to the panel surface. Drag and lift produced by the wind on the surface area of solar panels causes a torsional moment on the tracker structure, and may induce torsional instability in the structure due to wind galloping. The use of hydraulic dampers is recommended to reduce the intensity of galloping and reduce the potential risk of damage to the tracker structure and panels. In this paper, the vibration characteristics of the tracker structure are analysed (theoretically and numerically) with and without external dampers, using the lumped spring-mass-damper approach to identify the suitability of the selected dampers for the application.

Keywordsr Solar tracker · Damping

1 Introduction

Recent advances in technology, the international consensus on climate change action and relevant changes in national and international policies have made photo voltaic solar energy economical, and by implication the solar industry very competitive. The adoption of solar trackers has increased due to the added gain that the system provides. A solar tracker is an array of solar panels (referred to as modules) mounted on cold-formed steel mechanical structures (referred to as table) along with a prime-mover for tracking of modules. Due to the competitive landscape of the solar industry, designers increasingly rely on higher stronger grades of steel to reduce the tonnage of steel in the structure.

N. H. Watwe (✉)
Mahindra Susten Private Limited, Mumbai, India
e-mail: watwe.ninad@mahindra.com

V. Kartik
Department of Mechanical Engineering, Indian Institute of Technology Bombay, Mumbai, India
e-mail: vkartik@iitb.ac.in

D. Sen et al. (eds.), *Mechanism and Machine Science*, Lecture Notes in Mechanical Engineering, https://doi.org/10.1007/978-981-15-4477-4_36

This reduced tonnage due to stronger steel allows the system to keep its strength, but results in reduced stiffness of the system. Reduced structural stiffness of the system reduces the modal frequency of the system making it susceptible to excitation in its fundamental modes of vibration.

A common solution used in the industry is to add structural damping elements in the system using hydraulic dampers. There exists no systematic method for examining the suitability of dampers for performing their duty, and thus no method for sizing of dampers for use in solar trackers. Previous studies, such as [1] have focused on stowing logic for reduction of "galloping" in wind, where galloping is commonly referring to the phenomenon of excessive large-amplitude torsional movement. An attempt is made in this paper to study the velocity and displacement response of a tracker table under typical forcing with and without the damper. The tracker system is modelled as a lumped torsional spring-mass-damper system. Each of these system components is analysed individually. A forcing function is approximated based on the simplified 2-dimensional CFD model of a tracker structure. Using all the generated data, the vibration response of the structure is simulated and analysed to check for suitability of the elected dampers.

2 Tracker Structure and Working

A typical tracker structure is a cold-formed steel structure. Modules are mounted on slender cantilever beams called rails. These rails are in turn mounted on long steel pipes called torque tube. The torque tube is then suspended on vertical steel members using bearings. The vertical posts are grouted through concrete pile foundations in the ground. Figure 1 schematically represents the arrangement of a tracker table.

The torque tube is free to rotate in the bearing and is held in the middle by the linear actuator [2]. The extension and retraction of the linear actuator causes the lever-arm coupled to the torque tube to move. Thus, the extension and retraction of the linear actuator is used to rotate the torque tube around its longitudinal axis. Modules thus mounted in the torque tube can track the sun by controlling the length of the linear actuator. Based on the time of day, date and year, astronomical algorithms calculate the ideal azimuth angle to track the sun and execute tracking through electronic control [1].

The load on the system, apart from contact friction in the bearings, is due to the self-weight of the system, and moment due to impact of wind on the module surface. Each module contributes to torque which is generated due to the exposure to wind and connection to the torque tube. When analysed as a shaft, the centre post is assumed to be the torsional fixity and the rest of the torque tube is supported in bearings.

The torque tube is modelled as a doubly overhanging continuous beam with supports at every vertical post. The beam thus undergoes loading on its entire span due to the self-weight of the torque tube, apart from the load due to other members of the payload including rails and modules. Additional load is also due to wind impact on modules. This load may change depending on the wind conditions and angle of

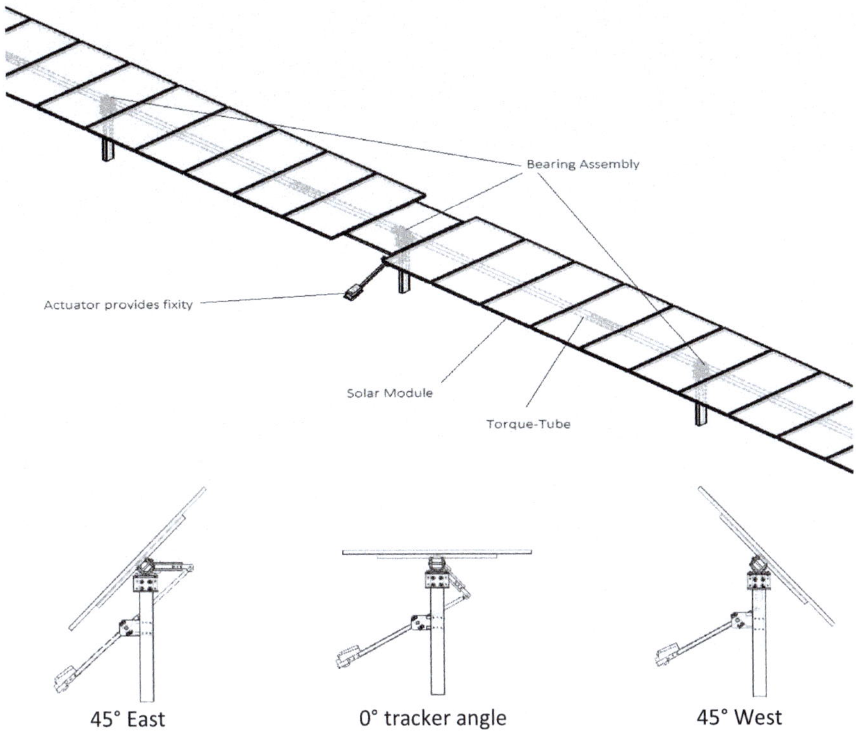

Fig. 1 Tracker table solid model and its configuration at various angles

attack of the wind on the modules. Simultaneously, this tube is also modelled as a shaft under torsion across its entire span. A simplified model of the torque diagram is given in Fig. 2. Through modal analysis done on the structural model, it is found that the first fundamental mode shape is torsional vibration about the torque tube as shown

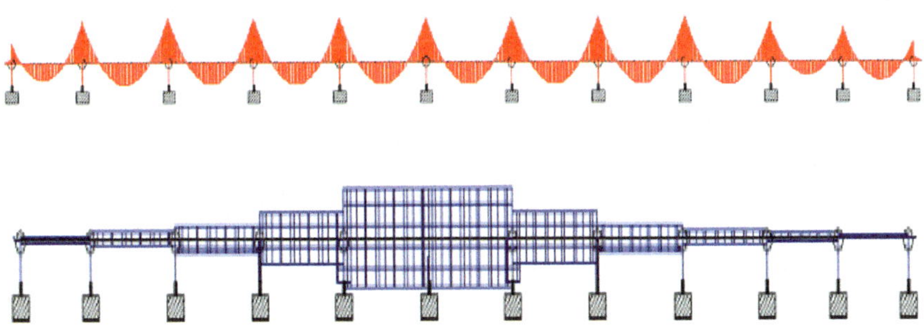

Fig. 2 Bending moment diagram of tracker (top) and torque diagram of torque on the torque tube due to wind (bottom). The height of the blue rectangle represents the magnitude of torque. Note that torque keeps on adding in each bay between vertical post and causes maximum torque to concentrate near the centre

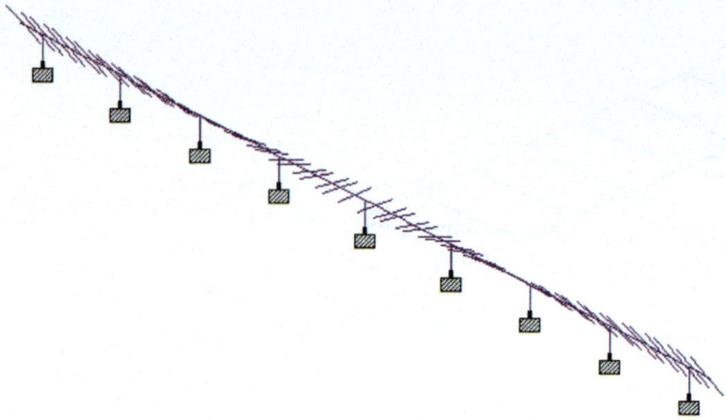

Fig. 3 First mode shape in the tracker showing twisting of torque tube that is clamped in the middle and free at one end

in Fig. 3. This is also corroborated from various reports of torsional galloping of the structure under load due to wind impact. Hence, in this paper, only this dominant mode of vibration is analysed (Fig. 4).

For reducing the magnitude of torsional vibration of the system, hydraulic dampers are added to the system by coupling ends of the dampers to radial levers on the torque tube and other ends fixed to the vertical posts as shown in Fig. 5. Thus, if the tracker structure gallops due to wind, the high linear velocity formed because of the damping shall cause a high resultant linear velocity in the damper plunger, resulting in a high damping force from the dampers. This then resists the vibration and reduces its magnitude.

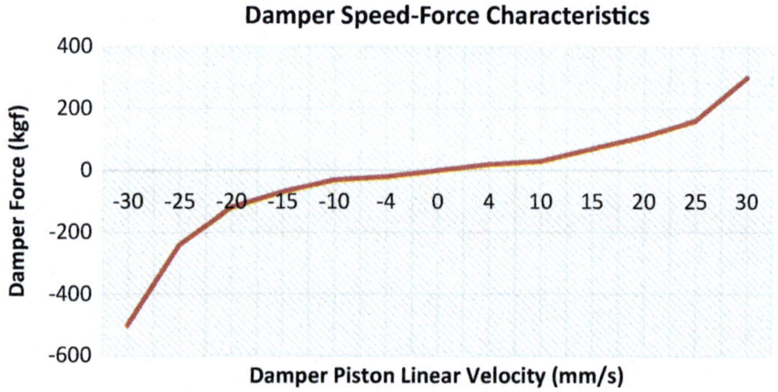

Fig. 4 Graph showing damping force versus linear velocity of the plunger of the damper

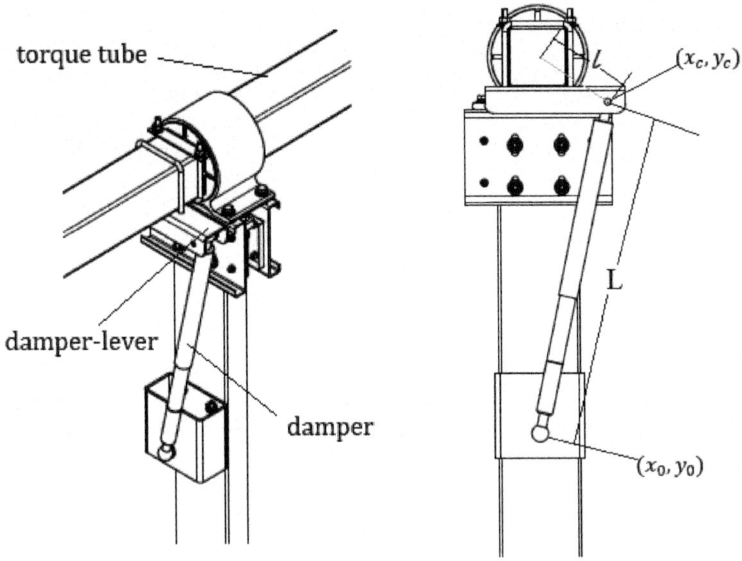

Fig. 5 Typical damper mounting done on tracker table

3 Analysis of Structural Properties

For the analysis of the vibrating structure, a simplified single degree of freedom lumped system model is constructed. The equivalent torsional stiffness of the torque tube and equivalent mass moment of inertia of the system are calculated for the full suspended system and lumped. The damping torque in the system is calculated based on the geometry of the dampers and mounting.

3.1 Stiffness

It may be noted that the torque tube spans the full length of the tracker. However, since the linear actuator in the centre is assumed to provide fixity to the shaft, the suspended system that is free to oscillate in the first torsional mode of vibration is only half of the length of the tracker table.

The torsional stiffness of non-circular hollow rectangular cross-section is given as [4]:

$$K_{\text{tube}} = \frac{T}{\theta} = \frac{kG}{L} \tag{1}$$

where variable k for non-circular rectangular hollow section is given as

Table 1 Parameters used for calculating the thickness of torque tube

Parameter	Value	Unit
t	2.5	mm
a	101	mm
G	79.73	GPa
L	30	m

$$k = \frac{2tt_1(a-t)^2(b-t_1)^2}{at + bt_1 - t^2 - t_1^2} \tag{2}$$

where a and b are lengths of the sides of rectangles and t and t_1 are the thicknesses of the corresponding sides. In this work, the cross-section of the torque tube is considered to be a hollow square with uniform thickness. Thus, the formula for k reduces to

$$k = t(a-t)^3 \tag{3}$$

and the torsional stiffness of the torque tube is obtained as

$$\mathrm{K_{tube}} = \frac{t(a-t)^3 G}{L} \tag{4}$$

The value for the given material and section size of a typical torque tube is calculated to be 6349.6 Nm/rad. Parameters used for calculating the value is given in Table 1.

3.2 Inertia

The rotational inertia of the system is given by the mass moment of inertia of the system about the axis of rotation of the tracker table. Mass moment of inertia is calculated for all individual components of the system using solid modelling CAD software. The software provides the numerical value of the mass moment of inertia about the centre of gravity of the component. Using the parallel axis theorem, the value of the moment of inertia about the axis of rotation is calculated. The typical value of the mass moment of inertia of the system is calculated as 249.6 kg m^2.

3.3 Damping

The damping in the system is assumed to be derived only from the hydraulic dampers added. The value of damping for various velocities is ascertained by methods prescribed in [5]. The typical values of damping at various linear plunger velocities are given in Table 2.

Compressive velocity is taken as negative, and extension as positive. A sixth-order polynomial is fit through the given data points to obtain the relation between velocity and damping force as shown in Fig. 4. This gives the response of the damper to linear speed of the plunger.

It should be noted that the damper may slope towards the windward side or away from it, as the direction of the wind may vary naturally. Based on the geometry of the mounting of the damper as shown in Fig. 5, the linear speed of the damper for the angular velocity of the tracker table is derived by taking the first derivative of the expression for the length of the damper.

$$L^2 = (x_c - x_0)^2 + (y_c - y_0)^2 \quad (5)$$

$$2.L.\dot{L} = 2.(x_c - x_0).(\dot{x_c}) + 2.(y_c - y_0).(\dot{y_c}) \quad (6)$$

Then the value of $(\dot{x_c})$ is simplified as $-\dot{l}.\dot{\theta}.\sin\theta$ and value of $(\dot{y_c})$ as $\dot{l}.\dot{\theta}.\cos\theta$, where θ is the tracker table angle. After simplification, the linear velocity of the plunger, as a function of the angular speed and angular position of the tracker, is derived as

$$\dot{L} = -\frac{l\dot{\theta}}{L}[(x_c - x_0)\sin\theta - (y_c - y_0)\cos\theta] \quad (7)$$

where $\dot{L}$ is the instantaneous linear velocity of the damper plunger at instantaneous damper length L and damper-lever length l is achieved at tracker table angle θ at an angular velocity of the tracker table $\dot{\theta}$. Here, (x_c, y_c) are the co-ordinates of the point of coupling of damper-lever and damper end, calculated for the instance from

Table 2 Damping force for various linear plunger velocities

Plunger velocity (m/s)	Force during extension (kgf)	Force during compression (kgf)
4	20	20
10	30	30
15	70	70
20	110	120
25	160	240
30	300	500

the tracker table angle, while (x_0, y_0) are the co-ordinates of points of coupling of the damper with vertical post, which are fixed.

The torque due to damping is calculated by cross-multiplying the damping force to the radial distance of coupling of the damper from the centre of rotation. It is noted that the damping torque will always act in the direction opposite to the direction of motion, as it will always resist motion. The value is calculated as

$$\vec{\mathrm{T}}_{\mathrm{damp}} = \vec{r} \times \vec{F} \tag{8}$$

where $\vec{T}_{\mathrm{damp}}$ is the damping torque vector, $\vec{F}$ is the damping force vector and $\vec{r}$ is the radius vector from the centre of rotation to the point of coupling of damper and lever. The damping torque is in the direction of the axis of the torque tube and changes direction based on the direction of the force, and always resists the motion of the torque tube.

3.4 Force Due to Wind

The design of the tracker is done with compliance to various wind design codes such as [6, 7] and as such may be used to approximate the force and moment caused by wind on the structures. Studies such as [8] have focused on wind moment and force on roof-top structures that typically have small aspect ratios. However, for values of the moment that are more relevant to the type of structure under analysis (small height, large aspect ratio), a separate steady-state 2-dimensional CFD analysis was conducted for tracker table at different angles to determine the coefficient of force and centres of pressures at different angles of the tracker table. A typical pressure contours of wind on a flat plate, simulating a tracker, is given in Table 3.

The coefficients of force and centre of pressure obtained are given in Table 4. The moment on tracker table is approximated as

$$M_{t\text{-wind}}(v) = \frac{1}{2}\rho_a.v^2.A_{\mathrm{mod}}.C_\alpha.(0.5 - p_c).t \tag{9}$$

where t is the tributary width, A_{mod} is the area of modules and ρ_a is the density of air.

Figure 6 shows the variation of the normalized wind moment with the angle of attack obtained using the above analysis. From the data, an 8th degree curve that closely tracks the data points in Fig. 6 for this range is given as below:

Table 3 Typical pressure contour plots of wind impact on a flat plate. It is observed that the total pressure on the plate increases with an increase in the angle of attack. It is also noted that the centre of pressure always acts on the windward side. The distance of the centre of pressure from the leading windward edge increases with an increase in angle. Red areas denote positive pressure, while blue areas denote negative pressure (suction)

Angle	Wind from left	Wind from right
45°		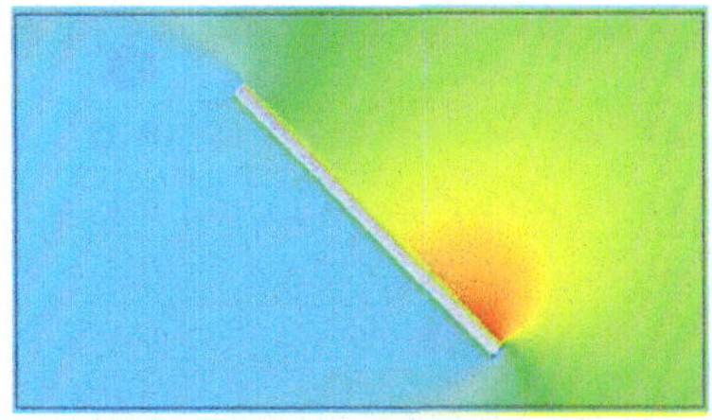
30°	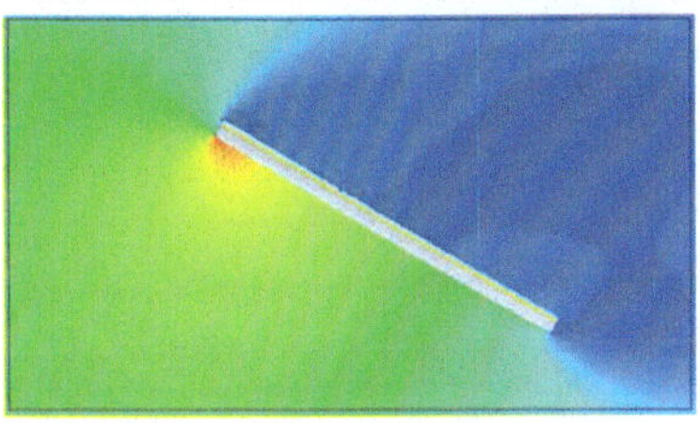	
20°	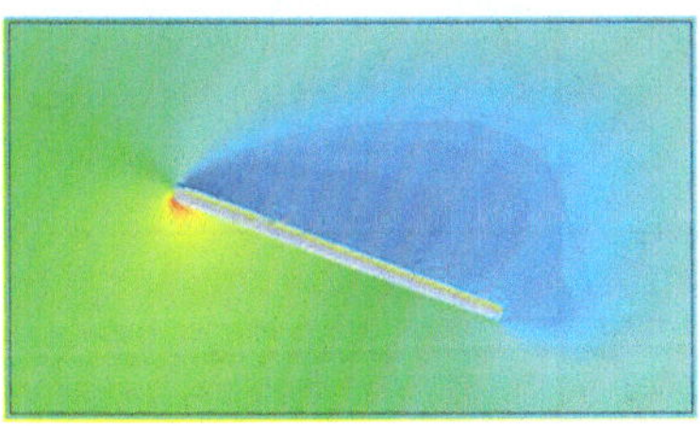	
10°	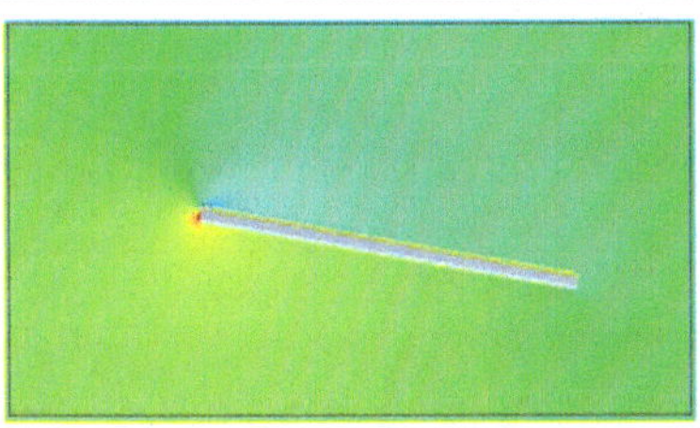	
0°	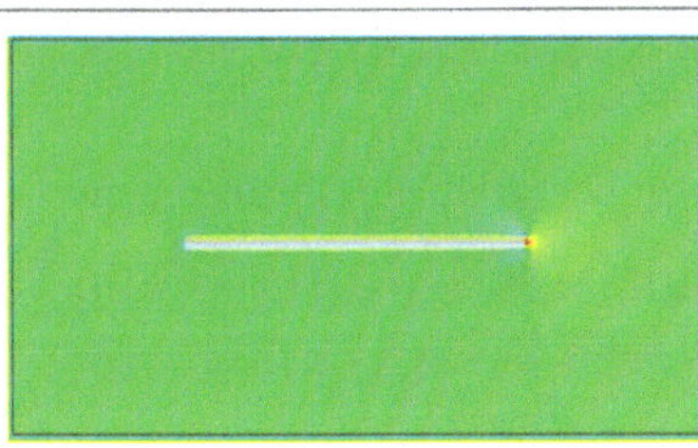	

Table 4 Coefficient of pressure and centre of pressure for various tracker table angles

Angle	Suction		Pressure	
	Suction coefficient (C_α)	Centre of pressure (p_c)	Pressure coefficient (C_α)	Centre of pressure (p_c)
0	−0.14	0.04	0.14	0.04
5	−0.62	0.3	0.65	0.3
10	−0.66	0.3	0.69	0.3
20	−1	0.35	1.1	0.3
30	−1.1	0.4	1.1	0.4
40	−1.24	0.4	1.21	0.4
45	−1.32	0.4	1.29	0.4
50	−1.41	0.4	1.3	0.4
60	−1.53	0.4	1.46	0.4

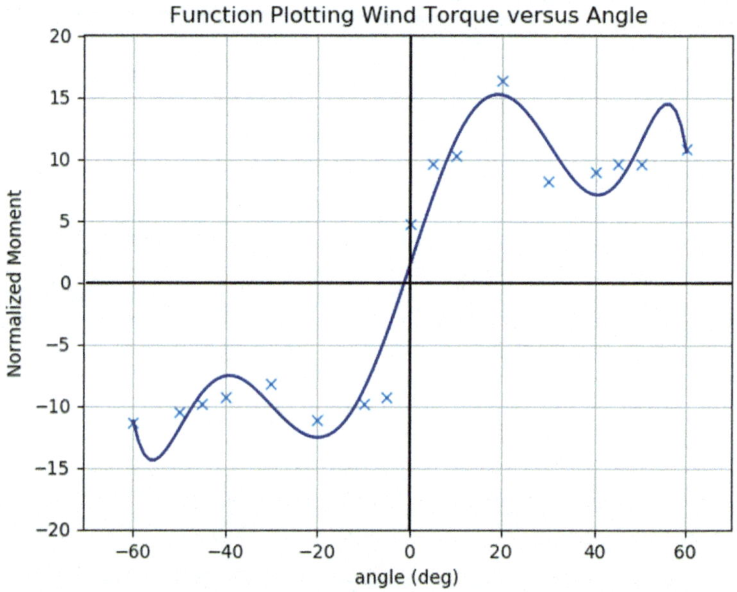

Fig. 6 Moment caused by wind at 1 m/s wind speed is calculated as per the formula given in Eq. (7) and plotted with 'x' markers, and a polynomial, bounded in the range of ±60° is plotted in blue

$$M_{t-\text{wind}}(\theta) = v^2 . \begin{pmatrix} -\ 1.98104260 \times 10^{-13} \times \theta^8 - 8.71827985 \times 10^{-11} \times \theta^7 \\ +\ 1.41949619 \times 10^{-9} \times \theta^6 + 6.20715384 \times 10^{-7} \times \theta^5 \\ -\ 2.89871114 \times 10^{-6} \times \theta^4 - 1.37341678 \times 10^{-3} \times \theta^3 \\ +\ 8.15112138 \times 10^{-4} \times \theta^2 + 1.14948143\theta + 1.40464709 \end{pmatrix} \tag{10}$$

where the angle θ is given in degrees and velocity v is given in metres per second.

3.5 Solution of Differential Equation

The equilibrium equation for linear vibration for a single degree of freedom system is given as

$$I\ddot{\theta} + C_t\dot{\theta} + \mathrm{K}_t\theta = \mathrm{T}(t) \tag{11}$$

However, since the system under consideration does not have linear damping, Eq. (11) is modified as follows:

$$I\ddot{\theta} + \mathrm{T}_{\mathrm{damp}} + \mathrm{K}_{\mathrm{tube}}\theta = M_{t-\mathrm{wind}}(\theta) \tag{12}$$

To obtain the numerical solution, the initial conditions are assumed to be that of a static slightly positively tilted tracker table, i.e. $\theta = 1^\circ$ and $\dot{\theta} = 0$. The equation is solved numerically using the Dormand–Prince method of the Runge Kutta family of ODE solvers.

The wind speed under consideration is taken to be an arbitrary value that is usually faced at site. As an indicative value, a constant wind speed of 17 m/s is assumed. Displacement and velocity response of the undamped system are shown in Fig. 7. Similar responses in the presence of damping when the damper is installed such that its slope is towards or away from the windward direction are shown in Fig. 8 and Fig. 9, respectively.

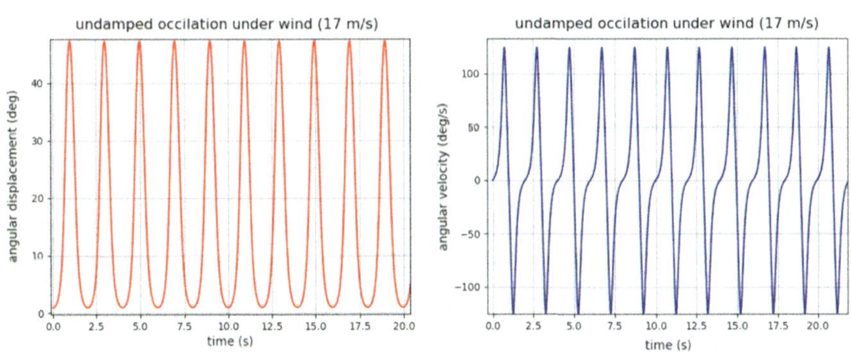

Fig. 7 Graphs showing the angular displacement (left) and velocity (right) of tracker table under wind forcing without damping

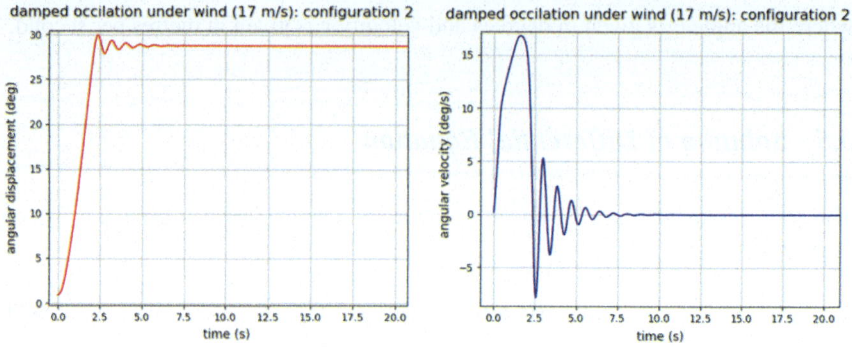

Fig. 8 Graphs showing angular displacement (left) and velocity (right) of the tracker table when the damper is installed with slope towards the windward side

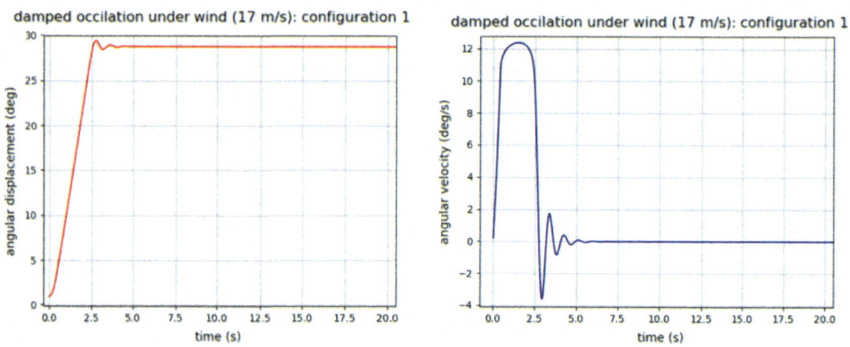

Fig. 9 Graphs showing the angular displacement (left) and velocity (right) of the tracker table when the damper is installed with slope away from the windward side

4 Summary

The tracker structure is a compliant structure under aero-dynamic forcing due to wind. The forcing due to wind is simulated through computational fluid dynamics. The oscillation of the tracker structure is modelled by approximating values of stiffness and inertia and forcing function. The response of the structure is plotted on graphs. It is observed from the graphs generated from simulation, that the dampers chosen are capable of arresting the oscillations of the table induced by wind, no matter the direction of wind, in under 10 s.

Acknowledgements The authors would like express gratitude to Mr. Abhimanyu Sable, Mr. Pankaj Sanap, Mr. Sandeep Bhosale, Mr. Apurav Jain, Mr. Basant Jain and the management of Mahindra Susten for the support and encouragement to carry out this work.

References

1. Christian R, Bourke PA, David B Torsional instability of single axis solar tracking systems. In: 14th international conference on wind engineering
2. Sanap PA, Bhosale SJ et al Arrangement for angularly displacing solar panels. US Patent US20170317640A1
3. Jain AP et al Solar tracking system and a method thereof. US Patent US20170070187A1
4. Young WC, Budyans RG, Sadgeh AM (2011) Rorak's formulas for stress and strain. McGraw Hill
5. Indian Standard automotive vehicle—shock absorbers—specifications "IS 5423: 1989"
6. Design loads (other than earthquakes) for buildings and structures—code of practice. Part 3 wind loads "IS 875 (Part3): 2015"
7. Minimum design loads for buildings and other structures "ASCE Standard ASCE/SEI 7–10."
8. Ortiz X, Rival D, Wood D (2015) Forces and moments on flat plates of small aspect ratio with application to PV wind loads and small wind turbine blades. Energies 8(4):2438–2453

Investigation on Hydrodynamic Characteristics of Textured Meso Scale Gas Bearing

Nilesh D. Hingawe and Skylab P. Bhore

Abstract In this paper, surface texturing has been adopted for meso scale gas journal bearing. Square texture shape is considered for the study. The compressible Reynolds equation of textured gas bearing is solved by the finite element method. For this, COMSOL 5.3 simulation software is used. The influence of partial texture on bearing hydrodynamic characteristics at variable process parameters such as eccentricity ratio, clearance, and journal speed has been studied. The hydrodynamic characteristics such as load capacity and friction coefficient are considered. In comparison with plain bearing, the significant improvement in hydrodynamic characteristics of textured bearing is observed. The texture arrangement in 0°–180° zone influences more to the load capacity and friction coefficient.

Keywords Meso scale gas bearing · Surface texture · Partial texture · Load capacity · Friction coefficient

1 Introduction

With the development in the technology of miniaturized rotating systems, the demand of meso scale bearings is widely increased. Meso scale bearings are the miniature type of bearings having dimensions in millimeters (dimensions lies between microscale and macroscale bearings). These bearings are the component of meso scale machines generally weighing up to 50 g [1]. Meso scale gas bearings have extensive applications in miniaturized rotating systems such as hard disk drive (HDD), laser scanner, dentist drill, miniature machine tools, meso scale gas turbine, gyroscope, etc. It has a simple design with high accuracy. The lubricant (gas/air) is clean to the environment and leakage problem is eliminated. Moreover, meso scale gas bearing can be operated at high speed with lesser frictional loss. Therefore, low noise and lesser heat generation take place in the bearing. In spite of these advantages, the gas journal

N. D. Hingawe (✉) · S. P. Bhore
Motilal Nehru National Institute of Technology Allahabad, Prayagraj, Uttar Pradesh 211004, India
e-mail: rme1603@mnnit.ac.in

S. P. Bhore
e-mail: skylabpbhore@mnnit.ac.in

D. Sen et al. (eds.), *Mechanism and Machine Science*, Lecture Notes in Mechanical Engineering, https://doi.org/10.1007/978-981-15-4477-4_37

bearings have relatively low load carrying capacity and dynamic stability. Thus, gas journal bearings are suitable only at low load operating conditions. These crucial issues limit its applications and further advancement. However, surface texturing has the potential to address these issues.

Surface texturing is a feasible and promising technique to improve the tribological performance of machine components in terms of load capacity, friction coefficient, wear, and power loss [2]. Adopting surface texturing in machine components such as mechanical seals, gas seals, journal bearings, thrust bearing, and piston ring improves the hydrodynamic characteristics viz. load capacity and friction coefficient [3]. Marian et al. [4] analyzed the square textured thrust bearing. They found that, square texture improves the hydrodynamic performance significantly. Moreover, Yu et al. [5] observed that square texture performs better than circular texture for friction reduction. However, Siripuram and Stephens [6] done the comparative analysis of different texture shapes viz. square, diamond, circle, triangle, and hexagonal for hydrodynamic lubrication. They found that the friction coefficient is less affected by texture shape and square texture is relatively easier to model.

Texture arrangement (partial texturing) influence the hydrodynamic performance of the bearing significantly. Tala-Ighil et al. [7] compared the plane oil lubricated bearing with full textured and partial textured bearing for maximization of pressure, minimization of film thickness, and frictional torque. They found that full texturing is ineffective compared to the partial texturing. Specifically, surface texturing in 185°–230° zone improves the bearing performance significantly. Also, they [8] investigated that journal speed influences on partial texturing for load capacity and friction coefficient. At lower journal speed, the surface texturing in 0°–180° region showed the best result. But, at average and high journal speed the 185°–230° texture zone obtained the best performance. Contradiction to these results, Liang et al. [9] observed that partial texturing may have a positive or negative effect on the bearing performance. Surface texturing at 60°–120° zone showed a negative influence on load capacity and friction coefficient. However, surface textured zone 0°–60° and 0°–48° have obtained maximum load capacity at low and high texture depth, respectively. But the minimum friction coefficient is achieved at 144°–180° zone. However, Meng et al. [10] evaluated the hydrodynamic characteristics of compound surface texture at 90°–180° texture zone. Shinde and Pawar [11] compared the influence of partial grooved texture at different zone (90°–180°, 90°–270°, and 90°–360°) on the bearing process parameters such as journal speed, eccentricity ratio, oil viscosity, and fluid film clearances. For these process parameters, they found that 90°–180° texture zone gives maximum load capacity and minimum frictional torque. Further, they [12] done multi-objective optimization of partial grooved bearing for maximization of load capacity and minimization of frictional torque. The bearing configurations such as 90°–155°, 90°–165°, and 90°–175° are considered for the analysis. Under the optimized condition, partial grooving at 90°–175° texture zone showed the best performance.

Meso scale gas journal bearings have relatively low hydrodynamic characteristics. Therefore, these bearings are suitable only at lower operating conditions. However, the above research work conforms that surface texturing has the potential to address these issues. So far, surface texturing has not been implemented in meso scale gas journal bearing. While, for macroscale oil lubricated bearings, the partial texturing is found to be more effective than full texturing for maximization of load capacity and minimization of friction coefficient. With this consideration, in the present work, partial surface texturing has been adopted in meso scale gas bearing to improve the hydrodynamic characteristics. Different texture zones such as 0°–150°, 0°–180°, 0°–210°, 0°–240°, and 0°–270° are considered for the analysis. This analysis is done at variable bearing parameters viz. eccentricity ratio, fluid film clearance, and journal speed. Finally, the texture zone is optimized for maximum load capacity and minimum friction coefficient.

2 Numerical Model

2.1 Compressible Reynolds Equation

The present study is based on numerical analysis. The gas journal bearing consist of shaft rotates in the sleeve. Eccentricity ratio and attitude angle are sufficient to describe the position of journal in the bearing. The load acting on the shaft is represented by eccentricity ratio. The load acting W' is balanced by bearing reaction force W (see Fig. 1).

Numerical analysis is done using COMSOL Multiphysics 5.3 fluid flow based thin film model. In the present model, fluid is considered to be isothermal. Pressure variation normal to the direction of flow is neglected. The Reynolds number (Re) is the function of fluid film parameters such as density, dynamic viscosity, clearance, and rotational speed. For miniaturized gas bearing the Reynolds number is in the order of unity or less [13]. For the present model, Re is given as,

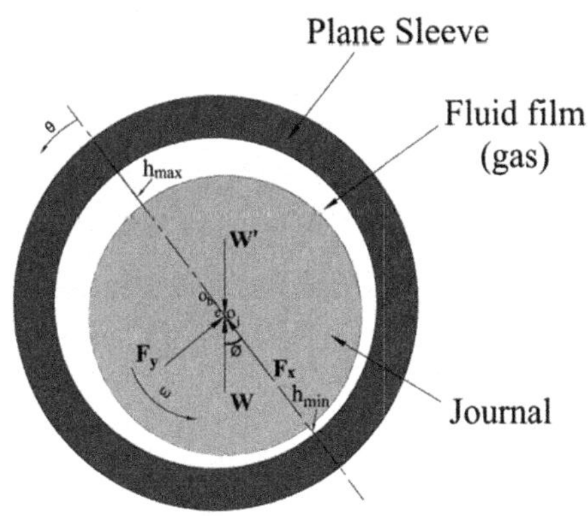

Fig. 1 Shaft position in the bearing

Table 1 Specification of air bearing system

Parameters	Values	Variable range
Bearing diameter (mm)	2	–
L/D ratio	1	–
Clearance (μm)	4	3–7
Eccentricity ratio, ε	0.8	0.3–0.8
Rotational speed (rpm)	8000	5000–25,000
Air viscosity (Pa s)	1.85×10^{-5}	–
Air density (kg/m^3)	1.225	
Texture dimensions (μm)	50 × 50	–
Texture depth (μm)	1	–

$$\mathrm{Re} = \frac{\rho \mathrm{v} \mathrm{C}}{\mu} \tag{1}$$

Where C is the diametric clearance and v is the journal speed in m/s. Considering Table 1, the Reynolds number is obtained to be 0.4439. This clearly justified that flow is laminar inside the bearing. At the bearing surface, no slip boundary condition is imposed. For a stationary loading, the steady-state two dimensional compressible Reynolds equation is written as

$$\frac{1}{r^2}\frac{\partial}{\partial \theta}\left(ph^3\frac{\partial p}{\partial \theta}\right) + \frac{\partial}{\partial z}\left(ph^3\frac{\partial p}{\partial z}\right) = 6\mu\,\omega\frac{\partial (ph)}{\partial \theta} \tag{2}$$

where p and h are the pressure and film thickness, respectively. The plane and square textured meso scale gas journal bearing are shown in Fig. 2 a and b, respectively. The film thickness relation in the textured bearing can be expressed as

$$h = c(1 + \varepsilon cos\theta) + \delta(\theta, z) \tag{3}$$

where

$$\delta(\theta, z) = \begin{cases} \Delta h\,(\theta, z) & \text{In the texture} \\ 0 & \text{Elsewhere} \end{cases} \tag{4}$$

The compressible Reynolds Eq. (2) is solved using finite element numerical discretization technique in COMSOL 5.3 with following boundary conditions

In axial direction:

$$p(z = 0, \theta) = p(z = L, \theta) = p_{\mathrm{atm}} \tag{5}$$

In circumferential direction:

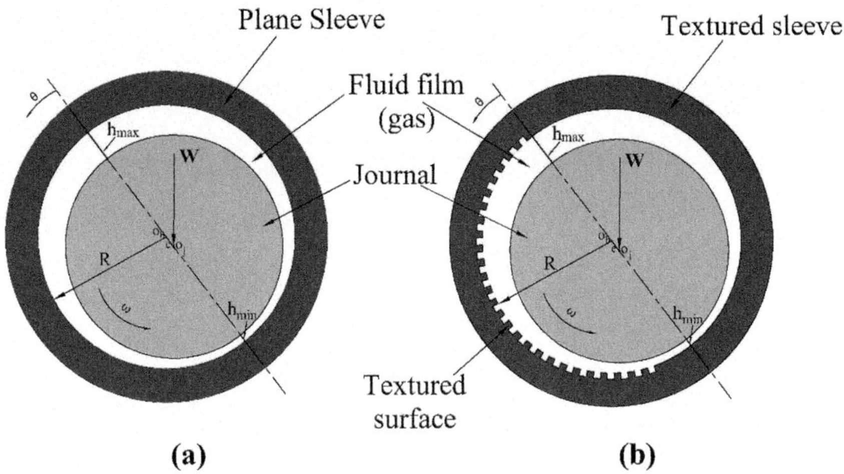

Fig. 2 Bearing coordinate system (**a**) Plane bearing (**b**) Textured bearing

$$p(z, \theta) = p(z, \theta + 2\pi) \tag{6}$$

Considering the above boundary conditions the pressure distribution is obtained. To estimate the accurate pressure following convergence criteria is used.

$$\sum\sum\left|\frac{(p_{i,j})_N - (p_{i,j})_{N-1}}{(p_{i,j})_N}\right| \leq 1 \times 10^{-5} \tag{7}$$

Based on this, the load components are obtained as follows:

$$F_X = \int_0^L \int_0^{2\pi} p\, r \cos\theta \mathrm{d}\theta \mathrm{d}z \tag{8}$$

$$F_Y = \int_0^L \int_0^{2\pi} p\, r \sin\theta \mathrm{d}\theta \mathrm{d}z \tag{9}$$

Using the above load components, the load capacity of the bearing is evaluated as

$$W = \sqrt{F_X^2 + F_Y^2} \tag{10}$$

Attitude angle is given as follows:

$$\phi = \tan^{-1}\frac{F_Y}{F_X} \tag{11}$$

Similarly, friction force can be evaluated as

$$F = \int_0^L \int_0^{2\pi} \tau r d\theta dz \tag{12}$$

Friction coefficient is the ratio of frictional force and load capacity as given below

$$\mu = \frac{F}{W} \tag{13}$$

2.2 *Mesh Sensitivity and Numerical Validation*

In the present study, the partial texturing is adopted for meso scale gas bearing. Texture dimensions are much smaller than the bearing dimensions (see Table 1). At the textured surface, extra-fine meshing is essential for accurate estimation of pressure distribution. But, at non-textured surface the coarse mesh results in accurate pressure. With this consideration, the physics controlled mesh is adopted for textured meso scale gas bearing. Singularities at the corner of the texture elements are removed by refining the triangular mesh. Further, to evaluate the accurate pressure, the mesh sensitivity analysis is done as shown in Fig. 3. Upon increasing the number of elements (fine meshing), load capacity first increases and then remains constant. To estimate the accurate result with the least computation time, the meshing parameters corresponding to 378043 elements are used for the analysis.

To validate the simulated model, the pressure distribution for the plain bearing is compared with Ref. [14]. Good agreement between the results is obtained as shown in Fig. 4. Hence, the present model can be employed for textured meso scale gas bearing.

3 Results and Discussion

In the present study, the influence of partial texture on load capacity and the friction coefficient is evaluated for meso scale gas bearing. Theses output responses are evaluated at variable process parameters such as eccentricity ratio, clearance, and journal speed. Bearing design parameters taken from Ref. [15] is given in Table 1. The pressure distribution of textured bearing is shown in Fig. 5. It shows the counters of pressure in the plainand textured zone.

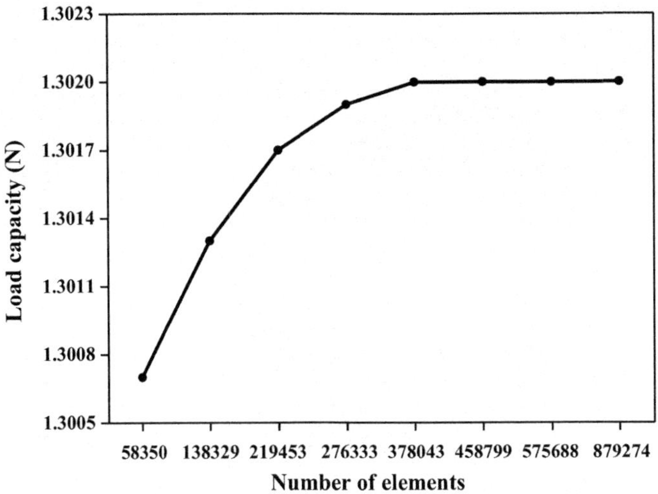

Fig. 3 Effect of number of elements on load carrying capacity

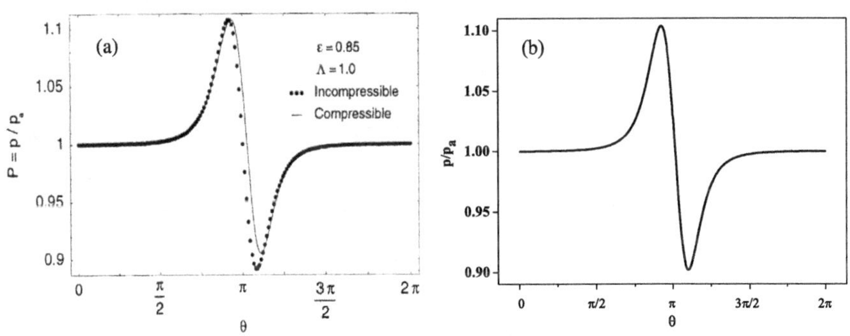

Fig. 4 Pressure distribution for plane bearing, (**a**) Orr [14] (**b**) present study

3.1 At Varied Eccentricity Ratio

Eccentricity ratio influences the bearing hydrodynamic characteristics such as load capacity and friction coefficient. Its effect on plain and partial textured bearing is analyzed at a variable range of 0.3–0.8. For this operating condition, the load capacity and friction coefficient are evaluated as shown in Fig. 6 a and b, respectively. A significant improvement in hydrodynamic characteristics is observed for textured bearing compared to the plain bearing. This is because surface texturing increases the fluid film thickness which results in increasing the volume of fluid accumulation in the bearing. However, surface texturing at 0°–150° zone is found to be least effective for both load capacity and friction coefficient. This emphasizes that surface texturing is effective in the minimum film thickness zone (active pressure zone). The

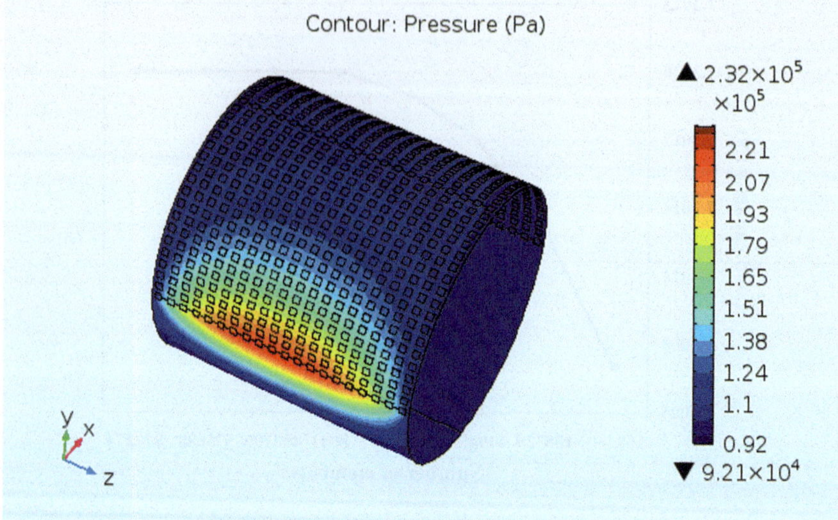

Fig. 5 Pressure distribution of partial textured (0°–180°) meso scale bearing

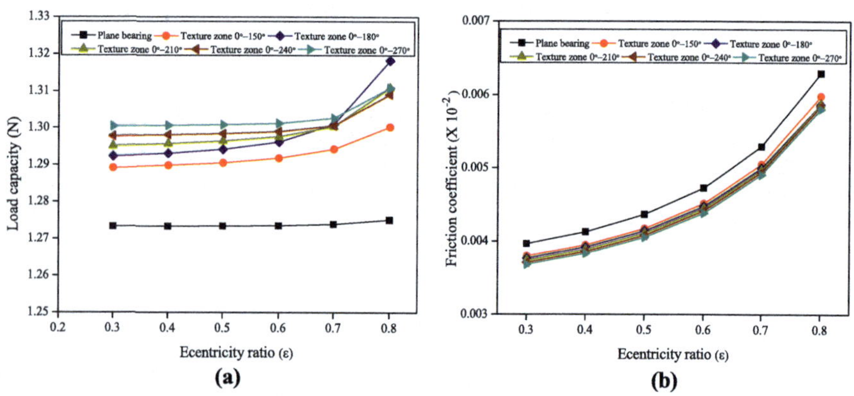

Fig. 6 Effect of eccentricity ratio on (**a**) load capacity (**b**) friction coefficient at different texture zones

maximum load capacity at a lower eccentricity ratio (0.3–0.6) is obtained at 0°–270° zone. However, at a higher eccentricity ratio (0.7–0.8) it is obtained at 0°–180° zone. This is because at higher eccentricity ratio, the pressure at the outlet of the active pressure zone decreases monotonously and thus compressibility at the outlet of active pressure zone in the texture is negligible.

The minimum friction coefficient is achieved at 0°–270° zone for all eccentricity ratios. This is because increasing film thickness decreases the velocity gradient and hence shear stress of the fluid, this results in a decrease of frictional force. The decrease in frictional force decreases the frictional coefficient. But the variation in frictional coefficient achieved at 0°–270° zone is not significant compared with 0°–180°, 0°–210°, and 0°–240° zone. Under the optimized condition, for maximization of load capacity and minimization of friction coefficient, at low and medium eccentricity ratio the surface texturing at 0°–270° zone should be considered. Whereas, at higher eccentricity ratio the surface texture should be employed at 0°–180° zone.

3.2 At Varied Fluid Film Clearance

Fluid film clearance influences the bearing hydrodynamic characteristics such as load capacity and friction coefficient substantially. These hydrodynamic characteristics are investigated at different textured zones (0°–150°, 0°–180°, 0°–210°, 0°–240°, and 0°–270°) and plane bearing at variable clearance as shown in Fig. 7. A significant improvement in load capacity and frictional coefficient is observed for textured bearing compared to plain bearing as shown in Fig. 7 a and b, respectively. However, surface texturing is found to be more beneficial at low clearance than at large clearance. This is because at low clearance the compression of fluid in the texture is more. Thus, the pressure developed is also more and hence the load capacity. But at high clearance, for the same mass of fluid, the volume is more. This decreases the pressure. Of all texture zone considered, the surface texturing at 0°–150° is found to be least effective for both load capacity and friction coefficient. At low clearance, the maximum load capacity is obtained at 0°–180° zone, but at large clearance it is obtained at 0°–270° zone. The minimum frictional coefficient is achieved at 0°–270° zone. Deviation in frictional coefficient obtained at 0°–270° zone is very less for other texture zones such as 0°–180°, 0°–210°, and 0°–240°. This is because adopting surface texturing at the outlet of the active pressure zone decreases the velocity gradient of the fluid. For maximization of load capacity and minimization of the frictional coefficient, the optimized textured zone is found to be 0°–180°.

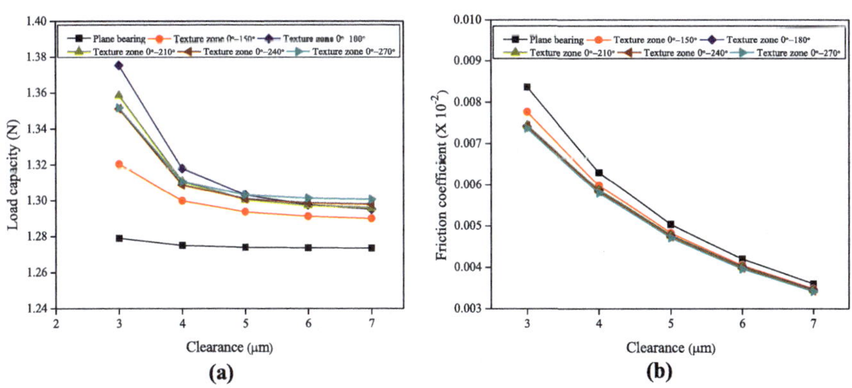

Fig. 7 Effect of fluid film clearance on (**a**) load capacity (**b**) friction coefficient at different texture zones

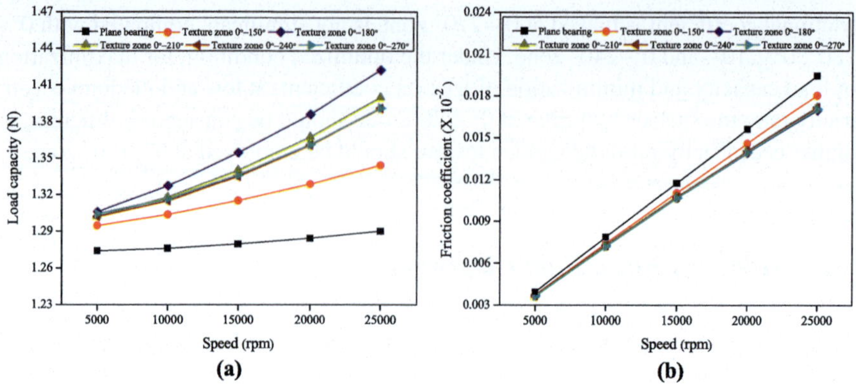

Fig. 8 Effect of journal speed on (**a**) load capacity (**b**) friction coefficient at different texture zones

3.3 At Varied Journal Rotational Speed

Journal speed influences the hydrodynamic characteristics such as load capacity and frictional coefficient significantly. The detailed analysis of plane and textured bearing is done at a variable speed range of 5000–25,000 rpm. The significant improvement in load capacity and the frictional coefficient is observed for textured bearing than plane bearing as shown in Fig. 8 a and b, respectively. However, surface texturing is found to be more effective at higher journal speed than lower. Of all textured zone considered, for maximization of load capacity the surface texturing at 0°–180° is found to most effective and 0°–150° zone is found to be least effective. Upon increasing journal speed the frictional coefficient is obviously increased. Under all operating speeds, the minimum friction coefficient is achieved at 0°–270° texture zone. Upon increasing the journal speed, the compressibility effect of the fluid increases in the convergent zone. However, it does not affect at divergence zone. Thus surface texturing in the convergent zone (active pressure zone) compresses more fluid. But the variation in friction coefficient obtained at 0°–270° zone is slight compared to other textured zones such as 0°–180°, 0°–210°, and 0°–240°. Therefore, for the maximization of load capacity and minimization of the frictional coefficient, the optimized textured zone is found to be 0°–180°.

4 Conclusions

In the present analysis, a square textured meso scale gas bearing is considered. To investigate the influence of partial texture on hydrodynamic characteristics such as load capacity and frictional coefficient, the different textured zone: 0°–150°, 0°–180°, 0°–210°, 0°–240°, and 0°–270° are considered. For this, bearing parameters namely eccentricity ratio, fluid film clearance, and journal rotational speed are varied. It is concluded that

1. Surface texturing is found to be an effective technique to improve the hydrodynamic characteristics such as load capacity and frictional coefficient of the meso scale gas bearing.
2. For the maximization of load capacity and minimization of frictional coefficient, textured zone of 0°–180° is found to be most effective and textured zone of 0°–150° to be least effective.
3. For the textured bearing, the process parameters such as eccentricity ratio, clearance, and journal speed are found to be more sensitive to load capacity than the frictional coefficient.

References

1. Salehi M, Heshmat H, Walton JF II, Tomaszewski M (2007) Operation of a mesoscopic gas turbine simulator at speeds inexcess of 700,000 rpm on foil bearings. J Eng Gas Turbines Power 129:170–176
2. Grooper D, Wang L, Harvey TJ (2016) Hydrodynamic lubrication of textured surfaces: a review of modeling techniques and key findings. Tribol Int 94:509–529
3. Etsion I (2013) Modeling of surface texturing in hydrodynamic lubrication. Friction 1(3):195–209
4. Marian VG, Kilian M, Scholz W (2007) Theoretical and experimental analysis of a partiallytextured thrust bearing with square dimples. IMechE Part J: J. Engineering Tribology 221:771–778
5. Yu H, Deng H, Huang W, Wang X (2011) The effect of dimple shapes on friction of parallel surfaces. Inst Mech Eng Part J J Eng Tribol 225:693–703
6. Siripuram RB, Stephens LS (2004) Effect of deterministic asperity geometry on hydrodynamic lubrication. J Tribol 126:527–534
7. Tala-Ighil N, Fillon M, Maspeyrot P (2011) Effect of textured area on the performances of a hydrodynamic journal bearing. Tribol Int. 44(2011): 211–219
8. Tala-Ighil N, Fillon M (2015) A numerical investigation of both thermal and texturing surface effects on the journal bearings static characteristics. Tribol Int. 90(2015): 228–239
9. Liang X, Liu Z, Wang H, Zhou X, Zhou X (2016) Hydrodynamic lubrication of partial texturedsliding journal bearing based on three-dimensional CFD. Ind Lubr Tribol 68(1):106–115
10. Meng FM, Zhang L, Liu Y (2015) Effect of compound dimple on tribological performances of journal bearing. Tribol Int 91(2015): 99–110
11. Shinde A, Pawar P (2017) Effect of partial grooving on the performanceof hydrodynamic journal bearing. Ind Lubric Tribol 69(4):574–584
12. Shinde A, Pawar P (2017) Multi-objective optimization of surface textured journal bearing by Taguchi based grey relational analysis. Tribol Int. 114(2017): 349–357
13. Maureau J, Sharatchandra MC, Sen M, Gad-el-Hak M (1997) Flow and load characteristics of microbearings with slip. J Micromechanics Microengineering 7:55–64
14. Orr DJ (1999) Macro-scale investigation of high Speed gas bearings for MEMS Devices. Ph.D. thesis, MIT Department of Aeronautics and Astronautics, USA
15. Chen CY, Liu CS, Li YC, Mou S (2015) Geometry optimization for asymmetrical herringbone grooves of miniature hydrodynamic journal bearings by using Taguchi technique. Proc Inst Mech Eng Part J J Eng Tribol 229(2):196–206

Underwater Glider Path Planning Using Partially Observable Markov Decision Processes

Wenwen Wang, Zhiliang Wu, Mengyuan Zhao, and Wei Ma

Abstract Underwater glider has an outstanding advantage of long endurance for ocean observation. This paper presents underwater glider path planning with uncertainties under the framework of partially observable Markov decision processes (POMDPs). The kinematics model and the sensor model of the underwater glider have been built, and the uncertainty of action has been taken into consideration. The priori probability distribution and posteriori probability distribution are obtained from the kinematic model and sensor model, respectively. Particle filtering has been used to combine the two probability distributions. Results show that integrating the uncertainties in state estimation can improve the accuracy of waypoint estimation. Obstacle avoidance is also presented in the same framework.

Keywords Underwater glider · Path planning · Partially observable markov decision process (POMDP) · Uncertainty · Particle filter

1 Introduction

Underwater glider is an effective tool for observing the marine environment. It uses horizontally fixed wings to convert gravity into a forward driving force, which greatly reduces power consumption with respect to conventional autonomous underwater

W. Wang · Z. Wu (✉) · W. Ma
Key Laboratory of Mechanism Theory and Equipment Design of Ministry of Education, School of Mechanical Engineering, Tianjin University, Tianjin, China
e-mail: zhlwu@tju.edu.cn

W. Wang
e-mail: 1329777897@qq.com

W. Ma
e-mail: maweiyezi1990@163.com

M. Zhao
China Railway Engineering Equipment Group Co., Ltd, Zhengzhou, China
e-mail: zhaomengyuan10@163.com

D. Sen et al. (eds.), *Mechanism and Machine Science*, Lecture Notes in Mechanical Engineering, https://doi.org/10.1007/978-981-15-4477-4_38

vehicles (AUVs) [1]. It also has the advantages of low cost and low noise, making it feasible for wide applications in marine scientific research, environmental monitoring, resource exploration, and military reconnaissance [2, 3].

Path planning for underwater gliders is indispensable for ocean observations, especially in complicated observing areas where ships are busy traveling on the water surface or the underwater environment is unknown or partially unknown. Underwater gliders follow a sawtooth trajectory in the water and surface intermittently for communication. Collision to ships or underwater obstacles may be fatal to the safety of underwater gliders. It is, therefore, important to generate a serial of waypoints and develop a navigation strategy following which the underwater glider can avoid obstacles while traveling in the water.

Conventional underwater glider path planning in practice usually focuses on waypoint setting on the water surface given the map, the goal location, and the underwater glider's desired attitudes. Optimal paths are generated by different objectives, such as the shortest distance and minimal energy assumption. Various path generation approaches have been developed, including A* algorithm [4, 5], RRT [6], iterative optimization [7], genetic algorithm [8], fuzzy algorithm [9], artificial potential field approach [10], and level set method [11]. The design of waypoints has been further improved by taking into account the dynamic environment on the water surface [12]. Potential collision to ships could be effectively avoided. Actually, uncertainties during observation result not only from the dynamic environment but also from the vehicle actuating mechanism. These uncertainties may affect the underwater glider's movement in the water, producing deviation from the planned trajectory and the preset waypoint and causing adjustment or even replanning of the waypoints [13].

In this paper, underwater glider path planning has been performed under the framework of a partially observable Markov decision process (POMDP). The action uncertainty of underwater glider and the uncertainty induced by ocean currents have been integrated in vehicle state estimation. Obstacle avoidance in the water is also presented with a sensor model.

The rest of this paper is organized as follows: Establishment of the kinematic model is presented in Sect. 2. In Sect. 3, the glider's sensor model and the map are built. The POMDP framework and the simulation results are presented in Sect. 4. Conclusions are given in the last section.

2 Kinematic Model

This section focuses on the underwater glider's kinematic model which describes the effect of control actions on the configuration of the glider. Usually, the configuration of a rigid mobile robot is described by six variables, three Cartesian coordinates and three Euler angles (roll, pitch, and yaw) relative to the external coordinate. In this kinematic model, the underwater glider is represented by its position (x, y, z) and the posture, pitch angle θ, roll angle φ, and yaw angle ψ, as illustrated in Fig. 1.

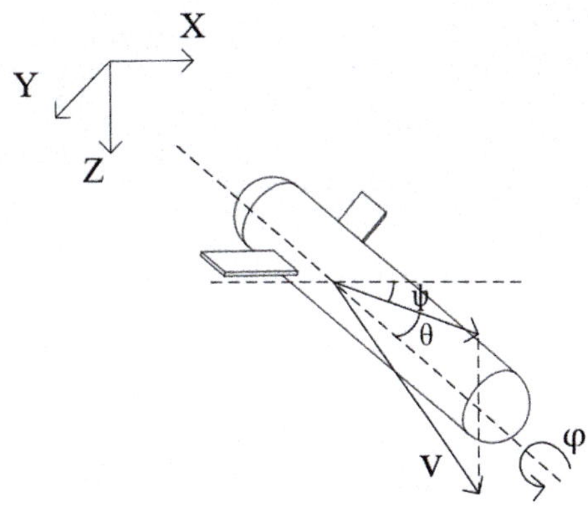

Fig. 1 The configuration of underwater glider, where v denotes glider's velocity and θ, φ and ψ denote pitch angle, roll angle, and yaw angle respectively

As the underwater glider travels in the water, it changes the pitch angle and the roll angle to change its gliding direction. In this work, the variation in the pitch angle and the yaw angle are taken as the control $u = (\Delta\theta, \Delta\psi)$ for convenience and the state of underwater glider is represented by $X = (x, y, z, \theta, \varphi, \psi)$. For discrete time intervals, the control actions and the state can be represented, respectively as $u_t = (\Delta\theta_t, \Delta\psi_t)$ and $X_t = (x_t, y_t, z_t, \theta_t, \varphi_t, \psi_t)$.

The kinematic model can be expressed as $X_t = G(X_{t-1}, u_t)$. Theoretically, the underwater glider's motion trail is helical, as shown in Fig. 2. But actually, the pitch angle does not change actively as the glider is gliding in the water. The kinematics model is then derived as

$$\begin{cases} \bar{x}_t = x_{t-1} + \frac{|\wedge\bar{z}_{t-1}|\left[\sin\left(\psi_{t-1}+\Delta\bar{\psi}_{t-1}\right)-\sin(\psi_{t-1})\right]}{\Delta\bar{\psi}_{t-1}\tan(\theta_{t-1})} \\ \bar{y}_t = y_{t-1} + \frac{|\Delta\bar{z}_{t-1}|\left[\cos(\psi_{t-1})-\cos\left(\psi_{t-1}+\Delta\bar{\psi}_{t-1}\right)\right]}{\Delta\bar{\psi}_{t-1}\tan(\theta_{t-1})} \\ \bar{z}_t = z_{t-1} + \Delta\bar{z}_{t-1}, \\ \bar{\psi}_t = \psi_{t-1} + \Delta\bar{\psi}_{t-1}, \\ \bar{\theta}_t = \theta_{t-1}, \end{cases} \tag{1}$$

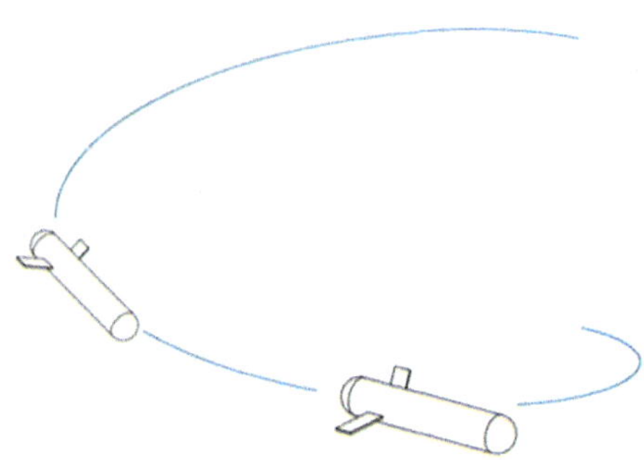

Fig. 2 Helical motion of underwater glider

Table 1 Particle filter algorithm

1:Algorithm particle filter $\left\{\hat{X}_{t-1}, p\left(\hat{X}_t \middle| u_t, \hat{X}_{t-1}\right), p(Z_t\vert X_t)\right\}$
2: for $v = 1$ to $n_{(x,y)}$ do
3:$w_t^v = p\left(\hat{X}_t^v \vert Z_{1:t}\right)$
4: end for
5:$r = rand\left(0 : n_{(x,y)}^{-1}\right)$
6:$c = w_t^1$
7:$i = 1$
8: for $m = 1$ to $n_{(x,y)}$
9:$U = r + (m-1)n_{(x,y)}^{-1}$
10: while $U > c$
11:$i = i + 1$
12:$c = c + w_t^i$
13: end while 14: add $\hat{X}_t^i$ to $\hat{X}_t$
15: end for
16: return $\hat{X}_t$

where $\bar{u}_{t-1} = \left(\Delta\bar{\theta}_{t-1}, \Delta\bar{\psi}_{t-1}\right)$ is the control at time $t-1$, and $\bar{X}_t = (\bar{x}_t, \bar{y}_t, \bar{z}_t, \bar{\theta}_t, \bar{\varphi}_t, \bar{\psi}_t)$ is the expected state of the underwater glider at time t.

The state transition probability $p(X_t\vert u_{t-1}, X_{t-1})$, which is the probability distribution of the state from the current time to the next moment, is associated with the actuation mechanisms. The uncertainty of action is represented by introducing Gaussian noise to the control variable and the estimate is

$$\hat{u}_t = (\Delta\theta_t + \mathbf{N}(0, \sigma_\theta), \Delta\psi_t + \mathbf{N}(0, \sigma_\psi)) \quad (2)$$

where $N(0, \sigma)$ is a normal distribution with a standard deviation of σ, which can be obtained by linear regression of the actual variance in the control variable. The underwater glider's state is also affected by the dynamic ocean environment. The state estimation is then updated according to measurements that are relevant to the environmental uncertainty. Table 1 illustrates the principal procedure of state estimation using particle filter algorithm. The posterior distribution is approximated using the most recent state, control, and measurement.

3 Sensor Model and Map

In this section, the sensor model is described as a sector area as shown in Fig. 3.

The measurement area can be discretized in the relative horizontal direction and vertical direction as

$$S = \left\{ S_{(1,1)},\ S_{(1,2)}, \ldots, S_{(n_\theta, n_\phi)} \right\} \tag{3}$$

where $S_{(i,j)}$, $\left(i = 1, 2, \ldots, n_\theta;\ j = 1, 2, \ldots, n_\phi\right)$, n_θ is the number of parts divided in the horizontal direction and n_ϕ is the number of parts divided in the vertical direction. When an obstacle is detected in the area $S_{(i,j)}$, θ_i, ϕ_j and distance between the obstacle and glider r will be returned. The position of the obstacle in the world coordinate is also presented in probability distribution, since the location of the glider $\hat{X}_t$ is not deterministic. And three-dimensional space is divided into grids, and the map can be expressed as $m = \left\{m_1, m_2, \ldots, m_{n_m}\right\}$. $m_i = 1$ means occupied and $m_i = 0$ means free. The natation $p(m_i = 1)$ or $p(m_i)$ refers to the probability that a grid cell is occupied. The relationship between the location information of obstacle and occupation of the grid is shown as

$$\begin{aligned} p\left(Z_t|\hat{X}_t\right) &= sensor\ model\left(\hat{X}_t,\ S_{(i,j)}\right) \\ &= \left(p'(m_1),\ p'(m_2), \ldots,\ p'\left(m_{N_m}\right)\right) \end{aligned} \tag{4}$$

The algorithm in Table 2 reflects the measurement information of the obstacle to the grid map. The map can be expressed as

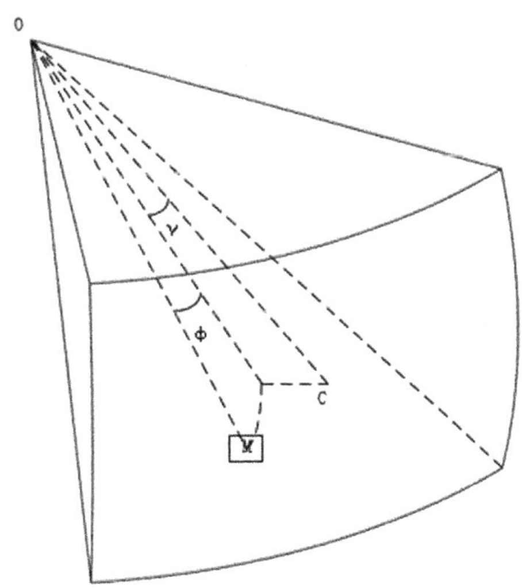

Fig. 3 Sensor model: line OC is the direction of the glider, and γ is the angle between line OM and line OC in the relative horizontal direction, and Φ is the angle in the relative vertical direction

Table 2 Sensor model algorithm

1: Algorithm sensor model $\left(\hat{X}_t, S_{(i,j)}\right)$
2: The position of obstacle $obstacle = (x_o, y_o, z_o)$ was presented by set of particle
$\left\{\left(x_o^1, y_o^1, z_o^1\right), \left(x_o^2, y_o^2, z_o^2\right), \ldots, \left(x_o^{n_{(x,y)}}, y_o^{n_{(x,y)}}, z_o^{n_{(x,y)}}\right)\right\}$
3: for $i = 1$ to $n_{(x,y)}$
4: $x_o^i = x_t^i + r\cos(\theta_t^i + \phi)\cos(\beta_t^i + \gamma)$ $y_o^i = y_t^i + r\cos(\theta_t^i + \phi)\sin(\beta_t^i + \gamma)$ $z_o^i = z_t^i + r\sin(\theta_t^i + \phi)$
5: end for
6: for $i = 1$ to n_m
7: $k = the\ number\ of\ particle\ belong\ to\ grid\ m_i$
8: $p'(m_i) = k/n_{(x,y)}$
9: end for
10: return $\left(p'(m_1), p'(m_2), \ldots, p'\left(m_{N_m}\right)\right)$

$$p\left(m|Z_{1:t}, \hat{X}_{1:t}\right) = \prod_i p\left(m_i|Z_{1:t}, \hat{X}_{1:t}\right) \tag{5}$$

$$p\left(m_i|Z_{1:t}, \hat{X}_{1:t}\right) = p\left(Z_t|\hat{X}_t\right)p\left(m_i|Z_{1:t-1}, \hat{X}_{1:t}\right) \tag{6}$$

where the measurement Z_t includes the information of the obstacle detected and the position of the glider. The mapping algorithm in Table 3 combines the information of the obstacle obtained at time t and the previous information. For convenience, the occupancy grid uses *log* form in the representation of occupancy:

Table 3 Mapping algorithm

1: Algorithm mapping $\left(p\left(m|Z_{1:t-1}, \hat{X}_{1:t}\right), p\left(Z_t|\hat{X}_t\right)\right)$
2: for $i = 1$ to n_m
3: $l_{t,i} = l_{t-1,i} + \ln\frac{p\left(m_i|Z_t,\hat{X}_t\right)}{1-p\left(m_i|Z_t,\hat{X}_t\right)}$
4: $p\left(m_i|Z_{1:t}, \hat{X}_{1:t}\right) = 1 - \left(1 + e^{l_{t,i}}\right)^{-1}$
5: end for
6: return $p\left(m|Z_{1:t}, \hat{X}_{1:t}\right)$

$$l_{t,i} = \ln \frac{p\left(m_i | Z_{1:t}, \hat{X}_{1:t}\right)}{1 - p\left(m_i | Z_{1:t}, \hat{X}_{1:t}\right)} \quad (7)$$

$$p\left(m_i | Z_{1:t}, \hat{X}_{1:t}\right) = 1 - \left(1 + e^{l_{t,i}}\right)^{-1} \quad (8)$$

The constant l_0 is the prior of occupancy represented as

$$l_0 = \ln \frac{p(m_i = 1)}{p(m_i = 0)} = \ln \frac{p(m_i)}{1 - p(m_i)} \quad (9)$$

4 Planning and Control

In this section, the Partially Observable Markov Decision Process (POMDP) framework is built. First of all, the control variable $u = \left(\Delta\bar{z}, \Delta\bar{\psi}\right)$ will be discretized as $u_1, u_2, \ldots, u_{n_u}$. The probability distribution of the glider's position is distributed to the grid on the map. The immediate payoff is defined as

$$r(x, u) = \begin{cases} |x - goal| - |g(x, u) - goal| & no\ obstacles\ \ encountered \\ -10000 & obstacles\ \ encountered \end{cases} \quad (10)$$

This equation illustrates a decrease in the distance between the glider and the goal after the action has been performed. The value function is

$$V_T\left(\hat{X}\right) = \lambda \max_u \left[r\left(\hat{X}, u\right) + \sum_x V_{T-1}\left(x'\right) p\left(x' | x, u\right) \right] \quad (11)$$

$$V_1(x) = \max_u r(x, u) \quad (12)$$

$$\pi_T\left(\hat{X}\right) = \arg\max_u \left[r\left(\hat{X}, u\right) + \sum_x V_{T-1}\left(x'\right) p\left(x' | x, u\right) \right] \quad (13)$$

where $\lambda \in (0, 1)$ is the discount factor, T represents the steps taken into account, $\pi_T\left(\hat{X}\right)$ is the strategy to obtain the maximum. Table 4 illustrates the algorithm to estimate the value by sampling from the probability distribution of the state $\hat{X}$.

The experimental data obtained by the underwater glider PETREL [2] is used to train the model to predict the next position which is on the surface of the water. The 11 continuous profiles' data have been used to get the parameter Eq. (2). Two cases of control of the yaw angle are obtained from the analysis of data. One of them is that $\Delta\bar{\psi} = 0.12°$ with a standard deviation of 1.93° when the gilder actively adjusts

Table 4 Planning algorithm

1: Algorithm planning
2: for all controls u
3: $Q(u) = 0$
4: for $i = 1$ to n
5: sample $x \sim \hat{X}$
6: $Q(u) = Q(u) + \frac{1}{n}\lambda\left[r(x, u) + V\left(\hat{X}'\right)\right]$
7: end for
8: $V\left(\hat{X}\right) = \max_u Q(u)$

the yaw angle. The other is $\Delta\bar{\psi} = -0.07°$ and the standard deviation is 1.03° when the glider's yaw angle is only affected by the current. The standard deviation of the pitch angle is 2.87°. In the simulation, the desired yaw angle is 199.6°, the destined depth is 610 m and the pitch angle is 20°. Comparison between the generated path, the ideal path, and the actual waypoint is shown in Fig. 4.

In the simulation, the position of the glider is estimated at each time interval. The green trajectory is obtained with action uncertainty and environmental uncertainty. The surface point is closer to the actual waypoint located by GPS, as indicated in Fig. 4.

The model is also used for obstacle avoidance in the water. The path simulated the real-time path planning of glider by making decisions after detecting a limited range of environments, as shown in Fig. 5.

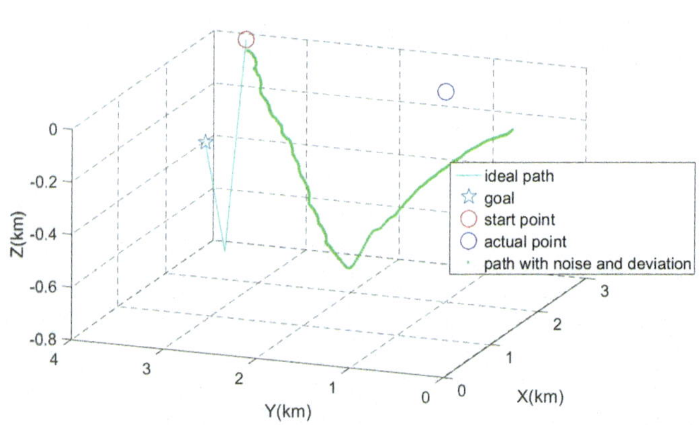

Fig. 4 Simulation result: underwater glider's trajectory

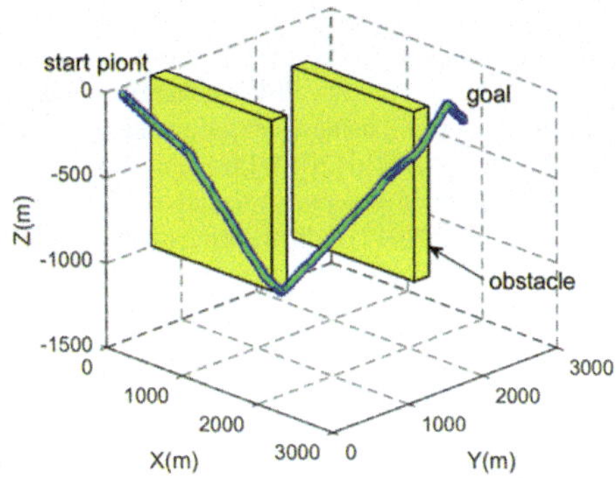

Fig. 5 A collision-free trajectory

5 Conclusion

This paper presents underwater glider navigation using partially observable Markov decision processes (POMDPs). A probability model has been built to estimate the glider's position. Uncertainties including the dynamic ocean environment and the action uncertainty coming from the actuating mechanism of the underwater glider have been integrated into state estimation. Simulation results have been presented. Comparison with experimental data extracted from sea trial indicates an improved accuracy of waypoint design using the presented approach. The approach is further extended for obstacle avoidance in the three-dimensional space.

References

1. Davis RE, Charles CE, Clayton PJ (2002) Autonomous buoyancy-driven underwater gliders. Technol Appl Auton Underw Veh (2002) 37–58
2. Wang S et al (2011) Dynamic modeling and motion simulation for a winged hybrid-driven underwater glider. China Ocean Eng 25(1):97–112
3. Zhao B (2015) Underwater glider path optimization based on improved particle swarm algorithm. Ship Sci Technol 8:140–145
4. Carroll KP, McClaran SR, Nelson EL, et al (1992) AUV path planning: an A* approach to path planning with consideration of variable vehicle speeds and multiple, overlapping, time-dependent exclusion zones. In: Proceedings of the 1992 symposium on autonomous underwater vehicle technology, pp 79–84
5. Isern-González J, Hernández-Sosa D, Fernández-Perdomo E et al (2012) Obstacle avoidance in underwater glider path planning. J Phys Agent 6(1):11–20
6. Urmson C, Simmons R, Approaches for heuristically biasing RRT growth. In Proceedings of 2003 IEEE/RSJ international conference on intelligent robots and systems, vol 2, pp 1178–1183
7. Isern-González J, Hernández-Sosa D, Fernández-Perdomo E, et al (2011) Path planning for underwater gliders using iterative optimization. In: Proceedings of 2011 IEEE international conference on robotics and automation, pp 1538–1543. IEEE
8. Alvarez A, Caiti A, Onken R (2004) Evolutionary path planning for autonomous underwater vehicles in a variable ocean. IEEE J Ocean Eng 29(2):418–429
9. Xu HL, Feng XS (2009) An AUV fuzzy obstacle avoidance method under event feedback supervision. Oceans

10. Witt J, Dunbabin M (2008) Go with the flow: optimal AUV path planning in coastal environments. In Proceedings of Australian conference on robotics and automation, vol 2
11. Lolla T, Lermusiaux PFJ, Uechermann MP, Haley PJJ (2014) Time-optimal path planning in dynamic flows using level set equations: theory and schemes. Ocean Dyn 64:1373–1397
12. Geoffrey AH et al (2016) Learning uncertainty in ocean current predictions for safe and reliable navigation of underwater vehicles. J Field Robot 33(1):47–66
13. Wu Z, et al (2017) Path planning for underwater gliders with motion constraints. In: Proceedings of 2016 IFToMM Asian conference on mechanism and machine science, vol 408, pp 3–10

Fabrication of Micro-Compliant Mechanisms Using Micro-Stereolithography

Ratnesh Bafna, Abhijit Tanksale, and Prasanna Gandhi

Abstract Compliant mechanisms provide friction-less and backlash-free smooth motion as it is obtained through deformation of flexible members. Micro-compliant mechanisms are widely used in many applications such as MEMS, micro-gripper, and tweezers for biological applications, precise motion of focusing lens. In the literature, various fabrication techniques such as wire EDM, micro-milling, Very-Large-Scale Integration (VLSI)-based lithography, and etching processes are used to fabricate micro-compliant mechanisms. These processes suffer from limitation that high speed, high resolution, and high aspect ratio (required for high performance mechanisms) all cannot be there in a single process. 3D micro-printing has the potential to meet these stringent requirements in a cost-effective manner. We demonstrate in this paper, development of high aspect ratio micro-mechanisms via 3D micro-printing based on indigenous technology. This 3D micro-printer has been developed based on micro-stereolithography technology with patented scanning. We report comparable short printing time, resolution, and ultra-high aspect ratios obtained using the proposed method.

Keywords Micro-gripper · 3D printing · On-Axis scanning · Cell manipulation · Double parallelogram-compliant mechanism

---

R. Bafna · A. Tanksale · P. Gandhi (✉)
Indian Institute of Technology Bombay, Mumbai, India
e-mail: gandhi@iitb.ac.in

R. Bafna
e-mail: ratnesh.bafna1@gmail.com

A. Tanksale
e-mail: 144100009@iitb.ac.in

D. Sen et al. (eds.), *Mechanism and Machine Science*, Lecture Notes in Mechanical Engineering, https://doi.org/10.1007/978-981-15-4477-4_39

1 Introduction

Compliant mechanism is a device which transmits force or motion by only means of deformation in flexible links. More often, compliant mechanisms are monolithic in design, which eliminates the assembly of different parts and relative frictional motion between parts, which increases the performance of device due to less wear and elimination of friction and backlash. Due to these advantages of compliant mechanisms over rigid body mechanisms, they are preferred for obtaining miniature motion with high precision and accuracy. The compliant mechanisms, which are used to obtain miniature motion in the range from few micron to nanometer, has their own dimensions in micron to few millimeters. These compliant mechanisms are called as micro-compliant mechanisms. Micro-compliant mechanisms have several applications such as MEMS for micro-sensor, micro-gripper, Bi-stable micro-mechanism [1], micro-air vehicles [2], and so on.

In the literature, common techniques used for fabrication of micro-compliant mechanisms are lithography and etching or vacuum casting [3], 3D printing, projection-micro-stereolithography (PMSL) [4], wire electro-discharge machining (WEDM) [5], etc. Conventional VLSI-based lithography and etching processes give excellent resolution, however they suffer from limitation on achievable aspect ratios and often need expensive tools such as Bosch process of Deep Reactive Ion Etching (DRIE) [6, 7]. Digital Light Processing (DLP) process uses pixel-by-pixel control over spatial exposure, where pixel sizes are around 10 μ. However the resolution in addition shows the effect of light diffraction and is seldom as small as pixel size. Conventional stereolithography (SLA) method is based on galvano mirror laser scanning, where laser spot intensity is not homogeneous and laser profile also vary with scanning distance. Table 1 shows a comparative study of common techniques used for fabrication of micro-compliant mechanisms.

To overcome the above-mentioned limitations of various fabrication techniques, we have proposed a novel micro-stereolithography MSL technique [8–10] achieving uniform focus spot over the entire range of scan. Another salient feature of our micro-

Table 1 Comparison of different 3D printing technique for fabrication of micro-compliant mechanisms

3D printing techniques	Aspect ratio (width:height)	Printing (resolution) (μm)	Printing speed
DLP (Rapid Shape S30E)	1:5	25	250–300 layer/h
SLA (Form Lab 2)	1:8	140	2000 mm/s
VLSI-based lithography	1:1–1:10	1	5–10 mm/s
MSL	1:12	9	8 mm/s

printer is that the motion stages are implemented using compliant mechanisms to get ultra-high precision in positioning at high speeds. We demonstrate using this printer fabrication of the proposed micro-mechanisms and characterize them further.

2 Micro-Stereolithography Setup

The micro-stereolithography is a technology developed in house based on compliant mechanism. In this technology, the device is printed using layer by layer photo-polymerization of photo-polymer resin. The involvement of compliant mechanism and novel on-axis laser scanning technique [9] gives high resolution, homogeneous composition for large size of the printed components. The process is highly repeatable and silent. The MSL setup is shown in Fig. 1. It consists of mainly three subassemblies (a) XY stage based on double parallelogram compliant mechanism (b) Z stage with substrate (c) Laser for photo-polymerization along with acousto-optic modulator (AOM) for switching the laser.

Double parallelogram mechanism in XY stage provides perfect straight-line guidance accuracy with zero backlash and friction. The position accuracy obtained in XY motion is less than 1 μm [11]. The constraint surface technique [12] is used to achieve constant layer thickness. Photo-polymer resin is squeezed between two substrates and constant layer thickness is maintained while printing. Minimum constant layer thickness of 10 um can be achieved. Thin transparent Teflon sheet is used as tank bottom substrate. Hence printing layer always adheres to top substrate during layer separation. The laser is exposed on printing surface from the bottom. Raster scanning

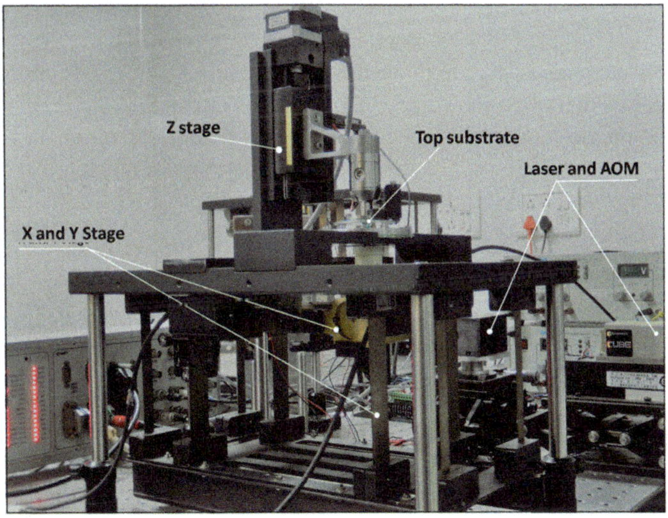

Fig. 1 Micro-stereolithography setup

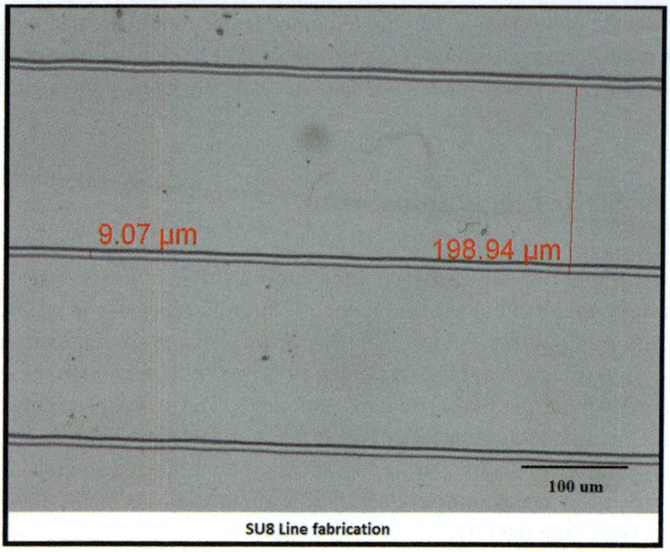

Fig. 2 Minimum printed line width

method is used for laser scanning [13]. The minimum printed line width is 9 µm as shown in Fig. 2. 1,6-hexanediol diacrylate (HDDA), trimethylolpropane triacrylate (TMPTA), SU8, resin has been used for printing micro-components.

3 Fabrication of Micro-compliant Mechanism

The various advantages like high accuracy, high aspect ratio, and homogeneity of cured material makes MSL a good choice for printing micro-compliant mechanisms. Two micro-compliant mechanisms (1) Parallelogram-compliant mechanism (2) Micro-gripper [3] are printed using the MSL setup. Manufactured examples are compliant mechanisms with distributed compliance solely. The dimensional accuracy and functioning of these printed micro-mechanisms are discussed below.

3.1 Parallelogram-Compliant Mechanism

Parallelogram-compliant mechanism is selected because the device has high aspect ratio with minimum feature dimension of 100 µm. The dimensions of compliant beam in device are 0.1 mm thickness × 5.5 mm length × 1.2 mm height. The device has possible application in optomechatronics and for miniaturizing compliant stage by introducing magnetic bead for actuation.

Table 2 Parameters of printing and specification of single parallelogram mechanism

Parameters	Values
Device dimension XYZ	$6 \times 9 \times 1.2$ mm
Minimum feature size	0.1 mm
Line spacing for scanning	20 μm
Layer thickness	25 μm
Scanning speed	8 mm/s
Printing time	5 h 30 min

Table 3 Dimensions inaccuracy of printed parallelogram-compliant mechanism

Critical model dimension (μm)	Measured dimension avg. (μm)	Dimensional inaccuracy avg. (%)
100	103.75	3.75
1000	1006.66	0.666

Resin used for printing is TMPTA 60% added with benzoin ethyl ether (BEE) by 4% in wt of TMPTA, and Digi-fab resin 40% by volume. Resin is acrylate-based plastic like material. Table 2 shows parameters that are used for printing.

Parallelogram-compliant mechanism is supported with appropriate support structure while printing which was removed after cleaning process. Mechanism is separated from substrate by gentle pricking using syringe needle and extra resin is flushed and cleaned using ethanol. The dimension of the printed mechanism is measured under OSLO olympus microscope. From Table 3, it is observed that dimensional inaccuracy of printed parallelogram mechanism is within 1–4% of the desired dimensions. The critical dimensions of desired CAD model and printed mechanism are shown in Fig. 3. As seen from Table 2, we can get the micro-mechanism prototype with high aspect ratio of 1:12 (width:height) in a single printing process (as against multiple lithography and etching in VLSI) and within short time of about 6 h. With further increase in the speed and range, the process can be viable for production level as well. One of the drawbacks of this process is that there are no easy ways to integrate electronic components into the structures thus fabricated. The printed micro-mechanism was tested by actuating it for few cycles as shown in Figs. 4 and 5.

3.2 *Micro-Gripper*

Micro-gripper is another micro-mechanism printed using the same MSL setup. Micro-gripper is largely used in biomedical applications. They are used for manipulation of cell [14], sperm insertion, cell trapping, etc. The other methods for cell manipulation using vacuum with micro-pipette do not have fair control over cell for

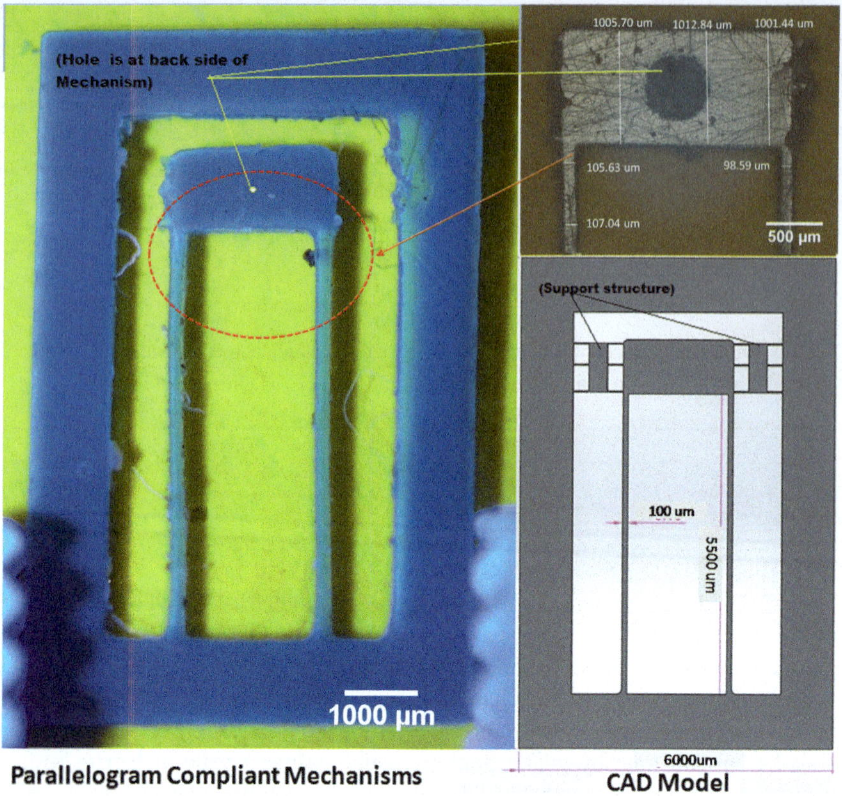

Fig. 3 Magnified image under microscope of parallelogram-compliant mechanism and CAD model

sorting and isolation. Micro-gripper can gain good control for cell manipulation with multifunctional capability of cell sorting, positions, and force sensing.

CAD model as shown in Fig. 6 was prepared with appropriate dimension for the micro-gripper found in the literature [3]. Resin used for printing is TMPTA 60% added with BEE by 4% in wt of TMPTA, and Digi-fab resin 40% by volume. The parameters used for printing of micro-gripper are given in Table 4.

The cleaning and separation process is the same as mentioned for the micro-parallelogram mechanism. The critical dimension of the printed micro-gripper and actual micro-gripper is shown in Fig. 6. From Table 4, the aspect ratio of printed component is 1:5.66 (width:height). From Table 5 the inaccuracy in the dimension of the fabricated micro-gripper is within 1–5% of the actual dimension of gripper.

Fig. 4 Parallelogram-compliant mechanisms

Fig. 5 Parallelogram-compliant mechanisms after actuation

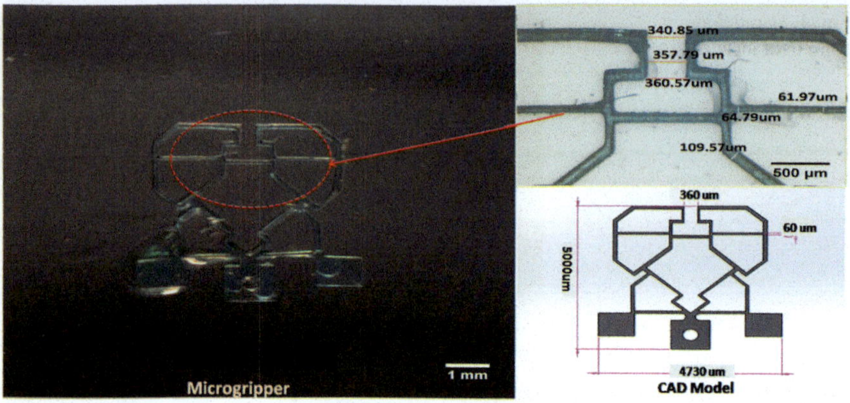

Fig. 6 Micro-gripper printed using MSL setup and CAD model

Table 4 Parameters of printing and specification of micro-gripper

Parameters	Values
Device dimension XYZ	5 × 5 × 0.34 mm
Minimum feature size	0.06 mm
Line spacing for scanning	10 µm
Layer thickness	25 µm
Scanning speed	8 mm/s
Printing time	55 min

Table 5 Dimensions inaccuracy of printed micro-gripper

Critical model dimension (µm)	Measured dimension avg. (µm)	Dimensional inaccuracy avg. (%)
360	353.07	1.92
111	109.57	1.28
60	63.38	5.63

3.3 Conclusion

The micro-mechanisms were successfully printed using micro-stereo-lithography technique. The overall dimensional accuracy of the printed micro-compliant mechanisms is very high and better than the fabrication techniques used in the literature. The MSL technique is also capable of printing structures with a high aspect ratio of 1:12. This shows that the proposed fabricating technique has good capability to print various micro-compliant mechanisms, micro-structures, and devices.

References

1. Jensen BD, Parkinson MB, Kurabayashi K, Howell LL, Baker MS (2001) Design optimization of a fully compliant bistable micro mechanism. In: ASME international mechanical engineering congress and exposition
2. Zhang C, Rossi C (2017) A review of compliant transmission mechanisms for bio-inspired flapping-wing micro air vehicles. Bioinspir Biomim 12:2
3. Reddy A, Maheshwari N, Sahu DK, Ananthasuresh GK (2010) Miniature compliant grippers with vision-based force sensing. IEEE Trans Robot 26(5):867–877
4. Alblalaihid K, Overton J, Lawes S, Peter Kinnell I (2017) A 3d-printed polymer micro-gripper with self-defined electrical tracks and thermal actuator. J Micromech Microeng 27:4
5. Thornell G, Bexell M, Schweitz J-A, Johansson S (1996) Design and fabrication of a gripping tool for micromanipulation. Elsevier Sensor Actuation A(53):428–433
6. Marty F, Rousseau L, Saadany B, Mercier B, Français O, Mita Y, Bourouina T (2005) Advanced etching of silicon based on deep reactive ion etching for silicon high aspect ratio microstructures and three-dimensional micro- and nanostructures. Microelectr J 36(7):673–677. European Micro and Nano Systems
7. Wasilik M, Chen N (2004) Deep reactive ion etch conditioning recipe, vol 5342, p 01
8. Deshmukh S, Gandhi P (2009) Optomechanical scanning systems for microstereolithography (MSL): analysis and experimental verification. J Mater Process Technol 209:1275–1285
9. Gandhii PS, Deshmuk S, Ramtekkar R, Bhole K, Barak A (2013) On-axis linear focused spot scanning microstereolithography system: optomechatronics design, analysis and development. J Adv Manuf Syst 12(1):43–68
10. Deshmukh S (2009) Design and development of microstereolithography system for 3D fabrication. PhD thesis, Indian Institute of Technology, Bombay, India
11. Gandhi PS, Deshmukh S (2009) A 2d optomechanical focused laser spot scanner: analysis and experimental results for microstereolithography. J Micromech Microeng 20:1275–1285
12. Bertsch A, Bernhard P, Renaud P (2001) Microstereolithography: concepts and applications. In: Emerging technologies and factory automation ETFA, vol 2, pp 289–298
13. Baraki A, Bhole K, Gandhi P (2010) Novel large-range, high-resolution 3d printing of micro-components using dSPACE. In: dSPACE user conference 2010, India
14. Honnatti M, Hughes G, Lee J (2006) Directed cellular manipulation using polymer microgrippers. Zyvex application note 9720, vol 2, pp 289–298

Development of a Powered Assistive Device for Patients with Lower Limb Muscle Weakness

Shishir Shah and Abhishek Gupta

Abstract Lower limb orthoses are wearable robotic devices used to assist people suffering from diseases such as paralysis, paraplegia, foot drop, muscle weakness, etc. People suffering from such diseases either lose their ability to walk or they walk in an asymmetric gait cycle with reduced speed and get exhausted easily by walking a short distance. The commercially available orthosis is very costly and moreover they do not provide complete rehabilitation. In the study, people suffering from lower limb muscle weakness are considered and a simulation study on orthosis design to assist them during swing phase of ground-level walking is described. The structural as well as functional aspect of a biological leg is considered to perform the simulation. Three-link (representing thigh, shank and foot) model is considered to perform the simulation of human lower limb in swing phase. Based on biomechanics, different muscles of lower limb are mimicked by using springs and series elastic actuator. A trajectory optimization problem is formulated to get human leg model's hip, knee and ankle joint trajectories in the range of normal human during swing phase by varying stiffness of different springs and biarticular actuator parameters. The simulation results showed that the model's different joint trajectories are well within the bounds of normal human by using only one biarticular series elastic actuator at the place of gastrocnemius muscle and passive elements such as springs. Thus, we can assist human during the swing phase of ground-level walking by using such a small actuator set.

Keywords Human gait cycle · Biomechanics of human during walking · Orthosis · Exoskeleton · Series elastic actuator

S. Shah (✉) · A. Gupta
Mechanical Department, IIT Bombay, Mumbai, India
e-mail: shishirshah9@gmail.com

A. Gupta
e-mail: abhi.gupta@iitb.ac.in

D. Sen et al. (eds.), *Mechanism and Machine Science*, Lecture Notes in Mechanical Engineering, https://doi.org/10.1007/978-981-15-4477-4_40

1 Introduction

People suffering from paralysis, paraplegia or lower limb muscle weakness have to rely on others to do the activity of daily living. They can perform some of the activities using wheel chair but it does not involve any physical movement. Due to the lack of mobility, they are at a greater risk of developing some serious problems with their urinary, respiratory, cardiovascular and digestive systems. On an emotional level, they cannot address people at eye level and cannot stand up and hug using a wheel chair [1]. All these problems cause psychological effects such as depression, anxiety, and grief.

The robotic exoskeleton/orthosis helps people suffering from paralysis, paraplegia to walk again. A small amount of exercise with an assistive device can improve the health of paraplegics. Some of the physical benefits of using an assistive device include increased bone density, decrease in body fat percentage, improved cardiovascular, bowel and bladder function, and improved sitting posture [1]. There are commercial exoskeletons available such as eLEGs [2], ReWalk [3] and HAL (Hybrid Assistive Leg) [4], which assist people with paraplegia or paralysis to walk. The orthosis such as Vanderbilt orthosis [5], Mina [6], CUHK-EXO [7], etc. are under research phase. All of these devices provide power at hip and knee joints in sagittal plane. The other degrees of freedom at hip are either locked or loaded with complaint element. The ankle joint is passive with spring loading except for HAL. HAL gives power to ankle joint also. Apart from this, all the devices have some additional unique features such as eLEGs has adjustable thigh and lower leg segments to accommodate large range of patients, HAL-5 uses EMG sensor to measure muscle activity and understand the human intentions, Vanderbilt orthosis have electrically controllable normally locked brakes at knee so the knee is locked during stance phase and free to rotate during swing phase and CUHK-EXO has smart crutches, smartphone application to monitor motion of the wearer. The review on control strategy used by these devices is described in [8]. All these devices try to provide the required torque at the joints but do not consider the structural aspects of the body. The biarticular muscles perform very important function during walking. They help in stability and also transfer energy from one part of body to the other [9]. None of these devices has explored the use of biarticular actuator.

The current study considers the use of biarticular series elastic actuator (SEA) to assist human during swing phase of ground-level walking. The SEA is considered at the place of gastrocnemius muscle as it performs some important functions as described in the next section. The current study also incorporates the use of passive elements such as springs to assist the actuator. The simulation study by considering the proposed actuator set to assist human during swing phase of ground-level walking is given in the subsequent sections.

2 Human Gait Cycle and Its Biomechanics During Walking

The systematic study of human motion is known as gait analysis. The human walking is a successive repetition of a basic movement known as the gait cycle. The gait cycle is divided into two phases, the stance phase and the swing phase. During stance phase, the foot stays in the contact with the ground while during swing phase, the leg swings in forward direction without touching the ground. Approximately 60% of the gait cycle is stance phase and remaining 40% is swing phase. An instant when the heel touches the ground is called heel strike, and the moment when the foot leaves the ground is known as toe off. The stance phase starts from the initial contact of one leg and ends at the toe off of the same leg, whereas swing phase starts at the toe off of the one leg and ends at the initial contact of the same leg.

Most of the human motion occurs in sagittal plane while walking and therefore in this study, only the movement in sagittal planes are considered. Sagittal plane is the plane which divides the body into left and right side. The motions in the sagittal plane are called flexion and extension. In case of the ankle they are called plantar flexion and dorsiflexion. The motions of hip, knee, and ankle joints in sagittal plane are shown in Fig. 1. Flexion/dorsiflexion is the motion in the sagittal plane, by which the angle between two parts of the body reduces. Extension/plantar flexion is the motion in the sagittal plane by which the angle between the parts increases.

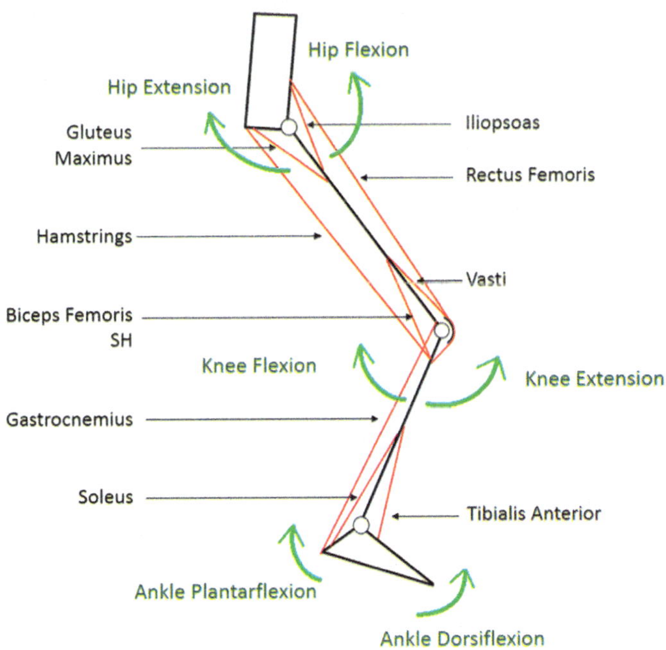

Fig. 1 Major muscles acting on lower limb [10]

Figure 1 shows the major muscles acting on the lower limb of the human. There are two types of muscles, namely uniarticular and biarticular muscles, which are important during walking. Uniarticular muscle spans only across one joint whereas biarticular muscle spans across two joints. Uniarticular muscles generate positive power at the joint whereas biarticular muscles help to transfer energy synergistically between the different segments of the leg. During stance phase, the whole body moves forward. In this phase, the soleus and gastrocnemius muscles of stance leg stores energy by contracting eccentrically. In eccentric contraction, the muscle extends from its neutral position and stores energy by acting as a spring. This stored energy in soleus and gastrocnemius muscles helps to lift the heel (push-off) of stance leg from the ground during pre-swing phase. The gastrocnemius muscle facilitates control falling of the body during stance phase and allows synergistic energy transfer from the upper body to the ankle joint. It prevents the hyperextension of knee by acting as a knee flexor and stabilizes it during the stance phase. It also assists in swing initiation [9].

During swing phase, the swing leg moves ahead of the stance leg. The positive power is generated at the hip during initial swing phase to accelerate the swinging leg in forward direction. The thigh and shank of swing leg moves because of the double pendulum effect. The iliopsoas muscle contracts concentrically to flex the hip. In concentric contraction, the muscle length decreases and it produces positive power. Vasti muscle is active in the initial part of swing phase to prevent the flexion of knee. The hamstring is active in the later part of the phase to prevent the hyperextension of the knee. The hamstring contracts eccentrically during the terminal swing to limit the extension of knee joint. Tibialis anterior is active during the whole swing phase to dorsiflex the ankle [9]. The tibialis anterior muscle stores energy by contracting eccentrically during plantar flexion motion of ankle, just before swing phase and this stored energy is utilized to dorsiflex the ankle during swing. Thus, hamstring and tibialis anterior muscles can be mimicked by a spring. Based on the biomechanics of human, different muscles are mimicked using springs and actuators in the present study. The next section describes the unique combination of actuator and springs used in swing phase of walking.

3 Modelling of Human Leg in Swing Phase

The model for the leg in swing phase is shown in Fig. 2. The three links from the top represents the thigh, shank and the foot of the swing leg, respectively. It has three revolute joints representing (from the top) the hip, knee and ankle joint of the leg. The foot link is taken as a triangular link so that it looks similar to human foot and helps in determining the heel and toe clearance. The hip joint coincides with the origin of reference frame taken. The model rotates about the hip joint.

Inspired from the biomechanics of the human leg in swing phase, two springs are added to mimic hamstring and tibialis anterior muscles. Also one biarticular series elastic actuator (SEA) is added at the place of gastrocnemius muscle to give positive

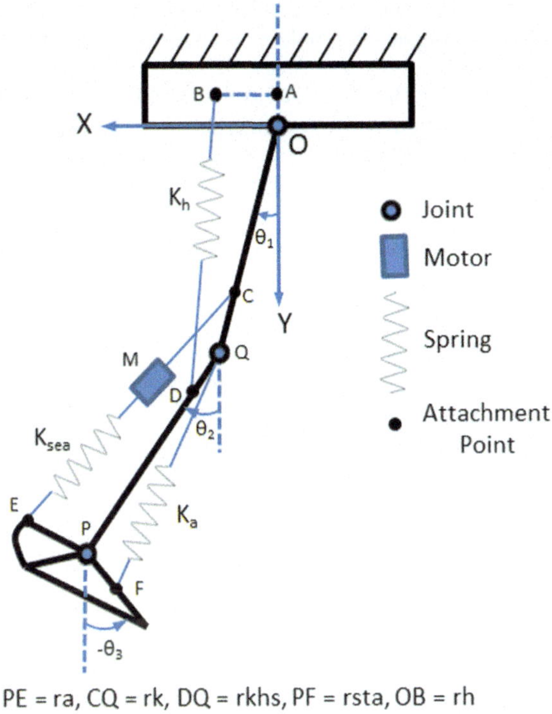

Fig. 2 Modelling of the leg in swing phase

power to the leg. The actuator is added at the place of gastrocnemius muscle because gastrocnemius muscle performs some important functions as mentioned in Sect. 2. Also, power can be provided to all the joints using actuator in this position. The equations of motion for the considered system are derived using Lagrange method. The anthropometric related data for human leg are taken from [11]. The system is under-actuated and to check the controllability of this system with the considered actuator set is very difficult since the system is highly non-linear. Thus instead of checking for controllability and then decide on the actuator set, it is decided to find the trajectory of all the states and control inputs of the system such that it satisfies the dynamic equations of motion. The direct collocation method is used to solve the trajectory search problem [12].

Direct collocation method is a numerical method to solve the trajectory optimization problem. It involves discretization of the dynamic system and giving it as a constraint to the optimization problem. Thus, it converts differential equation into algebraic equations. The optimization parameters are the states and the control input at discrete points. For any continuous dynamic system $\dot{x} = f(x(t), u(t), t)$, it approximates $\int f(x(t), u(t), t)dt$ to find the discrete version of the system by using numerical integration methods such as Trapezoidal, Hermit–Simpson and Range–Kutta. The optimization problem is formulated in MATLAB to find the optimum trajectory of states and control inputs. Objective function, bounds, and constraints of the optimization problem are described below.

1. Objective function: Our system can be written in a form, $\dot{x} = f(x, u)$. The only control input to the system is the motor force given by SEA. The objective function considered to solve the trajectory optimization problem is given by

$$\underset{x,u,p}{\text{minimize}} \int_{t_o}^{t_f} w(x, u, \tau)d\tau \tag{1}$$

   where
   t_0 and t_f are start and end time of the trajectory
   $w = u^2$
   u = force given by SEA
   $P = [ra, rk, rkhs, K_h, K_a, K_{\text{sea}}]$
   ra = distance between ankle joint and lower attachment point of SEA
   rk = distance between knee joint and upper attachment point of SEA
   $rkhs$ = distance between knee joint and lower attachment point of hamstring spring
   K_h, K_a, K_{sea} are spring constant of hamstring, tibia and SEA spring, respectively. The parameter vector P includes the spring stiffness, spring and actuator's attachment points as shown in Fig. 2. The value of parameters rh and $rsta$ are taken constant and are equal to 0.072 m and 0.116 m, respectively. rh represents the distance between upper attachment point of hamstring spring and hip joint whereas $rsta$ represents the distance between lower attachment point of tibia spring and ankle joint. The value of rh is taken by considering the geometry and space requirement to attach the spring. The lower attachment point of tibia is considered on the centre of gravity of the foot and the value 0.116 m represents the distance of centre of gravity of foot from ankle joint. According to the direct collocation method, objective function is discretized using Hermit–Simpson method [12] and supplied to the solver.

2. Constraints:

   (a) Dynamic equations constraints: The equations of motion are discretized using Hermit–Simpson method and supplied to the solver as constraints.
   (b) Path constraints: These are the constraints applicable throughout the trajectory of the system apart from the dynamic equations constraints. The constraints are given below.

$$\begin{aligned} \theta_1 - \theta_2 &\leqslant 0 \\ -Y_3 &\leqslant 0.95 \\ -Y_4 &\leqslant 0.95 \end{aligned} \tag{2}$$

Table 1 Values of optimization parameters

Parameter	Value
ra	0.04 m
rk	0.04 m
rkhs	0.06 m
K_h	19 N/mm
K_a	270 N/m
K_{sea}	35 N/mm

where
$\theta_1 - \theta_2$ is knee angle of leg
Y_3 and Y_4 are vertical distance of toe and heel, respectively, in the reference frame shown in Fig. 2.
The first constraint is to restrict the hyperextension of knee joint of the leg. The last two equations are added to ensure the toe and heel clearance of the leg from the ground, respectively. The number 0.95 m represents the distance between the ground and the origin. It is the sum of thigh, shank and foot length of the leg.

3. Bounds: The upper and lower bounds for angle trajectories of considered model are taken from [13], where mean and deviation of joint angles over one gait cycle are given for healthy ground-level walking. The lower and upper bound for velocities are taken as -10 and $+10$ rad/s, respectively. Arbitrarily high values are taken as the bounds for control input. The bounds on spring stiffness are considered based on the manufacturing possibility and that for springs and actuators attachment point distances are considered based on the available space and geometry.
4. Results: We found a local minima after optimizing the above described problem in MATLAB using fmincon solver and interior point method algorithm. The optimized parameter vector is given in Table 1. We also found the open-loop profile for control input (SEA force). But we cannot apply this to a real system without using a stabilizing controller. Thus, this open-loop force profile is made a closed loop by using impedance control as described in the next section.

4 Impedance Control for Swing Phase

Impedance control regulates the dynamic interaction between actuator and its environment. Thus using it, we can mimic spring or spring and damper behaviour through the actuator. The open-loop SEA force profile from direct collocation method is made a closed loop by approximating it using the control law given below.

Table 2 Impedance parameters for SEA

Sub-phase	Description	K (N/mm)	B (Ns/mm)	C (N)
1	Start to max knee flexion	11.36	0.24	38.4
2	Max knee flexion to max hip flexion	3.1	0.3	27.4
3	Max hip flexion to heel strike	4.09	0.2	35

$$F = K\delta + B\dot{\delta} + C \tag{3}$$

where,
δ is the deflection of series spring
K, B and C are impedance parameters.

Here, K and B can be considered as spring constant and damping coefficient term and C as feed-forward term. Thus by measuring the deflection of series spring, the force needed to apply can be determined. The values of K, B, and C are found by curve fitting the force profile using least square method. To use a single set of impedance parameters to approximate open-loop force profile for the whole phase is difficult. Therefore, the swing phase is divided into three subphases according to the movement of leg. Subphases are selected in such a way that it can be detected easily using sensors. The description of three subphases and impedance parameters in these subphases are given in Table 2. The knee and hip flexion mentioned in Table 2 is calculated by using knee ($\theta_1 - \theta_2$) and hip (θ_1) joint angle, respectively. The joint angle sensor such as rotary potentiometer can be used to determine the subphase in which the user is walking. After determining the subphase, a linear potentiometer can be used to calculate the deflection series spring. Finally, the force requirement of SEA can be calculated using sensor reading, Table 2, and Eq. 3. Figure 3 shows the

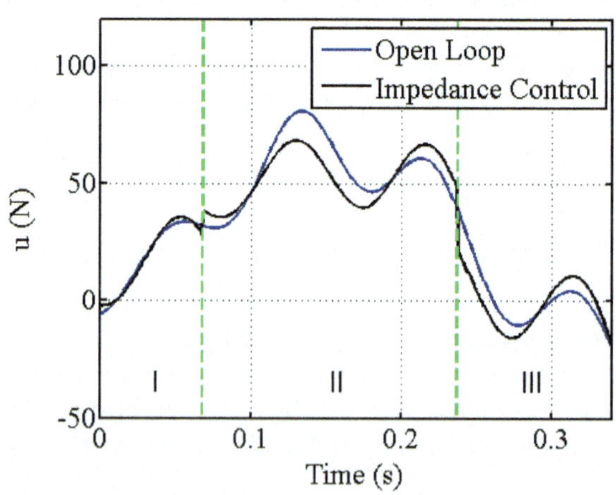

Fig. 3 Open- and closed-loop force profile of SEA, the green dashed line divides the total swing phase into different subphases as mentioned in Table 2

open-loop and closed-loop force profile of SEA. The vertical green dashed line in Fig. 3 divides the swing phase into different subphases as mentioned in Table 2. The number written in roman on Fig. 3 represents the corresponding subphase. The step in the graph of closed-loop force profile is observed when the value of impedance parameters changes from one subphase to other. The future work includes the design of controller with fast step response characteristics to take care of this.

5 Final Results and Discussion for Swing Phase

After making force profile a closed loop, system dynamics are solved piecewise using ode45 in MATLAB. The initial condition used is the same as gotten from direct collocation optimization. Figures 4, 5 and 6 gives the trajectory of θ_1 to θ_3. The bounds used in the below figures are of healthy humans walking on a level ground surface and taken from [13]. The vertical green dashed line in all figures divides the swing phase in different subphases as mention in Sect. 4. From the below figures, we can see that all angles of the mechanism are within the bound of healthy walking.

According to the sign convention taken, the positive force of SEA means it acts upward on the thigh segment of leg. From Fig. 3, the force applied by the actuator moves the thigh in anticlockwise direction (flexion) and the leg moves forward. The thigh reaches to its maximum flexion and then extends before the heel strike. Therefore, the force given by SEA is negative during the end part of the cycle. The hamstring spring is compressed at the start of swing phase due to the initial conditions and then expands and reaches to maximum deflection at the end of the phase. It helps the actuator at the beginning of the cycle to initiate the swing and then resists to store energy and prevent the hyperextension of knee as done by hamstring muscle.

Fig. 4 θ_1

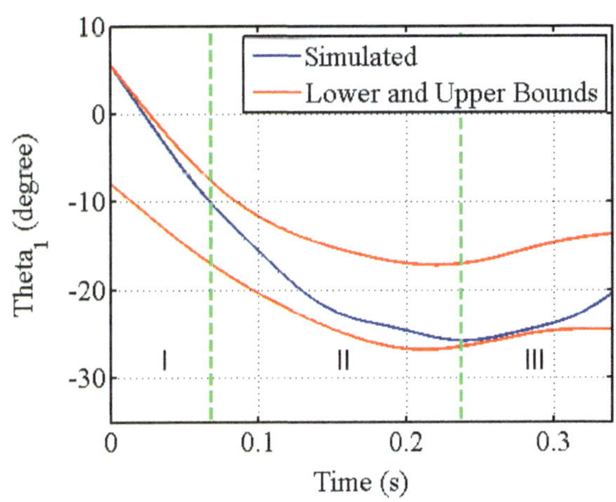

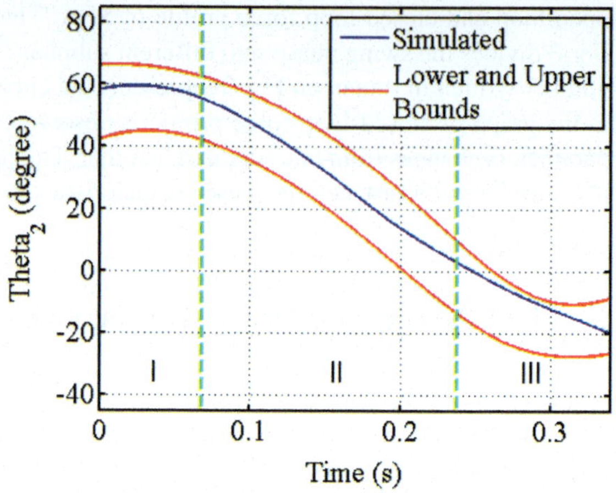

Fig. 5 θ_2

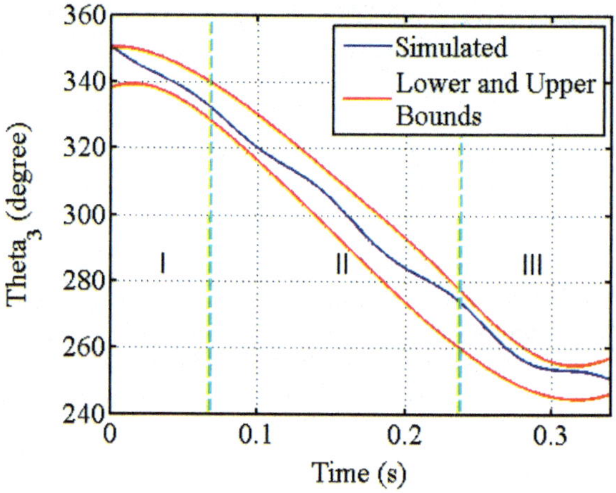

Fig. 6 θ_3

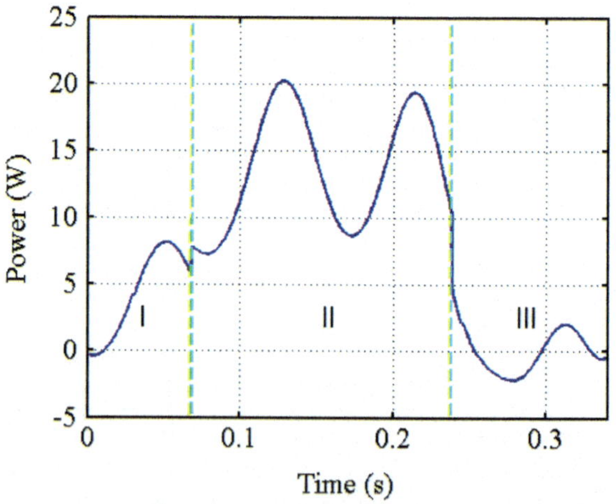

Fig. 7 Power given by SEA

Fig. 8 Deflection of hamstring spring

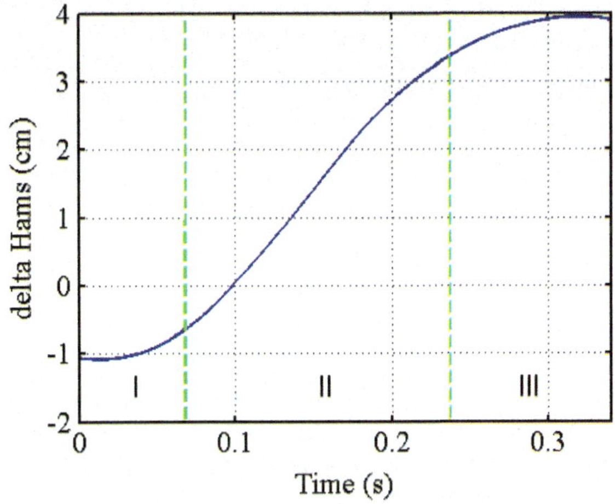

Fig. 9 Deflection of tibia spring

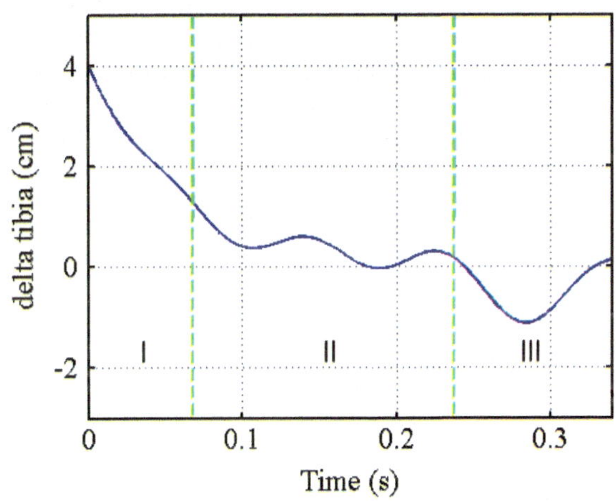

The tibia spring is extended at the start of this phase due to the initial conditions and it will be the case since the foot plantar flexes at the end of stance phase. The tibia spring applies torque on the foot in anticlockwise direction (dorsiflexion) to dorsiflex the ankle and come to zero deflection at the end of this phase. This is essentially the function of tibialis anterior muscle. Figures 8 and 9 give the deflection of hamstring and tibia springs, respectively. The negative deflection in Figs. 8 and 9 indicates that the spring is compressed. Figure 7 gives the power requirement of SEA. The RMS power used by SEA is only 10 W. Thus from the above simulation study, we can say that there is a possibility of assisting human during swing phase using only one actuator and two springs.

6 Conclusion and Future Work

The biomechanics of human walking on a level surface and actuation of different muscles during the gait cycle is studied thoroughly. The model of human lower extremity during swing phase is created and the equations of motion are derived using Lagrange formulation. Trajectory optimization problem is formulated for swing phase to get human leg model's different angle trajectories within the range of normal human by varying different parameters of the considered actuator set. The optimized open-loop SEA force profile is made closed-loop through impedance control by dividing the swing phase into different detectable subphases. The simulation results showed that the humans can be assisted during swing phase using only one biarticular series elastic actuator and two springs. The future work includes the dynamic simulation of both the legs together during the different phases of gait cycle using the same actuator set as used in the current study. It also includes the CAD design of an assistive device based on the simulation, its actual fabrication and testing.

References

1. Exoskeletons for paraplegics. https://sites.google.com/a/cortland.edu/exoskeletons-for-paraplegics/advantages. Accessed 17 May 2017
2. Strausser KA, Swift TA, Zoss AB, Kazerooni H, Bennett BC (2011) Mobile exoskeleton for spinal cord injury: development and testing. In : ASME 2011 dynamic systems and control conference and bath/ASME symposium on fluid power and motion control, DSCC 2011, vol 2, pp 419–425
3. Talaty M, Esquenazi A, Briceno JE (2013) Differentiating ability in users of the ReWalk^{TM} powered exoskeleton: an analysis of walking kinematics. In: 2013 IEEE international conference on rehabilitation robotics (ICORR). IEEE, pp 1–5
4. Suzuki K, Mito G, Kawamoto H, Hasegawa Y, Sankai Y (2007) Intention-based walking support for paraplegia patients with Robot Suit HAL. Adv Robot 21(12):1441–1469
5. Farris RJ, Quintero HA, Goldfarb M (2011) Preliminary evaluation of a powered lower limb orthosis to aid walking in paraplegic individuals. IEEE Trans Neural Syst Rehabil Eng 19(6):652–659
6. Raj AK, Neuhaus PD, Moucheboeuf AM, Noorden JH, Lecoutre DV (2011) Mina: a sensorimotor robotic orthosis formobility assistance. J Robot
7. Chen B, Ma H, Qin LY, Guan X, Chan KM, Law SW, Qin L, Liao WH (2015) Design of a lower extremity exoskeleton for motion assistance in paralyzed individuals. In: 2015 IEEE international conference on robotics and biomimetics, IEEE-ROBIO 2015
8. Yan T, Cempini M, Oddo CM, Vitiello N (2015) Review of assistive strategies in powered lower-limb orthoses and exoskeletons. Robot Auton Syst 64:120–136
9. Whittle MW (2007) Gait analysis-an introduction. Heidi Harrison

10. Joshi SD, Gupta A (2015) Conceptual design of an active transtibial prosthesis based on expected joint and muscle forces in a unilateral transtibial amputee: a modelling study. In: Proceedings of the ASME 2015 international mechanical engineering congress & exposition, IMECE 2015. Houston, Texas
11. Onyshko S, Winter DA (1980) A mathematical model for the dynamics of human locomotion. J Biomech 13(4):361–368
12. Kelly M (2017) An introduction to trajectory optimization: How to do your own direct collocation. SIAM Rev 59(4):849–904
13. Winter D (1987) The biomechanics and motor control of human gait waterloo

Simulation of Small Rigid Particle Transportation by an Inclined Linear Vibratory Conveyor

Pranav Lad and V. Kartik

Abstract The motion of a small rigid particle over a conveying tray that has oscillatory motion in an inclined direction with respect to horizontal is simulated and analyzed. Initially, when the amplitude of oscillation is small enough, the particle always remains in contact with the tray; however, it travels longitudinally due to differential friction in forward and backward directions which is produced due to vertical motion of the tray. As the oscillation amplitude is increased, the body loses contact and has periodic sliding and projectile phases of motion. The effect of friction coefficient, acceleration amplitude and vibration angle on the mean conveying velocity is examined. When the amplitude is further increased, the body starts bouncing on the tray after the impact and the motion appears to be chaotic. However, it is found that when a particle is dropped from a certain height, the periodic bouncing motion can be obtained with specific parameters.

Keywords Vibratory conveyor · High-throughput · Hopping motion

1 Introduction

Linear vibratory conveyors find many engineering applications such as in packaging lines of pharmaceutical, agricultural, and food processing industry to transport various granular rigid particles between processing machines. Such conveyors have simple construction and have unique advantage such as the suitability to handle abrasive and hot materials. However, such conveyors are generally used for limited transport distances. Also, they have noisy operation when their net harmonic acceleration is very high. In packaging lines, to improve the productivity, it is essential to increase the throughput of such conveyors. The basic understanding of the design of

P. Lad (✉) · V. Kartik
Department of Mechanical Engineering, Indian Institute of Techology Bombay, Mumbai, India
e-mail: pranavlad@iitb.ac.in

D. Sen et al. (eds.), *Mechanism and Machine Science*, Lecture Notes in Mechanical Engineering, https://doi.org/10.1007/978-981-15-4477-4_41

linear vibratory conveyors and the effect of different operating parameters on their mean conveying velocity is essential to achieve high-throughput of the conveyor.

The most common type of such conveyor consists of a tray mounted horizontally (or slightly inclined to the base) with the help of inclined leaf springs. The tray is excited periodically (usually due to electromagnetic force) in the direction inclined to the surface of the tray. Redford and Boothroyd [1] were probably the first to study the design of vibratory bowl feeder. Parameswaran and Ganapathy [2] reviewed the research in this area thereafter. However, most of the work was empirical or experimental in nature. Winkler [3] attempted theoretical analysis of vibrating conveyors especially when the velocity profile of the tray is triangular periodic. The particle was assumed to be always in contact with the tray. However, as the acceleration of tray is increased, at some point the particle loses contact and undergoes sliding and projectile phases of motion periodically. The analysis of such hopping phenomenon is presented later [4]. Sloot and Kruyt [5] studied the transport of granular materials by inclined conveyors theoretically and experimentally. However, large deviations are obtained between theoretical and experimental results at higher acceleration when material undergoes hopping motion. The error may be reduced by incorporating the change in the normal velocity of a particle after collision with the tray, which can be done by the introduction of coefficient of restitution during the impact.

In this paper, an algorithm simulating the motion of a point mass on an inclined vibratory tray is developed. It incorporates all phases of motion such as in-contact (sticking and sliding in both forward and backward direction relative to the tray) and out of contact (projectile motion) phase including the impact between the object and the tray.

2 Physical Model

Figure 1 shows the schematic representation of a vibratory conveyor. A flat tray inclined at an angle "θ" to the horizontal is mounted on the fixed base with the help of leaf springs. These springs make an angle "$\alpha + \theta$" to vertical so that the tray has periodic sinusoidal motion of amplitude A_0 at an angle α to its surface. The co-ordinate system assumed here is such that x- and y-axes are oriented parallel and normal to the surface of the tray, respectively. Following assumptions are made in this work:

- the tray has pure sinusoidal motion;
- particle is a rigid body considered to be a point mass with negligible rotation and its motion is independent of its shape;
- there is no tendency of rolling down for the body;
- friction between tray and body is only considered during in-contact phase; i.e., it is neglected during the impact;
- coefficients of static and dynamic friction are constant;
- the friction between point mass and side walls of the tray is negligible;

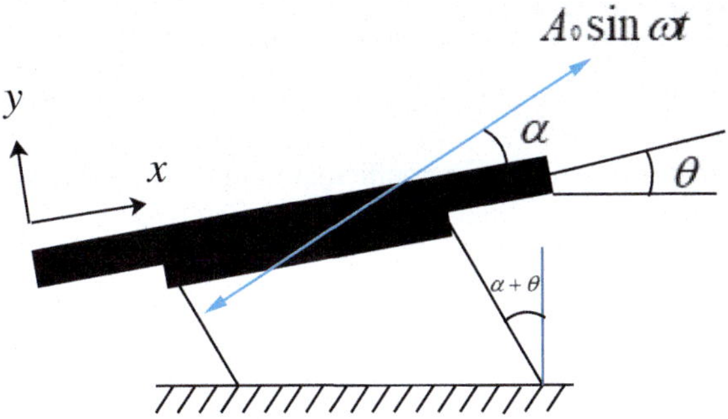

Fig. 1 Schematic representation of an inclined vibratory conveyor

- air drag during the flight phase is neglected;
- tray is rigid and a particle has negligible mass compared to the tray so that the velocity of tray after the impact remains unaffected.

Consider a particle placed on the tray horizontal to the base (i.e., $\theta = 0$). When a tray oscillates sinusoidally in the horizontal direction along its surface (i.e., $\alpha = 0$), the particle follows the track motion unless the amplitude and frequency of oscillation is such that the inertia force overcomes the static friction between the tray and itself. If this critical limit is crossed, particle slides back and forth relative to the surface of the tray, which motion is opposed by kinetic friction. However, a net forward or backward motion of a particle is zero. To achieve such a net forward motion, the kinetic friction during forward sliding of a particle is required to be less than that during backward sliding. This differential friction can either be achieved by variation in the coefficient of friction or that in the normal reaction between particle and the tray during both the strokes. It is usually impossible to control the coefficient of friction precisely; thus the second approach is used to obtain net motion of a particle from one end of the tray to another. The variation in the normal reaction is possible if the tray oscillates in inclined direction to its surface. During part of the cycle, the normal reaction experienced by a particle is the addition of self-weight (due to gravity) and inertia force (due to normal acceleration of the tray), whereas during remaining part of the cycle, it is the difference between the two. Thus, the frictional force experienced by a particle when it is sliding forward (relative to a tray) is less than that during backward sliding.

When a particle is placed on an inclined tray which is oscillating in a direction inclined to its surface, there exists a minimum limit on the magnitude and frequency of oscillation, so that inertial force overcomes the static friction and either slides forward or backward relative to the tray. To obtain these conditions, the magnitude of inertia force experienced by a particle due to the motion of tray can be resolved in parallel and normal components as $mA_0\omega^2 \cos\alpha$ and $mA_0\omega^2 \sin\alpha$ respectively, where "m" is the mass of a particle. Now, it can be seen that for forward sliding of a particle to occur,

$$mA_0\omega^2 \cos\alpha > mg \sin\theta + F_r \tag{1}$$

where friction force,

$$F_r = \mu_s N = \mu_s (mg\cos\theta - mA_0\omega^2 \sin\alpha) \tag{2}$$

where μ_s and N are coefficient of static friction and normal reaction, respectively. Thus, condition for forward sliding is given as

$$\frac{A_0\omega^2}{g} > \frac{\mu_s \cos\theta + \sin\theta}{\cos\alpha + \mu_s \sin\alpha} \tag{3}$$

By defining a dimensionless acceleration parameter as $A_n/g_n = A_0\omega^2 \sin\alpha/g\cos\theta$, Eq. 3 is modified to

$$\frac{A_n}{g_n} > \frac{\mu_s + \tan\theta}{\cot\alpha + \mu_s} \tag{4}$$

Similarly, the condition for a particle to slide backward relative to a tray is written as

$$\frac{A_n}{g_n} > \frac{\mu_s - \tan\theta}{\cot\alpha - \mu_s} \tag{5}$$

For sufficiently high amplitude of vibration, the normal reaction experienced by a particle becomes zero and it leaves the tray to hop forward. The condition for this to happen can be written as

$$\frac{A_0\omega^2}{g} > \frac{\cos\theta}{\sin\alpha} \tag{6}$$

or, in dimensionless form,

$$\frac{A_n}{g_n} > 1 \tag{7}$$

3 Different Phases of Motion of a Particle

Depending on various parameters, a particle may have different phases of motion such as in-contact (stationary/forward/backward sliding), projectile, or collision phase.

3.1 In-Contact Phase

Consider a particle is placed on a tray and it is at rest initially. Depending on the initial phase of sinusoidal motion, a particle will either remain stationary or immediately start to slide forward or backward. To obtain the forward sliding region in the cycle, we can write using Newton's law,

$$mA_0\omega^2 \cos\alpha\ \sin(\omega t_f) = mg\sin\theta + \mu\left(mg\cos\theta - mA_0\omega^2 \sin\alpha\ \sin(\omega t_f)\right) \quad (8)$$

where ωt_f represents the angular position in the cycle in which forward sliding occurs. Thus,

$$\sin(\omega t_f) = \frac{\mu\cos\theta + \sin\theta}{\frac{A_0\omega^2}{g}(\cos\alpha + \mu\sin\alpha)} \quad (9)$$

Note that Eq. 9 has two solutions in a cycle; each represents the start and the end of the forward sliding region, respectively. Similarly, for backward sliding,

$$mA_0\omega^2 \cos\alpha\ \sin(\omega t_b) = mg\sin\theta - \mu\left(mg\cos\theta - mA_0\omega^2 \sin\alpha\ \sin(\omega t_b)\right) \quad (10)$$

where ωt_b represents the angular position in the cycle during which a particle slides backward relative to the tray. Thus,

$$\sin(\omega t_b) = \frac{-\mu\cos\theta + \sin\theta}{\frac{A_0\omega^2}{g}(\cos\alpha - \mu\sin\alpha)} \quad (11)$$

Solutions of this equation denote the start and the end of the backward sliding region in a cycle, respectively. A particle remains stationary relative to the tray during the remaining part of the cycle (other than forward and backward sliding region), as it fails to overcome the static friction.

Figure 2a shows the limits of angular positions defining the forward and backward sliding regions. An illustrative diagram representing corresponding time instants is shown in Fig. 2b, for $A_n/g_n = 1.2, \theta = 2°$, and $\alpha = 20°$. If a particle is at rest and a tray starts oscillating suddenly at time t lying within the blue region, it will remain stationary until the cycle enters into red or black region in which it starts to slide forward or backward, respectively. However, if the tray suddenly starts at time t lying between red or black region, a particle will immediately start to slide forward or backward, respectively.

When $A_n/g_n < 1$, the velocity of a particle decreases (or increases) while sliding forward (or backward) due to kinetic friction, until it becomes equal to that of the tray. Depending on the region in which this time instant lies in the cycle, a particle either remains stationary or immediately slides backward (or forward) relative to the tray and the cycle continues periodically.

3.2 Projectile Motion

When $A_n/g_n > 1$, a particle leaves contact with the tray at some instant "t_1" in the cycle, which can be obtained by equating the normal reaction to zero,

$$N = g\cos\theta - A_0\omega^2 \sin\alpha\ \sin(\omega t_1) = 0 \quad (12)$$

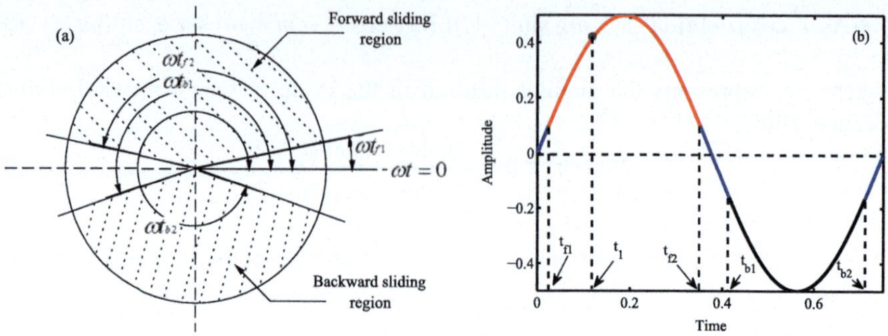

Fig. 2 Regions within which a particle remains stationary, slides forward or backward relative to a tray. Respective **a** angular positions, **b** time instants in the cycle

or,

$$\sin(\omega t_1) = \frac{g_n}{A_n} \tag{13}$$

For illustration, this instant is shown in Fig. 2b. Thus, when a tray starts to oscillate at $t = 0$, a particle at rest will stick to the tray until $t = t_{f1}$; it will start sliding forward thereafter relative to the tray until $t = t_1$. At this instant, it will lose contact with the tray and will undergo a projectile motion. Initial position at the start of this phase is equal to that of the tray and velocity is that at the end of forward sliding phase at $t = t_1$. The particle experiences accelerations "$-g \cos \theta$" and "$-g \sin \theta$" in the direction normal and parallel to the surface of the tray respectively, during the flight phase.

3.3 Collision Phase

A particle undergoes a projectile motion until it collides with the tray when its normal position becomes equal to that of the tray. If A_n/g_n is sufficiently low, the particle does not bounce back significantly after the impact, and the collision is assumed to be perfectly plastic. Thus, the normal component of velocity of a particle after the impact becomes equal to that of the tray. Most of the time, the parallel component of velocity of a particle at the time of impact is more than that of the tray. Thus, a particle slides forward until its velocity becomes equal to that of the tray. Now, depending on the region in which this time instant lies in the cycle, a particle either remains stationary or immediately slides backward relative to the tray; and the cycle continues periodically.

However, if A_n/g_n is very high, a particle starts to bounce back after the impact and the collision cannot be assumed perfectly plastic anymore. The normal component of velocity of a particle after impact can be obtained as

$$v_{py} = v_{cy}(t_i) - \epsilon(u_{py} - v_{cy}(t_i)) \tag{14}$$

where v_{py}, u_{py}, $v_{cy}(t_i)$, and ϵ denote normal velocities of a particle after and before impact, normal velocity of a tray at the time of impact, and coefficient of restitution respectively. Note that the effect of collision on the motion of a tray is assumed to be negligible and the horizontal velocity of the particle is assumed to be unaffected during the collision.

The particle will undergo projectile motion again with this velocity as an initial condition. The successive impacts will continue to occur until the normal velocity of a particle after the impact becomes less than or equal to that of the tray. Once this condition is reached, the cycle will repeat as explained earlier for the case of plastic collision.

The flow chart of the algorithm which incorporates all these phases of motion is shown in Appendix.

4 Results and Discussion

The effect of various parameters on the mean conveying velocity of a tray is discussed in this section. Using an algorithm in Appendix, the total parallel distance covered by a particle in a specific amount of time can be calculated. The mean conveying velocity is defined as the ratio of total distance covered by a particle in the direction parallel to the surface of the tray and the total amount of time. Initially, the coefficient of restitution ϵ is assumed to be zero representing the case of perfectly plastic collision.

Figure 3 shows the variation of mean conveying velocity of the tray relative to the frequency of oscillation for constant A_n/g_n. In this case, the tray is assumed to be horizontal with respect to the base ($\theta = 0°$) and it vibrates at an angle $\alpha = 20°$.

Fig. 3 Variation of mean conveying velocity of tray relative to the frequency of oscillation

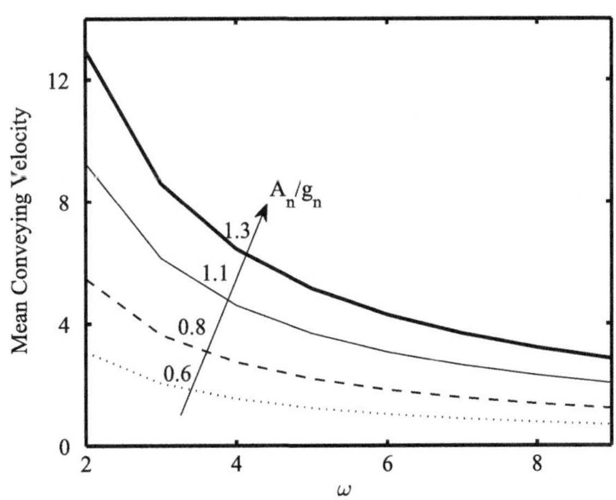

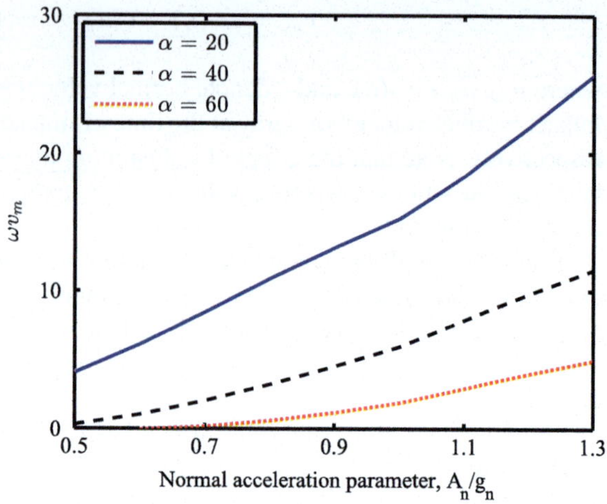

Fig. 4 Variation of conveying velocity with normal acceleration parameter A_n/g_n. Solid, dashed, and dotted curves correspond to $\alpha = 20°$, $40°$, and $60°$ respectively

It is observed that the conveying velocity decreases as the frequency of oscillation is increased for a constant normal acceleration parameter A_n/g_n, i.e., the conveying velocity is inversely proportional to the frequency; and the product "ωv_m" is approximately constant, where v_m is the mean conveying velocity. As A_n/g_n is increased, for same value of frequency, the conveying velocity increases.

The variation of the product of frequency and the conveying velocity with normal acceleration parameter A_n/g_n is shown in Fig. 4 for different values of vibration angle. As expected, the conveying velocity increases with A_n/g_n almost linearly. Note that if $A_n/g_n > 1$, a particle starts losing contact and undergoes projectile motion for the part of the cycle. It can be observed from Fig. 4 that there exists a small but sudden change in the slope of the curve between $A_n/g_n = 0.9$ and 1.1. Thus, it can be said that the projectile phase of motion helps to increase the throughput of the conveyor significantly.

According to Fig. 4, the mean conveying speed decreases as the vibration angle is increased. However, there exists an optimum vibration angle for which the conveying speed is maximum for the same parameters. To illustrate this, the variation of mean conveying speed is shown in Fig. 5 for different values of coefficient of friction. For this analysis, normal acceleration parameter A_n/g_n is assumed to be 0.8. Clearly, as the vibration angle increases, the conveying velocity first increases and then decreases after reaching maximum. This optimum value of vibration angle shifts to a lower side as the coefficient of friction increases.

Practically, the coefficient of friction between metal parts is generally found to be close to 0.2–0.3. However, there is a common practice in the industry to provide rubber-like coating on the conveying tray, which increases the coefficient of friction up to 0.8. Figure 6 shows the variation of mean conveying velocity of the tray with coefficient of friction for different values of normal acceleration parameter when vibration angle $\alpha = 20°$. For very low acceleration, there exists an optimum value of coefficient of friction for which the conveying speed is maximum. However, to

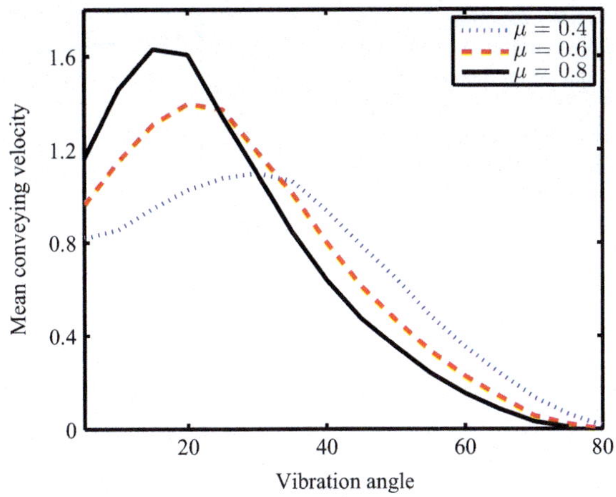

Fig. 5 Variation of mean conveying speed of the conveyor with the angle of vibration. Solid, dashed, and dotted curves correspond to the coefficient of friction $\mu = 0.8$, 0.6, and 0.4 respectively

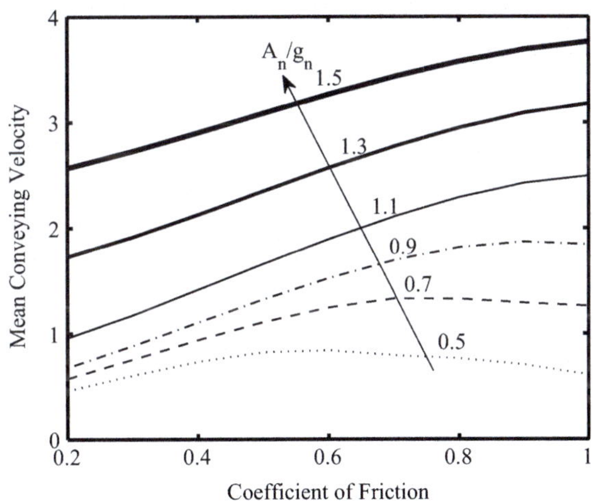

Fig. 6 Variation of mean conveying velocity with the coefficient of friction for different values of normal acceleration parameter

improve the throughput, A_n/g_n is normally kept higher than 1. For such a range of accelerations, the conveying speed increases with coefficient of friction monotonically. Thus, it is recommended to provide rubber-like coating to increase the coefficient of friction which in turn improves the throughput of the conveyor. This coating also helps to reduce the noisy operation at higher accelerations.

All the results discussed above correspond to the case when tray is perfectly horizontal with respect to the base, i.e., $\theta = 0°$ and when the impact is considered to be perfectly plastic. Figure 7 shows the variation of mean conveying velocity with the tray inclination angle for different acceleration parameters (note that negative value of θ indicates declination). As expected, the conveying velocity decreases monotonically as the inclination angle increases. Thus, it is beneficial to mount a tray with negative inclination angles; however, the declination angle should not be so

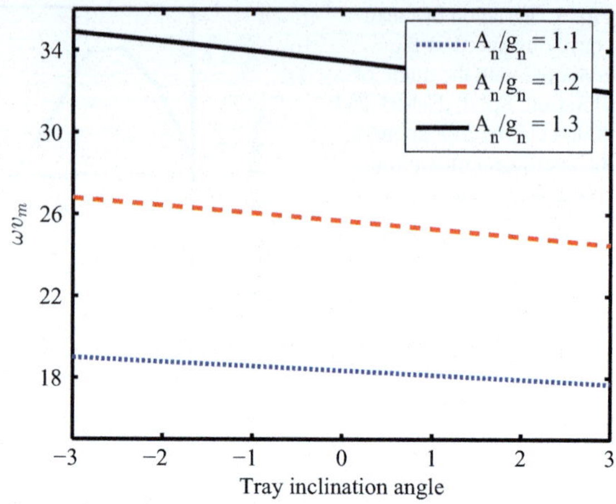

Fig. 7 Variation of mean conveying velocity with the angle of inclination of tray. Dotted, dashed, and solid lines correspond to $A_n/g_n = 1.1$, 1.2, and 1.3 respectively

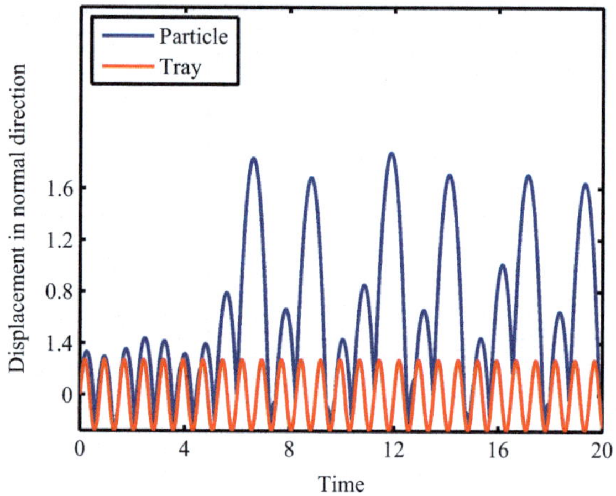

Fig. 8 Displacement of a particle in normal direction when it is initially at rest on the tray. Blue and red curves represent the displacement of a particle and that of a tray respectively

high that a particle starts to roll down, in which case, the control over the operation is lost.

As seen already, the conveying speed of a tray increases with the acceleration parameter. However, for sufficiently high acceleration parameter, a particle starts bouncing on the tray and the collision can no longer be considered to be perfectly plastic. To account for energy change during the impact, coefficient of restitution is introduced in the model. When a particle is at rest on the tray which suddenly starts oscillating with higher accelerations, the motion of a particle becomes chaotic or unstable. As an example, Fig. 8 shows the displacement of a particle in the normal direction when the acceleration parameter $A_n/g_n = 1.92$ and the coefficient of restitution $\epsilon = 0.6$. The tray is assumed to be horizontal relative to the base and the vibration angle is considered to be 20°. Such a chaotic motion limits the maximum acceleration at which a tray can oscillate in practical applications. However, if a particle is dropped on the tray from a certain height instead of being at rest initially, there exists a possibility of obtaining stable motion of a particle at higher conveying speed.

5 Stability of Motion of a Particle at High Acceleration Parameter

Holmes [6] was the first to systematically investigate the dynamics of seemingly simple nonlinear system of a bouncing ball on the vertically oscillating plate. Luo and Han [7] revisited the dynamics of this system using differential equations of motion. Hill et. al. [8] presented an elementary analysis of an elastic bouncing ball on vertically vibrating plate and predicted the oscillonic behavior for a ball initially at rest. Barroso et. al. [9] reviewed all the ramifications of this fundamentally simple problem of bouncing ball on the plate vibrating sinusoidally in the vertical direction. On the basis of discrete map of difference equations, the stability of periodic motion is also examined.

Following the work of Barroso et. al. [9], the set of conditions to obtain the periodic motion of subharmonic index n, for a particle dropped from initial height h_0 on the tray vibrating at an angle α to its surface with frequency ω, are written as

$$\omega t_i = \omega t_0 + n\pi \tag{15}$$

$$\frac{h_0 \omega^2 \sin\alpha}{g_n} = \frac{(n\pi)^2}{2} + \frac{A_n}{g_n}\sin(\omega t_i) \tag{16}$$

$$\frac{A_n}{g_n}\cos(\omega t_i) = n\pi\frac{1-\epsilon}{1+\epsilon} \tag{17}$$

where ωt_0 and ωt_i denote the initial phase of a periodic motion of a tray in the normal direction and that at the instant of impact, respectively. (Note that A_n/g_n is the dimensionless acceleration parameter mentioned earlier.) As long as $0 < \frac{A_n}{g_n}\sin(\omega t_i) < 1$, it is possible to obtain periodic motion for a particular value of coefficient of restitution, provided that the initial height from which the particle is dropped and the initial phase of oscillation of tray satisfy the above equations.

For example, if a tray vibrates at an angle $\alpha = 20°$ with coefficient of restitution $\epsilon = 0.4$ and $A_n/g_n = 1.5$, for period-1 mode of the motion, the angular phase of oscillation of the tray at the instant of impact can be obtained from Eq. 17. Initial height from which a particle should be dropped is obtained by substituting this angular phase in Eq. 16 and initial angular phase of the tray when a particle is dropped is found using Eq. 15. Figure 9a shows such a period-1 motion of a particle bouncing on the vibrating tray when these initial height h_0 and angular phase ωt_0 are used in the simulation. Figure 9b shows the periodic motion of a particle with subharmonic index

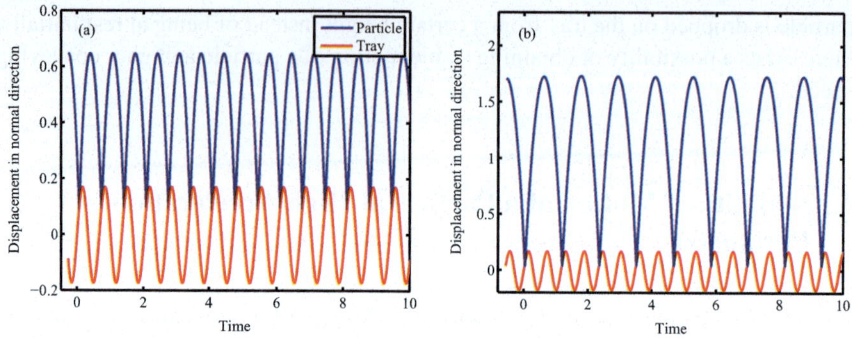

Fig. 9 Displacement of a particle in normal direction when it is dropped from h_0. Blue and red curves represent the displacement of a particle and that of a tray respectively. **a** $A_n/g_n = 1.5$, $h_0 = 1.9$, and $\epsilon = 0.4$ **b** $A_n/g_n = 2$, $h_0 = 5$, and $\epsilon = 0.52$

$n = 2$. This periodic motion is obtained when $A_n/g_n = 2$, $\epsilon = 0.52$, and $h_0 = 5$. Note that, in this case, a tray completes two cycles of oscillatory motion between successive impacts with a particle.

Thus, by using above set of conditions, it is possible to obtain maximum throughput of a conveying tray, without unstable or chaotic motion at higher acceleration parameters.

Summary

An algorithm is developed to simulate the motion of a particle on an inclined tray vibrating at an angle to its surface. Different phases of motion such as in-contact phase (stationary, forward or backward sliding), projectile phase, and collision phase are discussed. The effect of different parameters (such as frequency of oscillation, acceleration parameter, vibration angle, coefficient of friction, and inclination angle) on the mean conveying speed of the tray is examined. As the acceleration parameter is increased, the mean conveying speed of the tray increases; however, for sufficiently high acceleration, the motion of a particle becomes chaotic. Specific conditions to obtain periodic stable motion of a particle in such cases are discussed. The developed algorithm can be used to obtain optimum design parameters to improve the overall throughput of the conveyor.

Appendix

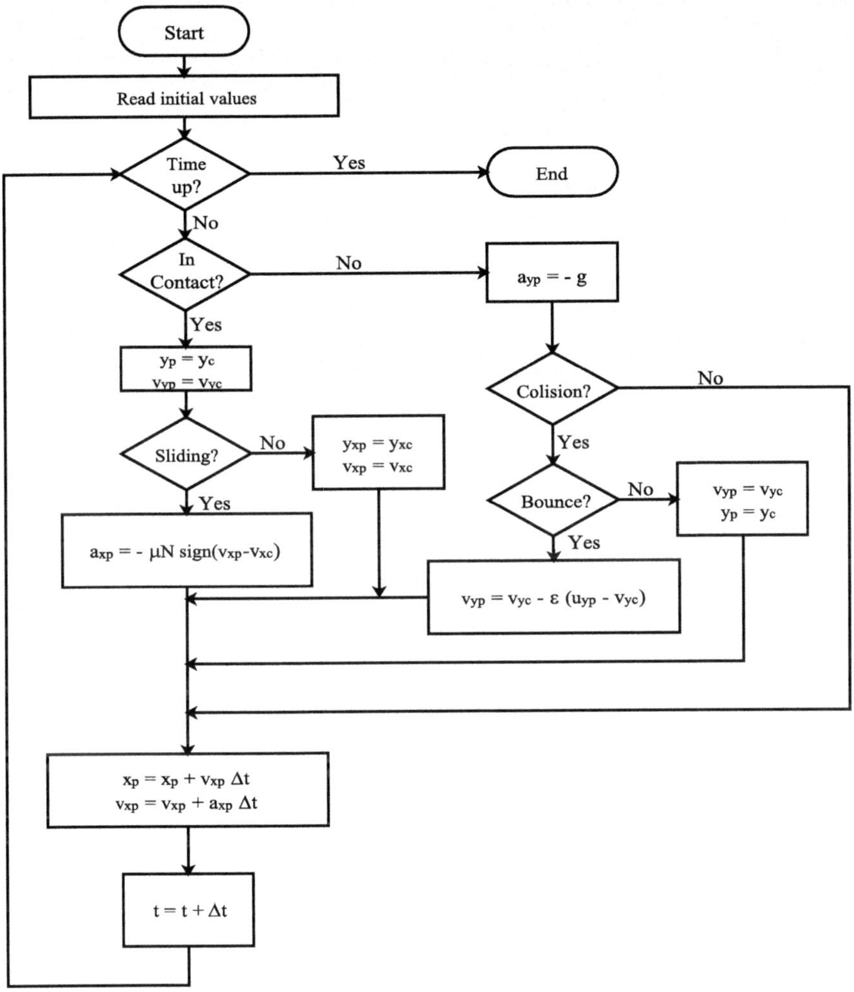

References

1. Redford A, Boothroyd G (1968) Vibratory feeding. Proc Inst Mech Eng 182(6):135–146
2. Parameswaran MA, Ganapathy S (1979) Vibratory conveying-analysis and design: a review. Mech Mach Theory 14:89–97
3. Winkler G (1978) Analysing the vibrating conveyor. Int J Mech Sci 20:561–570
4. Winkler G (1979) Analysing the hopping conveyor. Int J Mech Sci 21:651–658
5. Sloot EM, Kruyt NP (1996) Theoretical and experimental study of the transport of granular materials by inclined vibratory conveyors. Powder Technol 87:203–210

6. Holmes PJ (1982) The dynamics of repeated impacts with a sinusoidally vibrating table. J Sound Vib 84(2):173–189
7. Luo ACJ, Han RPS (1996) The dynamics of a bouncing ball with a sinusoidally vibrating table revisited. Nonlinear Dyn 10:1–18
8. Hill JM, Jennings MJ, To DV, Williams KA (2000) Dynamics of an elastic ball bouncing on an oscillating plane and the oscillon. Appl Math Modell 24:715–732
9. Barroso JJ, Carneiro MV, Macau EEN (2009) Bouncing ball problem: stability of the periodic modes. Phys Rev E 79(1–10):026206

Comparative Analysis of the Speed Coupling and Torque Coupling Hybrid Modes of a Parallel Hybrid Electric Mini Truck

Parth Joshi and V. Kartik

Abstract Two variants of the drivetrain of a post-transmission parallel hybrid electric mini truck namely SCHD and TCHD, operated in the speed coupling (SC) and torque coupling (TC) hybrid modes respectively are analyzed. The equivalent fuel consumption of the conventional and hybrid drivetrains was estimated for three urban drive cycles. Substantial improvement was achieved in the equivalent fuel consumption in comparison with the conventional vehicle for both the variants. TC provided better reduction (up to 55.9%) in the equivalent fuel consumption than SC (up to 34.2%). A metric ΔSFC is introduced to determine the mean deviation of the engine operating point relative to its optimum over a drive cycle. ΔSFC was evaluated over three drive cycles for the conventional and hybrid drivetrains and was least in case of TCHD, indicating that in TC, the IC engine is operated more efficiently than SC.

Keywords Commercial vehicles · Parallel hybrid · Post-transmission · Planetary gear · Equivalent fuel consumption

1 Introduction

In India, the annual diesel consumption of transport vehicles like buses and trucks is about 38% of the total [1]. Many of these vehicles operate in congested urban traffic conditions that are typically observed in India and several other developing countries. These vehicles usually run at low speeds and have to stop frequently over their route. This leads to poor fuel economy and high emissions. Therefore, a significant reduction in diesel consumption and tailpipe emissions could be achieved through hybridization and electrification of their drivetrains [2]. Mini trucks (small commercial vehicles such as Tata Ace, Mahindra Jeeto, etc.) are generally used for

P. Joshi (✉) · V. Kartik (✉)
Indian Institute of Technology Bombay, Mumbai, India
e-mail: parthsjoshi@iitb.ac.in; parthsjoshi@gmail.com

V. Kartik
e-mail: vkartik@iitb.ac.in

D. Sen et al. (eds.), *Mechanism and Machine Science*, Lecture Notes in Mechanical Engineering, https://doi.org/10.1007/978-981-15-4477-4_42

urban delivery in India. Reports show annual sales of more than 300,000 mini trucks which is about 45% of commerical vehicle units sold in India [3]. Gupta et al. [4] reported substantial benefits in terms of energy consumption and acceleration performance achieved by converting a light duty truck into a parallel hybrid vehicle with a post-transmission (P3) drivetrain architecture. They retrofitted an electric motor, a lead acid battery pack, and a planetary gear-based transmission that combined the mechanical output of the IC engine and the electric motor. In our previous work [5], the transmission gear ratios and gear selection logic and two parameters of the supervisory control logic for an identical P3 parallel hybrid mini truck were optimized. The optimal solution minimized the equivalent fuel consumption per kilometer over three drive cycles that represent congested urban driving conditions. The hybrid drivetrain is capable of operating in several modes such as pure engine, pure electric, regenerative braking, hybrid, and power-split. In the hybrid mode, the drivetrain can operate in two ways, namely, speed coupling (SC) and torque coupling (TC) [6]. In the present work, the same hybrid mini truck is studied. The aim of this work is to analyze which one of the two hybrid modes, SC and TC, provides a greater reduction in the energy consumption and to examine which one of them is more effective in maintaining IC engine operating point close to its optimum, irrespective of fluctuations in power demand, for low-speed urban drive cycles involving high acceleration and deceleration.

2 Modeling

The drivetrain architecture and the specifications of the conventional and hybrid vehicles examined in the present work are identical to those in the previous work. Figure 1 shows a block diagram of the drivetrain of the hybrid vehicle [5]. The hybrid vehicle is powered by a diesel engine and a three-phase induction motor with rated powers of 12 kW and 8 kW, respectively. A 72 V lead acid battery pack with a capacity of 215 Ah is used to drive the induction motor. A planetary gear-based mechanical power coupling device combines the power output of the IC engine and electric motor.

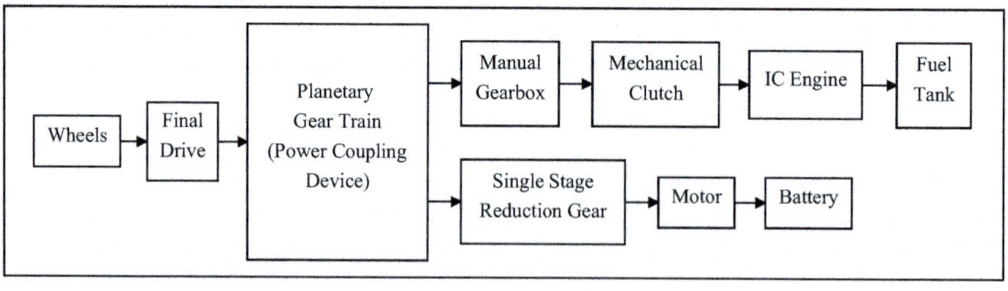

Fig. 1 Block diagram of the hybrid drivetrain [5]

A backward modeling approach [7] is applied for the drivetrains of the conventional and hybrid vehicles. The conventional and hybrid drivetrain models are used to estimate the energy consumption over a given drive cycle. The required output torque and speed of the IC engine and the electric motor are computed assuming that the drive cycle is exactly followed. The instantaneous tractive force required at the tire–road interface to drive the vehicle at required speed is calculated using the equations of longitudinal vehicle dynamics. The maximum tractive power that the drivetrain can supply is limited by the maximum allowable adhesive force at the contact between the tire and road. A peak value of tractive force coefficient of 0.85 at 16% tire slip is assumed for calculating the maximum allowable tractive force [6]. If the power required at any instant is greater than the maximum power that the drivetrain can deliver at the required speed, then the output torque and speed of the IC engine and electric motor are calculated using the maximum power that the drivetrain can provide at the required speed.

The fuel consumption rate of the IC engine is estimated by linearly interpolating between the data points of a map of brake-specific fuel consumption (BSFC) as a function of brake torque and speed. The BSFC data was obtained experimentally. Figure 2 shows the lines of optimum operation for the IC engine [5] in the hybrid modes SC and TC, described in Sect. 2.2. These lines represent the best IC engine operating points in the hybrid modes. The optimum operation lines are loci of the IC engine's operating points for which the BSFC is minimum for a given torque or a given speed. The efficiency of the electric motor is estimated by linearly interpolating

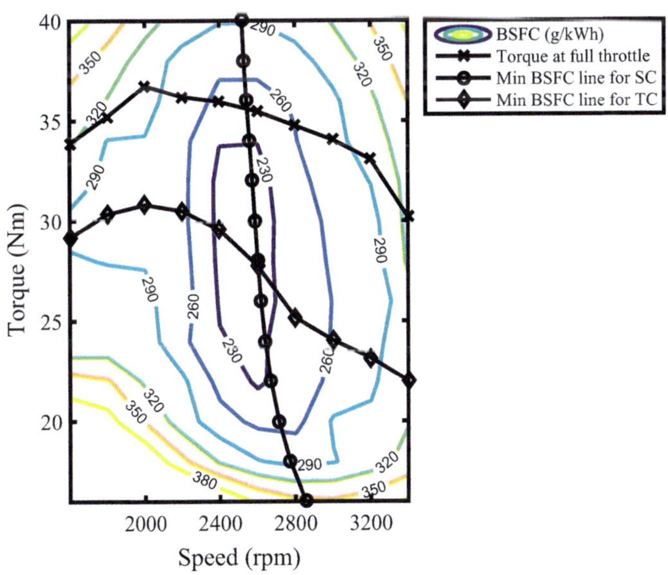

Fig. 2 Contours of constant BSFC and the lines of optimum IC engine operation in the hybrid modes SC and TC [5]

between the data points of an experimentally obtained map of efficiency as a function of the torque and speed. In these models, thermal effects and the inertia of rotating components are neglected. An equivalent circuit-based lead acid battery model is used [8–10]. The gear selection logic for the five-speed gearbox is based on maps of upshift and downshift speeds for each gear [11, 12].

2.1 Power Coupling Device

Figure 3 shows the architecture of the power coupling device [4, 5] used in the hybrid drivetrain. It is equipped with electromagnetically actuated brake (EMB) and clutch (EMC). The different modes of hybrid drivetrain operation are realized by appropriately engaging/disengaging the EMB or EMC. When EMB is engaged, the ring gear gets clamped to the housing. When EMC is engaged, the planet carrier and sun gear get locked together. Therefore, the relative motion between the sun gear, planet gears, and the ring gear gets constrained. At steady state, the equation for the energy balance for the planetary gear train is given by

$$T_1\omega_1 + T_2\omega_2 = T_3\omega_3 \tag{1}$$

where T_1, T_2, and T_3 are torques acting along the axes of the ring gear, sun gear, and the planet carrier, respectively, while ω_1, ω_2, and ω_3 are the angular speeds of the ring gear, sun gear, and the planet carrier respectively. The gears in the planetary gear set of the power coupling device accelerate and decelerate while rotating as the

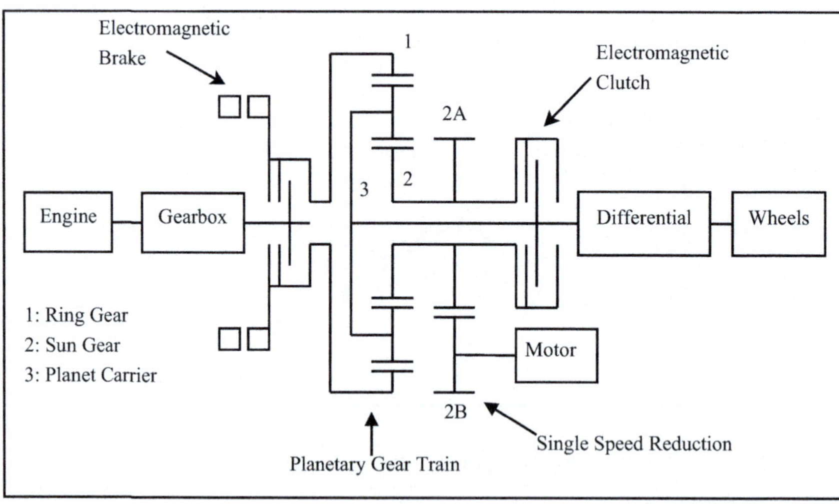

Fig. 3 Architecture of the power coupling device [4, 5]

speed of the vehicle changes with time. Power is consumed in overcoming the inertia of these rotating components. However, in the present work, the power required to overcome the inertia of these rotating components is assumed to be negligible, since the vehicle does not undergo very high acceleration or deceleration. The losses due to friction are also assumed to be negligible. Therefore, Eq. (1) for the addition of the power output of the IC engine and electric motor is assumed to hold good.

2.2 *Modes of Operation*

The five modes of operation of the hybrid drivetrain were described earlier. The pure electric mode is used for propulsion from idle up to the speed of 30 km/h. The hybrid mode is used when the electric motor or the IC engine alone is insufficient to meet the power demanded. When the power requirement is low, the power-split mode is applied. In this mode, the electric motor acts as a generator and the IC engine generates excess power for charging the battery in such a way that its operating point is close to optimum. A rule-based supervisory control logic [5] determines the appropriate mode of operation of the hybrid drivetrain in order to meet the instantaneous power demand as efficiently as possible. The control logic splits the power demand between the two power sources such that the IC engine provides the steady component of the power demand while the electric motor provides the fluctuating component of the power demand. It attempts to place the IC engine operating point close to its optimum [6].

2.2.1 Speed Coupling Hybrid Mode

During the operation of the hybrid drivetrain in SC mode, both EMB and EMC are disengaged, as shown in Fig. 4. In this mode, the planet carrier speed is a combination of the speeds of the sun gear and ring gear, while the torques are such that Eq. (1) is satisfied. The speed ω_3, and the torque T_3, at the output of the power coupling device are given by

$$\begin{aligned} \omega_3 &= \frac{\omega_1 R_1}{(1+R_1)} + \frac{\omega_2}{(1+R_1)} \\ T_3 &= \frac{T_1(1+R_1)}{R_1} = T_2(1+R_1) \\ \omega_1 &= \frac{\omega_e}{i_g},\ \omega_2 = \frac{\omega_m}{R_2} \\ T_1 &= T_e i_g,\ T_2 = T_m R_2 \\ R_1 &= \frac{Z_1}{Z_2} \\ R_2 &= \frac{Z_{2A}}{Z_{2B}} \end{aligned} \tag{2}$$

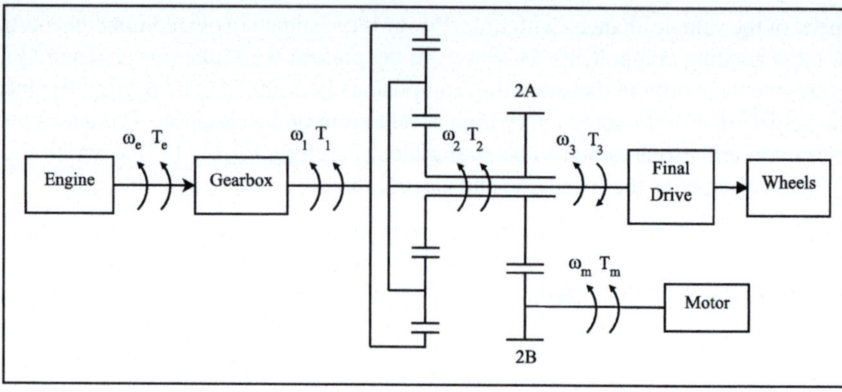

Fig. 4 Schematic diagram of hybrid drivetrain operation in the SC hybrid mode

where, ω_1 and ω_2 are the speeds and T_1 and T_2 are the torques transmitted by the ring gear and sun gear, respectively, T_e and T_m are the output torques and, ω_e and ω_m are the speeds of the IC engine and the electric motor, respectively, and i_g is the gear ratio of the manual gearbox, Z_1, Z_2, Z_{2A}, and Z_{2B} are the number of teeth of the ring gear, sun gear, gears 2 A and 2B, respectively. The vehicle speed v, and the tractive torque T_t, at the wheels for the speed coupling hybrid mode are given by

$$\begin{aligned} v &= \frac{r\omega_3}{i_0} = \frac{r\omega_e R_1}{(1+R_1)\, i_0 i_g} + \frac{r\omega_m}{(1+R_1)\, i_0 R_2} \\ T_t = i_0 T_3 &= \frac{T_e\,(1+R_1)\, i_0 i_g}{R_1} = T_m\,(1+R_1)\, i_0 R_2 \end{aligned} \tag{3}$$

where r is the tire radius and i_0 is the final drive gear ratio. These equations are used in the algorithm for SC for determining the operating point of the IC engine. The algorithm identifies the feasible range of speeds in which the IC engine and electric motor can be operated in order to meet the power demand. The IC engine operating point for which the BSFC is minimum over the feasible range of speeds is selected and the corresponding operating point of the electric motor is determined.

2.2.2 Torque Coupling Hybrid Mode

The schematic diagram of the hybrid drivetrain operated in the TC hybrid mode is shown in Fig. 5. In this mode, the EMC is engaged because of which the entire planetary gear set rotates as a single body and the engine directly drives the shaft labeled as 2. The electric motor power is transmitted though the single speed reduction (2A, 2B). In the TC hybrid mode, the power coupling device combines the torque of the two power sources with a constraint on the speeds in order to satisfy Eq. (1). The speed ω_3, and the torque T_3, at the output of the power coupling device are given by

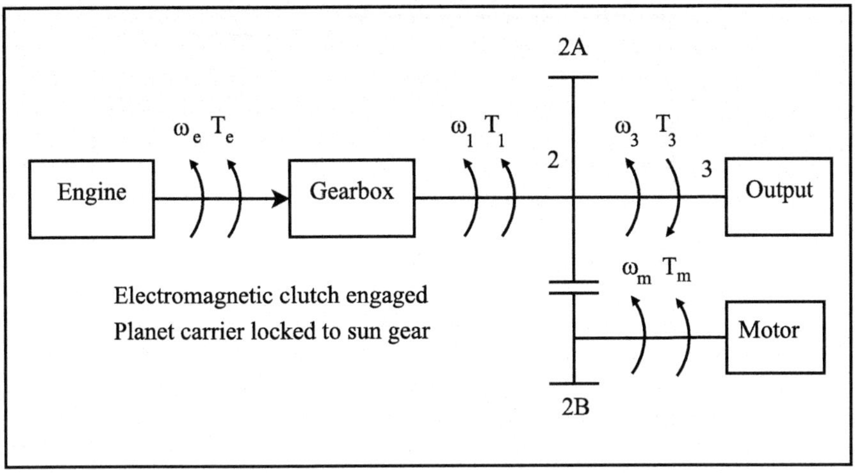

Fig. 5 Schematic diagram of the hybrid drivetrain operated in the TC hybrid mode

$$\begin{gathered} T_3 = T_1 + T_2 \\ \omega_3 = \omega_1 = \omega_2 \\ \omega_1 = \frac{\omega_e}{i_g}, \omega_2 = \frac{\omega_m}{R_2} \\ T_1 = T_e i_g, T_2 = T_m R_2. \end{gathered} \tag{4}$$

The vehicle speed v, and the torque T_t at wheels, for the TC hybrid mode are given by

$$\begin{gathered} v = \frac{r\omega_3}{i_0} = \frac{r\omega_e}{i_0 i_g} = \frac{r\omega_m}{i_0 R_2} \\ T_t = i_0 T_3 = i_0 \left(i_g T_e + R_2 T_m \right). \end{gathered} \tag{5}$$

These equations are used in the algorithm for TC for determining the IC engine's operating point. The algorithm identifies the feasible range of torques that the IC engine and the electric motor can supply in order to meet the power demand. The operating point of the IC engine for which the BSFC is minimum over the feasible range of torques is selected and the corresponding operating point of the electric motor is determined.

2.2.3 Algorithm for Operating Point Selection in the Hybrid Modes

Figure 6 shows the maximum torque available in each of the five gears of the manual transmission for the SC and TC hybrid modes. In the SC hybrid mode, for a torque T_t

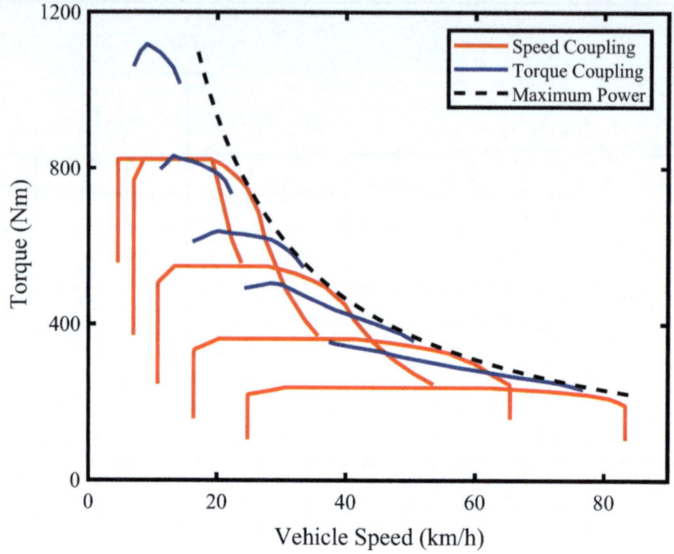

Fig. 6 Maximum tractive torque available in each of the five gears of the manual transmission for the SC and TC hybrid modes

required at wheels, the IC engine's output torque $T_{e,k}$ corresponding to the kth gear of the manual transmission and the electric motor's output torque T_m are obtained using Eq. (3) as follows:

$$\begin{aligned} T_{e,k} &= \frac{T_t R_1}{(1+R_1)\, i_0 i_{g,k}} \\ T_m &= \frac{T_t}{(1+R_1)\, i_0 R_2} \end{aligned} \tag{6}$$

where $i_{g,k}$ is the kth gear ratio of the manual transmission and $k = 1, 2, 3, 4, 5$. The ranges of the IC engine's speed $\omega_{e,k}$ corresponding to the kth gear of the manual transmission, the electric motor's speed ω_m, and the vehicle speed over which the drivetrain can provide the required tractive torque T_t in the SC hybrid mode are given by

$$\begin{aligned} \omega_{e,k}^{(\min)} &\le \omega_{e,k} \le \omega_{e,k}^{(\max)} \\ 0 &\le \omega_m \le \omega_m^{(\max)} \\ \frac{r\omega_{e,k}^{(\min)} R_1}{(1+R_1)\, i_0 i_{g,k}} \le v &\le \frac{r\omega_{e,k}^{(\max)} R_1}{(1+R_1)\, i_0 i_{g,k}} + \frac{r\omega_m^{(\max)}}{(1+R_1)\, i_0 R_2} \end{aligned} \tag{7}$$

where $\omega_{e,k}^{(\min)}$ and $\omega_{e,k}^{(\max)}$ are the minimum and maximum speeds of the IC engine for the kth gear of the manual transmission and $\omega_m^{(\max)}$ is the maximum speed of the electric motor. The maximum and minimum vehicle speeds for which the drivetrain can provide the required tractive torque are thus obtained for all the five gears of the manual transmission.

In the torque coupling hybrid mode, for a given vehicle speed v, the IC engine's speed $\omega_{e,k}$ corresponding to the kth gear of the manual transmission and the electric motor's speed ω_m are obtained using Eq. (5) as follows:

$$\begin{aligned} \omega_{e,k} &= \frac{v i_0 i_{g,k}}{r} \\ \omega_m &= \frac{v i_0 R_2}{r} \end{aligned} \tag{8}$$

where $i_{g,k}$ is the kth gear ratio of the manual transmission and $k = 1, 2, 3, 4, 5$. The maximum tractive torque $T_t^{(\max)}$ that the drivetrain can provide in the torque coupling hybrid mode in the kth gear of the manual transmission is given by

$$T_t^{(\max)} = i_0 \left(i_{g,k} T_{e,k}^{(\max)} + R_2 T_m^{(\max)} \right) \tag{9}$$

where $T_{e,k}^{(\max)}$ is the maximum IC engine torque at speed $\omega_{e,k}$ and $T_m^{(\max)}$ is the maximum motor torque at speed ω_m. The maximum available tractive torque at wheels at any vehicle speed is thus obtained for all the five gears of the manual transmission.

It is seen from Fig. 6 that the maximum speeds that the vehicle can achieve in the SC and TC hybrid modes are 83 km/h and 77 km/h, respectively. The hyperbolic curve represents the tractive torque available at a given vehicle speed if the hybrid drivetrain provides the combined maximum power of 19.3 kW that the IC engine and electric motor together can provide. However, it is seen that both the hybrid modes are incapable of delivering this maximum power for the first gear of the transmission. A major objective of a multi-gear manual transmission is to closely approximate the characteristics of an ideal power source so as to have the maximum power available over a large range of vehicle speeds. In this case, the hybrid modes together with the five-speed manual transmission enable the hybrid drivetrain to provide power close to the combined maximum available power over a wide range of vehicle speeds between 25 and 75 km/h.

3 Simulation Results

In order to investigate which one of the hybrid modes, SC and TC, provides more efficient IC engine operation, two variants of the hybrid drivetrain, denoted by SCHD and TCHD, that are operated in the hybrid modes SC and TC respectively, are analyzed. The performance of the conventional and hybrid drivetrains is simulated for the following drive cycles: Modified Indian Driving Cycle (MIDC) [13], New York

Table 1 Improvement obtained in the fuel consumption with the hybrid drivetrain over the conventional drivetrain

Drive cycle	SCHD (%)		TCHD (%)	
	FC_{ICE}	EFC	FC_{ICE}	EFC
MIDC	21.92	17.04	43.31	42.63
NYCC	75.40	34.16	77.55	57.45
MHBC	80.33	34.15	83.17	55.92

City Cycle (NYCC), and Manhattan Bus Cycle (MHBC) [14]. These drive cycles represent some of the characteristics of congested urban traffic. MIDC has lower frequency of stops, lower accelerations, and higher average speed than MHBC and NYCC. The methodology of simulation is identical to that of our previous work [5]. The equivalent fuel consumption (EFC) for the hybrid drivetrain is calculated as follows:

$$EFC = FC_{ICE} + FC_{ELEC} \tag{10}$$

where FC_{ICE} is the fuel consumed by the IC engine alone over the drive cycle, and FC_{ELEC} is the fuel consumed by the IC engine, run at its optimum operating point, for charging the battery through the motor/generator for restoring the battery SOC from the final value at the end of the drive cycle back to its initial SOC. Table 1 shows the % improvement obtained in EFC with the hybrid drivetrains SCHD and TCHD relative to the baseline (conventional drivetrain) for the drive cycles MIDC, NYCC, and MHBC. Substantial improvement in EFC is obtained for all the drive cycles, particularly in the case of NYCC and MHBC. TCHD exhibits significantly better improvement over the baseline than SCHD. The improvement in FC_{ICE} represents the benefit that could be obtained if electrical energy from the grid is used to charge the battery for restoring its SOC to the initial value.

Figures 7, 8, and 9 show the IC engine operating points for the hybrid drivetrains SCHD and TCHD for the drive cycles MHBC, NYCC, and MIDC, respectively. A significant number of the operating points lie within the region of optimum IC engine operation for both the hybrid drivetrains in the case of MIDC. It is seen that most of the operating points in the case of TCHD lie along the minimum BSFC line for TC for all the three drive cycles. On the other hand, very few operating points lie along the minimum BSFC line for SC and most of them lie outside the optimum BSFC region, in the case of SCHD for all the drive cycles. A large number of operating points, at which the pure engine mode is used, lie away from the optimum BSFC region in the case of MHBC and NYCC. In the case of SCHD for MIDC and NYCC, a significant number of operating points with low torque and high speeds close to the upper limit of engine speed are observed, where BSFC is quite high. From these qualitative results, it can be inferred that TC performs better than SC for all the three drive cycles.

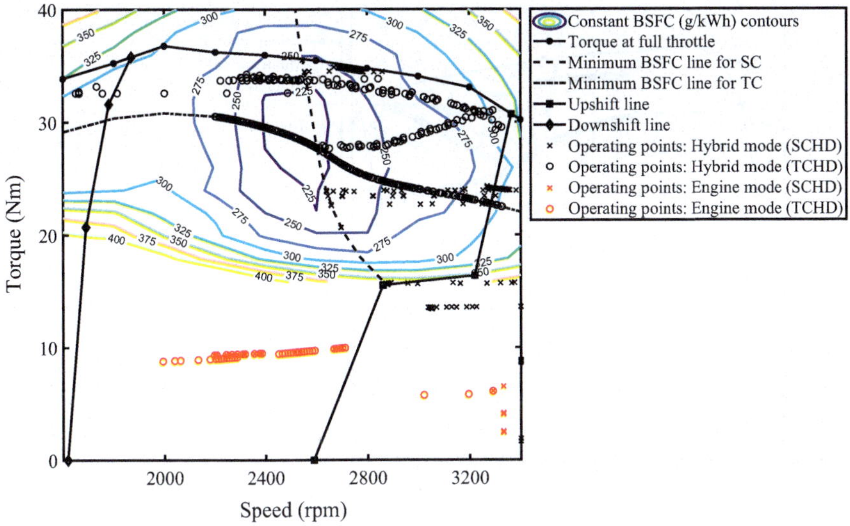

Fig. 7 Simulated engine operating points for the hybrid drivetrains SCHD and TCHD over Manhattan Bus Cycle

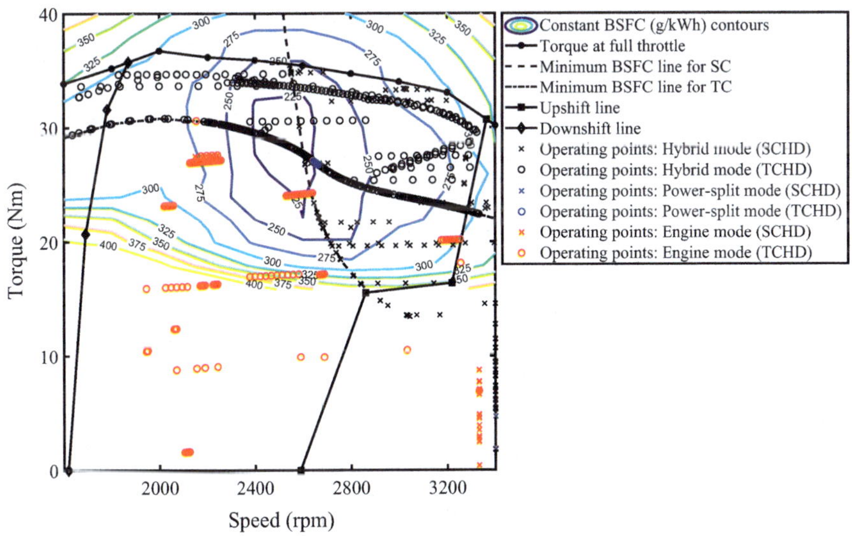

Fig. 8 Simulated engine operating points for the hybrid drivetrains SCHD and TCHD over New York City Cycle

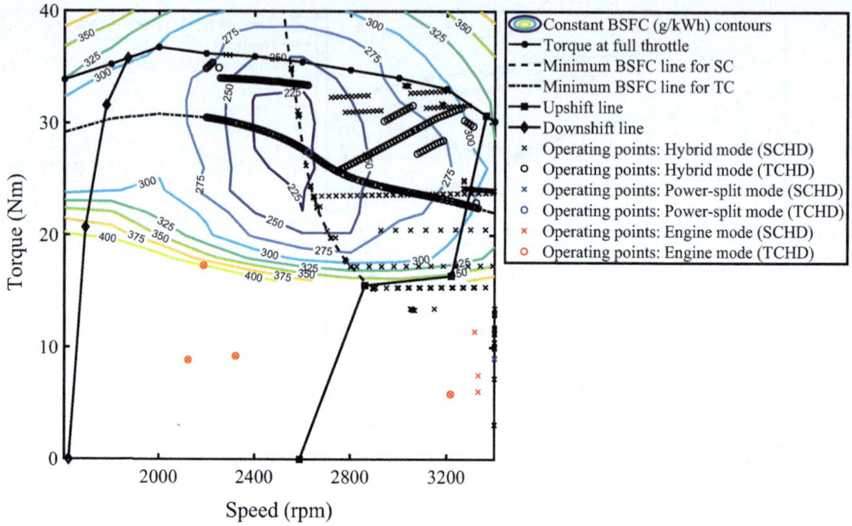

Fig. 9 Simulated engine operating points for the hybrid drivetrains SCHD and TCHD over Modified Indian Driving Cycle

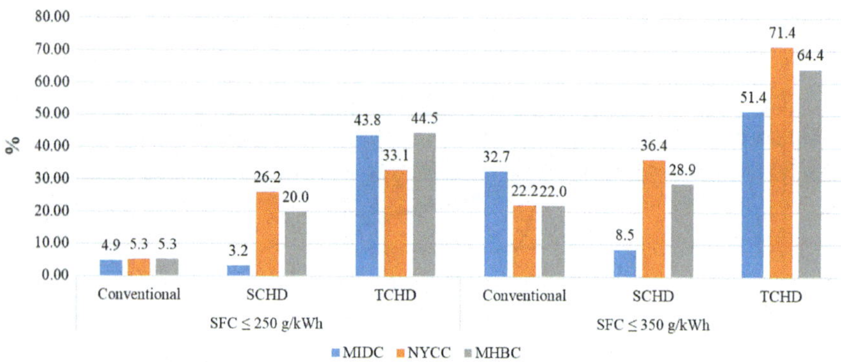

Fig. 10 Fraction of IC engine operating points that lie within the contours: BSFC = 250 and BSFC = 350 g/kWh for the conventional and hybrid drivetrains

In order to quantitatively characterize the deviation of IC engine operating points over a drive cycle, the fractions of operating points that lie within the optimum BSFC region are obtained for each drive cycle. Figure 10 shows the comparison of the fraction of IC engine operating points that lie inside the regions (a) $BSFC \leq 250$ g/kWh and (b) $BSFC \leq 350$ g/kWh for the conventional and hybrid drivetrains. The values of these fractions for TCHD are substantially higher than that of the conventional drivetrain and the hybrid drivetrain SCHD for all the three drive cycles.

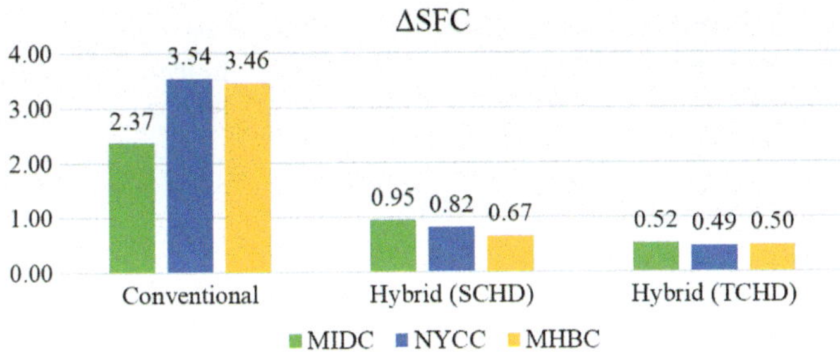

Fig. 11 Metric ΔSFC for the conventional and hybrid drivetrains SCHD and TCHD

The fraction of operating points inside the regions (a) and (b) provides a measure of the deviation of the operating points over a drive cycle. However, this approach has certain limitations. It cannot provide the cumulative deviation as well as the distribution of the operating points relative to the optimum. Therefore, it cannot be correlated to the fuel consumption. In order to quantify the mean deviation of the operating points of the IC engine over a drive cycle relative to its optimum operating point, a metric denoted as ΔSFC is introduced and is defined as follows:

$$\Delta SFC = \frac{1}{N}\frac{\sum_{j=1}^{N}\left|BSFC\left(t_j\right) - BSFC_{\min}\right|}{\left(BSFC_{\max} - BSFC_{\min}\right)} \quad (11)$$

where t_j denotes the time at the jth step in the simulation, $BSFC_{\max}$ and $BSFC_{\min}$ are the maximum and minimum values of the BSFC for the IC engine map. The advantage of this metric is that it is a dimensionless quantity and is independent of the particular engine BSFC map. Therefore, it can also be used to compare hybrid drivetrains with dissimilar engines. Figure 11 shows the comparison of ΔSFC values for the conventional and hybrid drivetrains for the drive cycles MHBC, NYCC, and MIDC.

The relative contribution of each of the five modes of operation of the hybrid drivetrain is examined. Table 2 shows the fraction of drive cycle time duration for which each mode of operation of the hybrid drivetrain is utilized for MHBC, NYCC, and MIDC. It is seen that the pure electric mode is used over a significant portion of all the three drive cycles, particularly in the case of NYCC and MHBC. The power-split and pure engine modes are used over a substantial portion of MIDC, while these modes are used only over a small portion of NYCC and MHBC. Therefore, the extensive use of the pure electric mode is one of the causes of significant reduction in the overall energy consumption of the hybrid drivetrain as compared to that of the conventional vehicle.

Table 2 Utilization of the modes of operation of the hybrid drivetrain

Drive cycle	Pure electric (%)	Hybrid (%)	Power-Split (%)	Regenerative braking (%)	Pure engine (%)
MIDC	34.77	9.200	21.07	6.360	28.62
NYCC	58.83	11.34	1.440	21.20	7.210
MHBC	68.57	10.60	0	15.98	4.870

4 Conclusion

The present work deals with a comparative analysis of two hybrid modes, viz., SC and TC. Two alternative models of the hybrid drivetrain, SCHD and TCHD in which the SC and TC hybrid modes respectively are enabled, have been simulated over three urban drive cycles. Substantial reduction in the equivalent fuel consumption relative to the baseline conventional vehicle was obtained for both the models. Among the two hybrid modes, TC provided a better reduction in the energy consumption. A metric ΔSFC is introduced in order to compute the mean deviation of the IC engine's operating points in comparison with its optimum over a drive cycle. ΔSFC was evaluated over three drive cycles for the conventional and hybrid drivetrains SCHD and TCHD. The value of ΔSFC was found to be the least in case of TCHD, indicating that TC provides a more efficient operation of the IC engine than SC. In the current scenario of traffic congestion in urban India, this methodology could prove to be a useful analytical tool for evaluating the benefits of hybridization and electrification of commerical vehicles.

References

1. All India Study on Sectoral Demand of Diesel & Petrol (2013) Technical report, Petroleum Planning and Analysis Cell, Ministry of Petroleum and Natural Gas, Government of India, New Delhi
2. Karali N, Gopal AR, Sharpe B, Delgado O, Bandivadekar A, Garg M (2017) Improved heavy-duty vehicle fuel efficiency in India: Benefits, costs and environmental impacts. Technical report, April, Lawrence Berkeley National Laboratory, The International Council on Clean Transportation
3. Thakkar K (2018) Mahindra & Mahindra drives past Tata Motors in race for mini-trucks sales. 03 July 2018
4. Gupta AK, Kartik V, Ramanarayanan CP (2016) Experimental study on performance of a parallel diesel hybrid vehicle retrofitted with a single planetary gear train as an add-on transmission. In: International mobility conference. SAE International, February 2016
5. Joshi P, Kartik V (June 2018) Transmission and control strategy optimization for a parallel hybrid electric micro-utility vehicle in urban drive cycles. In: Proceedings of the 2018 IEEE transportation electrification conference and expo (ITEC). Long Beach, CA, USA
6. Ehsani M, Gao Y, Emadi A (2010) Modern electric, hybrid electric and fuel cell vehicles—fundamentals, theory and design, 2nd edn. CRC Press, Boca Raton

7. Onori S, Serrao L, Rizzoni G (2016) Hybrid electric vehicles—energy management strategies, 1st edn. Springer, London
8. Larminie J, Lowry J (2003) Electric vehicle technology explained, 1st ed. Wiley, West Sussex, UK
9. Husain I (2003) Electric and hybrid vehicles—design fundamentals, 1st edn. CRC press, Boca Raton
10. Mi C, Masrur MA, Gao DW (2011) Hybrid electric vehicles—principles and applications with practical perspectives, 1st edn. Wiley, West Sussex, UK
11. Lee H-D, Sul S-K, Cho H-S, Lee J-M (2000) Advanced gear-shifting and clutching strategy for a parallel-hybrid vehicle. IEEE Ind Appl Mag 6:26–32
12. ADVISOR Vehicle Simulator—Gearbox Control. Accessed 23 February 2018
13. Sharma S, Goel A, Suresh R, Lakshmi CS, Mahtta R, Sundar S, Bansal G, German J, Posada F, Franco V (2013) Assessment of emission test driving cycles in India: a case for improving compliance, Technical report, The Energy and Resources Institute (TERI), Reviewed by The International Council for Clean Transportation, Octtober 2013
14. Kelly K, Prohaska R, Ragatz A, Konan A (2016) NREL DriveCAT—chassis dynamometer test cycles. Accessed 12 March 2018

[illegible]

A Passive Universal Grasping Mechanism Based on an Everting Shell

Mythra V. S. Balakuntala, Safvan Palathingal, and G. K. Ananthasuresh

Abstract A passive monolithic compliant grasping mechanism that works based on the eversion of an elastically deformable bistable shell is conceptualized. It comprises grasping arms made of beam segments that work in conjunction with the everting shell. The grasper is capable of picking up a stiff object of any shape up to a maximum size and weight. The bistable shell everts upon contact with the object to enable the grasping arms envelop the object forming an enclosure. The mechanism then stays in that configuration until it is actuated again to turn the shell back to its original configuration and thereby opening the enclosure to release the object. The stiffness of the arms decides the payload of the mechanism. The size of the arms decides the largest object that can be grasped and held. The arms have distributed compliance so that they can conform to the shape of the object without applying undue force on it.

Keywords Bistable shells · Compliant gripper · End-effector

1 Introduction

Universal passive grippers ought to grasp objects of varying size and shape. In comparison to active multi-fingered grippers, they do not need sophisticated grasping algorithms and thus are easier to use in practice [1, 2]. Existing passive grippers can be broadly classified as follows: grippers with soft contacts [3, 4], granular grippers [5–7], and underactuated grippers [8, 9]. In this work, we explore an alternative way

M. V. S. Balakuntala (✉) · S. Palathingal · G. K. Ananthasuresh
Indian Institute of Science, Bangalore, India
e-mail: mythra.vbs@gmail.com

S. Palathingal
e-mail: safvanp@iisc.ac.in

G. K. Ananthasuresh
e-mail: suresh@iisc.ac.in

D. Sen et al. (eds.), *Mechanism and Machine Science*, Lecture Notes in Mechanical Engineering, https://doi.org/10.1007/978-981-15-4477-4_43

to passively grip arbitrarily-shaped objects using a monolithic compliant grasping mechanism based on bistable shells, also known as *everting shells*.

Everting shells can maintain two force-free equilibrium states as shown in Fig. 1. The bistable shell considered here is stress-free in its as-fabricated state and stressed in its everted stable state. The ability to maintain two structural orientations without consuming power makes them ideal to be used in a passive gripper. A planar passive gripper using bistable arches was considered before by Nguyen and Wang [10]. We extend this concept by using everting shell as the critical element and attaching grasping arms to it to conceive a gripper that is passive as well as capable of picking up objects of a variety of shapes.

The gripper presented here has monolithic design that combines a switching mechanism, an everting shell, and grasping arms. The switching mechanism releases grasped objects by transmitting the input force to the everting shell. The bistable shell everts upon contact with the object to enable the grasping arms envelop the object. In Sect. 2, we explain the working principle of the gripper in detail. Furthermore, the analysis of the everting shell, switching mechanism, and grasping arms is presented. We use finite element analysis (FEA) in Abaqus [11] software to obtain force-displacement characteristics for the selected shape of the shell. The switching mechanism is designed by incorporating the force needed to evert the shell and the maximum actuation force available. The grasping arms are modeled as cantilever beams to obtain an approximate measure of payload. A 3D-printed prototype and results from preliminary grasping trials are given in Sect. 3. Section 4 includes the scope for future work.

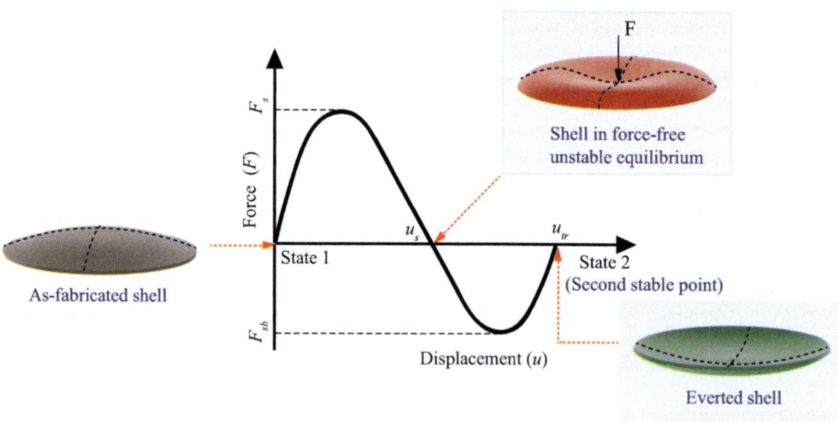

Fig. 1 A typical force-displacement curve of a bistable shell

2 Design

2.1 Working Principle

The everting compliant gripper consists of the switching mechanism, an everting shell, and grasping arms as shown in Fig. 2. The switching mechanism is the connecting element between the actuator and the everting shell. It switches the everting shell from its stress-free configuration to the second stable everted configuration of the shell. Figure 3 shows the grasping arms opening and closing in conjuction with the everting shell.

As illustrated in Fig. 4, three parts of the mechanism work in the following three steps to grasp and release objects: (1) the input force applied on the switching mechanism switches the everting shell and opens the grasping arms to grasp the object of interest (see Fig. 4a, b and c); (2) the everted shell, upon contact with the object, returns to the stress-free configuration causing the grasping arms to close around the

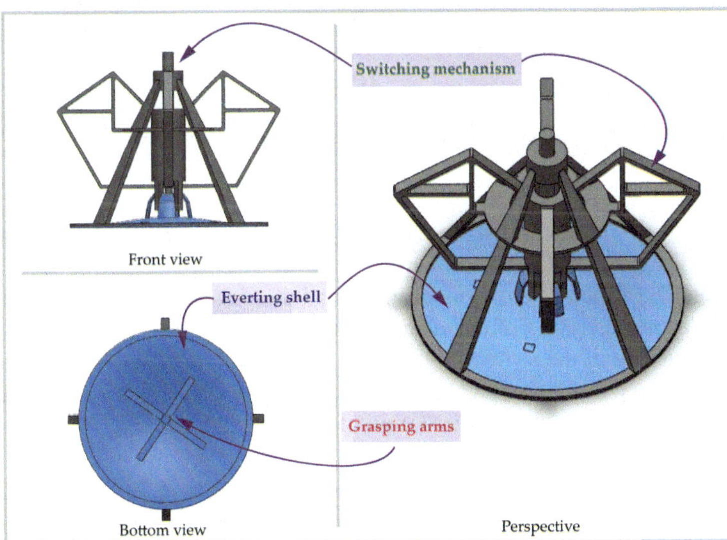

Fig. 2 The compliant monolithic grasping mechanism based on an everting shell. The mechanism consists of three parts, (1) switching mechanism, (2) everting shell, and (3) grasping arms

Fig. 3 Bistable shell with contracting mechanism and grasping arms attached

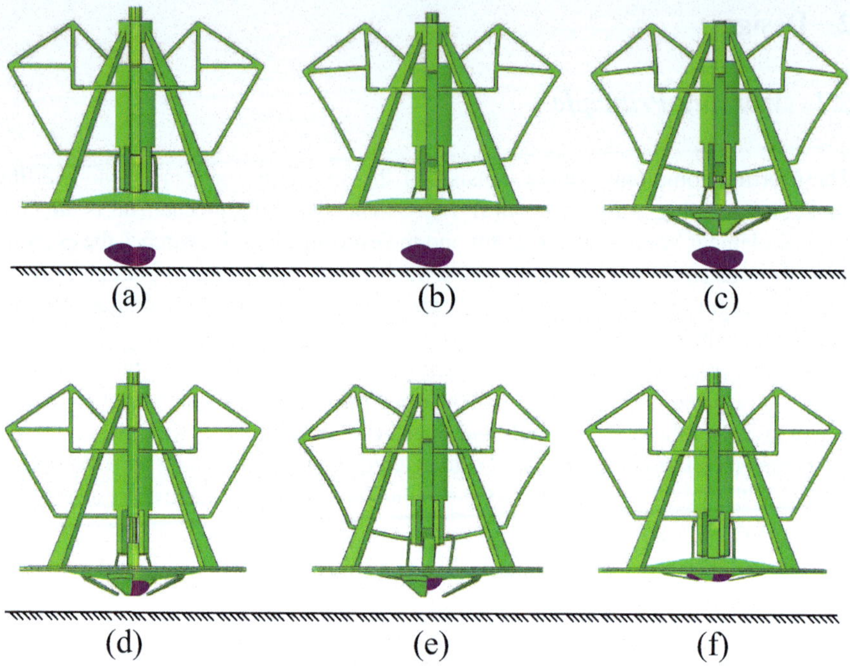

Fig. 4 Working of the grasping mechanism

object and grasp it (see Fig. 4d, e and f); (3) the switching mechanism is actuated again to release the object (not shown in Fig. 4).

The design of the switching mechanism and the grasping arms is dependent on the bistable characteristics of the everting shell. These characteristics determine the topology and size of the switching mechanism, and the dimensions of the grasping arms as explained in the next two subsections. We systematically approach the analysis of the gripper considering the design of an everting shell first, followed by the switching mechanism and the grasping arms.

2.2 Everting Shell

The two force-free stable equilibrium configurations of the everting shell correspond to the open and closed positions of the gripper. The critical static characteristics of the shell are switching (F_s) and switch-back (F_{sb}) forces, switching displacement (u_s), and the travel (u_{tr}). These characteristics are labeled in the force-displacement curve given in Fig. 1. F_s is the minimum force required by the shell to switch from its stress-free stable configuration to the stressed everted shape. F_{sb} is the maximum force that can be resisted by the shell in the everted state before switching back to

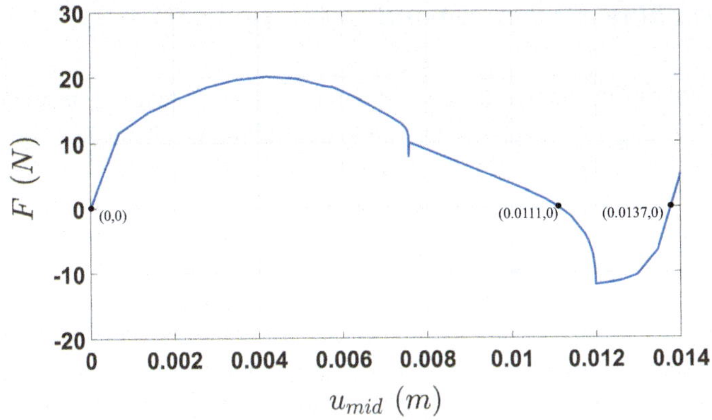

Fig. 5 Simulated force-displacement curve of the cosine-shaped bistable shell

its as-fabricated initial shape. u_s is the minimum displacement the shell needs to be actuated to initiate its eversion and u_{tr} is the total displacement of the point of interest between the two stable states. Note that the everting shell considered here is bistable solely because of its as-fabricated shape, not due to pre-stress. This is preferred as the gripper in its default closed configuration is stress-free.

In the case of planar arches, it is known that a single cosine profile does not show bistability for fixed-fixed boundary conditions [12]. Here, we show that a shell formed from a revolved cosine profile is bistable for fixed-periphery boundary conditions. The height of the shell at a distance r from the center, $h(r)$, is taken as

$$h(r) = \frac{h_{\text{mid}}}{2}\left[1 - \cos\left(2\pi - \frac{2\pi r}{R}\right)\right] \tag{1}$$

where h_{mid} is the height at the midpoint of the shell and R the radius of the platform. We analyzed a shell of uniform thickness, $t = 1$ mm, $h_{\text{mid}} = 7$ mm, and $R = 30$ mm using a quasi-static FEA in Abaqus. The dimensions were selected for realizing a prototype design that can be actuated by hand. Young's modulus and Poisson's ratio of Verowhite and TangloPlus mixture, a 3D printing material used in Object Conex 260 for prototyping, were taken as $E = 1.2$ GPa and $\nu = 0.3$, respectively. For a point load applied at the center of the shell with edge fixed, the force-displacement characteristics obtained is shown in Fig. 5. Note that u_{mid} is the deformation at the point of application of the load, F. The curve intersects the u_{mid} axis, $F = 0$, at three points corresponding to two stable and one in-between unstable states, indicating bistability. Two important design considerations for the prototype gripper were sufficiently high $\frac{F_{sb}}{F_s}$ ratio and low F_s. An ideal $\frac{F_{sb}}{F_s}$ ratio of one implies that the forces required to switch from one state to the other are identical. F_s is kept low so that the gripper can be actuated by hand. The dimensions of the shell were arrived at by considering multiple designs such that $\frac{F_{sb}}{F_s} > 0.5$ and $F_s < 35$ N.

2.3 Switching Mechanism and Grasping Arms

The switching mechanism transmits the force from the actuator to the center of the bistable shell. The parameters that affect the switching mechanism design are force and output displacement from the actuator, and switching force and displacement of the shell. For the everting shell considered here, $F_s = 20$ N and $u_s = 11.12$ mm. The switching mechanism attaches to the edge of the shell as shown in Fig. 2. The force is applied at the protruding part at the top of the central axis of the mechanism. This initiates a contact at the top of the shell. When the deformation at this point of contact exceeds u_s, the shell everts to the second stable state. For other cases where force or displacement amplification is required, the switching mechanism could be designed to be a force or displacement amplifying compliant mechanism without affecting the aforementioned grasping functionality [13].

Four symmetrically arranged rectangular beams that are attached to the bottom surface of the everting shell act as the grasping arms that hold objects. In the as-fabricated state of the arch, the grasping arms assume horizontal configuration. When the shell is in the second stable state, grasping arms protrude outside as shown in Fig. 3. The grasping arms should be wide apart when the shell is in the second stable state. This is important to hold objects of larger size. This determines the length of the arms and the position at which they are attached to the bottom surface of the shell. For the prototype considered here, beams of length 12 mm attached at a distance of 15 mm from the shell edge satisfy these constraints. When the everted shell comes in contact with stiff objects, it switches back to its initial state. In the process, the grasping arms come together and hold the object firmly. By taking the width of the beam as 5 mm and depth as 1 mm, grasping is compliant enough to wrap around the object, at the same time, stiff enough to support the weight of the object.

3 Results and Performance

The 3D-printed prototype of the everting shell and grasping arms in Object Conex 260 is shown in Fig. 6a, b, c, d, and e. The figure illustrates grasping of a range of objects such as (a) ring, (b) paper clip, (c) aluminum cylinder, (d) mini motor, and (e) plastic spoon. A complete 3D-printed prototype of the grasping mechanism is depicted in Fig. 6f, g, and h. In Fig. 6f, gripper with an everted shell comes in contact with a screwdriver, which acts as the object here. The grasping mechanism holds the screwdriver by everting the shell in Fig. 6g. In Fig. 6h, the gripper releases the screwdriver by actuating at the actuating point of the switching mechanism.

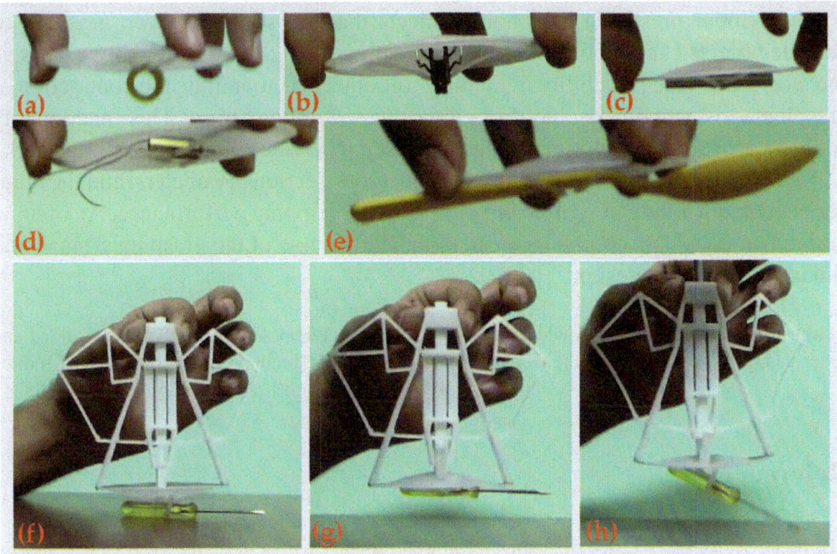

Fig. 6 Grasping arms can capture objects of varying size, shape, and weight: (**a**) ring (**b**) paper clip (**c**) aluminum cylinder (**d**) mini motor (**e**) plastic spoon. The grasping mechanism (**f**) grasps a screwdriver, (**g**) holds, and (**h**) releases it using the switching mechanism

It may be noted here that the gripper is able to grasp objects with length larger than the span of the platform of the shell; for example, the plastic spoon in Fig. 6e. This is possible due to the free space left in between the grasping arms by limiting their number to four. If one increases the number of grasping arms, the gripping performance improves but limits the size of graspable objects due to the reduction in the free space between the grasping arms.

4 Summary and Future Work

A monolithic compliant grasping mechanism based on the bistability of an everting shell is proposed and illustrated with the aid of a 3D-printed prototype. The salient features of the grasping mechanism include lower power consumption, ability to grasp objects of a variety of shapes, grasping of stiff objects initiated from contact, and a completely monolithic design. Scope for future work in the design of everting shell, switching mechanism, and grasping arms is discussed next.

Cosine profile is not the only shape that shows bistability with fixed-fixed boundary conditions. For example, selecting the fundamental buckling mode of a circular plate with fixed-fixed boundary conditions as the as-fabricated shape may also show bistability. Further, optimizing the arch profile that fits into design constraints can improve the grasper functionality.

The gripper presented here holds objects that are stiff. However, grasping of delicate objects cannot be effected by contacting them as it would apply excessive force on them. One possibility in holding soft objects is to initiate the contraction of grasping arms actively, i.e., not from the contact with the object. This functionality could be achieved by redesigning the switching mechanism and the everting shell.

Rational design of the profile, orientation, cross-section parameters, and the number of grasping arms can significantly improve the gripping performance. An optimal design that considers compliance, strength, and volume of the grasping arms would enhance the present design.

Acknowledgements We acknowledge the grant from the Technology Initiative for Disabled and Elderly (TIDE) programme of the Department of Science and Technology, Government of India.

References

1. Bicchi A, Kumar V (2000) Robotic grasping and contact: a review. In: ICRA, vol 348. Citeseer, p 353
2. Murray RM (2017) A mathematical introduction to robotic manipulation. CRC Press
3. Choi H, Koc M (2006) Design and feasibility tests of a flexible gripper based on inflatable rubber pockets. Int J Mach Tools Manuf 46(12–13):1350–1361
4. Maruyama R, Watanabe T, Uchida M (2013) Delicate grasping by robotic gripper with incompressible fluid-based deformable fingertips. In: 2013 IEEE/RSJ international conference on intelligent robots and systems (IROS). IEEE, pp 5469–5474
5. Amend JR, Brown E, Rodenberg N, Jaeger HM, Lipson H (2012) A positive pressure universal gripper based on the jamming of granular material. IEEE Trans Robot 28(2):341–350
6. Brown E, Rodenberg N, Amend J, Mozeika A, Steltz E, Zakin MR, Lipson H, Jaeger HM (2010) Universal robotic gripper based on the jamming of granular material. Proc Natl Acad Sci 107(44):18809–18814
7. Manti M, Cacucciolo V, Cianchetti M (2016) Stiffening in soft robotics: a review of the state of the art. IEEE Robot Autom Mag 23(3):93–106
8. Laliberté T, Birglen L, Gosselin C (2002) Underactuation in robotic grasping hands. Mach Intell Robot Control 4(3):1–11
9. Stavenuiter RA, Birglen L, Herder JL (2017) A planar underactuated grasper with adjustable compliance. Mech Mach Theory 112:295–306
10. Nguyen T-A, Wang D-A (2016) A gripper based on a compliant bitable mechanism for gripping and active release of objects. In: International conference on manipulation, automation and robotics at small scales (MARSS). IEEE, pp 1–4
11. Hibbett, Karlsson, Sorensen (1998) ABAQUS/standard: user's manual, vol 1. Karlsson & Sorensen, Hibbitt
12. Palathingal S, Ananthasuresh G (2017) Design of bistable arches by determining critical points in the force-displacement characteristic. Mech Mach Theory 117:175–188
13. Krishnan G, Ananthasuresh G (2008) Evaluation and design of displacement-amplifying compliant mechanisms for sensor applications. J Mech Des 130(10):102304

Dimensional Synthesis of Delta Manipulator Using Genetic Algorithm-Based Multi-objective Optimization

Anil K. Patidar, Sasanka S. Sinha, and S. Mukherjee

Abstract This paper deals in design synthesis of a three translational degree-of-freedom Delta manipulator by combining together two different approaches (average condition number and power of point). By doing so, shortcomings of both approaches are eliminated. To solve this, multi-objective problem is solved with GA-based multi-objective optimization which works on the principle of Pareto optimality. Different constraints used in optimization are also discussed. A numerical example is presented to observe the effect of different constraints and objective function on the design variables which in turn dictate the manipulator workspace. Further, we try to conclude with a general procedure to design a Delta manipulator without using optimization routines.

Keywords Power of points method · Manipulability · Reciprocal screws · Pareto optimal

1 Introduction

The Delta robot is a three degree-of-freedom (DoF) parallel manipulator which was developed by Reymond Clavel at the Institut de Microtechnique de l'Ecole Polytechnique Federale de Lausanne in 1990. It can translate in the 3-dimensional Euclidean space. It is generally used for pick and place operations, and has added advantages of accuracy, speed, and load-sharing compared to its serial counterparts. The parallel architecture evident in the form of several in-parallel legs supporting the end-effector allows us to use motors of similar size and power. Designing a Delta manipulator for

A. K. Patidar (✉) · S. S. Sinha · S. Mukherjee
IIT Delhi, New Delhi 110016, India
e-mail: mez158449@mech.iitd.ac.in

S. S. Sinha
e-mail: mez168275@mech.iitd.ac.in

S. Mukherjee
e-mail: sudipto@mech.iitd.ac.in

D. Sen et al. (eds.), *Mechanism and Machine Science*, Lecture Notes in Mechanical Engineering, https://doi.org/10.1007/978-981-15-4477-4_44

a prescribed workspace is a challenging task. There are multiple facets to this process out of which the one most discussed in the literature is dimensional synthesis which deals with the identification and evaluation of kinematic or geometric parameters like link lengths, platform, and base radii, etc. Besides, there is an interplay with other processes like dynamic analysis to determine the forces and moments followed by stress-strain analysis to decide material properties and link shapes. Researchers have worked a lot on creating design methodologies. In general, design variables for parallel manipulators are obtained via optimization.

In 1991, Gosselin and Angeles [1] defined global conditioning index (GCI) which gives a global estimate of the manipulability by averaging the condition number of the robot Jacobian within the specified workspace. It is a basic parameter used in dimensional synthesis. In 1997, Stampler et al. [2] used Monte Carlo method to determine the GCI for a Delta robot during optimization to determine the values of the manipulator design variables. Huang et al. [3] added an additional term along with a GCI-like term to describe the deviation between the maximum and minimum values of the condition number. This was demonstrated by designing a two degrees-of-freedom planer parallel manipulator. Yun et al. [4] compared to two methods of dimensional synthesis considering maximum inscribed workspace and GCI. Their results favored GCI.

Later, it became quite evident that optimizing manipulability alone gave designs which are large with small workspaces. Laribi et al. [5] optimized for space utilization of a Delta robot based on the mathematical concept of the power of a point. Stock and Miller [6] had earlier defined a new performance index called space utilization which reflects the ratio of the workspace size to the robot structure size. A bounding box enclosing the robot's structure gives the robot size. Based on this, they proposed a kinematic design method that offers a compromise between manipulability and space utilization. They applied their method to a Linear Delta robot. Apart from these, there exist interval analysis-based methods for the multi-criteria optimal design of parallel manipulators as well [7, 8].

For better performance during high-speed motion, it is desirable to take the dynamics or the inertia of movable components into account in the design process. In literature, isotropy of the inertia matrix termed as dynamic isotropy was proposed by Ma and Angeles [9] to take care of this. Zhang et al. [10] defined pressure/transmission angles for the Delta robot and proposed two new dynamic metrics related to the inertia and Coriolis/centrifugal components of the driving torque. Brinker et al. [11] gave formulations for the transmissibility and constrainability of the Delta robot based on power coefficients and pressure angles.

This paper is primarily concerned with dimensional synthesis of the Delta for a specified workspace and does not take into account the dynamic requirements. As a supplement to the work proposed in [6], the paper aims to combine workspace and GCI-based optimizations. From the results of optimization, the authors try to come up with a general procedure for the synthesis of a Delta manipulator which does away with the time-consuming optimization process and offers a design satisfying the desired objectives. The paper is divided into five sections. Section 2 deals with the kinematics and workspace analysis of the Delta manipulator. Section 3 describes the design and synthesis problem formulated as a constrained optimization problem. The constraints and the objectives are expressed in this section. Section 4 provides a comparative study of the various design optimization approaches presented in Sect. 3 via a numerical example. The steps for a general design procedure of the Delta manipulator are also developed and enumerated here. Section 5 summarizes the research findings and proposes future work.

2 Kinematics and Workspace Analysis

Kinematics is the study of the relations existing between the link geometries, joint variables, and workspace coordinates. It is essential to determine the workspace boundaries. The Delta robot considered in this study is fully actuated and holonomic. Thereby, zeroth-order kinematic relations can be derived.

2.1 *Inverse Kinematics*

The inverse kinematics for the Delta robot is well known in the literature. In this paper, we follow the approach presented in [12]. The Delta robot consists of 3 parallel legs as shown in Fig. 1a and each leg has two links—proximal (l_p) and distal (l_d). Each leg is oriented by φ_i angle from the base frame. The subscript i represents the ith leg and so i varies from 1 to 3. The proximal link is connected to the base by revolute joint $\mathbf{B}_i$.

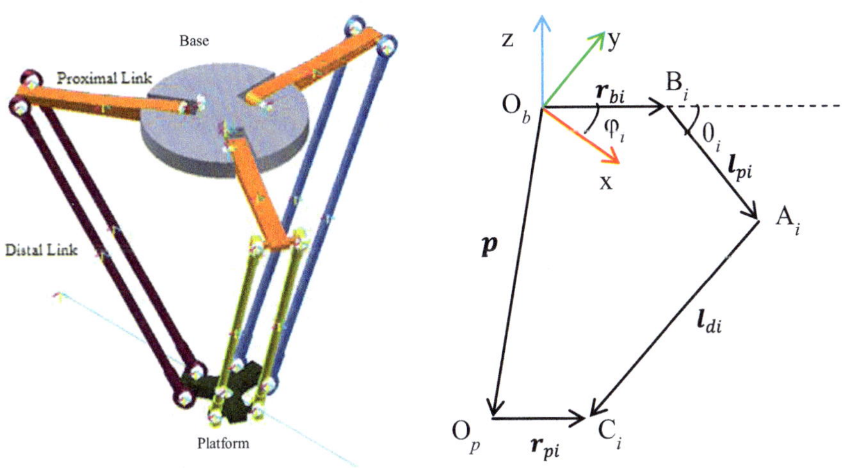

(a) CAD model (b) Line diagram representation

Fig. 1 The Delta manipulator

The distal link is connected to the proximal link and the mobile platform by spherical joints respectively at $\mathbf{A}_i$ and $\mathbf{C}_i$. Figure 1b is the line representation of a single leg of the manipulator in which $\mathbf{O}_p$ represents the platform center, $\mathbf{O}_b$ represents the base center, r_b and r_p are the base and platform radii, respectively. It will be seen later that the workspace depends on the difference, $R = r_b - r_p$, of the platform and the base radii. The relation between platform center coordinates (x, y, z) and joint coordinates θ_i can be obtained by writing the loop closure equation between $\mathbf{A}_i$ and $\mathbf{C}_i$:

$$\boldsymbol{p} + \boldsymbol{r}_p = \boldsymbol{r}_b + \boldsymbol{l}_d + \boldsymbol{l}_p. \tag{1}$$

The above equation in vector form can be decomposed into three scalar equations as

$$E_i \cos\theta_i + F_i \sin\theta_i + G_i = 0, \qquad i = 1, 2, 3 \tag{2}$$

where

$$\begin{aligned}
E_i &= 2L_i(R - x\cos\varphi_i - y\sin\varphi_i)\\
F_i &= -2zL_i\\
G_i &= R^2 + x^2 + y^2 + z^2 + L_i^2 - l_i^2 + 2R(x\cos\varphi_i + y\sin\varphi_i).
\end{aligned}$$

This equation can be solved by tangent half-angle substitution. By substituting $\cos\theta_i = (1 - t_i^2)/(1 + t_i^2)$ and $\sin\theta_i = 2t_i/(1 + t_i^2)$, where $t_i = \tan(\theta_i/2)$, Eq. (2) can be rewritten as

$$(G_i - E_i)t_i^2 + 2F_i t_i + (G_i + E_i) = 0. \tag{3}$$

Solution of this quadratic equation is $t_{1,2} = \frac{-F_i \pm 2\sqrt{E_i^2 + F_i^2 - G_i^2}}{G_i - E_i}$ which gives the values of the joint angles. Differentiating Eq. (2) with respect to time yields

$$\begin{aligned}
\mathbf{J}_a\dot{\boldsymbol{x}} &= \mathbf{J}_b\dot{\boldsymbol{\theta}}\\
\Rightarrow \dot{\boldsymbol{x}} &= \mathbf{J}\dot{\boldsymbol{\theta}}
\end{aligned} \tag{4}$$

where $\dot{\boldsymbol{x}} = [\dot{x}, \dot{y}, \dot{z}]^\mathsf{T}$ is the Cartesian velocity vector, $\dot{\boldsymbol{\theta}} = \left[\dot{\theta}_1, \dot{\theta}_2, \dot{\theta}_3\right]^\mathsf{T}$ is the joint velocity vector, $\mathbf{J}_a$ is a 3×3 matrix with ith row as

$$[\cos\varphi_i(r_b - r_p + l_p\cos\theta_i) - x,\ \sin\varphi_i(r_b - r_p + l_p\cos\theta_i) - y,\ l_p\sin\theta_i - z],$$

$\mathbf{J}_b$ is a 3×3 diagonal matrix with ith row diagonal entry as

$$l_p\sin\theta_i(r_p - r_b + x\cos\varphi_i + y\sin\varphi_i) - l_p z\cos\theta_i,$$

and $\mathbf{J} = \mathbf{J}_a^{-1}\mathbf{J}_b$ is the Jacobian matrix which relates the Cartesian velocities to the joint rates. The manipulability of a manipulator is defined as the condition number of the matrix $\mathbf{A} \equiv \mathbf{J}\mathbf{J}^\mathsf{T}$ [1, 3].

2.2 Singularity Analysis of the Delta Manipulator

The Delta manipulator can have both gain and loss type singularities. The loss of degrees of freedom occurs at the workspace boundaries while gain singularities occur within the workspace. The analysis of gain singularities is more important. As the Delta manipulator is a lower mobility manipulator [11, 13], the singularity visualization can be better done by screw theory. The Delta manipulator in the present study is R-2S-2S which can be represented by R-2U-2U (shown in Fig. 2) as one of the degrees of freedom of spherical joint is redundant along the distal link. The instantaneous twist of the moving platform is a zero pitch screw $\$_p = [\boldsymbol{\omega}^{\mathsf{T}}\ \mathbf{v}^{\mathsf{T}}]^{\mathsf{T}}$ which can be written as a linear combination of all the seven screws corresponding to the seven joints in the ith leg.

$$\$_p = \dot{\theta}_{j,i}\hat{\$}_{j,i}, \qquad j = 1, 2, ..., 7. \tag{5}$$

As $\theta_{3,i} = \theta_{4,i}$ and $\theta_{6,i} = \theta_{7,i}$, the above equation can be reduced as

$$\$_p = \dot{\theta}_{1,i}\hat{\$}_{1,i} + \dot{\theta}_{2,i}\hat{\$}_{2,i} + \dot{\theta}_{3,i}\hat{\$}_{3,i} + \dot{\theta}_{5,i}\hat{\$}_{5,i} + \dot{\theta}_{6,i}\hat{\$}_{6,i}. \tag{6}$$

The leg constraint screw system is a 1-system. The reciprocal screw to all the joint screws of the leg can be identified as an infinite pitch screw

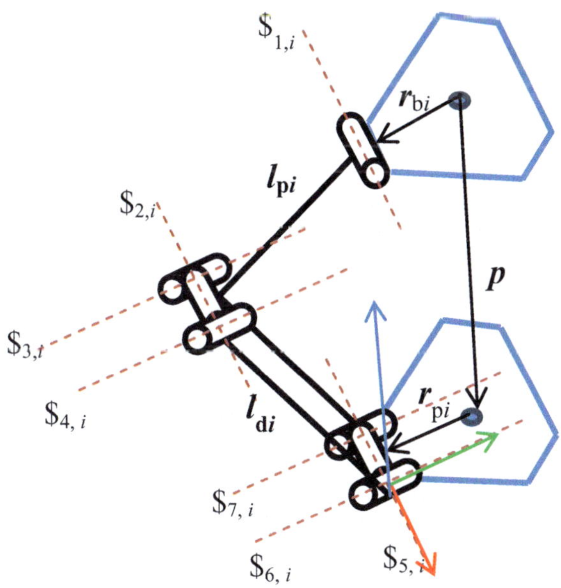

Fig. 2 Schematic of Delta robot showing screws of one leg

$$\hat{\$}_{r,1,i} = \begin{bmatrix} 0 \\ \boldsymbol{n}_i \end{bmatrix} \tag{7}$$

where $\boldsymbol{n}_i$ is a unit vector in the direction $s_{5,i} \times s_{6,i}$ ($s_{j,i}$ is the unit vector along the jth joint axis of the ith leg.) Taking the orthogonal product of both sides of Eq. (6) with $\hat{\$}_{r,1,i}$ and using reciprocity of screws, we have

$$\hat{\$}^{\mathsf{T}}_{r,1,i}\$_p = 0. \tag{8}$$

Writing for all the legs we have

$$\mathbf{J}_c\$_p = 0 \tag{9}$$

where $\mathbf{J}_c$ is the constraint Jacobian matrix given by

$$\mathbf{J}_c = \begin{bmatrix} \boldsymbol{n}_1^{\mathsf{T}} & 0 \\ \boldsymbol{n}_2^{\mathsf{T}} & 0 \\ \boldsymbol{n}_3^{\mathsf{T}} & 0 \end{bmatrix}.$$

To remove all the passive joints from Eq. (6), we need to find a reciprocal screw of all the passive joints of ith leg. From geometry, a zero pitch screw along the line passing through distal link is identified to satisfy this.

$$\hat{\$}_{r,2,i} = \begin{bmatrix} \hat{\boldsymbol{l}}_{d,i} \\ \hat{\boldsymbol{l}}_{d,i} \times \hat{\boldsymbol{r}}_{p,i} \end{bmatrix}. \tag{10}$$

Taking orthogonal product of the above screw with the twist in Eq. (6) gives

$$\hat{\$}^{\mathsf{T}}_{r,2,i}\$_p = \hat{\$}^{\mathsf{T}}_{r,2,i}\hat{\$}_{1,i}\dot{\theta}_{1,i}. \tag{11}$$

Collecting for all the three legs we get

$$\tilde{\mathbf{J}}_a\$_p = \tilde{\mathbf{J}}_b\dot{\boldsymbol{\theta}} \tag{12}$$

where $\tilde{\mathbf{J}}_a$ and $\tilde{\mathbf{J}}_b$ are given by

$$\tilde{\mathbf{J}}_a = \begin{bmatrix} (\hat{\boldsymbol{l}}_{d,1} \times \hat{\boldsymbol{r}}_{p,1})^{\mathsf{T}} & \hat{\boldsymbol{l}}^{\mathsf{T}}_{d,1} \\ (\hat{\boldsymbol{l}}_{d,2} \times \hat{\boldsymbol{r}}_{p,2})^{\mathsf{T}} & \hat{\boldsymbol{l}}^{\mathsf{T}}_{d,2} \\ (\hat{\boldsymbol{l}}_{d,3} \times \hat{\boldsymbol{r}}_{p,3})^{\mathsf{T}} & \hat{\boldsymbol{l}}^{\mathsf{T}}_{d,3} \end{bmatrix}$$

and

$$\tilde{\mathbf{J}}_b = \begin{bmatrix} \hat{\$}^{\mathsf{T}}_{r,2,1}\hat{\$}_{1,1} & 0 & 0 \\ 0 & \hat{\$}^{\mathsf{T}}_{r,2,2}\hat{\$}_{1,2} & 0 \\ 0 & 0 & \hat{\$}^{\mathsf{T}}_{r,2,3}\hat{\$}_{1,3} \end{bmatrix}.$$

The matrix $\tilde{\mathbf{J}}_a$ and the matrix $\mathbf{J}_a$ in Sect. 2.2 are equivalent provided $\boldsymbol{\omega} = 0$. Similarly, $\tilde{\mathbf{J}}_b$ and $\mathbf{J}_b$ are equivalent. Equation (12) can be further written as

$$\mathbf{J}_f \$_p = \dot{\boldsymbol{\theta}} \tag{13}$$

where $\mathbf{J}_f \equiv \tilde{\mathbf{J}}_b^{-1} \tilde{\mathbf{J}}_a$ is the force Jacobian. The rows of matrix $\mathbf{J}_c$ represent the three constraint moments on the moving platform. Similarly, the rows of $\mathbf{J}_f$ represent the three forces of actuation on the moving platform. The manipulator will be under constraint singularity when $\mathbf{J}_c$ is singular. Then the three constraint moments are coplanar or any two moments are parallel. The matrix $\mathbf{J}_f$ will be singular when $\tilde{\mathbf{J}}_a$ is singular. Then the actuation forces are coplanar or any two of them are parallel. Coincidentally, in the case of the Delta manipulator, both matrices $\mathbf{J}_f$ and $\mathbf{J}_c$ become singular simultaneously. So, manipulability of the matrix $\mathbf{J}$ in Sect. 2.2 is sufficient to take care of all types of singularities of the Delta manipulator.

2.3 Workspace Analysis of the Delta Manipulator

It is observed that the workspace of the Delta manipulator is geometrically the intersection of 3 tori centered at the revolute joints. Each leg of manipulator can move in a torus centered at $\mathbf{B}_i$ with l_d and l_p acting as minor and major radii, respectively. Mathematically, the equation of the torus for the ith leg can be obtained by setting the discriminant of Eq. (3) to be greater than or equal to zero:

$$F_i^2 + F_i^2 - G_i^2 \geq 0, \qquad i = 1, 2, 3. \tag{14}$$

The shape of the torus is dependent on the ratio of minor and major radii as shown in Fig. 3. Different torii give different shapes of the manipulator workspace such as conical, spherical, etc. So proper selection of link lengths is essential to design. It is found that the above equation is dependent on the quantity R which controls the rectilinear displacement of the torus towards the center of the fixed platform as seen in Fig. 4. This in turn affects the relative placement of the three tori which in turn determines the overlapping region which is the workspace of the Delta manipulator. When $R = 0$, maximum overlap occurs and the workspace volume is maximum for a given set of l_d and l_p. As the difference increases in magnitude, the overlap decreases continually until $R = l_d + l_p$ in which case the workspace degenerates to a single point (Fig. 4). Thereafter, when $|R| > l_d + l_p$, there is no overlapping and no feasible workspace exists. The workspace gets rotated by 180° about the z-axis of symmetry for the same value of R but with a negative sign. Lastly, the workspace is symmetric

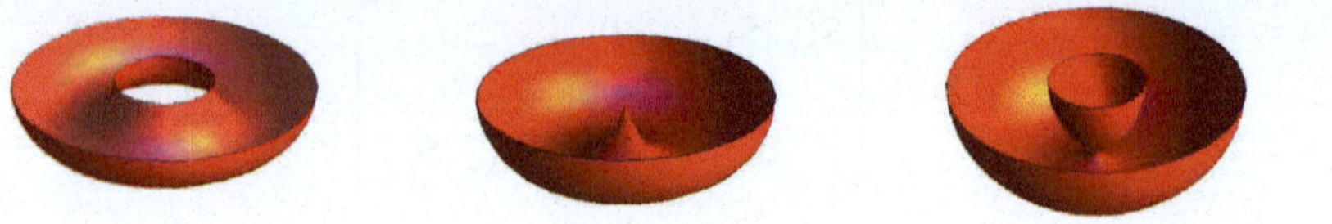

(a) Ring torus $l_p > l_d$ (b) Horn torus $l_p = l_d$ (c) Spindle torus $l_p < l_d$

Fig. 3 Different types of tori

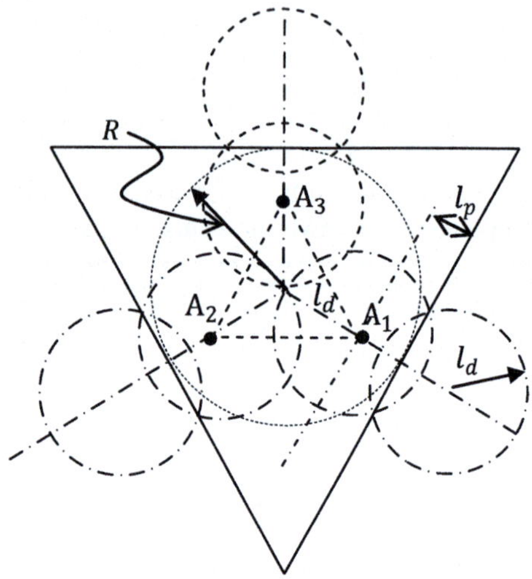

Fig. 4 Point workspace as seen perpendicular to the *xy*-plane

about the manipulator base and the manipulator structure does not allow the platform to cross the base. Therefore, only half of the workspace will be considered for further study.

3 Synthesis Problem Formulation

In general, four design variables can be identified for Delta manipulator—proximal link length (l_p), distal link length (l_d), difference between base and platform radii (R), and center height of workspace (H). So our design vector is $\boldsymbol{\chi} = \left[l_p, l_d, R, H\right]^{\mathsf{T}}$. These parameters are shown in Fig. 5. The desired workspace is cylindrical with cylinder height and radius as h and r, respectively. Only one-third of the workspace is considered because of the z-axis of symmetry. In our case, the problem is a highly nonlinear multivariable problem so that the values of these variables can be conveniently found by optimization. There are multiple optimization algorithms available to solve nonlinear optimization problems. In the present study, genetic algorithm (GA) is used because it can handle a wide range of problems from single objective to multi-objective. The other feature of GA is that it has multiple starts so that we can get to the global minimum rather than being stuck at a local minimum.

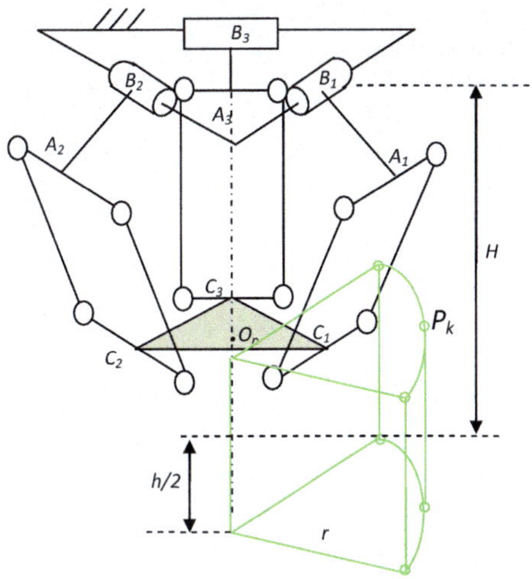

Fig. 5 Setup for the optimization algorithm

3.1 Objective Function

We are considering multiple objectives based on various standards of performance. One of them is the minimization of the weighted sum of the manipulability of the Delta robot and its variance. The variance was replaced by the ratio of the maximum and the minimum condition number $\tilde{\eta}$ in [3]. It is required to minimize the objective function

$$\eta \equiv \sqrt{\bar{\eta}^2 + w_\eta \tilde{\eta}^2} \tag{15}$$

where w_η is the weight. A kinematically good design has a homogeneous well-conditioned workspace and the above objective best serves this purpose. There is no universal criteria to select a favorable condition number for a desired workspace. So homogenization often leads to a situation where good manipulability region is confined to a small fraction of the total robot workspace which may result in large robot dimensions. It becomes important to limit the residual or unused workspace which is our second objective. Following the work by Laribi et al. [5], we used power of points method for six boundary points in one-third of the robot workspace as shown in Fig. 5. We then calculate the distances of these six points from the three tori and try to minimize their absolute summation or sum of absolute distances

$$\beta \equiv \sum_{k=1}^{6} \sum_{i=1}^{3} |f_i(\chi, P_k)| \tag{16}$$

where $f_i(\chi, P_k)$ is a function related to the distance of point P_k from the ith torus. This function value is found by substituting the point coordinates in the LHS of the corresponding torus equation as presented in Eq. (13). Thus, we were able to get the combined benefits of good manipulability zone and small robot size.

3.2 Pareto Efficiency

In a single-objective problem, the solution is best if it gives minimum value of cost function but for multi-objective problems it is not possible to get the minimum value of both cost functions for the same solution. Therefore a single cost function is created by assigning different weights to individual cost functions. The other important question is what weights to be assigned to different objective functions. An Italian engineer and economist Vilfredo Pareto came up with a different approach for his economics problems and named it as 'Pareto efficiency' or 'Pareto optimal.' According to the Pareto optimality, optimality is not achieved if there is a solution where improvements can be made by improving one objective without degrading another objective. The finding of new solution is called 'Pareto improvement.' If there is no further Pareto improvement possible then the solution is 'Pareto optimal.'

3.3 Constraints

The robot must be able to reach every point in the considered cylindrical workspace. This can be incorporated as an assembly criteria by studying the extremity conditions. A three-dimensional depiction of the robot in its extremities is difficult. A planar depiction is possible by viewing in a plane perpendicular to the base platform and containing the concerned leg as shown in Fig. 6. From Fig. 6a, b, one can respectively deduce

$$l_d - l_p - H + h/2 < 0 \tag{17}$$

$$\sqrt{(H + h/2)^2 + (R + r)^2} - l_d - l_p < 0. \tag{18}$$

Intersection of ring tori will give a roughly spherical workspace while intersection of spindle tori will lead to a conical-like workspace. Accordingly, the torus conditions can be selected as constraints. To accommodate the servomotors on the base,

$$R > R_{\min} \tag{19}$$

where $R_{\min}$ is of the order of motor dimensions. A maximum limit can also be imposed to constrain the mass of the base for portability of the robot. The other constraint used by Huang et al. [3] and Zhang et al. [10] to reduce robot footprint is machine to workspace volume ratio which is given as

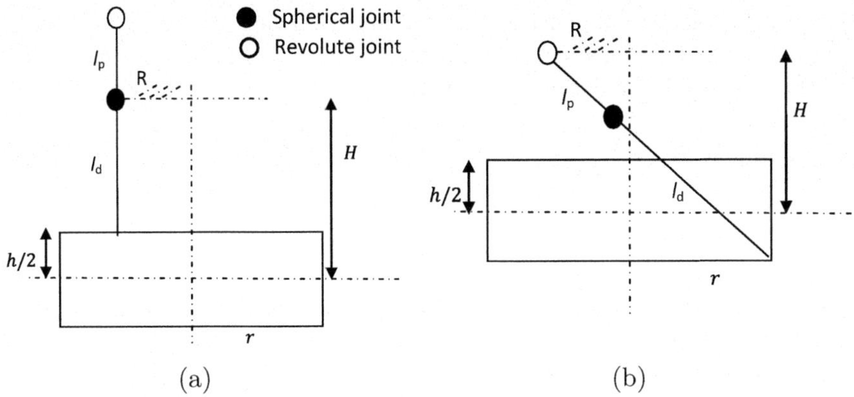

Fig. 6 Extreme robot assembly condition

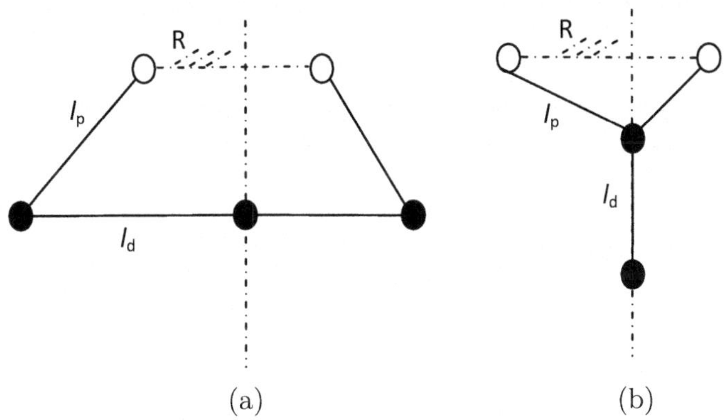

Fig. 7 Gain singularity configurations

$$1.1(R + l_p) < r. \tag{20}$$

This ensures that the outer end of the proximal link does not protrude out of the workspace region. The dimensions of the robot can be adjusted to take care of several gain singularity conditions as well. Figure 7a demonstrates a gain singularity in which the distal links and the platform lie in a single plane so that the platform has infinitesimal translations in the z-direction. To take care of this, the constraint to be imposed is

$$l_p - l_d + R < 0. \tag{21}$$

Similarly, the gain singularity in Fig. 7b can be prevented by using the constraint

$$H + h/2 - \sqrt{l_p^2 - R^2} - l_d < 0. \tag{22}$$

The planar depiction in Fig. 7 is possible from the viewing plane described above. The constraints for gain singularity are optional as one of the objective functions related to the well-conditioned Jacobian matrix is sufficient for our case as discussed in Sect. 2.2.

4 A Numerical Example

It is required to determine the design parameters for a Delta robot with a workspace enclosing a cylindrical volume of radius $r = 200$ mm and height $h = 100$ mm. Genetic algorithm toolbox in MATLAB 2015b was used to get the optimized results. The simulations were carried out on an Intel 4-core i7 processor with 3 GHz clock-speed and 8 GB RAM. The simulations required an average computation time of about 8475 s. 30 individuals were selected at evenly spaced points in the one-third cylindrical region shown in Fig. 5. The rest of the GA settings are enlisted in Table 1.

Figure 9 shows the influence of the objective functions on the optimized designs. The workspace for a design solely based on manipulability criteria is shown in Fig. 9a which clearly reveals that the region of interest occupies a central position in the entire workspace leading to a large unused volume. Minimizing the residual volume alone results in a workspace such that the boundaries of the cylindrical volume lie on the workspace boundary as seen in Fig. 9b. Using both the objectives together results in a design whose workspace is depicted in Fig. 9c. Although the residual volume surrounding the lateral surface of the cylinder has been minimized, there still remains a lot of unused space above and below the cylinder. This can be taken care by the inclusion of a third objective function which minimizes the distances of the top and bottom centers of the cylinder from the top and bottom points of the workspace. Similar procedure was carried out but with additional active constraint on machine to workspace volume ratio. The results in Fig. 10 reveal that the imposition of this constraint leads to marginal difference in workspaces for the various objective functions. This further suggests that the constraint is strict and dictates the values of the design parameters. The pareto optimal front for the case in Fig. 10c is depicted in Fig. 8 and the optimal solution selected is marked as a black star.

Table 1 GA settings

Lower bound	[100, 200, 60, −500]
Upper bound	[100, 200, 60, −500]
Population	200
Mutation	0.2
Reproduction	Random

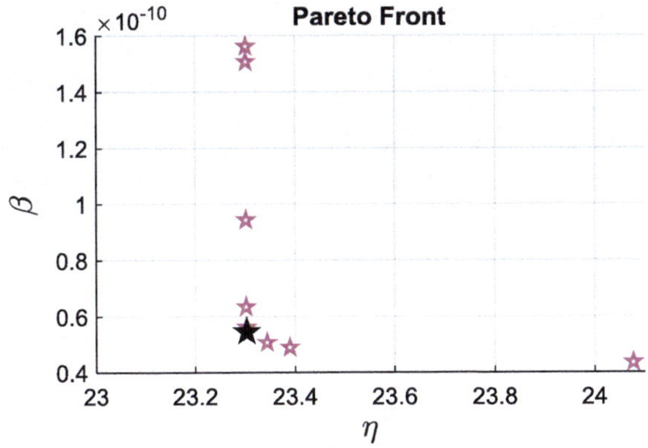

Fig. 8 Pareto optimal front

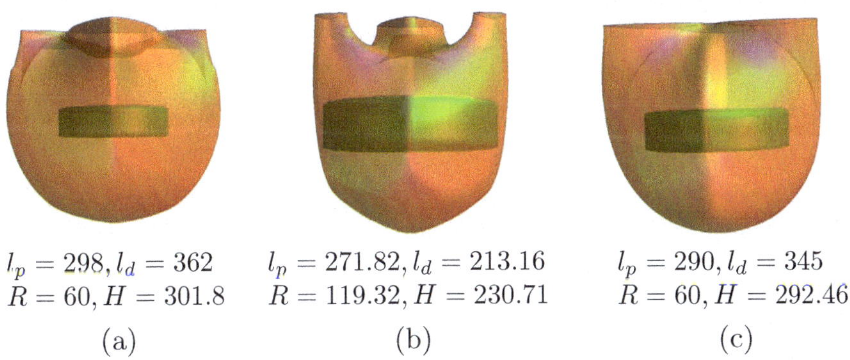

Fig. 9 Workspaces of optimized designs for different objective functions

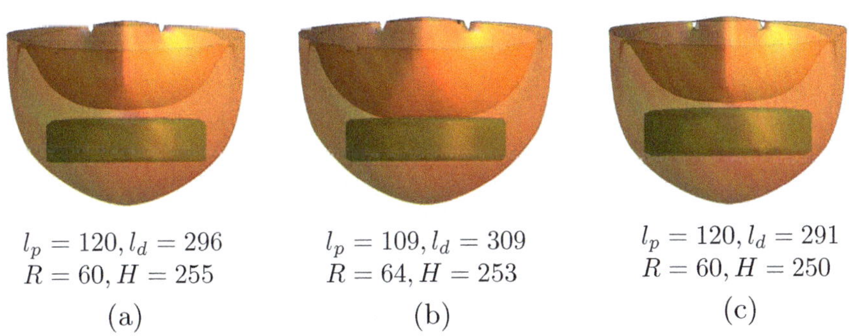

Fig. 10 Optimized designs showing the dominance of the constraint related to the machine to workspace volume ratio

4.1 Design Procedure

From the above observations, a general procedure can be developed for the design of a Delta manipulator without performing the optimization routines. In view of the workspace limitations and constraint in Eq. (10), with increase in R, both the workspace volume and manipulability decrease. So R should be minimum for good manipulability. The proximal link length can be calculated from Eq. (11). Now, constraints in Eqs. (8) and (9) can be rewritten as

$$\frac{l_d - l_p}{H - h/2} = R1$$

$$\frac{\sqrt{(H + h/2)^2 + (R + r)^2}}{l_d + l_p} = R2$$

where $R1$ and $R2$ are ratios which should be less than unity for assembly.

Figure 11 shows the variation of η with $R1$ at different values of $R2$. For good manipulability, a small $R1$ and a large $R2$ should be chosen as seen in Fig. 11. Finally, the guidelines are summarized as

1. $R = R_{\min}$.
2. $l_p = r/1.1 - R$.
3. $R2$ should be chosen between $\frac{R+r}{\sqrt{4L^2 - 4hL + (R+r)^2 + h^2}}$ to 1.
4. $R1$ should be the minimum value for assembly at a specified $R2$ which is

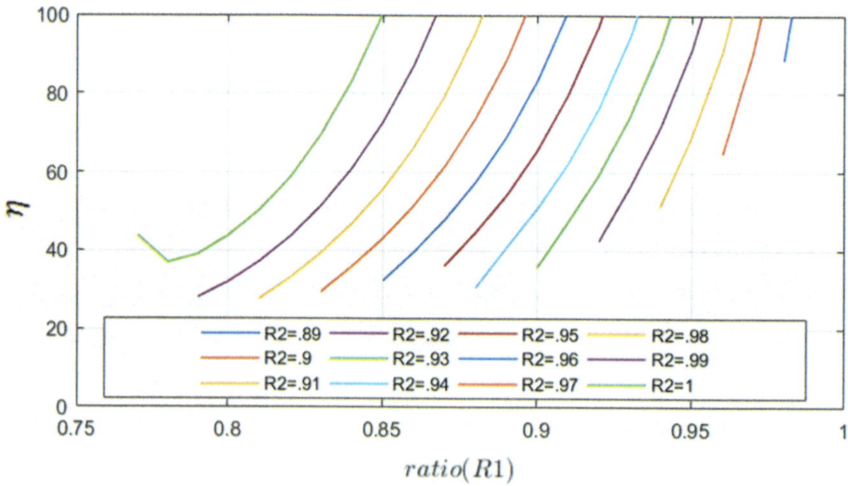

Fig. 11 Plot of η versus $R1$. Values taken are $R = 60$ mm, $r = 200$ mm, and $h = 100$ mm

$$\frac{2hL + \sqrt{(R+r)^4 + (R+r)^2h^2 - 4(R+r)^2L^2R2^2}}{(R+r)^2 + h^2}.$$

It is advised to avoid values of $R2$ close to 1 as the manipulability trend changes as seen in Fig. 11.

5 Conclusion and Future Work

Some interesting conclusions can be drawn from the present study. The use of multiple objectives enables us to satisfy multiple requirements simultaneously which is not possible in the single objective case and provide better design. The machine to workspace volume ratio dictates the optimization results to a large extent so much so that one can have a rough estimate of the values of the design variables. To avoid optimization for the design process, some empirical relations are also developed to design a Delta manipulator.

For future work, additional minimizing function concerning the robot dynamics can also be included as one of the objectives.

References

1. Gosselin C, Angeles J (1991) A global performance index for the kinematic optimization of robotic manipulators. J Mech Design 113:220
2. Stamper R, Tsai LW, Walsh G (1997) Optimization of a three DOF translational platform for well-conditioned workspace. Proceedings of international conference on robotics and automation 4:3250–3255
3. Huang T, Li Z, Li M, Chetwynd DG, Gosselin CM (2004) Conceptual design and dimensional synthesis of a novel 2-DOF translational parallel robot for pick-and-place operations. J Mech Design 126(3):449
4. Yun Y, Wang L, Guan L (2004) Dimensional synthesis of a 3-DOF parallel manipulator. In: 2004 IEEE international conference on systems, man and cybernetics, vol 6, pp 5317–5323
5. Laribi MA, Romdhane L, Zeghloul S (2007) Analysis and dimensional synthesis of the Delta robot for a prescribed workspace. Mech Mach Theory 42(7):859–870
6. Stock M, Miller K (2003) Optimal kinematic design of spatial parallel manipulators: application to linear delta robot. J Mech Design 125(2):292
7. Hao F, Merlet JP (2005) Multi-criteria optimal design of parallel manipulators based on interval analysis. Mech Mach Theory 40(2):157–171
8. Chablat D, Wenger P, Majou F, Merlet JP (2004) An interval analysis based study for the design and the comparison of three-degrees-of-freedom parallel kinematic machines. Int J Robot Res 23(6):615–624
9. Ma O, Angeles J (1993) Optimum design of manipulators under dynamic isotropy conditions. In: 1993 Proceedings IEEE international conference on robotics and automation
10. Zhang L, Mei J, Zhao X, Huang T (2012) Dimensional synthesis of the Delta robot using transmission angle constraints. Robotica 30(3):343–349
11. Brinker J, Corves B, Takeda Y (2018) Kinematic performance evaluation of high-speed delta parallel robots based on motion/force transmission indices. Mech Mach Theory 125:111–125

12. Williams II RL (2015) The delta parallel robot: kinematics solutions. http://www.ohio.edu/people/williar4/html/pdf/DeltaKin.pdf
13. Joshi SA, Tsai L-W (2002) Jacobian analysis of limited-DOF parallel manipulators. In: ASME 2002 international design engineering technical conferences and computers and information in engineering conference. American Society of Mechanical Engineers, pp 341–348

Dynamic Modeling of a Passive Flexible Link Biped for a Symmetric Periodic Gait

Saptarshi Jana and Abhishek Gupta

Abstract The aim of this study is to investigate the existence of a symmetric and periodic walking gait in a flexible passive dynamic biped for walking along an inclined plane. This model is a simple, two-link, 2-D passive compass gait biped. These two links are connected at the hip through a freely rotating revolute joint and one of the links (stance leg) can change its length when it is in contact with the ground. Hence, the robot has three degrees of freedom (DoF). Dynamic equations of the system during the swing phase have been determined using the Euler–Lagrange method. For the linearized model, we have analytically obtained a natural set of initial conditions to obtain a periodic gait with above constraints. These initial conditions were used to obtain the periodic cycle for the non-linear model and the corresponding time period through numerical optimization. It has been investigated the effect of leg stiffness on the gait parameters. We found a flexible biped which can produce symmetric walking gait and walks fast.

Keywords Compass gait · Passive dynamic walker · Flexible leg biped · Dynamic modeling

1 Introduction

The most sophisticated walking mechanism is the human walking. During the last four decades, various researchers have done plenty of research on the bipedal walking mechanism. The first initiative on the bipedal walking robot (WABOT) was taken by Professor Ichiro Kato, Waseda University in 1970 [1]. Some of the pioneer researchers, McMohan [2] and McGeer [3], have reported their study in the late twentieth century. McGeer has reported in his study on the transformation of a rimless wheel into a biped walker [3]. The primary objective of a biped is to achieve

S. Jana (✉) · A. Gupta
Department of Mechanical Engineering, Indian Institute of Technology, Mumbai, India
e-mail: saptarshi.jana@iitb.ac.in

A. Gupta
e-mail: abhi.gupta@iitb.ac.in

D. Sen et al. (eds.), *Mechanism and Machine Science*, Lecture Notes in Mechanical Engineering, https://doi.org/10.1007/978-981-15-4477-4_45

a dynamically balanced gait. Bipeds are widely classified into four categories based on their walking strategies. Passive dynamic walkers (PDW) are one of them which are fully uncontrolled and have no actuator [3]. It uses gravitational energy to compensate for the energy loss during the cycle [4]. PDWs are the most energy-efficient biped among all on a flat ground [5]. A numerical model has been established by M. Gomes which can walk on level ground with even no energy input [6]. Basically, all these models have the same goal, i.e., to achieve a stable dynamic gait with no regard to controllability or stability against external influences.

The addition of compliance at different positions of the biped could be beneficial to achieve different goals. A smooth walking pattern generated by adding torsional spring at the different joints of the biped [7, 8]. A combination of linear and torsional spring also sometimes used for attaining better stability in the walking [9]. It has been found that the consumption of energy can be reduced by adding compliant ankle [10]. Multiple linear springs between links can act as biarticular muscle which exhibits better human-like walking pattern [11, 12]. From these examples, it has always been found that compliant bipeds have advantages over the rigid link biped.

In this study, a 2-D biped robot comprises of two flexible legs, of variable length, connected by a revolute joint at the hip is analyzed. This point foot biped model is inspired by the simplest walking model of Garcia et al. [13]. The main objective of our study is to investigate a set of suitable initial conditions that can generate a symmetric walking gait biped for any particular model parameter when it is walking along an inclined plane. The effect of leg stiffness on the gait parameters is also investigated.

2 Analysis of the Model

A kinematic sketch of the model has shown in Fig. 1. This model resembles a two-link compass gait biped. This biped robot comprises of two equal length legs; each can change its length (d_1) when in contact with the ground (stance leg) otherwise it

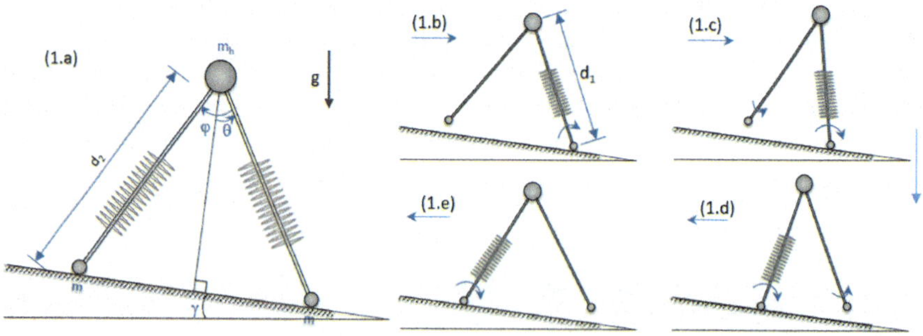

Fig. 1 **a** The kinematic sketch of a passive dynamic biped. **a–e** Shows a walking step through different positions

(swing leg) is of fixed length (d_2). Both the legs connected with a revolute joint at the hip. The swinging leg is modeled as the massless rigid link of constant length equal to that of the uncompressed leg, and the stance leg comprises of two massless links which are connected through a sliding pair and linear spring at the knee to model its flexibility. The stance leg can only change its length but it cannot twist or bend. This whole system has three point masses, two at the end of each leg (m) and one at the hip joint (m_h). These leg masses are very small in comparison with hip mass (i.e. $m_h \gg m$). Stance leg makes an angle θ with the perpendicular plane to the inclined plane. The angle between the stance and swing leg is ϕ. This mechanism consists of two phases—swing phase during which one of the legs is in contact with the ground and the other leg swings; another phase is the heel strike which is assumed to be instantaneous. The robot has three degrees of freedom (DoF) and it is a planar mechanism.

2.1 *Walking Step Cycle*

A walking step starts with heel strike (Fig. 1a) and ends with the next heel strike (Fig. 1a). Just after heel strike (Fig. 1a) stance leg starts rotating about the point of contact and length starts to change as well throughout the process. Just after the impact, the swing leg has no angular velocity (Fig. 1b). Then it starts rotating about the hip joint under the action of gravity when it loses contact with the ground (Fig. 1c). It rotates further due to the inertia (Fig. 1d) After a certain time, it stops rotating and the whole model falls under the action of gravity (Fig. 1e) and the heel of swing leg strikes (Fig. 1a). Thus, the walking cycle completes.

2.2 *Assumption*

- Legs and hip are considered as point masses and links are considered as massless rigid links,
- Foot masses (m) are very much smaller than the hip mass (m_h),
- The stance leg is rigidly connected to the ground so there are no slips and no bounce,
- Feet have point contact with ground,
- Collision is instantaneous and plastic.

2.3 Method of Solution

I. Dynamic equations of the system during the swing phase have been determined using the Euler–Lagrange method.
II. Keeping in mind that the double stance phase is considered to be instantaneous, conservation of angular momentum principle was used to obtain impact conditions.
III. Linearization of the non-linear model has been done around an equilibrium point and analytically obtained a natural set of initial conditions to obtain a symmetric period gait with some constraints.
IV. Numerical optimization is used to obtain the initial conditions for a symmetric gait using the non-linear model. The ICs found for the linearized case were used as initial guesses for the optimization.

3 Dynamics of the Model

As the double stance phase is instantaneous, swing phase prevails the whole period of a step cycle. Dynamics of the system controlled by gravity without any other external influences. Here is the coupled differential equation (Eq. 1). The equations of motion of the system during the swing phase have been determined using the Euler–Lagrange method. Equations are simplified using mass ratio, $\beta = m/m_h$ and length ratio, $d = d_1/d_2$. These equations are non-linear. In the above equations,g' is scaled gravity, $g' = g/d_2$; natural frequency of the stance leg's oscillation, $\omega_n = \sqrt{k/m_h}$; leg stiffness, k. The equations are further simplified with the partial suppression of β in the Sect. 3.1.

$$\begin{pmatrix} \beta(1+d^2)+d^2-2\beta d\cos\phi & -\beta\sin\phi & -(\beta-\beta d\cos\phi) \\ -\beta\sin\phi & 1+\beta & \beta\sin\phi \\ -\beta(1-d\cos\phi) & \beta\sin\phi & \beta \end{pmatrix}\begin{pmatrix} \ddot{\theta} \\ \ddot{d} \\ \ddot{\phi} \end{pmatrix}$$
$$+\begin{pmatrix} (2\beta d\sin\phi)\dot{\theta}\dot{\phi}+2d(1+\beta)\dot{d}\dot{\theta}-(\beta d\sin\phi)\dot{\phi}^2-(2\beta\cos\phi)\dot{d}\dot{\theta} \\ -(2\beta\cos\phi)\dot{\theta}\dot{\phi}+(\beta\cos\phi)\dot{\phi}^2+(\beta\cos\phi)\dot{\theta}^2-d\dot{\theta}^2+\omega_n^2(d-1)-\beta d\dot{\theta}^2 \\ -(d\sin\phi)\dot{\theta}^2+(2\cos\phi)\dot{d}\dot{\theta} \end{pmatrix}$$
$$+\begin{pmatrix} -(d\sin\theta+\beta\sin(\phi-\theta)+\beta d\sin\theta) \\ (1+\beta)\cos\theta \\ \beta\sin(\phi-\theta) \end{pmatrix}g' = \begin{pmatrix} 0 \\ 0 \\ 0 \end{pmatrix} \tag{1}$$

3.1 The Resulting Equations of Motion (EoM)

$$
\begin{aligned}
& d\,\ddot{\theta} + 2\dot{d}\,\dot{\theta} - g'\sin(\theta - \gamma) = 0 \\
& \ddot{d} + g'\cos(\theta - \gamma) - d\,\dot{\theta}^2 + \omega_n^2(d - 1) = 0 \\
& \ddot{\phi} - (1 - d\cos\phi)\ddot{\theta} + \sin\phi\,\ddot{d} - d\sin\phi\,\dot{\theta}^2 + 2\cos\phi\,\dot{d}\,\dot{\theta} + g'\sin(\phi - \theta + \gamma) = 0
\end{aligned}
\tag{2}
$$

We have these equations (Eq. 2) considering the γ slope and effect of mass ratio β, although β does not appear in the equations. The effect of swing leg mass is very less on the dynamics of stance leg and also on the oscillation of stance leg as the hip mass dominates overs other masses but in the swing leg dynamics it has an important contribution. So, β is taken as zero in the first two equations but not in the third equation.

3.2 Collision Condition

During heel strike, an impulsive force is generated at the ground contact point and passed through the hip mass. All other non-impulsive forces are negligible in comparison with the impulsive force. Conservation of angular momentum principle has been used to obtain the robot states after the impact of the swinging leg. Assuming that, the before impact states are $\theta^-,\ \dot{\theta}^-,\ d^-,\ \dot{d}^-,\ \phi^-,\ \dot{\phi}^-$ and the after impact states are denoted as $\theta^+,\ \dot{\theta}^+,\ d^+,\ \dot{d}^+,\ \phi^+,\ \dot{\phi}^+$. At the end of the swing phase (or the beginning of double stance phase), the swing leg stops rotating and the stance leg restores its neutral length before impact ($d_1 = d_2$). The radial velocity of hip mass along the stance leg length, thus, is zero at impact ($\dot{d}_1 = \dot{d}_2 = 0$). From the geometric relationship at the time of impact, we get $\phi = 2\theta$. During heel strike, the configuration of the mechanism remains unchanged. $\mathbf{q}^+ = \mathbf{A}\mathbf{q}^-$; Matrix $\mathbf{A}$ (Eq. 3) gives the relation between both the state, named "jump" matrix by Garcia et. al [13].

$$
\begin{bmatrix} \theta \\ \dot{\theta} \\ d \\ \dot{d} \\ \phi \\ \dot{\phi} \end{bmatrix}^+
=
\overbrace{\begin{bmatrix}
-1 & 0 & 0\,0\,0\,0 \\
0 & \cos 2\theta^- & 0\,0\,0\,0 \\
0 & 0 & 1\,0\,0\,0 \\
0 & 0 & 0\,1\,0\,0 \\
-2 & 0 & 0\,0\,0\,0 \\
0 & (1 - \cos 2\theta^-)\cos 2\theta^- & 0\,0\,0\,0
\end{bmatrix}}^{\mathbf{A}}
\begin{bmatrix} \theta \\ \dot{\theta} \\ d \\ \dot{d} \\ \phi \\ \dot{\phi} \end{bmatrix}^-
\tag{3}
$$

4 Solution Through Linearization

4.1 Linearized EoM

EoMs have been linearized in order to carry out analytical solution of the system. In the process angles of rotation are assumed to be sufficiently small. A suitable equilibrium point has been found about which equation of motions are linearized which is $\left[\Theta\ \dot{\Theta}\ D\ \dot{D}\ \Phi\ \dot{\Phi}\right]^{eq} = [\gamma\ 0\ (1-p)\ 0\ \gamma p/(1-p)\ 0]$; where variable $p = g'/\omega_n^2$ and γ is the ground slope. The linearized equations are:

$$\begin{aligned} &\ddot{\Theta} - \{g'/(1-p)\}\Theta = 0 \\ &\ddot{D} + \omega_n^2 D = 0 \\ &\ddot{\Phi} - \{g'/(1-p)\}\Theta + g'\Phi + \{\gamma g' p/(1-p)\}D = 0 \end{aligned} \tag{4}$$

These parameters, in capital, indicate the linearized parameters which are measured from the equilibrium point for the same. These equations will lead us to obtain a natural set of initial conditions (ICs) for stable gait with assumptions given in Sect. 2.2.

4.2 Initial Conditions

Initial conditions for solving these equations are $\left[\Theta_0\ \dot{\Theta}_0\ D_0\ \dot{D}_0\ \Phi_0\ \dot{\Phi}_0\right]$. Initially, that is just after heel strike, swing leg is almost stationary and the stance leg is uncompressed. Thus, the radial velocity of hip mass along the stance leg length is zero. According to our model's requirement we have few known values and relationships from the geometry. These are:

$$\begin{aligned} &\Phi_0 = 2\Theta_0 + 2\gamma - \{\gamma p/(1-p)\} \\ &\dot{\Phi}_0 = \dot{D}_0 = 0\,. \end{aligned} \tag{5}$$

4.3 Analytical Solution

Solutions for the liner ODE (Eq. 4) have been found using above initial conditions (Sect. 4.2).

$$
\begin{aligned}
\Theta &= \tfrac{1}{2}(e^{rt}+e^{-rt})\Theta_0 + \tfrac{1}{2r}(e^{rt}-e^{-rt})\dot{\Theta}_0\\
D &= D_0\cos(\omega_n t)\\
\Phi &= \left\{\tfrac{1}{2(2-p)}(e^{rt}+e^{-rt}) + \left(2-\tfrac{1}{(2-p)}\right)\cos(\sqrt{g'}t)\right\}\Theta_0\\
&+\left\{\tfrac{1}{2r(2-p)}(e^{rt}-e^{-rt}) + \tfrac{1}{\sqrt{g'}(2-p)}\sin(\sqrt{g'}t)\right\}\dot{\Theta}_0\\
&+\left\{\tfrac{p\gamma r^2(\cos(\omega_n t)-\cos(\sqrt{g'}t))}{(\omega_n^2-g')(1-p)}\right\}D_0 + \left(2\gamma-\tfrac{\gamma p}{(1-p)}\right)\cos(\sqrt{g'}t)
\end{aligned}
\tag{6}
$$

Where, $r = \sqrt{g'/(1-p)}$; $p = g'/\omega_n^2$; The solution is given in terms three independent initial conditions (i.e. Θ_0, $\dot{\Theta}_0$, D_0). As we have employed a relationship between Θ & Φ so Φ will not give us any independent relation. We have solved for Θ , $\dot{\Theta}$, D, $\dot{D}$. Rearranging it in a matrix form is given in Eq. (6).

$$
\begin{bmatrix}\Theta\\ \dot{\Theta}\\ D\\ \dot{D}\end{bmatrix} = \overbrace{\begin{bmatrix}\frac{1}{2}(e^{rt}+e^{-rt}) & \frac{1}{2r}(e^{rt}-e^{-rt}) & 0 & 0\\ \frac{r}{2}(e^{rt}-e^{-rt}) & \frac{1}{2}(e^{rt}+e^{-rt}) & 0 & 0\\ 0 & 0 & \cos(\omega_n t) & 0\\ 0 & 0 & -\omega_n\sin(\omega_n t) & 0\end{bmatrix}}^{\mathbf{B}}\begin{bmatrix}\Theta_0\\ \dot{\Theta}_0\\ D_0\\ \dot{D}_0\end{bmatrix}
\tag{7}
$$

For a particular initial condition, the above equation (Eq. 7) varies only with natural frequency (ω_n) of oscillation in the length of stance leg and time period (T). So we need to find a set of initial conditions which gives us a periodic value after every impact. Impact conditions need to be applied to understand the necessary changes to the variables. Under the assumption of small angles, the impact matrix (Eq. 3) becomes,

$$
\mathbf{A}_1 = \begin{bmatrix}-1\,0\,0\,0\\ 0\;\;1\,0\,0\\ 0\;\;0\,1\,0\\ 0\;\;0\,0\,1\end{bmatrix}
\tag{8}
$$

4.4 Eigen Vectors and Interrelations

After applying impact conditions, we get a relation (Eq. 9) between initial conditions of two consecutive gaits.$\mathbf{Q}^+$ in the Eq. (8) indicates the states just after impact, i.e,. the ICs for the next cycle. Eigen vectors for the Eigen value one of the matrix $\mathbf{C}$ gives us a relation between ICs. Eigen values are found to be $\lambda = (1,\ -1,\ \cos(\omega_n T))$.

$$\mathbf{Q}^+ = \begin{bmatrix} \Theta \\ \dot{\Theta} \\ D \\ \dot{D} \end{bmatrix}^+ = \overbrace{\mathbf{A_1 B}}^{\mathbf{C}} \begin{bmatrix} \Theta_0 \\ \dot{\Theta}_0 \\ D_0 \\ \dot{D}_0 \end{bmatrix}$$

$$or, \quad \mathbf{Q}^+ = \mathbf{CQ_0} \tag{9}$$

For the eigen value 1, it returns the original initial condition. Third eigen value, $\cos(\omega_n T)$, also returns the original ICs in a condition that $\cos(\omega_n T) = 1$ or, $\omega_n T = 2n\pi$, where n = 1,2,3... $\dot{D}_0$ be always zero for any value D_0 as it satisfies the above-mentioned relation. Relations we get are:

$$\begin{aligned} \lambda = 1, \quad & \dot{\Theta}_0 = r\left[\frac{1+e^{rT}}{1-e^{rT}}\right]\Theta_0 \\ \lambda = \cos(\omega_n \mathrm{T}), \quad & \dot{\mathrm{D}}_0 = -[\omega_n \tan(\omega_n T)] D_0 \end{aligned} \tag{10}$$

4.5 Step Period Finding

From Eq. (10), a set of ICs are obtained for a particular Θ_0 and D_0 but there are two more unknowns time period and natural frequency which varies with the leg stiffness. For a particular ω_n, it could possible to get few feasible solutions of time period. Substituting the above relations in the expression of Φ in Eq. (6) and using $\cos(\omega_n T) = 1$ or, $\omega_n T = 2n\pi$, where n = 1,2,3..., an expression for time period is found. From the symmetry of a stable gait it can be said $\Phi(T) = -\Phi_0$. We have used the expression for Φ_0 in Eq. (5). Note that the initial deflection of leg length equal to the static deflection from the equilibrium point for this linear case, i.e., $D_0 = g'$.

$$\begin{aligned} \Phi(T) = & \left\{\frac{1}{2(2-p)}(e^{rT} + e^{-rT}) + \left(2 - \frac{1}{(2-p)}\right)\cos(\sqrt{g'}T)\right\}\Theta_0 \\ & + \left\{\frac{1}{2r(2-p)}(e^{rT} - e^{-rT}) + \frac{1}{\sqrt{g'}(2-p)}\sin(\sqrt{g'}T)\right\} r\left[\frac{1+e^{rT}}{1-e^{rT}}\right]\Theta_0 \\ & + \left\{\frac{p\gamma r^2(\cos(\omega_n T) - \cos(\sqrt{g'}T))}{(\omega_n^2 - g')(1-p)}\right\} D_0 + \left(2\gamma - \frac{\gamma p}{(1-p)}\right)\cos(\sqrt{g'}T) \end{aligned} \tag{11}$$

MATLAB has been found helpful for solving this equation for minimum time period. In this case, we get time period *(T)* = 1.1966 s for $\omega_n = 21$ rps which is marked within a black box in Fig. 3. Now using these values in the Eq. (10) helps to find a natural set of ICs for periodically walking biped.

A biped with 1-m uncompressed leg ($d_2 = 1$ m), hip mass of 1 kg ($m_h = 1$ kg), leg oscillation frequency 21 rps ($\omega_n = 21$ rps), starts with an initial step of 0.2-rad stance leg angle ($\Theta_0 = 0.2$ rad) and the rest comes from the relations found before. A symmetric periodic walking cycle we have found with the above values of parameters over a slop of $\gamma = 0.009$ rad which is here in the Fig. 2. The pattern shows symmetric behavior and the repeatability of initial conditions at the beginning of each step.

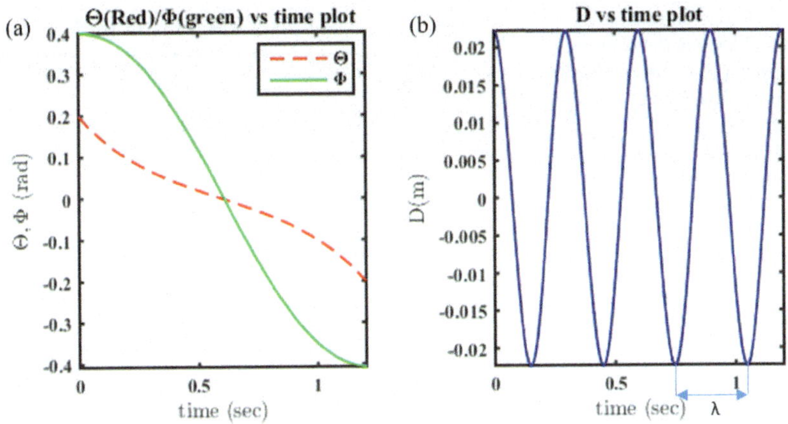

Fig. 2 This figure shows for linear case **a** Leg angles versus time for a single cycle **b** Change in length of stance leg over the time period at an angular frequency 21 rps. λ is one cycle of the oscillation

Where both the leg angle intersects is called the mid-stance position, the stance leg oscillates in a simple harmonic motion with an amplitude equal to static deflection as this is independent of other parameters as shown in the Fig. 2 b.

Effect of ω_n. We have investigated the effect of leg stiffness on the gait parameters. We get a relationship between all initial states which are dependent on time period and the stiffness of the stance leg. From the Eq. (11), we get a relation between these two (T and ω_n) which has been plotted in the Fig. 3 and also the relationship $\omega_n T = 2n\pi$ for $n = 1,2,3\ldots$is shown within a suitable range. The intersection points of these two are the possible solutions. It is possible that there may not be a solution for every ω_n and there also a possibility that we can see multiple solutions for some ω_n as well.

5 Solution Through Numerical Optimization

Analysis has been done in order to find a stable and periodic walking gait which gives us a set of suitable initial condition and time period through numerical optimization. MATLAB is used for numerical optimization. ODE45 is the solver for solving ODEs and *fminsearch* is the function, based on the Nelder–Mead simplex algorithm which can optimize a cost function of multiple parameters. Initial guess given based on the interrelation found through the analytical derivations and from the ω_n versus T relation. It gives the optimum value of the function around the initial guess within our region of interest. A tolerance of 1×10^{-13} is given to the function for high accuracy.

Non-linear equations of motion (Eq. 2) used for numerical optimization. The relations between variables in ICs have been given as input in the process. The

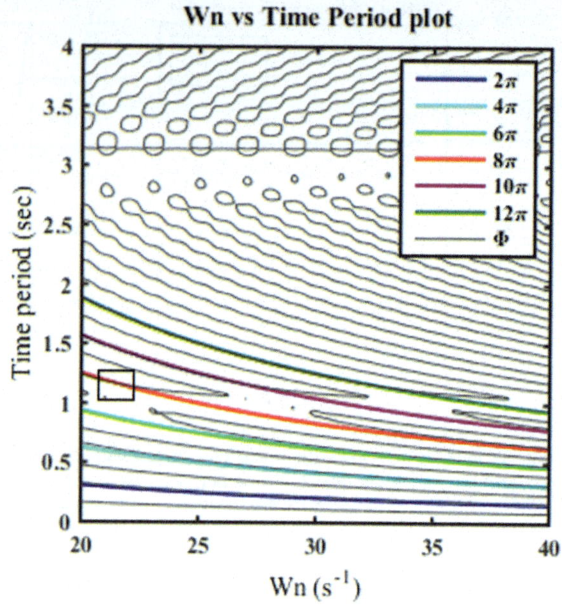

Fig. 3 Figure shows the relation between the natural frequency (ω_n) of oscillation of the stance leg and the time period (T) of a cycle. Stiffness of the leg is proportional with ω_n. So in the variation of stiffness has an effect on T. Intersection point within the black box shows one solution that we have found at $\omega_n = 21$ rps

collision condition (Eq. 3) applied to the system. All the ICs are considered in the cost function.

So, the difference between ICs of two successive cycles has been minimized with a minimum tolerance limit. The time period (T) and frequency (ω_n) are the parameters to be optimized. Results have been investigated with the same biped parameters as stated above for the linearized system. It has been found that the time period (T) to be 1.2043 s. Figure 4 shows the change in leg angles with respect to time in a cycle and the change in length of stance leg in the same time period. A periodic gait satisfies the conditions for producing a symmetric gait.

6 Results

6.1 Comparison of Results

Results for the both method has been compared in the Fig. 5 and in the Table 1. As we are not investigating foot scuffing in this study so no comparison of change in leg length has been made here. Both the leg angles in each case show motion of similar nature and returns to the initial condition at the end of the cycle. There are little differences in values for similar leg angles which is a result of linearization as some approximations made. As in the numerical optimization there are all coupled non-linear equations so the non-linearity of the plot depends on all variables unlike

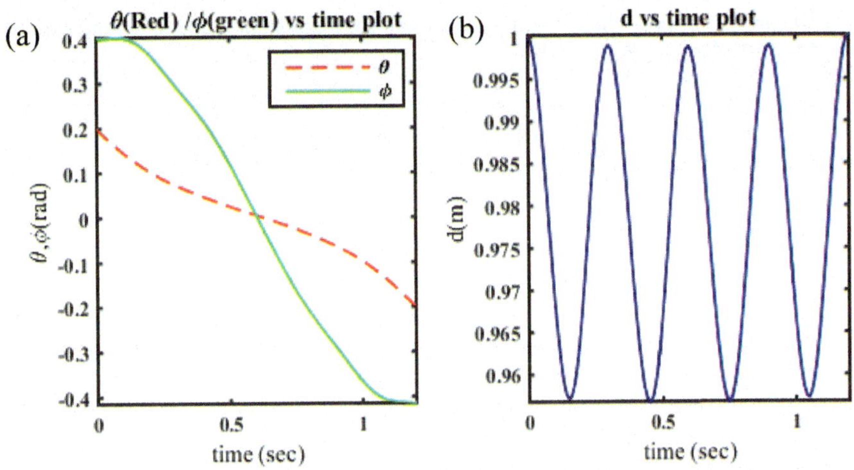

Fig. 4 This figure shows for non-linear case **a** Leg angles versus time for a single cycle **b** Change in length of stance leg over the time period at a frequency of 21 rps

Fig. 5 This figure compares both the results of Linear approximation of leg angles (Θ, Φ) and the non-linear leg angles (θ, ϕ) with time in one cycle

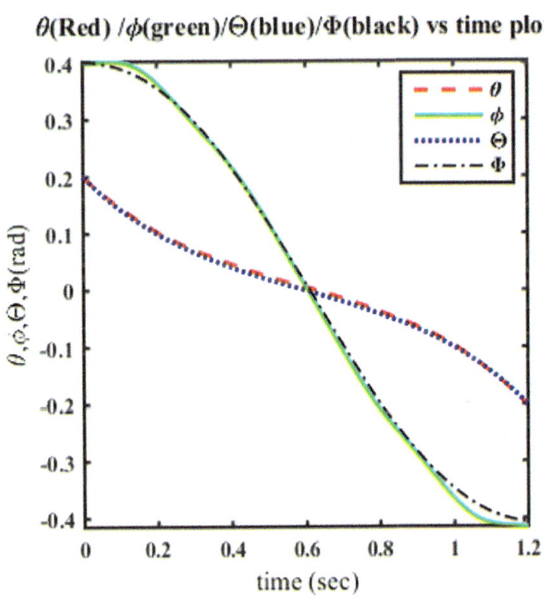

Table 1 Initial velocity and time period found for a periodic gait

Process	Init. velocity (rad/sec)	Time period (sec)
Linearization solution	−0.6624	1.1966
Numerical solution	−0.6085	1.2043

the linearized model. These results are shown for a constant slope and vary with the inclination of the walking surface.

7 Discussions and Conclusion

In this paper, we have presented a simplest 2-D biped robot which is inspired by the work of Garcia et. al. [13] and modified with two flexible legs. Dynamic equations of the system during the swing phase have been determined using the Euler–Lagrange method. These equations are non-dimensionalized, and the ratio of the foot mass to the hip mass is used to further reduce and simplify the equations. Conservation of angular momentum principle was used to obtain the robot states after the impact of the swinging leg. For the linearized model, we have analytically obtained a natural set of initial conditions to obtain a symmetric period gait with above constraints. These initial conditions were used to obtain the periodic cycle for the non-linear model and the corresponding time period. For the non-linear model, an analysis carried out to find a symmetric and periodic walking gait which gives us a set of suitable initial condition and time period through numerical optimization. Both the results confirm a possibility of getting a symmetric periodic gait when it is walking down a slope. The stiffness of the leg spring has an effect on the time period so it could further be helpful to make faster stable gait biped with proper foot model. The time period of the walking cycle indicates that the biped can walk reasonably fast. Four oscillations of the stance leg in a walking step is consistent for that ω_n though with one oscillation in a step could make the gait even more natural but it does not affect the periodic gait.

From the study, we have found a passive biped with periodic gait which can walk fast under only the action of gravitational force. Foot scuffing could have been an issue which we need to investigate further. It could be helpful in future for producing faster gait and reduction of impact energy dissipation with proper foot model due to its flexibility. We believe that this study will be helpful for better understanding in compass gait bipeds.

References

1. Kato I, Waseda University robotics research (https://www.waseda.jp/top/en-news/13858)
2. Mochon S (1980) Ballistic walking. J Biomech 13:49–57
3. Mcgeer T (1990) Passive dynamic walking. Int J Rob Res 9(2):62–82
4. Park JH, Kim KD (1998) Biped robot walking using gravity-compensated inverted pendulum mode and computed torque control. In: Proceedings of 1998 IEEE International Conference on Robotics and Automation (Cat. No.98CH36146), vol 4, no May, pp 3528–3533
5. Collins S, Ruina A, Tedrake R, Wisse M (2005) Efficient bipedal robots based on passive-dynamics walkers. Science(80) 307(5712): 1082–1085
6. Gomes M, Ruina A (2011) Walking model with no energy cost. Phys Rev E Stat Nonlinear Soft Matter Phys 83(3): 6–9

7. Li Z, Tsagarakis NG, Caldwell DG (2013) Walking pattern generation for a humanoid robot with compliant joints, pp. 1–14
8. Sarkar A, Kishore NN, Dutta A (2014) 8-DoF biped robot with compliant links. In: 2014 13th International Conference on Control, Automation, Robotics and Vision, ICARCV 2014, vol 2014, no December, pp 1195–1200
9. He J, Ren G (2018) On the stability of passive dynamic walker with flat foot and series ankle spring. Adv Mech Eng 10(3):1–12
10. Schauβ T, Scheint M, Sobotka M, Seiberl W, Buss M (2009) Effects of compliant ankles on bipedal locomotion. In: Proceedings of IEEE International Conference on Robotics and Automation, pp 2761–2766
11. Iida F, Rummel J, Seyfarth A (2008) Bipedal walking and running with spring-like biarticular muscles. J Biomech 41(3):656–667
12. Iida F, Minekawa Y, Rummel J, Seyfarth A (2009) Toward a human-like biped robot with compliant legs. Rob Auton Syst 57(2):139–144
13. Garcia M, Chatterjee A, Ruina A, Coleman M (1998) The simplest walking model: stability, complexity, and scaling. J Biomech Eng 120(2):281

Modelling and Analysis of 3-PSS Parallel Kinematic Mechanism

Thomas Mervin Joe, Mallick Gaurav, A. P. Sudheer, and M. L. Joy

Abstract Parallel Kinematic Mechanism's (PKM) are being used for machining complex and intricate shapes for précised applications with higher stiffness and rigidity. It is possible to achieve different geometries with variable poses of the tool. PKMs are also being used for various operations such as inspection, surgical, assembly, etc. This paper focuses on modelling and analysis of a novel three legged PSS PKM for precise applications. The available automated systems are expensive and bulky making it less suitable for small/medium-scale industries. The focus of this paper is to develop a simple and cost-effective 3-PSS PKM. Kinematic modelling is done using the geometrical approach. The workspace for this PKM is plotted using MATLAB and compared with the existing mechanisms. The optimum workspace is determined using Genetic Algorithm (GA). Further, the static structural analysis is done using ANSYS for the determination of maximum stresses and deformation induced in the robot on the application of static load. The dynamic analysis is carried out using the MSC ADAMS software. The dynamic torques, acceleration and forces acting at each joint are determined instantaneously by considering velocity variations across sliders and friction at all joints.

Keywords Parallel kinematic machinery (PKM) · Kinematic modelling · Dynamic modelling · Work space analysis · Genetic algorithm (GA)

T. Mervin Joe (✉) · M. Gaurav · A. P. Sudheer · M. L. Joy
National Institute of Technology, Calicut, India
e-mail: mervin_p170047me@nitc.ac.in

M. Gaurav
e-mail: gaurav_m160341me@nitc.ac.in

A. P. Sudheer
e-mail: apsudheer@nitc.ac.in

M. L. Joy
e-mail: mlj@nitc.ac.in

D. Sen et al. (eds.), *Mechanism and Machine Science*, Lecture Notes in Mechanical Engineering, https://doi.org/10.1007/978-981-15-4477-4_46

1 Introduction

Presently, a lot of researches are carried out in the development of parallel robotic systems. This is because of its high rigidity, stiffness and higher load carrying capability of the structure. These mechanisms are also capable of achieving higher precise and accurate motions. However, there exists a drawback of limited workspace for parallel configurations. Therefore, one of the major challenges of this work is to develop a design having more workspace. The idea of parallel manipulators is introduced first by Gough, followed by the development of Stewart platform. Nabat et al. [1] has explained the development of a parallel robot that is used for pick and place operations. The PKM is designed such that it facilitates SCARA motions at the least possible cycle time. It is basically a combination of the DELTA manipulator and a 3RP serial mechanism. Even though it offers high speed motions, this hybrid mechanism is expensive and difficult to control. Clavel [2] explained the complete synthesis and development of the DELTA robotic system. It basically consists of three legs with each leg having series of revolute joints. The drawback of this design is the inability to carry larger loads. Tsai et al. [3] has developed a translational platform which is a modification of the DELTA robotic system. In this configuration, the end effector platform is located at the top. In addition to this, one revolute joint in each leg is replaced with a translational joint and that design is used for flight simulation applications. Because of this modification, the workspace of this robot has reduced compared to that of the DELTA robot. The closed form forward and inverse kinematics of the mechanism was explained in detail. Li et al. [4] has explained the complete kinematics of a 3-PRS parallel manipulator. The workspace for the 3-PRS system is depicted in Table 3. The important consideration to be taken to ensure proper mobility and manoeuvrability of the end effector is detailed in Yoshikawa [5]. The mobility and kinematic analysis of redundantly actuated 2UPR–2PRU four limbed parallel manipulator is detailed in Xu et al. [6]. The singularity analysis is done to determine the singular positions using the Jacobian matrix. However, the dynamic modelling for the mechanism is not explained. Zhang et al. [7] explained the design and analysis of a novel three DoF parallel manipulator with decoupled motions. The kinematic relations were derived using the screw theory approach. The computational complexities involved in modelling were depicted in the paper.

It is observed from the comprehensive literature review that the parallel manipulators generally have limited workspace, bulky structure, difficult to control and relatively more expensive. Therefore, the primary focus of this paper is to propose a parallel manipulator that will overcome the challenges faced by such manipulators. To overcome the demerits of PKM, the proposed design must have a simpler design higher workspace and easier mobility to enable all possible motions to the platform attached. This paper also explains the modelling, analysis and workspace optimization of 3-PSS parallel manipulator. In the coming section, workspace of the 3-PSS manipulator is compared to other parallel manipulators having spherical joints with similar dimensions. This manipulator consists of three legs with each leg having one prismatic and two spherical joints. Geometric approach is used for the kinematic

Table 1 f_i and f_{id} values for slider, link, shaft and platform

Element	f_i	f_{id}	$\sum_{i=1}^{g} f_i - f_{id}$
Sliding blocks	1	3	−2
Connecting link	3	3	0
Lead screw	0	3	−3
Moving platform	6	0	6

modelling of the manipulator. This mechanism can be used for multiple applications depending upon the end effector used and is also suitable for surgical applications. The robot is modelled individually across each legs and finally combined together at the moving platform.

The number of degrees of freedom for the mechanism is calculated using the Gruebler's criterion, stated as follows:

$$DoF = b(n - g - 1) + \sum_{i=1}^{g} f_i - f_{id} + s \tag{1}$$

where 'b' is the number of DoF in three-dimensional space, 'n' is the number of elements, 'g' is the number of joints, 'f_i' is the number of DoF of the ith element, 'f_{id}' is the number of identical DoF and 's' is the number of passive joints. The values for 'b', 'n', 'g' and 's' in the above equation for this mechanism are 6, 10, 9 and 6, respectively. The values of '$\sum_{i=1}^{g} f_i - f_{id}$' for each element is listed in the Table 1.

On substituting the above values into the Gruebler's equation stated in 1, the number of DoF for this mechanism is found to be 3. The three degrees refers to one linear translation and two rotation of the moving platform. The linear motion of the platform is about the 'x' axis and rotation about the 'x' and 'y' axes (refer Fig. 1). The workspace for the mechanism is plotted using the kinematic relations obtained from the geometric model by the aid of MATLAB. The maximum possible workspace achievable is determined using the Genetic Algorithm (GA) for a specific range of joint parameters. The static structural analysis is done for the robot frame and structure. Analysis for bending, torsion and shear is considered for the static structural analysis. The maximum deformations and the stresses induced in the frame is analysed using ANSYS. The dynamic analysis for the robot is done using MSC ADAMS for a particular case study.

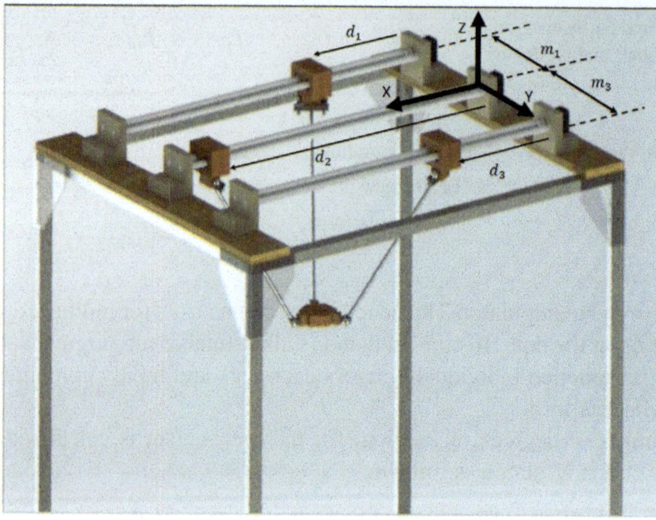

Fig. 1 Different variables assigned to the PKM

2 Kinematic Modelling of the 3-PSS Robot

There are many literatures that involve the modelling of mechanisms with prismatic, revolute or hook's joints. It is observed that the use of hook's joint instead of spherical joints will cause a reduction in mobility of the end effector. This would therefore reduce the available workspace. Kinematic modelling is done to obtain kinematic relations of the end effector with respect to a fixed point. These relations are further used to plot the workspace and in addition for the control of end effector to reach the desired positions. The mechanism is modelled by initially assigning frames to all necessary joints. The relationships between these frames can be represented by closed form mathematical equations using a suitable loop closure equation. The manipulator kinematics mainly studies how the position of the frame located at the end effector changes if the robot joints are subjected to motion. The kinematic relations are obtained in terms of the joint variables. Each leg of the PKM is modelled separately. The spherical joints in the mechanism are modelled such that it has three mutually perpendicular revolute joints acting along the same origin. This approximation is done to simplify the modelling of spherical joints. A coordinate system XYZ is initially assigned at the end of second leg as shown in Fig. 2. This frame is considered to be the base frame.

In Fig. 1, 'm_i' refers to the distance between two prismatic sliders measured along the y axis. The value of m_i will change depending upon the leg being modelled. Since there are three legs, 'i' ranges from 1 to 3. 'd_i' refers to the linear distance travelled by the slider measured along the X axis with respect to the origin axis of the assigned coordinate system XYZ. Each prismatic joint consists of a slider, sliding over smooth shafts and it is actuated by a motor-driven lead screw. The moving

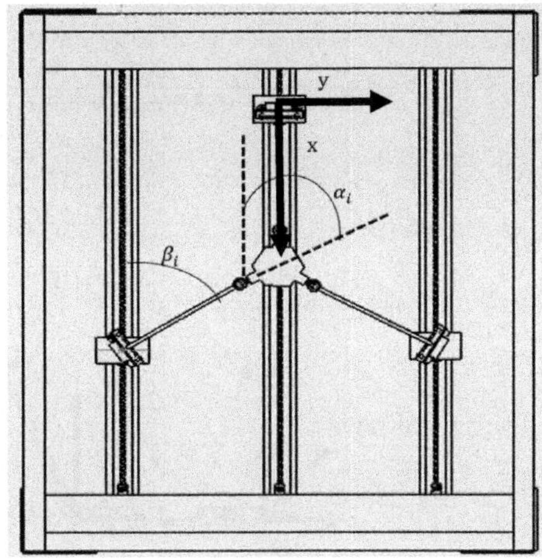

Fig. 2 Apparent joint angles assigned to the PKM

platform is connected to slider through a link of length '*l*' by means of spherical joints. The geometrical approach uses three angles to describe the orientation of link, obtained by the rotation of spherical joints. The three angles are the apparent angles of the link projected to two different planes. The link connecting the slider and the platform can be considered as a line inclined to multiple planes. α_i is the apparent angle made by the lower spherical joint axis with the x axis as shown in Fig. 2. Depending upon the rotation of the spherical joint, correspondingly α_i will alter. Similarly, β_i is the apparent angle made by the rigid link with the x axis. α_i and β_i are both measured by projecting onto the XY plane.

Similarly, γ_i is the apparent angle made by the link measured along the XZ plane. Let '*r*' be the distance measured from the centre of the platform to the spherical joint connected to the link. The distance from the prismatic joint's axis to the upper spherical joint is assigned 'd_{si}' and the distance from the downstream spherical joint to the end effector is assigned 'd_{ei}'. These two offsets are measured along the Z axis. Figure 3 shows γ_i, r, d_{si} and d_{ei}. The basic concepts of geometry are then used to obtain the position coordinates of the centre of moving platform. In this methodology, each leg is modelled individually like an open chain. Finally, the three legs are coupled together using a suitable loop closure equation to obtain the closed form kinematic relations of the mechanism. The point modelled across the three different legs is the same, which is the centre of moving platform. In this approach the mechanism is projected onto a two-dimensional plane and concepts of line geometry is used to model the mechanism. The pose of the end effector is defined with respect to the base frame through geometric modelling. The position coordinates of any *i*th leg can be represented as follows:

$$x_i = d_i + r\cos\alpha_i + l\cos(\alpha_i + \beta_i)\sin\gamma_i \tag{2}$$

$$y_i = m_i + r\sin\alpha_i + l\sin(\alpha_i + \beta_i)\sin\gamma_i \tag{3}$$

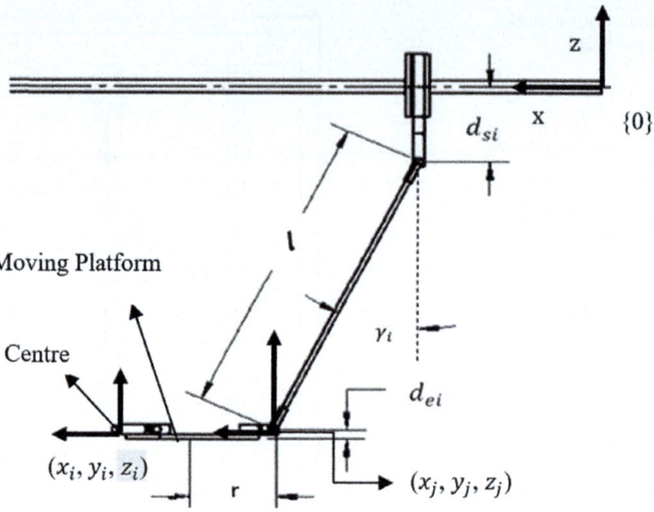

Fig. 3 Apparent angle assigned across the side view for the PKM

$$z_i = -d_{si} - d_{ei} - l\cos\gamma_i \tag{4}$$

The values r,l, d_{si} and d_{ei} remains a constant for a particular mechanism. The above three equations represent the open chain general expression for the centre of moving platform with respect to the global frame. The three legs are joined at the moving platform with the condition that the distance between any two consecutive spherical joint is a constant. The moving platform is taken to be an equilateral triangle with side 'k' as shown in Fig. 4. Let '$x_j^{'}$,' $y_j^{'}and^{'}z_j^{'}$ be the position coordinates of the spherical joint as depicted in the figure.

$$k^2 = \left(x_j - x_{j+1}\right)^2 + \left(y_j - y_{j+1}\right)^2 + \left(z_j - z_{j+1}\right)^2, j = 0, 1, 2 \tag{5}$$

where,

$$x_j = d_j + l\cos\left(\alpha_j + \beta_j\right)\sin\gamma_j \tag{6}$$

Fig. 4 Equilateral triangle shaped moving platform

$$y_j = m_j + l\sin(\alpha_j + \beta_j)\sin\gamma_j \tag{7}$$

$$z_j = -d_{sj} - d_{ej} - l\cos\gamma_j \tag{8}$$

Substituting Eqs. 6, 7 and 8 into Eq. 5 yields three different equations with distinct joint angles. The three equations are solved for obtaining the sliding distance 'd_j'for each leg by substituting the constraints values of Euler angles for each leg. From the inverse kinematic relations, the Euler angles corresponding to the position of 'd_j'can be determined. By re-arranging Eq. 3,

$$\gamma_i = \cos^{-1}\left(\frac{-d_{si} - d_{ei} - z_i}{l}\right) \tag{9}$$

Substituting Eq. 9 into Eqs. 2 and 3, the expression for the slider distance from the origin along the x axis can be derived as,

$$d_i = (x_i - r\cos\alpha_i) \pm \sqrt{(l\sin\gamma_i)^2 - (y_i - m_i - r\sin\alpha_i)^2} \tag{10}$$

This value then obtained by solving the above equation is substituted into Eqs. 2, 3 and 4 to obtain the position coordinates of the end effector with respect to the global frame. The workspace for the manipulator is plotted by substituting the link dimensions and range of angles listed in Table 2 into Eqs. 2, 3 and 4. On doing this, the final work volume plot for the mechanism is obtained. Table 3 shows workspace plot of 3-PSS and few other parallel manipulator architectures with spherical joints and its corresponding geometry for which the dimensions of the frames were taken to be similar to that of the 3-PSS manipulator for comparing the work volumes. The respective work volume for each mechanism is also given in the table for all the mechanisms considered having spherical joints.

Genetic Algorithm is used to determine the optimised work volume possible with the selected dimensions of the various links in the mechanism. GA is a very robust algorithm and proved efficient for optimisation in engineering applications [10]. The workspace plot of the 3-PSS parallel mechanism is approximated to be cuboidal in shape. The objective will be to maximize the work volume. However, the actual work volume will be slightly lesser than the calculated value since it has a slight tapering towards the bottom as seen in the workspace plot for the 3-PSS in Table 3. On comparing the workspace plots obtained for the 3-PSS manipulator with that of the other parallel robots having spherical joints listed in Table 3, it is observed that the work space for this mechanism is more. The general trend observed for parallel manipulators having spherical joints is that the work volume obtained is comparatively less. However, in the proposed design having spherical joints, it is seen that the work volume obtained is higher. The reason for using spherical joints is to enable all possible mobility for the moving platform. However, in this preliminary work, the singularity analysis and trajectory planning is not included. This mechanism is also suitable for full body surgical/scanning operations. This

Table 2 Link dimensions and angle constraints

	m_i (m)	r (m)	l (m)	d_{si} (m)	d_{ei} (m)	d_i/(m)	α_i °	β_i°	γ_i°
Leg1	−0.2	0.05	0.450	0.025	0.025	0–0.7	0–30	0–30	−30–30
Leg2	0	0.05	0.450	0.025	0.025	0–0.7	0–30	0–30	−30–30
Leg3	0.2	0.05	0.450	0.025	0.025	0–0.7	0–30	0–30	−30–30

Table 3 Workspace plots for 3-SPS [8], 3-PRS [4], 6-SPS [9] and 3-PSS manipulator

S.no	Configuration	Sliding Distance (m)	Workspace Plot
1	3-SPS Robot [8] 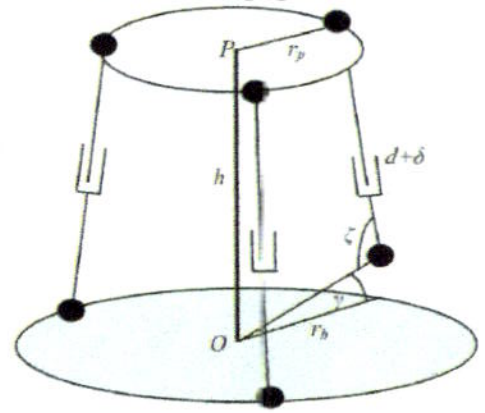	0.7 (3nos)	Approx. Work Volume = 1.32×10^{-5} m^3 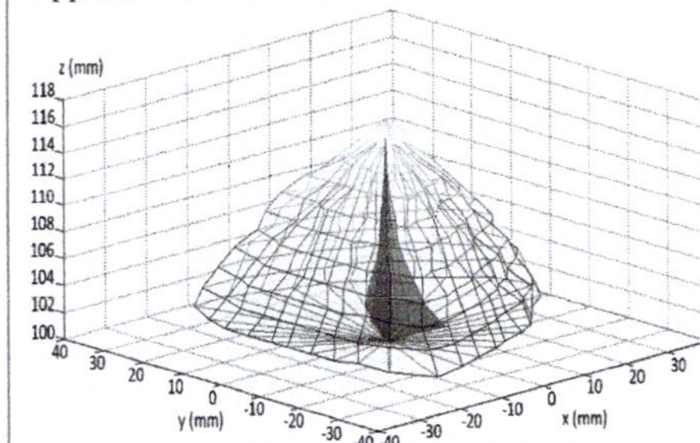
2	3-PRS Robot [4] 	0.8 (3nos)	Approx. Work Volume = 3.2×10^{-4} m^3 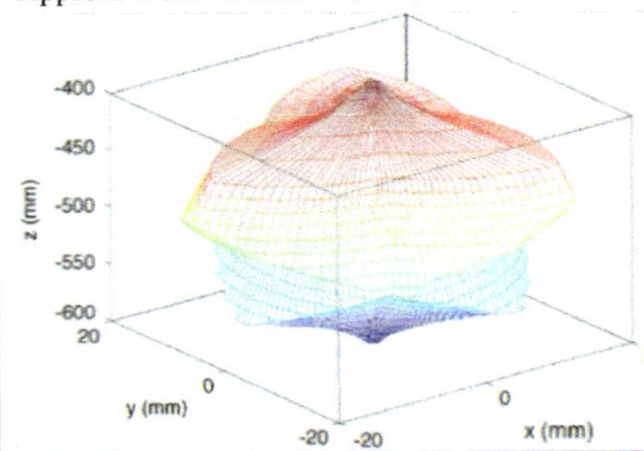

(continued)

Table 3 (continued)

S.no	Configuration	Sliding Distance (m)	Workspace Plot
3	6-PSS Manipulator [9] 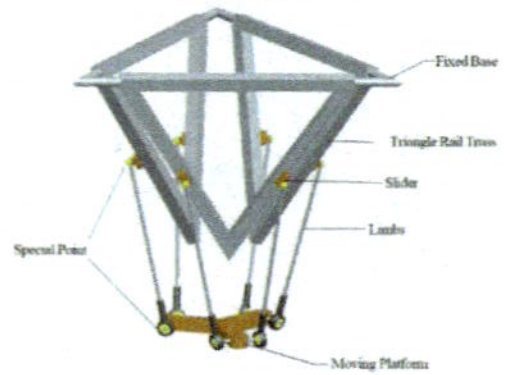	0.8 (6nos)	Approx. Work Volume = $6.28 \times 10^{-3}\ m^3$ 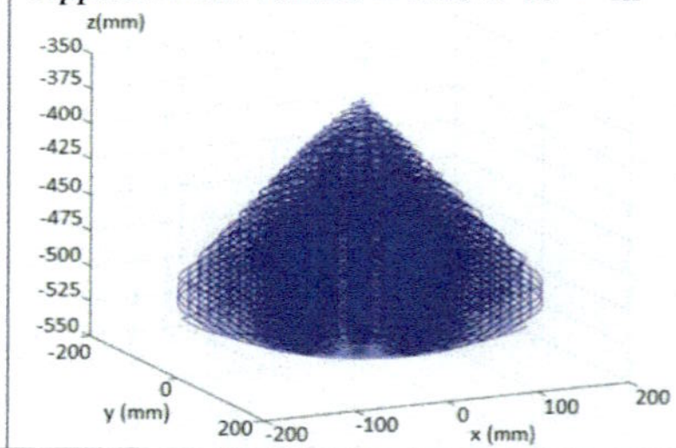
4	3-PSS Manipulator 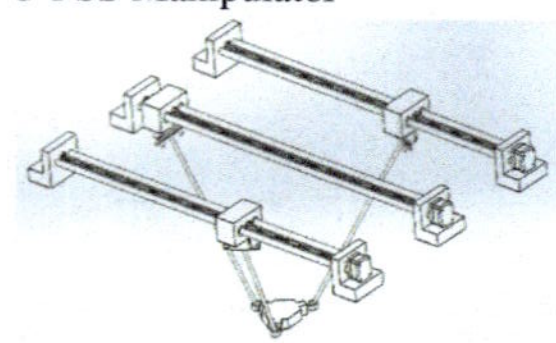	0.7 (3nos)	Work Volume = $1.388 \times 10^{-2}\ m^3$ 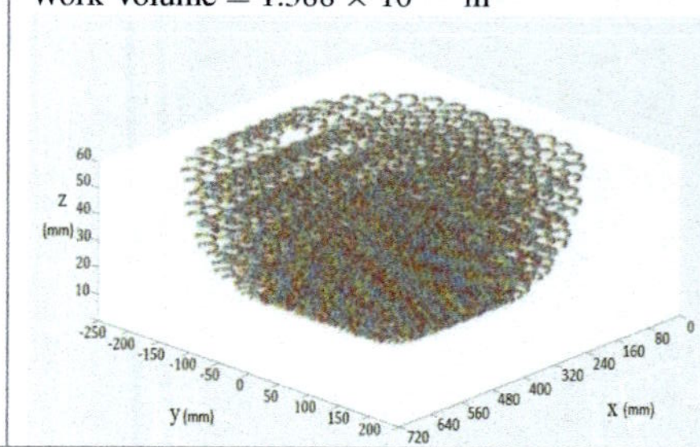

Table 4 GA Parameters

S.no	GA parameters	Value
1	Crossover	Single point crossover
2	Population	100
3	Generations	500
4	Crossover rate	0.9
5	Mutation	Bit-flip mutation with fixed mutation probability 0.1

can be made possible by increasing the longitudinal workspace by changing the sliding distance which is not practical in other mechanisms. By solving the coupled equations (refer Eq. 4), the joint variables are determined and correspondingly the constraints are formed for the optimisation of work volume using GA. The volume of the workspace is maximised by considering the volume equation of a cuboid. Hence, the objective function is to:

$$\text{Maximise}\,V = \text{Max}\int\limits_{x}\int\limits_{y}\int\limits_{z} \Delta x \Delta y \Delta z \tag{11}$$

With respect to the constraints:

- $0 \leq d_j \leq 1$ m
- $0 \leq \alpha_j \leq 30^{\text{o}}$
- $0 \leq \beta_j \leq 30^{\text{o}}$
- $0 \leq \gamma_j \leq 30^{\text{o}}$

The various parameters used to solve GA problem is shown in Table 4.

During the simulation run, the population and generations are kept constant. The cross-over rates are changed from 0.8 to 0.95 and the mutation rates varied from 0.01 to 1. The optimised work volume is calculated by using the GA parameters listed in Table 4. The optimisation is carried out in an Intel Core i7 processor, 8 GB RAM, and the maximum time taken for getting the optimised result is 3.2 s. Based on this optimisation problem, the maximum work volume for the considered 3-PSS manipulator is 0.01388 m^3.

3 Static Structural Analysis

The static structural analysis is done to compute the effects of steady loading conditions on the robot structure. It is done to determine the maximum deformations induced in the system, stresses and strains induced at extreme loading conditions on the robotic structure. The simulation results can be further used to identify the weaker parts and make necessary modifications to avoid the failure of the mechanism. The analysis is done using ANSYS finite element software. The materials are assigned accordingly from the material library available in the ANSYS. Table 5 shows the name of the part and the corresponding material assigned for the analysis.

The analysis is done by applying a static load of 100 N at the centre of moving platform. The corresponding deflections induced in the system are determined from the simulation results. The dimensions of certain parts like the smooth shafts, lead screws, etc. is changed accordingly to avoid excessive bendings or deflections. Excessive deflections can interfere into the smooth motions of the sliders along the linear smooth shafts. Figure 5 shows the simulation results for the considered conditions. The maximum deflection induced in the shaft is 0.2622×10^{-3} m. This deflection happening on the linear smooth shaft will not hinder the motion of the prismatic sliders which was verified experimentally. The deflections induced at all other parts are small and are within safer limits. The analysis is done for that instantaneous position such that the moving platform is at the exact middle. At this position, the deflection acting in the linear shafts will be maximum. The two ends of the smooth shaft can be assumed to be fixed. The maximum stress (von Mises) induced for the same load and position is 28.976 MPa. Figure 6 shows the simulation results for stresses induced in the structure. The stress induced is maximum on the linear shafts. However, this stress is far below the yield stress of the material. Therefore, the present design is safe for the maximum load that is applied.

Table 5 Part name and material assigned for the ANSYS simulation

S.no.	Part	Material assigned	Properties
1	Lead screw	Alloy Steel 6061	Wear resistant, high hardness, toughness, corrosion resistance
2	Threaded Fasteners	Mild Steel 1091	High toughness, wear resistant, high tensile and impact strength
3	Frame	Aluminium Alloy 6061	Light weight, high toughness, tensile strength, weldability, corrosion resistance
4	Smooth Shafts	Stainless Steel	High strength, wear resistant, corrosion resistance, lower maintenance
5	Rigid Links	Aluminium Alloy 6061	Light weight, high toughness
6	Sliders	Galvanised Steel	High toughness, high strength, higher shear strength
7	Platform	Aluminium Alloy 6061	Light weight, high toughness
8	Spherical joints	Stainless steel	High strength, wear resistant, higher fatigue strength

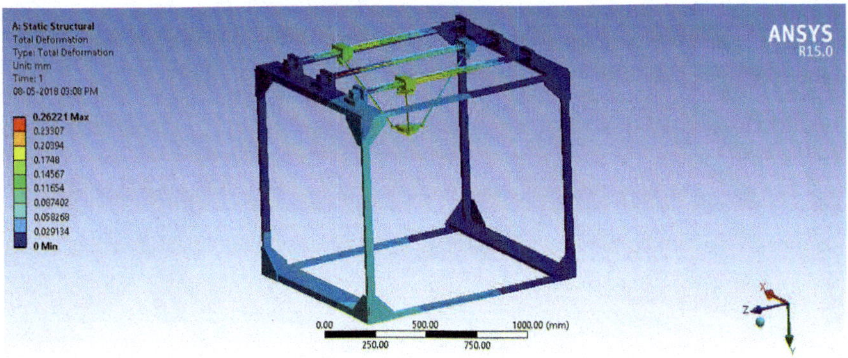

Fig. 5 Static structural analysis for deflection of PKM using ANSYS

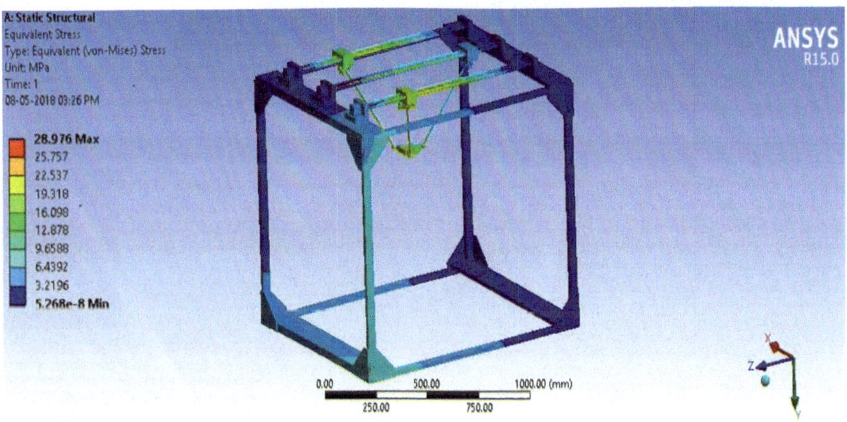

Fig. 6 Static structural analysis for maximum stress induced using ANSYS

4 Dynamic Analysis Using MSC ADAMS

The MSC ADAMS multi-physics simulation software is used to determine the dynamic actuator torques, velocities and acceleration acting along the links and joints. Figure 7a and b shows the 3-PSS mechanism modelled in the MSC ADAMS software with all the constraints and joints specified. The slider and the lead screw are connected using a screw pair. The motor and the lead screw are connected via a revolute joint. Therefore, the rotation of the lead screw enables a linear motion to the slider. Spherical joints are defined for the rigid links connected to the sliders and moving platform. The materials assigned to the parts are similar to that given for the ANSYS simulation as listed in Table 5.

In this paper, a case study simulating pick and place operations for a maximum payload of 40 N is explained. The simulation is programmed to complete the task in

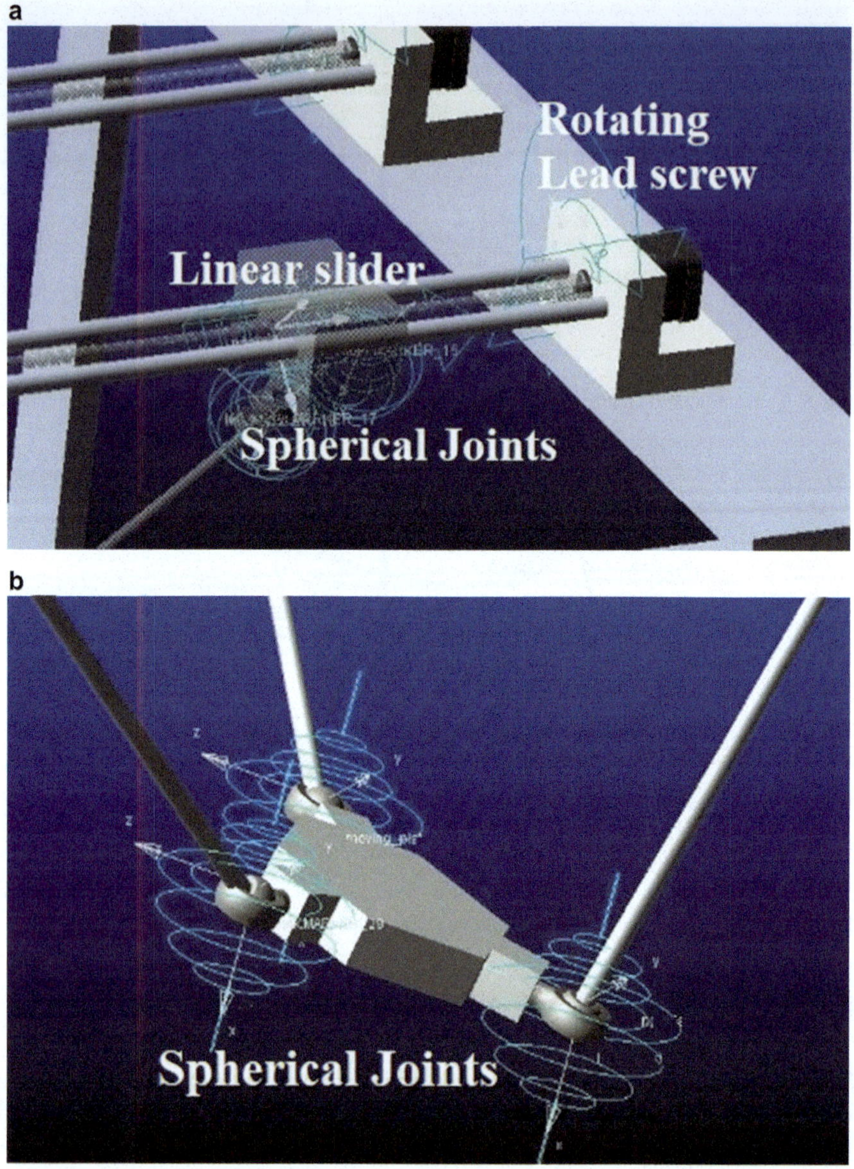

Fig. 7 Different joints and boundary conditions assigned to PKM in MSC ADAMS

1 second. Within this time span, the sliders move by a distance of 0.1 m. Simultaneously, the dynamic torques acting along the active joints are computed. Figures 8, 9 and 10 show the torque variations acting along the revolute joints in the first, second and third legs which rotate the lead screw.

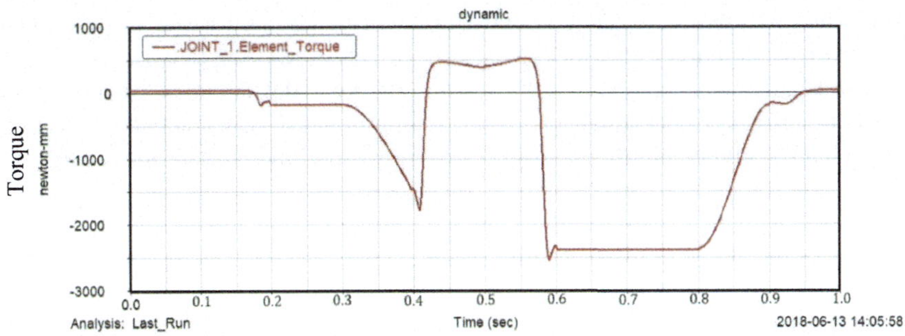

Fig. 8 Torque variations of first active joint with respect to time

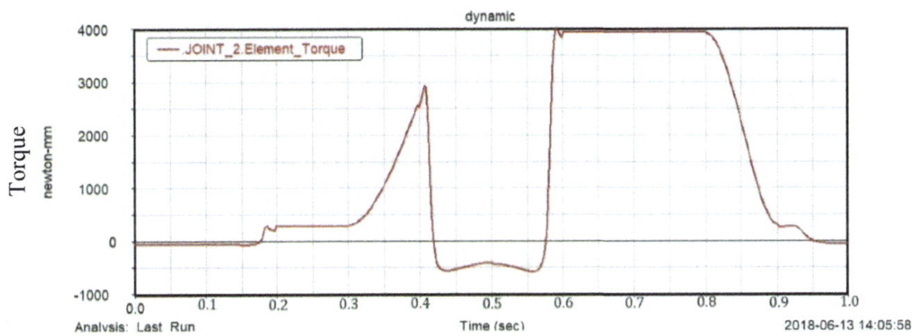

Fig. 9 Torque variations of second active joint with respect to time

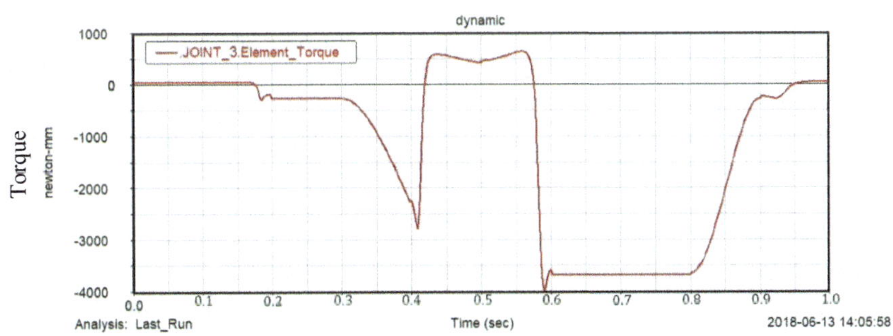

Fig. 10 Torque variations of third active joint with respect to time

The simulation is done to determine the maximum torque necessary at the active joint to rotate the lead screw. The lead screw rotation will further enable linear motions for the sliders. The sudden peaks in the graph denote jerks that will be induced in the system. This is normally not preferred for the smooth working of the mechanism. Therefore, to avoid the jerk motions it is required to define a trapezoidal velocity profile that will have a gradual ascend and descend. The jerk is due to the sudden change in acceleration of the mechanism.

It is seen from the analysis that the maximum torque developed at the active joint is 4 Nm. Each figure corresponds to the torque induced at respective active joint for each leg while lifting the payload. Depending upon the kind of application, the mechanism is to be used; accordingly the load acting at the end effector will change. This would demand the appropriate choice of the active joint to facilitate the smooth motions of the slider and successfully complete the task. The maximum theoretical torque obtained from the plot of individual leg is required to manipulate a payload of 40 N. The actual torque can be calculated by considering a safety factor of 1.5. This actual torque is used for the selection of stepper motors from the actuator data sheet.

5 Results and Discussion

In this paper, a novel 3-PSS parallel manipulator is proposed for which the kinematic modelling, workspace analysis and dynamic modelling using MSC ADAMS has been explained. The mechanism involves lesser complexity and has three active joints to control the motions of the moving platform. The number of degrees for the mechanism according to the Gruebler's criterion is 3. The platform has one linear and two rotations. The kinematic modelling is done using geometric approach based on which the workspace is plotted and compared with that of 3-SPS, 3-PRS and 6-PSS parallel manipulators. It is observed that the work volume is highest for the 3-PSS parallel robotic mechanism. The optimised work volume is calculated by the GA. The maximum volume possible for the considered dimension is calculated to be 0.01388 m^3. The vertical linear motion of the end effector for the proposed design is approximately 0.065 m with maximum planar area of 0.55×0.4 m^2. The static structural analysis is done to determine the appropriate link dimensions to withstand the loads and stresses acting. The maximum deflection induced in the robot is 0.2622×10^{-3} m and maximum stress induced is 28.976 MPa. Since, both the values are within the safer limit, the corresponding dimensions of the links are fixed for the actual operation of the robot. The dynamic torque required by the active joints to carry a payload of 4 kg is analysed using the MSC ADAMS software. The actuator to be chosen at the active joint is calculated by considering a safety factor to the results obtained from ADAMS. Therefore, the actuator to be chosen for pick and place operation is 6 Nm. It is also observed that the vertical motion of the mechanism is comparatively lesser.

6 Conclusions and Outlook

This paper focuses on design of a novel three leg PSS PKM which is suitable for pick and place operation and electronic assembly. Based on the kinematic model, the workspace is plotted for the mechanism. The three legs are coupled together considering the distance between two adjacent corners of the platform to be a constant. This consideration imparts the constraint motion to the platform attached to the three independent legs. The maximum workspace possible for the current dimensions of the link is also quantified using the GA. The results obtained from the static structural analysis done in ANSYS are explained in this paper. The results showed that the considered materials and link dimensions are within the safer limits. This work also depicts the results obtained from the MSC ADAMS software to dynamically model the mechanism for pick and place operation and electronic component assembly. The payload and controllability factors for the proposed mechanism are on the upper side when compared to the 3-RRR (DELTA) and 3-RUU parallel mechanisms. The workspace can be further improvised by adding more degrees of freedom to the system. The vertical motion of the proposed mechanism is limited. This can be increased by changing the planar motion of the sliders at varying heights. This additional motions to the end effector improve the work volume further. This mechanism can be used for surgical application by integrating the 3-PSS parallel manipulator with two-dimension planar mechanism of the serial manipulator. This would impart more intricate and precise motions to the end effector for various similar applications.

References

1. Nabat V, Krut O, Poignet F, Pierrot (2008) On the design of a fast parallel robot based on its dynamic model. J Exp Robot 39:409–419
2. Clavel R (1988) DELTA, a fast robot with parallel geometry. Proceedings of 18th international symposium on industrial robots. Lausanne, pp 91–100
3. Tsai LW, Walsh GC (1996) Kinematics of a novel three DOF translational platform. Proc IEEE Int Conf Robotics Autom 4(4):3446–3451
4. Yangmin L, Xu Q (2007) Kinematic analysis of a 3-PRS parallel manipulator. J Robot Comput Integr Manuf 23:395–408
5. Yoshikawa T (1985) Manipulability of robotic mechanisms, Int J Robot Res 4(2):3 9
6. Lingmun X, Li Q (2017) Mobility, kinematic analysis and dimensional optimization of a new three DOF parallel robot using actuator redundancy. ASME 9(041008):1–11
7. Zhang D, Su X, Gao Z, Qian J (2008) Design, analysis and fabrication of a novel three degrees of freedom parallel robotic manipulator with decoupled motions. Int J Robot Autom 7(2):16
8. Khalid A, Mekid S (2017) Intelligent spherical joints based tri-actuated spatial parallel manipulator for precision applications. Robot Comput Integr Manuf, 1–12
9. Xu W, Li Y (2013) Kinematics and workspace analysis for a novel 6-PSS parallel manipulator. Proceeding of the IEEE, International Conference on Robotics and Biomimetics (ROBIO). Shenzhen, China
10. Haupt RL, Haupt SE (2004) Practical genetic algorithm, 2nd edn. John Wiley & Sons, Inc Habuken, New Jersey

6 Conclusions and Outlook

Development of a Climbing Robot Based on Multi-suction Cups Mounted on Timing Belt Mechanism

Ravindra Singh Bisht, Pushparaj Mani Pathak, and Soraj Kumar Panigrahi

Abstract Climbing robot requires adhesion mechanism for holding the robot on to the surface and locomotion mechanism for moving the robot along the wall surface in order to perform various tasks such as glass façade cleaning and remote surface inspection of civil structures. Pneumatic adhesion force control has a prominent role for safe locomotion of the climbing robot. In case of insufficient adhesion force or airflow leakage, the robot may not work properly while carrying payload such as cleaning device, multiple sensors, and power systems, etc. Actuation and de-actuation (open and close modes) of adhesion force control of multi-suction cups based on the traditional design restricts its application particularly in climbing on wall structures where the weight and electrical power have a major concern. In the present study, pneumatic adhesion mechanism based on flat suction cup has been chosen for its performance characteristics analysis, while operation under rough and smooth wall surface conditions. Based on these experimental investigations and stability analysis, pneumatic adhesion and timing belt mechanisms for simultaneous operations of adhesion and locomotion have been developed. The developed concept for multi-suction cup-based timing belt mechanism is implemented in a mobile robot for climbing on a vertical/inclined wall surface. The manufacturing trials for simultaneous locomotion and adhesion required for surface climbing have also been conducted under static and dynamic situations.

Keywords Climbing robot · Adhesion mechanism · Guiderail mechanism · Multi-suction cup · Stability analysis

R. S. Bisht · P. M. Pathak (✉)
Mechanical and Industrial Engineering Department, IIT Roorkee, Roorkee, India
e-mail: pushparaj.pathak@me.iitr.ac.in

R. S. Bisht
e-mail: rbisht@me.iitr.ac.in; rsbisht@cbri.res.in

R. S. Bisht · S. K. Panigrahi
Acoustics Instrumentation and Mechanical Systems Group, CSIR-CBRI, Roorkee, India
e-mail: skpanigrahi@cbri.res.in

D. Sen et al. (eds.), *Mechanism and Machine Science*, Lecture Notes in Mechanical Engineering, https://doi.org/10.1007/978-981-15-4477-4_47

1 Introduction

Pneumatic adhesion technique is most suitable for developing high payload carrying climbing robot, particularly for glass façade cleaning and remote surface inspection tasks. Climbing robot requires adhesion mechanism for holding the robot on to the surface and locomotion mechanism for moving the robot along the surface in order to reach various locations of the desired structures. Various types of pneumatic adhesion techniques for wall climbing robot have been explored a lot in the recent past [1–4]. The pneumatic adhesion mechanisms can be further categorized into three groups, based on the method used for generating negative pressure, such as active pneumatic suction [1, 2], passive pneumatic suction [3], and vortex aerodynamic suction [4].

Active pneumatic adhesion technique generates adhesion force using external vacuum supply unit with airflow valve connected to single/multiple suction cups [2, 5] for creating negative pressure to smoothly hold the climbing robot on the desired wall surface, whereas passive suction adhesion mechanism works on attachment and detachment of multi-suction cups by a mechanical press to evacuate the inside air for holding the robot without requiring any vacuum supply unit [3]. The third method of pneumatic adhesion mechanisms is to use impeller mounted on a motor arrangement with an enclosed suction chamber to create adhesion force using aerodynamic principle. Here the impeller motor is driven by DC/AC power supply with pressure sensor based feedback controller to control the inside negative pressure of enclosed chamber by varying the rotational speed of the used impeller [6]. All these pneumatic techniques have been used in combination with wheel [6], leg [7], track [1, 2], and translation [8, 9] locomotion mechanisms. All these pneumatic techniques for wall climbing robot have their own advantages and disadvantages [10].

Pneumatic adhesion force control has a prominent role for safe locomotion of climbing robot. In case of insufficient adhesion force, the robot may not work properly while carrying a payload such as roller brush, sensors and power systems etc. Generally, active pneumatic adhesion mechanisms are used to generate vacuum force by actuation and de-actuation of airflow inside the single-/multiple suction cups in order to smoothly hold and move the climbing robot on the desired wall surface. Actuation and de-actuation of multi-suction cups are generally performed by individual solenoid valves connected through pneumatic tubes emanating from the vacuum supply unit. The devised solenoid valves on suction cups increase the overall weight of the system as well as electrical power requirement [8]. Adhesion force open and close mode control of multi-suction cups based on the traditional design of climbing robot may restrict its application particularly in climbing on wall structures where the weight and electrical power has a major concern. Efforts have also been made previously to improve the pneumatic actuate techniques with different concepts for climbing robot adhesion [1, 2, 11] and position control [9].

It is revealed from the literature that most of the previous research work was involved in the development of various types of pneumatic adhesion mechanisms for laboratory prototypes of climbing robot as stated above. So far, active pneumatic adhesion techniques require perfect sealing at the contact interface. This technique can be used more appropriately for adhesion force control using multi-suction cups [1, 2] or multiple vacuum chambers [12] for achieving the required adhesion force to hold the desired payload. Thus, climbing robot application may not be limited to the smooth and nonporous material surface. Current focus [13, 14] on pneumatic adhesion-based mechanism is to develop adhesion gripper mechanism for climbing robot to be implemented in the rough and uneven concrete wall surface. To tackle this issue, in this paper, open and close mode actuation by mechanical means of flat multi-suction cup-based timing belt mechanism has been developed after that it has been implemented in a climbing robot. The proposed mechanism not only improves the overall payload carrying capacity of the climbing robot but also provides continuous locomotion for efficient coverage required, particularly, in glass façade cleaning as well as inspection of a large vertical wall surface of civil structures.

The next section gives experimental characteristics of flat suction cup chosen for the development of pneumatic adhesion mechanism. In the third section, force analysis of the climbing robot is presented along with robot design parameters. The fourth section shows the design and development of timing belt mechanism for simultaneous locomotion and adhesion of the climbing robot. Experimental trials on developed mechanisms implemented in the climbing robot are described in the fifth section. The last section of this paper highlights conclusions derived from this work.

2 Experimental Testing of Flat Suction Cup

Experiments for measuring adhesion force generated by flat suction cups have been performed. The adhesion force produced by the flat suction cup is due to the negative pressure created using a vacuum generator. Vacuum level inside the flat suction cup is controlled through a pressure-regulated valve of Filter, Regulator, and Lubricator (FRL) unit. Air compressor is used to supply air through the FRL unit to the vacuum generator. Experimental setup using a motorized test stand (ESM301LE) with digital force gauge (M5-200E) shown in Fig. 1 was used to measure the adhesion force.

As seen from the Fig. 2, adhesion force of flat suction cup is increasing as we are increasing different vacuum levels ranging from 200 to 600 mbar. For different levels of vacuum inside the flat suction cup, the generated forces are 43.8 N, 83.2 N, and 121.0 N, respectively.

Results in Fig. 3 indicate that the generated adhesion force is more in smooth surface 1 compared to relatively rough cement surface 2. Therefore, climbing robot design based on active suction cup can work more effectively on smooth cement/tile surface, wood surface, and glass façade surface of multi-story buildings.

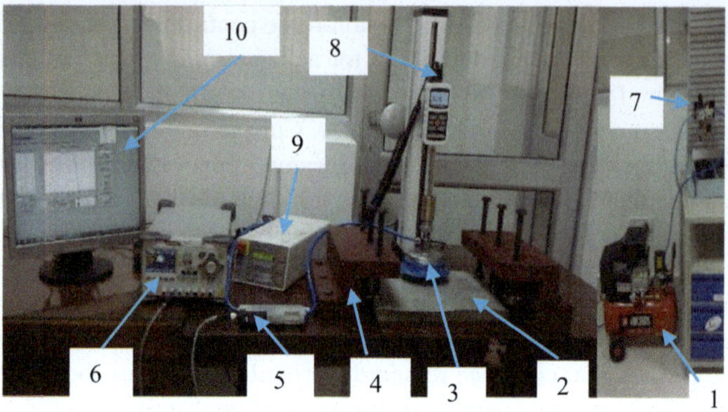

Fig. 1 Experimental setup for testing and measuring adhesion force of suction cup. 1 Air compressor, 2 Cement surface, 3 Suction cup, 4 Test surface fixture, 5 Vacuum generator, 6 DC power supply, 7 FRL unit, 8 Digital force gauge mounted on motorized actuator, 9 Controller, 10 Data visualization software (MESUR^{TM} gauge) for analysis

Fig. 2 Pneumatic force characteristics of flat suction cup for different vacuum pressure versus actuator speed

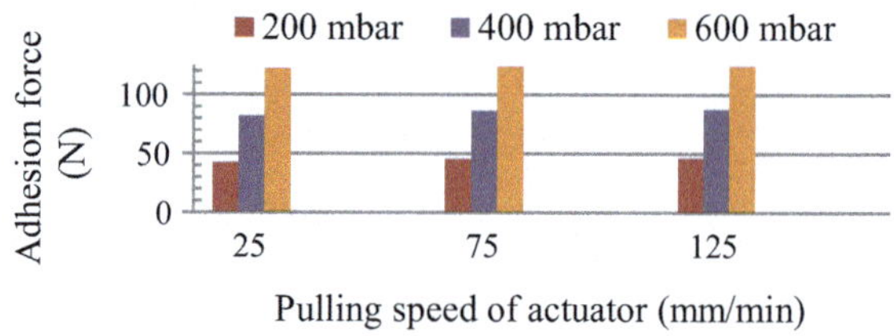

Fig. 3 Comparison of test results of flat suction cup for different surface conditions

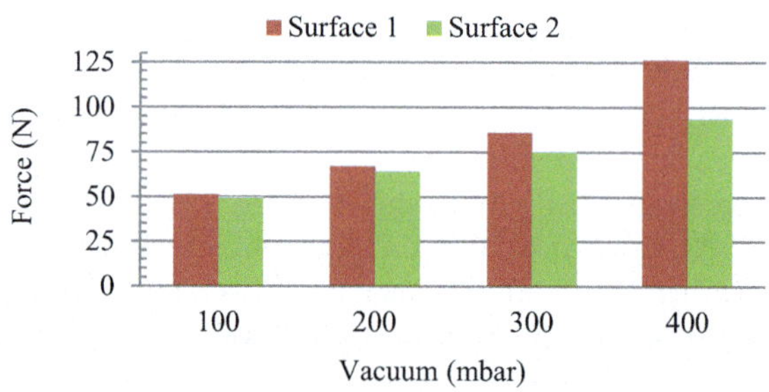

3 Analysis of Forces Due to Multi-suction Cup Based Timing Belt Mechanism of Climbing Robot

Climbing robot has to work safely while climbing in any orientation of the target wall surface. To avoid any mode of failures such as slipping and toppling from the inclined/vertical wall surface, stability analysis has been performed to obtain minimum pneumatic adhesion force (F_i) required at each flat suction cup of the climbing robot with a payload like a cleaning roller brush.

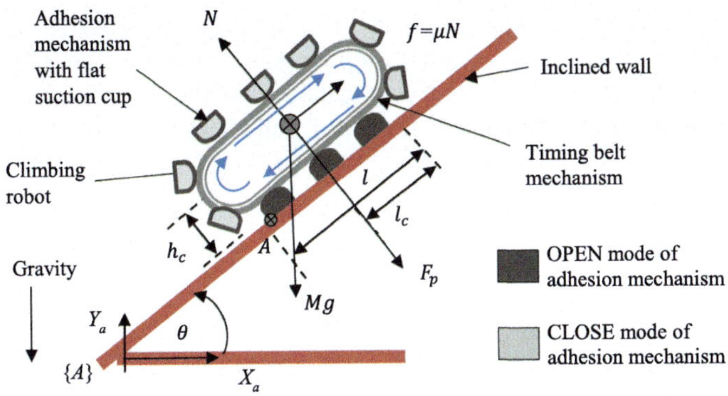

Fig. 4 Forces on the climbing robot due to multi-suction cups mounted on timing belt mechanism for wall inclined angle ($0° < \theta \leq 180°$)

Table 1 Design parameters chosen for the climbing robot used in the stability analysis

Design parameters	Mass M (kg)	Length l (m)	Width b (m)	Length l_c (m)	CG Height h_c (m)	Max. suction cups
Value	20	0.30	0.55	0.15	0.05–0.15	20

Where in Fig. 4, $N = \sum_{i=1}^{n} N_i$ is net reaction force acting on the climbing robot body, N_i is the normal reaction on robot body by each flat suction cup attached with the wall surface, $F_p = \sum_{i=1}^{n} F_i$ is net pneumatic adhesion force acting on the robot body, F_i is adhesion force generated by each suction cup, and f is net wall interface contact friction force. The robot physical parameters used in the stability analysis are shown in Table 1.

For the inclined wall orientation, the net pneumatic adhesion force required at each suction cup to avoid climbing robot from slipping is derived as:

$$F_i \geq \left\{ \frac{F_g}{n} \left(\frac{\sin\theta}{\mu_s} - \cos\theta \right) \right\}, \quad \text{for } 0^\circ < \theta \leq 180^\circ \tag{1}$$

where $F_g = Mg$ is the total weight of the climbing robot including additional payload considered as roller brush, μ_s is coefficient of static friction between rubber suction cup and wall surface, n is the total number of suction cups which are in contact with the wall surface at a time and θ is the wall orientation. Net adhesion force required at each flat suction cup to avoid climbing robot from toppling about a contact point A is given by:

$$F_i \geq \left\{ \frac{F_g}{n} \left(\frac{h_c}{l_c} \sin\theta - \cos\theta \right) \right\}, \quad \text{for } 0^\circ < \theta \leq 180^\circ \tag{2}$$

where h_c is the height of the climbing robot center of gravity (CG) measured from the wall surface and l_c is the length of tipping point measured from the robot CG. From above Eqs. (1) and (2), a stability criterion for the climbing robot to avoid sliding and toppling from the inclined/vertical wall is expressed as:

$$F_i = \left\{ \max \left\{ \frac{F_g}{n}\left(\frac{\sin\theta}{\mu_s} - \cos\theta\right), \frac{F_g}{n}\left(\frac{h_c}{l_c}\sin\theta - \cos\theta\right) \right\} \right\}, \quad \text{for } 0^\circ < \theta \le 180^\circ \tag{3}$$

The results evaluated analytically in Fig. 5 indicate that the required adhesion force for avoiding toppling is comparatively lower than the sliding. Hence, for given design parameters in Table 1 and $\left(\frac{h_c}{l_c} = 0.60\right)$ for manufacturing trials of the climbing robot will have no toppling as long as the climbing robot is able to avoid slipping, hence the more stable design of the climbing can be developed. From the above results shown in Fig. 5 and Table 2 for 20-kg robot weight including payload, if $0.4 \le \mu_{\text{static}} \le 0.8$, the adhesion force requirement in each suction cup of the robot is $53.00\,\text{N} \le F_i \le 134.50\,\text{N}$ for avoiding any modes of failure in any orientation of test surface. By comparing both the results obtained from force analysis of robot and experiments on flat suction cup, it is found that the (100–600 mbar) vacuum levels of operating range are most suitable to operate the prototype-climbing robot. DC motor driving torque required at each timing belt mechanism is derived as, $\tau_{tb} = F_t r_{tb}$, where F_t is a traction force is derived from Fig. 4, r_{tb} is the turning radius of the timing belt mechanism. Maximum necessary driving torque of 12.80 Nm for each timing belt mechanism is evaluated from analysis when the 20-kg climbing robot weight is working under most critical condition ($\theta = \theta_{cr}$). Hence, geared DC motor (10 RPM, 24V DC) with maximum of 15 Nm continuous torque capacity is selected for the climbing robot development.

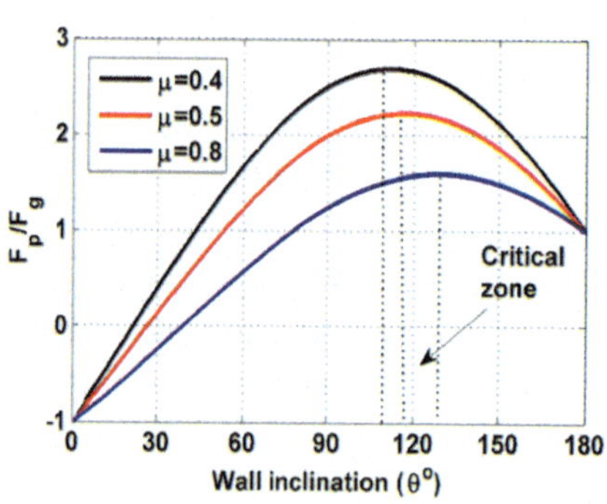

(a)

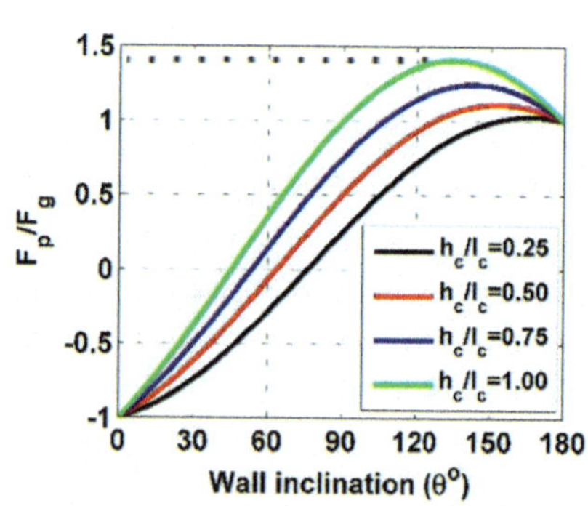

(b)

Fig. 5 Relation between adhesion force to robot weight ratio $\left(\frac{F_p}{F_g}\right)$ to wall inclination angle (θ°) for avoiding robot **a** sliding and **b** toppling on inclined/vertical wall

Table 2 Minimum adhesion force and critical wall conditions for prototype climbing robot on inclined wall ($0° < \theta \leq 180°$)

Friction	F_g (kg)	Critical wall angle ($\theta = \theta_{cr}$)	F_p (N)	F_p (N)	F_i (N) (if $n = 6$)	F_i (N) (if $n = 4$)
$\mu_s = 0.4$ (rough cement [15])	20	110°	$F_p \geq 2.69F_g$	538.0	89.67	134.50
$\mu_s = 0.5$ (glass surface [16])	20	115°	$F_p \geq 2.23F_g$	446.0	74.34	111.50
$\mu_s = 0.8$ (smooth cement [15])	20	125°	$F_p \geq 1.59F_g$	318.0	53.00	79.50

4 Development of Mechanisms for Climbing Robot

Climbing robot requires simultaneous operation of locomotion and adhesion mechanisms, where locomotion mechanism for moving along the surface, and adhesion mechanisms for holding the robot on to the surface, respectively. Climbing robot should have an appropriate mechanism for its proper functioning while climbing. It is found from the above laboratory experiments that the flat suction cup based design is suitable for developing mechanisms to be used in high payload carrying climbing robot for vertical/inclined surface cleaning and inspection tasks.

4.1 *Pneumatic Adhesion Mechanism*

Firstly, pneumatic adhesion mechanism has been designed. Then the prototype of the adhesion mechanism has been manufactured using high-strength and lightweight material (Nylon–Ertalon) as weight is very critical for climbing robot design. Open and close mode operations of the mechanism are performed with the help of reciprocating piston motion, where (δ_d) is net displacement of the piston by subjecting up and down force. In the open mode (ON) as shown in Fig. 6a, when the displacement is δ_d, the net vacuum pressure ($p_i = \delta P$) inside the suction cup will be produced. While in the closed mode (OFF) shown in Fig. 6b, when ($\delta_d = 0$), there will be no vacuum pressure inside the suction cup. Mechanical components used for the effective functioning of the mechanism are explained in the assembly CAD design as well as fabricated unit shown in Fig. 6a–c, where arrows indicate relative motions of parts as shown in adhesion mechanism.

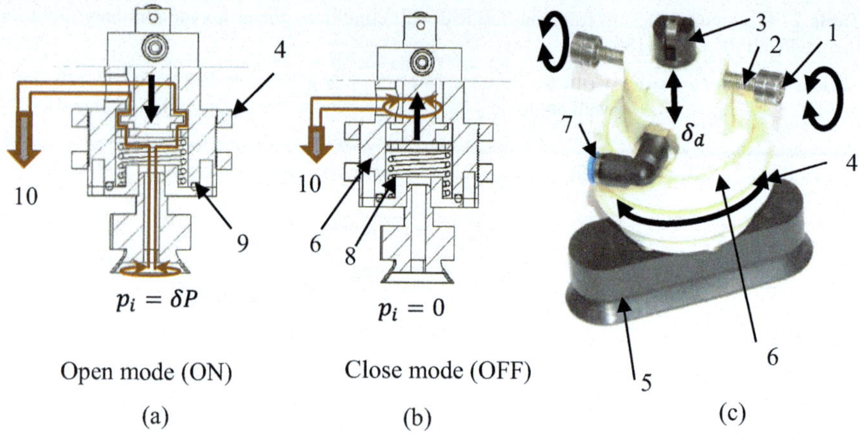

Fig. 6 **a** Inside view of open and **b** close mode of pneumatic adhesion mechanism **c** Developed prototype of pneumatic adhesion mechanism for climbing robot. 1 Roller ball bearing, 2 Shaft, 3 Piston, 4 Chuck nut for rubber track mounting, 5 Flat suction cup, 6 Pneumatic cylinder, 7 Airflow valve, 8 Helical spring, 9 Rubber sealing ring, 10 Connection to vacuum pump through airflow valve

4.2 Mechanism for Simultaneous Operation of Adhesion and Locomotion

Here, mechanism required for simultaneous adhesion and locomotion has been further divided into two parts, namely timing belt mechanism and guiderail mechanism. Guiderail mechanism is a part of timing belt mechanism. Guiderail mechanism has been further divided into three parts A, B and C shown in Fig. 7a, where parts A and C for holding the n number of adhesion mechanism units and a guided path is provided for roller motion of the adhesion units. Part B has a cam profile, where cam profile height is designed $\geq \delta_d$, to actuate the adhesion mechanisms which are in contact with the wall surface (Fig. 7b). The timing belt mechanism consists of n number of adhesion mechanisms, a rubber track, a pair of track wheel system, and guiderail mechanism. Where the n number of adhesion mechanisms can be decided based on the required robot payload capacity. When the pneumatic adhesion mechanism shown in Sect. 4.1 integrated with the guiderail of timing belt mechanism, it shows various motions as indicated by arrows in Fig. 7. These motions are taking place while the mechanism is under driving operation. With the help of these mechanisms, the flat suction cups which are attached with the wall will be in ON ($\delta_d = 5$ mm height of cam profile) operation depicted in Fig. 6a, hence, vacuum will create inside these suction cups. Those are moving away from the wall surface will get OFF ($\delta_d = 0$ mm) or de-actuated as depicted in Fig. 6b. If design condition $\left(l_{cp} \leq l_g - 2d\right)$ satisfies then suction cups which are in open mode will be perpendicular to the wall surface, where l_{cp} effective length of the cam profile, l_g length

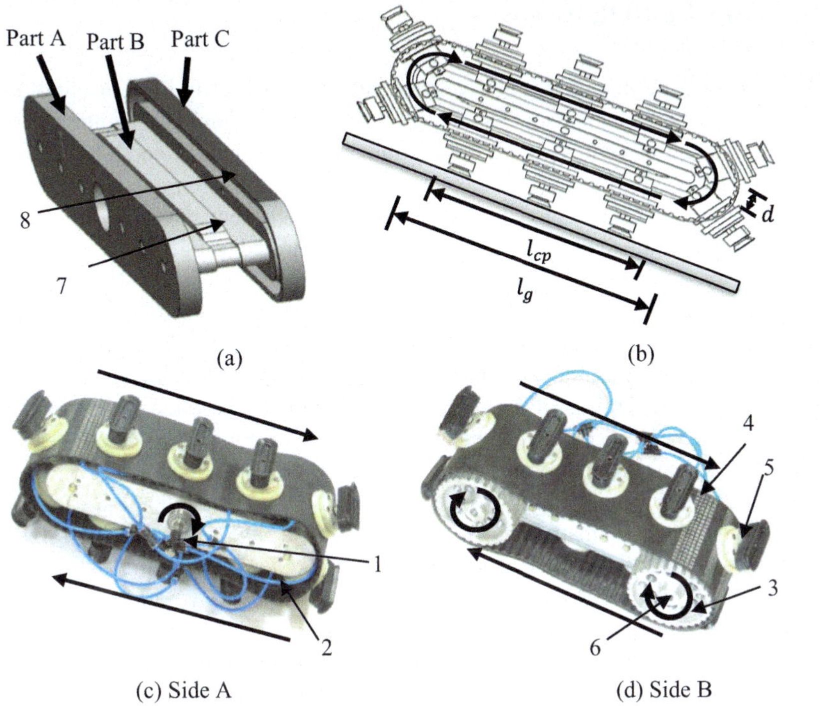

Fig. 7 Timing belt mechanism shows simultaneous operation of locomotion and adhesion mechanism **a** Guiderail mechanism **b** Inside view of timing belt mechanism (where $n = 10$ number of suction cups mounted on timing belt mechanism, $p = p_i$ for $n = 1$–3 and $p = 0$ for $n = 4$–10), **c** and **d** Prototype of developed mechanism. 1 Rotary joint, 2 Piping connections, 3 Track-wheel, 4 Rubber track belt, 5 Adhesion mechanism, 6 Connection to geared DC motor, 7 Cam profile, 8 Guided path for roller motion, (where arrows indicate direction of motion)

of the guide rail or center distance of two track wheel of timing belt, and d corner profile diameter of Part B of guiderail mechanism.

To drive the timing belt mechanism, a pair of front and rear track wheels is mounted at the Side B of timing belt mechanism. A vacuum supply unit may be provided either by vacuum generator along with air compressor for laboratory trials of climbing robot or portable vacuum pump for the autonomous mode of climbing robot for the field trials. A rotary joint connected through vacuum source is provided at Side A of the timing belt mechanism to avoid tangling of pneumatic tubes connecting the multi-suction cups.

5 Experimental Trials of Developed Mechanisms and Climbing Robot

Two sets of experimental trials have been performed to verify the concept of developed timing belt mechanism. The first experiment shown in Fig. 8 confirms that the adhesion force generates due to the vacuum created inside the suction cups, which are in contact with the wall surface, and rest other suction cups are in close mode due to the guiderail mechanism. In the second experiment shown in Fig. 9a–f, the motion of the timing belt mechanism has been confirmed using a DC motor mounted with the track wheel and pillow block bearing support system.

A schematic diagram is shown in Fig. 10 clearly shows the control methodology and connection flow diagram of multi-suction cup-based timing belt mechanism for simultaneous locomotion and adhesion of the climbing robot. Vacuum level inside the multi-suction cups of a timing belt mechanism is controlled through the flow-regulated valve of FRL unit of pneumatic circuit. Air compressor is used to supply air to the vacuum generator for obtaining output vacuum. Vacuum level inside the suction cup can be monitored from the vacuum indicator mounted at each side of the timing belt mechanism. A tee junction flow-control valve is used to bifurcate the vacuum supply to each side of the timing belt mechanism. All the connections of various used pneumatic components are to be airtight for avoiding any loss in the required vacuum level to support the robot weight on vertical wall motion. To confirm this, several experimental trials have been conducted to verify the robot design, so that it can safely climb to the vertical/inclined wall surface without falling.

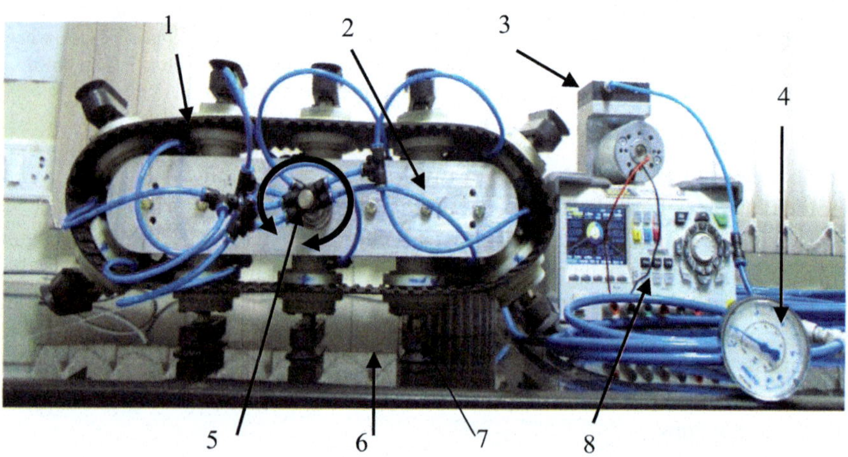

Fig. 8 Test setup to confirm open and close mode of operations of multi-suction cups mounted on a timing belt mechanism. 1 Timing belt mechanism, 2 Guiderail mechanism, 3 Vacuum pump supply, 4 Vacuum indicator, 5 Rotary joint, 6 Glossy test surface, 7 Flat suction cup, 8 Power supply (static condition)

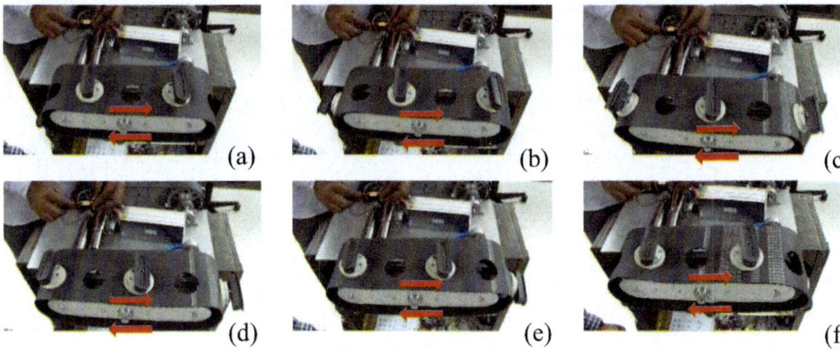

Fig. 9 Motion trials of a single timing belt mechanism of climbing robot, motion sequence of timing belt motion shown from (**a**) to (**f**)

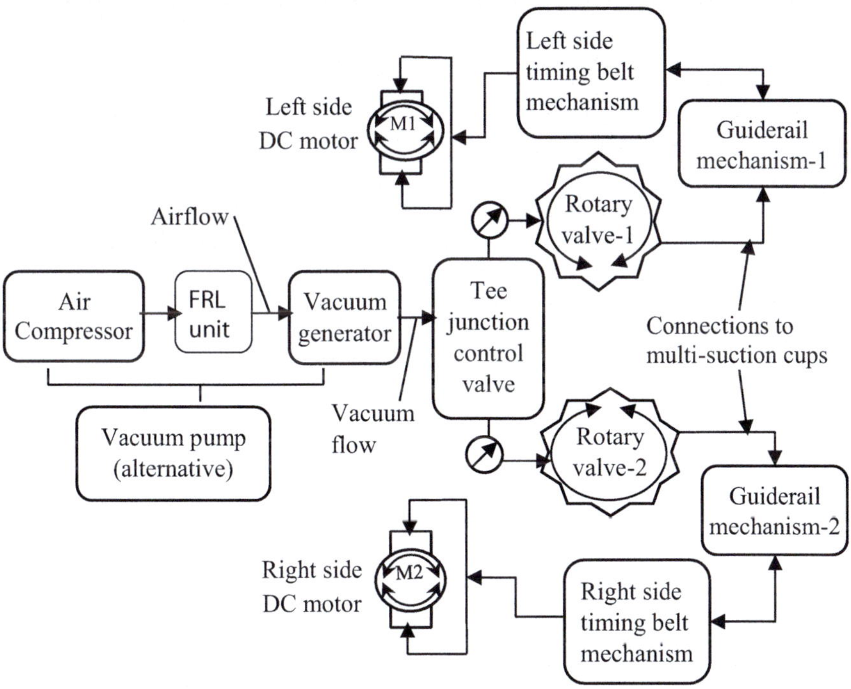

Fig. 10 Control methodology and connection flow diagram of multi-suction cup based timing belt mechanism for simultaneous locomotion and adhesion of climbing robot

The developed prototype of the climbing robot shown in Fig. 11a is based on a pair of timing belt mechanisms for simultaneous locomotion and adhesion on a vertical/inclined wall. Two front wheels of the timing belt are connected with individual DC geared motor, and the two rear wheels of the timing belt are ideal for free rotation provided by Plummer block bearing support system with the robot frame.

Fig. 11 **a** Developed prototype climbing robot, **b** Working trials for climbing action up to 60° wall inclination of glass surface, where when $n = 6$ suction cups mounted on each timing belt mechanism

PWM technique is used in the control algorithm for varying speed as well as direction control of the timing belt DC Motors. In Fig. 11b, the working of the climbing robot based on the laboratory trials has been performed up to 60° wall inclination of glass surface. Further work is required to test the robot in the laboratory on critical vertical wall orientations, along with roller brush assembly with the improved climbing robot design for additional payload testing. The improved climbed robot design will be tested in a proper wall surface along with roof assisted mounting system which is to be needed so that the robot can effectively be deployed in various types of vertical wall orientations safely.

6 Conclusion

This paper has investigated the working of pneumatic adhesion and timing belt mechanisms using multi-suction cups for simultaneous operations of adhesion and locomotion of climbing robot. Stability analysis on different wall orientations of the proposed climbing robot has been performed to obtain suitable design parameters for manufacturing trials. This work has revealed that the proposed mechanism not only improves the overall payload carrying capacity of the climbing robot but also provides continuous locomotion for efficient coverage required, particularly, in glass façade cleaning as well as inspection of a large vertical wall surface of civil structures. For the proposed design of a timing belt mechanism, a maximum ($n = 10$) number of adhesion mechanism units can be integrated. However, under dynamic trials of the climbing robot, it was found that ($n < 10$) adhesion mechanism units are needed to avoid any interference due to manufacturing tolerances. Further experimental investigations are needed to estimate the performance of the climbing robot on critical vertical wall orientations. A proper wall surface along with roof-assisted mounting system is to be needed so that the robot can effectively be deployed in various types of vertical wall orientations safely.

Acknowledgments The authors are thankful to the Director CSRI-CBRI for providing the opportunity to carry out this research work. The research work is financially supported by XII 5-year plan network projects: ESC0102 and ESC0112 of CSIR-CBRI, Roorkee. Thanks to all the team members of XII 5-year plan project and supporting staff of AIMS Group of CSIR-CBRI, Roorkee for their cooperation and support while conducting laboratory trials.

References

1. Kim H, Kim D, Yang H, Lee K, Seo K, Chang D, Kim J (2008) Development of a wall-climbing robot using a tracked wheel mechanism. J Mech Sci Technol 22(8):1490–1498
2. Giuk L, Kim H, Seo K, Kim J, Kim HS (2015) MultiTrack: a multi-linked track robot with suction adhesion for climbing and transition. Robot Auton Syst 72:207–216
3. Ge D, Ren C, Matsuno T, Ma S (2016) Guide rail design for a passive suction cup based wall-climbing robot. In: 2016 IEEE/RSJ international conference on intelligent robots and systems (IROS). IEEE, pp 5776–5781
4. Zhou Q, Li X (2018) Experimental investigation on climbing robot using rotation-flow adsorption unit. Robot Auton Syst 105:112–120
5. Liu J, Tanaka K, Bao LM, Yamaura I (2006) Analytical modelling of suction cups used for window-cleaning robots. Vacuum 80(6):593–598
6. Koo IM, Trong TD, Lee YH, Moon H, Koo J, Park SK, Choi HR (2013) Development of wall climbing robot system by using impeller type adhesion mechanism. J Intell Robot Syst 72(1):57–72
7. Akutsu A, Sasaki E, Takeya K, Kobayashi Y, Suzuki K, Tamura H (2017) A comprehensive study on development of a small-sized self-propelled robot for bridge inspection. Struct Infrastruct Eng 13(8):1056–1067
8. Zhang H, Zhang J, Wang W, Liu R, Zong G (2007) A series of pneumatic glass-wall cleaning robots for high-rise buildings. Ind Robot Int J 34(2):150–160
9. Zhang HX, Wang W, Zhang JW (2009) High stiffness pneumatic actuating scheme and improved position control strategy realization of a pneumatic climbing robot. In: ROBIO 2008 IEEE international conference on robotics and biomimetics, pp 1806–1811
10. Bisht RS, Alexander SJ (2013) Mobile robots for periodic maintenance and inspection of civil infrastructure: a review. In: International conference on machines and mechanisms (iNaCoMM), pp 1050–1057
11. Xiao J, Li B, Ushiroda K, Song Q (2015) Rise-Rover: a wall-climbing robot with high reliability and load-carrying capacity. In: 2015 IEEE international conference on robotics and biomimetics (ROBIO). IEEE, pp 2072–2077
12. Hillenbrand C, Schmidt D, Berns K (2008) CROMSCI: development of a climbing robot with negative pressure adhesion for inspections. Ind Robot Int J 35(3):228–237
13. Huston D, Burns D, Gardner-Morse J, Montane P, Angola E (2014) Dual-durometer soft suction foot robot for concrete inspection. In: Health monitoring of structural and biological systems, international society for optics and photonics, (9064), 90642Q
14. Fujita M, Ikeda S, Fujimoto T, Shimizu T, Ikemoto S, Miyamoto T (2018) Development of universal vacuum gripper for wall-climbing robot. Adv Robot 32(6):283–296
15. El-Sherbiny YM, Hasouna AT, Ali WY (2012) Friction coefficient of rubber sliding against flooring materials. J Eng Appl Sci 7(1):121–126
16. Tuononen AJ (2016) Onset of frictional sliding of rubber–glass contact under dry and lubricated conditions. Sci Rep 6:27951

mounting system is to be needed so that the robot can effectively be deployed in various types of vertical wall orientations safely.

Acknowledgements. The authors are thankful to the Director, CSIR-[illegible] for providing the opportunity [illegible] work. The research work [illegible] supported by [illegible]

Multi-patch Isogeometric Analysis of Planar Compliant Mechanisms

Sagar Bodkhe and G. K. Ananthasuresh

Abstract Multi-patch isogeometric analysis (IGA) of planar compliant mechanism, based on the planar Timoshenko beam theory, is presented. IGA is a new computational framework that seamlessly integrates geometric modelling and deformation analysis. The multi-patch technique enables analysis of multi-segmented beam structures with bifurcation points and star junctions. Methods to handle various Dirichlet and Neumann boundary conditions in the framework of IGA are also discussed. Displacement and section force distributions obtained from IGA are compared with the results obtained from finite element analysis (FEA) and analytical solutions. IGA is particularly advantageous in the problems involving geometries with large curvatures, such as curved beams, curved thin plates, etc. Such curved geometries can be used to model various interconnected members in a compliant mechanism. The multi-patch IGA approach discussed in this paper can be used to improve the efficiency of shape optimization process of planar compliant mechanisms, by providing better control over the shape using fewer design variables and by reducing errors arising due to geometric approximation.

Keywords Isogeometric analysis · Complaint mechanism · Multi-patch IGA · Finite element analysis

1 Introduction

Isogeometric analysis (IGA) is a computational framework that integrates computer-aided design (CAD) with finite element analysis (FEA). In FEA, actual geometry is approximated by a mesh of finite elements. In contrast, in IGA, the geometric descriptions are imported directly from the exact CAD model, which reduces the errors arising due to geometric approximation [1]. IGA is also based on isoparametric formulation used in FEA. The non-uniform rational B-spline (NURBS) basis

S. Bodkhe · G. K. Ananthasuresh (✉)
Indian Institute of Science, Bengaluru, India
e-mail: suresh@iisc.ac.in

S. Bodkhe
e-mail: sagarbodkhe13@gmail.com

D. Sen et al. (eds.), *Mechanism and Machine Science*, Lecture Notes in Mechanical Engineering, https://doi.org/10.1007/978-981-15-4477-4_48

functions are used to interpolate both geometry and the displacement field. In addition, the convergence rates are found to be higher in IGA than FEA because of higher continuity across the element boundaries. In IGA, a separate algorithm for mesh refinement is not required and thus it is not necessary to update the finite element models after geometrical changes. The integration of modelling (CAD) and analysis (FEA) in a single framework reduces the total design time, which makes IGA an arguably superior alternative to traditional FEA, especially in shape optimization. In this paper, we present only the analysis framework that is suitable for compliant mechanisms.

Fully compliant mechanisms are elastic continua that are used to transmit or transform force and motion. The advantages of these mechanisms over rigid-body mechanisms are due to the absence of rigid-body kinematic joints [2]. These mechanisms consist of various interconnected straight and curved segments, which can be modelled as beams. The use of IGA in design and optimization of compliant mechanisms has the potential to improve the overall efficiency of the process by reducing the computational cost and modelling errors by providing easier control over the shape of the geometry.

Analysis of curved beams with a constant-curvature finite element is explained in [3, 4]. Using these elements, only the geometries having constant curvatures can be exactly represented. On the other hand, using IGA, curves with arbitrarily varying curvatures can be represented exactly at all the levels of mesh refinement. IGA of single-patch curved beams based on Timoshenko theory is discussed by Cazzani et al. [5]. Only simple geometries, obtained from a single curve, can be studied using the single-patch formulation discussed in [5]. As an extension, we present the multi-patch approach, which can be used to analyse complicated topologies. This study is particularly useful in design and optimization of compliant mechanisms made of slender planar segments.

Numerical locking may occur in a thick Timoshenko beam solved by displacement-based IGA [6]. In addition to shear locking, curvature locking also exists in curved beams. Echter and Bischoff [7] explain discrete shear gap (DSG) method for static IGA of straight beams. Improved selective reduced integration schemes are used by Adam et al. [6] to overcome the locking effect in IGA of curved beams. However, with higher order NURBS basis functions (degree more than three), a locking-free behaviour is observed [7]. Utilizing this fact, in this paper, higher order NURBS basis functions obtained using p-refinement are used in the analysis to alleviate the effects of locking.

Isogeometric formulation of curved Timoshenko beam is described in Sect. 2. In Sect. 3, imposition of various Dirichlet and Neumann boundary conditions is discussed; this is an important aspect of IGA. In Sect. 4, we extend the ideas developed to multi-patch formulation. In Sect. 5, we present several problems and compare the results with those obtained from traditional FEA and analytical solutions.

2 Isogeometric Formulation of Curved Timoshenko Beams

2.1 Non-uniform Rational B-Splines (NURBS)

NURBS is a standard technique used for geometric representation in various CAD packages. NURBS is a superset of B-splines (basis-spline). A B-spline is defined using a set of control points. The B-spline basis functions, which are piecewise polynomials, are defined on a knot vector Ξ. A knot vector is a non-decreasing sequence of numbers. Each element of this vector is called a knot in the parametric space. A typical knot vector is represented as

$$\Xi = \left\{\xi_1, \xi_2 \ldots \ldots \xi_{n+p+1}\right\},$$

where n is the total number of control is points and p the degree of the basis functions. The interval $[\xi_i, \xi_{i+1})$ is called a knot span. If $\xi_i \neq \xi_{i+1}$ then the knot span represents an element in the physical space. The knots not only subdivide the curve but also control the continuity of the basis functions, if a knot is repeated s times then the basis functions are C^{p-s} continuous at that knot. A knot vector is said to be uniform if its knots are uniformly spaced and non-uniform otherwise. If the first $p+1$ knots are repeated (C^{-1} continuous) then the curve is clamped at the first and last control points. In this paper, we will consider only uniform, clamped knot vectors.

NURBS curve in $\mathbb{R}^d$ is simply the projection of B-spline curve on a unit plane defined in $\mathbb{R}^{d+1}$. A point on the NURBS curve in $\mathbb{R}^d$ is given by [8]

$$C(\xi) = \sum_{i=1}^{n} R_i^p(\xi) Q_i, \tag{1}$$

$$R_i^p(\xi) = \frac{N_i^p(\xi) w_i}{\sum_{i=1}^{n} N_i^p(\xi) w_i} \tag{2}$$

where $Q_i s$ are the control points in $\mathbb{R}^d$, $R_i^p s$ are the NURBS basis functions of degree p, $N_i^p s$ are the B-spline basis functions of degree p and $w_i s$ are the weight-associated control points.

Some of the important properties of the NURBS basis function are [8]:

1. Partition of unity: For a particular ξ the sum of all basis function is always 1.

$$\sum_{i=1}^{n} R_i^p(\xi) = 1,$$

2. Non-negativity: NURBS basis functions are always non-negative.
3. Local Support: In a particular knot span, only few basis functions are non-zero. The non-zero or effective basis function in ith knot span are $\left[R_{i-p}^p \ldots R_i^p\right]_{1\times(p+1)}$.

2.2 IGA Model for the Curved Beams

A local (n–t) coordinate system used in the curved beam model is shown in Fig. 1. Radius of curvature at any point is given by R. The curve is parametrized by arc length. The equations used to derive the finite element formulation are described by Lim et al. [9].

The compatibility equations are:

$$\varepsilon = \frac{du}{ds} + \frac{w}{R}, \quad \gamma = \frac{dw}{ds} - \frac{u}{R} + \varphi, \quad \chi = \frac{d\varphi}{ds} \tag{3}$$

where ε, γ and χ are the generalized axial, shear and bending strains. Displacements in the tangential direction are denoted by u and w, respectively, while the rotations are denoted by φ. (See [5] for detailed discussion.) By minimizing the potential energy with respect to u, w and φ, we get:

$$u, w, \varphi = \arg \min_{u,w,\varphi} \left[\frac{1}{2} \int_0^L \left(EA\varepsilon^2 + GA\kappa\gamma^2 + EI\chi^2 \right) ds - \int_0^L (q_t u + q_n w + m\varphi) ds \right] \tag{4}$$

where Young's modulus is represented by E, shear modulus by G, cross-sectional area by A and the second moment of area by I; and, κ is the shear correction factor.

By using Eqs. (1) and (3) in (4), we get a discrete set of governing equations, where unknowns are the displacements at the control points. To carry out the integration in a physical domain, two transformations are required as opposed to a single transformation in FEA, as shown in Fig. 2. The physical domain is mapped to a parametric domain, which in turn is mapped to a parent domain, where integrations are performed using Gaussian quadrature. However, NURBS interpolation is different from polynomial interpolation, this implies that the integration using Gaussian quadrature is not as accurate as in case of polynomial interpolations. In "half-point rule" suggested by Hughes et al. in [10], the total Gauss points are taken to be half of the total degrees of freedom. The strain displacement matrix $\boldsymbol{B}$, stiffness matrix

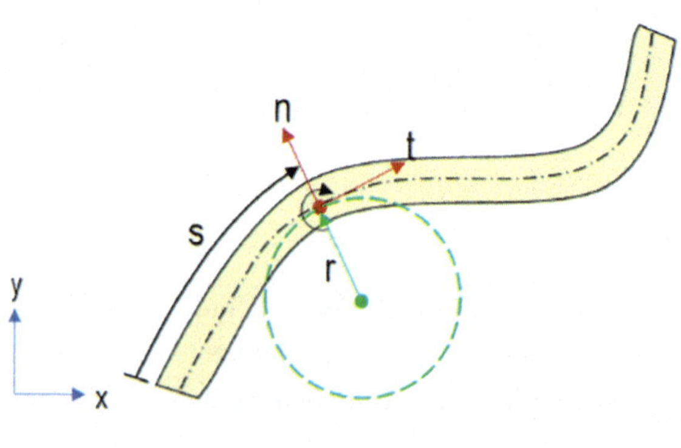

Fig. 1 In-plane model for curved beam

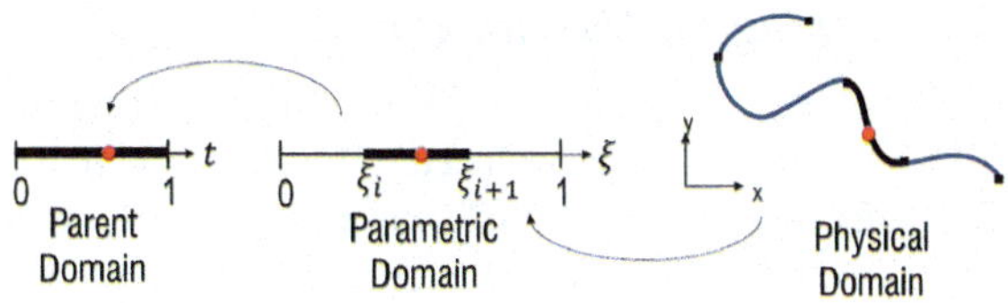

Fig. 2 Domain mappings in IGA

$\boldsymbol{K}$ and load vector $\boldsymbol{f}$ are obtained from Eq. (3). The variable R_i denotes the shape functions, whereas r and s are the radius and coordinate on the centroid line, respectively (see Fig. 1). The Jacobian J maps between parameter and physical space, the variable ξ is the coordinate in parametric space as depicted in Fig. 2. The variable $\boldsymbol{q}$ denotes distributed load.

$$\boldsymbol{B} = \begin{bmatrix} \frac{dR_1}{ds} & \frac{R_1}{r} & 0 & \dots & \frac{dR_n}{ds} & \frac{R_n}{r} & 0 \\ -\frac{R_1}{r} & \frac{dR_1}{ds} & R_1 & \dots & -\frac{R_n}{r} & \frac{dR_n}{ds} & R_n \\ 0 & 0 & \frac{dR_1}{ds} & \dots & 0 & 0 & \frac{dR_n}{ds} \end{bmatrix}, \quad \boldsymbol{K} = \int_{\xi} \boldsymbol{B}^T \begin{bmatrix} EA & 0 & 0 \\ 0 & GA\kappa & 0 \\ 0 & 0 & EI \end{bmatrix} \boldsymbol{B} J d\xi,$$

$$\boldsymbol{f} = \int_{\xi} \boldsymbol{R}^T \boldsymbol{q} J d\xi \tag{5}$$

There are three types of refinement strategies possible in IGA [1]:

(1) h-refinement: Knot insertion without changing the curve geometrically or parametrically;
(2) p-refinement: The order of basis functions may be increased and the continuities of elements remain unchanged;
(3) k-refinement: Order elevation followed by knot insertion, which can achieve the continuity of C^{p-1}.

We have used k-refinement strategy in this work because the polynomial order and smoothness of the NURBS basis functions are simultaneously increased, which results in higher order inter element continuity.

3 Boundary Conditions

The degrees of freedom in IGA are the displacements of the control points, which does not, in general, hold any physical meaning related to the problem. Due to this, the imposition of boundary condition is not straightforward. In this section, multi-freedom constraint (MFC) [11] approach is used to impose various Dirichlet boundary condition imposed in the interior of the curve for which $\xi \in (0, 1)$. A technique to apply a point load at any point on the curve is also explained.

3.1 Fixing an Interior Point

To fix an interior point, we need to find the effective control points in the knot span. By "effective control points", we mean all the control points corresponding to the non-zero basis function in the active knot span. Once the knot span index i, in which the point to be fixed lies, and the degree of the basis function p, are known, the control point indices which are effective are from $i - p$ to i.

Consider the problem shown in Fig. 3. The geometry is quarter of a circle defined using four control points, the degree of the basis function is two. The point corresponding to $\xi = 0.5$ is to be fixed. The knot vector for the curve is $\Xi(\xi) = \{0, 0, 0, 0.5, 1, 1, 1\}$, the knot index for $\xi = 0.5$ is $i = 4$. The effective control points are thus,

$$(i - p : i) = (4 - 2 : 4) = \{2, 3, 4\},$$

Since the number of DoF per node (control point) is 3, the effective DoFs are 4, 5, 6, 7, 8, 9 and 10, 11, 12.

By using the NURBS basis function, the displacements at the control points can be interpolated to get the displacement at any point on the curve. Tangential (u_t), normal (u_n) and rotational (u_θ) displacement at $\xi = 0.5$ are given by

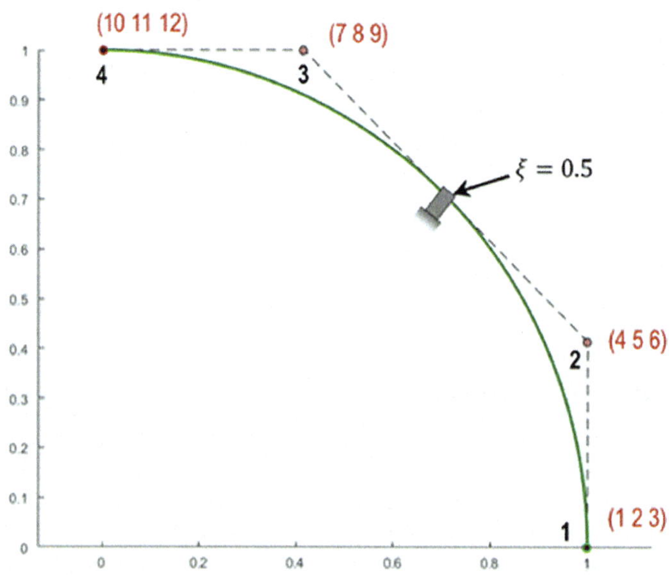

Fig. 3 Curved beam with fixed interior point

$$\begin{bmatrix} u_{t1} \\ u_{n1} \\ u_{\theta 1} \end{bmatrix} = \begin{bmatrix} u_4 R_2(0.5) + u_7 R_3(0.5) + u_{10} R_4(0.5) \\ u_5 R_2(0.5) + u_8 R_3(0.5) + u_{11} R_4(0.5) \\ u_6 R_2(0.5) + u_9 R_3(0.5) + u_{12} R_4(0.5) \end{bmatrix} \tag{6}$$

where R_2, R_3 and R_4 are the non-zero NURBS basis function in the knots-span $(i = 4)$. It should be noted that only $p + 1 = 3$ basis functions are non-zero in the knot-span. This follows directly from the property 3 discussed in Sect. 2.2. u_4, u_7 and u_{10} are the tangential displacements, u_5, u_8 and u_{11} are the radial displacements, u_6, u_9 and u_{12} are the angular displacements at the control points 2, 3 and 4, respectively. To fix a point, we set the displacement in Eq. (6) to be zero.

$$\begin{bmatrix} u_4 R_2(0.5) + u_7 R_3(0.5) + u_{10} R_4(0.5) \\ u_5 R_2(0.5) + u_8 R_3(0.5) + u_{11} R_4(0.5) \\ u_6 R_2(0.5) + u_9 R_3(0.5) + u_{12} R_4(0.5) \end{bmatrix} = \begin{bmatrix} 0 \\ 0 \\ 0 \end{bmatrix} \tag{7}$$

The above constraint Eq. (7) can be written in matrix form as

$$\boldsymbol{A}\boldsymbol{u} = \boldsymbol{0},$$
$$\text{where } \boldsymbol{u} = [u_1 u_2 \ldots\ldots u_{10}]^T,$$
$$\boldsymbol{A} = \begin{bmatrix} 0\,0\,0\,1\,0\,0\,1\,0\,0\,1\,0\,0 \\ 0\,0\,0\,0\,1\,0\,0\,1\,0\,0\,1\,0 \\ 0\,0\,0\,0\,0\,1\,0\,0\,1\,0\,0\,1 \end{bmatrix}.$$

By augmenting the matrix $\boldsymbol{A}$ with the stiffness matrix and introducing Lagrangian variables (λ) as additional variables for each constrain equation, we get the final set of governing equation as

$$\begin{bmatrix} \boldsymbol{K}_{12\times12} & \boldsymbol{A}^T_{12\times3} \\ \boldsymbol{A}_{3\times12} & 0_{3\times3} \end{bmatrix} \begin{bmatrix} \boldsymbol{U}_{12\times1} \\ \boldsymbol{\lambda}_{3\times1} \end{bmatrix} = \begin{bmatrix} \boldsymbol{F}_{n\times1} \\ 0_{3\times1} \end{bmatrix}. \tag{8}$$
$$\hat{\boldsymbol{K}}\hat{\boldsymbol{u}} = \hat{\boldsymbol{F}}.$$

3.2 Simply Supported Interior Point

By keeping the third degree of freedom (rotational) un-constrained, we get a simply supported connection at any desired point on the curve as depicted in Fig. 4. It can be considered as a special case of Sect. 3.1. The last row of the $\boldsymbol{A}$ matrix represents constraint on the angular displacement; hence, it need not be considered.

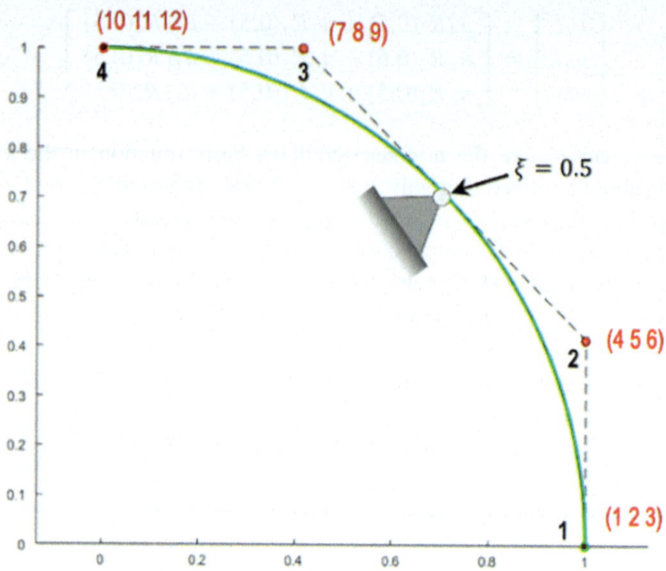

Fig. 4 Simply supporter internal joint

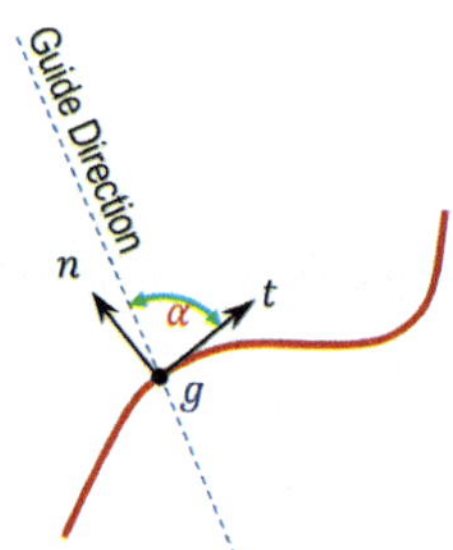

Fig. 5 Slider joint in a curved beam

3.3 *Slider Joint at Any Interior Point*

The case of an internal slider joint, guided in any arbitrary direction as shown in Fig. 6, is considered in this section. Consider point q in Fig. 5. For q to slide only in some specific direction, the required constraint is given by,

$$u_t \cos\alpha + u_n \sin\alpha = 0.$$

By proceeding similar to the method discussed in Sect. 3.1, the constraint equation can be written as following,

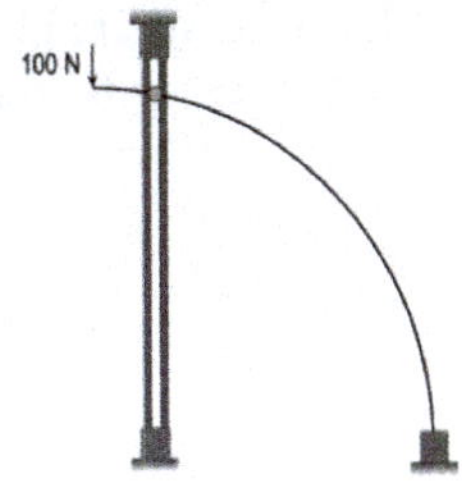

Fig. 6 Curved beam with slider joint

$$\left(u_k R_j + u_{k+3} R_{j+1} + \cdots + u_{k+3p} R_{j+p}\right)\cos\alpha + \left(u_{k+1} R_j + u_{k+4} R_{j+1} + \cdots + u_{k+3p+1} R_{j+p}\right)\sin\alpha = 0 \quad (9)$$

$$k = 3(i - p) - 2,$$

where $R_{j+p}s$ are the non-zero basis functions in the active knot span (i). u_{k+3p} and u_{k+3p+1} are the tangential and normal displacements at the active control points, respectively, and p is the degree of the basis function. Equation (9) can be written in the form of $\boldsymbol{A}$ matrix and then the augmented governing equations can be obtained as before.

4 Multi-patch IGA of Curved Beams

In the previous sections, we discussed the IGA of single-patch curved beam. Geometries with multiple interconnected beams can be studied using a multi-patch approach. We classify the connection in a multi-patch problem in three different types as shown in Fig. 7.

1. Junction at the ends of the curves (Type 1)
2. Junction at the end of one curve and on the interior of another (Type 2)
3. Junction on the interior of both the curves (Type 3).

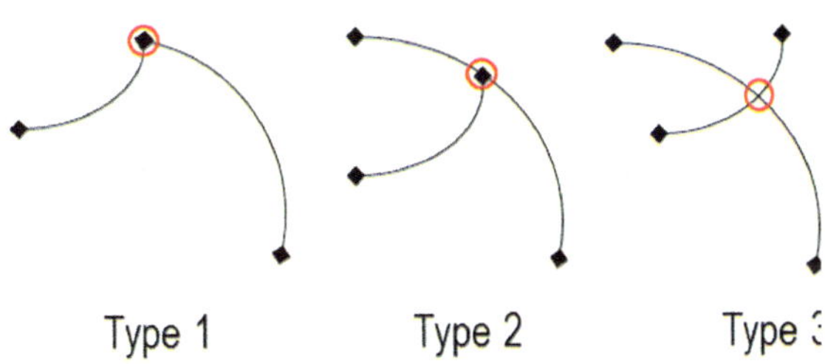

Fig. 7 Types of junctions in a multi-patch problem

It can be inferred from Fig. 7 that the junction types 1 and 2 are the special cases of type 3. We will discuss junction type 3 in the next section.

4.1 Type 3 Junction

Consider the problem shown in Fig. 8. The junction point g clearly lies in the interior of both the patches. The curve in each patch is defined by three controls points, while the degree of the basis is two. The knot vector for both the patches is $\Xi = \{0, 0, 0, 1, 1, 1\}$. The DoF indices are shown in blue colour for patch 1 and in red colour for patch 2.

Let the parameter values at the junction g for patch 1 and patch 2 be ζ_{j1} and ζ_{j2}, respectively. The next step is to find out the knot span index i_{j1} and i_{j2} for the evaluated parameter values.

The displacement at the junction, evaluated on patch 1, can be calculated using the linear combination of the effective NURBS basis and displacements of the control points in the active knot span. The indices for active control point and bases are $i_{j1} - p_1$ to i_{j1}. Let $i_{j1} = 3$ and $p_1 = 2$, the active indices are then 1, 2 and 3. The displacement is thus given by

$$\begin{bmatrix} u_{tj}^1 \\ u_{nj}^1 \\ u_{\theta j}^1 \end{bmatrix} = \begin{bmatrix} u_1 S_1(\zeta_{j1}) + u_4 S_2(\zeta_{j1}) + u_7 S_3(\zeta_{j1}) \\ u_2 S_1(\zeta_{j1}) + u_5 S_2(\zeta_{j1}) + u_8 S_3(\zeta_{j1}) \\ u_3 S_1(\zeta_{j1}) + u_6 S_2(\zeta_{j1}) + u_9 S_3(\zeta_{j1}) \end{bmatrix} \tag{10}$$

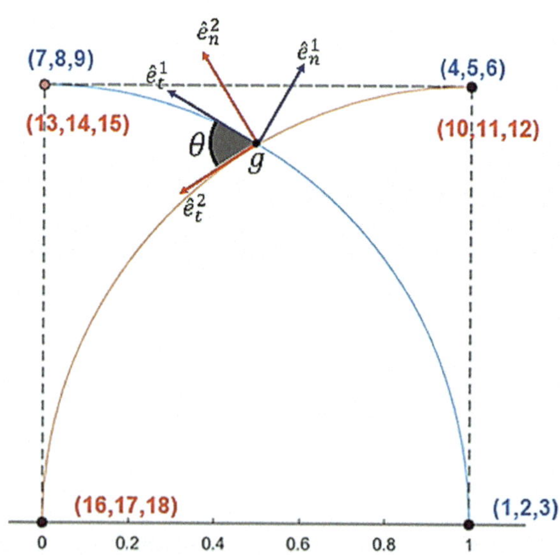

Fig. 8 Coordinate transformation between two patches

where S_1, S_2 and S_3 are the non-zero NURBS basis functions for patch 1 in the knot span index i_{j1}, evaluated at $\zeta = \zeta_{j1}$.

Similarly, for patch 2, $i_{j2} = 3$ and $p_2 = 2$, the displacement at the junction is

$$\begin{bmatrix} u_{tj}^2 \\ u_{nj}^2 \\ u_{\theta j}^2 \end{bmatrix} = \begin{bmatrix} u_{10}R_1(\zeta_{j2}) + u_{13}R_2(\zeta_{j2}) + u_{16}R_3(\zeta_{j2}) \\ u_{11}R_1(\zeta_{j2}) + u_{14}R_2(\zeta_{j2}) + u_{17}R_3(\zeta_{j2}) \\ u_{12}R_1(\zeta_{j2}) + u_{15}R_2(\zeta_{j2}) + u_{18}R_3(\zeta_{j2}) \end{bmatrix} \tag{11}$$

where R_1, R_2 and R_3 are the non-zero NURBS basis functions for patch 2 in the knot span index i_{j2}, evaluated at $\zeta = \zeta_{j2}$.

By transforming displacement on patch 2 to the local coordinate system of patch 1 using the transformation matrix $\boldsymbol{T}$ and equating the two displacements in Eqs. (10) and (11), we get

$$\begin{bmatrix} u_{tj}^1 \\ u_{nj}^1 \\ u_{\theta j}^1 \end{bmatrix} = -\boldsymbol{T}^T \begin{bmatrix} u_{tj}^2 \\ u_{nj}^2 \\ u_{\theta j}^2 \end{bmatrix}$$

$$\begin{bmatrix} u_1S_1(\zeta_{j1}) + u_4S_2(\zeta_{j1}) + u_7S_3(\zeta_{j1}) \\ u_2S_1(\zeta_{j1}) + u_5S_2(\zeta_{j1}) + u_8S_3(\zeta_{j1}) \\ u_3S_1(\zeta_{j1}) + u_6S_2(\zeta_{j1}) + u_9S_3(\zeta_{j1}) \end{bmatrix} = -T^T \begin{bmatrix} u_{10}R_1(\zeta_{j2}) + u_{13}R_2(\zeta_{j2}) + u_{16}R_3(\zeta_{j2}) \\ u_{11}R_1(\zeta_{j2}) + u_{14}R_2(\zeta_{j2}) + u_{17}R_3(\zeta_{j2}) \\ u_{12}R_1(\zeta_{j2}) + u_{15}R_2(\zeta_{j2}) + u_{18}R_3(\zeta_{j2}) \end{bmatrix} \tag{12}$$

The preceding constraint equation can be written in matrix form as,

$$\boldsymbol{A}\boldsymbol{u} = 0$$
$$\text{where } \boldsymbol{u} = [u_1 u_2 \ldots\ldots u_{18}]^T$$

$$A = \left[\overset{1\;2\;3}{S_1 I_{3\times 3}} \quad \overset{4\;5\;6}{S_2 I_{3\times 3}} \quad \overset{7\;8\;9}{S_3 I_{3\times 3}} \quad \overset{10\;11\;12}{-R_1 T^T_{3\times 3}} \quad \overset{13\;14\;15}{-R_2 T^T_{3\times 3}} \quad \overset{16\;17\;18}{-R_3 T^T_{3\times 3}} \right] \tag{13}$$

The final equation is as follows.

$$\begin{bmatrix} \boldsymbol{K}_{3n\times 3n} & \boldsymbol{\alpha A}^T_{3n\times 3} \\ \boldsymbol{\alpha A}_{3\times 3n} & 0_{3\times 3} \end{bmatrix} \begin{bmatrix} \boldsymbol{U}_{3n\times 1} \\ \boldsymbol{\lambda}_{3\times 1} \end{bmatrix} = \begin{bmatrix} \boldsymbol{F}_{3n\times 1} \\ 0_{3\times 1} \end{bmatrix} \tag{14}$$

$$\hat{\boldsymbol{K}}\hat{\boldsymbol{u}} = \hat{\boldsymbol{F}}.$$

Type 1 and type 2 junctions can be obtained from the method discussed in this section. For type 1 junction, the parameter value for both the patches are either 1 or 0. For type 2 junction, the parameter value for one patch is either 0 or 1 and, for the other patch, parameter can take any value between 0 and 1.

4.2 Geometry with Multiple Junctions

For a multiple junction problem (see Fig. 9), each intersection yields three sets of constraint equation. This leads to stacking of previously calculated $\boldsymbol{A}$ matrix in Eq. 13 over one another. The individual $\boldsymbol{A}$ matrices can be evaluated using the method explained in Sect. 4.1.

The $\boldsymbol{A}$ Matrix for a typical multi-junction problem is of the following form,

$$A = \begin{bmatrix} \cdots & s_1 I_{3\times3} & s_2 I_{3\times3} & s_3 I_{3\times3} & -R_1 T^T_{3\times3} & -R_2 T^T_{3\times3} & -R_3 T^T_{3\times3} & \cdots & & \\ & \cdots & s_1 I_{3\times3} & s_2 I_{3\times3} & s_3 I_{3\times3} & -R_1 T^T_{3\times3} & -R_2 T^T_{3\times3} & -R_3 T^T_{3\times3} & \cdots & \\ \cdots & s_1 I_{3\times3} & s_2 I_{3\times3} & s_3 I_{3\times3} & -R_1 T^T_{3\times3} & -R_2 T^T_{3\times3} & -R_3 T^T_{3\times3} & \cdots & & \end{bmatrix}$$

The final governing equation can be written as,

$$\begin{bmatrix} \boldsymbol{K}_{3n\times3n} & \alpha \boldsymbol{A}^T_{3n\times3m} \\ \alpha \boldsymbol{A}_{3m\times3n} & 0_{3m\times3m} \end{bmatrix} \begin{bmatrix} \boldsymbol{U}_{3n\times1} \\ \boldsymbol{\lambda}_{3m\times1} \end{bmatrix} = \begin{bmatrix} \boldsymbol{F}_{3n\times1} \\ 0_{3m\times1} \end{bmatrix} \tag{15}$$

$$\hat{\boldsymbol{K}}\hat{\boldsymbol{u}} = \hat{\boldsymbol{F}}$$

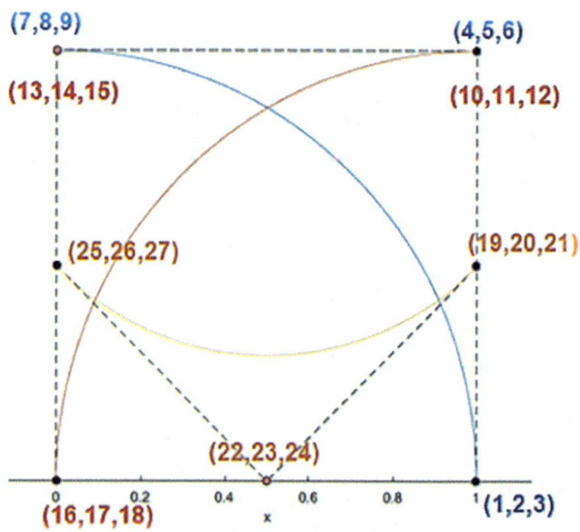

Fig. 9 Multi-patch geometry with multiple junctions

where m is the number of intersection points and n the number of control points.

4.3 Beams with Hinged Connection

Beams with hinged connection can be easily handled using ideas developed in the previous sections. The constraint on the rotational DoF needs to be removed. Consequently, the last row of the $\boldsymbol{A}$ matrix is no more required. This operation ensures that the junction will have equal x and y direction displacement and zero moment at the junction. For an intermediate hinge in the beam, the beam can be sub-divided at the hinge location into two different patches and then it can be handled in the similar fashion as a multi-patch problem.

5 Examples

5.1 A Compliant Gripper

IGA of a compliant gripper shown in Fig. 10a is considered. A one-half model of the problem is analyzed utilizing the symmetry. The geometry is constructed using six patches. The cross-section is square with edge length 0.2 m, Young's modulus is taken to be 80 GPa and Poisson's ratio is 0.2. The displacement result is plotted in Fig. 10c. The section force contours are plotted in Fig. 11. The convergence study

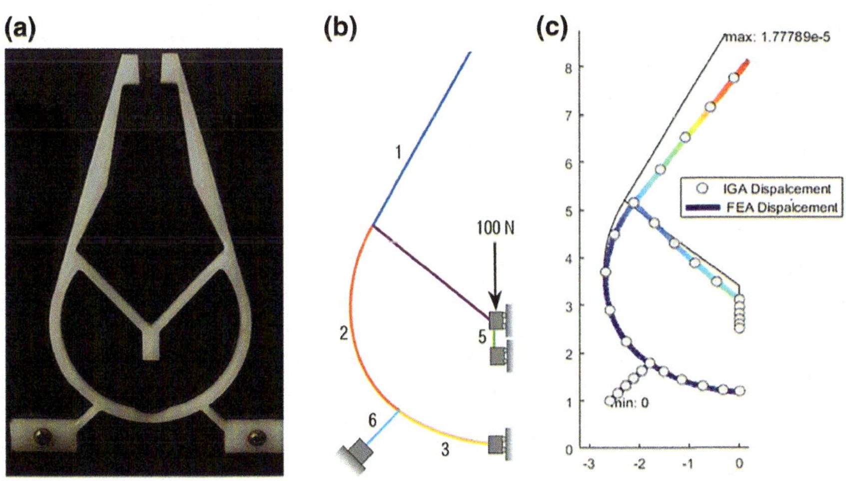

Fig. 10 A complaint gripper. **a** Actual Mechanism, **b** Schematic and Patch numbering, **c** IGA versus FEA Displacement

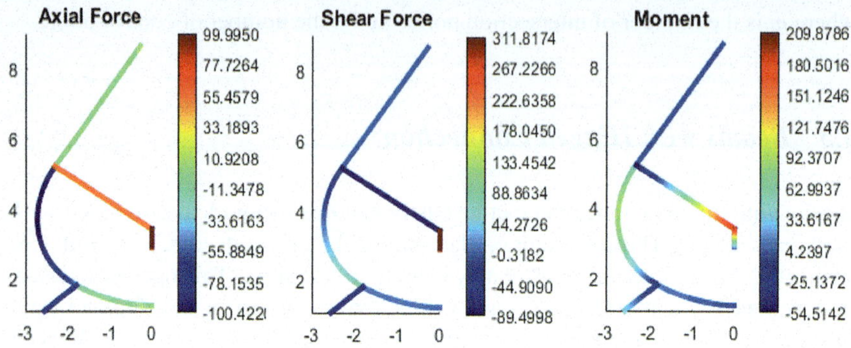

Fig. 11 Section forces results for Problem 5.1

for displacement and strain energy is shown in Figs. 12 and 13, respectively. The convergence rates are found to be higher in case of a higher degree basis. Also, better results are obtained at lower level of h-refinement when higher degree basis functions are used.

As pointed out earlier, Gaussian quadrature is not efficient for NURBS basis functions. It can be seen from the data presented in Table 1 that the section forces have a strong dependence on the degree of the basis and the number of gauss points used. The FEA results are also presented for comparison. In FEA, quadratic polynomial basis are used as the shape function and three gauss points are used for integral evaluation. The comparison between IGA and FEA for tip displacement and strain

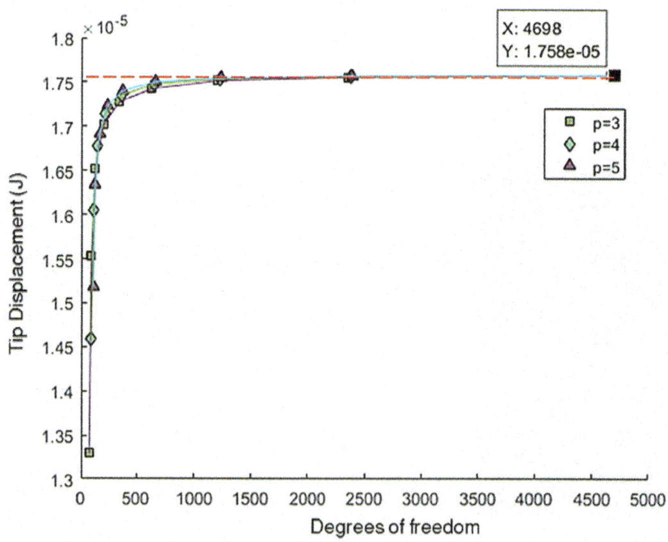

Fig. 12 Displacement convergence study for compliant gripper (Problem 5.1)

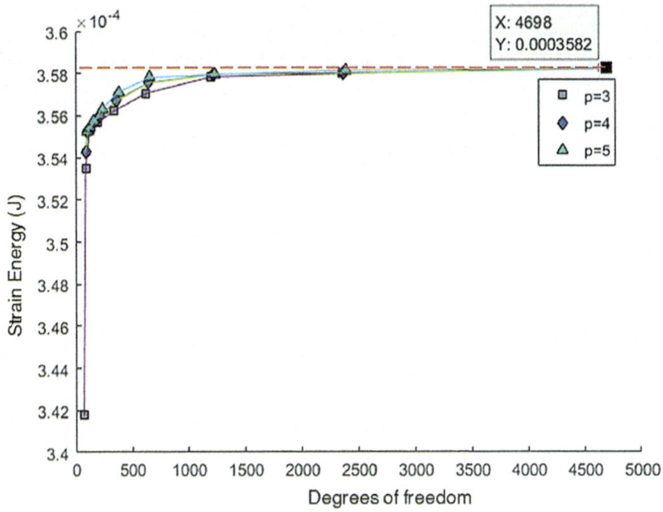

Fig. 13 Strain energy convergence study for compliant gripper (Problem 5.1)

energy is presented in Table 2. The small discrepancy in the result can be attributed to difference in the nature basis functions and the geometric description used. Exact geometry is used in IGA, while linearly interpolated geometry is used for analysis in FEA. The FEA results are considered as reference for the problems for which analytical results are unavailable. For the comparison, the DoF used in both FEA and IGA models are taken to be of the same order.

5.2 A Structure Subjected to Distributed Load

A structure with curved members, subjected to a distributed load is considered. The problem is shown in Fig. 14, it is composed of two curved members triangulated using straight beams. The structure is subjected to a distributed load. The geometry is made up of seven patches. The two ends of the curved beams are fixed. All the members have a rectangular cross section. The displacement and section force distribution is shown in Figs. 15 and 16, respectively.

The tip displacement and stain energy comparison between FEA and IGA result is presented in Table 3.

Table 1 Section forces variation with degree of the basis and number of gauss points

Section force		Degree (p) = 2			Degree (p) = 3			Degree (p) = 3			FEA
		GP = 3	GP = 5	GP = 7	GP = 3	GP = 5	GP = 7	GP = 3	GP = 5	GP = 7	
Axial force (N)	Max	100.00	100.00	100.00	100.00	100.00	100.00	100.00	100.00	100.00	100.00
	Min	−100.53	−100.77	−100.85	−100.24	−100.25	−100.25	−100.24	−100.25	−100.25	−100.36
Shear force (N)	Max	82.20	94.78	97.12	83.37	94.96	92.71	87.52	87.78	86.13	87.05
	Min	− 84.32	− 90.78	− 89.47	− 84.36	− 86.07	− 84.44	− 83.49	− 83.70	− 83.45	− 83.98
Moment (Nm)	Max	179.01	210.15	217.02	182.81	211.02	206.18	187.80	194.87	191.73	194.34
	Min	−48.95	−49.20	−49.28	−48.97	−49.22	−49.30	−48.97	−49.22	−49.30	−50.92

Table 2 Discrepancy between tip displacement and strain energy

Parameter	FEA (4317 DoFs)	IGA (4698 DoFs)	Discrepancy (%)
Tip displacement [$\times 10^{-5}$] (M)	1.775	1.758	0.95
Strain energy [$\times 10^{-4}$] (J)	3.611	3.582	0.80

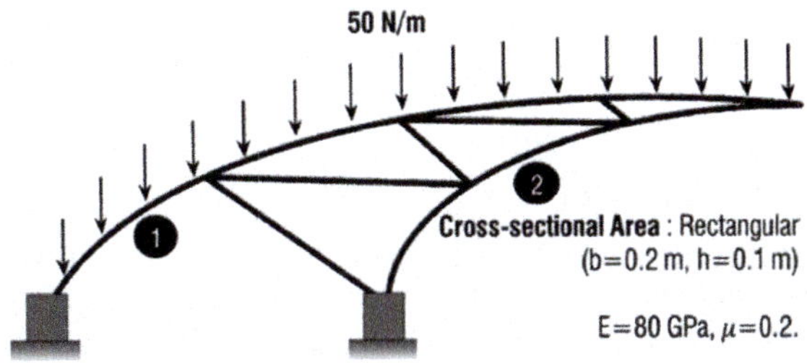

Fig. 14 Structure under distributed load

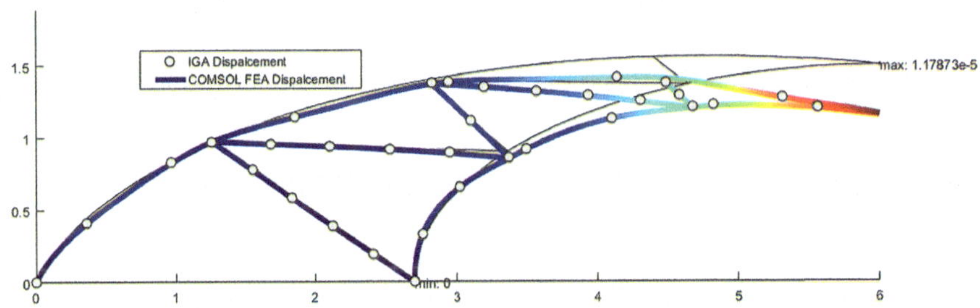

Fig. 15 IGA versus FEA displacement comparison (Problem 5.2)

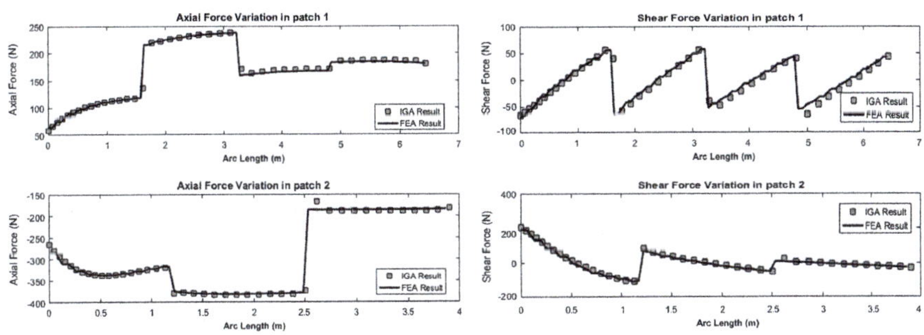

Fig. 16 Axial force and shear force variation in patch 1 and 2

Table 3 Discrepancy between tip displacement and strain energy

Parameter	FEA (2577 DoFs)	IGA (2751 DoFs)	Discrepancy (%)
Tip displacement [$\times 10^{-5}$] (M)	1.1787	1.1739	0.40
Strain energy [$\times 10^{-5}$] (J)	5.0482	4.9984	0.98

5.3 *Beam with a Slider Joint*

Consider the problem shown in Fig. 17. A curved beam with a slider joint at $\xi = 0.8$ is subjected to a horizontal load of 100 N in x-direction. One end of the beam is fixed. The geometry consists of only single patch. Technique to handle the slider joint is explained in Sect. 3.3.

The displacement results are shown in the Fig. 18. For verification, comparison between IGA and FEA results are shown in Figs. 19, 20 and 21. The small discrepancy in the results can be attributed to the nature of basis of functions used, difference in geometric representation and interpolation technique used in section force evaluation.

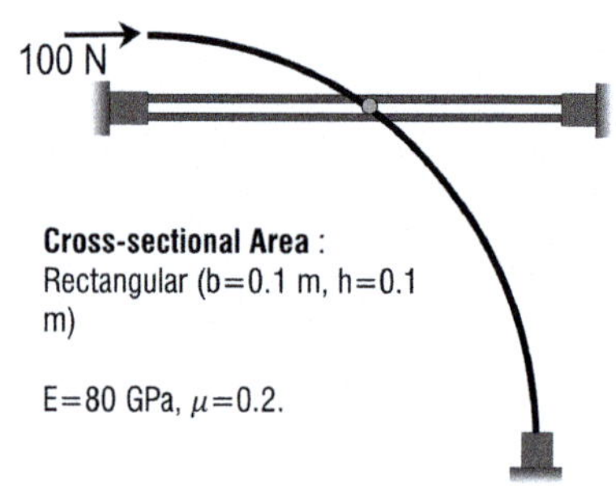

Fig. 17 Beam with a slider joint

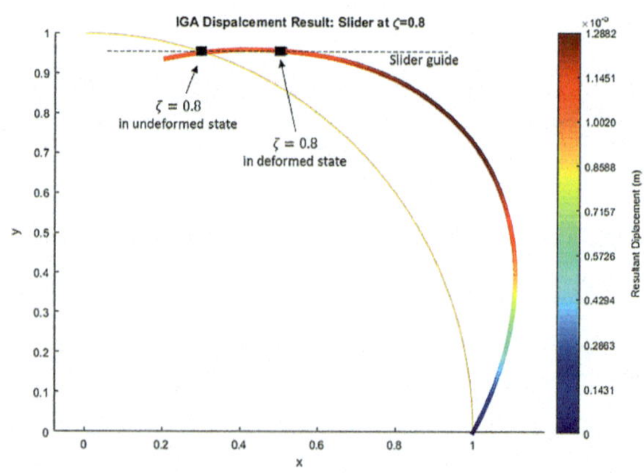

Fig. 18 IGA displacement result

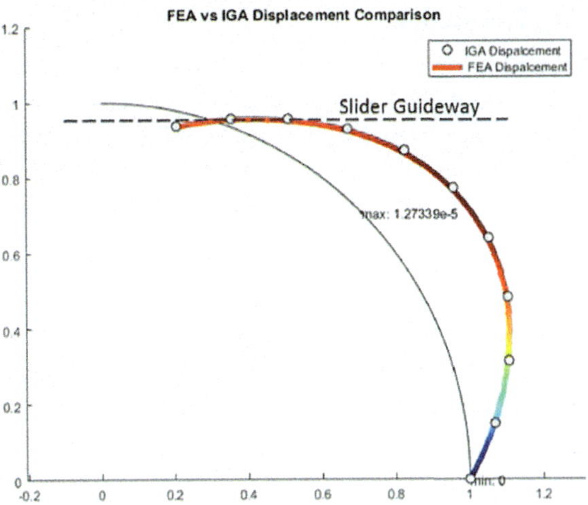

Fig. 19 IGA versus FEA displacement comparison

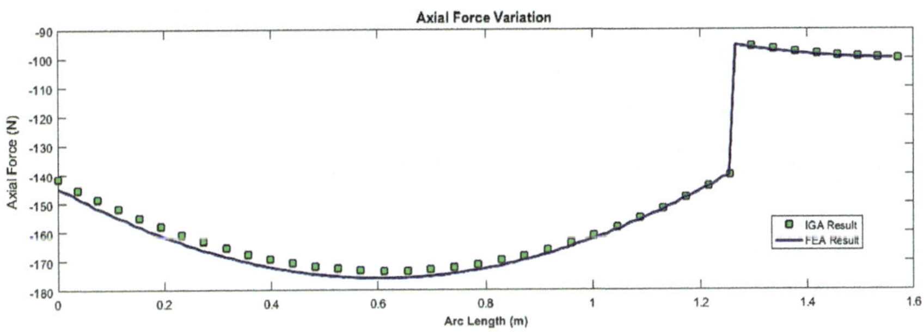

Fig. 20 Axial force variation in the beam

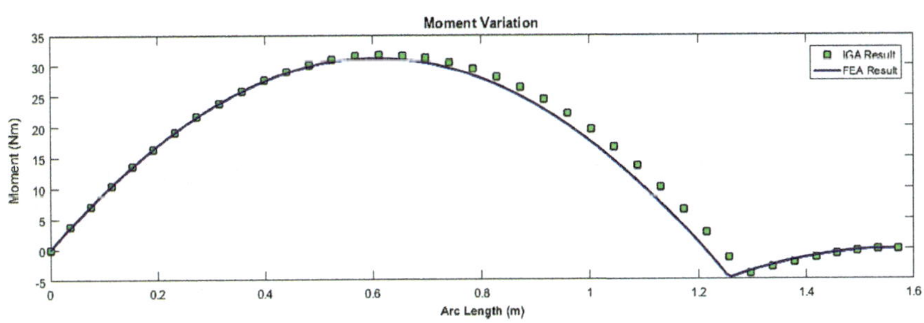

Fig. 21 Moment variation in the beam

6 Concluding Remarks

In the study, we have developed an approach for multi-patch curved beam models to be used in compliant mechanisms. Also, we have discussed techniques to handle Dirichlet and Neumann boundary conditions in the framework of IGA. This includes fixed joints, pinned joints and slider joints at any location on the curve. It may be noted that modelling pinned and slider joints is easier in IGA as compared to traditional FEA. The results obtained are found to be in good agreement with those obtained from commercial FEA software. However, several other issues such as locking in the Timoshenko beam, inefficiency of the Gaussian quadrature in IGA, etc., are not discussed. In future, the ideas can be extended to spatial multi-patch beams. More importantly, using the IGA framework in shape optimization [12] is the next step of this work.

References

1. Hughes T, Cottrell J, Bazilevs Y (2005) Isogeometric analysis: CAD, finite elements, NURBS, exact geometry and mesh refinement
2. Kota S, Ananthasuresh GK (1995) Designing compliant mechanisms. Mech Eng CIME 117(11):93–97
3. Litewka P, Jerzy R (1997) An efficient curved beam finite element. Int J Numer Methods Eng 40(14):2629–2652
4. Yang F, Sedaghati R, Esmailzadeh E (2008) Free in-plane vibration of general curved beams using finite element method. J Sound Vib 318(4–5):850–867
5. Cazzani A, Marcello M, Emilio T (2016) Isogeometric analysis of plane-curved beams. Math Mech Solids 21(5):562–577
6. Adam C, Bouabdallah S, Zarroug M, Maitournam H (2014) Improved numerical integration for locking treatment in isogeometric structural elements, Part I: Beams. Comput Methods Appl Mech Eng 279:1–28
7. Echter R, Bischoff M (2010) Numerical efficiency, locking and unlocking of NURBS finite elements. Comput Methods Appl Mech Eng 199(5–8):374–382
8. Piegl L, Tiller W (1995) The NURBS book. Springer, London, UK
9. Lim C, Wan C, Kitipornchai S (1997) Timoshenko curved beam bending solutions in terms of Euler-Bernoulli solutions. Arch Appl Mech 67:179–190
10. Hughes TJ, Reali A, Sangalli G (2010) Efficient quadrature for NURBS-based isogeometric analysis. Comput Methods Appl Mech Eng 199:301–313
11. Felippa IC (2017) MultiFreedom Constraints II. University of Colorado. http://www.colorado.edu/engineering/CAS/courses.d/IFEM.d/
12. Xu D, Ananthasuresh GK (2003) Freeform skeletal shape optimization of compliant mechanisms. J Mech Design 125(2):253–261

Design and Development of a Contact-Aided Compliant Flapping Wing for Micro Air Vehicle

M. Masruddin, Raunak Singh Rana, Deepak Kumar Patel, and Narayana Reddy

Abstract In this paper, a contact-aided compliant flapping wing for micro air vehicle (MAV) is presented. The design consists of a perforated plate, covered with flaps which allow motion in one direction only. In order to arrest the motion of flaps in the other direction, a contact stopper is provided. When the wing (i.e., perforated plate along with flaps) is making a flapping motion in a fluid environment, it experiences different drag forces in forward and reverse strokes. The wing is made with acrylic (Poly (methyl methacrylate) or PMMA) sheet having 2 mm thick. The flaps are connected to perforated plate by compliant joints that act as torsional springs. Furthermore, these compliant joints aid in moving flaps from open to a closed position. Experiments were carried out in both air and water medium for various plate configurations. It is observed that only 14.71% of input power is saved during a reverse stroke of four-flap perforated plate while flapping in air. Hence, the advantage is less in air, wing thickness needs to be reduced substantially to get an advantage in air medium. But in a water medium, input power is saved by 32.80% for the current dimensions. This increment is due to the fact that the density of water is more compared to air. It is also noticed that there is no advantage beyond 65 and 70% of area reduction in air and water mediums respectively as it is the minimum power required to drive the servomotor.

Keywords Torsional spring · Perforated plate · Acrylic sheet

M. Masruddin (✉) · R. S. Rana · D. K. Patel · N. Reddy
Department of Mechanical Engineering, IIT-Guwahati, Guwahati, India
e-mail: masruddin@iitg.ac.in

R. S. Rana
e-mail: raunak123rana@gmail.com

D. K. Patel
e-mail: deepakpateljnv@gmail.com

N. Reddy
e-mail: annem@iitg.ac.in

D. Sen et al. (eds.), *Mechanism and Machine Science*, Lecture Notes in Mechanical Engineering, https://doi.org/10.1007/978-981-15-4477-4_49

1 Introduction

Compliant mechanisms grab a lot of attention from the past few years. It is mainly due to the huge advantages over rigid body mechanisms. Major advantages are reduction in cost, less weight, less movable joints, and hence less requirement of lubrication. These benefits eliminate the backlash and increase the efficiency of the mechanism. Such approach is adopted while designing high precision instruments [1, 2]. Also, injection moulding is used in manufacturing most of the compliant mechanisms as a single piece [3, 4].

Studies on contact-aided compliant mechanism (CCM) had been carried out in the early 2000s. Mankame and Ananthasuresh were the first to explore the concept of contact-aided compliant mechanisms by designing displacement contact-aided compliant gripper which consists of a single contact point [5]. After that, research on CCMs took several turns; one of the interesting applications is the implementation of CCMs in the design of a flapping wing micro air vehicle (MAV) [6–8]. Some interesting attempts were made in the construction of insect thorax involving CCMs [9–11] and also wing actuation using compliant elements [12–15]. Many studies were proposed for the design of compliant transmission mechanisms for MAVs [16–19]. Studies suggest that such implementation of CCMs in MAVs reduces input power requirement by 30% compared to MAVs made with rigid body mechanisms [12, 20].

This paper presents a novel wing design for MAV using CCMs. The advantage of this design is reduction in input power requirement during reverse stroke of wing.

2 Wing Design

The design of the wing consists of a perforated plate covered with flaps which allows motion in only one direction as shown in Fig. 1. The flap is connected to the perforated plate by means of a compliant hinge. 50 μm thin steel strip is used as a compliant hinge. In order to arrest the motion of flaps in the other direction, a contact stopper is provided.

When the wing (i.e, perforated plate along with flaps) moves in the forward direction, the flap is in the closed position at the beginning of the stroke. By the end of the forward stroke the flap fully opens. Hence, at the beginning of reverse stroke, the flap will be in open position. This reduces the effective area of the wing during

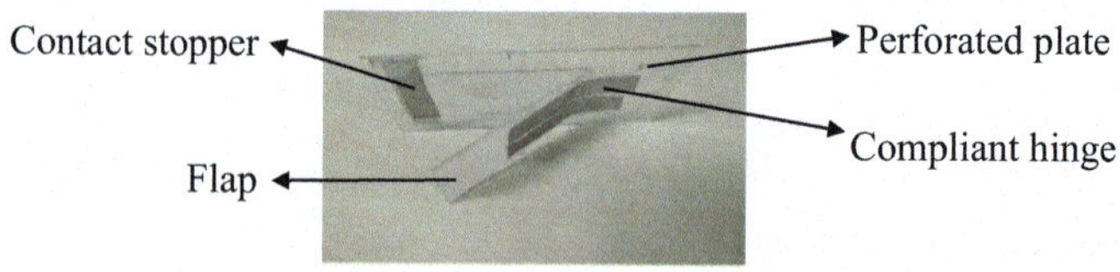

Fig. 1 Single-flap perforated plate

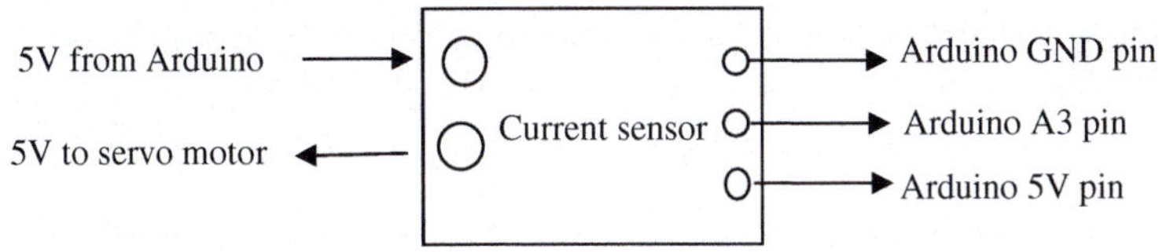

Fig. 2 Circuit diagram

reverse stroke. By the end of the reverse stroke, flap tries to come back to its original position. When forward stroke is about to start, compliant hinge acts as a torsional spring and restores the flap to its original position. Thus the perforated plate acts as a solid plate in forward stroke. Clearly, drag force is more when flaps are closed and less when flaps are open. The advantage of this particular phenomenon is considered in the current design (Fig. 2).

2.1 *Construction and Working*

The experimental setup consists of a rectangular base plate made with 8 mm thick acrylic sheet. Servomotor is mounted on the base plate as presented in Fig. 3. A shaft of 10 mm diameter made with acrylic material is used to flap the wing. Pillow block bearings of the same size are used to support the shaft in the horizontal position. One end of the shaft is fixed to the servomotor with the help of a shaft coupler made with nylon material. The wing is fixed to the other end of the shaft by using aluminium strips. Arduino microcontroller was used to supply power to the servomotor. ACS 712 Hall effect current sensor was connected to sense the current in the circuit. Arduino supplies 5 V to the current sensor and this voltage is received by the servomotor through the current sensor as shown in Fig. 2.

The power consumed to flap the wing is calculated from the product of voltage supply and current. Firstly, the experiments were conducted in air medium for

Fig. 3 Experimental setup

different dimensions of wings. Further, experiments have been carried out in water environment to observe the drag force dominance over inertia force.

3 Torque Calculations

Firstly, time taken by the wing to complete one cycle is measured, as presented in Table 1. Then angular speed is calculated using Eq. (1).

$$t = \frac{2\pi}{\omega} \tag{1}$$

From Fig. 3, it is clear that the shaft is subjected to pure shear. Hence, from elementary strength of materials principal strain is calculated as follows:

Maximum principal stress

$$\sigma_1 = +\tau \tag{2}$$

And minimum principal stress

$$\sigma_2 = -\tau \tag{3}$$

Then from Hooke's law Maximum principal strain is given by

$$\varepsilon_1 = \frac{\sigma_1}{E} - (v)\frac{\sigma_2}{E} \tag{4}$$

By substituting Eqs. (2) and (3) in Eq. (4), then

$$\varepsilon_1 = \left(\frac{1+v}{E}\right)\left(\frac{16T}{\pi d^3}\right) \tag{5}$$

Table 1 Variation of parameters with respect to different configurations of the wing

Plate configuration		Time period (s)	Strain (μ)	Torque (N-m)	Power (mW)
Single-flap perforated plate	In air	5.60	95.65	0.03478	39.00
	In water	6.04	132.69	0.04825	50.16
Four-flap perforated plate	In air	5.64	98.49	0.03581	39.87
	In water	6.23	140.76	0.05118	51.59
Solid plate	In air	6.01	109.23	0.03972	41.50
	In water	6.62	157.03	0.05710	54.16

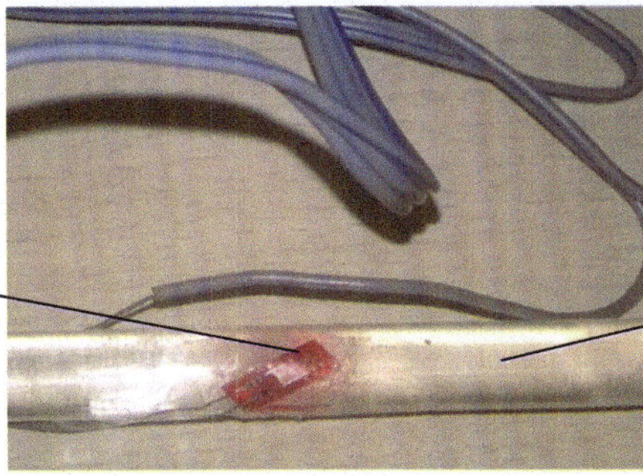

Fig. 4 Strain gauge attached at 45° to the acrylic shaft

where

- ε_1 Strain in principal direction
- ν Poisson's ratio
- E Young's modulus
- T Input torque required to rotate the shaft
- d Diameter of the shaft
- t Time taken by the wing to complete one cycle (forward and reverse strokes)
- ω Angular speed of the shaft

For acrylic material $\nu = 0.35$, $E = 2.5$ Gpa, and $d = 10$ mm.

Strain gauge is attached to the shaft at 45° as shown in Fig. 4 to measure the strain in the principal direction. Strain buster is used to capture the strain from the strain gauge. By knowing strain, corresponding Torque and Power are calculated and presented in Table 1.

4 Results and Discussions

Experiments were carried out in air and water medium to observe required advantage of input power. It is observed that 45.147 mW of input power is required in forward stroke with time period of 5.60 s in air for a single-flap perforated plate. But in return stroke, the input power of 37.995 mW is required. The power requirement in return stroke is reduced because the flaps are fully opened at the beginning of return stroke and this reduces the effective area during return stroke.

But for a four-flap perforated plate, 45.594 mW of power is required during forward stroke having time period of 5.64 s. In return stroke 38.889 mW of power is consumed in air medium as given in Fig. 5. The shape of graph is sinusoidal for single-flap and four-flap perforated plates due to the decrement of power in return stroke. But for solid plate, 46.488 mW required in air medium with time period of 6.01 s and the shape is a straight line because the same amount of power is required in both forward and reverse strokes.

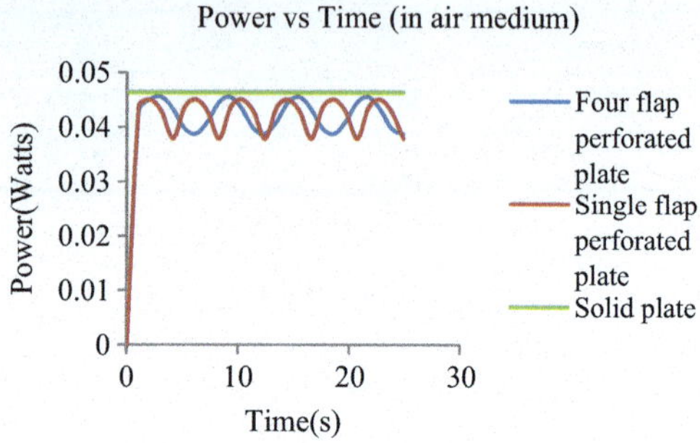

Fig. 5 Variation of power with respect to time in air medium

Figure 6 shows that the perforated plate with four flaps consumes 56.769 mW of power in forward stroke and 45.150 mW in return stroke in water with a time period of 6.23 s. But 55.875 mW are required during forward stroke and 37.548 mW in return stroke for single-flap perforated plate in water medium having time period of 6.04 s. The input power requirement for a perforated plate with four flaps is more than that of a single-flap perforated plate. This is due to the fact that effective area of four-flapped perforated plate is more compared to that of the single-flap perforated plate. Only 25% of the area is reduced from the solid plate to design a four-flap perforated plate. On the other hand, 40% of the area is cut out from the solid plate to form a single-flap perforated plate.

For solid plate 59.004 mW of power is required in water medium having time period of 6.62 s, as usual there is no variation in power during forward and backward stroke in water, because the effective area is the same in both strokes.

Figures 7 and 8 represent torque variation is similar to power in air and water medium with respect to time. It is observed that maximum torque of 0.0623 N-m is required in the case of solid plate flapping in water with an error of 8.3% from the strain gauge readings. Of all plate configurations, 0.04027 N-m is the minimum torque observed in case of single-flap perforated plate flapping in air with an error of 13.63% from the torque calculated using strain gauge readings.

The drop in torque is due to higher density of water compared to air. Also, the shape of graph is a straight line for the solid plate because the same amount of input

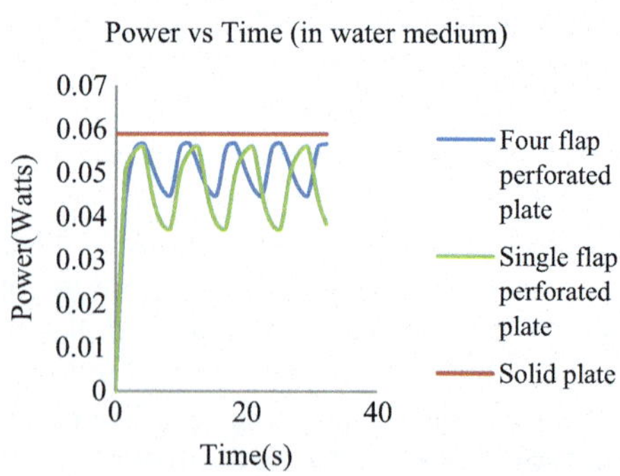

Fig. 6 Variation of power with respect to time in water medium

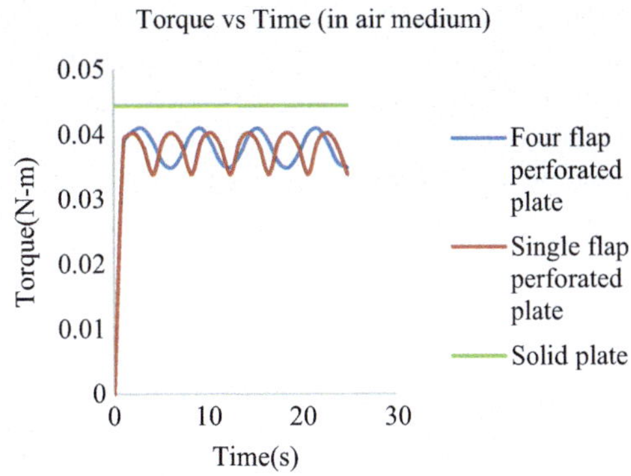

Fig. 7 Variation of Torque with respect to time in air medium

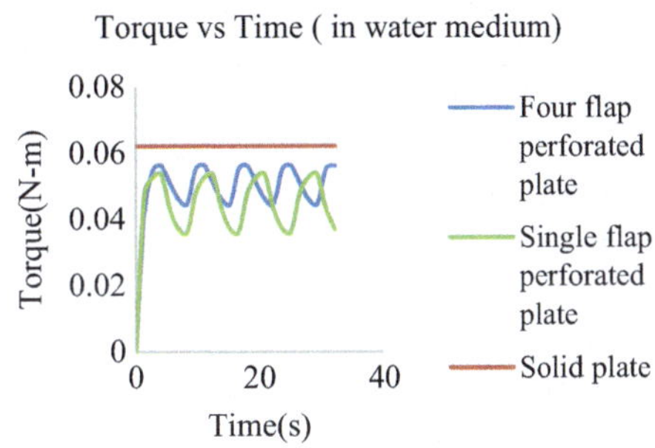

Fig. 8 Variation of Torque with respect to time in air medium

torque is required in both forward and reverse strokes. For four-flap and single-flap perforated plate, the effective area reduces in reverse stroke. Due to this, drop in input torque is observed and hence, the shape of the graph is sinusoidal in both air and water medium.

It is observed that the advantage of input power reduction increases with the increase in area reduction as presented in Fig. 9. Four-flap perforated plate is designed by reducing 25% of area from the solid plate. In air, for a four-flap perforated plate 45.6 mW of power is required in forward stroke and 38.89 mW in return stroke. Hence, 14.71% of power is saved in air medium. But in water 20.45% of

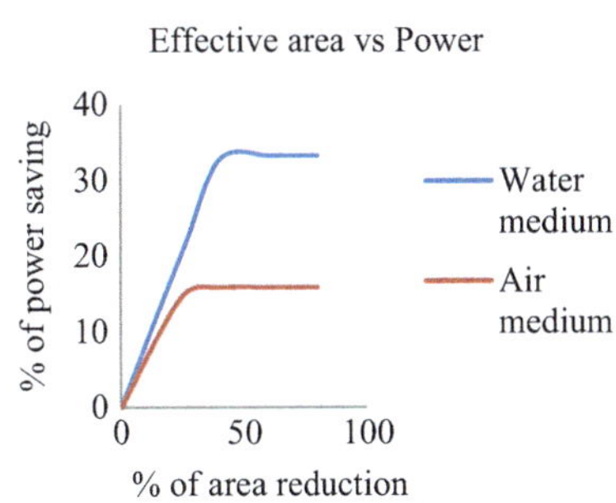

Fig. 9 Variation of % power with respect to % of area reduction

input power is saved from forward to return strokes. The advantage in water is more than in air as water is denser than air. Later, 40% of area is reduced from the solid plate to form a single-flap perforated plate. For this configuration, 15.85 and 32.80% of power is saved in air and water medium, respectively. But it came to be noticed that, there is no advantage beyond 65% of area reduction in air and 70% that in water because it is the minimum power required to drive the servomotor.

5 Conclusions

The wing is designed as a plate with slots, covered by flaps. The flaps are connected to the plate by compliant hinges. The wing is made with acrylic (Poly (methyl methacrylate) or PMMA) sheet having 2 mm thick. Motion is provided to the wing by Arduino microcontroller and servomotor. Hall effect current sensor is used to sense the current in the circuit. The simulations have been carried out for various plate configurations with different physical and geometrical parameters. Experiments are carried out in both air and water environment to observe the required advantage. Torque measurement is done by attaching strain gauge at an angle of 45° on the acrylic shaft. There is not much advantage observed in air medium due to high inertia forces. Wing thickness needs to be reduced substantially to get an advantage in air medium. However, the current dimensions show advantage in water medium.

Design of the wing presented in this paper has the advantage of decreasing power in reverse stroke. This is achieved by reducing the effective area of the wing in reverse stroke. It is observed that 14.71% of input power is saved from forward to reverse strokes in the case of a single-flap perforated plate flapping in air. But 32.80% of input power is saved, when a four-flap perforated plate flaps in water. The advantage in case of the single-flap perforated plate is more compared to the four-flap perforated plate because 40% of the effective area is reduced in case of the single-flap perforated plate and only 25% of the effective area is reduced in the four-flap perforated plate. It also came to be noticed that, time period of flapping the wing is more in water than in air for all configurations of the wing. This is due to the fact that the density of water is more compared to the density of air.

6 Future Work

The implementation of this novel approach in the design of wings for a micro air vehicle (MAV) leads to the reduction in input power requirement during backward stroke. This idea can be explored in the area of wing actuation mechanism and thorax construction of MAVs. Hence, in future it is possible to develop MAVs with very low power source.

Acknowledgements This paper is dedicated to the memory of our mentor Dr. A. Narayana Reddy.

References

1. Motsinger RN (1964) Flexural devices in measurement systems. In: Stein PK (ed) Measurement engineering. Stein Engineering Services, Phoenix
2. Tuttle SB (1967) Semi-fixed flexural mechanisms. In: Mechanisms for engineering design. Wiley, New York
3. Roach GM, Lyon SM, Howell LL (1998) A compliant, over-running Ratchet and Pawl clutch with centrifugal throw-out. In: Proceedings of the ASME design engineering technical conferences, DETC98/MECH-5819
4. Roach GM, Lyon SM, Howell LL (1998) A compliant, over-running Ratchet and Pawl clutch with centrifugal throw-out. U. S. patent 6,148,979, 21 Nov 1998
5. Mankame ND, Ananthasuresh GK (2002) Contact-aided compliant mechanisms: concept and preliminaries. In: Proceedings of ASME international design engineering technical conferences. Montreal, Canada, DETC2002/MECH-34211
6. Wood RJ (2004) Liftoff of a 60 mg flapping-wing MAV. In: IEEE/RSJ IROS international conference on robotics and systems, pp 1889–1894, October, 2004
7. Wood RJ, Avadhanula S, Fearing RS (2003) Micro-robotics using composite materials; the micromechanical flying insect thorax. In: IEEE international conference on robotics and automation, Taipei, Taiwan
8. Yan J, Wood RJ, Avadhanula S, Sitti M, Fearing RS (2001) Towards flapping wing control for micro-mechanical flying insect. In: IEEE international conference on robotics and automation, pp 3901–3908, Seoul, Korea
9. Wootton RJ (2009) Springy shells, pliant plates and minimal motors: abstracting the insect thorax to drive a micro-air vehicle flying insects and robots. Berlin, pp 207–217
10. Gullan PJ, Cranston PS (2009) The insects: an outline of entomology. Wiley, New York
11. Nation JL (2008) Insect physiology and biochemistry. CRC Press, Boca Raton, FL
12. Madangopal R, Khan ZA, Agrawal SK (2005) Biologically inspired design of small flapping wing air vehicles using four-bar mechanisms and quasi-steady aerodynamics. J Mech Des 127:809–16
13. Khan ZA, Agrawal SK (2007) Design and optimization of a biologically inspired flapping mechanism for flapping wing micro air vehicles. In: IEEE international conference on robotics and automation, pp 373–378
14. Khan ZA, Agrawal SK (2011) Optimal hovering kinematics of flapping wings for micro air vehicles. AIAA J 49:257–68
15. Park JH, Agrawal SK (2014) Dynamic effects of asymmetric in-phase flapping (AIF) on forward flight. In: IEEE international conference on robotics and automation, pp 3550–3555
16. Keennon M, Klingebiel K, Won H (2012) Development of the nano hummingbird: a tailless flapping wing micro air vehicle. In: AIAA aerospace sciences meeting, p 588
17. Jeon J, Cho H, Kim Y, Lee J, Shin JS, Kim C, Kim S (2016) Design and analysis of the link mechanism in the flapping wing MAV using flexible multi-body dynamic analysis. In: 24th AIAA/AHS adaptive structures conference, p 0819
18. Nguyen TA, Phan HV, Au TKL, Park HC (2016) Experimental study on thrust and power of flapping-wing system based on rack-pinion mechanism bio-inspiration bio-mimetics, 11 046001
19. Liu CH, Chen CK (2015) Kinematic analysis of a flapping wing micro-aerial-vehicle with Watt straight-line linkage. J Appl Sci Eng 18:355–62
20. Baek SS, Ma KY, Fearing RS (2009) Efficient resonant drive of flapping-wing robots. In: IEEE/RSJ international conference on intelligent robots and systems, pp 2854–60

References

Nonlinear Behavior of Planar Compliant Tensegrity Mechanism with Variable Free-Lengths

Tyler Rhodes, Miranda Tanouye, and Vishesh Vikas

Abstract Traditional tensegrity mechanisms comprise compressive (rigid rods) and tensile members (cables). Compliant tensegrity mechanisms (CoTM) include springs alongside cables. Introduction of spring elements allows these structures to be more adaptable and robust. The kinematic and stability analyses of such mechanisms will facilitate better understanding of their behaviors for developing control and design methodologies. The analysis of CoTM often involves making the zero free-length (ZFL) assumption, i.e., the free-length of the spring is zero, which disqualifies the analysis for most real-word applications. The paper illustrates the drastic increase in computational complexity for finding static solutions as the assumption of ZFL for spring elements are relaxed for a simple planar compliant tensegrity mechanism comprising two rigid triangular platforms connected by a compressive member and two spring elements. The resulting nonlinear behavior of obtained static solutions shows intersecting manifolds of equilibrium orientation angles where the number of solutions vary from minimum of 4 to beyond 10 as the spring free-lengths are varied.

Keywords Tensegrity · Compliant mechanisms · Non-zero free-length

1 Introduction

Tensegrity mechanisms are made up of compressive and tensile members where no pair of struts touch and each end is connected to three non-coplanar ties. This provides some unique features: strain is distributed through deformation of the structure, they expand in all axes at once, and can be built from one another [1]. These unique

T. Rhodes · M. Tanouye · V. Vikas (✉)
Agile Robotics Lab, University of Alabama, Tuscaloosa, AL, USA
e-mail: vvikas@ua.edu

T. Rhodes
e-mail: trhodes1@crimson.ua.edu

M. Tanouye
e-mail: mtanouye@crimson.ua.edu

D. Sen et al. (eds.), *Mechanism and Machine Science*, Lecture Notes in Mechanical Engineering, https://doi.org/10.1007/978-981-15-4477-4_50

properties render them useful in the fields of robotics [2, 3], space applications [4], bridges [5], and biological modeling [6]. Several natural structures can be modeled as a tensegrity, such as the human spine [7] and DNA supermolecules [8]. Their ability to withstand large amounts of strain and efficient packing (occupying less volume for storage) make them ideal for deployment applications [9].

Static analysis, dynamic and stability analysis of traditional tensegrity mechanism provide insight into their behavior [10–14]. The direct static problem yields equilibrium equations which are nonlinear in the angle, thus, yielding multiple equilibrium solutions. The analytical solution structure of the equilibrium equations is of special interest as they are helpful in providing the bound on the number of possible solutions. Some analytical solutions for prestressed tensegrity structures have been proposed and explored [15, 16].

Compliant tensegrity mechanisms (CoTM) augment traditional tensegrity structures by adding spring elements that can compress and as well as extend. Moreover, the mechanism orientation may be controlled through its compliance, such as by indirectly varying free-lengths of the spring elements. Such modifications have been experimentally introduced to build tensegrity robots, however, their kinematic analysis has been limited [17, 18]. The presented research analyzes the effect of variation of spring free-lengths on the static equilibrium and stability of a planar tensegrity mechanism.

Commonly, the static and dynamic analysis of these structures assumes zero free-lengths (ZFL). For a ZFL spring, the unstretched (free) length of a linear spring is zero and a general ZFL spring is not an existing product [19], however, it can be realized through different methodologies including pretension [20] and geometric manipulation [21]. This ZFL assumption greatly simplifies the analysis of the system but is difficult to implement in practice. The paper presents detailed analysis for an example planar CoTM where ZFL assumption is gradually relaxed where (1) both springs have ZFL, (2) one spring has ZFL, and (3) no springs have ZFLs. The application of the presented analysis will be instrumental in control, modeling of deployable CoTM structures, and tensegrity-based biomechanical modeling.

2 Problem Statement and Definition

The proposed mechanism consists of two triangular rigid bodies connected by a rigid rod and two spring members as shown in Fig. 1 where the relative distance between points 1, 4 is L_3. The spring free-lengths and spring constants are denoted by L_{0i}, k_i where $i = 1, 2$. The objective is to find all of its stable equilibrium positions, as defined by γ_1 and γ_2, for a given set of parameters.

As illustrated in Fig. 1, let the coordinate system B be fixed on the bottom rigid body with origin at point 1, x-axis between points 1, 2, and z-axis out of the plane of paper. Similarly, the coordinate system T is fixed on the top rigid body with origin at point 4, x-axis between points 4, 6, and z-axis out of the plane of paper. Let

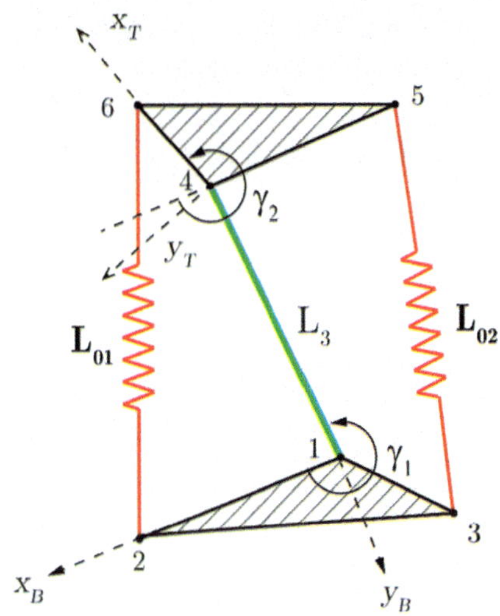

Fig. 1 The proposed planar compliant tensegrity mechanism (CoTM) comprises two triangular rigid bodies connected by a rigid rod (green) and two spring members (red)

${}^{A}\boldsymbol{P}_i$, $\boldsymbol{P}_{i\to j}$ represent a point i in coordinate system A and vector between points i and j, respectively. Consequently, the points are

$$ {}^{B}\boldsymbol{P}_2 = [p_{2x}, 0, 0]^T \ , \quad {}^{B}\boldsymbol{P}_3 = \left[p_{3x}, p_{3y}, 0\right]^T \tag{1} $$

$$ {}^{T}\boldsymbol{P}_5 = [p_{5x}, 0, 0]^T \ , \quad {}^{T}\boldsymbol{P}_6 = \left[p_{6x}, p_{6y}, 0\right]^T \tag{2} $$

The transformation matrix ${}^{T}_{B}\boldsymbol{T}$ between coordinate system T and B is written as

$$ {}^{T}_{B}\boldsymbol{T} = \begin{bmatrix} c_2 & -s_2 & 0 & L_3 c_1 \\ s_2 & c_2 & 0 & L_3 s_1 \\ 0 & 0 & 1 & 0 \\ 0 & 0 & 0 & 1 \end{bmatrix} \tag{3} $$

where c_i, s_i are $\cos\gamma_i$, $\sin\gamma_i$ corresponding to angles γ_1, γ_2 as shown in Fig. 1. Hence, for $i = 4, 5$

$$ \begin{bmatrix} {}^{B}\boldsymbol{P}_i \\ 1 \end{bmatrix} = {}^{T}_{B}\boldsymbol{T} \begin{bmatrix} {}^{T}\boldsymbol{P}_i \\ 1 \end{bmatrix} \tag{4} $$

The superscript will be dropped for the remaining section of the paper as all the calculations will be performed in the B coordinate system. Let f_3 be the unknown force along the rigid bar and d_i, $f_i = k_i(d_i - L_{0i})\ \forall i = 1, 2$ be the length and forces in spring elements. Using screw theory notation, for static equilibrium, sum of the three wrenches along the corresponding tensegrity elements is zero

$$ \frac{f_1}{d_1}\begin{bmatrix} \boldsymbol{P}_{2\to 6} \\ \boldsymbol{P}_{2\to 6} \times \boldsymbol{P}_{1\to 2} \end{bmatrix} + \frac{f_2}{d_2}\begin{bmatrix} \boldsymbol{P}_{3\to 5} \\ \boldsymbol{P}_{3\to 5} \times \boldsymbol{P}_{1\to 3} \end{bmatrix} + f_3 \begin{bmatrix} \boldsymbol{P}_{1\to 4} \\ 0 \end{bmatrix} = 0 \tag{5} $$

The analysis results in four equations of interest—two equilibrium equations coupled with two constraint equations

$$k_1\left(1-\frac{L_{01}}{d_1}\right)(\boldsymbol{P}_{2\to6}\times\boldsymbol{P}_{1\to4})+k_2\left(1-\frac{L_{02}}{d_2}\right)(\boldsymbol{P}_{3\to5}\times\boldsymbol{P}_{1\to4})=0 \tag{6a}$$

$$k_1\left(1-\frac{L_{01}}{d_1}\right)(\boldsymbol{P}_{2\to6}\times\boldsymbol{P}_{1\to2})+k_2\left(1-\frac{L_{02}}{d_2}\right)(\boldsymbol{P}_{3\to5}\times\boldsymbol{P}_{1\to3})=0 \tag{6b}$$

$$d_1^2-\|\boldsymbol{P}_{2\to6}\|^2=0 \tag{7a}$$

$$d_2^2-\|\boldsymbol{P}_{3\to5}\|^2=0 \tag{7b}$$

3 Equilibrium and Stability Analysis

3.1 Case 1: Both Springs Have Zero Free-Length

Equilibrium Analysis For the first case, both springs are assumed to have ZFL, i.e., $L_{01}=L_{02}=0$. The variables d_1 and d_2 do not appear in the equilibrium equations as a result of ZFL assumption, hence, Eqs. (6a), (6b) are decoupled from Eqs. (7a), (7b). However, these equations are nonlinear in the desired equilibrium angles with constrained solution space $c_i, s_i \in [-1, 1]\ \forall i = 1, 2$. Tan-half angle identities are instrumental in converting these equations to polynomial form with unconstrained solution space where $x_i \in [-\infty, \infty]$

$$c_i=\frac{1-x_i^2}{1+x_i^2},\ s_i=\frac{2x_i}{1+x_i^2}\quad\text{where}\quad x_i=\tan\frac{\gamma_i}{2}\forall i=1,2 \tag{8}$$

The resulting polynomial form of Eqs. (6a), (6b) is

$$f_1(x_1,x_2)=C_1(x_2)x_1^2+C_2(x_2)x_1+C_3(x_2)=0 \tag{9a}$$

$$f_2(x_1,x_2)=D_1(x_2)x_1^2+D_2(x_2)x_1+D_3(x_2)=0 \tag{9b}$$

where the coefficients $C_i(x_2), D_i(x_2) \forall i = 1, 2, 3$ are second-order polynomials in x_2 and are listed in Appendix 6.1. The solution of this system of equations can be obtained through polynomial elimination by constructing a Sylvester matrix [22, 23]. The coefficients for x_1 in Eqs. (9a), (9b) are combined to create a linear system of the form

$$\begin{bmatrix} C_1 & C_2 & C_3 & 0 \\ 0 & C_1 & C_2 & C_3 \\ D_1 & D_2 & D_3 & 0 \\ 0 & D_1 & D_2 & D_3 \end{bmatrix}\begin{bmatrix} x_1^3 \\ x_1^2 \\ x_1 \\ 1 \end{bmatrix}=\begin{bmatrix} 0 \\ 0 \\ 0 \\ 0 \end{bmatrix} \tag{10}$$

As these system of equations have a non-trivial solution, the determinant of the 4×4 Sylvester matrix needs to be zero. This determinant will result in a sixth (6) degree polynomial in variable x_2 which can be numerically solved.

Uniqueness There is a one-to-one mapping between x_1 and x_2, i.e., for a given solution of x_2, there exists a unique x_1. The unique x_1 can be calculated by rearranging Eq. (10) and substituting x_2 into the coefficients.

$$\begin{bmatrix} x_1^3 \\ x_1^2 \\ x_1 \end{bmatrix} = \begin{bmatrix} C_1(x_2) & C_2(x_2) & C_3(x_2) \\ 0 & C_1(x_2) & C_2(x_2) \\ D_1(x_2) & D_2(x_2) & D_3(x_2) \end{bmatrix}^{-1} \begin{bmatrix} 0 \\ -C_3(x_2) \\ 0 \end{bmatrix} \tag{11}$$

As evident from Eq. (8), x_1 and x_2 correspond to unique γ_1 and γ_2.

Stability Analysis. The interest of this study is not only evaluating all possible solutions of a given system, but determining the stability of these solutions by observing the eigenvalues of the Jacobian $\boldsymbol{J}_{L_{01}=0, L_{02}=0}$. The Jacobian was constructed from the force and torque equations with respect to the dependent variables x_1 and x_2

$$\boldsymbol{J}_{L_{01}=0, L_{02}=0} = \begin{bmatrix} f_{1,x_1} & f_{1,x_2} \\ f_{2,x_1} & f_{2,x_2} \end{bmatrix} \tag{12}$$

where $f_{i,x} = \dfrac{\partial f_i}{\partial x}$. The corresponding solution is described as stable if the real parts of all the eigenvalues are negative.

3.2 *Case 2: One Spring Has Zero Free-Length*

Equilibrium Analysis For the scenario when only one of the springs is assumed to have ZFL, i.e., $L_{01} \neq 0$, $L_{02} - 0$, the equilibrium Eqs. (7a), (7b) are coupled with the constraint Eq. (6a) through a third variable d_1. Polynomial conversion results in

$$g_1(x_1, x_2, d_1) = (E_{1,1}d_1 + E_{1,2})x_1^2 + (E_{2,1}d_1 + E_{2,2})x_1 + (E_{3,1}d_1 + E_{3,2}) \tag{13a}$$

$$g_2(x_1, x_2, d_1) = (F_{1,1}d_1 + F_{1,2})x_1^2 + (F_{2,1}d_1 + F_{2,2})x_1 + (F_{3,1}d_1 + F_{3,2}) \tag{13b}$$

$$g_3(x_1, x_2, d_1) = \left[(1 + x_2^2)\right](1 + x_1^2)d_1^2 - (G_1 x_1^2 + G_2 x_1 + G_3) \tag{13c}$$

where $E_{i,j}(x_2)$, $F_{i,j}(x_2)$, and $G_{i,j}(x_2)$ coefficients are listed in Appendix 6.2. The system is reduced to two equations in variables x_1, x_2 by substituting for the linear solution of d_1 from Eqs. (13a) and (13b) into Eq. (13c).

$$g_4(x_1, x_2) = J_6 x_1^6 + J_5 x_1^5 + J_4 x_1^4 + J_3 x_1^3 + J_2 x_1^2 + J_1 x_1 + J_0 \tag{14a}$$

$$g_5(x_2, x_2) = K_6 x_1^6 + K_5 x_1^5 + K_4 x_1^4 + K_3 x_1^3 + K_2 x_1^2 + K_1 x_1 + K_0 \tag{14b}$$

where J_i and K_i are 6th degree polynomials in x_2. The resulting 12×12 Sylvester matrix is formulated similar to Eq. (10) and is dependent only on x_2. The determinant of this matrix results in a 72-degree polynomial in x_2.

Uniqueness. It is easy to observe that for a given solution of x_2, unique $x_1, d_1, \gamma_1, \gamma_2$ can be obtained by rearranging the Sylvester matrix and Eqs. (13c) and (8).

Stability Analysis. The Jacobian $\boldsymbol{J}_{L_{01} \neq 0, L_{02}=0}$ is constructed using Eqs. (13a), (13b), and (13c) and the system is termed stable when the real part of all the eigenvalues are negative.

$$\boldsymbol{J}_{L_{01} \neq 0, L_{02}=0} = \begin{bmatrix} g_{1,x_1} & g_{1,x_2} & g_{1,d_1} \\ g_{2,x_1} & g_{2,x_2} & g_{2,d_1} \\ g_{3,x_1} & g_{3,x_2} & g_{3,d_1} \end{bmatrix} \tag{15}$$

3.3 *Case 3: None of the Springs Have Zero Free-Lengths*

Equilibrium Analysis When the ZFL constraint for both the springs are relaxed, the complexity increases even further as the equilibrium and both constraint equations are coupled through variables d_1, d_2.

$$h_1(x_1, x_2, d_1, d_2) = L_1 x_1^2 + L_2 x_1 + L_3 \tag{16a}$$
$$h_2(x_1, x_2, d_1, d_2) = M_1 x_1^2 + M_2 x_1 + M_3 \tag{16b}$$
$$h_3(x_1, x_2, d_1, d_2) = (1 + x_1^2)(1 + x_2^2)d_1^2 - (N_1 x_1^2 + N_2 x_1 + N_3) \tag{16c}$$
$$h_4(x_1, x_2, d_1, d_2) = (1 + x_1^2)(1 + x_2^2)d_2^2 - (O_1 x_1^2 + O_2 x_1 + O_3) \tag{16d}$$

The coefficients $L_i(x_2, d_1, d_2)$, $M_i(x_2, d_1, d_2)$, $N_i(x_2)$, $O_i(x_2) \forall i = 1, 2, 3$ are listed in Appendix 6.3. This scenario presents computational challenge as the polynomial elimination algorithms become inefficient [22] but can be numerically approached using homotopy-based numerical techniques.

Stability Analysis Again, the Jacobian $\boldsymbol{J}_{L_{01} \neq 0, L_{02} \neq 0}$ is constructed using Eqs. (16a)–(16d) and the system is stable when the real part of all the eigenvalues are negative.

$$\boldsymbol{J}_{L_{01} \neq 0, L_{02} \neq 0} = \begin{bmatrix} h_{1,x_1} & h_{1,x_2} & h_{1,d_1} & h_{1,d_2} \\ h_{2,x_1} & h_{2,x_2} & h_{2,d_1} & h_{2,d_2} \\ h_{3,x_1} & h_{3,x_2} & h_{3,d_1} & h_{3,d_2} \\ h_{4,x_1} & h_{4,x_2} & h_{4,d_1} & h_{4,d_2} \end{bmatrix} \tag{17}$$

4 Numerical Examples and Discussion

Case 1: For an example mechanisms with parameters as listed in Table 1 where both springs have ZFL, the roots of the sixth-order polynomial resulting from polynomial elimination are tabulated in Table 2 where only the third solution is stable. The four real solutions are illustrated in Fig. 2, with the stable solution in bold.

Case 2: One spring has ZFL The ZFL constraint was relaxed for the CoTM example with $L_{01} = 11.5$m. The analytical solution resulted in a 72-degree polynomial with

Table 1 CoTM parameters

Parameter	Value
L_3	9 m
k_1, k_2	2.81N/m
p_{2x}	3.19 m
p_{3x}, p_{3y}	−2.55, 4.93 m
p_{6x}	4.25 m
p_{5x}, p_{5y}	−0.3, −3.45 m

Table 2 Solutions for Case 1

Solution	γ_1 (deg)	γ_2 (deg)
1	75.034	289.04
2	276.53	330.79
3 (stable)	**252.35**	**123.74**
4	93.67	124.97
5, 6	−149.80	−160.71
	$\pm i85.41$	$\pm i167.38$

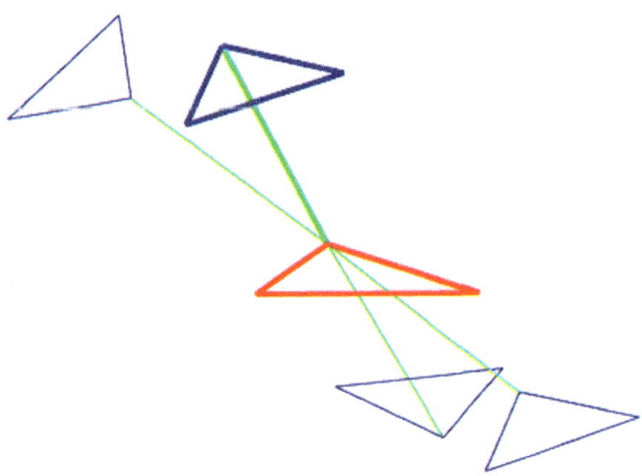

Fig. 2 Visualization of solutions when both springs have ZFL. The four real solutions are indicated where the fixed bottom rigid body, the top rigid body, and the compressive rigid member are marked in red, blue, and green, respectively. The single stable solution is marked in bold

Table 3 Solutions for Case 2 where $L_{01} = 11.5$ m. Stable solutions in bold

	γ_1(deg)	γ_2(deg)	d_1(m)
1	121.69	−170.30	13.95
2	−53.70	−137.78	10.16
3	−134.63	−118.07	−15.35
4	109.42	−113.03	9.08
5	43.84	−111.10	−2.88
6	108.06	−79.05	6.78
7	115.19	−5.97	−8.19
8	**−49.06**	**13.59**	**−8.97**
9	121.04	41.45	11.50
10	−121.20	64.23	−7.14
11	−57.21	65.86	5.03
12	**35.41**	**67.86**	**−10.81**
13	−55.72	94.46	−3.56
14	**129.99**	**112.18**	**−15.14**

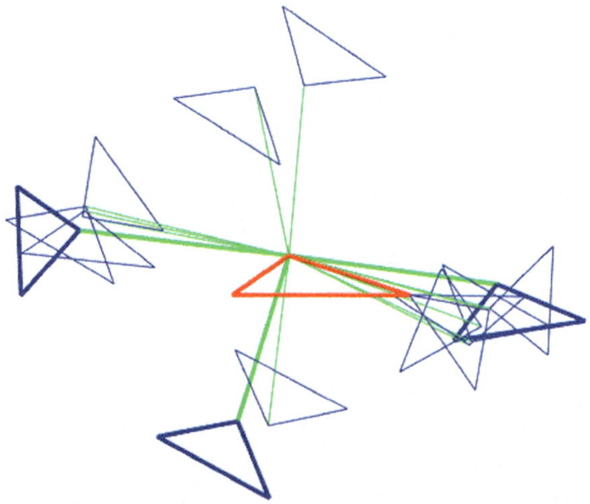

Fig. 3 Fourteen real solutions were obtained when the ZFL assumption was relaxed for one spring element of which three were stable (marked in bold)

14 real solutions. These solutions are listed in Table 3 and illustrated in Fig. 3 where the 8th, 12th, and 14th solutions were stable. The Fig. 4 highlights how the stability of these solutions varies with change in the free-length L_{01}.

Case 3: No springs have ZFLs Solving four polynomial equations in four becomes unfeasible for traditional polynomial elimination methods [22]. Bertini software [24] was used to numerically obtain 88 solutions for $L_{01} = L_{02} = 11.5$ m. Thirty of these solutions were real of which five were found to be stable as collated in Table 4 and Fig. 5, respectively.

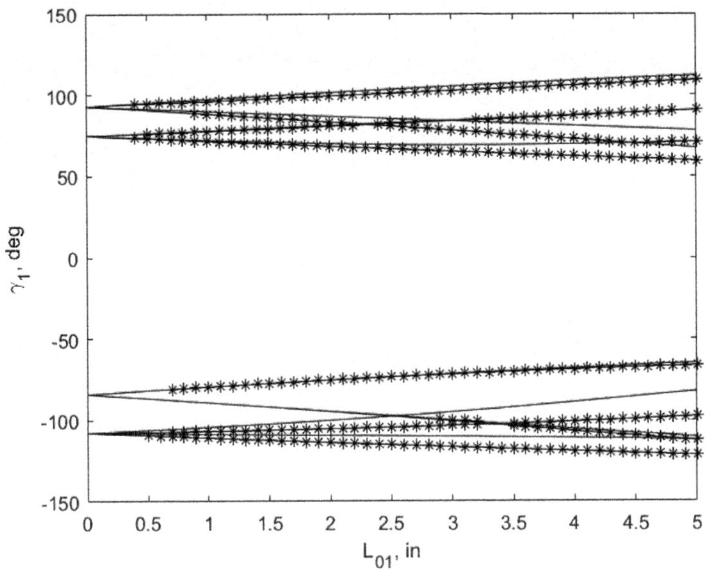

Fig. 4 Solutions of γ_1 as L_{01} is varied with stable solutions marked with '*'

Table 4 Solutions for Case 3 ($L_{01} = L_{02} = 11.5$ m). Stable ones are highlighted

	γ_1(deg)	γ_2(deg)	d_1	d_2
1	118.75	35.52	11.13	0.06
2	1.86	11.31	−10.04	14.45
3	123.69	48.35	−11.93	0.001
4	158.85	147.91	−16.15	3.86
5	27.01	−135.42	2.11	8.56
6	81.58	−60.61	−5.20	−2.62
7	114.38	24.21	10.39	−0.03
8	113.32	148.87	14.74	−6.26
9	158.45	−18.37	−7.78	8.66
10	121.96	43.85	−11.65	−0.001
11	−16.11	152.17	1.78	13.86
12	1.55	−177.34	−1.56	11.77
13	116.20	−141.81	11.83	6.92
14	**−99.16**	**157.70**	11.23	11.05
15	**−178.95**	**−69.89**	11.50	11.50
16	−109.34	109.14	−8.79	−12.91
17	−60.03	49.95	6.08	−17.95
18	−119.62	−94.00	14.44	13.33
19	−82.25	−172.07	11.34	10.96
20	−92.63	−46.78	−12.11	−16.10

(continued)

Table 4 (continued)

	γ_1(deg)	γ_2(deg)	d_1	d_2
21	−173.90	5.31	7.92	11.33
22	−1.03	−33.82	9.68	12.22
23	−59.06	−143.69	10.42	−11.06
24	111.37	16.39	−9.87	−0.00
25	52.62	85.56	11.68	11.61
26	**56.10**	**108.90**	11.50	11.50
27	118.69	−152.47	12.74	−6.92
28	**48.02**	**60.96**	11.50	11.50
29	**−60.20**	**−120.65**	11.50	11.50
30	113.25	21.29	−10.20	0.001

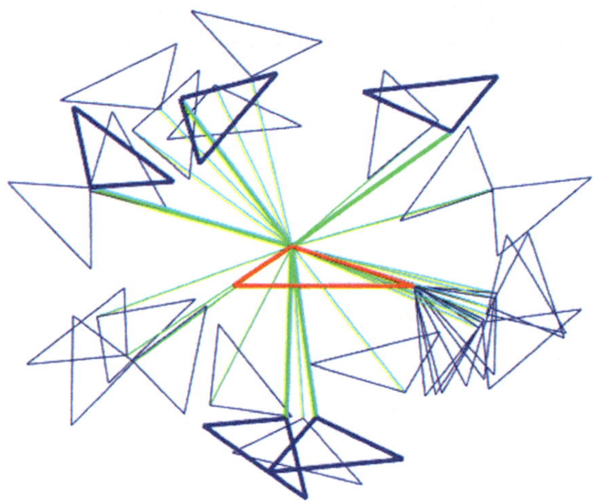

Fig. 5 Example solutions when the ZFL assumptions are completely relaxed. Five stable solutions (bold) were observed out of the resulting 30 real solutions

4.1 Discussion

Complexity As the ZFL assumption was relaxed, the overall computational complexity of this simple planar example increased exponentially such that the analytical solution could not be obtained in the third case. Moving forward to a 3D CoTM example is only expected to increase complexity further. Finding better methods for solving these complex polynomials will be a continuous problem in this study. The Bertini software was used instead to solve the solutions, as a result, the degree of the solution polynomial remains unknown and correspondingly, the total number of possible solutions.

Nonlinear behavior with change in free-length The behavior of the mechanism as a function of free-lengths is of special interest due to possible applications in robust control of deployable structure and tensegrity-based biomechanical modeling. Figure 4 shows the change in orientation of the mechanism (γ_1) as the free-length L_{01} is varied. The behavior of CoTM orientation is nonlinear—4 real solutions for ZFL assumption bifurcate to as many as 10. This implies that it is possible to move from one orientation to another by slowly shortening or lengthening the free-length of L_{01}, thus, controlling the mechanism's movement. This can be done by altering the length of the string on the cable-spring series combination of the tensile element of the mechanism. It also shows that as free-length increases, the behavior of the system changes and the number of stable equilibrium solutions vary. The top set and bottom set of solutions are very similar in terms of pattern—this may imply that for some solutions there is a mirrored configuration. Stable solutions are marked with an asterisk and start to appear when $L_{01} = 0.4$ m and tend to follow one bifurcation line. Where these lines intersect, the system loses stability except for around $(4.5\text{m}, 77.35°)$ where the stable solutions start to follow the line they intersected with. Figure 6 shows intersecting solution manifolds along which the mechanism can be maneuvered for obtaining robust, energy efficient control of the example mechanism.

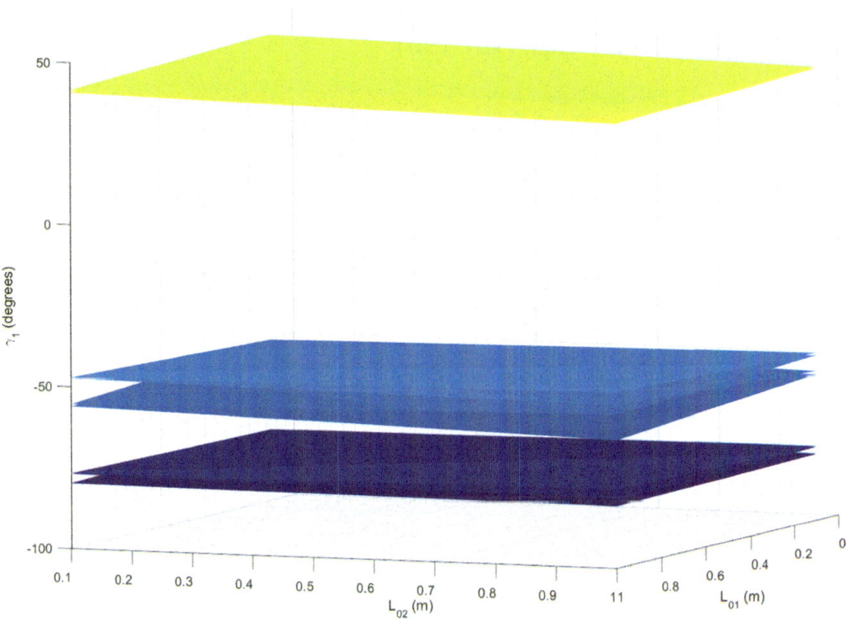

Fig. 6 The intersecting solution manifolds for γ_2

5 Conclusion

This paper presented a method for determining the stable equilibrium solutions for a planar compliant tensegrity mechanism that is comprised of two rigid bodies connected by a compressive member and two spring members. Complexity of the problem increased dramatically from assuming (a) both ZFL springs—sixth degree polynomial corresponding to equilibrium orientations; (b) only one spring ZFL—72-degree polynomial; (c) no ZFL spring—88 numerical solutions for example problem but unknown degree of polynomial using homotopy-based Bertini numerical software. The resulting behavior of the mechanism as ZFL assumption for one spring is relaxed and varied is also observed. Here, the stable solutions show how to control the configuration of the mechanism through compliance of the springs by shortening or lengthening its free-length. This control is possible by varying L_{01} to move between desired γ_1 solutions.

6 Appendix

6.1 *Coefficients for Eqs.* (9a), (9b)

$$C_i(x_2) = A_{i,2}x_2^2 + A_{i,1}x_2 + A_{i,0} \tag{18a}$$

$$D_i(x_2) = B_{i,2}x_2^2 + B_{i,1}x_2 + B_{i,0} \qquad \forall i = 1, 2 \tag{18b}$$

$$A_{1,2} = L_3(k_2 p_{3y} + k_2 p_{5y}) \qquad B_{1,2} = L_3 k_2 p_{3y} - k_2 p_{3x} p_{5y} + k_2 p_{5x} p_{3y}$$

$$A_{1,1} = -L_3(2k_1 p_{6x} + 2k_2 p_{5x}) \qquad B_{1,1} = 2k_1 p_{2x} p_{6x} + 2k_2 p_{3x} p_{5x} + 2k_2 p_{3y} p_{5y}$$

$$A_{1,0} = L_3(k_2 p_{3y} - k_2 p_{5y}) \qquad B_{1,0} = L_3 k_2 p_{3y} + k_2 p_{3x} p_{5y} - k_2 p_{5x} p_{3y}$$

$$A_{2,2} = L_3(2k_1 p_{2x} + 2k_2 p_{3x} + 2k_1 p_{6x} + 2k_2 p_{5x}) \qquad B_{2,2} = 2L_3 k_1 p_{2x} + 2L_3 k_2 p_{3x}$$

$$A_{2,1} = 4L_3 k_2 p_{5y} \qquad B_{2,1} = 0$$

$$A_{2,0} = L_3(2k_1 p_{2x} + 2k_2 p_{3x} - 2k_1 p_{6x} - 2k_2 p_{5x}) \qquad B_{2,0} = 2L_3 k_1 p_{2x} + 2L_3 k_2 p_{3x}$$

$$A_{3,2} = -L_3(k_2 p_{3y} + k_2 p_{5y}) \qquad B_{3,2} = k_2 p_{5x} p_{3y} - k_2 p_{3x} p_{5y} - L_3 k_2 p_{3y}$$

$$A_{3,1} = L_3(2k_1 p_{6x} + 2k_2 p_{5x}) \qquad B_{3,1} = 2k_1 p_{2x} p_{6x} + 2k_2 p_{3x} p_{5x} + 2k_2 p_{3y} p_{5y}$$

$$A_{3,0} = -L_3(k_2 p_{3y} - k_2 p_{5y}) \qquad B_{3,0} = k_2 p_{3x} p_{5y} - l_3 k_2 p_{3y} - k_2 p_{5x} p_{3y}$$

6.2 *Coefficients for Eqs.* (13a)–(13c)

$$E_1 = (L_3(d_1k_2p_{3y} + d_1k_2p_{5y}))x_2^2 + (2L_3(L_{01}k_1p_{6x} - d_1k_1p_{6x} - d_1k_2p_{5x}))x_2 + L_3(d_1k_2p_{3y} - d_1k_2p_{5y})$$

$$E_2 = (L_3(2d_1k_1p_{2x} - 2L_{01}k_1p_{6x} - 2L_{01}k_1p_{2x} + 2d_1k_2p_{3x} + 2d_1k_1p_{6x} + 2d_1k_2p_{5x}))x_2^2 + (4L_3d_1k_2p_{5y})x_2 - L_3(2L_{01}k_1p_{2x} - 2L_{01}k_1p_{6x} - 2d_1k_1p_{2x} - 2d_1k_2p_{3x} + 2d_1k_1p_{6x} + 2d_1k_2p_{5x})$$

$$E_3 = (-L_3(d_1k_2p_{3y} + d_1k_2p_{5y}))x_2^2 + (L_3(2d_1k_1p_{6x} - 2L_{01}k_1p_{6x} + 2d_1k_2p_{5x}))x_2 - L_3(d_1k_2p_{3y} - d_1k_2p_{5y})$$

$$F_1 = d_1k_2(L_3p_{3y} - p_{3x}p_{5y} + p_{5x}p_{3y})x_2^2 + (2d_1k_1p_{2x}p_{6x} + 2d_1k_2p_{3x}p_{5x} + 2d_1k_2p_{3y}p_{5y} - 2L_{01}k_1p_{2x}p_{6x})x_2 + d_1k_2(L_3p_{3y} + p_{3x}p_{5y} - p_{5x}p_{3y})$$

$$F_2 = 2L_3(x_2^2 + 1)(d_1k_1p_{2x} - L_{01}k_1p_{2x} + d_1k_2p_{3x})$$

$$F_3 = -d_1k_2(L_3p_{3y} + p_{3x}p_{5y} - p_{5x}p_{3y})x_2^2 + (2d_1k_1p_{2x}p_{6x} + 2d_1k_2p_{3x}p_{5x} + 2d_1k_2p_{3y}p_{5y} - 2L_{01}k_1p_{2x}p_{6x})x_2 - d_1k_2(L_3p_{3y} - p_{3x}p_{5y} + p_{5x}p_{3y})$$

$$G_1 = (L_3 + p_{2x} + p_{6x})^2x_2^2 + (L_3 + p_{2x} - p_{6x})^2$$

$$G_2 = (8L_3p_{6x})x_2$$

$$G_3 = (p_{2x} - L_3 + p_{6x})^2x_2^2 + (L_3 - p_{2x} + p_{6x})^2$$

6.3 *Coefficient Definitions for Eqs.* (16a), (16d)

$$L_1 = -L_3d_1k_2(p_{3y} + p_{5y})(L_{02} - d_2)x_2^2 - L_3(2d_1d_2k_1p_{6x} + 2d_1d_2k_2p_{5x} - 2L_{01}d_2k_1p_{6x} - 2L_{02}d_1k_2p_{5x})x_2 - L_3d_1k_2(L_{02} - d_2)(p_{3y} - p_{5y})$$

$$L_2 = L_3(2d_1d_2k_1p_{2x} + 2d_1d_2k_2p_{3x} + 2d_1d_2k_1p_{6x} + 2d_1d_2k_2p_{5x} - 2L_{01}d_2k_1p_{2x} - 2L_{02}d_1k_2p_{3x} - 2L_{01}d_2k_1p_{6x} - 2L_{02}d_1k_2p_{5x})x_2^2 - 4L_3d_1k_2p_{5y}(L_{02} - d_2)x_2 + L_3(2d_1d_2k_1p_{2x} + 2d_1d_2k_2p_{3x} - 2d_1d_2k_1p_{6x} - 2d_1d_2k_2p_{5x} - 2L_{01}d_2k_1p_{2x} - 2L_{02}d_1k_2p_{3x} + 2L_{01}d_2k_1p_{6x} + 2L_{02}d_1k_2p_{5x})$$

$$L_3 = L_3d_1k_2(p_{3y} + p_{5y})(L_{02} - d_2)x_2^2 + L_3(2d_1d_2k_1p_{6x} + 2d_1d_2k_2p_{5x} - 2L_{01}d_2k_1p_{6x} - 2L_{02}d_1k_2p_{5x})x_2 + L_3d_1k_2(L_{02} - d_2)(p_{3y} - p_{5y})$$

$$M_1 = -d_1k_2(L_{02} - d_2)(L_3p_{3y} - p_{3x}p_{5y} + p_{5x}p_{3y})x_2^2 +$$
$$(2d_1d_2k_1p_{2x}p_{6x} - 2L_{02}d_1k_2p_{3x}p_{5x} - 2L_{02}d_1k_2p_{3y}p_{5y} -$$
$$2L_{01}d_2k_1p_{2x}p_{6x} + 2d_1d_2k_2p_{3x}p_{5x} + 2d_1d_2k_2p_{3y}p_{5y})x_2 -$$
$$d_1k_2(L_{02} - d_2)(L_3p_{3y} + p_{3x}p_{5y} - p_{5x}p_{3y})$$
$$M_2 = 2L_3(x_2^2 + 1)(d_1d_2k_1p_{2x} + d_1d_2k_2p_{3x} -$$
$$L_{01}d_2k_1p_{2x} - L_{02}d_1k_2p_{3x})$$
$$M_3 = d_1k_2(L_{02} - d_2)(L_3p_{3y} + p_{3x}p_{5y} - p_{5x}p_{3y})x_2^2 +$$
$$(2d_1d_2k_1p_{2x}p_{6x} - 2L_{02}d_1k_2p_{3x}p_{5x} - 2L_{02}d_1k_2p_{3y}p_{5y} -$$
$$2L_{01}d_2k_1p_{2x}p_{6x} + 2d_1d_2k_2p_{3x}p_{5x} + 2d_1d_2k_2p_{3y}p_{5y})x_2 +$$
$$d_1k_2(L_{02} - d_2)(L_3p_{3y} - p_{3x}p_{5y} + p_{5x}p_{3y})$$
$$N_i = G_i \quad \forall i = 1, 2, 3$$

$$O_1 = (L_3^2 + 2L_3p_{3x} + 2L_3p_{5x} + p_{3x}^2 + 2p_{3x}p_{5x} + p_{5x}^2 + p_{3y}^2 +$$
$$2p_{3y}p_{5y} + p_{5y}^2)x_2^2 + (4L_3p_{5y} + 4p_{3x}p_{5y} - 4p_{5x}p_{3y})x_2 +$$
$$L_3^2 + 2L_3p_{3x} - 2L_3p_{5x} + p_{3x}^2 - 2p_{3x}p_{5x} + p_{5x}^2 + p_{3y}^2 -$$
$$2p_{3y}p_{5y} + p_{5y}^2$$
$$O_2 = (-4L_3p_{3y} - 4L_3p_{5y})x_2^2 + (8L_3p_{5x})x_2 + 4L_3p_{5y} - 4L_3p_{3y}$$
$$O_3 = (L_3^2 - 2L_3p_{3x} - 2L_3p_{5x} + p_{3x}^2 + 2p_{3x}p_{5x} + p_{5x}^2 + p_{3y}^2 +$$
$$2p_{3y}p_{5y} + p_{5y}^2)x_2^2 + (4p_{3x}p_{5y} - 4L_3p_{5y} - 4p_{5x}p_{3y})x_2 +$$
$$L_3^2 - 2L_3p_{3x} + 2L_3p_{5x} + p_{3x}^2 - 2p_{3x}p_{5x} + p_{5x}^2 + p_{3y}^2 -$$
$$2p_{3y}p_{5y} + p_{5y}^2$$

References

1. Pugh A (1976) An introduction to tensegrity. University of California Press
2. Shibata M, Saijyo F, Hirai S (2009) Crawling by body deformation of tensegrity structure robots. In: 2009 IEEE international conference on robotics and automation, pp. 4375–4380, May 2009
3. Iscen A, Agogino A, SunSpiral V, Tumer K (2014) Flop and roll: learning robust goal-directed locomotion for a tensegrity robot. In: IEEE/RSJ international conference on intelligent robots and systems, pp. 2236–2243, Sept 2014
4. Tibert G (2002) Deployable tensegrity structures for space applications. PhD Thesis, KTH Royal Institute of Technology, Sweden, 2002
5. Veuve N, Dalil Safaei S, Smith IFC (2016) Active control for mid-span connection of a deployable tensegrity footbridge. Eng Struct 112:245–255
6. Ingber DE (1998) The architecture of life. Sci Am 278(1):48–57
7. Levin SM (2002) The tensegrity-truss as a model for spine mechanics: biotensegrity. J Mech Med Biol 2(03–04):375–388

8. Liedl T, Hgberg B, Tytell J, Ingber DE, Shih WM (2010) Self-assembly of three-dimensional prestressed tensegrity structures from DNA. Nat Nanotechnol 5:520–524
9. Sultan C, Skelton RT (1998) Tendon control deployment of tensegrity structures. In: Smart structures and materials 1998: mathematics and control in smart structures. International society for optics and photonics, vol 3323, pp 455–467
10. Juan SH, Mirats Tur JM (2008) Tensegrity frameworks: static analysis review. Mech Mach Theory 43:859–881
11. Zhang JY, Ohsaki M (2007) Stability conditions for tensegrity structures. Int J Solids Struct 44:3875–3886
12. Murakami H (2001) Static and dynamic analyses of tensegrity structures. Part 1. Nonlinear equations of motion. Int J Solids Struct 38:3599–3613
13. Connelly R, Whiteley W (1996) Second-order rigidity and prestress stability for tensegrity frameworks. SIAM J Discret Math 9:453–491
14. Arsenault M, Gosselin CM (2006) Kinematic, static and dynamic analysis of a planar 2-dof tensegrity mechanism. Mech Mach Theory 41:1072–1089
15. Sultan C, Corless M, Skelton RE (2001) The prestressability problem of tensegrity structures: some analytical solutions. Int. J. Solids Struct. 38:5223–5252
16. Pellegrino S (1990) Analysis of prestressed mechanisms. Int J Solids Struct 26:1329–1350
17. Vasquez RE, Correa JC (2007) Kinematics, dynamics and control of a planar 3-DOF tensegrity robot manipulator. In: ASME 2007 international design engineering technical conferences and computers and information in engineering conference. American society of mechanical engineers, pp 855–866
18. Vasquez RE, Crane CD, Correa JC (2014) Analysis of a planar tensegrity mechanism for ocean wave energy harvesting. J Mech Robot 6:031015–031015
19. Delissen AATM, Radaelli G, Herder JL (2017) Design and optimization of a general planar zero free length spring. Mech Mach Theory 117:56–77
20. Herder JL (2001) Energy-free systems. Theory, conception and design of statically, vol 2
21. Deepak SR, Ananthasuresh GK (2012) Perfect static balance of linkages by addition of springs but not auxiliary bodies. J Mech Robot 4:021014–021014
22. Kapur D (1995) Algorithmic elimination methods. In: Tutorial notes, international symposium on symbolic and algebraic computation (ISSAC), Montreal
23. Sylvester JJ (1853) On a theory of the syzygetic relations of two rational integral functions, comprising an application to the theory of sturm's functions, and that of the greatest algebraical common measure. Philos Trans R Soc Lond 143:407–548
24. Bates DJ, Hauenstein JD, Sommese AJ, Wampler CW (2006) Bertini: software for numerical algebraic geometry

Underactuated Control of Serial Manipulator

Aditya Jain, Rajesh Kumar, Indra Narayan Kar, and Subir Kumar Saha

Abstract Serial manipulators have limited purpose in the area where quick motion is a prerequisite. Parallel manipulators are known to be advantageous due to their low actuation inertia. However, the dynamics of parallel manipulators are highly coupled and nonlinear, and control structures require de-coupling modules, if quick motion is required with direct drive systems. Underactuated (base actuated) serial manipulators use the benefit of the two and serve the purpose of quick motion. The proposed method is different from the available "*flatness*"-based methods of underactuated control. Additionally, the requirement of passive actuation in the joints is evaded. In this paper, we propose a control algorithm for swing up control of Double Inverted Pendulum on a Moving Cart (DIPMC) and Single Inverted Pendulum on a Moving Cart (SIPMC). The control algorithm is divided into two parts: Feedforward controller and Feedback controller. In feedforward controller, a force predictor input is applied from the base to ensure the manipulator achieves a pose in the required domain and then a subsequent feedback-based control is implied on the system.

Keywords Underactuated serial manipulator · Dynamics-based control · Linear Quadratic Regulator (LQR)

A. Jain (✉) · R. Kumar
Indian Institute of Technology, Delhi, India
e-mail: aj.mechiitd@gmail.com

R. Kumar
e-mail: rajeshkr96@gmail.com

I. N. Kar
Department of Electrical Engineering, Indian Institute of Technology, Delhi, India
e-mail: ink@ee.iitd.ac.in

S. K. Saha
Department of Mechanical Engineering, Indian Institute of Technology, Delhi, India
e-mail: saha@mech.iitd.ac.in

D. Sen et al. (eds.), *Mechanism and Machine Science*, Lecture Notes in Mechanical Engineering, https://doi.org/10.1007/978-981-15-4477-4_51

1 Introduction

The double inverted pendulum on a moving cart (DIPMC) is an underactuated system consisting of two passive revolute joints and moving cart as a prismatic joint as shown in Fig. 1. Swinging up DIPMC to an unstable equilibrium position (upright position) has been a benchmark problem for the testing of equilibrium control-related algorithms and techniques. The DIPMC has been implemented in various forms over the years especially in the category of *two-wheeled inverted pendulum* robots. Ha et al. [1] developed *Yamabico Kurara* robot which has two active actuators for controlling wheels and the freely moving upper body of the robot with feedback from a gyro sensor. Shiroma et al. [2] used the underactuation of wheeled inverted pendulum robot for transportation of external load via cooperation between robots at the same time maintaining their agility and stability. The commercial success of DIPMC category robotics occurred in Segway PT series robots as self-balancing robots designed for traversing in rough terrain, assisting in attending medical emergency, and in warehouses for transportation of goods. Several control techniques have been implemented and tested on DIPMC and its variants. Knut Graichen et al. [3] implemented control strategy under finite-time transition between stationary setpoints with free parameters to solve two-point *Boundary Value Problem* (BVP) for generating feedforward input. Henrik et al. [4] implemented a passive fault-tolerant controller for early detection of faults in a closed-loop system and taking corresponding action for reconfiguration of the system. Bui et al. [5] implemented a controller based on Lyapunov stability for position and velocity of a welding robot which was modeled on the concept of two-wheeled inverted pendulum following a setpoint trajectory. Additionally, there exist differential flatness-based system methods which have been instrumental in motion planning for generating and following fast trajectories with high accuracy positioning [6]. The drawback of differential flatness is that only some specific *flat output* trajectories are available for a manipulator to reach the desired target. This method is best preferable for trajectory control but for setpoint control and stabilization control, alternate non-differential flatness-based methods can be used, which do not require closed-form discovery of a trajectory leading to a flat

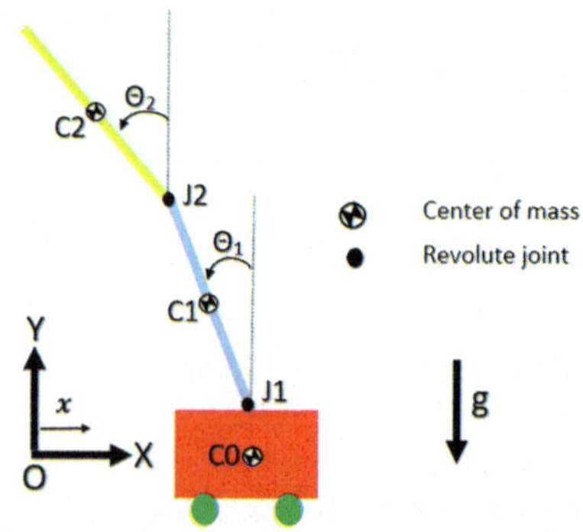

Fig. 1 Schematic diagram of DIPMC

output. Our algorithm presented is based on setpoint control where we are not giving any specific trajectory to the system, therefore no such specific flat trajectory for achieving the stabilization of DIPMC and SIPMC is required .

We present a new control algorithm for the stabilization of DIPMC and SIPMC in an inverted position. First, the Euler Lagrange model of DIPMC and SIPMC has been designed. For implementing the controller, we categorized it into two parts which are feedforward controller and feedback controller. In feedforward controller, an initial force estimator of sinusoidal variant is applied on the cart in the horizontal direction and then for feedback controller, the LQR method has been implemented. We demonstrated the working of a controller via simulations and trajectories of DIPMC and SIPMC.

2 Mathematical Model

We wish to control an underactuated set of serial manipulators using a base actuation of the prismatic joint. A traditional serial manipulator usually consists of mass matrix (***M***), quadratic velocity product terms (***V***), and gravity compensation vectors (***G***) to be nonlinear functions of the state vectors (***X***). The non-linearized equations have been derived for DIPMC and SIPMC using Euler Lagrange equations as shown below.

- DIPMC Euler Lagrange Formulation

The double inverted pendulum consisting of two links and a cart is presented in Fig. 1.

We have assumed in our model that all revolute and prismatic joints are ideal, the center of mass of link 1, link 2, and cart lies at their respective Center of Geometry, ***g*** is constant, no friction is acting in joints, and cart (that is, prismatic joint between link1 and ground) is constrained to move in only one dimension (here X direction as shown in Fig. 1). The equations of motion as per the "*Lagrangian Formulation*" with the generalized coordinates being "θ_1", "θ_2", and "x" where J_1 = Revolute joint between cart and link 1, J_2 = Revolute joint between link 1 and link 2, C_0 = Center of mass of cart, C_1 = Center of mass of link 1, and C_2 = Center of mass of link 2 (Table 1).

Formulating Euler Lagrange equation of motion from a system's kinetic energy and potential energy, we get

$$L = T - V$$

$$\frac{\mathrm{d}}{\mathrm{d}t}\left(\frac{\partial L}{\partial \dot{q}_i}\right) - \frac{\partial L}{\partial q_i} = \phi_i \tag{1}$$

Table 1 The expressions for mechanical energy components of DIPMC

	Kinetic energy	Potential energy
Cart	$T_1 = \frac{1}{2}M\dot{x}^2$	$V_1 = 0$
Link 1	$T_3 = \frac{1}{2}I_1\dot{\theta}_1^2 + \frac{1}{2}m_1v_1^2$	$V_2 = m_1 g r_1 \sin\theta_1$
Link 2	$T_3 = \frac{1}{2}I_2\dot{\theta}_2^2 + \frac{1}{2}m_2v_2^2$	$V_3 = m_2 g(2r_2 \sin\theta_1 + r_2 \sin(\theta_1 + \theta_2))$

where M = Mass of cart, $\dot{x}$ = Velocity of cart, m_1 = Mass of link 1, $r_1 = J_1C_1$, I_1 = Moment of Inertia of link 1 about C_1, θ_1 = Angle of link 1 made with respect to the horizontal axis (in plane of slider), $\dot{\theta}_1$ = Angular velocity of link 1, v_1 = Velocity of C_1 with respect to ground frame, m_2 = Mass of link 2, $r_2 = J_2C_2$, I_2 = Moment of Inertia of link 2 about C_2, θ_2 = Angle of link 2 made with reference to the axis of link 1, $\dot{\theta}_2$ = Angular velocity of link 2, and v_2 = Velocity of C_2 with respect to ground frame

where $T = \sum_{i=1}^{3} T_i$, $V = \sum_{i=1}^{3} V_i$, and $q_i = x, \theta_1 and \theta_2$.

$$\boldsymbol{M}(\boldsymbol{q})\ddot{\boldsymbol{q}} + \boldsymbol{V}(\boldsymbol{q}, \dot{\boldsymbol{q}}) + \boldsymbol{G}(\boldsymbol{q}) = \boldsymbol{\tau} \tag{2}$$

The system of equations reduces to the following set of matrices—

$$\boldsymbol{M}(\boldsymbol{q}) = \begin{bmatrix} I_1 + m_1r_1^2 + 4m_2r_1^2 & 2m_2r_1r_2\cos(\theta_1 - \theta_2) & -(m_1r_1\cos\theta_1 + 2m_2r_1\cos\theta_1) \\ 2m_2r_1r_2\cos(\theta_1 - \theta_2) & I_2 + m_2r_2^2 & -m_2r_2\cos(\theta_2) \\ -(m_1r_1\cos\theta_1 + 2m_2r_1\cos\theta_1) & -m_2r_2\cos(\theta_2) & M + m_1 + m_2 - b \end{bmatrix} \tag{3}$$

$$\boldsymbol{V}(\boldsymbol{q}, \dot{\boldsymbol{q}}) = \begin{bmatrix} 2m_2r_1r_2\sin(\theta_1 - \theta_2)\dot{\theta}_2^2 \\ -2m_2r_1r_2\sin(\theta_1 - \theta_2)\dot{\theta}_1^2 \\ ((m_1 + 2m_2)r_1 sin\theta_1)\dot{\theta}_1^2 + m_2r_2\sin(\theta_2)\dot{\theta}_2^2 \end{bmatrix} \tag{4}$$

$$\boldsymbol{G}(\boldsymbol{q}) = \begin{bmatrix} -gm_1r_1\sin(\theta_1) - 2gm_2r_1\sin(\theta_1) \\ -gm_2r_2\sin(\theta_2) \\ 0 \end{bmatrix} \tag{5}$$

$$\boldsymbol{\tau} = ForceVector = \begin{bmatrix} 0 & 0 & F \end{bmatrix}^T \tag{6}$$

where $\boldsymbol{q} = Generalisedcoordinateset = \begin{bmatrix} \theta_1 & \theta_2 & x \end{bmatrix}^T$.

- SIPMC Euler Lagrange Formulation

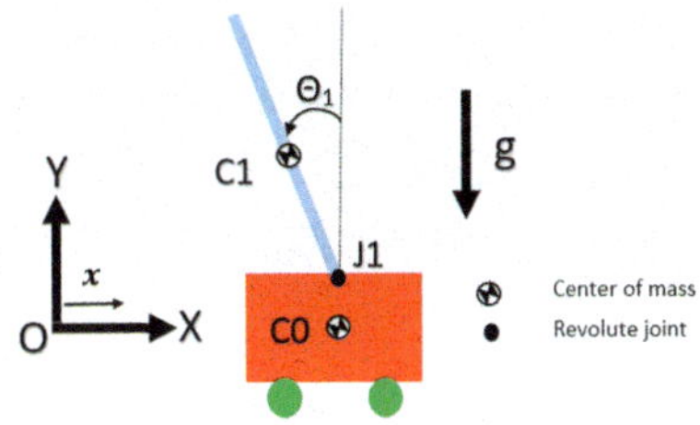

Fig. 2 Schematic diagram of a single inverted pendulum on a moving cart

The equation of motion for single inverted pendulum consisting of two generalized coordinates are shown in Eqs. (8)–(11) (Fig. 2).

$$\boldsymbol{M}(\boldsymbol{q})\ddot{\boldsymbol{q}} + \boldsymbol{V}(\boldsymbol{q}, \dot{\boldsymbol{q}}) + \boldsymbol{G}(\boldsymbol{q}) = \boldsymbol{\tau} \tag{7}$$

$$\boldsymbol{M} = \begin{matrix} I_1 + m_1 r_1^2 & -m_1 r_1 \cos\theta \\ -m_1 r_1 \cos\theta & m + m_1 - b \end{matrix} \tag{8}$$

$$\boldsymbol{V} = \begin{bmatrix} 0 \\ m_1 r_1 sin\theta\dot{\theta}^2 \end{bmatrix} \tag{9}$$

$$\boldsymbol{G} = \begin{matrix} -g m_1 r_1 sin\theta \\ 0 \end{matrix} \tag{10}$$

where $\boldsymbol{q} = Generalisedcoordinateset = \begin{bmatrix} \theta & x \end{bmatrix}^T$, m = Mass of cart, $\dot{x}$ = Velocity of cart, m_1 = Mass of link 1, $r_1 = J_1C_1$, I_1 = Moment of Inertia of link 1 about C_1, θ_1 = Angle of link 1 made with respect to the horizontal axis (in plane of slider), $\dot{\theta}$ = Angular velocity of link 1, and v_1 = velocity of C_1 with respect to ground frame.

3 Controller Design

Equilibrium pose is the state where the derivatives of the state vectors is a null vector [7]. The controller has been designed in two parts: Feedforward controller and Feedback controller. To linearize the governing equation with respect to the equilibrium points, we use the assumption that the system is near its required equilibrium state. However, to ensure that the assumption of linearization around equilibrium pose is true, we need to bring the system close to the equilibrium pose. Hence, a quick impulse is applied to the cart to ensure that the end effectors are close to the equilibrium state during feedforward controller and then the controller switches to feedback controller which uses the *Linear Quadratic Regulator* (LQR) method [8] for stabilization around equilibrium points. A sinusoidal force term is used as an enforcing function in a feedforward controller as the output of the end effector is a combination of multiple sinusoids. So, a first-hand assumption a three-parameter-based sinusoidal function of the form "$Fcos(\omega t + \phi)$" is used. The three parameters

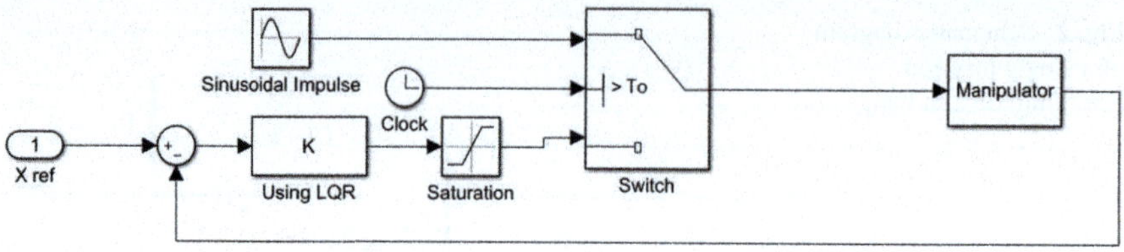

Fig. 3 Control block diagram

are defined by ensuring smoothness of the state vector at all times besides being differentiable throughout the trajectory except during a controller shift. During the change from a feedforward to feedback controller, the *system required* acceleration is limited which is presented via examples in Sect. 4. In the feedforward controller, value of force magnitude, frequency, and phase difference are dependent on system parameters and initial configuration of the system. Once linearized with respect to the equilibrium state vector, the controller equation can be represented in Fig. 3. In order to formulate the state vector ($\boldsymbol{X}$) from the Eq. (11),

$$\dot{\boldsymbol{X}} = \boldsymbol{A}\boldsymbol{X} + \boldsymbol{B}\boldsymbol{U} \tag{11}$$

The LQR minimization functional is presented in Eq. (12).

$$\boldsymbol{J} = \frac{1}{2}(\boldsymbol{X}_f^T \boldsymbol{S}_f \boldsymbol{X}_f) + \frac{1}{2}\int_{t_o}^{t_f} (\boldsymbol{X}^T \boldsymbol{Q}\boldsymbol{X} + \boldsymbol{U}^T \boldsymbol{R}\boldsymbol{U})\mathrm{d}t \tag{12}$$

where $\boldsymbol{S}_f$ is nonnegative-definite symmetric matrix and $\boldsymbol{Q}$, $\boldsymbol{R}$ are positive definite matrices.

In Appendix, the value of $\boldsymbol{K}$, linearized form of Euler Lagrange, and the corresponding $\boldsymbol{A}$ and $\boldsymbol{B}$ have been described for DIPMC and SIPMC.

4 Simulations and Graphs of DIPMC and SIPMC

(a) Double Inverted Pendulum Control

Slider is the only actuated system in the structure. The mechanical properties of the DIPMC are as presented in Table 2. The mechanical parameters M, m_1, m_2, r_1, and r_2 are taken suitably for simulation. However, I_1 and I_2 of links have been calculated using the standard formulae for moment of inertia of rod of length l and mass m around its center of mass as $I = \frac{ml^2}{12}$.

The link 1 has been designed in such a way to support the revolute joint and encoder for measuring the joint angle along the rotation axis while maintaining the symmetricity of center of mass of link 1. Applying Linear Quadratic Regulator (LQR) function on $\boldsymbol{A}$, $\boldsymbol{B}$, $\boldsymbol{Q} and \boldsymbol{R}$ with

Table 2 Mechanical properties of the DIPMC used for simulation

Mass of cart, M	0.5 kg
Mass of link 1, m_1	0.2 kg
Mass of link 2, m_2	0.1 kg
Length of link 1, $2r_1$	0.30 m
Length of link 2, $2r_2$	0.30 m
Moment of inertia of link 1, I_1	0.006 kgm^2
Moment of inertia of link 2, I_2	0.003 kgm^2

$$Q = \begin{bmatrix} 100 & 1 & 0 & 0 & 0 & 0 \\ 0 & 0.1 & 0 & 0 & 0 & 0 \\ 0 & 0 & 100 & 0 & 0 & 0 \\ 0 & 0 & 0 & 0.1 & 0 & 0 \\ 0 & 0 & 0 & 0 & 100 & 0 \\ 0 & 0 & 0 & 0 & 0 & 0.1 \end{bmatrix}$$

and $R = 100$ gives the gain $\boldsymbol{K}$.

$\boldsymbol{K} = [89.03629.6917 - 102.0109 - 20.74291.00002.0564]Z.$

Substituting, $F = -\boldsymbol{K}\boldsymbol{X}$ in Eq (16), we get $\dot{\boldsymbol{X}} = (\boldsymbol{A} - \boldsymbol{B}\boldsymbol{K})\boldsymbol{X}$ and then subsequently feedback control loop was implemented. Also, F has been saturated to $24N$ in feedback control as shown in Fig. 8 so that the torque applied by the actuator is within permissible limits. To control the structure, an initial impulsive force, $f = F_o\cos(\omega t)$ was applied to the cart. After doing several iterations, by varying F_o and ω, we were able to stabilize the system from the initial conditions $X_{\text{initial}} = \begin{bmatrix} \pi & 0 & \pi & 0 & 0 & 0 \end{bmatrix}$ at $F_o = 24N \ and \ \omega = 2.8\pi \ rad \, s^{-1}$.

The system behavior is presented in Fig. 4. Till $t = 0.55$ s, the feedforward controller is implemented onto the system and then the feedback controller.

The variation of generalized coordinates with respect to time is presented in Fig. 5. It is evident from the figures that derivatives of the state vectors (Figs. 6 and 7) are approaching 0 thus signifying that the system is going toward equilibrium. Also, they are smooth and continuous throughout the period except at the kink due to change in shift from feedforward controller to feedback controller. This kink, however, in practical aspect has a negligible effect on the system's trajectory as the time constant for rise in current is comparable to the mechanical time constant. The nature of the cart's position with respect to time (Fig. 5) shows that the force estimator used before the control shift is satisfying with the nature of the cart's trajectory. Additionally, the estimator helps to limit the acceleration of the cart ($\ddot{x}$) during control change (Fig. 7) which is achievable by a motor in practical means (Fig. 8).

(b) Single Inverted Pendulum Control

The mechanical properties of the pendulum developed is presented in Table 3. Similar to DIPMC, the mechanical parameters m, m_1, and r_1 are taken randomly for

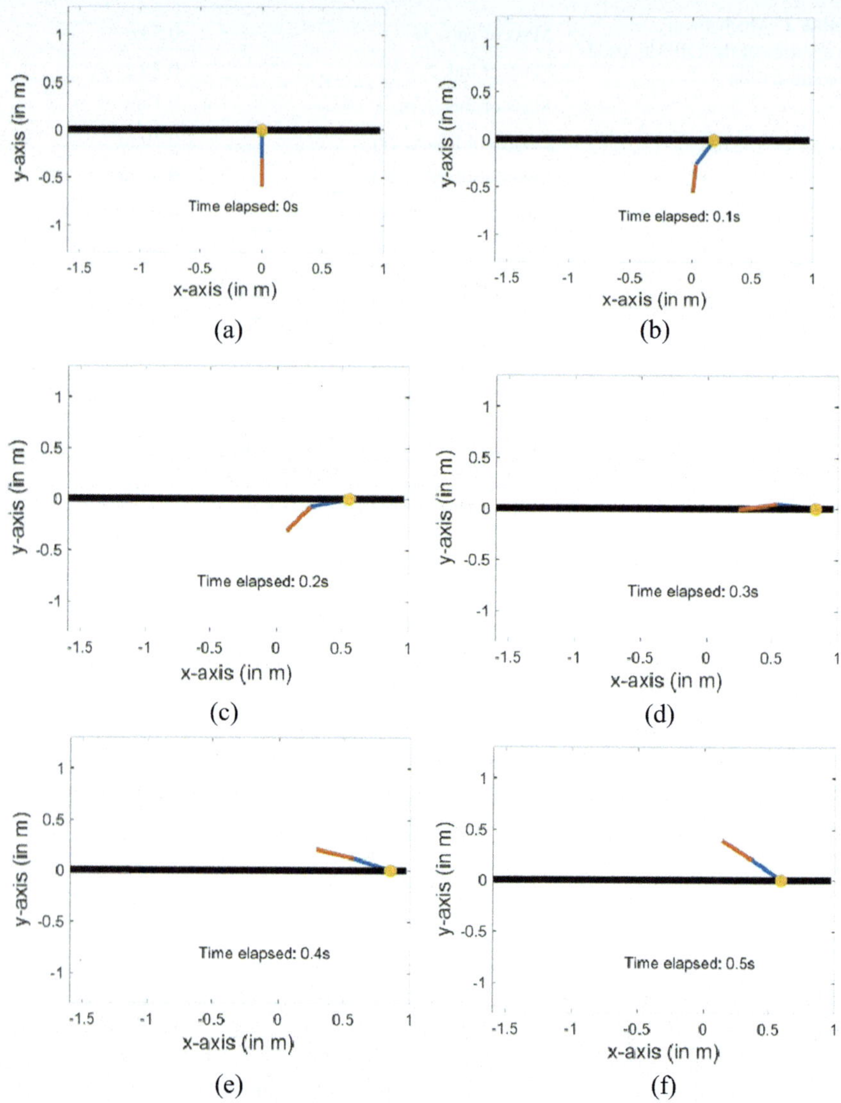

Fig. 4 Simulation results of DIPMC. It takes around 5 s to stabilize in an inverted upright position from its initial vertical stable position

simulation. However, I_1 of link 1 has been calculated using the standard formulae for moment of inertia of rod of length l and mass m around its center of mass as $I = \frac{ml^2}{12}$.

In the feedback controller, the system was reaching its inverted state equilibrium position ($\theta_1 = 0$) only when $|\theta_1| \leq \frac{2\pi}{5}$. The initial estimator force applied on the

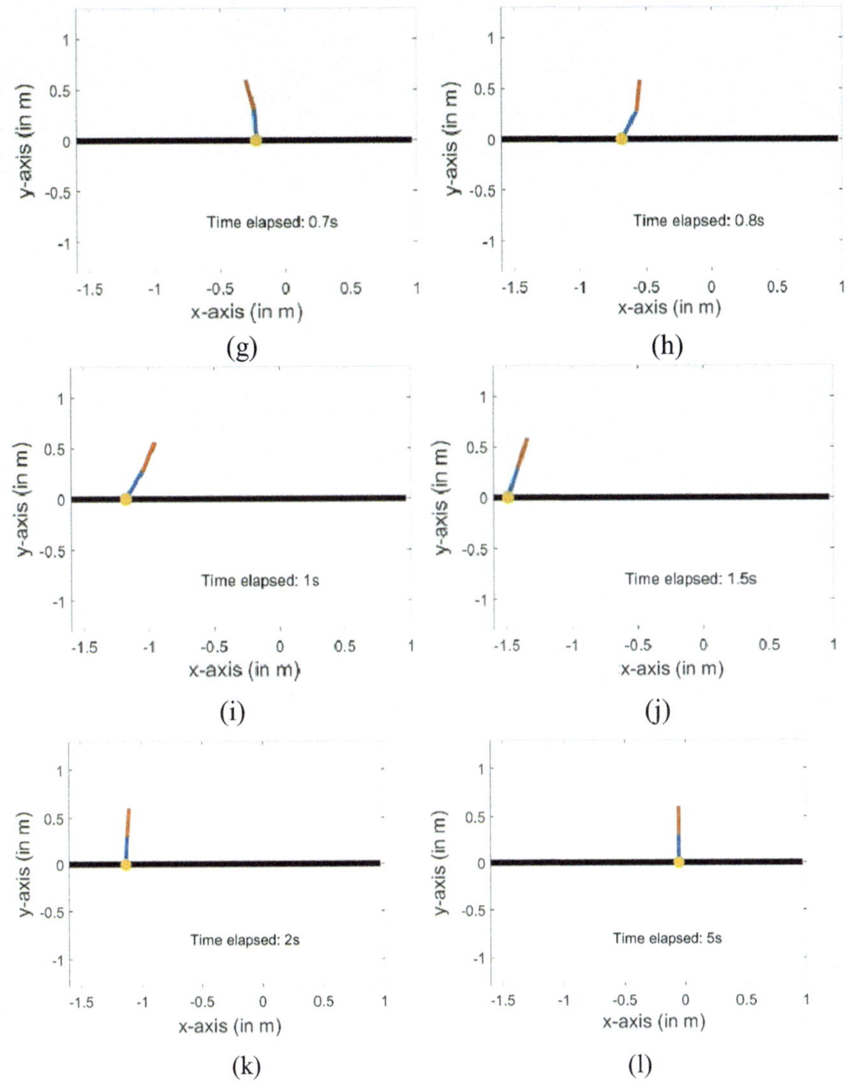

Fig. 4 (continued)

cart is of the form, $f = F_o \sin(\omega t)$. Applying Linear Quadratic Regulator (LQR) function on A, B, $Q\, and R$ with $Q = 0.01 I_4$ and $R = 0.001$ gives the gain $\boldsymbol{K}$.

The LQR solution for single inverted pendulum results in the gain of $\boldsymbol{K} = \begin{bmatrix} 23.38 & 4.60 & -3.16 & -4.82 \end{bmatrix}$. Therefore, substituting $F = \boldsymbol{KX}$, we get the control law. Unlike, after doing several iterations, we were able to stabilize the system from the initial conditions $\boldsymbol{X} = \begin{bmatrix} \pi & 0 & 0 & 0 \end{bmatrix}$ at $F_o = 15 N and \omega = 5\pi rads^{-1}$. The

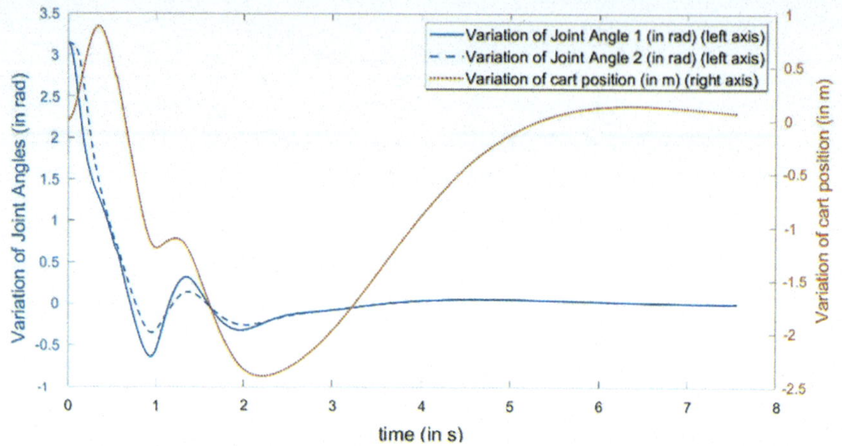

Fig. 5 Change in generalized coordinates with respect to time. Prismatic joint needs to be related to the right vertical axis whereas the revolute joint coordinates need to be related to the right vertical axis

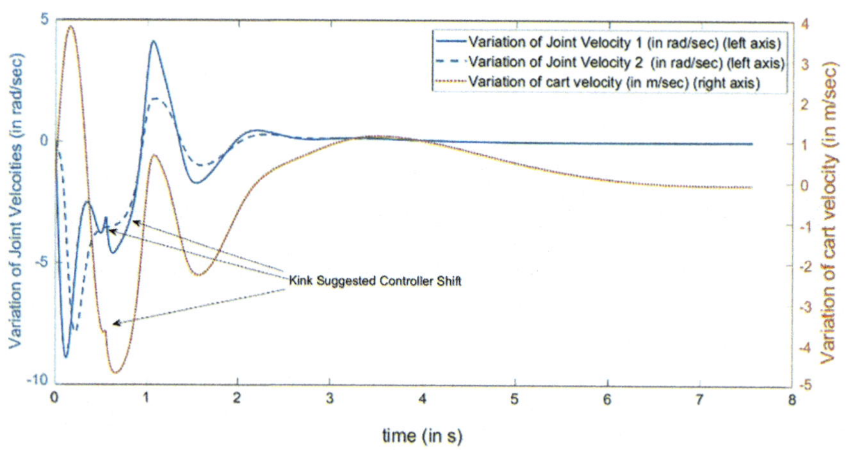

Fig. 6 Change in derivatives of generalized coordinates with respect to time

system behavior as simulation snips is presented in Fig. 9. Around $t = 0.55$ s, the system shifts from feedforward controller to feedback controller.

The behavior of the generalized coordinates for a single inverted pendulum is presented in Fig. 10. The joint angle stabilizes to 0° which was a requisite for the controller. The derivative of state vector for single inverted pendulum reduces to a null vector, as after around 3.5 s, the joint velocity and cart velocity (Fig. 11) stabilize to zero. Also, the joint acceleration and the cart acceleration (Fig. 12) reduce to none. Hence, the derivative of the state vector reduces to zero which is indicative of the equilibrium.

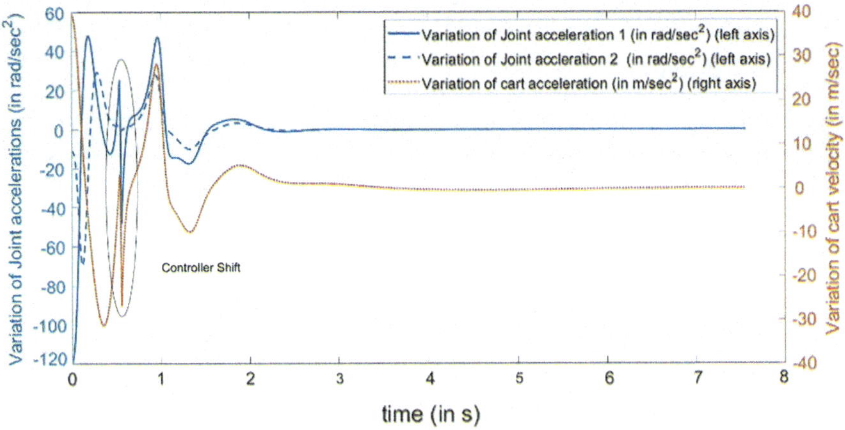

Fig. 7 Second derivatives of generalized coordinates with respect to time

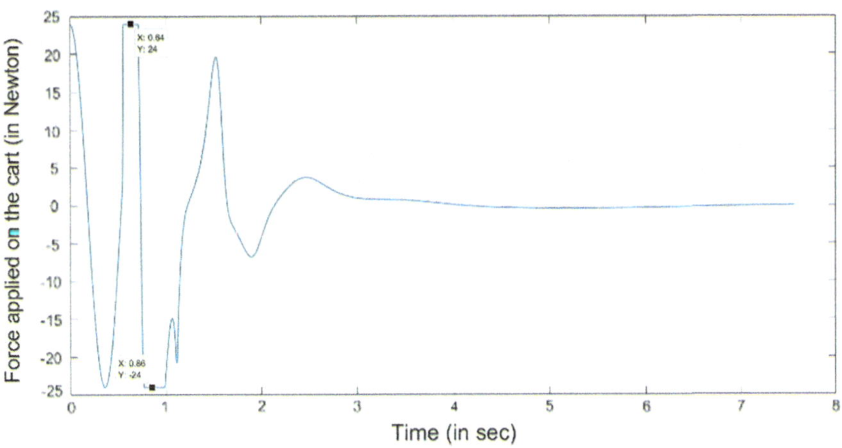

Fig. 8 External force acting on the cart in horizontal direction

Table 3 Mechanical properties of the single inverted pendulum used for simulation

Mass of cart, m	0.2 kg
Mass of link 1, m_1	0.1 kg
Length of link 1, $2r_1$	0.30 m
Moment of inertia of link 1, I_1	0.001 kgm^2

5 Conclusion

The paper aims to formulating a methodology to ensure stabilization control of under-actuated serial arms. An open-loop controlling sinusoidal input is used to bring the

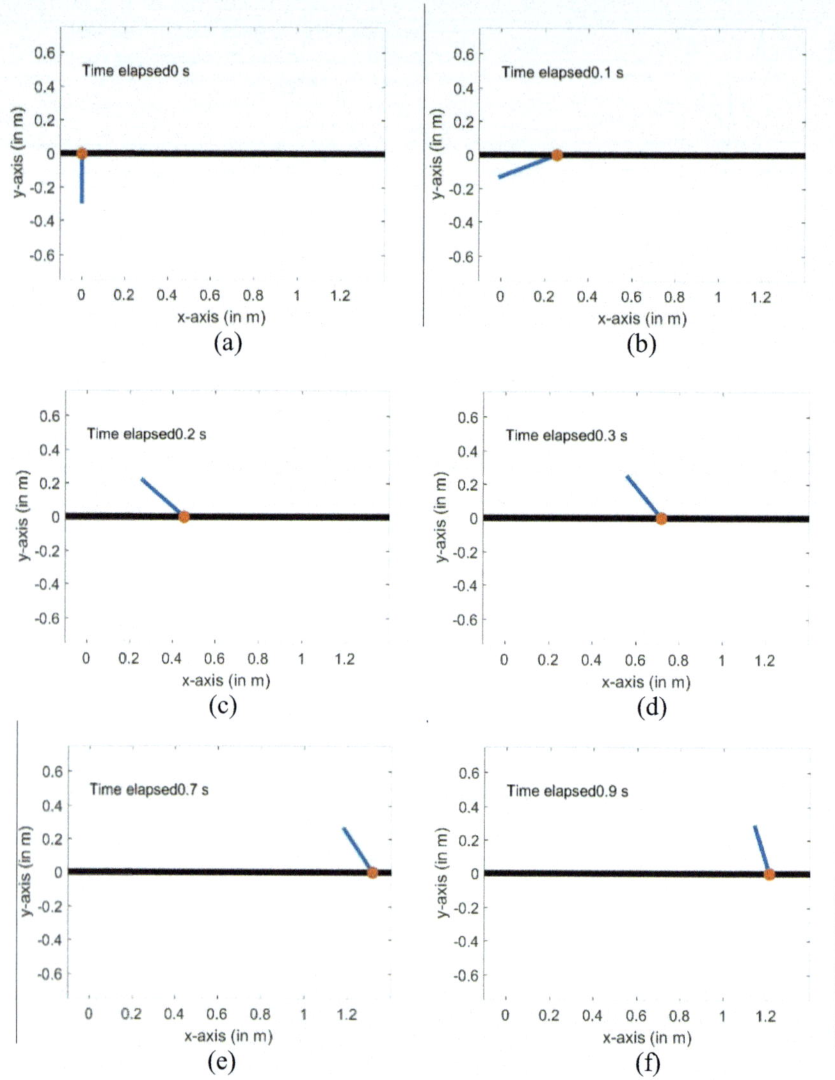

Fig. 9 Simulation results of stabilization of SIPMC. It takes around 3.5 s for SIPMC to stabilize in an upright vertical position from its initial stable equilibrium position

system around the region of equilibrium and then closed loop is applied with the system being linearized around the point of equilibrium. The input ensures that the net trajectory of the generalized coordinates is smooth and differentiable at all times. Additionally, the requisite acceleration to shift from a feedforward term to a feedback controller is within practical limits by using such a method. The methodology proposed is applied to a single inverted pendulum and a double inverted pendulum on a moving cart.

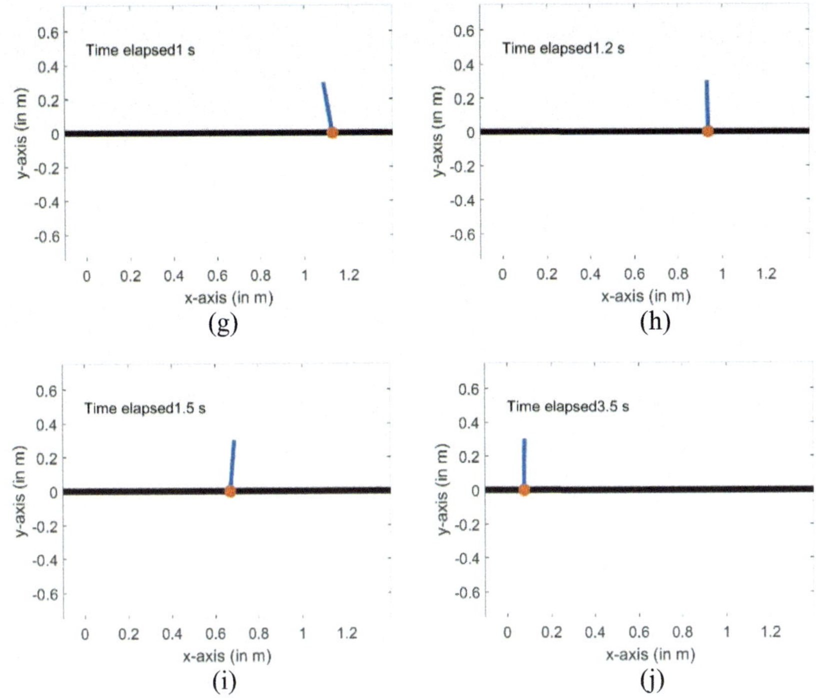

Fig. 9 (continued)

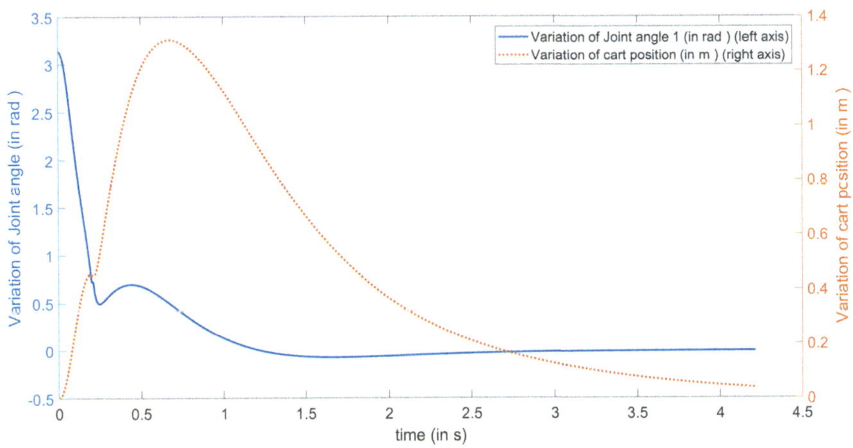

Fig. 10 Variation of the generalized coordinates of a single inverted pendulum with time

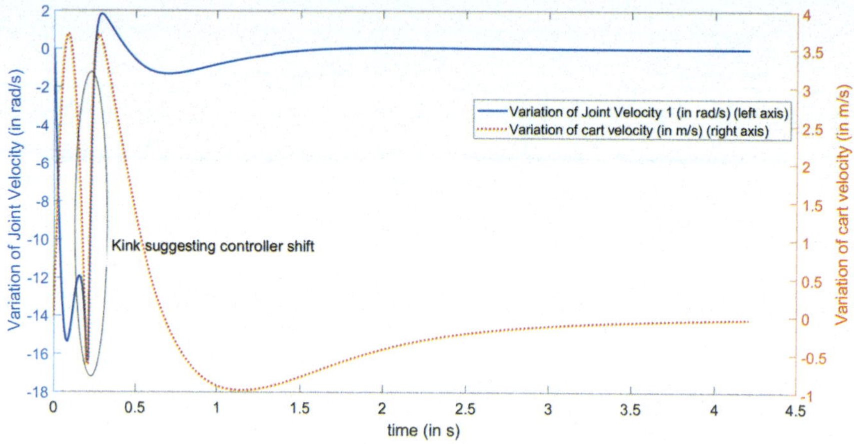

Fig. 11 First derivatives of generalized coordinates of a single inverted pendulum with respect to time

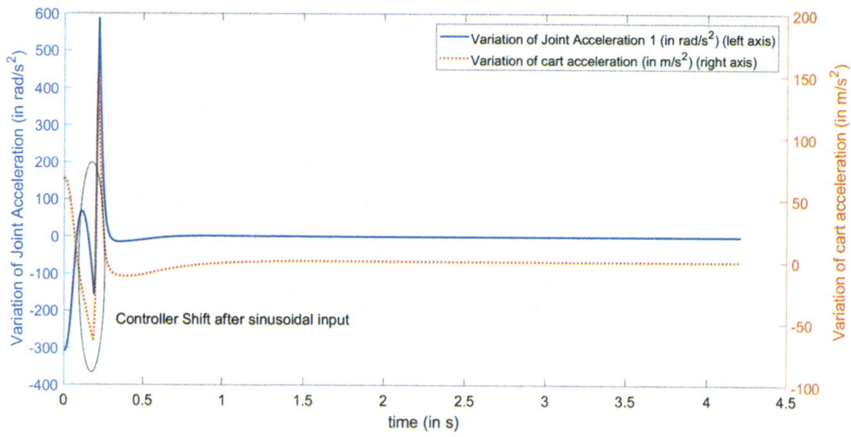

Fig. 12 Second derivatives of generalized coordinates of a single inverted pendulum with respect to time

Appendix

1. DIPMC

The initial input puts the system into a state such that $\theta_1 and \theta_2$ are into the region close to $\theta_1 = 0$ and $\theta_2 = 0$ and ignoring higher order terms. Therefore, applying approximations to the joint angles, we get $\cos\theta_1 = 1$, $\sin\theta_1 = \theta_1$, $\cos\theta_2 = 1$, and $\sin\theta_2 = \theta_2$. The equation of motion now relates to the linearized matrices and vectors.

$$\widetilde{M(q)}\ddot{q} + \widetilde{V(q,\dot{q})} + \widetilde{G(q)} = \tau \tag{13}$$

$$\Rightarrow \ddot{q} = \widetilde{M(q)}^{-1}\left(\tau - \widetilde{V(q,\dot{q})} + \widetilde{G(q)}\right) \tag{14}$$

where $\widetilde{M(q)}$, $\widetilde{V(q,\dot{q})}$, *and* $\widetilde{G(q)}$ are linearized forms of $M(q)$, $V(q,\dot{q})$, *and* $G(q)$, respectively.

$$\widetilde{M(q)} = \begin{bmatrix} I_1 + m_1 r_1^2 + 4m_2 r_1^2 & 2m_2 r_1 r_2 & -(m_1 r_1 + 2m_2 r_1) \\ 2m_2 r_1 r_2 & I_2 + m_2 r_2^2 & -m_2 r_2 \\ -(m_1 r_1 + 2m_2 r_1) & -m_2 r_2 & M + m_1 + m_2 - b \end{bmatrix} \tag{15}$$

$$\widetilde{V(q,\dot{q})} = \begin{bmatrix} 0\,0\,0 \end{bmatrix}^T \tag{16}$$

$$\widetilde{G(q)} = \begin{bmatrix} h_1\theta_1 \; h_2\theta_2 \; 0 \end{bmatrix}^T \tag{17}$$

where $h_1 = -(gm_1 r_1 + 2gm_2 r_1)$, $h_2 = -gm_2$.

Rewriting Eq (14) in state-space form of

$$\dot{X} = AX + Bu$$

$$X = \begin{bmatrix} \theta_1 \; \dot{\theta}_1 \;\; \theta_2 \; \dot{\theta}_2 \;\; x \; \dot{x} \end{bmatrix}^T \tag{18}$$

$$A = \begin{bmatrix} 0 & 1 & 0 & 0\,0\,0 \\ -\widetilde{M(q)}^{-1}_{(1,1)} h_1 & 0 & -\widetilde{M(q)}^{-1}_{(2,1)} h_2 & 0\,0\,0 \\ 0 & 0 & 0 & 1\,0\,0 \\ -\widetilde{M(q)}^{-1}_{(2,1)} h_1 & 0 & -\widetilde{M(q)}^{-1}_{(2,2)} h_2 & 0\,0\,0 \\ 0 & 0 & 0 & 0\,0\,1 \\ -\widetilde{M(q)}^{1}_{(3,1)} h_1 & 0 & -\widetilde{M(q)}^{-1}_{(3,2)} h_2 & 0\,0\,0 \end{bmatrix} \tag{19}$$

$$B = \begin{bmatrix} 0 \; \widetilde{M(q)}^{-1}_{(1,3)} \; 0 \; \widetilde{M(q)}^{-1}_{(2,3)} \; 0 \; \widetilde{M(q)}^{-1}_{(3,3)} \end{bmatrix}^T \tag{20}$$

$$u = [00000F]^T \tag{21}$$

2. SIPMC

The linearized equation of motion around the unstable equilibrium is formulated and the equation of motion by replacing "sin(θ)" by "θ", "cos(θ)" by 1 and ignoring higher order terms is presented below.

$$\underbrace{\begin{bmatrix} I_1 + m_1 r_1^2 & -m_1 r_1 \\ -m_1 r_1 & m + m_1 - b \end{bmatrix}}_{\widetilde{M(q)}'} \begin{bmatrix} \ddot{\theta} \\ \ddot{x} \end{bmatrix} + \underbrace{\begin{bmatrix} -g m_1 r_1 \theta \\ 0 \end{bmatrix}}_{\widetilde{G(q)}'} = \tau \tag{22}$$

Therefore, the state-space form of SIPMC

$$\dot{X} = AX + Bu$$

where

$$X = \begin{bmatrix} \theta \ \dot{\theta} \ \ x \ \dot{x} \end{bmatrix}^T \tag{23}$$

$$A = \begin{bmatrix} 0 & 1 & 0\,0 \\ -\widetilde{M(q)}'^{-1}_{(1,1)} h & 0 & 0\,0 \\ 0 & 0 & 0\,1 \\ -\widetilde{M(q)}'^{-1}_{(2,1)} h & 0 & 0\,0 \end{bmatrix} \tag{24}$$

where $h = -g m_1 r_1$

$$B = \begin{bmatrix} 0 \ \widetilde{M(q)}'^{-1}_{(1,2)} \ \ 0 \ \widetilde{M(q)}'^{-1}_{(2,2)} \end{bmatrix}^T \tag{25}$$

$$u = \begin{bmatrix} 0\,0 \ \ 0 \ F \end{bmatrix}^T \tag{26}$$

References

1. Ha YS (1995) Trajectory tracking control for navigation of the inverse pendulum type self-contained mobile robot. Intell Robot Syst 637–655
2. Shiroma N, Matsumoto O, Tani K (1999) Cooperative behavior of a mechanically unstable mobile robot for object transportation. JSME Int J Ser C Mech Syst Machine Elements and Manuf 42(4):965–973
3. Graichen K, Hagenmeyer V, Zeitz M (2005) A new approach to inversion-based feedforward control design for nonlinear systems. Automatica 41(12):2033–2041
4. Niemann H, Stoustrup J (2005) Passive fault tolerant control of a double inverted pendulum—a case study. Control Eng Pract 13(8):1047–1059
5. Bui TH, Nguyen TT, Chung TL, Kim SB (2003) A simple nonlinear control of a two-wheeled welding mobile robot. Int J Control Autom Syst 1(1):35–42
6. Fliess M, Lévine J, Martin P, Rouchon P (1995) Flatness and defect of non-linear systems: introductory theory and examples. Int J Control 61(6): 1327–1361
7. Meirovitch L (2010) Methods of analytical dynamics, pp 170–208. Courier Corporation
8. Kwakernaak H, Sivan R (1972) Linear optimal control systems, vol 1. Wiley, New York, pp 201–209

Soft Hand Exoskeleton for Adaptive Grasping Using a Compact Differential Mechanism

Ajay Bajaj, Vishal Jain, Prabhat Kumar, Aynur Unal, and Anupam Saxena

Abstract The work proposes an underactuated design of a glove-like soft hand exoskeleton for grasping and lifting objects of varied shapes and sizes. Strings are used to flex all finger joints, assuming that finger anatomy of an impaired hand is intact. A pulley-based differential mechanism is designed to actuate all four fingers via a single motor to allow adaptive grasping. Two DC motors are used, one for flexion of all four fingers and the second for thumb flexion. Finger extension is passively achieved via elastic bands on the dorsal side. A prototype of the hand exoskeleton, weighing 300 g without battery is fabricated that occupies $100 \times 56\,\text{mm}^2$ space over the palmer side of the forearm. Novelty in the design lies in reducing the required length of the existing pulley-based differential mechanism from 20 cm to 10 cm. Lightweight, and compactness make the device portable. Performance of the soft exoskeleton is demonstrated via testing it on a healthy subject.

Keywords Soft hand exoskeleton · Finger flexion-extension · Differential mechanism size reduction · Artificial tendon pulley system

---

A. Bajaj · V. Jain · A. Saxena (✉)
Department of Mechanical Engineering, Indian Institute of Technology Kanpur, Kanpur, India
e-mail: anupams@iitk.ac.in

A. Bajaj
e-mail: kajaybajaj.2222@gmail.com

V. Jain
e-mail: vishaljain1216@gmail.com

P. Kumar
Faculty of 3mE, Department of Precision and Microsystems Engineering, Delft University of Technology, Delft, The Netherlands
e-mail: prabhatkumar.rns@gmail.com

A. Unal
Digital Monozukuri, Palo Alto, CA, USA
e-mail: aynurunal@alumni.stanford.edu

D. Sen et al. (eds.), *Mechanism and Machine Science*, Lecture Notes in Mechanical Engineering, https://doi.org/10.1007/978-981-15-4477-4_52

1 Introduction

Post spinal cord/brain injuries or stroke, normal functioning of the hand may be partially/totally impaired. In such cases, anatomy of a human hand remains intact while only motor control is lost.

For such patients, this makes it comparatively easier to develop a lightweight compact device, which can enable them to perform diurnal tasks. Only an actuation mechanism needs to be designed.

Grasping and lifting common objects are some of the most basic tasks one needs in daily living. In literature, several designs for hand exoskeletons exist, aiming for actuation of all finger joints. This full acutation is essential for tasks wherein, independent joint control is important, such as fine motor tasks like, writing, painting etc. However, that makes the design complex, and requires intricate control system for even the tasks such as grasping and lifting, which a healthy person performs easily. Complexity exists because, the control system has to adjust with varying shapes and sizes of the objects user wants to grasp. Underactuation, i.e. not actuating all joints independently, helps overcome this problem. When less actuators are used, than the degrees of freedom present in the fingers, there is redundancy. Consequently, during flexion, fingers can adapt to the shape/size of the object to be grasped. Some designs exist in literature [1, 2] (related to implants and robotic hands) which demonstrate this capability, by use of differential mechanism in their actuation schemes. Differential mechanism allows the user to flex some fingers, even if others are held stationary. This helps in making successful contact with objects of different shapes and sizes, even though not all fingers make contact simultaneously. Grasp force can be increased thereafter.

Biomimetic exoskeleton designs are inspired by the human hand anatomy, with the aim to mimic the tendon pulley systems in the hand, which are responsible for finger flexion and extension (Fig. 1). Lee et al. [3] propose a biomimetic hand exosksleton, in which they use seven DC motors to actuate tendons and employ springs as a differential mechanism. However, springs limit the shape adaptability, and also, they may behave differently after few iterations if permanent extension occurs. In et al. [4, 5] and Kang et al. [6] introduce differential mechanisms which enable adaptive grasping of various objects having different shapes/sizes. Two DC motors are used, one for index and middle figures, and another for the thumb. They use a passive brake mechanism [7, 8] to prevent slacking in tendons and ensure the efficient working of mechanism. Kang et al. [6] use 'U' shape differential mechanism only for index and middle fingers. This mechanism cannot be employed for all four fingers as it generates large frictional force leading to a high motor load.

The present work employs a pulley-based differential mechanism and inextensible strings to design a soft hand exoskeleton for adaptive grasping. The exoskeleton is developed for lightweight, compactness, portability and less friction. All four fingers are flexed using a single actuator, through a differential mechanism. The notion is similar to the one described in [9] wherein the differential pulley mechanism is employed on the palm of a prosthetic hand. The design therein uses four pulleys

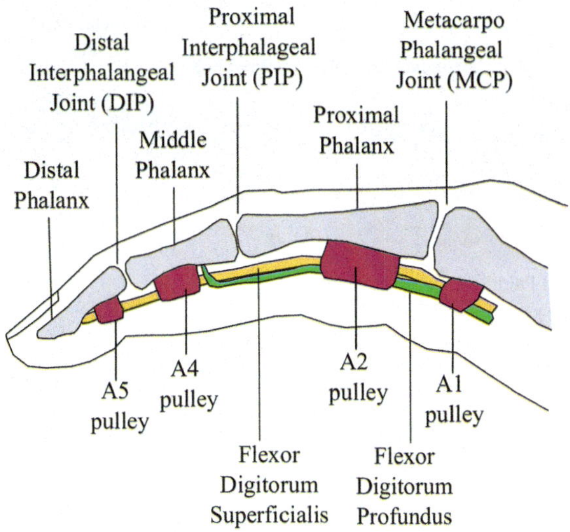

Fig. 1 Anatomy of a human finger: Tendon pulley system for finger flexion

whereas in the one proposed for an orthotic soft hand exoskeleton, seven moving pulleys and seven idlers are employed in multiple layers to reduce space over the forearm. The novelty is in the size reduction of the differential mechanism to obtain a compact design. In Sect. 2, design of a glove with tendon-pulley system, and a differential mechanism, CAD model, actuation and control are presented. The proposed design is tested and validated on a healthy subject (Sect. 3). Design aspects are discussed and lastly, conclusions are drawn in Sect. 4.

2 Exoskeleton Design

The presented design mimics the tendons responsible for finger flexion and extension via inextensible strings and elastic bands respectively. These strings are referred to as artificial tendons (Fig. 2a) in this paper. Pulleys responsible for maintaining these tendons close to hand are also replicated via inextensible string loops, referred to as artificial finger and palm pulleys. A woolen glove is used as an interface for the artificial tendon-pulley system. Tendons are actuated by a differential mechanism mounted on the forearm. Detailed explanation of the design is provided in the following subsections.

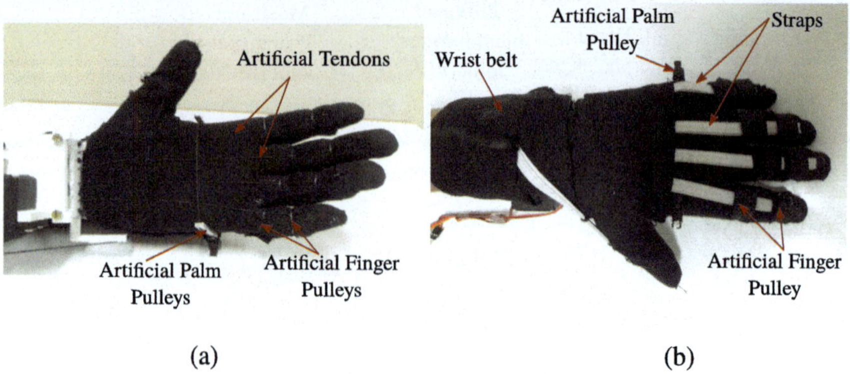

Fig. 2 Glove with **a** artificial tendon pulley system on palmer side for finger flexion, and **b** elastic straps on dorsal side for finger extension

2.1 Glove with Extensor and Flexor Tendon Pulley System

Artificial tendons and artificial finger pulleys are arranged like the flexor digitorum profundus (FDP) tendon pulley system found in biological fingers (Fig. 1). These tendons are attached to the distal phalanges, on their palmer side. Pulling these tendons flexes the three joints of a finger namely, the distal interphalangeal joint (DIP), proximal interphalageal joint (PIP) and metacarpophalangeal joint (MCP) (Figs. 1 and 2).

In case of the thumb, the artificial tendon actuates its MCP and IP joints.

Artificial tendons are woven in a flexible glove such as a woolen one. If these tendons are pulled as they are, they shift away from finger joints (phenomenon called bowstringing, Fig. 3). Consequently, the glove stretches and lifts away from the hand. This is undesirable as it interferes with the objects while gripping and may damage the glove. To avoid bowstringing, two sets of artificial pulleys are employed: (i) around finger, and (ii) on palm. Artificial finger pulleys are formed at middle and proximal phalanges of fingers, and proximal phalanx of thumb, using closed loops of inextensible string (Fig. 2). String loops are closed by securing Velcro straps attached at its ends (Fig. 4a). The straps make it easy to wear the glove, as well as, help in adjusting to different finger girths.

Implementation of a similar pulley (string loop) around palm, to prevent bowstringing near MCP joints, proves ineffective. This happens because, as these tendons are pulled, this pulley (string) also gets pulled away from hand, causing significant palm squeezing, as well as, bowstringing. To prevent this, a U shaped aluminum bracket is used around the palm, which eliminates contact between string and palm sides (Figs. 2 and 4b).

When artificial tendons are directly attached to the glove at distal phalanx, the glove stretches in proximal direction (Fig. 5a). Consequently, it interferes during grasping, and, damages the glove too. These problem is circumvented by using another pulley in form of string loop around distal phalanges, woven in the glove (Fig. 5b).

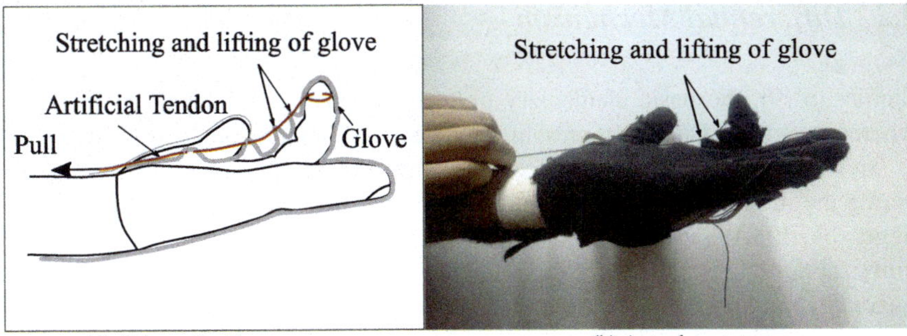

(a) Schematic diagram (b) Actual prototypes

Fig. 3 **a** Schematic diagram and **b** Actual prototype showing lifting/stretching of the glove away from fingers and palm when AT is pulled (towards left)

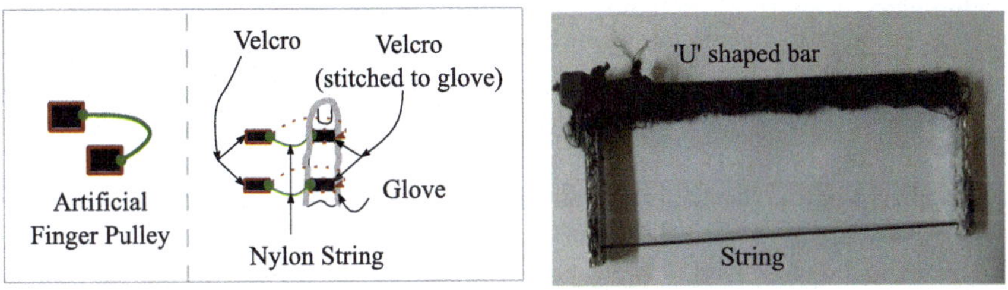

(a) (b)

Fig. 4 Artificial Pulley design: **a** Pulleys around finger phalanges for preventing bowstring near PIP and DIP joints, **b** Pulley around palm for preventing bowstringing near MCP joint

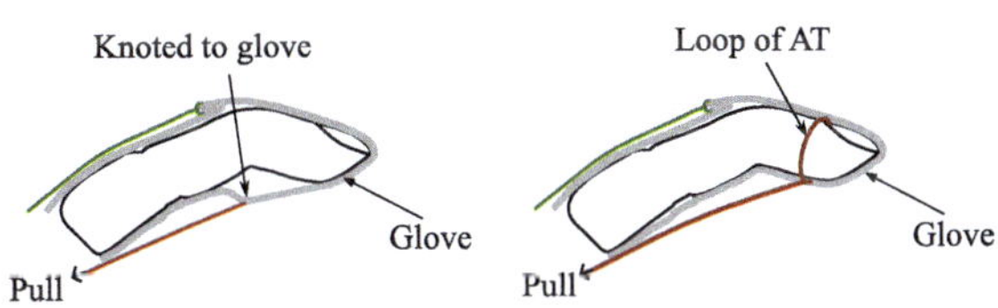

Fig. 5 Tendon attachment on distal phalanx: Attaching to a point on the glove leads to stretching. Solution is to wind a loop of the tendon (string) on the phalanx

Finger/thumb extension is performed passively by elastic bands. One band is attached to each of the five distal phalanges on their dorsal side. These bands are sewn onto the glove to terminate on the back of the hand (Fig. 2b). They are in relaxed state, in the fully extended configuration of the hand. Post flexion, when extension is desired, artificial flexor tendons are slackened, and, band elasticity helps the digits regain their fully extended states.

For finger flexion, the tendons are actuated via a differential mechanism discussed next. Thumb flexion is performed by directly actuating the corresponding tendon via a separate actuator.

2.2 Differential Mechanism

Similar to [9], the basic elementary unit of this mechanism is shown in Fig. 6a, wherein blocks and strings represent fingers and artificial tendons, respectively.

This unit has two degrees of freedom. Therefore, while moving the pulley, hindering the motion of one finger (block) does not prevent the motion of the other finger (block). At a given time instant, only the finger which offers lower resistance moves (Fig. 6b), when pulling force is just sufficient to cause uniform motion. In finger flexion, the tendon force (tendon tension) keeps increasing with flexion, in all fingers [10]. This means, as the finger with lower resistance flexes by some amount, its resistance increases, and can become higher than that of the other one. Then, the other finger starts flexing. This switch keeps happening, and both fingers get flexed consequently step by step. In practice, it is difficult to have identical resistance in both fingers. Therefore, simultaneous flexion does not happen. However, switching between fingers happens very fast, as only a very small differential in resistance is required to cause the switch. Thus, fingers appear to move simultaneously. In case fingers are being flexed to grasp an object, if a finger gets into contact with the object, it stops flexing further. Then, the other finger starts flexing faster, until this finger also makes contact. Thereafter, the grasping force starts increasing, if pulley is being exerted further force from the actuator. Based on these observations, this elementary unit can be employed to flex a pair of fingers for adaptive grasping via a single actuator.

To design the exoskeleton for actuating all four fingers for adaptive grasping, a pair of elementary units is connected via a third pulley (Fig. 7). This arrangement becomes a two-stage differential mechanism, such that, when the third pulley is pulled by an actuator, the pulley among P_1 and P_2 offering lower resistance moves at a time. Consequently, the finger offering least resistance moves at a time, as discussed earlier. All fingers keep flexing turn by turn, until some finger gets into contact with the object to be grasped. Rest of the fingers keep flexing, until each makes contact with the object. Then, the grasp force can be increased, by further increasing actuator input. In this way, fingers can grasp objects of varying shapes and sizes (adaptive grasping). Grasp force is also expected to be sufficient for diurnal activities, by use of an appropriately powerful actuator.

To flex a finger completely using the above design, strings directly attached to fingers, i.e. the blocks, should be pulled by around $d = 10\,\text{cm}$ for an adult human. Therefore, to close a fist, one needs a minimum $2d = 20\,\text{cm}$ space, as shown in Fig. 7. This excludes the pulley sizes. Yet, it is significantly large and will cover most of the forearm. Therefore, a strategy is adopted to reduce the longitudinal size of the differential mechanism in three stages.

In stage-1 (Fig. 8), two idler pulleys A_1, and A_2, are introduced to reverse the direction of motion of the pulley P_3. This reduces the required size to $3d/2 = 15\,\text{cm}$, excluding the pulley sizes. In stage-2 (Fig. 9), instead of connecting fingers directly to pulleys P_1 and P_2, intermediate pulleys P_4–P_7 are introduced. One end of the strings (artifical tendons) connecting to the fingers is fixed. This arrangement reduces the

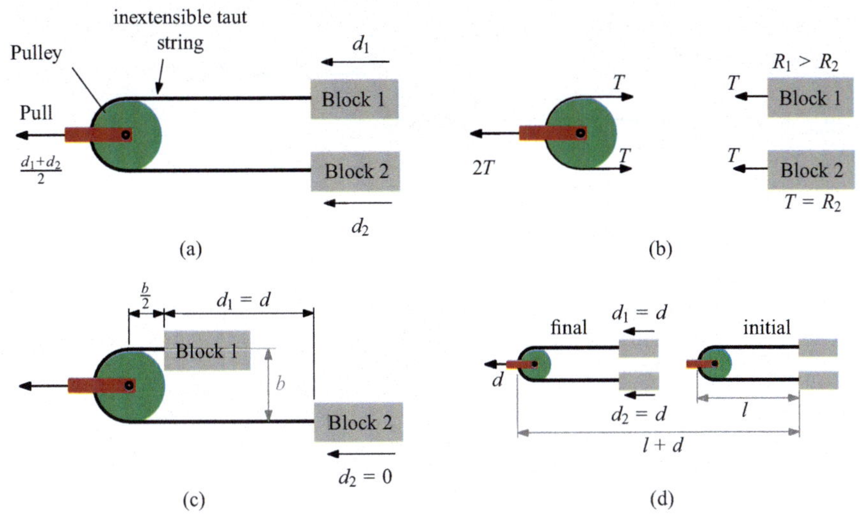

Fig. 6 Pulley-block system as a differential mechanism: **a** Kinematics: Two degrees of freedom: Out of the three elements (two blocks and one pulley), any two can move independently at a time, while maintaining the string taut. Pulley displacement has to be the average of block displacements d_1 and d_2. If one block stops moving, the other will move twice as fast as the pulley. **b** Statics: Starting from rest, if the blocks offer unequal resistances R_1, R_2 to motion, the string tension T increases until the less resistant block starts moving. Pulling force is therefore, twice the lower resistance value. Assumptions: (i) Pulley axis friction is negligible. (ii) Rolling friction (no slip) between string and pulley is negligible compared to string tension T. (iii) String is very light to be considered massless. (iv) Pulley inertia force (either mass or acceleration) small compared to string tension T. Under these assumptions, string tension T remains uniform over its entire length, and the pulling force is twice the tension T. **c** Design: Configuration with minimum distance between pulley and a block: If $d_1, d_2 \leq d$, the minimum string length required is $d + b + \pi b$. **d** Design: Configuration with maximum displacement: Length $l = d/2 + b$. Hence, minimum space needed for the assembly $= l + d = 3d/2 + b$

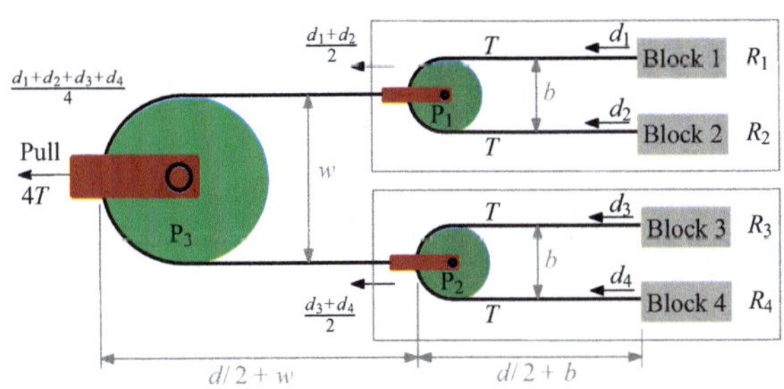

Fig. 7 Differential Mechanism for four fingers: Two pulley-block elements act as two separate blocks for the bigger pulley. Displacement of the bigger pulley is the average of all four block displacements. At a given time, only the block with lowest resistance will move, just like in Fig. 6, provided that pulley-3 is pulled with just enough force to cause motion. The pulling force will be four times the lowest resistance among the four. If $d_1, d_2, d_3, d_4 \leq d$, then, the maximum displacement of the smaller pulleys $= d$. Therefore, similar to Fig. 6d, the minimum longitudinal space needed for the entire assembly $= l + d = 2d + b + w$, where $l = d + b + w$. Lateral space required $= b + w$

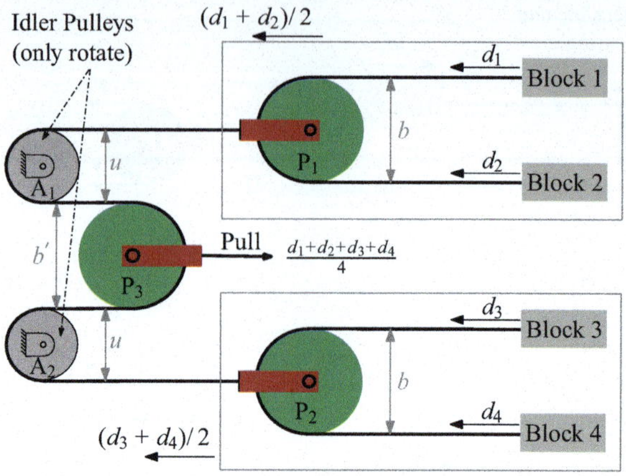

Fig. 8 Length reduction of the differential mechanism in Fig. 7: Stage-1 of the reduction: Two idler pulleys A_1, A_2 used to reverse the direction of the pulley P_3. Each of the two pulley-block subassemblies on the right, needs minimum space of $3d/2 + b$, as shown in Fig. 6, for block displacements $d_1, d_2, d_3, d_4 \leq d$. The pulley P_3 has the maximum displacement $= d$. Thus, the maximum longitudinal space need for the entire assembly $= 3d/2 + b + u$. Lateral space required $= b + b' + 2u$

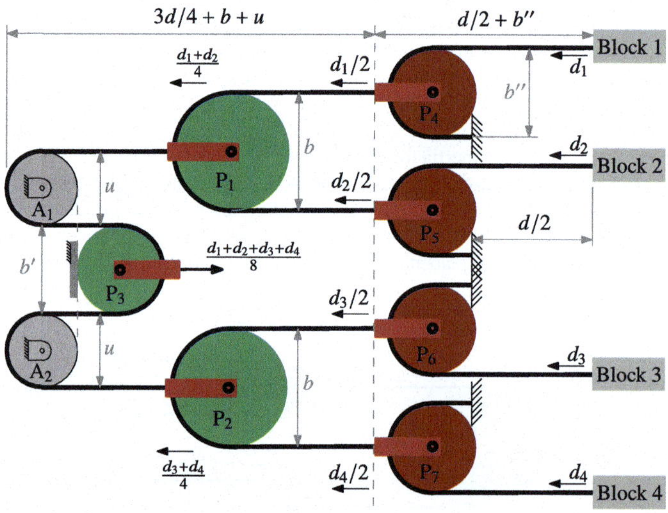

Fig. 9 Length reduction of the differential mechanism in Fig. 7: Stage-2 of the reduction: Each block in Fig. 8 is replaced by a pulley-block element with one end of the string fixed. These new pulleys P_4 - P_7 have maximum displacement $= d/2$ for block displacements $d_1, d_2, d_3, d_4 \leq d$. Other pulleys, therefore, have their maximum displacements halved to $d/2$. Comparing to Fig. 8, the logitudinal space requirement is reduced to $3d/4 + b + u$. However, addition of new pulleys increases it by $= d/2 + b''$. Hence, total longitudinal space needed $= 5d/4 + b + b'' + u$. Lateral space $= b + b' + b'' + 2u$

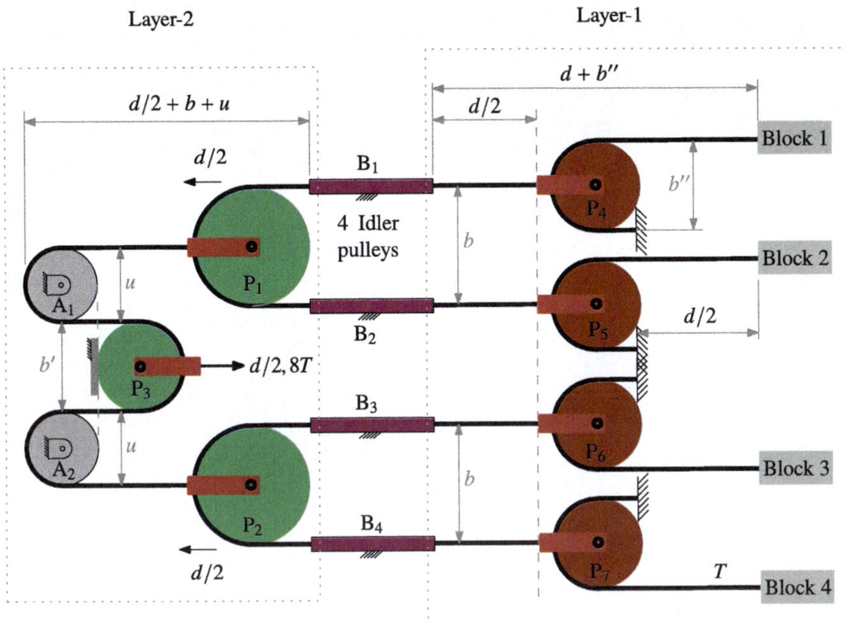

Fig. 10 Length reduction of the differential mechanism in Fig. 7: Stage-3 of the reduction: The design in Fig. 9 is split into two halves. Each half is stacked vertically in two layers. Four vertical idler pulleys B_1–B_4 are used to connect the two layers. This figure shows the two layers in single plane for visual clarity. All pulleys P_1–P_7 have maximum displacement $= d/2$ for block displacements $d_1, d_2, d_3, d_4 \leq d$. Logitudinal space requirement for the layer-1 is $d + b''$, which is more than that of the layer-2 ($= d/2 + b + u$). Hence, total longitudinal space needed by the entire assembly is $d + b'' + u'$, where u' is the diameter of idler pulleys B_1–B_4. Lateral space $= b + b' + b'' + 2u$. Note that, pulling force has doubled, as compared to the original design in Fig. 7. Longitudinal space is approximately halved ($d \gg b, b', b'', u, u', w$). Input pulley P_3 displacement is also halved

longitudinal size to $5d/4 = 12.5$ cm. However, it doubles the actuator force requirement, and halves actuator amplitude. To take advantage of this arrangement, four idler pulleys B_1–B_4 are introduced in stage-3 (Fig. 10). The design is split into two halves which are then, stacked over one another in two layers. This brings down the longitudinal size to $d - 10$ cm from the original 20 cm (ignoring pulley sizes). The compromizes required are, (i) increase in the lateral width and the height of the assembly, (ii) actuator force required is doubled, i.e. eight times the lowest resisting finger force, (iii) thirteen pulleys required, which might increase friction considerably compared to just three in the original design, if not properly designed.

To accommodate thumb motion, a hole is made in the frame of the differential mechanism, to pass the string (artificial tendon) coming from the thumb of the glove. The tendon is connected to motor directly. This allows the thumb to flex in the direction perpendicular to the axis of the hole.

Motors are mounted on the third layer of the assembly, and connected to the pulleys via strings passing over an idler pulley-C (Figs. 11 and 12). Figure 11 depicts

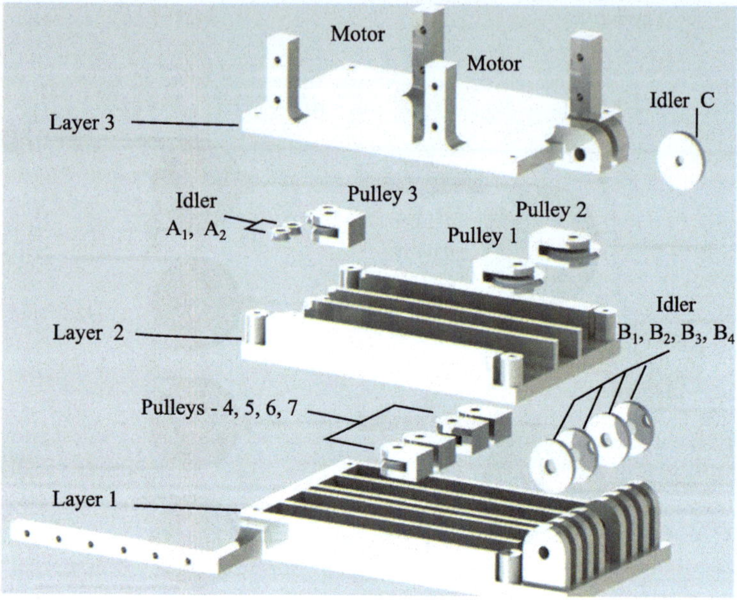

Fig. 11 CAD model: Exploded view, showing three layers. Layers 1, and 2 house the mechanism, while the actuators (motors) are assembled in the third (topmost) layer

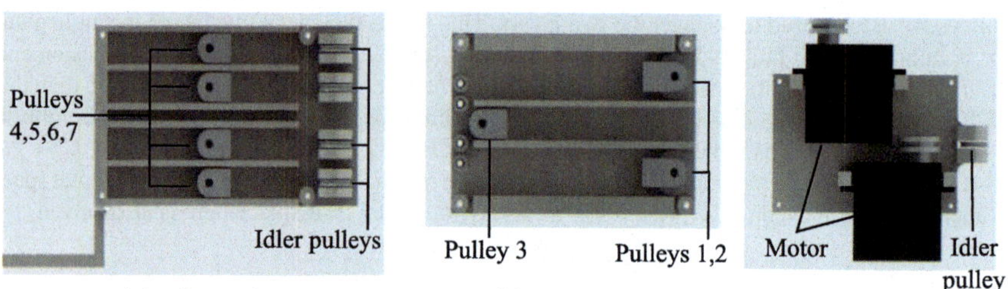

(a) Layer 1 (b) Layer 2 (c) Layer 3

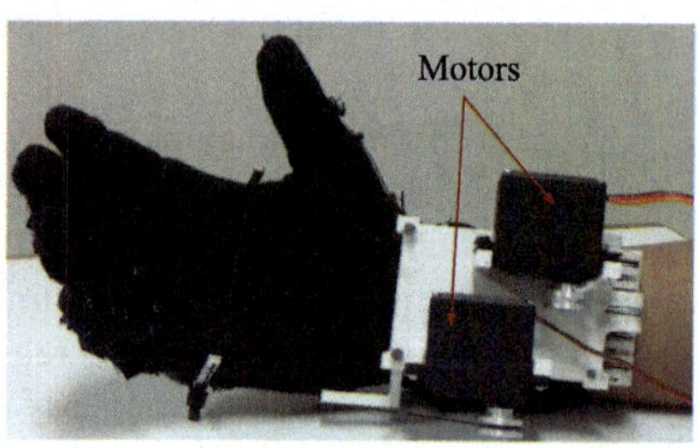

(d)

Fig. 12 Prototype of the soft hand exoskeleton: **a**, **b**, **c** show different layers (top view) in the differential mechanism. **d** shows the exosekleton in use the CAD model of the final differential mechanism. A prototype of the final model, is fabricated which weighs 300 g and occupies $100 \times 56\,\text{mm}^2$ on forearm (Fig. 12b).

2.3 *Motor Selection and Control*

Most daily tasks performed by hand require 10.5 N of force [11]. In et al. [4] proposed an empirical relation between the normal force applied by fingertip F_n and artificial tendon tension F_t using an experiment as:

$$F_\text{n} = 0.38 F_\text{t} \tag{1}$$

The maximum force exerted by the flexor digitorum profundus (FDP) muscle is approximately 140 N [12]. Therefore, the actuator which can pull the artificial tendons with 140 N (FDP muscle) is chosen. As per Eq. 1, we expect our soft hand exoskeleton to deliver maximum normal fingertip force of $0.38 \times 140 = 53$ N, which is well above the magnitude needed for daily tasks.

Motors are manually controlled by double pole double throw (DPDT) switches, which is expected to be operated by user's healthy hand (if feasible).

3 Tests and Validation

The exoskeleton is tested on a healthy 22 year old male subject. The subject is instructed to keep his hands free and loose, during all tests. The exoskeleton is mounted on his left hand, and is operated by the subject himself using his right hand (Fig. 13a). The subject is able to make fist (Fig. 13b), as well as, grasp and lift objects of different shapes and sizes (Fig. 13e–h). The adaptability is evident from the fact, that, blocking motion of some fingers, does not hinder other fingers (Fig. 13c, d). As control of ab/ad-duction is absent, thumb position is adjusted manually to hold the objects properly. In these tests, forearm, elbow and shoulder are controlled by the subject himself. Subject experiences uneasiness at distal phalanx, where tendon attachment loops apply squeezing pressure (Fig. 14). Similar uneasiness is experienced at the contact of wrist with the mechanism box, and at contact of palm bracket on dorsal side of hand, as circled in Fig. 14.

4 Discussion

The proposed soft hand exoskeleton uses two actuators to control fourteen degrees of freedom in hand responsible for finger/thumb flexion. The device weighs 300 g and occupies $100 \times 56\,\text{mm}^2$ on forearm. It is lighter than most previous soft hand

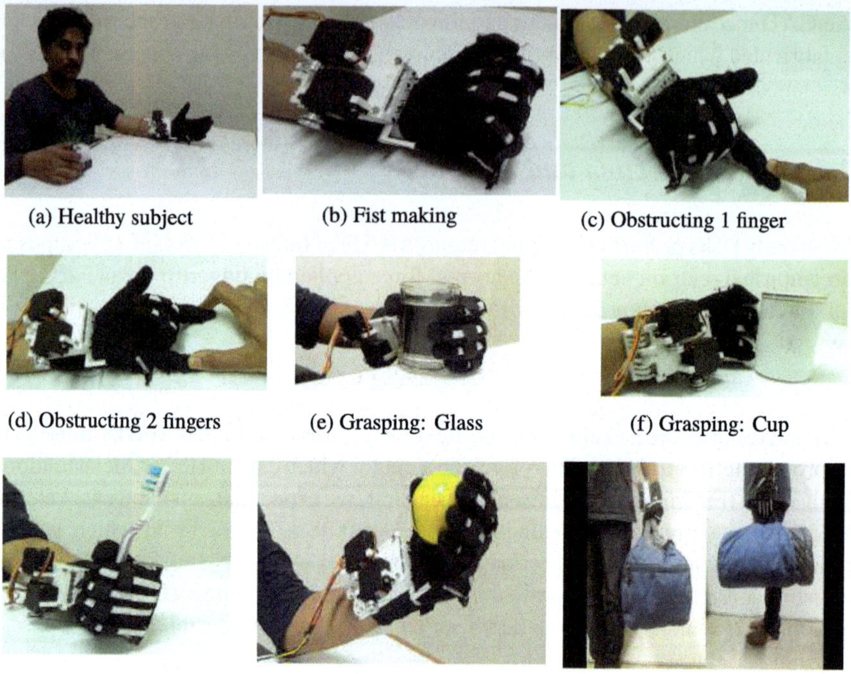

(a) Healthy subject (b) Fist making (c) Obstructing 1 finger

(d) Obstructing 2 fingers (e) Grasping: Glass (f) Grasping: Cup

(g) Grasping: Brush (h) Grasping: Ball (i) Lifting Bag (>1kg)

Fig. 13 Various tests performed by a healthy subject: Subject is instructed to keep his digits loose, and, operate the DPDT switch by his other hand

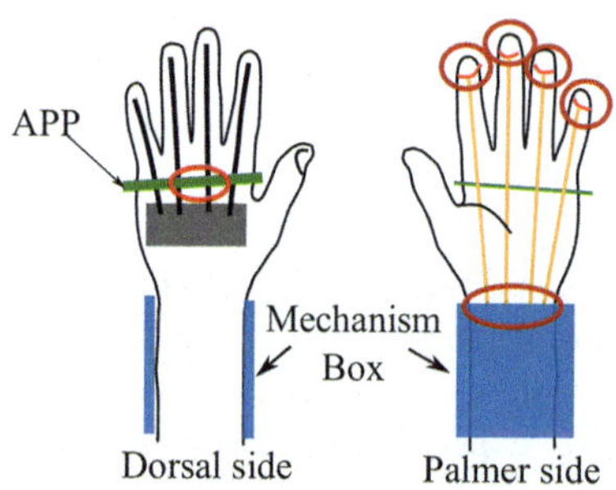

Fig. 14 Regions (circled) of uncomfortable contact forces

exoskeletons, e.g., 1 kg [3], 194 g (only glove part) [5]. Lightweight and compactness of the exoskeleton makes it easily portable, comfortable to wear and use. The device has low friction, as evident from the fact that all four fingers are actuated by a single motor. It was not possible in the design presented by Kang [5, 6] wherein 'U' shaped tubes were used which increased friction abruptly. In addition, one can further reduce the frictional force in this design by using pulleys with bearings.

The size of the existing pulley based differential mechanism, has been reduced considerably from 20 cm of length to 10 cm. Without the reduction, the corresponding mechanism would occupy most of the forearm area. The proposed reduction makes the device compact, and portable. The motor rotation required for the same flexion is reduced, whereas, motor torque requirement gets doubled. Number of pulleys has significantly increased from three to fourteen, which might increase friction significantly, if lubrication/bearings are not used in pulleys.

There are some limitations of this design. User can operate the exoskeleton using his/her healthy hand, if feasible. This is difficult, if both hands are disfunctional. As mechanism is underactuated, different independent control of fingers/joints is not possible. As a consequence, many activities cannot be performed, such as, squeezing of a spherical sponge ball, which requires flexion of DIP and MIP joint independent of PIP joint. Effective gripping of bigger spherical ball, which demands ab/ad-duction of fingers is also not possible. Grasping of objects, via fingers, of varied shapes and sizes is well achieved by using a single motor actuated mechanism. Implemented mechanism for flexion of thumb does not cover the complete range of motion required for gripping all type of objects. Similar to flexion, ab/ad-duction of thumb could be controlled via a pair of an elastic strap and an artificial tendon, connecting digit of thumb to an additional actuator. This would enable exoskeleton to grip a wide range of objects.

Assistance would be needed by an impaired subject to wear the glove. As the glove is close and opaque, it would be difficult for the subject to insert his/her impaired fingers into corresponding fingers of glove on his own. Semi-closed glove could be designed, enhancing visibility and reachablity of fingers, thus easing wearability.

When intentionally/accidentally finger(s) or/and wrist is/are flexed by subject himself or gets flexed by any external means, other than motor, the artificial tendons slack and become prone to slip off any mechanical pulley employed in mechanism. Provision of cover over pulley might resolve this issue.

The tendon lengths are subjected to user's hand. Instead of knotting the tendons to glove at distal phalanx, a better way should be explored, to tune the tendon lengths as per the user's finger length. User experiences uncomfortable contact forces at multiple locations (Fig. 14). Cushioning at these regions would enhance the user experience.

The presented soft hand exoskeleton is an attempt in the direction to provide a user friendly solution to people with impaired hand(s). The device must be tested on many patients to validate the feasiblity and to explore the unencountered challenges associated with it. Some possible modifications in this regard could be (i) if both hand are impaired, device could be controlled by some other means e.g. voice or leg, (ii) for people with wrist or/and elbow or/and shoulder impairment, the device could be coupled with other assisting devices. Use of elastic straps for extension of fingers and thumb is not a reliable solution in long run. Some alternative extensors may be explored.

5 Conclusion

Size reduction of the pulley-based differential mechanism allows us to develop a compact lightweight soft hand exoskeleton. The entire assembly occupies much smaller space (10 cm of length on forearm) than, without the reduction which would cover most of the forearm (atleast 20 cm of length). While the design provides a proof of concept for a wearable hand exoskeleton that exhibits adaptive grasping, further research is required before the proposed design becomes useful and amenable to the beneficiaries.

Acknowledgements Shyam Sunder Nishad and Vitthal Khatik, Ph.D. students at IIT Kanpur, are sincerely acknowledged for critically reviewing and significantly revising this manuscript.

References

1. Dollar AM, Howe RD (2010) The highly adaptive sdm hand: design and performance evaluation. Int J Robot Res 29(5):585–597
2. Mardula KL, Balasubramanian R, Allan CH (2015) Implanted passive engineering mechanism improves hand function after tendon transfer surgery: a cadaver-based study. Hand 10(1):116–122
3. Lee S, Landers KA, Park H-S (2014) Development of a biomimetic hand exotendon device (biomhed) for restoration of functional hand movement post-stroke. IEEE Trans Neural Syst Rehabil Eng 22(4):886–898
4. In H, Cho K-J, Kim K, Lee B (2011) Jointless structure and under-actuation mechanism for compact hand exoskeleton. In 2011 IEEE international conference on rehabilitation robotics (ICORR). IEEE, pp 1–6
5. In H, Kang BB, Sin M, Cho K-J (2015) Exo-glove: a wearable robot for the hand with a soft tendon routing system. IEEE Robot Autom Mag 22(1):97–105
6. Kang BB, Lee H, In H, Jeong U, Chung J, Cho K-J (2016) Development of a polymer-based tendon-driven wearable robotic hand. In 2016 IEEE international conference on robotics and automation (ICRA). IEEE, pp 3750–3755
7. Kang S, In H, Cho K-J (2012) Design of a passive brake mechanism for tendon driven devices. Int J Precis Eng Manuf 13(8):1487–1490
8. In H, Kang S, Cho K-J (2012) Capstan brake: passive brake for tendon-driven mechanism. In: 2012 IEEE/RSJ international conference on intelligent robots and systems (IROS). IEEE, pp 2301–2306
9. Birglen L, Laliberté T, Gosselin CM (2007) In: Siciliano B, Khatib O, Groen F (eds) Underactuated robotic hands in Springer tracts in advanced robotics, vol 40. Springer, Berlin, Heidelberg
10. Kamper DG, Hornby TG, Rymer WZ (2002) Extrinsic flexor muscles generate concurrent flexion of all three finger joints. J Biomech 35(12):1581–1589
11. Smaby N, Johanson ME, Baker B, Kenney DE et al (2004) Identification of key pinch forces required to complete functional tasks. J Rehabil Res Dev 41(2):215
12. Pollard NS, Gilbert RC (2002) Tendon arrangement and muscle force requirements for human-like force capabilities in a robotic finger. In: Proceedings of the IEEE international conference on robotics and automation, 2002 (ICRA'02), vol 4. IEEE, pp 3755–3762

Angle-Bound Smoothing with Applications in Kinematics

Josep M. Porta, Soheil Sarabandi, and Federico Thomas

Abstract Distance-bound smoothing is a method to characterize point configurations satisfying a given set of constraints on the pairwise ranges of distances between the points. This technique essentially consists in the iterative application of a set of algebraic conditions to reduce the given initial distance ranges. Although originally developed in a Computational Chemistry context, it has successfully been applied to problems of planar point configurations arising in Robotics, which includes the position analysis of planar robots. In this paper, this technique is generalized to deal with points on a sphere so that the planar case can now be deduced from the one presented here as a limit case. To exemplify the interest of this technique, it is applied to solve the position analysis of multiloop spherical linkages.

Keywords Spherical geometry · Bound smoothing · Spherical mechanisms

1 Introduction

Let us consider all pairwise distances between a set of points. If one of these distances is unknown, we can assign it the range $[0, \infty)$. If its value is estimated using a sensor, we can assign it the range $[d - \delta, d + \delta]$, where d is the measured distance and δ accounts for possible errors. If it corresponds to a perfectly calibrated distance, we can assign it a single-point range $[d, d]$. Since, in general, all these distances are not independent, their redundancy can be used to obtain better bounds for the unknown and estimated distances.

J. M. Porta (✉) · S. Sarabandi · F. Thomas
Institut de Robòtica i Informàtica Industrial (IRI), CSIC-UPC,
Llorens i Artigas 4-6, 08028 Barcelona, Spain
e-mail: porta@iri.upc.edu

S. Sarabandi
e-mail: ssarabandi@iri.upc.edu

F. Thomas
e-mail: f.thomas@csic.es

D. Sen et al. (eds.), *Mechanism and Machine Science*, Lecture Notes in Mechanical Engineering, https://doi.org/10.1007/978-981-15-4477-4_53

Pairwise distances between a set of points must satisfy constraints that depend on the space in which the points are embedded. For example, if they are embedded in a line, the distances between all subsets of three points must satisfy the triangular identity. In this case, we can easily come up with an iterative process in which the triangular identity is applied to improve the bounds between all subsets of three points until no further improvement is obtained. This *bound-smoothing* process can be extended to higher dimensions where the distances between the points must satisfy more sophisticated constraints [10]. For instance, it has been extended to the plane where sets of four points must satisfy the tetrangular identity, and also to the case in which orientation constraints on the formed triangles have to be considered [14]. Nevertheless, the only available generalizations to three dimensions [11, 12] are partial since complex constraints involving subsets of up to five points must be considered [6]. In this paper, we consider the case of points embedded on the surface of a sphere and hence the term *angle-bound smoothing* because the distances between points on a sphere are measured as lengths of arcs of great circles.

To our knowledge, no previous works exist for solving the stated problem. Nevertheless, the problem of propagating ranges of motion in multiloop spherical robots bears some resemblance with it. Some methods have been proposed in [1, 3] to perform this kind of propagations. They are based on processing a set of independent loop equations that completely characterize the kinematics of the analyzed robot. While expressing the problems tackled in these works as angle-bound smoothing problems is straightforward, expressing a general angle-bound smoothing problem so that it can be solved using the techniques presented in them is just not obvious, if even possible.

To have a rapid glimpse on how the method presented in this paper works for a problem in Robotics, let us consider the problem of solving the forward kinematics of the 3-DOF 3-$\underline{R}$RR spherical parallel manipulator appearing in Fig. 1(left) [7]. That is, the problem of locating the triangle $P_7P_8P_9$ with respect to the base $P_1P_2P_3$ as a function of the motor angles α_1, α_2, and α_3. Since triangle $P_4P_5P_6$ is located with respect to $P_1P_2P_3$ once these angles are known, the problem can be reformulated as that of finding all possible configurations for the set of points $\{P_4, \ldots, P_9\}$ on the unit sphere up to global rotations. The location of these points are constrained because some pairwise distances between them are known: those connected by a segment in Fig. 1(right). These constraints can be exploited to reduce the ranges of the rest of distances. Moreover, this smoothing process can be used in a branch-and-bound search scheme to isolate all the solutions [8, 9, 13]. In such scheme, the initial ranges define the search space including all the solutions of interest. This space is reduced as much as possible and then split and recursively processed until all solutions are bounded with a desired accuracy. This is in contrast with local Newton-like approaches which approximate a single solution.

In a branch-and-prune process, the split is trivial and, thus the relevant issue is to define a smoothing method that rapidly converges to the solution set. This paper proposes an efficient smoothing method for points on the sphere. This method is based on the monotonicity analysis of the trigonometric expression relating the distances between four points on a sphere, since the distances between them cannot

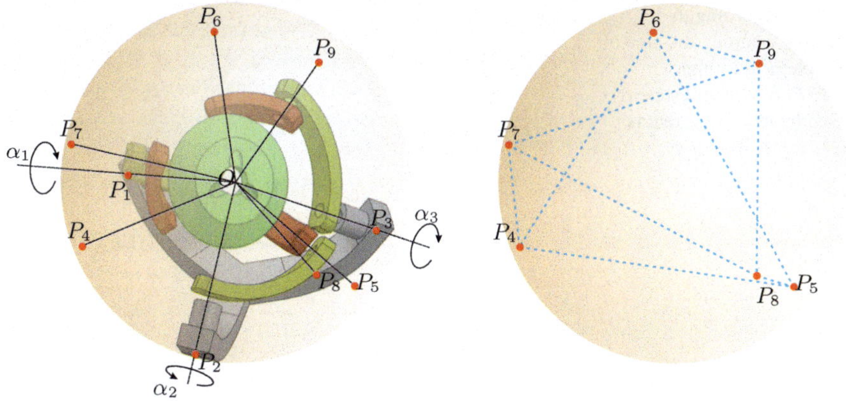

Fig. 1 A 3-DOF 3-RRR spherical parallel manipulator (left). The nine rotation axes intersect in the center of the mechanism. The corresponding semiaxes going through the revolute joints intersect with the unit sphere at $P_1, \ldots, P_9$. Once the rotation angles α_1, α_2, and α_3 about the motor axes are known, the location of the triangle $P_4 P_5 P_6$ is determined with respect to $P_1 P_2 P_3$. This implies that the forward kinematics of this robot essentially reduces to find all possible configurations for $\{P_4, \ldots, P_9\}$ given the shown pairwise distances (right)

be arbitrarily assigned. Fortunately, as we will see, we do not need to considered all possible subsets of four points. This means that only the bounds of a subset of unknown distances are progressively reduced, but they are enough to finally obtain the corresponding point configuration by trivial sequences of bilaterations.

This paper is structured as follows. Section 2 reviews some basic facts on spherical geometry that are needed in Sect. 3, where some new results on a family of Gramians—which can be seen as the spherical counterpart of Cayley–Menger determinants—are presented. Section 4 exploits these results to solve the position analysis of two multiloop spherical linkages. Finally, Sect. 5 summarizes the main results and points to directions deserving further attention.

2 Background

In spherical geometry, the shortest distance between two points, P_i and P_j, also known as the geodesic distance, is the length of an arc of a great circle containing both points. This great circle is the result of intersecting the sphere with the plane defined by the center of the sphere and the two points. Then, on the unit sphere the distance is $d(P_i, P_j) = \cos^{-1}\langle \mathbf{p}_i, \mathbf{p}_j \rangle = \theta_{ij}$, the angle between the vectors from the center to points P_i and P_j. Here $\langle \mathbf{p}_i, \mathbf{p}_j \rangle$ is the standard Euclidean inner product.

A spherical triangle is a three-sided polygon on a sphere whose sides are arcs of great circles. The spherical excess E of the spherical triangle $P_1 P_2 P_3$ (see Fig. 2) is given by

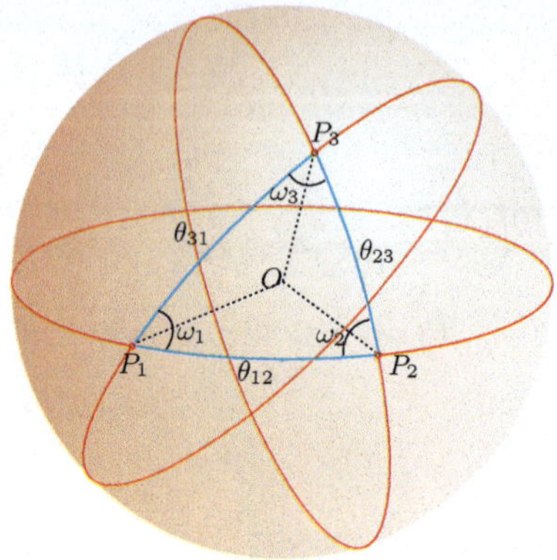

Fig. 2 According to Girard's theorem, the area of the spherical triangle $P_1 P_2 P_3$ on the unit sphere is given by its excess, that is $\omega_1 + \omega_2 + \omega_3 - \pi$

$$E = \omega_1 + \omega_2 + \omega_3 - \pi, \tag{1}$$

which, according to Girard's theorem, coincides with the area of the triangle for unit spheres [15, § 127].

The triangle inequality between three points on a sphere holds provided that the distance between any two points is the lowest of the two arcs of great circle connecting them [5, p. 194]. If we exclude from our analysis those cases in which (a) two points are antipodal because, in this case, there are infinite arcs of great circle realizing the minimum distance; and (b) two points are coincident, the considered distances will always lie in the range $(0, \pi)$. In what follows, independently of the number of points characterizing a problem, we will always work with subsets of four points which will be limited in the following way.

Remark 1 We will only consider sets of four points whose pairwise distances lie in the range $(0, \pi)$.

As a consequence of this limitation, the correspondence between θ_{ij} and $\cos\theta_{ij}$ becomes one-to-one. Thus, as a matter of convenience, we will indistinctly use θ_{ij} or $\cos\theta_{ij}$ when referring to the distance between P_i and P_j.

3 New Results on Gramians on the Unit Sphere

Given the location vectors, $\mathbf{p}_1, \ldots, \mathbf{p}_n$, of points $P_1, \ldots, P_n$, the Gram determinant, or Gramian, is the determinant defined as

$$G(1, \ldots, n) = \begin{vmatrix} \langle \mathbf{p}_1, \mathbf{p}_1 \rangle & \langle \mathbf{p}_1, \mathbf{p}_2 \rangle & \ldots & \langle \mathbf{p}_1, \mathbf{p}_n \rangle \\ \langle \mathbf{p}_2, \mathbf{p}_1 \rangle & \langle \mathbf{p}_2, \mathbf{p}_2 \rangle & \ldots & \langle \mathbf{p}_2, \mathbf{p}_n \rangle \\ \vdots & \vdots & \ddots & \vdots \\ \langle \mathbf{p}_n, \mathbf{p}_1 \rangle & \langle \mathbf{p}_n, \mathbf{p}_2 \rangle & \ldots & \langle \mathbf{p}_n, \mathbf{p}_n \rangle \end{vmatrix}, \tag{2}$$

which, in the particular case in which all points lie on the unit sphere, reduces to

$$G(1,\ldots,n) = \begin{vmatrix} 1 & \cos\theta_{12} & \ldots & \cos\theta_{1n} \\ \cos\theta_{12} & 1 & \ldots & \cos\theta_{2n} \\ \vdots & \vdots & \ddots & \vdots \\ \cos\theta_{1n} & \cos\theta_{2n} & \ldots & 1 \end{vmatrix}. \quad (3)$$

Gramians are zero if, and only if, the involved coordinate vectors are linearly dependent, and strictly positive otherwise [4, p. 251]. Negative Gramians only arise if the given distances do not correspond to any configuration of real points. Since we limit our analysis to a sphere in three dimensions, a Gramian of four points necessarily vanishes. For the same reason, a Gramian of three points lying on a great circle will also vanish. Next, we analyze the three-and four-point cases in more detail.

Lemma 1 *The Gramian of three points, P_1, P_2, and P_3, on the unit sphere can be expressed as*

$$G(1,2,3) = \sin(p)\sin(p-\theta_{12})\sin(p-\theta_{13})\sin(p-\theta_{13}), \quad (4)$$

where p is the semiperimeter of the triangle they define, that is,

$$p = \frac{\theta_{12}+\theta_{23}+\theta_{13}}{2}.$$

Proof 1 The direct expansion of the Gramian yields:

$$G(1,2,3) = 2\cos\theta_{12}\cos\theta_{13}\cos\theta_{23} \\ -\cos^2\theta_{12} - \cos^2\theta_{13} - \cos^2\theta_{23} + 1.$$

Then, expressing the cosine functions in terms of sines, and simplifying the result, the lemma follows.

Observe the resemblance of (4) with Heron's formula for the computation of the area of a triangle given its edge lengths. This suggests a connection between the Gramian of three points on the unit sphere and the area of the spherical triangle they define. Indeed, using Cagnoli's theorem [15, § 132], it is straightforward to conclude that

$$G(1,2,3) = 4\cos^2\frac{\theta_{12}}{2}\cos^2\frac{\theta_{23}}{2}\cos^2\frac{\theta_{13}}{2}\sin^2\frac{E}{2}, \quad (5)$$

where E is the excess of the triangle or, equivalently for a sphere of unit radius, its area.

Since the Gramian of four points vanishes, any of the pairwise distances between these four points can be expressed in terms of the others. The explicit formula is given in the next lemma.

Lemma 2 *Given four points, P_1, P_2, P_3, and P_4, on the unit sphere, the distance between P_3 and P_4 can be expressed in terms of all other distances between these four points as*

$$\cos\theta_{34}^{\pm} = \frac{-B_{34} \pm \sqrt{B_{34}^2 + A_{34}C_{34}}}{A_{34}}, \tag{6}$$

where

$$A_{34} = \begin{vmatrix} 1 & \cos\theta_{12} \\ \cos\theta_{12} & 1 \end{vmatrix},$$

$$B_{34} = \begin{vmatrix} 1 & \cos\theta_{12} & \cos\theta_{13} \\ \cos\theta_{12} & 1 & \cos\theta_{23} \\ \cos\theta_{14} & \cos\theta_{24} & 0 \end{vmatrix},$$

$$C_{34} = \begin{vmatrix} 1 & \cos\theta_{12} & \cos\theta_{13} & \cos\theta_{14} \\ \cos\theta_{12} & 1 & \cos\theta_{23} & \cos\theta_{24} \\ \cos\theta_{13} & \cos\theta_{23} & 1 & 0 \\ \cos\theta_{14} & \cos\theta_{24} & 0 & 1 \end{vmatrix}.$$

Proof 2 It can be verified that

$$G(1, 2, 3, 4) = -A_{34}\cos^2\theta_{34} - 2B_{34}\cos\theta_{34} + C_{34}. \tag{7}$$

Then, since in our case the Gramian of four points vanishes, the lemma directly follows.

The sign of the square root in (6) accounts for the two possible solutions for θ_{34}, $\theta_{3,4}^{+}$ and $\theta_{3,4}^{-}$ (see Fig. 3). While the negative sign corresponds to the configuration in which P_3 and P_4 lie on the same hemisphere with respect to the great circle defined by P_1 and P_2 (*cis* configuration), the positive one corresponds to the configuration in which P_3 and P_4 lie on different hemispheres (*trans* configuration).

To simplify the presentation of the current version of our algorithm, we also introduce the following limitation.

Remark 2 We will not consider orientation constraints for spherical triangles.

As a consequence of this, any two triangles sharing an edge could indistinctly be in cis or trans configuration.

Observe that $\theta_{3,4}^{+} = \theta_{3,4}^{-}$ if the discriminant in (6) vanishes. Clearly, this only happens when any of the two triangles $P_1P_2P_4$ and $P_1P_2P_3$ degenerate (their areas are null). This suggests, according to (5), that this discriminant is directly connected to $G(1, 2, 4)$ and $G(1, 2, 3)$. This connection is established in the following lemma.

Lemma 3 *The discriminant in* (6) *can be expressed as*

$$D_{34} := B_{34}^2 + A_{34}C_{34} = G(1, 2, 4)\, G(1, 2, 3). \tag{8}$$

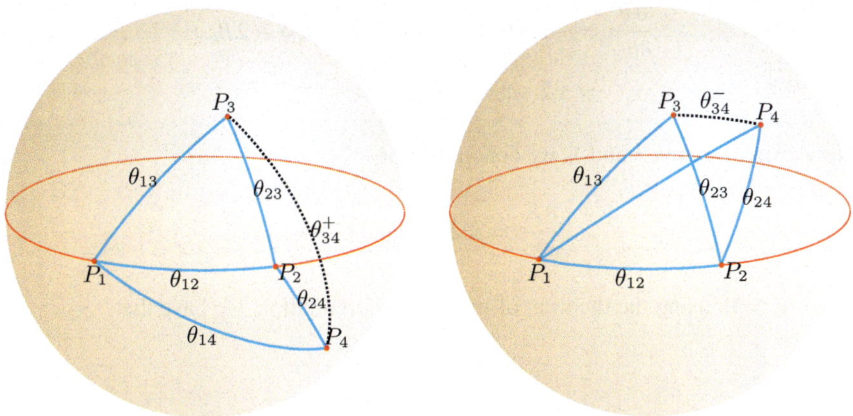

Fig. 3 Given four points on the unit sphere, P_1, P_2, P_3, and P_4, the distance between, say (without loss of generality) P_3 and P_4, is determined by all other distances. In general, two solutions are possible which correspond to the cis (left drawing) and trans configuration (right drawing)

Proof 3 The proof can be obtained factorizing the discriminant using a computer algebra system.

The analysis of the monotonicity of either $\theta_{3,4}^{+}$ or $\theta_{3,4}^{-}$ by computing the derivatives of (6) is an uphill task. Fortunately, the following theorem gives us the clue to simplify this analysis.

Theorem 1 *Given points P_1, P_2, P_3, and P_4 on the unit sphere, then*

$$\frac{\partial\theta_{kl}}{\partial\theta_{ij}} = \pm\frac{\sin\theta_{ij}}{\sin\theta_{kl}}\sqrt{\frac{G_i\ G_j}{G_k\ G_l}}$$

for $i, j, k, l \in \{1, 2, 3, 4\}$ *and where* $G_1 = G(2, 3, 4)$, $G_2 = G(1, 3, 4)$, $G_3 = G(1, 2, 4)$, *and* $G_4 = G(1, 2, 3)$.

Proof 4 Let us denote $F := G(1, 2, 3, 4)$. Then, using the chain rule, the derivative of F with respect to θ_{kl} can be expressed as

$$\frac{\partial F}{\partial\theta_{kl}} = \frac{\partial F}{\partial\cos\theta_{kl}}\ \frac{\partial\cos\theta_{kl}}{\partial\theta_{kl}},$$

which, after computing the derivatives, according to expression (7), reads as

$$\frac{\partial F}{\partial\theta_{kl}} = -\sin\theta_{kl}(-2A_{kl}\cos\theta_{kl} - 2B_{kl}).$$

Then, using Lemma 2, it can be rewritten as

$$\frac{\partial F}{\partial \theta_{kl}} = -\sin\theta_{kl}[-2(-B_{kl} \pm \sqrt{D_{kl}}) - 2B_{kl}]$$
$$= \pm 2\ \sin\theta_{kl}\sqrt{D_{kl}}.$$

Now, using Lemma 3 and 1, we conclude that

$$\frac{\partial F}{\partial \theta_{kl}} = \pm 2\ \sin\theta_{kl}\sqrt{G_k\, G_l}.$$

Since $F = 0$, using the theorem of implicit differentiation, we have that

$$\frac{\partial \theta_{kl}}{\partial \theta_{ij}} = -\frac{\partial F/\partial \theta_{ij}}{\partial F/\partial \theta_{kl}}, \tag{9}$$

from which the theorem follows.

Corollary 1 *Given points P_i, P_j, P_k, and P_l on the unit sphere, the distance between P_k and P_l, θ_{kl}, is monotonic with respect to all other distances between these four points provided that no alignment on the sphere is generated (no subset of three points lie on a great circle).*

Proof 5 Since, according to Remark 1, $\sin\theta_{ij}$ and $\sin\theta_{kl}$ are always positive, then according to Lemma 1

$$G_i = G(j, k, l)$$
$$= \sin(p_{jkl})\sin(p_{jkl}-\theta_{jk})\sin(p_{jkl}-\theta_{jl})\sin(p_{jkl}-\theta_{kl}),$$

with $p_{jkl} = (\theta_{jk} + \theta_{kl} + \theta_{jl})/2$ the semiperimeter of the triangle $P_j P_k P_l$. Whenever any of the arguments of the sine functions in G_i includes the value $r\,\pi$ with $r \in \mathbb{Z}$, there is a change in the sign of the derivative $\partial\theta_{kl}/\partial\theta_{ij}$. Thus, the change in the sign of the derivative occurs when

$$p_{jkl} = \pi \Rightarrow \theta_{jk} + \theta_{jl} + \theta_{kl} = 2\pi,$$
$$p_{jkl} - \theta_{jk} = 0 \Rightarrow -\theta_{jk} + \theta_{jl} + \theta_{kl} = 0,$$
$$p_{jkl} - \theta_{jl} = 0 \Rightarrow \theta_{jk} - \theta_{jl} + \theta_{kl} = 0,$$
$$p_{jkl} - \theta_{kl} = 0 \Rightarrow \theta_{jk} + \theta_{jl} - \theta_{kl} = 0.$$

These conditions correspond to alignments of P_i, P_j, and P_k. The same holds for G_j, G_k, and G_l.

Remark 3 We assume that in the considered subsets of four points no three points can get aligned in the given distance domains.

This limitation is introduced to simplify the presentation of the technique described in the next section to compute exact bounds on the pairwise distances between four points.

4 Examples

The angle-bound smoothing approach introduced in this paper has been evaluated on the two spherical mechanisms shown in Fig. 4. The first one is a 3-DOF 3-RRR, the robot described in the introduction. This robot has been used in [1] and [3] to evaluate previous approaches dealing with spherical robots. The second test case is a four-loop kinematic linkage and it has been considered before in [2]. In this case, previous approaches relying on interval propagations over kinematic loops [1, 3] may have difficulties converging to the solution due to the number of loops in the problem. Actually, such approaches have been only validated on mechanisms involving up to two loops.

Since due to Remark 3, we are interested in proving the convergence properties of the angle smoothing procedure without considering degeneracies, i.e., arc alignments which typically appear in symmetric robots and configurations, the problems are defined at random. For each of them, the required number of points are sampled from a uniform distribution on a unit-radius sphere. Such set of points is a valid solution for the problem at hand and the variable arc lengths are defined as intervals around this solution. The size of such intervals is as large as possible, but without activating any of the alignment conditions. In this way, the initial search space is a box in the space of cosines of the variable arc lengths where all the functions (6) are monotonic.

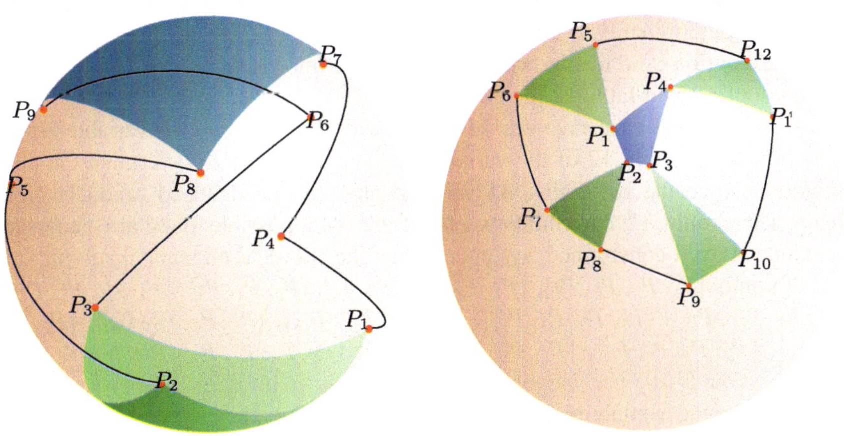

Fig. 4 The topology of the two mechanisms used to evaluate the angle-bound smoothing approach introduced in this paper. Left: A 3-DOF 3-RRR spherical robot. Right: A four-loop spherical linkage

If a function is continuous and locally monotonic with respect to each argument in a given domain, its extrema are in the boundary of the domain. Moreover, if the domain is an axis-aligned box, finding exact bounds without overestimation simply reduces to evaluate this function in all the corners of the box [14]. Thus, the range of one variable can be exactly computed taking the extremes of the evaluation of (6) in the corners of the box defined by the rest of variables involved in the determinant. This typically reduces the bounds for the variable of interest. The iteration of this procedure for all the variables in the considered determinants reduces the initial search box until the solution is bounded with the required accuracy.

The spherical 3-RRR problem involves 6 points, 9 fix-length arcs and 6 arcs whose length vary with the configuration, but one of them can be deduced uniquely once the other 5 are fixed. In this example, the values for the unknowns are enlarged with an amplitude of about 10–15° around the randomly generated solution. Fifteen determinants involving 4 points can be defined over 6 points, but only 5 of them are necessary to obtain a well-constrained system. Note that using a small set of determinants not only has performance advantages, but it also increases the size of the monotonic areas since only the alignments between the points in the considered determinants are relevant. The determinants that we used in this case are the ones involving the sets of points $\{P_4, P_5, P_7, P_8\}$, $\{P_4, P_7, P_8, P_9\}$, $\{P_4, P_6, P_7, P_9\}$, $\{P_6, P_7, P_8, P_9\}$, and $\{P_5, P_6, P_7, P_9\}$. Note that each determinant includes points fulfilling Remark 1. The first determinant involves two variable arc lengths, θ_{48} and θ_{57}, the next three determinants introduce one new variable each (θ_{49}, θ_{67}, and θ_{69}, respectively), and the last determinant is defined over previously introduced variables. In this way, the uncertainty in θ_{48} is used to reduce the range of the rest of variables in sequence until the last determinant, which closes the cycle and reduces the range for θ_{48}. The iteration of this procedure reduces the error in the solution approximation as shown in Fig. 5. The error is computed as the sum of the variable ranges in the problem and each iteration corresponds to the application of the range reduction with the considered determinants. It can be seen that in few steps the error is canceled. The overall execution time for the whole angle bound smoothing process is about 0.3 s in a non-optimized Matlab implementation.

The second test case involves 12 points with 20 fixed and 46 variable-length arcs. In this case only 12 of the variable arcs are necessary to determine the configuration since the remaining 34 variable arcs can be deduced uniquely from them. Then, only 12 determinants out of the 495 possible ones are necessary to define a well-constrained system. Using the 12 determinants involving the sets of points $\{P_1, P_2, P_6, P_7\}$, $\{P_1, P_2, P_5, P_6\}$, $\{P_1, P_2, P_4, P_5\}$, $\{P_1, P_4, P_5, P_{12}\}$, $\{P_1, P_4, P_{11}, P_{12}\}$, $\{P_1, P_3, P_4, P_{11}\}$, $\{P_3, P_4, P_{10}, P_{11}\}$, $\{P_3, P_4, P_9, P_{10}\}$, $\{P_2, P_3, P_4, P_9\}$, $\{P_2, P_3, P_8, P_9\}$, $\{P_2, P_3, P_7, P_8\}$, and $\{P_1, P_2, P_3, P_7\}$, the initial error, which is about 5–10° per variable, is reduced as shown in Fig. 6. This process takes about 0.9 s in the Matlab implementation. As in the previous case, the error rapidly reduces and in a few iterations the solution included in the initial box is accurately bounded. Figure 6-bottom shows that the convergence ratio is higher than the one for the previous test case (see Fig. 5-bottom), which is remarkable. This indicates a good scalability of the proposed angle-bound smoothing technique. The shape of

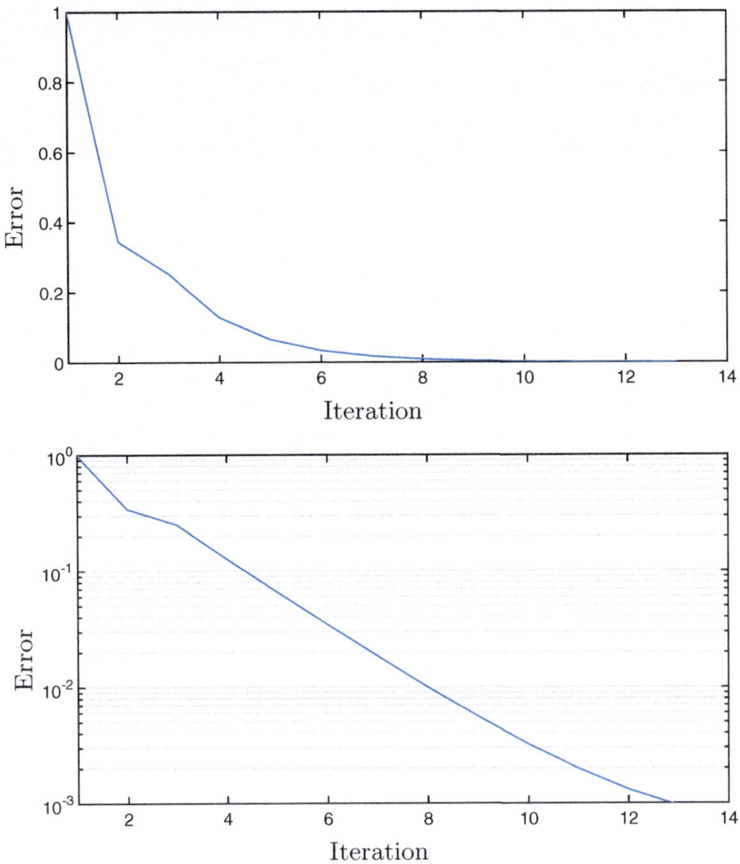

Fig. 5 Error reduction in the 3-RRR robot. Top: In linear scale. Bottom: In logarithmic scale to better appreciate the convergence ratio

the triangles involved in each propagation step seem to have influence in the convergence ratio, i.e., the closer to an alignment the slower the convergence. However, the detailed analysis of the factors that have influence on the convergence is left as an open issue.

5 Conclusions

This paper extends the distance-bound smoothing technique to the case of points on a sphere. It introduces the necessary tools to generate constraints between pairwise distances involving sets of four points on a sphere. When iteratively applied, such constraints allow bounding the solutions with the required accuracy. Moreover, the

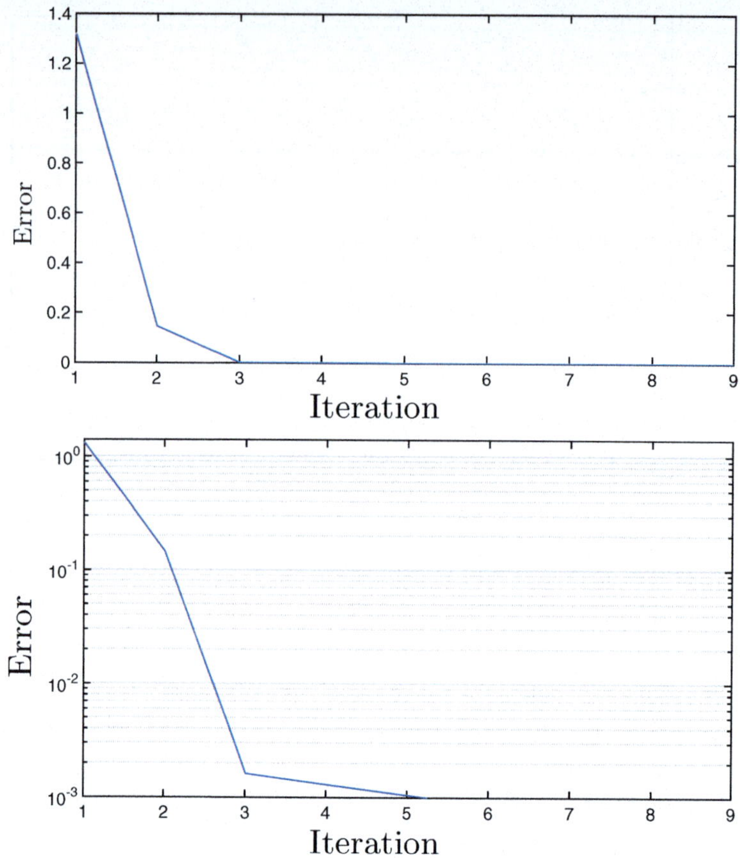

Fig. 6 Error reduction in the four-loop spherical linkage. Top: In linear scale. Bottom: In logarithmic scale to better appreciate the convergence ratio

results presented here characterize the monotonicity regions for the given constraints, which provides a simpler method to determine tight ranges for a given variable arc.

From this starting point, several extensions are possible. The tools introduced in this paper can be used to implement a full monotonicity analysis parallel to the one described in [14] for the planar case. This essentially requires to exploit the equivalent of Stewart's theorem for the sphere. Moreover, the inclusion of orientation constraints is also feasible by considering the geometric interpretation of the determinants introduced in this paper. Finally, the presented approach gives some clues on how to extend the proposed technique to points in 3D. In this way a 3D-bound smoothing method with orientation constraints could be derived. This is a fundamental, but still missing, tool in the field.

Acknowledgements This work has been partially supported by the Spanish Ministry of Economy and Competitiveness through projects DPI2014-57220-C2-2-P, DPI2017-88282-P, and MDM-2016-0656.

References

1. Bombín C, Ros L, Thomas F (2001) On the computation of the direct kinematics of parallel spherical mechanisms using Bernstein polynomials. IEEE International Conference on Robotics and Automation, pp 3332–3337
2. Borrás J, Di Gregorio R (2008) Direct position analysis of a large family of spherical and planar parallel manipulators with four loops. Second International Workshop on Fundamental Issues and Future Research Directions for Parallel Mechanisms and Manipulators. pp 19–29
3. Celaya E (2002) Interval propagation for solving parallel spherical mechanisms. In: Lenarcic J, Thomas F (eds) Advances in robot kinematics. Springer, pp 415–422
4. Gantmacher FR (1984) The theory of matrices, vol 1. Chelsea Pub Co
5. Han L, Rudolph L (2008) Simplex-tree based kinematics of foldable objects as multi-body systems involving loops. Robotics: Science and Systems IV
6. Havel T (1995) Distance geometry. Wiley, New York
7. Kong X, Gosselin CM (2004) Type synthesis of 3-DOF spherical parallel manipulators based on screw theory. ASME J. Mech Des 126(1):101–108
8. Porta JM, Thomas F, Ros L, Torras C (2003) A branch-and-prune algorithm for solving systems of distance constraints. IEEE International Conference on Robotics and Automation. pp 342–348
9. Porta JM (2005) CuikSlam: a kinematics-based approach to SLAM. IEEE international conference on robotics and automation. pp 2425–2431
10. Porta JM, Ros L, Thomas F, Corcho F, Cantó J, Pérez JJ (2007) Complete maps of molecular loop conformational spaces. J Comput Chem 28(13):2170–2189
11. Porta JM, Rull A, Thomas F (2016) Sensor localization from distance and orientation constraints. Sensors
12. Porta JM, Thomas F (2017) Closed-Form position analysis of variable geometry trusses. Mech Mach Theory 109:14–21
13. Rosales C, Porta JM, Suarez R, Ros L (2008) Finding all valid hand configurations for a given precision grasp. IEEE International Conference on Robotics and Automation. pp 1634–1640
14. Rull A, Porta JM, Thomas F (2014) Distance-bound smoothing under orientation constraints. IEEE International Conference on Robotics and Automation. pp 1431–1436
15. Todhunter I (1914) Spherical trigonometry. McMillan & Co, London

A New Sitting-Type Lower-Limb Rehabilitation Robot Based on a Spatial Parallel Kinematic Machine

J. K. Mohanta, S. Mohan, P. Wenger, and C. Chevallereau

Abstract This paper proposes a lying/sitting type lower-limb rehabilitation robot based on a three degrees of freedom spatial parallel kinematic manipulator namely "The Orthoglide" along with an actuated rotational degree of freedom at the end-effector for the purpose of lower-limb rehabilitation treatments. This rehabilitation robot is an end-effector/foot-plate based mechanism which controls the ankle-joint movements to assist physiotherapists in performing therapeutic treatments along with a passive orthosis (supporting system). The conceptual and detail design of the rehabilitation device is presented along with its kinematics. Further, the functional validation of the proposed robot in terms of the kinematic motion capabilities to provide physiological motions like knee flexion-extension, hip adduction-abduction and hip flexion-extension are also investigated. The required safety features for the proposed rehabilitation robot in terms of mechanical constraints and non-actuated joints are also discussed.

Keywords Lower limb rehabilitation robot · Orthoglide · Foot-plate based · Parallel manipulator · Passive orthosis

J. K. Mohanta (✉)
Department of Electrical Engineering, IIT Kanpur, Kanpur, India
e-mail: jkmjayant@gmail.com

S. Mohan
Department of Mechanical Engineering, IIT Palakkad, Palakkad, India
e-mail: santhakumar@iitpkd.ac.in

P. Wenger · C. Chevallereau
Laboratoire des Sciences du Numérique de Nantes (LS2N), CNRS, Nantes, France
e-mail: Philippe.Wenger@ls2n.fr

C. Chevallereau
e-mail: Christine.Chevallereau@ls2n.fr

D. Sen et al. (eds.), *Mechanism and Machine Science*, Lecture Notes in Mechanical Engineering, https://doi.org/10.1007/978-981-15-4477-4_54

1 Introduction

Lower limb treatments using mechanized systems has improved the speed of recovery of patients from the bedridden state [1]. The lower limb impairments are caused by motor nerves dysfunction, which are generally caused by spinal cord injury (SCI), stroke and accidental disability. Continuous physical motion of the impaired limb, known as motor rehabilitation, required for this process [2]. These are done in numerous ways for example with a body weight support (BWS), using supportive orthosis, using foot-pate motion and limb motion in sitting/lying posture. The sitting/lying treatments are appropriate for the bedridden patients. Example of sitting/lying type systems are the ANKUR-LL2 [8], an improved Lambda [6], the Supine [4], Physiotherabot [3], the Lambda [5] and the ANKUR-LL1 [7].

Lower limb mechanisms with high mechanical advantage joints, outfits the system with a self-locking capability, which helps in case of emergency stop [9]. Structures of the mechanisms consist of bulky actuators and system appears dynamically heavy. To avoid such difficulty a passive light-weight orthosis, with no actuators, is used along with joint feedback sensors [6]. Here the orthosis is indirectly controlled.

Rehabilitation devices are generally having motion in sagittal plane [1]. Although many therapies need off-planar movements but complexity of the arrangement and safety of the patient is major concern. In this direction a device namely Hipbot [10] is designed to give off-planar motions to the lower limbs, when the patient is in lying posture. Still Hipbot is having limitation that it can give adduction-abduction away from the body not towards the body because of physical limitations. To address this issue a new manipulator configuration is proposed. The major difficulty in off-planar movement is movement of the actuator assembly as they are mounted in single plane. So, here we have used a parallel manipulator based mechanism with passive orthosis which can render four degrees of freedom to the lower limb.

The rest of the paper is organized as follows: Sect. 2 discusses the architecture of the proposed rehabilitation robot and its geometrical aspects. Section 3 describes the system kinematics, Sect. 4 demonstrates the physiological motion validation of the proposed manipulator and Sect. 5 concludes the paper with safety aspects and scope of future works.

2 Proposed Manipulator's Architecture

The proposed rehabilitation robot shown in Fig. 1 consists of a driving mechanism and a passive orthosis. The driving mechanism is made using the concept of a parallel manipulator namely "The Orthoglide" [11]. Instead of an overhead structure a modified version with actuators on ground is taken for usage in rehabilitation purpose. Here three linear joints are placed orthogonally and three links with each leg having configuration PUU is connecting to the end-effector. The orthosis consists of three links corresponding to the thigh, crus and foot of the limb with a vertical

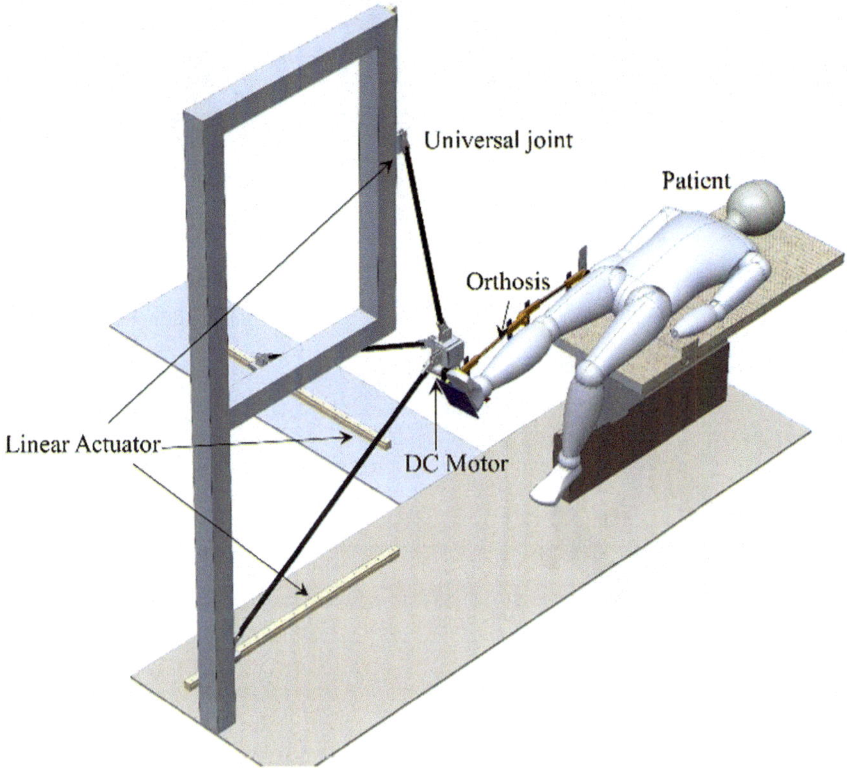

Fig. 1 Proposed lower limb rehabilitation robot

hip joint. These three links are joined in series at the hip, knee and ankle positions, with rotary joint. The solid model of the proposed orthosis is shown in Fig. 2. The passive orthosis can be seen in Fig. 1. Forward kinematics for the passive orthosis is as follows:

$$\begin{aligned} x &= \left(L_{thigh}\cos\phi_1 + L_{crus}\cos(\phi_1 - \phi_2)\right)\cos\phi_3 \\ y &= \left(L_{thigh}\cos\phi_1 + L_{crus}\cos(\phi_1 - \phi_2)\right)\sin\phi_3 \\ z &= L_{thigh}\sin\phi_1 + L_{crus}\sin(\phi_1 - \phi_2) \end{aligned} \quad (1)$$

Here ϕ_1, ϕ_2 and ϕ_3 are the hip flexion-extension, knee flexion-extension and hip adduction-abduction angles, which are shown in Fig. 2. L_{thigh} and L_{crus} are the length of the thigh support and crus support. Where (x, y, z) are the coordinate location of the ankle joint. The required workspace based on the physiological motion specifications are plotted and shown in Fig. 3. Here the reachable workspace is shown by considering the hip flexion-extension joint as origin. The possible joint movements on the sagittal plane for the human limbs are as: $-30^\circ \leq \phi_1 \leq 60^\circ$,$0^\circ \leq \varphi_2 \leq 105^\circ$

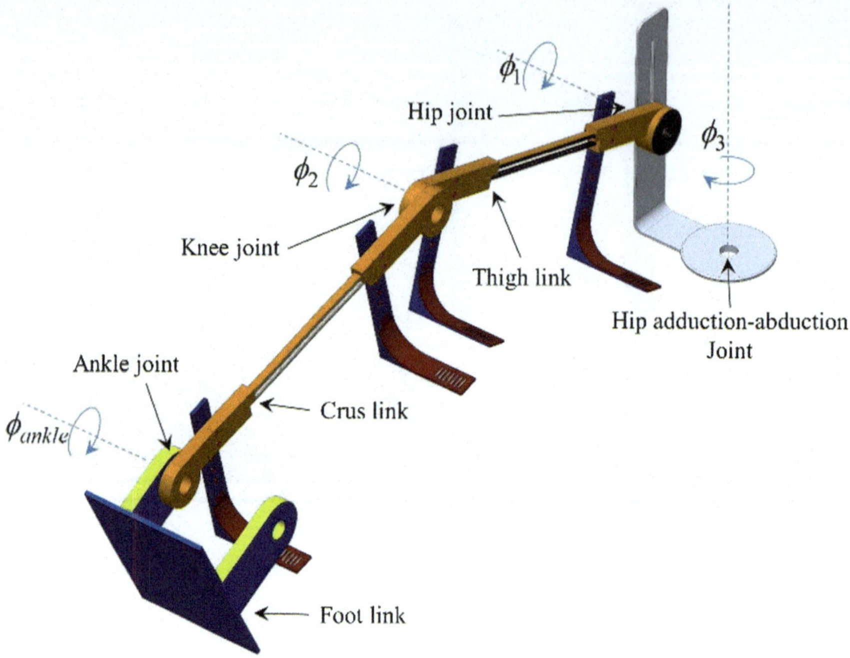

Fig. 2 Passive orthosis

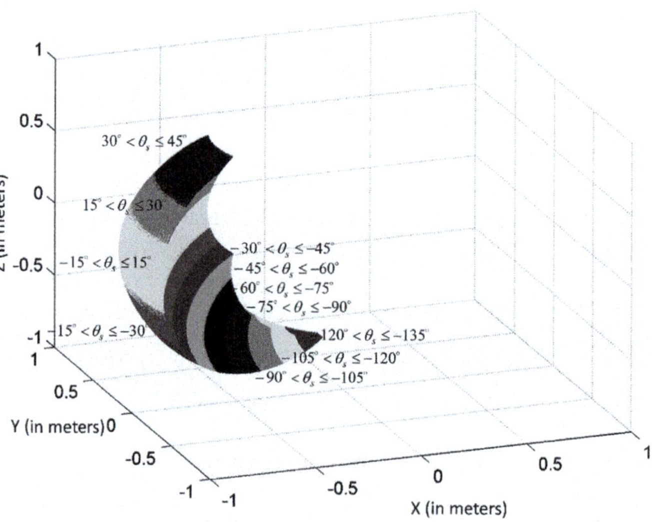

Fig. 3 Distribution of the required workspace area for ankle joint

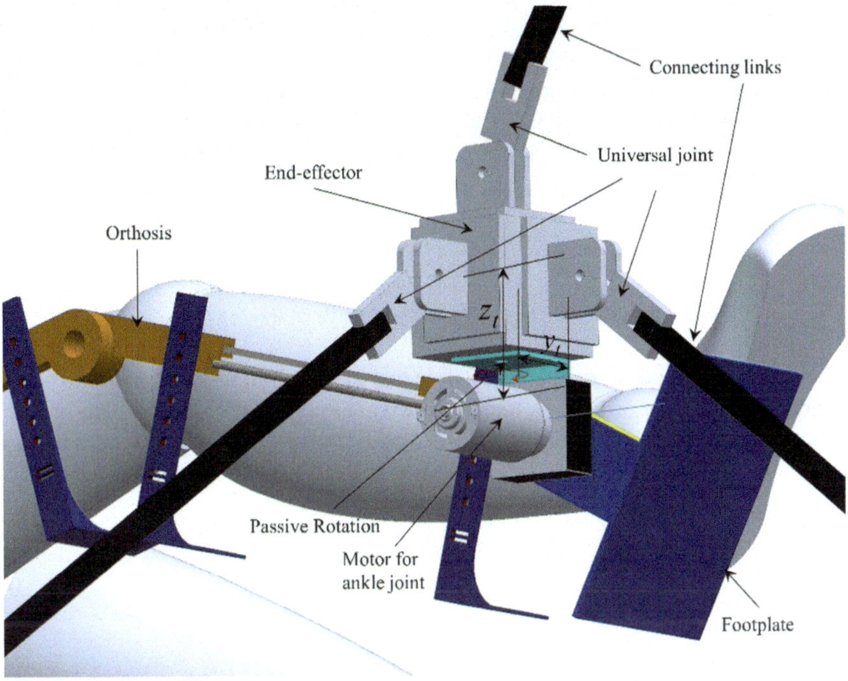

Fig. 4 End-effector orientation arrangement

and $-15^\circ \leq \phi_3 \leq 15^\circ$. In the Fig. 3, $\theta_S = \phi_1 - \phi_2$, shows orientation of the crus link at the ankle joint. Here, the ankle rotation is controlled by an independent actuator which is placed at the ankle joint. The angle requirement for the foot link at the end effector is. $\theta_s + \phi_{ankle}$.

Figure 4 shows the junction end-effector arrangement for the system, where the orthosis ankle joint and Orthoglide's end-effector is connected. The orthosis ankle joint is mounted on a body. This body has a passive vertical rotary joint at the bottom of the Orthoglide's end-effector's cube. This enable the orthosis not to hinder the hip adduction-abduction motion.

3 Manipulator Kinematics

For the Orthoglide kinematic solution can be determined by the following relations

$$
\begin{aligned}
(x - x_1)^2 + y^2 + z^2 &= l^2 \\
x^2 + (y - y_2)^2 + z^2 &= l^2 \\
x^2 + y^2 + (z - z_3)^2 &= l^2
\end{aligned} \tag{2}
$$

Here, $x_1 = r_1 - d_1$, $y_2 = r_2 + d_2$ and $z_3 = r_3 + d_3$. Where r_1, r_2 and r_3 are the joint space variable. Hence our joint space vector consist of $[r_1 r_2 r_3 \theta_{ankle}]^{\mathrm{T}}$ and task space consists of $[\phi_1 \phi_2 \phi_3 \phi_{ankle}]^{\mathrm{T}}$

Inverse kinematics for the Orthoglide part can be written as follows:

$$r_1 = \sqrt{l^2 - y_o^2 - z_o^2} + x_o + d_1$$
$$r_2 = \sqrt{l^2 - x_o^2 - z_o^2} - y_o - d_2$$
$$r_3 = \sqrt{l^2 - x_o^2 - y_o^2} + z_o - d_3 \quad (3)$$

where the x_o, y_o and z_o are the Orthoglide end-effector point (Fig. 5).

Kinematics of the orthosis system is shown in Eq. (1) relation between the orthosis coordinates to the to the end-effector coordinates of the Orthoglide can be written as

$$x_o = x - p_x - y_t \sin\phi_3$$
$$y_o = y - p_y - y_t \cos\phi_3$$
$$z_o = z + p_z - z_t \quad (4)$$

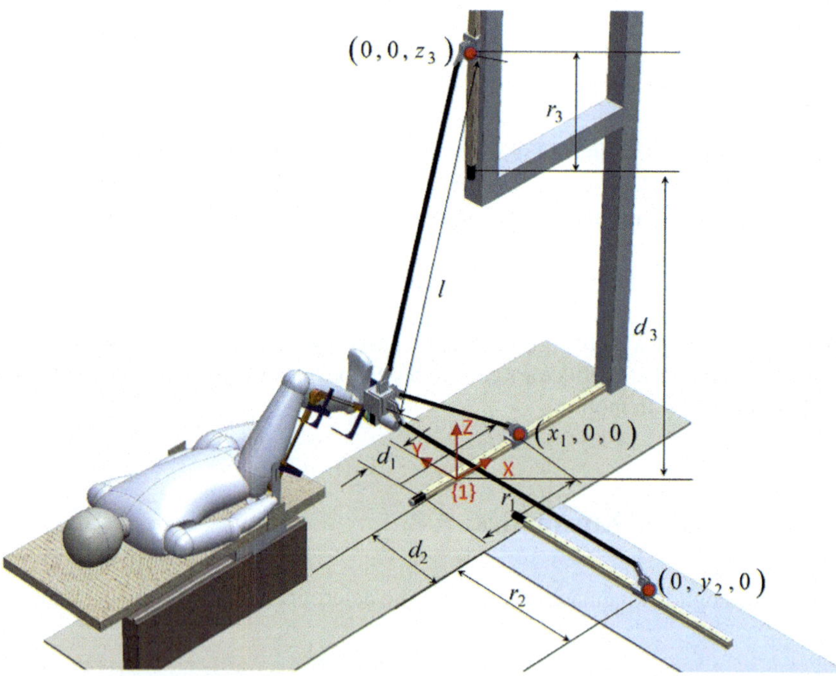

Fig. 5 Kinematic parameters of Orthoglide

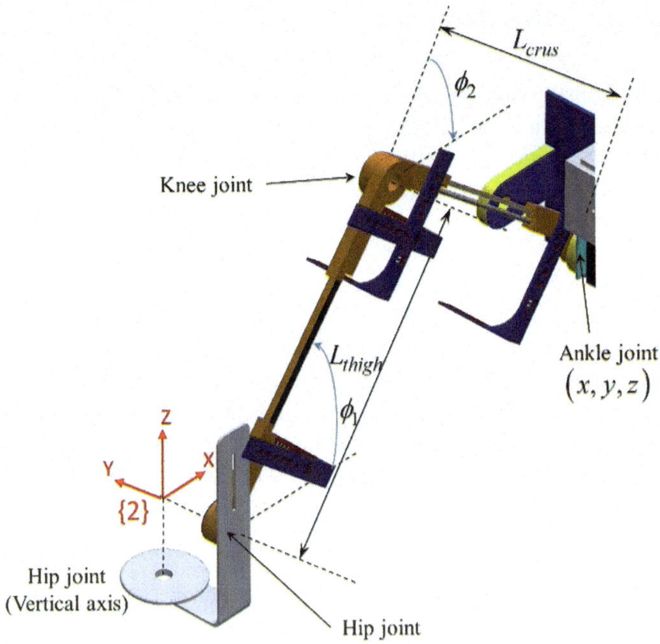

Fig. 6 Passive orthosis

By substituting Eqs. (4) and (1) in Eq. (3) we can obtain complete inverse kinematics relation for the assembly. The parameters x_t and z_t are shown in Fig. 4. The point (x, y, z) is with respect to (w.r.t.) frame {2} (Fig. 6), while (x_o, y_o, z_o) is w.r.t. frame {1} (Fig. 5). To bring all the points to one coordinates system linear transformation is applied to the frames, the distances $\left[\, p_x,\ p_y,\ p_z \,\right]$ are the linear transformation for the coordinates. In motion assembly, the parameters $\left[r_1\ r_2\ r_3 \right] = f(\phi_1, \phi_2, \phi_3)$ while the ankle joint is having independent actuator whose actuation is represented by ϕ_{ankle} which is a function of hip and knee adduction-abduction angle, $\theta_{ankle} = f(\phi_1, \phi_2, \phi_{ankle})$.

4 Physiological Motions

Physiological motions are decided by the therapist according to rehabilitation status of the patient. Here in this section basic physiological motions are shown to validate the kinematics of the proposed system. Each individual joint are given motion according to the angular limits used in therapeutic treatments [12].

4.1 Hip Flexion-Extension

Hip flexion-extension motion refers to motion of hip joint about the sagittal plane. To visualize this motion the values input is shown in Table 1.

To validate the motion possibility each joint is simulated for a 30 s motion trajectory. Here in this case the hip flexion-extension, $-15° < \phi_1 < 45°$ is simulated the geometric parameters are taken as:$L_{thigh} = 0.42$ m, $L_{crus} = 0.4$ m, $l = 1.4$ m, $d_1 = 0.28$ m, $d_2 = 0.4$ m, $d_3 = 1.6$ m, $p_x = 1.2$ m, $p_y = 0.14$ m,$p_z = 0.917$ m, $y_t = 0.05$ m and $z_t = 0.1$ m.

The joint-space movement shown in Fig. 7 shows that the motion is possible and mechanism is not encountering any singularities while performing the task. Parallel singularity can be avoided by regulation the parameter l, higher values are more stable, which is the length of the connecting link from the actuator to end end-effector with universal joints. Furthermore, optimization can be done as this parameter. The stroke length of the linear actuator is taken as 1.2 m. Hence, during the motion system is not overruling the limits.

Table 1 Task-space for validation of hip flexion-extension motion

Joint	ϕ_1	ϕ_2	ϕ_3	ϕ_{ankle}
Value	$\left(\frac{60(168.75-(t-15)^2)}{225}\right)^\circ$	10°	0°	0°

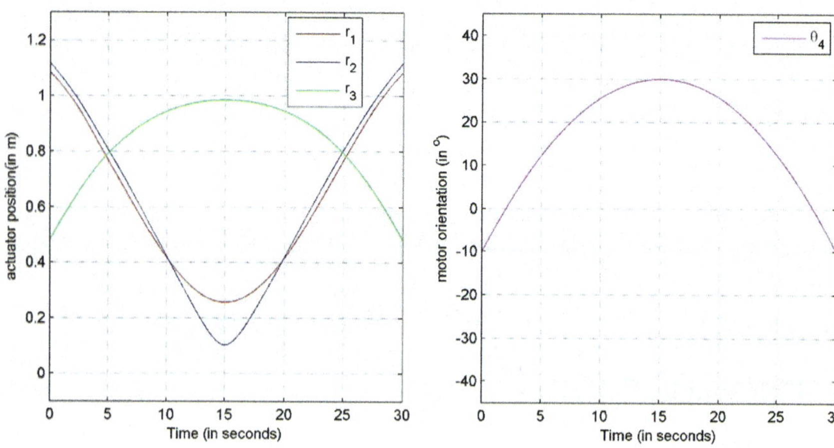

Fig. 7 Hip flexion-extension joint space movement

Table 2 Task-space for validation of hip adduction-abduction motion

Joint	ϕ_1	ϕ_2	ϕ_3	ϕ_{ankle}
Value	30°	20°	$15\sin(12t)$°	0°

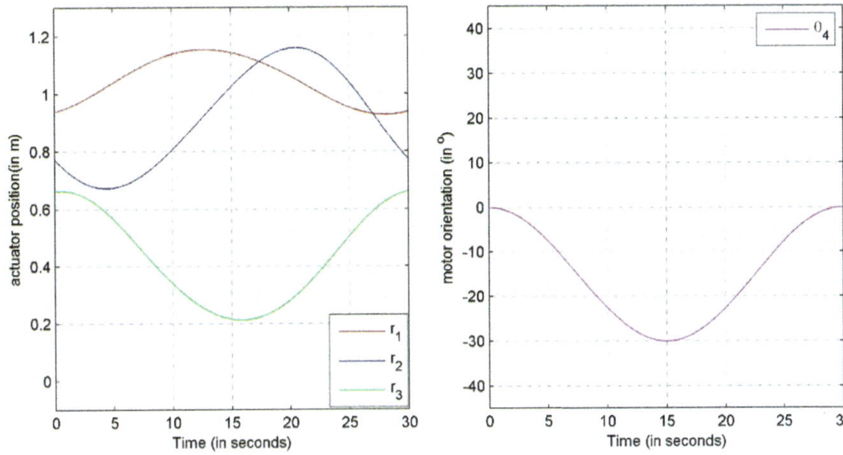

Fig. 8 Hip adduction-abduction joint space movement

4.2 *Hip Adduction-Abduction*

The trajectory taken for hip adduction-abduction is shown in Table 2. Here we have given motion to ϕ_3 as, $-15° < \phi_3 < 15°$ using as sinusoidal function with a time period of 30 s. The trajectory would be a circle in transverse plane as the patient is in lying posture.

This joint-space in Fig. 8 clearly shows that mechanism is able to provide bidirectional adduction and abduction which was not possible in case of Hipbot. Only one condition is to be verified that the lower limb under treatment is should not hit the other limb which is not being treated.

4.3 *Knee Flexion-Extension*

For knee flexion-extension the ϕ_2 angle is to be varied, this angle cannot be less than zero as it is constrained by human body constraints, also a safety feature in the knee joint is also required which stops the orthosis to go less than zero physically, for this simulation the task-space is shown in Table 3 and the joint-space output is shown in Fig. 9. Here we have taken $0 < \phi_2 < 40°$ using a polynomial equation.

All the joint must be protected by physical restriction and numerical restriction through integrated software. Other than this two emergency switches must be

Table 3 Task-space for validation of knee flexion-extension motion

Joint	ϕ_1	ϕ_2	ϕ_3	ϕ_{ankle}
Value	10°	$\left(\frac{40(225-(t-15)^2)}{225}\right)^\circ$	0°	0°

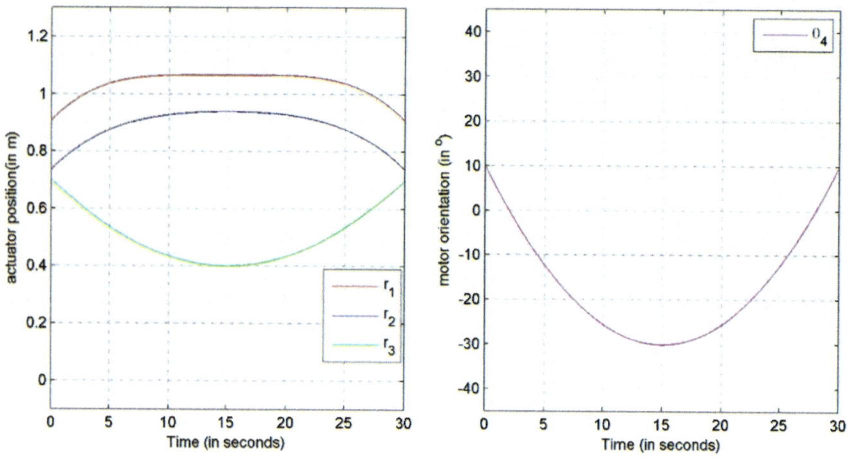

Fig. 9 Knee flexion-extension joint space movement

installed. One near patient's hand and another with the physiotherapist. These treatments must be provided in presence of a therapist. This devices are supportive devices which increases the efficiency of the therapist to give long duration treatments without making the therapist tired.

5 Conclusions

Lower limb rehabilitation techniques are designed in order to provide continuous therapy for a longer period of time at a stretch. For such purpose the mechanisms should perform continuous operation without facing any kind of singularity and joint space limitations. Also, it should be safe for the patient and the therapist. The proposed system is designed to provide and addition dimension to the therapists that is hip adduction-abduction which is not present with proper features in other devices. The design and kinematics is explained in order to realize the system. The basic physiological motions are kinematically simulated to verify the manipulator behavior towards the joint input. This architecture has potential to provide wide range of rehabilitation treatments. In future we would like to find dynamic based control schemes and real-time validation on human subject.

References

1. Diaz I, Gil JJ, S´anchez E (2011) Lower-limb robotic rehabilitation: literature review and challenges. J Robot 2011, Article ID 759764
2. Johnson MJ, Schmidt H (2009) Robot assisted neurological rehabilitation at home: motivational aspects and concepts for tele-rehabilitation. Public Health Forum, 17(4):8e1–8e4
3. Mohan S, Mohanta JK, Kurtenbach S, Paris J, Corves B, Huesing M (2017) Design, development and control of a 2PRP-2PPR planar parallel manipulator for lower limb rehabilitation therapies. Mech Mach Theory 112:272–294
4. Mohanta JK, Saxena C, Gupta G, Santhakumar M (2015) Kinematic analysis of a passive sitting/lying type lower limb rehabilitation robot. In: Proceedings of the 2nd international conference on machines and mechanisms, Kanpur, India, pp 1–12
5. Lim FM, Foong R, Yu H (2014) A supine gait training device for stroke rehabilitation. J Med Devices 8 (2)
6. Akdoğan E, Adli MA (2011) The design and control of a therapeutic exercise robot for lower limb rehabilitation: Physiotherabot. Mechatronics 21(3):509–522
7. Bouri M, Le Gall B, Clavel R (2009) A new concept of parallel robot for rehabilitation and fitness: the Lambda. In: IEEE international conference on robotics and biomimetics (ROBIO), pp 2503–2508
8. Mohanta JK, Mohan S, Deepasundar P, Kiruba-Shankar R (2018) Development and control of a new sitting-type lower limb rehabilitation robot. Comput Electr Eng 67:330–347
9. Wang W, Hou Z, Tong L, Zhang F, Chen Y, Tan MH (2014) A novel leg orthosis for lower limb rehabilitation robots of the sitting/lying type. Mech Mach Theory 74:337–353
10. Guzmán-Valdivia CH, Blanco-Ortega A, Oliver-Salazar MA, Gómez-Becerra FA, Carrera-Escobedo JL (2015) HipBot–The design, development and control of a therapeutic robot for hip rehabilitation. Mechatronics 30:55–64
11. Chablat D, Wenger P (2003) Architecture optimization of a 3-DOF translational parallel mechanism for machining applications, the Orthoglide. IEEE Trans Robot Autom 19(3):403–410
12. Stansfield BW, Hillman SJ, Hazlewood ME, Robb JE (2006) Regression analysis of gait parameters with speed in normal children walking at self-selected speeds. Gait & Posture 23:288–294

References

Design of Single Degree-of-Freedom Mechanisms for Hand Neurorehabilitation

Aravind Nehrujee, Sandeep Guguloth, Reethajanetsurekha, Samuelkamaleshkumar Selvaraj, Sujatha Srinivasan, and Sivakumar Balasubramanian

Abstract Hand rehabilitation requires intensive training of various gross and fine movements. Robotic devices have been developed and tested for implementing intense training of different hand functions in the current literature. Current hand rehabilitation robots can be broadly grouped into two categories: (a) simple robots with one or two degrees-of-freedom (DOF) that train only one or two hand functions; or (b) complex robots with several DOF capable of training a wide range of functions. Thus, to train different hand functions, we either need a set of simple robots or one complex robot, both of which are not economically viable solutions. A potential solution is to have a single DOF robot that can train a wide range of functions, i.e. a robot with a single actuator, along with a set of easily pluggable passive mechanisms for different hand functions. In this paper, we present the design and analysis of various kinematic mechanisms for a single DOF hand rehabilitation robot plug and train hand rehabilitation robot (PLUTO). Three different kinematic mechanisms have been designed and analyzed to train four different functions: wrist flexion/extension, wrist ulnar and radial deviation, forearm pronation/supination, and gross hand opening-

A. Nehrujee (✉) · S. Guguloth · S. Srinivasan
Department of Mechanical Engineering, Indian Institute of Technology Madras, Chennai, Tamil Nadu, India
e-mail: me17d600@smail.iitm.ac.in

S. Guguloth
e-mail: me15s063@smail.iitm.ac.in

S. Srinivasan
e-mail: sujsrcc@iitm.ac.in

Reethajanetsurekha · S. Selvaraj
Department of Physical Medicine and Rehabilitation, Christian Medical College, Vellore, Tamil Nadu, India
e-mail: reets_jan84@yahoo.co.in

S. Selvaraj
e-mail: stopatsams@cmcvellore.ac.in

A. Nehrujee · S. Balasubramanian
Department of Bioengineering, Christian Medical College, Vellore, Tamil Nadu, India
e-mail: siva82kb@cmcvellore.ac.in

D. Sen et al. (eds.), *Mechanism and Machine Science*, Lecture Notes in Mechanical Engineering, https://doi.org/10.1007/978-981-15-4477-4_55

closing. Proper alignment of the robot and human joints is essential for safe and appreciate interaction with a human. Misalignment in the robot and human joints can cause undesired forces on the human joint, makes the interaction potentially unsafe. The problem of misalignment was addressed in the proposed mechanisms through the use of appropriate passive DOFs between the robot and human joints. This paper presents the design and analysis of the three mechanisms for the different hand functions. The analysis of the mechanisms were carried out by considering variations in human hand anthropometry, and uncertainty in the robot-human axis alignments.

Keywords Robotics · Mechanisms · Neurorehabilitation · Hand · Kinematics

1 Introduction

Hand functions are one of the most commonly affected functions following a stroke [1], which severely affects a person's ability to carry out activities of daily living (ADL) and thus reduces quality of life. Increasing stroke population, a reducing healthcare budget, lack of trained therapists and poor access to rehabilitation facilities mean that a vast majority of stroke patients do not receive the recommended type and amount of therapy following a stroke [2]. Additionally, given the focus on balance and mobility early after a stroke, the hand is often neglected and, in general, has poor recovery.

There has been a lot of work in the last 15–20 years in the development and clinical evaluation of robotic devices for hand rehabilitation [3]. Hand rehabilitation involves the training of different types of movements to recover hand functions, e.g. wrist flexion/extension, pronation/supination, power grasp, pinch and lateral grasp, individuated finger control, etc. In the current literature, hand rehabilitation robots fall into two categories:

1. **Simple devices** are robots with 1 or 2 degrees-of-freedom (DOFs) designed to assist and train one of two specific hand movements (e.g. wrist flexion-extension, forearm pronation-supination, etc.) [4–6]. Such devices tend to be compact and affordable, and are more likely to be adopted for clinical use. However, training the different types of hand functions would require a set of such simple robots designed for training different hand function.
2. **Complex devices** have more than 2 DOFs and can train a wide range of functions [7–11], but these robot are likely to be expensive because of the use of multiple actuators for the different DOFs.

Thus, both simple and complex robots, in their current form, do not present an economically viable solution for training a wide range of hand functions. A possible solution to address this problem is to develop a simple (1DOF) robot capable of training a wide range of 1DOF hand movements. Such a robot will use a single actuator along with a set of pluggable passive kinematic mechanisms (which can be

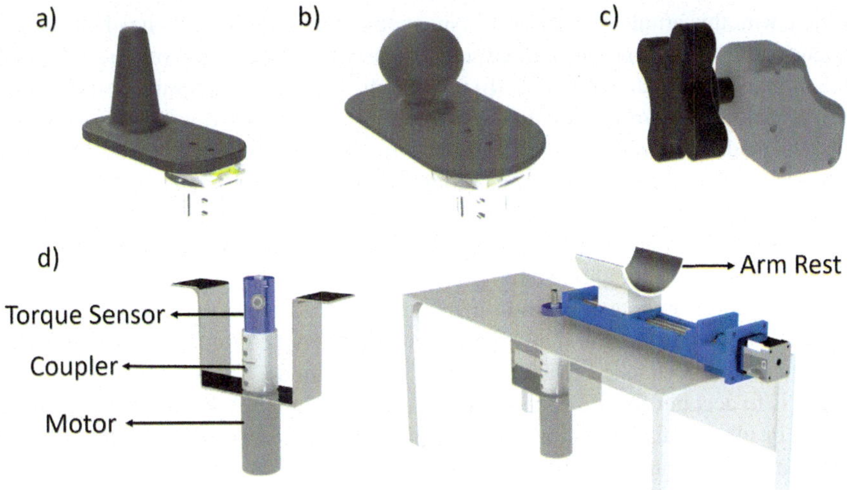

Fig. 1 Design of PLUTO, which consists of a single actuator and a set of pluggable passive mechanisms for training different hand tasks. Mechanisms for training the different hand functions **a** wrist flexion/extension, **b** wrist ulnar-radial deviation, **c** pronation/supination, and **d** Robot housing which includes the torque sensor coupled to the actuator and the passive mechanism The arm is mounted on a powered linear rail for adjustment of the arm position

easily attached/detached from the actuator) designed to train specific hand functions. We are currently working toward the development of such a robot, called the plug-and-train hand rehabilitation robot (PLUTO) for hand rehabilitation (see Fig. 1).

In this paper, we present the design and analysis of four kinematic mechanisms for PLUTO to train (a) wrist flexion/extension (WFE); (b) wrist ulnar/radial (WUR) deviation; (c) forearm pronation/supination (FPS); and (d) hand opening/closing (HOC). In this paper we only focus on the design and analysis of the pluggable mechanism for PLUTO, and do not present any other details about the robot's actuators, sensors, control, or the plug-in/plug-out mechanism.

The paper starts with a discussion of the requirements for the different mechanisms in Sect. 3. Section 4 provides the details of the kinematic design and analysis of the different mechanisms. This is followed by a brief discussion on the work presented along with details of our current and future work on the realization and testing of these mechanisms (Sect. 5).

2 PLUTO- Hardware Design

The primary design requirement was to make the rehabilitation kit portable and compact. To best serve this we choose a compact brushless DC motor (Maxon, Brushless EC Flat45) with a planetary gear head with a speed reduction of 26:1. The maximum torque with this arrangement was 3.5 Nm at 384 rpm. Since the gear head introduced friction, we coupled a rotary torque sensor (Forsentek, FYF: max torque 5 Nm) with the motor shaft (as shown in Fig. 1b) to implement admittance control and make the setup back drivable/transparent. The DC motor was combined with an incremental optical quadrature encoder (1024 counts per revolution) and a Hall sensor. The motor was connected to the PC via a microcontroller (Arduino DUE). The second design feature was to have simple and fast way to mount and unmount the passive mechanisms from the actuator. To do this, we used a universal mounting hub, the advantage of using a universal mounting hub is that its commercially available/easy to fabricate and it requires fastening of two nuts to mount. Figure 1a shows the plugin mechanism for wrist flexion extension that would be mounted on the torque sensors shaft. The plug in mount is attached to a handle with a linear guide. The purpose of the linear guide is to account for offset between the axis of rotation of the wrist and the actuator. For ulnar/radial deviation, the same setup as wrist flexion would be used just by changing the handle. A bevel gear with is used to train for pronation/supination (Fig. 1c).

3 Pluggable Mechanisms for PLUTO

Unlike most robots, PLUTO has a slightly unconventional design, where the robot's kinematic mechanism and its actuation system are fully decoupled. The robot contains a base-unit which houses the robot's single actuator. Passive single DOF mechanisms can be attached to the shaft of the robot for assisting training of different functions. The topology of the kinematic mechanism attached to the motor shaft will determine the nature of the human–machine interaction.

A schematic of the physical connection and interaction between the robot actuator and the human, through the passive mechanism, is shown in Fig. 2. The passive pluggable kinematic mechanism acts as a power transmission system between the robot actuator and the human hand. The interface between the actuator and the mechanism (left dashed line in Fig. 2) is formed by the physical connection between the actuator shaft and the plug-in connector of the mechanism, while the interface between the mechanism and the human hand is through an appropriately designed handle that the subject holds to interact with the robot. Overall, the system shown in Fig. 2 forms a closed kinematic chain between the robot and the human, consisting of the several joints: (a) the single revolute joint of the robot actuator; (b) the different joints of the mechanism; and (c) the relevant joint or DOF of the human hand. The closed chain can allow the robot to impose movements or apply force/torque to the human hand, and vice versa. This is shown through the double-sided arrows in Fig. 2, which indicate that power can either flow from the robot to the human hand (when the robot assisted the human), or from the human to the robot (when the subject voluntarily moves the actuator). The effort and flow variables associated with the

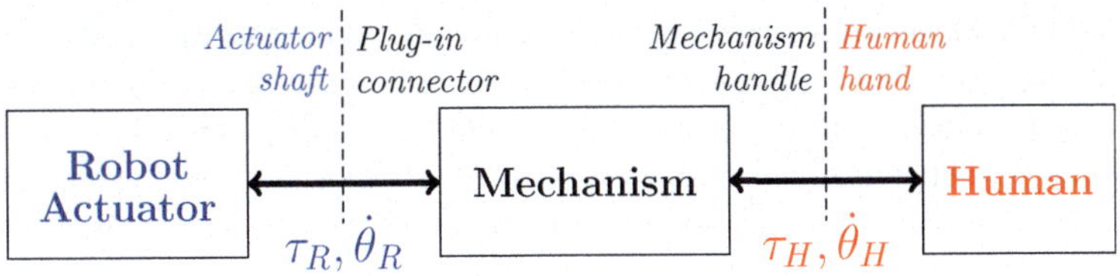

Fig. 2 Nature of the physical interaction between the robot, mechanisms, and human subject

robot actuator and the human are $(\tau_R, \dot{\theta}_R)$ and $(\tau_H, \dot{\theta}_H)$; an ideal mechanism would transmit power without any loss of power in either direction, i.e. $\tau_R \dot{\theta}_R = \tau_H \dot{\theta}_H$.

In the case of PLUTO, there are three important requirements that a mechanism must satisfy for the safe and efficient transmission of torque between the robot and the human.

1. **Safety** is the most important requirement when connecting a human limb to a robot. The robot must only apply force/toques in desired directions on the human limb, and must not apply any forces/torques in unintended directions. The problem of unintended forces on the human limb can result because of the differences in the kinematics of the robot and the human (e.g. misalignments between the robot and human joint axes) [12]. Jarrassé and Morel recommend the use of passive DOFs between the robot and human kinematic chains to address this problem to ensure safety [12].
2. **Efficiency** of the torque transmission between the robot and the human, i.e. the mechanism must allow faithful transmission of torque from the robot to the human and vice versa ($\tau_H \approx \tau_R$). This will ensure that movement of the actuator $\dot{\theta}_R$ and its torque (τ_R) are a close reflection of the movement (θ_H) and torque (τ_H) of the human joint.
3. The mechanism must not have **redundancy**, i.e. the human joint must move when the robot is moved and vice versa. This would imply that $\dot{\theta}_H \neq 0 \implies \dot{\theta}_R \neq 0$, although this condition can be violated at specific configurations of the closed chain (points of movement reversal).

Different mechanisms have been designed and analyzed to satisfy these three requirements. The details of these mechanisms for PLUTO are described in the following section.

4 Design and Analysis of PLUTO Mechanisms

Three different mechanisms have been designed for PLUTO to train four hand movements:

1. **Wrist mechanism** for training WFE and WUR movements.
2. **Prono-supination mechanism** for training FPS movements.
3. **Power grasp mechanism** for training HOC movements.

Two main issues these mechanisms have to address are the variations in human hand sizes, and the misalignments between the robot and human joint axes when a subject is connected to the robot. This is achieved through the addition of extra passive DOFs in the mechanism that connects the robot actuator to the human hand. This section presents the details of the structure of the three mechanisms, along with their kinematic analysis. The kinematic analysis was carried out to evaluate the torque transmission of the mechanisms for different human hand sizes and different levels of misalignments between the robot and human joint axes. For all kinematic analysis, the modified Denavit–Hartenberg convention [13] is followed to assign coordinate frames to the different links in the robot and human chains.

4.1 Wrist Mechanism

This mechanism is used for training wrist flexion/extension, and wrist ulnar/radial deviation movements, which is a single DOF movement. Kinematically, mechanisms that train the wrist flexion/extension and ulnar/radial deviation are equivalent. Consider the single DOF robot and human (wrist) kinematic chain shown in Fig. 3a. When the human is connected to the robot, as in Fig. 3b, the closed chain will move only when $l^r = l^h$, and the robot and human revolute joints are aligned (**R** and **H** are in the same location). When both of these conditions are not satisfied, then the mobility of this closed chain is zero (as shown in Fig. 3b). However, if a passive prismatic and a revolute joint are added to the robot, then the closed chain involving the human and the robot has a mobility of 1, where changes in θ_R will result in changes in θ_H (except for some pathological configurations). The additional revolute and prismatic joints in the robot will allow changes in l^r and ϕ to allow the human wrist to move when the robot is moved and vice versa. The wrist mechanism designed for PLUTO uses this particular kinematic design, consisting of a passive prismatic joint and a revolute joint. One end of the prismatic joint houses the plug-in connector, which attaches the mechanism to the shaft of the robot's actuator, and the revolute joint of the mechanism consists of the handle that the human holds.

The detailed kinematic structure of the closed chain formed between the robot, wrist mechanism, and the human wrist joint are shown in Fig. 4. As a first step in the analysis, the robot and human kinematic chains are considered to be two separate open chains, with their corresponding endpoints at the point **E** shown in Fig. 4. The global base-frame is (x_0, y_0, z_0), where z_0 is pointing perpendicularly out of the sheet. The robot links and their associated coordinate frames are shown in red, while that of the human hand are shown in blue. The robot has three DOF (RPR) and the human wrist has a single DOF (R), each with an associated coordinate frame; the axes for the different robot coordinate frames are represented with a superscript 'r', while that of the human have a superscript 'h'. The DH-parameters associated with the robot and human open chains are also shown in Fig. 4.

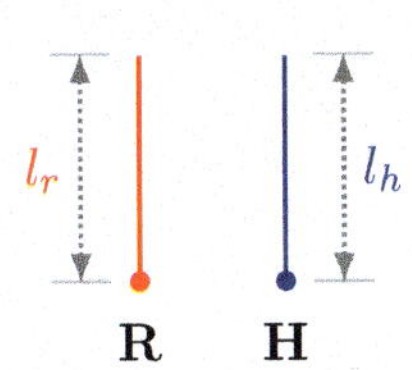

(a) Robot and human 1DOF kinematic chains.

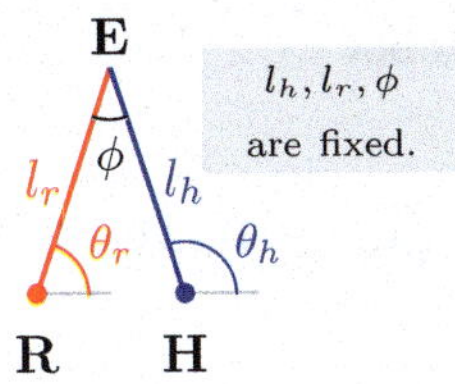

(b) Robot and Human kinematic chains joined together at **E** without any additional DOF.

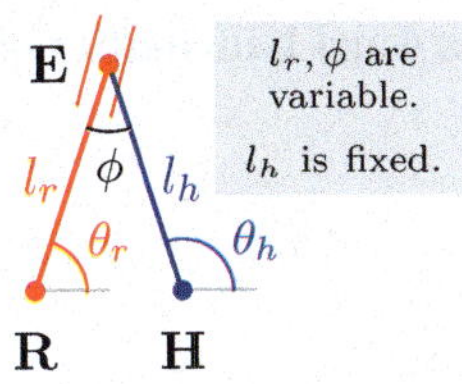

(c) Robot kinematic chain with an additional prismatic and revolute joints. The closed robot-human chain now has a mobility of 1.

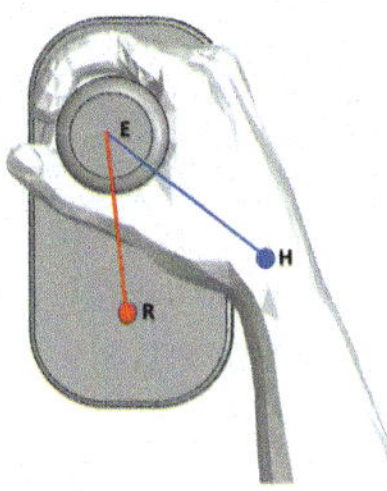

(d) Misalignment of the wrist flexion/extension axis (H) and the motor axis (R).

Fig. 3 Demonstration of the problems of connection a the single DOF robot and human kinematic chains, and the use of additional passive mechanisms to address this issue

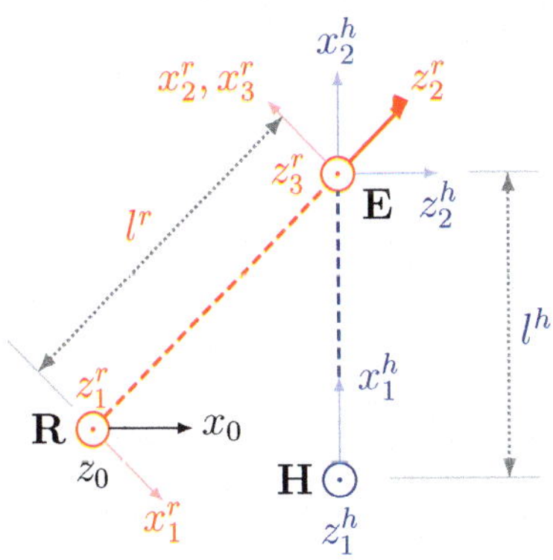

Robot DH parameters				
i^{th} joint	a_{i-1}	α_{i-1}	d_i	θ_i
1	0	0	0	θ_r
2	0	$+\frac{\pi}{2}$	l^r	0
3	0	$-\frac{\pi}{2}$	0	θ_3

Human DH parameters				
i^{th} joint	a_{i-1}	α_{i-1}	d_i	θ_i
1	l^h	0	0	θ_h

Fig. 4 Details of the wrist flexion-extension mechanism. **R** and **H** represent the point where the robot's and the human's kinematic chain start, respectively. The different frames attached to robot and human kinematic chains are shown in red and blue, respectively; the global reference frame (x_0, y_0, z_0) is shown in black. It should be noted that the robot and human chains are considered as open serial chains, with their endpoints connected at **E**, which imposes a position and orientation constraint.The global reference frame is (x_0, y_0, y_0)

Forward Kinematics of the Robot

$$ {}^{3}_{0}\mathbf{H}_r = {}^{1}_{0}\mathbf{H}_r {}^{2}_{1}\mathbf{H}_r {}^{3}_{2}\mathbf{H}_r $$

where ${}^{j}_{i}\mathbf{H}_r$ is the homogeneous transformation matrix representing frame j in frame i for the robot.

$$ {}^{3}_{0}\mathbf{p}_r = \begin{bmatrix} l_r s_r \\ -l_r c_r \\ 0 \end{bmatrix} \qquad {}^{3}_{0}\mathbf{R}_r = {}^{1}_{0}\mathbf{R}_r {}^{2}_{1}\mathbf{R}_r {}^{3}_{2}\mathbf{R}_r \tag{1} $$

$$ {}^{1}_{0}\mathbf{R}_r = \begin{bmatrix} c_r & -s_r & 0 \\ s_r & c_r & 0 \\ 0 & 0 & 1 \end{bmatrix} \quad {}^{2}_{1}\mathbf{R}_r = \begin{bmatrix} 1 & 0 & 0 \\ 0 & 0 & 1 \\ 0 & -1 & 0 \end{bmatrix} \quad {}^{3}_{2}\mathbf{R}_r = \begin{bmatrix} c_3 & 0 & -s_3 \\ 0 & 1 & 0 \\ -s_3 & 0 & c_3 \end{bmatrix} \tag{2} $$

where ${}^{3}_{0}\mathbf{p}_r$ and ${}^{3}_{0}\mathbf{R}_r$ are the position and orientation of the robot's coordinate frame 3 with respect to the global frame, respectively. The terms s_i and c_i stand for $\sin\theta_i$ and $\cos\theta_i$.

Forward Kinematics of the Human

$$ {}^{2}_{0}\mathbf{H}_h = {}^{1}_{0}\mathbf{H}_h {}^{2}_{1}\mathbf{H}_r \qquad {}^{1}_{0}\mathbf{H}_h = \begin{bmatrix} 0 & -1 & 0 & d_x \\ 1 & 0 & 0 & d_y \\ 0 & 0 & 1 & 0 \\ 0 & 0 & 0 & 1 \end{bmatrix} $$

where ${}^{j}_{i}\mathbf{H}_h$ is the homogeneous transformation matrix representing frame j in frame i for the human. d_x, d_y are the x and y coordinates of the misalignment between the robot actuator and human wrist axes.

$$ {}^{3}_{0}\mathbf{p}_h = \begin{bmatrix} -l_h c_h + d_x \\ l_h s_h + d_y \\ 0 \end{bmatrix} \qquad {}^{3}_{0}\mathbf{R}_h = \begin{bmatrix} c_h & s_h & 0 \\ -s_h & c_h & 0 \\ 0 & 0 & 1 \end{bmatrix} $$

where ${}^{3}_{0}\mathbf{p}_r$ and ${}^{3}_{0}\mathbf{R}_r$ are the position and orientation of the robot's coordinate frame 3 with respect to the global frame, respectively. The terms s_i and c_i stand for $\sin\theta_i$ and $\cos\theta_i$.

Transmission Properties of the Mechanism

In order to understand the transmission properties of the wrist mechanism, the relation between θ_r and θ_h has to be established. This can be done by solving the kinematic equations of the closed chain for a particular value of l_h, d_x, and d_y. The following procedure is followed in solving the closed chain kinematics,

1. Choosing a particular value for θ_h to carry out the forward kinematics of human chain to obtain ${}_0^2\mathbf{p}_h$ and ${}_0^2\mathbf{R}_h$.
2. Determine θ_r, l_r by solving the inverse kinematics problem for the robot chain by imposing the constraints ${}_0^3\mathbf{p}_r = {}_0^2\mathbf{p}_h$ and ${}_0^3\mathbf{R}_r = {}_0^2\mathbf{R}_h$. Also, $\theta_3 = \theta_h$, thus we have from Eqs. 1 to 2, ${}_0^1\mathbf{R}_r = {}_0^2\mathbf{R}_h \left({}_1^2\mathbf{R}_r {}_2^3\mathbf{R}_r\right)^T$. This results in the following,

$$\theta_r = \text{atan2}(l_h c_h - d_x, l_h s_h + d_y) \tag{3}$$

$$l_r = -\sqrt{(l_h c_h + d_x)^2 + (l_h c_h + d_y)^2}$$

The transmission ratio of the mechanism is defined as the ratio of the torque in the human joint τ_h to that of the robot's joint τ_r, which can be obtained from the corresponding joint velocities $\dot{\theta}_r$ and $\dot{\theta}_h$,

$$R = \frac{\tau_h}{\tau_r} = \frac{\dot{\theta}_r}{\dot{\theta}_h} = \frac{\partial \theta_r}{\partial \theta_h}$$

After obtaining the values of θ_r for different values of θ_h, $\frac{\partial \theta_r}{\partial \theta_h}$ is determined numerically to obtain the transmission ratio for different values of θ_h.

Effect of Different Hand Sizes and Misalignments

In order to understand the effect of different hand sizes (l_h) and different robot-human misalignments (d_x, d_y), the transmission ratio R was determined for the wrist mechanism by varying θ_h between $\pm 60°$, for 2 different hand sizes ($l_h = (4, 8)$) and 9 different misalignments corresponding to three different values for x and y misalignments $(d_x, d_y \in \{-3, 0, 3\})$; the values of θ_r, l_r were also determined for different θ_h for each combination of (l_h, d_x, d_y).

The result of this analysis is shown in Fig. 5, which shows the transmission ratio for over the ranges of wrist movements $\pm 60°$ for different hand sizes and misalignment. A transmission ratio of 1 is obtained when there is no misalignment. As expected, when the misalignment is large the transmission ratio varies about 1 for the range of wrist movements. For extreme misalignments $(d_x = 3, d_y = 3)$, the transmission ratio approximately varies between 0.4 and 1.6. It could be seen that even misalignments of the same magnitude (distance) have different transmission ratios, this is because the analysis is done only at $\pm 60°$. If the analysis was done from 0 to 360°, the transmission ratio versus angle would trace the same curve with a phase shift.

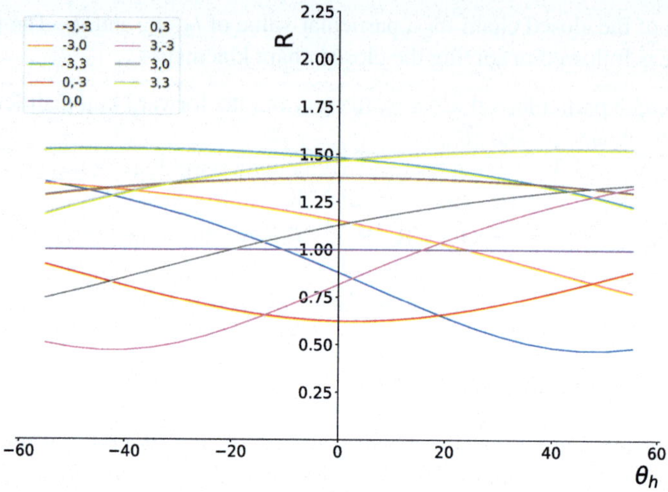

Fig. 5 Wrist mechanism transmission ratio for different range of wrist movements for different different misalignments for $l_h = 8\,\text{cm}$

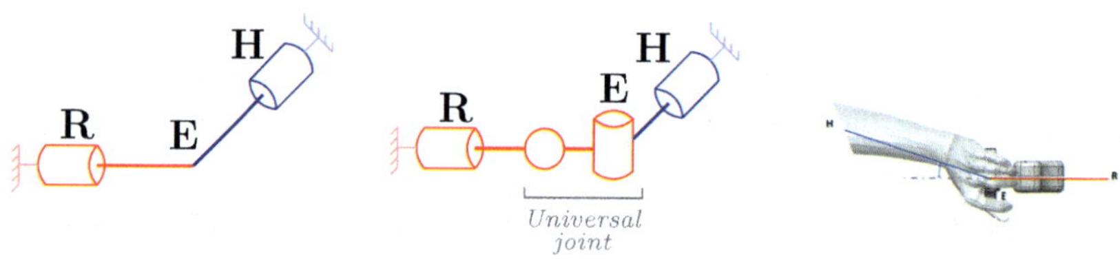

(a) Robot and human 1DOF kinematic chains.

(b) Robot and Human kinematic chains joined together at **E** without any additional DOFs.

(c) Misalignment of the forearm pronation-supination axis (H) and the axis of rotation of the mechanism (R).

Fig. 6 Demonstration misalignment problem for the forearm pronation-supination movement. Addition of a universal joint between the robot and the human kinematic chains will allow the transmission of movements and torques between the robot and the human

4.2 *Prono-Supination Mechanism*

This mechanism is used for training forearm pronation and supination movements with PLUTO. The problem due to joint angle misalignment and its resolution through the use of a universal joint is shown in Fig. 6. When there is misalignment between the robot actuator's rotation axis and that of the human's forearm, there is no rotation possible for the closed chain shown in Fig. 6a. However, this issue can be addressed through the use of universal joint, which now allows the movements and torque from the robot's actuator to be transmitted to the human forearm axis (Fig. 6b).

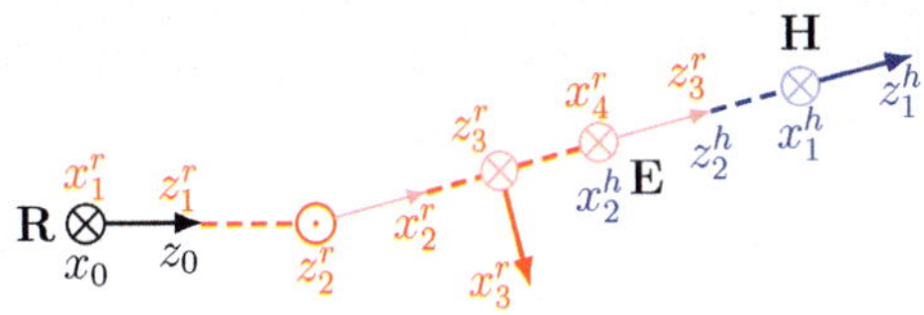

Robot DH parameters				
i^{th} joint	a_{i-1}	α_{i-1}	d_i	θ_i
1	0	0	0	θ_r
2	0	$\frac{\pi}{2}$	0	θ_2
3	0	$\frac{\pi}{2}$	0	θ_3
4	0	$\frac{\pi}{2}$	0	0

Human DH parameters				
i^{th} joint	a_{i-1}	α_{i-1}	d_i	θ_i
1	l^h	0	0	θ_h

Fig. 7 Details of the pronation-supination mechanism. **R** and **H** represent the points where the robot's and the human's kinematic chain start, respectively. The different frames attached to robot and human kinematic chains are shown in red and blue, respectively; the global reference frame (x_0, y_0, z_0) is shown in black. It should be noted that the robot and human chains are considered as open serial chains, with their endpoints connected at **E**, which imposes a position and orientation constraint. The global reference frame is (x_0, y_0, y_0)

The detailed kinematic structure, the different coordinate frames attached to the robot and human kinematic chains, and the corresponding DH parameters are shown in Fig. 7. For this mechanism, the position of **E** does not change, so we only need to consider the orientation kinematics to analyze the mechanism.

Forward Kinematics of the Robot

$$ {}^4_0\mathbf{R}_r = {}^1_0\mathbf{R}_r {}^2_1\mathbf{R}_r {}^3_2\mathbf{R}_r {}^4_3\mathbf{R}_r \tag{4}$$

$$ {}^1_0\mathbf{R}_r = \begin{bmatrix} c_r & -s_r & 0 \\ s_r & c_r & 0 \\ 0 & 0 & 1 \end{bmatrix} \quad {}^2_1\mathbf{R}_r = \begin{bmatrix} -s_2 & 0 & -c_2 \\ 0 & 0 & 1 \\ c_2 & -s_2 & 0 \end{bmatrix}$$

$$ {}^3_2\mathbf{R}_r = \begin{bmatrix} -s_3 & -c_3 & 0 \\ 0 & 0 & -1 \\ -c_3 & -s3 & 0 \end{bmatrix} \quad {}^4_3\mathbf{R}_r = \begin{bmatrix} 0 & -1 & 0 \\ 0 & 0 & -1 \\ 1 & 0 & 0 \end{bmatrix} \tag{5}$$

The conventions followed are the same as the one used for the wrist mechanism.

4.2.1 Forward Kinematics of the Human

The misalignment in orientation between robot (z_1^r) and human axis (z_1^h) is assumed $\alpha°$ about the x_1^r axis. Thus the overall forward kinematic equation for the human is given by ${}_0^2\mathbf{R}_h = {}_0^1\mathbf{R}_h {}_1^2\mathbf{R}_h$.

$$
{}_0^1\mathbf{R}_r = \begin{bmatrix} 1 & 0 & 0 \\ 0 & c_\alpha & -s_\alpha \\ 0 & c_\alpha & s_\alpha \end{bmatrix} \quad {}_1^2\mathbf{R}_r = \begin{bmatrix} c_r & -s_r & 0 \\ s_r & c_r & 0 \\ 0 & 0 & 1 \end{bmatrix}
$$

$$
{}_0^2\mathbf{R}_h = \begin{bmatrix} c_h & -s_h & 0 \\ s_h c_\alpha & c_h c_\alpha & -s_\alpha \\ s_h s_\alpha & c_h s_\alpha & c_\alpha \end{bmatrix} \tag{6}
$$

Equating ${}_0^4\mathbf{R}_r$ and ${}_0^2\mathbf{R}_h$, we obtain the relationship between the robot angle (θ_r) and human angle (θ_h),

$$
\theta_1^r = \text{atan2}(\tan(\theta_1^h), c_\alpha) \tag{7}
$$

Effect of Different Hand Sizes and Misalignments

The transmission properties of this mechanism was analyzed for four different misalignment angles $\alpha \in [0, 10, 20, 30]°$ between the robot and the human joint axes, over a range of $\pm 90°$ for θ_h. The effect of the transmission ratio for the different misalignments for the range of θ_h is shown in Fig. 8.

4.3 *Grasp Mechanism*

The grasp mechanism is designed to support and assist hand opening and closing movements, involving the coordinated flexion/extension of the fingers and the thumb. The power grasp movement involves the coordinated extension of the finger and the thumb to open the hand, followed by coordinated flexion of the fingers and the thumb. For a power grasp movement, the fingers flex and extend about the metacarpophalangeal joint, and the thumb flexes and extends about the carpometacarpal joint [14].

A simple schematic of the closed kinematic chain of the robot, the grasp mechanism, and the human hand is depicted in Fig. 9. The grasp mechanism consists of two revolute joints corresponding to the fingers ($\mathbf{R}_F$) and the thumb ($\mathbf{R}_T$). The joint corresponding to the finger is driven by the robot's actuator, which in turn drives the thumb joint in the opposite direction through a gear with gear ratio 1:2; thus, the mechanism's thumb joint will rotate at half the speed of the finger joint in the grasp mechanism $\dot{\theta}_{r_F} = 2\dot{\theta}_{r_T}$. Individually, the mechanisms for the fingers and the thumb have the same kinematic structure as the wrist mechanism described earlier.

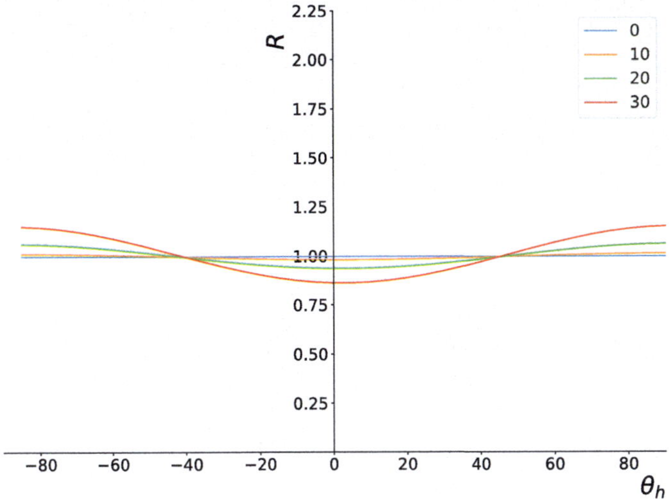

Fig. 8 Pronation-supination mechanism transmission ratio for different the orientation misalignment for pronation-supination movement range of ±90

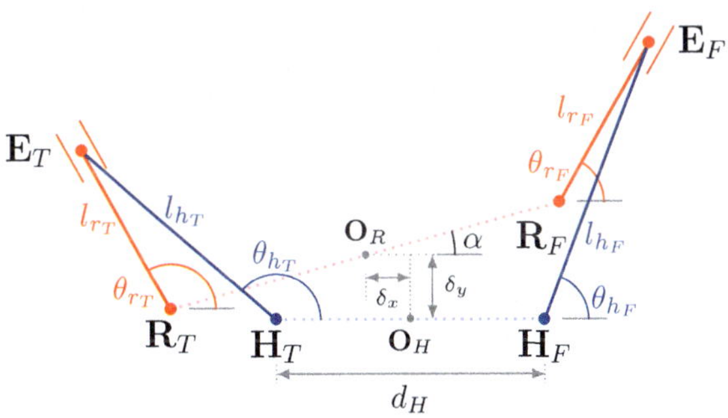

Fig. 9 Closed kinematic chain of the robot with the grasp mechanism and the human hand. The grasp mechanisms consists of two closed loops similar to that of the wrist mechanism. The two closed chains are constrained as their two robot joints corresponding to the thumb $\mathbf{R}_T$ and the fingers $\mathbf{R}_F$ are driven by gears with gear ratio 1:2, i.e. $\dot{\theta}_{r_F} = 2\dot{\theta}_{r_T}$. The points $\mathbf{O}_R$ and $\mathbf{O}_H$ represent the center point between the finger and thumb joints of the robot and human, respectively. The misalignment between $\mathbf{O}_R$ and $\mathbf{O}_H$ is the positional misalignment between the robot and the human hand. The angle α is the angle between the lines $\mathbf{H}_T$-$\mathbf{H}_F$ and $\mathbf{R}_T$-$\mathbf{R}_F$, which represents the misalignment in orientation between the robot and the human

Effect of Different Hand Sizes and Misalignments

The kinematic analysis of this mechanism is exactly the same as the wrist mechanism. Because there are two coupled closed chains, we will have two transmission ratios for each link, corresponding to the fingers R_F and the thumb R_T. We evaluated the transmission ratio of this mechanism for different hand sizes ($d_H \in \{6, 7, 8\}$ units), different positional misalignments ($\delta_x, \delta_y \in \{-1, 0, 1\}$ units), and different angular

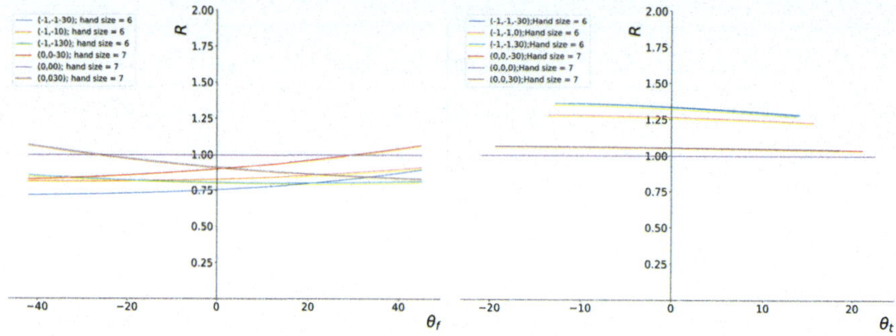

(a) Grasp mechanism transmission ratio for the fingers for different hand orientation misalignments.

(b) Grasp mechanism transmission ratio for the thumb for different hand orientation misalignments.

Fig. 10 Transmission ratio for the grasp mechanism

misalignments ($\alpha \in \{0, 30\}°$) of the human hand. The distance between $\mathbf{R}_T$ and $\mathbf{R}_F$ was set to 7 units in this analysis. The transmission ratios for the finger and the thumb as a function of the human finger joint angle and the thumb angle are shown in Fig. 10a and 10b, respectively. In general the overall transmission ratios were in and around 1 for the hand sizes and misalignments considered in the analysis.

5 Discussion

We presented a preliminary design and analysis of simple single DOF mechanisms for PLUTO to train different hand functions. The work presented in the current paper only focused on the kinematic analysis of three different mechanisms for training four different hand movements. The focus here was to design mechanisms that can allow assisted training of hands of different sizes, and also allow for minor misalignments in the rotational axes of the robot's actuator and the human joint.

For the wrist mechanisms, we only considered misalignments in position for the analysis. The results shown in Fig. 5 are for maximum misalignments of ±3 cm. As expected we find that the transmission ratio is 1 for all wrist angles θ_h when there is no misalignment between the robot and the human. However, when the misalignment increases, the transmission ratio varies significantly approximately between 0.4 and 1.6. This analysis was also used for determining the range of l_r for the robot for the different hand sizes and misalignments (results not shown). These values will determine the length of the sliding joint used for implementing the wrist mechanism.

In the case of the prono-supination mechanism, only angular misalignment was considered, which is parametrized by the α representing the rotation about the global x axis. Because of the symmetry in the mechanism, misalignment of α about the x-axis is the same as misalignment of the same magnitude about any axis in the

xy-plane. Thus all human axis falling on a particular cone about the global z-axis will have the same transmission properties. The result in Fig. 8 for three different α values indicate that they have good transmission ratio for this mechanism over the entire range of motion of the human hand.

The power grasp mechanism is in essence just a two-couple wrist mechanism, where the robot's finger and thumb joints are constrained to move at fixed rate relative to each other. Because of the kinematic coupling, both the human finger and thumb might not have appropriate range of motion due to changes in hand sizes and joint misalignments. This is seen in Fig. 10b, where the range of thumb movements for the human hand are different (extent of the graphs along the x-axis). This is because in the analysis the movements of the human fingers are fixed to $\pm 60°$, which results in varying ranges of movements for the thumb. If we had fixed the movement of the thumb, the range of movements of the fingers would have varied. The maximum misalignment in position was assumed 1 cm, and resulted in almost constant transmission ratio for the finger and the thumb.

5.1 Future Work

The work presented focused only on the kinematic analysis of the different mechanisms. Our future work will focus on the following:

1. Complete kinostatic analysis of all the mechanisms to understand the different forces acting on the human joint for the entire range of movements.
2. Selection of mechanism parameters to optimize range of movement, transmission ratio, mechanism size, and forces acting on the human joint.
3. Realization and testing the ergonomic properties of the different mechanisms for use by healthy subjects and subjects with hand impairments.

Acknowledgements We would like to thank the Department of Biotechnology, Government of India for funding this project.

References

1. Lai S-M, Studenski S, Duncan PW, Perera S (2002) Persisting consequences of stroke measured by the stroke impact scale. Stroke 33(7):1840–1844
2. Clarke DJ, Burton L-J, Tyson SF, Rodgers H, Drummond A, Palmer R, Hoffman A, Prescott M, Tyrrell P, Brkic L, Grenfell K, Forster A "Why do stroke survivors not receive recommended amounts of active therapy? findings from the react study, a mixed-methods case-study evaluation in eight stroke units. Clin Rehabil 0, no 0:0269215518765329, 0. PMID: 29582712
3. Balasubramanian S, Klein J, Burdet E (2010) Robot-assisted rehabilitation of hand function. Curr Opin Neurol 23(6):661–670
4. Dovat L, Lambercy O, Ruffieux Y, Chapuis D, Gassert R, Bleuler H, Teo CL, Burdet E (2006) A haptic knob for rehabilitation of stroke patients. In: 2006 IEEE/RSJ international conference on intelligent robots and systems, pp 977–982

5. Allington J, Spencer SJ, Klein J, Buell M, Reinkensmeyer DJ, Bobrow J (2011) Supinator extender (sue): a pneumatically actuated robot for forearm/wrist rehabilitation after stroke. In: 2011 annual international conference of the IEEE engineering in medicine and biology society, pp 1579–1582, Aug 2011
6. Cempini M, Cortese M, Vitiello N (2015) A powered finger-thumb wearable hand exoskeleton with self-aligning joint axes. IEEE/ASME Trans Mechatron 20:705–716
7. Omarkulov N, Telegenov K, Zeinullin M, Tursynbek I, Shintemirov A (2016) Preliminary mechanical design of nu-wrist: a 3-dof self-aligning wrist rehabilitation robot. In: 2016 6th IEEE international conference on biomedical robotics and biomechatronics (BioRob), pp 962–967, June 2016
8. Taheri H, Rowe JB, Gardner D, Chan V, Gray K, Bower C, Reinkensmeyer DJ, Wolbrecht ET (2014) Design and preliminary evaluation of the finger rehabilitation robot: controlling challenge and quantifying finger individuation during musical computer game play. J Neuroeng Rehabil 11:10 Feb
9. Schabowsky CN, Godfrey SB, Holley RJ, Lum PS (2010) Development and pilot testing of hexorr: hand exoskeleton rehabilitation robot. J Neuroeng Rehabil 7:36 Jul
10. Takahashi CD, Der-Yeghiaian L, Le V, Cramer SC (2005) A robotic device for hand motor therapy after stroke. In: 9th international conference on rehabilitation robotics, 2005. ICORR 2005. IEEE, pp 17–20
11. Li J, Zhang Z, Tao C, Ji R (2017) A number synthesis method of the self-adapting upper-limb rehabilitation exoskeletons. Int J Adv Robot Syst 14(3):1729881417710796
12. Jarrassé N, Morel G (2012) Connecting a human limb to an exoskeleton. IEEE Trans Robot 28(3):697–709
13. Craig JJ (2005) Introduction to robotics: mechanics and control, vol 3. Pearson/Prentice Hall Upper Saddle River, NJ, USA
14. Jones LA, Lederman SJ (2006) Human hand function. Oxford University Press

A Review on Designs of Various Ankle Foot Orthosis (AFO) Used to Treat Drop Foot Disease

Prashanth R. Kubasad, Somasekhara Rao Todeti, and Yogeesh D. Kamat

Abstract Compared to the last few decades, there is an increase in prevalence of neuromuscular diseases like stroke, multiple sclerosis, and cerebral palsy. These diseases cause lower limb disability like drop foot. The main reason for drop foot is weakness in dorsiflexor muscles. Drop foot results in 'toe drag during swing phase' and 'foot slap during heel contact'. Ankle foot orthosis (AFO) is a mechanical device, which is used to treat drop foot. Based on usage of sensors, actuators, and control systems, there are three types of AFOs: Semi active, Active, and Passive AFOs. Semi active and Active AFOs contain sensors, actuators, control systems, and onboard power source. Passive AFOs do not contain electrical boards but contain mechanical elements to control relative motion between foot part and shank part of the AFOs. Based on relative motion between foot and shank parts of AFOs, AFOs are also classified into two types: Non-articulated (or Fixed) and Articulated AFOs. Non-articulated AFOs are single piece devices having no relative motion between foot part and shank part of the device. Articulated AFOs are two-piece devices, having relative motion between foot part and shank part of the device, and the relative motion is controlled by passive and active actuators. In this paper, different working principles, advantages, and disadvantages of the existing AFOs are presented.

Keywords Drop foot · AFO · Articulated joints · Gait

P. R. Kubasad · S. R. Todeti (✉)
Department of Mechanical Engineering, National Institute of Technology-Karnataka, Surathkal 575025, India
e-mail: ssrao@nitk.edu.in

P. R. Kubasad
e-mail: prashanthrk431@gmail.com

Y. D. Kamat
KMC, Mangalore 575001, India
e-mail: yogeesh.kamat@manipalhospitals.com

D. Sen et al. (eds.), *Mechanism and Machine Science*, Lecture Notes in Mechanical Engineering, https://doi.org/10.1007/978-981-15-4477-4_56

1 Introduction

Walking is a special characteristic of locomotion by human beings. Each individual has a unique walking style [1]. Walking is a coordinated activity involving brain, spinal cord, muscles, peripheral nerves, bones and joints (Hip, Knee, and Ankle) [2]. But millions of people in the world are unable to experience a normal walking pattern because of gait disabilities [3]. These gait disabilities are due to neurological diseases (such as stroke, multiple sclerosis, cerebral palsy), muscular deficiencies (such as dorsiflexor or plantar flexor muscle weakness) or injuries (such as trauma or partial spinal cord injuries). These individuals (with gait disabilities) need assistance by assistive devices or by rehabilitation to overcome their limb pathologies. An orthosis is a device that supports body parts of a human externally. One of the orthoses, which supports ankle and foot of the patients to overcome the gait disability, is Ankle Foot Orthosis (AFO). In this paper, working principles of AFOs, their advantages, and limitations are presented.

Based on usage of sensors, actuators, and control systems, there are three types of AFOs: Semi active, Active, and Passive. Semi active and Active AFOs contain sensors, actuators, control systems, and onboard power source. Passive AFOs do not contain electrical boards, but contain mechanical elements (like dampers or springs) to control relative motion between foot and shank parts of the AFOs. Based on relative motion between foot and shank parts of AFOs, AFOs are also classified into two types: Non-articulated (or Fixed) and Articulated AFOs. Non-articulated AFOs are single piece devices made of lightweight thermoformable material like polypropylene. Articulated AFOs are two-piece (foot part and shank part) devices made of lightweight carbon composite or thermoplastic connected by joints. The foot and shank parts of the AFOs are connected by using hinges, springs, dampers, or flexion stops. Usually patients prefer passive AFOs, which are compact in size and can be used for daily life. Active AFOs are bulky in size and require external power supply. So, majority of the patients are not using these Active AFOs.

2 Normal Gait and Pathological Gait

Gait describes the manner of walking. A gait cycle starts from heel contact of one leg to the corresponding heel contact of the same leg. The gait cycle can be broadly divided into two phases: (1) Stance phase (the leg is in contact with the ground) and (2) Swing phase (the leg is in the air). The gait cycle can be further divided into heel contact, foot flat, mid stance, heel off, toe off, mid swing, and again heel contact as shown in Fig. 1. Usually in a gait cycle, stance phase is of 60% and swing phase is of 40% [2].

A leg contacts the ground through the heel and the ankle plantarflexes leading to a foot flat phase. At mid stance phase, the ankle motion results in propulsion of the leg. Then the ankle dorsiflexes by movement of the body over the foot. At heel off

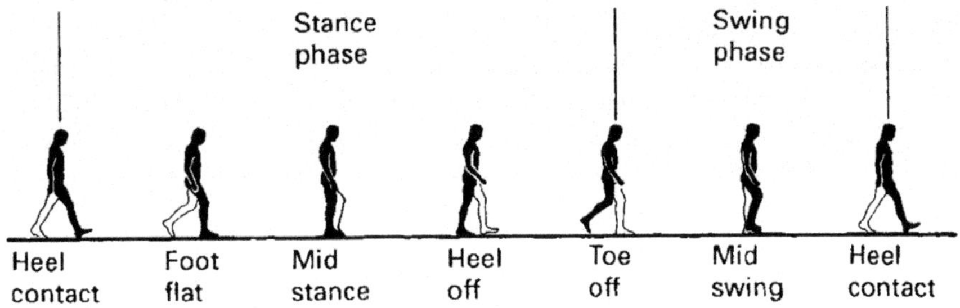

Fig. 1 Position of right leg from heel contact to heel contact [2]

phase, due to the heel rise, the ankle dorsiflexes to a maximum value. At the end of the heel off phase, due to motion of the calf muscle, the ankle plantarflexes to give push off. In toe off phase, the ankle dorsiflexes due to the pretibial muscle. During next heel contact, again the pretibial muscle controls motion of the ankle [4, 5].

Weakness in plantar flexor and dorsiflexor muscles are main reasons for gait disabilities. Plantar flexor muscle is located at posterior side of an ankle joint and includes soleus, gastrocnemius, and tibialis posterior muscles. Damage to any one of the above muscles causes weakness in plantar flexors. Weakness in plantar flexor muscle causes—'a reduction in push off during the toe off phase' and 'an increase in energy cost' [6]. Dorsiflexor muscle is located at anterior side of an ankle joint, and includes extensor digitorum longus, tibialis posterior, extensor hallucius longus muscles, and common peroneal nerve. Damage to any one of the above muscles or nerve causes weakness in dorsiflexor muscle and affects the gait during stance and swing phases. Patients having weakness in dorsiflexor muscle can not lift their foot through the ankle. The foot will fall uncontrollably after the heel contacts the ground. This condition is called as 'slap foot'. Also, during the swing phase, the toe is dragged due to lack of dorsiflexion. This condition is called as—'toe drag'. Drop foot (or a foot drop) is a disease, which include Slap foot and Toe drag. Drop foot causes imbalance of the patients while walking and the patients may fall down. The treatment for drop foot includes therapeutic, surgical, nerve stimulation [7], and orthotic. The research presented in this paper focuses on the orthotic.

Usually dorsiflexion muscles suffer more weakness compared to plantar flexion muscles [8]. For this reason, most of the AFOs are designed to prevent drop foot. The AFOs reduce the energy consumption during walking [9]. The AFOs secure the push off and provide progression during the stance phase of the gait [10].

3 Desirable Requirements of an AFO

Desirable requirements of an AFO [11] are: (i) AFO provides assist in inversion and eversion motions, (ii) AFO allows the initial ankle angle to be changed from 0° to 10° in dorsiflexion, (iii) AFO does not generate plantar flexion assist moment during dorsiflexion, (iv) AFO generates dorsiflexion assist moment during plantar flexion, (v) AFO coincides the mechanical ankle joint with the ankle joint of the patient, (vi) AFO easily be fitted into a shoe, (vii) AFO is aesthetic, (viii) AFO is lighter in weight, and (ix) AFO is easily handled by the patients. The comparison of different parameters of Passive, Semi active and Active AFOs is shown in Table 1.

4 Existing AFOs

4.1 Passive AFOs

4.1.1 Polypropylene AFO

Polypropylene AFO, shown in Fig. 2a, has less weight, good aesthetics, and can be worn within a shoe. This AFO is usually custom molded to fit for the patient's lower limb. When weight of the patient is applied on the foot part of this AFO, the energy storing and releasing capacity of the AFO provides resistance to plantar flexion during initial contact. From the mid stance to the terminal stance, this AFO permits dorsiflexion of ankle with little resistance. During the swing phase, the AFO holds the foot from dropping and prevents toe drag. The stiffness of the AFO depends on material properties, shape of flexible region and thickness. But fabrication of the AFO is carried out by trial and error method, which results in negative effect on the knee while walking. If this AFO is too stiff, it may delay the loading response and the knee can be more flexed. If this AFO is less stiff, it affects the patient by excessive extension of his knee. So polypropylene AFO should be designed to provide minimum ankle stiffness [12].

4.1.2 Type I AFO and Type II AFO

Yamamoto et al. [13] developed an AFO for hemiplegic patients in which initial ankle angle and magnitude of assist moments could be adjusted easily. They designed two AFOs and evaluated these AFOs by carrying out gait analysis for different gait parameters. Between these two AFOs, Type 1 AFO consisted of both dorsiflexion and plantar flexion assist moments as shown in Fig. 2b. Foot part of the AFO was made of plastic material connected by two aluminium uprights with Klenzak joints and two springs (one spring at anterior and another spring at posterior side of the ankle joint). Dorsiflexion assist moment was generated by the anterior side spring and plantar flexion assist moment by the posterior side spring. These two assist moments and the initial ankle angle could be adjusted separately by varying lengths of the springs. Gait analysis was carried out, on 33 hemiplegic patients wearing this AFO, by taking 10 gait cycles on each patient. From this gait analysis, they concluded that-for hemiplegic patients, plantar flexion assist moment was not necessary because

Table 1 Comparison of resistive moments, weight, advantages and disadvantages of reviewed AFOs

Type of AFO	Resistive/assistive	Resistive moment	Weight	Advantages	Disadvantages
DACS AFO [11]	Resistive	2–17 Nm	300 gm	Compact, untethered, adjustability of initial angle	Heavy, resistive force cannot be varied
AFO with oil damper [14]	Resistive	5–14 Nm	400 gm	Stiffness can be varied, untethered	Resistive force cannot be varied
AFO with four bar linkage [16]	Assistive	–	625 gm	Simple structure, light weight, untethered	Foot slap cannot be prevented, uncomfortable
PhAFO [17]	Resistive	–	1 kg	Untethered, variable motion control	Heavy, bulky in size
Six bar linkage orthotic device [18]	Assistive	–	–	Untethered, simple	Initial angle cannot be varied
AFO with MR brake [20]	Resistive	24 Nm	1.6 kg	Stiffness can be varied, compact	Tethered, heavy
AAFO with SEA [22]	Assistive	–	2.6 kg	Assist both plantar flexion and dorsiflexion	Tethered, heavy
Two DOF AFO [23]	Assistive	–	2.5 kg	Provide two DOF, assist both plantar flexion and dorsiflexion	Tethered, heavy
DAFO [24]	Assistive	–	–	Assist both plantar flexion and dorsiflexion, provide two DOF	Tethered, complicated mechanism
MACCEPA AFO [26]	Assistive	–	1.7 kg	Assist both plantar flexion and dorsiflexion, can be adjusted for any length of foot	Tethered, complicated design
ADJUST [28]	Assistive	–	–	Assist both plantar flexion and dorsiflexion	Tethered, heavy, bulky in size

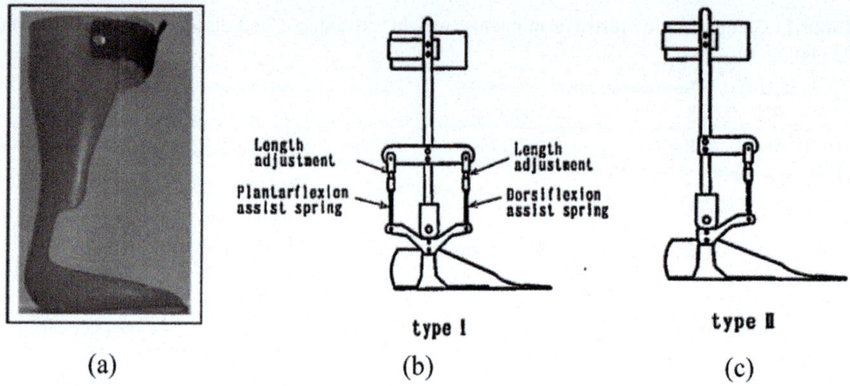

Fig. 2 **a** Polypropylene AFO [12]. **b** Type I AFO with both plantar flexor and dorsiflexor assist springs. **c** Type II AFO with only dorsiflexor assist springs [13]

the patients were uncomfortable due to presence of the posterior spring. So, Type 2 AFO, shown in Fig. 2c, was developed without the posterior spring so that the ankle could rotate freely in dorsiflexion. Gait analysis of a patient, wearing this AFO, was carried out. Results from this analysis showed—(1) reduction in the knee flexion moment and (2) absence of the quick plantar flexion.

4.1.3 Dorsiflexion Assist Controlled by Spring AFO

Based on observations from testing of Type 1 and Type 2 AFOs, Yamamoto et al. [11] developed a new AFO called as 'Dorsiflexion Assist Controlled by Spring (DACS) AFO'. DACS AFO had two parts, namely foot and shank parts, made of plastic, as shown in Fig. 3a. These foot and shank parts were connected at the ankle joint. The AFO contained an assistive device with a spring at the shank part. The spring developed dorsiflexion assist moment of about 2–17 Nm per 10° of plantar flexion.

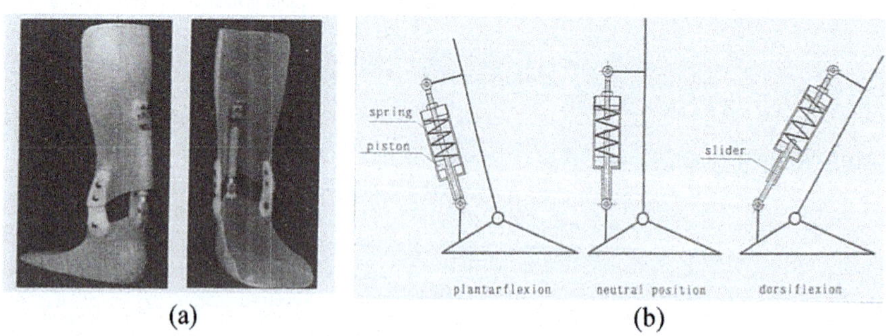

Fig. 3 **a** DACS AFO. **b** Kinematic Diagram of DACS AFO for plantar flexion and dorsiflexion [11]

But the spring did not reach the ideal value of 20 Nm of dorsiflexion assist moment. DACS AFO also had 4 springs of different spring coefficients to adjust the dorsiflexion assist moments. The DACS AFO's initial ankle angle could be adjusted by varying length of the assistive device.

When the ankle rotated to plantar flexion, the piston compressed the spring and developed assist moment proportional to the plantar flexion angle. In reverse direction, when the ankle rotated to dorsiflexion, foot part of the AFO rotated freely due to the motion of slider with less friction. So, the AFO only developed dorsiflexion assist moment and did not develop plantar flexion assist moment. Kinematic diagram of DACS AFO is shown in Fig. 3b. This DACS AFO reduced knee hypertension and improved the delayed progression of hip joint and increased the walking speed. The feedback of the patients were also positive. The limitations of this AFO were it had large spring units and bulky in size.

4.1.4 An AFO with Oil Damper

Yamamoto et al. [14, 15] designed a new AFO which consisted of an oil damper (weighing 100 gm, fitted at lateral side of ankle joint of the AFO), a metal plate, ring portion (to fit the ankle of the patient), and single axis joint (at medial side of the AFO) as shown in Fig. 4a.

When foot part of the AFO moved to plantar flexion during the initial contact, the piston in the oil damper moved upward with the help of a hook located on the metal plate. Due to movement of the piston, oil in the cylinder forced out through

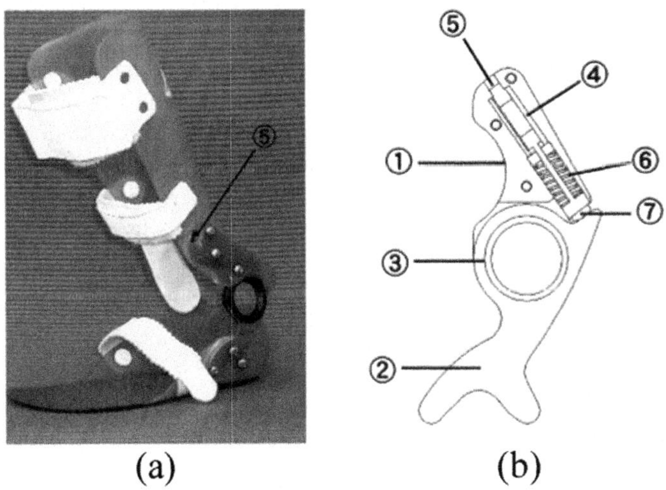

Fig. 4 **a** AFO with oil damper. **b** Oil damper (where 1: Oil damper unit, 2: Metal Part, 3: Ring Portion, 4: Hydraulic cylinder, 5: Adjustable screw, 6: Spring to assist dorsiflexion, 7: Rod cap to set initial ankle angle) [14]

an orifice with resistance. The resistance for oil flow through the orifice could be increased or decreased by varying the diameter of the orifice. It could be carried out by rotating the adjustable screw even while using this AFO. If the diameter of the orifice was more, the oil squeezed out without much resistance. So that resistance to the plantar flexion was less and vice versa. The oil coming out of the cylinder went back through check valves provided in the piston without resistance. This results in free rotation of foot part of the AFO in dorsiflexion direction. The arrangement of an oil damper is shown in Fig. 4b. It also contained a spring, which helped the piston to return to its initial position. The initial ankle angle could also be adjusted to 0°, 5°, and 8° dorsiflexion. The total mass of the AFO was approximately 400 g.

The plantar flexion resistance moment was induced between 5 and 14 Nm per 10° of plantar flexion. Evaluation of the AFO was carried out, on two patients with hemiplegia (Note: These two patients were using polypropylene AFOs before being tested with the AFO with oil damper), by 3D gait analysis. Evaluation with this AFO (with oil damper) revealed that- 'speed of walking of the patients increased' and 'cadence of the patients also increased'. Both patients were more comfortable using AFO with oil damper, when compared to using the polypropylene AFO. It was also easy to wear this AFO inside a shoe.

4.1.5 An AFO with Four Bar Linkage

Peter et al. [16] developed a new passive AFO, shown in Fig. 5a, which had four bar rigid links coupled together with flexion of knee and ankle of the patient. It was designed to provide necessary assistive force to lift the foot of the patient during the swing phase. The link joints were pivoted and made of aluminium bars. Mass of the AFO was 625 gm.

When the knee was flexed at an angle between 5 and 20°, the linkage and thigh part would come into contact. This resulted in lifting of the foot by assistive force provided by the thigh part of the linkage. When the patient was in stationary position and the knee was not flexed, assistive force would not act on his foot. The standing and walking positions of the patient with the AFO are shown in Fig. 5b and c respectively. Gait analysis (by plotting hip, knee and ankle angle graphs) of a healthy person

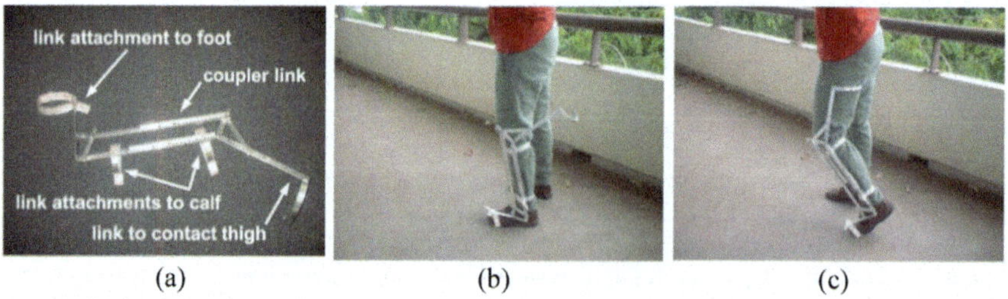

Fig. 5 **a** Four bar linkage. **b** Standing linkage position. **c** Walking linkage position [16]

wearing this prototype was carried out. The graphs showed lift of the foot during the swing phase and prevented toe drag. The main drawbacks of this prototype were the following: (1) it was not comfortable to wear, (2) it was bulky in appearance, (3) it prevented only toe drag, and could not prevent slap foot.

4.1.6 Pneumatic harvested AFO

Pneumatic harvested AFO (PhAFO) was developed by Chin et al. [17] consisted of two components, called foot plate and tibial upright. These two parts were made of carbon composite laminate. The AFO had a heel of 1 cm. Sole of the AFO had a bellow pump, two check valves, an actuator, and a pressure release valve as shown in Fig. 6. It also comprised of a cam lock mechanism attached at the lateral side of the AFO. The actuator consisted of a roller follower, a guide rail, and a linear cylinder. The total mass of the AFO was 1 kg. At the heel contact, the compressed air from the cylinder released to atmosphere through the release valve. The spring in the cylinder would come to the initial position leading to unlock the cam and this resulted in plantar flexion of the foot. During the stance phase, the bellow pump compressed and the release valve closed. This resulted in extension of roller follower and due to the cam design, free dorsiflexion was allowed. During the swing phase, cam was locked due to the follower and prevented the toe drag.

Gait analysis was carried out on a healthy person with the AFO on one leg and shoe on the other leg. Kinetic and kinematics data showed the clearance of the toe during the swing phase. But during the mid-swing, excessive dorsiflexion occurred. Testing of AFO was not conducted on any patient.

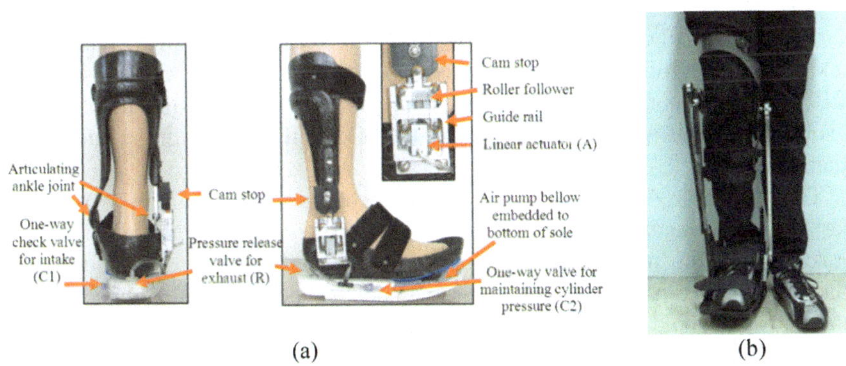

Fig. 6 **a** Posterior and lateral view of power harvesting AFO [17]. **b** Six-bar linkage orthotics [18]

4.1.7 Six Bar Linkage Orthotic Device

Ghosh et al. [18, 19] developed an orthotic device, which contained a six-bar mechanical linkage to be attached to the thigh of the patients. At first, the walking data (knee and ankle angles, and toe movement) were obtained from a healthy person and set that as a standard data for their studies. They also set 11 task points (obtained from the standard data) for path synthesis of toe movement for the normal walking. By using Mathematica Software and the standard data obtained from the healthy person, they developed a six-bar linkage orthotic device based on RR anthropometric backbone chain as shown in Fig. 6b. Evaluation of the orthotic device was carried out by simulation (using OpenSim software to get knee and ankle angles of the orthotic device). It was observed that there was $\pm 10^\circ$ error between the simulation data (i.e. knee and ankle angles) of the orthotic device and the standard data (i.e. knee and ankle angles of the healthy person measured above).

Gait analysis (to measure knee and ankle angles, and toe movement of the wearer) was carried out on a healthy person wearing the orthotic device. Observations from the gait analysis showed $\pm 12\%$ error in the knee and ankle angles and the toe was missing a single task point, out of the 11 task points of the standard data, which were defined earlier for the toe movement of the normal walking. This was one of the major drawbacks of this device.

4.2 Semi Active AFOs

4.2.1 AFO with MR Brake

Furusho et al. [20] developed an AFO with Magnetorheological (MR) brake at the ankle joint to control the ankle angle. The MR brake consisted of MR fluid in it. The MR fluid was a non-colloidal solution, that contained magnetic metal particles as a solute and oil as a solvent. Viscosity of the MR fluid varied with the intensity of magnetic field.

The MR brake AFO was made by using a compressive force sensor at its sole part and a bending moment sensor at its shank part to control the timing of the MR brake. During the swing phase, the toe was cleared off from the ground due to its plantar flexion resistance. During the heel contact, shock due to the impact of heel on the ground could be absorbed by the MR brake. They designed two prototypes of AFO with MR brake. First prototype (see Fig. 7a) was made of a rotor and a stator of 0.2 mm thick, MR fluid layer of 50 μm thick, number of coils were 44 turns, current flow of 1 A, magnetic flux of 0.3 T, and shear stress developed was 10 kPa. The torque obtained at ankle of the AFO was only 0.71 Nm (theoretical torque was 2.2 Nm). They designed a second prototype (see Fig. 7b) by increasing the number of coils to get a magnetic flux of 0.5 T, increasing the number of disk and outer diameter, and piston mechanism was introduced to prevent fluid leakage. The total weight of the AFO was 1.6 kg. The AFO also contained a linkage mechanism (see Fig. 7c)

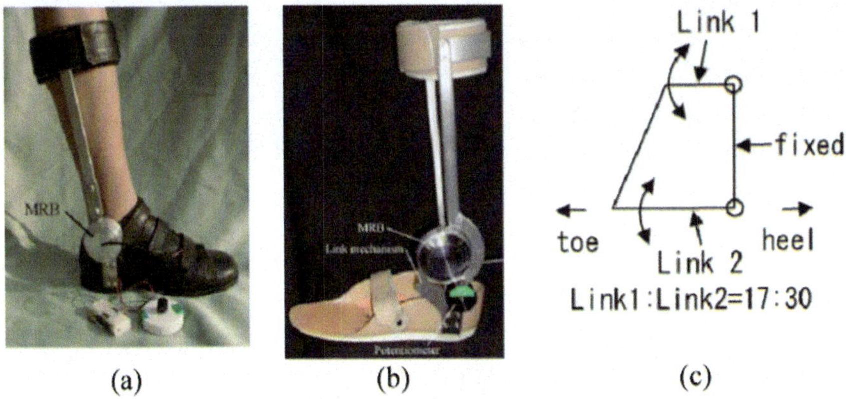

Fig. 7 **a** First prototype with MR damper. **b** Second prototype with MR damper. **c** Linkage mechanism [20]

(to amplify the torque up to 24 Nm), a potentiometer at its ankle joint (to measure the ankle angle), a six axis force torque sensor at the bottom (to measure GRF) and bending moment sensor at the shank (to measure bending moment around the ankle). The AFO worked according to control algorithm in four stages. *Stage 1*: During the heel contact, depending on the ankle angular velocity, braking torque was applied to prevent the slap foot. *Stage 2*: During the stance phase, the AFO allowed free ankle motion with zero braking torque. *Stage 3*: During the toe off phase, constant torque was applied to prevent the toe drag. *Stage 4*: From the toe off to heel contact, torque turned down for the next heel contact. This AFO was evaluated for gait analysis on a patient and his gait improved. The limitations of the AFO were its heavy weight and require external power supply.

4.2.2 An AFO with SMA Alloys

Mataee et al. [21] proposed an AFO with shape memory alloys (SMA). The SMA alloy used for AFO was super elastic NiTi (Nickel-Titanium) element. This SMA element at ankle joint of the AFO, stored and released energy to assist the ankle joint. They designed two models of the AFO. In the first model, a NiTi rod of length 25 mm and diameter 6.4 mm was used along with an actuator at the ankle joint of the AFO. The NiTi rod was axially loaded using the actuator on one side and torsional load by motion of the ankle of wearer on another side. These loading conditions were divided into four stages followed by (i) axial preload during mid stance by the actuator, (ii) rotational (or torsional) load by motion of the ankle during terminal stance, (iii) unloading partial load during pre-swing, and (iv) unloading remaining load during swing phase to prevent the toe drag. The stiffness of NiTi rod could be varied by controlling the preload and partial axial load recovery. So, this AFO could

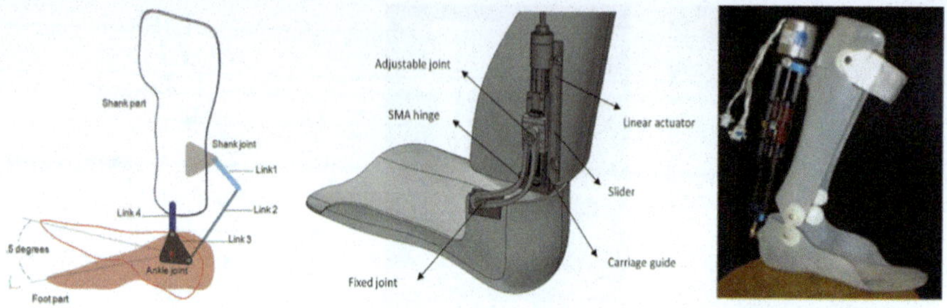

(a) (b) (c)

Fig. 8 **a** Four link transmission mechanism (first model) [21]. **b** AFO with SMA hinge second model) [21]. **c** AFO with series elastic actuator [22]

be used for slow, normal, and fast walking speeds. The transmission mechanism contained four bar link is shown in Fig. 8a.

In the second model of the AFO, the NiTi element was used as an adjustable hinge (at the ankle joint of the AFO). This AFO contained a linear actuator, a slider, and a carriage guide as shown in Fig. 8b. Here, by adjusting the slider with the help of the actuator, active length of the hinge could be varied. Bending stiffness and the corresponding ankle stiffness of the AFO depended on the active length of the hinge. By varying the active length of the hinge, stiffness of the device could be adjusted for slow, normal, and fast walking speed.

They analysed these two models of the AFOs numerically. Numerical results showed that the first model (tension-torsion load) gave better performance in ankle stiffness variation than the second model (bending stiffness model). They concluded that for slow walking speed, both models did not give satisfactory results to exactly mimic the motion of the ankle. Physical models or prototypes were not developed for study purpose.

4.3 Active AFOs

4.3.1 AAFO with SEA Actuator

Blaya and Herr [22] presented a variable impedance active ankle foot orthosis (AAFO). They used SEA (Series Elastic Actuator) to control stiffness of the ankle joint according to the plantar flexion and dorsiflexion movement. The AAFO contained an articulated AFO (made of polypropylene) with mechanical hinge, a potentiometer (to determine ankle angle), and a GRF (Ground Reaction Force) sensor (at the bottom of the AFO to measure GRF) as shown in Fig. 8c. The SEA was controlled by PC with codes written in C and based on control algorithm. They divided the control algorithm into three stages. First stage: from the heel strike to midstance, the SEA automatically adjusted the stiffness to avoid the slap foot. Second stage: from

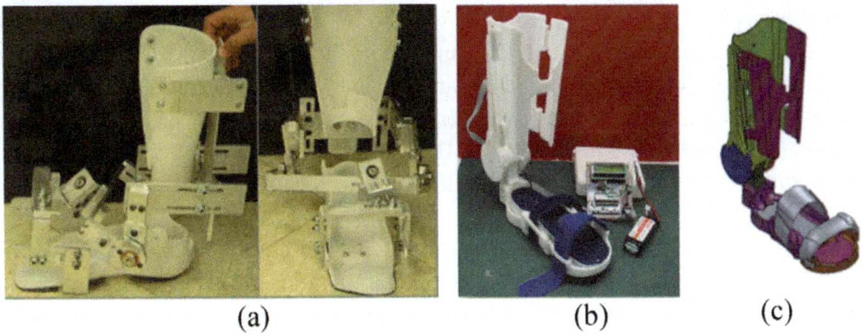

Fig. 9 **a** AFO with 2 DOF [23]. **b** DAFO [24]. **c** DAFO CAD model [24]

midstance to toe off, the SEA force was set to zero to allow free ankle movement by minimizing the ankle joint impedance. Third stage: during swing phase, the SEA was set to constant impedance to prevent the toe drag.

Gait analysis was carried out by using two drop foot patients and three healthy persons. Feedback from the patients were good and they preferred to use the AFO in future, if its weight could be reduced (actual weight-2.6 kg).

4.3.2 Two DOF AFO

Agrawal et al. [23] developed an AFO, shown in Fig. 9a, which had two degrees of freedom (plantar flexion-dorsiflexion and inversion-eversion motions). Among these two motions, plantar flexion-dorsiflexion motion was controlled by an active DC servomotor. The inversion-eversion motion was controlled by a spring and a damper. At first, the kinematic and dynamic models of the ankle were designed. The AFO was developed with help of the kinematic model of the ankle. Limbs of the AFO were made of aluminium. The plantar flexion-dorsiflexion axis and subtalar joints were designed using cubes obtained from the kinematic model of the ankle. Limbs were made in telescopic structure, due to the fitting for different foot dimensions of the patients. The inversion eversion motions could be controlled by using the torsional spring. The plantar flexion-dorsiflexion joint was actuated by DC servomotor by using set-point control law for point to point motion. The drawback of this AFO was its weight (2.5 kg). Also, this AFO was designed for anthropometric data from average person. To change the kinematic angles to fit for other patients, cubes could be replaced for specific patient.

4.3.3 Dynamic AFO

Patar et al. [24] developed a Dynamic Ankle Foot Orthosis (DAFO) manufactured using rapid prototype method and Fuse Deposition Method. The DAFO was made of polypropylene material. The DAFO worked on a DC Servomotor actuator. The prototype was designed to get two DOF motion including plantar flexion/dorsiflexion and inversion/eversion motions. The plantar flexion/dorsiflexion motion worked by the DC Servomotor actuator and the inversion/eversion motion by passive spring and damper (see Fig. 9b and c respectively). The bottom of sole of the DAFO contains two limit switches (at the front and rear position of the sole). These switches controlled rotation of the DC motor.

Based on signals provided by the switches, servomotor rotated clockwise and counterclockwise directions to get the plantar flexion and dorsiflexion motions. DAFO worked according to 11 sequences. The gait cycle started from the standing position and lead to heel off (Sequence 1,2), toe off (Sequence 3,4), heel strike (Sequence 5,6), stance phase (Sequence 7,8), again heel off (Sequence 9,10), toe off (Sequence 11), and heel strike and corresponding motor rotated to 22° Counter clockwise (CCW), 34° CW, 12° Clockwise (CW), 0°, 22° CCW, 34° CCW and 12° CW respectively to perform one complete cycle. This DAFO could be used for different walking speeds by changing rotation speed of the motor. They only concentrated on the cost reduction to produce the prototype and achieved. But no report was found on evaluation of the AFO being carried out on any patient.

4.3.4 Three DOF AFO

Zhang et al. [25] proposed an AFO having three DOF including plantar flexion/dorsiflexion, abduction/adduction, and internal/external rotation of the ankle. This AFO consisted of an electromagnetic clutch, a DC motor, a bidirectional power transmission ball screw, a steel wire, an angle sensor, and a force sensor (see Fig. 10a).

The AFO worked on engagement and disengagement of the electromagnetic clutch. During the loading response, the electromagnetic clutch disengaged and spring was compressed to store energy. In heel off phase, the electromagnetic clutch

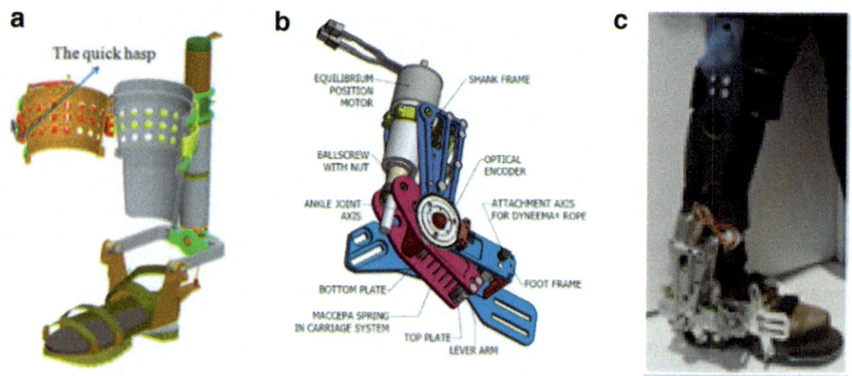

Fig. 10 **a** Three DOF AFO [25]. **b** CAD Model of AFO. **c** Prototype of AFO [26]

engaged to run the DC motor. The rotary motion of the DC motor was converted to linear motion by using the ball screw. This ball screw pulled the steel wire to prevent foot drop during swing phase. Abduction/adduction motion of the AFO was provided by using resilient rubber. Internal/external rotation of the AFO was provided by using double tier beleaguered structure. Only biomechanical analysis of the ankle joint of the AFO was carried out and its prototype has not been tested.

4.3.5 MACCEPA AFO

Maltedo et al. [26] presented an AFO having a compliant actuator, called as MACCEPA (Mechanically Adjustable Compliance and Controllable Equilibrium Position Actuator). The AFO assisted both plantar flexion and dorsiflexion moments. The AFO consisted of an equilibrium motor, a ball screw, a SODEMANN compression spring, a toe plate, a heel plate, a foot frame, and a shank frame. The rotary motion of the equilibrium motor was transformed to translation of nut in ball screw, which lead to rotary motion of lever arm around the joint axis. Spring lied between top and bottom plates (see Fig. 10b). The bottom plate connected to the spring by using a rope and the rope was wrapped to the foot frame. When the foot frame rotated due to the actuator, the rope either released or wrapped. This lead to changes in pre-compression of the spring and torque around the ankle. The AFO could be adjusted for different foot length. The total mass of this device was 1.7 kg. Testing of the AFO was not conducted on any patient. Prototype of the AFO is shown in Fig. 10c.

4.3.6 AFO with ServoTube

Gmerek et al. [27] developed a kinematic and dynamic model of an AFO with ServoTube actuator. It consists of shank and foot parts connected by a hinge. The hinge consists of a ball bearing which helps the user to control tilting of the body. The AFO contains a force sensor, a position sensor, a load cell, sliders, and an actuator at the rear portion of the shank (Fig. 11a).

This AFO works by controller, which has two different feedback loops, an outer loop (which operates on tracking error), and a inner loop (based on dynamic model of the AFO). Evaluation of the AFO was carried out in virtual environment which would reduce the selection of hardware and control design complications. But this AFO was not tested experimentally on any patient.

4.3.7 ADJUST

Wilk et al. [28] developed an AFO, called as ADJUST which controlled both plantar flexion and dorsiflexion motions independently. ADJUST consisted of two mechanical hinges (at the lateral and medial side of the ankle), a foot plate, and a shank cover (see Fig. 11b). Each hinge contained a pawl and ratchet (see Fig. 11c), a solenoid, an Arduino, a battery, two FSR (force sensing resistors) (at the bottom of the sole),

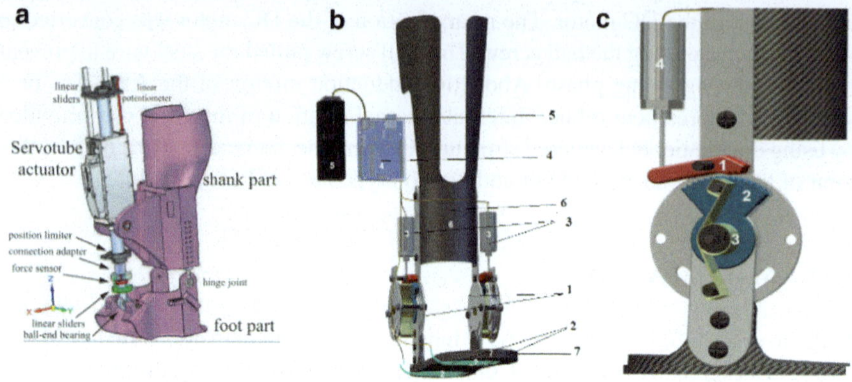

Fig. 11 **a** Active AFO with ServoTube [27]. **b** ADJUST (1: Mechanical Hinge, 2: Force Sensing Resistor, 3: Arduino, 4: Solenoids, 5: Battery, 6: Shank Cover, 7: Foot Plate). **c** Mechanical hinge of ADJUST (1: Pawl, 2: Ratchet, 3: Leaf Spring, 4: Solenoid) [28]

and two springs. The lateral side springs controlled the plantar flexor function, and the medial side springs controlled the dorsiflexor function. Signals from the FSR activated and deactivated the Solenoid. When the Solenoid was activated, ratchet unlocks the pawl, and the hinge rotated freely. When the Solenoid was deactivated, pawl and ratchet locked, and rattling rotation occured only in one direction. During the stance phase, both the Solenoids were deactivated, due to pressing of the FSR which locked both the hinges. During heel contact to foot flat phase, the medial hinge stored energy, and the lateral hinge allowed rattling rotation to plantar flexion. During foot flat to heel off phase, the lateral hinge stored energy, and the medial hinge allowed rattling rotation to the dorsiflexion. During heel off to toe off phase, the lateral hinge released partial energy. During swing phase, the Solenoid activation lead to locking of the ratchet and pawl. The lateral and medial hinges released energy and lifted the foot plate preventing the toe drag.

Gait analysis was carried out on a patient having both plantar flexor and dorsiflexor muscle paralysis in five different conditions. Initial one with his own AFO and remaining with different stiffness of the ADJUST. (1. Both medial and lateral spring to stiff, 2. Medial stiff and lateral flexible, 3. Both medial and lateral spring to flexible, 4. Medial flexible and lateral stiff). Patient preferred first two conditions in the ADJUST over the other two conditions, where the foot slap was noticed. But the patient was uncomfortable with the weight and bulky size of the AFO.

5 Discussion

Though many AFOs are reported in the literature, only Polypropylene AFO [12] and AFO with oil damper [14] are commercially available for daily use. Other AFOs are still in research and development phase. AFO with oil damper shows improvement in walking speed, step length, peak negative power, and peak plantar flexor moment [15, 29]. Only one AFO in the literature is observed to have change of stiffness for slow, normal and fast walking conditions [21]. Remaining AFOs are designed for normal walking speed only and none of these AFOs are suitable for running purpose. Five AFOs [22–24, 26, 28] in the literature assist both the plantar flexion and dorsiflexion motions. Other AFOs either assist or resist only in one direction. Except the three AFOs [23–25], other AFOs have only one DOF of motion. i.e., plantar flexion/dorsiflexion. Lack of inversion/eversion and abduction/adduction motions may lead to unnatural gait. Except Polypropylene AFO [12] and six bar linkage orthotic device [18], all the AFOs are able to change their initial ankle angle.

Passive AFOs are less in weight compared to active and semi active AFOs. Passive AFOs successfully restrict the unnecessary motion of the ankle during the swing phase. But these AFOs will not provide much push off during the toe off phase. Additionally, frictional forces reduce their functions. Passive AFOs are not able to adjust their stiffness according to walking conditions. While active AFOs, using different sensors, can change their stiffness during walking.

Active and Semi Active AFOs overcome some of the limitations of passive AFOs by changing the stiffness according to walking conditions. Active and Semi Active AFOs provide assist moments for dorsiflexion during the stance phase and propulsive force during the toe off phase. Due to lack of assist torque, their utilization for daily use is limited. These AFOs use external power supply and are bulky in size. So, these AFOs are limited to experimentation in clinical laboratories only.

Among active AFOs, AFO with SMA [21] changes the stiffness of ankle joint by minimum actuation. The property of restoring and releasing energy of SMA element makes it possible. But slow response for applied load is a drawback. In future, by considering other design parameters, there is a chance to improve the AFO with SMA.

Researchers face many challenges like light weight actuators, power supply, and efficient transmission to design compact AFOs. Designing an AFO that is suitable for every patient is still challenging, due to the nature of disability that varies from one patient to another patient.

6 Conclusion and Future Work

Here, a literature review was carried out on various designs of the existing AFOs used to treat the drop foot. From this review, we can conclude that, there is a need for design of compact, lightweight, untethered, and efficient AFO that could fit for daily life usage. Compared to passive AFOs, active and semi active AFOs give better rehabilitation for patients with the drop foot. Active and semi active AFOs provide assistance to dorsiflexion during the swing phase and allows free rotation of ankle during the stance phase. There is still a challenge to develop an AFO which meets all the above said characteristics.

References

1. Rose J, Gamble JG (2006) Human walking, 3rd edn. Lippincott Williams and Wilkins, Philadelphia
2. Whittle MW (1991) Gait analysis: an introduction, 1st edn. Butterworth-Heinemann
3. Birch R (2008) A history of limb amputation. J Bone Jt Surg Br 90–B(10):1276–1277
4. Gage J (1990) An overview of normal walking. Instructional course lectures, vol 39. University of Waterloo Press
5. Perry J (1992) Gait analysis: normal and pathological function. Slack Incorporated
6. Meinders M, Gitter A, Czerniecki JM (1998) The role of ankle plantar flexor muscle work during walking. Scand J Rehabil Med 30(1):39–46
7. Durham S, Eve L, Stevens C, Ewins D (2004) Effect of functional electrical stimulation on asymmetries in gait of children with hemiplegic cerebral palsy. Physiotherapy 90(2):82–90
8. Stewart JD (2008) Foot drop: where, why and what to do? Pract Neurol 8(3):158–169
9. Bregman DJJ, Harlaar J, Meskers CGM, De Groot V (2012) Spring-like ankle foot orthoses reduce the energy cost of walking by taking over ankle work. Gait Posture 35(1):148–153
10. Bhadane Deshpande M (2012) Towards a shape memory alloy based variable stiffness ankle foot orthosis. PhD thesis, The University of Toledo
11. Yamamoto S, Ebina M, Kubo S, Hayashi T, Akita Y, Hayakawa Y (1999) Development of an ankle foot orthosis with dorsiflexion assit, part 2: structure and evaluation. J Prosthet Orthot 11(2)
12. Ramsey JA (2011) Development of a method for fabricating polypropylene non-articulated dorsiflexion assist ankle foot orthoses with predetermined stiffness. Prosthet Orthot Int
13. Yamamoto S, Ebina M, Miyazaki S, Kawai H, Kubota T (1997) Development of a new ankle-foot orthoses with dorsiflexion assist, part 1—desirable characteristics of ankle-foot orthoses for hemiplegic patients. J Prosthet Orthot 9(4)
14. Yamamoto S, Hagiwara A, Mizobe T, Yokoyama O, Yasui T (2005) Development of an ankle-foot orthosis with an oil damper. Prosthet Orthot Int 29(3):209–219
15. Yokoyama O, Sashika H, Hagiwara A, Yamamoto S, Yasui T (2005) Kinematic effects on gait of a newly designed ankle-foot orthosis with oil damper resistance: a case series of 2 patients with hemiplegia. Arch Phys Med Rehabil 86(1):162–166
16. Berkelman P, Rossi P, Lu T, Ma J (2007) Passive orthosis linkage for locomotor rehabilitation. In: 2007 IEEE 10th international conference on rehabilitation robotics ICORR'07, vol 00, no c, pp 425–431
17. Chin R et al (2009) A pneumatic power harvesting ankle-foot orthosis to prevent foot-drop. J Neuroeng Rehabil 6(1):1–11
18. Ghosh S, Robson N, McCarthy JM (2017) Design of wearable lower leg orthotic based on six-bar linkage. IDETC/CIE, pp 1–10
19. Ghosh S, Robson N, McCarthy JM (2018) Development of customized orthotics based on lower-leg anthropometric data and task. In: International conference on applied human factors and ergonomics, vol 608
20. Furusho J, Kikuchi T, Tokuda M, Kakehashi T, Ikeda K, Morimoto S, Hashimoto Y, Tomiyama H, Nakagawa A, Akazawa Y (2007) Development of shear type compact MR brake for the intelligent ankle-foot orthosis and its control. In: IEEE 10th international conference on rehabilitation robotics ICORR'07, vol 00, no c, pp 89–94
21. Mataee MG, Andani MT, Elahinia M (2015) Adaptive ankle-foot orthoses based on superelasticity of shape memory alloys. J Intell Mater Syst Struct 26(6):639–651
22. Blaya J, Herr H (2004) Adaptive control of a variable-impedance ankle-foot orthosis to assist drop foot gait. Neural Syst Rehabil Eng IEEE Trans 12(1):24–31
23. Agrawal A, Banala SK, Agrawal SK, Binder-Macleod SA (2005) Design of a two degree-of-freedom ankle-foot orthosis for robotic rehabilitation. In: IEEE 9th International Conference *on* Rehabilitation Robotics, pp 41–44
24. Patar A, Jamlus N, Makhtar K, Mahmud J, Komeda T (2012) Development of dynamic ankle foot orthosis for therapeutic application. Int Symp Robot Intell Sensors 41(Iris):1432–1440

25. Zhang C, Zhu Y, Fan J, Zhao J, Yu H (2015) Design of a quasi-passive 3 DOFs ankle-foot wearable rehabilitation orthosis. Biomed Mater Eng 26(2):647–654
26. Moltedo M, Bacek T, Junius K, Vanderborght B, Lefeber D (2016) Mechanical design of a lightweight compliant and adaptable active ankle foot orthosis. IEEE RAS/EMBS Int Conf Biomed Robot Biomech 1224–1229
27. Gmerek A, Davoodi M, Meskin N, Tehrani ES, Kearney RE (2017) A new ankle foot orthosis: modeling and control. Int Conf Control Decis Inf Technol 1:1066–1071
28. Wilk DVD, Reints R, Postema K, Gort T, Harlaar J, Hijmans JM, Verkere GJ (2018) Development of an ankle-foot orthosis that provides support for flaccid paretic plantarflexor and dorsiflexor muscles. IEEE Trans Neural Syst Rehabil Eng 26(5):1036–1045
29. Yamamoto S, Ibayashi S, Fuchi M, Yasui T (2015) Immediate-term effects of use of an ankle–foot orthosis with an oil damper on the gait of stroke patients when walking without the device. Prosthet Orthot Int 39(2):140–149

Kinematic and Dynamic Analyses of Lower Body Exoskeleton Mechanism Using Adams

Shubham S. Kawale and M. Sreekumar

Abstract This paper illustrates the validation of a mathematical model developed for the kinematic and dynamic analyses of the lower body exoskeleton mechanism using Adams software. This framework is designed to assist individuals in the shipbuilding industry for lifting heavy payload with reduced physical fatigue. It contains modeling, simulation, and validation of the proposed lower body exoskeleton mechanism. An Adams software-based simulation environment is developed to validate the results obtained from the kinematic and dynamic analyses of the mechanism. Further, this work also extends to actuator and valve selection, sizing by utilizing clinical gait data (CGA). Torque versus gait cycle curves obtained for all active exoskeleton joints shows that the design has adequate torque to perform tasks with a payload of 30 kg. The design also ensures a minimum consumption of energy at the same time. Future work will be to build a prototype and test simulation results.

Keywords Exoskeleton mechanism · Wearable mechanisms · Clinical gait analysis

1 Introduction

The exoskeleton mechanism is an external framework attached to the human operator. It works parallel with the operator. It acts as an amplifier that augments the reinforce or restore human performances [1]. Simply, it provides physical assistance to a human body part for specific purposes such as reducing physical fatigue and providing physical support to body parts.

Exoskeletons are majorly used for military, medical and industrial purposes. Military exoskeleton improves physical capabilities and used in a combat environment. Whereas medical exoskeletons are majorly used for restoring human natural capabilities and assisting patients with physical limitations. Industrial exoskeletons are

S. S. Kawale (✉) · M. Sreekumar
Department of Mechanical Engineering, Indian Instituete of Information Technology, Design and Manufacturing, Kancheepuram, Chennai 600127, India
e-mail: mds16m006@iiitdm.ac.in

M. Sreekumar
e-mail: msk@iiitdm.ac.in

D. Sen et al. (eds.), *Mechanism and Machine Science*, Lecture Notes in Mechanical Engineering, https://doi.org/10.1007/978-981-15-4477-4_57

developed for fully abled personal to prevent muscular diseases and reduce physical fatigue as a result of various operational tasks. One of the demanding fields is the shipbuilding industry. Because of its complex, unrepeated tasks and various space constraints limit automation [2]. Whereas exoskeleton technology can be offered as an alternative solution in such problems. A concept mechanism design is already presented to offer an optimized solution [3]. This work presents the concept, kinematic and dynamic analysis of lower body exoskeleton for shipbuilding application. After presenting the kinematics and the dynamics of the exoskeleton, this work focuses on the optimization of the positions of the anchor points of the linear actuators. Further, the proposed exoskeleton design is compared with NTU wearable exoskeleton [4]. It is shown that the proposed planar kinematics improves existing solutions by reducing the power consumption of the actuators. In short, the novelty introduced by this work is the use of linear actuators at optimized anchor points to obtain an energy-efficient system (Fig. 1).

The current work will be focusing on the evaluation of kinematic and dynamic analyses of the lower body exoskeleton mechanism using Adams. The results obtained are compared to a mathematical model using the Lagrangian method and Adams. This work also shows linear actuator design and valve selection and sizing.

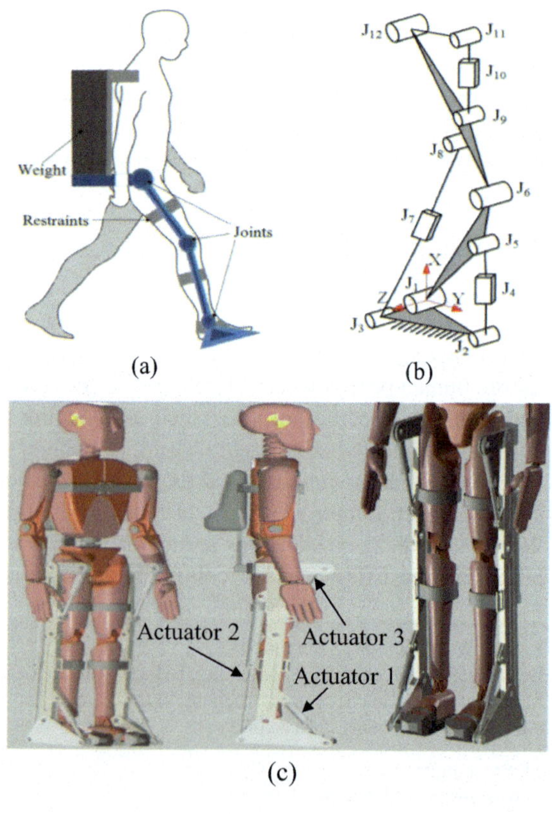

Fig. 1 **a** Typical exoskeleton mechanism, **b** Kinematic configuration of the proposed exoskeleton mechanism, **c** CAD models of the proposed lower body exoskeleton mechanism

Further, it shows the use of clinical gait analyses (CGA) to decide the dimensions of actuators, which have sufficient force to walk with minimum energy consumption. This hydraulic system design is inspired by BLEEX [5, 6]. The objective function of the current work is to validate the mathematical modeling using Adams software and to design hydraulic system components for minimum power consumption. This paper is composed mainly of three sections. The actuator selection and valve sizing are presented in Sect. 2, the simulation and its study of results which is explained in Sect. 3 and finally, the work is concluded in the subsequent section and followed by references.

1.1 Contributions

This work provides a cost-efficient lower body exoskeleton mechanism. It shows the evaluation of the kinematic and dynamic parameters of exoskeleton using Adams model. This paper also shows the use of CGA parameters to fix the actuator dimensions which ultimately yield to low weight and cost-effective design.

2 Actuator Selection and Valve Sizing

2.1 Requirement of the Actuators

One of the important factors in the design and development of industrial exoskeleton is the selection of actuators and sizing. The ideal actuators should have a high ratio of power to weight, which ensures the availability of sufficient torque to drive the body's movement while lifting a specific payload. During normal walking, the flexion and extension of the hip, knee, and ankle require a bilateral driving mechanism. The driving torque of the knee and ankle joint are about 230 Nm and 15 Nm respectively [5, 7]. Looking toward the requirement of the proposed exoskeleton, the hydraulic power system is selected for further design.

2.2 Selection of the Actuator and Sizing

To determine the optimum size of actuators, an analytical relationship must be established between required torque output and pressure input. A linear actuator is producing torque about the ankle joint as shown in Fig. 2 a. can relate to varying moment arm $R(\theta)$, supply pressure, and size of the actuator. Whereas, the hydraulic circuit diagram is shown in Fig. 2 b.

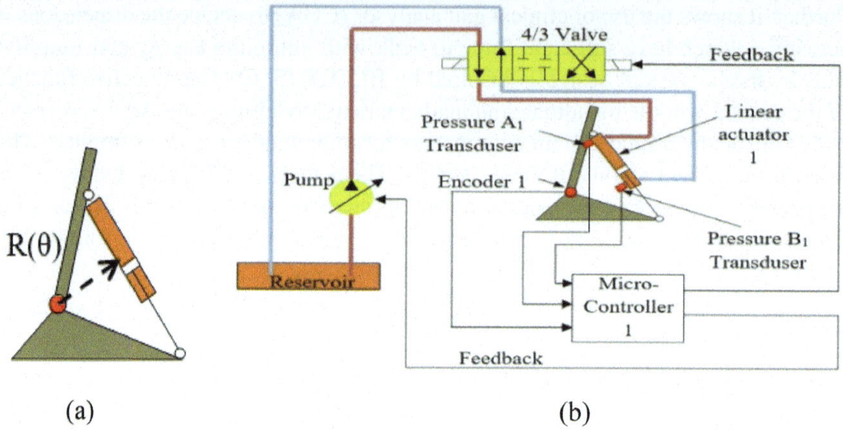

Fig. 2 **a** Arrangement of a linear actuator about an ankle joint, **b** Hydraulic circuit for the proposed exoskeleton system

$$T_{\text{max.push}} = P_s \pi D_{\text{Bore}}^2 R(\theta)/4 \quad (1)$$

$$T_{\text{max.pull}} = P_s \pi (D_{\text{Bore}}^2 - D_{\text{rod}}^2) R(\theta)/4 \quad (2)$$

where T is the torque acting on exoskeleton joint, D_{Bore} is bore diameter of the actuator, D_{rod} is rod diameter of the actuator, P_s is the input/supply pressure, and $R(\theta)$ is a moment arm which is the perpendicular distance between rotary joint and action of a force. In short, the output torque is a result of linear actuator force times of the moment arm [5]. The actuator sizing analyses is performed to determine the best possible combination of actuator's standard bore size, actuator pivot location, and input pressure. It ultimately provides the torque which is greater than the minimum required one and has the least hydraulic power consumption. The minimum required torque is based on the CGA data at normal walking speed. Additionally, the actuator must provide the intended range of motion with ergonomically suitable velocity and torque. In general, there is not a unique solution, but many feasible possibilities. To determine the best possible actuator's arrangement, an actuator's standard dimensions (bore diameter, rod diameter, and stroke length), input pressures and actuator's pivot location were assumed for each joint. Then the available output torque at exoskeleton joints has calculated using (Eqs. 1 and 2) and compared with the least required torque at human body joints which is determined from CGA data. Multiple iterations were conducted with different combinations of assumed parameters for each joint and the best possible arrangement was found out which has least hydraulic power consumption and enough torque to assist the operator. Figure 3 a–c shows the comparison between the calculated available torque and minimum required torque versus the gait cycle for each joint. This arrangement utilizes the smallest available actuator dimensions (bore diameter of 20 mm, and rod end diameter of 10 mm) and 3.5 MPa of supply pressure for all actuators. This arrangement yields a feasible

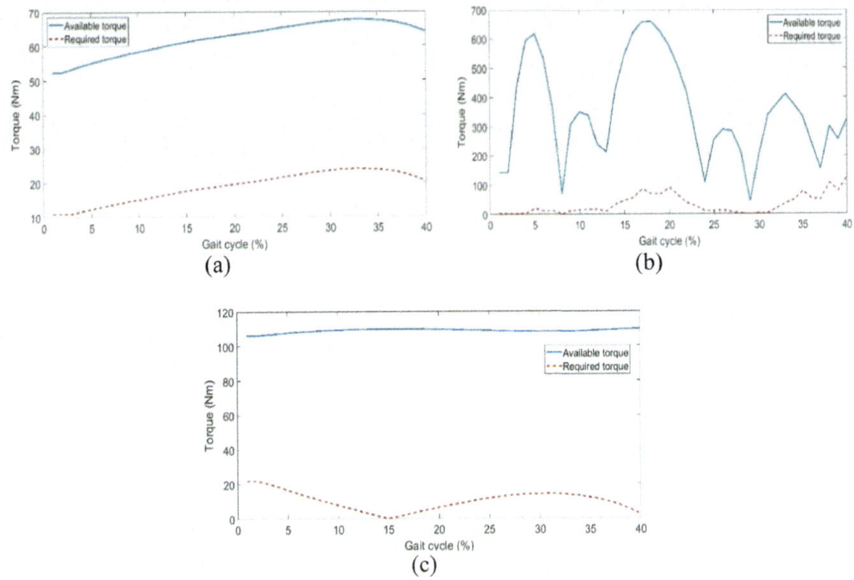

Fig. 3 Torque versus Gait cycle compared with CGA data **a** Hip, **b** Knee, **c** Ankle

solution but not to the optimized one. There is a large scope available to optimize the hydraulic system in order to reduce the gap between available and required torque. This optimization can be achieved by customizing input pressure, actuator sizing, etc.

As shown in Fig. 3 a, available torque is far more than the required torque. This arrangement for the hip joint uses 3.5 MPa supply pressure and a 20 mm bore diameter of actuator provides output torque five times than the required one. Whereas for the knee joint, the configuration uses a supply pressure of 3.5 MPa and a bore diameter of 25 mm. At the end of the support phase (near about 35–40% of the gait cycle), the available torque is slightly more than the required torque. Figure 3 c, shows the ankle joint has sufficient torque with the same supply pressure and a bore diameter of 20 mm. These parameters (actuator size and supply pressure) of the hydraulic system are found to be best according to the feasibility and requirement of the exoskeleton mechanism.

2.3 *Valve Sizing*

After finalizing the actuator size and supply pressure, servo valves can be chosen. It should be as small as possible and able to supply the required flow rates. For simplification, the valve equations (Eqs. 3 and 4), assuming no leakages in the valves, incompressible fluid, constant flow rate while entering and leaving the valve, and negligible return pressure.

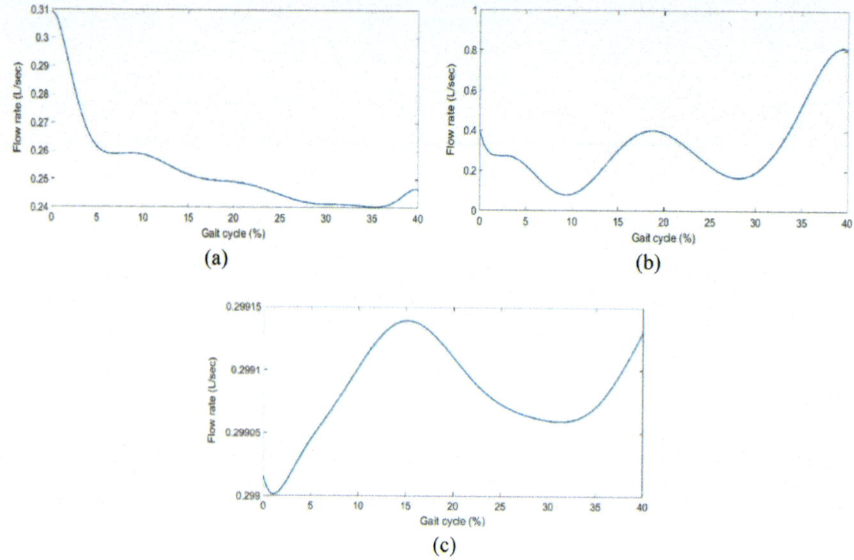

Fig. 4 Maximum valve flow rate of **a** Hip joint, **b** Knee joint, **c** Ankle joint

$$Q_{\text{push}} = R(\theta)\frac{\pi (D_{\text{Bore}})^2}{4}\left(\frac{\mathrm{d}\theta}{\mathrm{d}t}\right) \tag{3}$$

$$Q_{\text{pull}} = R(\theta)\frac{\pi (D_{\text{Bore}}^2 - D_{\text{rod}}^2)}{4}\left(\frac{\mathrm{d}\theta}{\mathrm{d}t}\right) \tag{4}$$

where, $Q_{\text{push/pull}}$ is the required flow rate through the valve to get required angular velocity and $\mathrm{d}\theta/\mathrm{d}t$ is the angular velocity of human body joint based on CGA data.

The maximum positive flow rate through the valves (using Eq. 3) is plotted against the gait cycle as shown in Fig. 4 a–c. The above plots can be referred to for selecting valve size.

3 Simulation and Study

In this section, exoskeleton mechanism simulation is performed in Adams and MATLAB software environment. Adams software has given CGA-based data as input for joint movement. Whereas MATLAB uses values obtained from the mathematical model of kinematic and dynamic analysis as an input. As shown in Fig. 5 a, the various constraints are applied to the exoskeleton model in Adams software environment. Exoskeleton links dimensions are suitable for an average adult with 75 kg of weight and 1.68 m of height [8]. The mechanism has three active and two passive degrees of freedom (DoFs) for each leg.

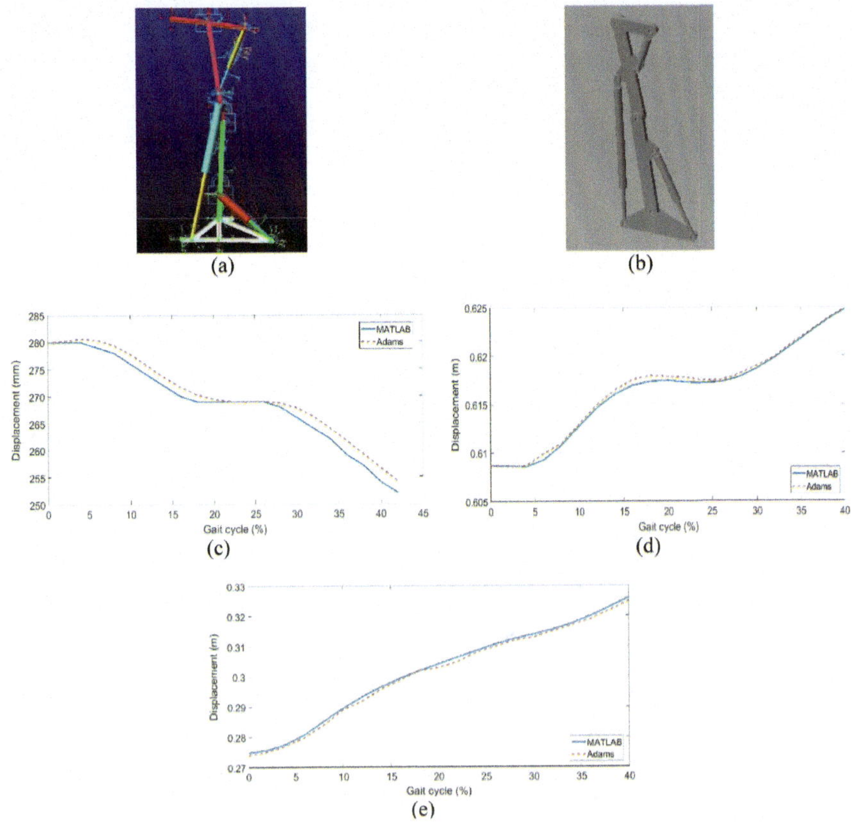

Fig. 5 **a** Adams model of the proposed exoskeleton, **b** MATLAB model of the proposed exoskeleton, Comparison of displacement plots of: **c** Actuator 1, **d** Actuator 2, **e** Actuator 3

Figure 5 b shows the MATLAB model of the proposed exoskeleton. The payload of 30 kg is applied to the exoskeleton's structural members. The kinematic and dynamic simulation has been performed in Adams and MATLAB. This simulation is performed only on a single leg of the exoskeleton. Figures 5 c–e and 6 a–c shows validation of the displacement and velocity plots of the actuators, respectively. Displacement plots using MATLAB are almost matching with plots using Adams and hence mathematical model for kinematics analyses is validated by comparing with Adams software results. Figure 6 d–f shows the plots of the actuator's force. The actuator's force will be zero while the swing phase. For actuator 1, there are little variations between MATLAB and Adams results. Results obtained from the mathematical model are quite lesser than that of Adams model. Whereas for actuators 2 and 3, these variations are quiet lesser. And hence it is validating the dynamic mathematical model. The variations between these two results are due to the difference in the inertia values. Adams calculates the inertia values of links automatically. Whereas, in MATLAB theses inertia values are calculated manually and given as input.

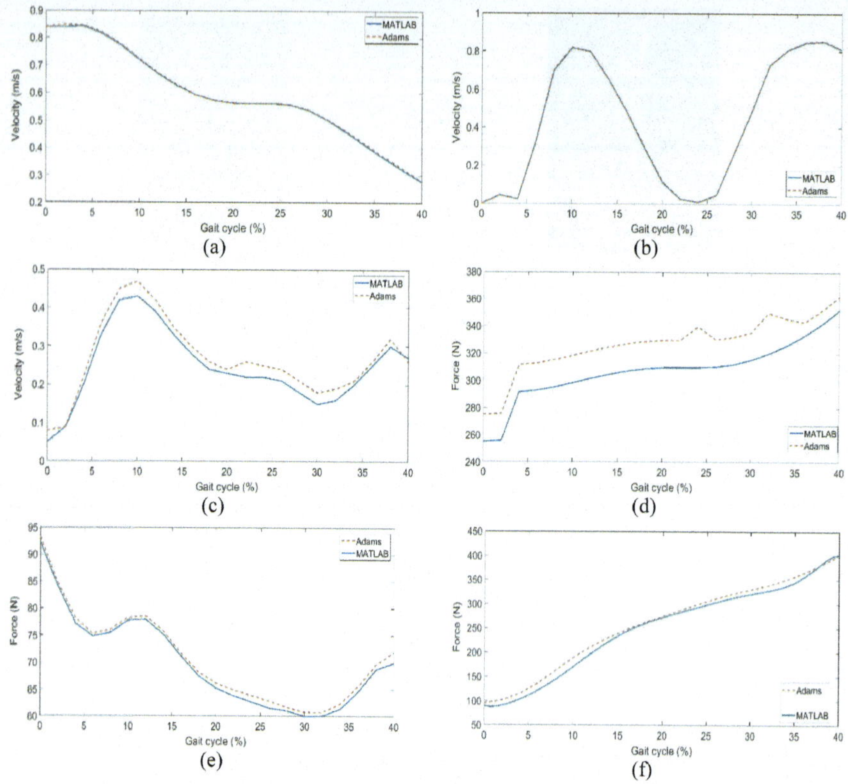

Fig. 6 Validation of velocity plots: **a** Actuator 1, **b** Actuator 2, **c** Actuator 3 and Validation of force plots: **d** Actuator 1, **e** Actuator 2, **f** Actuator 3

4 Conclusion

The validation of kinematic and dynamic analyses of the lower body exoskeleton mechanism has been performed using Adams software simulation. The actuator sizes and servo valves were selected using CGA data. It is shown that the CGA-based design has sufficient force to walk with minimum hydraulic power consumption. Future work will be concentrated on the prototyping of the proposed exoskeleton for validating theoretical results.

References

1. Pons JL (2008) Wearable robots: bio mechatronic exoskeletons, vol 70. Wiley Online Library
2. Na Li F, Lei Yan S, Hua Qian T (2015) Review on lower extremity exoskeleton robot. J Open Autom control Syst 7:441–453

3. Shubham S, Kawale M, Sreekumar (2018) Design of a wearable lower body exoskeleton mechanism for shipbuilding industry. Proc Comput Sci 133(2018): 1027–1028
4. Low KH, Liu X, Yu H (2006) Design and implementation of NTU wearable exoskeleton as an enhancement and assistive device. ABBI 3(3): 209–225
5. Zoss AB, Kazerooni H (2005) Architecture and hydraulics of a lower extremity exoskeleton. In: ASME 2005 international mechanical engineering congress and exposition, 1447–1455. Orlando, Florida, USA
6. Chu A, Kazerooni H, Zoss AB (2005) On the biomimetic design of the Berkeley lower extremity exoskeleton (BLEEX). In: IEEE international conference on robotics and automation, vol 43, pp 45–52. IEEE, Barcelona, Spain
7. Zoss AB, Kazerooni H, Chu A (2006) Biomechanical design of the Berkeley lower extremity exoskeleton (BLEEX). IEEE ASME Trans Mechatron 11:28–38
8. Ph360 homepage. Accessed 22 Jan 2018 https://ph360.me/ph360/anthropometry/

Motion Control of Omnidirectional Mobile Robot Using Bond Graph and Flatness-Based Controller

Saumya Ranjan Sahoo and Shital S. Chiddarwar

Abstract Omnidirectional mobile robots are widely used in various fields like warehouses, hospitals, military and nuclear power plants. Due to independent actuation of each wheel, it facilitates the highest order maneuverability and mobility. However, it being a complex mechatronics system, dynamic modeling and control is a challenging task. In this paper, an attempt is made to model a four-wheel omnidirectional mobile robot using highly intuitive bond graph technique. The dynamic model thus obtained is used to derive the control law for the robot. Control of the robot along desired trajectory is attained using principle of differential flatness theory. The validation of the approach is done in real time by performing simulation and experimental trials. These trials revealed effectiveness of bond graph technique for dynamic modeling of robot and flatness-based controller for smooth and accurate trajectory tracking.

Keywords Bond graph · Flatness-based controller · Omnidirectional robot

1 Introduction

Omnidirectional wheeled mobile robot offers advantages over conventional wheeled robots like better mobility, ability to turn in any direction and capability to move in confined space. These robots are being used in applications like automatic wheel chairs, service robots, fork lifter, mobile manipulators, etc. [1]. Among omnidirectional robots, four-wheeled omnidirectional mobile robots are unique holonomic systems those have three degree of freedom on a horizontal plane [2]. However, due to independent actuation of the four wheels, the modeling and control of motion becomes challenging, for which many researchers have contributed [3]. Based on the holonomy, the dynamics of different kinds of mobile robots are presented by Compion et al. [4]. Canceiocao et al. [5] proposed a practical approach for modelling and

S. R. Sahoo (✉) · S. S. Chiddarwar
Mechanical Department, Visvesvaraya National Institute of Technology, Nagpur 440010, India
e-mail: saumyasynergy@gmail.com

D. Sen et al. (eds.), *Mechanism and Machine Science*, Lecture Notes in Mechanical Engineering, https://doi.org/10.1007/978-981-15-4477-4_58

parameter estimation of an omnidirectional robot. Tlale [6] developed a mathematical model of mobile robot with Mecanum wheel to achieve omnidirectionality in the platform.

Due to the complex mechatronics structure of the mobile robots, recently there has been a noteworthy increase in use of computational dynamics for design, development, analysis, simulation and control of these robotic systems. Bond graph being more intuitive among available computational dynamic systems, it has great potential to deal with complex mechatronics systems. The application of bond graph for different disciplines of science has been reported by Troncate [7]. On the similar lines, Van and Peter [8] illustrated the importance of bond graph for modelling physical system from control aspects. Kannopp et al. [9] demonstrated the role of bond graph in modelling and simulation of mechatronics systems. Using bond graph method, the modelling of complex mechatronics system like space platform, hexapod, robotic leg mechanism are also spotted in literature [10–12].

The slippage and other energy losses have a critical influence on the robot dynamics and may lead to poor tracking performance. In recent years, a large number of control algorithms have been proposed by researchers [13, 14]. The concept of flatness was proposed by Fliess et al. [15], and has received attention because of its impressive application in trajectory tracking and motion planning. The application of differential flatness technique in various mechanical systems was illustrated by Murry [16].

In the present work, the dynamic model of the 4-WMR (Four-wheel Mecanum robot) is derived using bond graph technique. A generalized equation of motion is estimated by implementing bond and junction properties on the developed bond graph. This equation of motion is converted into control law to track desired trajectory with less variation in control input and minimum trajectory tracking error. The efficacy of the proposed controller is established using circular and S-shaped trajectories for simulation and experimental runs.

The outline of the paper is as follows: In Sect. 2, the kinematics and dynamics of the robot using bond graph is illustrated. Section 3 discusses about the differential flatness analysis of the robot and design of flatness based controller. In Sect. 4, a brief description about the simulation and experimental study is given. Section 5 discusses about the results obtained from simulation and experimental study. Conclusions drawn from this study are presented in Sect. 6.

2 Kinematics and Dynamics of the Robot

The robot consists of four independently actuated Mecanum wheels with peripheral roller inclined at 45°. Because of peripheral rollers, the wheel moves freely at an angle 45° with respect to the driven direction. It is assumed that the motion of the robot is on a smooth horizontal plane. Figure 1 shows the robot having four wheels with coordinate frames $O_{cwi}(i = 1, 2, 3, 4)$. The angular velocity $v_{cwi} = \left[\dot{x}_{cwi}, \dot{y}_{cwi}, \dot{\theta}_{cwi}\right]$ of the wheel 'i' has three components:

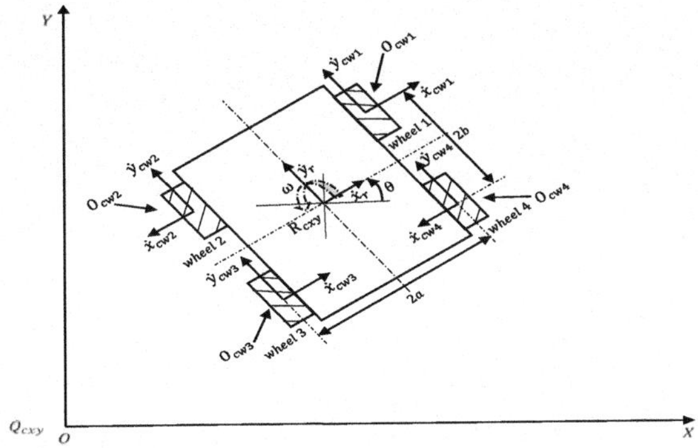

Fig. 1 Kinematic structure of the robot

- $\dot{\theta}_{ix}(i = 1, 2, 3, 4)$: Rotational speed of the roller around the hub.
- $\dot{\theta}_{ir}(i = 1, 2, 3, 4)$: Rotational speed of the roller.
- $\dot{\theta}_{iz}(i = 1, 2, 3, 4)$: Rotational speed of the wheel around the point of contact with surface.

2.1 Kinematics

The wheel velocity vector $v_{cwi} = \left[\dot{x}_{cwi}, \dot{y}_{cwi}, \dot{\theta}_{cwi}\right]$ in $O_{cwi}(i = 1, 2, 3, 4)$ coordinates is given as

$$\begin{bmatrix} \dot{x}_{cwi} \\ \dot{y}_{cwi} \\ \dot{\theta}_{cwi} \end{bmatrix} = \begin{bmatrix} 0 & r_i \sin(\alpha_i) & 0 \\ R_i & -r_i \cos(\alpha_i) & 0 \\ 0 & 0 & 1 \end{bmatrix} \begin{bmatrix} \dot{\theta}_{ix} \\ \dot{\theta}_{ir} \\ \dot{\theta}_{iz} \end{bmatrix} \tag{1}$$

Where R_i: Radius of the wheel and r_i: Radius of the roller. The robot velocity vector $\dot{R}_{cxy} = [\dot{x}_r, \dot{y}_r, \dot{\theta}_r]^T$ in the R_{cxy} coordinate frame (moving frame) is given by

$$\dot{R}_{cxy} = \begin{bmatrix} \dot{x}_r \\ \dot{y}_r \\ \dot{\theta}_r \end{bmatrix} = \begin{bmatrix} \cos(\phi^r_{ri}) & -\sin(\phi^r_{ri}) & d^r_{riy} \\ \sin(\phi^r_{ri}) & \cos(\phi^r_{ri}) & d^r_{rix} \\ 0 & 0 & 1 \end{bmatrix} \begin{bmatrix} \dot{x}_{cwi} \\ \dot{y}_{cwi} \\ \dot{\theta}_{cwi} \end{bmatrix} \tag{2}$$

Where $\phi^r_{ri}(i = 1, 2, 3, 4)$ is the angle between $O_{cwi}(i = 1, 2, 3, 4)$ and R_{cxy} frames. Here d^r_{rix} and d^r_{riy} are the longitudinal and lateral distances between these two coordinate frames. From the kinematic structure of the robot, $\phi^r_{ri} = 0$. Hence, Eq. (2) can be written as follows:

$$\dot{R}_{cxy} = \begin{bmatrix} \dot{x}_r \\ \dot{y}_r \\ \dot{\theta}_r \end{bmatrix} = \begin{bmatrix} 1 & 0 & d^r_{riy} \\ 0 & 1 & d^r_{rix} \\ 0 & 0 & 1 \end{bmatrix} \begin{bmatrix} \dot{x}_{cwi} \\ \dot{y}_{cwi} \\ \dot{\theta}_{cwi} \end{bmatrix} \tag{3}$$

Introducing Eq. (3) into Eq. (1), we get

$$\begin{bmatrix} \dot{x}_r \\ \dot{y}_r \\ \dot{\theta}_r \end{bmatrix} = \begin{bmatrix} 0 & r_i \sin(\alpha_i) & d^r_{riy} \\ R_i & -r_i \cos(\alpha_i) & d^r_{rix} \\ 0 & 0 & 1 \end{bmatrix} \begin{bmatrix} \dot{\theta}_{ix} \\ \dot{\theta}_{ir} \\ \dot{\theta}_{iz} \end{bmatrix} \tag{4}$$

$J_i = \begin{bmatrix} 0 & r_i \sin(\alpha_i) & d^r_{riy} \\ R_i & -r_i \cos(\alpha_i) & d^r_{rix} \\ 0 & 0 & 1 \end{bmatrix}$ is the Jacobian matrix of wheel 'i', which is square and invertible matrix.

Form the above equations, the inverse kinematic solution is obtained as

$$\begin{bmatrix} \dot{x}_r \\ \dot{y}_r \\ \dot{\theta}_r \end{bmatrix} = \frac{R}{4} \begin{bmatrix} -1 & 1 & -1 & 1 \\ 1 & 1 & 1 & 1 \\ \frac{1}{a+b} & -\frac{1}{a+b} & -\frac{1}{a+b} & \frac{1}{a+b} \end{bmatrix} \begin{bmatrix} \dot{\theta}_{cw1} \\ \dot{\theta}_{cw2} \\ \dot{\theta}_{cw3} \\ \dot{\theta}_{cw4} \end{bmatrix} \tag{5}$$

Where $\dot{\theta}_{cwi} (i = 1,\ 2,\ 3,\ 4)$ is the angular velocity of each wheel.

2.2 *Dynamics*

The Mecanum wheel robot consists of steel chassis equipped with four independently actuated Mecanum wheels. Each wheel is actuated by a toothed geared DC motor. The motor is powered by the battery. The different inertial forces from the geared DC motor provide motion to the robot. To develop the dynamic model of such mechatronics system, bond graph provides a better platform, which uses energy flow inside the system. Using the energy flow in the system, the bond graph models of subsystems like motors, gears, the Mecanum wheel, longitudinal, lateral, and yaw motion are developed. These submodels are integrated to obtain the global bond graph structure for the robot as shown in Fig. 2.

On applying the bond and junction properties to the global bond graph, the following mathematical model is extracted:

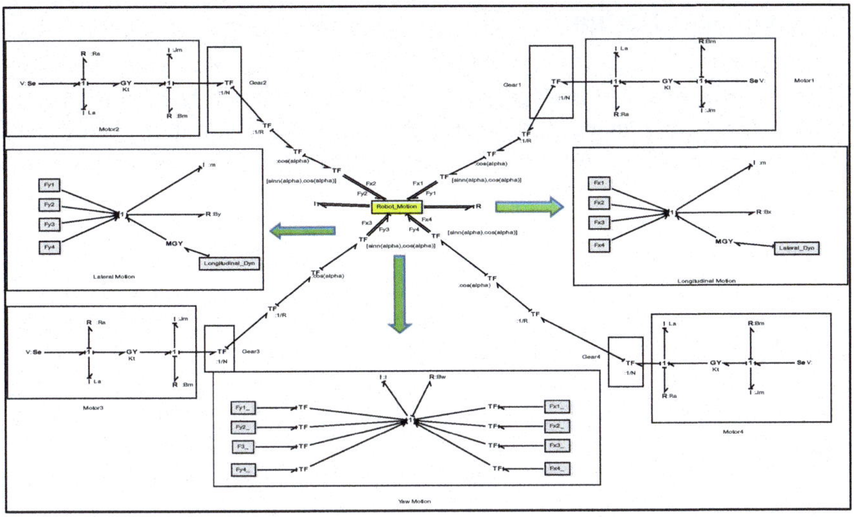

Fig. 2 Bond graph of the robot

$$\ddot{x}_q = \frac{\begin{array}{r}\tau_1(\cos\theta - \sin\theta) + \tau_2(\cos\theta + \sin\theta)\\ +\tau_3(\cos\theta - \sin\theta) + \tau_4(\cos\theta - \sin\theta)\\ -\dot{x}_q(2R\beta_x\cos^2\theta + 2R\beta_y\cos^2\theta)\\ -\dot{y}_q(2R\beta_x\cos\theta\sin\theta - 2R\beta_y\cos\theta\sin\theta)\end{array}}{2mR} \tag{6}$$

$$\ddot{y}_q = \frac{\begin{array}{r}\tau_w(\cos\theta + \sin\theta) + \tau_2(-\cos\theta + \sin\theta)\\ +\tau_w(\cos\theta + \sin\theta) + \tau_w - (\cos\theta + \sin\theta)\\ -\dot{y}_q(2R\beta_y\cos^2\theta + 2R\beta_x\cos^2\theta)\\ -\dot{x}_q(2R\beta_x\cos\theta\sin\theta - 2R\beta_y\cos\theta\sin\theta)\end{array}}{2mR} \tag{7}$$

$$\ddot{\theta} = \frac{(\tau_w - \tau_w - \tau_w + \tau_{4w})}{(2IR)}(a+b) - \frac{\beta_v\dot{\theta}}{I} \tag{8}$$

$$V = L_a\frac{di}{dt} + R_a i + k_b\omega;\ J_m\frac{dw}{dt} = T - b_m\omega - T_l \tag{9}$$

Where m: mass of the robot, I: inertia of the robot, $\beta_x,\ \beta_y,\ \beta_\omega$: viscous friction along longitudinal, lateral and yaw direction, R_a: armature resistance, L_a: inductance of the motor, J_m: motor inertia and b_m: motor friction. The model extracted from bond graph is used to derive the control law for trajectory tracking by the robot.

3 Controller Design

The main objective here is to develop a control law that will be easy to interface with the bond graph model and capable to track the desired trajectory efficiently. Moreover, it should be noted that the trajectory followed by the robot is known a priori.

In a nonlinear system $\dot{x} = f(x, u)$, if the state variable $x = [x_1, x_2 \ldots\ldots x_n]^T$ and control input $u = [u_1, u_2 \ldots\ldots u_n]$ can be parameterized by a set of output $y = [y_1, y_2 \ldots\ldots y_m]^T$ and its derivatives, then the system is said to be differentially flat having flat output $y = [y_1, y_2 \ldots\ldots y_m]^T$ [15]. In view of this, the state variables and control variables can be written as follows:

$$x = \psi(y, \dot{y}, \ddot{y}, \ldots y^k); \qquad x = \beta(y, \dot{y}, \ddot{y}, \ldots y^{k+1});$$

As in the current system, the number of inputs is 4 and the states of the system are 3, virtual control inputs $u_i(i = 1, 2, 3)$ need to be considered.

Hence, $u_i(i = 1, 2, 3)$ can be redefined as

$$\begin{bmatrix} u_1 \\ u_2 \\ u_3 \end{bmatrix} = \begin{bmatrix} \frac{1}{2mR} & 0 & \frac{1}{2mR} & 0 \\ 0 & \frac{1}{2mR} & 0 & \frac{1}{2mR} \\ \frac{1}{2IR} & -\frac{1}{2IR} & -\frac{1}{2IR} & \frac{1}{2IR} \end{bmatrix} \begin{bmatrix} \tau_1 \\ \tau_2 \\ \tau_3 \\ \tau_4 \end{bmatrix} \tag{10}$$

where $\tau_i(i = 1, 2, 3, 4)$ is the torque generated by the motors considered as a real control input to the robotic system.

The system dynamics (Eqs. (6)–(9)) can be represented in the control affine form as $\dot{x} = f(x) + g(x)u$; where $x = \left[x_q, y_q, \theta_q, \dot{x}_q, \dot{y}_q, \dot{\theta}_q\right]^T$ is the state and $u_i(i = 1, 2, 3)$ is the control input.

Now, the complete set of the flat output $y = [y_1\ y_2\ y_3]^T = \left[x_q\ y_q\ \theta_q\right]^T$ is chosen in such a way that it make the entire system differentially flat. The three flat outputs have parameterized all the states and inputs variables of the system. The states and control inputs can be represented as

States:

$x = \left[x_q\ y_q\ \theta_q\ \dot{x}_q\ \dot{y}_q\ \dot{\theta}_q\right]^T = [y_1\ \ y_2\ \ y_3\ \ \dot{y}_1\ \ \dot{y}_2\ \ \dot{y}_3]^T$

Control inputs:

$$\begin{bmatrix} u_1 \\ u_2 \\ u_3 \end{bmatrix} = \begin{bmatrix} (\cos\ y_3 - sin\ y_3)/2 & (\cos\ y_3 + sin\ y_3)/2 & 0 \\ (\cos\ y_3 + sin\ y_3)/2 & (-\cos\ y_3 + sin\ y_3)/2 & 0 \\ 0 & 0 & 1 \end{bmatrix} \begin{bmatrix} \ddot{y}_1 \\ \ddot{y}_2 \\ \ddot{y}_3 \end{bmatrix} + \begin{bmatrix} \frac{2\beta(\cos y_3 - \sin y_3)}{2m} & \frac{2\beta(\cos y_3 + \sin y_3)}{2m} & 0 \\ \frac{2\beta(\cos y_3 + \sin y_3)}{2m} & \frac{2\beta(-\cos y_3 + \sin y_3)}{2m} & 0 \\ 0 & 0 & \frac{-\beta}{I} \end{bmatrix} \begin{bmatrix} \dot{y}_1 \\ \dot{y}_2 \\ \dot{y}_3 \end{bmatrix} \tag{11}$$

Now, the system is differentially flat and the mapping of state space to the flat output space indicates the full state controllability of the system. Based on the flatness of the system, the control law for the virtual input of the desired trajectory in the form of flat output can be defined as

$$\begin{bmatrix} \ddot{y}_1 \\ \ddot{y}_2 \\ \ddot{y}_3 \end{bmatrix} = \begin{bmatrix} \ddot{y}_1^d \\ \ddot{y}_2^d \\ \ddot{y}_3^d \end{bmatrix} + \begin{bmatrix} k_{p1} & 0 & 0 \\ 0 & k_{p2} & 0 \\ 0 & 0 & k_{p3} \end{bmatrix} \begin{bmatrix} e_{y1} \\ e_{y2} \\ e_{y3} \end{bmatrix} + \begin{bmatrix} k_{i1} & 0 & 0 \\ 0 & k_{i2} & 0 \\ 0 & 0 & k_{i3} \end{bmatrix} \begin{bmatrix} \int e_{y1} dt \\ \int e_{y2} dt \\ \int e_{y3} dt \end{bmatrix} + \begin{bmatrix} k_{d1} & 0 & 0 \\ 0 & k_{d2} & 0 \\ 0 & 0 & k_{d3} \end{bmatrix} \begin{bmatrix} \dot{e}_{y1} \\ \dot{e}_{y2} \\ \dot{e}_{y3} \end{bmatrix} \tag{12}$$

where $[k_p]_{3\times3}$,$[k_i]_{3\times3}$,$[k_d]_{3\times3}$ are the controller gain parameters and $\begin{bmatrix} e_{y1} \\ e_{y2} \\ e_{y3} \end{bmatrix} = \begin{bmatrix} y_1^d - y_1 \\ y_2^d - y_2 \\ y_3^d - y_3 \end{bmatrix}$ is the error vector.

The corresponding error dynamics of the system can be written as

$$[\ddot{e}_y]_{3\times1} + [k_p]_{3\times3} [e_y]_{3\times1} + [k_i]_{3\times3} \left[\int e_y \, dt \right]_{3\times1} + [k_d]_{3\times3} [\dot{e}_y]_{3\times1} = 0 \tag{13}$$

Finally, $[k_p]_{3\times3}$,$[k_i]_{3\times3}$,$[k_d]_{3\times3}$ are selected in such a way that the roots of Eq. (13) lie on the left half of the s-plane and the tracking error $[e_y]_{3\times1}$ exponentially converges to zero.

4 Simulation and Experimental Analysis

To demonstrate the effectiveness of the proposed dynamic model and the controller, simulation and experimentations are performed using MATLAB Simulink 2015a. The robot use for experimentation is having $m = 6$ kg, $I = 0.0945$ kgm^{2,} R = 0.05 m.

To check the performance of the robot, two trajectories are taken into consideration.

Trajectory I: Circular Trajectory

$$\begin{aligned} x(t) &= 0.6\sin(t) \\ y(t) &= 0.6\cos(t) \qquad t \geq 0 \\ \theta(t) &= 0 \end{aligned}$$

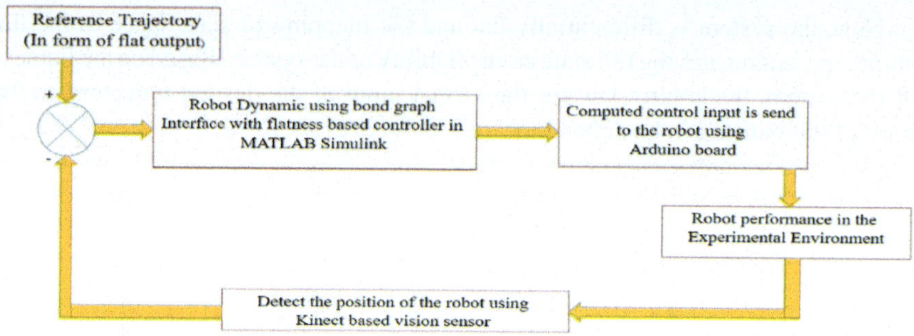

Fig. 3 Block diagram for real-time implementation of control law

where '*t* 'is the time period and [0, 0, 0] is the initial posture vector of the mobile robot.

Trajectory II: S Trajectory (With orientation change)

$$\begin{aligned} x(t) &= t \\ y(t) &= \sin t \qquad\qquad t \geq 0 \\ \theta(t) &= \tfrac{\pi}{26} t \end{aligned}$$

In Trajectory I, emphasis is on position control, however, for applications like motion of robot inside curved duct, tunnels, etc., it is necessary to change the position and orientation simultaneously. In such scenario, to check the capability of the controller to track complex trajectory with desired orientation, Trajectory II is selected. In most of the reported work, either position or orientation tracing of a Mecanum wheel robot is attained. However, in this work emphasis is given on simultaneous position and orientation control of the Mecanum wheel robot. First, simulations are carried out in MATLAB Simulink and then real-time trajectory tracking is performed with robot under consideration in experimental environment (Fig. 3).

5 Result and Discussion

The selection of the proper controller parameters to minimize the tracking error with minimal control effort is the most crucial aspect from controller performance point of view. For this, extensive simulations were carried out in MATLAB to get the proper controller parameters using non-linear least square optimization technique.

The controller parameters are estimated as

$$[k_p]_{3\times3} = \begin{bmatrix} 1.2 & 0 & 0 \\ 0 & 1.2 & 0 \\ 0 & 0 & 25.11 \end{bmatrix} [k_i]_{3\times3} = \begin{bmatrix} 0.3 & 0 & 0 \\ 0 & 0.3 & 0 \\ 0 & 0 & 0.003 \end{bmatrix}, [k_d]_{3\times3} = \begin{bmatrix} 5 & 0 & 0 \\ 0 & 5 & 0 \\ 0 & 0 & 27.63 \end{bmatrix}$$

Figures 4a and 5a show the real-time implementation of Trajectory I and II, respectively. The tracking error of about 3–4 cm was observed for Trajectory I and 8–10 cm for Trajectory II during experimentation. The tracking error is on higher side for Trajectory II because controller tries to achieve the desired angular orientation at the cost of x,y position. Another reason might be the slippage between the driving wheel and the surface, nature of surface, and unevenness of the surface. The simulation results are comparable with experimental results even though the factors like internal friction, backlash in gear and other energy losses are not considered during dynamic modelling. Otherwise, the error might have further reduce. The novelty of this work lies in successful control of robot in terms position as well as orientation.

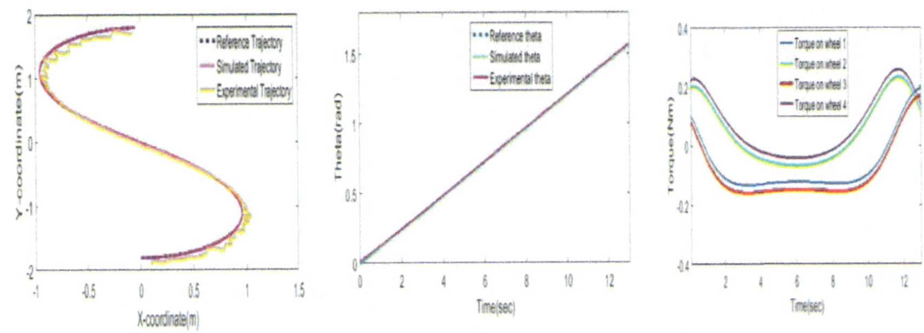

(a) (b) (c)

Fig. 4 Simulated and experimental trajectory **a** Position **b** orientation **c** control input

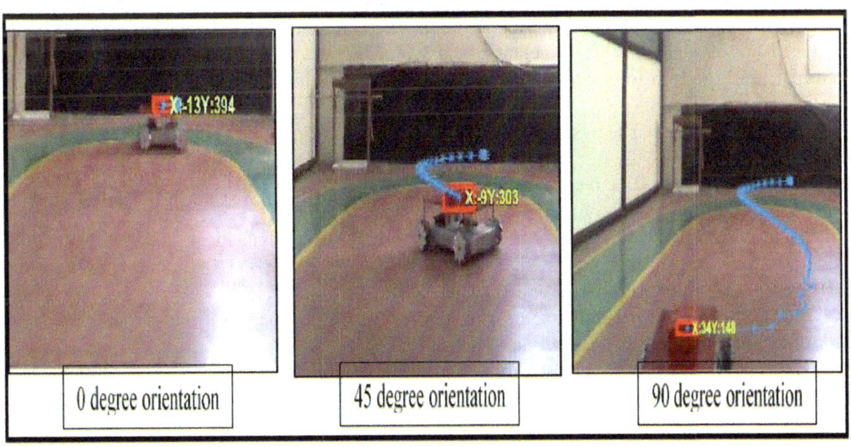

Fig. 5 Real-time trajectory tracking by the Kinect **a** 0° orientation **b** 45° orientation **c** 90° orientation

6 Conclusion

In this paper, the bond graph approach with a flatness-based controller for controlling the Mecanum wheel robot along the desired trajectory is proposed. The bond graph model gave an accurate dynamic model for the robot. This helped in obtaining a flatness-based control law to navigate the robot along the desired trajectories pertaining to different complexities. Asymptotical stability of the proposed controller was achieved using classical pole placement technique. The optimal controller parameters were found using non-linear least square optimization technique. The proposed approach was capable to control robot with minimum error. It was able to control the position and orientation of the robot for simple as well as complex trajectory. The performance of the proposed approach can be improved by considering uncertainties and disturbances during dynamic modelling and making it adaptive.

References

1. Ye C, Ma S, Hui L (2011) An omnidirectional mobile robot. Sci China Inf Sci 54(12):2631–2638
2. Pin Fg, Killough SM (1994) A new family of omnidirectional and holonomic wheeled platforms for mobile robots. IEEE Trans Robot Autom 10(4): 480–489. (Harold A. Rothbart (2003) Cam design handbook: dynamics and accuracy, McGraw-Hill Professional)
3. Tzafestas SG (2013) Introduction to mobile robot control. Elsevier
4. Campion G, Bastin G, Dandrea-Novel B (1996) Structural properties and classification of kinematic and dynamic models of wheeled mobile robots. IEEE Trans Robot Autom 12(1):47–62
5. Conceicao AS, Moreira AP, Costa PJ (2009) Practical approach of modeling and parameters estimation for omnidirectional mobile robots. IEEE/ASME Trans Mechatron 14(3):377–381
6. Tla N, de Villiers M (2008) Kinematics and dynamics modelling of a mecanum wheeled mobile platform. In: 15th International conference on mechatronics and machine vision in practice, M2VIP 2008, pp 657–662. IEEE
7. Troncale L (1988) The systems sciences: What are they? are they one, or many? Eur J Oper Res 37(1):8–33
8. Van Amerongen J, Breedveld P (2003) Modelling of physical systems for the design and control of mechatronic systems. Annual Reviews in Control 27(1):87–117
9. Karnopp DC, Margolis DL, Rosenberg RC. (2012) System dynamics: modeling, simulation, and control of mechatronic systems. Wiley. (Ye S, Ma, Hui L (2011) An omnidirectional mobile robot. Sci China Inf Sci 54(12): 2631–2638)
10. Pathak PM, Kumar RP, Mukherjee A, Dasgupta A (2008) A scheme for robust trajectory control of space robots. Simul Model Pract Theory 16(9):1337–1349
11. Ragusila V, Emami MR (2014) Modelling of a robotic leg using bond graphs. Simul Model Pract Theory 40:132–143
12. Damic V, Cohodar M (2015) Dynamic analysis of Stewart platform by bond graphs. Procedia Eng 100:226–233
13. Alakshendra V, Chiddarwar SS (2017) Adaptive robust control of Mecanum-wheeled mobile robot with uncertainties. Nonlinear Dyn 87(4):2147–2169
14. Sahoo SR, Chiddarwar SS, Alakshendra V (2017) Intuitive dynamic modeling and flatness-based nonlinear control of a mobile robot. SIMULATION, p 0037549717741192

15. Fliess M, Lévine J, Martin P, Rouchon P, Flatness and defect of non-linear systems: introductory theory and examples
16. Murray RM, Rathinam M, and Sluis W (1995) Differential flatness of mechanical control systems: A catalog of prototype systems. In: ASME international mechanical engineering congress and exposition

Innovative Low-Cost Stair Climbing Mechanism

Girish Sudhir Modak and Manmohan Manikrao Bhoomkar

Abstract This paper presents elaborated optimization process for lobed petals of a novel affordable setup capable of climbing staircases. Completely innovative method of specifically built wheels is already presented by the authors in their previous research paper. This setup uses driving wheels similar to lobe cam. The outline of each cam is conjugate to the staircase. Number of petals was an independent parameter. So it was important to select suitable number of petals for the projected setup. Standard process of 'Design of Experiments' is adopted for deciding the appropriate number of petals. Last section presents results of the study.

Keywords Stair climbing chassis · Petal wheels · Matching profile · Design of experiments

List of Abbreviations

θ	Angle of slope of the staircase
Ø	Wheel rotation angle when contact is with Going
r_0	Minimum radius of Logarithmic Spiral
$Ø_0$	Angle of wheel radius w.r.t min. radius position
H	Step height
D	Step depth
F	Ratio (D/H) of the step
n	Number of lobes of the wheel
Hyp	Hypotenuse of the right-angled triangle bounded by Rise and Go of the step

G. S. Modak (✉) · M. M. Bhoomkar
Pune Vidyarthi Griha's College of Engineering and Technology, S.P. Pune University, Pune, India
e-mail: gsmodak@yahoo.com

D. Sen et al. (eds.), *Mechanism and Machine Science*, Lecture Notes in Mechanical Engineering, https://doi.org/10.1007/978-981-15-4477-4_59

1 Introduction

Old constructions from densely populated regions do not have elevators fitted thereto. It becomes difficult to the disabled citizens to climb the high going staircases. To solve their problems, an innovative setup that uses wheels conjugate to the stairs was proposed [1] earlier. It is nothing but chassis. Body is built on chassis that can transform it into 'Staircase Climbing Wheelchair' or 'Stair Climbing Trolley'. The driving wheels are similar to the cams perfectly matching with the steps on which the setup is supposed to climb. Design of Experiments method is used to decide the most appropriate number of petals applicable for this use.

2 Conjugate Profile Generation Process

Matching cam is generated using standard method of 'Rack Shift'. Staircase is considered as 'Rack' with a combination of linear profile. Figure 1 shows 'Wheel Blank' in relation to steps in the form of rack teeth.

Figure 2 shows thr process of transferring profile of teeth on blank of wheel. This process is repeated by moving the rack forward and by tilting blank up to next mark. After one complete rotation of wheel blank, completed lobbed gear is obtained as shown in Fig. 3. If the blank size is increased appropriately, we get wheel with four petals as shown in Fig. 4.

Petals can be any in number, but their minimum number is two. DOE analysis can be used to decide optimum number of petals for the wheel.

Longer profile will be in contact with horizontal surface of step while climbing. Shorter curve should touch vertical surface but that is not possible. So the shorter profile can be conveniently neglected and the petal wheel is relieved in the corresponding part.

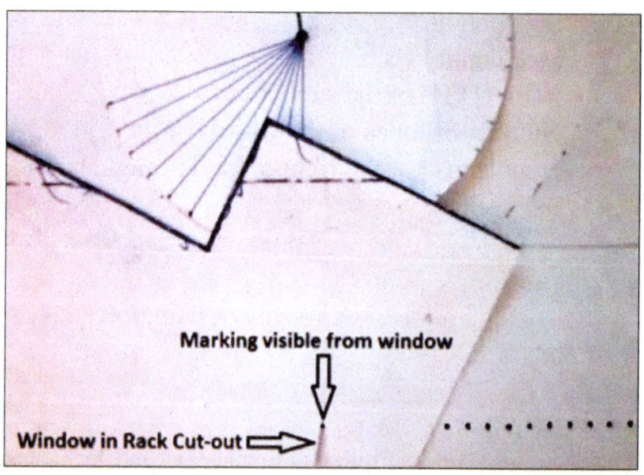

Fig. 1 Wheel blank with steps as rack teeth

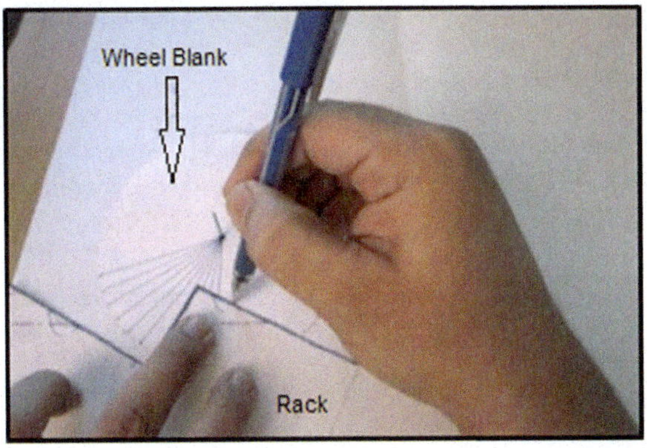

Fig. 2 Copying the profile of rack on blank of wheel

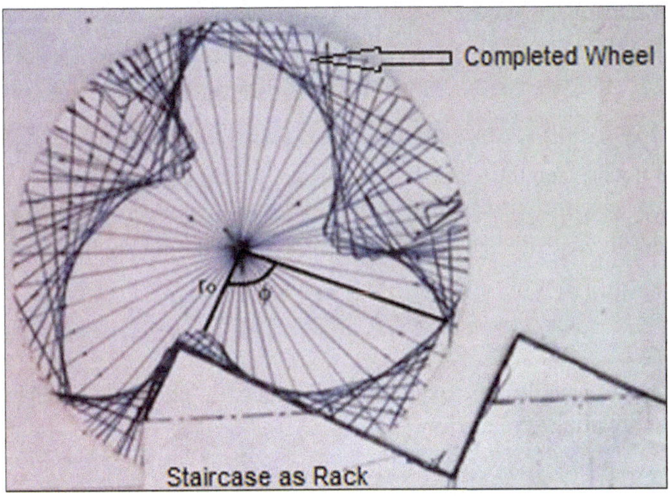

Fig. 3 Cam wheel consisting of three petals

3 Mathematically Derived Equation of the Profile

Cam formation process shows that the obtained profile is applicable for particular step array. If step proportion or Height to Depth ratio is changed, the outline of wheel changes. So developing of mathematical model was an important task.

Basic mathematics is used to arrive at universal expression of the curve which is given below [1]:

$$r = r_0 \, .e^{\tan\theta.(\emptyset)} \tag{1}$$

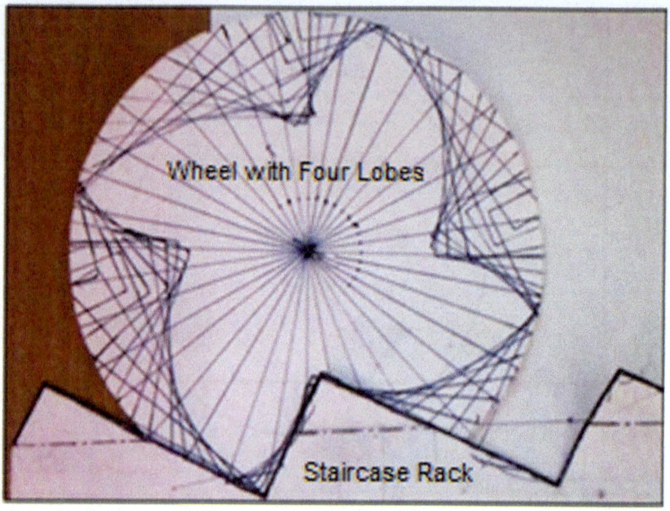

Fig. 4 Cam wheel consisting of four petals

r_0 and (Ø) are 'Least Radius' and 'Rotation Angle of wheel during rise'. Decided number of petals, step ratio and dimensions are governing parameters.

Defined steps are used to find decide the two quantities for the any arbitrary staircase. Deep mathematical procedure is used to find least radius [1]:

$$\boldsymbol{r_0} = \frac{\mathbf{D} * \mathbf{H}}{\boldsymbol{Hyp}(\boldsymbol{e}^{\tan\theta.(\emptyset)} - 1)} \tag{2}$$

Wheel rotation angle corresponding to contact with 'Going' (Ø) can be calculated by using the relation given below [1]:

$$\emptyset = \frac{2\pi.\boldsymbol{F}^2}{\boldsymbol{n}\left(\boldsymbol{F}^2 + 1\right)} \text{ (\textbf{Angle in Radians})} \tag{3}$$

For angles in degrees then the expression (3) changes to

$$\emptyset = \frac{360.\boldsymbol{F}^2}{\boldsymbol{n}\left(\boldsymbol{F}^2 + 1\right)} \text{ (\textbf{Angle in Degrees})} \tag{4}$$

Values of $\boldsymbol{r_0}$ and **Ø** can be found out from Eqs. (1) to (4). In these relations, the wheel is assumed to have ***'n'*** petals. Staircase is assumed to have step height **'H'** while Depth to Height ratio of steps is assumed to be ***'F'***

4 Use of Design of Experiments Technique for Optimization

Torque required to drive the setup is also derived as given below [1]:

$$T = (\mathbf{W}\sin\theta)(r_{\max}) \tag{5}$$

If Eqs. (2), (3) and (4) are substituted in (1),

$$r_{\max} = \left(\frac{\mathbf{D} * \mathbf{H}}{Hyp\,(e^{\tan\theta.\left(\frac{2\pi.F^2}{n(F^2+1)}\right)} - 1)}\right).e^{\tan\theta.\left(\frac{2\pi.F^2}{n(F^2+1)}\right)} \tag{6}$$

Equations (6) and (7) give

$$T = (\mathbf{W}\sin\theta)\left(\frac{\mathbf{D} * \mathbf{H}}{Hyp(e^{\tan\theta.\left(\frac{2\pi.F^2}{n(F^2+1)}\right)} - 1)}\right).e^{\tan\theta.\left(\frac{2\pi.F^2}{n(F^2+1)}\right)} \tag{7}$$

One petal of the wheel climbs one step so if wheel is made up of more number of petals and if wish to maintain constant rate of climb, wheel must be rotated slower. So as the petals increase, angular velocity (ω) reduces accordingly. As the requirement of power is the governing criterion and can be calculated by using the following equation [1]:

$$P = T\,.\,\omega \tag{8}$$

Substituting Eq. (7) in (8), we get

$$P = (\mathbf{W}\sin\theta)\left(\frac{\mathbf{D} * \mathbf{H}}{Hyp(e^{\tan\theta.\left(\frac{2\pi.F^2}{n(F^2+1)}\right)} - 1)}\right).e^{\tan\theta.\left(\frac{2\pi.F^2}{n(F^2+1)}\right)} \times \omega \tag{9}$$

Weight (W) is constant and angular velocity (ω) need be adjusted based on required number of petals (n). Other variables (Hyp) (F) and (θ) are based on (H) and (D). Thus the important parameters are (n), (H) and (D).

Formula (9) calculates power required for raising the setup and is the response for above-mentioned variables. It is must to select optimum values of these parameters.

Standards from USA [2], UK [3], India [4] and Australia [5] are studied in depth to decide standard ranges of (D) and (H).

Building Code of Australia (BCA) gives the top and bottom limit for 2(H) + (D) as 735 and 470, respectively, while other standards specify the same highest value for step height as 190 mm and minimum value of depth as 250 mm while ranges of (D) and (H) are comparatively wider.

Based on all these standards, ranges for 'Height' and 'Depth' are as given in Table 1.

Middle values for these parameters are selected midway and the number of petals is selected as 3, 4 and 5. So this is three-factor and three-level problem.

MS Excel worksheet is designed for $3^3 = 27$ values of power for various permutations of (D), (H) and (n) (Fig. 5).

One petal of the wheel climbs one step so angular speed (ω) need to be adjusted for various numbers of petals so as to maintain speed of ascend of the chassis.

Required power to climb the steps is treated as output based on the three parameters. Excel sheet is modified to include power (in kW) just like torque and the obtained values are stored separately. Design of Experiments (DOE) technique is used to complete the full factorial design.

Table 1 Reference parameters of staircases

Step height (H)		Step depth (D)		2(H) + (D)	
Maximum	Minimum	Maximum	Minimum	Maximum	Minimum
190	115	355	240	735	470

	A	B	C	D	E	F	G	H	I
1			Min	Med	Max				
2	W in N	1200	1200	1200	1200	1200	1200	1200	1200
3	RPM	10							
4	Going	D	240	297.5	355		240	240	240
5	Rise	H	115	152.5	190		115	115	115
6	No	N	3	4	5		3	4	5
7	Factor	F					2.086956522	2.086956522	2.086956522
8									
9	D SQ						57600	57600	57600
10	H SQ						13225	13225	13225
11	SUM						70825	70825	70825
12	SQRT	S					266.1296676	266.1296676	266.1296676
13									
14	Phi deg	ϕ d					97.59265796	73.19449347	58.55559478
15	Phi rad	ϕ c					1.70399879	1.277999092	1.022399274
16									
17	Delta deg	δ d					120	90	72
18	Alfa deg	α d					22.40734204	16.80550653	13.44440522
19									
20	tan θ						0.479166667	0.479166667	0.479166667
21	(tan θ).ϕ						0.81649942	0.612374565	0.489899652
22	e^tan θ. ϕ						2.262565666	1.844806816	1.632152428
23	coeff of rmin						1.262565666	0.844806816	0.632152428
24	DH/S						103.7088433	103.7088433	103.7088433
25	rmin						82.14134605	122.7604245	164.0567032
26	rmax						185.8501893	226.4692678	267.7655465
27									
28	TORQUE						96371.54083	117434.329	138848.2755

Fig. 5 Excel output giving the required parameters

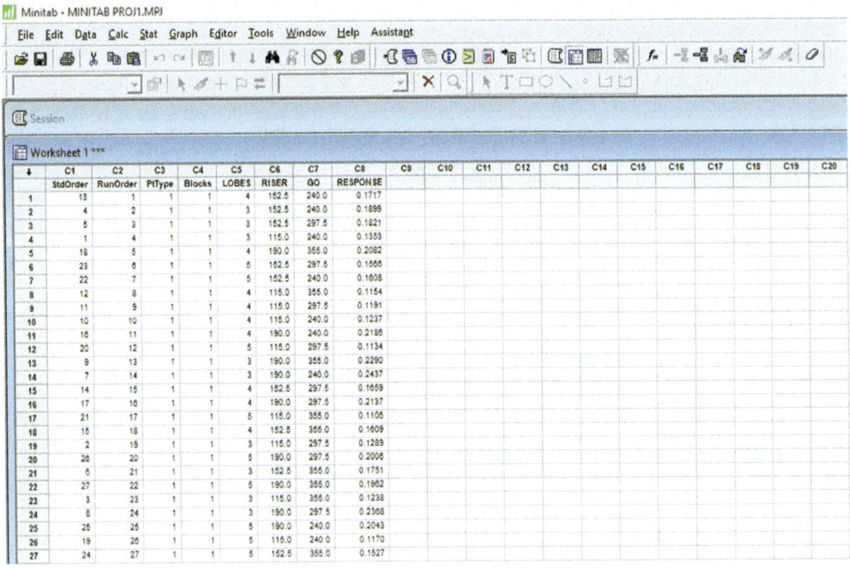

↓	C1 StdOrder	C2 RunOrder	C3 PtType	C4 Blocks	C5 LOBES	C6 RISER	C7 GO	C8 RESPONSE
1	13	1	1	1	4	152.5	240.0	0.1717
2	4	2	1	1	3	152.5	240.0	0.1899
3	5	3	1	1	3	152.5	297.5	0.1821
4	1	4	1	1	3	115.0	240.0	0.1353
5	18	5	1	1	4	190.0	355.0	0.2082
6	23	6	1	1	5	152.5	297.5	0.1566
7	22	7	1	1	5	152.5	240.0	0.1608
8	12	8	1	1	4	115.0	355.0	0.1154
9	11	9	1	1	4	115.0	297.5	0.1191
10	10	10	1	1	4	115.0	240.0	0.1237
11	16	11	1	1	4	190.0	240.0	0.2186
12	20	12	1	1	5	115.0	297.5	0.1134
13	9	13	1	1	3	190.0	355.0	0.2290
14	7	14	1	1	3	190.0	240.0	0.2437
15	14	15	1	1	4	152.5	297.5	0.1659
16	17	16	1	1	4	190.0	297.5	0.2137
17	21	17	1	1	5	115.0	355.0	0.1106
18	15	18	1	1	4	152.5	355.0	0.1609
19	2	19	1	1	3	115.0	297.5	0.1289
20	26	20	1	1	5	190.0	297.5	0.2006
21	6	21	1	1	3	152.5	355.0	0.1751
22	27	22	1	1	5	190.0	355.0	0.1962
23	3	23	1	1	3	115.0	355.0	0.1238
24	8	24	1	1	3	190.0	297.5	0.2368
25	25	25	1	1	5	190.0	240.0	0.2043
26	19	26	1	1	5	115.0	240.0	0.1170
27	24	27	1	1	5	152.5	355.0	0.1527

Fig. 6 Input screen of Minitab as applied for the above study

Pennsylvania State University has developed a statistics package called 'MINITAB' for completing statistical analysis. Various parameters are used to obtain 'Main Effects' and 'Interactions' plots with reference to their response.

This package is efficiently used for understanding the inter-relations among different variables from the specified problem. Input pallet of this software, as applied for the above study, is shown in Fig. 6.

Figure 7 shows 'Main effects plot' for Power versus Number of lobes:

Above graph reveals that the requirement of power is inversely proportional to number of petals, just like the rate of reduction in power requirement.

Figure 8 gives interactions graphs for step height (H), step depth (D) and number of petals (n) versus power

We can conclude that requirement of power reduces with increase in the step height (H) irrespective of the number of petals while the requirement of power is least for five petals.

Requirement of power increases with increase in height of step. Overall increase in requirement of power is less if the petals are more.

Analysis of Means (ANOM) is shown in Fig 9

For highest number of petals, the requirement of power for climbing the stairs is less than the mean.

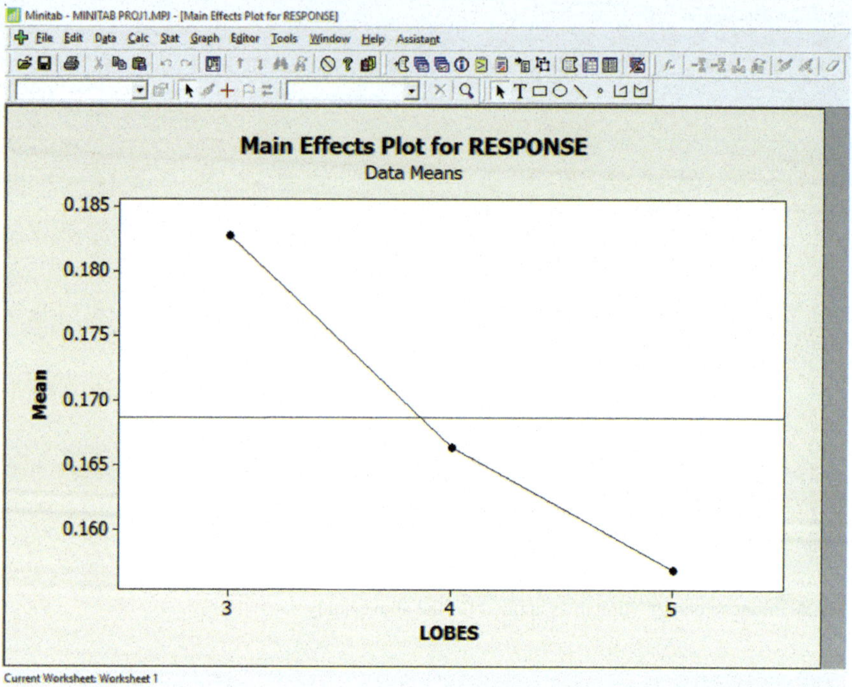

Fig. 7 Main effects graph showing Power against number of petals

5 Experimental Verification

To confirm the viability of the setup, a full-scale platform is developed [1] as shown in Fig. 10. Two drive shafts are coupled by using chain sprocket. The drive being positive, lobes of driving wheel rotate in phase. Electric motor is used as prime mover for the setup and gearbox provides required speed reduction and torque magnification.

Components of the configuration:-

1. Frame
2. Plummer blocks
3. Drive shafts
4. Sprockets
5. Roller Chain
6. Gear reducer
7. Electric motor
8. On–off Switches (Not shown)
9. Wires (Not shown)
10. Drive wheels with designed profile
11. Supporting hubs
12. Nuts and Bolts

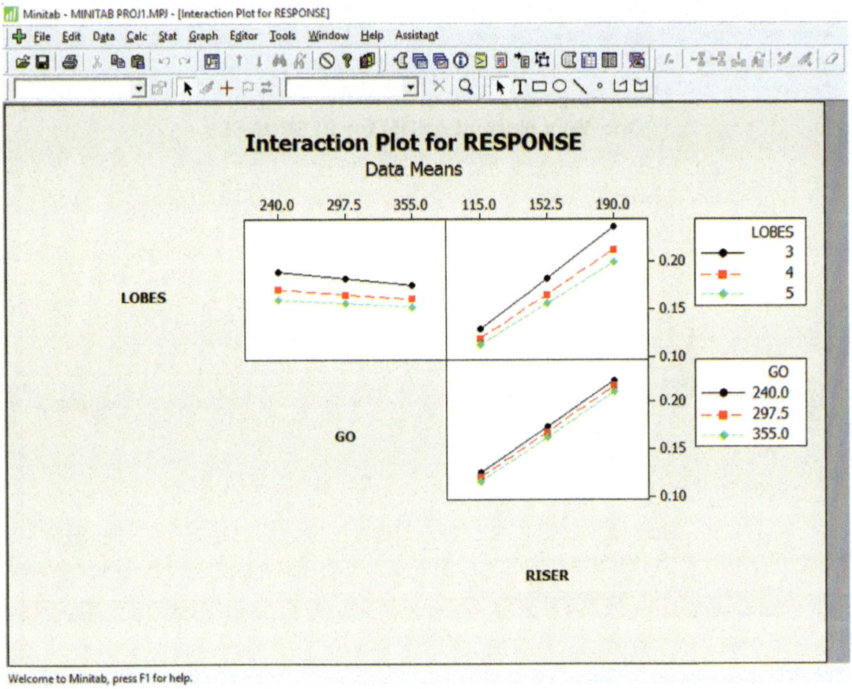

Fig. 8 Interactions graph showing effect of power against (D), (H) and (n)

Other parameters are as given below:

- Frame dimensions (cm):- 115 × 60
- Bare Weight (in kg):- 40
- Pay load (in kg):- 95
- Motor capacity (in HP):- 0.75
- Timc to climb (in sec):- 36
- Speed of shaft (RPM):- 10.

It is seen that the setup moves perfectly in the path parallel to 'pitch line'.

Force perpendicular to the torque arm at farthest point of contact is determined with the help of load cells. It is found that the said force is 40 kgf ≈ 392 N.

Preliminary analysis [1] projects that the design with three petals is ideal [1] but the said inference was completely based on the requirement of torque.

To compare the behaviour of various setups, rate of ascend should be kept same. So the wheel with lesser petals should be rotated faster as compared with the wheel with more petals.

Table 2 gives a comparison of different parameters for drive speed as 10 and 6 RPM.

Table 3 shows comparative observations as well as calculations of different parameters, ultimately leading to power required to drive the setup.

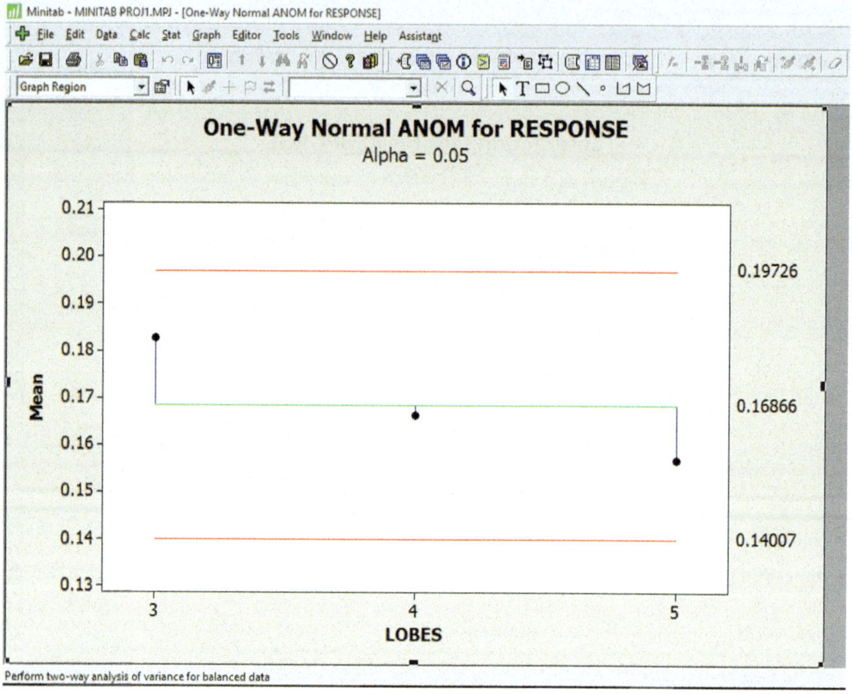

Fig. 9 Graph showing interactions of power with (D), (H) and (n)

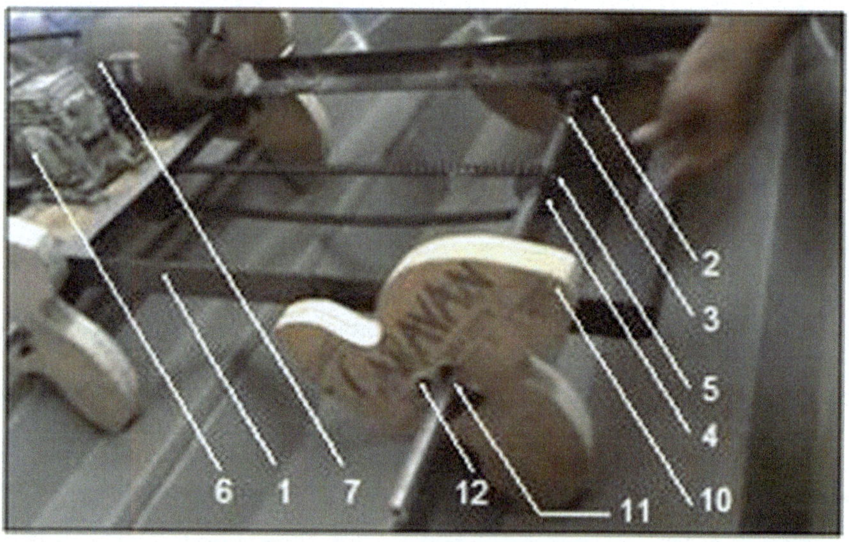

Fig. 10 Working of proposed setup

Table 2 Comparison of parameters corresponding to two different drive speeds

Parameter	For three lobe design	For five lobe design
Motor RPM	1440	1440
Reduction in pulley	1:2	1:2
Reduction in gearbox	1:60	1:80
Reduction in sprocket	5:6	2:3
Shaft speed in RPM	$1440 \times \frac{1}{2} \times \frac{1}{60} \times \frac{5}{6} = 10$	$1440 \times \frac{1}{2} \times \frac{1}{80} \times \frac{2}{3} = 6$

Table 3 Relative calculations and observations for designs with three and five petals

Parameter (Unit)	Value for setup with three lobes	Value for setup with five lobes
Force (N)	392 N	398 N
Longest torque arm (m)	0.18 m	0.22 m
Torque (Nm)	81.6 Nm	101.5 Nm
Shaft speed (RPM and Rad/s)	10 RPM = 1.047 rad/s	06 RPM = 0.6283 rad/s
Power required for 1 wheel (W)	85.58 W	63.13 W
Total power requirement (W)	343.82 W	258.02 W

Torque required for the drive wheel with laser number of petals is less than the one for wheel with more number of petals. But the angular velocity for the latter is less than the former. Combined effect of these variables is seen in the overall requirement of power.

6 Conclusion

Effect of increase in the requirement of torque by increasing the number of petals is less that the effect of corresponding decrease in the angular velocity.

Accordingly, based on the 'Requirement of power for raising the staircase climbing platform', it is observed that the requirement of power is minimum is for design with five petals, even though the corresponding requirement of torque is more.

References

1. Modak GS, Bhoomkar MM (2016) Front Mech Eng 11:299. https://doi.org/10.1007/s11465-016-0374-1
2. 2006 Stair Building Code (1997) As a supplement of visual interpretation of the international residential code 1997. Stairway Manufacturers' Association, USA

3. Stairs, ladders and walkways design for straight stairs. British Standard BS5395 Part 1, 1977
4. Design and construction of stairs for houses. Indian Standard IS 1634:1992 (Revised 2002), Bureau of Indian Standards, 2002
5. Building code of Australia. Australian Building Codes Board under Australian Standards, 2015

Forward Dynamics of the Double-Wishbone Suspension Mechanism Using the Embedded Lagrangian Formulation

Vyankatesh Ashtekar and Sandipan Bandyopadhyay

Abstract The double-wishbone (DWB) is a popular suspension system, particularly in the high-end automobiles. Simulation of its kinematics and dynamics are vital elements in the process of analysis and design of these complex mechanical systems. However, the simulation of the DWB suspension can be computationally demanding, due to the large number of nonlinear, coupled ordinary differential equations (ODEs) that arise from the motion of the multiple links present in the system. In this paper, a less common approach to the Lagrangian formulation is adopted, which is known as the embedded formulation or the actuator space formulation. In this formulation, the number of ODEs to be solved come down to only two-a number that equals the degree-of-freedom of the system. The joint variables associated with the unactuated links are computed through the forward kinematics of the system. This method has the advantage of reducing the computational burden in the numerical solution of the ODEs significantly, as the number of ODEs comes down to two, as opposed to eleven in the more commonly used configuration space formulation. Furthermore, the unactuated variables are determined from the kinematic constraints, as opposed to being computed from the numerical solutions to ODEs, which make them more accurate. The formulation is illustrated via numerical examples implemented in the Computer Algebra System (CAS), Mathematica. It is believed that such a formulation would aid in the understanding of the dynamics of the suspension systems, and help in the process of their design.

Keywords Double wishbone suspension · Forward dynamics · Lagrangian formulation · Root-tracking

V. Ashtekar · S. Bandyopadhyay (✉)
Indian Institute of Technology Madras, Chennai, India
e-mail: sandipan@iitm.ac.in

V. Ashtekar
e-mail: vyankatesh0811@gmail.com

D. Sen et al. (eds.), *Mechanism and Machine Science*, Lecture Notes in Mechanical Engineering, https://doi.org/10.1007/978-981-15-4477-4_60

1 Introduction

Dynamic modelling and simulation of an automotive suspension system such as the double wishbone is a difficult task, due to the number of loops involved, and the number of variables contained therein. Kinematic as well as dynamic analysis of DWB suspension is presented in [1, pp. 202–247], as a case study. The analyses are configuration dependent, i.e., presented with numbers only, and hence, are of little help in plotting the response of the suspension to varying inputs. Often, certain parts of it are omitted in the analysis to make the problem simpler, and amenable to mathematical analysis. For example, the vertical motion aspect can be studied in a manner decoupled from the steering motion (albeit approximately), as shown in [2]. Spatial models of the DWB mechanism are not very common in literature. In Cherian et al. [3] use the constrained Lagrangian formulation to study the mechanism in its entirety. Attia [4] presents a formulation in the joint-space, in which the dynamic model consists of 11 coupled, non-linear ordinary differential equations. Uchida and McPhee [5] uses the *embedded formulation* (see, e.g., [6, pp. 196–200]), of the Lagrangian equation of motion. This formulation has the advantage of bringing down the number of ODEs to be solved to the minimum possible, i.e., the degree-of-freedom of the mechanism (which is just 2 in this case). However, the embedded formulation needs to solve the forward kinematics of the mechanism at each integration step, which in itself, is a very hard task. In [5] the authors needed to resort to a number of compromises to achieve this. Firstly, the inputs included in the model were not the actual physical inputs, namely, the steering rack displacement and the road input, in terms of the vertical displacement of the wheel-type contact point. The kinematic constraint equations were solved using a numerical procedure, whose convergence was not guaranteed. Finally, a stabilising scheme had to be used to keep the error accrual during the numerical integration under check.

This paper uses the embedded formulation to reduce the dynamic model to only two coupled ODEs. However, it alleviates the caveats mentioned in [5], by using an analytical model of kinematics reported in [7]. Therefore the inputs to the dynamic model are the *actual* physical inputs, and the solutions are guaranteed at each step. Further, due to the accurate nature of the kinematic solutions, the numerical integration does not need any special treatments while returning highly accurate results. The results of the simulation are analysed further for their accuracy in terms of kinematics and dynamics (i.e., work-energy principle, in the differential or "power" form), their veracity is established in a quantitative manner.

The rest of the paper is organised as follows: in Sect. 2, the formulation of the kinematic constraints and the process of their solution are discussed briefly. Equations of motion are derived and the procedure for the time integration of the same is presented in Sect. 3. The forward dynamic analysis of the DWB is illustrated with a numerical example in Sect. 4. The paper is concluded in Sect. 5.

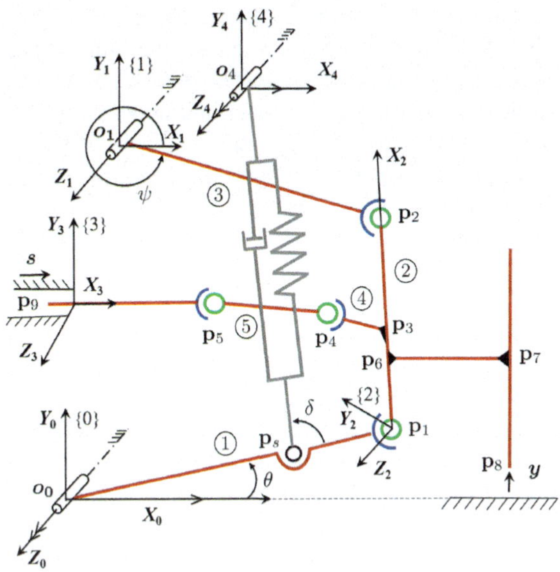

Fig. 1 Schematic of the DWB suspension system

2 Kinematics of the Double Wishbone Suspension

The forward kinematic problem of the DWB suspension is defined and two methods for the solution of the same are discussed in this section. First method is adopted from [7], which gives all possible solutions to the forward kinematics problem. The second method, based on the tracking of the roots of a polynomial equation is discussed in Sect. 2.4.

2.1 Geometry of the Suspension

A kinematic sketch of the DWB suspension is depicted in Fig. 1. Kinematics and dynamics analyses of the DWB suspension system are performed with reference to the vehicle body (frame {0}), which serves as a stationary frame of reference. The suspension mechanism is a combination of a spatial four-bar loop, $\boldsymbol{o}_0\boldsymbol{p}_1\boldsymbol{p}_2\boldsymbol{o}_1\boldsymbol{o}_0$, which accommodates the jounce and the rebound movement of the wheel, coupled with a spatial five-bar loop, $\boldsymbol{o}_0\boldsymbol{p}_1\boldsymbol{p}_4\boldsymbol{p}_5\boldsymbol{p}_9\boldsymbol{o}_0$, enabling the steering motion of the wheel. The king-pin (link ② in Fig. 1) is the only link coupling the two loops together. Link ② is a rigid assembly of the king-pin, i.e., the wheel hub and the steering link, and the wheel.[1]

[1] The wheel actually *rotates* on the hub axle, but that is irrelevant to the kinematic model.

Three frames are used to describe the configuration of the DWB and one for specifying the pivot of the spring and the damper. The global frame of reference, $\{0\}$ is attached to $\boldsymbol{o}_0$ and its Z axis is (denoted by Z_0) is aligned with the axis of hinges of the lower A-arm. The lower and upper A-arms are equivalently represented as single bars/beams, (link ① and ③, respectively, in Fig. 1). Link ① moves in the X_0Y_0-plane. Two different body-fixed frames of reference, namely, $\{1\}$ and $\{2\}$, are attached to $\boldsymbol{o}_1$ and $\boldsymbol{p}_1$, respectively. The Z_1-axis of $\{1\}$ is aligned with the axis of the hinges of the upper A-arm. link ③ moves in X_1Y_1 plane. The X_2-axis of 2 is along the link vector $\boldsymbol{l}_2 = \boldsymbol{p}_2 - \boldsymbol{p}_1$, and the plane X_2Y_2 contains the steering link, i.e., link ④.

The damper is connected between the vehicle body (at $\boldsymbol{o}_4$) and the lower A-arm (at $\boldsymbol{p}_s$). The damper is attached via rod-end bearings, which are kinematically equivalent to the spherical joints, to either of the pivots.[2] The spring is attached between the barrel and the rod of the damper. Often, in real life, the spring is in contact with lower A-arm or chassis directly through a circular seat. In that case, the spring undergoes slight buckling in addition to its longitudinal motion, during its compression. However, in this paper, it is assumed that the spring undergoes longitudinal motion only. The spring and the damper undergo planar motion as the origin of frame $\{4\}$ lies in the plane X_0Y_0, i.e., the plane of motion of link ①.

Any vector in its own the body-fixed frame of reference can be transformed to the global frame of reference by pre-multiplying with an appropriate rotation matrix. The rotation matrix ${}^2_0\boldsymbol{R}$ describes the orientation of $\{2\}$ w.r.t. $\{0\}$, i.e., the orientation of king-pin axis (KPA) w.r.t. the global frame of reference. In this work, ${}^2_0\boldsymbol{R}$ is expressed in terms of the Rodrigues' parameters, $\boldsymbol{c} = [c_1, c_2, c_3]^\top \in \mathbb{R}^3$.

2.2 Formulation of the Loop-Closure Equations

In the following, for the sake of brevity, vectors in $\{0\}$ are written simply as $\boldsymbol{p}, \boldsymbol{l}$, etc., i.e., without the leading superscript. The steering/rack input is given to point $\boldsymbol{p}_9$, and the road profile input is applied to $\boldsymbol{p}_8$. It may be noted that the road profile input y is defined as the Y-coordinate of $\boldsymbol{p}_8$ w.r.t. $\{0\}$. Link ② is defined by points $\boldsymbol{p}_1$ and $\boldsymbol{p}_2$. They can be expressed as:

$$\boldsymbol{p}_1 = \boldsymbol{o}_0 + \boldsymbol{R}_{Z_0}(\theta)[l_1, 0, 0]^\top, \text{ and} \tag{1}$$

$$\boldsymbol{p}_2 = \boldsymbol{o}_1 + {}^1_0\boldsymbol{R}\boldsymbol{R}_{Z_1}(\psi)[l_3, 0, 0]^\top, \tag{2}$$

respectively, where ${}^1_0\boldsymbol{R} = \boldsymbol{R}_X(\alpha_1)\boldsymbol{R}_Y(\alpha_2)\boldsymbol{R}_Z(\alpha_3)$, and α_i are the XYZ Euler angles, relating the orientation of $\{1\}$ w.r.t. $\{0\}$. The vector $\boldsymbol{l}_2$ is given by:

[2]Figure 1 depicts the pivot of the damper at $\boldsymbol{o}_4$ as a pin joint because the other two degrees-of-freedom of the spherical joint are irrelevant to the analysis. See Sect. 2.2 for more details.

$$\boldsymbol{l}_2 = \boldsymbol{p}_2 - \boldsymbol{p}_1, \text{ and also as} \tag{3}$$

$$\boldsymbol{l}_2 = {}^2_0\boldsymbol{R}[l_2, 0, 0]^\top. \tag{4}$$

From Eqs. (1) to (4) one can write:

$$(\boldsymbol{o}_1 + {}^1_0\boldsymbol{R}\boldsymbol{R}_{Z_1}(\psi)[l_3, 0, 0]^\top) - (\boldsymbol{o}_0 + \boldsymbol{R}_{Z_0}(\theta)[l_1, 0, 0]^\top) - {}^2_0\boldsymbol{R}[l_2, 0, 0]^\top = \boldsymbol{0}. \tag{5}$$

Equation (5) models the kinematics of the four-bar loop, $\boldsymbol{o}_0\boldsymbol{p}_1\boldsymbol{p}_2\boldsymbol{o}_1\boldsymbol{o}_0$. The LHS of Eq. (5) is written compactly as:

$$\boldsymbol{\eta}_1 = [\eta_{1x}, \eta_{1y}, \eta_{1z}]^\top. \tag{6}$$

The explicit form of Eq. (5) is available in [7], and is therefore skipped here for brevity. End-points of link ⑤ are given by:

$$\boldsymbol{p}_4 = \boldsymbol{p}_1 + {}^2_0\boldsymbol{R}[rl_2, 0, 0]^\top + {}^2_0\boldsymbol{R}\,{}^2\boldsymbol{l}_4, \text{ and} \tag{7}$$

$$\boldsymbol{p}_5 = [p_{5x} + s_x, p_{5y} + s_y, p_{5z} + s_z]^\top \text{ respectively,} \tag{8}$$

where rl_2 is the distance between $\boldsymbol{p}_1$ and $\boldsymbol{p}_3$; ${}^2\boldsymbol{l}_4 = [l_{4x}, l_{4y}, l_{4z}]^\top$ (expressed in $\{2\}$), and $\boldsymbol{s} = [s_x, s_y, s_z]^\top$ is the rack input vector in $\{0\}$, given by:

$$\boldsymbol{s} = {}^3_0\boldsymbol{R}[s, 0, 0]^\top, \text{ where}$$
$${}^3_0\boldsymbol{R} = \boldsymbol{R}_X(\beta_1)\boldsymbol{R}_Y(\beta_2)\boldsymbol{R}_Z(\beta_3),$$

and β_1, β_2 and β_3 are the XYZ Euler angles relating the orientation of $\{3\}$ w.r.t. $\{0\}$. The link vector $\boldsymbol{l}_5$ can be expressed as:

$$\boldsymbol{l}_5 = \boldsymbol{p}_5 - \boldsymbol{p}_4.$$

The kinematic constraint arising out of the spatial five-bar loop, $\boldsymbol{o}_0\boldsymbol{p}_1\boldsymbol{p}_4\boldsymbol{p}_5\boldsymbol{p}_9\boldsymbol{o}_0$, is captured in a single scalar equation $\eta_2 = 0$, where:

$$\eta_2 = (\boldsymbol{p}_5 \quad \boldsymbol{p}_4) \cdot (\boldsymbol{p}_5 - \boldsymbol{p}_4) - l_5^2. \tag{9}$$

The spatial four-bar loop is connected with the road input y through the equation:

$$\boldsymbol{p}_8 + [x, y, z]^\top - {}^2_0\boldsymbol{R}\,{}^2\boldsymbol{l}_6 - {}^2_0\boldsymbol{R}[\rho l_2, 0, 0]^\top - \boldsymbol{p}_1 = \boldsymbol{0}, \tag{10}$$

where x, z are horizontal components of the displacement of $\boldsymbol{p}_8$ w.r.t. $\{0\}$, as a consequence of s and y departing from their initial reference (i.e., zero) values; ${}^2\boldsymbol{l}_6 = [l_{6x}, l_{6y}, l_{6z}]^\top = {}^2\boldsymbol{p}_8 - {}^2\boldsymbol{p}_6$; ρl_2 is the distance between points $\boldsymbol{p}_6$ and $\boldsymbol{p}_1$. The expression on the LHS of Eq. (10) is written compactly as:

$$\boldsymbol{\eta}_3 = [\eta_{3x}, \eta_{3y}, \eta_{3z}]^\top. \tag{11}$$

The full form of Eq. (11) can be found in [7]. The seven scalar equations, namely, Eqs.(6), (9), and (11) define the kinematics of the DWB mechanism. In addition to those, equations for the calculation of the spring and the damper displacement are given by:

$$\boldsymbol{p}_s + l_s[\cos(\theta + \delta), \sin(\theta + \delta), 0]^\top - \boldsymbol{o}_4 = \boldsymbol{0}, \tag{12}$$

where δ is the angle between the damper and link ①, as illustrated in Fig. 1, and l_s is the distance between $\boldsymbol{o}_4$ and $\boldsymbol{p}_s$. The spring and the damper undergo planar motion in the plane of link ①. Hence, the Z-component of Eq. (12) is trivial. The expression on the LHS of Eq. (12) is written compactly as:

$$\boldsymbol{\eta}_4 = [\eta_{4x}, \eta_{4y}]^\top, \text{ where} \tag{13}$$

$$\eta_{4x} = o_{0x} + l_1 \cos(\theta) + l_s \cos(\theta + \delta) - o_{4x}, \text{ and} \tag{14}$$

$$\eta_{4y} = o_{0y} + l_1 \sin(\theta) + l_s \sin(\theta + \delta) - o_{4y}. \tag{15}$$

Equations (14)–(15) and Eqs. (6), (9), and (10) are dependent upon eleven joint-space variables, namely:

$$\boldsymbol{q} = [s, y, c_1, c_2, c_3, \theta, \psi, x, z, l_s, \delta]^\top, \tag{16}$$

Put together, the loop-closure equations are written as:

$$\boldsymbol{\eta}(\boldsymbol{q}) = [\eta_{1x}, \eta_{1y}, \eta_{1z}, \eta_2, \eta_{3x}, \eta_{3y}, \eta_{3z}, \eta_{4x}, \eta_{4y}]^\top = \boldsymbol{0}. \tag{17}$$

The procedure to solve for the unknowns, given the steering input s and the road input y is discussed in the following section.

2.3 *Zeroth-Order Kinematics*

The problem of computing the orientations of the king-pin axis (defined in terms of $[c_1, c_2, c_3]^\top$) given s and y, is defined as the forward kinematics problem in this context. The solution to the forward kinematics problem, in the general case, is presented in [7] comprehensively. The vectors of the known variables, $\boldsymbol{\theta}$, and the unknown variables, $\boldsymbol{\phi}$, are defined as, respectively:

$$\boldsymbol{\theta} = [s, y]^\top, \ \boldsymbol{\phi} = [c_1, c_2, c_3, \theta, \psi, x, z, l_s, \delta]^\top. \tag{18}$$

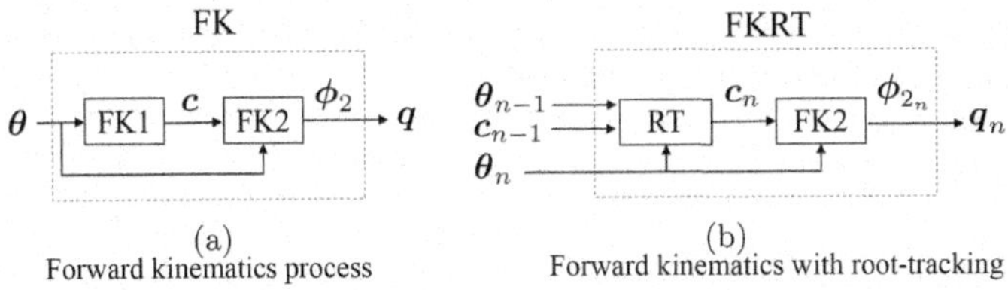

Fig. 2 Block-diagram representation of the two solution schemes. The subscript n signifies the nth step of the numerical integration (see Sect. 3.2)

Hence, $\boldsymbol{q} = [\boldsymbol{\theta}^\top, \boldsymbol{\phi}^\top]^\top$. The unknowns $\boldsymbol{\phi}$ are split further: $\boldsymbol{\phi} = [\boldsymbol{c}^\top, \boldsymbol{\phi_2}^\top]^\top$, where:

$$\boldsymbol{c} = [c_1, c_2, c_3]^\top, \ \boldsymbol{\phi}_2 = [\theta, \psi, x, z, l_s, \delta]^\top. \tag{19}$$

Solution for $\boldsymbol{\phi}$ is obtained through the following steps.

1. Successive elimination of the unknowns in $\boldsymbol{\phi}_2$ from the loop-closure constraints (Eq. 17) leads to three *quartic* equations in $\boldsymbol{c}$:

$$\boldsymbol{f}(\boldsymbol{c}) = \boldsymbol{f}(c_1, c_2, c_3) = [f_1, f_2, f_3]^\top = \mathbf{0}. \tag{20}$$

   The coefficients in Eq. (20) are functions of only the architecture parameters, and the input $\boldsymbol{\theta}$. The Bézout limit on the number of complex solutions of Eq. (20) is 64. It has been observed that, all 64 solutions are finite for the architecture parameters[3] listed in Table 3, out of which authors have found not more than four to be real. This step is depicted as the block "FK1" in Fig. 2a.
2. The variables $\{\theta, \psi, x, z\}$ are computed from the known values of $\boldsymbol{c}$.
3. The variables describing the spring and the damper pose, i.e., l_s and δ are calculated from θ alone, from Eq. (14) to (15). Second step combined with this step, are depicted in the block "FK2" in Fig. 2a.

These steps together form the "FK" block in Fig. 2a.

2.4 Tracking the Roots of a Set of Polynomial Equations

The solution method discussed above is computationally expensive, since it computes all the roots of Eq. (20). However, only one of the real branches of the solution has to be followed in a dynamic simulation. Hence, the complete set of solutions is obtained *only once* at the initial state of the simulation. Then a method to track a particular solution branch is used to compute the passive variables at every integration step. A two-step process is followed for root tracking. Firstly, the increment vector, $\Delta\boldsymbol{c}$

[3]Positions, lengths are measured in mm, the angles are measured in degrees, masses are measured in kg and inertiæ in kg-m^2 throughout the paper unless mentioned otherwise.

at any step, is found through the linearisation of Eq. (20) about the solution at the current step, from the following equations:

$$\Delta \boldsymbol{f} = \boldsymbol{J}_{fc}\,\Delta \boldsymbol{c} + \boldsymbol{J}_{f\theta}\,\Delta\boldsymbol{\theta}, \text{ where } \boldsymbol{J}_{fc} = \frac{\partial \boldsymbol{f}}{\partial \boldsymbol{c}}, \text{ and } \boldsymbol{J}_{f\theta} = \frac{\partial \boldsymbol{f}}{\partial \boldsymbol{\theta}}.$$
$$\Rightarrow \Delta\boldsymbol{c} = -\boldsymbol{J}_{fc}^{-1}\boldsymbol{J}_{f\theta}\,\Delta\boldsymbol{\theta}, \text{ since } \Delta\boldsymbol{f} = \boldsymbol{0}, \text{ and assuming } \boldsymbol{J}_{fc} \text{ is invertible.} \quad (21)$$

The matrix $\boldsymbol{J}_{fc}$ is not invertible when $\boldsymbol{f}$ has root(s) of multiplicity of two or more. However, such cases are outside the purview of this paper. Numerical computations are aborted when the *condition number* of the matrix $\boldsymbol{J}_{fc}$ reaches a predefined upper limit κ_M, which is taken to be 10^7 in this work. The new vector $\boldsymbol{c}$ is then used as the *initial guess* in a Newton-Raphson procedure, to improve the accuracy of the root. The procedure is implemented using the internal function `FindRoot` of the CAS `Mathematica` with an `AccuracyGoal` (denoted by ϵ_f) of 10 significant digits and `PrecisionGoal` (denoted by ϵ_X) of 10 significant digits. The Jacobian matrices are updated at every iteration. The maximum number of iterations is set to $n_I = 100$. This procedure of tracking the roots is condensed in the block "RT" in Fig. 2b. The block-diagram for the second method of forward kinematics with root tracking is shown in Fig. 2b and denoted as "FKRT" block. The entire process of root-tracking is illustrated through a flowchart in Fig. 3. In case of a computational failure or lack of convergence in the FKRT block, the control is returned to the FK block, to reinitialise the tracking process.

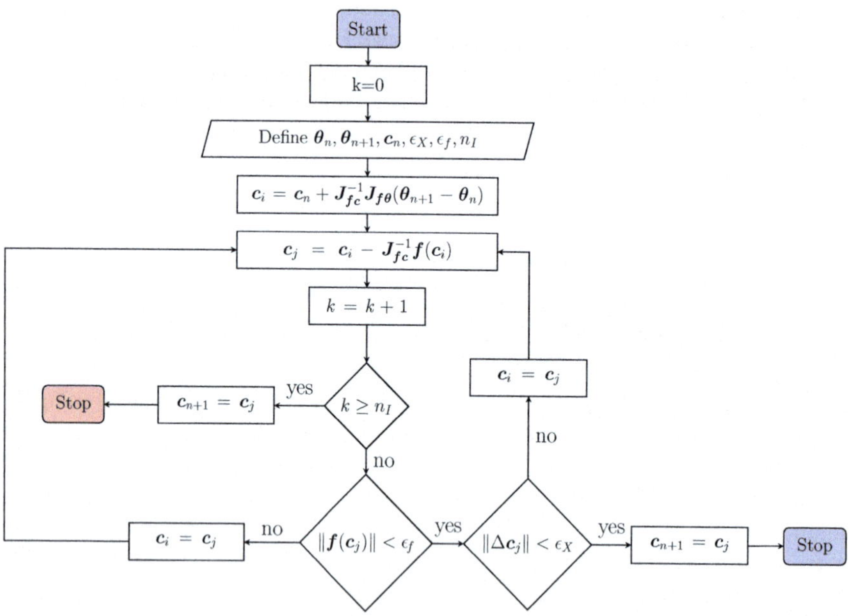

Fig. 3 Flowchart for tracking the roots of the system of polynomials $\boldsymbol{f}$

2.5 Velocity Kinematics

This section discusses the calculation of the passive joint rates, $\dot{\boldsymbol{\phi}}$, from the actuator joint rates, $\dot{\boldsymbol{\theta}}$. The passive joint rates are calculated from the loop-closure constraints as follows. Restating Eq. (17) as $\boldsymbol{\eta}(\boldsymbol{q}) = \boldsymbol{\eta}(\boldsymbol{\theta}, \boldsymbol{\phi})$ and differentiating it w.r.t. time, one obtains:

$$\frac{\partial \boldsymbol{\eta}}{\partial \boldsymbol{\theta}} \dot{\boldsymbol{\theta}} + \frac{\partial \boldsymbol{\eta}}{\partial \boldsymbol{\phi}} \dot{\boldsymbol{\phi}} = \mathbf{0}; \tag{22}$$

$$\Rightarrow \dot{\boldsymbol{\phi}} = -\left(\frac{\partial \boldsymbol{\eta}}{\partial \boldsymbol{\phi}}\right)^{-1} \frac{\partial \boldsymbol{\eta}}{\partial \boldsymbol{\theta}} \dot{\boldsymbol{\theta}}, \text{ assuming } \det\left(\frac{\partial \boldsymbol{\eta}}{\partial \boldsymbol{\phi}}\right) \neq 0. \tag{23}$$

Writing $\boldsymbol{J}_{\eta\theta} = \frac{\partial \boldsymbol{\eta}}{\partial \boldsymbol{\theta}}$ and $\boldsymbol{J}_{\eta\phi} = \frac{\partial \boldsymbol{\eta}}{\partial \boldsymbol{\phi}}$, Eq. (23) is re-written as:

$$\dot{\boldsymbol{\phi}} = \boldsymbol{J}_{\phi\theta}\dot{\boldsymbol{\theta}}, \text{ where } \boldsymbol{J}_{\phi\theta} = -\boldsymbol{J}_{\eta\phi}^{-1}\boldsymbol{J}_{\eta\theta} \text{ and } \det(\boldsymbol{J}_{\eta\phi}) \neq 0. \tag{24}$$

Equation (24) gives an expression for the passive joint rates in terms of the active joint rates. Similarly, $\ddot{\boldsymbol{\phi}}$ can be obtained by differentiating Eq. (24). Further, Eq. (24) leads to a linear map between $\dot{\boldsymbol{q}}$ and the *input rates*, $\dot{\boldsymbol{\theta}}$:

$$\dot{\boldsymbol{q}} = \boldsymbol{J}_{q\theta}\dot{\boldsymbol{\theta}}, \text{ where, } \boldsymbol{J}_{q\theta} = \begin{bmatrix} \boldsymbol{I}_2 \\ \boldsymbol{J}_{\theta\phi} \end{bmatrix}, \tag{25}$$

where $\boldsymbol{I}_2 \in \Re^{2\times 2}$ is an *identity matrix*. Consequently, the velocity of *any* point on the mechanism, say $\boldsymbol{p}_i$, can be expressed in terms of the actuator joint rates alone. For example, if $\boldsymbol{p}_i = \boldsymbol{p}_i(\boldsymbol{q})$, then its velocity is given by:

$$\dot{\boldsymbol{p}}_i = \frac{\partial \boldsymbol{p}_i}{\partial \boldsymbol{q}} \dot{\boldsymbol{q}} = \boldsymbol{J}_{p_i q}\dot{\boldsymbol{q}} = \boldsymbol{J}_{p_i q}\boldsymbol{J}_{q\theta}\dot{\boldsymbol{\theta}}, \text{ where } \boldsymbol{J}_{p_i q} = \frac{\partial \boldsymbol{p}_i}{\partial \boldsymbol{q}}. \tag{26}$$

Equation (25) facilitates the conversion of the equation of motion of the constrained mechanical system only in term of the actuator accelerations. The velocity kinematics computations are represented as first order forward kinematics (FoFK) block in Fig. 4.

Fig. 4 Block representation of the velocity kinematics

3 Dynamics of the Double Wishbone Suspension

The main objective of this paper is to reduce the number of simultaneous ODEs to be solved in the forward dynamic simulation to the smallest possible, i.e., the degree-of-freedom of the mechanism. This is done by following the embedded formulation the of constrained Lagrangian dynamics.

3.1 Derivation of the Equation of Motion

Equation of motion of a constrained mechanical system can be written as:

$$\boldsymbol{M}(\boldsymbol{q})\ddot{\boldsymbol{q}} + \boldsymbol{C}(\boldsymbol{q}, \dot{\boldsymbol{q}})\dot{\boldsymbol{q}} + \boldsymbol{G}(\boldsymbol{q}) = \boldsymbol{\tau}^{\mathrm{nc}} + \boldsymbol{\tau}^{\mathrm{c}}, \tag{27}$$

where $\boldsymbol{M}$ represents the generalised mass matrix, $\boldsymbol{C}$ contains the centripetal and Coriolis forces, $\boldsymbol{G}$ is the vector of potential forces and $\boldsymbol{\tau}^{\mathrm{nc}}$ is the vector of external forces. The constraint forces term, $\boldsymbol{\tau}^{\mathrm{c}}$ in Eq. (27) can be written as $\boldsymbol{J}_{\eta q}^{\top}\boldsymbol{\lambda}$ (see, e.g., [8, pp. 187–192]), where $\boldsymbol{\lambda}$ is the vector of the Lagrange multipliers. Substituting $\dot{\boldsymbol{q}}$ and $\ddot{\boldsymbol{q}}$ in Eq. (27), using Eq. (25) and its derivative w.r.t. time, and pre-multiplying with $\boldsymbol{J}_{q\theta}^{\top}$, following equation is obtained:

$$\boldsymbol{J}_{q\theta}^{\top}\boldsymbol{M}\boldsymbol{J}_{q\theta}\ddot{\boldsymbol{\theta}} + \boldsymbol{J}_{q\theta}^{\top}\big(\boldsymbol{M}\dot{\boldsymbol{J}}_{q\theta} + \boldsymbol{C}\boldsymbol{J}_{q\theta}\big)\dot{\boldsymbol{\theta}} + \boldsymbol{J}_{q\theta}^{\top}\boldsymbol{G} = \boldsymbol{J}_{q\theta}^{\top}\boldsymbol{\tau}^{\mathrm{nc}} + \boldsymbol{J}_{q\theta}^{\top}\boldsymbol{J}_{\eta q}^{\top}\boldsymbol{\lambda}. \tag{28}$$

With $\boldsymbol{J}_{q\theta}^{\top}\boldsymbol{J}_{\eta q}^{\top}\boldsymbol{\lambda} = \boldsymbol{0}$, [9], Eq. (28) is rewritten as:

$$\boldsymbol{M}_{\theta}\ddot{\boldsymbol{\theta}} + \boldsymbol{C}_{\theta}\dot{\boldsymbol{\theta}} + \boldsymbol{G}_{\theta} = \boldsymbol{\tau}_{\theta}, \tag{29}$$

where $\boldsymbol{M}_{\theta} = \boldsymbol{J}_{q\theta}^{\top}\boldsymbol{M}\boldsymbol{J}_{q\theta}$, $\boldsymbol{C}_{\theta} = \boldsymbol{J}_{q\theta}^{\top}\big(\boldsymbol{M}\dot{\boldsymbol{J}}_{q\theta} + \boldsymbol{C}\boldsymbol{J}_{q\theta}\big)$ and $\boldsymbol{G}_{\theta} = \boldsymbol{J}_{q\theta}^{\top}\boldsymbol{G}$ and $\boldsymbol{\tau}_{\theta} = \boldsymbol{J}_{q\theta}^{\top}\boldsymbol{\tau}^{\mathrm{nc}}$ is the active generalised force vector. It should be noted that Eq. (29) is valid only when $\det(\boldsymbol{J}_{\eta\phi}) \neq 0$. Detailed derivation of Eq. (29) can be found in [9].

3.2 Integration of the Equation of Motion

Since the free response of the DWB suspension is studied in this paper, the damping force is the only non-conservative/external force applied to the system. The term $\boldsymbol{\tau}_{\theta}$ is calculated as:

$$\boldsymbol{\tau}_{\theta} = \boldsymbol{J}_{q\theta}^{\top}\boldsymbol{J}_{\boldsymbol{v}_{P_s}}^{\top}\boldsymbol{F}_{d}, \text{ where } \boldsymbol{F}_{d} = -C_{d}\dot{l}_{s}. \tag{30}$$

In Eq. (30) C_d is the damping coefficient, $\boldsymbol{F}_d$ is the damping force generated in the hydraulic damper and $\boldsymbol{J}_{v_{p_s}}$ is the velocity Jacobian matrix corresponding to the point of application (point $\boldsymbol{p}_s$, see Fig. 1) of the damping force. Equation (29) is rearranged to facilitate the forward dynamic simulation:

$$\ddot{\boldsymbol{\theta}} = \boldsymbol{M_\theta}^{-1}(\boldsymbol{\tau}_\theta - \boldsymbol{C}_\theta\dot{\boldsymbol{\theta}} - \boldsymbol{G}_\theta). \quad (31)$$

Equation (31) is a set of non-linear second order ordinary differential equations (ODEs). In order to convert them to a set of first order ODEs, the state-space vector $\boldsymbol{X}$ is defined as $\boldsymbol{X} = [\boldsymbol{\theta}^\top, \dot{\boldsymbol{\theta}}^\top]^\top$. The derivative of $\boldsymbol{X}$ is computed as:

$$\dot{\boldsymbol{X}} = \begin{pmatrix} \dot{\boldsymbol{\theta}} \\ \ddot{\boldsymbol{\theta}} \end{pmatrix} = \begin{pmatrix} \dot{\boldsymbol{\theta}} \\ \boldsymbol{M_\theta}^{-1}(\boldsymbol{\tau}_\theta - \boldsymbol{C}_\theta\dot{\boldsymbol{\theta}} - \boldsymbol{G}_\theta) \end{pmatrix} = \boldsymbol{f}_d(\boldsymbol{X}). \quad (32)$$

Similarly, for the configuration space, one can define $\boldsymbol{Y} = [\boldsymbol{q}^\top, \dot{\boldsymbol{q}}^\top]^\top$. The vector $\boldsymbol{Y}$ is computed from $\boldsymbol{X}$ using methods of forward kinematics (represented as the block "KIN" in Fig. 6). Equation (32) is depicted in the form of a block, "DYN" in Fig. 5. The above set of first order ODEs are integrated using the fourth-order Runge-Kutta algorithm (depicted as the block "RK4" in Fig. 6) implemented in computer algebra system (CAS) `Mathematica`. The complete process of numerical integration is depicted as a block-diagram in Fig. 6.

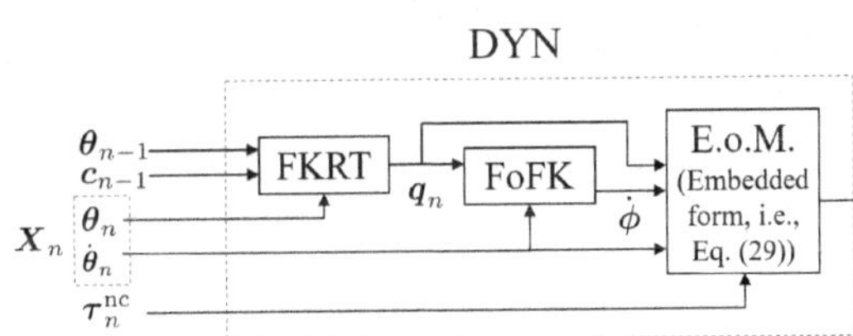

Fig. 5 Block-diagram representation of Eq. (32): $\boldsymbol{X}_n$ represents the value of $\boldsymbol{X}$ at the nth step of integration. The equation of motion is condensed in the "E.o.M" block

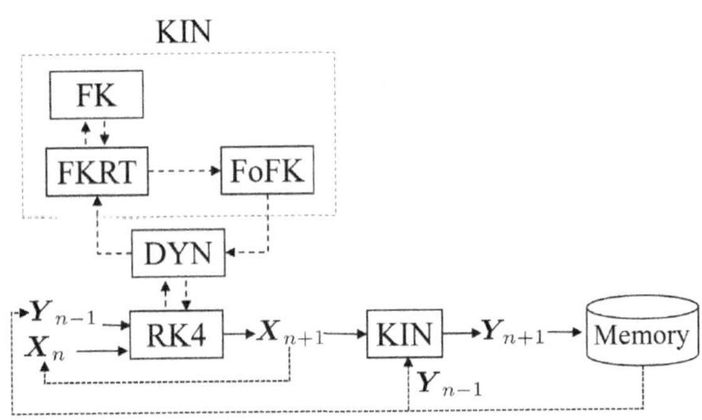

Fig. 6 Integration process

Table 1 Solutions and the residuals of Eq. (17)

c	ϕ_2	$\|\eta\|$ ($\times 10^{-15}$, m)
(0.089, 0.013, 2.259)	(−0.440, −1.051, −0.001, −0.035, 0.622, 2.186)	1.932
(0.162, 0.145, 1.423)	(−0.359, −0.689, −0.048, −0.031, 0.603, 2.125)	0.828
(−1.795, −2.857, 1.359)	(−0.408, −0.808, −0.004, 0.008, 0.614, 2.162)	6.818
(−2.157, −4.092, 1.546)	(−0.445, −0.958, 0.007, 0.0122, 0.622, 2.189)	14.501

(a) (b) (c) (d)

Fig. 7 Real initial configurations, at $s = 0$ and $y = -150$. Configurations **a**, **b**, **c**, and **d** correspond to the four rows of the Table 1, in that order

4 Numerical Results

Free response of the DWB system is simulated with an initial condition corresponding to a real-life scenario in which the wheel has encountered a ditch, and is at its lower-most position, with no initial deviation in the steering angle (see Fig. 7c). This condition is translated into its mathematical form by choosing $s = 0$ and $y = -150$. The initial condition is taken to be stationary, i.e., $\dot{s} = 0$ and $\dot{y} = 0$. Numerical values of the architecture parameters are given in Table 3 in Appendix 6. Mass, inertia, position of the centre of mass of links and the spring and the damper characteristics are listed in Table 4 in Appendix 6.

4.1 *Initial Configuration of the DWB Mechanism*

All the possible solutions of Eq. (20) are computed at the given initial configuration using the `NSolve` routine of the CAS `Mathematica`, with the default working precision of 15 significant digits. The results[4] contain 4 real solutions, which are presented in Table 1, along with the residues produced when the solutions are substituted back in the LHS of Eq. (17). Solution number 3, shown in Fig. 7c, corresponds to a real-life DWB configuration and is chosen as the solution branch to be followed in the root-tracking and numerical integration.

[4]Lengths are in meters and the angles are in radians.

4.2 Numerical Results and Their Validation

The simulation was carried out for 1.5 s with a step size of 0.002 s. It consumed 324 s of CPU-time on an Intel®Core™i7-7500U CPU running at 3.46 GHz. The trajectories obtained are shown in Fig. 8. The time rates of y and s are plotted in Fig. 9. It can be observed that y settles down with time, but s does not. This observation shows that the damper has a little influence on the phenomenon of "bump-steer". The forward dynamics simulation is validated using the following checks:

1. Zeroth-order check on the loop-closure equations, quantified as the error in satisfaction of the loop-closure constraints in Eq. (17): $e_0 = \|\boldsymbol{\eta}(\boldsymbol{q})\|$.
2. Mechanical power check, to ensure a balance between the rate of change of the total energy of the system and the net energy influx, leading to the measure of error: $e_1 = \frac{\mathrm{d}E}{\mathrm{d}t} - \boldsymbol{\tau}^{\mathrm{nc}} \cdot \dot{\boldsymbol{q}}$.

Table 2 lists the absolute maximum and the RMS errors e_0 and e_1 during the simulation. Since the DWB system under consideration is damped and there is no other external input, total mechanical energy of the system falls monotonically with time, as expected, reaching a minimum energy in the steady state (see Fig. 11). After a "sufficiently long" time, all input accelerations vanish. This is observed from the plots of $\ddot{s}$ and $\ddot{y}$ shown in Fig. 10.

The error measures and plots of the results confirm the correctness of the results from a numerical as well as physical perspectives.

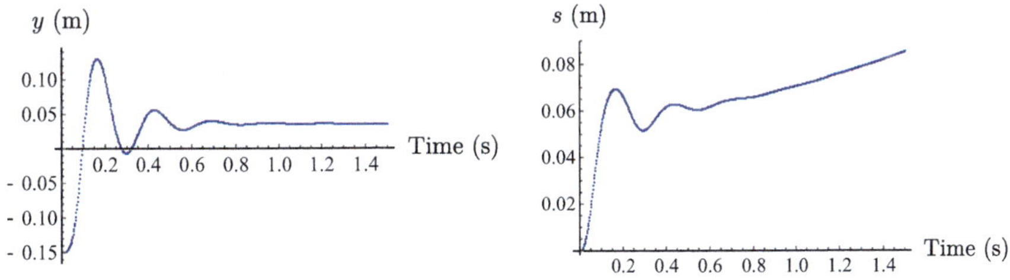

(a) y vs. time

(b) s vs. time

Fig. 8 Trajectories of y and s obtained from the simulation

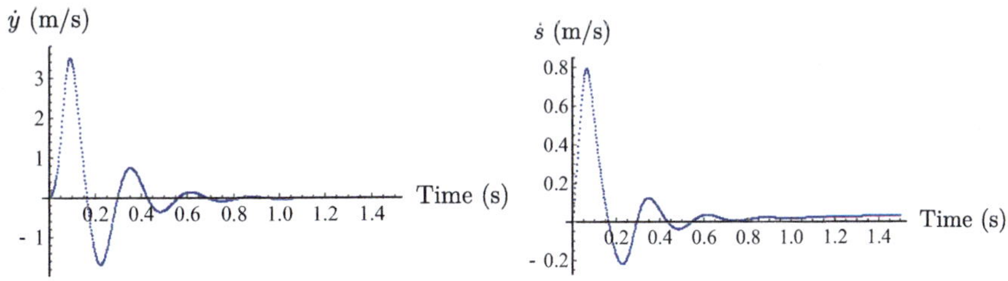

(a) $\dot{y}$ vs. time

(b) $\dot{s}$ vs. time

Fig. 9 Variation in the values of $\dot{y}$ and $\dot{s}$ w.r.t. time

Table 2 Summary of the errors in the simulation

Error measure	$e_0(\times 10^{-15})$ (m)	$e_1(\times 10^{-12})(W)$
Maximum($\|e_i\|$)	29.3	37.3
RMS(e_i)	8.9	3.6

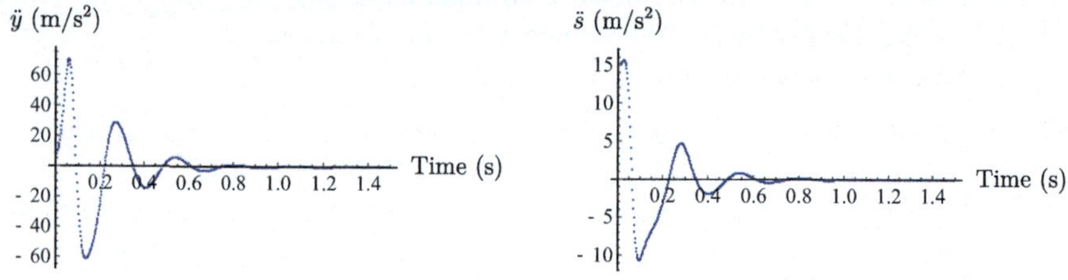

(a) $\ddot{y}$ vs. time (b) $\ddot{s}$ vs. time

Fig. 10 Plot of acceleration of the active variables

Fig. 11 Variation of the total mechanical energy as a function of time

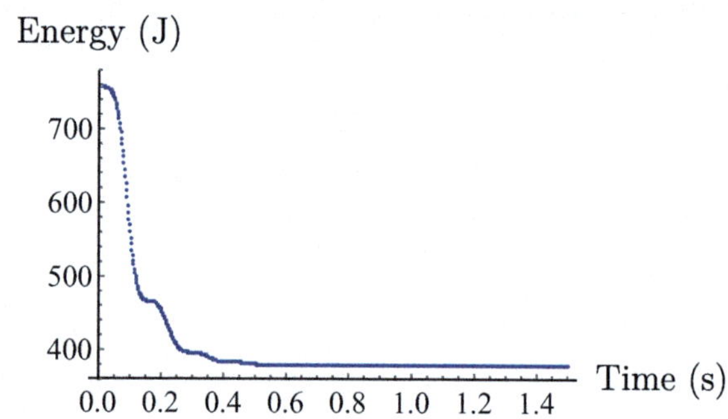

5 Conclusions

This paper presents a method to simulate the dynamics of the DWB suspension system. Using a procedure to solve the kinematics of the system, its dynamic model is reduced to just two ODEs, as opposed to eleven reported in existing literature. Moreover, since the kinematic constraints are being solved in a direct, algebraic manner and not via a larger system of ODEs, the constraints are satisfied to a high level of accuracy. The method is supposed to speed up the numerical simulation significantly. However, the numerical efficiency is not reflected in the present work because of the use of a general purpose CAS. In the future, these algorithms would be implemented in `C/C++`, which is expected to bring forth the computational advantages of the proposed method.

Table 3 Architecture parameters of the DWB mechanism

Parameter	Symbol	Value
Equivalent length of lower A-arm	l_1	420
Length of the KPA	l_2	163
Equivalent length of upper A-arm	l_3	260
Length of tie-rod	l_5	520
Euler angles relating {1} to {0}	$(\alpha_1, \alpha_2, \alpha_3)$	$(-2, -1, 3)$
Euler angles relating {3} to {0}	$(\beta_1, \beta_2, \beta_3)$	$(2, 1, -3)$
Link (4) vector w.r.t. {2}	(l_{4x}, l_{4y}, l_{4z})	(0, 103, 0)
Position of the origin of {0}	$\boldsymbol{o}_0$	(0, 0, 0)
Position of the origin of {1}	$\boldsymbol{o}_1$	(130, 160, 0)
Initial position of $\boldsymbol{p}_5$	(p_{5x}, p_{5y}, p_{5z})	$(10, 20, -400)$
Initial position of $\boldsymbol{p}_7$	(p_{7x}, p_{7y}, p_{7z})	(480, 50, 0)
Initial position of $\boldsymbol{p}_8$	(p_{8x}, p_{8y}, p_{8z})	$(480, -150, -4.801)$
${}^2\boldsymbol{l}_6$ vector w.r.t. {2}	(l_{6x}, l_{6y}, l_{6z})	$(-209, 7, -30)$
$\|\boldsymbol{p}_1 - \boldsymbol{p}_3\|$ as fraction of $\boldsymbol{l}_2$	r	0.473
$\|\boldsymbol{p}_1 - \boldsymbol{p}_6\|$ as fraction of $\boldsymbol{l}_2$	ρ	0.312
$\|\boldsymbol{p}_1 - \boldsymbol{p}_s\|$ as fraction of $\boldsymbol{l}_2$	σ	0.642

Appendix

6 Simulation Data

Some of the architecture parameters of the DWB suspension system are adopted from [7], and are listed in Table 3. The values are transformed suitably, as per the reference frames used in this paper. Some of the mass and inertia parameters are adopted from [4]. The parameters not available there are calculated on the basis of certain reasonable assumptions, e.g., the tie-rod inertia is calculated by assuming it to be a cylinder of uniform diameter. The weight of the steering rack is assumed appropriately. All the inertia[5] parameters are listed in Table 4. The positions of the centres of mass of the links in their own body fixed-frames are mentioned in the same table.

The spring and the damper characteristics from [4] did not fit directly to this problem, as the dynamic models under consideration are different. In [4], a three-degree-of-freedom quarter car model is studied. Hence, the spring stiffness and the damping coefficients are reduced from the values reported therein to suit the dynamics of the problem under consideration in this paper. These parameters are listed in Table 4.

---

[5] If not mentioned otherwise, the inertiæ are measured about the respective centres of mass of the links, and in the reference frames formed by the respective principal axes.

Table 4 Centres of mass, mass and inertia parameters, and the spring and the damper characteristics

Parameter	Symbol	Value
Mass of lower A-arm	m_1	1
Mass of link of upper A-arm	m_3	1
Mass of knuckle and steering link	m_2	8
Mass of wheel	m_w	22
Mass of tie-rod	m_5	1
Mass of steering rack	m_6	6.5/2
Mass of spring and damper combined	m_{sd}	4
Inertia of lower A-arm	(I_{1x}, I_{1y}, I_{1z})	(0.028, 0.002, 0.03)
Inertia of upper A-arm	(I_{3x}, I_{3y}, I_{3z})	(0.028, 0.002, 0.03)
Inertia of knuckle and steering link	(I_{2x}, I_{2y}, I_{2z})	(1.6, 1.6, 1.6)
Inertia of wheel	(I_{wx}, I_{wy}, I_{wz})	(2, 1.35, 2)
Inertia of tie-rod (about one of the ends)	(I_{5x}, I_{5y}, I_{5z})	$(0, \frac{m_5 l_5^2}{3}, \frac{m_5 l_5^2}{3})$
Coordinates of c.o.m. of link ①	$(rc_{1x}, rc_{1y}, rc_{1z})$	(0.14, 0, 0)
Coordinates of c.o.m. of link ②	$(rc_{2x}, rc_{2y}, rc_{2z})$	(0.0815, 0.0035, −0.01)
Coordinates of c.o.m. of link ③	$(rc_{3x}, rc_{3y}, rc_{3z})$	(0.13, 0, 0)
Coordinates of c.o.m. of link ⑤	$(rc_{5x}, rc_{5y}, rc_{5z})$	$(l_5/2, 0, 0)$
Coordinates of c.o.m. of the wheel	(rcw_x, rcw_y, rcw_z)	(0.0815, −0.0035, 0)
Spring constant (N/m)	K	5.11×10^4
Coefficient of damping (N-s/m)	C_d	0.1×10^4
Free length of the spring	l_s	482

References

1. Blundell M, Harty D (2004) The multibody systems approach to vehicle dynamics, 1st edn. Butterworth-Heinemann, Oxford
2. Balike KP (2010) Kineto-dynamic analyses of vehicle suspension for optimal synthesis. PhD thesis, Concordia University, September 2010
3. Cherian V, Jalili N, Ayglon V (2009) Modelling, simulation, and experimental verification of the kinematics and dynamics of a double wishbone suspension configuration. Proc Inst Mech Eng Part D: J Automob Eng 223(10):1239–1262
4. Attia HA (2002) Dynamic modelling of the double wishbone motor-vehicle suspension system. Eur J Mech A/Solids 21(1):167–174
5. Uchida T, McPhee J (2012) Driving simulator with double-wishbone suspension using efficient block-triangularized kinematic equations. Multibody Syst Dyn 28(4):331–347
6. Shabana AA (2009) Computational dynamics, 2nd edn. Wiley, New York
7. Reddy KV, Kodati M, Chatra K, Bandyopadhyay S (2016) A comprehensive kinematic analysis of the double wishbone and MacPherson strut suspension systems. Mech Mach Theory 105:441–470
8. Ghosal A (2006) Robotics: fundamental concepts and analysis, 1st edn. Oxford University Press, New Delhi

9. Muralidharan V, Mamidi TK, Guptasarma S, Nag A, Bandyopadhyay S (2018) A comparative study of the configuration-space and actuator-space formulations of the lagrangian dynamics of parallel manipulators and the effects of kinematic singularities on these. Mech Mach Theory 130:403–434

A Mechanical OR Gate Using Pinned-Pinned Bistable Arches

Deepayan Banik, Safvan Palathingal, G. K. Ananthasuresh, and Amitabha Ghosh

Abstract Mechanical logic gates, whose embodiments were demonstrated in the past with rigid bodies connected with kinematic joints, are revisited in this work using compliant mechanisms. In particular, an OR gate is designed using pinned-pinned bistable arches. The compliant OR gate consists of five bistable arches arranged in such a way that a central arch acts as the output with 0 and 1 stable states while two pairs of arches, with their own 0 and 1 states, act as inputs. The arch-profiles of all arches are designed so that the forces of switching and switching back between the two stable states and the travel between the two stable configurations are as per the desired functionality. An analytical method, which is based on linear superposition of buckling mode shapes of a straight pinned-pinned column, was used to design the arch-profiles. Finite element analysis was used to validate the OR logic behaviour of the entire setup. A macroscale prototype was made using 3D printing to demonstrate the working of the OR logic. Since mechanical gates are useful at the micro-scale, using fixed-fixed arches instead of pinned-pinned arches is beneficial from the viewpoint of microfabrication.

Keywords Compliant mechanisms · Mechanism design · Micro/nano mechanisms · Systems

D. Banik · A. Ghosh (✉)
Indian Institute of Engineering Science and Technology, Shibpur, India
e-mail: amitabha@iitk.ac.in

D. Banik
e-mail: banik@iitk.ac.in

S. Palathingal · G. K. Ananthasuresh
Indian Institute of Science, Bangalore, India
e-mail: safvanp@iisc.ac.in

G. K. Ananthasuresh
e-mail: suresh@iisc.ac.in

D. Sen et al. (eds.), *Mechanism and Machine Science*, Lecture Notes in Mechanical Engineering, https://doi.org/10.1007/978-981-15-4477-4_61

1 Introduction

The earliest logic gates were mechanical. Charles Babbage in 1837 designed a gear-based system called the Analytical Engine, capable of performing logical operations [1]. Fluidic logic gates have undergone some development too. Other kinds of mechanical logic gates used flip-flops as in Digi-Comp I and II, which were commercially available educational toys during the 1960s. Despite all developments, mechanical logic gates remained reasonably large in size, comprising revolute joints. Being applicable to the construction of logical circuits requires them to be sufficiently small so as to accommodate several of them in a small area. Since manufacturing at small scales essentially requires structures to be monolithic, all kinematic joints should be avoided using compliant design. This brings us to the subject of our project which is to design logic gates using compliant mechanisms. We have used bistable arches to achieve this. From the point of view of designing at micro- or nanoscales, one needs to be able to do the same at larger scales first, because, manufacturing micromechanisms is not only expensive but also time-consuming. Here, we have tried to achieve the design on the macroscale first. Because of this fact, we are allowed to use pinned-pinned arches, which cannot be used otherwise. At microscale, we ought to use fixed-fixed arches that can be easily realized using photolithography.

Arches with two force-free stable equilibrium positions are called bistable arches. They are conceptually advanced versions of pre-stressed beams or columns that have two similar stable states. Applying the correct amount of pre-stress is a difficult task during manufacturing in large quantities or at the microscale. Alternately, bistability may be achieved by fabricating an arch in a shape that is a linear combination of buckling modes of the corresponding straight column [2, 3]. Analysis of sine-curved arches, the fundamental mode shape of a straight column with hinged-hinged boundary conditions, has been studied in a pioneering work [3].

Figure 1 shows the as-fabricated unstressed state as State 1 and the stressed force-free equilibrium state as State 2. The minimum force required to switch from State 1 to State 2 is the switch force, F_s; the minimum force required to switch back to

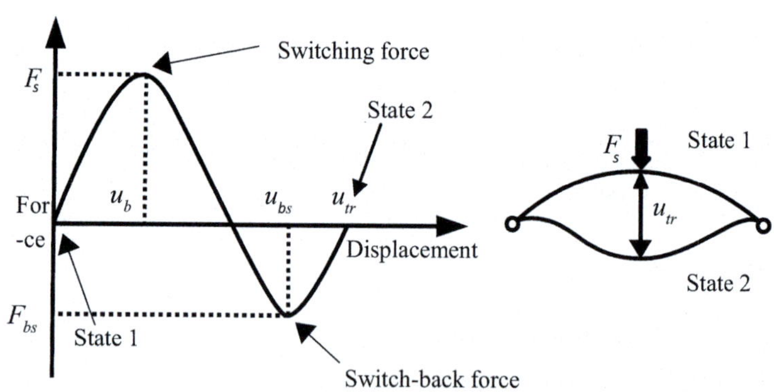

Fig. 1 Typical force-displacement characteristics of a bistable arch

State 1 from State 2 is the switch-back force, F_{sb}; and distance of the midpoint of the arch moves between the two stable states is the travel $u_{tr.}$ Bistability of an arch can be concluded from the values of F_s and F_{sb}. Other important parameters for design are the switch and switch-back threshold displacements D_s and D_{sb}.

In Sect. 2, we describe an analytical method used to design the arches based on their bistable characteristics. The method is based on the minimization of total potential energy (PE) in order to determine the possible stable configurations of a deformed arch. The initial force-free equilibrium shape of the arch consists of certain buckling modes. Upon applying forces at certain points on the arches, they deform and store potential energy. The deformed shape is assumed to be composed of the same modes as the original shape for which the mode weights are unknown. These are determined by solving the equations obtained by minimizing the PE. Bistability is concluded from the second-order derivatives of the TPE. This method has been dealt with extensive mathematical detail in [4, 5].

The next step is carrying out Finite Element Analysis (FEA) for the individual arches so as to obtain the force-displacement characteristic. This is presented in Sect. 3. This gives us the travel required at specific points on the arches to cause switching from State 1 to State 2. The output arch is analysed first, resulting in parameters required for the design of the two input arches along with the ones necessary to bring about the logical operation. Here we have worked on the OR logic gate. Finally, the arches have been 3D-printed at the macroscale to demonstrate the working of the OR logic.

2 Analytical Method for Determination of Bistability

A brief outline of the analytical method used for the determination of bistability is given in this section.

2.1 Buckling Analysis

Figure 2 shows an unstressed pinned-pinned column with torsion springs at the joints acted upon by an axial load, P.

The buckling modes of this system may be determined by applying suitable boundary conditions to the governing differential equation given by [6].

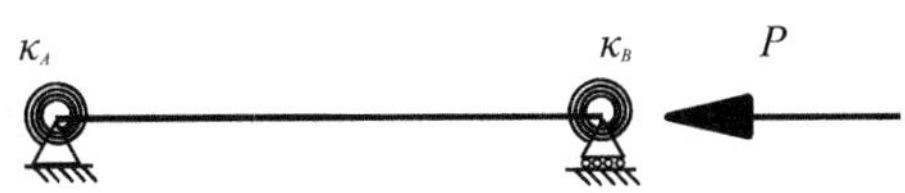

Fig. 2 An unstressed pinned-pinned with torsion springs beam acted upon by axial load P

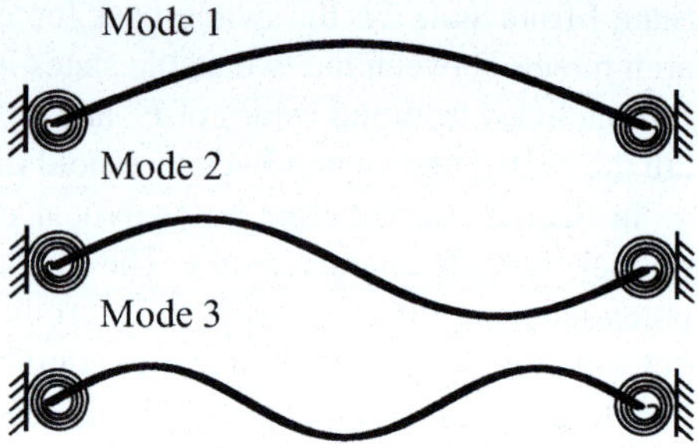

Fig. 3 First three buckling modes

$$EI\frac{d^4w}{dx^4} + P\frac{d^2w}{dx^2} = 0 \tag{1}$$

$$EI\frac{d^2w}{dx^2}(\text{at } x = 0) = \kappa_A\frac{dw}{dx}(\text{at } x = 0) \tag{2}$$

$$EI\frac{d^2w}{dx^2}(\text{at } x = L) = -\kappa_B\frac{dw}{dx}(\text{at } x = L) \tag{3}$$

where P is the axial load, w the transverse displacement of the beam perpendicular to the axial force, E the Young's Modulus, I the second moment of area of the cross section, κ_A the stiffness of torsional spring at the left end and κ_B the stiffness of torsional spring at the right end. By using the transcendental equation for the non-trivial buckled configuration derived in [5], the buckling modes look like those in Fig. 3. The following is the equation:

$$\frac{4\kappa_A}{M^3EI}\sin^2\left(\frac{ML}{2}\right)-\left[\frac{\kappa_B}{\kappa_A M^2} - \frac{\kappa_B L}{M^2EI} - \frac{EIL}{\kappa_A} - \frac{1}{M^2}\right]\sin(ML) = \left[\frac{L}{M} - \frac{\kappa_B L}{\kappa_A M}\right]\cos(ML) \tag{4}$$

$$M^2 = \frac{N^2}{L^2} \quad \text{and} \quad N^2 = \frac{PL^2}{EI}$$

These buckling modes correspond to sine waves of different wavelengths between the ends.

2.2 *Post-buckling Analysis with Pinned Ends*

Bistability has been proved for the first buckling mode with the fixed-fixed boundary condition used as the as-fabricated, stress-free geometric profile in [7]. In [2], the

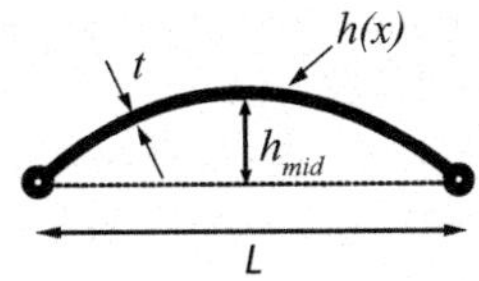

Fig. 4 Model of a pinned-pinned bistable arch

same has been done for a sine-curved arch which is the first mode for the pinned-pinned boundary condition. We consider a linear combination of the buckling modes for a given boundary condition as the as-fabricated shape as done in [3].

Let $h(x)$ denote the arch-profile (Fig. 4). Let w_i indicate the ith buckling mode shape of the straight beam of the same boundary conditions. Since it is likely to get a bistable behaviour from a linear combination of w_is, we write

$$h(x) = \sum_{i=1}^{n} a_i w_i(x) \tag{5}$$

where a_i is the weight for $w_i(x)$ and n is the number of buckling modes considered. Let a transverse load applied at the mid-point of the arch deform it to a shape given by $w(x)$. Since buckling mode shapes form a basis set, we approximate $w(x)$ as follows:

$$w(x) = \sum_{i=1}^{m} A_i w_i(x) \tag{6}$$

where A_i is the weight for $w_i(x)$ for each buckling mode shape in the deformed shape and m is the number of mode shapes used for approximating the deformed profile. It may be noted that m should not be smaller than n because to capture the as-fabricated shape itself m should be equal to n.

2.3 Equilibrium Equations

As we cause the arch to deform, it gains potential energy in terms of strain energy due to bending in the beam (SE_b), strain energy due to compression along the length of the arch (SE_c), and the work potential (WP) due to the external transverse force F applied at the mid-point. The normalized potential energy for a bistable arch of length L, depth or thickness t, moment of inertia I, height at the mid-point h_{mid}, and Young's modulus E, is given by

$$PE = SE_b + SE_c + WP$$

$$= \frac{1}{2}\int_0^1 \left(\frac{\mathrm{d}^2 H}{\mathrm{d}X^2} - \frac{\mathrm{d}^2 W}{\mathrm{d}X^2}\right)\mathrm{d}X + 6Q^2\left\{\int_0^1 \left[\frac{1}{2}\left(\frac{\mathrm{d}H}{\mathrm{d}X}\right)^2 - \frac{1}{2}\left(\frac{\mathrm{d}W}{\mathrm{d}X}\right)^2\right]\mathrm{d}X\right\}^2 \tag{7}$$

Table 1 List of normalized quantities

Parameter	Normalized quantity	Normalizing factor
x	X	$1/L$
$h(x)$, $w(x)$	$H(X)$, $W(X)$	$1/h_{\mathrm{mid}}$
f	F	$L3/EIh_{\mathrm{mid}}$
–	Q	h_{mid}/t

$$-F(H_{\mathrm{mid}} - W_{\mathrm{mid}})$$

where the normalized quantities are described in Table 1.

The equilibrium equations obtained from the principle of minimum potential energy are

$$\frac{\partial PE}{\partial A_i} = 0 \quad \text{where} \quad i = 1, 2, 3, \ldots m \tag{8}$$

Solving these equations gives the values of A_i, which constitutes the second (toggled) state of the arch.

2.4 Shape Relations

The standard expressions for obtaining the toggled shape are derived in [5]. The steps are given as follows:
Step 1. The coefficient of the fundamental mode (A_1) is calculated by solving the following equation in terms of the coefficients or weights (a_i) of all the modes in the as-fabricated shape using a numerical technique.

$$3Q^2 \sum_{i=1}^{m} \left\{ \frac{\frac{1}{i^2}\left(1 - \frac{a_1}{A_1}\right) - 2}{\left[1 - \frac{1}{i^2}\left(1 - \frac{a_1}{A_1}\right)\right]^2} \right\} a_i^2 = 1 \tag{9}$$

Step 2. The remaining weights are obtained using

$$A_i = \frac{a_i}{1 - \frac{1}{i^2}\left(1 - \frac{a_1}{A_1}\right)} \tag{10}$$

It is to be noted that the final or toggled shape has only those modes that are present in the initial or as-fabricated shape of the arch.

Step 3. Compute the Hessian matrix. This is a matrix consisting of second-order derivatives of the total potential energy in any deformed shape.

Diagonal elements:

$$H_{ii} = \frac{M_i^4}{2} - \frac{3Q^2 M_i^2 C}{2} - 3Q^2 A_i^2 M_i^4 \tag{11}$$

Non-diagonal elements:

$$H_{ij} = 3Q^2 A_i A_j M_i^2 M_j^2 \tag{12}$$

where $M_i = i\pi$.

Step 4. Check for positive definiteness of the Hessian matrix by checking if all the eigenvalues of the Hessian matrix are positive. This is the condition for bistability of the arch.

3 The Design for the Mechanical OR Gate

By applying the analytical technique described earlier, we check the bistability of a few standard combinations of modes and their mode weights. For our final design of the output of the OR gate, the first and the fifth buckling modes are used, as the shape provides for two valleys (locations 1 and 2) to accept inputs. The analytical method offers an efficient alternative for the determination of bistability as compared to a Finite Element Analysis (FEA) for the structure. However, after the design, it is helpful to use FEA of the continuum model of the arch to get the complete force-displacement curve.

The length of the arch is 100 mm, thickness t is 1 mm, weight of the first and the fifth modes are 10 and 2.5, respectively, while the in-plane depth is 5 mm. The shape of the output arch can be seen in Fig. 5 and is given by the following expression:

$$h(x) = 10 \sin\left(\frac{\pi x}{100}\right) + 2.5 \sin\left(\frac{5\pi x}{100}\right) \tag{13}$$

From the method described in the previous section, we obtain the toggled shape of the arch as given in Fig. 5; the corresponding expression is

$$w(x) = -8.25885 \sin\left(\frac{\pi x}{100}\right) + 2.74253 \sin\left(\frac{5\pi x}{100}\right) \tag{14}$$

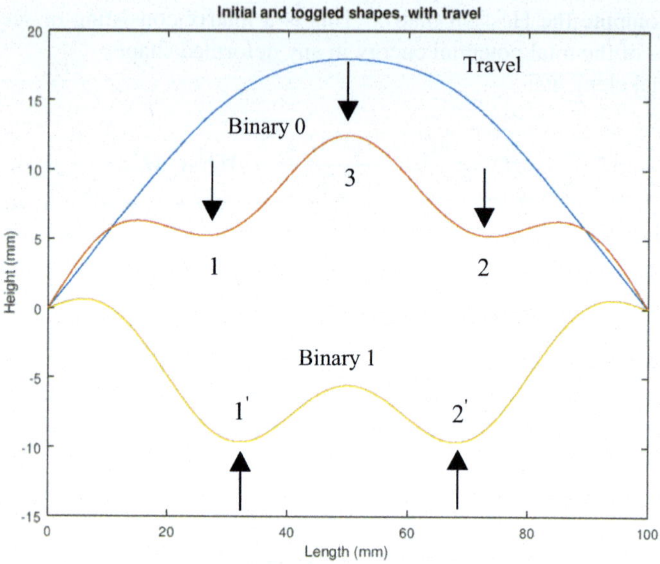

Fig. 5 The as-fabricated (middle) and toggled (bottom-most) shapes with travel (top-most). The arrows represent the direction of actuation

3.1 *The OR Logic*

According to the OR logic, when either of the two inputs is 1, the output must also be 1. The shape corresponding to State 1 in Fig. 1 represents the binary digit 0, while State 2 corresponds to binary digit 1. By using FEA, we perform parametric sweeps for displacements at location 3 (50 mm along the length of the arch) given in Fig. 5 to recheck bistability from the force-displacement curve. The same is done at location 1 (or location 2 at 25 mm along the length of the arch from either ends) to obtain the threshold displacement required at that particular location for switching from 0 to 1. This displacement in turn determines the travel (say T_1) of the input arches (the schematic is given in Fig. 6). Thus, if T_1 displacement is given at location 1 or 2 the output will switch to its less-stable second state (binary 1) from its as-fabricated state

Fig. 6 Schematic arrangement of the input and output arches

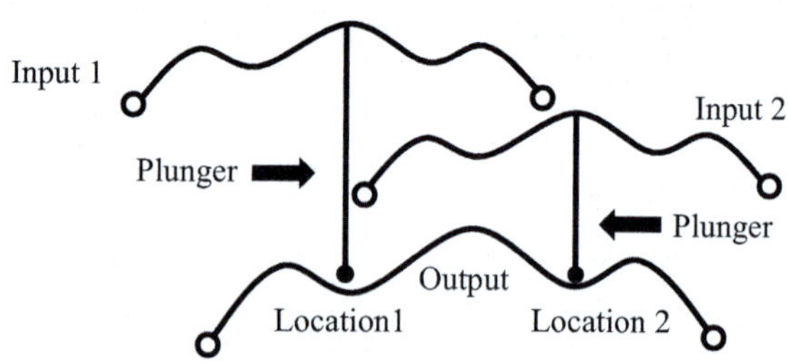

Table 2 Truth table for the OR gate

Input 1	Input 2	Output
0	0	0
0	1	1
1	0	1
1	1	1

(binary 0). When both the inputs are actuated, the output will still be in the second state, exhibiting the OR logic. The truth table for the OR gate is given in Table 2.

3.2 *Force-Displacement (F-D) Characteristics*

In this section, we present the relationship between force and displacement applied at various points on the output and input arches. Note that the displacement is assumed to be positive in the downward direction while force is assumed to be negative in the same direction.

The output arch. From Fig. 7a, we find that the arch concerned has reasonable bistable properties. This is understood from the fact that there are three zero-force or equilibrium shapes for the arch. The first and the third shapes with displacements 0 and 18 mm (at the mid-point or location 3) are stable equilibria because a slight displacement in any direction results in a restoring force in the opposite direction. The second shape with a displacement of 13 mm is unstable because a slight displacement in any direction results in a force in the same direction. From the figure, it is also evident that the as-fabricated shape is more stable than the toggled shape as the energy required for switching from State 1 to State 2, i.e., the area under the F-D curve till the position for the unstable equilibrium, is more than that required for switch-back.

In Fig. 7b, we see the force-displacement characteristic for actuation along location 1 which is positioned eccentrically for receiving inputs. The abscissa at the point where the curve cuts the x-axis for the second time (position for unstable equilibrium) gives us the travel required for switching from binary 0 to binary 1. From the graph, this, denoted by T_1 (defined earlier), comes out to be 13.25 mm. This means that actuating any one or both of the inputs given in Fig. 6 would result in switching of the output from 0 to 1, which is the intent of the OR logic.

When both the locations 1 and 2 are actuated, the switching is perfectly symmetric. We know that in cases of symmetric switching the force-displacement curve traces itself back when the direction of actuation is reversed and the arch switches from State 2 to State 1. Figure 7a and c both demonstrate cases of symmetric switching. Therefore, once the arch has been switched to State 1, we need to apply a displacement, say T_2 at locations $1'$ and $2'$ (Fig. 5) together to switch it back to 0. The reason we need to analyse the switch-back case is that we want the output arch to return to

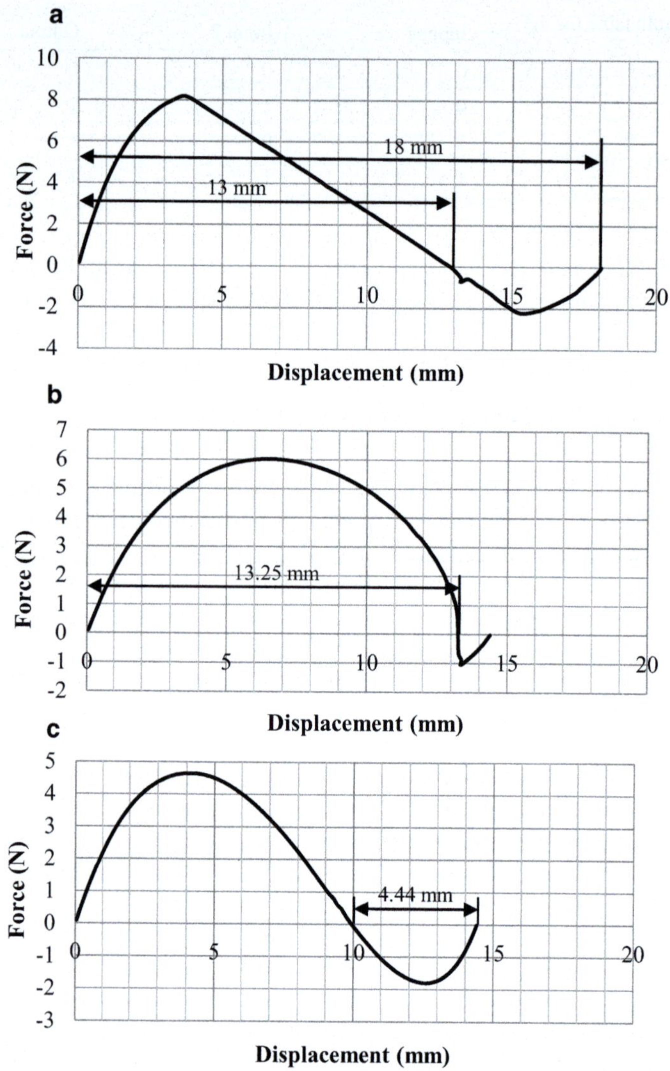

Fig. 7 **a** F-D curve for actuation at mid-point (location 3). **b** F-D curve for eccentric actuation (location 1 or 2). **c** F-D curve for symmetric eccentric actuation (location 1 and 2)

State 1 (binary 0) when both the inputs are reverted back to their as-fabricated shapes (binary 0). This is necessary for an OR gate.

From Fig. 7c the value of T_2 is found to be 4.4 mm. It would not have been possible to determine this from the plot of the forward switch had the switch not been perfectly symmetric. It would be necessary to perform a sweep in the opposite direction otherwise.

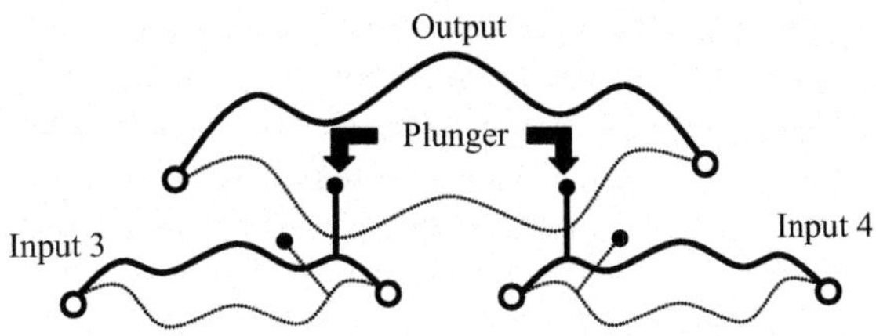

Fig. 8 Schematic arrangement of the input and output arches for reverse actuation

It is now clear that, in order to revert the output arch to State 1, the input arches 1 and 2, shown in Fig. 6 will not be sufficient. We need to use two more to give the actuation in the reverse direction. Figure 8 shows the input arches 3 and 4 required for the purpose. Locations 1′ and 2′ (Fig. 5) receive the actuations given by the inputs 3 and 4, respectively. Note that, according to OR logic, both the inputs 3 and 4 must be made 0 for the output arch to return to 0. This makes T_2 very important from the point of designing. The displacement provided by either of the arches 3 and 4 needs to be slightly greater than T_2 so that just any one of them is not sufficient for switching the output from State 2 to State 1. To be able to do so by singular efforts, the displacement that has to be provided must be significantly greater than T_2 and hence it need not be calculated.

Input arches. The input arches are composed of the same modes as the output arch because it provides a greater choice for travel and has better bistable properties compared to simple sine arches. Another reason is that input arches would be outputs for other inputs in a complete logical circuit. For inputs 1 and 2, we need to have a travel of at least T_1 (13.25 mm) at the mid-point for switching of the output. The length of these arches is 70 mm, thickness t is 0.7 mm, weight of the first and the fifth modes are 8 and 2, respectively, and in-plane depth is 5 mm. Using the analytical method described in Sect. 2, these arches are found to be bistable. This is evident from Fig. 9 as well. Figure 9 also tells us that the travel at the mid-point of such arches is around 14.25 mm, which satisfies the criterion for switching of the output arch.

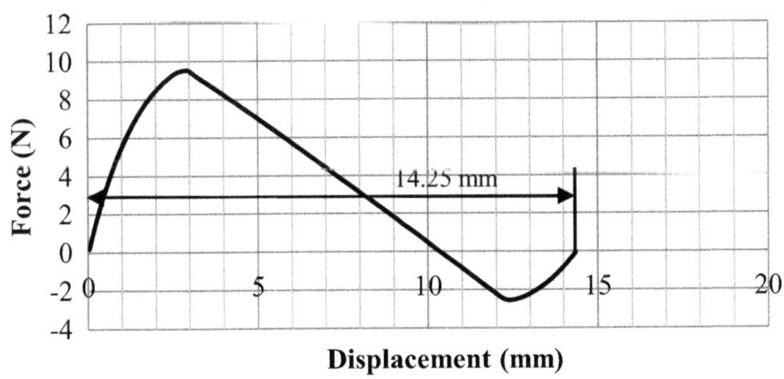

Fig. 9 F-D curve for actuation at mid-point for forward inputs 1 and 2

In case of input arches 3 and 4, we need a travel slightly more than T_2 (4.44 mm). If we try to keep such a small travel at the mid-point, then the entire size of the arch has to be reduced to such an extent that it would result in a loss of bistability and also structural strength. This is why the plungers in these arches are located eccentrically (8.6 mm from the corresponding end), which allows for smaller travel without jeopardizing the aforementioned factors. The length of these arches is 50 mm, thickness t is 0.5 mm, weight of the first and the fifth modes are 5 and 1.25, respectively, and in-plane depth is 5 mm. By using the analytical method, we find the travel at this location to be 4.65 mm that is slightly greater than T_2 (4.44 mm). In Fig. 10, the force-displacement curve shows that the arch is bistable.

In Fig. 11, the different configurations of the entire assembly have been displayed. The models for all the arches were developed and then 3D-printed using Acrylonitrile Butadiene Styrene (ABS) having a Poisson ratio, Young's modulus, and density similar to those of Polypropylene that was assumed in FEA.

It must be noted here that inputs 1 and 3 have to be actuated simultaneously and the same is true for inputs 2 and 4. Figure 11a shows the initial un-deformed state of the assembly where both the inputs (1 and 3 are considered as a single input now and so are 2 and 4) and the outputs are corresponding to binary 0. In Fig. 11b, one of the inputs has been actuated to 1 causing the output to switch to 1 as well. When both the inputs are actuated, as in Fig. 11d, the output remains in the toggled state. Finally, in Fig. 11c, we see that, as one of the inputs is reverted back to 0, the output arch does not switch back to 0, just as it should, to maintain the OR logic. Figure 11a also represents the state when both the inputs have been switched back to 0 causing the output to return to 0 as well. A video clip of the operation of the mechanism can be accessed at http://www.mecheng.iisc.ac.in/~suresh/ORGate.mpg.

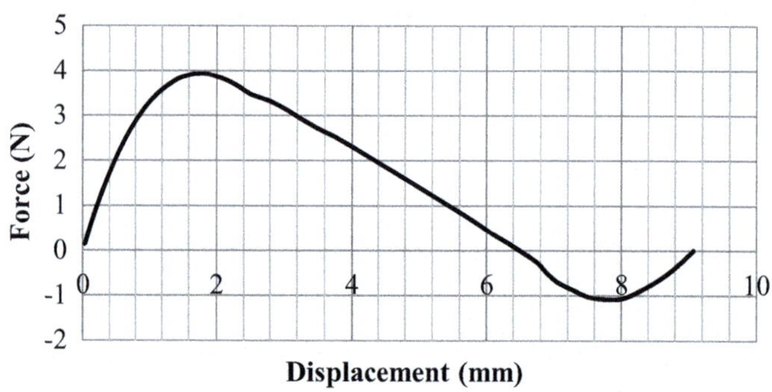

Fig. 10 F-D curve for actuation at mid-point for backward inputs 3 and 4

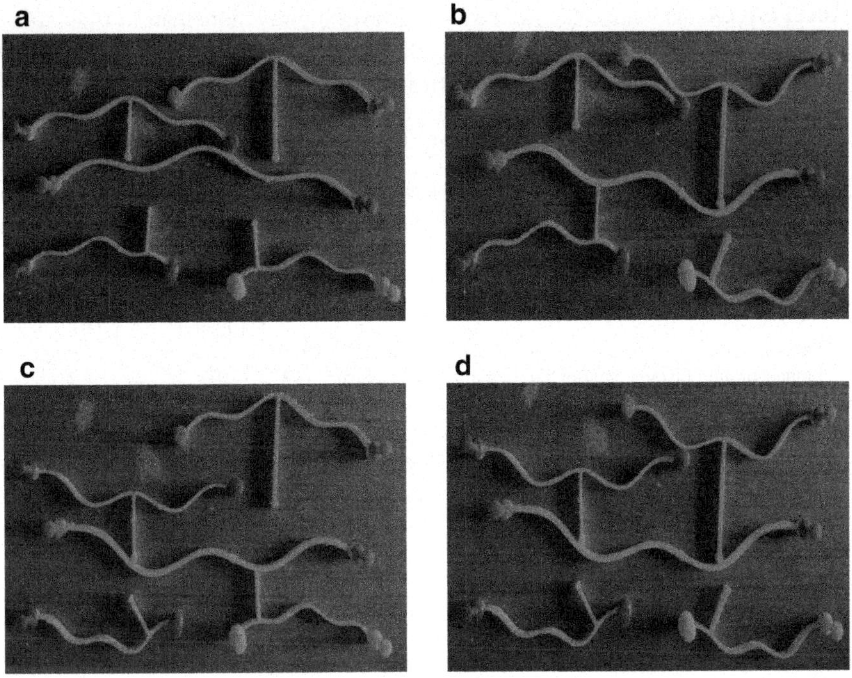

Fig. 11 A fabricated set of arches that work in tandem to execute the OR logic based on the design described so far. The inputs are **a** 0 OR 0 **b** 0 OR 1 **c** 1 OR 0 **d** 1 OR 1. The corresponding outputs are **a** 0 **b** 0 **c** 0 **d** 1

4 Conclusion

The design presented here is preliminary and a lot of work lies further in this direction. The way logic has been demonstrated for the OR gate in this work paves way for designing AND, NAND, and NOR gates which could result from modifications of the current design itself. This mechanism has to be designed at microscales because that is where the real utility of the system lies. Mechanisms can be miniaturized to a significantly small scale today. Compliant mechanisms are better for this purpose as they may be manufactured from monolithic structures using photolithography. Once this is achieved, the way to link two or more gates together to make a circuit has to be figured out. Being mechanical gates, these would have the advantage of sustenance through a higher range of temperatures given that a suitable material is used. In addition, mechanical logic does not use electrons, which is why they could find potential application in systems which are exposed to a considerable amount of electromagnetic waves that interfere with the electrons. This work serves as a proof of concept for the feasibility of such systems.

Acknowledgements Support from a grant from the Technology Initiative for Disabled and Elderly (TIDE) programme of the Department of Science and Technology is gratefully acknowledged.

References

1. Babbage's analytical engine, 1834–1871. (Trial model). Science Museum. Accessed 23 August 2017
2. Qiu J, Lang JH, Slocum AH (2004) A curved-beam bistable mechanism. J Microelectromech Syst 13(2):137–146
3. Fung Y-C, Kaplan A (1952) Buckling of low arches or curved beams of small curvature. National Advisory Committee for Aeronautics; Washington, DC, United States, Technical Note 2840, p 78
4. Palathingal S, Ananthasuresh GK (2017) Design of bistable pinned-pinned arches with torsion springs by determining critical points. In: Mechanism and machine science: proceedings of ASIAN MMS 2016 & CCMMS 2016, December, Guangzhou. Springer, China, pp 677–688
5. Palathingal S, Ananthasuresh GK (2018) A bilateral relationship between stable profiles of pinned-pinned bistable shallow arches. Int J Solids Struct 143:183–193
6. Palathingal S, Sarojini D, Katti AN, Ananthasuresh GK (2015) A generalised method to investigate the bistability of curved beams using buckling analysis. In: 2nd international and 17th national conference on machines and mechanisms, iNaCoMM, p 73
7. Simitses GJ, Hodges DH (2006) Fundamentals of structural stability. Butterworth-Heinemann

The manufacturer's authorised representative in the EU is Springer Nature Customer Service Centre GmbH, Europaplatz 3, 69115 Heidelberg, Germany. If you have any concerns regarding our products, please contact ProductSafety@springernature.com

Printed and bound by CPI Group (UK) Ltd, Croydon, CR0 4YY
01/10/2026
02226525-0002